Fundamentals of Aerodynamics

Seventh Edition

John D. Anderson, Jr.

Christopher P. Cadou

McGRAW HILL SERIES IN AERONAUTICAL AND AEROSPACE ENGINEERING

The Wright brothers invented the first practical airplane in the first decade of the twentieth century. Along with this came the rise of aeronautical engineering as an exciting, new, distinct discipline. College courses in aeronautical engineering were offered as early as 1914 at the University of Michigan and at MIT. Michigan was the first university to establish an aeronautics department with a four-year degree-granting program in 1916; by 1926 it had graduated over one hundred students. The need for substantive textbooks in various areas of aeronautical engineering became critical. Rising to this demand, McGraw Hill became one of the first publishers of aeronautical engineering textbooks, starting with *Airplane Design and Construction* by Ottorino Pomilio in 1919, and the classic and definitive text *Airplane Design: Aerodynamics* by the iconic Edward P. Warner in 1927. Warner's book was a watershed in aeronautical engineering textbooks.

Since then, McGraw Hill has become the time-honored publisher of books in aeronautical engineering. With the advent of high-speed flight after World War II and the space program in 1957, aeronautical and aerospace engineering grew to new heights. There was, however, a hiatus that occurred in the 1970s when aerospace engineering went through a transition, and virtually no new books in the field were published for almost a decade by anybody. McGraw Hill broke this hiatus with the foresight of its Chief Engineering Editor, B.J. Clark, who was instrumental in the publication of *Introduction to Flight* by John Anderson. First published in 1978, *Introduction to Flight* is now in its 8th edition. Clark's bold decision was followed by McGraw Hill riding the crest of a new wave of students and activity in aerospace engineering, and it opened the flood-gates for new textbooks in the field.

In 1988, McGraw Hill initiated its formal series in Aeronautical and Aerospace Engineering, gathering together under one roof all its existing texts in the field, and soliciting new manuscripts. This author is proud to have been made the consulting editor for this series, and to have contributed some of the titles. Starting with eight books in 1988, the series now embraces 24 books covering a broad range of discipline in the field. With this, McGraw Hill continues its tradition, started in 1919, as the premier publisher of important textbooks in aeronautical and aerospace engineering.

John D. Anderson, Jr.

Fundamentals of Aerodynamics

Seventh Edition

John D. Anderson, Jr.
Curator of Aerodynamics
National Air and Space Museum
Smithsonian Institution
and
Professor Emeritus
University of Maryland

Christopher P. Cadou
Professor of Aerospace
Engineering
and
Keystone Professor,
University of Maryland

FUNDAMENTALS OF AERODYNAMICS

3 4 5 6 7 8 9 LCR 28 27 26 25

ISBN 978-1-266-07644-2
MHID 1-266-07644-1

Cover Image: *U.S. Navy photo by Mass Communication Specialist 2nd Class Kristopher Wilson*

John D. Anderson, Jr., was born in Lancaster, Pennsylvania, on October 1, 1937. He attended the University of Florida, graduating in 1959 with high honors and a bachelor of aeronautical engineering degree. From 1959 to 1962, he was a lieutenant and task scientist at the Aerospace Research Laboratory at Wright-Patterson Air Force Base. From 1962 to 1966, he attended the Ohio State University under the National Science Foundation and NASA Fellowships, graduating with a PhD in aeronautical and astronautical engineering. In 1966, he joined the U.S. Naval Ordnance Laboratory as Chief of the Hypersonics Group. In 1973, he became Chairman of the Department of Aerospace Engineering at the University of Maryland, and since 1980 has been professor of Aerospace Engineering at the University of Maryland. In 1982, he was designated a Distinguished Scholar/Teacher by the University. During 1986–1987, while on sabbatical from the University, Dr. Anderson occupied the Charles Lindbergh Chair at the National Air and Space Museum of the Smithsonian Institution. He continued with the Air and Space Museum one day each week as their Special Assistant for Aerodynamics, doing research and writing on the history of aerodynamics. In addition to his position as professor of aerospace engineering, in 1993, he was made a full faculty member of the Committee for the History and Philosophy of Science and in 1996 an affiliate member of the History Department at the University of Maryland. In 1996, he became the Glenn L. Martin Distinguished Professor for Education in Aerospace Engineering. In 1999, he retired from the University of Maryland and was appointed Professor Emeritus. He is currently the Curator for Aerodynamics at the National Air and Space Museum, Smithsonian Institution.

Dr. Anderson has published 11 books: *Gasdynamic Lasers: An Introduction,* Academic Press (1976), and under McGraw Hill, *Introduction to Flight* (1978, 1984, 1989, 2000, 2005, 2008, 2012, 2016), *Modern Compressible Flow* (1982, 1990, 2003), *Fundamentals of Aerodynamics* (1984, 1991, 2001, 2007, 2011), *Hypersonic and High Temperature Gas Dynamics* (1989), *Computational Fluid Dynamics: The Basics with Applications* (1995), *Aircraft Performance and Design* (1999), *A History of Aerodynamics and Its Impact on Flying Machines,* Cambridge University Press (1997 hardback, 1998 paperback), *The Airplane: A History of Its Technology,* American Institute of Aeronautics and Astronautics (2003), *Inventing Flight,* Johns Hopkins University Press (2004), and X-15, The World's Fastest Rocket Plane and the Pilots Who Ushered in the Space Age, with co-author Richard Passman, Zenith Press in conjunction with the Smithsonian Institution (2014). He is the author of over 120 papers on radiative gasdynamics, reentry aerothermodynamics, gasdynamic and chemical lasers, computational fluid dynamics, applied aerodynamics, hypersonic flow, and the history of aeronautics. Dr. Anderson is a member of the National Academy of Engineering, and is in *Who's Who in America*. He is an Honorary Fellow of the American Institute of Aeronautics and Astronautics (AIAA). He is also a fellow of the Royal Aeronautical Society, London. He is a member of Tau Beta Pi, Sigma Tau, Phi Kappa Phi, Phi Eta Sigma, The American Society for Engineering Education, the History of Science Society, and the Society for the History of Technology. In 1988, he was elected as Vice President of the AIAA for Education. In 1989, he was awarded the John Leland Atwood Award jointly by the American Society for Engineering Education and the American Institute of Aeronautics and Astronautics "for the lasting influence of his recent contributions to aerospace engineering education." In 1995, he was awarded the AIAA Pendray Aerospace Literature Award "for writing undergraduate and graduate textbooks in aerospace engineering which have received worldwide acclaim for their readability and clarity of presentation, including historical content." In 1996, he was elected Vice President of the AIAA for Publications. He has recently been honored by the AIAA with its 2000 von Karman Lectureship in Astronautics.

From 1987 to the present, Dr. Anderson has been the senior consulting editor on the McGraw Hill Series in Aeronautical and Astronautical Engineering.

Christopher P. Cadou earned his undergraduate degrees (BS in Mechanical Engineering and a BA in History) from Cornell University in 1989 and went on to graduate school at the University of California Los Angeles where he studied fluid mechanics and acoustically excited flames using laser-induced fluorescence imaging techniques. Dr. Cadou graduated with a PhD in 1996 and moved to the California Institute of Technology as a post-doctoral scholar where he studied boundary layer instabilities associated with SCRAMJet inlet unstart using time-resolved Schlieren imaging and taught a course in air breathing propulsion at UCLA. This was an important time because he learned that he liked both research and teaching. Two years later he took another post-doctoral position at the Massachusetts Institute of Technology to help develop the combustor and turbomachinery of a tiny gas turbine engine being constructed using the same technologies used to make computer chips. However, his interest in teaching remained and he left MIT in 2000 to join the University of Maryland's Aerospace Engineering Department as an Assistant Professor. He has taught compressible flow at the graduate and undergraduate levels to more than 1,000 students, authored one book on micro-scale energy conversion systems, and published more than 100 scholarly papers in the areas of micro-scale combustion, supersonic film cooling, laser diagnostics, smart materials-based actuators, and hybrid turbine/fuel cell energy conversion systems. He is a fellow of the American Society of Mechanical Engineers and an Associate Fellow of the American Institute of Aeronautics and Astronautics.

CONTENTS

PART 2
Inviscid, Incompressible Flow   207

Chapter 3
Fundamentals of Inviscid, Incompressible Flow   209

PREFACE TO THE SEVENTH EDITION

FUNDAMENTALS OF AERODYNAMICS

This book follows in the same tradition as the previous editions: it is for students—to be read, understood, and enjoyed. It is consciously written in a clear, informal, and direct style to talk to the reader and gain their immediate interest in the challenging and yet beautiful discipline of aerodynamics. The explanation of each topic is carefully constructed to make sense to the reader. Moreover, the structure of each chapter is highly organized to keep the reader aware of where we are, where we were, and where we are going with the flow of new and important ideas and concepts.

This edition continues with the same instructional and learning features introduced in the previous editions, such as preview boxes at the beginning of each chapter, road maps to keep the reader focused on the flow of new ideas and concepts, and end-of-chapter integrated work challenges that help to consolidate the important concepts in the minds of the readers. It also continues with such features as an introduction to computational fluid dynamics as an integral part of the modern study of aerodynamics, a chapter devoted entirely to hypersonic aerodynamics which has applications to new vehicle designs, and historical notes placed at the end of many of the chapters (a unique tradition that started with the first edition of this book, and that has carried on through all of the subsequent editions). Due to the extremely favorable comments from readers and users of the first six editions, virtually all the content of the earlier editions has been carried over intact to the present edition.

The major new feature of this edition is the addition of a valuable co-author, Dr. Christopher Cadou, Keystone Professor of Engineering at the University of Maryland. Dr. Cadou has contributed new worked examples and many new end-of-chapter homework problems which constitute most of the new content to this Seventh Edition, and which in fact greatly enhances the learning power of the new edition.

This book is organized along classical lines. It deals first with inviscid incompressible flow, then progresses to inviscid compressible flow, and then viscous flow in sequence. The material nicely divides into a two semester course, with Parts 1 and 2 in the first semester and Parts 3 and 4 in the second semester. The entire book has been used in a fast-paced first semester graduate course intended to introduce the fundamentals of aerodynamics to new graduate students who have not had this material as part of their undergraduate education. The book works well in such a mode.

Thanks go to the McGraw Hill editorial and production staff for their excellent help in producing this book, and to the legions of students over the years for many stimulating discussions that have influenced the development of this book. Special thanks go to both families of the two authors; families who have been patient and understanding about the time devoted to the preparation of the book.

As a final comment, aerodynamics is a subject of intellectual beauty, composed and drawn by many great minds over the centuries. Fundamentals of Aerodynamics is intended to portray and convey this beauty. Do you feel challenged and interested by these thoughts? If so, then read on, and enjoy.

John D. Anderson, Jr.
Christopher P. Cadou

ACKNOWLEDGMENTS

The authors wish to thank the reviewers of the seventh edition text for their suggestions and valued feedback:

H. Pat Artis
Virginia Tech AOE

Mark Nathaniel Callender
Middle Tennessee State University

Jose Camberos
University of Dayton, Ohio

Russell Cummings
United States Air Force Academy

Frederick Ferguson
North Carolina A&T State University

Christopher Griffin
West Virginia University

Dr. Sidaard Gunasekaran
University of Dayton, Ohio

David Guo
Southern New Hampshire University

Adam Huang
University of Arkansas

Gabriel Karpouzian
United States Naval Academy

Randall Kent
Weber State University

John Lee
John Brown University

David Miklosovic
United States Naval Academy

Anupam Sharma
Iowa State University

David B Stringer
Kent State University

Hui Wan
University of Colorado, Colorado Springs

Adam Wickenheiser
University of Delaware

ADDITIONAL RESOURCES

Proctorio

Remote Proctoring & Browser-Locking Capabilities

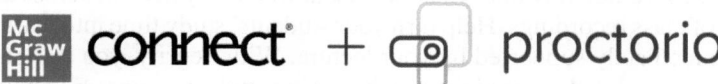

Remote proctoring and browser-locking capabilities, hosted by Proctorio within Connect, provide control of the assessment environment by enabling security options and verifying the identity of the student.

Seamlessly integrated within Connect, these services allow instructors to control the assessment experience by verifying identification, restricting browser activity, and monitoring student actions.

Instant and detailed reporting gives instructors an at-a-glance view of potential academic integrity concerns, thereby avoiding personal bias and supporting evidence-based claims.

 ReadAnywhere®

Read or study when it's convenient for you with McGraw Hill's free ReadAnywhere® app. Available for iOS or Android smartphones or tablets, ReadAnywhere gives users access to McGraw Hill tools including the eBook and SmartBook® 2.0 or Adaptive Learning Assignments in Connect. Take notes, highlight, and complete assignments offline—all of your work will sync when you open the app with Wi-Fi access. Log in with your McGraw Hill Connect username and password to start learning—anytime, anywhere!

OLC-Aligned Courses

Implementing High-Quality Instruction and Assessment Through Preconfigured Courseware

In consultation with the Online Learning Consortium (OLC) and our certified Faculty Consultants, McGraw Hill has created pre-configured courseware using OLC's quality scorecard to align with best practices in online course delivery. This turnkey courseware contains a combination of formative assessments, summative assessments, homework, and application activities, and can easily be customized to meet an individual instructor's needs and desired course outcomes. For more information, visit https://www.mheducation.com/highered/olc.

Tegrity: Lectures 24/7

Tegrity in Connect is a tool that makes class time available 24/7 by automatically capturing every lecture. With a simple one-click start-and-stop process, you capture all computer screens and corresponding audio in a format that is easy to search, frame by frame. Students can replay any part of any class with easy-to-use, browser-based viewing on a PC, Mac, or mobile device.

Educators know that the more students can see, hear, and experience class resources, the better they learn. In fact, studies prove it. Tegrity's unique search feature helps students efficiently find what they need, when they need it, across an entire semester of class recordings. Help turn your students' study time into learning moments immediately supported by your lecture. With Tegrity, you also increase intent listening and class participation by easing students' concerns about note-taking. Using Tegrity in Connect will make it more likely you will see students' faces, not the tops of their heads.

Writing Assignment

Available within Connect and Connect Master, the Writing Assignment tool delivers a learning experience to help students improve their written communication skills and conceptual understanding. As an instructor, you can assign, monitor, grade, and provide feedback on writing more efficiently and effectively.

Create
Your Book, Your Way

McGraw Hill's Content Collections Powered by Create® is a self-service website that enables instructors to create custom course materials—print and eBooks—by drawing upon McGraw Hill's comprehensive, cross-disciplinary content. Choose what you want from our high-quality textbooks, articles, and cases. Combine it with your own content quickly and easily, and tap into other rights-secured, third-party content such as readings, cases, and articles. Content can be arranged in a way that makes the most sense for your course, and you can include the course name and information as well. Choose the best format for your course: color print, black-and-white print, or eBook. The eBook can be included in your Connect course and is available on the free ReadAnywhere® app for smartphone or tablet access as well. When you are finished customizing, you will receive a free digital copy to review in just minutes! Visit McGraw Hill Create®—www.mcgrawhillcreate.com—today and begin building!

Reflecting the Diverse World Around Us

McGraw Hill believes in unlocking the potential of every learner at every stage of life. To accomplish that, we are dedicated to creating products that reflect, and are accessible to, all the diverse, global customers we serve. Within McGraw Hill, we foster a culture of belonging, and we work with partners who share our commitment to equity, inclusion, and diversity in all forms. In McGraw Hill Higher Education, this includes, but is not limited to, the following:

- Refreshing and implementing inclusive content guidelines around topics including generalizations and stereotypes, gender, abilities/disabilities, race/ethnicity, sexual orientation, diversity of names, and age.

- Enhancing best practices in assessment creation to eliminate cultural, cognitive, and affective bias.

- Maintaining and continually updating a robust photo library of diverse images that reflect our student populations.

- Including more diverse voices in the development and review of our content.

- Strengthening art guidelines to improve accessibility by ensuring meaningful text and images are distinguishable and perceivable by users with limited color vision and moderately low vision.

Instructors
The Power of Connections

A complete course platform

Connect enables you to build deeper connections with your students through cohesive digital content and tools, creating engaging learning experiences. We are committed to providing you with the right resources and tools to support all your students along their personal learning journeys.

65%
Less Time Grading

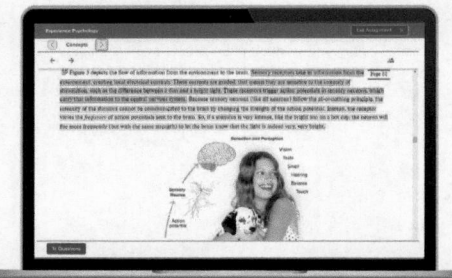

Laptop: Getty Images; Woman/dog: George Doyle/Getty Images

Every learner is unique

In Connect, instructors can assign an adaptive reading experience with SmartBook® 2.0. Rooted in advanced learning science principles, SmartBook 2.0 delivers each student a personalized experience, focusing students on their learning gaps, ensuring that the time they spend studying is time well-spent.
mheducation.com/highered/connect/smartbook

Affordable solutions, added value

Make technology work for you with LMS integration for single sign-on access, mobile access to the digital textbook, and reports to quickly show you how each of your students is doing. And with our Inclusive Access program, you can provide all these tools at the lowest available market price to your students. Ask your McGraw Hill representative for more information.

Solutions for your challenges

A product isn't a solution. Real solutions are affordable, reliable, and come with training and ongoing support when you need it and how you want it. Visit **supportateverystep.com** for videos and resources both you and your students can use throughout the term.

Students
Get Learning that Fits You

Effective tools for efficient studying

Connect is designed to help you be more productive with simple, flexible, intuitive tools that maximize your study time and meet your individual learning needs. Get learning that works for you with Connect.

Study anytime, anywhere

Download the free ReadAnywhere® app and access your online eBook, SmartBook® 2.0, or Adaptive Learning Assignments when it's convenient, even if you're offline. And since the app automatically syncs with your Connect account, all of your work is available every time you open it. Find out more at **mheducation.com/readanywhere**

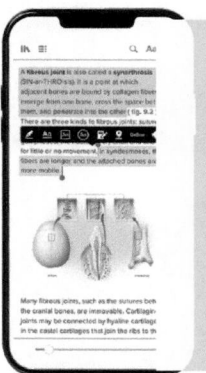

"I really liked this app—it made it easy to study when you don't have your text-book in front of you."

- Jordan Cunningham, Eastern Washington University

iPhone: Getty Images

Everything you need in one place

Your Connect course has everything you need—whether reading your digital eBook or completing assignments for class—Connect makes it easy to get your work done.

Learning for everyone

McGraw Hill works directly with Accessibility Services Departments and faculty to meet the learning needs of all students. Please contact your Accessibility Services Office and ask them to email accessibility@mheducation.com, or visit **mheducation.com/about/accessibility** for more information.

Fundamental Principles

I n Part 1, we cover some of the basic principles that apply to aerodynamics in
general. These are the pillars on which all of aerodynamics is based. ■

CHAPTER 1

Aerodynamics: Some Introductory Thoughts

The term "aerodynamics" is generally used for problems arising from flight and other topics involving the flow of air.
  Ludwig Prandtl, 1949

Aerodynamics: The dynamics of gases, especially atmospheric interactions with moving objects.
  The American Heritage
  Dictionary of the English
  Language, 1969

PREVIEW BOX

Why learn about aerodynamics? For an answer, just take a look at the following five photographs showing a progression of airplanes over the past 70 years. The Douglas DC-3 (Figure 1.1), one of the most famous aircraft of all time, is a low-speed subsonic transport designed during the 1930s. Without a knowledge of low-speed aerodynamics, this aircraft would have never existed. The Boeing 707 (Figure 1.2) opened high-speed subsonic flight to millions of passengers beginning in the late 1950s. Without a knowledge of high-speed subsonic aerodynamics, most of us would still be relegated to ground transportation.

Figure 1.1 Douglas DC-3 (*NASA*).

Figure 1.2 Boeing 707 (*CSU Archives/Everett Collection/Alamy Stock Photo*).

Figure 1.3 Bell X-1 (*Library of Congress, Prints & Photographs Division [LC-USZ6-1658]*).

Figure 1.4 Lockheed F-104 (*Library of Congress, Prints & Photographs Division [LC-USZ62-94416]*).

Figure 1.5 Lockheed-Martin F-22 (*U.S. Air Force Photo/Staff Sgt. Vernon Young Jr.*).

Figure 1.6 Blended wing body (*NASA*).

The Bell X-1 (Figure 1.3) became the first piloted airplane to fly faster than sound, a feat accomplished with Captain Chuck Yeager at the controls on October 14, 1947. Without a knowledge of transonic aerodynamics (near, at, and just above the speed of sound), neither the X-1, nor any other airplane, would have ever broken the sound barrier. The Lockheed F-104 (Figure 1.4) was the first supersonic airplane point-designed to fly at twice the speed of sound, accomplished in the 1950s. The Lockheed-Martin F-22 (Figure 1.5) is a modern fighter aircraft designed for sustained supersonic flight. Without a knowledge of supersonic aerodynamics, these supersonic airplanes would not exist. Finally, an example of an innovative new vehicle concept for high-speed subsonic flight is the blended wing body shown in Figure 1.6. At the time of writing, the blended-wing-body promises to carry from 400 to 800 passengers over long distances with almost 30 percent less fuel per seat-mile than a conventional jet transport. This would be a "renaissance" in long-haul transport. The salient design aspects of this exciting new concept are discussed in Section 11.10. The airplanes in Figures 1.1–1.6 are six good reasons to learn about aerodynamics. The major purpose of this book is to help you do this. As you continue to read this and subsequent chapters, you will progressively learn about low-speed aerodynamics, high-speed subsonic aerodynamics, transonic aerodynamics, supersonic aerodynamics, and more.

Airplanes are by no means the only application of aerodynamics. The air flow over an automobile, the gas flow through the internal combustion engine powering an automobile, weather and storm prediction, the flow through a windmill, the production of thrust by gas turbine jet engines and rocket engines, and the movement of air through building heater and air-conditioning systems are just a few other examples of the application of aerodynamics. The material in this book is powerful stuff—important stuff. Have fun reading and learning about aerodynamics.

To learn a new subject, you simply have to start at the beginning. This chapter is the beginning of our study of aerodynamics; it weaves together a series of introductory thoughts, definitions, and concepts essential to our discussions in subsequent chapters. For example, how does nature reach out and grab hold of an airplane in flight—or any other object immersed in a flowing fluid—and exert an aerodynamic force on the object? We will find out here. The resultant aerodynamic force is frequently resolved into two components defined as lift and drag; but rather than dealing with the lift and drag forces themselves, aerodynamicists deal instead with lift and drag *coefficients*. What is so magic about lift and drag coefficients? We will see. What is a Reynolds number? Mach number? Inviscid flow? Viscous flow? These rather mysterious sounding terms will be demystified in the present chapter. They and others constitute the language of aerodynamics, and as we all know, to do anything useful you have to know the language. Visualize this chapter as a beginning language lesson, necessary to go on to the exciting aerodynamic applications in later chapters. There is a certain enjoyment and satisfaction in learning a new language. Take this chapter in that spirit, and move on.

1.1 IMPORTANCE OF AERODYNAMICS: HISTORICAL EXAMPLES

On August 8, 1588, the waters of the English Channel churned with the gyrations of hundreds of warships. The Spanish Armada had arrived to carry out an invasion of Elizabethan England and was met head-on by the English fleet under the command of Sir Francis Drake. The Spanish ships were large and heavy; they were packed with soldiers and carried formidable cannons that fired 50 lb round shot that could devastate any ship of that era. In contrast, the English ships were smaller and lighter; they carried no soldiers and were armed with lighter, shorter-range cannons. The balance of power in Europe hinged on the outcome of this naval encounter. King Philip II of Catholic Spain was attempting to squash Protestant England's rising influence in the political and religious affairs of Europe; in turn, Queen Elizabeth I was attempting to defend the very existence of England as a sovereign state. In fact, on that crucial day in 1588, when the English floated six fire ships into the Spanish formation and then drove headlong into the ensuing confusion, the future history of Europe was in the balance. In the final outcome, the heavier, sluggish, Spanish ships were no match for the faster, more maneuverable, English craft, and by that evening the Spanish Armada lay in disarray, no longer a threat to England. This naval battle is of particular importance because it was the first in history to be fought by ships on both sides powered completely by sail (in contrast to earlier combinations of oars and sail), and it taught the world that political power was going to be synonymous with naval power. In turn, naval

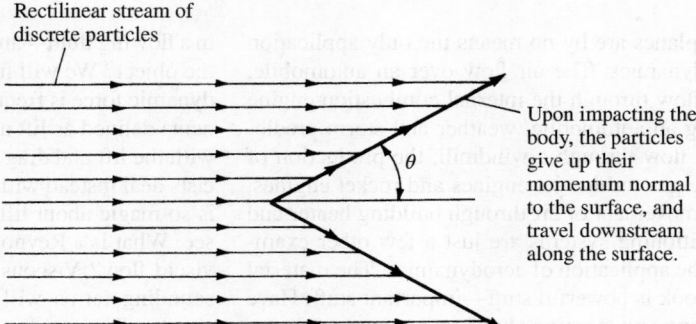

Rectilinear stream of
discrete particles

θ

Upon impacting the
body, the particles
give up their
momentum normal
to the surface, and
travel downstream
along the surface.

Figure 1.7 Isaac Newton's model of fluid flow in the year 1687. This
model was widely adopted in the seventeenth and eighteenth centuries
but was later found to be conceptually inaccurate for most fluid flows.

power was going to depend greatly on the speed and maneuverability of ships. To
increase the speed of a ship, it is important to reduce the resistance created by the water
flow around the ship's hull. Suddenly, the drag on ship hulls became an engineering
problem of great interest, thus giving impetus to the study of fluid mechanics.

This impetus hit its stride almost a century later, when, in 1687, Isaac Newton
(1642–1727) published his famous *Principia,* in which the entire second book
was devoted to fluid mechanics. Newton encountered the same difficulty as others
before him, namely, that the analysis of fluid flow is conceptually more difficult
than the dynamics of solid bodies. A solid body is usually geometrically well defined, and its motion is therefore relatively easy to describe. On the other hand, a
fluid is a "squishy" substance, and in Newton's time it was difficult to decide even
how to qualitatively model its motion, let alone obtain quantitative relationships.
Newton considered a fluid flow as a uniform, rectilinear stream of particles, much
like a cloud of pellets from a shotgun blast. As sketched in Figure 1.7, Newton
assumed that upon striking a surface inclined at an angle θ to the stream, the
particles would transfer their normal momentum to the surface but their tangential momentum would be preserved. Hence, after collision with the surface, the
particles would then move along the surface. This led to an expression for the
hydrodynamic force on the surface which varies as $\sin^2 \theta$. This is Newton's famous sine-squared law (described in detail in Chapter 14). Although its accuracy
left much to be desired, its simplicity led to wide application in naval architecture. Later, in 1777, a series of experiments was carried out by Jean LeRond
d'Alembert (1717–1783), under the support of the French government, in order
to measure the resistance of ships in canals. The results showed that "the rule that
for oblique planes resistance varies with the sine square of the angle of incidence
holds good only for angles between 50 and 90° and must be abandoned for lesser
angles." Also, in 1781, Leonhard Euler (1707–1783) pointed out the physical inconsistency of Newton's model (Figure 1.7) consisting of a rectilinear stream of
particles impacting without warning on a surface. In contrast to this model, Euler

noted that the fluid moving toward a body "*before* reaching the latter, bends its direction and its velocity so that when it reaches the body it flows past it along the surface, and exercises no other force on the body except the pressure corresponding to the single points of contact." Euler went on to present a formula for resistance that attempted to take into account the shear stress distribution along the surface, as well as the pressure distribution. This expression became proportional to $\sin^2 \theta$ for large incidence angles, whereas it was proportional to $\sin \theta$ at small incidence angles. Euler noted that such a variation was in reasonable agreement with the ship-hull experiments carried out by d'Alembert.

This early work in fluid dynamics has now been superseded by modern concepts and techniques. (However, amazingly enough, Newton's sine-squared law has found new application in very high-speed aerodynamics, to be discussed in Chapter 14.) The major point here is that the rapid rise in the importance of naval architecture after the sixteenth century made fluid dynamics an important science, occupying the minds of Newton, d'Alembert, and Euler, among many others. Today, the modern ideas of fluid dynamics, presented in this book, are still driven in part by the importance of reducing hull drag on ships.

Consider a second historical example. The scene shifts to Kill Devil Hills, 4 mi south of Kitty Hawk, North Carolina. It is summer of 1901, and Wilbur and Orville Wright are struggling with their second major glider design, the first being a stunning failure the previous year. The airfoil shape and wing design of their glider are based on aerodynamic data published in the 1890s by the great German aviation pioneer Otto Lilienthal (1848–1896) and by Samuel Pierpont Langley (1834–1906), secretary of the Smithsonian Institution—the most prestigious scientific position in the United States at that time. Because their first glider in 1900 produced no meaningful lift, the Wright brothers have increased the wing area from 165 to 290 ft² and have increased the wing camber (a measure of the airfoil curvature—the larger the camber, the more "arched" is the thin airfoil shape) by almost a factor of 2. But something is still wrong. In Wilbur's words, the glider's "lifting capacity seemed scarcely one-third of the calculated amount." Frustration sets in. The glider is not performing even close to their expectations, although it is designed on the basis of the best available aerodynamic data. On August 20, the Wright brothers despairingly pack themselves aboard a train going back to Dayton, Ohio. On the ride back, Wilbur mutters that "nobody will fly for a thousand years." However, one of the hallmarks of the Wrights is perseverance, and within weeks of returning to Dayton, they decide on a complete departure from their previous approach. Wilbur later wrote that "having set out with absolute faith in the existing scientific data, we were driven to doubt one thing after another, until finally after two years of experiment, we cast it all aside, and decided to rely entirely upon our own investigations." Since their 1901 glider was of poor aerodynamic design, the Wrights set about determining what constitutes good aerodynamic design. In the fall of 1901, they design and build a 6 ft long, 16 in square wind tunnel powered by a two-bladed fan connected to a gasoline engine. A replica of the Wrights' tunnel is shown in Figure 1.8a. In their wind tunnel they test over 200 different wing and airfoil shapes, including flat plates,

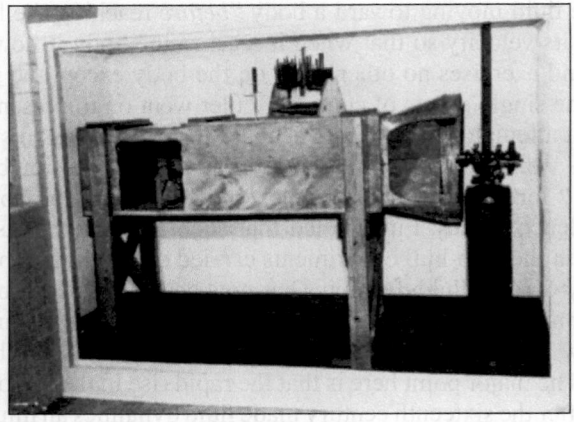

(a)

(b)

Figure 1.8 (a) Replica of the wind tunnel designed, built, and used by the Wright brothers in Dayton, Ohio, during 1901–1902. (b) Wing models tested by the Wright brothers in their wind tunnel during 1901–1902. [(a) NASA; (b) Courtesy of John Anderson.]

curved plates, rounded leading edges, rectangular and curved planforms, and various monoplane and multiplane configurations. A sample of their test models is shown in Figure 1.8b. The aerodynamic data are taken logically and carefully. Armed with their new aerodynamic information, the Wrights design a new glider in the spring of 1902. The airfoil is much more efficient; the camber is reduced considerably, and the location of the maximum rise of the airfoil is moved closer to the front of the wing. The most obvious change, however, is that the ratio of the length of the wing (wingspan) to the distance from the front to the rear of the airfoil (chord length) is increased from 3 to 6. The success of this glider during the summer and fall of 1902 is astounding; Orville and Wilbur accumulate over a thousand flights during this period. In contrast to the previous year, the Wrights return to Dayton flushed with success and devote all their subsequent efforts to powered flight. The rest is history.

The major point here is that good aerodynamics was vital to the ultimate success of the Wright brothers and, of course, to all subsequent successful airplane designs up to the present day. The importance of aerodynamics to successful manned flight goes without saying, and a major thrust of this book is to present the aerodynamic fundamentals that govern such flight.

Consider a third historical example of the importance of aerodynamics, this time as it relates to rockets and space flight. High-speed, supersonic flight had become a dominant feature of aerodynamics by the end of World War II. By this time, aerodynamicists appreciated the advantages of using slender, pointed body shapes to reduce the drag of supersonic vehicles. The more pointed and slender the body, the weaker the shock wave attached to the nose, and hence the smaller the wave drag. Consequently, the German V-2 rocket used during the last stages of World War II had a pointed nose, and all short-range rocket vehicles flown during the next decade followed suit. Then, in 1953, the first hydrogen bomb was exploded by the United States. This immediately spurred the development of long-range intercontinental ballistic missiles (ICBMs) to deliver such bombs. These vehicles were designed to fly outside the region of the earth's atmosphere for distances of 5000 mi or more and to reenter the atmosphere at suborbital speeds of from 20,000 to 22,000 ft/s. At such high velocities, the aerodynamic heating of the reentry vehicle becomes severe, and this heating problem dominated the minds of high-speed aerodynamicists. Their first thinking was conventional—a sharp-pointed, slender reentry body. Efforts to minimize aerodynamic heating centered on the maintenance of laminar boundary layer flow on the vehicle's surface; such laminar flow produces far less heating than turbulent flow (discussed in Chapters 15 and 19). However, nature much prefers turbulent flow, and reentry vehicles are no exception. Therefore, the pointed-nose reentry body was doomed to failure because it would burn up in the atmosphere before reaching the earth's surface.

However, in 1951, one of those major breakthroughs that come very infrequently in engineering was created by H. Julian Allen at the NACA (National Advisory Committee for Aeronautics) Ames Aeronautical Laboratory—he introduced the concept of the *blunt* reentry body. His thinking was paced by the following concepts. At the beginning of reentry, near the outer edge of the atmosphere, the vehicle has a large amount of kinetic energy due to its high velocity and a large amount of potential energy due to its high altitude. However, by the time the vehicle reaches the surface of the earth, its velocity is relatively small and its altitude is zero; hence, it has virtually no kinetic or potential energy. Where has all the energy gone? The answer is that it has gone into (1) heating the body and (2) heating the airflow around the body. This is illustrated in Figure 1.9. Here, the shock wave from the nose of the vehicle heats the airflow around the vehicle; at the same time, the vehicle is heated by the intense frictional dissipation within the boundary layer on the surface. Allen reasoned that if more of the total reentry

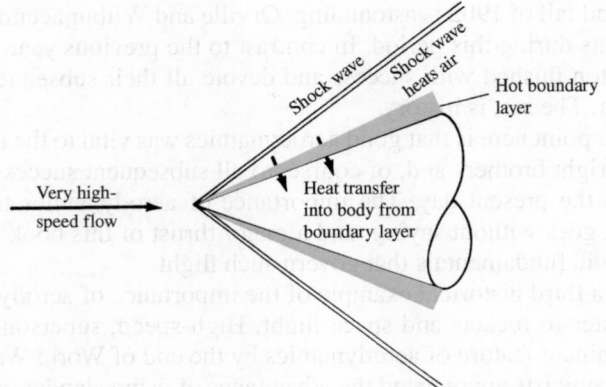

Figure 1.9 Energy of reentry goes into heating both the body and the air around the body.

energy could be dumped into the airflow, then less would be available to be transferred to the vehicle itself in the form of heating. In turn, the way to increase the heating of the airflow is to create a stronger shock wave at the nose (i.e., to use a blunt-nosed body). The contrast between slender and blunt reentry bodies is illustrated in Figure 1.10. This was a stunning conclusion—to minimize aerodynamic heating, you actually want a blunt rather than a slender body. The result was so important that it was bottled up in a secret government document. Moreover, because it was so foreign to contemporary intuition, the blunt-reentry-body concept was accepted only gradually by the technical community. Over the next few years, additional aerodynamic analyses and experiments confirmed the validity of blunt reentry bodies. By 1955, Allen was publicly recognized for his work, receiving the Sylvanus Albert Reed Award of the Institute of the Aeronautical Sciences (now the American Institute of Aeronautics and Astronautics). Finally, in 1958, his work was made available to the public in the pioneering document NACA Report 1381 entitled "A Study of the Motion and Aerodynamic Heating of Ballistic Missiles Entering the Earth's Atmosphere at High Supersonic Speeds." Since Harvey Allen's early work, all successful reentry bodies, from the first Atlas ICBM to the manned Apollo lunar capsule, have been blunt. Incidentally, Allen went on to distinguish himself in many other areas, becoming the director of the NASA Ames Research Center in 1965, and retiring in 1970. His work on the blunt reentry body is an excellent example of the importance of aerodynamics to space vehicle design.

In summary, the purpose of this section has been to underscore the importance of aerodynamics in historical context. The goal of this book is to introduce the fundamentals of aerodynamics and to give the reader a much deeper insight to many technical applications in addition to the few described above. Aerodynamics is also a subject of intellectual beauty, composed and drawn by many great minds over the centuries. If you are challenged and interested by these thoughts, or even the least bit curious, then read on.

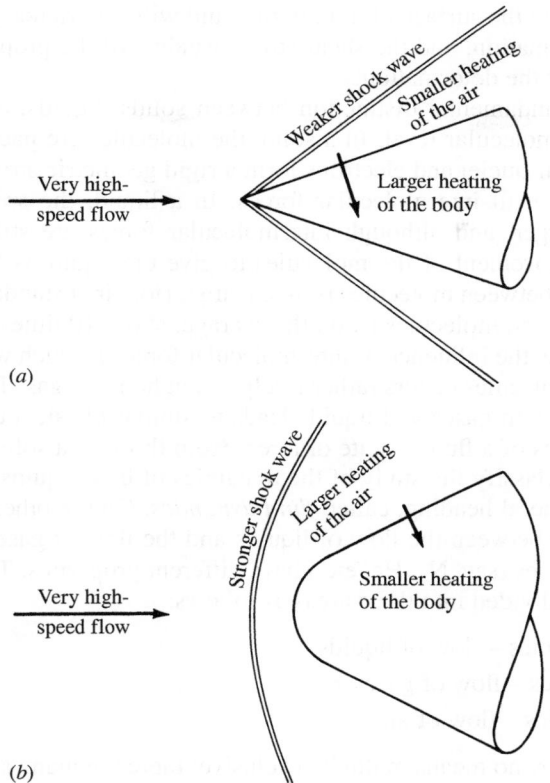

Figure 1.10 Contrast of aerodynamic heating for slender and blunt reentry vehicles. (*a*) Slender reentry body. (*b*) Blunt reentry body.

1.2 AERODYNAMICS: CLASSIFICATION AND PRACTICAL OBJECTIVES

A distinction between solids, liquids, and gases can be made in a simplistic sense as follows. Put a solid object inside a larger, closed container. The solid object will not change; its shape and boundaries will remain the same. Now put a liquid inside the container. The liquid will change its shape to conform to that of the container and will take on the same boundaries as the container up to the maximum depth of the liquid. Now put a gas inside the container. The gas will completely fill the container, taking on the same boundaries as the container.

The word *fluid* is used to denote either a liquid or a gas. A more technical distinction between a solid and a fluid can be made as follows. When a force is applied tangentially to the surface of a solid, the solid will experience a *finite* deformation, and the tangential force per unit area—the shear stress—will usually be proportional to the amount of deformation. In contrast, when a tangential shear stress is applied to the surface of a fluid, the fluid will experience a *continuously increasing* deformation, and the shear stress usually will be proportional to the rate of change of the deformation.

The most fundamental distinction between solids, liquids, and gases is at the atomic and molecular level. In a solid, the molecules are packed so closely together that their nuclei and electrons form a rigid geometric structure, "glued" together by powerful intermolecular forces. In a liquid, the spacing between molecules is larger, and although intermolecular forces are still strong, they allow enough movement of the molecules to give the liquid its "fluidity." In a gas, the spacing between molecules is much larger (for air at standard conditions, the spacing between molecules is, on the average, about 10 times the molecular diameter). Hence, the influence of intermolecular forces is much weaker, and the motion of the molecules occurs rather freely throughout the gas. This movement of molecules in both gases and liquids leads to similar physical characteristics, the characteristics of a fluid—quite different from those of a solid. Therefore, it makes sense to classify the study of the dynamics of both liquids and gases under the same general heading, called *fluid dynamics*. On the other hand, certain differences exist between the flow of liquids and the flow of gases; also, different species of gases (say, N_2, He, etc.) have different properties. Therefore, fluid dynamics is subdivided into three areas as follows:

Hydrodynamics—flow of liquids

Gas dynamics—flow of gases

Aerodynamics—flow of air

These areas are by no means mutually exclusive; there are many similarities and identical phenomena between them. Also, the word "aerodynamics" has taken on a popular usage that sometimes covers the other two areas. As a result, this author tends to interpret the word *aerodynamics* very liberally, and its use throughout this book does *not* always limit our discussions just to air.

Aerodynamics is an applied science with many practical applications in engineering. No matter how elegant an aerodynamic theory may be, or how mathematically complex a numerical solution may be, or how sophisticated an aerodynamic experiment may be, all such efforts are usually aimed at one or more of the following practical objectives:

1. The prediction of forces and moments on, and heat transfer to, bodies moving through a fluid (usually air). For example, we are concerned with the generation of lift, drag, and moments on airfoils, wings, fuselages, engine nacelles, and most importantly, whole airplane configurations. We want to estimate the wind force on buildings, ships, and other surface vehicles. We are concerned with the hydrodynamic forces on surface ships, submarines, and torpedoes. We need to be able to calculate the aerodynamic heating of flight vehicles ranging from the supersonic transport to a planetary probe entering the atmosphere of Jupiter. These are but a few examples.

Figure 1.11 A CO_2-N_2 gas-dynamic laser, circa 1969 (*Courtesy of John Anderson*).

2. Determination of flows moving internally through ducts. We wish to calculate and measure the flow properties inside rocket and air-breathing jet engines and to calculate the engine thrust. We need to know the flow conditions in the test section of a wind tunnel. We must know how much fluid can flow through pipes under various conditions. A recent, very interesting application of aerodynamics is high-energy chemical and gas-dynamic lasers (see Reference 1), which are nothing more than specialized wind tunnels that can produce extremely powerful laser beams. Figure 1.11 is a photograph of an early gas-dynamic laser designed in the late 1960s.

The applications in item 1 come under the heading of *external aerodynamics* since they deal with external flows over a body. In contrast, the applications in item 2 involve *internal aerodynamics* because they deal with flows internally within ducts. In external aerodynamics, in addition to forces, moments, and aero-dynamic heating associated with a body, we are frequently interested in the details of the flow field away from the body. For example, the communication blackout experienced by the space shuttle during a portion of its reentry trajectory is due to a concentration of free electrons in the hot shock layer around the body. We need to calculate the variation of electron density throughout such flow fields. Another example is the propagation of shock waves in a supersonic flow; for instance, does the shock wave from the wing of a supersonic airplane impinge upon and interfere with the tail surfaces? Yet another example is the flow associated with the strong vortices trailing downstream from the wing tips of large subsonic airplanes such as the Boeing 747. What are the properties of these vortices, and how do they affect smaller aircraft which happen to fly through them?

The above is just a sample of the myriad applications of aerodynamics. One purpose of this book is to provide the reader with the technical background necessary to fully understand the nature of such practical aerodynamic applications.

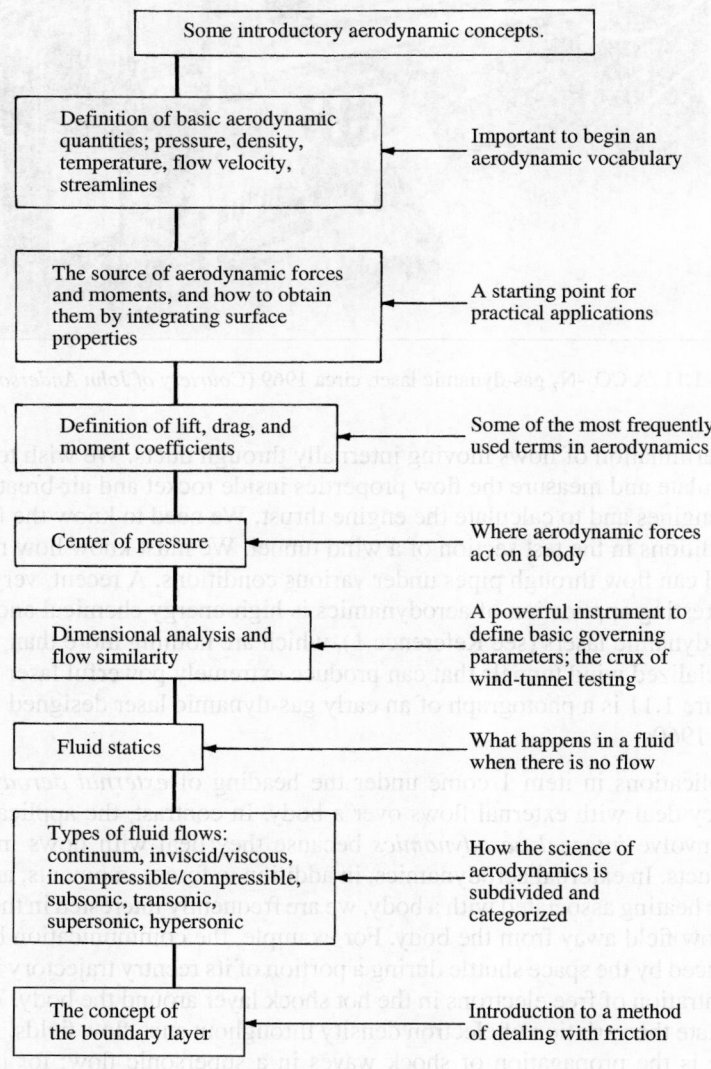

Figure 1.12 Road map for Chapter 1.

1.3 ROAD MAP FOR THIS CHAPTER

When learning a new subject, it is important for you to know where you are, where you are going, and how you can get there. Therefore, at the beginning of each chapter in this book, a road map will be given to help guide you through the material of that chapter and to help you obtain a perspective as to how the material fits within the general framework of aerodynamics. For example, a road map for Chapter 1 is given in Figure 1.12. You will want to frequently refer back to these road maps as you progress through the individual chapters. When you reach the end of each chapter, look back over the road map to see where you started, where you are now, and what you learned in between.

1.4 SOME FUNDAMENTAL AERODYNAMIC VARIABLES

A prerequisite to understanding physical science and engineering is simply learning the vocabulary used to describe concepts and phenomena. Aerodynamics is no exception. Throughout this book, and throughout your working career, you will be adding to your technical vocabulary list. Let us start by defining four of the most frequently used words in aerodynamics: *pressure, density, temperature,* and *flow velocity.*[1]

Consider a surface immersed in a fluid. The surface can be a real, solid surface such as the wall of a duct or the surface of a body; it can also be a free surface which we simply imagine drawn somewhere in the middle of a fluid. Also, keep in mind that the molecules of the fluid are constantly in motion. *Pressure* is the normal force per unit area exerted on a surface due to the time rate of change of momentum of the gas molecules impacting on (or crossing) that surface. It is important to note that even though pressure is defined as force "per unit area," you do not need a surface that is exactly 1 ft^2 or 1 m^2 to talk about pressure. In fact, pressure is usually defined at a *point* in the fluid or a *point* on a solid surface and can vary from one point to another. To see this more clearly, consider a point B in a volume of fluid. Let

dA = elemental area at B

dF = force on one side of dA due to pressure

Then, the pressure at point B in the fluid is defined as

$$p = \lim \left(\frac{dF}{dA} \right) \qquad dA \rightarrow 0$$

The pressure p is the limiting form of the force per unit area, where the area of interest has shrunk to nearly zero at the point B.[2] Clearly, you can see that pressure

---

[1] A basic introduction to these quantities is given on pages 56–61 of Reference 2.

[2] Strictly speaking, dA can never achieve the limit of zero, because there would be no molecules at point B in that case. The above limit should be interpreted as dA approaching a very small value, near zero in terms of our macroscopic thinking, but sufficiently larger than the average spacing between molecules on a microscopic basis.

is a *point property* and can have a different value from one point to another in the fluid.

Another important aerodynamic variable is *density,* defined as the mass per unit volume. Analogous to our discussion on pressure, the definition of density does not require an actual volume of 1 ft³ or 1 m³. Rather, it is a *point property* that can vary from point to point in the fluid. Again, consider a point B in the fluid. Let

$$dv = \text{elemental volume around } B$$

$$dm = \text{mass of fluid inside } dv$$

Then, the density at point B is

$$\rho = \lim \frac{dm}{dv} \qquad dv \to 0$$

Therefore, the density ρ is the limiting form of the mass per unit volume, where the volume of interest has shrunk to nearly zero around point B. (Note that dv cannot achieve the value of zero for the reason discussed in the footnote concerning dA in the definition of pressure.)

Temperature takes on an important role in high-speed aerodynamics (introduced in Chapter 7). The temperature T of a gas is directly proportional to the average kinetic energy of the molecules of the fluid. In fact, if KE is the mean molecular kinetic energy, then temperature is given by $KE = \frac{3}{2}kT$, where k is the Boltzmann constant. Hence, we can qualitatively visualize a high-temperature gas as one in which the molecules and atoms are randomly rattling about at high speeds, whereas in a low-temperature gas, the random motion of the molecules is relatively slow. Temperature is also a point property, which can vary from point to point in the gas.

The principal focus of aerodynamics is fluids in motion. Hence, flow velocity is an extremely important consideration. The concept of the velocity of a fluid is slightly more subtle than that of a solid body in motion. Consider a solid object in translational motion, say, moving at 30 m/s. Then all parts of the solid are simultaneously translating at the same 30 m/s velocity. In contrast, a fluid is a "squishy" substance, and for a fluid in motion, one part of the fluid may be traveling at a different velocity from another part. Hence, we have to adopt a certain perspective, as follows. Consider the flow of air over an airfoil, as shown in Figure 1.13. Lock your eyes on a specific, infinitesimally small element of mass in the gas, called

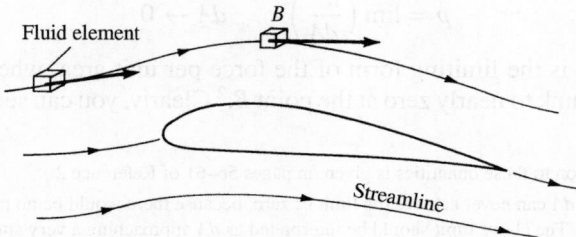

Figure 1.13 Illustration of flow velocity and streamlines.

a *fluid element,* and watch this element move with time. Both the speed and direction of this fluid element can vary as it moves from point to point in the gas. Now fix your eyes on a specific fixed point in space, say, point B in Figure 1.13. *Flow velocity* can now be defined as follows: The velocity of a flowing gas at any fixed point B in space is the velocity of an infinitesimally small fluid element as it sweeps through B. The flow velocity $\mathbf{V}$ has both magnitude and direction; hence, it is a vector quantity. This is in contrast to p, ρ, and T, which are scalar variables. The scalar magnitude of $\mathbf{V}$ is frequently used and is denoted by V. Again, we emphasize that velocity is a point property and can vary from point to point in the flow.

Referring again to Figure 1.13, a moving fluid element traces out a fixed path in space. As long as the flow is steady (i.e., as long as it does not fluctuate with time), this path is called a *streamline* of the flow. Drawing the streamlines of the flow field is an important way of visualizing the motion of the gas; we will frequently be sketching the streamlines of the flow about various objects. A more rigorous discussion of streamlines is given in Chapter 2.

Finally, we note that friction can play a role internally in a flow. Consider two adjacent streamlines a and b as sketched in Figure 1.14. The streamlines are an infinitesimal distance, dy, apart. At point 1 on streamline b, the flow velocity is V; at point 2 on streamline a, the flow velocity is slightly higher, $V + dV$. You can imagine that streamline a is rubbing against streamline b and, due to friction, exerts a force of magnitude dF_f on streamline b acting tangentially toward the right. Furthermore, imagine this force acting on an elemental area dA, where dA is perpendicular to the y axis and tangent to the streamline b at point 1. The local *shear* stress, τ, at point 1 is

$$\tau = \lim \left(\frac{dF_f}{dA} \right) \qquad dA \to 0$$

The shear stress τ is the limiting form of the magnitude of the frictional force per unit area, where the area of interest is perpendicular to the y axis and has shrunk

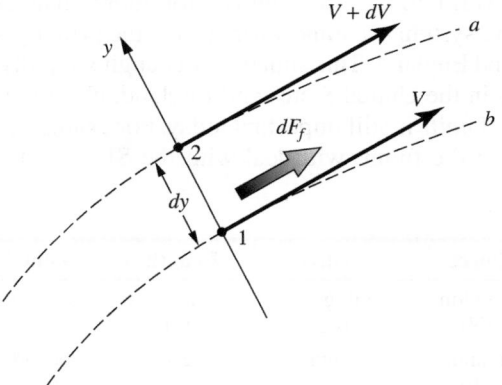

Figure 1.14 Generation of frictional force due to a velocity gradient in a flow.

to nearly zero at point 1. Shear stress acts tangentially along the streamline. For the type of gases and liquids of interest in aerodynamic applications, the value of the shear stress at a point on a streamline is proportional to the spatial rate of change of velocity normal to the streamline at that point (i.e., for the flow illustrated in Figure 1.14, $\tau \propto dV/dy$). The constant of proportionality is defined as the *viscosity coefficient, μ*. Hence,

$$\tau = \mu \frac{dV}{dy}$$

where dV/dy is the velocity gradient. In reality, μ is not really a constant; it is a function of the temperature of the fluid. We will discuss these matters in more detail in Section 1.11. From the above equation, we deduce that in regions of a flow field where the velocity gradients are small, τ is small and the influence of friction locally in the flow is small. On the other hand, in regions where the velocity gradients are large, τ is large and the influence of friction locally in the flow can be substantial.

1.4.1 Units

Two consistent sets of units will be used throughout this book, SI units (Systeme International d'Unites) and the English engineering system of units. The basic units of force, mass, length, time, and absolute temperature in these two systems are given in Table 1.1.

For example, units of pressure and shear stress are lb/ft^2 or N/m^2, units of density are slug/ft^3 or kg/m^3, and units of velocity are ft/s or m/s. When a consistent set of units is used, physical relationships are written without the need for conversion factors in the basic formulae; they are written in the pure form intended by nature. Consistent units will always be used in this book. For an extensive discussion on units and the significance of consistent units versus nonconsistent units, see pages 65–70 of Reference 2.

The SI system of units (metric units) is the standard system of units throughout most of the world today. In contrast, for more than two centuries the English engineering system (or some variant) was the primary system of units in the United States and England. This situation is changing rapidly, especially in the aerospace industry in the United States and England. Nevertheless, a familiarity with both systems of units is still important today. For example, even though most engineering work in the future will deal with the SI units, there exists a huge

Table 1.1

	Force	Mass	Length	Time	Temp.
SI Units	Newton (N)	kilogram (kg)	meter (m)	second (s)	Kelvin (K)
English Engineering Units	pounds (lb)	slug	feet (ft)	second (s)	deg. Rankine (°R)

bulk of present and past engineering literature written in the English engineering system of units, literature that will be used well into the future. The modern engineering student must be bilingual in these units, and must feel comfortable with both systems. For this reason, although many of the worked examples and end-of-the-chapter problems in this book are in the SI units, some are in the English engineering system of units. You are encouraged to join this bilingual spirit and to work to make yourself comfortable in both systems.

1.5 AERODYNAMIC FORCES AND MOMENTS

At first glance, the generation of the aerodynamic force on a giant Boeing 747 may seem complex, especially in light of the complicated three-dimensional flow field over the wings, fuselage, engine nacelles, tail, etc. Similarly, the aerodynamic resistance on an automobile traveling at 55 mi/h on the highway involves a complex interaction of the body, the air, and the ground. However, in these and all other cases, the aerodynamic forces and moments on the body are due to only two basic sources:

1. *Pressure distribution* over the body surface
2. *Shear stress distribution* over the body surface

No matter how complex the body shape may be, the aerodynamic forces and moments on the body are due entirely to the above two basic sources. The *only* mechanisms nature has for communicating a force to a body moving through a fluid are pressure and shear stress distributions on the body surface. Both pressure p and shear stress τ have dimensions of force per unit area (pounds per square foot or newtons per square meter). As sketched in Figure 1.15, p acts *normal* to the surface, and τ acts *tangential* to the surface. Shear stress is due to the "tugging action" on the surface, which is caused by friction between the body and the air (and is studied in great detail in Chapters 15 to 20).

The net effect of the p and τ distributions integrated over the complete body surface is a resultant aerodynamic force R and moment M on the body, as sketched in Figure 1.16. In turn, the resultant R can be split into components, two sets of

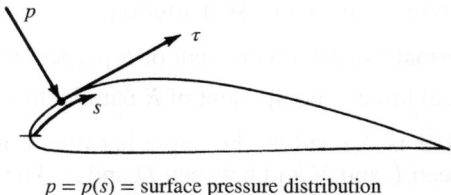

$p = p(s)$ = surface pressure distribution
$\tau = \tau(s)$ = surface shear stress distribution

Figure 1.15 Illustration of pressure and shear stress on an aerodynamic surface.

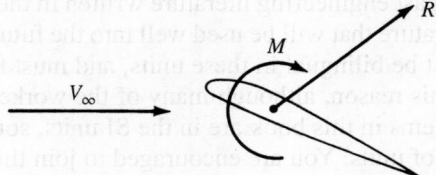

Figure 1.16 Resultant aerodynamic force
and moment on the body.

1.5 AERODYNAMIC FORCES AND MOMENTS

At first glance, the generation of the aerodynamic force on a giant Boeing 747 may
seem complex, especially in light of the complicated three-dimensional flow field
over the vehicle, that is, the airplane nacelles, etc. Similarly, the aerodynamic resistance on an automobile traveling at 55 mi/h on the highway involves a complex
interaction of the body, the ground, and the air around. However, for these and all other
cases, the aerodynamic forces and moments on the body are due to only two basic
sources:

1. Pressure distribution over the body surface
2. Shear stress distribution over the body surface

No matter how complex the body shape may be, the aerodynamic forces and
moments on the body are due entirely to the above two basic sources. The only
mechanisms nature has for communicating a force to a body moving through a
fluid are pressure and shear stress distributions on the body surface. Both pressure
p and shear stress τ have dimensions of force per unit area (pounds per square foot

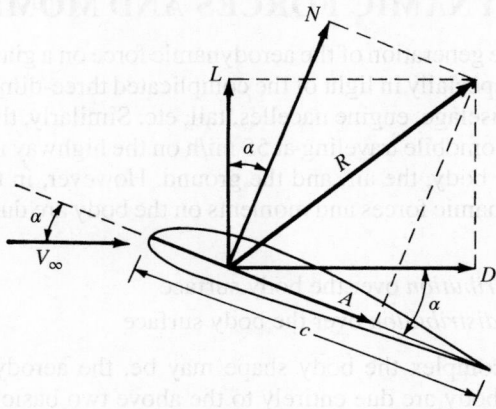

Figure 1.17 Resultant aerodynamic force and the
components into which it splits.

which are shown in Figure 1.17. In Figure 1.17, V_∞ is the *relative wind,* defined as
the flow velocity far ahead of the body. The flow far away from the body is called
the *freestream,* and hence V_∞ is also called the freestream velocity. In Figure 1.17,
by definition,

$$L \equiv \text{lift} \equiv \text{component of } R \text{ perpendicular to } V_\infty$$

$$D \equiv \text{drag} \equiv \text{component of } R \text{ parallel to } V_\infty$$

The chord c is the linear distance from the leading edge to the trailing edge of
the body. Sometimes, R is split into components perpendicular and parallel to the
chord, as also shown in Figure 1.17. By definition,

$$N \equiv \text{normal force} \equiv \text{component of } R \text{ perpendicular to } c$$

$$A \equiv \text{axial force} \equiv \text{component of } R \text{ parallel to } c$$

The angle of attack α is defined as the angle between c and V_∞. Hence, α is
also the angle between L and N and between D and A. The geometrical relation
between these two sets of components is, from Figure 1.17,

$$L = N \cos \alpha - A \sin \alpha \tag{1.1}$$

$$D = N \sin \alpha + A \cos \alpha \tag{1.2}$$

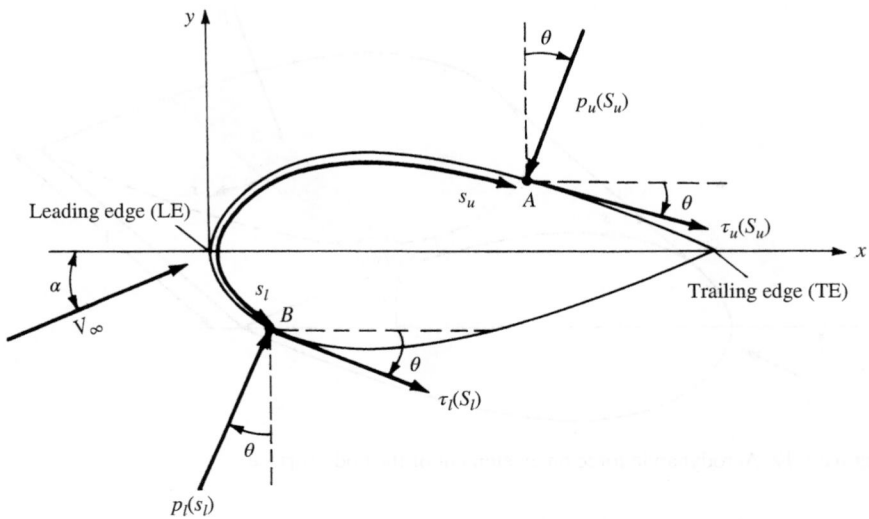

Figure 1.18 Nomenclature for the integration of pressure and shear stress distributions over a two-dimensional body surface.

Let us examine in more detail the integration of the pressure and shear stress distributions to obtain the aerodynamic forces and moments. Consider the two-dimensional body sketched in Figure 1.18. The chord line is drawn horizontally, and hence the relative wind is inclined relative to the horizontal by the angle of attack α. An xy coordinate system is oriented parallel and perpendicular, respectively, to the chord. The distance from the leading edge measured along the body surface to an arbitrary point A on the upper surface is s_u; similarly, the distance to an arbitrary point B on the lower surface is s_l. The pressure and shear stress on the upper surface are denoted by p_u and τ_u, both p_u and τ_u are functions of s_u. Similarly, p_l and τ_l are the corresponding quantities on the lower surface and are functions of s_l. At a given point, the pressure is normal to the surface and is oriented at an angle θ relative to the perpendicular; shear stress is tangential to the surface and is oriented at the same angle θ relative to the horizontal. In Figure 1.18, the sign convention for θ is positive when measured *clockwise* from the vertical line to the direction of p and from the horizontal line to the direction of τ. In Figure 1.18, all thetas are shown in their positive direction. Now consider the two-dimensional shape in Figure 1.18 as a cross section of an infinitely long cylinder of uniform section. A unit span of such a cylinder is shown in Figure 1.19. Consider an elemental surface area dS of this cylinder, where $dS = (ds)(1)$ as shown by the shaded area in Figure 1.19. We are interested in the contribution to the total normal force N' and the total axial force A' due to the pressure and shear stress on the elemental area dS. The primes on N' and A' denote force per unit span. Examining both Figures 1.18 and 1.19, we see that

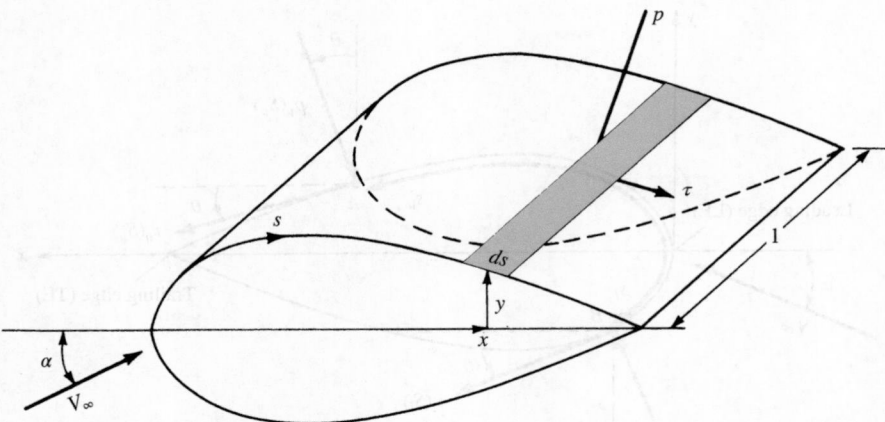

Figure 1.19 Aerodynamic force on an element of the body surface.

the elemental normal and axial forces acting on the elemental surface dS on the *upper* body surface are

$$dN'_u = -p_u ds_u \cos\theta - \tau_u ds_u \sin\theta \qquad (1.3)$$

$$dA'_u = -p_u ds_u \sin\theta + \tau_u ds_u \cos\theta \qquad (1.4)$$

On the *lower* body surface, we have

$$dN'_l = p_l ds_l \cos\theta - \tau_l ds_l \sin\theta \qquad (1.5)$$

$$dA'_l = p_l ds_l \sin\theta + \tau_l ds_l \cos\theta \qquad (1.6)$$

In Equations (1.3) to (1.6), the positive directions of N' and A' are those shown in Figure 1.17. In these equations, the positive clockwise convention for θ must be followed. For example, consider again Figure 1.18. Near the leading edge of the body, where the slope of the upper body surface is positive, τ is inclined upward, and hence it gives a positive contribution to N'. For an upward inclined τ, θ would be counterclockwise, hence negative. Therefore, in Equation (1.3), $\sin\theta$ would be negative, making the shear stress term (the last term) a positive value, as it should be in this instance. Hence, Equations (1.3) to (1.6) hold in general (for both the forward and rearward portions of the body) as long as the above sign convention for θ is consistently applied.

The total normal and axial forces *per unit span* are obtained by integrating Equations (1.3) to (1.6) from the leading edge (LE) to the trailing edge (TE):

$$N' = -\int_{LE}^{TE} (p_u \cos\theta + \tau_u \sin\theta)\, ds_u + \int_{LE}^{TE} (p_l \cos\theta - \tau_l \sin\theta)\, ds_l \quad (1.7)$$

$$A' = \int_{LE}^{TE} (-p_u \sin\theta + \tau_u \cos\theta)\, ds_u + \int_{LE}^{TE} (p_l \sin\theta + \tau_l \cos\theta)\, ds_l \quad (1.8)$$

Figure 1.20 Sign convention for aerodynamic moments.

In turn, the total lift and drag per unit span can be obtained by inserting Equations (1.7) and (1.8) into (1.1) and (1.2); note that Equations (1.1) and (1.2) hold for forces on an arbitrarily shaped body (unprimed) and for the forces per unit span (primed).

The aerodynamic moment exerted on the body depends on the point about which moments are taken. Consider moments taken about the leading edge. By convention, moments that tend to increase α (pitch up) are positive, and moments that tend to decrease α (pitch down) are negative. This convention is illustrated in Figure 1.20. Returning again to Figures 1.18 and 1.19, the moment per unit span about the leading edge due to p and τ on the elemental area dS on the upper surface is

$$dM'_u = (p_u \cos\theta + \tau_u \sin\theta)x \, ds_u + (-p_u \sin\theta + \tau_u \cos\theta)y \, ds_u \quad (1.9)$$

On the bottom surface,

$$dM'_l = (-p_l \cos\theta + \tau_l \sin\theta)x \, ds_l + (p_l \sin\theta + \tau_l \cos\theta)y \, ds_l \quad (1.10)$$

In Equations (1.9) and (1.10), note that the same sign convention for θ applies as before and that y is a positive number above the chord and a negative number below the chord. Integrating Equations (1.9) and (1.10) from the leading to the trailing edges, we obtain for the moment about the leading edge per unit span

$$M'_{LE} = \int_{LE}^{TE} [(p_u \cos\theta + \tau_u \sin\theta)x - (p_u \sin\theta - \tau_u \cos\theta)y] \, ds_u$$

$$(1.11)$$

$$+ \int_{LE}^{TE} [(-p_l \cos\theta + \tau_l \sin\theta)x + (p_l \sin\theta + \tau_l \cos\theta)y] \, ds_l$$

In Equations (1.7), (1.8), and (1.11), θ, x, and y are known functions of s for a given body shape. Hence, if p_u, p_l, τ_u, and τ_l are known as functions of s (from theory or experiment), the integrals in these equations can be evaluated. Clearly, Equations (1.7), (1.8), and (1.11) demonstrate the principle stated earlier, namely, *the sources of the aerodynamic lift, drag, and moments on a body are the pressure and shear stress distributions integrated over the body.* A major goal of theoretical aerodynamics is to calculate $p(s)$ and $\tau(s)$ for a given body shape and freestream conditions, thus yielding the aerodynamic forces and moments via Equations (1.7), (1.8), and (1.11).

As our discussions of aerodynamics progress, it will become clear that there are quantities of an even more fundamental nature than the aerodynamic forces and moments themselves. These are *dimensionless force and moment coefficients,* defined as follows. Let ρ_∞ and V_∞ be the density and velocity, respectively, in the freestream, far ahead of the body. We define a dimensional quantity called the freestream *dynamic pressure* as

Dynamic pressure: $\qquad\qquad q_\infty \equiv \frac{1}{2}\rho_\infty V_\infty^2$

The dynamic pressure has the units of pressure (i.e., pounds per square foot or newtons per square meter). In addition, let S be a reference area and l be a reference length. The dimensionless force and moment coefficients are defined as follows:

Lift coefficient: $\qquad\qquad C_L \equiv \dfrac{L}{q_\infty S}$

Drag coefficient: $\qquad\qquad C_D \equiv \dfrac{D}{q_\infty S}$

Normal force coefficient: $\qquad C_N \equiv \dfrac{N}{q_\infty S}$

Axial force coefficient: $\qquad C_A \equiv \dfrac{A}{q_\infty S}$

Moment coefficient: $\qquad\quad C_M \equiv \dfrac{M}{q_\infty Sl}$

In the above coefficients, the reference area S and reference length l are chosen to pertain to the given geometric body shape; for different shapes, S and l may be different things. For example, for an airplane wing, S is the planform area, and l is the mean chord length, as illustrated in Figure 1.21a. However, for a sphere, S is the cross-sectional area, and l is the diameter, as shown in Figure 1.21b.

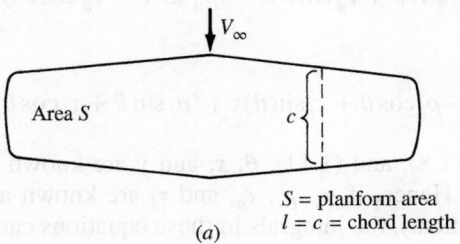

Area S $\qquad\qquad\qquad\qquad c$

S = planform area
$l = c$ = chord length
(a)

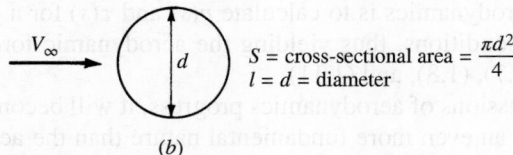

S = cross-sectional area $= \dfrac{\pi d^2}{4}$
$l = d$ = diameter
(b)

Figure 1.21 Some reference areas and lengths.

The particular choice of reference area and length is not critical; however, when using force and moment coefficient data, you must always know what reference quantities the particular data are based upon.

The symbols in capital letters listed above (i.e., C_L, C_D, C_M, and C_A) denote the force and moment coefficients for a complete three-dimensional body such as an airplane or a finite wing. In contrast, for a two-dimensional body, such as given in Figures 1.18 and 1.19, the forces and moments are per unit span. For these two-dimensional bodies, it is conventional to denote the aerodynamic coefficients by lowercase letters; for example,

$$c_l \equiv \frac{L'}{q_\infty c} \quad c_d \equiv \frac{D'}{q_\infty c} \quad c_m \equiv \frac{M'}{q_\infty c^2}$$

where the reference area $S = c(1) = c$.

Two additional dimensionless quantities of immediate use are

Pressure coefficient: $\qquad C_p \equiv \dfrac{p - p_\infty}{q_\infty}$

Skin friction coefficient: $\qquad c_f \equiv \dfrac{\tau}{q_\infty}$

where p_∞ is the freestream pressure.

The most useful forms of Equations (1.7), (1.8), and (1.11) are in terms of the dimensionless coefficients introduced above. From the geometry shown in Figure 1.22,

$$dx = ds \cos \theta \tag{1.12}$$

$$dy = -(ds \sin \theta) \tag{1.13}$$

$$S = c(1) \tag{1.14}$$

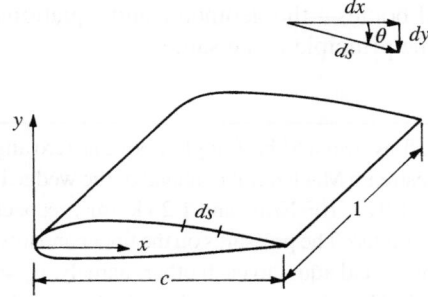

Figure 1.22 Geometrical relationship of differential lengths.

Substituting Equations (1.12) and (1.13) into Equations (1.7), (1.8), and (1.11); dividing by q_∞; and further dividing by S in the form of Equation (1.14), we obtain the following integral forms for the force and moment coefficients:

$$c_n = \frac{1}{c} \left[\int_0^c (C_{p,l} - C_{p,u}) \, dx + \int_0^c \left(c_{f,u} \frac{dy_u}{dx} + c_{f,l} \frac{dy_l}{dx} \right) dx \right] \tag{1.15}$$

$$c_a = \frac{1}{c} \left[\int_0^c \left(C_{p,u} \frac{dy_u}{dx} - C_{p,l} \frac{dy_l}{dx} \right) dx + \int_0^c (c_{f,u} + c_{f,l}) \, dx \right] \tag{1.16}$$

$$c_{m_{\text{LE}}} = \frac{1}{c^2} \left[\int_0^c (C_{p,u} - C_{p,l})x \, dx - \int_0^c \left(c_{f,u} \frac{dy_u}{dx} + c_{f,l} \frac{dy_l}{dx} \right) x \, dx \right.$$
$$\left. + \int_0^c \left(C_{p,u} \frac{dy_u}{dx} + c_{f,u} \right) y_u \, dx + \int_0^c \left(-C_{p,l} \frac{dy_l}{dx} + c_{f,l} \right) y_l \, dx \right] \tag{1.17}$$

The simple algebraic steps are left as an exercise for the reader. When evaluating these integrals, keep in mind that y_u is directed above the x axis, and hence is positive, whereas y_l is directed below the x axis, and hence is negative. Also, dy/dx on both the upper and lower surfaces follow the usual rule from calculus (i.e., positive for those portions of the body with a positive slope and negative for those portions with a negative slope).

The lift and drag coefficients can be obtained from Equations (1.1) and (1.2) cast in coefficient form:

$$c_l = c_n \cos \alpha - c_a \sin \alpha \tag{1.18}$$

$$c_d = c_n \sin \alpha + c_a \cos \alpha \tag{1.19}$$

Integral forms for c_l and c_d are obtained by substituting Equations (1.15) and (1.16) into (1.18) and (1.19).

It is important to note from Equations (1.15) through (1.19) that the aerodynamic force and moment coefficients can be obtained by integrating the pressure and skin friction coefficients over the body. This is a common procedure in both theoretical and experimental aerodynamics. In addition, although our derivations have used a two-dimensional body, an analogous development can be presented for three-dimensional bodies—the geometry and equations only get more complex and involved—the principle is the same.

EXAMPLE 1.1

Consider the supersonic flow over a 5° half-angle wedge at zero angle of attack, as sketched in Figure 1.23a. The freestream Mach number ahead of the wedge is 2.0, and the freestream pressure and density are 1.01×10^5 N/m² and 1.23 kg/m³, respectively (this corresponds to standard sea level conditions). The pressures on the upper and lower surfaces of the wedge are constant with distance s and equal to each other, namely, $p_u = p_l = 1.31 \times 10^5$ N/m², as shown in Figure 1.23b. The pressure exerted on the base of the wedge is equal to p_∞. As seen in Figure 1.23c, the shear stress varies over both the upper and lower surfaces as $\tau_w = 431s^{-0.2}$. The chord length, c, of the wedge is 2 m. Calculate the drag coefficient for the wedge.

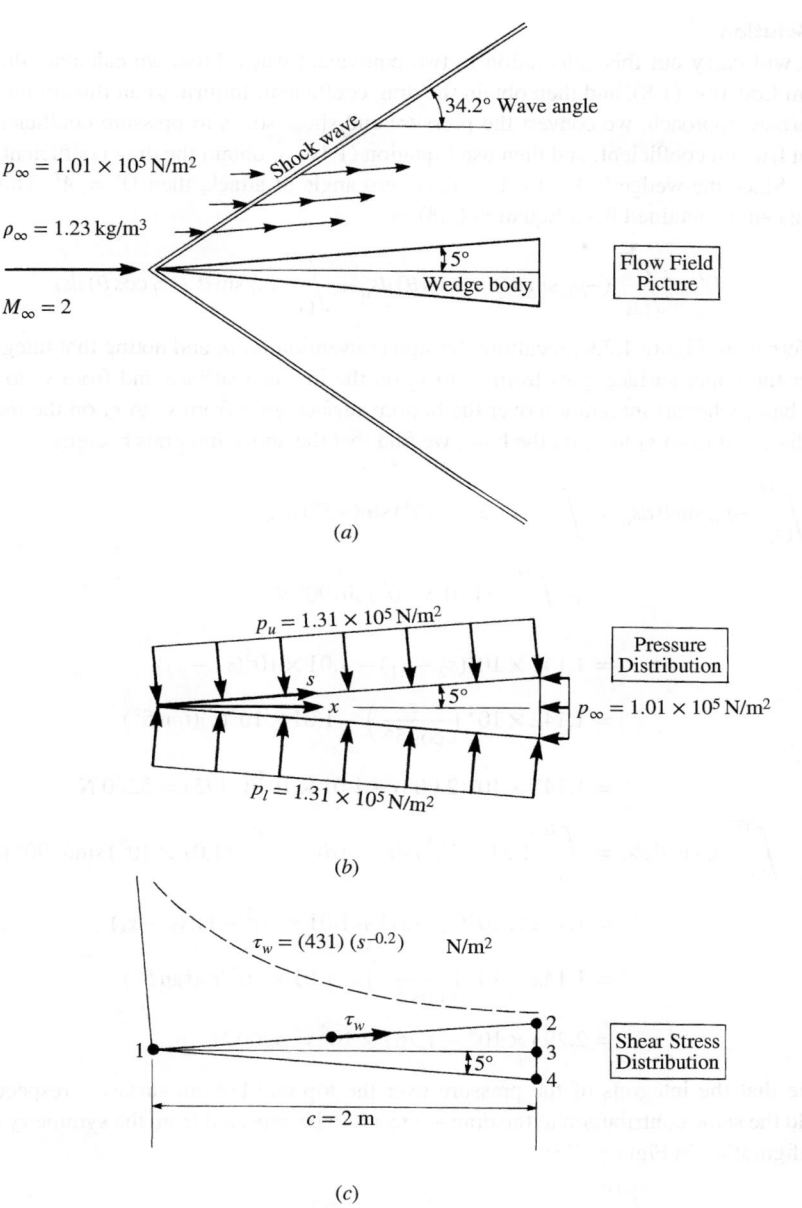

Figure 1.23 Illustration for Example 1.1.

■ Solution

We will carry out this calculation in two equivalent ways. First, we calculate the drag from Equation (1.8), and then obtain the drag coefficient. In turn, as an illustration of an alternate approach, we convert the pressure and shear stress to pressure coefficient and skin friction coefficient, and then use Equation (1.16) to obtain the drag coefficient.

Since the wedge in Figure 1.23 is at zero angle of attack, then $D' = A'$. Thus, the drag can be obtained from Equation (1.8) as

$$D' = \int_{LE}^{TE} (-p_u \sin\theta + \tau_u \cos\theta)\, ds_u + \int_{LE}^{TE} (p_l \sin\theta + \tau_l \cos\theta)\, ds_l$$

Referring to Figure 1.23c, recalling the sign convention for θ, and noting that integration over the upper surface goes from s_1 to s_2 on the inclined surface and from s_2 to s_3 on the base, whereas integration over the bottom surface goes from s_1 to s_4 on the inclined surface and from s_4 to s_3 on the base, we find that the above integrals become

$$\int_{LE}^{TE} -p_u \sin\theta\, ds_u = \int_{s_1}^{s_2} -(1.31 \times 10^5) \sin(-5°)\, ds_u$$

$$+ \int_{s_2}^{s_3} -(1.01 \times 10^5) \sin 90°\, ds_u$$

$$= 1.142 \times 10^4 (s_2 - s_1) - 1.01 \times 10^5 (s_3 - s_2)$$

$$= 1.142 \times 10^4 \left(\frac{c}{\cos 5°} \right) - 1.01 \times 10^5 (c)(\tan 5°)$$

$$= 1.142 \times 10^4 (2.008) - 1.01 \times 10^5 (0.175) = 5260 \text{ N}$$

$$\int_{LE}^{TE} p_l \sin\theta\, ds_l = \int_{s_1}^{s_4} (1.31 \times 10^5) \sin(5°)\, ds_l + \int_{s_4}^{s_3} (1.01 \times 10^5) \sin(-90°)\, ds_l$$

$$= 1.142 \times 10^4 (s_4 - s_1) + 1.01 \times 10^5 (-1)(s_3 - s_4)$$

$$= 1.142 \times 10^4 \left(\frac{c}{\cos 5°} \right) - 1.01 \times 10^5 (c)(\tan 5°)$$

$$= 2.293 \times 10^4 - 1.767 \times 10^4 = 5260 \text{ N}$$

Note that the integrals of the pressure over the top and bottom surfaces, respectively, yield the same contribution to the drag—a result to be expected from the symmetry of the configuration in Figure 1.23:

$$\int_{LE}^{TE} \tau_u \cos\theta\, ds_u = \int_{s_1}^{s_2} 431 s^{-0.2} \cos(-5°)\, ds_u$$

$$= 429 \left(\frac{s_2^{0.8} - s_1^{0.8}}{0.8} \right)$$

$$= 429 \left(\frac{c}{\cos 5°} \right)^{0.8} \frac{1}{0.8} = 936.5 \text{ N}$$

$$\int_{LE}^{TE} \tau_l \cos\theta \, ds_l = \int_{s_1}^{s_4} 431 s^{-0.2} \cos(-5°) \, ds_l$$

$$= 429\left(\frac{s_4^{0.8} - s_1^{0.8}}{0.8}\right)$$

$$= 429\left(\frac{c}{\cos 5°}\right)^{0.8} \frac{1}{0.8} = 936.5 \text{ N}$$

Again, it is no surprise that the shear stresses acting over the upper and lower surfaces, respectively, give equal contributions to the drag; this is to be expected due to the symmetry of the wedge shown in Figure 1.23. Adding the pressure integrals, and then adding the shear stress integrals, we have for total drag

$$D' = \underbrace{1.052 \times 10^4}_{\substack{\text{pressure}\\\text{drag}}} + \underbrace{0.1873 \times 10^4}_{\substack{\text{skin friction}\\\text{drag}}} = \boxed{1.24 \times 10^4 \text{ N}}$$

Note that, for this rather slender body, but at a supersonic speed, most of the drag is pressure drag. Referring to Figure 1.23a, we see that this is due to the presence of an oblique shock wave from the nose of the body, which acts to create pressure drag (sometimes called *wave drag*). In this example, only 15 percent of the drag is skin friction drag; the other 85 percent is the pressure drag (wave drag). This is typical of the drag of slender supersonic bodies. In contrast, as we will see later, the drag of a slender body at subsonic speed, where there is no shock wave, is mainly skin friction drag.

The drag coefficient is obtained as follows. The velocity of the freestream is twice the sonic speed, which is given by

$$a_\infty = \sqrt{\gamma R T_\infty} = \sqrt{(1.4)(287)(288)} = 340.2 \text{ m/s}$$

See Chapter 8 for a derivation of this expression for the speed of sound.) Note that, in the above equation, the standard sea level temperature of 288 K is used. Hence, $V_\infty = 2(340.2) = 680.4$ m/s. Thus,

$$q_\infty = \tfrac{1}{2}\rho_\infty V_\infty^2 = (0.5)(1.23)(680.4)^2 = 2.847 \times 10^5 \text{ N/m}^2$$

Also,
$$S = c(1) = 2.0 \text{ m}^2$$

Hence,
$$c_d = \frac{D'}{q_\infty S} = \frac{1.24 \times 10^4}{(2.847 \times 10^5)(2)} = \boxed{0.022}$$

An alternate solution to this problem is to use Equation (1.16), integrating the pressure coefficients and skin friction coefficients to obtain directly the drag coefficient. We proceed as follows:

$$C_{p,u} = \frac{p_u - p_\infty}{q_\infty} = \frac{1.31 \times 10^5 - 1.01 \times 10^5}{2.847 \times 10^5} = 0.1054$$

On the lower surface, we have the same value for C_p, that is,

$$C_{p,l} = C_{p,u} = 0.1054$$

Also,

$$c_{f,u} = \frac{\tau_w}{q_\infty} = \frac{431s^{-0.2}}{q_\infty} = \frac{431}{2.847 \times 10^5} \left(\frac{x}{\cos 5°}\right)^{-0.2} = 1.513 \times 10^{-3} x^{-0.2}$$

On the lower surface, we have the same value for c_f, that is,

$$c_{f,l} = 1.513 \times 10^{-3} x^{-0.2}$$

Also,

$$\frac{dy_u}{dx} = \tan 5° = 0.0875$$

and

$$\frac{dy_l}{dx} = -\tan 5° = -0.0875$$

Inserting the above information into Equation (1.16), we have

$$c_d = c_a = \frac{1}{c} \int_0^c \left(C_{p,u}\frac{dy_u}{dx} - C_{p,l}\frac{dy_l}{dx}\right) dx + \frac{1}{c}\int_0^c (c_{f,u} + c_{f,l})\, dx$$

$$= \frac{1}{2}\int_0^2 [(0.1054)(0.0875) - (0.1054)(-0.0875)]\, dx$$

$$+ \frac{1}{2}\int_0^2 2(1.513 \times 10^{-3})x^{-0.2}\, dx$$

$$= 0.009223x \Big|_0^2 + 0.00189x^{0.8} \Big|_0^2$$

$$= 0.01854 + 0.00329 = \boxed{0.022}$$

This is the same result as obtained earlier.

EXAMPLE 1.2

Consider a cone at zero angle of attack in a hypersonic flow. (Hypersonic flow is very high-speed flow, generally defined as any flow above a Mach number of 5; hypersonic flow is further defined in Section 1.10.) The half-angle of the cone is θ_c, as shown in Figure 1.24. An approximate expression for the pressure coefficient on the surface of a hypersonic body is given by the newtonian sine-squared law (to be derived in Chapter 14):

$$C_p = 2\sin^2\theta_c$$

Note that C_p, hence, p, is constant along the inclined surface of the cone. Along the base of the body, we assume that $p = p_\infty$. Neglecting the effect of friction, obtain an expression for the drag coefficient of the cone, where C_D is based on the area of the base S_b.

■ **Solution**

We cannot use Equations (1.15) to (1.17) here. These equations are expressed for a two-dimensional body, such as the airfoil shown in Figure 1.22, whereas the cone in Figure 1.24 is a shape in three-dimensional space. Hence, we must treat this three-dimensional body as follows. From Figure 1.24, the drag force on the shaded strip of surface area is

$$(p\sin\theta_c)(2\pi r)\frac{dr}{\sin\theta_c} = 2\pi rp\, dr$$

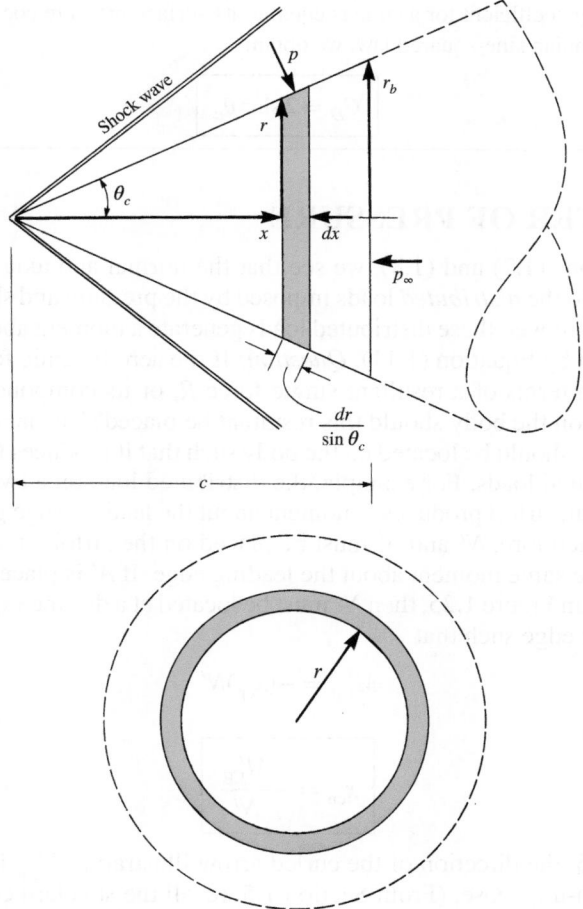

Figure 1.24 Illustration for Example 1.2.

The total drag due to the pressure acting over the total surface area of the cone is

$$D = \int_0^{r_b} 2\pi r p \, dr - \int_0^{r_b} r 2\pi p_\infty \, dr$$

The first integral is the horizontal force on the inclined surface of the cone, and the second integral is the force on the base of the cone. Combining the integrals, we have

$$D = \int_0^{r_b} 2\pi r (p - p_\infty) \, dr = \pi (p - p_\infty) r_b^2$$

Referenced to the base area, πr_b^2, the drag coefficient is

$$C_D = \frac{D}{q_\infty \pi r_b^2} = \frac{\pi r_b^2 (p - p_\infty)}{\pi r_b^2 q_\infty} = C_p$$

(*Note:* The drag coefficient for a cone is equal to its surface pressure coefficient.) Hence, using the newtonian sine-squared law, we obtain

$$C_D = 2 \sin^2 \theta_c$$

1.6 CENTER OF PRESSURE

From Equations (1.7) and (1.8), we see that the normal and axial forces on the body are due to the *distributed* loads imposed by the pressure and shear stress distributions. Moreover, these distributed loads generate a moment about the leading edge, as given by Equation (1.11). *Question:* If the aerodynamic force on a body is specified in terms of a resultant single force R, or its components such as N and A, *where* on the body should this resultant be placed? The answer is that the resultant force should be located on the body such that it produces the same effect as the distributed loads. For example, the distributed load on a two-dimensional body such as an airfoil produces a moment about the leading edge given by Equation (1.11); therefore, N' and A' must be placed on the airfoil at such a location to generate the same moment about the leading edge. If A' is placed on the chord line as shown in Figure 1.25, then N' must be located at a distance x_{cp} downstream of the leading edge such that

$$M'_{LE} = -(x_{cp})N'$$

$$x_{cp} = -\frac{M'_{LE}}{N'} \qquad (1.20)$$

In Figure 1.25, the direction of the curled arrow illustrating M'_{LE} is drawn in the positive (pitch-up) sense. (From Section 1.5, recall the standard convention that aerodynamic moments are positive if they tend to increase the angle of attack.) Examining Figure 1.25, we see that a positive N' creates a negative (pitch-down) moment about the leading edge. This is consistent with the negative sign in Equation (1.20). Therefore, in Figure 1.25, the actual moment about the leading edge is negative, and hence is in a direction opposite to the curled arrow shown.

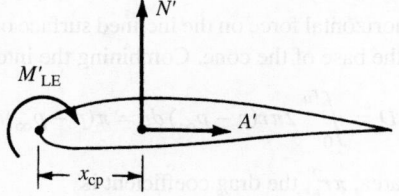

Figure 1.25 Center of pressure for an airfoil.

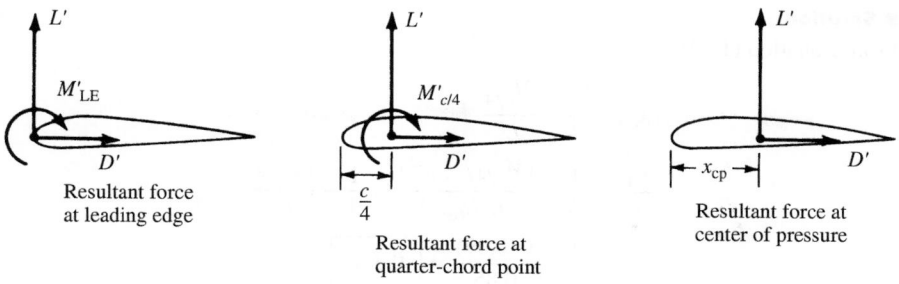

Figure 1.26 Equivalent ways of specifying the force-and-moment system on an airfoil.

In Figure 1.25 and Equation (1.20), x_{cp} is defined as the *center of pressure*. It is the location where the resultant of a distributed load effectively acts on the body. If moments were taken about the center of pressure, the integrated effect of the distributed loads would be zero. Hence, an alternate definition of the center of pressure is that point on the body about which the aerodynamic moment is zero.

In cases where the angle of attack of the body is small, $\sin \alpha \approx 0$ and $\cos \alpha \approx 1$; hence, from Equation (1.1), $L' \approx N'$. Thus, Equation (1.20) becomes

$$x_{cp} \approx -\frac{M'_{LE}}{L'} \tag{1.21}$$

Examine Equations (1.20) and (1.21). As N' and L' decrease, x_{cp} increases. As the forces approach zero, the center of pressure moves to infinity. For this reason, the center of pressure is not always a convenient concept in aerodynamics. However, this is no problem. To define the force-and-moment system due to a distributed load on a body, the resultant force can be placed at *any* point on the body, as long as the value of the moment about that point is also given. For example, Figure 1.26 illustrates three equivalent ways of specifying the force-and-moment system on an airfoil. In the left figure, the resultant is placed at the leading edge, with a finite value of M'_{LE}. In the middle figure, the resultant is placed at the quarter-chord point, with a finite value of $M'_{c/4}$. In the right figure, the resultant is placed at the center of pressure, with a zero moment about that point. By inspection of Figure 1.26, the quantitative relation between these cases is

$$M'_{LE} = -\frac{c}{4}L' + M'_{c/4} = -x_{cp}L' \tag{1.22}$$

EXAMPLE 1.3

In low-speed, incompressible flow, the following experimental data are obtained for an NACA 4412 airfoil section at an angle of attack of $4°$: $c_l = 0.85$ and $c_{m,c/4} = -0.09$. Calculate the location of the center of pressure.

■ Solution

From Equation (1.22),

$$x_{cp} = \frac{c}{4} - \frac{M'_{c/4}}{L'}$$

$$\frac{x_{cp}}{c} = \frac{1}{4} - \frac{(M_{c/4}/q_\infty c^2)}{(L'/q_\infty c)} = \frac{1}{4} - \frac{c_{m,c/4}}{c_l}$$

$$= \frac{1}{4} - \frac{(-0.09)}{0.85} = \boxed{0.356}$$

(*Note:* In Chapter 4, we will learn that, for a thin, symmetrical airfoil, the center of pressure is at the quarter-chord location. However, for the NACA 4412 airfoil, which is not symmetric, the center-of-pressure location is behind the quarter-chord point.)

EXAMPLE 1.4

Consider the DC-3 shown in Figure 1.1. Just outboard of the engine nacelle, the airfoil chord length is 15.4 ft. At cruising velocity (188 mi/h) at sea level, the moments per unit span at this airfoil location are $M'_{c/4} = -1071$ ft lb/ft and $M'_{LE} = -3213.9$ ft lb/ft. Calculate the lift per unit span and the location of the center of pressure on the airfoil.

■ Solution

From Equation (1.22),

$$\frac{c}{4}L' = M'_{c/4} - M'_{LE} = -1071 - (-3213.9) = 2142.9$$

At this airfoil location on the wing, $\dfrac{c}{4} = \dfrac{15.4}{4} = 3.85$ ft.

Thus,

$$L' = \frac{2142.9}{3.85} = \boxed{556.6 \text{ lb/ft}}$$

Returning to Equation (1.22),

$$-x_{cp}L' = M'_{LE}$$

$$-x_{cp} = -\frac{M'_{LE}}{L'} = -\frac{(-3213.9)}{556.6} = \boxed{5.774 \text{ ft}}$$

Note: In this section, we have shown that the force and moment system acting on an airfoil is uniquely specified by giving the lift acting at any point on the airfoil and the moment about that point. Analogously, this example proves that the force and moment system is also uniquely specified by giving the moments acting about any two points on the airfoil.

1.7 DIMENSIONAL ANALYSIS: THE BUCKINGHAM PI THEOREM

The aerodynamic forces and moments on a body, and the corresponding force and moment coefficients, have been defined and discussed in Section 1.5. *Question:* What physical quantities determine the variation of these forces and moments?

The answer can be found from the powerful method of *dimensional analysis,* which is introduced in this section.[3]

Consider a body of given shape at a given angle of attack (e.g., the airfoil sketched in Figure 1.17). The resultant aerodynamic force is R. On a physical, intuitive basis, we expect R to depend on:

1. Freestream velocity V_∞.
2. Freestream density ρ_∞.
3. Viscosity of the fluid. We have seen that shear stress τ contributes to the aerodynamic forces and moments, and that τ is proportional to the velocity gradients in the flow. For example, if the velocity gradient is given by $\partial u/\partial y$, then $\tau = \mu \partial u/\partial y$. The constant of proportionality is the viscosity coefficient μ. Hence, let us represent the influence of viscosity on aerodynamic forces and moments by the freestream viscosity coefficient μ_∞.
4. The size of the body, represented by some chosen reference length. In Figure 1.17, the convenient reference length is the chord length c.
5. The compressibility of the fluid. The technical definition of compressibility is given in Chapter 7. For our present purposes, let us just say that compressibility is related to the *variation* of density throughout the flow field, and certainly the aerodynamic forces and moments should be sensitive to any such variation. In turn, compressibility is related to the speed of sound a in the fluid, as shown in Chapter 8.[4] Therefore, let us represent the influence of compressibility on aerodynamic forces and moments by the freestream speed of sound, a_∞.

In light of the above, and without any a priori knowledge about the variation of R, we can use common sense to write

$$R = f(\rho_\infty, V_\infty, c, \mu_\infty, a_\infty) \qquad (1.23)$$

Equation (1.23) is a general functional relation, and as such is not very practical for the direct calculation of R. In principle, we could mount the given body in a wind tunnel, incline it at the given angle of attack, and then systematically measure the variation of R due to variations of ρ_∞, V_∞, c, μ_∞, and a_∞, taken one at a time. By cross-plotting the vast bulk of data thus obtained, we might be able to extract a precise functional relation for Equation (1.23). However, it would be hard work, and it would certainly be costly in terms of a huge amount of required wind-tunnel time. Fortunately, we can simplify the problem and considerably reduce our time and effort by first employing the method of dimensional analysis.

[3] For a more elementary treatment of dimensional analysis, see Chapter 5 of Reference 2.

[4] Common experience tells us that sound waves propagate through air at some finite velocity, much slower than the speed of light; you see a flash of lightning in the distance, and hear the thunder moments later. The speed of sound is an important physical quantity in aerodynamics and is discussed in detail in Section 8.3.

This method will define a set of dimensionless parameters that governs the aerodynamic forces and moments; this set will considerably reduce the number of independent variables as presently occurs in Equation (1.23).

Dimensional analysis is based on the obvious fact that in an equation dealing with the real physical world, each term must have the same dimensions. For example, if

$$\psi + \eta + \zeta = \phi$$

is a physical relation, then ψ, η, ζ, and ϕ must have the same dimensions. Otherwise, we would be adding apples and oranges. The above equation can be made dimensionless by dividing by any one of the terms, say, ϕ:

$$\frac{\psi}{\phi} + \frac{\eta}{\phi} + \frac{\zeta}{\phi} = 1$$

These ideas are formally embodied in the Buckingham pi theorem, stated below without derivation. (See Reference 3, pages 21–28, for such a derivation.)

Buckingham Pi Theorem

Let K represent the number of fundamental dimensions required to describe the physical variables. (In mechanics, all physical variables can be expressed in terms of the dimensions of *mass, length,* and *time;* hence, $K = 3$.) Let P_1, $P_2, \ldots, P_N$ represent N physical variables in the physical relation

$$f_1(P_1, P_2, \ldots, P_N) = 0 \qquad (1.24)$$

Then, the physical relation Equation (1.24) may be reexpressed as a relation of $(N - K)$ dimensionless products (called Π products),

$$f_2(\Pi_1, \Pi_2, \ldots, \Pi_{N-K}) = 0 \qquad (1.25)$$

where each Π product is a dimensionless product of a set of K physical variables plus one other physical variable. Let $P_1, P_2, \ldots, P_K$ be the selected set of K physical variables. Then,

$$\Pi_1 = f_3(P_1, P_2, \ldots, P_K, P_{K+1}) \qquad (1.26)$$

$$\Pi_2 = f_4(P_1, P_2, \ldots, P_K, P_{K+2})$$

$$\cdots\cdots\cdots\cdots\cdots\cdots\cdots\cdots\cdots$$

$$\Pi_{N-K} = f_5(P_1, P_2, \ldots, P_K, P_N)$$

The choice of the repeating variables, $P_1, P_2, \ldots, P_K$ should be such that they include *all* the K dimensions used in the problem. Also, the dependent variable [such as R in Equation (1.23)] should appear in only one of the Π products.

Returning to our consideration of the aerodynamic force on a given body at a given angle of attack, Equation (1.23) can be written in the form of Equation (1.24):

$$g(R, \rho_\infty, V_\infty, c, \mu_\infty, a_\infty) = 0 \qquad (1.27)$$

Following the Buckingham pi theorem, the fundamental dimensions are

$$m = \text{dimension of mass}$$
$$l = \text{dimension of length}$$
$$t = \text{dimension of time}$$

Hence, $K = 3$. The physical variables and their dimensions are

$$[R] = mlt^{-2}$$
$$[\rho_\infty] = ml^{-3}$$
$$[V_\infty] = lt^{-1}$$
$$[c] = l$$
$$[\mu_\infty] = ml^{-1}t^{-1}$$
$$[a_\infty] = lt^{-1}$$

Hence, $N = 6$. In the above, the dimensions of the force R are obtained from Newton's second law, force = mass × acceleration; hence, $[R] = mlt^{-2}$. The dimensions of μ_∞ are obtained from its definition [e.g., $\mu = \tau/(\partial u/\partial y)$], and from Newton's second law. (Show for yourself that $[\mu_\infty] = ml^{-1}t^{-1}$.) Choose ρ_∞, V_∞, and c as the arbitrarily selected sets of K physical variables. Then, Equation (1.27) can be reexpressed in terms of $N - K = 6 - 3 = 3$ dimensionless Π products in the form of Equation (1.25):

$$f_2(\Pi_1, \Pi_2, \Pi_3) = 0 \qquad (1.28)$$

From Equation (1.26), these Π products are

$$\Pi_1 = f_3(\rho_\infty, V_\infty, c, R) \qquad (1.29a)$$
$$\Pi_2 = f_4(\rho_\infty, V_\infty, c, \mu_\infty) \qquad (1.29b)$$
$$\Pi_3 = f_5(\rho_\infty, V_\infty, c, a_\infty) \qquad (1.29c)$$

For the time being, concentrate on Π_1, from Equation (1.29a). Assume that

$$\Pi_1 = \rho_\infty^d V_\infty^b c^e R \qquad (1.30)$$

where d, b, and e are exponents to be found. In dimensional terms, Equation (1.30) is

$$[\Pi_1] = (ml^{-3})^d (lt^{-1})^b (l)^e (mlt^{-2}) \qquad (1.31)$$

Because Π_1 is dimensionless, the right side of Equation (1.31) must also be dimensionless. This means that the exponents of m must add to zero, and similarly for the exponents of l and t. Hence,

For m: $d + 1 = 0$
For l: $-3d + b + e + 1 = 0$
For t: $-b - 2 = 0$

Solving the above equations, we find that $d = -1$, $b = -2$, and $e = -2$. Substituting these values into Equation (1.30), we have

$$\Pi_1 = R\rho_\infty^{-1} V_\infty^{-2} c^{-2} \tag{1.32}$$

$$= \frac{R}{\rho_\infty V_\infty^2 c^2}$$

The quantity $R/\rho_\infty V_\infty^2 c^2$ is a dimensionless parameter in which c^2 has the dimensions of an area. We can replace c^2 with any reference area we wish (such as the planform area of a wing S), and Π_1 will still be dimensionless. Moreover, we can multiply Π_1 by a pure number, and it will still be dimensionless. Thus, from Equation (1.32), Π_1 can be redefined as

$$\Pi_1 = \frac{R}{\frac{1}{2}\rho_\infty V_\infty^2 S} = \frac{R}{q_\infty S} \tag{1.33}$$

Hence, Π_1 is a force coefficient C_R, as defined in Section 1.5. In Equation (1.33), S is a reference area germane to the given body shape.

The remaining Π products can be found as follows. From Equation (1.29b), assume

$$\Pi_2 = \rho_\infty V_\infty^h c^i \mu_\infty^j \tag{1.34}$$

Paralleling the above analysis, we obtain

$$[\Pi_2] = (ml^{-3})(lt^{-1})^h(l)^i(ml^{-1}t^{-1})^j$$

Hence,

For m: $1 + j = 0$
For l: $-3 + h + i - j = 0$
For t: $-h - j = 0$

Thus, $j = -1$, $h = 1$, and $i = 1$. Substitution into Equation (1.34) gives

$$\Pi_2 = \frac{\rho_\infty V_\infty c}{\mu_\infty} \tag{1.35}$$

The dimensionless combination in Equation (1.35) is defined as the freestream *Reynolds number* $\mathrm{Re} = \rho_\infty V_\infty c/\mu_\infty$. The Reynolds number is physically a measure of the ratio of inertia forces to viscous forces in a flow and is one of the most powerful parameters in fluid dynamics. Its importance is emphasized in Chapters 15 to 20.

Returning to Equation (1.29c), assume

$$\Pi_3 = V_\infty \rho_\infty^k c^r a_\infty^s \tag{1.36}$$

$$[\Pi_3] = (lt^{-1})(ml^{-3})^k(l)^r(lt^{-1})^s$$

For m: $\qquad\qquad\qquad k = 0$

For l: $\qquad\qquad\qquad 1 - 3k + r + s = 0$

For t: $\qquad\qquad\qquad -1 - s = 0$

Hence, $k = 0$, $s = -1$, and $r = 0$. Substituting Equation (1.36), we have

$$\Pi_e = \frac{V_\infty}{a_\infty} \tag{1.37}$$

The dimensionless combination in Equation (1.37) is defined as the freestream *Mach number* $M = V_\infty/a_\infty$. The Mach number is the ratio of the flow velocity to the speed of sound; it is a powerful parameter in the study of gas dynamics. Its importance is emphasized in subsequent chapters.

The results of our dimensional analysis may be organized as follows. Inserting Equations (1.33), (1.35), and (1.37) into (1.28), we have

$$f_2\left(\frac{R}{\frac{1}{2}\rho_\infty V_\infty^2 S}, \frac{\rho_\infty V_\infty c}{\mu_\infty}, \frac{V_\infty}{a_\infty}\right) = 0$$

or $\qquad\qquad\qquad f_2(C_R, \text{Re}, M_\infty) = 0$

or $\qquad\qquad\qquad \boxed{C_R = f_6(\text{Re}, M_\infty)} \tag{1.38}$

This is an important result! Compare Equations (1.23) and (1.38). In Equation (1.23), R is expressed as a general function of five independent variables. However, our dimensional analysis has shown that:

1. R can be expressed in terms of a dimensionless force coefficient, $C_R = R/\frac{1}{2}\rho_\infty V_\infty^2 S$.
2. C_R is a function of only Re and M_∞, from Equation (1.38).

Therefore, by using the Buckingham pi theorem, we have reduced the number of independent variables from five in Equation (1.23) to two in Equation (1.38). Now, if we wish to run a series of wind-tunnel tests for a given body at a given angle of attack, we need only to vary the Reynolds and Mach numbers in order to obtain data for the direct formulation of R through Equation (1.38). With a small amount of analysis, we have saved a huge amount of effort and wind-tunnel time. More importantly, we have defined two dimensionless parameters, Re and M_∞, which govern the flow. They are called *similarity parameters,* for reasons to be discussed in the following section. Other similarity parameters are introduced as our aerodynamic discussions progress.

Since the lift and drag are components of the resultant force, corollaries to Equation (1.38) are

$$C_L = f_7(\text{Re}, M_\infty) \tag{1.39}$$

$$C_D = f_8(\text{Re}, M_\infty) \tag{1.40}$$

Moreover, a relation similar to Equation (1.23) holds for the aerodynamic moments, and dimensional analysis yields

$$C_M = f_9(\text{Re}, M_\infty) \tag{1.41}$$

Keep in mind that the above analysis was for a given body shape at a given angle of attack α. If α is allowed to vary, then C_L, C_D, and C_M will in general depend on the value of α. Hence, Equations (1.39) to (1.41) can be generalized to

$$\boxed{\begin{aligned} C_L &= f_{10}(\text{Re}, M_\infty, \alpha) \\ C_D &= f_{11}(\text{Re}, M_\infty, \alpha) \\ C_M &= f_{12}(\text{Re}, M_\infty, \alpha) \end{aligned}} \tag{1.42} \tag{1.43} \tag{1.44}$$

Equations (1.42) to (1.44) assume a given body shape. Much of theoretical and experimental aerodynamics is focused on obtaining explicit expressions for Equations (1.42) to (1.44) for specific body shapes. This is one of the practical applications of aerodynamics mentioned in Section 1.2, and it is one of the major thrusts of this book.

For mechanical problems that also involve thermodynamics and heat transfer, the temperature, specific heat, and thermal conductivity of the fluid, as well as the temperature of the body surface (wall temperature), must be added to the list of physical variables, and the unit of temperature (say, kelvin or degree Rankine) must be added to the list of fundamental dimensions. For such cases, dimensional analysis yields additional dimensionless products such as heat transfer coefficients, and additional similarity parameters such as the ratio of specific heat at constant pressure to that at constant volume c_p/c_v, the ratio of wall temperature to freestream temperature T_w/T_∞, and the Prandtl number $\text{Pr} = \mu_\infty c_p/k_\infty$, where k_∞ is the thermal conductivity of the freestream.[5] Thermodynamics is essential to the study of compressible flow (Chapters 7 to 14), and heat transfer is part of the study of viscous flow (Chapters 15 to 20). Hence, these additional similarity parameters will be emphasized when they appear logically in our subsequent discussions. For the time being, however, the Mach and Reynolds numbers will suffice as the dominant similarity parameters for our present considerations.

---

[5] The *specific heat* of a fluid is defined as the amount of heat added to a system, δq, per unit increase in temperature; $c_v = \delta q/dT$ if δq is added at constant volume, and similarly, for c_p if δq is added at constant pressure. Specific heats are discussed in detail in Section 7.2. The thermal conductivity relates heat flux to temperature gradients in the fluid. For example, if $\dot{q}_x$ is the heat transferred in the x direction per second per unit area and dT/dx is the temperature gradient in the x direction, then thermal conductivity k is defined by $\dot{q}_x = -k(dT/dx)$. Thermal conductivity is discussed in detail in Section 15.3.

1.8 FLOW SIMILARITY

Consider two different flow fields over two different bodies. By definition, different flows are *dynamically similar* if:

1. The streamline patterns are geometrically similar.
2. The distributions of V/V_∞, p/p_∞, T/T_∞, etc., throughout the flow field are the same when plotted against common nondimensional coordinates.
3. The force coefficients are the same.

Actually, item 3 is a consequence of item 2. If the nondimensional pressure and shear stress distributions over different bodies are the same, then the nondimensional force coefficients will be the same.

   The definition of dynamic similarity was given above. *Question:* What are the *criteria* to ensure that two flows are dynamically similar? The answer comes from the results of the dimensional analysis in Section 1.7. Two flows will be dynamically similar if:

1. The bodies and any other solid boundaries are geometrically similar for both flows.
2. The similarity parameters are the same for both flows.

So far, we have emphasized two parameters, Re and M_∞. For many aerodynamic applications, these are by far the dominant similarity parameters. Therefore, in a limited sense, but applicable to many problems, we can say that flows over geometrically similar bodies at the same Mach and Reynolds numbers are dynamically similar, and hence the lift, drag, and moment coefficients will be identical for the bodies. This is a key point in the validity of wind-tunnel testing. If a scale model of a flight vehicle is tested in a wind tunnel, the measured lift, drag, and moment coefficients will be the same as for free flight as long as the Mach and Reynolds numbers of the wind-tunnel test-section flow are the same as for the free-flight case. As we will see in subsequent chapters, this statement is not quite precise because there are other similarity parameters that influence the flow. In addition, differences in freestream turbulence between the wind tunnel and free flight can have an important effect on C_D and the maximum value of C_L. However, direct simulation of the free-flight Re and M_∞ is the primary goal of many wind-tunnel tests.

EXAMPLE 1.5

Consider the flow over two circular cylinders, one having four times the diameter of the other, as shown in Figure 1.27. The flow over the smaller cylinder has a freestream density, velocity, and temperature given by ρ_1, V_1, and T_1, respectively. The flow over the larger cylinder has a freestream density, velocity, and temperature given by ρ_2, V_2, and T_2, respectively, where $\rho_2 = \rho_1/4$, $V_2 = 2V_1$, and $T_2 = 4T_1$. Assume that both μ and a are proportional to $T^{1/2}$. Show that the two flows are dynamically similar.

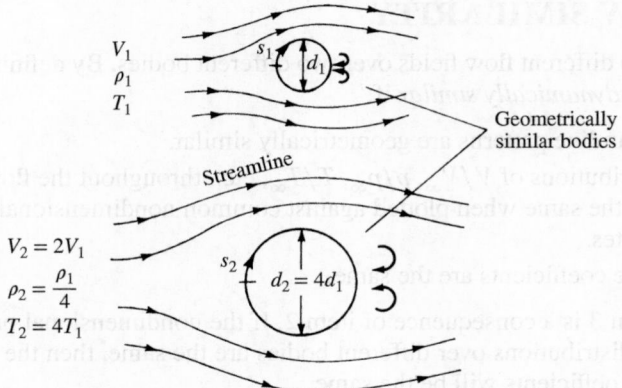

V_1
ρ_1
T_1

$V_2 = 2V_1$
$\rho_2 = \dfrac{\rho_1}{4}$
$T_2 = 4T_1$

Geometrically
similar bodies

Streamline

Figure 1.27 Example of dynamic flow similarity. Note that as
part of the definition of dynamic similarity, the streamlines
(lines along which the flow velocity is tangent at each point)
are geometrically similar between the two flows.

■ **Solution**

Since $\mu \propto \sqrt{T}$ and $a \propto \sqrt{T}$, then

$$\frac{\mu_2}{\mu_1} = \sqrt{\frac{T_2}{T_1}} = \sqrt{\frac{4T_1}{T_1}} = 2$$

and

$$\frac{a_2}{a_1} = \sqrt{\frac{T_2}{T_1}} = 2$$

By definition,

$$M_1 = \frac{V_1}{a_1}$$

and

$$M_2 = \frac{V_2}{a_2} = \frac{2V_1}{2a_1} = \frac{V_1}{a_1} = M_1$$

Hence, the Mach numbers are the same. Basing the Reynolds number on the diameter d
of the cylinder, we have by definition,

$$\text{Re}_1 = \frac{\rho_1 V_1 d_1}{\mu_1}$$

and

$$\text{Re}_2 = \frac{\rho_2 V_2 d_2}{\mu_2} = \frac{(\rho_1/4)(2V_1)(4d_1)}{2\mu_1} = \frac{\rho_1 V_1 d_1}{\mu_1} = \text{Re}_1$$

Hence, the Reynolds numbers are the same. Since the two bodies are geometrically similar and M_∞ and Re are the same, we have satisfied all the criteria; the two flows are dynamically similar. In turn, as a consequence of being similar flows, we know from the definition that:

1. The streamline patterns around the two cylinders are geometrically similar.
2. The nondimensional pressure, temperature, density, velocity, etc., distributions are the same around two cylinders. This is shown schematically in Figure 1.28, where

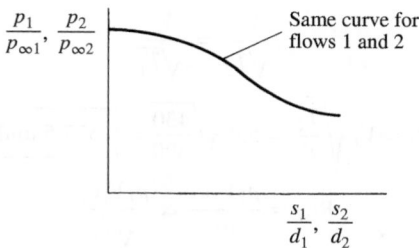

Figure 1.28 One aspect of the definition of dynamically similar flows. The nondimensional flow variable distributions are the same.

the nondimensional pressure distribution p/p_∞ is shown as a function of the nondimensional surface distance s/d. It is the same curve for both bodies.

3. The drag coefficients for the two bodies are the same. Here, $C_D = D/q_\infty S$, where $S = \pi d^2/4$. As a result of the flow similarity, $C_{D1} = C_{D2}$. (*Note:* Examining Figure 1.27, we see that the lift on the cylinders is zero because the flow is symmetrical about the horizontal axis through the center of the cylinder. The pressure distribution over the top is the same as over the bottom, and they cancel each other in the vertical direction. Therefore, drag is the only aerodynamic force on the body.)

Consider a Boeing 747 airliner cruising at a velocity of 550 mi/h at a standard altitude of 38,000 ft, where the freestream pressure and temperature are 432.6 lb/ft^2 and 390°R, respectively. A one-fiftieth scale model of the 747 is tested in a wind tunnel where the temperature is 430°R. Calculate the required velocity and pressure of the test airstream in the wind tunnel such that the lift and drag coefficients measured for the wind-tunnel model are the same as for free flight. Assume that both μ and a are proportional to $T^{1/2}$.

■ **Solution**

Let subscripts 1 and 2 denote the free-flight and wind-tunnel conditions, respectively. For $C_{L,1} = C_{L,2}$ and $C_{D,1} = C_{D,2}$, the wind-tunnel flow must be dynamically similar to free flight. For this to hold, $M_1 = M_2$ and $\text{Re}_1 = \text{Re}_2$:

$$M_1 = \frac{V_1}{a_1} \propto \frac{V_1}{\sqrt{T_1}}$$

and

$$M_2 = \frac{V_2}{a_2} \propto \frac{V_2}{\sqrt{T_2}}$$

Hence,
$$\frac{V_2}{\sqrt{T_2}} = \frac{V_1}{\sqrt{T_1}}$$

or
$$V_2 = V_1\sqrt{\frac{T_2}{T_1}} = 550\sqrt{\frac{430}{390}} = \boxed{577.5 \text{ mi/h}}$$

$$\text{Re}_1 = \frac{\rho_1 V_1 c_1}{\mu_1} \propto \frac{\rho_1 V_1 c_1}{\sqrt{T_1}}$$

and
$$\text{Re}_2 = \frac{\rho_2 V_2 c_2}{\mu_2} \propto \frac{\rho_2 V_2 c_2}{\sqrt{T_2}}$$

Hence,
$$\frac{\rho_1 V_1 c_1}{\sqrt{T_1}} = \frac{\rho_2 V_2 c_2}{\sqrt{T_2}}$$

or
$$\frac{\rho_2}{\rho_1} = \left(\frac{V_1}{V_2}\right)\left(\frac{c_1}{c_2}\right)\sqrt{\frac{T_2}{T_1}}$$

However, since $M_1 = M_2$, then
$$\frac{V_1}{V_2} = \sqrt{\frac{T_1}{T_2}}$$

Thus,
$$\frac{\rho_2}{\rho_1} = \frac{c_1}{c_2} = 50$$

The equation of state for a perfect gas is $p = \rho RT$, where R is the specific gas constant. Thus,

$$\frac{p_2}{p_1} = \frac{\rho_2}{\rho_1}\frac{T_2}{T_1} = (50)\left(\frac{430}{390}\right) = 55.1$$

Hence,
$$p_2 = 55.1 p_1 = (55.1)(432.6) = \boxed{23{,}836 \text{ lb/ft}^2}$$

Since 1 atm = 2116 lb/ft^2, then $p_2 = 23{,}836/2116 = \boxed{11.26 \text{ atm}}$.

In Example 1.6, the wind-tunnel test stream must be pressurized far above atmospheric pressure in order to simulate the proper free-flight Reynolds number. However, most standard subsonic wind tunnels are not pressurized as such, because of the large extra financial cost involved. This illustrates a common difficulty in wind-tunnel testing, namely, the difficulty of simulating both Mach number and Reynolds number simultaneously in the same tunnel. It is interesting to note that the NACA (National Advisory Committee for Aeronautics, the predecessor of NASA) in 1922 began operating a pressurized wind tunnel at the NACA Langley Memorial Laboratory in Hampton, Virginia. This was a subsonic wind tunnel contained entirely inside a large tank pressurized to as high as 20 atm. Called the variable density tunnel (VDT), this facility was used in the 1920s and

Figure 1.29 The NACA variable density tunnel (VDT). Authorized in March of 1921, the VDT was operational in October 1922 at the NACA Langley Memorial Laboratory at Hampton, Virginia. It is essentially a large, subsonic wind tunnel entirely contained within an 85-ton pressure shell, capable of 20 atm. This tunnel was instrumental in the development of the various families of NACA airfoil shapes in the 1920s and 1930s. In the early 1940s, it was decommissioned as a wind tunnel and used as a high-pressure air storage tank. In 1983, due to its age and outdated riveted construction, its use was discontinued altogether. Today, the VDT remains at the NASA Langley Research Center; it has been officially designated as a National Historic Landmark. (*NASA*)

1930s to provide essential data on the NACA family of airfoil sections at the high Reynolds numbers associated with free flight. A photograph of the NACA variable density tunnel is shown in Figure 1.29; notice the heavy pressurized shell in which the wind tunnel is enclosed. A cross section of the VDT inside the pressure cell is shown in Figure 1.30. These figures demonstrate the extreme measures sometimes taken in order to simulate simultaneously the free-flight values of the important similarity parameters in a wind tunnel. Today, for the most part, we do not attempt to simulate all the parameters simultaneously; rather, Mach number simulation is achieved in one wind tunnel, and Reynolds number simulation in another tunnel. The results from both tunnels are then analyzed and correlated to obtain reasonable values for C_L and C_D appropriate for free flight. In any event, this example serves to illustrate the difficulty of full free-flight simulation in a given wind tunnel and underscores the importance given to dynamically similar flows in experimental aerodynamics.

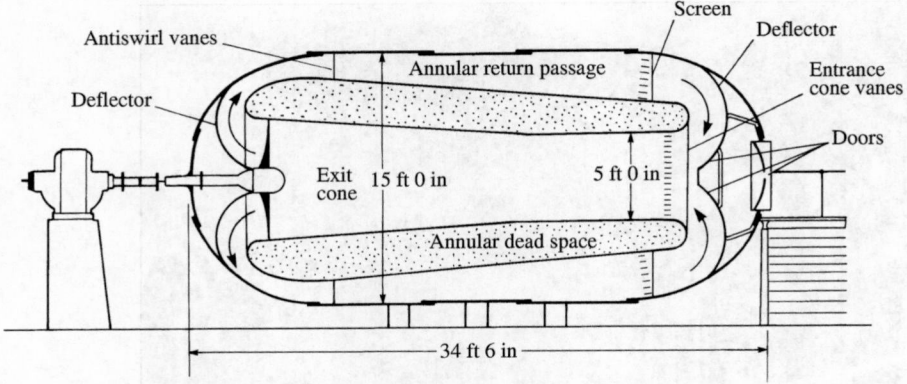

Figure 1.30 Schematic of the variable density tunnel (*From Baals, D. D. and Carliss, W. R., Wind Tunnels of NASA, NASA SP-440, 1981*).

DESIGN BOX

Lift and drag coefficients play a strong role in the preliminary design and performance analysis of airplanes. The purpose of this design box is to enforce the importance of C_L and C_D in aeronautical engineering; they are much more than just the conveniently defined terms discussed so far—they are fundamental quantities, which make the difference between intelligent engineering and simply groping in the dark.

Consider an airplane in steady, level (horizontal) flight, as illustrated in Figure 1.31. For this case, the weight W acts vertically downward. The lift L acts vertically upward, perpendicular to the relative wind

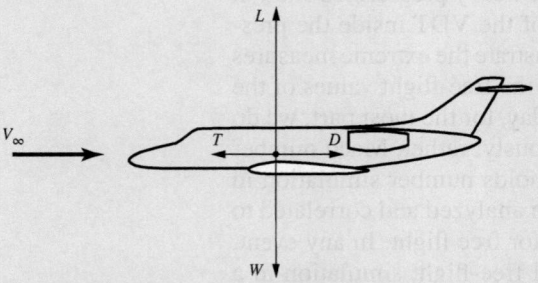

Figure 1.31 The four forces acting on an airplane in flight.

V_∞ (by definition). In order to sustain the airplane in level flight,

$$L = W$$

The thrust T from the propulsive mechanism and the drag D are both parallel to V_∞. For steady (unaccelerated) flight,

$$T = D$$

Note that for most conventional flight situations, the magnitude of L and W is much larger than the magnitude of T and D, as indicated by the sketch in Figure 1.31. Typically, for conventional cruising flight, $L/D \approx 15$ to 20.

For an airplane of *given shape,* such as that sketched in Figure 1.31, at given Mach and Reynolds number, C_L and C_D are simply functions of the angle of attack, α of the airplane. This is the message conveyed by Equations (1.42) and (1.43). It is a simple and basic message—part of the beauty of nature—that the actual values of C_L and C_D for a given body shape just depend on the orientation of the body in the flow (i.e., angle of attack). Generic variations for C_L and C_D versus α are sketched in Figure 1.32. Note that C_L increases linearly with α until an angle of

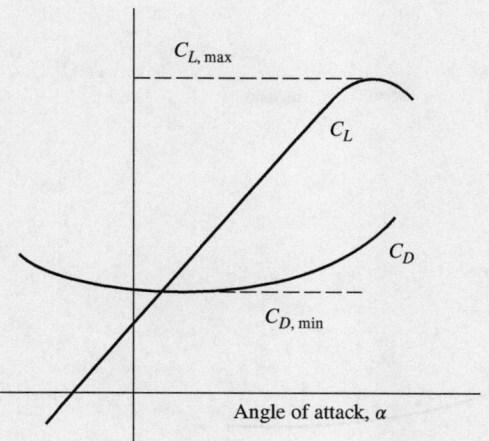

Figure 1.32 Schematic of lift and drag coefficients versus angle of attack; illustration of maximum lift coefficient and minimum drag coefficient.

attack is reached when the wing stalls, the lift coefficient reaches a peak value, and then drops off as α is further increased. The maximum value of the lift coefficient is denoted by $C_{L,\max}$, as noted in Figure 1.32.

The lowest possible velocity at which the airplane can maintain steady level flight is the stalling velocity, V_{stall}; it is dictated by the value of $C_{L,\max}$, as follows.[6] From the definition of lift coefficient given in Section 1.5, applied for the case of level flight where $L = W$, we have

$$C_L = \frac{L}{q_\infty S} = \frac{W}{q_\infty S} = \frac{2W}{\rho_\infty V_\infty^2 S} \qquad (1.45)$$

Solving Equation (1.45) for V_∞,

$$V_\infty = \sqrt{\frac{2W}{\rho_\infty S C_L}} \qquad (1.46)$$

For a given airplane flying at a given altitude, W, ρ, and S are fixed values; hence from Equation (1.46), each value of velocity corresponds to a specific value

of C_L. In particular, V_∞ will be the smallest when C_L is a maximum. Hence, the stalling velocity for a given airplane is determined by $C_{L,\max}$ from Equation (1.46)

$$V_{\text{stall}} = \sqrt{\frac{2W}{\rho_\infty S C_{L,\max}}} \qquad (1.47)$$

For a given airplane, without the aid of any artificial devices, $C_{L,\max}$ is determined purely by nature, through the physical laws for the aerodynamic flowfield over the airplane. However, the airplane designer has some devices available that artificially increase $C_{L,\max}$ beyond that for the basic airplane shape. These mechanical devices are called *high-lift devices;* examples are flaps, slats, and slots on the wing which, when deployed by the pilot, serve to increase $C_{L,\max}$, and hence decrease the stalling speed. High-lift devices are usually deployed for landing and take-off; they are discussed in more detail in Section 4.12.

On the other extreme of flight velocity, the maximum velocity for a given airplane with a given maximum thrust from the engine is determined by the value of minimum drag coefficient, $C_{D,\min}$, where $C_{D,\min}$ is marked in Figure 1.32. From the definition of drag coefficient in Section 1.5, applied for the case of steady level flight where $T = D$, we have

$$C_D = \frac{D}{q_\infty S} = \frac{T}{q_\infty S} = \frac{2T}{\rho_\infty V_\infty^2 S} \qquad (1.48)$$

Solving Equation (1.48) for V_∞,

$$V_\infty = \sqrt{\frac{2T}{\rho_\infty S C_D}} \qquad (1.49)$$

For a given airplane flying at maximum thrust $T_{\max}$ and a given altitude, from Equation (1.49) the maximum value of V_∞ corresponds to flight at $C_{D,\min}$

$$V_{\max} = \sqrt{\frac{2T_{\max}}{\rho_\infty S C_{D,\min}}} \qquad (1.50)$$

From the above discussion, it is clear that the aerodynamic *coefficients* are important engineering

---

[6] The lowest velocity may instead be dictated by the power required to maintain level flight exceeding the power available from the powerplant. This occurs on the "back side of the power curve." The velocity at which this occurs is usually less than the stalling velocity, so is of academic interest only. See Anderson, *Aircraft Performance and Design,* McGraw Hill, 1999, for more details.

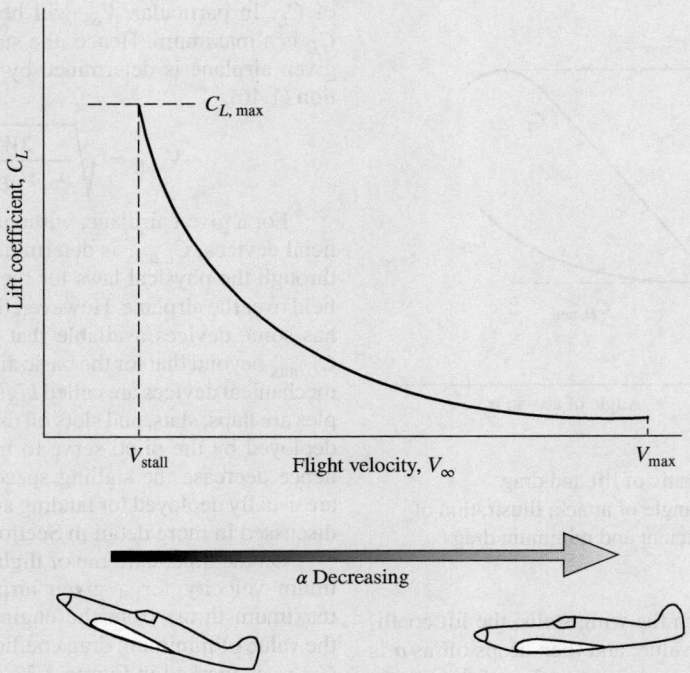

Figure 1.33 Schematic of the variation of lift coefficient with flight velocity for level flight.

quantities that dictate the performance and design of airplanes. For example, stalling velocity is determined in part by $C_{L,\max}$, and maximum velocity is determined in part by $C_{D,\min}$.

Broadening our discussion to the whole range of flight velocity for a given airplane, note from Equation (1.45) that each value of V_∞ corresponds to a specific value of C_L. Therefore, over the whole range of flight velocity from V_{stall} to $V_{\max}$, the airplane lift coefficient varies as shown generically in Figure 1.33. The values of C_L given by the curve in Figure 1.33 are *what are needed* to maintain level flight over the whole range of velocity at a given altitude. The airplane designer must design the airplane to *achieve* these values of C_L for an airplane of given weight and wing area. Note that the required values of C_L decrease as V_∞ increases. Examining the lift coefficient variation with angle of attack shown in Figure 1.33, note that as the airplane flies faster, the angle of attack must be

smaller, as also shown in Figure 1.33. Hence, at high speeds, airplanes are at low α, and at low speeds, airplanes are at high α; the specific angle of attack which the airplane must have at a specific V_∞ is dictated by the specific value of C_L required at that velocity.

Obtaining raw lift on a body is relatively easy—even a barn door creates lift at angle of attack. The name of the game is to obtain the necessary lift with as *low* a drag as possible. That is, the values of C_L required over the entire flight range for an airplane, as represented by Figure 1.33, can sometimes be obtained even for the least effective lifting shape—just make the angle of attack high enough. But C_D also varies with V_∞, as governed by Equation (1.48); the generic variation of C_D with V_∞ is sketched in Figure 1.34. A poor aerodynamic shape, even though it generates the necessary values of C_L shown in Figure 1.33, will have inordinately high values of C_D (i.e., the C_D curve in Figure 1.34 will ride high on the

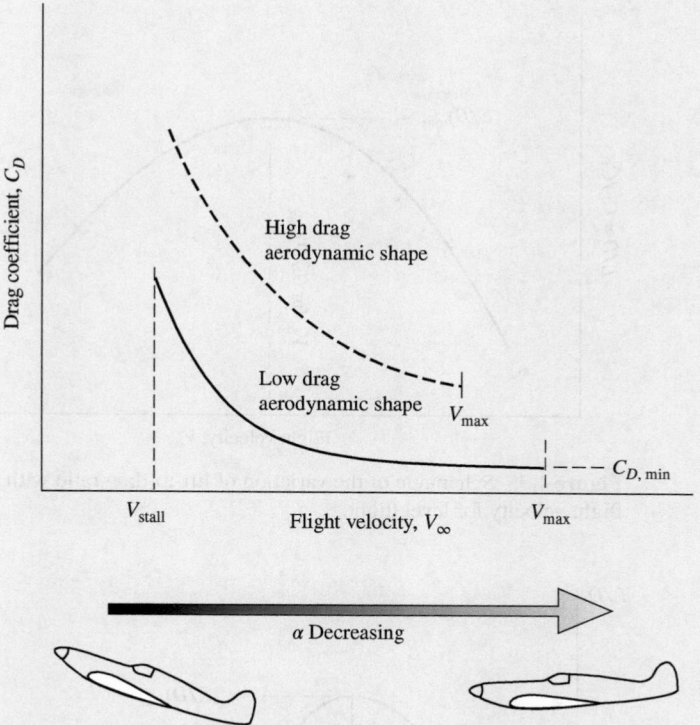

Figure 1.34 Schematic of the variation of drag coefficient with flight velocity for level flight. Comparison between high and low drag aerodynamic bodies, with the consequent effect on maximum velocity.

graph), as denoted by the dashed curve in Figure 1.34. An aerodynamically efficient shape, however, will produce the requisite values of C_L prescribed by Figure 1.33 with much lower drag, as denoted by the solid curve in Figure 1.34. An undesirable by-product of the high-drag shape is a lower value of the maximum velocity for the same maximum thrust, as also indicated in Figure 1.34.

Finally, we emphasize that a true measure of the aerodynamic efficiency of a body shape is its *lift-to-drag ratio,* given by

$$\frac{L}{D} = \frac{q_\infty S C_L}{q_\infty S C_D} = \frac{C_L}{C_D} \qquad (1.51)$$

Since the value of C_L necessary for flight at a given velocity and altitude is determined by the airplane's weight and wing area (actually, by the *ratio* of W/S, called the *wing loading*) through the relationship given by Equation (1.45), the value of L/D at this velocity is controlled by C_D, the denominator in Equation (1.51). At any given velocity, we want L/D to be as high as possible; the higher is L/D, the more aerodynamically efficient is the body. For a given airplane at a given altitude, the variation of L/D as a function of velocity is sketched generically in Figure 1.35. Note that, as V_∞ increases from a low value, L/D first increases, reaches a maximum at some intermediate velocity, and then decreases. Note that, as V_∞ increases, the angle of attack of the airplane decreases, as explained earlier. From a strictly aerodynamic consideration, L/D for a given body shape depends on angle of attack. This can be seen from Figure 1.32,

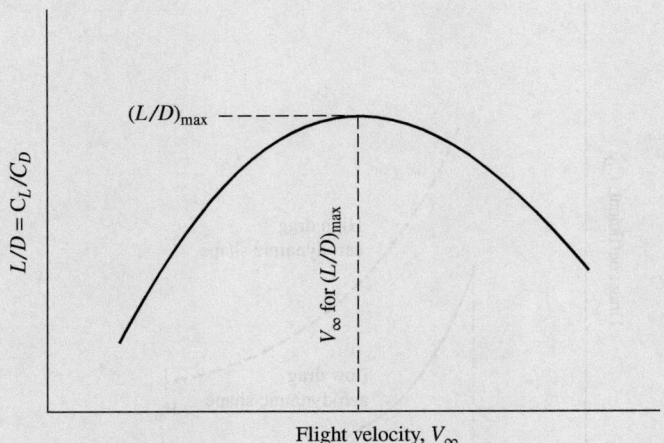

Figure 1.35 Schematic of the variation of lift-to-drag ratio with flight velocity for level flight.

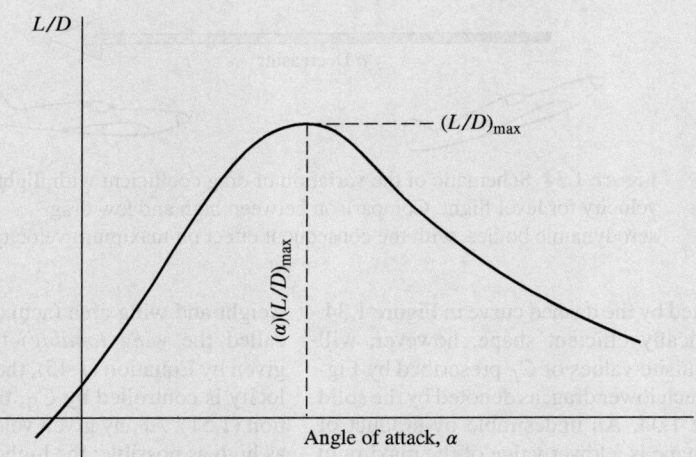

Figure 1.36 Schematic of the variation of lift-to-drag ratio with angle of attack.

where C_L and C_D are given as a function of α. If these two curves are ratioed, the result is L/D as a function of angle of attack, as sketched generically in Figure 1.36. The relationship between Figure 1.35 and Figure 1.36 is that, when the airplane is flying at the velocity that corresponds to $(L/D)_{\max}$ as shown in Figure 1.35, it *is* at the angle of attack for $(L/D)_{\max}$ as shown in Figure 1.36.

In summary, the purpose of this design box is to emphasize the important role played by the aerodynamic *coefficients* in the performance analysis and design of airplanes. In this discussion, what has been important is not the lift and drag per se, but rather C_L and C_D. These coefficients are a wonderful intellectual construct that helps us to better understand the aerodynamic characteristics of a body, and to make reasoned, intelligent calculations. Hence, they are more than just conveniently defined quantities as might first appear when introduced in Section 1.5.

For more insight to the engineering value of these coefficients, see Anderson, *Aircraft Performance and Design,* McGraw Hill, 1999, and Anderson and Bowden, *Introduction to Flight,* 9th edition, McGraw Hill, 2022. Also, homework Problem 1.15 at the end of this chapter gives you the opportunity to construct specific curves for C_L, C_D, and L/D versus velocity for an actual airplane so that you can obtain a feel for some real numbers that have been only generically indicated in the figures here. (In our present discussion, the use of generic figures has been intentional for pedagogic reasons.) Finally, an historical note on the origins of the use of aerodynamic coefficients is given in Section 1.14.

EXAMPLE 1.7

Consider an executive jet transport patterned after the Cessna 560 Citation V shown in three views in Figure 1.37. The airplane is cruising at a velocity of 492 mph at an altitude of 33,000 ft, where the ambient air density is 7.9656×10^{-4} slug/ft^3. The weight and wing planform areas of the airplane are 15,000 lb and 342.6 ft^2, respectively. The drag coefficient at cruise is 0.015. Calculate the lift coefficient and the lift-to-drag ratio at cruise.

■ **Solution**

The units of miles per hour for velocity are not consistent units. In the English engineering system of units, feet per second are consistent units for velocity (see Section 2.4 of Reference 2). To convert between mph and ft/s, it is useful to remember that 88 ft/s = 60 mph.

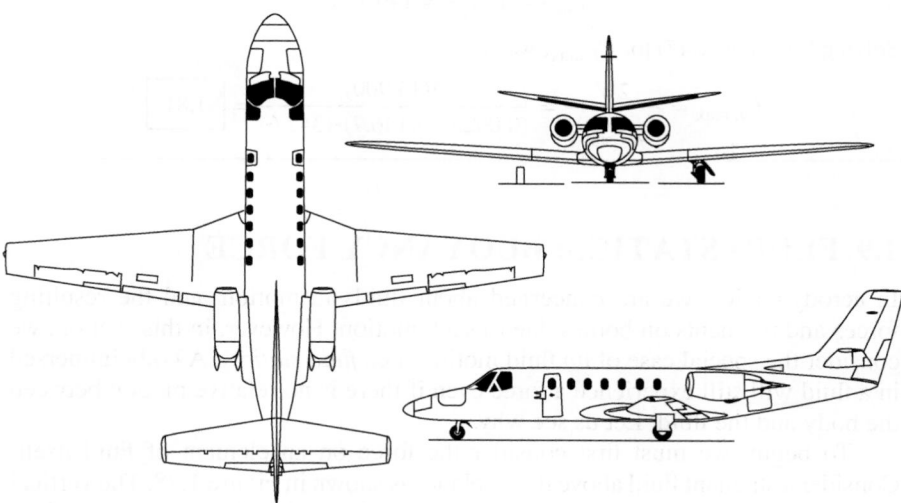

Figure 1.37 Cessna 560 Citation V.

For the present example,

$$V_\infty = 492\left(\frac{88}{60}\right) = 721.6 \text{ ft/s}$$

From Equation (1.45),

$$C_L = \frac{2W}{\rho_\infty V_\infty^2 S} = \frac{2(15,000)}{(7.9659 \times 10^{-4})(721.6)^2(342.6)} = \boxed{0.21}$$

From Equation (1.51),

$$\frac{L}{D} = \frac{C_L}{C_D} = \frac{0.21}{0.015} = \boxed{14}$$

Remarks: For a conventional airplane such as shown in Figure 1.37, almost all the lift at cruising conditions is produced by the wing; the lift of the fuselage and tail are very small by comparison. Hence, the wing can be viewed as an aerodynamic "lever." In this example, the lift-to-drag ratio is 14, which means that for the expenditure of one pound of thrust to overcome one pound of drag, the wing is lifting 14 pounds of weight—quite a nice leverage.

EXAMPLE 1.8

The same airplane as described in Example 1.7 has a stalling speed at sea level of 100 mph at the maximum take-off weight of 15,900 lb. The ambient air density at standard sea level is 0.002377 slug/ft^3. Calculate the value of the maximum lift coefficient for the airplane.

■ **Solution**

Once again we have to use consistent units, so

$$V_{\text{stall}} = 100\frac{88}{60} = 146.7 \text{ ft/s}$$

Solving Equation (1.47) for $C_{L,\text{max}}$, we have

$$C_{L,\text{max}} = \frac{2W}{\rho_\infty V_{\text{stall}}^2 S} = \frac{2(15,900)}{(0.002377)(146.7)^2(342.6)} = \boxed{1.81}$$

1.9 FLUID STATICS: BUOYANCY FORCE

In aerodynamics, we are concerned about fluids in motion, and the resulting forces and moments on bodies due to such motion. However, in this section, we consider the special case of *no* fluid motion (i.e., *fluid statics*). A body immersed in a fluid will still experience a force even if there is no relative motion between the body and the fluid. Let us see why.

To begin, we must first consider the force on an element of fluid itself. Consider a stagnant fluid above the *xz* plane, as shown in Figure 1.38. The vertical direction is given by *y*. Consider an infinitesimally small fluid element with sides of length *dx*, *dy*, and *dz*. There are two types of forces acting on this fluid element: pressure forces from the surrounding fluid exerted on the surface of the element,

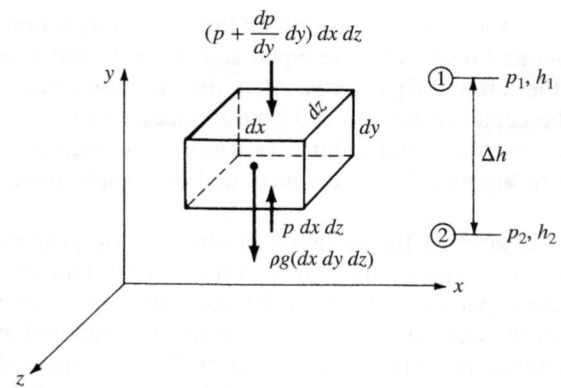

Figure 1.38 Forces on a fluid element in a stagnant fluid.

and the gravity force due to the weight of the fluid inside the element. Consider forces in the y direction. The pressure on the bottom surface of the element is p, and hence the force on the bottom face is $p(dx\,dz)$ in the upward direction, as shown in Figure 1.38. The pressure on the top surface of the element will be slightly different from the pressure on the bottom because the top surface is at a different location in the fluid. Let dp/dy denote the rate of change of p with respect to y. Then the pressure exerted on the top surface will be $p + (dp/dy)\,dy$, and the pressure force on the top of the element will be $[p + (dp/dy)\,dy](dx\,dz)$ in the downward direction, as shown in Figure 1.38. Hence, letting upward force be positive, we have

$$\text{Net pressure force} = p(dx\,dz) - \left(p + \frac{dp}{dy}dy\right)(dx\,dz)$$

$$= -\frac{dp}{dy}(dx\,dy\,dz)$$

Let ρ be the mean density of the fluid element. The total mass of the element is $\rho(dx\,dy\,dz)$. Therefore,

$$\text{Gravity force} = -\rho(dx\,dy\,dz)g$$

where g is the acceleration of gravity. Since the fluid element is stationary (in equilibrium), the sum of the forces exerted on it must be zero:

$$-\frac{dp}{dy}(dx\,dy\,dz) - g\rho(dx\,dy\,dz) = 0$$

or

$$\boxed{dp = -g\rho\,dy} \tag{1.52}$$

Equation (1.52) is called the *Hydrostatic equation;* it is a differential equation which relates the change in pressure dp in a fluid with a change in vertical height dy.

The net force on the element acts only in the vertical direction. The pressure forces on the front and back faces are equal and opposite and hence cancel; the same is true for the left and right faces. Also, the pressure forces shown in Figure 1.38 act at the center of the top and bottom faces, and the center of gravity is at the center of the elemental volume (assuming the fluid is homogeneous); hence, the forces in Figure 1.38 are colinear, and as a result, there is no moment on the element.

Equation (1.52) governs the variation of atmospheric properties as a function of altitude in the air above us. It is also used to estimate the properties of other planetary atmospheres such as for Venus, Mars, and Jupiter. The use of Equation (1.52) in the analysis and calculation of the "standard atmosphere" is given in detail in Reference 2; hence, the details will not be repeated here. Appendices D and E, however, contain a tabulation of the properties of the 1959 ARDC model atmosphere for earth as compiled by the U.S. Air Force. These standard atmosphere tables are included in this book for use in solving certain worked examples and some end-of-chapter homework problems. Moreover, Example 1.10 at the end of this section illustrates how the Hydrostatic equation is used to obtain some of the entries in Appendices D and E.

Let the fluid be a liquid, for which we can assume ρ is constant. Consider points 1 and 2 separated by the vertical distance Δh as sketched on the right side of Figure 1.38. The pressure and y locations at these points are (p_1, h_1) and (p_2, h_2) respectively. Integrating Equation (1.52) between points 1 and 2, we have

$$\int_{p_1}^{p_2} dp = -\rho g \int_{h_1}^{h_2} dy$$

or
$$p_2 - p_1 = -\rho g(h_2 - h_1) = \rho g \, \Delta h \qquad (1.53)$$

where $\Delta h = h_1 - h_2$. Equation (1.53) can be more conveniently expressed as

$$p_2 + \rho g h_2 = p_1 + \rho g h_1$$

or
$$\boxed{p + \rho g h = \text{constant}} \qquad (1.54)$$

Note that in Equations (1.53) and (1.54), increasing values of h are in the positive (upward) y direction.

A simple application of Equation (1.54) is the calculation of the pressure distribution on the walls of a container holding a liquid, and open to the atmosphere at the top. This is illustrated in Figure 1.39, where the top of the liquid is at a height h_1. The atmospheric pressure p_a is impressed on the top of the liquid; hence, the pressure at h_1 is simply p_a. Applying Equation (1.54) between the top (where $h = h_1$) and an arbitrary height h, we have

$$p + \rho g h = p_1 + \rho g h_1 = p_a + \rho g h_1$$

or
$$p = p_a + \rho g(h_1 - h) \qquad (1.55)$$

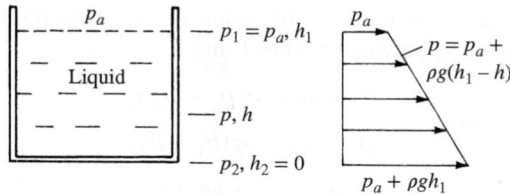

Figure 1.39 Hydrostatic pressure distribution on the walls of a container.

Equation (1.55) gives the pressure distribution on the vertical sidewall of the container as a function of h. Note that the pressure is a *linear* function of h as sketched on the right of Figure 1.39, and that p increases with depth below the surface.

Another simple and very common application of Equation (1.54) is the liquid-filled U-tube manometer used for measuring pressure differences, as sketched in Figure 1.40. The manometer is usually made from hollow glass tubing bent in the shape of the letter U. Imagine that we have an aerodynamic body immersed in an airflow (such as in a wind tunnel), and we wish to use a manometer to measure the surface pressure at point b on the body. A small pressure orifice (hole) at point b is connected to one side of the manometer via a long (usually flexible) pressure tube. The other side of the manometer is open to the atmosphere, where the pressure p_a is a known value. The U tube is partially filled with a liquid of known density ρ. The tops of the liquid on the left and right sides of the U tube are at points 1 and 2, with heights h_1 and h_2, respectively. The body surface pressure p_b is transmitted through the pressure tube and impressed on the top of the liquid at point 1. The atmospheric pressure p_a is impressed on the top of the liquid at point 2. Because in general $p_b \neq p_a$, the tops of the liquid will be at different heights (i.e., the two sides of the manometer will show a displacement $\Delta h = h_1 - h_2$ of the fluid). We wish to obtain the value of the surface pressure

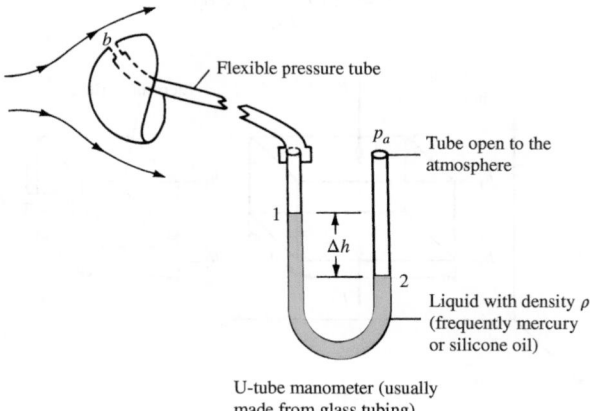

Figure 1.40 The use of a U-tube manometer.

at point b on the body by reading the value of Δh from the manometer. From Equation (1.54) applied between points 1 and 2,

$$p_b + \rho g h_1 = p_a + \rho g h_2$$

or

$$p_b = p_a - \rho g(h_1 - h_2)$$

or

$$p_b = p_a - \rho g \, \Delta h \tag{1.56}$$

In Equation (1.56), p_a, ρ, and g are known, and Δh is read from the U tube, thus allowing p_b to be measured.

At the beginning of this section, we stated that a solid body immersed in a fluid will experience a force even if there is no relative motion between the body and the fluid. We are now in a position to derive an expression for this force, henceforth called the *buoyancy force*. We will consider a body immersed in either a stagnant gas or liquid, hence ρ can be a variable. For simplicity, consider a rectangular body of unit width, length l, and height $(h_1 - h_2)$, as shown in Figure 1.41. Examining Figure 1.41, we see that the vertical force F on the body due to the pressure distribution over the surface is

$$F = (p_2 - p_1)l(1) \tag{1.57}$$

There is no horizontal force because the pressure distributions over the vertical faces of the rectangular body lead to equal and opposite forces which cancel each other. In Equation (1.57), an expression for $p_2 - p_1$ can be obtained by integrating the hydrostatic equation, Equation (1.52), between the top and bottom faces:

$$p_2 - p_1 = \int_{p_1}^{p_2} dp = -\int_{h_1}^{h_2} \rho g \, dy = \int_{h_2}^{h_1} \rho g \, dy$$

Substituting this result into Equation (1.57), we obtain the buoyancy force

$$F = l(1) \int_{h_2}^{h_1} \rho g \, dy \tag{1.58}$$

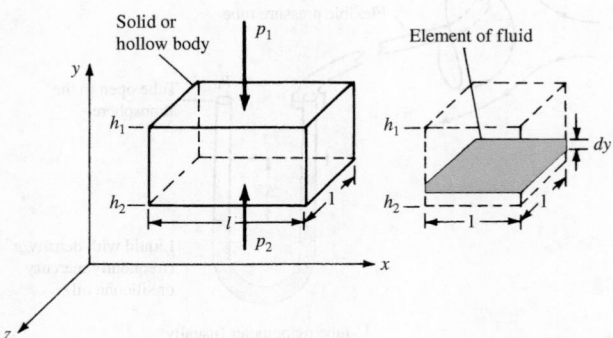

Figure 1.41 Source of the buoyancy force on a body immersed in a fluid.

Consider the physical meaning of the integral in Equation (1.58). The weight of a small element of fluid of height dy and width and length of unity as shown at the right of Figure 1.41 is $\rho g\, dy\,(1)(1)$. In turn, the weight of a column of fluid with a base of unit area and a height $(h_1 - h_2)$ is

$$\int_{h_2}^{h_1} \rho g\, dy$$

which is precisely the integral in Equation (1.58). Moreover, if we place l of these fluid columns side by side, we would have a volume of fluid equal to the volume of the body on the left of Figure 1.41, and the *weight* of this total volume of *fluid* would be

$$l \int_{h_2}^{h_1} \rho g\, dy$$

which is precisely the right-hand side of Equation (1.58). Therefore, Equation (1.58) states in words that

$$\frac{\text{Buoyancy force}}{\text{on body}} = \frac{\text{weight of fluid}}{\text{displaced by body}}$$

We have just proved the well-known *Archimedes principle,* first advanced by the Greek scientist, Archimedes of Syracuse (287–212 B.C.). Although we have used a rectangular body to simplify our derivation, the Archimedes principle holds for bodies of any general shape. (See Problem 1.14 at the end of this chapter.) Also, note from our derivation that the Archimedes principle holds for both gases and liquids and does not require that the density be constant.

The density of liquids is usually several orders of magnitude larger than the density of gases (e.g., for water $\rho = 10^3$ kg/m^3, whereas for air $\rho = 1.23$ kg/m^3). Therefore, a given body will experience a buoyancy force a thousand times greater in water than in air. Obviously, for naval vehicles buoyancy force is all important, whereas for airplanes it is negligible. On the other hand, lighter-than-air vehicles, such as blimps and hot-air balloons, rely on buoyancy force for sustentation; they obtain sufficient buoyancy force simply by displacing huge volumes of air. For most problems in aerodynamics, however, buoyancy force is so small that it can be readily neglected.

EXAMPLE 1.9

A hot-air balloon with an inflated diameter of 30 ft is carrying a weight of 800 lb, which includes the weight of the hot air inside the balloon. Calculate (*a*) its upward acceleration at sea level the instant the restraining ropes are released and (*b*) the maximum altitude it can achieve. Assume that the variation of density in the standard atmosphere is given by $\rho = 0.002377(1 - 7 \times 10^{-6}h)^{4.21}$, where h is the altitude in feet and ρ is in slug/ft^3.

■ **Solution**

(a) At sea level, where $h = 0$, $\rho = 0.002377$ slug/ft^3. The volume of the inflated balloon is $\frac{4}{3}\pi(15)^3 = 14{,}137$ ft^3. Hence,

$$\text{Buoyancy force} = \text{weight of displaced air}$$

$$= g\rho\mathcal{V}$$

where g is the acceleration of gravity and $\mathcal{V}$ is the volume.

$$\text{Buoyancy force} \equiv B = (32.2)(0.002377)(14{,}137) = 1082 \text{ lb}$$

The net upward force at sea level is $F = B - W$, where W is the weight. From Newton's second law,

$$F = B - W = ma$$

where m is the mass, $m = \frac{800}{32.2} = 24.8$ slug. Hence,

$$a = \frac{B - W}{m} = \frac{1082 - 800}{24.8} = \boxed{11.4 \text{ ft/s}^2}$$

(b) The maximum altitude occurs when $B = W = 800$ lb. Since $B = g\rho\mathcal{V}$, and assuming the balloon volume does not change,

$$\rho = \frac{B}{g\mathcal{V}} = \frac{800}{(32.2)(14{,}137)} = 0.00176 \text{ slug/ft}^3$$

From the given variation of ρ with altitude, h,

$$\rho = 0.002377(1 - 7 \times 10^{-6}h)^{4.21} = 0.00176$$

Solving for h, we obtain

$$h = \frac{1}{7 \times 10^{-6}}\left[1 - \left(\frac{0.00176}{0.002377}\right)^{1/4.21}\right] = \boxed{9842 \text{ ft}}$$

EXAMPLE 1.10

The purpose of this example is to show how the standard altitude tables in Appendices D and E are constructed with the use of the Hydrostatic equation. A complete discussion on the construction and use of the standard altitude tables is given in Chapter 3 of Reference 2.

From sea level to an altitude of 11 km, the standard altitude is based on a linear variation of temperature with altitude, h, where T decreases at a rate of -6.5 K per kilometer (the lapse rate). At sea level, the standard pressure, density, and temperature are 1.01325×10^5 N/m^2, 1.2250 kg/m^3, and 288.16 K, respectively. Calculate the pressure, density, and temperature at a standard altitude of 5 km.

■ **Solution**

Repeating Equation (1.52),

$$dp = -g\rho \, dy = -g\rho \, dh$$

The equation of state for a perfect gas is given in Chapter 7 as Equation (7.1),

$$p = \rho RT$$

where R is the specific gas constant. Dividing Equation (1.52) by (7.1), we have

$$\frac{dp}{p} = -\frac{g}{R}\frac{dh}{T} \tag{E1.1}$$

Denoting the lapse rate by a, we have by definition

$$a \equiv \frac{dT}{dh}$$

or

$$dh = \frac{dT}{a} \tag{E1.2}$$

Substituting (E1.2) into (E1.1), we obtain

$$\frac{dp}{p} = -\frac{g}{aR}\frac{dT}{T} \tag{E1.3}$$

Integrate Equation (E1.3) from sea level where the standard values of pressure and temperature are denoted by p_s and T_s, respectively, and a given altitude h where the values of pressure and temperature are p and T, respectively.

$$\int_{p_s}^{p} \frac{dp}{p} = -\int_{T_s}^{T} \frac{g}{aR}\frac{dT}{T}$$

or

$$\ln\frac{p}{p_s} = -\int_{T_s}^{T} \frac{g}{aR}\frac{dT}{T} \tag{E1.4}$$

In Equation (E1.4), a and R are constants, but the acceleration of gravity, g, varies with altitude. The integral in Equation (E1.4) is simplified by assuming that g is constant with altitude, equal to its value at sea level, g_s. With this assumption, Equation (E1.4) becomes

$$\ln\frac{p}{p_s} = -\frac{g_s}{aR}\ln\frac{T}{T_s}$$

or

$$\frac{p}{p_s} = \left(\frac{T}{T_s}\right)^{-g_s/aR} \tag{E1.5}$$

Note: Here we must make a distinction between the geometric altitude, h_G, which is the actual "tape measure" altitude above sea level, and the geopotential altitude, h, which is a slightly fictitious altitude consistent with the assumption of a constant value of g. That is, when we write the Hydrostatic equation as

$$dp = -g\rho\, dh_G$$

we are treating g as a variable with altitude and hence the altitude is the actual geometric altitude, h_G. On the other hand, when we assume that g is constant, say equal to its value at sea level, g_s, the Hydrostatic equation is

$$p = -g_s\rho\, dh \tag{E1.6}$$

where h is denoted as the geopotential altitude, consistent with the assumption of constant g. For reasonable altitudes associated with conventional atmospheric flight, the difference between h_G and h is very small. Observe from Appendix D that altitude is listed under two columns, the first being the geometric altitude, h_G, and the second being the geopotential altitude, h. For the current example, we are calculating the properties at an altitude of 5 km. This is the real "tape measure" altitude, h_G. The corresponding value of geopotential altitude, h, is 4.996 km, only a 0.08 percent difference. The calculation of h for a given h_G is derived in Reference 2; it is not important to our discussions here. What is important, however, is that when we use Equation (E1.5), or any other such equation assuming a constant value of g, we must use the geopotential altitude. For the calculations in this example, where we are calculating properties at a geometric altitude of 5 km, we must use the value of the geopotential altitude, 4.996 km, in the equations.

Equation (E1.5) explicitly gives the variation of pressure with temperature, and implicitly the variation with altitude because temperature is a known function of altitude via the given lapse rate $a = dT/dh = -6.5$ K/km. Specifically, because T varies linearly with altitude for the altitude region under consideration here, we have

$$T - T_s = ah \tag{E1.7}$$

In Equation (E1.7), h is the geopotential altitude. The given value of $a = -6.5$ K/km $= -0.0065$/m is based on the change in geopotential altitude. Thus, from Equation (E1.7), we have at the specified geometric altitude of 5 km,

$$T - 288.16 = -(0.0065)(4996) = -32.474$$

$$T = 288.16 - 32.474 = \boxed{255.69 \text{ K}}$$

Note that this value of T is precisely the value entered in Appendix D for a geometric altitude of 5000 m.

In Equation (E1.5), for air, $R = 287$ J/kg K. Thus, the exponent is

$$\frac{-g_s}{aR} = -\frac{(9.80)}{(-0.0065)(287)} = 5.25328$$

and

$$\frac{p}{p_s} = \left(\frac{T}{T_s}\right)^{-g_s/aR} = \left(\frac{255.69}{288.16}\right)^{5.25328} = 0.53364$$

$$p = 0.53364 p_s = 0.53364(1.01325 \times 10^5)$$

$$p = \boxed{5.407 \times 10^4 \text{ N/m}^2}$$

This value of pressure agrees within 0.04 percent with the value entered in Appendix D. The very slight difference is due to the value of $R = 287$ J/(kg)(K) used here, which depends on the molecular weight of air, which in turn varies slightly from one source to another.

Finally, the density can be obtained from the equation of state

$$\rho = \frac{p}{RT} = \frac{5.407 \times 10^4}{(287)(255.69)} = \boxed{0.7368 \text{ kg/m}^3}$$

which agrees within 0.05 percent with the value entered in Appendix D.

EXAMPLE 1.11

Consider a U-tube mercury manometer oriented vertically. One end is completely sealed with a total vacuum above the column of mercury. The other end is open to the atmosphere where the atmospheric pressure is that for standard sea level. What is the displacement height of the mercury in centimeters, and in which end is the mercury column the highest? The density of mercury is 1.36×10^4 kg/m^3.

■ **Solution**

Examining Figure 1.40, consider the sealed end with the total vacuum to be on the left, where $p_b = 0$, and the height of the mercury column is h_1. This is balanced by the right column of mercury with height h_2 plus the atmospheric pressure p_a exerted on the top of the column. Clearly, when these two columns of mercury are balanced, the left column must be higher to account for the finite pressure being exerted at the top of the right column, i.e., $h_1 > h_2$. From Equation (1.56), we have

$$p_b = p_a - \rho g \, \Delta h$$

where

$$\Delta h = h_1 - h_2$$

Thus,

$$\Delta h = \frac{p_a}{\rho g}$$

From Appendix D, at sea level $p_a = 1.013 \times 10^5$ N/m^2. Hence,

$$\Delta h = \frac{p_a}{\rho g} = \frac{1.013 \times 10^5}{\left(1.36 \times 10^4\right)(9.8)} = 0.76 \text{ m} = \boxed{76 \text{ cm}}$$

Note: Since 2.54 cm = 1 in, then $\Delta h = 76/2.54 = 29.92$ in. The above calculation explains why, on a day when the atmospheric pressure happens to be that for standard sea level, the meteorologist on television will usually say that the pressure is now "76 cm, or 760 mm, or 29.92 in." You rarely hear the pressure quoted in terms of "pounds per square foot," or "pounds per square inch," and hardly ever in "newtons per square meter."

1.10 TYPES OF FLOW

An understanding of aerodynamics, like that of any other physical science, is obtained through a "building-block" approach—we dissect the discipline, form the parts into nice polished blocks of knowledge, and then later attempt to reassemble the blocks to form an understanding of the whole. An example of this process is the way that different types of aerodynamic flows are categorized and visualized. Although nature has no trouble setting up the most detailed and complex flow with a whole spectrum of interacting physical phenomena, we must attempt to understand such flows by modeling them with less detail, and neglecting some of the (hopefully) less significant phenomena. As a result, a study of aerodynamics has evolved into a study of numerous and distinct types of flow. The purpose of this section is to itemize and contrast these types of flow, and to briefly describe their most important physical phenomena.

1.10.1 Continuum Versus Free Molecule Flow

Consider the flow over a body, say, for example, a circular cylinder of diameter d. Also, consider the fluid to consist of individual molecules, which are moving about in random motion. The mean distance that a molecule travels between collisions with neighboring molecules is defined as the *mean-free path* λ. If λ is orders of magnitude smaller than the scale of the body measured by d, then the flow appears to the body as a continuous substance. The molecules impact the body surface so frequently that the body cannot distinguish the individual molecular collisions, and the surface feels the fluid as a continuous medium. Such flow is called *continuum flow*. The other extreme is where λ is on the same order as the body scale; here the gas molecules are spaced so far apart (relative to d) that collisions with the body surface occur only infrequently, and the body surface can feel distinctly each molecular impact. Such flow is called *free molecular flow*. For manned flight, vehicles such as the space shuttle encounter free molecular flow at the extreme outer edge of the atmosphere, where the air density is so low that λ becomes on the order of the shuttle size. There are intermediate cases, where flows can exhibit some characteristics of both continuum and free molecule flows; such flows are generally labeled "low-density flows" in contrast to continuum flow. By far, the vast majority of practical aerodynamic applications involve continuum flows. Low-density and free molecule flows are just a small part of the total spectrum of aerodynamics. Therefore, in this book we will always deal with continuum flow; that is, we will always treat the fluid as a continuous medium.

1.10.2 Inviscid Versus Viscous Flow

A major facet of a gas or liquid is the ability of the molecules to move rather freely, as explained in Section 1.2. When the molecules move, even in a very random fashion, they obviously transport their mass, momentum, and energy from one location to another in the fluid. This transport on a molecular scale gives rise to the phenomena of mass diffusion, viscosity (friction), and thermal conduction.

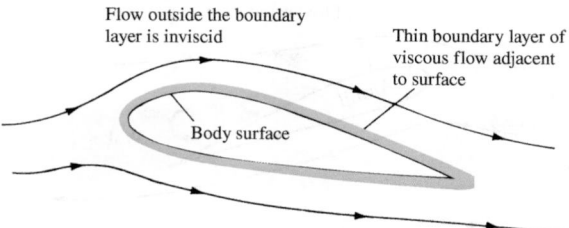

Figure 1.42 The division of a flow into two regions: (1) the thin viscous boundary layer adjacent to the body surface and (2) the inviscid flow outside the boundary layer.

Such "transport phenomena" will be discussed in detail in Chapter 15. For our purposes here, we need only to recognize that all real flows exhibit the effects of these transport phenomena; such flows are called *viscous flows*. In contrast, a flow that is assumed to involve no friction, thermal conduction, or diffusion is called an *inviscid flow*. Inviscid flows do not truly exist in nature; however, there are many practical aerodynamic flows (more than you would think) where the influence of transport phenomena is small, and we can *model* the flow as being inviscid. For this reason, more than 70 percent of this book (Chapters 3 to 14) deals primarily with inviscid flows.

Theoretically, inviscid flow is approached in the limit as the Reynolds number goes to infinity (to be proved in Chapter 15). However, for practical problems, many flows with high but finite Re can be assumed to be inviscid. For such flows, the influence of friction, thermal conduction, and diffusion is limited to a very thin region adjacent to the body surface called the boundary layer, and the remainder of the flow outside this thin region is essentially inviscid. This division of the flow into two regions is illustrated in Figure 1.42. Hence, most of the material discussed in Chapters 3 to 14 applies to the flow outside the boundary layer. For flows over slender bodies, such as the airfoil sketched in Figure 1.42, inviscid theory adequately predicts the pressure distribution and lift on the body and gives a valid representation of the streamlines and flow field away from the body. However, because friction (shear stress) is a major source of aerodynamic drag, inviscid theories by themselves cannot adequately predict total drag.

In contrast, there are some flows that are dominated by viscous effects. For example, if the airfoil in Figure 1.42 is inclined to a high incidence angle to the flow (high angle of attack), then the boundary layer will tend to separate from the top surface, and a large wake is formed downstream. The separated flow is sketched at the top of Figure 1.43; it is characteristic of the flow field over a "stalled" airfoil. Separated flow also dominates the aerodynamics of blunt bodies, such as the cylinder at the bottom of Figure 1.43. Here, the flow expands around the front face of the cylinder, but separates from the surface on the rear face, forming a rather fat wake downstream. The types of flow illustrated in Figure 1.43 are dominated by viscous effects; no inviscid theory can independently predict the

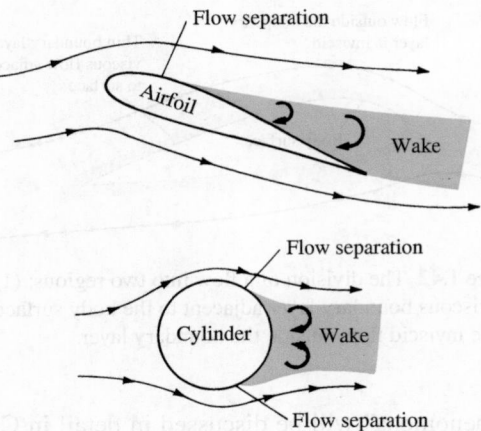

Figure 1.43 Examples of viscous-dominated flow.

aerodynamics of such flows. They require the inclusion of viscous effects, to be presented in Part 4.

1.10.3 Incompressible Versus Compressible Flows

A flow in which the density ρ is constant is called *incompressible*. In contrast, a flow where the density is variable is called *compressible*. A more precise definition of compressibility will be given in Chapter 7. For our purposes here, we will simply note that all flows, to a greater or lesser extent, are compressible; truly incompressible flow, where the density is precisely constant, does not occur in nature. However, analogous to our discussion of inviscid flow, there are a number of aerodynamic problems that can be modeled as being incompressible without any detrimental loss of accuracy. For example, the flow of homogeneous liquids is treated as incompressible, and hence most problems involving hydrodynamics assume $\rho = $ constant. Also, the flow of gases at a low Mach number is essentially incompressible; for $M < 0.3$, it is always safe to assume $\rho = $ constant. (We will prove this in Chapter 8.) This was the flight regime of all airplanes from the Wright brothers' first flight in 1903 to just prior to World War II. It is still the flight regime of most small, general aviation aircraft of today. Hence, there exists a large bulk of aerodynamic experimental and theoretical data for incompressible flows. Such flows are the subject of Chapters 3 to 6. On the other hand, high-speed flow (near Mach 1 and above) must be treated as compressible; for such flows ρ can vary over wide latitudes. Compressible flow is the subject of Chapters 7 to 14.

1.10.4 Mach Number Regimes

Of all the ways of subdividing and describing different aerodynamic flows, the distinction based on the Mach number is probably the most prevalent. If M is the local Mach number at an arbitrary point in a flow field, then by definition the flow is locally:

> *Subsonic* if $M < 1$
>
> *Sonic* if $M = 1$
>
> *Supersonic* if $M > 1$

Looking at the whole field simultaneously, four different speed regimes can be identified using Mach number as the criterion:

1. *Subsonic flow* ($M < 1$ everywhere). A flow field is defined as *subsonic* if the Mach number is less than 1 at every point. Subsonic flows are characterized by smooth streamlines (no discontinuity in slope), as sketched in Figure 1.44a. Moreover, since the flow velocity is everywhere less than the speed of sound, disturbances in the flow (say, the sudden deflection of the trailing edge of the airfoil in Figure 1.44a) propagate both upstream and downstream, and are felt throughout the entire flow field. Note that a freestream Mach number M_∞ less than 1 does not guarantee a totally subsonic flow over the body. In expanding over an aerodynamic shape, the flow velocity increases above the freestream value, and if M_∞ is close enough to 1, the local Mach number may become supersonic in certain regions of the flow. This gives rise to a rule of thumb that $M_\infty < 0.8$ for subsonic flow over slender bodies. For blunt bodies, M_∞ must be even lower to ensure totally subsonic flow. (Again, emphasis is made that the above is just a loose rule of thumb and should not be taken as a precise quantitative definition.) Also, we will show later that incompressible flow is a special limiting case of subsonic flow where $M \to 0$.

2. *Transonic flow* (mixed regions where $M < 1$ and $M > 1$). As stated above, if M_∞ is subsonic but is near unity, the flow can become locally supersonic ($M > 1$). This is sketched in Figure 1.44b, which shows pockets of supersonic flow over both the top and bottom surfaces of the airfoil, terminated by weak shock waves behind which the flow becomes subsonic again. Moreover, if M_∞ is increased slightly above unity, a bow shock wave is formed in front of the body; behind this shock wave, the flow is locally subsonic, as shown in Figure 1.44c. This subsonic flow subsequently expands to a low supersonic value over the airfoil. Weak shock waves are usually generated at the trailing edge, sometimes in a "fishtail" pattern as shown in Figure 1.44c. The flow fields shown in Figure 1.44b and c are characterized by mixed subsonic-supersonic flows and are dominated by the physics of both types of flow. Hence, such flow fields are called *transonic flows*. Again, as a rule of thumb for slender bodies, transonic flows occur for freestream Mach numbers in the range $0.8 < M_\infty < 1.2$.

3. *Supersonic flow* ($M > 1$ everywhere). A flow field is defined as *supersonic* if the Mach number is greater than 1 at every point. Supersonic flows are frequently characterized by the presence of shock waves across which the flow properties and streamlines change discontinuously (in contrast to the

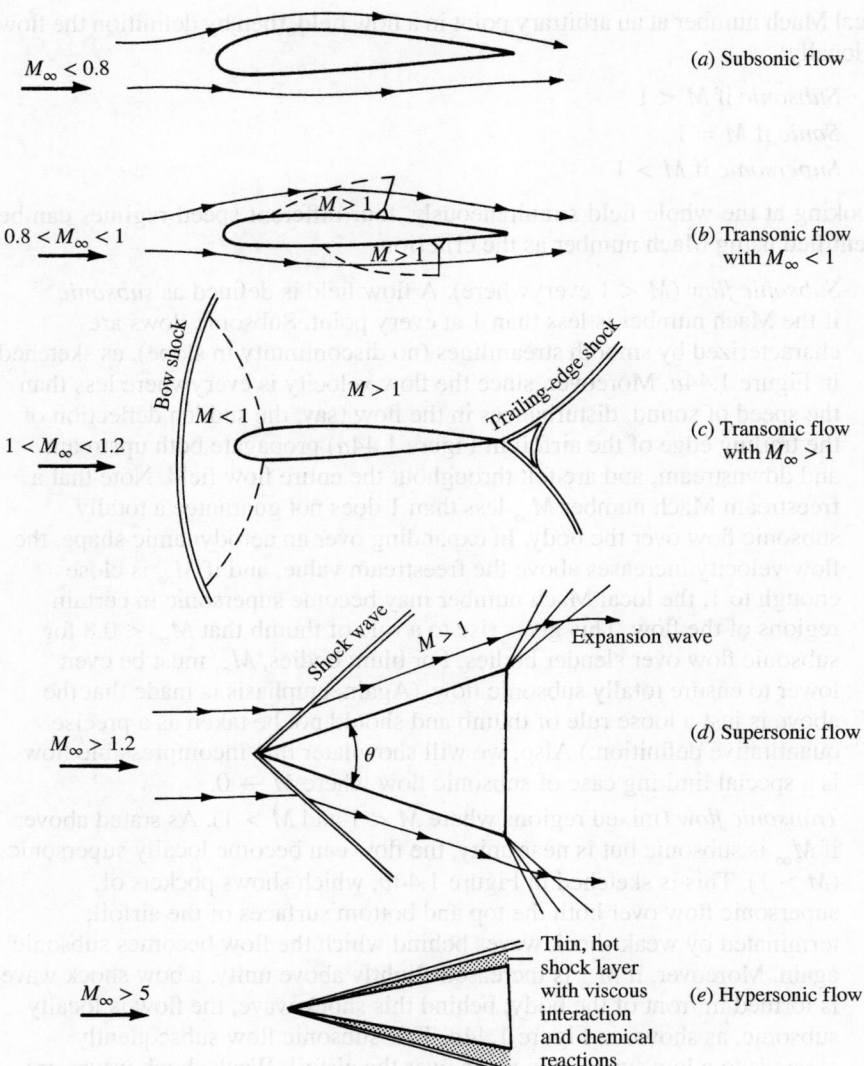

$M_\infty < 0.8$

(a) Subsonic flow

$0.8 < M_\infty < 1$

$M > 1$

$M > 1$

(b) Transonic flow
with $M_\infty < 1$

Bow shock

$1 < M_\infty < 1.2$

$M > 1$

$M < 1$

Trailing-edge shock

(c) Transonic flow
with $M_\infty > 1$

Shock wave

$M > 1$

Expansion wave

$M_\infty > 1.2$

θ

(d) Supersonic flow

$M_\infty > 5$

Thin, hot
shock layer
with viscous
interaction
and chemical
reactions

(e) Hypersonic flow

Figure 1.44 Different regimes of flow.

smooth, continuous variations in subsonic flows). This is illustrated in
Figure 1.44*d* for supersonic flow over a sharp-nosed wedge; the flow
remains supersonic behind the oblique shock wave from the tip. Also
shown are distinct expansion waves, which are common in supersonic
flow. (Again, the listing of $M_\infty > 1.2$ is strictly a rule of thumb. For
example, in Figure 1.44*d*, if θ is made large enough, the oblique shock wave
will detach from the tip of the wedge and will form a strong, curved bow
shock ahead of the wedge with a substantial region of subsonic flow behind
the wave. Hence, the totally supersonic flow sketched in Figure 1.44*d* is

destroyed if θ is too large for a given M_∞. This shock detachment phenomenon can occur at any value of $M_\infty > 1$, but the value of θ at which it occurs increases as M_∞ increases. In turn, if θ is made infinitesimally small, the flow field in Figure 1.44d holds for $M_\infty \geq 1.0$. These matters will be considered in detail in Chapter 9. However, the above discussion clearly shows that the listing of $M_\infty > 1.2$ in Figure 1.44d is a very tenuous rule of thumb and should not be taken literally.) In a supersonic flow, because the local flow velocity is greater than the speed of sound, disturbances created at some point in the flow *cannot* work their way upstream (in contrast to subsonic flow). This property is one of the most significant physical differences between subsonic and supersonic flows. It is the basic reason why shock waves occur in supersonic flows, but do not occur in steady subsonic flow. We will come to appreciate this difference more fully in Chapters 7 to 14.

4. *Hypersonic flow* (very high supersonic speeds). Refer again to the wedge in Figure 1.44d. Assume θ is a given, fixed value. As M_∞ increases above 1, the shock wave moves closer to the body surface. Also, the strength of the shock wave increases, leading to higher temperatures in the region between the shock and the body (the shock layer). If M_∞ is sufficiently large, the shock layer becomes very thin, and interactions between the shock wave and the viscous boundary layer on the surface occur. Also, the shock layer temperature becomes high enough that chemical reactions occur in the air. The O_2 and N_2 molecules are torn apart; that is, the gas molecules dissociate. When M_∞ becomes large enough such that viscous interaction and/or chemically reacting effects begin to dominate the flow (Figure 1.44e), the flow field is called *hypersonic*. (Again, a somewhat arbitrary but frequently used rule of thumb for hypersonic flow is $M_\infty > 5$.) Hypersonic aerodynamics received a great deal of attention during the period 1955–1970 because atmospheric entry vehicles encounter the atmosphere at Mach numbers between 25 (ICBMs) and 36 (the Apollo lunar return vehicle). Again during the period 1985–1995, hypersonic flight received a great deal of attention with the concept of air-breathing supersonic-combustion ramjet-powered transatmospheric vehicles to provide single-stage-to-orbit capability. Today, hypersonic aerodynamics is just part of the whole spectrum of realistic flight speeds. Some basic elements of hypersonic flow are treated in Chapter 14.

In summary, we attempt to organize our study of aerodynamic flows according to one or more of the various categories discussed in this section. The block diagram in Figure 1.45 is presented to help emphasize these categories and to show how they are related. Indeed, Figure 1.45 serves as a road map for this entire book. All the material to be covered in subsequent chapters fits into these blocks, which are lettered for easy reference. For example, Chapter 2 contains discussions of some fundamental aerodynamic principles and equations which fit into both blocks C and D. Chapters 3 to 6 fit into blocks D and E, Chapter 7 fits into blocks D and F, etc. As we proceed with our development of aerodynamics,

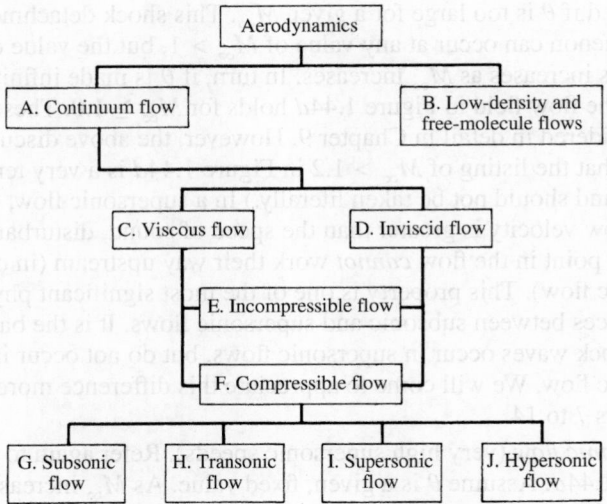

Figure 1.45 Block diagram categorizing the types of aerodynamic flows.

we will frequently refer to Figure 1.45 in order to help put specific, detailed material in proper perspective relative to the whole of aerodynamics.

1.11 VISCOUS FLOW: INTRODUCTION TO BOUNDARY LAYERS

Section 1.10.2 speaks to the problem of friction in an aerodynamic flow. Frictional shear stress is defined in Section 1.4; shear stress, τ, exists at any point in a flow where there is a velocity gradient across streamlines. For most problems in aerodynamics, the local shear stress has a meaningful effect on the flow only where the velocity gradients are substantial. Consider, for example, the flow over the body shown in Figure 1.42. For the vast region of the flow field away from the body, the velocity gradients are relatively small, and friction plays virtually no role. For the thin region of the flow adjacent to the surface, however, the velocity gradients are large, and friction plays a defining role. This natural division of the flow into two regions, one where friction is much more important than the other, was recognized by the famous German fluid dynamicist Ludwig Prandtl in 1904. Prandtl's concept of the boundary layer created a breakthrough in aerodynamic analysis. Since that time, theoretical analyses of most aerodynamic flows have treated the region away from the body as an *inviscid flow* (i.e., no dissipative effects due to friction, thermal conduction, or mass diffusion), and the thin region immediately adjacent to the body surface as a *viscous flow* where these dissipative effects are included. The thin viscous region adjacent to the body is called the *boundary layer;* for most aerodynamic problems of interest, the boundary layer is very thin compared to the extent of the rest of the flow. But what an effect this thin boundary layer has! It is the source of the friction drag on an aerodynamic body. The friction drag on an Airbus 380 jumbo jet, for example, is generated by the boundary layer that wets the entire surface of the airplane in flight. The phenomena of flow separation, as sketched in Figure 1.43, is associated with the presence of the boundary layer; when the flow separates from the surface, it dramatically changes the pressure distribution over the surface resulting in a large increase in drag called *pressure drag*. So this thin viscous boundary layer adjacent to the body surface, although small in extent compared to the rest of the flow, is extremely important in aerodynamics.

Parts 2 and 3 of this book deal primarily with inviscid flows; viscous flow is the subject of Part 4. However, at the end of some of the chapters in Parts 2 and 3, you will find a "viscous flow section" (such as the present section) for the benefit of those readers who are interested in examining the practical impact of boundary layers on the inviscid flows studied in the given chapters. These viscous flow sections are stand-alone sections and do not break the continuity of the inviscid flow discussion in Parts 2 and 3; they simply complement those discussions.

Why are the velocity gradients inside the boundary layer so large? To help answer this question, first consider the purely inviscid flow over the airfoil shape in Figure 1.46. By definition there is no friction effect, so the streamline that is right on the surface of the body *slips* over the surface; for example, the flow velocity at point *b* on the surface is a finite value, unhindered by the effect of friction. In actuality, due to friction the infinitesimally thin layer of air molecules immediately adjacent to the body surface sticks to the surface; thus it has *zero* velocity relative to the surface. This is the *no-slip* condition, and it is the cause of the large velocity gradients within the boundary layer. To see why, consider the flow illustrated in Figure 1.47; here the boundary layer is shown greatly magnified in thickness

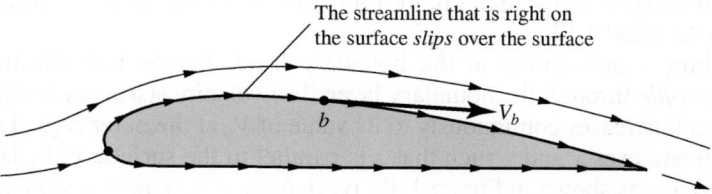

Figure 1.46 Inviscid (frictionless) flow.

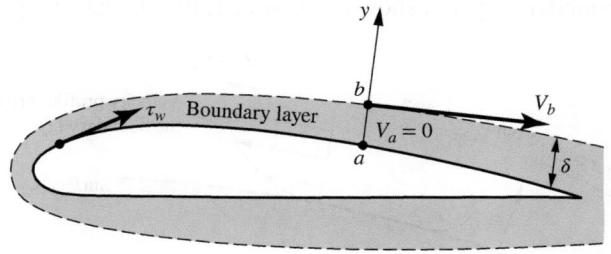

Figure 1.47 Flow in real life, with friction. The thickness of the boundary layer is greatly overemphasized for clarity.

for clarity. The flow velocity at point *a* on the body surface is zero due to the no-slip condition. Above point *a*, the velocity increases until it reaches the value of V_b at point *b* at the outer edge of the boundary layer. Because the boundary layer is so thin, V_b at point *b* in Figure 1.47 is assumed to be the same as V_b at point *b* on the body in the inviscid flow shown in Figure 1.46. Conventional boundary layer analysis assumes that the flow conditions at the outer edge of the boundary layer are the same as the surface flow conditions from an inviscid flow analysis. Examining Figure 1.47, because the flow velocity inside the boundary layer increases from zero at point *a* to a significant finite velocity at point *b*, and this increase takes place over a very short distance because the boundary layer is so thin, then the velocity gradients, the local values of dV/dy, are large. Hence the boundary layer is a region of the flow where frictional effects are dominant.

Also identified in Figure 1.47 is the shear stress at the wall, τ_w, and the boundary layer thickness, δ. Both τ_w and δ are important quantities, and a large part of boundary layer theory is devoted to their calculation.

It can be shown experimentally and theoretically that the pressure through the boundary layer in a direction perpendicular to the surface is constant. That is, letting p_a and p_b be the pressures at points *a* and *b*, respectively, in Figure 1.47, where the *y*-axis is perpendicular to the body at point *a*, then $p_a = p_b$. This is an important phenomenon. This is why the surface pressure distribution calculated for an inviscid flow (Figure 1.46) gives accurate results for the real-life surface pressures; it is because the inviscid calculations give the correct pressure at the outer edge of the thin boundary layer (point *b* in Figure 1.47), and these pressures are impressed without change through the boundary layer right down to the surface (point *a*). The preceding statements are reasonable for thin boundary layers that remain attached to the body surface; they do not hold for regions of separated flow such as those sketched in Figure 1.43. Such separated flows are discussed in Sections 4.12 and 4.13.

Looking more closely at the boundary layer, Figure 1.48 illustrates the *velocity profile* through the boundary layer. The velocity starts out at zero at the surface and increases continuously to its value of V_b at the outer edge. Let us set up coordinate axes *x* and *y* such that *x* is parallel to the surface and *y* is normal to the surface, as shown in Figure 1.48. By definition, a *velocity profile* gives the variation of velocity in the boundary layer as a function of *y*. In general, the velocity profiles at different *x* stations are different. Similarly, the *temperature profile* through the boundary layer is shown in Figure 1.49. The gas temperature at the

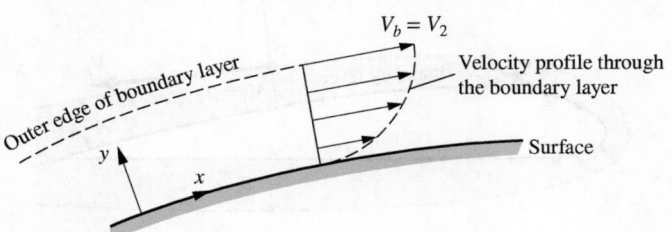

Figure 1.48 Velocity profile through a boundary layer.

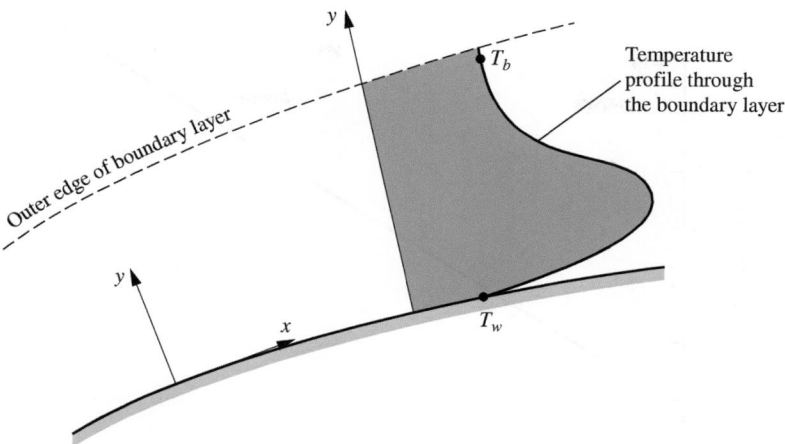

Figure 1.49 Temperature profile through a boundary layer.

wall (which is the same as the surface temperature of the wall itself—a kind of "no slip" condition on temperature) is T_w, and the temperature at the outer edge of the boundary layer is T_b. As before, the value of T_b is the same as the gas temperature at the surface calculated from an inviscid flow analysis. By definition, a *temperature profile* gives the variation of temperature in the boundary layer as a function of y. In general, the temperature profiles at different x stations are different. The temperature inside the boundary layer is governed by the combined mechanisms of thermal conduction and frictional dissipation. Thermal conduction is the transfer of heat from a hotter region to a colder region by random molecular motion. Frictional dissipation, in a very simplistic sense, is the local heating of the gas due to one streamline rubbing over another, somewhat analogous to warming your hands by vigorously rubbing them together. A better explanation of frictional dissipation is that, as a fluid element moves along a streamline inside the boundary layer, it slows down due to frictional shear stress acting on it, and some of the original kinetic energy it had before it entered the boundary layer is converted to internal energy inside the boundary layer, hence increasing the gas temperature inside the boundary layer.

The slope of the velocity profile at the wall is of particular importance because it governs the wall shear stress. Let $(dV/dy)_{y=0}$ be defined as the velocity gradient at the wall. Then the shear stress at the wall is given by

$$\tau_w = \mu \left(\frac{dV}{dy} \right)_{y=0} \tag{1.59}$$

where μ is the *absolute viscosity coefficient* (or simply the *viscosity*) of the gas. The viscosity coefficient has dimensions of mass/(length)(time), as can be verified from Equation (1.59) combined with Newton's second law. It is a physical property of the fluid; μ is different for different gases and liquids. Also, μ varies with T. For liquids, μ decreases as T increases (we all know that oil gets "thinner" when the temperature is increased). But for gases, μ increases as T increases (air

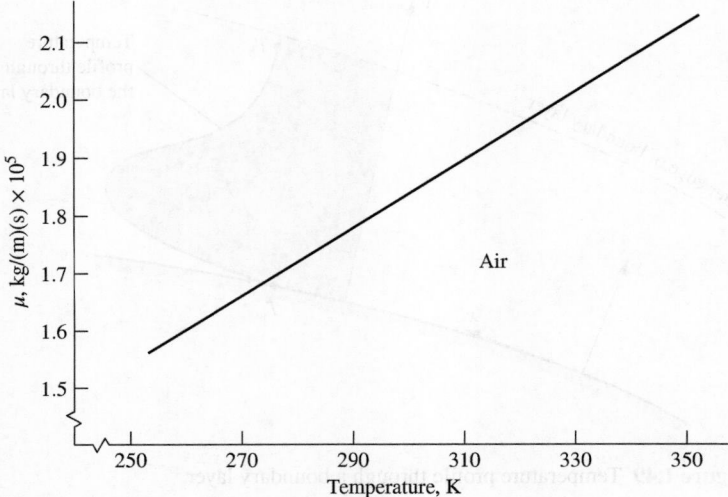

Figure 1.50 Variation of viscosity coefficient with temperature.

gets "thicker" when the temperature is increased). For air at standard sea-level temperature,

$$\mu = 1.7894 \times 10^{-5} \text{kg/(m)(s)} = 3.7373 \times 10^{-7} \text{slug/(ft)(s)}$$

The temperature variation of μ for air over a small range of interest is given in Figure 1.50.

Similarly, the slope of the temperature profile at the wall is very important; it dictates the aerodynamic heating to or from the wall. Let $(dT/dy)_{y=0}$ be defined as the temperature gradient at the wall. Then the aerodynamic heating rate (energy per second per unit area) at the wall is given by

$$\dot{q}_w = -k \left(\frac{dT}{dy} \right)_{y=0} \tag{1.60}$$

where k is the *thermal conductivity* of the gas, and the minus sign connotes that heat is conducted from a warm region to a cooler region, in the opposite direction as the temperature gradient. That is, if the temperature gradient in Equation (1.60) is positive (temperature increases in the direction above the wall), the heat transfer is from the gas *into* the wall, opposite to the direction of increasing temperature. If the temperature gradient is negative (temperature decreases in the direction above the wall), the heat transfer is *away* from the wall into the gas, again opposite to the direction of increasing temperature. Frequently, the heating or cooling of a wall by a flow over the wall is called "convective heat transfer," although from Equation (1.60) the actual mechanism by which heat is transferred between the gas and the wall is thermal *conduction*. In this book, we will label the heat transfer taking place between the boundary layer and the wall as *aerodynamic heating*. Aerodynamic heating is important in high-speed flows, particularly supersonic

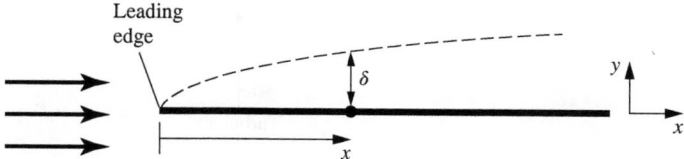

Figure 1.51 Growth of the boundary layer thickness.

flows, and it becomes absolutely dominant in hypersonic flows. Finally, we note that k in Equation (1.60), like μ in Equation (1.59), is a physical property of the fluid, and is a function of temperature. For air at standard sea-level temperature,

$$k = 2.53 \times 10^{-2} \text{J}/(\text{m})(\text{s})(\text{K}) = 3.16 \times 10^{-3} \text{lb}/(\text{s})(°\text{R})$$

Thermal conductivity is essentially proportional to the viscosity coefficient (i.e., $k = (\text{constant}) \times \mu$), so the temperature variation of k is proportional to that shown in Figure 1.49 for μ.

Sections 1.7 and 1.8 introduced the Reynolds number as an important similarity parameter. Consider the development of a boundary layer on a surface, such as the flat plate sketched in Figure 1.51. Let x be measured from the leading edge, that is, from the tip of the plate. Let V_∞ be the flow velocity far upstream of the plate. The *local* Reynolds number at a local distance x from the leading edge is defined as

$$\text{Re}_x = \frac{\rho_\infty V_\infty x}{\mu_\infty} \tag{1.61}$$

where the subscript ∞ is used to denote conditions in the freestream ahead of the plate. The local values of τ_w and δ are functions of Re_x; this is shown in Chapter 4, and in more detail in Part 4 of this book. The Reynolds number has a powerful influence over the properties of a boundary layer, and it governs the nature of viscous flows in general. We will encounter the Reynolds number frequently in this book.

Up to this point in our discussion, we have considered flow streamlines to be smooth and regular curves in space. However, in a viscous flow, and particularly in boundary layers, life is not quite so simple. There are two basic types of viscous flow:

1. *Laminar flow,* in which the streamlines are smooth and regular and a fluid element moves smoothly along a streamline.
2. *Turbulent flow,* in which the streamlines break up and a fluid element moves in a random, irregular, and tortuous fashion.

If you observe smoke rising from a lit cigarette, as sketched in Figure 1.52, you see first a region of smooth flow—laminar flow—and then a transition to irregular, mixed-up flow—turbulent flow. The differences between laminar and turbulent flow are dramatic, and they have a major impact on aerodynamics. For example, consider the velocity profiles through a boundary layer, as sketched in Figure 1.53. The profiles are different, depending on whether the flow is laminar

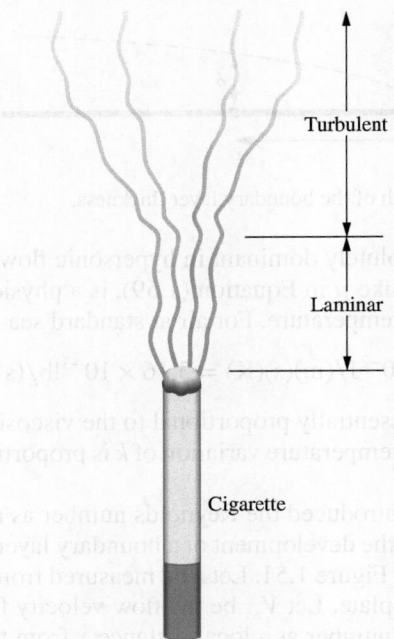

Figure 1.52 Smoke pattern illustrating transition from laminar to turbulent flow.

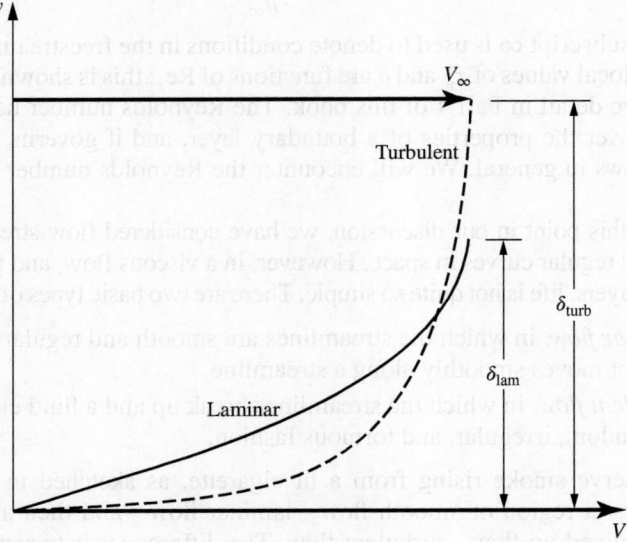

Figure 1.53 Velocity profiles for laminar and turbulent boundary layers. Note that the turbulent boundary layer thickness is larger than the laminar boundary layer thickness.

or turbulent. The turbulent profile is fuller than the laminar profile. For the turbulent profile, from the outer edge to a point near the surface, the velocity remains reasonably close to the freestream velocity; it then rapidly decreases to zero at the surface. In contrast, the laminar velocity profile gradually decreases to zero from the outer edge to the surface. Now consider the velocity gradient at the wall, $(dV/dy)_{y=0}$, which is the reciprocal of the slope of the curves shown in Figure 1.52 evaluated at $y = 0$. From Figure 1.53, it is clear that

$$\left(\frac{dV}{dy}\right)_{y=0} \text{ for laminar flow} < \left(\frac{dV}{dy}\right)_{y=0} \text{ for turbulent flow}$$

Recalling Equation (1.59) for τ_w leads us to the fundamental and highly important fact that *laminar shear stress is less than turbulent shear stress:*

$$(\tau_w)_{\text{laminar}} < (\tau_w)_{\text{turbulent}}$$

This obviously implies that the skin friction exerted on an airplane wing or body will depend on whether the boundary layer on the surface is laminar or turbulent, with laminar flow yielding the smaller skin friction drag.

The same trends hold for aerodynamic heating. We have for the temperature gradients at the wall:

$$\left(\frac{dT}{dy}\right)_{y=0} \text{ for laminar flow} < \left(\frac{dT}{dy}\right)_{y=0} \text{ for turbulent flow}$$

Recalling Equation (1.60) for q_w, we see that turbulent aerodynamic heating is larger than laminar aerodynamic heating, sometimes *considerably* larger. At hypersonic speeds, turbulent heat transfer rates can be almost a factor of 10 larger than laminar heat transfer rates—a showstopper in some hypersonic vehicle designs. We will have a lot to say about the effects of turbulent versus laminar flows in subsequent sections of this book.

In summary, in this section we have presented some introductory thoughts about friction, viscous flows, and boundary layers, in keeping with the introductory nature of this chapter. In subsequent chapters we will expand on these thoughts, including discussions on how to calculate some of the important practical quantities such as τ_w, q_w, and δ.

1.12 APPLIED AERODYNAMICS: THE AERODYNAMIC COEFFICIENTS—THEIR MAGNITUDES AND VARIATIONS

With the present section, we begin a series of special sections in this book under the general heading of "applied aerodynamics." The main thrust of this book is to present the *fundamentals* of aerodynamics, as is reflected in the book's title. However, *applications* of these fundamentals are liberally sprinkled throughout the book, in the text material, in the worked examples, and in the homework problems. The term *applied aerodynamics* normally implies the application of aerodynamics to the practical evaluation of the aerodynamic characteristics of real configurations such as airplanes, missiles, and space vehicles moving through an atmosphere (the earth's, or that of another planet). Therefore, to enhance the reader's appreciation of such applications, sections on applied aerodynamics will appear near the end of many of the chapters. To be specific, in this section, we address the matter of the aerodynamic coefficients defined in Section 1.5; in particular, we focus on lift, drag, and moment coefficients. These nondimensional coefficients are the primary language of applications in external aerodynamics (the distinction between external and internal aerodynamics was made in Section 1.2). It is important for you to obtain a feeling for typical values of the aerodynamic coefficients. (For example, do you expect a drag coefficient to be as low as 10^{-5}, or maybe as high as 1000—does this make sense?) The purpose of this section is to begin to provide you with such a feeling, at least for some common aerodynamic body shapes. As you progress through the remainder of this book, make every effort to note the typical magnitudes of the aerodynamic coefficients that are discussed in various sections. Having a realistic feel for these magnitudes is part of your technical maturity.

Question: What are some typical drag coefficients for various aerodynamic configurations? Some basic values are shown in Figure 1.54. The dimensional analysis described in Section 1.7 proved that $C_D = f(M, \text{Re})$. In Figure 1.54, the drag-coefficient values are for low speeds, essentially incompressible flow; therefore, the Mach number does not come into the picture. (For all practical purposes, for an incompressible flow, the Mach number is theoretically zero, not because the velocity goes to zero, but rather because the speed of sound is infinitely large. This will be made clear in Section 8.3.) Thus, for a low-speed flow, the aerodynamic coefficients for a fixed shape at a fixed orientation to the flow are functions of just the Reynolds number. In Figure 1.54, the Reynolds numbers are listed at the left and the drag-coefficient values at the right. In Figure 1.54a, a flat plate is oriented perpendicular to the flow; this configuration produces the largest possible drag coefficient of any conventional configuration, namely, $C_D = D'/q_\infty S = 2.0$, where S is the frontal area per unit span, i.e., $S = (d)(1)$, where d is the height of the plate. The Reynolds number is based on the height d; that is, $\text{Re} = \rho_\infty V_\infty d/\mu_\infty = 10^5$. Figure 1.54b illustrates flow over a circular cylinder of diameter d; here, $C_D = 1.2$, considerably smaller than the vertical plate value in Figure 1.54a. The drag coefficient can be reduced dramatically by streamlining the body, as shown in Figure 1.54c. Here, $C_D = 0.12$; this is an order of magnitude smaller than the circular cylinder in Figure 1.54b. The Reynolds numbers for Figure 1.54a, b, and c are all the same value, based on d (diameter). The drag coefficients are all defined the same, based on a reference area per unit span of $(d)(1)$. Note that the flow fields over the configurations in Figure 1.54a, b, and c show a wake downstream of the body; the wake is caused by the flow separating from the body surface, with a low-energy, recirculating flow inside the wake. The phenomenon of flow separation will be discussed in detail in Part 4 of this book, dealing with viscous flows. However, it is clear that the wakes diminish in size as we progressively go from Figure 1.54a, b, and c. The fact that

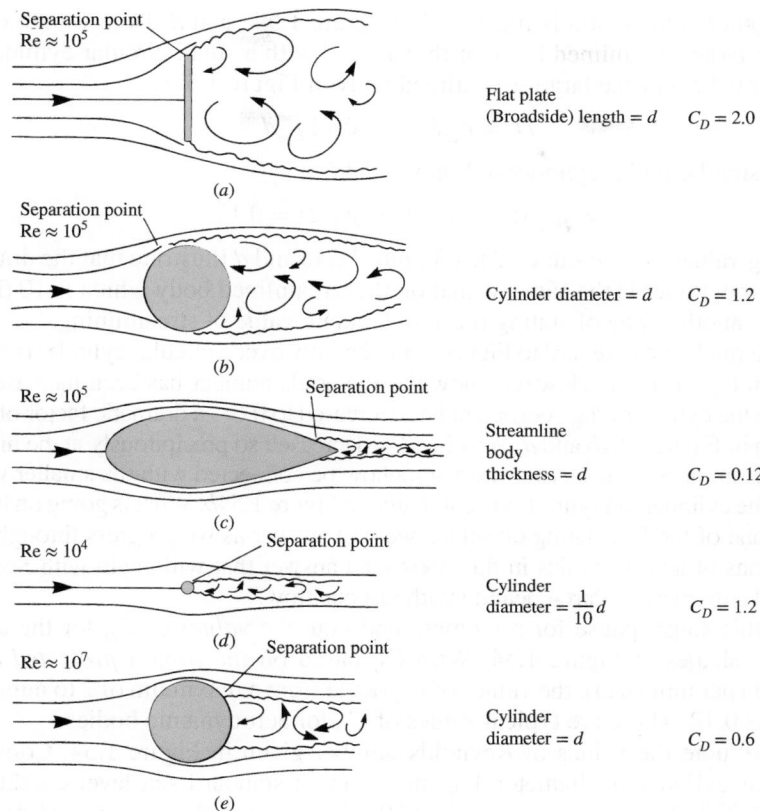

Figure 1.54 Drag coefficients for various aerodynamic shapes. (*Source:* Talay, T. A., *Introduction to the Aerodynamics of Flight,* NASA SP-367, 1975.)

C_D also diminishes progressively from Figure 1.54*a*, *b*, and *c* is no accident—it is a direct result of the regions of separated flow becoming progressively smaller. Why is this so? Simply consider this as one of the many interesting questions in aerodynamics—a question that will be answered in due time in this book. Note, in particular that the physical effect of the streamlining in Figure 1.54*c* results in a very small wake, hence a small value for the drag coefficient.

Consider Figure 1.54*d*, where once again a circular cylinder is shown, but of much smaller diameter. Since the diameter here is 0.1*d*, the Reynolds number is now 10^4 (based on the same freestream V_∞, ρ_∞, and μ_∞ as Figure 1.54*a*, *b*, and *c*). It will be shown in Chapter 3 that C_D for a circular cylinder is relatively independent of Re between Re $= 10^4$ and 10^5. Since the body *shape* is the same between Figure 1.54*d* and *b*, namely, a circular cylinder, then C_D is the same value of 1.2 as shown in the figure. However, since the drag is given by $D' = q_\infty S C_D$, and S is one-tenth smaller in Figure 1.54*d*, then the *drag force* on the small cylinder in Figure 1.54*d* is one-tenth smaller than that in Figure 1.54*b*.

Another comparison is illustrated in Figure 1.54c and d. Here we are comparing a large streamlined body of thickness d with a small circular cylinder of diameter 0.1d. For the large streamlined body in Figure 1.54c,

$$D' = q_\infty S C_D = 0.12 q_\infty d$$

For the small circular cylinder in Figure 1.54d,

$$D' = q_\infty S C_D = q_\infty (0.1d)(1.2) = 0.12 q_\infty d$$

The drag values are the same! Thus, Figure 1.54c and d illustrate that the drag on a circular cylinder is the same as that on the streamlined body which is 10 times thicker—another way of stating the aerodynamic value of streamlining.

As a final note in regard to Figure 1.54, the flow over a circular cylinder is again shown in Figure 1.54e. However, now the Reynolds number has been increased to 10^7, and the cylinder drag coefficient has decreased to 0.6—a dramatic factor of two less than in Figure 1.54b and d. Why has C_D decreased so precipitously at the higher Reynolds number? The answer must somehow be connected with the smaller wake behind the cylinder in Figure 1.54e compared to Figure 1.54b. What is going on here? This is one of the fascinating questions we will answer as we progress through our discussions of aerodynamics in this book—an answer that will begin with Section 3.18 and culminate in Part 4 dealing with viscous flow.

At this stage, pause for a moment and note the *values* of C_D for the aerodynamic shapes in Figure 1.54. With C_D based on the *frontal projected area* ($S = d(1)$ per unit span), the values of C_D range from a maximum of 2 to numbers as low as 0.12. These are typical values of C_D for aerodynamic bodies.

Also, note the values of Reynolds number given in Figure 1.54. Consider a circular cylinder of diameter 1 m in a flow at standard sea level conditions ($\rho_\infty = 1.23$ kg/m^3 and $\mu_\infty = 1.789 \times 10^{-5}$ kg/m $\cdot$ s) with a velocity of 45 m/s (close to 100 mi/h). For this case,

$$\mathrm{Re} = \frac{\rho_\infty V_\infty d}{\mu_\infty} = \frac{(1.23)(45)(1)}{1.789 \times 10^{-5}} = 3.09 \times 10^6$$

Note that the Reynolds number is over 3 million; values of Re in the millions are typical of practical applications in aerodynamics. Therefore, the large numbers given for Re in Figure 1.54 are appropriate.

Let us examine more closely the nature of the drag exerted on the various bodies in Figure 1.54. Since these bodies are at zero angle of attack, the drag is equal to the axial force. Hence, from Equation (1.8) the drag per unit span can be written as

$$D' = \underbrace{\int_{\mathrm{LE}}^{\mathrm{TE}} -p_u \sin\theta \, ds_u + \int_{\mathrm{LE}}^{\mathrm{TE}} p_l \sin\theta \, ds_l}_{\text{pressure drag}}$$

$$+ \underbrace{\int_{\mathrm{LE}}^{\mathrm{TE}} \tau_u \cos\theta \, ds_u + \int_{\mathrm{LE}}^{\mathrm{TE}} \tau_l \cos\theta \, ds_l}_{\text{skin friction drag}} \qquad (1.62)$$

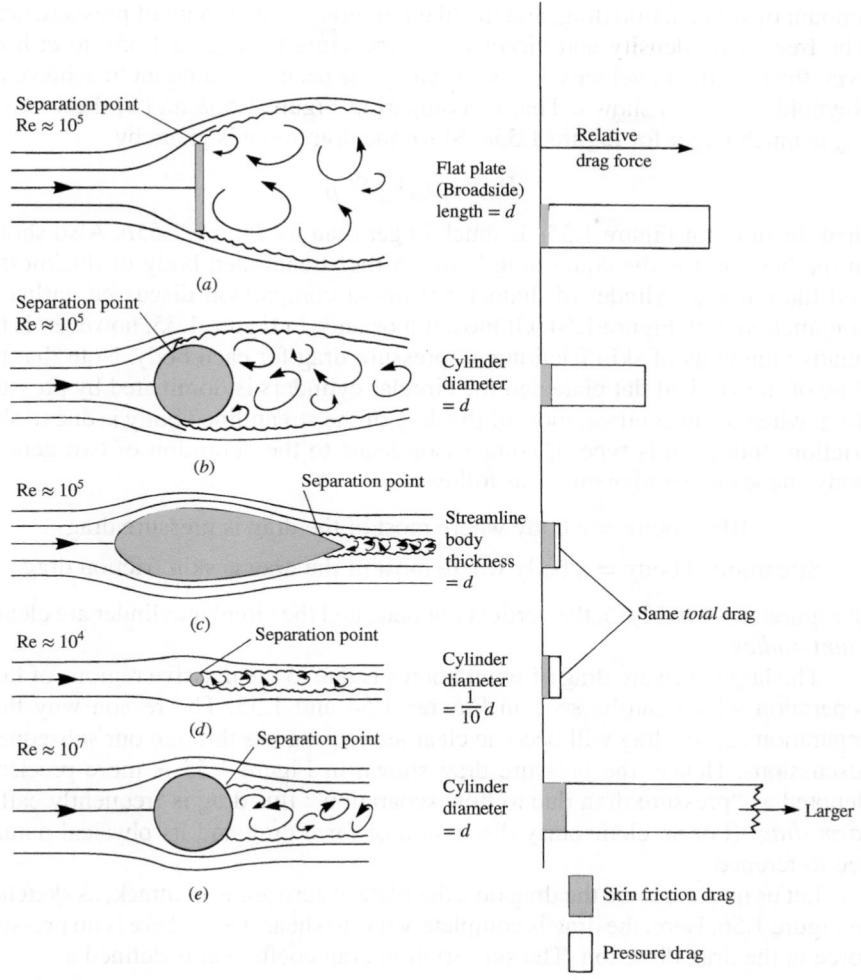

Figure 1.55 The relative comparison between skin friction drag and pressure drag for various aerodynamic shapes. (*Source:* Talay, T. A., *Introduction to the Aerodynamics of Flight,* NASA SP-367, 1975.)

That is, the drag on any aerodynamic body is composed of pressure drag and skin friction drag; this is totally consistent with our discussion in Section 1.5, where it is emphasized that the only two basic sources of aerodynamic force on a body are the pressure and shear stress distributions exerted on the body surface. The division of total drag onto its components of pressure and skin friction drag is frequently useful in analyzing aerodynamic phenomena. For example, Figure 1.55 illustrates the comparison of skin friction drag and pressure drag for the cases shown in Figure 1.54. In Figure 1.55, the bar charts at the right of the figure give the relative drag force on each body; the cross-hatched region denotes the amount of skin friction drag, and the blank region is the amount of pressure drag. The freestream density and viscosity are the same for Figure 1.55a to e; however, the freestream velocity V_∞ is varied by the necessary amount to achieve the Reynolds numbers shown. That is, comparing Figure 1.55b and e, the value of V_∞ is much larger for Figure 1.55e. Since the drag force is given by

$$D' = \tfrac{1}{2}\rho_\infty V_\infty^2 S C_D$$

then the drag for Figure 1.55e is much larger than for Figure 1.55b. Also shown in the bar chart is the equal drag between the streamlined body of thickness d and the circular cylinder of diameter $0.1d$—a comparison discussed earlier in conjunction with Figure 1.54. Of most importance in Figure 1.55, however, is the relative amounts of skin friction and pressure drag for each body. Note that the drag of the vertical flat plate and the circular cylinders is dominated by pressure drag, whereas, in contrast, most of the drag of the streamlined body is due to skin friction. Indeed, this type of comparison leads to the definition of two generic body shapes in aerodynamics, as follows:

Blunt body = a body where most of the drag is pressure drag

Streamlined body = a body where most of the drag is skin friction drag

In Figures 1.54 and 1.55, the vertical flat plate and the circular cylinder are clearly *blunt bodies*.

The large pressure drag of blunt bodies is due to the massive regions of flow separation which can be seen in Figures 1.54 and 1.55. The reason why flow separation causes drag will become clear as we progress through our subsequent discussions. Hence, the pressure drag shown in Figure 1.55 is more precisely denoted as "pressure drag due to flow separation"; this drag is frequently called *form drag*. (For an elementary discussion of form drag and its physical nature, see Reference 2.)

Let us now examine the drag on a flat plate at zero angle of attack, as sketched in Figure 1.56. Here, the drag is completely due to shear stress; there is no pressure force in the drag direction. The skin friction drag coefficient is defined as

$$C_f = \frac{D'}{q_\infty S} = \frac{D'}{q_\infty c(1)}$$

where the reference area is the *planform* area per unit span, that is, the surface area as seen by looking down on the plate from above. C_f will be discussed further in Chapters 4 and 16. However, the purpose of Figure 1.56 is to demonstrate that:

1. C_f is a strong function of Re, where Re is based on the chord length of the plate, $\text{Re} = \rho_\infty V_\infty c / \mu_\infty$. Note that C_f decreases as Re increases.
2. The value of C_f depends on whether the flow over the plate surface is laminar or turbulent, with the turbulent C_f being higher than the laminar C_f at the same Re. What is going on here? What is laminar flow? What is turbulent flow? Why does it affect C_f? The answers to these questions will be addressed in Chapters 4, 15, 17, and 18.

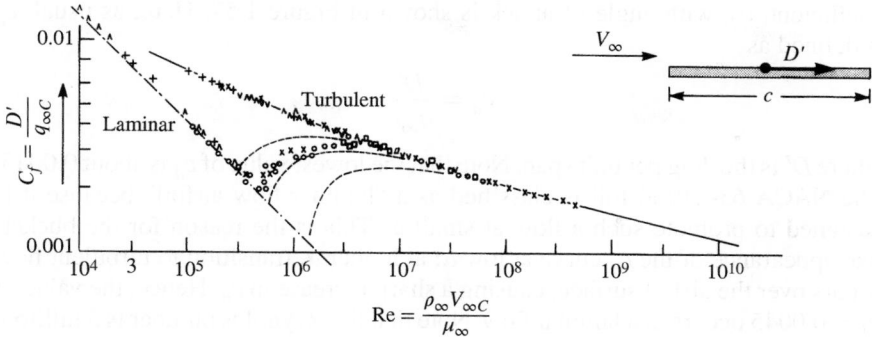

Figure 1.56 Variation of laminar and turbulent skin friction coefficient for a flat plate as a function of Reynolds number based on the chord length of the plate. The intermediate dashed curves are associated with various transition paths from laminar flow to turbulent flow.

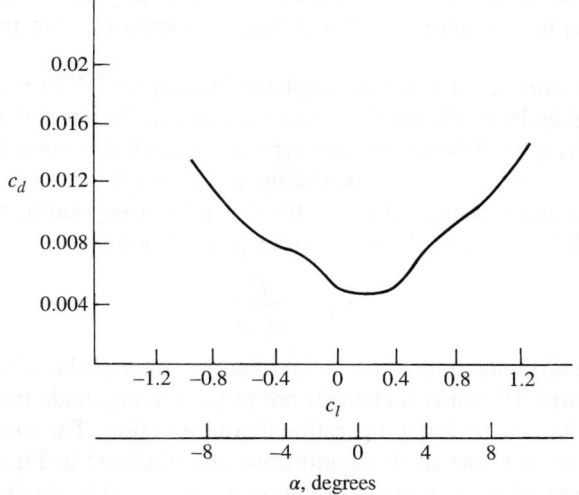

Figure 1.57 Variation of section drag coefficient for an NACA 63-210 airfoil. $\text{Re} = 3 \times 10^6$.

3. The magnitudes of C_f range typically from 0.001 to 0.01 over a large range of Re. These numbers are considerably smaller than the drag coefficients listed in Figure 1.54. This is mainly due to the different reference areas used. In Figure 1.54, the reference area is a cross-sectional area normal to the flow; in Figure 1.56, the reference area is the *planform* area.

A flat plate is not a very practical aerodynamic body—it simply has no volume. Let us now consider a body with thickness, namely, an airfoil section. An NACA 63-210 airfoil section is one such example. The variation of the drag coefficient, c_d, with angle of attack is shown in Figure 1.57. Here, as usual, c_d is defined as

$$c_d = \frac{D'}{q_\infty c}$$

where D' is the drag per unit span. Note that the lowest value of c_d is about 0.0045. The NACA 63-210 airfoil is classified as a "laminar-flow airfoil" because it is designed to promote such a flow at small α. This is the reason for the bucket-like appearance of the c_d curve at low α; at higher α, transition to turbulent flow occurs over the airfoil surface, causing a sharp increase in c_d. Hence, the value of $c_d = 0.0045$ occurs in a laminar flow. Note that the Reynolds number is 3 million. Once again, a reminder is given that the various aspects of laminar and turbulent flows will be discussed in Part 4. The main point here is to demonstrate that typical airfoil drag-coefficient values are on the order of 0.004 to 0.006. As in the case of the streamlined body in Figures 1.54 and 1.55, most of this drag is due to skin friction. However, at higher values of α, flow separation over the top surface of the airfoil begins to appear and pressure drag due to flow separation (form drag) begins to increase. This is why c_d increases with increasing α in Figure 1.57.

Let us now consider a complete airplane. In Chapter 3, Figure 3.2 is a photograph of the Seversky P-35, a typical fighter aircraft of the late 1930s. Figure 1.58 is a detailed drag breakdown for this type of aircraft. Configuration 1 in Figure 1.58 is the stripped-down, aerodynamically cleanest version of this aircraft; its drag coefficient (measured at an angle of attack corresponding to a lift coefficient of $C_L = 0.15$) is $C_D = 0.0166$. Here, C_D is defined as

$$C_D = \frac{D}{q_\infty S}$$

where D is the airplane drag and S is the planform area of the wing. For configurations 2 through 18, various changes are progressively made in order to bring the aircraft to its conventional, operational configuration. The incremental drag increases due to each one of these additions are tabulated in Figure 1.58. Note that the drag coefficient is increased by more than 65 percent by these additions; the value of C_D for the aircraft in full operational condition is 0.0275. This is a typical airplane drag-coefficient value. The data shown in Figure 1.58 were obtained in the full-scale wind tunnel at the NACA Langley Memorial Laboratory just prior to World War II. (The full-scale wind tunnel has test-section dimensions of 30 by 60 ft, which can accommodate a whole airplane—hence the name "full-scale.")

The values of drag coefficients discussed so far in this section have applied to low-speed flows. In some cases, their variation with the Reynolds number has been illustrated. Recall from the discussion of dimensional analysis in Section 1.7 that drag coefficient also varies with the Mach number. *Question:* What is the effect of increasing the Mach number on the drag coefficient of an airplane?

Airplane condition

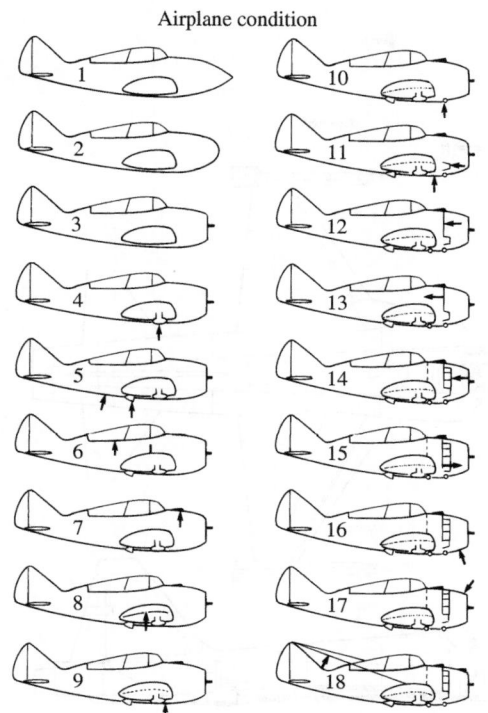

Condition number	Description	C_D ($C_L = 0.15$)	ΔC_D	ΔC_D, %[a]
1	Completely faired condition, long nose fairing	0.0166		
2	Completely faired condition, blunt nose fairing	0.0169		
3	Original cowling added, no airflow through cowling	0.0186	0.0020	12.0
4	Landing-gear seals and fairing removed	0.0188	0.0002	1.2
5	Oil cooler installed	0.0205	0.0017	10.2
6	Canopy fairing removed	0.0203	−0.0002	−1.2
7	Carburetor air scoop added	0.0209	0.0006	3.6
8	Sanded walkway added	0.0216	0.0007	4.2
9	Ejector chute added	0.0219	0.0003	1.8
10	Exhaust stacks added	0.0225	0.0006	3.6
11	Intercooler added	0.0236	0.0011	6.6
12	Cowling exit opened	0.0247	0.0011	6.6
13	Accessory exit opened	0.0252	0.0005	3.0
14	Cowling fairing and seals removed	0.0261	0.0009	5.4
15	Cockpit ventilator opened	0.0262	0.0001	0.6
16	Cowling venturi installed	0.0264	0.0002	1.2
17	Blast tubes added	0.0267	0.0003	1.8
18	Antenna installed	0.0275	0.0008	4.8
	Total		0.0109	

[a]Percentages based on completely faired condition with long nose fairing.

Figure 1.58 The breakdown of various sources of drag on a late 1930s airplane, the Seversky XP-41 (derived from the Seversky P-35 shown in Figure 3.2). [*Source:* Experimental data from Coe, Paul J., "Review of Drag Cleanup Tests in Langley Full-Scale Tunnel (From 1935 to 1945) Applicable to Current General Aviation Airplanes," NASA TN-D-8206, 1976.]

Consider the answer to this question for a Northrop T-38A jet trainer, shown in Figure 1.59. The drag coefficient for this airplane is given in Figure 1.60 as a function of the Mach number ranging from low subsonic to supersonic. The aircraft is at a small negative angle of attack such that the lift is zero; hence the C_D in Figure 1.60 is called the *zero-lift drag coefficient*. Note that the value of C_D is relatively constant from $M = 0.1$ to about 0.86. Why? At Mach numbers of about 0.86, the C_D rapidly increases. This large increase in C_D near Mach one is typical of all flight vehicles. Why? Stay tuned; the answers to these questions will become clear in Part 3 dealing with compressible flow. Also, note in Figure 1.60 that at low subsonic speeds, C_D is about 0.015. This is considerably lower than the 1930s-type airplane illustrated in Figure 1.58; of course, the T-38 is a more modern, sleek, streamlined airplane, and its drag coefficient should be smaller.

We now turn our attention to lift coefficient and examine some typical values. As a complement to the drag data shown in Figure 1.57 for an NACA 63-210

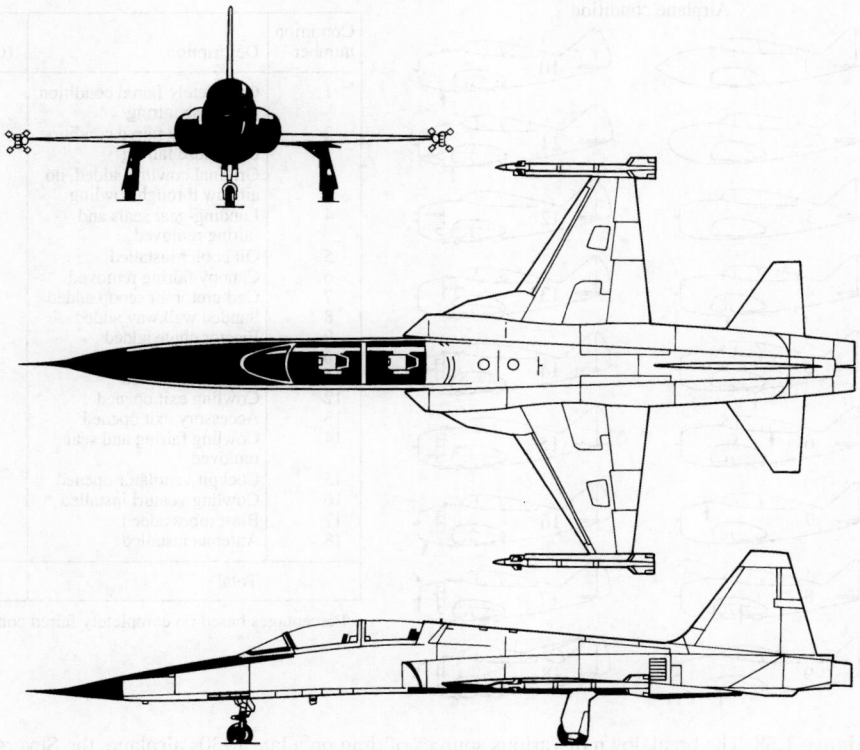

Figure 1.59 Three-view of the Northrop T-38 jet trainer (*Courtesy of the U.S. Air Force*).

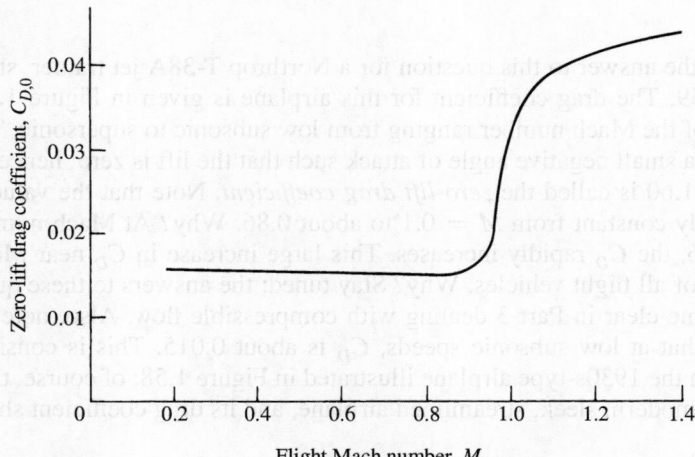

Figure 1.60 Zero-lift drag coefficient variation with Mach number for the T-38. (*Courtesy of the U.S. Air Force.*)

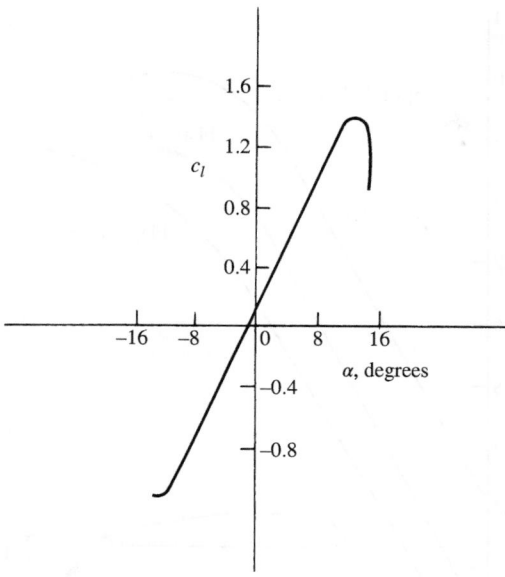

Figure 1.61 Variation of section lift coefficient for an NACA 63-210 airfoil. $Re = 3 \times 10^6$. No flap deflection.

airfoil, the variation of lift coefficient versus angle of attack for the same airfoil is shown in Figure 1.61. Here, we see c_l increasing linearly with α until a maximum value is obtained near $\alpha = 14°$, beyond which there is a precipitous drop in lift. Why does c_l vary with α in such a fashion—in particular, what causes the sudden drop in c_l beyond $\alpha = 14°$? An answer to this question will evolve over the ensuing chapters. For our purpose in the present section, observe the *values* of c_l; they vary from about -1.0 to a maximum of 1.5, covering a range of α from -12 to $14°$. Conclusion: For an airfoil, the magnitude of c_l is about a factor of 100 larger than c_d. A particularly important figure of merit in aerodynamics is the *ratio* of lift to drag, the so-called L/D ratio; many aspects of the flight performance of a vehicle are directly related to the L/D ratio (see, e.g., Reference 2). Other things being equal, a higher L/D means better flight performance. For an airfoil—a configuration whose primary function is to produce lift with as little drag as possible—values of L/D are large. For example, from Figures 1.57 and 1.61, at $\alpha = 4°$, $c_l = 0.6$ and $c_d = 0.0046$, yielding $L/D = \frac{0.6}{0.0046} = 130$. This value is much larger than those for a complete airplane, as we will soon see.

To illustrate the lift coefficient for a complete airplane, Figure 1.62 shows the variation of C_L with α for the T-38 in Figure 1.59. Three curves are shown, each for a different flap deflection angle. (Flaps are sections of the wing at the trailing edge which, when deflected downward, increase the lift of the wing. See Section 5.17 of Reference 2 for a discussion of the aerodynamic properties of flaps.) Note that at a given α, the deflection of the flaps increases C_L. The values of C_L shown

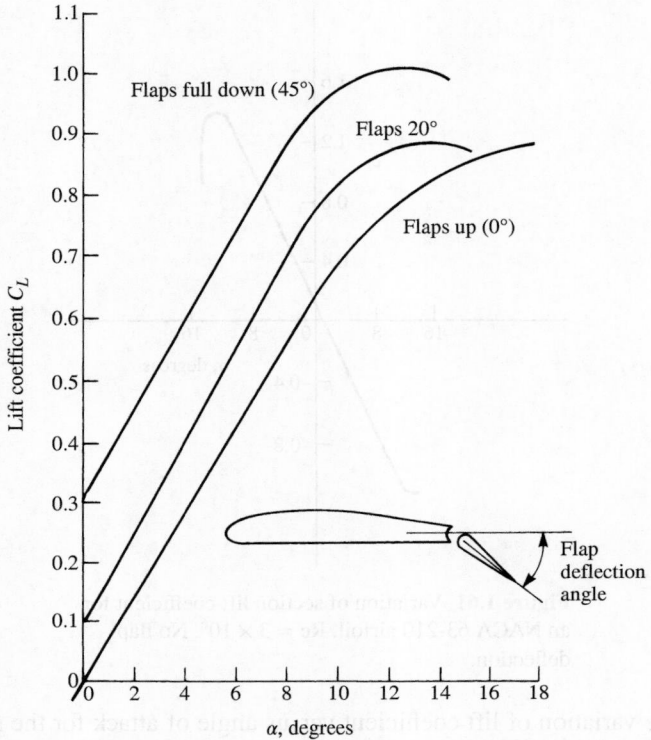

Figure 1.62 Variation of lift coefficient with angle of attack for the T-38. Three curves are shown corresponding to three different flap deflections. Freestream Mach number is 0.4. (*Courtesy of the U.S. Air Force.*)

in Figure 1.62 are about the same as that for an airfoil—on the order of 1. On the other hand, the maximum L/D ratio of the T-38 is about 10—considerably smaller than that for an airfoil alone. Of course, an airplane has a fuselage, engine nacelles, etc., which are elements with other functions than just producing lift, and indeed produce only small amounts of lift while at the same time adding a lot of drag to the vehicle. Thus, the L/D ratio for an airplane is expected to be much less than that for an airfoil alone. Moreover, the wing on an airplane experiences a much higher pressure drag than an airfoil due to the adverse aerodynamic effects of the wing tips (a topic for Chapter 5). This additional pressure drag is called induced drag, and for short, stubby wings, such as on the T-38, the induced drag can be large. (We must wait until Chapter 5 to find out about the nature of induced drag.) As a result, the L/D ratio of the T-38 is fairly small as most airplanes go. For example, the maximum L/D ratio for the Boeing B-52 strategic bomber is 21.5 (see Reference 45). However, this value is still considerably smaller than that for an airfoil alone.

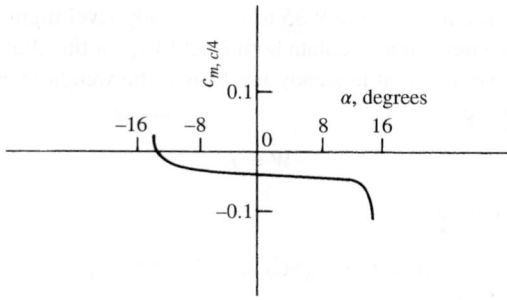

Figure 1.63 Variation of section moment coefficient about the quarter chord for an NACA 63-210 airfoil. Re $= 3 \times 10^6$.

Finally, we turn our attention to the values of moment coefficients. Figure 1.63 illustrates the variation of $c_{m,c/4}$ for the NACA 63-210 airfoil. Note that this is a *negative* quantity; all conventional airfoils produce negative, or "pitch-down," moments. (Recall the sign convention for moments given in Section 1.5.) Also, notice that its value is on the order of -0.035. This value is typical of a moment coefficient—on the order of hundredths.

With this, we end our discussion of typical values of the aerodynamic coefficients defined in Section 1.5. At this stage, you should reread this section, now from the overall perspective provided by a first reading, and make certain to fix in your mind the typical values discussed—it will provide a useful "calibration" for our subsequent discussions.

EXAMPLE 1.12

Consider the Seversky P-35 shown in Figure 3.2. Assume that the drag breakdown given in Figure 1.58 for the XP-41 applies also to the P-35. Note that the data given in Figure 1.58 apply for the specific condition where $C_L = 0.15$. The wing planform area and the gross weight of the P-35 are 220 ft^2 and 5599 lb, respectively. Calculate the horsepower required for the P-35 to fly in steady level flight with $C_L = 0.15$ at standard sea level.

■ **Solution**

From basic mechanics, if **F** is a force exerted on a body moving with a velocity **V**, the power generated by this system is $P = \mathbf{F} \cdot \mathbf{V}$. When **F** and **V** are in the same direction, then the dot product becomes $P = FV$, where F and V are the scalar magnitudes of force and velocity, respectively. When the airplane is in steady level flight (no acceleration) the thrust obtained from the engine exactly counteracts the drag, i.e., $T = D$. Hence the power required for the airplane to fly at a given velocity V_∞ is

$$P = TV_\infty = DV_\infty \qquad \text{(E1.12.1)}$$

To obtain the power required for the P-35 to fly in steady level flight with $C_L = 0.15$, at standard sea level, we must first calculate both D and V_∞ for this flight condition.

To obtain V_∞, we note that in steady level flight the weight is exactly balanced by the aerodynamic lift; i.e.,

$$W = L \tag{E1.12.2}$$

From Section 1.5, we have

$$W = L = q_\infty S C_L = 1/2\rho_\infty V_\infty^2 S C_L \tag{E1.12.3}$$

where S is the planform area of the wing. Solving Equation (E1.12.3) for V_∞, we have

$$V_\infty = \sqrt{\frac{2W}{\rho_\infty S C_L}} \tag{E1.12.4}$$

At standard sea level, from Appendix E (using English engineering units, consistent with the units used in this example), $\rho_\infty = 0.002377$ slug/ft^3. Also, for the P-35 in this example, $S = 220$ ft^2, $W = 5599$ lb, and $C_L = 0.15$. Hence, from Equation (E1.12.3), we have

$$V_\infty = \sqrt{\frac{2(5599)}{(0.002377)(220)(0.15)}} = 377.8 \text{ ft/s}$$

This is the flight velocity when the airplane is flying at standard sea level such that its lift coefficient is 0.15. We note that to fly in steady level flight at any other velocity the lift coefficient would have to be different; to fly slower C_L must be larger and to fly faster C_L must be smaller. Recall that C_L for a given airplane is a function of angle of attack, so our flight condition in this example with $C_L = 0.15$ corresponds to a specific angle of attack of the airplane.

Noting that 88 ft/s = 60 mi/h, V_∞ in miles per hour is $(377.8)\left(\frac{60}{88}\right) = 257.6$ mi/h. In the reference book *The American Fighter* by Enzo Angelucci and Peter Bowers, Orion Books, New York, 1985, the cruising speed of the Seversky P-35 is given as 260 mi/h. Thus, for all practical purposes, the value of $C_L = 0.15$ pertains to cruise velocity at sea level, and this explains why the drag data given in Figure 1.58 was given for a lift coefficient of 0.15.

To complete the calculation of power required, we need the value of D. From Figure 1.58, the drag coefficient for the airplane in full configuration is $C_D = 0.0275$. For the calculated flight velocity, the dynamic pressure is

$$q_\infty = 1/2\rho_\infty V_\infty^2 = 1/2(0.002377)(377.8)^2 = 169.6 \text{ lb/ft}^2$$

Thus,

$$D = q_\infty S C_D = (169.6)(220)(0.0275) = 1026 \text{ lb}$$

From Equation (E1.12.1),

$$P = DV_\infty = (1026)(377.8) = 3.876 \times 10^5 \text{ ft lb/s}$$

Note that 1 horsepower is 550 ft lb/s. Thus, in horsepower,

$$P = \frac{3.876 \times 10^5}{550} = \boxed{704 \text{ hp}}$$

The P-35 was equipped with a Pratt & Whitney R-1830-45 engine rated at 1050 hp. The power required at cruising velocity calculated here is 704 hp, consistent with this engine throttled back for efficient cruise conditions.

The purpose of this worked example is to illustrate typical values of C_L and C_D for a real airplane flying at real conditions, consistent with the subtitle of this section "Applied Aerodynamics: The Aerodynamic Coefficients—Their Magnitudes and Variations." Moreover, we have shown how these coefficients are used to calculate useful aerodynamic performance characteristics for an airplane, such as cruising velocity and power required for steady level flight. This example also underscores the importance and utility of aerodynamic *coefficients*. We made these calculations for a given airplane knowing only the values of the lift and drag *coefficients*, thus again reinforcing the importance of the dimensional analysis given in Section 1.7 and the powerful concept of flow similarity discussed in Section 1.8.

1.13 HISTORICAL NOTE: THE ILLUSIVE CENTER OF PRESSURE

The center of pressure of an airfoil was an important matter during the development of aeronautics. It was recognized in the nineteenth century that, for a heavier-than-air machine to fly at stable, equilibrium conditions (e.g., straight-and-level flight), the moment about the vehicle's center of gravity must be zero (see Chapter 7 of Reference 2). The wing lift acting at the center of pressure, which is generally a distance away from the center of gravity, contributes substantially to this moment. Hence, the understanding and prediction of the center of pressure was felt to be absolutely necessary in order to design a vehicle with proper equilibrium. On the other hand, the early experimenters had difficulty measuring the center of pressure, and much confusion reigned. Let us examine this matter further.

The first experiments to investigate the center of pressure of a lifting surface were conducted by the English engineer George Cayley (1773–1857) in 1808. Cayley was the inventor of the modern concept of the airplane, namely, a vehicle with fixed wings, a fuselage, and a tail. He was the first to separate conceptually the functions of lift and propulsion; prior to Cayley, much thought had gone into ornithopters—machines that flapped their wings for both lift and thrust. Cayley rejected this idea, and in 1799, on a silver disk now in the collection of the Science Museum in London, he inscribed a sketch of a rudimentary airplane with all the basic elements we recognize today. Cayley was an active, inventive, and long-lived man, who conducted numerous pioneering aerodynamic experiments and fervently believed that powered, heavier-than-air, manned flight was inevitable.

(See Chapter 1 of Reference 2 for an extensive discussion of Cayley's contributions to aeronautics.)

In 1808, Cayley reported on experiments of a winged model which he tested as a glider and as a kite. His comments on the center of pressure are as follows:

> By an experiment made with a large kite formed of an hexagon with wings extended from it, all so constructed as to present a hollow curve to the current, I found that when loaded nearly to 1 lb to a foot and 1/2, it required the center of gravity to be suspended so as to leave the anterior and posterior portions of the surface in the ratio of 3 to 7. But as this included the tail operating with a double leverage behind, I think such hollow surfaces relieve about an equal pressure on each part, when they are divided in the ratio of 5 to 12, 5 being the anterior portion. It is really surprising to find so great a difference, and it obliges the center of gravity of flying machines to be much forwarder of the center of bulk (the centroid) than could be supposed a priori.

Here, Cayley is saying that the center of pressure is 5 units from the leading edge and 12 units from the trailing edge (i.e., $x_{cp} = 5/17c$). Later, he states in addition: "I tried a small square sail in one plane, with the weight nearly the same, and I could not perceive that the center-of-resistance differed from the center of bulk." That is, Cayley is stating that the center of pressure in this case is $1/2c$.

There is no indication from Cayley's notes that he recognized that center of pressure moves when the lift, or angle of attack, is changed. However, there is no doubt that he was clearly concerned with the location of the center of pressure and its effect on aircraft stability.

The center of pressure on a flat surface inclined at a small angle to the flow was studied by Samuel P. Langley during the period 1887–1896. Langley was the secretary of the Smithsonian at that time, and devoted virtually all his time and much of the Smithsonian's resources to the advancement of powered flight. Langley was a highly respected physicist and astronomer, and he approached the problem of powered flight with the systematic and structured mind of a scientist. Using a whirling arm apparatus as well as scores of rubber-band powered models, he collected a large bulk of aerodynamic information with which he subsequently designed a full-scale aircraft. The efforts of Langley to build and fly a successful airplane resulted in two dismal failures in which his machine fell into the Potomac River—the last attempt being just 9 days before the Wright brothers' historic first flight on December 17, 1903. In spite of these failures, the work of Langley helped in many ways to advance powered flight. (See Chapter 1 of Reference 2 for more details.)

Langley's observations on the center of pressure for a flat surface inclined to the flow are found in the *Langley Memoir on Mechanical Flight, Part I, 1887 to 1896,* by Samuel P. Langley, and published by the Smithsonian Institution in 1911—5 years after Langley's death. In this paper, Langley states:

> The center-of-pressure in an advancing plane in soaring flight is always in advance of the center of figure, and moves forward as the angle-of-inclination of the sustaining

surfaces diminishes, and, to a less extent, as horizontal flight increases in velocity. These facts furnish the elementary ideas necessary in discussing the problem of equilibrium, whose solution is of the most vital importance to successful flight.

The solution would be comparatively simple if the position of the center-of-pressure could be accurately known beforehand, but how difficult the solution is may be realized from a consideration of one of the facts just stated, namely, that the position of the center-of-pressure in horizontal flight shifts with velocity of the flight itself.

Here, we see that Langley is fully aware that the center of pressure moves over a lifting surface, but that its location is hard to pin down. Also, he notes the correct variation for a flat plate, namely, x_{cp} moves forward as the angle of attack decreases. However, he is puzzled by the behavior of x_{cp} for a curved (cambered) airfoil. In his own words:

> Later experiments conducted under my direction indicate that upon the curved surfaces I employed, the center-of-pressure moves forward with an increase in the angle of elevation, and backward with a decrease, so that it may lie even behind the center of the surface. Since for some surfaces the center-of-pressure moves backward, and for others forward, it would seem that there might be some other surface for which it will be fixed.

Here, Langley is noting the totally opposite behavior of the travel of the center of pressure on a cambered airfoil in comparison to a flat surface, and is indicating ever so slightly some of his frustration in not being able to explain his results in a rational scientific way.

Three-hundred-fifty miles to the west of Langley, in Dayton, Ohio, Orville and Wilbur Wright were also experimenting with airfoils. As described in Section 1.1, the Wrights had constructed a small wind tunnel in their bicycle shop with which they conducted aerodynamic tests on hundreds of different airfoil and wing shapes during the fall, winter, and spring of 1901–1902. Clearly, the Wrights had an appreciation of the center of pressure, and their successful airfoil design used on the 1903 Wright Flyer is a testimonial to their mastery of the problem. Interestingly enough, in the written correspondence of the Wright brothers, only one set of results for the center of pressure can be found. This appears in Wilbur's notebook, dated July 25, 1905, in the form of a table and a graph. The graph is shown in Figure 1.64—the original form as plotted by Wilbur. Here, the center of pressure, given in terms of the percentage of distance from the leading edge, is plotted versus angle of attack. The data for two airfoils are given, one with large curvature (maximum height to chord ratio = 1/12) and one with more moderate curvature (maximum height to chord ratio = 1/20). These results show the now familiar travel of the center of pressure for a curved airfoil, namely, x_{cp} moves forward as the angle of attack is increased, at least for small to moderate values of α. However, the most forward excursion of x_{cp} in Figure 1.64 is 33 percent behind the leading edge—the center of pressure is always behind the quarter-chord point.

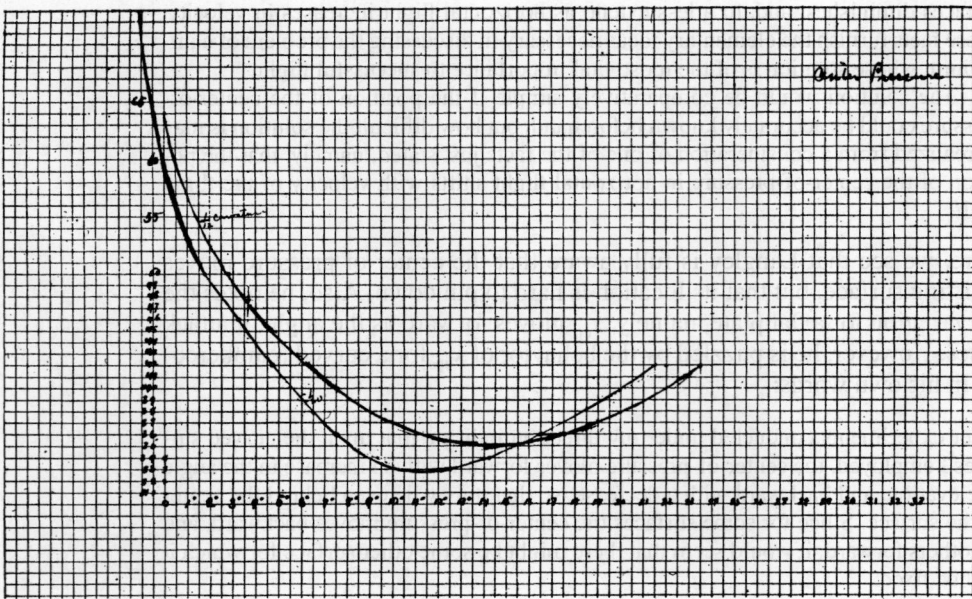

Figure 1.64 Wright brothers' measurements of the center of pressure as a function of angle of attack for a curved (cambered) airfoil. Center of pressure is plotted on the ordinate in terms of percentage distance along the chord from the leading edge. This figure shows the actual data as hand-plotted by Wilbur Wright, which appears in Wilbur's notebook dated July 25, 1905.

The first practical airfoil theory, valid for thin airfoils, was developed by Ludwig Prandtl and his colleagues at the University of Göttingen, Germany, during the period just prior to and during World War I. This thin airfoil theory is described in detail in Chapter 4. The result for the center of pressure for a curved (cambered) airfoil is given by Equation (4.66), and shows that x_{cp} moves forward as the angle of attack (hence c_l) increases, and that it is always behind the quarter-chord point for finite, positive values of c_l. This theory, in concert with more sophisticated wind-tunnel measurements that were being made during the period 1915–1925, finally brought the understanding and prediction of the location of the center of pressure for a cambered airfoil well into focus.

Because x_{cp} makes such a large excursion over the airfoil as the angle of attack is varied, its importance as a basic and practical airfoil property has diminished. Beginning in the early 1930s, the National Advisory Committee for Aeronautics (NACA), at its Langley Memorial Aeronautical Laboratory in Virginia, measured the properties of several systematically designed families of airfoils—airfoils which became a standard in aeronautical engineering. These NACA airfoils are discussed in Sections 4.2 and 4.3. Instead of giving the airfoil data in terms of lift, drag, and center of pressure, the NACA chose the alternate systems of reporting lift, drag, and moments about either the quarter-chord point or the aerodynamic center. These are totally appropriate alternative methods of defining the force-and-moment system on an airfoil, as discussed in Section 1.6 and illustrated in Figure 1.26. As a result, the center of pressure is rarely given as part of modern airfoil data. On the other hand, for three-dimensional bodies, such as slender projectiles and missiles, the location of the center of pressure still remains an important quantity, and modern missile data frequently include x_{cp}. Therefore, a consideration of center of pressure still retains its importance when viewed over the whole spectrum of flight vehicles.

1.14 HISTORICAL NOTE: AERODYNAMIC COEFFICIENTS

In Section 1.5, we introduced the convention of expressing aerodynamic force in terms of an aerodynamic *coefficient,* such as

$$L = \tfrac{1}{2}\rho_\infty V_\infty^2 S C_L$$

and

$$D = \tfrac{1}{2}\rho_\infty V_\infty^2 S C_D$$

where L and D are lift and drag, respectively, and C_L and C_D are the lift coefficient and drag coefficient, respectively. This convention, expressed in the form shown above, dates from about 1920. But the use of some type of aerodynamic coefficients goes back much further. In this section, let us briefly trace the genealogy of aerodynamic coefficients. For more details, see the author's recent book, *A History of Aerodynamics and Its Impact on Flying Machines* (Reference 58).

The first person to define and use aerodynamic force coefficients was Otto Lilienthal, the famous German aviation pioneer at the end of the nineteenth century. Interested in heavier-than-flight from his childhood, Lilienthal carried out the first definitive series of aerodynamic force measurements on cambered (curved) airfoil shapes using a whirling arm. His measurements were obtained over a period of 23 years, culminating in the publication of his book *Der Vogelflug als Grundlage der Fliegekunst (Birdflight as the Basis of Aviation)* in 1889. Many of the graphs in his book are plotted in the form that today we identify as a drag polar, that is, a plot of drag coefficient versus lift coefficient, with the different data points being measured at angles of attack ranging from below zero to 90°. Lilienthal had a degree in Mechanical Engineering, and his work reflected a technical professionalism greater than most at that time. Beginning in 1891, he put his research into practice by designing several gliders, and executing over 2000 successful glider flights before his untimely death in a crash on August 9, 1896. At the time of his death, Lilienthal was working on the design of an engine to power his machines. Had he lived, there is some conjecture that he would have beaten the Wright brothers in the race for the first heavier-than-air, piloted, and powered flight.

In his book, Lilienthal introduced the following equations for the normal and axial forces, which he denoted by N and T, respectively (for normal and "tangential")

$$N = 0.13\eta FV^2 \tag{1.63}$$

and

$$T = 0.13\theta FV^2 \tag{1.64}$$

where, in Lilienthal's notation, F was the reference planform area of the wing in m^2, V is the freestream velocity in m/s, and N and T are in units of kilogram force (the force exerted on one kilogram of mass by gravity at sea level). The number 0.13 is Smeaton's coefficient, a concept and quantity stemming from measurements made in the eighteenth century on flat plates oriented perpendicular to the flow. Smeaton's coefficient is proportional to the density of the freestream; its use is archaic, and it went out of favor at the beginning of the twentieth century. By means of Equations (1.63) and (1.64) Lilienthal introduced the "normal" and "tangential" coefficients, η and θ versus angle of attack. A copy of this table, reproduced in a paper by Octave Chanute in 1897, is shown in Figure 1.65. This became famous as the "Lilienthal Tables," and was used by the Wright brothers for the design of their early gliders. It is proven in Reference 58 that Lilienthal did not use Equations (1.63) and (1.64) explicitly to reduce his experimental data to coefficient form, but rather determined his experimental values for η and θ by dividing the experimental measurements for N and T by his measured force on the wing at 90° angle of attack. In so doing, he divided out the influence of uncertainties in Smeaton's coefficient and the velocity, the former being particularly important because the classical value of Smeaton's coefficient of 0.13 was in error by almost 40 percent. (See Reference 58 for more details.) Nevertheless, we have Otto Lilienthal to thank for the concept of aerodynamic force coefficients, a tradition that has been followed in various modified forms to the present time.

Following on the heals of Lilienthal, Samuel Langley at the Smithsonian Institution published whirling arm data for the resultant aerodynamic force R on a flat plate as a function of angle of attack, using the following equation:

$$R = kSV^2F(\alpha) \tag{1.65}$$

where S is the planform area, k is the *more accurate* value of Smeaton's coefficient (explicitly measured by Langley on his whirling arm), and $F(\alpha)$ was the corresponding force coefficient, a function of angle of attack.

The Wright brothers preferred to deal in terms of lift and drag, and used expressions patterned after Lilienthal and Langley to define lift and drag coefficients:

$$L = kSV^2C_L \tag{1.66}$$

$$D = kSV^2C_D \tag{1.67}$$

The Wrights were among the last to use expressions written explicitly in terms of Smeaton's coefficient k. Gustave Eiffel in 1909 defined a "unit force coefficient" K_i as

$$R = K_iSV^2 \tag{1.68}$$

TABLE OF NORMAL AND TANGENTIAL PRESSURES

Deduced by Lilienthal from the diagrams on Plate VI., in his book "Bird-flight as the Basis of the Flying Art."

α Angle.	η Normal.	ϑ Tangential.	α Angle.	η Normal.	ϑ Tangential.
−9°	0.000	+0.070	16°	0.909	−0.075
−8°	0.040	+0.067	17°	0.915	−0.073
−7°	0.080	+0.064	18°	0.919	−0.070
−6°	0.120	+0.060	19°	0.921	−0.065
−5°	0.160	+0.055	20°	0.922	−0.059
−4°	0.200	+0.049	21°	0.923	−0.053
−3°	0.242	+0.043	22°	0.924	−0.047
−2°	0.286	+0.037	23°	0.924	−0.041
−1°	0.332	+0.031	24°	0.923	−0.036
0°	0.381	+0.024	25°	0.922	−0.031
+1°	0.434	+0.016	26°	0.920	−0.026
+2°	0.489	+0.008	27°	0.918	−0.021
+3°	0.546	0.000	28°	0.915	−0.016
+4°	0.600	−0.007	29°	0.912	−0.012
+5°	0.650	−0.014	30°	0.910	−0.008
+6°	0.696	−0.021	32°	0.906	0.000
+7°	0.737	−0.028	35°	0.896	+0.010
+8°	0.771	−0.035	40°	0.890	+0.016
+9°	0.800	−0.042	45°	0.888	+0.020
10°	0.825	−0.050	50°	0.888	+0.023
11°	0.846	−0.058	55°	0.890	+0.026
12°	0.864	−0.064	60°	0.900	+0.028
13°	0.879	−0.070	70°	0.930	+0.030
14°	0.891	−0.074	80°	0.960	+0.015
15°	0.901	−0.076	90°	1.000	0.000

Figure 1.65 The Lilienthal Table of normal and axial force coefficients. This is a facsimile of the actual table that was published by Octave Chanute in an article entitled "Sailing Flight," *The Aeronautical Annual,* 1897, which was subsequently used by the Wright brothers.

In Equation (1.68), Smeaton's coefficient is nowhere to be seen; it is buried in the direct measurement of K_i. (Eiffel, of Eiffel Tower fame, built a large wind tunnel in 1909, and for the next 14 years reigned as France's leading aerodynamicist until his death in 1923.) After Eiffel's work, Smeaton's coefficient was never used in the aerodynamic literature—it was totally passé.

Gorrell and Martin, in wind tunnel tests, carried out in 1917 at MIT on various airfoil shapes, adopted Eiffel's approach, giving expressions for lift and drag:

$$L = K_y A V^2 \tag{1.69}$$

$$D = K_x A V^2 \tag{1.70}$$

where A denoted planform area and K_y and K_x were the lift and drag coefficients, respectively. For a short period, the use of K_y and K_x became popular in the United States.

However, also by 1917, the density ρ began to appear explicitly in expressions for force coefficients. In NACA Technical Report No. 20, entitled "Aerodynamic Coefficients and Transformation Tables," we find the following expression:

$$F = C\rho S V^2$$

where F is the total force acting on the body, ρ is the freestream density, and C is the force coefficient, which was described as "an abstract number, varying for a given airfoil with its angle of incidence, independent of the choice of units, provided these are consistently used for all four quantities (F, ρ, S, and V)."

Finally, by the end of World War I, Ludwig Prandtl at Gottingen University in Germany established the nomenclature that is accepted as standard today. Prandtl was already famous by 1918 for his pioneering work on airfoil and wing aerodynamics, and for his conception and development of boundary layer theory. (See Section 5.8 for a biographical description of Prandtl.) Prandtl reasoned that the dynamic pressure, $\frac{1}{2}\rho_\infty V_\infty^2$ (he called it "dynamical pressure"), was well suited to describe aerodynamic force. In his 1921 English-language review of works performed at Gottingen before and during World War I (Reference 59), he wrote for the aerodynamic force,

$$W = cFq \tag{1.71}$$

where W is the force, F is the area of the surface, q is the dynamic pressure, and c is a "pure number" (i.e., the force coefficient). It was only a short, quick step to express lift and drag as

$$L = q_\infty S C_L \tag{1.72}$$

and

$$D = q_\infty S C_D \tag{1.73}$$

where C_L and C_D are the "pure numbers" referred to by Prandtl (i.e., the lift and drag coefficients). And this is the way it has been ever since.

1.15 SUMMARY

Refer again to the road map for Chapter 1 given in Figure 1.11. Read again each block in this diagram as a reminder of the material we have covered. If you feel uncomfortable about some of the concepts, go back and reread the pertinent sections until you have mastered the material.

This chapter has been primarily qualitative, emphasizing definitions and basic concepts. However, some of the more important quantitative relations are summarized below:

The normal, axial, lift, drag, and moment coefficients for an aerodynamic body can be obtained by integrating the pressure and skin friction coefficients over the body surface from the leading to the trailing edge. For a two-dimensional body,

$$c_n = \frac{1}{c} \left[\int_0^c (C_{p,l} - C_{p,u}) \, dx + \int_0^c \left(c_{f,u} \frac{dy_u}{dx} + c_{f,l} \frac{dy_l}{dx} \right) dx \right] \qquad (1.15)$$

$$c_a = \frac{1}{c} \left[\int_0^c \left(C_{p,u} \frac{dy_u}{dx} - C_{p,l} \frac{dy_l}{dx} \right) dx + \int_0^c (c_{f,u} + c_{f,l}) \, dx \right] \qquad (1.16)$$

$$c_{m_{\mathrm{LE}}} = \frac{1}{c^2} \left[\int_0^c (C_{p,u} - C_{p,l}) x \, dx - \int_0^c \left(c_{f,u} \frac{dy_u}{dx} + c_{f,l} \frac{dy_l}{dx} \right) x \, dx \right.$$

$$\left. + \int_0^c \left(C_{p,u} \frac{dy_u}{dx} + c_{f,u} \right) y_u \, dx + \int_0^c \left(-C_{p,l} \frac{dy_l}{dx} + c_{f,l} \right) y_l \, dx \right] \qquad (1.17)$$

$$c_l = c_n \cos \alpha - c_a \sin \alpha \qquad (1.18)$$

$$c_d = c_n \sin \alpha + c_a \cos \alpha \qquad (1.19)$$

The center of pressure is obtained from

$$x_{\mathrm{cp}} = -\frac{M'_{\mathrm{LE}}}{N'} \approx -\frac{M'_{\mathrm{LE}}}{L'} \qquad (1.20) \text{ and } (1.21)$$

The criteria for two or more flows to be dynamically similar are:

1. The bodies and any other solid boundaries must be geometrically similar.
2. The similarity parameters must be the same. Two important similarity parameters are Mach number $M = V/a$ and Reynolds number $\mathrm{Re} = \rho V c / \mu$.

If two or more flows are dynamically similar, then the force coefficients C_L, C_D, etc., are the same.

In fluid statics, the governing equation is the hydrostatic equation:

$$dp = -g\rho\, dy \tag{1.52}$$

For a constant density medium, this integrates to

$$p + \rho g h = \text{constant} \tag{1.54}$$

or $\quad\quad\quad\quad\quad\quad\quad p_1 + \rho g h_1 = p_2 + \rho g h_2$

Such equations govern, among other things, the operation of a manometer, and also lead to Archimedes' principle that the buoyancy force on a body immersed in a fluid is equal to the weight of the fluid displaced by the body.

1.16 INTEGRATED WORK CHALLENGE: FORWARD-FACING AXIAL AERODYNAMIC FORCE ON AN AIRFOIL—CAN IT HAPPEN AND, IF SO, HOW?

Note: This section is the first of a number of "Integrated Work Challenges" that appear at the end of a chapter. In contrast to the numerous worked examples that appear at the end of specific sections and which bear on the material just in that specific section, the concept of the "Integrated Work Challenge" is to provide an extended worked example that bears on the chapter as a whole—that serves to integrate and use material from many sections of the chapter.

Concept: Imagine that you place yourself on the chord line of an airfoil that is oriented at a given angle of attack in a flow, such as shown in Figure 1.17. The airfoil is feeling an aerodynamic force generated by the net pressure and shear stress distributions exerted on every square centimeter of the exposed surface of the airfoil (Figure 1.18). There is no propulsion device (no jet engine, rocket, propeller, etc.) on the airfoil, so there is no source of a forward-facing thrust on the airfoil (at least you do not think so). Looking forward along the chord line, common sense (sometimes called "intuition") tells you that the aerodynamic force must always be in the opposite direction, i.e., always "dragging" you back.

Surprise: Your intuition is wrong. There is a situation where the net aerodynamic force results in a forward-facing component along the chord line, i.e., a forward-facing axial force acting on the airfoil. In Figure 1.17, such a forward-facing axial force would be *negative*.

This situation was first noticed by the German engineer Otto Lilienthal during his whirling arm experiments before 1889. (See Anderson, *A History of Aerodynamics*, Cambridge University Press, 1998, pp. 138–164.) In his book *Birdflight as the Basis of Aviation*, page 69, he contrasts the aerodynamic force generated on a cambered (curved) airfoil compared to that on a flat plate: "Under these conditions the characteristic difference of curved surfaces against planes appears still more striking; not only does the direction of the air pressures closely approach that of the perpendicular to the surface, but for certain angles it actually passes beyond it to the other side, converting the usual restraining component into a propelling component." Indeed, turning to Figure 1.65 in Section 1.14, you quickly see negative "tangentials," i.e., a negative forward-facing axial force, tabulated for a wide range of angle of attack. This situation was treated as a "curiosity"; such a forward-facing axial force is counterintuitive. It is real, however.

Challenge: Explain this forward-facing axial force, and find the conditions under which it occurs.

Solution: The solution to this problem is based on geometry. It requires a close examination of the aerodynamic force diagramed in Figure 1.17 and an understanding of the meaning and significance of resolving the resultant aerodynamic force R into its components normal and parallel to the chord (the normal and axial components N and A, respectively) and alternatively resolving R into its components normal and parallel to the freestream velocity (the lift L and drag D, respectively). Since we are interested in the case of a forward-facing axial force, i.e., a negative value of A, let us first examine what Figure 1.17 would look like if the axial force were *zero*. This special case is sketched in Figure 1.66a. Here, because the axial force is zero, the normal force N is the resultant force R on the airfoil. Since, by definition, N is perpendicular to the chord line, then R is perpendicular to the chord line; i.e., the only way for the axial force to be zero is for the resultant aerodynamic force to be perpendicular to the chord line. Since the angle between the chord line and the freestream is, by definition, the angle of attack α, then from Figure 1.66a, where L is perpendicular to the freestream and R is perpendicular to the chord line, the angle between the lift L and the resultant R is also equal to α. From the trigonometry of Figure 1.66,

$$\frac{D}{L} = \tan \alpha$$

or,

$$\frac{L}{D} = \frac{1}{\tan \alpha} = \cot \alpha$$

Consider the case when the drag D is actually *smaller* than that shown in Figure 1.66a, as sketched in Figure 1.66b. In this case, when R is resolved into its normal and axial components, the axial component is facing forward, i.e., A is now facing *forward* along the chord line. Because D in Figure 1.66b is smaller than in Figure 1.66a,

$$\frac{D}{L} < \tan \alpha$$

or,

$$\boxed{\frac{L}{D} > \cot \alpha}$$

This is the answer. When the lift-to-drag ratio for the airfoil is larger than the cotangent of the angle of attack, then there exists a forward-facing axial force on the airfoil, the "propelling component," in the words of Otto Lilienthal quoted earlier.

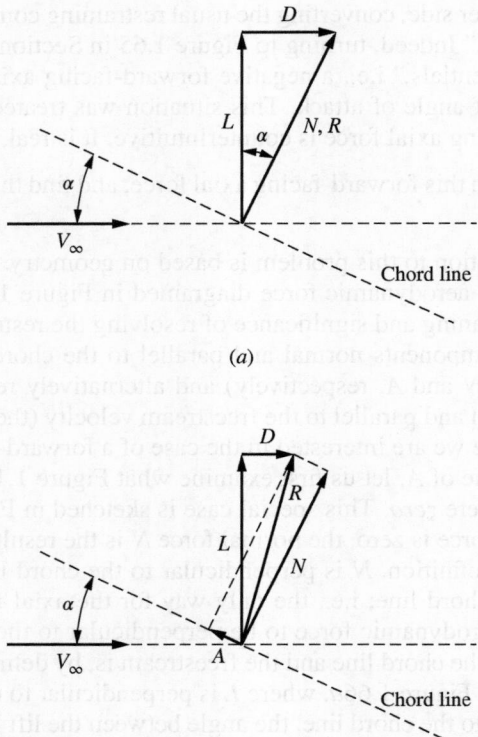

Figure 1.66 Illustration of the condition for a forward-facing axial force. (*a*) Resultant force along the normal. (*b*) Resultant force ahead of the normal.

For example, take the case of a particular airfoil, the NACA 2412 airfoil, with lift and drag data given in Figures 4.10 and 4.11, respectively. (The fact that we are leaping ahead several chapters is irrelevant; we are just extracting some experimental data for an airfoil.) At $\alpha = 6°$, from Figure 4.10, we see that the airfoil lift coefficient is $c_\ell = 0.88$, and from Figure 4.11 the drag coefficient is $c_d = 0.008$. The lift-to-drag ratio is

$$\frac{c_\ell}{c_d} = \frac{0.88}{0.008} = 110$$

In comparison,

$$\cot 6° = 9.52$$

Hence, in this case the airfoil lift-to-drag ratio is much larger than the cot α, and the condition

$$\frac{L}{D} > \cot \alpha$$

is satisfied; the airfoil experiences a forward-facing axial force.

Note: The lift-to-drag ratios for airfoils can be much higher than the lift-to-drag ratios for whole airplanes. Compare the value of 110 obtained here for the NACA 2412 airfoil with the value of 14 for the executive jet transport in Example 1.7. One of the reasons for this is that the whole airplane has a greater surface area contributed by parts of the airplane that do not produce lift, with a consequently much larger skin-friction drag.

Also note: As explained in Chapter 4, by convention, the lift and drag coefficients for an airfoil are expressed by lowercase letters, whereas those for a three-dimensional body, such as the airplane in Example 1.7, are expressed by uppercase letters. We have followed this convention in our present discussion.

Finally, return to the quote from Otto Lilienthal given at the beginning of this section. He makes a distinction between "curved surfaces against planes," meaning that his measurements revealed a forward-facing axial force, i.e., a "propelling component," only for the cambered (curved) airfoils, but not for flat plates ("planes"). This begs the question: Can a forward-facing axial force occur on a flat plate? Answering this question is the essence of Problem 1.20 at the end of this chapter.

1.17 PROBLEMS

1.1 For most gases at standard or near standard conditions, the relationship among pressure, density, and temperature is given by the perfect gas equation of state: $p = \rho RT$, where R is the specific gas constant. For air at near standard conditions, $R = 287$ J/(kg · K) in the International System of Units and $R = 1716$ ft · lb/(slug · °R) in the English Engineering System of Units. (More details on the perfect gas equation of state are given in Chapter 7.) Using the above information, consider the following two cases:

    a. At a given point on the wing of a Boeing 727, the pressure and temperature of the air are 1.9×10^4 N/m^2 and 203 K, respectively. Calculate the density at this point.

    b. At a point in the test section of a supersonic wind tunnel, the pressure and density of the air are 1058 lb/ft^2 and 1.23×10^{-3} slug/ft^3, respectively. Calculate the temperature at this point.

1.2 Starting with Equations (1.7), (1.8), and (1.11), derive in detail Equations (1.15), (1.16), and (1.17).

1.3 Consider an infinitely thin flat plate of chord c at an angle of attack α in a supersonic flow. The pressures on the upper and lower surfaces are different but constant over each surface; that is, $p_u(s) = c_1$ and $p_l(s) = c_2$, where c_1 and c_2 are constants and $c_2 > c_1$. Ignoring the shear stress, calculate the location of the center of pressure.

1.4 Consider an infinitely thin flat plate with a 1 m chord at an angle of attack of 10° in a supersonic flow. The pressure and shear stress distributions on the upper and lower surfaces are given by $p_u = 4 \times 10^4(x-1)^2 + 5.4 \times 10^4$, $p_l = 2 \times 10^4(x-1)^2 + 1.73 \times 10^5$, $\tau_u = 288x^{-0.2}$, and $\tau_l = 731x^{-0.2}$, respectively, where x is the distance from the leading edge in meters and p and τ are in newtons per square meter. Calculate the normal and axial forces, the lift and drag, moments about the leading edge, and moments about the quarter chord, all per unit span. Also, calculate the location of the center of pressure.

1.5 Consider an airfoil at 12° angle of attack. The normal and axial force coefficients are 1.2 and 0.03, respectively. Calculate the lift and drag coefficients.

1.6 Consider an NACA 2412 airfoil (the meaning of the number designations for standard NACA airfoil shapes is discussed in Chapter 4). The following is a tabulation of the lift, drag, and moment coefficients about the quarter chord for this airfoil, as a function of angle of attack.

α (degrees)	c_l	c_d	$c_{m,c/4}$
−2.0	0.05	0.006	−0.042
0	0.25	0.006	−0.040
2.0	0.44	0.006	−0.038
4.0	0.64	0.007	−0.036
6.0	0.85	0.0075	−0.036
8.0	1.08	0.0092	−0.036
10.0	1.26	0.0115	−0.034
12.0	1.43	0.0150	−0.030
14.0	1.56	0.0186	−0.025

From this table, plot on graph paper the variation of x_{cp}/c as a function of α.

1.7 The drag on the hull of a ship depends in part on the height of the water waves produced by the hull. The potential energy associated with these waves therefore depends on the acceleration of gravity g. Hence, we can state that the wave drag on the hull is $D = f(\rho_\infty, V_\infty, c, g)$, where c is a length scale associated with the hull, say, the maximum width of the hull. Define the drag coefficient as $C_D \equiv D/q_\infty c^2$. Also, define a similarity parameter called the *Froude number*, $\mathrm{Fr} = V/\sqrt{gc}$. Using Buckingham's pi theorem, prove that $C_D = f(\mathrm{Fr})$.

1.8 The shock waves on a vehicle in supersonic flight cause a component of drag called supersonic wave drag D_w. Define the wave-drag coefficient as $C_{D,w} = D_w/q_\infty S$, where S is a suitable reference area for the body. In supersonic flight, the flow is governed in part by its thermodynamic properties, given by the specific heats at constant pressure c_p and at constant volume c_v. Define the ratio $c_p/c_v \equiv \gamma$. Using Buckingham's pi theorem, show that $C_{D,w} = f(M_\infty, \gamma)$. Neglect the influence of friction.

1.9 Consider two different flows over geometrically similar airfoil shapes, one airfoil being twice the size of the other. The flow over the smaller airfoil has freestream properties given by $T_\infty = 200$ K, $\rho_\infty = 1.23$ kg/m^3, and $V_\infty = 100$ m/s. The flow over the larger airfoil is described by $T_\infty = 800$ K, $\rho_\infty = 1.739$ kg/m^3, and $V_\infty = 200$ m/s. Assume that both μ and a are proportional to $T^{1/2}$. Are the two flows dynamically similar?

1.10 Consider a Lear jet flying at a velocity of 250 m/s at an altitude of 10 km, where the density and temperature are 0.414 kg/m^3 and 223 K, respectively. Consider also a one-fifth scale model of the Lear jet being tested in a wind tunnel in the laboratory. The pressure in the test section of the wind tunnel is 1 atm $= 1.01 \times 10^5$ N/m^2. Calculate the necessary velocity, temperature, and density of the airflow in the wind-tunnel test section such that the lift and drag coefficients are the same for the wind-tunnel model and the actual airplane in flight. *Note:* The relation among pressure, density, and temperature is given by the equation of state described in Problem 1.1.

1.11 A U-tube mercury manometer is used to measure the pressure at a point on the wing of a wind-tunnel model. One side of the manometer is connected to the model, and the other side is open to the atmosphere. Atmospheric pressure and the density of liquid mercury are 1.01×10^5 N/m^2 and 1.36×10^4 kg/m^3, respectively. When the displacement of the two columns of mercury is 20 cm, with the high column on the model side, what is the pressure on the wing?

1.12 The German Zeppelins of World War I were dirigibles with the following typical characteristics: volume $= 15,000$ m^3 and maximum diameter $= 14.0$ m. Consider a Zeppelin flying at a velocity of 30 m/s at a standard altitude of 1000 m (look up the corresponding density in Appendix D). The Zeppelin is at a small angle of attack such that its lift coefficient is 0.05 (based on the maximum cross-sectional area). The Zeppelin is flying in straight-and-level flight with no acceleration. Calculate the total weight of the Zeppelin.

1.13 Consider a circular cylinder in a hypersonic flow, with its axis perpendicular to the flow. Let ϕ be the angle measured between radii drawn to the leading edge (the stagnation point) and to any arbitrary point on the cylinder. The pressure coefficient distribution along the cylindrical surface is given by $C_p = 2\cos^2\phi$ for $0 \le \phi \le \pi/2$ and $3\pi/2 \le \phi \le 2\pi$ and $C_p = 0$ for $\pi/2 \le \phi \le 3\pi/2$. Calculate the drag coefficient for the cylinder, based on projected frontal area of the cylinder.

1.14 Derive Archimedes' principle using a body of general shape.

1.15 Consider a light, single-engine, propeller-driven airplane similar to a Cessna Skylane. The airplane weight is 2950 lb and the wing reference area is 174 ft^2. The drag coefficient of the airplane C_D is a function of the lift coefficient C_L for reasons that are given in Chapter 5; this function for the given airplane is $C_D = 0.025 + 0.054C_L^2$.

a. For a steady level flight at sea level, where the ambient atmospheric density is 0.002377 slug/ft^3, plot on a graph the variation of C_L, C_D, and the lift-to-drag ratio L/D with flight velocity ranging between 70 ft/s and 250 ft/s.

b. Make some observations about the variation of these quantities with velocity.

1.16 Consider a flat plate at zero angle of attack in a hypersonic flow at Mach 10 at standard sea level conditions. At a point 0.5 m downstream from the leading edge, the local shear stress at the wall is 282 N/m^2. The gas temperature at the wall is equal to standard sea level temperature. At this point, calculate the velocity gradient at the wall normal to the wall.

1.17 Consider the Space Shuttle during its atmospheric entry at the end of a mission in space. At the altitude where the Shuttle has slowed to Mach 9, the local heat transfer at a given point on the lower surface of the wing is 0.03 MW/m^2. Calculate the normal temperature gradient in the air at this point on the wall, assuming the gas temperature at the wall is equal to the standard sea-level temperature.

1.18 The purpose of this problem is to give you a feel for the magnitude of Reynolds number appropriate to real airplanes in actual flight.

a. Consider the DC-3 shown in Figure 1.1. The wing root chord length (distance from the front to the back of the wing where the wing joins the fuselage) is 14.25 ft. Consider the DC-3 flying at 200 miles per hour at sea level. Calculate the Reynolds number for the flow over the wing root chord. (This is an important number, because as we will see later, it governs the skin-friction drag over that portion of the wing.)

b. Consider the F-22 shown in Figure 1.5, and also gracing the cover of this book. The chord length where the wing joins the center body is 21.5 ft. Consider the airplane making a high-speed pass at a velocity of 1320 ft/s at sea level (Mach 1.2). Calculate the Reynolds number at the wing root.

1.19 For the design of their gliders in 1900 and 1901, the Wright brothers used the Lilienthal Table given in Figure 1.65 for their aerodynamic data. Based on these data, they chose a design angle of attack of 3 degrees, and made all their calculations of size, weight, etc., based on this design angle of attack. Why do you think they chose three degrees?

Hint: From the table, calculate the ratio of lift to drag, *L/D*, at 3 degrees angle of attack, and compare this with the lift-to-drag ratio at other angles of attack. You might want to review the design box at the end of Section 1.8, especially Figure 1.36, for the importance of *L/D*.

1.20 Consider the existence of a forward-facing axial aerodynamic force on an airfoil, as discussed in Section 1.16. Can a forward-facing axial force exist on a flat plate at an angle of attack in a flow? Thoroughly explain your answer.

1.21 Derive $P = \rho RT$ from $PV = nR_uT$, where n is the number of moles.

1.22 Calculate R for a 20 percent mixture of H_2 in N_2.

1.23 Calculate the lift force generated by a NACA 1408 airfoil with a chord of 1.5 m and a span of 15 m operating at 50 m/s and an angle of attack of 8 degrees in air at a temperature of 20°C and a pressure of 1 atm. The kinematic viscosity of air at 20°C is $15.11 \times 10^{-6} \, m^2/s$.

1.24 Starting with Equations (1.1) and (1.2), use the definitions of the force coefficients to show that:

$$C_L = C_N \cos \alpha - C_A \sin \alpha \text{ and } C_D = C_N \sin \alpha + C_A \cos \alpha$$

1.25 Derive the expression for the moment coefficient about the leading edge [Equation (1.17)] from the expression for the moment about the leading edge [Equation (1.11)].

1.26 You work for a company that is developing a new family of propeller for fixed-wing UAVs. Your task is to design and test the propellers in a wind tunnel. The key performance parameters are the propeller's thrust T and its power consumption P. Assume that these will be determined by the free stream density ρ_∞, velocity U_∞, propeller diameter D, and propeller speed ω, use dimensional analysis to derive the main nondimensional parameters associated with the performance of the propeller. How should you use these parameters in the conduct of your job?

1.27 Consider two airfoils tested in two different environments where Airfoil 2 is a three times scale replica of Airfoil 1.

Factors	Airfoil 1	Airfoil 2
ρ	1.28 kg/m^3	1.01 kg/m^3
V_∞	100 m/s	300 m/s
T	200 K	1800 K

a. Determine if these flows are dynamically similar assuming that both μ and a are proportional to $T^{\frac{1}{2}}$.

b. If the flows are similar, explain why. If the flows are not similar, what one parameter would you change in Test 2 to make them similar and what would its new value be?

c. Now assume that Airfoil 2 is no longer a three times scale replica of Airfoil 1. Airfoil 2 still has a chord that is three times larger than that of Airfoil 1, but now Airfoil 1 is symmetric whereas Airfoil 2 has positive camber. Are these two flows dynamically similar (assuming both μ and a are proportional to $T^{\frac{1}{2}}$) or could they be made similar by changing one experimental parameter as in b?

1.28 Consider an airplane in level flight at 578 MPH at an altitude of 38,000 ft. It has a rectangular wing 25 ft long and 4 ft wide with a drag coefficient of 0.012. Find the weight of the airplane if its lift to drag ratio is 15.

1.29 Derive an expression for dynamic pressure in terms of pressure and the Mach number.

CHAPTER 2

Aerodynamics: Some Fundamental Principles and Equations

There is so great a difference between a fluid and a collection of solid particles that the laws of pressure and of equilibrium of fluids are very different from the laws of the pressure and equilibrium of solids.
Jean Le Rond d'Alembert, 1768

The principle is most important, not the detail.
Theodore von Karman, 1954

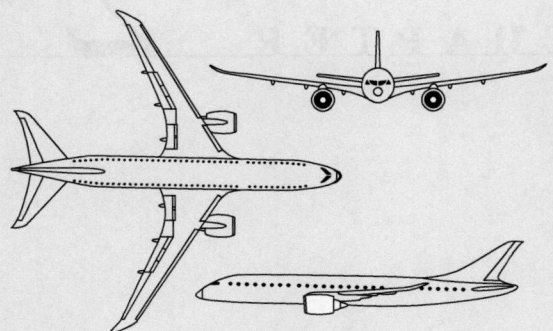

Figure 2.2 Boeing 787 jet airliner.

Boeing aerodynamicists first come up with possible aerodynamic shapes for this airplane? Answer: By using fundamental principles and equations.

This chapter is all about fundamental principles and equations in aerodynamics. The material discussed here is essential to the development of your understanding and appreciation of aerodynamics. Study this material with an open mind. Although this chapter is full of trees, see the forest as well. And this will not be the only chapter in which you will be studying fundamental concepts and equations; as we progress through this book, we will be expanding our inventory of fundamental concepts and equations as necessary to deal with the exciting engineering applications presented in subsequent chapters. So view your study of this chapter as building your foundation in aerodynamics from which you can spring to a lot of interesting applications later on. With this mind set, I predict that you will find this chapter to be intellectually enjoyable.

Central to this chapter is the derivation and discussion of the three most important and fundamental equations in aerodynamics: the continuity, momentum, and energy equations. The continuity equation is a mathematical statement of the fundamental principle that mass is conserved. The momentum equation is a mathematical statement of Newton's second law. The energy equation is a mathematical statement of energy conservation (i.e., the first law of thermodynamics). Nothing in aerodynamics is more fundamental than these three physical principles, and no equations in aerodynamics are more basic than the continuity, momentum, and energy equations. Make these three equations and fundamental principles your constant companions as you travel through this book—indeed, as you travel through all your study and work in aerodynamics, however far that may be.

2.1 INTRODUCTION AND ROAD MAP

To be a good craftsperson, one must have good tools and must know how to use them effectively. Similarly, a good aerodynamicist must have good aerodynamic tools and must know how to use them for a variety of applications. The purpose of this chapter is "tool-building"; we develop some of the concepts and equations that are vital to the study of aerodynamic flows. However, please be cautioned: A craftsperson usually derives his or her pleasure from the works of art created with the *use* of the tools; the actual building of the tools themselves is sometimes considered a mundane chore. You may derive a similar feeling here. As we proceed to build our aerodynamic tools, you may wonder from time to time why such tools are necessary and what possible value they may have in the solution of practical problems. Rest assured, however, that every aerodynamic tool we develop in this and subsequent chapters is important for the analysis and understanding of practical problems to be discussed later. So, as we move through this chapter, do not get lost or disoriented; rather, as we develop each tool, simply put it away in the store box of your mind for future use.

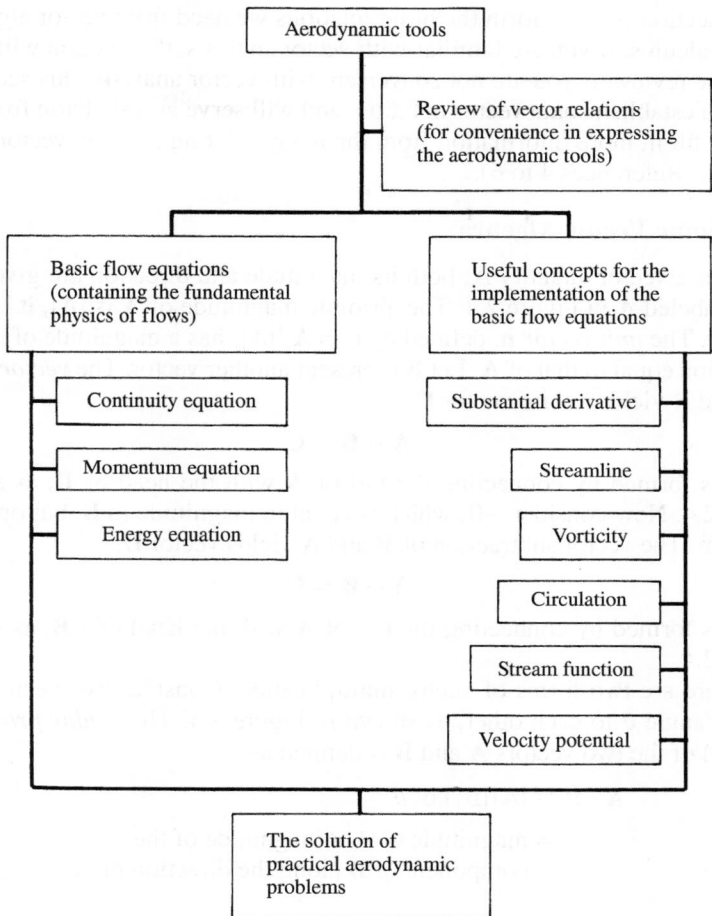

Figure 2.3 Road map for Chapter 2.

To help you keep track of our tool building, and to give you some orientation, the road map in Figure 2.3 is provided for your reference. As we progress through each section of this chapter, use Figure 2.3 to help you maintain a perspective of our work. You will note that Figure 2.3 is full of strange-sounding terms, such as "substantial derivative," "circulation," and "velocity potential." However, when you finish this chapter, and look back at Figure 2.3, all these terms will be second nature to you.

2.2 REVIEW OF VECTOR RELATIONS

Aerodynamics is full of quantities that have both magnitude and direction, such as force and velocity. These are *vector quantities,* and as such, the mathematics of aerodynamics is most conveniently expressed in vector notation. The purpose of this section is to set forth the basic relations we need from vector algebra and vector calculus. If you are familiar with vector analysis, this section will serve as a concise review. If you are not conversant with vector analysis, this section will help you establish some vector notation, and will serve as a skeleton from which you can fill in more information from the many existing texts on vector analysis (see, e.g., References 4 to 6).

2.2.1 Some Vector Algebra

Consider a vector quantity $\mathbf{A}$; both its magnitude and direction are given by the arrow labeled $\mathbf{A}$ in Figure 2.4. The absolute magnitude of $\mathbf{A}$ is $|\mathbf{A}|$, it is a scalar quantity. The *unit vector* $\mathbf{n}$, defined by $\mathbf{n} = \mathbf{A}/|\mathbf{A}|$, has a magnitude of unity and a direction equal to that of $\mathbf{A}$. Let $\mathbf{B}$ represent another vector. The *vector addition* of $\mathbf{A}$ and $\mathbf{B}$ yields a third vector $\mathbf{C}$,

$$\mathbf{A} + \mathbf{B} = \mathbf{C} \tag{2.1}$$

which is formed by connecting the tail of $\mathbf{A}$ with the head of $\mathbf{B}$, as shown in Figure 2.4. Now consider $-\mathbf{B}$, which is equal in magnitude to $\mathbf{B}$, but opposite in direction. The vector subtraction of $\mathbf{B}$ and $\mathbf{A}$ yields vector $\mathbf{D}$,

$$\mathbf{A} - \mathbf{B} = \mathbf{D} \tag{2.2}$$

which is formed by connecting the tail of $\mathbf{A}$ with the head of $-\mathbf{B}$, as shown in Figure 2.4.

There are two forms of vector multiplication. Consider two vectors $\mathbf{A}$ and $\mathbf{B}$ at an angle θ to each other, as shown in Figure 2.4. The *scalar product* (dot product) of the two vectors $\mathbf{A}$ and $\mathbf{B}$ is defined as

$$\mathbf{A} \cdot \mathbf{B} \equiv |\mathbf{A}||\mathbf{B}| \cos \theta \tag{2.3}$$

$$= \text{magnitude of } \mathbf{A} \times \text{magnitude of the}$$
$$\text{component of } \mathbf{B} \text{ along the direction of } \mathbf{A}$$

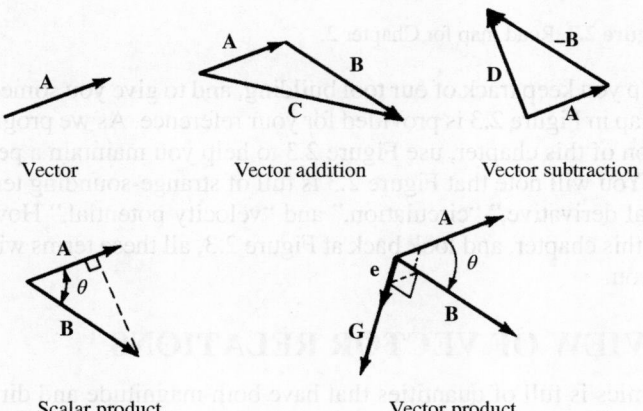

Vector   Vector addition   Vector subtraction

Scalar product   Vector product

Figure 2.4 Vector diagrams.

Note that the scalar product of two vectors is a scalar. In contrast, the *vector product* (cross product) of the two vectors **A** and **B** is defined as

$$\mathbf{A} \times \mathbf{B} \equiv (|\mathbf{A}||\mathbf{B}| \sin \theta)\mathbf{e} = \mathbf{G} \qquad (2.4)$$

where **G** is perpendicular to the plane of **A** and **B** and in a direction which obeys the "right-hand rule." (Rotate **A** into **B**, as shown in Figure 2.4. Now curl the fingers of your right hand in the direction of the rotation. Your right thumb will be pointing in the direction of **G**.) In Equation (2.4), **e** is a unit vector in the direction of **G**, as also shown in Figure 2.4. Note that the vector product of two vectors is a vector.

2.2.2 Typical Orthogonal Coordinate Systems

To describe mathematically the flow of fluid through three-dimensional space, we have to prescribe a three-dimensional coordinate system. The geometry of some aerodynamic problems best fits a rectangular space, whereas others are mainly cylindrical in nature, and yet others may have spherical properties. Therefore, we have interest in the three most common orthogonal coordinate systems: cartesian, cylindrical, and spherical. These systems are described below. (An orthogonal coordinate system is one where all three coordinate directions are mutually perpendicular. It is interesting to note that some modern numerical solutions of fluid flows utilize nonorthogonal coordinate spaces; moreover, for some numerical problems the coordinate system is allowed to evolve and change during the course of the solution. These so-called adaptive grid techniques are beyond the scope of this book. See Reference 7 for details.)

A *cartesian coordinate system* is shown in Figure 2.5a. The x, y, and z axes are mutually perpendicular, and **i**, **j**, and **k** are unit vectors in the x, y, and z directions, respectively. An arbitrary point P in space is located by specifying

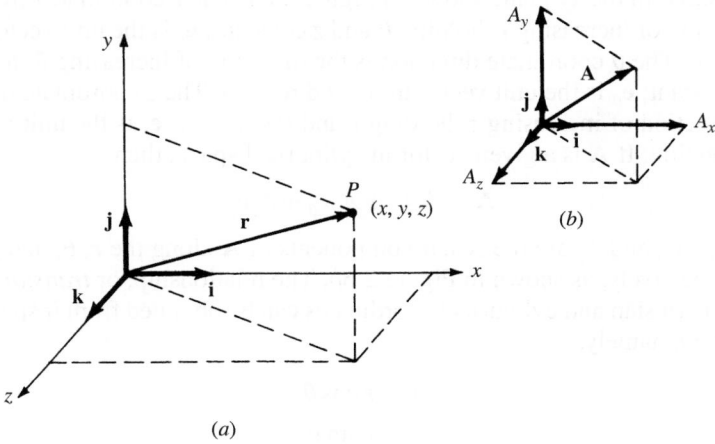

Figure 2.5 Cartesian coordinates.

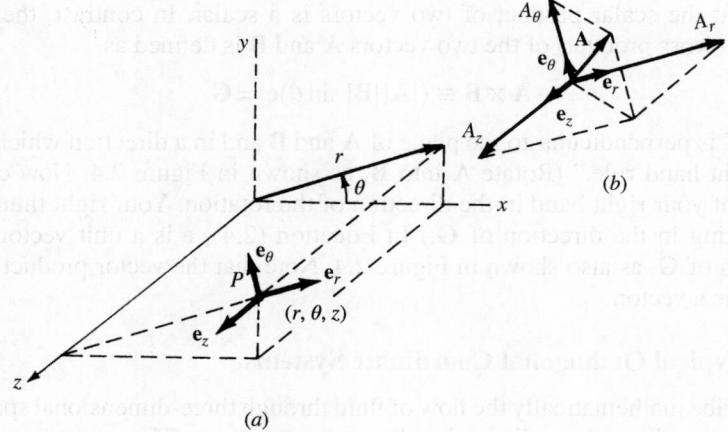

Figure 2.6 Cylindrical coordinates.

the three coordinates (x, y, z). The point can also be located by the *position vector* $\mathbf{r}$, where

$$\mathbf{r} = x\mathbf{i} + y\mathbf{j} + z\mathbf{k}$$

If $\mathbf{A}$ is a given vector in cartesian space, it can be expressed as

$$\mathbf{A} = A_x\mathbf{i} + A_y\mathbf{j} + A_z\mathbf{k}$$

where A_x, A_y, and A_z are the scalar components of $\mathbf{A}$ along the x, y, and z directions, respectively, as shown in Figure 2.5*b*.

A *cylindrical coordinate system* is shown in Figure 2.6*a*. A "phantom" cartesian system is also shown with dashed lines to help visualize the figure. The location of point P in space is given by three coordinates (r, θ, z), where r and θ are measured in the xy plane shown in Figure 2.6*a*. The r coordinate direction is the direction of increasing r, holding θ and z constant; $\mathbf{e}_r$ is the unit vector in the r direction. The θ coordinate direction is the direction of increasing θ, holding r and z constant; $\mathbf{e}_\theta$ is the unit vector in the θ direction. The z coordinate direction is the direction of increasing z, holding r and θ constant; $\mathbf{e}_z$ is the unit vector in the z direction. If $\mathbf{A}$ is a given vector in cylindrical space, then

$$\mathbf{A} = A_r\mathbf{e}_r + A_\theta\mathbf{e}_\theta + A_z\mathbf{e}_z$$

where A_r, A_θ, and A_z are the scalar components of $\mathbf{A}$ along the r, θ, and z directions, respectively, as shown in Figure 2.6*b*. The relationship, or *transformation*, between cartesian and cylindrical coordinates can be obtained from inspection of Figure 2.6*a*, namely,

$$\begin{aligned} x &= r\cos\theta \\ y &= r\sin\theta \\ z &= z \end{aligned} \tag{2.5}$$

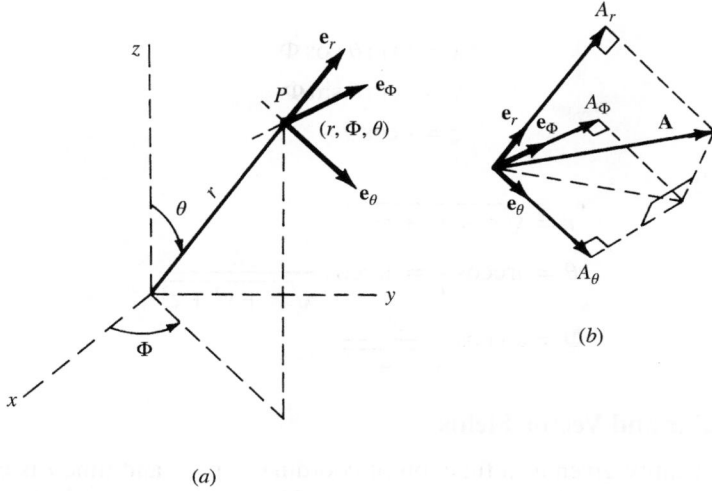

Figure 2.7 Spherical coordinates.

or inversely,

$$r = \sqrt{x^2 + y^2}$$
$$\theta = \arctan \frac{y}{x}$$
$$z = z$$
(2.6)

A *spherical coordinate system* is shown in Figure 2.7a. Once again, a phantom cartesian system is shown with dashed lines. (However, for clarity in the picture, the z axis is drawn vertically, in contrast to Figures 2.5 and 2.6.) The location of point P in space is given by the three coordinates (r, θ, Φ), where r is the distance of P from the origin, θ is the angle measured from the z axis and is in the rz plane, and Φ is the angle measured from the x axis and is in the xy plane. The r coordinate direction is the direction of increasing r, holding θ and Φ constant; $\mathbf{e}_r$ is the unit vector in the r direction. The θ coordinate direction is the direction of increasing θ, holding r and Φ constant; $\mathbf{e}_\theta$ is the unit vector in the θ direction. The Φ coordinate direction is the direction of increasing Φ, holding r and θ constant; $\mathbf{e}_\Phi$ is the unit vector in the Φ direction. The unit vectors $\mathbf{e}_r$, $\mathbf{e}_\theta$, and $\mathbf{e}_\Phi$ are mutually perpendicular. If $\mathbf{A}$ is a given vector in spherical space, then

$$\mathbf{A} = A_r \mathbf{e}_r + A_\theta \mathbf{e}_\theta + A_\Phi \mathbf{e}_\Phi$$

where A_r, A_θ, and A_Φ are the scalar components of $\mathbf{A}$ along the r, θ, and Φ directions, respectively, as shown in Figure 2.7b. The transformation between cartesian and spherical coordinates is obtained from inspection of Figure 2.7a, namely,

$$x = r \sin \theta \cos \Phi$$
$$y = r \sin \theta \sin \Phi \qquad (2.7)$$
$$z = r \cos \theta$$

or inversely,

$$r = \sqrt{x^2 + y^2 + z^2}$$
$$\theta = \arccos \frac{z}{r} = \arccos \frac{z}{\sqrt{x^2 + y^2 + z^2}} \qquad (2.8)$$
$$\Phi = \arccos \frac{x}{\sqrt{x^2 + y^2}}$$

2.2.3 Scalar and Vector Fields

A scalar quantity given as a function of coordinate space and time t is called a *scalar field*. For example, pressure, density, and temperature are scalar quantities, and

$$p = p_1(x, y, z, t) = p_2(r, \theta, z, t) = p_3(r, \theta, \Phi, t)$$
$$\rho = \rho_1(x, y, z, t) = \rho_2(r, \theta, z, t) = \rho_3(r, \theta, \Phi, t)$$
$$T = T_1(x, y, z, t) = T_2(r, \theta, z, t) = T_3(r, \theta, \Phi, t)$$

are scalar fields for pressure, density, and temperature, respectively. Similarly, a vector quantity given as a function of coordinate space and time is called a *vector field*. For example, velocity is a vector quantity, and

$$\mathbf{V} = V_x \mathbf{i} + V_y \mathbf{j} + V_z \mathbf{k}$$

where
$$V_x = V_x(x, y, z, t)$$
$$V_y = V_y(x, y, z, t)$$
$$V_z = V_z(x, y, z, t)$$

is the vector field for $\mathbf{V}$ in cartesian space. Analogous expressions can be written for vector fields in cylindrical and spherical space. In many theoretical aerodynamic problems, the above scalar and vector fields are the unknowns to be obtained in a solution for a flow with prescribed initial and boundary conditions.

2.2.4 Scalar and Vector Products

The scalar and vector products defined by Equations (2.3) and (2.4), respectively, can be written in terms of the components of each vector as follows.

Cartesian Coordinates Let

$$\mathbf{A} = A_x \mathbf{i} + A_y \mathbf{j} + A_z \mathbf{k}$$
and
$$\mathbf{B} = B_x \mathbf{i} + B_y \mathbf{j} + B_z \mathbf{k}$$
Then
$$\mathbf{A} \cdot \mathbf{B} = A_x B_x + A_y B_y + A_z B_z \qquad (2.9)$$

and

$$\mathbf{A} \times \mathbf{B} = \begin{bmatrix} \mathbf{i} & \mathbf{j} & \mathbf{k} \\ A_x & A_y & A_z \\ B_x & B_y & B_z \end{bmatrix} = \mathbf{i}(A_y B_z - A_z B_y) + \mathbf{j}(A_z B_x - A_x B_z)$$
$$+ \mathbf{k}(A_x B_y - A_y B_x) \tag{2.10}$$

Cylindrical Coordinates   Let

$$\mathbf{A} = A_r \mathbf{e}_r + A_\theta \mathbf{e}_\theta + A_z \mathbf{e}_z$$

and

$$\mathbf{B} = B_r \mathbf{e}_r + B_\theta \mathbf{e}_\theta + B_z \mathbf{e}_z$$

Then

$$\mathbf{A} \cdot \mathbf{B} = A_r B_r + A_\theta B_\theta + A_z B_z \tag{2.11}$$

and

$$\mathbf{A} \times \mathbf{B} = \begin{vmatrix} \mathbf{e}_r & \mathbf{e}_\theta & \mathbf{e}_z \\ A_r & A_\theta & A_z \\ B_r & B_\theta & B_z \end{vmatrix} \tag{2.12}$$

Spherical Coordinates   Let

$$\mathbf{A} = A_r \mathbf{e}_r + A_\theta \mathbf{e}_\theta + A_\Phi \mathbf{e}_\Phi$$

and

$$\mathbf{B} = B_r \mathbf{e}_r + B_\theta \mathbf{e}_\theta + B_\Phi \mathbf{e}_\Phi$$

Then

$$\mathbf{A} \cdot \mathbf{B} = A_r B_r + A_\theta B_\theta + A_\Phi B_\Phi \tag{2.13}$$

and

$$\mathbf{A} \times \mathbf{B} = \begin{vmatrix} \mathbf{e}_r & \mathbf{e}_\theta & \mathbf{e}_\Phi \\ A_r & A_\theta & A_\Phi \\ B_r & B_\theta & B_\Phi \end{vmatrix} \tag{2.14}$$

2.2.5 Gradient of a Scalar Field

We now begin a review of some elements of vector calculus. Consider a scalar field

$$p = p_1(x, y, z) = p_2(r, \theta, z) = p_3(r, \theta, \Phi)$$

The *gradient* of p, ∇p, at a given point in space is defined as a vector such that:

1. Its magnitude is the maximum rate of change of p per unit length of the coordinate space at the given point.
2. Its direction is that of the maximum rate of change of p at the given point.

For example, consider a two-dimensional pressure field in cartesian space as sketched in Figure 2.8. The solid curves are lines of constant pressure (i.e., they connect points in the pressure field which have the same value of p). Such lines are called *isolines*. Consider an arbitrary point (x, y) in Figure 2.8. If we move away from this point in an arbitrary direction, p will, in general, change because we are moving to another location in space. Moreover, there will be some direction from this point along which p changes *the most* over a unit length in that direction. This defines the *direction of the gradient* of p and is identified in Figure 2.8. The magnitude of ∇p is the rate of change of p per unit length in that

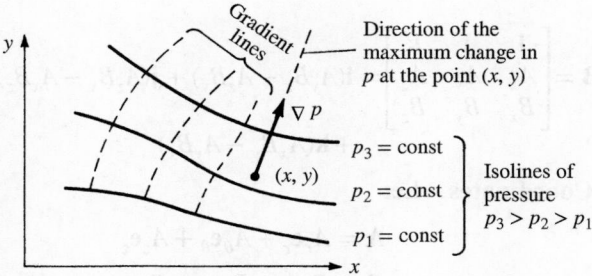

Figure 2.8 Illustration of the gradient of a scalar field.

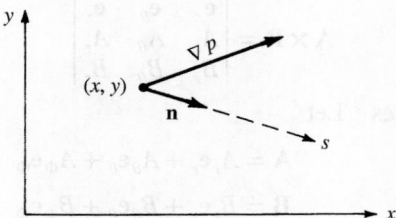

Figure 2.9 Sketch for the directional derivative.

direction. Both the magnitude and direction of ∇p will change from one point to another in the coordinate space. A line drawn in this space along which ∇p is tangent at every point is defined as a *gradient line,* as sketched in Figure 2.8. The gradient line and isoline through any given point in the coordinate space are perpendicular.

Consider ∇p at a given point (x, y) as shown in Figure 2.9. Choose some arbitrary direction s away from the point, as also shown in Figure 2.9. Let **n** be a unit vector in the s direction. The rate of change of p per unit length in the s direction is

$$\frac{dp}{ds} = \nabla p \cdot \mathbf{n} \tag{2.15}$$

In Equation (2.15), dp/ds is called the *directional derivative* in the s direction. Note from Equation (2.15) that the rate of change of p in any arbitrary direction is simply the component of ∇p in that direction.

Expressions for ∇p in the different coordinate systems are given below:

Cartesian: $p = p(x, y, z)$

$$\nabla p = \frac{\partial p}{\partial x}\mathbf{i} + \frac{\partial p}{\partial y}\mathbf{j} + \frac{\partial p}{\partial z}\mathbf{k} \tag{2.16}$$

Cylindrical:
$$p = p(r, \theta, z)$$

$$\boxed{\nabla p = \frac{\partial p}{\partial r}\mathbf{e}_r + \frac{1}{r}\frac{\partial p}{\partial \theta}\mathbf{e}_\theta + \frac{\partial p}{\partial z}\mathbf{e}_z}$$
(2.17)

Spherical:
$$p = p(r, \theta, \Phi)$$

$$\boxed{\nabla p = \frac{\partial p}{\partial r}\mathbf{e}_r + \frac{1}{r}\frac{\partial p}{\partial \theta}\mathbf{e}_\theta + \frac{1}{r\sin\theta}\frac{\partial p}{\partial \Phi}\mathbf{e}_\Phi}$$
(2.18)

2.2.6 Divergence of a Vector Field

Consider a vector field

$$\mathbf{V} = \mathbf{V}(x, y, z) = \mathbf{V}(r, \theta, z) = \mathbf{V}(r, \theta, \Phi)$$

In the above, $\mathbf{V}$ can represent any vector quantity. However, for practical purposes, and to aid in physical interpretation, consider $\mathbf{V}$ to be the flow velocity. Also, visualize a small fluid element of fixed mass moving along a streamline with velocity $\mathbf{V}$. As the fluid element moves through space, its volume will, in general, change. In Section 2.3, we prove that the time rate of change of the volume of a moving fluid element of fixed mass per unit volume of that element, is equal to the *divergence* of $\mathbf{V}$, denoted by $\nabla \cdot \mathbf{V}$. The divergence of a vector is a scalar quantity; it is one of two ways that the derivative of a vector field can be defined. In different coordinate systems, we have

Cartesian:
$$\mathbf{V} = \mathbf{V}(x, y, z) = V_x\mathbf{i} + V_y\mathbf{j} + V_z\mathbf{k}$$

$$\boxed{\nabla \cdot \mathbf{V} = \frac{\partial V_x}{\partial x} + \frac{\partial V_y}{\partial y} + \frac{\partial V_z}{\partial z}}$$
(2.19)

Cylindrical:
$$\mathbf{V} = \mathbf{V}(r, \theta, z) = V_r\mathbf{e}_r + V_\theta\mathbf{e}_\theta + V_z\mathbf{e}_z$$

$$\boxed{\nabla \cdot \mathbf{V} = \frac{1}{r}\frac{\partial}{\partial r}(rV_r) + \frac{1}{r}\frac{\partial V_\theta}{\partial \theta} + \frac{\partial V_z}{\partial z}}$$
(2.20)

Spherical:
$$\mathbf{V} = \mathbf{V}(r, \theta, \Phi) = V_r\mathbf{e}_r + V_\theta\mathbf{e}_\theta + V_\Phi\mathbf{e}_\Phi$$

$$\boxed{\nabla \cdot \mathbf{V} = \frac{1}{r^2}\frac{\partial}{\partial r}(r^2 V_r) + \frac{1}{r\sin\theta}\frac{\partial}{\partial \theta}(V_\theta \sin\theta) + \frac{1}{r\sin\theta}\frac{\partial V_\Phi}{\partial \Phi}}$$
(2.21)

2.2.7 Curl of a Vector Field

Consider a vector field

$$\mathbf{V} = \mathbf{V}(x, y, z) = \mathbf{V}(r, \theta, z) = \mathbf{V}(r, \theta, \Phi)$$

Although $\mathbf{V}$ can be any vector quantity, again consider $\mathbf{V}$ to be the flow velocity. Once again visualize a fluid element moving along a streamline. It is possible for this fluid element to be rotating with an angular velocity ω as it translates along the streamline. In Section 2.9, we prove that ω is equal to one-half of the *curl* of $\mathbf{V}$, where the curl of $\mathbf{V}$ is denoted by $\nabla \times \mathbf{V}$. The curl of $\mathbf{V}$ is a vector quantity; it is the alternate way that the derivative of a vector field can be defined, the first being $\nabla \cdot \mathbf{V}$ (see Section 2.2.6, Divergence of a Vector Field). In different coordinate systems, we have

Cartesian:
$$\mathbf{V} = V_x\mathbf{i} + V_y\mathbf{j} + V_z\mathbf{k}$$

$$\nabla \times \mathbf{V} = \begin{vmatrix} \mathbf{i} & \mathbf{j} & \mathbf{k} \\ \dfrac{\partial}{\partial x} & \dfrac{\partial}{\partial y} & \dfrac{\partial}{\partial z} \\ V_x & V_y & V_z \end{vmatrix} = \mathbf{i}\left(\frac{\partial V_z}{\partial y} - \frac{\partial V_y}{\partial z} \right) + \mathbf{j}\left(\frac{\partial V_x}{\partial z} - \frac{\partial V_z}{\partial x} \right) \\ + \mathbf{k}\left(\frac{\partial V_y}{\partial x} - \frac{\partial V_x}{\partial y} \right) \quad (2.22)$$

Cylindrical:
$$\mathbf{V} = V_r\mathbf{e}_r + V_\theta\mathbf{e}_\theta + V_z\mathbf{e}_z$$

$$\nabla \times \mathbf{V} = \frac{1}{r}\begin{vmatrix} \mathbf{e}_r & r\mathbf{e}_\theta & \mathbf{e}_z \\ \dfrac{\partial}{\partial r} & \dfrac{\partial}{\partial \theta} & \dfrac{\partial}{\partial z} \\ V_r & rV_\theta & V_z \end{vmatrix} \quad (2.23)$$

Spherical:
$$\mathbf{V} = V_r\mathbf{e}_r + V_\theta\mathbf{e}_\theta + V_\Phi\mathbf{e}_\Phi$$

$$\nabla \times \mathbf{V} = \frac{1}{r^2 \sin\theta}\begin{vmatrix} \mathbf{e}_r & r\mathbf{e}_\theta & (r\sin\theta)\mathbf{e}_\Phi \\ \dfrac{\partial}{\partial r} & \dfrac{\partial}{\partial \theta} & \dfrac{\partial}{\partial \Phi} \\ V_r & rV_\theta & (r\sin\theta)V_\Phi \end{vmatrix} \quad (2.24)$$

2.2.8 Line Integrals

Consider a vector field

$$\mathbf{A} = \mathbf{A}(x, y, z) = \mathbf{A}(r, \theta, z) = \mathbf{A}(r, \theta, \Phi)$$

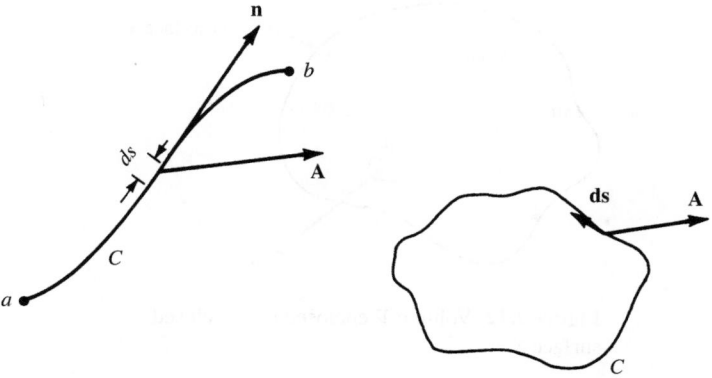

Figure 2.10 Sketch for line integrals.

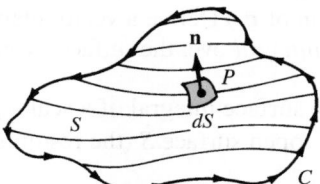

Figure 2.11 Sketch for surface integrals. The three-dimensional surface area S is bounded by the closed curve C.

Also, consider a curve C in space connecting two points a and b as shown on the left side of Figure 2.10. Let ds be an elemental length of the curve, and $\mathbf{n}$ be a unit vector tangent to the curve. Define the vector $\mathbf{ds} = \mathbf{n}\,ds$. Then, the *line integral* of $\mathbf{A}$ along curve C from point a to point b is

$$\int_a^b \mathbf{A} \cdot \mathbf{ds}$$

If the curve C is closed, as shown at the right of Figure 2.10, then the line integral is given by

$$\oint_C \mathbf{A} \cdot \mathbf{ds}$$

where the *counterclockwise* direction around C is considered positive. (The positive direction around a closed curve is, by convention, that direction you would move such that the area enclosed by C is always on your left.)

2.2.9 Surface Integrals

Consider an open surface S bounded by the closed curve C, as shown in Figure 2.11. At point P on the surface, let dS be an elemental area of the surface and $\mathbf{n}$ be a unit vector normal to the surface. The orientation of $\mathbf{n}$ is in the

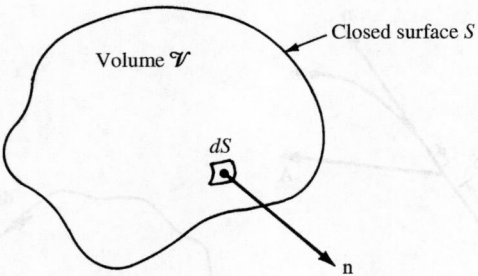

Figure 2.12 Volume $\mathcal{V}$ enclosed by the closed surface S.

direction according to the right-hand rule for movement along C. (Curl the fingers of your right hand in the direction of movement around C; your thumb will then point in the general direction of **n**.) Define a vector elemental area as $\mathbf{dS} = \mathbf{n}\,dS$. In terms of $\mathbf{dS}$, the *surface integral* over the surface S can be defined in three ways:

$$\iint_S p\,\mathbf{dS} = \text{surface integral of a scalar } p \text{ over the}$$
$$\text{open surface } S \text{ (the result is a vector)}$$

$$\iint_S \mathbf{A} \cdot \mathbf{dS} = \text{surface integral of a vector } \mathbf{A} \text{ over the}$$
$$\text{open surface } S \text{ (the result is a scalar)}$$

$$\iint_S \mathbf{A} \times \mathbf{dS} = \text{surface integral of a vector } \mathbf{A} \text{ over the}$$
$$\text{open surface } S \text{ (the result is a vector)}$$

If the surface S is *closed* (e.g., the surface of a sphere or a cube), **n** points out of the surface, away from the enclosed volume, as shown in Figure 2.12. The surface integrals over the closed surface are

$$\oiint_S p\,\mathbf{dS} \qquad \oiint_S \mathbf{A} \cdot \mathbf{dS} \qquad \oiint_S \mathbf{A} \times \mathbf{dS}$$

2.2.10 Volume Integrals

Consider a volume $\mathcal{V}$ in space. Let ρ be a scalar field in this space. The *volume integral* over the volume $\mathcal{V}$ of the quantity ρ is written as

$$\oiiint_{\mathcal{V}} \rho\,d\mathcal{V} = \text{volume integral of a scalar } \rho \text{ over the}$$
$$\text{volume } \mathcal{V} \text{ (the result is a scalar)}$$

Let $\mathbf{A}$ be a vector field in space. The volume integral over the volume $\mathcal{V}$ of the quantity $\mathbf{A}$ is written as

$$\oiiint_{\mathcal{V}} \mathbf{A}\,d\mathcal{V} = \text{volume integral of a vector } \mathbf{A} \text{ over the}$$
$$\text{volume } \mathcal{V} \text{ (the result is a vector)}$$

2.2.11 Relations Between Line, Surface, and Volume Integrals

Consider again the open area S bounded by the closed curve C, as shown in Figure 2.11. Let $\mathbf{A}$ be a vector field. The line integral of $\mathbf{A}$ over C is related to the surface integral of $\mathbf{A}$ over S by *Stokes' theorem:*

$$\oint_C \mathbf{A} \cdot \mathbf{ds} = \iint_S (\nabla \times \mathbf{A}) \cdot \mathbf{dS} \tag{2.25}$$

Consider again the volume $\mathcal{V}$ enclosed by the closed surface S, as shown in Figure 2.12. The surface and volume integrals of the vector field $\mathbf{A}$ are related through the *divergence theorem:*

$$\oiint_S \mathbf{A} \cdot \mathbf{dS} = \oiiint_\mathcal{V} (\nabla \cdot \mathbf{A}) \, d\mathcal{V} \tag{2.26}$$

If p represents a scalar field, a vector relationship analogous to Equation (2.26) is given by the *gradient theorem:*

$$\oiint_S p \, \mathbf{dS} = \oiiint_\mathcal{V} \nabla p \, d\mathcal{V} \tag{2.27}$$

2.2.12 Summary

This section has provided a concise review of those elements of vector analysis that we will use as tools in our subsequent discussions. Make certain to review these tools until you feel comfortable with them, especially the relations in boxes.

2.3 MODELS OF THE FLUID: CONTROL VOLUMES AND FLUID ELEMENTS

Aerodynamics is a fundamental science, steeped in physical observation. As you proceed through this book, make every effort to gradually develop a "physical feel" for the material. An important virtue of all successful aerodynamicists (indeed, of all successful engineers and scientists) is that they have good "physical intuition," based on thought and experience, which allows them to make reasonable judgments on difficult problems. Although this chapter is full of equations and (seemingly) esoteric concepts, now is the time for you to start developing this physical feel.

With this section, we begin to build the basic equations of aerodynamics. There is a certain philosophical procedure involved with the development of these equations, as follows:

1. Invoke three fundamental physical principles that are deeply entrenched in our macroscopic observations of nature, namely,
    a.   Mass is conserved (i.e., mass can be neither created nor destroyed).
    b.   Newton's second law: force = mass × acceleration.
    c.   Energy is conserved; it can only change from one form to another.
2. Determine a suitable *model* of the fluid. Remember that a fluid is a squishy substance, and therefore it is usually more difficult to describe than a well-defined solid body. Hence, we have to adopt a reasonable model of the fluid to which we can apply the fundamental principles stated in item 1.
3. Apply the fundamental physical principles listed in item 1 to the model of the fluid determined in item 2 in order to obtain mathematical equations which properly describe the physics of the flow. In turn, use these fundamental equations to analyze any particular aerodynamic flow problem of interest.

In this section, we concentrate on item 2; namely, we ask the question: What is a suitable model of the fluid? How do we visualize this squishy substance in order to apply the three fundamental physical principles to it? There is no single answer to this question; rather, three different models have been used successfully throughout the modern evolution of aerodynamics. They are (1) finite control volume, (2) infinitesimal fluid element, and (3) molecular. Let us examine what these models involve and how they are applied.

2.3.1 Finite Control Volume Approach

Consider a general flow field as represented by the streamlines in Figure 2.13. Let us imagine a closed volume drawn within a *finite* region of the flow. This volume defines a *control volume* $\mathcal{V}$, and a *control surface S* is defined as the closed surface

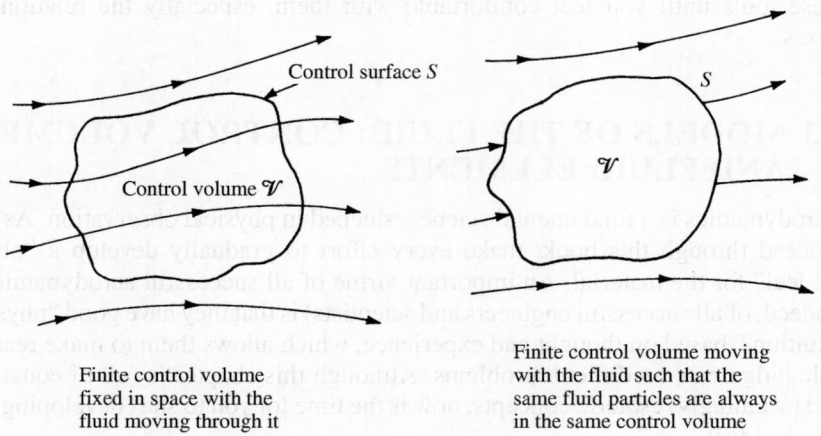

Finite control volume
fixed in space with the
fluid moving through it

Finite control volume moving
with the fluid such that the
same fluid particles are always
in the same control volume

Figure 2.13 Finite control volume approach.

which bounds the control volume. The control volume may be fixed in space with the fluid moving through it, as shown at the left of Figure 2.13. Alternatively, the control volume may be moving with the fluid such that the same fluid particles are always inside it, as shown at the right of Figure 2.13. In either case, the control volume is a reasonably large, finite region of the flow. The fundamental physical principles are applied to the fluid inside the control volume, and to the fluid crossing the control surface (if the control volume is fixed in space). Therefore, instead of looking at the whole flow field at once, with the control volume model we limit our attention to just the fluid in the finite region of the volume itself.

2.3.2 Infinitesimal Fluid Element Approach

Consider a general flow field as represented by the streamlines in Figure 2.14. Let us imagine an infinitesimally small fluid element in the flow, with a differential volume dV. The fluid element is infinitesimal in the same sense as differential calculus; however, it is large enough to contain a huge number of molecules so that it can be viewed as a continuous medium. The fluid element may be fixed in space with the fluid moving through it, as shown at the left of Figure 2.14. Alternatively, it may be moving along a streamline with velocity $\mathbf{V}$ equal to the flow velocity at each point. Again, instead of looking at the whole flow field at once, the fundamental physical principles are applied to just the fluid element itself.

2.3.3 Molecular Approach

In actuality, of course, the motion of a fluid is a ramification of the mean motion of its atoms and molecules. Therefore, a third model of the flow can be a microscopic approach wherein the fundamental laws of nature are applied directly to the atoms and molecules, using suitable statistical averaging to define the resulting fluid properties. This approach is in the purview of *kinetic theory,* which is a very

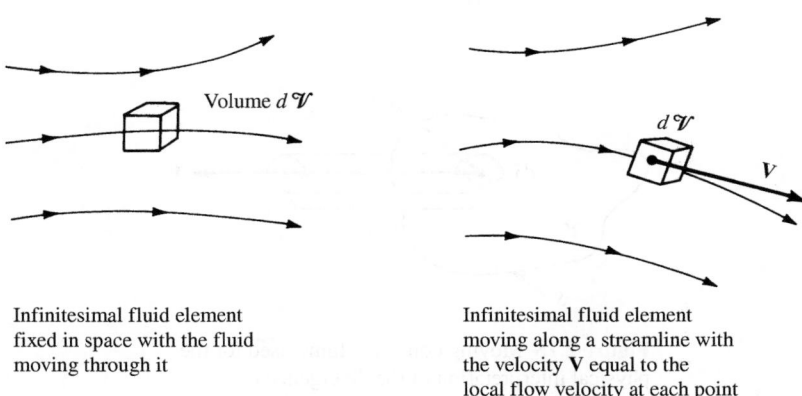

Infinitesimal fluid element
fixed in space with the fluid
moving through it

Infinitesimal fluid element
moving along a streamline with
the velocity **V** equal to the
local flow velocity at each point

Figure 2.14 Infinitesimal fluid element approach.

elegant method with many advantages in the long run. However, it is beyond the scope of the present book.

In summary, although many variations on the theme can be found in different texts for the derivation of the general equations of fluid flow, the flow model can usually be categorized under one of the approaches described above.

2.3.4 Physical Meaning of the Divergence of Velocity

In the equations to follow, the divergence of velocity, $\nabla \cdot \mathbf{V}$, occurs frequently. Before leaving this section, let us prove the statement made earlier (Section 2.2) that $\nabla \cdot \mathbf{V}$ is physically the time rate of change of the volume of a moving fluid element of fixed mass per unit volume of that element. Consider a control volume moving with the fluid (the case shown on the right of Figure 2.13). This control volume is always made up of the same fluid particles as it moves with the flow; hence, its mass is fixed, invariant with time. However, its volume $\mathcal{V}$ and control surface S are changing with time as it moves to different regions of the flow where different values of ρ exist. That is, this moving control volume of fixed mass is constantly increasing or decreasing its volume and is changing its shape, depending on the characteristics of the flow. This control volume is shown in Figure 2.15 at some instant in time. Consider an infinitesimal element of the surface dS moving at the local velocity $\mathbf{V}$, as shown in Figure 2.15. The change in the volume of the control volume $\Delta \mathcal{V}$, due to just the movement of dS over a time increment Δt, is, from Figure 2.15, equal to the volume of the long, thin cylinder with base area dS and altitude $(\mathbf{V}\Delta t) \cdot \mathbf{n}$; that is,

$$\Delta \mathcal{V} = [(\mathbf{V}\Delta t) \cdot \mathbf{n}]dS = (\mathbf{V}\Delta t) \cdot \mathbf{dS} \qquad (2.28)$$

Over the time increment Δt, the total change in volume of the whole control volume is equal to the summation of Equation (2.28) over the total control surface. In the limit as $dS \to 0$, the sum becomes the surface integral

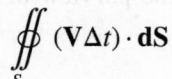

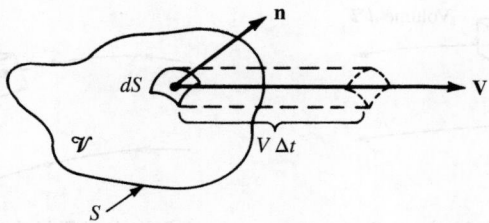

Figure 2.15 Moving control volume used for the physical interpretation of the divergence of velocity.

If this integral is divided by Δt, the result is physically the time rate of change of the control volume, denoted by $D\mathcal{V}/Dt$; that is,

$$\frac{D\mathcal{V}}{Dt} = \frac{1}{\Delta t} \oiint_S (\mathbf{V}\Delta t) \cdot d\mathbf{S} = \oiint_S \mathbf{V} \cdot d\mathbf{S} \qquad (2.29)$$

(The significance of the notation D/Dt is revealed in Section 2.9.) Applying the divergence theorem, Equation (2.26), to the right side of Equation (2.29), we have

$$\frac{D\mathcal{V}}{Dt} = \oiiint_{\mathcal{V}} (\nabla \cdot \mathbf{V}) d\mathcal{V} \qquad (2.30)$$

Now let us imagine that the moving control volume in Figure 2.15 is shrunk to a very small volume $\delta\mathcal{V}$, essentially becoming an infinitesimal moving fluid element as sketched on the right of Figure 2.14. Then Equation (2.30) can be written as

$$\frac{D(\delta\mathcal{V})}{Dt} = \oiiint_{\delta\mathcal{V}} (\nabla \cdot \mathbf{V}) d\mathcal{V} \qquad (2.31)$$

Assume that $\delta\mathcal{V}$ is small enough such that $\nabla \cdot \mathbf{V}$ is essentially the same value throughout $\delta\mathcal{V}$. Then the integral in Equation (2.31) can be approximated as $(\nabla \cdot \mathbf{V})\delta\mathcal{V}$. From Equation (2.31), we have

$$\frac{D(\delta\mathcal{V})}{Dt} = (\nabla \cdot \mathbf{V})\delta\mathcal{V}$$

or

$$\boxed{\nabla \cdot \mathbf{V} = \frac{1}{\delta\mathcal{V}} \frac{D(\delta\mathcal{V})}{Dt}} \qquad (2.32)$$

Examine Equation (2.32). It states that $\nabla \cdot \mathbf{V}$ is physically the *time rate of change of the volume of a moving fluid element per unit volume*. Hence, the interpretation of $\nabla \cdot \mathbf{V}$, first given in Section 2.2.6, Divergence of a Vector Field, is now proved.

2.3.5 Specification of the Flow Field

In Section 2.2.3, we defined both scalar and vector *fields*. We now apply this concept of a field more directly to an aerodynamic flow. One of the most straightforward ways of describing the details of an aerodynamic flow is simply to visualize the flow in three-dimensional space, and to write the variation of the aerodynamic properties as a function of space and time. For example, in cartesian coordinates the equations

$$p = p(x, y, z, t) \qquad (2.33a)$$

$$\rho = \rho(x, y, z, t) \qquad (2.33b)$$

$$T = T(x, y, z, t) \qquad (2.33c)$$

and

$$\mathbf{V} = u\mathbf{i} + v\mathbf{j} + w\mathbf{k} \qquad (2.34a)$$

where

$$u = u(x, y, z, t) \qquad (2.34b)$$

$$v = v(x, y, z, t) \qquad (2.34c)$$

$$w = w(x, y, z, t) \qquad (2.34d)$$

represent the *flow field*. Equation (2.33a–c) give the variation of the scalar flow field variables pressure, density, and temperature, respectively. [In equilibrium thermodynamics, the specification of two state variables, such as p and ρ, uniquely defines the values of all other state variables, such as T. In this case, one of Equations (2.33) can be considered redundant.] Equation (2.34a–d) give the variation of the vector flow field variable velocity $\mathbf{V}$, where the scalar components of $\mathbf{V}$ in the x, y, and z directions are u, v, and w, respectively.

Figure 2.16 illustrates a given fluid element moving in a flow field specified by Equations (2.33) and (2.34). At the time t_1, the fluid element is at point 1, located at (x_1, y_1, z_1) as shown in Figure 2.16.

At this instant, its velocity is $\mathbf{V}_1$ and its pressure is given by

$$p = p(x_1, y_1, z_1, t_1)$$

and similarly for its other flow variables.

By definition, an *unsteady* flow is one where the flow field variables at any given point are changing with time. For example, if you lock your eyes on point 1 in Figure 2.16, and keep them fixed on point 1, if the flow is unsteady you will

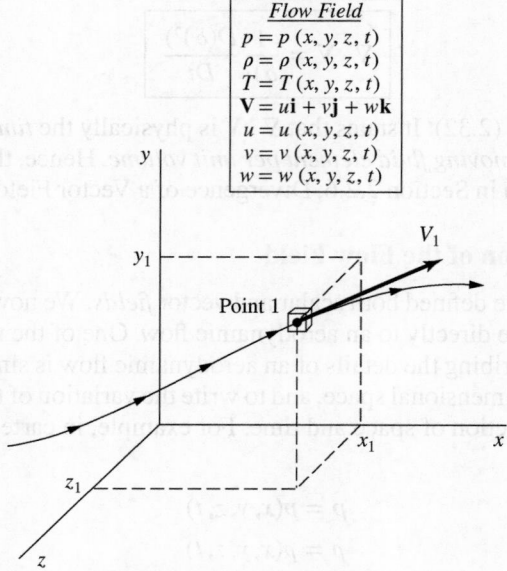

Flow Field
$p = p(x, y, z, t)$
$\rho = \rho(x, y, z, t)$
$T = T(x, y, z, t)$
$\mathbf{V} = u\mathbf{i} + v\mathbf{j} + w\mathbf{k}$
$u = u(x, y, z, t)$
$v = v(x, y, z, t)$
$w = w(x, y, z, t)$

Figure 2.16 A fluid element passing through point 1 in a flow field.

observe p, ρ, etc., fluctuating with time. Equations (2.33) and (2.34) describe an unsteady flow field because time t is included as one of the independent variables. In contrast, a *steady* flow is one where the flow field variables at any given point are invariant with time, that is, if you lock your eyes on point 1 you will continuously observe the same constant values for p, ρ, $\mathbf{V}$, etc., for all time. A steady flow field is specified by the relations

$$p = p(x, y, z)$$
$$\rho = \rho(x, y, z)$$
$$\text{etc.}$$

The concept of the flow field, and a specified fluid element moving through it as illustrated in Figure 2.16, will be revisited in Section 2.9, where we define and discuss the concept of the substantial derivative.

The subsonic compressible flow over a cosine-shaped (wavy) wall is illustrated in Figure 2.17. The wavelength and amplitude of the wall are l and h, respectively, as shown in Figure 2.17. The streamlines exhibit the same qualitative shape as the wall, but with diminishing amplitude as distance above the wall increases. Finally, as $y \rightarrow \infty$, the

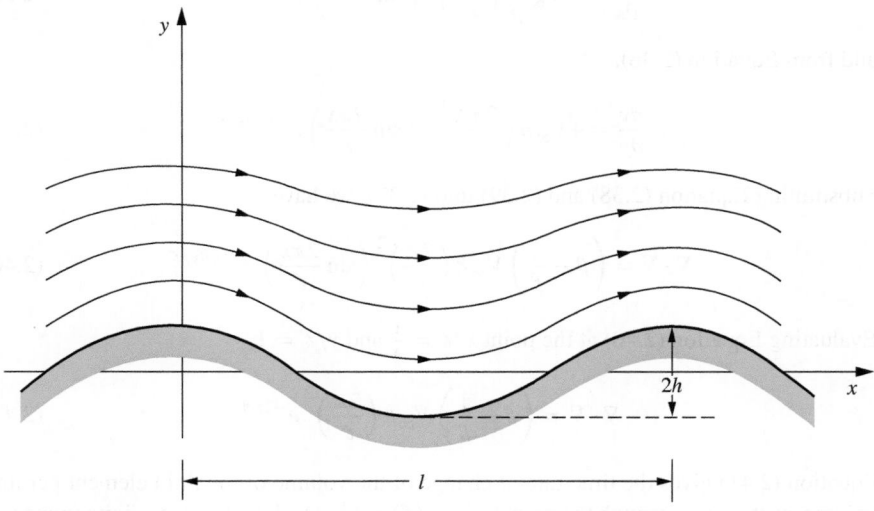

Figure 2.17 Subsonic compressible flow over a wavy wall; the streamline pattern.

streamline becomes straight. Along this straight streamline, the freestream velocity and Mach number are V_∞ and M_∞, respectively. The velocity field in cartesian coordinates is given by

$$u = V_\infty \left[1 + \frac{h}{\beta} \frac{2\pi}{\ell} \left(\cos \frac{2\pi x}{\ell} \right) e^{-2\pi\beta y/\ell} \right] \qquad (2.35)$$

and

$$v = -V_\infty h \frac{2\pi}{\ell} \left(\sin \frac{2\pi x}{\ell} \right) e^{-2\pi\beta y/\ell} \qquad (2.36)$$

where

$$\beta \equiv \sqrt{1 - M_\infty^2}$$

Consider the particular flow that exists for the case where $\ell = 1.0$ m, $h = 0.01$ m, $V_\infty = 240$ m/s, and $M_\infty = 0.7$. Also, consider a fluid element of fixed mass moving along a streamline in the flow field. The fluid element passes through the point $(x/\ell, y/\ell) = (\frac{1}{4}, 1)$. At this point, calculate the time rate of change of the volume of the fluid element per unit volume.

■ **Solution**

From Section 2.3.4, we know that the time rate of change of the volume of a moving fluid element of fixed mass per unit volume, is given by the divergence of the velocity $\nabla \cdot \mathbf{V}$. In cartesian coordinates, from Equation (2.19), we have

$$\nabla \cdot \mathbf{V} = \frac{\partial u}{\partial x} + \frac{\partial v}{\partial y} \qquad (2.37)$$

From Equation (2.35),

$$\frac{\partial u}{\partial x} = -V_\infty \frac{h}{\beta} \left(\frac{2\pi}{\ell} \right)^2 \left(\sin \frac{2\pi x}{\ell} \right) e^{-2\pi\beta y/\ell} \qquad (2.38)$$

and from Equation (2.36),

$$\frac{\partial v}{\partial y} = +V_\infty h \left(\frac{2\pi}{\ell} \right)^2 \beta \left(\sin \frac{2\pi x}{\ell} \right) e^{-2\pi\beta y/\ell} \qquad (2.39)$$

Substituting Equation (2.38) and (2.39) into (2.37), we have

$$\nabla \cdot \mathbf{V} = \left(\beta - \frac{1}{\beta} \right) V_\infty h \left(\frac{2\pi}{\ell} \right)^2 \left(\sin \frac{2\pi x}{\ell} \right) e^{-2\pi\beta y/\ell} \qquad (2.40)$$

Evaluating Equation (2.40) at the point $x/\ell = \frac{1}{4}$ and $y/\ell = 1$,

$$\nabla \cdot \mathbf{V} = \left(\beta - \frac{1}{\beta} \right) V_\infty h \left(\frac{2\pi}{\ell} \right)^2 e^{-2\pi\beta} \qquad (2.41)$$

Equation (2.41) gives the time rate of change of the volume of the fluid element per unit volume, as it passes through the point $(x/\ell, y/\ell) = (\frac{1}{4}, 1)$. Note that it is a finite (nonzero) value; the volume of the fluid element is changing as it moves along the streamline. This is consistent with the definition of a compressible flow, where the density is a variable and hence the volume of a fixed mass must also be variable. Note from Equation (2.40) that $\nabla \cdot \mathbf{V} = 0$ only along vertical lines denoted by $x/\ell = 0, \frac{1}{2}, 1, 1\frac{1}{2}, \dots$, where the $\sin(2\pi x/\ell)$ goes to zero. This is a peculiarity associated with the cyclical nature of the flow field over the cosine-shaped wall. For the particular flow considered here, where $\ell = 1.0$ m, $h = 0.01$ m, $V_\infty = 240$ m/s, $M_\infty = 0.7$, and

$$\beta = \sqrt{1 - M_\infty^2} = \sqrt{1 - (0.7)^2} = 0.714$$

Equation (2.41) yields

$$\nabla \cdot \mathbf{V} = \left(0.714 - \frac{1}{0.714}\right)(240)(0.01)\left(\frac{2\pi}{1}\right)e^{-2\pi(0.714)} = \boxed{-0.7327 \text{ s}^{-1}}$$

The physical significance of this result is that, as the fluid element is passing through the point $(\frac{1}{4}, 1)$ in the flow, it is experiencing a 73 percent rate of *decrease* of volume per second (the negative quantity denotes a decrease in volume). That is, the density of the fluid element is increasing. Hence, the point $(\frac{1}{4}, 1)$ is in a compression region of the flow, where the fluid element will experience an increase in density. Expansion regions are defined by values of x/ℓ which yield negative values of the sine function in Equation (2.40), which in turn yields a positive value for $\nabla \cdot \mathbf{V}$. This gives an increase in volume of the fluid element, hence a decrease in density. Clearly, as the fluid element continues its path through this flow field, it experiences cyclical increases and decreases in density, as well as the other flow field properties.

2.4 CONTINUITY EQUATION

In Section 2.3, we discussed several models which can be used to study the motion of a fluid. Following the philosophy set forth at the beginning of Section 2.3, we now apply the fundamental physical principles to such models. Unlike the above derivation of the physical significance of $\nabla \cdot \mathbf{V}$, wherein we used the model of a moving finite control volume, we now employ the model of a *fixed* finite control volume as sketched on the left side of Figure 2.13. Here, the control volume is fixed in space, with the flow moving through it. Unlike our previous derivation, the volume $\mathcal{V}$ and control surface $\mathbf{S}$ are now constant with time, and the mass of fluid contained within the control volume can change as a function of time (due to unsteady fluctuations of the flow field).

Before starting the derivation of the fundamental equations of aerodynamics, we must examine a concept vital to those equations, namely, the concept of *mass flow*. Consider a given area A arbitrarily oriented in a flow field as shown in Figure 2.18. In Figure 2.18, we are looking at an edge view of area A. Let A be small enough such that the flow velocity $\mathbf{V}$ is uniform across A. Consider the fluid elements with velocity $\mathbf{V}$ that pass through A. In time dt after crossing A, they have moved a distance $V\, dt$ and have swept out the shaded volume shown in Figure 2.18. This volume is equal to the base area A times the height of the

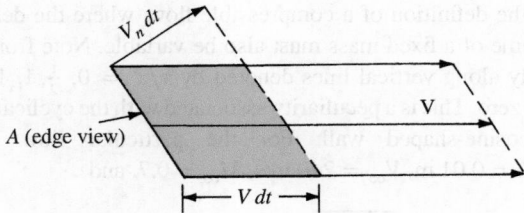

Figure 2.18 Sketch for discussion of mass flow through area A in a flow field.

cylinder $V_n\,dt$, where V_n is the component of velocity normal to A; that is,

$$\text{Volume} = (V_n\,dt)A$$

The mass inside the shaded volume is therefore

$$\text{Mass} = \rho(V_n\,dt)A \tag{2.42}$$

This is the mass that has swept past A in time dt. By definition, the *mass flow* through A is the mass crossing A per second (e.g., kilograms per second, slugs per second). Let $\dot{m}$ denote mass flow. From Equation (2.42),

$$\dot{m} = \frac{\rho(V_n\,dt)A}{dt}$$

or

$$\boxed{\dot{m} = \rho V_n A} \tag{2.43}$$

Equation (2.43) demonstrates that mass flow through A is given by the product

$$\boxed{\text{Area} \times \text{density} \times \text{component of flow velocity } \textit{normal} \text{ to the area}}$$

A related concept is that of *mass flux,* defined as the mass flow *per unit area.*

$$\boxed{\text{Mass flux} = \frac{\dot{m}}{A} = \rho V_n} \tag{2.44}$$

Typical units of mass flux are kg/(s · m^2) and slug/(s · ft^2).

The concepts of mass flow and mass flux are important. Note from Equation (2.44) that mass flux across a surface is equal to the product of density times the component of velocity perpendicular to the surface. Many of the equations of aerodynamics involve products of density and velocity. For example, in cartesian coordinates, $\mathbf{V} = V_x\mathbf{i} + V_y\mathbf{j} + V_z\mathbf{k} = u\mathbf{i} + v\mathbf{j} + w\mathbf{k}$, where u, v, and w denote the x, y, and z components of velocity, respectively. (The use of u, v, and w rather than V_x, V_y, and V_z to symbolize the x, y, and z components of velocity is quite common in aerodynamic literature; we henceforth adopt the u, v, and w notation.) In many of the equations of aerodynamics, you will find the products ρu, ρv, and ρw; always remember that these products are the mass fluxes in the x, y, and z

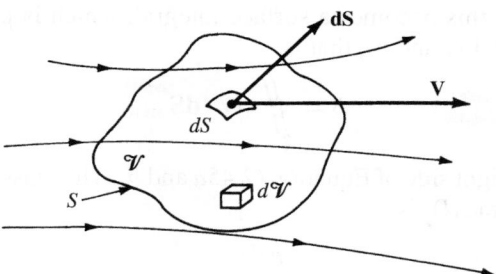

Figure 2.19 Finite control volume fixed in space.

directions, respectively. In a more general sense, if V is the magnitude of velocity in an arbitrary direction, the product ρV is physically the mass flux (mass flow per unit area) across an area oriented perpendicular to the direction of V.

We are now ready to apply our first physical principle to a finite control volume fixed in space.

Physical principle   Mass can be neither created nor destroyed.

Consider a flow field wherein all properties vary with spatial location and time; for example, $\rho = \rho(x, y, z, t)$. In this flow field, consider the fixed finite control volume shown in Figure 2.19. At a point on the control surface, the flow velocity is $\mathbf{V}$ and the vector elemental surface area is $\mathbf{dS}$. Also $d\mathcal{V}$ is an elemental volume inside the control volume. Applied to this control volume, the above physical principle means

$$
\begin{array}{cc}
\text{Net mass flow } \textit{out} \text{ of control} \\
\text{volume through surface } S
\end{array}
=
\begin{array}{c}
\text{time rate of decrease of} \\
\text{mass inside control volume } \mathcal{V}
\end{array}
\qquad (2.45a)
$$

or
$$
B = C \qquad\qquad (2.45b)
$$

where B and C are just convenient symbols for the left and right sides, respectively, of Equation (2.45a). First, let us obtain an expression for B in terms of the quantities shown in Figure 2.19. From Equation (2.43), the elemental mass flow across the area dS is

$$
\rho V_n \, dS = \rho \mathbf{V} \cdot \mathbf{dS}
$$

Examining Figure 2.19, note that by convention, $\mathbf{dS}$ always points in a direction *out* of the control volume. Hence, when $\mathbf{V}$ also points out of the control volume (as shown in Figure 2.19), the product $\rho \mathbf{V} \cdot \mathbf{dS}$ is *positive*. Moreover, when $\mathbf{V}$ points out of the control volume, the mass flow is physically leaving the control volume (i.e., it is an *outflow*). Hence, a positive $\rho \mathbf{V} \cdot \mathbf{dS}$ denotes an outflow. In turn, when $\mathbf{V}$ points into the control volume, $\rho \mathbf{V} \cdot \mathbf{dS}$ is *negative*. Moreover, when $\mathbf{V}$ points inward, the mass flow is physically entering the control volume (i.e., it is an *inflow*). Hence, a negative $\rho \mathbf{V} \cdot \mathbf{dS}$ denotes an *inflow*. The *net* mass flow *out* of the entire control surface S is the summation over S of the elemental mass flows. In the limit, this becomes a surface integral, which is physically the left side of Equation (2.45a and b); that is,

$$B = \oiint_S \rho \mathbf{V} \cdot d\mathbf{S} \tag{2.46}$$

Now consider the right side of Equation (2.45a and b). The mass contained within the elemental volume $d\mathcal{V}$ is

$$\rho \, d\mathcal{V}$$

Hence, the total mass inside the control volume is

$$\iiint_\mathcal{V} \rho \, d\mathcal{V}$$

The time rate of *increase* of mass inside $\mathcal{V}$ is then

$$\frac{\partial}{\partial t} \iiint_\mathcal{V} \rho \, d\mathcal{V}$$

In turn, the time rate of *decrease* of mass inside $\mathcal{V}$ is the negative of the above; that is

$$-\frac{\partial}{\partial t} \iiint_\mathcal{V} \rho \, d\mathcal{V} = C \tag{2.47}$$

Thus, substituting Equations (2.46) and (2.47) into (2.45b), we have

$$\oiint_S \rho \mathbf{V} \cdot d\mathbf{S} = -\frac{\partial}{\partial t} \iiint_\mathcal{V} \rho \, d\mathcal{V}$$

or

$$\frac{\partial}{\partial t} \iiint_\mathcal{V} \rho \, d\mathcal{V} + \oiint_S \rho \mathbf{V} \cdot d\mathbf{S} = 0 \tag{2.48}$$

Equation (2.48) is the final result of applying the physical principle of the conservation of mass to a finite control volume fixed in space. Equation (2.48) is called the *continuity equation*. It is one of the most fundamental equations of fluid dynamics.

Note that Equation (2.48) expresses the continuity equation in integral form. We will have numerous opportunities to use this form; it has the advantage of relating aerodynamic phenomena over a finite region of space without being concerned about the details of precisely what is happening at a given distinct point in the flow. On the other hand, there are many times when we are concerned with the details of a flow and we want to have equations that relate flow properties at a *given point*. In such a case, the integral form as expressed in Equation (2.48) is not particularly useful. However, Equation (2.48) can be reduced to another form that does relate flow properties at a given point, as follows. To begin with, since the control volume used to obtain Equation (2.48) is fixed in space, the limits of integration are also fixed. Hence, the time derivative can be placed inside the volume integral and Equation (2.48) can be written as

$$\iiint_{\mathcal{V}} \frac{\partial \rho}{\partial t} \, d\mathcal{V} + \iint_{S} \rho \mathbf{V} \cdot \mathbf{dS} = 0 \qquad (2.49)$$

Applying the divergence theorem, Equation (2.26), we can express the right-hand term of Equation (2.49) as

$$\iint_{S} (\rho \mathbf{V}) \cdot \mathbf{dS} = \iiint_{\mathcal{V}} \nabla \cdot (\rho \mathbf{V}) \, d\mathcal{V} \qquad (2.50)$$

Substituting Equation (2.50) into (2.49), we obtain

$$\iiint_{\mathcal{V}} \frac{\partial \rho}{\partial t} \, d\mathcal{V} + \iiint_{\mathcal{V}} \nabla \cdot (\rho \mathbf{V}) \, d\mathcal{V} = 0$$

or
$$\iiint_{\mathcal{V}} \left[\frac{\partial \rho}{\partial t} + \nabla \cdot (\rho \mathbf{V}) \right] d\mathcal{V} = 0 \qquad (2.51)$$

Examine the integrand of Equation (2.51). If the integrand were a finite number, then Equation (2.51) would require that the integral over part of the control volume be equal and opposite in sign to the integral over the remainder of the control volume, such that the net integration would be zero. However, the finite control volume is *arbitrarily* drawn in space; there is no reason to expect cancellation of one region by the other. Hence, the only way for the integral in Equation (2.51) to be zero for an arbitrary control volume is for the integrand to be zero at *all* points within the control volume. Thus, from Equation (2.51), we have

$$\boxed{\frac{\partial \rho}{\partial t} + \nabla \cdot (\rho \mathbf{V}) = 0} \qquad (2.52)$$

Equation (2.52) is the continuity equation in the form of a partial differential equation. This equation relates the flow field variables at a *point in the flow,* as opposed to Equation (2.48), which deals with a finite space.

It is important to keep in mind that Equations (2.48) and (2.52) are equally valid statements of the physical principle of conservation of mass. They are mathematical representations, but always remember that they speak words—they say that mass can be neither created nor destroyed.

Note that in the derivation of the above equations, the only assumption about the nature of the fluid is that it is a continuum. Therefore, Equations (2.48) and (2.52) hold in general for the three-dimensional, unsteady flow of any type of fluid, inviscid or viscous, compressible or incompressible. (*Note:* It is important to keep track of all assumptions that are used in the derivation of any equation because they tell you the limitations on the final result, and therefore prevent you from using an equation for a situation in which it is not valid. In all our future derivations, develop the habit of noting all assumptions that go with the resulting equations.)

134 PART 1 Fundamental Principles

It is important to emphasize the difference between unsteady and steady flows. In an *unsteady* flow, the flow-field variables are a function of both spatial location and time; for example,

$$\rho = \rho(x, y, z, t)$$

This means that if you lock your eyes on one fixed point in space, the density at that point will change with time. Such unsteady fluctuations can be caused by time-varying boundaries (e.g., an airfoil pitching up and down with time or the supply valves of a wind tunnel being turned off and on). Equations (2.48) and (2.52) hold for such unsteady flows. On the other hand, the vast majority of practical aerodynamic problems involve *steady* flow. Here, the flow-field variables are a function of spatial location only, for example,

$$\rho = \rho(x, y, z)$$

This means that if you lock your eyes on a fixed point in space, the density at that point will be a fixed value, invariant with time. For steady flow, $\partial/\partial t = 0$, and hence Equations (2.48) and (2.52) reduce to

$$\oiint_S \rho \mathbf{V} \cdot \mathbf{dS} = 0 \qquad (2.53)$$

and

$$\nabla \cdot (\rho \mathbf{V}) = 0 \qquad (2.54)$$

2.5 MOMENTUM EQUATION

Newton's second law is frequently written as

$$\mathbf{F} = m\mathbf{a} \qquad (2.55)$$

where $\mathbf{F}$ is the force exerted on a body of mass m and $\mathbf{a}$ is the acceleration. However, a more general form of Equation (2.55) is

$$\mathbf{F} = \frac{d}{dt}(m\mathbf{V}) \qquad (2.56)$$

which reduces to Equation (2.55) for a body of constant mass. In Equation (2.56), $m\mathbf{V}$ is the momentum of a body of mass m. Equation (2.56) represents the second fundamental principle upon which theoretical fluid dynamics is based.

Physical principle   Force = time rate of change of momentum

We will apply this principle [in the form of Equation (2.56)] to the model of a finite control volume fixed in space as sketched in Figure 2.19. Our objective is to obtain expressions for both the left and right sides of Equation (2.56) in terms of the familiar flow-field variables p, ρ, $\mathbf{V}$, etc. First, let us concentrate on the left side of Equation (2.56) (i.e., obtain an expression for **F**, which is the force exerted on the fluid as it flows through the control volume). This force comes from two sources:

1. *Body forces:* gravity, electromagnetic forces, or any other forces which "act at a distance" on the fluid inside $\mathcal{V}$.
2. *Surface forces:* pressure and shear stress acting on the control surface S.

Let **f** represent the net body force per unit mass exerted on the fluid inside $\mathcal{V}$. The body force on the elemental volume $d\mathcal{V}$ in Figure 2.19 is therefore

$$\rho \mathbf{f} \, d\mathcal{V}$$

and the total body force exerted on the fluid in the control volume is the summation of the above over the volume $\mathcal{V}$:

$$\text{Body force} = \iiint_{\mathcal{V}} \rho \mathbf{f} \, d\mathcal{V} \tag{2.57}$$

The elemental surface force due to pressure acting on the element of area dS is

$$-p \, \mathbf{dS}$$

where the negative sign indicates that the force is in the direction opposite of **dS**. That is, the control surface is experiencing a pressure force that is directed into the control volume and which is due to the pressure from the surroundings, and examination of Figure 2.19 shows that such an inward-directed force is in the direction opposite of **dS**. The complete pressure force is the summation of the elemental forces over the entire control surface:

$$\text{Pressure force} = -\oiint_{S} p \, \mathbf{dS} \tag{2.58}$$

In a viscous flow, the shear and normal viscous stresses also exert a surface force. A detailed evaluation of these viscous stresses is not warranted at this stage of our discussion. Let us simply recognize this effect by letting $\mathbf{F}_{\text{viscous}}$ denote the total viscous force exerted on the control surface. We are now ready to write an expression for the left-hand side of Equation (2.56). The total force experienced by the fluid as it is sweeping through the fixed control volume is given by the sum of Equations (2.57) and (2.58) and $\mathbf{F}_{\text{viscous}}$:

$$\mathbf{F} = \iiint_{\mathcal{V}} \rho \mathbf{f} \, d\mathcal{V} - \oiint_{S} p \, \mathbf{dS} + \mathbf{F}_{\text{viscous}} \tag{2.59}$$

Now consider the right side of Equation (2.56). The time rate of change of momentum of the fluid as it sweeps through the fixed control volume is the sum of two terms:

$$\begin{array}{c} \text{Net flow of momentum } out \\ \text{of control volume across surface } S \end{array} \equiv \mathbf{G} \tag{2.60a}$$

and

$$\text{Time rate of change of momentum due to} \atop \text{unsteady fluctuations of flow properties inside } \mathcal{V} \equiv \mathbf{H} \qquad (2.60b)$$

Consider the term denoted by **G** in Equation (2.60a). The flow has a certain momentum as it enters the control volume in Figure 2.19, and, in general, it has a different momentum as it leaves the control volume (due in part to the force **F** that is exerted on the fluid as it is sweeping through $\mathcal{V}$). The *net* flow of momentum *out* of the control volume across the surface S is simply this outflow minus the inflow of momentum across the control surface. This change in momentum is denoted by **G**, as noted above. To obtain an expression for **G**, recall that the mass flow across the elemental area dS is $(\rho \mathbf{V} \cdot \mathbf{dS})$; hence, the flow of momentum per second across dS is

$$(\rho \mathbf{V} \cdot \mathbf{dS})\mathbf{V}$$

The net flow of momentum out of the control volume through S is the summation of the above elemental contributions, namely,

$$\mathbf{G} = \oiint_S (\rho \mathbf{V} \cdot \mathbf{dS})\mathbf{V} \qquad (2.61)$$

In Equation (2.61), recall that positive values of $(\rho \mathbf{V} \cdot \mathbf{dS})$ represent mass flow out of the control volume, and negative values represent mass flow into the control volume. Hence, in Equation (2.61) the integral over the whole control surface is a combination of positive contributions (outflow of momentum) and negative contributions (inflow of momentum), with the resulting value of the integral representing the net outflow of momentum. If **G** has a positive value, there is more momentum flowing out of the control volume per second than flowing in; conversely, if **G** has a negative value, there is more momentum flowing into the control volume per second than flowing out.

Now consider **H** from Equation (2.60b). The momentum of the fluid in the elemental volume $d\mathcal{V}$ shown in Figure 2.19 is

$$(\rho\, d\mathcal{V})\mathbf{V}$$

The momentum contained at any instant inside the control volume is therefore

$$\iiint_{\mathcal{V}} \rho \mathbf{V}\, d\mathcal{V}$$

and its time rate of change due to unsteady flow fluctuations is

$$\mathbf{H} = \frac{\partial}{\partial t} \iiint_{\mathcal{V}} \rho \mathbf{V}\, d\mathcal{V} \qquad (2.62)$$

Combining Equations (2.61) and (2.62), we obtain an expression for the total time rate of change of momentum of the fluid as it sweeps through the fixed control volume, which in turn represents the right-hand side of Equation (2.56):

$$\frac{d}{dt}(m\mathbf{V}) = \mathbf{G} + \mathbf{H} = \oiint_S (\rho\mathbf{V} \cdot \mathbf{dS})\mathbf{V} + \frac{\partial}{\partial t}\iiint_\mathcal{V} \rho\mathbf{V} \, d\mathcal{V} \qquad (2.63)$$

Hence, from Equations (2.59) and (2.63), Newton's second law,

$$\frac{d}{dt}(m\mathbf{V}) = \mathbf{F}$$

applied to a fluid flow is

$$\boxed{\frac{\partial}{\partial t}\iiint_\mathcal{V} \rho\mathbf{V} \, d\mathcal{V} + \oiint_S (\rho\mathbf{V} \cdot \mathbf{dS})\mathbf{V} = -\oiint_S p \, \mathbf{dS} + \iiint_\mathcal{V} \rho\mathbf{f} \, d\mathcal{V} + \mathbf{F}_{\text{viscous}}}$$

$$(2.64)$$

Equation (2.64) is the momentum equation in integral form. Note that it is a vector equation. Just as in the case of the integral form of the continuity equation, Equation (2.64) has the advantage of relating aerodynamic phenomena over a finite region of space without being concerned with the details of precisely what is happening at a given distinct point in the flow. This advantage is illustrated in Section 2.6.

From Equation (2.64), we now proceed to a partial differential equation which relates flow-field properties at a point in space. Such an equation is a counterpart to the differential form of the continuity equation given in Equation (2.52). Apply the gradient theorem, Equation (2.27), to the first term on the right side of Equation (2.64):

$$-\oiint_S p \, \mathbf{dS} = -\iiint_\mathcal{V} \nabla p \, d\mathcal{V} \qquad (2.65)$$

Also, because the control volume is fixed, the time derivative in Equation (2.64) can be placed inside the integral. Hence, Equation (2.64) can be written as

$$\iiint_\mathcal{V} \frac{\partial(\rho\mathbf{V})}{\partial t} d\mathcal{V} + \oiint_S (\rho\mathbf{V} \cdot \mathbf{dS})\mathbf{V} = -\iiint_\mathcal{V} \nabla p \, d\mathcal{V} + \iiint_\mathcal{V} \rho\mathbf{f} \, d\mathcal{V} + \mathbf{F}_{\text{viscous}}$$

$$(2.66)$$

Recall that Equation (2.66) is a vector equation. It is convenient to write this equation as three scalar equations. Using cartesian coordinates, where

$$\mathbf{V} = u\mathbf{i} + v\mathbf{j} + w\mathbf{k}$$

the x component of Equation (2.66) is

$$\iiint_\mathcal{V} \frac{\partial(\rho u)}{\partial t} d\mathcal{V} + \oiint_S (\rho\mathbf{V} \cdot \mathbf{dS})u = -\iiint_\mathbf{V} \frac{\partial p}{\partial x} d\mathcal{V} + \iiint_\mathcal{V} \rho f_x \, d\mathcal{V} + (F_x)_{\text{viscous}}$$

$$(2.67)$$

[*Note:* In Equation (2.67), the product $(\rho\mathbf{V}\cdot\mathbf{dS})$ is a scalar, and therefore has no components.] Apply the divergence theorem, Equation (2.26), to the surface integral on the left side of Equation (2.67):

$$\oiint_S (\rho\mathbf{V}\cdot\mathbf{dS})u = \oiint_S (\rho u\mathbf{V})\cdot\mathbf{dS} = \oiiint_{\mathcal{V}} \nabla\cdot(\rho u\mathbf{V})\,d\mathcal{V} \qquad (2.68)$$

Substituting Equation (2.68) into Equation (2.67), we have

$$\oiiint_{\mathcal{V}} \left[\frac{\partial(\rho u)}{\partial t} + \nabla\cdot(\rho u\mathbf{V}) + \frac{\partial p}{\partial x} - \rho f_x - (\mathcal{F}_x)_{\text{viscous}} \right] d\mathcal{V} = 0 \qquad (2.69)$$

where $(\mathcal{F}_x)_{\text{viscous}}$ denotes the proper form of the x component of the viscous shear stresses when placed inside the volume integral (this form will be obtained explicitly in Chapter 15). For the same reasons as stated in Section 2.4, the integrand in Equation (2.69) is identically zero at all points in the flow; hence,

$$\boxed{\frac{\partial(\rho u)}{\partial t} + \nabla\cdot(\rho u\mathbf{V}) = -\frac{\partial p}{\partial x} + \rho f_x + (\mathcal{F}_x)_{\text{viscous}}} \qquad (2.70a)$$

Equation (2.70a) is the x component of the momentum equation in differential form. Returning to Equation (2.66), and writing the y and z components, we obtain in a similar fashion

$$\boxed{\frac{\partial(\rho v)}{\partial t} + \nabla\cdot(\rho v\mathbf{V}) = -\frac{\partial p}{\partial y} + \rho f_y + (\mathcal{F}_y)_{\text{viscous}}} \qquad (2.70b)$$

and

$$\boxed{\frac{\partial(\rho w)}{\partial t} + \nabla\cdot(\rho w\mathbf{V}) = -\frac{\partial p}{\partial z} + \rho f_z + (\mathcal{F}_z)_{\text{viscous}}} \qquad (2.70c)$$

where the subscripts y and z on f and $\mathcal{F}$ denote the y and z components of the body and viscous forces, respectively. Equation (2.70a to c) are the scalar x, y, and z components of the momentum equation, respectively; they are partial differential equations that relate flow-field properties at any point in the flow.

Note that Equations (2.64) and (2.70a to c) apply to the unsteady, three-dimensional flow of any fluid, compressible or incompressible, viscous or inviscid. Specialized to a steady ($\partial/\partial t \equiv 0$), inviscid ($\mathbf{F}_{\text{viscous}} = 0$) flow with no body forces ($\mathbf{f} = 0$), these equations become

$$\boxed{\oiint_S (\rho\mathbf{V}\cdot\mathbf{dS})\mathbf{V} = -\oiint_S p\,\mathbf{dS}} \qquad (2.71)$$

and

$$\nabla \cdot (\rho u \mathbf{V}) = -\frac{\partial p}{\partial x}$$ (2.72a)

$$\nabla \cdot (\rho v \mathbf{V}) = -\frac{\partial p}{\partial y}$$ (2.72b)

$$\nabla \cdot (\rho w \mathbf{V}) = -\frac{\partial p}{\partial z}$$ (2.72c)

Since most of the material in Chapters 3 through 14 assumes steady, inviscid flow with no body forces, we will have frequent occasion to use the momentum equation in the forms of Equations (2.71) and (2.72a to c).

The momentum equations for an inviscid flow [such as Equation (2.72a to c)] are called the *Euler equations*. The momentum equations for a viscous flow [such as Equation (2.70a to c)] are called the *Navier-Stokes equations*. We will encounter this terminology in subsequent chapters.

2.6 AN APPLICATION OF THE MOMENTUM EQUATION: DRAG OF A TWO-DIMENSIONAL BODY

We briefly interrupt our orderly development of the fundamental equations of fluid dynamics in order to examine an important application of the integral form of the momentum equation. During the 1930s and 1940s, the National Advisory Committee for Aeronautics (NACA) measured the lift and drag characteristics of a series of systematically designed airfoil shapes (discussed in detail in Chapter 4). These measurements were carried out in a specially designed wind tunnel where the wing models spanned the entire test section (i.e., the wing tips were butted against both sidewalls of the wind tunnel). This was done in order to establish two-dimensional (rather than three-dimensional) flow over the wing, thus allowing the properties of an airfoil (rather than a finite wing) to be measured. The distinction between the aerodynamics of airfoils and that of finite wings is made in Chapters 4 and 5. The important point here is that because the wings were mounted against both sidewalls of the wind tunnel, the NACA did not use a conventional force balance to measure the lift and drag. Rather, the lift was obtained from the pressure distributions on the *ceiling and floor* of the tunnel (above and below the wing), and the drag was obtained from measurements of the flow velocity *downstream* of the wing. These measurements may appear to be a strange way to measure the aerodynamic force on a wing. Indeed, how are these measurements related to lift and drag? What is going on here? The answers to these questions are addressed in this section; they involve an application of the fundamental momentum equation in integral form, and they illustrate a basic technique that is frequently used in aerodynamics.

Consider a two-dimensional body in a flow, as sketched in Figure 2.20a. A control volume is drawn around this body, as given by the dashed lines in

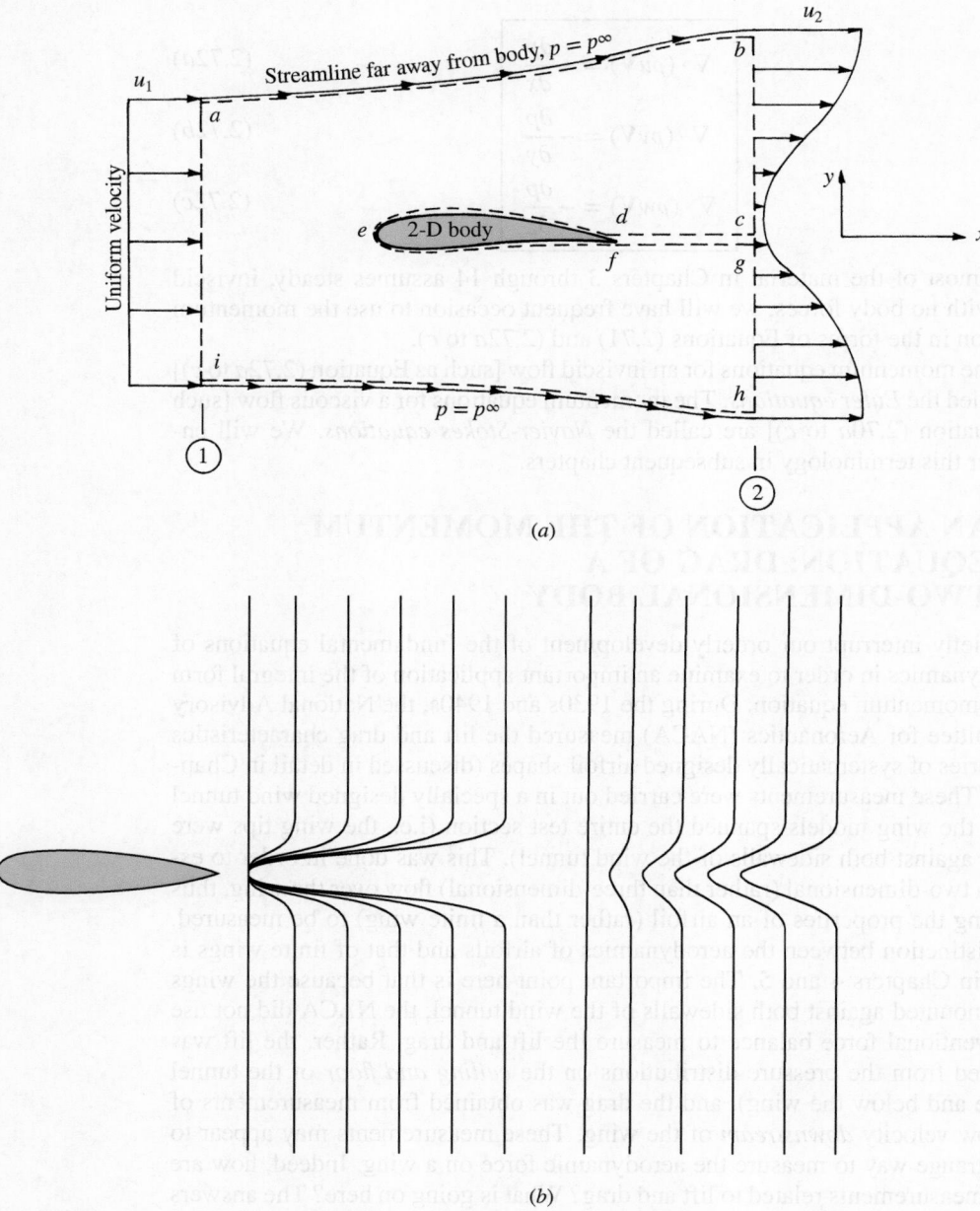

(a)

(b)

Figure 2.20 (a) Control volume for obtaining drag on a two-dimensional body.
(b) Rendering of the velocity profiles downstream of an airfoil. The profiles are made
visible in water flow by pulsing a voltage through a straight wire perpendicular to the flow, thus
creating small bubbles of hydrogen that subsequently move downstream with the flow.

Figure 2.20*a*. The control volume is bounded by:

1. The upper and lower streamlines far above and below the body (*ab* and *hi*, respectively).
2. Lines perpendicular to the flow velocity far ahead of and behind the body (*ai* and *bh*, respectively).
3. A cut that surrounds and wraps the surface of the body (*cdefg*).

The entire control volume is *abcdefghia*. The width of the control volume in the *z* direction (perpendicular to the page) is unity. Stations 1 and 2 are inflow and outflow stations, respectively.

Assume that the contour *abhi* is far enough from the body such that the pressure is same everywhere on *abhi* and equal to the freestream pressure $p = p_\infty$. Also, assume that the inflow velocity u_1 is uniform across *ai* (as it would be in a freestream, or a test section of a wind tunnel). The outflow velocity u_2 is *not* uniform across *bh*, because the presence of the body has created a wake at the outflow station. However, assume that both u_1 and u_2 are in the *x* direction; hence, $u_1 =$ constant and $u_2 = f(y)$.

An actual photograph of the velocity profiles in a wake downstream of an airfoil is shown in Figure 2.20*b*.

Consider the surface forces on the control volume shown in Figure 2.20*a*. They stem from *two* contributions:

1. The pressure distribution over the surface *abhi*,

$$-\iint\limits_{abhi} p\,\mathbf{dS}$$

2. The surface force on *def* created by the presence of the body

In the above list, the surface shear stress on *ab* and *hi* has been neglected. Also, note that in Figure 2.20*a* the cuts *cd* and *fg* are taken adjacent to each other; hence, any shear stress or pressure distribution on one is equal and opposite to that on the other (i.e., the surface forces on *cd* and *fg* cancel each other). Also, note that the surface force on *def* is the *equal and opposite reaction* to the shear stress and pressure distribution created by the flow over the surface of the body. To see this more clearly, examine Figure 2.21. On the left is shown the flow over the body. As explained in Section 1.5, the moving fluid exerts pressure and shear stress distributions over the body surface which create a resultant aerodynamic force per unit span $\mathbf{R}'$ on the body. In turn, by Newton's third law, the body exerts equal and opposite pressure and shear stress distributions on the flow (i.e., on the part of the control surface bounded by *def*). Hence, the body exerts a force $-\mathbf{R}'$ on the control surface, as shown on the right of Figure 2.21. With the above in mind, the total surface force on the entire control volume is

$$\text{Surface force} = -\iint\limits_{abhi} p\,\mathbf{dS} - \mathbf{R}' \qquad (2.73)$$

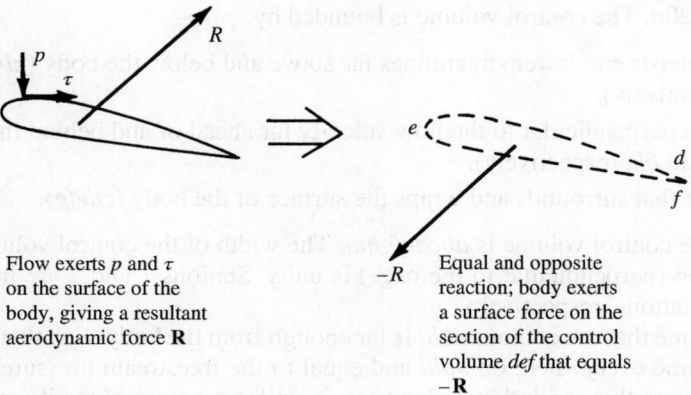

Flow exerts p and τ
on the surface of the
body, giving a resultant
aerodynamic force **R**

Equal and opposite
reaction; body exerts
a surface force on the
section of the control
volume *def* that equals
$-\mathbf{R}$

Figure 2.21 Equal and opposite reactions on a body and adjacent
section of control surface.

Moreover, this is the *total* force on the control volume shown in Figure 2.20*a*
because the volumetric body force is negligible.

Consider the integral form of the momentum equation as given by Equation (2.64). The right-hand side of this equation is physically the force on the
fluid moving through the control volume. For the control volume in Figure 2.20*a*,
this force is simply the expression given by Equation (2.73). Hence, using Equation (2.64), with the right-hand side given by Equation (2.73), we have

$$\frac{\partial}{\partial t} \iiint_{\mathcal{V}} \rho \mathbf{V}\, d\mathcal{V} + \oiint_{S} (\rho \mathbf{V} \cdot \mathbf{dS})\mathbf{V} = -\iint_{abhi} p\, \mathbf{dS} - \mathbf{R}' \qquad (2.74)$$

Assuming steady flow, Equation (2.74) becomes

$$\mathbf{R}' = -\oiint_{S} (\rho \mathbf{V} \cdot \mathbf{dS})\mathbf{V} - \iint_{abhi} p\, \mathbf{dS} \qquad (2.75)$$

Equation (2.75) is a vector equation. Consider again the control volume in Figure 2.20*a*. Take the x component of Equation (2.75), noting that the inflow and
outflow velocities u_1 and u_2 are in the x direction and the x component of **R**′ is
the aerodynamic drag per unit span D':

$$D' = -\oiint_{S} (\rho \mathbf{V} \cdot \mathbf{dS})u - \iint_{abhi} (p\, dS)_x \qquad (2.76)$$

In Equation (2.76), the last term is the component of the pressure force in the
x direction. [The expression $(p\, dS)_x$ is the x component of the pressure force
exerted on the elemental area dS of the control surface.] Recall that the boundaries
of the control volume *abhi* are chosen far enough from the body such that p is

constant along these boundaries. For a constant pressure,

$$\iint\limits_{abhi} (p \, dS)_x = 0 \tag{2.77}$$

because looking along the x direction in Figure 2.20a, the pressure force on *abhi* pushing toward the right exactly balances the pressure force pushing toward the left. This is true no matter what the shape of *abhi* is, as long as p is constant along the surface (for proof of this statement, see Problem 2.1). Therefore, substituting Equation (2.77) into (2.76), we obtain

$$D' = - \oiint\limits_{S} (\rho \mathbf{V} \cdot \mathbf{dS}) u \tag{2.78}$$

Evaluating the surface integral in Equation (2.78), we note from Figure 2.20a that:

1. The sections *ab*, *hi*, and *def* are streamlines of the flow. Since by definition $\mathbf{V}$ is parallel to the streamlines and $\mathbf{dS}$ is perpendicular to the control surface, along these sections $\mathbf{V}$ and $\mathbf{dS}$ are perpendicular vectors, and hence $\mathbf{V} \cdot \mathbf{dS} = 0$. As a result, the contributions of *ab*, *hi*, and *def* to the integral in Equation (2.78) are zero.
2. The cuts *cd* and *fg* are adjacent to each other. The mass flux out of one is identically the mass flux into the other. Hence, the contributions of *cd* and *fg* to the integral in Equation (2.78) cancel each other.

As a result, the only contributions to the integral in Equation (2.78) come from sections *ai* and *bh*. These sections are oriented in the y direction. Also, the control volume has unit depth in the z direction (perpendicular to the page). Hence, for these sections, $dS = dy(1)$. The integral in Equation (2.78) becomes

$$\oiint\limits_{S} (\rho \mathbf{V} \cdot \mathbf{dS}) u = - \int_i^a \rho_i u_1^2 \, dy + \int_h^b \rho_2 u_2^2 \, dy \tag{2.79}$$

Note that the minus sign in front of the first term on the right-hand side of Equation (2.79) is due to $\mathbf{V}$ and $\mathbf{dS}$ being in opposite directions along *ai* (station 1 is an inflow boundary); in contrast, $\mathbf{V}$ and $\mathbf{dS}$ are in the same direction over *hb* (station 2 is an outflow boundary), and hence the second term has a positive sign.

Before going further with Equation (2.79), consider the integral form of the continuity equation for steady flow, Equation (2.53). Applied to the control volume in Figure 2.20a, Equation (2.53) becomes

$$-\int_i^a \rho_1 u_1 \, dy + \int_h^b \rho_2 u_2 \, dy = 0$$

or

$$\int_i^a \rho_1 u_1 \, dy = \int_h^b \rho_2 u_2 \, dy \tag{2.80}$$

Multiplying Equation (2.80) by u_1, which is a constant, we obtain

$$\int_i^a \rho_1 u_1^2 \, dy = \int_h^b \rho_2 u_2 u_1 \, dy \tag{2.81}$$

Substituting Equation (2.81) into Equation (2.79), we have

$$\oiint_S (\rho \mathbf{V} \cdot \mathbf{dS})u = -\int_h^b \rho_2 u_2 u_1 \, dy + \int_h^b \rho_2 u_2^2 \, dy$$

or

$$\oiint_S (\rho \mathbf{V} \cdot \mathbf{dS})u = -\int_h^b \rho_2 u_2 (u_1 - u_2) \, dy \tag{2.82}$$

Substituting Equation (2.82) into Equation (2.78) yields

$$D' = \int_h^b \rho_2 u_2 (u_1 - u_2) \, dy \tag{2.83}$$

Equation (2.83) is the desired result of this section; it expresses the drag of a body in terms of the known freestream velocity u_1 and the flow-field properties ρ_2 and u_2, across a vertical station downstream of the body. These downstream properties can be measured in a wind tunnel, and the drag per unit span of the body D' can be obtained by evaluating the integral in Equation (2.83) numerically, using the measured data for ρ_2 and u_2 as a function of y.

Examine Equation (2.83) more closely. The quantity $u_1 - u_2$ is the velocity decrement at a given y location. That is, because of the drag on the body, there is a wake that trails downstream of the body. In this wake, there is a loss in flow velocity $u_1 - u_2$. The quantity $\rho_2 u_2$ is simply the mass flux; when multiplied by $u_1 - u_2$, it gives the decrement in momentum. Therefore, the integral in Equation (2.83) is physically the decrement in momentum flow that exists across the wake, and from Equation (2.83), this wake momentum decrement is equal to the drag on the body.

For incompressible flow, $\rho = $ constant and is known. For this case, Equation (2.83) becomes

$$D' = \rho \int_h^b u_2 (u_1 - u_2) \, dy \tag{2.84}$$

Equation (2.84) is the answer to the questions posed at the beginning of this section. It shows how a measurement of the velocity distribution across the wake of a body can yield the drag. These velocity distributions are conventionally measured with a Pitot rake, as shown in Figure 2.22. This is nothing more than a series of Pitot tubes attached to a common stem, which allows the simultaneous measurement of velocity across the wake. (The principle of the Pitot tube as a velocity-measuring instrument is discussed in Chapter 3. See also pages 188–210 of Reference 2 for an introductory discussion on Pitot tubes.)

The result embodied in Equation (2.84) illustrates the power of the integral form of the momentum equation; it relates drag on a body located at some position in the flow to the flow-field variables at a completely different location.

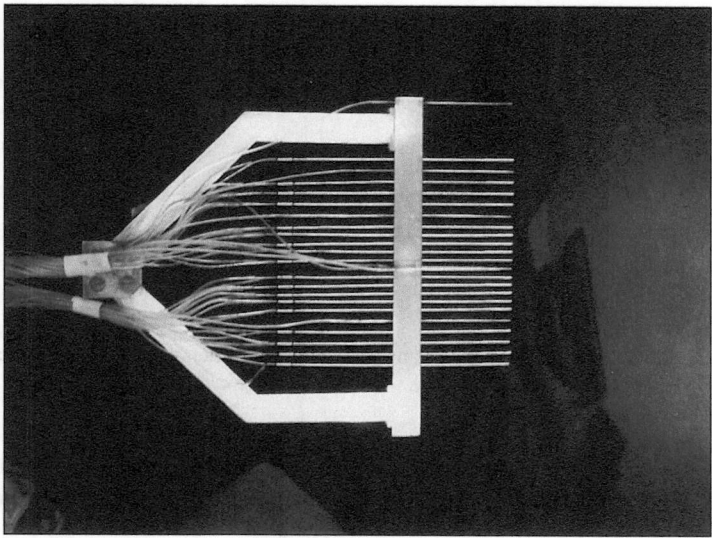

Figure 2.22 A Pitot rake for wake surveys (*Courtesy of John Anderson and the University of Maryland Aerodynamic Laboratory*).

At the beginning of this section, it was mentioned that lift on a two-dimensional body can be obtained by measuring the pressures on the ceiling and floor of a wind tunnel, above and below the body. This relation can be established from the integral form of the momentum equation in a manner analogous to that used to establish the drag relation; the derivation is left as a homework problem.

EXAMPLE 2.2

Consider an incompressible flow, laminar boundary layer growing along the surface of a flat plate, with chord length c, as sketched in Figure 2.23. The definition of a boundary layer was discussed in Sections 1.10 and 1.11. For an incompressible, laminar, flat plate boundary layer, the boundary-layer thickness δ at the trailing edge of the plate is

$$\frac{\delta}{c} = \frac{5}{\sqrt{Re_c}}$$

and the skin friction drag coefficient for the plate is

$$C_f \equiv \frac{D'}{q_\infty c(1)} = \frac{1.328}{\sqrt{Re_c}}$$

where the Reynolds number is based on chord length

$$Re_c = \frac{\rho_\infty V_\infty c}{\mu_\infty}$$

[*Note:* Both δ/c and C_f are functions of the Reynolds number—just another demonstration of the power of the similarity parameters. Since we are dealing with a low-speed,

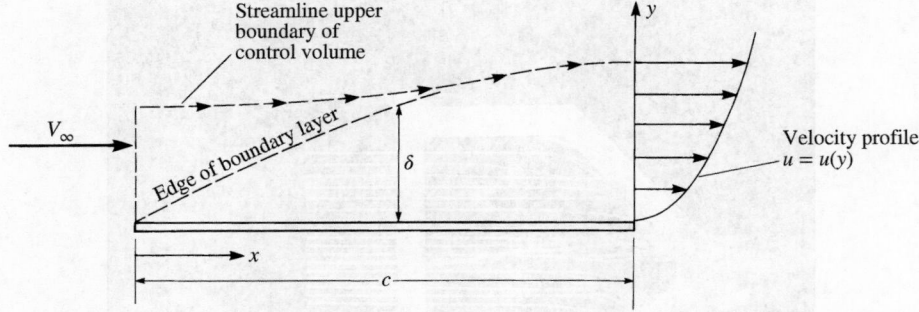

Figure 2.23 Sketch of a boundary layer and the velocity profile at $x = c$. The boundary-layer thickness δ is exaggerated here for clarity.

incompressible flow, the Mach number is not a relevant parameter here.] Let us *assume* that the velocity profile through the boundary layer is given by a power-law variation

$$u = V_\infty \left(\frac{y}{\delta} \right)^n$$

Calculate the value of n, consistent with the information given above.

■ Solution

From Equation (2.84)

$$C_f = \frac{D'}{q_\infty c} = \frac{\rho_\infty}{\frac{1}{2} \rho_\infty V_\infty^2 c} \int_0^\delta u_2(u_1 - u_2)\, dy$$

where the integral is evaluated at the trailing edge of the plate. Hence,

$$C_f = 2 \int_0^{\delta/c} \frac{u_2}{V_\infty} \left(\frac{u_1}{V_\infty} - \frac{u_2}{V_\infty} \right) d\left(\frac{y}{c} \right)$$

However, in Equation (2.84), applied to the control volume in Figure 2.23, $u_1 = V_\infty$. Thus,

$$C_f = 2 \int_0^{\delta/c} \frac{u_2}{V_\infty} \left(1 - \frac{u_2}{V_\infty} \right) d\left(\frac{y}{c} \right)$$

Inserting the laminar boundary-layer result for C_f as well as the assumed variation of velocity, both given above, we can write this integral as

$$\frac{1.328}{\sqrt{\text{Re}_c}} = 2 \int_0^{\delta/c} \left[\left(\frac{y/c}{\delta/c} \right)^n - \left(\frac{y/c}{\delta/c} \right)^{2n} \right] d\left(\frac{y}{c} \right)$$

Carrying out the integration, we obtain

$$\frac{1.328}{\sqrt{\text{Re}_c}} = \frac{2}{n+1} \left(\frac{\delta}{c} \right) - \frac{2}{2n+1} \left(\frac{\delta}{c} \right)$$

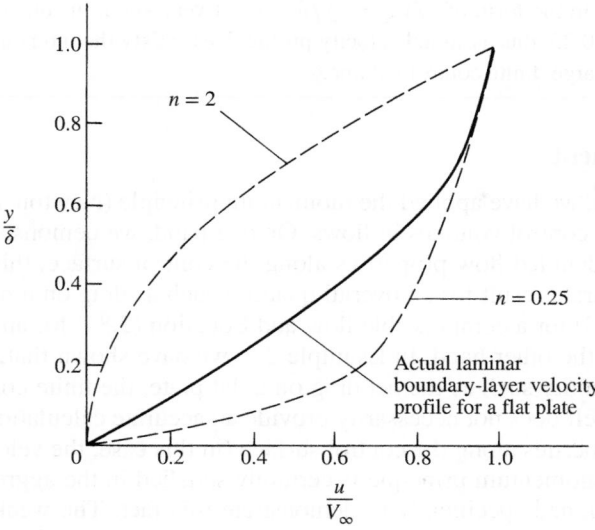

Figure 2.24 Comparison of the actual laminar boundary-layer profile with those calculated from Example 2.2.

Since $\delta/c = 5/\sqrt{\text{Re}_c}$, then

$$\frac{1.328}{\sqrt{\text{Re}_c}} = \frac{10}{n+1}\left(\frac{1}{\sqrt{\text{Re}_c}}\right) - \frac{10}{2n+1}\left(\frac{1}{\sqrt{\text{Re}_c}}\right)$$

or

$$\frac{1}{n+1} - \frac{1}{2n+1} = \frac{1.328}{10}$$

or

$$0.2656n^2 - 0.6016n + 0.1328 = 0$$

Using the quadratic formula, we have

$$n = 2 \quad \text{or} \quad 0.25$$

By *assuming* a power-law velocity profile in the form of $u/V_\infty = (y/\delta)^n$, we have found two different velocity profiles that satisfy the momentum principle applied to a finite control volume. Both of these profiles are shown in Figure 2.24 and are compared with an exact velocity profile obtained by means of a solution of the incompressible, laminar boundary-layer equations for a flat plate. (This boundary-layer solution is discussed in Chapter 18.) Note that the result $n = 2$ gives a concave velocity profile that is essentially nonphysical when compared to the convex profiles always observed in boundary layers. The result $n = 0.25$ gives a convex velocity profile that is qualitatively physically correct. However, this profile is quantitatively inaccurate, as can be seen in comparison to the exact profile. Hence, our original *assumption* of a power-law velocity profile for the laminar boundary layer in the form of $u/V_\infty = (y/\delta)^n$ is not very good, in spite of the fact that when $n = 2$ or 0.25, this assumed velocity profile does satisfy the momentum principle, applied over a large, finite control volume.

2.6.1 Comment

In this section, we have applied the momentum principle (Newton's second law) to large, fixed control volumes in flows. On one hand, we demonstrated that, by knowing the detailed flow properties along the control surface, this application led to an accurate result for an overall quantity such as drag on a body, namely, Equation (2.83) for a compressible flow and Equation (2.84) for an incompressible flow. On the other hand, in Example 2.2, we have shown that, by knowing an overall quantity such as the net drag on a flat plate, the finite control volume concept by itself does not necessarily provide an accurate calculation of detailed flow-field properties along the control surface (in this case, the velocity profile), although the momentum principle is certainly satisfied in the aggregate. Example 2.2 is designed specifically to demonstrate this fact. The weakness here is the need to *assume* some form for the variation of flow properties over the control surface; in Example 2.2, the assumption of the particular power-law profile proved to be unsatisfactory.

2.7 ENERGY EQUATION

For an incompressible flow, where ρ is constant, the primary flow-field variables are p and $\mathbf{V}$. The continuity and momentum equations obtained earlier are two equations in terms of the two unknowns p and $\mathbf{V}$. Hence, for a study of incompressible flow, the continuity and momentum equations are sufficient tools to do the job.

However, for a compressible flow, ρ is an additional variable, and therefore we need an additional fundamental equation to complete the system. This fundamental relation is the energy equation, to be derived in this section. In the process, two additional flow-field variables arise, namely, the internal energy e and temperature T. Additional equations must also be introduced for these variables, as will be mentioned later in this section.

The material discussed in this section is germane to the study of compressible flow. For those readers interested only in the study of incompressible flow for the time being, you may bypass this section and return to it at a later stage.

Physical principle   Energy can be neither created nor destroyed; it can only change in form.

This physical principle is embodied in the first law of thermodynamics. A brief review of thermodynamics is given in Chapter 7. Thermodynamics is essential to the study of compressible flow; however, at this stage, we will only introduce the first law, and we defer any substantial discussion of thermodynamics until Chapter 7, where we begin to concentrate on compressible flow.

Consider a fixed amount of matter contained within a closed boundary. This matter defines the *system*. Because the molecules and atoms within the system are constantly in motion, the system contains a certain amount of energy. For simplicity, let the system contain a unit mass; in turn, denote the internal energy per unit mass by e.

The region outside the system defines the *surroundings*. Let an incremental amount of heat δq be added to the system from the surroundings. Also, let δw be the work done on the system by the surroundings. (The quantities δq and δw are discussed in more detail in Chapter 7.) Both heat and work are forms of energy, and when added to the system, they change the amount of internal energy in the system. Denote this change of internal energy by de. From our physical principle that energy is conserved, we have for the system

$$\delta q + \delta w = de \qquad (2.85)$$

Equation (2.85) is a statement of the first law of thermodynamics.

Let us apply the first law to the fluid flowing through the fixed control volume shown in Figure 2.19. Let

B_1 = rate of heat added to fluid inside control volume from surroundings

B_2 = rate of work done on fluid inside control volume

B_3 = rate of change of energy of fluid as it flows through control volume

From the first law,

$$B_1 + B_2 = B_3 \qquad (2.86)$$

Note that each term in Equation (2.86) involves the *time rate* of energy change; hence, Equation (2.86) is, strictly speaking, a *power* equation. However, because it is a statement of the fundamental principle of conservation of energy, the equation is conventionally termed the "energy equation." We continue this convention here.

First, consider the rate of heat transferred to or from the fluid. This can be visualized as volumetric heating of the fluid inside the control volume due to absorption of radiation originating outside the system or the local emission of radiation by the fluid itself, if the temperature inside the control volume is high enough. In addition, there may be chemical combustion processes taking place inside the control volume, such as fuel-air combustion in a jet engine. Let this volumetric rate of heat addition per unit mass be denoted by $\dot{q}$. Typical units for $\dot{q}$ are J/s · kg or ft · lb/s · slug. Examining Figure 2.19, the mass contained within an elemental volume is $\rho\, d\mathcal{V}$; hence, the rate of heat addition to this mass is $\dot{q}(\rho\, d\mathcal{V})$. Summing over the complete control volume, we obtain

$$\text{Rate of volumetric heating} = \iiint_{\mathcal{V}} \dot{q}\rho\, d\mathcal{V} \qquad (2.87)$$

In addition, if the flow is viscous, heat can be transferred into the control volume by means of thermal conduction and mass diffusion across the control surface.

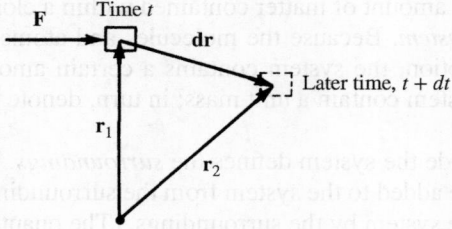

Figure 2.25 Schematic for the rate of doing work by a force **F** exerted on a moving body.

At this stage, a detailed development of these viscous heat-addition terms is not warranted; they are considered in detail in Chapter 15. Rather, let us denote the rate of heat addition to the control volume due to viscous effects simply by $\dot{Q}_{\text{viscous}}$. Therefore, in Equation (2.86), the total rate of heat addition is given by Equation (2.87) plus $\dot{Q}_{\text{viscous}}$:

$$B_1 = \iiint_{\mathcal{V}} \dot{q}\rho \, d\mathcal{V} + \dot{Q}_{\text{viscous}} \tag{2.88}$$

Before considering the rate of work done on the fluid inside the control volume, consider a simpler case of a solid object in motion, with a force **F** being exerted on the object, as sketched in Figure 2.25. The position of the object is measured from a fixed origin by the radius vector **r**. In moving from position $\mathbf{r}_1$ to $\mathbf{r}_2$ over an interval of time dt, the object is displaced through **dr**. By definition, the work done on the object in time dt is $\mathbf{F} \cdot \mathbf{dr}$. Hence, the time rate of doing work is simply $\mathbf{F} \cdot \mathbf{dr}/dt$. However, $\mathbf{dr}/dt = \mathbf{V}$, the velocity of the moving object. Hence, we can state that

$$\text{Rate of doing work on moving body} = \mathbf{F} \cdot \mathbf{V}$$

In words, the rate of work done on a moving body is equal to the product of its velocity and the component of force in the direction of the velocity.

This result leads to an expression for B_2, as follows. Consider the elemental area dS of the control surface in Figure 2.19. The pressure force on this elemental area is $-p\,\mathbf{dS}$. From the above result, the rate of work done on the fluid passing through dS with velocity **V** is $(-p\,\mathbf{dS}) \cdot \mathbf{V}$. Hence, summing over the complete control surface, we have

$$\begin{array}{c}\text{Rate of work done on fluid inside} \\ \mathcal{V} \text{ due to pressure force on } S\end{array} = - \oiint_{S} (p\,\mathbf{dS}) \cdot \mathbf{V} \tag{2.89}$$

In addition, consider an elemental volume $d\mathcal{V}$ inside the control volume, as shown in Figure 2.19. Recalling that **f** is the body force per unit mass, the rate of work done on the elemental volume due to the body force is $(\rho \mathbf{f}\, d\mathcal{V}) \cdot \mathbf{V}$. Summing over the complete control volume, we obtain

$$\begin{matrix} \text{Rate of work done on fluid} \\ \text{inside } \mathcal{V} \text{ due to body forces} \end{matrix} = \iiint\limits_{\mathcal{V}} (\rho \mathbf{f}\, d\mathcal{V}) \cdot \mathbf{V} \qquad (2.90)$$

If the flow is viscous, the shear stress on the control surface will also perform work on the fluid as it passes across the surface. Once again, a detailed development of this term is deferred until Chapter 15. Let us denote this contribution simply by $\dot{W}_{\text{viscous}}$. Then the total rate of work done on the fluid inside the control volume is the sum of Equations (2.89) and (2.90) and $\dot{W}_{\text{viscous}}$:

$$B_2 = - \oiint\limits_{S} p\mathbf{V} \cdot d\mathbf{S} + \iiint\limits_{\mathcal{V}} \rho(\mathbf{f} \cdot \mathbf{V})\, d\mathcal{V} + \dot{W}_{\text{viscous}} \qquad (2.91)$$

To visualize the energy inside the control volume, recall that in the first law of thermodynamics as stated in Equation (2.85), the internal energy e is due to the random motion of the atoms and molecules inside the system. Equation (2.85) is written for a stationary system. However, the fluid inside the control volume in Figure 2.19 is not stationary; it is moving at the local velocity $\mathbf{V}$ with a consequent kinetic energy per unit mass of $V^2/2$. Hence, the energy per unit mass of the moving fluid is the sum of both internal and kinetic energies $e + V^2/2$. This sum is called the *total energy* per unit mass.

We are now ready to obtain an expression for B_3, the rate of change of total energy of the fluid as it flows through the control volume. Keep in mind that mass flows into the control volume of Figure 2.19 bringing with it a certain total energy; at the same time mass flows out of the control volume taking with it a generally different amount of total energy. The elemental mass flow across dS is $\rho \mathbf{V} \cdot d\mathbf{S}$, and therefore the elemental flow of total energy across dS is $(\rho \mathbf{V} \cdot d\mathbf{S})(e + V^2/2)$. Summing over the complete control surface, we obtain

$$\begin{matrix} \text{Net rate of flow of total} \\ \text{energy across control surface} \end{matrix} = \oiint\limits_{S} (\rho \mathbf{V} \cdot d\mathbf{S}) \left(e + \frac{V^2}{2} \right) \qquad (2.92)$$

In addition, if the flow is unsteady, there is a time rate of change of total energy inside the control volume due to the transient fluctuations of the flow-field variables. The total energy contained in the elemental volume $d\mathcal{V}$ is $\rho(e + V^2/2)\, d\mathcal{V}$, and hence the total energy inside the complete control volume at any instant in time is

$$\iiint\limits_{\mathcal{V}} \rho \left(e + \frac{V^2}{2} \right) d\mathcal{V}$$

Therefore,

$$\begin{matrix} \text{Time rate of change of total energy} \\ \text{inside } \mathcal{V} \text{ due to transient variations} \\ \text{of flow-field variables} \end{matrix} = \frac{\partial}{\partial t} \iiint\limits_{\mathcal{V}} \rho \left(e + \frac{V^2}{2} \right) d\mathcal{V} \qquad (2.93)$$

In turn, B_3 is the sum of Equations (2.92) and (2.93):

$$B_3 = \frac{\partial}{\partial t} \iiint\limits_{\mathcal{V}} \rho \left(e + \frac{V^2}{2} \right) d\mathcal{V} + \iint\limits_{S} (\rho \mathbf{V} \cdot \mathbf{dS}) \left(e + \frac{V^2}{2} \right) \qquad (2.94)$$

Repeating the physical principle stated at the beginning of this section, the rate of heat added to the fluid plus the rate of work done on the fluid is equal to the rate of change of total energy of the fluid as it flows through the control volume (i.e., *energy is conserved*). In turn, these words can be directly translated into an equation by combining Equations (2.86), (2.88), (2.91), and (2.94):

$$\boxed{\begin{aligned} \iiint\limits_{\mathcal{V}} \dot{q}\rho \, d\mathcal{V} &+ \dot{Q}_{\text{viscous}} - \iint\limits_{S} p\mathbf{V} \cdot \mathbf{dS} + \iiint\limits_{\mathcal{V}} \rho(\mathbf{f} \cdot \mathbf{V}) \, d\mathcal{V} + \dot{W}_{\text{viscous}} \\ &= \frac{\partial}{\partial t} \iiint\limits_{\mathcal{V}} \rho \left(e + \frac{V^2}{2} \right) d\mathcal{V} + \iint\limits_{S} \rho \left(e + \frac{V^2}{2} \right) \mathbf{V} \cdot \mathbf{dS} \end{aligned}} \qquad (2.95)$$

Equation (2.95) is the energy equation in integral form; it is essentially the first law of thermodynamics applied to a fluid flow.

For the sake of completeness, note that if a shaft penetrates the control surface in Figure 2.19, driving some power machinery located inside the control volume (say, a compressor of a jet engine), then the rate of work delivered by the shaft, $\dot{W}_{\text{shaft}}$, must be added to the left side of Equation (2.95). Also note that the potential energy does not appear explicitly in Equation (2.95). Changes in potential energy are contained in the body force term when the force of gravity is included in $\mathbf{f}$. For the aerodynamic problems considered in this book, shaft work is not treated, and changes in potential energy are always negligible.

Following the approach established in Sections 2.4 and 2.5, we can obtain a partial differential equation for total energy from the integral form given in Equation (2.95). Applying the divergence theorem to the surface integrals in Equation (2.95), collecting all terms inside the same volume integral, and setting the integrand equal to zero, we obtain

$$\boxed{\begin{aligned} \frac{\partial}{\partial t} \left[\rho \left(e + \frac{V^2}{2} \right) \right] + \nabla \cdot \left[\rho \left(e + \frac{V^2}{2} \right) \mathbf{V} \right] &= \rho \dot{q} - \nabla \cdot (p\mathbf{V}) + \rho(\mathbf{f} \cdot \mathbf{V}) \\ &+ \dot{Q}'_{\text{viscous}} + \dot{W}'_{\text{viscous}} \end{aligned}}$$

$$(2.96)$$

where $\dot{Q}'_{\text{viscous}}$ and $\dot{W}'_{\text{viscous}}$ represent the proper forms of the viscous terms, to be obtained in Chapter 15. Equation (2.96) is a partial differential equation which relates the flow-field variables at a given point in space.

If the flow is steady ($\partial/\partial t = 0$), inviscid ($\dot{Q}_{\text{viscous}} = 0$ and $\dot{W}_{\text{viscous}} = 0$), adiabatic (no heat addition, $\dot{q} = 0$), and without body forces ($\mathbf{f} = 0$), then Equations (2.95) and (2.96) reduce to

$$\boxed{\iint\limits_{S} \rho \left(e + \frac{V^2}{2} \right) \mathbf{V} \cdot \mathbf{dS} = - \iint\limits_{S} p\mathbf{V} \cdot \mathbf{dS}} \qquad (2.97)$$

and

$$\nabla \cdot \left[\rho \left(e + \frac{V^2}{2} \right) \mathbf{V} \right] = -\nabla \cdot (p\mathbf{V}) \qquad (2.98)$$

Equations (2.97) and (2.98) are discussed and applied at length beginning with Chapter 7.

With the energy equation, we have introduced another unknown flow-field variable e. We now have three equations, continuity, momentum, and energy, which involve four dependent variables, $\rho, p, \mathbf{V}$, and e. A fourth equation can be obtained from a thermodynamic state relation for e (see Chapter 7). If the gas is calorically perfect, then

$$e = c_v T \qquad (2.99)$$

where c_v is the specific heat at constant volume. Equation (2.99) introduces temperature as yet another dependent variable. However, the system can be completed by using the perfect gas equation of state

$$p = \rho R T \qquad (2.100)$$

where R is the specific gas constant. Therefore, the continuity, momentum, and energy equations, along with Equations (2.99) and (2.100), are five independent equations for the five unknowns, $\rho, p, \mathbf{V}, e$, and T. The matter of a perfect gas and related equations of state are reviewed in detail in Chapter 7; Equations (2.99) and (2.100) are presented here only to round out our development of the fundamental equations of fluid flow.

2.8 INTERIM SUMMARY

At this stage, let us pause and think about the various equations we have developed. Do not fall into the trap of seeing these equations as just a jumble of mathematical symbols that, by now, might look all the same to you. Quite the contrary, these equations speak words: for example, Equations (2.48), (2.52), (2.53), and (2.54) all say that mass is conserved; Equations (2.64), (2.70a to c), (2.71), and (2.72a to c) are statements of Newton's second law applied to a fluid flow; Equations (2.95) to (2.98) say that energy is conserved. It is very important to be able to see the physical principles behind these equations. When you look at an equation, try to develop the ability to see past a collection of mathematical symbols and, instead, to read the physics that the equation represents.

The equations listed above are fundamental to all of aerodynamics. Take the time to go back over them. Become familiar with the way they are developed, and make yourself comfortable with their final forms. In this way, you will find our subsequent aerodynamic applications that much easier to understand.

Also, note our location on the road map shown in Figure 2.3. We have finished the items on the left branch of the map—we have obtained the basic flow equations containing the fundamental physics of fluid flow. We now start with the branch on the right, which is a collection of useful concepts helpful in the application of the basic flow equations.

2.9 SUBSTANTIAL DERIVATIVE

Consider a small fluid element moving through a flow field, as shown in Figure 2.26. This figure is basically an extension of Figure 2.16, in which we introduced the concept of a fluid element moving through a specified flow field. The velocity field is given by $V = u\mathbf{i} + v\mathbf{j} + w\mathbf{k}$, where

$$u = u(x, y, z, t)$$
$$v = v(x, y, z, t)$$
$$w = w(x, y, z, t)$$

In addition, the density field is given by

$$\rho = \rho(x, y, z, t)$$

At time t_1, the fluid element is located at point 1 in the flow (see Figure 2.26), and its density is

$$\rho_1 = \rho(x_1, y_1, z_1, t_1)$$

At a later time t_2 the same fluid element has moved to a different location in the flow field, such as point 2 in Figure 2.26. At this new time and location, the density of the fluid element is

$$\rho_2 = \rho(x_2, y_2, z_2, t_2)$$

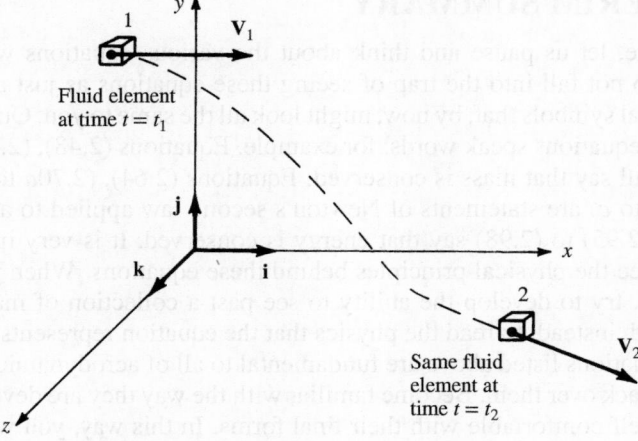

Figure 2.26 Fluid element moving in a flow field—illustration for the substantial derivative.

Since $\rho = \rho(x, y, z, t)$, we can expand this function in a Taylor series about point 1 as follows:

$$\rho_2 = \rho_1 + \left(\frac{\partial \rho}{\partial x}\right)_1 (x_2 - x_1) + \left(\frac{\partial \rho}{\partial y}\right)_1 (y_2 - y_1) + \left(\frac{\partial \rho}{\partial z}\right)_1 (z_2 - z_1)$$

$$+ \left(\frac{\partial \rho}{\partial t}\right)_1 (t_2 - t_1) + \text{higher-order terms}$$

Dividing by $t_2 - t_1$, and ignoring the higher-order terms, we have

$$\frac{\rho_2 - \rho_1}{t_2 - t_1} = \left(\frac{\partial \rho}{\partial x}\right)_1 \frac{x_2 - x_1}{t_2 - t_1} + \left(\frac{\partial \rho}{\partial y}\right)_1 \left(\frac{y_2 - y_1}{t_2 - t_1}\right) + \left(\frac{\partial \rho}{\partial z}\right)_1 \frac{z_2 - z_1}{t_2 - t_1} + \left(\frac{\partial \rho}{\partial t}\right)_1$$

$$(2.101)$$

Consider the physical meaning of the left side of Equation (2.101). The term $(\rho_2 - \rho_1)/(t_2 - t_1)$ is the *average* time rate of change in density of the fluid element as it moves from point 1 to point 2. In the limit, as t_2 approaches t_1, this term becomes

$$\lim_{t_2 \to t_1} \frac{\rho_2 - \rho_1}{t_2 - t_1} = \frac{D\rho}{Dt}$$

Here, $D\rho/Dt$ is a symbol for the *instantaneous* time rate of change of density of the fluid element as it moves through point 1. By definition, this symbol is called the *substantial derivative D/Dt*. Note that $D\rho/Dt$ is the time rate of change of density of a *given fluid element* as it moves through space. Here, our eyes are locked on the fluid element as it is moving, and we are watching the density of the element change as it moves through point 1. This is different from $(\partial \rho/\partial t)_1$, which is physically the time rate of change of density at the *fixed* point 1. For $(\partial \rho/\partial t)_1$, we fix our eyes on the stationary point 1, and watch the density change due to transient fluctuations in the flow field. Thus, $D\rho/Dt$ and $\partial \rho/\partial t$ are physically and numerically different quantities.

Returning to Equation (2.101), note that

$$\lim_{t_2 \to t_1} \frac{x_2 - x_1}{t_2 - t_1} \equiv u$$

$$\lim_{t_2 \to t_1} \frac{y_2 - y_1}{t_2 - t_1} \equiv v$$

$$\lim_{t_2 \to t_1} \frac{z_2 - z_1}{t_2 - t_1} \equiv w$$

Thus, taking the limit of Equation (2.101) as $t_2 \to t_1$, we obtain

$$\frac{D\rho}{Dt} = u\frac{\partial \rho}{\partial x} + v\frac{\partial \rho}{\partial y} + w\frac{\partial \rho}{\partial z} + \frac{\partial \rho}{\partial t} \qquad (2.102)$$

Examine Equation (2.102) closely. From it, we can obtain an expression for the substantial derivative in cartesian coordinates:

$$\frac{D}{Dt} \equiv \frac{\partial}{\partial t} + u\frac{\partial}{\partial x} + v\frac{\partial}{\partial y} + w\frac{\partial}{\partial z} \qquad (2.103)$$

Furthermore, in cartesian coordinates, the vector operator ∇ is defined as

$$\nabla \equiv \mathbf{i}\frac{\partial}{\partial x} + \mathbf{j}\frac{\partial}{\partial y} + \mathbf{k}\frac{\partial}{\partial z}$$

Hence, Equation (2.103) can be written as

$$\boxed{\frac{D}{Dt} \equiv \frac{\partial}{\partial t} + (\mathbf{V} \cdot \nabla)} \qquad (2.104)$$

Equation (2.104) represents a definition of the substantial derivative in vector notation; thus, it is valid for any coordinate system.

Focusing on Equation (2.104), we once again emphasize that D/Dt is the substantial derivative, which is physically the time rate of change following a moving fluid element; $\partial/\partial t$ is called the *local derivative,* which is physically the time rate of change at a fixed point; $\mathbf{V} \cdot \nabla$ is called the *convective derivative,* which is physically the time rate of change due to the movement of the fluid element from one location to another in the flow field where the flow properties are spatially different. The substantial derivative applies to any flow-field variable (e.g., Dp/Dt, DT/Dt, Du/Dt). For example,

$$\underset{\substack{\text{local} \\ \text{derivative}}}{\frac{DT}{Dt} \equiv \frac{\partial T}{\partial t}} + \underset{\substack{\text{convective} \\ \text{derivative}}}{(\mathbf{V} \cdot \nabla)T} \equiv \frac{\partial T}{\partial t} + u\frac{\partial T}{\partial x} + v\frac{\partial T}{\partial y} + w\frac{\partial T}{\partial z} \qquad (2.105)$$

Again, Equation (2.105) states physically that the temperature of the fluid element is changing as the element sweeps past a point in the flow because at that point the flow-field temperature itself may be fluctuating with time (the local derivative) and because the fluid element is simply on its way to another point in the flow field where the temperature is different (the convective derivative).

Consider an example that will help to reinforce the physical meaning of the substantial derivative. Imagine that you are hiking in the mountains, and you are about to enter a cave. The temperature inside the cave is cooler than outside. Thus, as you walk through the mouth of the cave, you feel a temperature decrease—this is analogous to the convective derivative in Equation (2.105). However, imagine that, at the same time, a friend throws a snowball at you such that the snowball hits you just at the same instant you pass through the mouth of the cave. You will feel an additional, but momentary, temperature drop when the snowball hits you—this is analogous to the local derivative in Equation (2.105). The net temperature drop you feel as you walk through the mouth of the cave is therefore a combination of both the act of moving into the cave, where it is cooler, and being struck by the snowball at the same instant—this net temperature drop is analogous to the substantial derivative in Equation (2.105).

EXAMPLE 2.3

Return to the subsonic compressible flow over a wavy wall treated in Example 2.1. In that example, we calculated the time rate of change of the volume of a fluid element per unit volume at the point $(x/\ell, y/\ell) = (\frac{1}{4}, 1)$ to be -0.7327 s^{-1}. That is, at the instant the fluid element was passing through this point, its volume was experiencing a rate of *decrease* of 73 percent per second, a substantial instantaneous rate of change. Moreover, we noted in Example 2.1 that, because the volume was decreasing and hence the density increasing, the point $(\frac{1}{4}, 1)$ must be in a compression region. This would imply that the fluid element is slowing down as it passes through point $(\frac{1}{4}, 1)$; i.e., it is experiencing a *deceleration*. Calculate the value of the deceleration at this point.

■ Solution

Acceleration (or deceleration) is physically the time rate of change of velocity. The time rate of change of velocity of a moving fluid element is, from the physical meaning of the substantial derivative, the substantial derivative of the velocity. Let us deal in terms of cartesian coordinates. For the two-dimensional flow considered in Example 2.1, the x and y components of acceleration are denoted by a_x and a_y, respectively, where

$$a_x = \frac{Du}{Dt} = u\frac{\partial u}{\partial x} + v\frac{\partial u}{\partial y} \tag{E2.1}$$

and

$$a_y = \frac{Dv}{Dt} = u\frac{\partial v}{\partial x} + v\frac{\partial v}{\partial y} \tag{E2.2}$$

Equations for u and v are given in Example 2.1 as

$$u = V_\infty \left[1 + \frac{h}{\beta}\frac{2\pi}{\ell}\left(\cos\frac{2\pi x}{\ell}\right) e^{-2\pi\beta y/\ell} \right] \tag{2.35}$$

and

$$v = -V_\infty h\frac{2\pi}{\ell}\left(\sin\frac{2\pi x}{\ell}\right) e^{-2\pi\beta y/\ell} \tag{2.36}$$

where

$$\beta = \sqrt{1 - M_\infty^2}$$

From Equation (2.35),

$$\frac{\partial u}{\partial x} = -\frac{V_\infty h}{\beta}\left(\frac{2\pi}{\ell}\right)^2 \left(\sin\frac{2\pi x}{\ell}\right) e^{-2\pi\beta y/\ell} \tag{E2.3}$$

and

$$\frac{\partial u}{\partial y} = -V_\infty h\left(\frac{2\pi}{\ell}\right)^2 \left(\cos\frac{2\pi x}{\ell}\right) e^{-2\pi\beta y/\ell} \tag{E2.4}$$

From Equation (2.36),

$$\frac{\partial v}{\partial x} = -V_\infty h \left(\frac{2\pi}{\ell} \right)^2 \left(\cos \frac{2\pi x}{\ell} \right) e^{-2\pi\beta y/\ell} \qquad \text{(E2.5)}$$

and

$$\frac{\partial v}{\partial y} = V_\infty h \beta \left(\frac{2\pi}{\ell} \right)^2 \left(\cos \frac{2\pi x}{\ell} \right) e^{-2\pi\beta y/\ell} \qquad \text{(E2.6)}$$

For the wavy wall in Example 2.1, the wavelength is $\ell = 1.0$ m and the amplitude $h = 0.01$ m (see Figure 2.17). Also given in Example 2.1 is $V_\infty = 240$ m/s and $M_\infty = 0.7$. For these conditions, and remembering that we are making the calculation for a fluid element as it passes through $(x, y) = (\frac{1}{4}, 1)$,

$$\frac{2\pi}{\ell} = \frac{2\pi}{1.0} = 6.283$$

$$\beta = \sqrt{1 - M_\infty^2} = \sqrt{1 - (0.7)^2} = 0.714$$

$$\frac{2\pi\beta y}{\ell} = 6.283\,(0.714)(1.0) = 4.486$$

$$e^{-2\pi\beta y/\ell} = e^{-4.486} = 0.01126$$

$$\sin \frac{2\pi x}{\ell} = \sin \frac{2\pi}{4} = \sin \frac{\pi}{2} = 1$$

$$\cos \frac{2\pi x}{\ell} = \cos \frac{\pi}{2} = 0$$

From Equation (2.35),

$$u = V_\infty \left[1 + \frac{h}{\beta} \frac{2\pi}{\ell} \left(\cos \frac{2\pi x}{\ell} \right) e^{-2\pi\beta y/\ell} \right]$$

$$u = V_\infty = 240 \text{ m/s}$$

From Equation (2.36),

$$v = -V_\infty h \frac{2\pi}{\ell} \left(\sin \frac{2\pi x}{\ell} \right) e^{-2\pi\beta y/\ell}$$

$$v = -(240)(0.01)(6.283)(1)(0.01126)$$

$$v = -0.1698 \text{ m/s}$$

From Equation (E2.3),

$$\frac{\partial u}{\partial x} = -\frac{V_\infty h}{\beta} \left(\frac{2\pi}{\ell} \right)^2 \left(\sin \frac{2\pi x}{\ell} \right) e^{-2\pi\beta y/\ell}$$

$$\frac{\partial u}{\partial x} = -\frac{(240)(0.01)}{0.714}(6.283)^2(1)(0.01126)$$

$$\frac{\partial u}{\partial x} = -1.494 \, \text{s}^{-1}$$

From Equation (E2.4),

$$\frac{\partial u}{\partial y} = -V_\infty h \left(\frac{2\pi}{\ell}\right)^2 \left(\cos \frac{2\pi x}{\ell}\right) e^{-2\pi\beta y/\ell}$$

$$\frac{\partial u}{\partial y} = 0$$

From Equation (E2.5),

$$\frac{\partial v}{\partial x} = -V_\infty h \left(\frac{2\pi}{\ell}\right)^2 \left(\cos \frac{2\pi x}{\ell}\right) e^{-2\pi\beta y/\ell}$$

$$\frac{\partial v}{\partial x} = 0$$

From Equation (E2.6),

$$\frac{\partial v}{\partial y} = V_\infty h \beta \left(\frac{2\pi}{\ell}\right)^2 \left(\sin \frac{2\pi x}{\ell}\right) e^{-2\pi\beta y/\ell}$$

$$\frac{\partial v}{\partial y} = (240)(0.01)(0.714)(6.283)^2(1)(0.01126)$$

$$\frac{\partial v}{\partial y} = 0.7617 \, \text{s}^{-1}$$

Substituting the above values into Equation (E2.1), we have

$$a_x = u\frac{\partial u}{\partial x} + v\frac{\partial u}{\partial y} = (240)(-1.494) - (0.1698)(0)$$

$$a_x = -358.56 \, \text{m/s}^2$$

From Equation (E2.2),

$$a_y = u\frac{\partial u}{\partial x} + v\frac{\partial v}{\partial y}$$

$$a_y = (240)(0) - (0.1698)(0.7617) = -0.129 \, \text{m/s}^2$$

The absolute magnitude of the acceleration is

$$|a| = \sqrt{a_x^2 + a_y^2} = \sqrt{(-358.56)^2 + (-0.129)^2}$$

$$|a| = 358.6 \, \text{m/s}^2$$

Note, however, that both a_x and a_y are negative, and hence the acceleration is *negative*, i.e., the fluid element is *decelerating* as it passes through point $(\frac{1}{4}, 1)$, with a value of

$$\boxed{\text{Deceleration} = 358.6 \text{ m/s}^2}$$

Note also that, by far, the deceleration is greatest in the x direction, with the deceleration in the y direction being very small.

The acceleration of gravity at sea level on earth is 9.8 m/s^2. Observe that the fluid element in this example is locally experiencing a deceleration with an absolute magnitude that is 36.6 times that of the acceleration of gravity; i.e., the fluid element as it passes through point $(\frac{1}{4}, 1)$ is experiencing a large deceleration of 36.6 g. This is totally consistent with the result from Example 2.1 that the fluid element is simultaneously experiencing a very rapid change in volume of 73 percent per second. (To relate to human experience, a human being can tolerate only up to 10 g acceleration or deceleration, and that for only a few seconds before life-threatening bodily injury.) The flow field shown in Figure 2.17 and treated here and in Example 2.1 is relatively benign; indeed, it is a flow involving only small perturbations from a uniform flow. Subsonic small perturbation flows are treated in Chapter 11. Yet, from this example we deduce that a given fluid element, even though it is moving through a rather calm flow field, gets rather drastically pushed around. The force that is pushing around the fluid element is supplied by the pressure gradients in the flow, as discussed in Section 2.5.

2.10 FUNDAMENTAL EQUATIONS IN TERMS OF THE SUBSTANTIAL DERIVATIVE

In this section, we express the continuity, momentum, and energy equations in terms of the substantial derivative. In the process, we make use of the following vector identity:

$$\nabla \cdot (\rho \mathbf{V}) \equiv \rho \nabla \cdot \mathbf{V} + \mathbf{V} \cdot \nabla \rho \tag{2.106}$$

In words, this identity states that the divergence of a scalar times a vector is equal to the scalar times the divergence of the vector plus the dot product of the vector and the gradient of the scalar.

First, consider the continuity equation given in the form of Equation (2.52):

$$\frac{\partial \rho}{\partial t} + \nabla \cdot (\rho \mathbf{V}) = 0 \tag{2.52}$$

Using the vector identity given by Equation (2.106), Equation (2.52) becomes

$$\frac{\partial \rho}{\partial t} + \mathbf{V} \cdot \nabla \rho + \rho \nabla \cdot \mathbf{V} = 0 \tag{2.107}$$

However, the sum of the first two terms of Equation (2.107) is the substantial derivative of ρ [see Equation (2.104)]. Thus, from Equation (2.107),

$$\boxed{\frac{D\rho}{Dt} + \rho\nabla \cdot \mathbf{V} = 0} \tag{2.108}$$

Equation (2.108) is the form of the continuity equation written in terms of the substantial derivative.

Next, consider the x component of the momentum equation given in the form of Equation (2.70a):

$$\frac{\partial(\rho u)}{\partial t} + \nabla \cdot (\rho u \mathbf{V}) = -\frac{\partial p}{\partial x} + \rho f_x + (\mathcal{F}_x)_{\text{viscous}} \tag{2.70a}$$

The first terms can be expanded as

$$\frac{\partial(\rho u)}{\partial t} = \rho\frac{\partial u}{\partial t} + u\frac{\partial \rho}{\partial t} \tag{2.109}$$

In the second term of Equation (2.70a), treat the scalar quantity as u and the vector quantity as $\rho\mathbf{V}$. Then the term can be expanded using the vector identity in Equation (2.106):

$$\nabla \cdot (\rho u \mathbf{V}) \equiv \nabla \cdot [u(\rho\mathbf{V})] = u\nabla \cdot (\rho\mathbf{V}) + (\rho\mathbf{V}) \cdot \nabla u \tag{2.110}$$

Substituting Equations (2.109) and (2.110) into (2.70a), we obtain

$$\rho\frac{\partial u}{\partial t} + u\frac{\partial \rho}{\partial t} + u\nabla \cdot (\rho\mathbf{V}) + (\rho\mathbf{V}) \cdot \nabla u = -\frac{\partial p}{\partial x} + \rho f_x + (\mathcal{F}_x)_{\text{viscous}}$$

or

$$\rho\frac{\partial u}{\partial t} + u\left[\frac{\partial \rho}{\partial t} + \nabla \cdot (\rho\mathbf{V})\right] + (\rho\mathbf{V}) \cdot \nabla u = -\frac{\partial p}{\partial x} + \rho f_x + (\mathcal{F}_x)_{\text{viscous}} \tag{2.111}$$

Examine the two terms inside the square brackets; they are precisely the left side of the continuity equation, Equation (2.52). Since the right side of Equation (2.52) is zero, the sum inside the square brackets is zero. Hence, Equation (2.111) becomes

$$\rho\frac{\partial u}{\partial t} + \rho\mathbf{V} \cdot \nabla u = -\frac{\partial p}{\partial x} + \rho f_x + (\mathcal{F}_x)_{\text{viscous}}$$

or

$$\rho\left(\frac{\partial u}{\partial t} + \mathbf{V} \cdot \nabla u\right) = -\frac{\partial p}{\partial x} + \rho f_x + (\mathcal{F}_x)_{\text{viscous}} \tag{2.112}$$

Examine the two terms inside the parentheses in Equation (2.112); their sum is precisely the substantial derivative Du/Dt. Hence, Equation (2.112) becomes

$$\boxed{\rho\frac{Du}{Dt} = -\frac{\partial p}{\partial x} + \rho f_x + (\mathcal{F}_x)_{\text{viscous}}} \tag{2.113a}$$

In a similar manner, Equation (2.70b and c) yield

$$\rho \frac{Dv}{Dt} = -\frac{\partial p}{\partial y} + \rho f_y + (\mathcal{F}_y)_{\text{viscous}} \qquad (2.113b)$$

$$\rho \frac{Dw}{Dt} = -\frac{\partial p}{\partial z} + \rho f_z + (\mathcal{F}_z)_{\text{viscous}} \qquad (2.113c)$$

Equation (2.113a to c) are the x, y, and z components of the *momentum equation* written in terms of the substantial derivative. Compare these equations with Equation (2.70a to c). Note that the right sides of both sets of equations are unchanged; only the left sides are different.

In an analogous fashion, the energy equation given in the form of Equation (2.96) can be expressed in terms of the substantial derivative. The derivation is left as a homework problem; the result is

$$\rho \frac{D(e + V^2/2)}{Dt} = \rho \dot{q} - \nabla \cdot (p\mathbf{V}) + \rho(\mathbf{f} \cdot \mathbf{V}) + \dot{Q}'_{\text{viscous}} + \dot{W}'_{\text{viscous}} \qquad (2.114)$$

Again, the right-hand sides of Equations (2.96) and (2.114) are the same; only the form of the left sides is different.

In modern aerodynamics, it is conventional to call the form of Equations (2.52), (2.70a to c), and (2.96) the *conservation* form of the fundamental equations (sometimes these equations are labeled as the *divergence* form because of the divergence terms on the left side). In contrast, the form of Equations (2.108), (2.113a to c), and (2.114), which deals with the substantial derivative on the left side, is called the *nonconservation* form. Both forms are equally valid statements of the fundamental principles, and in most cases, there is no particular reason to choose one form over the other. The nonconservation form is frequently found in textbooks and in aerodynamic theory. However, for the numerical solution of some aerodynamic problems, the conservation form sometimes leads to more accurate results. Hence, the distinction between the conservation form and the nonconservation form has become important in the modern discipline of computational fluid dynamics. (See Reference 7 for more details.)

2.11 PATHLINES, STREAMLINES, AND STREAKLINES OF A FLOW

In addition to knowing the density, pressure, temperature, and velocity fields, in aerodynamics we like to draw pictures of "where the flow is going." To accomplish this, we construct diagrams of pathlines and/or streamlines of the flow. The distinction between pathlines and streamlines is described in this section.

Consider an unsteady flow with a velocity field given by $\mathbf{V} = \mathbf{V}(x, y, z, t)$. Also, consider an infinitesimal fluid element moving through the flow field, say, element A as shown in Figure 2.27a. Element A passes through point 1. Let us

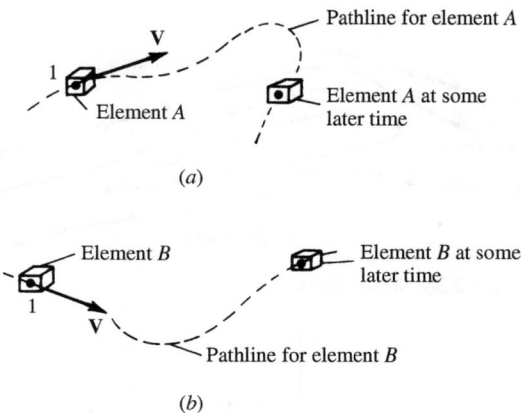

Pathline for element A

V

1

Element A

Element A at some
later time

(a)

Element B

Element B at some
later time

1

V

Pathline for element B

(b)

Figure 2.27 Pathlines for two different fluid elements
passing through the same point in space: unsteady flow.

trace the path of element A as it moves downstream from point 1, as given by
the dashed line in Figure 2.27a. Such a path is defined as the *pathline* for element A. Now, trace the path of another fluid element, say, element B as shown
in Figure 2.27b. Assume that element B also passes through point 1, but at some
different time from element A. The pathline of element B is given by the dashed
line in Figure 2.27b. Because the flow is unsteady, the velocity at point 1 (and at
all other points of the flow) changes with time. Hence, the pathlines of elements
A and B are different curves in Figure 2.27a and b. In general, for *unsteady* flow,
the pathlines for different fluid elements passing through the same point are not
the same.

In Section 1.4, the concept of a streamline was introduced in a somewhat
heuristic manner. Let us be more precise here. By definition, a *streamline* is a
curve whose tangent at any point is in the direction of the velocity vector at that
point. Streamlines are illustrated in Figure 2.28. The streamlines are drawn such
that their tangents at every point along the streamline are in the same direction as
the velocity vectors at those points. If the flow is unsteady, the streamline pattern
is different at different times because the velocity vectors are fluctuating with time
in both magnitude and direction.

In general, streamlines are different from pathlines. You can visualize a pathline as a time-exposure photograph of a given fluid element, whereas a streamline
pattern is like a single frame of a motion picture of the flow. In an unsteady
flow, the streamline pattern changes; hence, each "frame" of the motion picture
is different.

However, for the case of *steady flow* (which applies to most of the applications in this book), the magnitude and direction of the velocity vectors at all points
are fixed, invariant with time. Hence, the pathlines for different fluid elements going through the same point are the same. Moreover, the pathlines and streamlines
are identical. Therefore, in steady flow, there is no distinction between pathlines

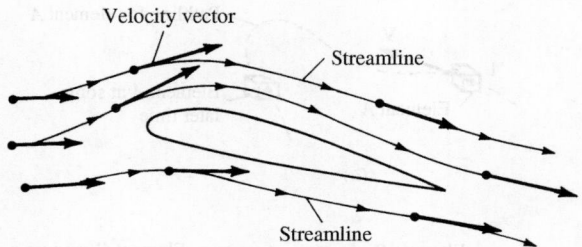

Figure 2.28 Streamlines.

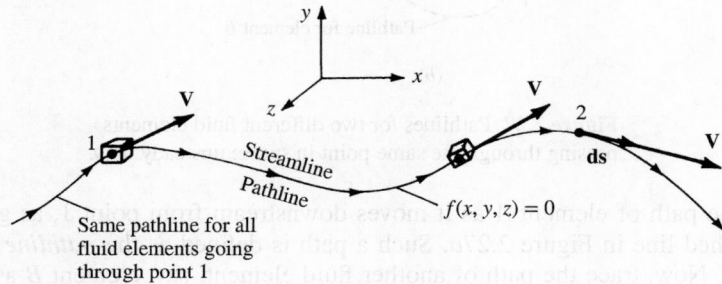

Figure 2.29 For steady flow, streamlines and pathlines are the same.

and streamlines; they are the same curves in space. This fact is reinforced in Figure 2.29, which illustrates the fixed, time-invariant streamline (pathline) through point 1. In Figure 2.29, a given fluid element passing through point 1 traces a pathline downstream. All subsequent fluid elements passing through point 1 at later times trace the same pathline. Since the velocity vector is tangent to the pathline at all points on the pathline for all times, the pathline is also a streamline. For the remainder of this book, we deal mainly with the concept of streamlines rather than pathlines; however, always keep in mind the distinction described above.

Question: Given the velocity field of a flow, how can we obtain the mathematical equation for a streamline? Obviously, the streamline illustrated in Figure 2.29 is a curve in space, and hence it can be described by the equation $f(x, y, z) = 0$. How can we obtain this equation? To answer this question, let **ds** be a directed element of the streamline, such as shown at point 2 in Figure 2.29. The velocity at point 2 is **V**, and by definition of a streamline, **V** is parallel to **ds**. Hence, from the definition of the vector cross product [see Equation (2.4)],

$$\boxed{\mathbf{ds} \times \mathbf{V} = 0} \tag{2.115}$$

Equation (2.115) is a valid equation for a streamline. To put it in a more recognizable form, expand Equation (2.115) in cartesian coordinates:

$$\mathbf{ds} = dx\mathbf{i} + dy\mathbf{j} + dz\mathbf{k}$$
$$\mathbf{V} = u\mathbf{i} + v\mathbf{j} + w\mathbf{k}$$

$$\mathbf{ds} \times \mathbf{V} = \begin{vmatrix} \mathbf{i} & \mathbf{j} & \mathbf{k} \\ dx & dy & dz \\ u & v & w \end{vmatrix}$$

$$= \mathbf{i}(w\,dy - v\,dz) + \mathbf{j}(u\,dz - w\,dx) + \mathbf{k}(v\,dx - u\,dy) = 0 \qquad (2.116)$$

Since the vector given by Equation (2.116) is zero, its components must each be zero:

$$w\,dy - v\,dz = 0 \qquad (2.117a)$$
$$u\,dz - w\,dx = 0 \qquad (2.117b)$$
$$v\,dx - u\,dy = 0 \qquad (2.117c)$$

Equation (2.117a to c) are differential equations for the streamline. Knowing u, v, and w as functions of x, y, and z, Equation (2.117a to c) can be integrated to yield the equation for the streamline: $f(x, y, z) = 0$.

To reinforce the physical meaning of Equation (2.117a to c), consider a streamline in two dimensions, as sketched in Figure 2.30a. The equation of this streamline is $y = f(x)$. Hence, at point 1 on the streamline, the slope is dy/dx. However, $\mathbf{V}$ with x and y components u and v, respectively, is tangent to the streamline at point 1. Thus, the slope of the streamline is also given by v/u, as shown in Figure 2.30. Therefore,

$$\frac{dy}{dx} = \frac{v}{u} \qquad (2.118)$$

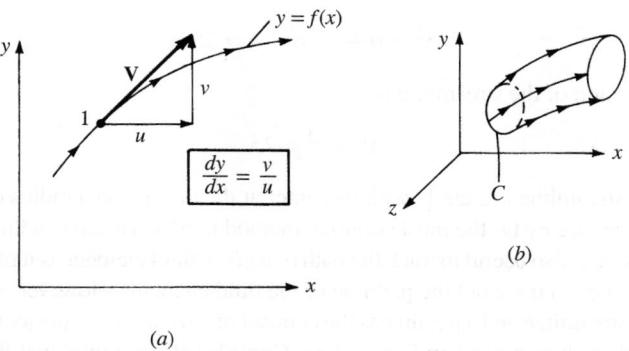

Figure 2.30 (a) Equation of a stream in two-dimensional cartesian space. (b) Sketch of a streamtube in three-dimensional space.

Equation (2.118) is a differential equation for a streamline in two dimensions. From Equation (2.118),

$$v\,dx - u\,dy = 0$$

which is precisely Equation (2.117c). Therefore, Equations (2.117a to c) and (2.118) simply state mathematically that the velocity vector is tangent to the streamline.

A concept related to streamlines is that of a streamtube. Consider an arbitrary closed curve C in three-dimensional space, as shown in Figure 2.30b. Consider the streamlines which pass through all points on C. These streamlines form a tube in space as sketched in Figure 2.30b; such a tube is called a *streamtube*. For example, the walls of an ordinary garden hose form a streamtube for the water flowing through the hose. For a steady flow, a direct application of the integral form of the continuity equation [Equation (2.53)] proves that the mass flow across all cross sections of a streamtube is constant. (Prove this yourself.)

EXAMPLE 2.4

Consider the velocity field given by $u = y/(x^2 + y^2)$ and $v = -x/(x^2 + y^2)$. Calculate the equation of the streamline passing through the point $(0, 5)$.

■ **Solution**

From Equation (2.118), $dy/dx = v/u = -x/y$, and

$$y\,dy = -x\,dx$$

Integrating, we obtain

$$y^2 = -x^2 + c$$

where c is a constant of integration.

For the streamline through $(0, 5)$, we have

$$5^2 = 0 + c \quad \text{or} \quad c = 25$$

Thus, the equation of the streamline is

$$x^2 + y^2 = 25$$

Note that the streamline is a circle with its center at the origin and a radius of 5 units.

Streamlines are by far the most common method used to visualize a fluid flow. In an unsteady flow, it is also useful to track the path of a *given* fluid element as it moves through the flow field (i.e., to trace out the pathline of the fluid element). However, separate from the ideas of a streamline and a pathline is the concept of a *streakline*. Consider a fixed point in a flow field, such as point 1 in Figure 2.31. Consider all the individual fluid elements that have passed through point 1 over a given time interval $t_2 - t_1$. These fluid elements, shown in Figure 2.31, are connected with each other, like a string of elephants connected trunk-to-tail. Element A is the fluid element that passed through point 1 at time t_1.

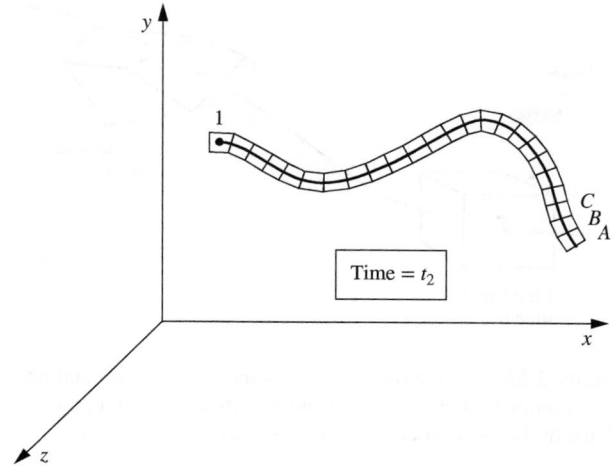

Figure 2.31 Illustration of a streakline through point 1.

Element B is the next element that passed through point 1, just behind element A. Element C is the element that passed through point 1 just behind element B, and so forth. Figure 2.31 is an illustration, made at time t_2, which shows all the fluid elements that have earlier passed through point 1 over the time interval $(t_2 - t_1)$. The line that connects all these fluid elements is, by definition, a *streakline*. We can more concisely define a streakline as the locus of fluid elements that have earlier passed through a prescribed point. To help further visualize the concept of a streakline, imagine that we are constantly injecting dye into the flow field at point 1. The dye will flow downstream from point 1, forming a curve in the x, y, z space in Figure 2.31. This curve *is* the streakline shown in Figure 2.31. A photograph of a streakline in the flow of water over a circular cylinder is shown in Figure 3.48. The white streakline is made visible by white particles that are constantly formed by electrolysis near a small anode fixed on the cylinder surface. These white particles subsequently flow downstream forming a streakline.

For a steady flow, pathlines, streamlines, and streaklines are all the same curves. Only in an unsteady flow are they different. So for steady flow, which is the type of flow mainly considered in this book, the concepts of a pathline, streamline, and streakline are redundant.

2.12 ANGULAR VELOCITY, VORTICITY, AND STRAIN

In several of our previous discussions, we made use of the concept of a fluid element moving through the flow field. In this section, we examine this motion more closely, paying particular attention to the orientation of the element and its change in shape as it moves along a streamline. In the process, we introduce the concept of vorticity, one of the most powerful quantities in theoretical aerodynamics.

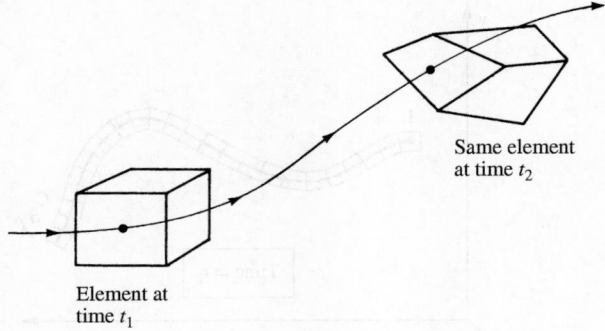

Figure 2.32 The motion of a fluid element along a streamline is a combination of translation and rotation; in addition, the shape of the element can become distorted.

Consider an infinitesimal fluid element moving in a flow field. As it translates along a streamline, it may also *rotate,* and in addition its shape may become *distorted* as sketched in Figure 2.32. The amount of rotation and distortion depends on the velocity field; the purpose of this section is to quantify this dependency.

Consider a two-dimensional flow in the *xy* plane. Also, consider an infinitesimal fluid element in this flow. Assume that at time *t* the shape of this fluid element is rectangular, as shown at the left of Figure 2.33. Assume that the fluid element is moving upward and to the right; its position and shape at time $t + \Delta t$ are shown at the right in Figure 2.33. Note that during the time increment Δt, the sides AB and AC have rotated through the angular displacements $-\Delta\theta_1$ and $\Delta\theta_2$, respectively. (Counterclockwise rotations by convention are considered positive; since line AB

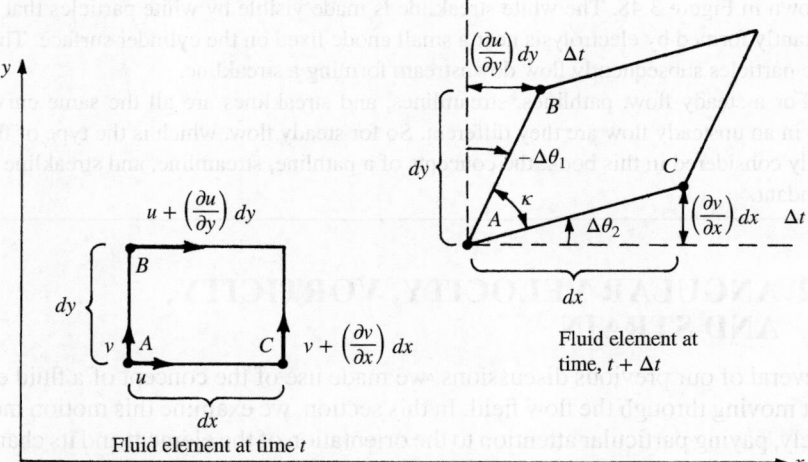

Figure 2.33 Rotation and distortion of a fluid element.

is shown with a clockwise rotation in Figure 2.33, the angular displacement is negative, $-\Delta\theta_1$.) At present, consider just the line AC. It has rotated because during the time increment Δt, point C has moved differently from point A. Consider the velocity in the y direction. At point A at time t, this velocity is v, as shown in Figure 2.33. Point C is a distance dx from point A; hence, at time t the vertical component of velocity of point C is given by $v + (\partial v/\partial x)\,dx$. Hence,

$$\frac{\text{Distance in } y \text{ direction that } A \text{ moves}}{\text{during time increment } \Delta t} = v\Delta t$$

$$\frac{\text{Distance in } y \text{ direction that } C \text{ moves}}{\text{during time increment } \Delta t} = \left(v + \frac{\partial v}{\partial x}dx\right)\Delta t$$

$$\frac{\text{Net displacement in } y \text{ direction}}{\text{of } C \text{ relative to } A} = \left(v + \frac{\partial v}{\partial x}dx\right)\Delta t - v\Delta t$$

$$= \left(\frac{\partial v}{\partial x}dx\right)\Delta t$$

This net displacement is shown at the right of Figure 2.33. From the geometry of Figure 2.33,

$$\tan \Delta\theta_2 = \frac{[(\partial v/\partial x)\,dx]\,\Delta t}{dx} = \frac{\partial v}{\partial x}\Delta t \tag{2.119}$$

Since $\Delta\theta_2$ is a small angle, $\tan \Delta\theta_2 \approx \Delta\theta_2$. Hence, Equation (2.119) reduces to

$$\Delta\theta_2 = \frac{\partial v}{\partial x}\Delta t \tag{2.120}$$

Now consider line AB. The x component of the velocity at point A at time t is u, as shown in Figure 2.33. Because point B is a distance dy from point A, the horizontal component of velocity of point B at time t is $u + (\partial u/\partial y)\,dy$. By reasoning similar to that above, the net displacement in the x direction of B relative to A over the time increment Δt is $[(\partial u/\partial y)\,dy]\,\Delta t$, as shown in Figure 2.33. Hence,

$$\tan(-\Delta\theta_1) = \frac{[(\partial u/\partial y)\,dy]\,\Delta t}{dy} = \frac{\partial u}{\partial y}\Delta t \tag{2.121}$$

Since $-\Delta\theta_1$ is small, Equation (2.121) reduces to

$$\Delta\theta_1 = -\frac{\partial u}{\partial y}\Delta t \tag{2.122}$$

Consider the angular velocities of lines AB and AC, defined as $d\theta_1/dt$ and $d\theta_2/dt$, respectively. From Equation (2.122), we have

$$\frac{d\theta_1}{dt} = \lim_{\Delta t \to 0}\frac{\Delta\theta_1}{\Delta t} = -\frac{\partial u}{\partial y} \tag{2.123}$$

From Equation (2.120), we have

$$\frac{d\theta_2}{dt} = \lim_{\Delta t \to 0} \frac{\Delta \theta_2}{\Delta t} = \frac{\partial v}{\partial x} \qquad (2.124)$$

By definition, the angular velocity of the fluid element as seen in the xy plane is the average of the angular velocities of lines AB and AC. Let ω_z denote this angular velocity. Therefore, by definition,

$$\omega_z = \frac{1}{2}\left(\frac{d\theta_1}{dt} + \frac{d\theta_2}{dt}\right) \qquad (2.125)$$

Combining Equations (2.123) to (2.125) yields

$$\omega_z = \frac{1}{2}\left(\frac{\partial v}{\partial x} - \frac{\partial u}{\partial y}\right) \qquad (2.126)$$

In the above discussion, we have considered motion in the xy plane only. However, the fluid element is generally moving in three-dimensional space, and its angular velocity is a vector $\boldsymbol{\omega}$ that is oriented in some general direction, as shown in Figure 2.34. In Equation (2.126), we have obtained only the component of $\boldsymbol{\omega}$ in the z direction; this explains the subscript z in Equations (2.125) and (2.126). The x and y components of $\boldsymbol{\omega}$ can be obtained in a similar fashion. The resulting angular velocity of the fluid element in three-dimensional space is

$$\boldsymbol{\omega} = \omega_x \mathbf{i} + \omega_y \mathbf{j} + \omega_z \mathbf{k}$$

$$\boxed{\boldsymbol{\omega} = \frac{1}{2}\left[\left(\frac{\partial w}{\partial y} - \frac{\partial v}{\partial z}\right)\mathbf{i} + \left(\frac{\partial u}{\partial z} - \frac{\partial w}{\partial x}\right)\mathbf{j} + \left(\frac{\partial v}{\partial x} - \frac{\partial u}{\partial y}\right)\mathbf{k}\right]} \qquad (2.127)$$

Equation (2.127) is the desired result; it expresses the angular velocity of the fluid element in terms of the velocity field, or more precisely, in terms of derivatives of the velocity field.

The angular velocity of a fluid element plays an important role in theoretical aerodynamics, as we shall see soon. However, the expression $2\boldsymbol{\omega}$ appears

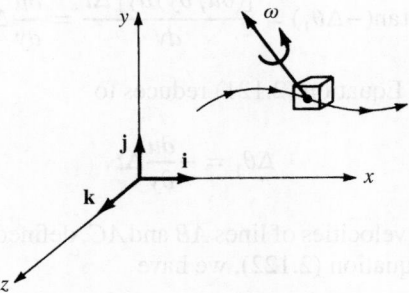

Figure 2.34 Angular velocity of a fluid element in three-dimensional space.

frequently, and therefore we define a new quantity, *vorticity*, which is simply twice the angular velocity. Denote vorticity by the vector ξ:

$$\xi \equiv 2\boldsymbol{\omega}$$

Hence, from Equation (2.127),

$$\xi = \left(\frac{\partial w}{\partial y} - \frac{\partial v}{\partial z}\right)\mathbf{i} + \left(\frac{\partial u}{\partial z} - \frac{\partial w}{\partial x}\right)\mathbf{j} + \left(\frac{\partial v}{\partial x} - \frac{\partial u}{\partial y}\right)\mathbf{k} \qquad (2.128)$$

Recall Equation (2.22) for $\nabla \times \mathbf{V}$ in cartesian coordinates. Since u, v, and w denote the x, y, and z components of velocity, respectively, note that the right sides of Equations (2.22) and (2.128) are identical. Hence, we have the important result that

$$\xi = \nabla \times \mathbf{V} \qquad (2.129)$$

In a velocity field, the curl of the velocity is equal to the vorticity.

The above leads to two important definitions:

1. If $\nabla \times \mathbf{V} \neq 0$ at every point in a flow, the flow is called *rotational*. This implies that the fluid elements have a finite angular velocity.
2. If $\nabla \times \mathbf{V} = 0$ at every point in a flow, the flow is called *irrotational*. This implies that the fluid elements have no angular velocity; rather, their motion through space is a pure translation.

The case of rotational flow is illustrated in Figure 2.35. Here, fluid elements moving along two different streamlines are shown in various modes of rotation. In contrast, the case of irrotational flow is illustrated in Figure 2.36. Here, the upper streamline shows a fluid element where the angular velocities of its sides are zero. The lower streamline shows a fluid element where the angular velocities of two intersecting sides are finite but equal and opposite to each other, and so their sum is identically zero. In both cases, the angular velocity of the fluid element is zero (i.e., the flow is irrotational).

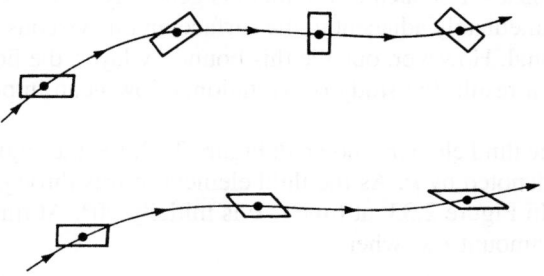

Figure 2.35 Fluid elements in a rotational flow.

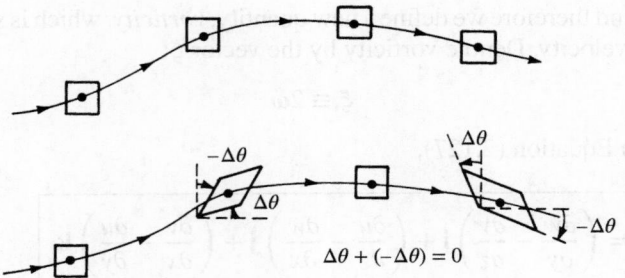

Figure 2.36 Fluid elements in an irrotational flow.

If the flow is two-dimensional (say, in the xy plane), then from Equation (2.128),

$$\xi = \xi_z \mathbf{k} = \left(\frac{\partial v}{\partial x} - \frac{\partial u}{\partial y} \right) \mathbf{k} \tag{2.130}$$

Also, if the flow is irrotational, $\xi = 0$. Hence, from Equation (2.130),

$$\boxed{\frac{\partial v}{\partial x} - \frac{\partial u}{\partial y} = 0} \tag{2.131}$$

Equation (2.131) is the *condition of irrotationality for two-dimensional flow*. We will have frequent occasion to use Equation (2.131).

Why is it so important to make a distinction between rotational and irrotational flows? The answer becomes blatantly obvious as we progress in our study of aerodynamics; we find that irrotational flows are much easier to analyze than rotational flows. However, irrotational flow may at first glance appear to be so special that its applications are limited. Amazingly enough, such is not the case. There are a large number of practical aerodynamic problems where the flow field is essentially irrotational, for example, the subsonic flow over airfoils, the supersonic flow over slender bodies at small angle of attack, and the subsonic-supersonic flow through nozzles. For such cases, there is generally a thin boundary layer of viscous flow immediately adjacent to the surface; in this viscous region, the flow is highly rotational. However, outside this boundary layer, the flow is frequently irrotational. As a result, the study of irrotational flow is an important aspect of aerodynamics.

Return to the fluid element shown in Figure 2.33. Let the angle between sides AB and AC be denoted by κ. As the fluid element moves through the flow field, κ will change. In Figure 2.33, at time t, κ is initially $90°$. At time $t + \Delta t$, κ has changed by the amount $\Delta \kappa$, where

$$\Delta \kappa = -\Delta \theta_2 - (-\Delta \theta_1) \tag{2.132}$$

By definition, the *strain* of the fluid element as seen in the xy plane is the change in κ, where positive strain corresponds to a *decreasing* κ. Hence, from Equation (2.132),

$$\text{Strain} = -\Delta\kappa = \Delta\theta_2 - \Delta\theta_1 \qquad (2.133)$$

In viscous flows (to be discussed in Chapters 15 to 20), the time rate of strain is an important quantity. Denote the time rate of strain by ε_{xy}, where in conjunction with Equation (2.133)

$$\varepsilon_{xy} \equiv -\frac{d\kappa}{dt} = \frac{d\theta_2}{dt} - \frac{d\theta_1}{dt} \qquad (2.134)$$

Substituting Equations (2.123) and (2.124) into (2.134), we have

$$\varepsilon_{xy} = \frac{\partial v}{\partial x} + \frac{\partial u}{\partial y} \qquad (2.135a)$$

In the yz and zx planes, by a similar derivation the strain is, respectively,

$$\varepsilon_{yz} = \frac{\partial w}{\partial y} + \frac{\partial v}{\partial z} \qquad (2.135b)$$

and

$$\varepsilon_{zx} = \frac{\partial u}{\partial z} + \frac{\partial w}{\partial x} \qquad (2.135c)$$

Note that angular velocity (hence, vorticity) and time rate of strain depend solely on the velocity derivatives of the flow field. These derivatives can be displayed in a matrix as follows:

$$\begin{bmatrix} \dfrac{\partial u}{\partial x} & \dfrac{\partial u}{\partial y} & \dfrac{\partial u}{\partial z} \\[2ex] \dfrac{\partial v}{\partial x} & \dfrac{\partial v}{\partial y} & \dfrac{\partial v}{\partial z} \\[2ex] \dfrac{\partial w}{\partial x} & \dfrac{\partial w}{\partial y} & \dfrac{\partial w}{\partial z} \end{bmatrix}$$

The sum of the diagonal terms is simply equal to $\nabla \cdot \mathbf{V}$, which from Section 2.3 is equal to the time rate of change of volume of a fluid element; hence, the diagonal terms represent the *dilatation* of a fluid element. The off-diagonal terms are cross derivatives which appear in Equations (2.127), (2.128), and (2.135a to c). Hence, the off-diagonal terms are associated with rotation and strain of a fluid element.

In summary, in this section, we have examined the rotation and deformation of a fluid element moving in a flow field. The angular velocity of a fluid element and the corresponding vorticity at a point in the flow are concepts that are useful in the analysis of both inviscid and viscous flows; in particular, the absence of vorticity—irrotational flow—greatly simplifies the analysis of the flow, as we will see. We take advantage of this simplification in much of our treatment of inviscid flows in subsequent chapters. On the other hand, we do not make use of the time rate of strain until Chapter 15.

EXAMPLE 2.5

For the velocity field given in Example 2.4, calculate the vorticity.

■ **Solution**

$$\boldsymbol{\xi} = \nabla \times \mathbf{V} = \begin{vmatrix} \mathbf{i} & \mathbf{j} & \mathbf{k} \\ \dfrac{\partial}{\partial x} & \dfrac{\partial}{\partial y} & \dfrac{\partial}{\partial z} \\ u & v & w \end{vmatrix} = \begin{vmatrix} \mathbf{i} & \mathbf{j} & \mathbf{k} \\ \dfrac{\partial}{\partial x} & \dfrac{\partial}{\partial y} & \dfrac{\partial}{\partial z} \\ \dfrac{y}{x^2+y^2} & \dfrac{-x}{x^2+y^2} & 0 \end{vmatrix}$$

$$= \mathbf{i}[0-0] - \mathbf{j}[0-0]$$

$$+ \mathbf{k}\left[\frac{(x^2+y^2)(-1)+x(2x)}{(x^2+y^2)^2} - \frac{(x^2+y^2)-y(2y)}{(x^2+y^2)^2}\right]$$

$$= 0\mathbf{i} + 0\mathbf{j} + 0\mathbf{k} = \mathbf{0}$$

The flow field is irrotational at every point except at the origin, where $x^2 + y^2 = 0$.

EXAMPLE 2.6

Consider the boundary-layer velocity profile used in Example 2.2, namely, $u/V_\infty = (y/\delta)^{0.25}$. Is this flow rotational or irrotational?

■ **Solution**

For a two-dimensional flow, the irrotationality condition is given by Equation (2.131), namely

$$\frac{\partial v}{\partial x} - \frac{\partial u}{\partial y} = 0.$$

Does this relation hold for the viscous boundary-layer flow in Example 2.2? Let us examine this question. From the boundary-layer velocity profile given by

$$\frac{u}{V_\infty} = \left(\frac{y}{\delta}\right)^{0.25}$$

we obtain

$$\frac{\partial u}{\partial y} = 0.25 \frac{V_\infty}{\delta}\left(\frac{y}{\delta}\right)^{-0.75} \tag{E2.7}$$

What can we say about $\partial v/\partial x$? In Example 2.2, the flow was incompressible. From the continuity equation for a steady flow given by Equation (2.54), repeated below,

$$\nabla \cdot (\rho \mathbf{V}) = \frac{\partial(\rho u)}{\partial x} + \frac{\partial(\rho v)}{\partial y} = 0$$

we have for an incompressible flow, where $\rho = $ constant,

$$\frac{\partial u}{\partial x} + \frac{\partial v}{\partial y} = 0 \tag{E2.8}$$

Equation (E2.8) will provide an expression for v as follows:

$$\frac{\partial u}{\partial x} = \frac{\partial}{\partial x}\left[V_\infty \left(\frac{y}{\delta}\right)^{0.25}\right] \qquad \text{(E2.9)}$$

However, from Example 2.2, we stated that

$$\frac{\delta}{c} = \frac{5}{\sqrt{\text{Re}_c}}$$

This equation holds at any x station along the plate, not just at $x = c$. Therefore, we can write

$$\frac{\delta}{x} = \frac{5}{\sqrt{\text{Re}, x}}$$

where

$$\text{Re}, x = \frac{\rho_\infty V_\infty x}{\mu_\infty}$$

Thus, δ is a function of x given by

$$\delta = 5\sqrt{\frac{\mu_\infty x}{\rho_\infty V_\infty}}$$

and

$$\frac{d\delta}{dx} = \frac{5}{2}\sqrt{\frac{\mu_\infty}{\rho_\infty V_\infty}} x^{-1/2}$$

Substituting into Equation (E2.9), we have

$$\frac{\partial u}{\partial x} = \frac{\partial}{\partial x}\left[V_\infty \left(\frac{y}{\delta}\right)^{0.25}\right] = V_\infty y^{0.25}(-0.25)\delta^{-1.25}\frac{d\delta}{dx}$$

$$= -V_\infty y^{0.25}\delta^{-1.25}\left(\frac{5}{8}\right)\sqrt{\frac{\mu_\infty}{\rho_\infty V_\infty}} x^{-1/2}$$

$$= -\frac{5}{8}V_\infty y^{0.25}\left(\frac{1}{5}\right)^{1.25}\left(\frac{\mu_\infty}{\rho_\infty V_\infty}\right)^{-1/8} x^{-9/8}$$

Hence,

$$\frac{\partial u}{\partial x} = -Cy^{1/4}x^{-9/8}$$

where C is a constant. Inserting this into Equation (E2.8), we have

$$\frac{\partial v}{\partial y} = C_y^{1/4}x^{-9/8}$$

Integrating with respect to y, we have

$$v = C_1 y^{5/4}x^{-9/8} + C_2 \qquad \text{(E2.10)}$$

where C_1 is a constant and C_2 can be a function of x. Evaluating Equation (E2.10) at the wall, where $v = 0$ and $y = 0$, we obtain $C_2 = 0$. Hence,

$$v = C_1 y^{5/4}x^{-9/8}$$

In turn, we obtain by differentiation

$$\frac{\partial v}{\partial x} = C_3 y^{5/4} x^{-17/8} \tag{E2.11}$$

(*Note:* v is finite inside a boundary layer; the streamlines within a boundary are deflected upward. However, this "displacement" effect is usually small compared to the running length in the x direction, and v is of small magnitude in comparison to u. Both of these statements will be verified in Chapters 17 and 18.) Recasting Equation (E2.7) in the same general form as Equation (E2.11), we have

$$\frac{\partial u}{\partial y} = 0.25 V_\infty y^{-0.75} \left(\frac{1}{\delta}\right)^{0.25}$$

$$= 0.25 V_\infty y^{-0.75} \left(\frac{1}{5\sqrt{\mu_\infty x / \rho_\infty V_\infty}}\right)^{0.25}$$

Hence,

$$\frac{\partial u}{\partial y} = C_4 y^{-3/4} x^{-1.8} \tag{E2.12}$$

From Equations (E2.11) and (E2.12), we can write

$$\frac{\partial v}{\partial x} - \frac{\partial u}{\partial y} = C_3 y^{5/4} x^{-17/8} - C_4 y^{-3/4} x^{-1/9} \neq 0$$

Therefore, the irrotationality condition does *not* hold; the flow is *rotational*.

In Example 2.6, we demonstrated a basic result that holds in general for viscous flows, namely, viscous flows are *rotational*. This is almost intuitive. For example, consider an infinitesimally small fluid element moving along a streamline, as sketched in Figure 2.37. If this is a viscous flow, and assuming that the velocity increases in the upward direction (i.e., the velocity is higher on the neighboring streamline above and lower on the neighboring streamline below), then the shear stresses on the upper and lower faces of the fluid element will be in the directions shown. Such shear stresses will be discussed at length in Chapter 15. Examining Figure 2.37, we see clearly that the shear stresses exert a rotational moment about the center of the element, thus providing a mechanism for setting

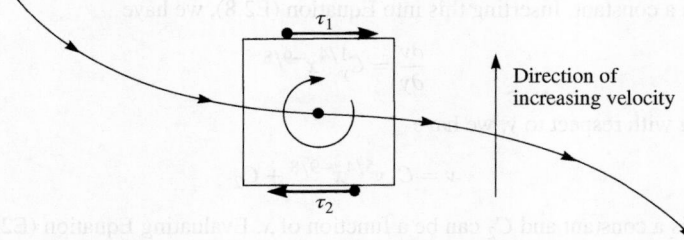

Figure 2.37 Shear stress and the consequent rotation of a fluid element.

the fluid element into rotation. Although this picture is overly simplistic, it serves to emphasize that viscous flows are rotational flows. On the other hand, as stated earlier in this section, there are numerous inviscid flow problems that are irrotational, with the attendant simplifications to be explained later. Some inviscid flows are rotational, but there exist such a large number of practical aerodynamic problems described by inviscid, irrotational flows that the distinction between rotational and irrotational flow is an important consideration.

EXAMPLE 2.7

Prove that the inviscid subsonic compressible flow over the wavy wall shown in Figure 2.17 and discussed in Example 2.1 is irrotational.

■ **Solution**
This flow is two-dimensional. From Equations (2.128) and (2.129),

$$\nabla \times \mathbf{V} = \left(\frac{\partial v}{\partial x} - \frac{\partial u}{\partial y} \right) \mathbf{k}$$

The velocity field is given by Equations (2.35) and (2.36) as

$$u = V_\infty \left[1 + \frac{h}{\beta} \frac{2\pi}{\ell} \left(\cos \frac{2\pi x}{\ell} \right) e^{-2\pi \beta y / \ell} \right] \tag{2.35}$$

and

$$v = -V_\infty h \frac{2\pi}{\ell} \left(\sin \frac{2\pi x}{\ell} \right) e^{-2\pi \beta y / \ell} \tag{2.36}$$

Differentiating Equation (2.36), we have

$$\frac{\partial v}{\partial x} = -V_\infty h \left(\frac{2\pi}{\ell} \right)^2 \left(\cos \frac{2\pi x}{\ell} \right) e^{-2\pi \beta y / \ell}$$

Differentiating Equation (2.35), we have

$$\frac{\partial u}{\partial y} = V_\infty \frac{h}{\beta} \frac{2\pi}{\ell} \left(\cos \frac{2\pi x}{\ell} \right) e^{-2\pi \beta y / \ell} \left(-\frac{2\pi \beta}{\ell} \right)$$

$$= -V_\infty h \left(\frac{2\pi}{\ell} \right)^2 \left(\cos \frac{2\pi x}{\ell} \right) e^{-2\pi \beta y / \ell}$$

Thus,

$$\frac{\partial v}{\partial x} - \frac{\partial u}{\partial y} = -V_\infty h \left(\frac{2\pi}{\ell} \right)^2 \left(\cos \frac{2\pi x}{\ell} \right) e^{-2\pi \beta y / \ell}$$

$$- \left[-V_\infty h \left(\frac{2\pi}{\ell} \right)^2 \left(\cos \frac{2\pi x}{\ell} \right) e^{-2\pi \beta y / \ell} \right]$$

$$= 0$$

Hence,

$$\nabla \times \mathbf{V} = \left(\frac{\partial v}{\partial x} - \frac{\partial u}{\partial y} \right) \mathbf{k} = 0$$

Conclusion: The inviscid subsonic compression flow over a wavy wall is *irrotational*.

Example 2.7 demonstrates another basic result. Return to Figure 2.17 and examine it closely. Here we have an inviscid flow where the freestream (shown far above the wall) is a uniform flow with velocity V_∞. In a uniform flow, $\partial u/\partial y = \partial v/\partial x = 0$. Therefore, a uniform flow is irrotational. We can view the original source of the flow shown in Figure 2.17 to be the uniform flow shown far above it, and this original flow is irrotational. Moreover, the whole flow field is inviscid, that is, there is no internal friction and no shear stress at the wall to introduce vorticity in the flow. The flow shown in Figure 2.17, on a physical basis, must therefore remain irrotational throughout. Of course, Example 2.7 proves *mathematically* that the flow is irrotational throughout. However, this is just an example of a broader concept: A flow field that is originally irrotational, without any internal mechanisms such as frictional shear stress to generate vorticity, will remain irrotational throughout. This makes sense, does it not?

2.13 CIRCULATION

You are reminded again that this is a tool-building chapter. Taken individually, each aerodynamic tool we have developed so far may not be particularly exciting. However, taken collectively, these tools allow us to obtain solutions for some very practical and exciting aerodynamic problems.

In this section, we introduce a tool that is fundamental to the calculation of aerodynamic lift, namely, *circulation*. This tool was used independently by Frederick Lanchester (1878–1946) in England, Wilhelm Kutta (1867–1944) in Germany, and Nikolai Joukowski (1847–1921) in Russia to create a breakthrough in the theory of aerodynamic lift at the turn of the twentieth century. The relationship between circulation and lift and the historical circumstances surrounding this breakthrough are discussed in Chapters 3 and 4. The purpose of this section is only to define circulation and relate it to vorticity.

Consider a closed curve C in a flow field, as sketched in Figure 2.38. Let $\mathbf{V}$ and $\mathbf{ds}$ be the velocity and directed line segment, respectively, at a point on C. The circulation, denoted by Γ, is defined as

$$\boxed{\Gamma \equiv - \oint_C \mathbf{V} \cdot \mathbf{ds}} \qquad (2.136)$$

The circulation is simply the negative of the line integral of velocity around a closed curve in the flow; it is a kinematic property depending only on the velocity field and the choice of the curve C. As discussed in Section 2.2.8, by mathematical

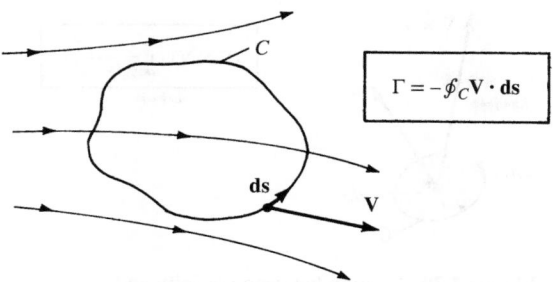

Figure 2.38 Definition of circulation.

convention, the positive sense of the line integral is counterclockwise. However, in aerodynamics, it is convenient to consider a positive circulation as being clockwise. Hence, a minus sign appears in the definition given by Equation (2.136) to account for the positive-counterclockwise sense of the integral and the positive-clockwise sense of circulation.[1]

The use of the word *circulation* to label the integral in Equation (2.136) may be somewhat misleading because it leaves a general impression of something moving around in a loop. Indeed, according to the *American Heritage Dictionary of the English Language,* the first definition given to the word "circulation" is "movement in a circle or circuit." However, in aerodynamics, circulation has a very precise technical meaning, namely, Equation (2.136). It does *not* necessarily mean that the fluid elements are moving around in circles within this flow field—a common early misconception of new students of aerodynamics. Rather, when circulation exists in a flow, it simply means that the line integral in Equation (2.136) is finite. For example, if the airfoil in Figure 2.28 is generating lift, the circulation taken around a closed curve enclosing the airfoil will be finite, although the fluid elements are by no means executing circles around the airfoil (as clearly seen from the streamlines sketched in Figure 2.28).

Circulation is also related to vorticity as follows. Refer back to Figure 2.11, which shows an open surface bounded by the closed curve C. Assume that the surface is in a flow field and the velocity at point P is $\mathbf{V}$, where P is any point on the surface (including any point on curve C). From Stokes' theorem [Equation (2.25)],

$$\Gamma \equiv -\oint_C \mathbf{V} \cdot \mathbf{ds} = -\iint_S (\nabla \times \mathbf{V}) \cdot \mathbf{dS} \qquad (2.137)$$

Hence, the circulation about a curve C is equal to the vorticity integrated over any open surface bounded by C. This leads to the immediate result that if the flow is

---

[1] Some books do not use the minus sign in the definition of circulation. In such cases, the positive sense of both the line integral and Γ is in the same direction. This causes no problem as long as the reader is aware of the convention used in a particular book or paper.

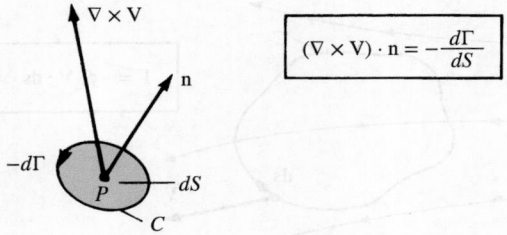

Figure 2.39 Relation between vorticity and circulation.

irrotational everywhere within the contour of integration (i.e., if $\nabla \times \mathbf{V} = 0$ over any surface bounded by C), *then* $\Gamma = 0$. A related result is obtained by letting the curve C shrink to an infinitesimal size, and denoting the circulation around this infinitesimally small curve by $d\Gamma$. Then, in the limit as C becomes infinitesimally small, Equation (2.137) yields

$$d\Gamma = -(\nabla \times \mathbf{V}) \cdot d\mathbf{S} = -(\nabla \times \mathbf{V}) \cdot \mathbf{n} \, dS$$

or

$$\boxed{(\nabla \times \mathbf{V}) \cdot \mathbf{n} = -\frac{d\Gamma}{dS}} \tag{2.138}$$

where dS is the infinitesimal area enclosed by the infinitesimal curve C. Referring to Figure 2.39, Equation (2.138) states that at a point P in a flow, the component of vorticity normal to dS is equal to the negative of the "circulation per unit area," where the circulation is taken around the boundary of dS.

EXAMPLE 2.8

For the velocity field given in Example 2.4, calculate the circulation around a circular path of radius 5 m. Assume that u and v given in Example 2.4 are in units of meters per second.

■ Solution

Since we are dealing with a circular path, it is easier to work this problem in polar coordinates, where $x = r\cos\theta$, $y = r\sin\theta$, $x^2 + y^2 = r^2$, $V_r = u\cos\theta + v\sin\theta$, and $V_\theta = -u\sin\theta + v\cos\theta$. Therefore,

$$u = \frac{y}{x^2 + y^2} = \frac{r\sin\theta}{r^2} = \frac{\sin\theta}{r}$$

$$v = -\frac{x}{x^2 + y^2} = -\frac{r\cos\theta}{r^2} = -\frac{\cos\theta}{r}$$

$$V_r = \frac{\sin\theta}{r}\cos\theta + \left(-\frac{\cos\theta}{r}\right)\sin\theta = 0$$

$$V_\theta = -\frac{\sin\theta}{r}\sin\theta + \left(-\frac{\cos\theta}{r}\right)\cos\theta = -\frac{1}{r}$$

$$\mathbf{V} \cdot \mathbf{ds} = (V_r \mathbf{e}_r + V_\theta \mathbf{e}_\theta) \cdot (dr \, \mathbf{e}_r + r \, d\theta \, \mathbf{e}_\theta)$$

$$= V_r \, dr + r V_\theta \, d\theta = 0 + r \left(-\frac{1}{r} \right) d\theta = -d\theta$$

Hence,
$$\Gamma = -\oint_C \mathbf{V} \cdot \mathbf{ds} = -\int_0^{2\pi} -d\theta = \boxed{2\pi \; \text{m}^2/\text{s}}$$

Note that we never used the 5-m diameter of the circular path; in this case, the value of Γ is independent of the diameter of the path.

2.14 STREAM FUNCTION

In this section, we consider two-dimensional steady flow. Recall from Section 2.11 that the differential equation for a streamline in such a flow is given by Equation (2.118), repeated below

$$\frac{dy}{dx} = \frac{v}{u} \tag{2.118}$$

If u and v are known functions of x and y, then Equation (2.118) can be integrated to yield the algebraic equation for a streamline:

$$f(x, y) = c \tag{2.139}$$

where c is an arbitrary constant of integration, with different values for different streamlines. In Equation (2.139), denote the function of x and y by the symbol $\bar{\psi}$. Hence, Equation (2.139) is written as

$$\bar{\psi}(x, y) = c \tag{2.140}$$

The function $\bar{\psi}(x, y)$ is called the *stream function*. From Equation (2.140) we see that the equation for a streamline is given by *setting the stream function equal to a constant* (i.e., c_1, c_2, c_3, etc.). Two different streamlines are illustrated in Figure 2.40; streamlines ab and cd are given by $\bar{\psi} = c_1$ and $\bar{\psi} = c_2$, respectively.

There is a certain arbitrariness in Equations (2.139) and (2.140) via the arbitrary constant of integration c. Let us define the stream function more precisely in order to reduce this arbitrariness. Referring to Figure 2.40, let us define the

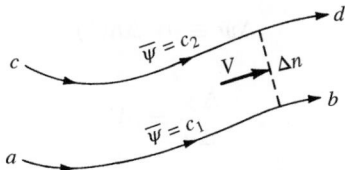

Figure 2.40 Different streamlines are given by different values of the stream function.

numerical value of $\bar{\psi}$ such that the *difference* $\Delta\bar{\psi}$ between $\bar{\psi} = c_2$ for stream-line *cd* and $\bar{\psi} = c_1$ for streamline *ab* is equal to the *mass* flow between the two streamlines. Since Figure 2.40 is a two-dimensional flow, the mass flow between two streamlines is defined *per unit depth perpendicular to the page*. That is, in Figure 2.40 we are considering the mass flow inside a streamtube bounded by streamlines *ab* and *cd*, with a rectangular cross-sectional area equal to Δn times a unit depth perpendicular to the page. Here, Δn is the normal distance between *ab* and *cd*, as shown in Figure 2.40. Hence, mass flow between streamlines *ab* and *cd* per unit depth perpendicular to the page is

$$\Delta\bar{\psi} = c_2 - c_1 \tag{2.141}$$

The above definition does not completely remove the arbitrariness of the constant of integration in Equations (2.139) and (2.140), but it does make things a bit more precise. For example, consider a given two-dimensional flow field. Choose one streamline of the flow, and give it an arbitrary value of the stream function, say, $\bar{\psi} = c_1$. Then, the value of the stream function for any other streamline in the flow, say, $\bar{\psi} = c_2$, is fixed by the definition given in Equation (2.141). Which streamline you choose to designate as $\bar{\psi} = c_1$ and what numerical value you give c_1 usually depend on the geometry of the given flow field, as we see in Chapter 3.

The equivalence between $\bar{\psi} =$ constant designating a streamline, and $\Delta\bar{\psi}$ equaling mass flow (per unit depth) between streamlines, is natural. For a steady flow, the mass flow inside a given streamtube is constant along the tube; the mass flow across any cross section of the tube is the same. Since by definition $\Delta\bar{\psi}$ is equal to this mass flow, then $\Delta\bar{\psi}$ itself is constant for a given streamtube. In Figure 2.40, if $\bar{\psi}_1 = c_1$ designates the streamline on the bottom of the streamtube, then $\bar{\psi}_2 = c_2 = c_1 + \Delta\bar{\psi}$ is also constant along the top of the streamtube. Since by definition of a streamtube (see Section 2.11) the upper boundary of the streamtube is a streamline itself, then $\psi_2 = c_2 =$ constant must designate this streamline.

We have yet to develop the most important property of the stream function, namely, derivatives of $\bar{\psi}$ yield the flow-field velocities. To obtain this relation-ship, consider again the streamlines *ab* and *cd* in Figure 2.40. Assume that these streamlines are close together (i.e., assume Δn is small), such that the flow ve-locity V is a constant value across Δn. The mass flow through the streamtube per unit depth perpendicular to the page is

$$\Delta\bar{\psi} \equiv \rho V \, \Delta n(1)$$

or

$$\frac{\Delta\bar{\psi}}{\Delta n} = \rho V \tag{2.142}$$

Consider the limit of Equation (2.142) as $\Delta n \to 0$:

$$\rho V = \lim_{\Delta n \to 0} \frac{\Delta\bar{\psi}}{\Delta n} \equiv \frac{\partial\bar{\psi}}{\partial n} \tag{2.143}$$

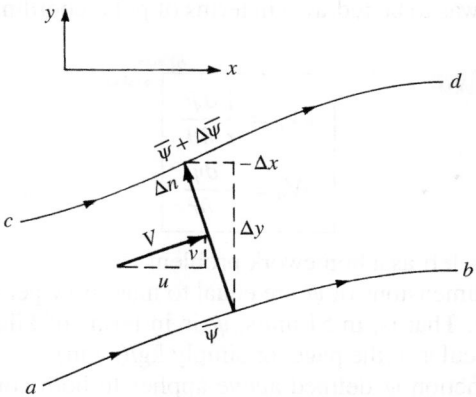

Figure 2.41 Mass flow through Δn is the sum of the mass flows through Δy and $-\Delta x$.

Equation (2.143) states that if we know $\bar{\psi}$, then we can obtain the product (ρV) by differentiating $\bar{\psi}$ in the direction *normal* to V. To obtain a practical form of Equation (2.143) for cartesian coordinates, consider Figure 2.41. Notice that the directed normal distance Δn is equivalent first to moving upward in the y direction by the amount Δy and then to the left in the negative x direction by the amount $-\Delta x$. Due to conservation of mass, the mass flow through Δn (per unit depth) is equal to the sum of the mass flows through Δy and $-\Delta x$ (per unit depth):

$$\text{Mass flow} = \Delta\bar{\psi} = \rho V \Delta n = \rho u \Delta y + \rho v(-\Delta x) \qquad (2.144)$$

Letting cd approach ab, Equation (2.144) becomes in the limit

$$d\bar{\psi} = \rho u \, dy - \rho v \, dx \qquad (2.145)$$

However, since $\bar{\psi} = \bar{\psi}(x, y)$, the chain rule of calculus states

$$d\bar{\psi} = \frac{\partial\bar{\psi}}{\partial x}dx + \frac{\partial\bar{\psi}}{\partial y}dy \qquad (2.146)$$

Comparing Equations (2.145) and (2.146), we have

$$\rho u = \frac{\partial\bar{\psi}}{\partial y} \qquad (2.147a)$$

$$\rho v = -\frac{\partial\bar{\psi}}{\partial x} \qquad (2.147b)$$

Equation (2.147a and b) are important. If $\bar{\psi}(x, y)$ is known for a given flow field, then at any point in the flow the products ρu and ρv can be obtained by differentiating $\bar{\psi}$ in the directions normal to u and v, respectively.

If Figure 2.41 was to be redrawn in terms of polar coordinates, then a similar derivation yields

$$\rho V_r = \frac{1}{r}\frac{\partial \bar{\psi}}{\partial \theta} \qquad (2.148a)$$

$$\rho V_\theta = -\frac{\partial \bar{\psi}}{\partial r} \qquad (2.148b)$$

Such a derivation is left as a homework problem.

Note that the dimensions of $\bar{\psi}$ are equal to mass flow per unit depth perpendicular to the page. That is, in SI units, $\bar{\psi}$ is in terms of kilograms per second per meter perpendicular to the page, or simply kg/(s · m).

The stream function $\bar{\psi}$ defined above applies to both compressible and incompressible flow. Now consider the case of incompressible flow only, where $\rho = $ constant. Equation (2.143) can be written as

$$V = \frac{\partial(\bar{\psi}/\rho)}{\partial n} \qquad (2.149)$$

We define a new stream function, for incompressible flow only, as $\psi \equiv \bar{\psi}/\rho$. Then Equation (2.149) becomes

$$V = \frac{\partial \psi}{\partial n}$$

and Equations (2.147) and (2.148) become

$$u = \frac{\partial \psi}{\partial y} \qquad (2.150a)$$

$$v = -\frac{\partial \psi}{\partial x} \qquad (2.150b)$$

and

$$V_r = \frac{1}{r}\frac{\partial \psi}{\partial \theta} \qquad (2.151a)$$

$$V_\theta = -\frac{\partial \psi}{\partial r} \qquad (2.151b)$$

The incompressible stream function ψ has characteristics analogous to its more general compressible counterpart $\bar{\psi}$. For example, since $\bar{\psi}(x,y) = c$ is the equation of a streamline, and since ρ is a constant for incompressible flow, then $\psi(x,y) \equiv \bar{\psi}/\rho = $ constant is also the equation for a streamline (for incompressible flow only). In addition, since $\Delta\bar{\psi}$ is mass flow between two streamlines (per unit depth perpendicular to the page), and since ρ is mass per unit volume, then physically $\Delta\psi = \Delta\bar{\psi}/\rho$ represents the *volume flow* (per unit depth) between two streamlines. In SI units, $\Delta\psi$ is expressed as cubic meters per second per meter perpendicular to the page, or simply m²/s.

In summary, the concept of the stream function is a powerful tool in aerodynamics for two primary reasons. Assuming that $\bar{\psi}(x, y)$ [or $\psi(x, y)$] is known through the two-dimensional flow field, then

1. $\bar{\psi}$ = constant (or ψ = constant) gives the equation of a streamline.
2. The flow velocity can be obtained by differentiating $\bar{\psi}$ (or ψ), as given by Equations (2.147) and (2.148) for compressible flow and Equations (2.150) and (2.151) for incompressible flow. We have not yet discussed how $\bar{\psi}(x, y)$ [or $\psi(x, y)$] can be obtained in the first place; we are assuming that it is known. The actual determination of the stream function for various problems is discussed in Chapter 3.

2.15 VELOCITY POTENTIAL

Recall from Section 2.12 that an irrotational flow is defined as a flow where the vorticity is zero at every point. From Equation (2.129), for an irrotational flow,

$$\xi = \nabla \times \mathbf{V} = 0 \tag{2.152}$$

Consider the following vector identity: if ϕ is a scalar function, then

$$\nabla \times (\nabla \phi) = 0 \tag{2.153}$$

that is, the curl of the gradient of a scalar function is identically zero. Comparing Equations (2.152) and (2.153), we see that

$$\boxed{\mathbf{V} = \nabla \phi} \tag{2.154}$$

Equation (2.154) states that for an *irrotational* flow, there exists a scalar function ϕ such that the velocity is given by the gradient of ϕ. We denote ϕ as the *velocity potential*. ϕ is a function of the spatial coordinates; that is, $\phi = \phi(x, y, z)$, or $\phi = \phi(r, \theta, z)$, or $\phi = \phi(r, \theta, \Phi)$. From the definition of the gradient in cartesian coordinates given by Equation (2.16), we have, from Equation (2.154),

$$u\mathbf{i} + v\mathbf{j} + w\mathbf{k} = \frac{\partial \phi}{\partial x}\mathbf{i} + \frac{\partial \phi}{\partial y}\mathbf{j} + \frac{\partial \phi}{\partial z}\mathbf{k} \tag{2.155}$$

The coefficients of like unit vectors must be the same on both sides of Equation (2.155). Thus, in cartesian coordinates,

$$\boxed{u = \frac{\partial \phi}{\partial x} \quad v = \frac{\partial \phi}{\partial y} \quad w = \frac{\partial \phi}{\partial z}} \tag{2.156}$$

In a similar fashion, from the definition of the gradient in cylindrical and spherical coordinates given by Equations (2.17) and (2.18), we have, in cylindrical coordinates,

$$V_r = \frac{\partial \phi}{\partial r} \quad V_\theta = \frac{1}{r}\frac{\partial \phi}{\partial \theta} \quad V_z = \frac{\partial \phi}{\partial z} \tag{2.157}$$

and in spherical coordinates,

$$V_r = \frac{\partial \phi}{\partial r} \quad V_\theta = \frac{1}{r}\frac{\partial \phi}{\partial \theta} \quad V_\Phi = \frac{1}{r\sin\theta}\frac{\partial \phi}{\partial \Phi} \tag{2.158}$$

The velocity potential is analogous to the stream function in the sense that derivatives of ϕ yield the flow-field velocities. However, there are distinct differences between ϕ and $\bar{\psi}$ (or ψ):

1. The flow-field velocities are obtained by differentiating ϕ in the same direction as the velocities [see Equations (2.156) to (2.158)], whereas $\bar{\psi}$ (or ψ) is differentiated normal to the velocity direction [see Equations (2.147) and (2.148), or Equations (2.150) and (2.151)].
2. The velocity potential is defined for irrotational flow only. In contrast, the stream function can be used in either rotational or irrotational flows.
3. The velocity potential applies to three-dimensional flows, whereas the stream function is defined for two-dimensional flows only.[2]

When a flow field is irrotational, hence allowing a velocity potential to be defined, there is a tremendous simplification. Instead of dealing with the velocity components (say, u, v, and w) as unknowns, hence requiring three equations for these three unknowns, we can deal with the velocity potential as one unknown, therefore requiring the solution of only one equation for the flow field. Once ϕ is known for a given problem, the velocities are obtained directly from Equations (2.156) to (2.158). This is why, in theoretical aerodynamics, we make a distinction between irrotational and rotational flows and why the analysis of irrotational flows is simpler than that of rotational flows.

Because irrotational flows can be described by the velocity potential ϕ, such flows are called *potential flows*.

In this section, we have not yet discussed how ϕ can be obtained in the first place; we are assuming that it is known. The actual determination of ϕ for various problems is discussed in Chapters 3, 6, 11, and 12.

EXAMPLE 2.9

Calculate the velocity potential for the flow field over a wavy wall given in Example 2.1.

■ **Solution**
From Example 2.1, the equations for u and v are given by

$$u = V_\infty\left[1 + \frac{h}{\beta}\frac{2\pi}{\ell}\left(\cos\frac{2\pi x}{\ell}\right)e^{-2\pi\beta y/\ell}\right] \tag{2.35}$$

---

[2] $\bar{\psi}$ (or ψ) can be defined for axisymmetric flows, such as the flow over a cone at zero degrees angle of attack. However, for such flows, only two spatial coordinates are needed to describe the flow field (see Chapter 6).

and

$$v = -V_\infty h \frac{2\pi}{\ell} \left(\sin \frac{2\pi x}{\ell} \right) e^{-2\pi\beta y/\ell} \tag{2.36}$$

From Equation (2.156),

$$u = \frac{\partial \phi}{\partial x} \quad \text{and} \quad v = \frac{\partial \phi}{\partial y}$$

To find an expression for ϕ, we first integrate u with respect to x and then integrate v with respect to y, as follows

$$\phi = \int \frac{\partial \phi}{\partial x} dx = \int u \, dx \tag{E2.13}$$

Substituting Equation (2.35) into (E2.13), we have

$$\phi = \int \left[V_\infty + \frac{V_\infty h}{\beta} \frac{2\pi}{\ell} \left(\cos \frac{2\pi x}{\ell} \right) e^{-2\pi\beta y/\ell} \right] dx$$

$$\phi = V_\infty x + \frac{V_\infty h}{\beta} \frac{2\pi}{\ell} \left(\sin \frac{2\pi x}{\ell} \right) \left(\frac{2\pi}{\ell} \right)^{-1} e^{-2\pi\beta y/\ell}$$

$$\phi = V_\infty x + \frac{V_\infty h}{\beta} \left(\sin \frac{2\pi x}{\ell} \right) e^{-2\pi\beta y/\ell} + f(y) \tag{E2.14}$$

The function of $f(y)$ is added to Equation (E2.14) in the spirit of a "constant of integration" since Equation (E2.13) is an integral just with respect to x. Indeed, since $u = \frac{\partial \phi}{\partial x}$, when Equation (E2.14) is differentiated with respect to x, the derivative of $f(y)$ with respect to x is zero, and we recover Equation (2.35) for u. Also, we have

$$\phi = \int \frac{\partial \phi}{\partial y} dy = \int v \, dy \tag{E2.15}$$

Substituting Equation (2.36) into (E2.15), we have

$$\phi = \int \left[-V_\infty h \frac{2\pi}{\ell} \left(\sin \frac{2\pi x}{\ell} \right) e^{-2\pi\beta y/\ell} \right] dy$$

$$\phi = -V_\infty h \frac{2\pi}{\ell} \left(\sin \frac{2\pi x}{\ell} \right) e^{-2\pi\beta y/\ell} \left(-\frac{\ell}{\partial\pi\beta} \right)$$

$$\phi = V_\infty \frac{h}{\beta} \left(\sin \frac{2\pi x}{\ell} \right) e^{-2\pi\beta y/\ell} + g(x) \tag{E2.16}$$

where $g(x)$ in Equation (E2.16) is again a "constant of integration" since Equation (E2.15) is an integral just with respect to y. When Equation (E2.16) is differentiated with respect to y, the derivative of $g(x)$ with respect to y is zero, and we recover Equation (2.36) for v.

Compare Equations (E2.14) and (E2.16). These are two equations for the same velocity potential. Hence, in Equation (E2.14), $f(y) = 0$ and in Equation (E2.16), $g(x) = V_\infty x$. Thus, the velocity potential for the flow over the wavy wall in Example 2.1 is

$$\phi = V_\infty x + \frac{V_\infty h}{\beta} \left(\sin \frac{2\pi x}{\ell} \right) e^{-2\pi\beta y/\ell} \tag{E2.17}$$

Recall that a velocity potential exists for any irrotational flow. In Example 2.7, we proved that the flow field over the wavy wall shown in Figure 2.17 as specified in Example 2.1 is irrotational. Hence, a velocity potential exists for that flow. Equation (E2.17) is the velocity potential.

2.16 RELATIONSHIP BETWEEN THE STREAM FUNCTION AND VELOCITY POTENTIAL

In Section 2.15, we demonstrated that for an irrotational flow $\mathbf{V} = \nabla\phi$. At this stage, take a moment and review some of the nomenclature introduced in Section 2.2.5 for the gradient of a scalar field. We see that a line of constant ϕ is an isoline of ϕ; since ϕ is the velocity potential, we give this isoline a specific name, *equipotential line*. In addition, a line drawn in space such that $\nabla\phi$ is tangent at every point is defined as a gradient line; however, since $\nabla\phi = \mathbf{V}$, this gradient line is a *streamline*. In turn, from Section 2.14, a streamline is a line of constant $\bar{\psi}$ (for a two-dimensional flow). Because gradient lines and isolines are perpendicular (see Section 2.2.5, Gradient of a Scalar Field), then equipotential lines ($\phi = $ constant) and streamlines ($\bar{\psi} = $ constant) are mutually perpendicular.

To illustrate this result more clearly, consider a two-dimensional, irrotational, incompressible flow in cartesian coordinates. For a streamline, $\psi(x, y) = $ constant. Hence, the differential of ψ along the streamline is zero; that is,

$$d\psi = \frac{\partial\psi}{\partial x}dx + \frac{\partial\psi}{\partial y}dy = 0 \tag{2.159}$$

From Equation (2.150a and b), Equation (2.159) can be written as

$$d\psi = -v\,dx + u\,dy = 0 \tag{2.160}$$

Solve Equation (2.160) for dy/dx, which is the slope of the $\psi = $ constant line, that is, the slope of the streamline:

$$\left(\frac{dy}{dx}\right)_{\psi\,=\,\text{const}} = \frac{v}{u} \tag{2.161}$$

Similarly, for an equipotential line, $\phi(x, y) = $ constant. Along this line,

$$d\phi = \frac{\partial\phi}{\partial x}dx + \frac{\partial\phi}{\partial y}dy = 0 \tag{2.162}$$

From Equation (2.156), Equation (2.162) can be written as

$$d\phi = u\,dx + v\,dy = 0 \tag{2.163}$$

Solving Equation (2.163) for dy/dx, which is the slope of the $\phi = $ constant line (i.e., the slope of the equipotential line), we obtain

$$\left(\frac{dy}{dx}\right)_{\phi\,=\,\text{const}} = -\frac{u}{v} \tag{2.164}$$

Combining Equations (2.161) and (2.164), we have

$$\left(\frac{dy}{dx}\right)_{\psi=\text{const}} = -\frac{1}{(dy/dx)_{\phi=\text{const}}} \qquad (2.165)$$

Equation (2.165) shows that the slope of a $\psi = $ constant line is the negative reciprocal of the slope of a $\phi = $ constant line (i.e., streamlines and equipotential lines are mutually perpendicular).

2.17 HOW DO WE SOLVE THE EQUATIONS?

This chapter is full of mathematical equations—equations that represent the basic physical fundamentals that dictate the characteristics of aerodynamic flow fields. For the most part, the equations are either in partial differential form or integral form. These equations are powerful and by themselves represent a sophisticated intellectual construct of our understanding of the fundamentals of a fluid flow. However, the equations by themselves are not very practical. They must be *solved* in order to obtain the actual flow fields over specific body shapes with specific flow conditions. For example, if we are interested in calculating the flow field around a Boeing 777 jet transport flying at a velocity of 800 ft/s at an altitude of 30,000 ft, we have to obtain a *solution* of the governing equations for this case—a solution that will give us the results for the dependent flow-field variables p, ρ, $\mathbf{V}$, etc., as a function of the independent variables of spatial location and time. Then we have to squeeze this solution for extra practical information, such as lift, drag, and moments exerted on the vehicle. How do we do this? The purpose of the present section is to discuss two philosophical answers to this question. As for practical solutions to specific problems of interest, there are literally hundreds of different answers to this question, many of which make up the content of the rest of this book. However, all these solutions fall under one or the other of the two philosophical approaches described next.

2.17.1 Theoretical (Analytical) Solutions

Students learning any field of physical science or engineering at the beginning are usually introduced to nice, neat analytical solutions to physical problems that are simplified to the extent that such solutions are possible. For example, when Newton's second law is applied to the motion of a simple, frictionless pendulum, students in elementary physics classes are shown a closed-form analytical solution for the time period of the pendulum's oscillation, namely,

$$T = 2\pi\sqrt{\ell/g}$$

where T is the period, ℓ is the length of the pendulum, and g is the acceleration of gravity. However, a vital assumption behind this equation is that of *small amplitude* oscillations. Similarly, in studying the motion of a freely falling body in a gravitational field, the distance y through which the body falls in time t after release is given by

$$y = \tfrac{1}{2}gt^2$$

However, this result neglects any aerodynamic drag on the body as it falls through the air. The above examples are given because they are familiar results from elementary physics. They are examples of theoretical, closed-form solutions—straightforward algebraic relations.

The governing equations of aerodynamics, such as the continuity, momentum, and energy equations derived in Sections 2.4, 2.5, and 2.7, respectively, are highly nonlinear, partial differential, or integral equations; to date, no general analytical solution to these equations has been obtained. In lieu of this, two different philosophies have been followed in obtaining useful solutions to these equations. One of these is the theoretical, or analytical, approach, wherein the physical nature of certain aerodynamic applications allows the governing equations to be simplified to a sufficient extent that analytical solutions of the simplified equations can be obtained. One such example is the analysis of the flow across a normal shock wave, as discussed in Chapter 8. This flow is one-dimensional, that is, the changes in flow properties across the shock take place only in the flow direction. For this case, the y and z derivatives in the governing continuity, momentum, and energy equations from Sections 2.4, 2.5, and 2.7 are identically zero, and the resulting one-dimensional equations, which are still *exact* for the one-dimensional case being considered, lend themselves to a direct analytical solution, which is indeed an exact solution for the shock wave properties. Another example is the compressible flow over an airfoil considered in Chapters 11 and 12. If the airfoil is thin and at a small angle of attack, and if the freestream Mach number is not near one (not transonic) nor above five (not hypersonic), then many of the terms in the governing equations are small compared to others and can be neglected. The resulting simplified equations are linear and can be solved analytically. This is an example of an *approximate* solution, where certain simplifying assumptions have been made in order to obtain a solution.

The history of the development of aerodynamic theory is in this category—the simplification of the full governing equations apropos a given application so that analytical solutions can be obtained. Of course this philosophy works for only a limited number of aerodynamic problems. However, classical aerodynamic theory is built on this approach and, as such, is discussed at some length in this book. You can expect to see a lot of closed-form analytical solutions in the subsequent chapters, along with detailed discussions of their limitations due to the approximations necessary to obtain such solutions. In the modern world of aerodynamics, such classical analytical solutions have three advantages:

1. The act of developing these solutions puts you in intimate contact with all the physics involved in the problem.

2. The results, usually in closed form, give you direct information on what are the important variables, and how the answers vary with increases or

decreases in these variables. For example, in Chapter 11 we will obtain a simple equation for the compressibility effects on lift coefficient for an airfoil in high-speed subsonic flow. Equation (11.52) tells us that the high-speed effect on lift coefficient is governed by just M_∞ alone, and that as M_∞ increases, the lift coefficient increases. Moreover, the equation tells us *in what way* the lift coefficient increases, namely, inversely with $(1 - M_\infty^2)^{1/2}$. This is powerful information, albeit approximate.

3. Finally, the results in closed form provide simple tools for rapid calculations, making possible the proverbial "back of the envelope calculations" so important in the preliminary design process and in other practical applications.

2.17.2 Numerical Solutions—Computational Fluid Dynamics (CFD)

The other general approach to the solution of the governing equations is numerical. The advent of the modern high-speed digital computer in the last third of the twentieth century has revolutionized the solution of aerodynamic problems and has given rise to a whole new discipline—computational fluid dynamics. Recall that the basic governing equations of continuity, momentum, and energy derived in this chapter are either in integral or partial differential form. In Anderson, *Computational Fluid Dynamics: The Basics with Applications,* McGraw Hill, 1995, computational fluid dynamics is defined as "the art of replacing the integrals or the partial derivatives (as the case may be) in these equations with discretized algebraic forms, which in turn are solved to obtain *numbers* for the flow field values at discrete points in time and/or space." The end product of computational fluid dynamics (frequently identified by the acronym CFD) is indeed a collection of numbers, in contrast to a closed-form analytical solution. However, in the long run, the objective of most engineering analyses, closed form or otherwise, is a quantitative description of the problem (i.e., *numbers*).

The beauty of CFD is that it can deal with the full nonlinear equations of continuity, momentum, and energy, in principle, without resorting to any geometrical or physical approximations. Because of this, many complex aerodynamic flow fields have been solved by means of CFD which had never been solved before. An example of this is shown in Figure 2.42. Here we see the unsteady, viscous, turbulent, compressible, separated flow field over an airfoil at high angle of attack (14° in the case shown), as obtained from Reference 53. The freestream Mach number is 0.5, and the Reynolds number based on the airfoil chord length (distance from the front to the back edges) is 300,000. An instantaneous streamline pattern that exists at a certain instant in time is shown, reflecting the complex nature of the separated, recirculating flow above the airfoil. This flow is obtained by means of a CFD solution of the two-dimensional, unsteady continuity, momentum, and energy equations, including the full effects of viscosity and thermal conduction, as developed in Sections 2.4, 2.5, and 2.7, *without* any further geometrical or physical simplifications. The equations with all the viscosity and thermal conduction terms explicitly shown are developed in Chapter 15; in this form, they are frequently

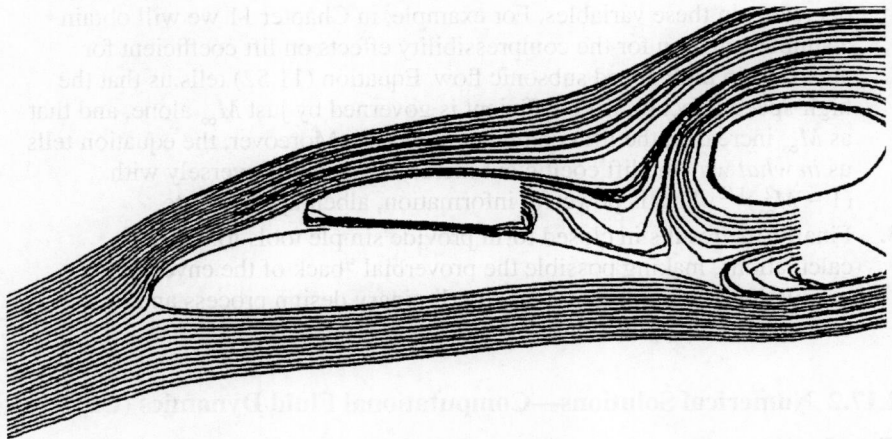

Figure 2.42 Calculated streamline pattern for separated flow over an airfoil. Re = 300,000, $M_\infty = 0.5$, angle of attack = 14°.

labeled as the *Navier-Stokes equations*. There is no analytical solution for the flow shown in Figure 2.42; the solution can only be obtained by means of CFD.

Let us explore the basic philosophy behind CFD. Again, keep in mind that CFD solutions are completely *numerical solutions,* and a high-speed digital computer must be used to carry them out. In a CFD solution, the flow field is divided into a number of *discrete* points. Coordinate lines through these points generate a *grid,* and the discrete points are called *grid points*. The grid used to solve the flow field shown in Figure 2.42 is given in Figure 2.43; here, the grid is wrapped around the airfoil, which is seen as the small white speck in the center-left of the figure, and the grid extends a very large distance out from the airfoil. This large extension of the grid into the main stream of the flow is necessary for a subsonic flow, because disturbances in a subsonic flow physically feed out large distances away from the body. We will learn why in subsequent chapters. The black region near the airfoil is simply the computer graphics way of showing a very large number of closely spaced grid points near the airfoil, for better definition of the viscous flow near the airfoil. The flow field properties, such as p, ρ, u, v, etc., are calculated just at the discrete grid points, and nowhere else, by means of the numerical solution of the governing flow equations. This is an inherent property that distinguishes CFD solutions from closed-form analytical solutions. Analytical solutions, by their very nature, yield closed-form equations that describe the flow as a function of continuous time and/or space. So we can pick any one of the infinite number of points located in this continuous space, feed the coordinates into the closed-form equations, and obtain the flow variables at that point. Not so with CFD, where the flow-field variables are calculated only at discrete grid points. For a CFD solution, the partial derivatives or the integrals, as the case may be, in the governing flow equations are *discretized,* using the flow-field variables at grid points only.

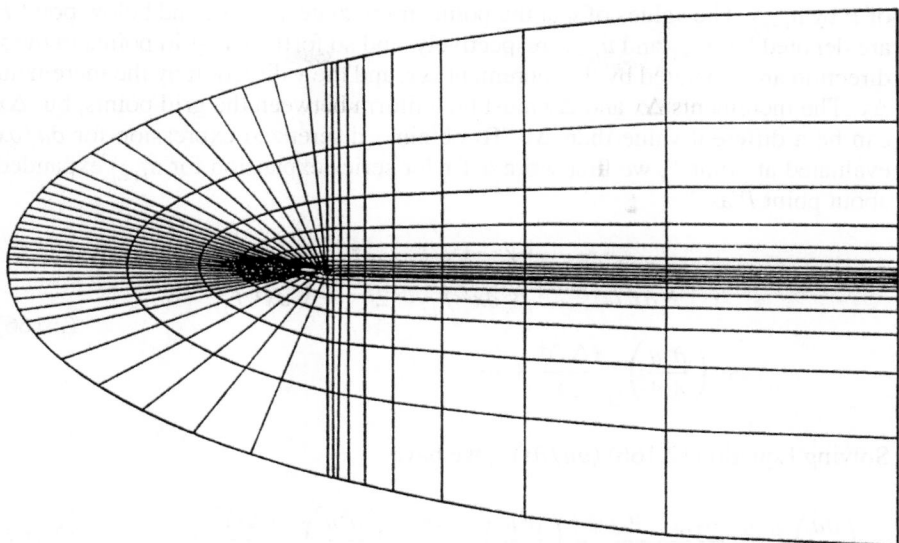

Figure 2.43 The grid used for the numerical solution of the flow over the airfoil in Figure 2.40. The airfoil is the small speck in the middle-left of the figure.

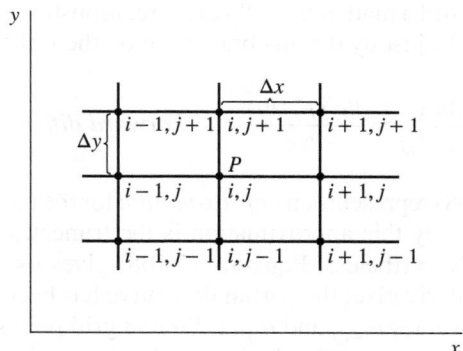

Figure 2.44 An array of grid points in a uniform, rectangular grid.

How is this discretization carried out? There are many answers to this equation. We will look at just a few examples, to convey the ideas.

Let us consider a partial derivative, such as $\partial u / \partial x$. How do we discretize this partial derivative? First, we choose a uniform rectangular array of grid points as shown in Figure 2.44. The points are identified by the index i in the x direction, and the index j in the y direction. Point P in Figure 2.44 is identified as point (i,j). The value of the variable u at point j is denoted by $u_{i,j}$. The value of u at the point immediately to the right of P is denoted by $u_{i+1,j}$ and that immediately to the left of P by $u_{i-1,j}$. The values of u at the points immediately above and below point P are denoted by $u_{i,j+1}$ and $u_{i,j-1}$, respectively, and so forth. The grid points in the x direction are separated by the increment Δx, and the y direction by the increment Δy. The increments Δx and Δy must be uniform between the grid points, but Δx can be a different value than Δy. To obtain a discretized expression for $\partial u/\partial x$ evaluated at point P, we first write a Taylor series expansion for $u_{i+1,j}$ expanded about point P as:

$$u_{i+1,j} = u_{i,j} + \left(\frac{\partial u}{\partial x}\right)_{i,j}\Delta x + \left(\frac{\partial^2 u}{\partial x^2}\right)_{i,j}\frac{(\Delta x)^2}{2} + \left(\frac{\partial^3 u}{\partial x^3}\right)_{i,j}\frac{(\Delta x)^3}{6}$$
$$+ \left(\frac{\partial^4 u}{\partial x^4}\right)_{i,j}\frac{(\Delta x)^4}{24} + \cdots \qquad (2.166)$$

Solving Equation (2.166) $(\partial u/\partial x)_{i,j}$, we have

$$\left(\frac{\partial u}{\partial x}\right)_{i,j} = \underbrace{\frac{u_{i+1,j} - u_{i,j}}{\Delta x}}_{\text{Forward difference}} \underbrace{- \left(\frac{\partial^2 u}{\partial x^2}\right)_{i,j}\frac{\Delta x}{2} - \left(\frac{\partial^3 u}{\partial x^3}\right)_{i,j}\frac{(\Delta x)^2}{6} + \cdots}_{\text{Truncation error}} \qquad (2.167)$$

Equation (2.167) is still a mathematically exact relationship. However, if we choose to represent $(\partial u/\partial x)_{i,j}$ just by the algebraic term on the right-hand side, namely,

$$\left(\frac{\partial u}{\partial x}\right)_{i,j} = \frac{u_{i+1,j} - u_{i,j}}{\Delta x} \qquad \textit{Forward difference} \qquad (2.168)$$

then Equation (2.168) represents an *approximation* for the partial derivative, where the error introduced by this approximation is the truncation error, identified in Equation (2.167). Nevertheless, Equation (2.168) gives us an algebraic expression for the partial derivative; the partial derivative has been *discretized* because it is formed by the values $u_{i+1,j}$ and $u_{i,j}$ at *discrete* grid points. The algebraic difference quotient in Equation (2.168) is called a *forward difference,* because it uses information ahead of point (i,j), namely $u_{i+1,j}$. Also, the forward difference given by Equation (2.168) has *first-order accuracy* because the leading term of the truncation error in Equation (2.167) has Δx to the first power.

Equation (2.168) is not the only discretized form for $(\partial u/\partial x)_{i,j}$. For example, let us write a Taylor series expansion for $u_{i-1,j}$ expanded about point P as

$$u_{i-1,j} = u_{i,j} + \left(\frac{\partial u}{\partial x}\right)_{i,j}(-\Delta x) + \left(\frac{\partial^2 u}{\partial x^2}\right)_{i,j}\frac{(-\Delta x)^2}{2}$$
$$+ \left(\frac{\partial^3 u}{\partial x^3}\right)_{i,j}\frac{(-\Delta x)^3}{6} + \cdots \qquad (2.169)$$

Solving Equation (2.169) for $(\partial u/\partial x)_{i,j}$, we have

$$\left(\frac{\partial u}{\partial x}\right)_{i,j} = \underbrace{\frac{u_{i,j} - u_{i-1,j}}{\Delta x}}_{\text{Rearward difference}} + \underbrace{\left(\frac{\partial^2 u}{\partial x^2}\right)_{i,j} \frac{\Delta x}{2} - \left(\frac{\partial^3 u}{\partial x^3}\right)_{i,j} \frac{(\Delta x)^2}{6} + \cdots}_{\text{Truncation error}} \qquad (2.170)$$

Hence, we can represent the partial derivative by the rearward difference shown in Equation (2.170), namely,

$$\left(\frac{\partial u}{\partial x}\right)_{i,j} = \frac{u_{i,j} - u_{i-1,j}}{\Delta x} \qquad \textit{Rearward difference} \qquad (2.171)$$

Equation (2.171) is an approximation for the partial derivative, where the error is given by the truncation error labeled in Equation (2.170). The rearward difference given by Equation (2.171) has first-order accuracy because the leading term in the truncation error in Equation (2.170) has Δx to the first power. The forward and rearward differences given by Equations (2.168) and (2.171), respectively, are equally valid representations of $(\partial u/\partial x)_{i,j}$, each with first-order accuracy.

In most CFD solutions, first-order accuracy is not good enough; we need a discretization of $(\partial u/\partial x)_{i,j}$ that has at least second-order accuracy. This can be obtained by subtracting Equation (2.169) from Equation (2.166), yielding

$$u_{i+1,j} - u_{i-1,j} = 2\left(\frac{\partial u}{\partial x}\right)_{i,j} \Delta x + \left(\frac{\partial^3 u}{\partial x^3}\right)_{i,j} \frac{(\Delta x)^3}{3} + \cdots \qquad (2.172)$$

Solving Equation (2.172) for $(\partial u/\partial x)_{i,j}$, we have

$$\left(\frac{\partial u}{\partial x}\right)_{i,j} = \underbrace{\frac{u_{i+1,j} - u_{i-1,j}}{2\Delta x}}_{\text{Central difference}} - \underbrace{\left(\frac{\partial^3 u}{\partial x^3}\right)_{i,j} \frac{(\Delta x)^2}{3} + \cdots}_{\text{Truncation error}} \qquad (2.173)$$

Hence, we can represent the partial derivative by the *central difference* shown in Equation (2.173), namely

$$\left(\frac{\partial u}{\partial x}\right)_{i,j} = \frac{u_{i+1,j} - u_{i-1,j}}{2\Delta x} \qquad \textit{Central difference} \qquad (2.174)$$

Examining Equation (2.173), we see that the central difference expression given in Equation (2.174) has *second-order accuracy,* because the leading term in the truncation error in Equation (2.173) has $(\Delta x)^2$. For most CFD solutions, second-order accuracy is sufficient.

So this is how it works—this is how the partial derivatives that appear in the governing flow equations can be discretized. There are many other possible discretized forms for the derivatives; the forward, rearward, and central differences obtained above are just a few. Note that Taylor series have been used to obtain these discrete forms. Such Taylor series expressions are the basic foundation of *finite-difference* solutions in CFD. In contrast, if the integral form of the governing flow equations are used, such as Equations (2.48) and (2.64), the individual integral terms can be discretized, leading again to algebraic equations that are the basic foundation of *finite-volume* solutions in CFD.

EXAMPLE 2.10

Consider a one-dimensional, unsteady flow, where the flow-field variables such as ρ, u, etc., are functions of distance x and time t. Consider the grid shown in Figure 2.45, where grid points arrayed in the x direction are denoted by the index i. Two rows of grid points are shown, one at time t and the other at the later time $t + \Delta t$. In particular, we are interested in calculating the unknown density at grid point i at time $t + \Delta t$, denoted by $\rho_i^{t+\Delta t}$. Set up the calculation of this unknown density.

■ **Solution**

Note in Figure 2.45 that the dashed loop (called the computational module) contain the grid points $i - 1$, i, and $i + 1$ at time t, where the flow field is known, and the grid point i at time $t + \Delta t$, where the flow field is unknown. From the continuity equation, Equation (2.52), repeated below

$$\frac{\partial \rho}{\partial t} + \nabla \cdot (\rho \mathbf{V}) = 0 \qquad (2.52)$$

written for unsteady, one-dimensional flow, we have

$$\frac{\partial \rho}{\partial t} + \frac{\partial (\rho u)}{\partial x} = 0 \qquad (2.175)$$

Rearranging Equation (2.175),

$$\frac{\partial \rho}{\partial t} = -\frac{\partial (\rho u)}{\partial x}$$

or

$$\frac{\partial \rho}{\partial t} = -\rho \frac{\partial u}{\partial x} - u \frac{\partial \rho}{\partial x} \qquad (2.176)$$

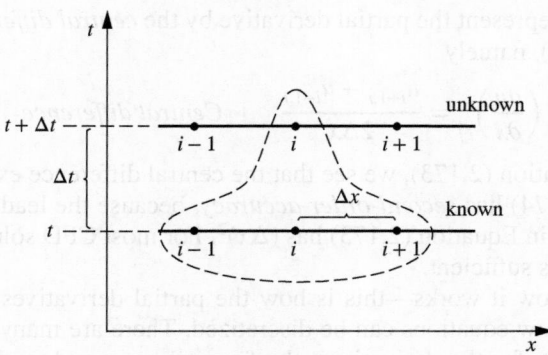

Figure 2.45 Computational module. The calculation of unknown properties at point i at time $(t + \Delta t)$ from known properties at points $i - 1$, i, $i + 1$ at time t.

In Equation (2.176), replace $\partial\rho/\partial t$ with a forward difference in time, and $\partial u/\partial x$ and $\partial\rho/\partial x$ with central differences in space, centered around grid point i

$$\frac{\rho_i^{t+\Delta t} - \rho_i^t}{\Delta t} = -\rho_i^t\left(\frac{u_{i+1}^t - u_{i-1}^t}{2\Delta x}\right) - u_i^t\left(\frac{\rho_{i+1}^t - \rho_{i-1}^t}{2\Delta x}\right) \qquad (2.177)$$

Equation (2.177) is called a *difference equation;* it is an approximate representation of the original partial differential equation, Equation (2.176), where the error in the approximation is given by the sum of the truncation errors associated with each of the finite differences used to obtain Equation (2.177). Solving Equation (2.177) for $\rho_i^{t+\Delta t}$

$$\rho_i^{t+\Delta t} = \rho_i^t - \frac{\Delta t}{2\Delta x}\left(\rho_i^t u_{i+1}^t - \rho_i^t u_{i-1}^t + u_i^t \rho_{i+1}^t - u_i^t \rho_{i-1}^t\right) \qquad (2.178)$$

In Equation (2.178), all quantities on the right-hand side are known values at time t. Hence, Equation (2.178) allows the direct calculation of the unknown value, $\rho_i^{t+\Delta t}$, at time $t + \Delta t$.

This example is a simple illustration of how a CFD solution to a given flow can be set up, in this case for an unsteady, one-dimensional flow. Note that the unknown velocity and internal energy at grid point i at time $t + \Delta t$ can be calculated in the same manner, writing the appropriate difference equation representations for the x component of the momentum equation, Equation (2.113a), and the energy equation, Equation (2.114).

The above example looks very straightforward, and indeed it is. It is given here only as an illustration of what is meant by a CFD technique. However, do not be misled. Computational fluid dynamics is a sophisticated and complex discipline. For example, we have said nothing here about the accuracy of the final solutions, whether or not a certain computational technique will be stable (some attempts at obtaining numerical solutions will go unstable—blow up—during the course of the calculations), and how much computer time a given technique will require to obtain the flow-field solution. Also, in our discussion we have given examples of some relatively simple grids. The generation of an appropriate grid for a given flow problem is frequently a challenge, and grid generation has emerged as a subdiscipline in its own right within CFD. For these reasons, CFD is usually taught only in graduate-level courses in most universities. However, in an effort to introduce some of the basic ideas of CFD at the undergraduate level, I have written a book, Reference 7, intended to present the subject at the most elementary level. Reference 7 is intended to be read *before* students go on to the more advanced books on CFD written at the graduate level. In the present book, we will, from time to time, discuss some applications of CFD as part of the overall fundamentals of aerodynamics. However, this book is not about CFD; Reference 7 is.

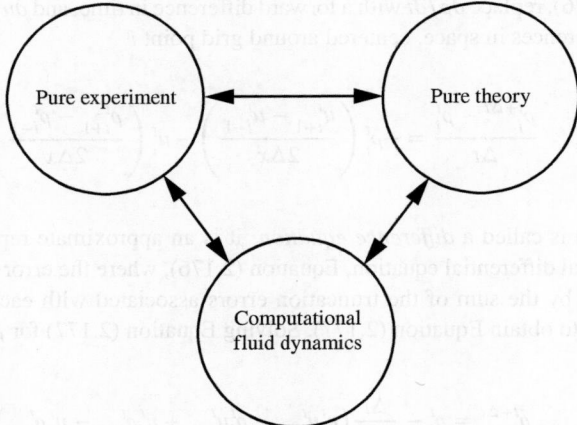

Figure 2.46 The three equal partners of modern
aerodynamics.

2.17.3 The Bigger Picture

The evolution of our intellectual understanding of aerodynamics is over 2500 years
old, going all the way back to ancient Greek science. The aerodynamics you are
studying in this book is the product of this evolution. (See Reference 68 for an in-
depth study of the history of aerodynamics.) Relevant to our current discussion
is the development of the experimental tradition in fluid dynamics, which took
place in the middle of the seventeenth century, principally in France, and the
introduction of rational analysis in mechanics pioneered by Isaac Newton toward
the end of the same century. Since that time, up until the middle of the twentieth
century, the study and practice of fluid dynamics, including aerodynamics, has
dealt with pure experiment on one hand and pure theory on the other. If you were
learning aerodynamics as recently as, say 1960, you would have been operating
in the "two-approach world" of theory and experiment. However, computational
fluid dynamics has revolutionized the way we study and practice aerodynamics
today. As sketched in Figure 2.46, today CFD is an *equal* partner with pure theory
and pure experiment in the analysis and solution of aerodynamic problems. This
is no flash in the pan—CFD will continue to play this role indefinitely, for as long
as our advanced human civilization exists. Also, the double arrows in Figure 2.46
imply that today each of the equal partners constantly interact with each other—
they do not stand alone, but rather help each other to continue to resolve and better
understand the "big picture" of aerodynamics.

2.18 SUMMARY

Return to the road map for this chapter, as given in Figure 2.3. We have now
covered both the left and right branches of this map and are ready to launch into
the solution of practical aerodynamic problems in subsequent chapters. Look at

each block in Figure 2.3; let your mind flash over the important equations and concepts represented by each block. If the flashes are dim, return to the appropriate sections of this chapter and review the material until you feel comfortable with these aerodynamic tools.

For your convenience, the most important results are summarized below:

Basic Flow Equations

Continuity equation

$$\frac{\partial}{\partial t} \iiint_\mathcal{V} \rho \, d\mathcal{V} + \oiint_S \rho \mathbf{V} \cdot d\mathbf{S} = 0 \tag{2.48}$$

or

$$\frac{\partial \rho}{\partial t} + \nabla \cdot (\rho \mathbf{V}) = 0 \tag{2.52}$$

or

$$\frac{D\rho}{Dt} + \rho \nabla \cdot \mathbf{V} = 0 \tag{2.108}$$

Momentum equation

$$\frac{\partial}{\partial t} \iiint_\mathcal{V} \rho \mathbf{V} \, d\mathcal{V} + \oiint_S (\rho \mathbf{V} \cdot d\mathbf{S})\mathbf{V} = - \oiint_S p \, d\mathbf{S} + \iiint_\mathcal{V} \rho \mathbf{f} \, d\mathcal{V} + \mathbf{F}_{\text{viscous}} \tag{2.64}$$

or

$$\frac{\partial (\rho u)}{\partial t} + \nabla \cdot (\rho u \mathbf{V}) = -\frac{\partial p}{\partial x} + \rho f_x + (\mathcal{F}_x)_{\text{viscous}} \tag{2.70a}$$

$$\frac{\partial (\rho v)}{\partial t} + \nabla \cdot (\rho v \mathbf{V}) = -\frac{\partial p}{\partial y} + \rho f_y + (\mathcal{F}_y)_{\text{viscous}} \tag{2.70b}$$

$$\frac{\partial (\rho w)}{\partial t} + \nabla \cdot (\rho w \mathbf{V}) = -\frac{\partial p}{\partial z} + \rho f_z + (\mathcal{F}_z)_{\text{viscous}} \tag{2.70c}$$

or

$$\rho \frac{Du}{Dt} = -\frac{\partial p}{\partial x} + \rho f_x + (\mathcal{F}_x)_{\text{viscous}} \tag{2.113a}$$

$$\rho \frac{Dv}{Dt} = -\frac{\partial p}{\partial y} + \rho f_y + (\mathcal{F}_y)_{\text{viscous}} \tag{2.113b}$$

$$\rho \frac{Dw}{Dt} = -\frac{\partial p}{\partial z} + \rho f_z + (\mathcal{F}_z)_{\text{viscous}} \tag{2.113c}$$

(continued)

Energy equation

$$\frac{\partial}{\partial t} \iiint\limits_{\mathcal{V}} \rho \left(e + \frac{V^2}{2} \right) d\mathcal{V} + \oiint\limits_{S} \rho \left(e + \frac{V^2}{2} \right) \mathbf{V} \cdot \mathbf{dS} \qquad (2.95)$$

$$= \iiint\limits_{\mathcal{V}} \dot{q}\rho \, d\mathcal{V} + \dot{Q}_{\text{viscous}} - \oiint\limits_{S} p\mathbf{V} \cdot \mathbf{dS}$$

$$+ \iiint\limits_{\mathcal{V}} \rho(\mathbf{f} \cdot \mathbf{V}) \, d\mathcal{V} + \dot{W}_{\text{viscous}}$$

or

$$\frac{\partial}{\partial t} \left[\rho \left(e + \frac{V^2}{2} \right) \right] + \nabla \cdot \left[\rho \left(e + \frac{V^2}{2} \right) \mathbf{V} \right] = \rho\dot{q} - \nabla \cdot (p\mathbf{V}) + \rho(\mathbf{f} \cdot \mathbf{V})$$

$$+ \dot{Q}'_{\text{viscous}} + \dot{W}'_{\text{viscous}} \qquad (2.96)$$

or

$$\rho\frac{D(e + V^2/2)}{Dt} = \rho\dot{q} - \nabla \cdot (p\mathbf{V}) + \rho(\mathbf{f} \cdot \mathbf{V}) + \dot{Q}'_{\text{viscous}} + \dot{W}'_{\text{viscous}} \qquad (2.114)$$

Substantial derivative

$$\frac{D}{Dt} \equiv \underbrace{\frac{\partial}{\partial t}}_{\substack{\text{local} \\ \text{derivative}}} + \underbrace{(\mathbf{V} \cdot \nabla)}_{\substack{\text{convective} \\ \text{derivative}}} \qquad (2.109)$$

A streamline is a curve whose tangent at any point is in the direction of the velocity vector at that point. The equation of a streamline is given by

$$\mathbf{ds} \times \mathbf{V} = 0 \qquad (2.115)$$

or, in cartesian coordinates,

$$w \, dy - v \, dz = 0 \qquad (2.117a)$$

$$u \, dz - w \, dx = 0 \qquad (2.117b)$$

$$v \, dx - u \, dy = 0 \qquad (2.117c)$$

The vorticity ξ at any given point is equal to twice the angular velocity of a fluid element $\boldsymbol{\omega}$, and both are related to the velocity field by

$$\xi = 2\boldsymbol{\omega} = \nabla \times \mathbf{V} \qquad (2.129)$$

When $\nabla \times \mathbf{V} \neq 0$, the flow is rotational. When $\nabla \times \mathbf{V} = 0$, the flow is irrotational.

Circulation Γ is related to lift and is defined as

$$\Gamma \equiv -\oint_C \mathbf{V} \cdot \mathbf{ds} \qquad (2.136)$$

Circulation is also related to vorticity via

$$\Gamma \equiv -\oint_C \mathbf{V} \cdot \mathbf{ds} = -\iint_S (\nabla \times \mathbf{V}) \cdot \mathbf{dS} \qquad (2.137)$$

or

$$(\nabla \times \mathbf{V}) \cdot \mathbf{n} = -\frac{d\Gamma}{dS} \qquad (2.138)$$

The stream function $\bar{\psi}$ is defined such that $\bar{\psi}(x, y) = $ constant is the equation of a streamline, and the difference in the stream function between two streamlines $\Delta\bar{\psi}$ is equal to the mass flow between the streamlines. As a consequence of this definition, in cartesian coordinates,

$$\rho u = \frac{\partial\bar{\psi}}{\partial y} \qquad (2.147a)$$

$$\rho v = -\frac{\partial\bar{\psi}}{\partial x} \qquad (2.147b)$$

and in cylindrical coordinates,

$$\rho V_r = \frac{1}{r}\frac{\partial\bar{\psi}}{\partial\theta} \qquad (2.148a)$$

$$\rho V_\theta = -\frac{\partial\bar{\psi}}{\partial r} \qquad (2.148b)$$

For incompressible flow, $\psi \equiv \bar{\psi}/\rho$ is defined such that $\psi(x, y) = $ constant denotes a streamline and $\Delta\psi$ between two streamlines is equal to the volume flow between these streamlines. As a consequence of this definition, in cartesian coordinates,

$$u = \frac{\partial\psi}{\partial y} \qquad (2.150a)$$

$$v = -\frac{\partial\psi}{\partial x} \qquad (2.150b)$$

and in cylindrical coordinates,

$$V_r = \frac{1}{r}\frac{\partial\psi}{\partial\theta} \qquad (2.151a)$$

$$V_\theta = -\frac{\partial\psi}{\partial r} \qquad (2.151b)$$

The stream function is valid for both rotational and irrotational flows, but it is restricted to two-dimensional flows only.

The velocity potential ϕ is defined for irrotational flows only, such that

$$\mathbf{V} = \nabla\phi \tag{2.154}$$

In cartesian coordinates,

$$u = \frac{\partial\phi}{\partial x} \quad v = \frac{\partial\phi}{\partial y} \quad w = \frac{\partial\phi}{\partial z} \tag{2.156}$$

In cylindrical coordinates,

$$V_r = \frac{\partial\phi}{\partial r} \quad V_\theta = \frac{1}{r}\frac{\partial\phi}{\partial\theta} \quad V_z = \frac{\partial\phi}{\partial z} \tag{2.157}$$

In spherical coordinates,

$$V_r = \frac{\partial\phi}{\partial r} \quad V_\theta = \frac{1}{r}\frac{\partial\phi}{\partial\theta} \quad V_\Phi = \frac{1}{r\sin\theta}\frac{\partial\phi}{\partial\Phi} \tag{2.158}$$

An irrotational flow is called a potential flow.

A line of constant ϕ is an equipotential line. Equipotential lines are perpendicular to streamlines (for two-dimensional irrotational flows).

2.19 PROBLEMS

2.1 Consider a body of arbitrary shape. If the pressure distribution over the surface of the body is constant, prove that the resultant pressure force on the body is zero. [Recall that this fact was used in Equation (2.77).]

2.2 Consider an airfoil in a wind tunnel (i.e., a wing that spans the entire test section). Prove that the lift per unit span can be obtained from the pressure distributions on the top and bottom walls of the wind tunnel (i.e., from the pressure distributions on the walls above and below the airfoil).

2.3 Consider a velocity field where the x and y components of velocity are given by $u = cx/(x^2 + y^2)$ and $v = cy/(x^2 + y^2)$, where c is a constant. Obtain the equations of the streamlines.

2.4 Consider a velocity field where the x and y components of velocity are given by $u = cy/(x^2 + y^2)$ and $v = -cx/(x^2 + y^2)$, where c is a constant. Obtain the equations of the streamlines.

2.5 Consider a velocity field where the radial and tangential components of velocity are $V_r = 0$ and $V_\theta = cr$, respectively, where c is a constant. Obtain the equations of the streamlines.

2.6 Consider a velocity field where the x and y components of velocity are given by $u = cx$ and $v = -cy$, where c is a constant. Obtain the equations of the streamlines.

2.7 The velocity field given in Problem 2.3 is called *source flow,* which will be discussed in Chapter 3. For source flow, calculate:

a. The time rate of change of the volume of a fluid element per unit volume.

b. The vorticity.

Hint: It is simpler to convert the velocity components to polar coordinates and deal with a polar coordinate system.

2.8 The velocity field given in Problem 2.4 is called *vortex flow,* which will be discussed in Chapter 3. For vortex flow, calculate:

a. The time rate of change of the volume of a fluid element per unit volume.

b. The vorticity.

Hint: Again, for convenience use polar coordinates.

2.9 Is the flow field given in Problem 2.5 irrotational? Prove your answer.

2.10 Consider a flow field in polar coordinates, where the stream function is given as $\psi = \psi(r, \theta)$. Starting with the concept of mass flow between two streamlines, derive Equation (2.148a and b).

2.11 Assuming the velocity field given in Problem 2.6 pertains to an incompressible flow, calculate the stream function and velocity potential. Using your results, show that lines of constant ϕ are perpendicular to lines of constant ψ.

2.12 Consider a length of pipe bent into a U-shape. The inside diameter of the pipe is 0.5 m. Air enters one leg of the pipe at a mean velocity of 100 m/s and exits the other leg at the same magnitude of velocity, but moving in the opposite direction. The pressure of the flow at the inlet and exit is the ambient pressure of the surroundings. Calculate the magnitude and direction of the force exerted on the pipe by the airflow. The air density is 1.23 kg/m^3.

2.13 Consider the subsonic compressible flow over the wavy wall treated in Example 2.1. Derive the equation for the velocity potential for this flow as a function of x and y.

2.14 In Example 2.1, the statement is made that the streamline an infinite distance above the wall is straight. Prove this statement.

2.15 A pressure field is defined by $P(x, y) = \sqrt{x^2 + y^2} + 2$. Here, x and y are in units of meters, and the pressure P has units of Pa. Calculate the resulting force that acts on the top of a circle of radius 2 m centered at the origin.

2.16 A jet engine undergoing ground testing vents its exhaust through a smokestack. Investigators set up two probes to measure the dispersion of the exhaust into the atmosphere. Probe A is set up at a fixed location near the exit of the smokestack. Probe B is attached to a balloon that is carried along by the local air current which, for the purposes of this analysis, will assumed to be steady. Define what type of derivatives these probes represent and identify which probe corresponds to which type of derivative.

2.17 If $\vec{V} = (3x + 2y)\mathbf{i} + 4(x + y + 2z)\mathbf{j} + (x - 3z)\mathbf{k}$ please:

    a.   Compute the curl of **V** and explain what the curl signifies physically.

    b.   Compute the divergence of **V** and explain what the divergence signifies physically.

2.18 Write the expression for the substantial derivative and explain its physical significance.

2.19 Show how the differential form of the x-component momentum equation [Equation (2.70a)] can be expressed using the substantial derivative:

$$\frac{\partial(\rho u)}{\partial t} + \nabla \circ (\rho u V) = -\frac{\partial p}{\partial x} + \rho f_x + (F_x)_{viscous}$$

Next, show how your result above can be generalized for the full momentum conservation equation (in differential form):

$$\frac{\partial(\rho V)}{\partial t} + \nabla \circ (\rho V V) = -\nabla p + \rho f + F_{viscous}$$

2.20 Consider steady flow through the duct with circular cross section illustrated below.

Use mass conservation to find U_2 in terms of U_1 assuming $D_2/D_1 = 2$ and:

    a.   The flow is steady, incompressible, and uniform across the entrance and exit of the duct.

    b.   The flow is steady, incompressible, with a parabolic distribution across the duct exit.

    c.   The flow is steady, compressible, and uniform across the entrance and exit of the duct with $T_2/T_1 = 5$ and $P_2/P_1 = 2$.

2.21 RP-1 flows steadily through the 90° reducing elbow shown in the illustration. The absolute pressure at the inlet to the elbow is 107.3 kPa and the cross-sectional area is 0.01 m². The cross-sectional area at the outlet is 0.0025 m² and the velocity there is 16 m/s. The elbow discharges to the atmosphere. Find the velocity of the propellant entering the elbow and the force required to hold the elbow in place. Assume that the outside pressure is 101 kPa, the density of RP-1 is constant at 810 kg/m³, that the volume of the elbow is 11, and that gravity points in the $-z$ direction. Neglect the weight of the elbow itself.

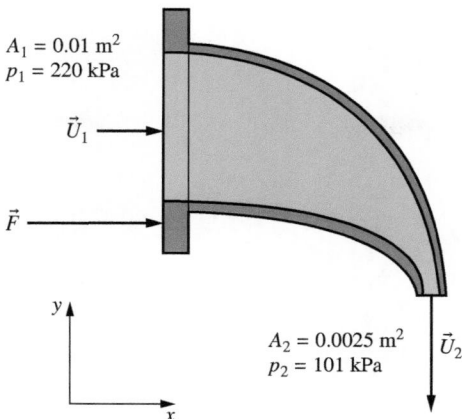

$A_1 = 0.01 \text{ m}^2$
$p_1 = 220 \text{ kPa}$

$\vec{U}_1 \longrightarrow$

$\vec{F} \longrightarrow$

y

$A_2 = 0.0025 \text{ m}^2$ $\vec{U}_2$
$p_2 = 101 \text{ kPa}$

x

2.22 The space shuttle main engine consumes approximately 140.3 kg/s of liquid hydrogen and 374.2 kg/s of liquid oxygen at full power. The overall ratio of hydrogen to oxygen on a mole-basis is approximately 6:1 so the combustion products are a gaseous mixture of water and excess hydrogen. The latter is used to cool the engine. The nozzle expands the hot, high pressure, gaseous combustion products to produce thrust—a process we will learn a lot more about later. If the temperature and pressure at the nozzle's exit are 12.7 kPa and 390 K, respectively, and the effective diameter of the nozzle exit is 2.16 m, find the speed of the gas mixture at the nozzle exit.

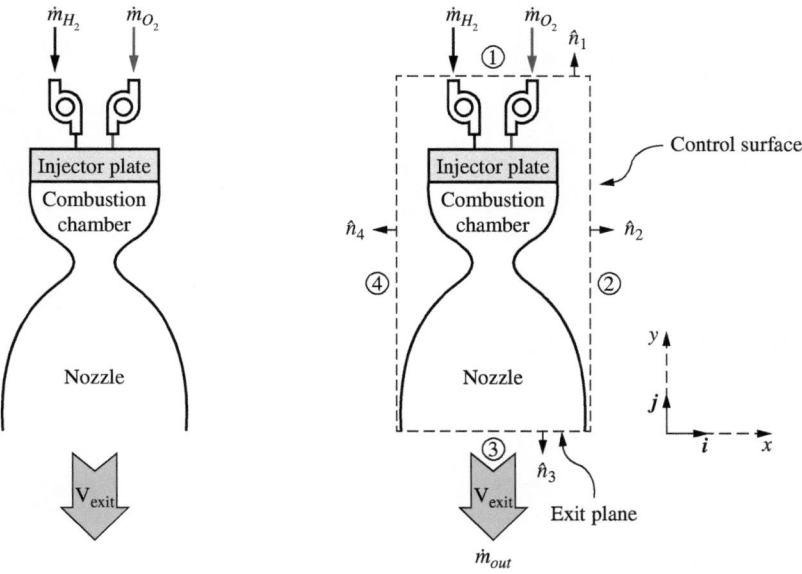

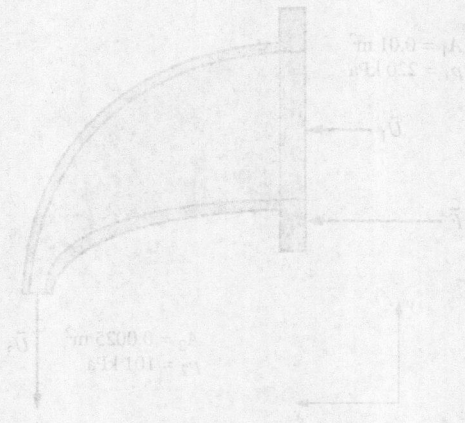

$A_2 = 0.01$ m²
$p_2 = 250$ kPa

$A_1 = 0.003$ m²
$p_1 = 101$ kPa

2.22 The space shuttle main engine consumes approximately 1340.3 kg/s of liquid hydrogen and 374.2 kg/s of liquid oxygen at full power. The overall ratio of hydrogen to oxygen on a mole basis is approximately 6:1 so the combustion products are a gaseous mixture of water and excess hydrogen. The latter is used to cool the engine. The nozzle expands the hot, high-pressure, gaseous combustion products to produce thrust—a process we will learn a lot more about later. If the temperature and pressure at the nozzle's exit are 1277 kPa and 390 K respectively, and the effective diameter of the nozzle exit is 2.16 m, find the speed of the gas mixture at the nozzle exit.

Inviscid, Incompressible Flow

I n Part 2, we deal with the flow of a fluid that has constant density—incompressible flow. This applies to the flow of liquids, such as water flow, and to low-speed flow of gases. The material covered here is applicable to low-speed flight through the atmosphere—flight at a Mach number of about 0.3 or less. ∎

CHAPTER 3

Fundamentals of Inviscid, Incompressible Flow

Theoretical fluid dynamics, being a difficult subject, is for convenience, commonly divided into two branches, one treating of frictionless or perfect fluids, the other treating of viscous or imperfect fluids. The frictionless fluid has no existence in nature, but is hypothesized by mathematicians in order to facilitate the investigation of important laws and principles that may be approximately true of viscous or natural fluids.

**Albert F. Zahm, 1912
(Professor of aeronautics, and
developer of the first aeronautical
laboratory in a U.S. university,
The Catholic University of America)**

PREVIEW BOX

Here we go again—fundamentals, and yet more fundamentals. However, in this chapter we focus on the fundamentals of a specific class of flow—inviscid, incompressible flow. Actually, such flow is a myth on two accounts. First, in real life there is always friction to some greater or lesser degree, so in nature there is, strictly speaking, no inviscid flow. Second, every flow is compressible to some greater or lesser degree, so in nature there is, strictly speaking, no incompressible flow. In engineering, however, if we were always "strictly speaking" about everything, we would not

be able to analyze anything. As you progress in your studies, you will find that virtually every engineering analysis involves approximations about the physics involved.

In regard to the material in this chapter, there are a whole host of aerodynamic applications that are *so close* to being inviscid and incompressible that we readily make that assumption and obtain amazingly accurate results. Here, you will find out how low-speed wind tunnels work. You will discover how to measure the velocity of a low-speed flow using a basic instrument called a *Pitot tube*. Pitot tubes are installed on virtually all airplanes to measure their flight velocity—something as important and basic as having the speedometer in your automobile. You will discover why a baseball curves when the pitcher puts spin on it, and why a golfball will sometimes hook or slice when a golfer hits it.

And there is a lot more. In the category of fundamentals, you will be introduced to Bernoulli's equation, far and away the most famous equation in fluid dynamics; it relates velocity and pressure from one point to another in an inviscid, incompressible flow. You will learn how to calculate the flow over a circular body shape, and how to calculate and plot the precise streamline shapes for this flow—something that is kind of fun to do, and that has importance further down the line in our study of aerodynamics.

So this chapter is a mixture of fundamentals and applications. Add these fundamentals to your growing inventory of basic concepts, and enjoy the applications.

3.1 INTRODUCTION AND ROAD MAP

The world of practical aviation was born on December 17, 1903, when, at 10:35 A.M., and in the face of cold, stiff, dangerous winds, Orville Wright piloted the Wright Flyer on its historic 12-s, 120-ft first flight. Figure 3.1 shows a photograph of the Wright Flyer at the instant of lift-off, with Wilbur Wright running along the right side of the machine, supporting the wing tip so that it will not drag the sand. This photograph is the most important picture in aviation history; the event it depicts launched the profession of aeronautical engineering into the mainstream of the twentieth century.[1]

The flight velocity of the Wright Flyer was about 30 mi/h. Over the ensuing decades, the flight velocities of airplanes steadily increased. By means of more powerful engines and attention to drag reduction, the flight velocities of airplanes rose to approximately 300 mi/h just prior to World War II. Figure 3.2 shows a typical fighter airplane of the immediate pre-World War II era. From an aerodynamic point of view, at air velocities between 0 and 300 mi/h the air density remains essentially constant, varying by only a few percent. Hence, the aerodynamics of the family of airplanes spanning the period between the two photographs shown in Figures 3.1 and 3.2 could be described by *incompressible flow*. As a result, a huge bulk of experimental and theoretical aerodynamic results was acquired over the 40-year period beginning with the Wright Flyer—results that applied to incompressible flow. Today, we are still very interested in incompressible aerodynamics because most modern general aviation aircraft still fly at speeds below 300 mi/h; a typical light general aviation airplane is shown in Figure 3.3. In addition to low-speed aeronautical applications, the principles of incompressible flow apply to the flow of fluids, for example, water flow through pipes, the motion of submarines and ships through the ocean, the design of wind turbines (the modern term for windmills), and many other important applications.

---

[1] See Reference 2 for historical details leading to the first flight by the Wright brothers.

Figure 3.1 Historic photograph of the first successful heavier-than-air powered crewed flight, achieved by the Wright brothers on December 17, 1903 (*NASA*).

Figure 3.2 The Seversky P-35 (*U.S. Air Force Photo*).

For all the above reasons, the study of incompressible flow is as relevant today as it was at the time of the Wright brothers. Therefore, Chapters 3 to 6 deal exclusively with incompressible flow. Moreover, for the most part, we ignore any effects of friction, thermal conduction, or diffusion; that is, we deal with *inviscid* incompressible flow in these chapters.[2] Looking at our spectrum of aerodynamic

[2] An inviscid, incompressible fluid is sometimes called an *ideal fluid,* or *perfect fluid.* This terminology will not be used here because of the confusion it sometimes causes with "ideal gases" or "perfect gases" from thermodynamics. This author prefers to use the more precise descriptor "inviscid, incompressible flow," rather than ideal fluid or perfect fluid.

Figure 3.3 Cessna 340 (*Philip Nealey/Photodisc/Getty Images*).

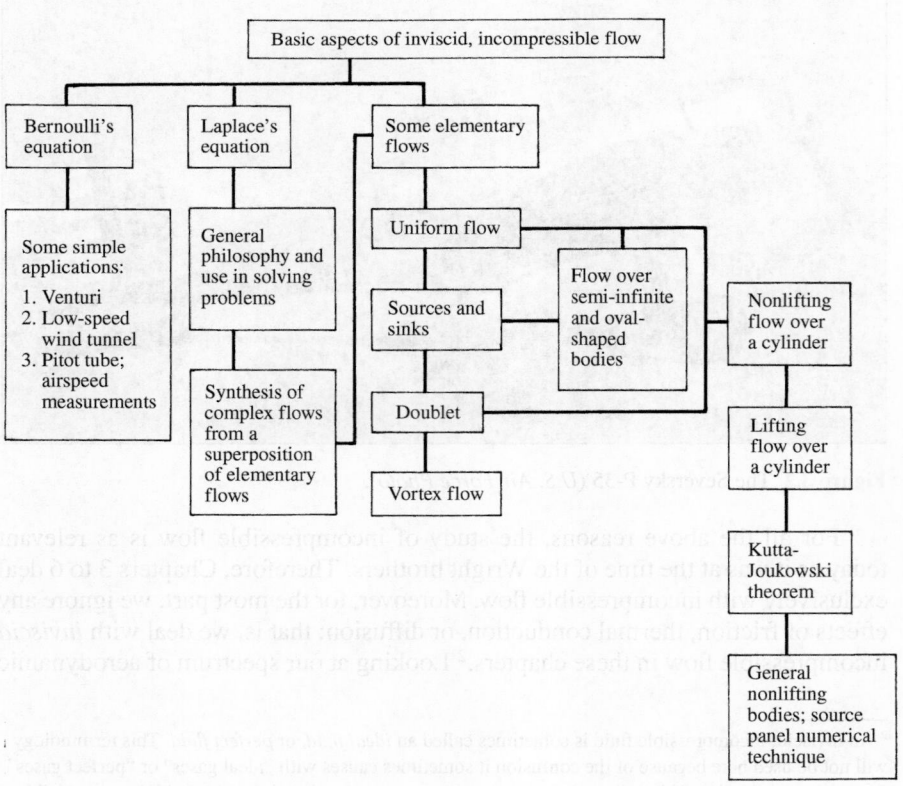

Figure 3.4 Road map for Chapter 3.

flows as shown in Figure 1.44, the material contained in Chapters 3 to 6 falls within the combined blocks D and E.

The purpose of this chapter is to establish some fundamental relations applicable to inviscid, incompressible flows and to discuss some simple but important flow fields and applications. The material in this chapter is then used as a launching pad for the airfoil theory of Chapter 4 and the finite wing theory of Chapter 5.

A road map for this chapter is given in Figure 3.4. There are three main avenues: (1) a development of Bernoulli's equation, with some straightforward applications; (2) a discussion of Laplace's equation, which is the governing equation for inviscid, incompressible, irrotational flow; and (3) the presentation of some elementary flow patterns, how they can be superimposed to synthesize both the nonlifting and lifting flow over a circular cylinder, and how they form the basis of a general numerical technique, called the *panel technique,* for the solution of flows over bodies of general shape. As you progress through this chapter, occasionally refer to this road map so that you can maintain your orientation and see how the various sections are related.

3.2 BERNOULLI'S EQUATION

As will be portrayed in Section 3.19, the early part of the eighteenth century saw the flowering of theoretical fluid dynamics, paced by the work of Johann and Daniel Bernoulli and, in particular, by Leonhard Euler. It was at this time that the relation between pressure and velocity in an inviscid, incompressible flow was first understood. The resulting equation is

$$p + \tfrac{1}{2}\rho V^2 = \text{const}$$

This equation is called *Bernoulli's equation,* although it was first presented in the above form by Euler (see Section 3.19). Bernoulli's equation is probably the most famous equation in fluid dynamics, and the purpose of this section is to derive it from the general equations discussed in Chapter 2.

Consider the x component of the momentum equation given by Equation (2.113a). For an inviscid flow with no body forces, this equation becomes

$$\rho \frac{Du}{Dt} = -\frac{\partial p}{\partial x}$$

or
$$\rho \frac{\partial u}{\partial t} + \rho u \frac{\partial u}{\partial x} + \rho v \frac{\partial u}{\partial y} + \rho w \frac{\partial u}{\partial z} = -\frac{\partial p}{\partial x} \tag{3.1}$$

For steady flow, $\partial u/\partial t = 0$. Equation (3.1) is then written as

$$u \frac{\partial u}{\partial x} + v \frac{\partial u}{\partial y} + w \frac{\partial u}{\partial z} = -\frac{1}{\rho} \frac{\partial p}{\partial x} \tag{3.2}$$

Multiply Equation (3.2) by dx:

$$u\frac{\partial u}{\partial x}dx + v\frac{\partial u}{\partial y}dx + w\frac{\partial u}{\partial z}dx = -\frac{1}{\rho}\frac{\partial p}{\partial x}dx \tag{3.3}$$

Consider the flow along a streamline in three-dimensional space. The equation of a streamline is given by Equation (2.117a to c). In particular, substituting

$$u\,dz - w\,dx = 0 \tag{2.117b}$$

and

$$v\,dx - u\,dy = 0 \tag{2.117c}$$

into Equation (3.3), we have

$$u\frac{\partial u}{\partial x}dx + u\frac{\partial u}{\partial y}dy + u\frac{\partial u}{\partial z}dz = -\frac{1}{\rho}\frac{\partial p}{\partial x}dx \tag{3.4}$$

or

$$u\left(\frac{\partial u}{\partial x}dx + \frac{\partial u}{\partial y}dy + \frac{\partial u}{\partial z}dz\right) = -\frac{1}{\rho}\frac{\partial p}{\partial x}dx \tag{3.5}$$

Recall from calculus that given a function $u = u(x, y, z)$, the differential of u is

$$du = \frac{\partial u}{\partial x}dx + \frac{\partial u}{\partial y}dy + \frac{\partial u}{\partial z}dz$$

This is exactly the term in parentheses in Equation (3.5). Hence, Equation (3.5) is written as

$$u\,du = -\frac{1}{\rho}\frac{\partial p}{\partial x}dx$$

or

$$\frac{1}{2}d(u^2) = -\frac{1}{\rho}\frac{\partial p}{\partial x}dx \tag{3.6}$$

In a similar fashion, starting from the y component of the momentum equation given by Equation (2.113b), specializing to an inviscid, steady flow, and applying the result to flow along a streamline, Equation (2.117a and c), we have

$$\frac{1}{2}d(v^2) = -\frac{1}{\rho}\frac{\partial p}{\partial y}dy \tag{3.7}$$

Similarly, from the z component of the momentum equation, Equation (2.113c), we obtain

$$\frac{1}{2}d(w^2) = -\frac{1}{\rho}\frac{\partial p}{\partial z}dz \tag{3.8}$$

Adding Equations (3.6) through (3.8) yields

$$\frac{1}{2}d(u^2 + v^2 + w^2) = -\frac{1}{\rho}\left(\frac{\partial p}{\partial x}dx + \frac{\partial p}{\partial y}dy + \frac{\partial p}{\partial z}dz\right) \tag{3.9}$$

However,

$$u^2 + v^2 + w^2 = V^2 \tag{3.10}$$

and

$$\frac{\partial p}{\partial x}dx + \frac{\partial p}{\partial y}dy + \frac{\partial p}{\partial z}dz = dp \tag{3.11}$$

Substituting Equations (3.10) and (3.11) into Equation (3.9), we have

$$\frac{1}{2}d(V^2) = -\frac{dp}{\rho}$$

or

$$\boxed{dp = -\rho V\,dV} \tag{3.12}$$

Equation (3.12) is called *Euler's equation*. It applies to an inviscid flow with no body forces, and it relates the change in velocity along a streamline dV to the change in pressure dp along the same streamline.

Equation (3.12) takes on a very special and important form for incompressible flow. In such a case, $\rho = $ constant, and Equation (3.12) can be easily integrated between any two points 1 and 2 along a streamline. From Equation (3.12), with $\rho = $ constant, we have

$$\int_{p_1}^{p_2} dp = -\rho \int_{V_1}^{V_2} V\,dV$$

or

$$p_2 - p_1 = -\rho\left(\frac{V_2^2}{2} - \frac{V_1^2}{2}\right)$$

or

$$\boxed{p_1 + \frac{1}{2}\rho V_1^2 = p_2 + \frac{1}{2}\rho V_2^2} \tag{3.13}$$

Equation (3.13) is *Bernoulli's equation,* which relates p_1 and V_1 at point 1 on a streamline to p_2 and V_2 at another point 2 on the same streamline. Equation (3.13) can also be written as

$$\boxed{p + \frac{1}{2}\rho V^2 = \text{const} \qquad \text{along a streamline}} \tag{3.14}$$

In the derivation of Equations (3.13) and (3.14), no stipulation has been made as to whether the flow is rotational or irrotational—these equations hold along a streamline in either case. For a general, rotational flow, the value of the constant in Equation (3.14) will change from one streamline to the next. However, if the flow is irrotational, then Bernoulli's equation holds between *any* two points in the flow, not necessarily just on the same streamline. For an irrotational flow, the constant in Equation (3.14) is the same for all streamlines, and

$$\boxed{p + \frac{1}{2}\rho V^2 = \text{const} \qquad \text{throughout the flow}} \tag{3.15}$$

The proof of this statement is given as Problem 3.1.

The physical significance of Bernoulli's equation is obvious from Equations (3.13) to (3.15); namely, *when the velocity increases, the pressure decreases, and when the velocity decreases, the pressure increases.*

Note that Bernoulli's equation was derived from the momentum equation; hence, it is a statement of Newton's second law for an inviscid, incompressible flow with no body forces. However, note that the dimensions of Equations (3.13) to (3.15) are energy per unit volume ($\frac{1}{2}\rho V^2$ is the kinetic energy per unit volume). Hence, Bernoulli's equation is also a relation for mechanical energy in an incompressible flow; it states that the work done on a fluid by pressure forces is equal to the change in kinetic energy of the flow. Indeed, Bernoulli's equation can be derived from the general energy equation, such as Equation (2.114). This derivation is left to the reader. The fact that Bernoulli's equation can be interpreted as either Newton's second law or an energy equation simply illustrates that the energy equation is redundant for the analysis of inviscid, incompressible flow. For such flows, the continuity and momentum equations suffice. (You may wish to review the opening comments of Section 2.7 on this same subject.)

The strategy for solving most problems in inviscid, incompressible flow is as follows:

1. Obtain the velocity field from the governing equations. These equations, appropriate for an inviscid, incompressible flow, are discussed in detail in Sections 3.6 and 3.7.

2. Once the velocity field is known, obtain the corresponding pressure field from Bernoulli's equation.

However, before treating the general approach to the solution of such flows (Section 3.7), several applications of the continuity equation and Bernoulli's equation are made to flows in ducts (Section 3.3) and to the measurement of airspeed using a Pitot tube (Section 3.4).

EXAMPLE 3.1

Consider an airfoil in a flow at standard sea level conditions with a freestream velocity of 50 m/s. At a given point on the airfoil, the pressure is 0.9×10^5 N/m^2. Calculate the velocity at this point.

■ **Solution**

At standard sea level conditions, $\rho_\infty = 1.23$ kg/m^3 and $p_\infty = 1.01 \times 10^5$ N/m^2. Hence,

$$p_\infty + \frac{1}{2}\rho V_\infty^2 = p + \frac{1}{2}\rho V^2$$

$$V = \sqrt{\frac{2(p_\infty - p)}{\rho} + V_\infty^2} = \sqrt{\frac{2(1.01 - 0.9) \times 10^5}{1.23} + (50)^2}$$

$$\boxed{V = 142.8 \text{ m/s}}$$

EXAMPLE 3.2

Consider the inviscid, incompressible flow of air along a streamline. The air density along the streamline is 0.002377 slug/ft^3, which is standard atmospheric density at sea level. At point 1 on the streamline, the pressure and velocity are 2116 lb/ft^2 and 10 ft/s, respectively. Further downstream, at point 2 on the streamline, the velocity is 190 ft/s. Calculate the pressure at point 2. What can you say about the relative change in pressure from point 1 to point 2 compared to the corresponding change in velocity?

■ **Solution**
From Equation (3.13),

$$p_1 + \frac{1}{2}\rho V_1^2 = p_2 + \frac{1}{2}\rho V_2^2$$

Hence,

$$p_2 = p_1 + \frac{1}{2}\rho\left(V_1^2 - V_2^2\right)$$

$$p_2 = 2116 + \frac{1}{2}(0.002377)[(10)^2 - (190)^2]$$

$$= 2116 + \frac{1}{2}(0.002377)(100 - 36{,}100)$$

$$= 2116 - 42.8 = \boxed{2073.2 \text{ lb/ft}^2}$$

Note: As the flow velocity increases from 10 ft/s to 190 ft/s along the streamline, the pressure decreases from 2116 lb/ft^2 to 2073.1 lb/ft^2. This is a factor of 19 increase in velocity, for only a factor of 42.8/2116, or 0.02 decrease in pressure. Stated another way, only a 2 percent decrease in the pressure creates a 1900 percent increase in the flow velocity. This is an example of a general characteristic of low-speed flows—only a small change in pressure is required to obtain a large change in velocity. You can sense this in the weather around you. Only a small barometric change from one location to another can create a strong wind.

3.3 INCOMPRESSIBLE FLOW IN A DUCT: THE VENTURI AND LOW-SPEED WIND TUNNEL

Consider the flow through a duct, such as that sketched in Figure 3.5. In general, the duct will be a three-dimensional shape, such as a tube with elliptical or rectangular cross sections which vary in area from one location to another. The flow through such a duct is three-dimensional and, strictly speaking, should be analyzed by means of the full three-dimensional conservation equations derived in Chapter 2. However, in many applications, the variation of area $A = A(x)$ is moderate, and for such cases it is reasonable to *assume* that the flow-field properties are uniform across any cross section, and hence vary only in the x direction. In Figure 3.5, uniform flow is sketched at station 1, and another but different uniform flow is shown at station 2. Such flow, where the area changes as a function

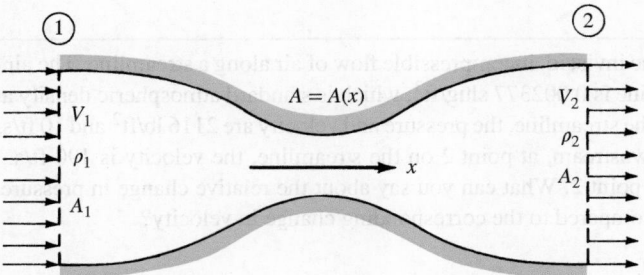

Figure 3.5 Quasi-one-dimensional flow in a duct.

of x and all the flow-field variables are assumed to be functions of x only, that is, $A = A(x)$, $V = V(x)$, $p = p(x)$, etc., is called *quasi-one-dimensional flow*. Although such flow is only an approximation of the truly three-dimensional flow in ducts, the results are sufficiently accurate for many aerodynamic applications. Such quasi-one-dimensional flow calculations are frequently used in engineering. They are the subject of this section.

Consider the integral form of the continuity equation written below:

$$\frac{\partial}{\partial t} \iiint\limits_{V} \rho \, d\mathcal{V} + \oiint\limits_{S} \rho \mathbf{V} \cdot \mathbf{dS} = 0 \qquad (2.48)$$

For steady flow, this becomes

$$\oiint\limits_{S} \rho \mathbf{V} \cdot \mathbf{dS} = 0 \qquad (3.16)$$

Apply Equation (3.16) to the duct shown in Figure 3.5, where the control volume is bounded by A_1 on the left, A_2 on the right, and the upper and lower walls of the duct. Hence, Equation (3.16) is

$$\iint\limits_{A_1} \rho \mathbf{V} \cdot \mathbf{dS} + \iint\limits_{A_2} \rho \mathbf{V} \cdot \mathbf{dS} + \iint\limits_{\text{wall}} \rho \mathbf{V} \cdot \mathbf{dS} = 0 \qquad (3.17)$$

Along the walls, the flow velocity is tangent to the wall. Since by definition $\mathbf{dS}$ is perpendicular to the wall, then along the wall, $\mathbf{V} \cdot \mathbf{dS} = 0$, and the integral over the wall surface is zero; that is, in Equation (3.17),

$$\iint\limits_{\text{wall}} \rho \mathbf{V} \cdot \mathbf{dS} = 0 \qquad (3.18)$$

At station 1, the flow is uniform across A_1. Noting that $\mathbf{dS}$ and $\mathbf{V}$ are in opposite directions at station 1 ($\mathbf{dS}$ always points *out* of the control volume by definition),

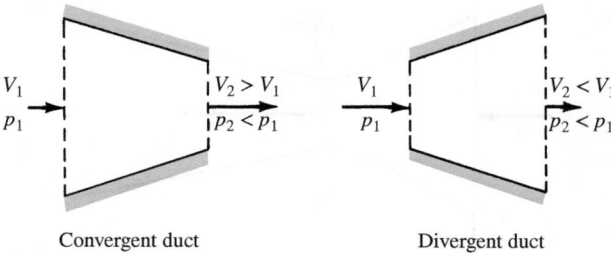

Convergent duct   Divergent duct

Figure 3.6 Incompressible flow in a duct.

we have in Equation (3.17)

$$\iint_{A_1} p\mathbf{V} \cdot \mathbf{dS} = -\rho_1 A_1 V_1 \qquad (3.19)$$

At station 2, the flow is uniform across A_2, and since $\mathbf{dS}$ and $\mathbf{V}$ are in the same direction, we have, in Equation (3.17),

$$\iint_{A_2} \rho\mathbf{V} \cdot \mathbf{dS} = \rho_2 A_2 V_2 \qquad (3.20)$$

Substituting Equations (3.18) to (3.20) into (3.17), we obtain

$$-\rho_1 A_1 V_1 + \rho_2 A_2 V_2 + 0 = 0$$

or

$$\boxed{\rho_1 A_1 V_1 = \rho_2 A_2 V_2} \qquad (3.21)$$

Equation (3.21) is the quasi-one-dimensional continuity equation; it applies to both compressible and incompressible flow.[3] In physical terms, it states that the mass flow through the duct is constant (i.e., what goes in must come out). Compare Equation (3.21) with Equation (2.43) for mass flow.

Consider *incompressible* flow only, where $\rho = $ constant. In Equation (3.21), $\rho_1 = \rho_2$, and we have

$$\boxed{A_1 V_1 = A_2 V_2} \qquad (3.22)$$

Equation (3.22) is the quasi-one-dimensional continuity equation for incompressible flow. In physical terms, it states that the volume flow (cubic feet per second or cubic meters per second) through the duct is constant. From Equation (3.22), we see that if the area decreases along the flow (convergent duct), the velocity increases; conversely, if the area increases (divergent duct), the velocity decreases. These variations are shown in Figure 3.6; they are fundamental consequences of

---

[3] For a simpler, more rudimentary derivation of Equation (3.21), see Chapter 4 of Reference 2. In the present discussion, we have established a more rigorous derivation of Equation (3.21), consistent with the general integral form of the continuity equation.

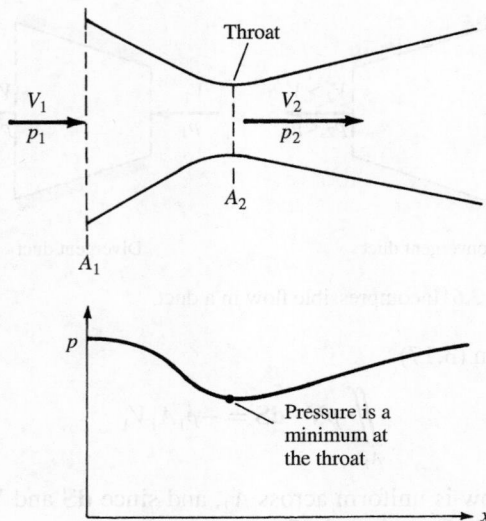

Figure 3.7 Flow through a venturi.

the incompressible continuity equation, and you should fully understand them. Moreover, from Bernoulli's equation, Equation (3.15), we see that when the velocity increases in a convergent duct, the pressure decreases; conversely, when the velocity decreases in a divergent duct, the pressure increases. These pressure variations are also shown in Figure 3.6.

Consider the incompressible flow through a convergent-divergent duct, shown in Figure 3.7. The flow enters the duct with velocity V_1 and pressure p_1. The velocity increases in the convergent portion of the duct, reaching a maximum value V_2 at the minimum area of the duct. This minimum area is called the *throat.* Also, in the convergent section, the pressure decreases, as sketched in Figure 3.7. At the throat, the pressure reaches a minimum value p_2. In the divergent section downstream of the throat, the velocity decreases and the pressure increases. The duct shown in Figure 3.7 is called a *venturi;* it is a device that finds many applications in engineering, and its use dates back more than a century. Its primary characteristic is that the pressure p_2 is lower at the throat than the ambient pressure p_1 outside the venturi. This pressure difference $p_1 - p_2$ is used to advantage in several applications. For example, in the carburetor of an automobile engine, there is a venturi through which the incoming air is mixed with fuel. The fuel line opens into the venturi at the throat. Because p_2 is less than the surrounding ambient pressure p_1, the pressure difference $p_1 - p_2$ helps to force the fuel into the airstream and mix it with the air downstream of the throat.

In an application closer to aerodynamics, a venturi can be used to measure airspeeds. Consider a venturi with a given inlet-to-throat area ratio A_1/A_2, as shown in Figure 3.7. Assume that the venturi is inserted into an airstream that has an unknown velocity V_1. We wish to use the venturi to measure this velocity. With regard to the venturi itself, the most direct quantity that can be measured is the pressure difference $p_1 - p_2$. This can be accomplished by placing a small hole (a pressure tap) in the wall of the venturi at both the inlet and the throat and connecting the pressure leads (tubes) from these holes across a differential pressure gage, or to both sides of a U-tube manometer (see Section 1.9). In such a fashion, the pressure difference $p_1 - p_2$ can be obtained directly. This measured pressure difference can be related to the unknown velocity V_1 as follows. From Bernoulli's equation, Equation (3.13), we have

$$V_1^2 = \frac{2}{\rho}(p_2 - p_1) + V_2^2 \tag{3.23}$$

From the continuity equation, Equation (3.22), we have

$$V_2 = \frac{A_1}{A_2}V_1 \tag{3.24}$$

Substituting Equation (3.24) into (3.23), we obtain

$$V_1^2 = \frac{2}{\rho}(p_2 - p_1) + \left(\frac{A_1}{A_2}\right)^2 V_1^2 \tag{3.25}$$

Solving Equation (3.25) for V_1, we obtain

$$V_1 = \sqrt{\frac{2(p_1 - p_2)}{\rho[(A_1/A_2)^2 - 1]}} \tag{3.26}$$

Equation (3.26) is the desired result; it gives the inlet air velocity V_1 in terms of the measured pressure difference $p_1 - p_2$ and the known density ρ and area ratio A_1/A_2. In this fashion, a venturi can be used to measure airspeeds. Indeed, historically the first practical airspeed indicator on an airplane was a venturi used by the French Captain A. Eteve in January 1911, more than 7 years after the Wright brothers' first powered flight. Today, the most common airspeed-measuring instrument is the Pitot tube (to be discussed in Section 3.4); however, the venturi is still found on some general aviation airplanes, including home-built and simple experimental aircraft.

Another application of incompressible flow in a duct is the low-speed wind tunnel. The desire to build ground-based experimental facilities designed to produce flows of air in the laboratory which simulate actual flight in the atmosphere dates back to 1871, when Francis Wenham in England built and used the first wind tunnel in history.[4] From that date to the mid-1930s, almost all wind tunnels

[4] For a discussion on the history of wind tunnels, see Chapter 4 of Reference 2.

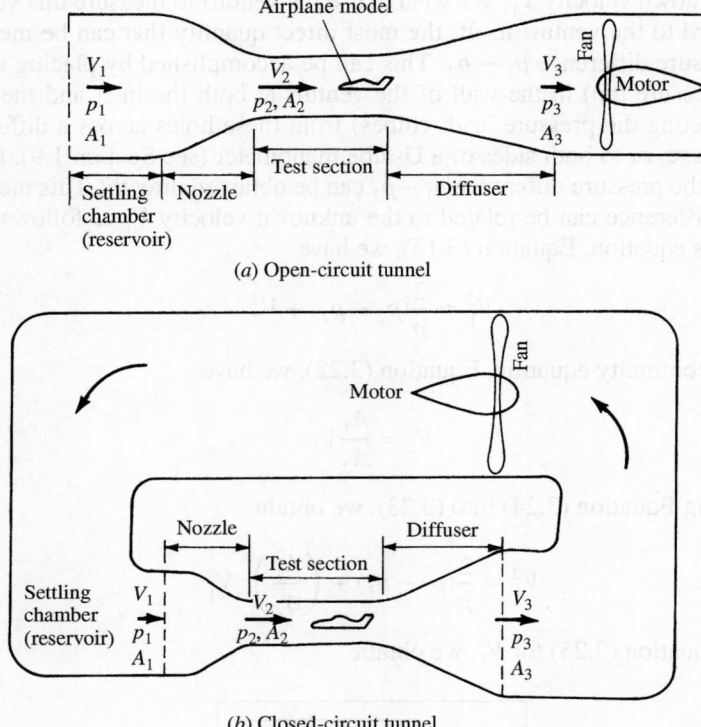

Airplane model

V_1 p_1 A_1

Settling chamber (reservoir)

Nozzle

Test section

V_2 p_2, A_2

Diffuser

V_3 p_3 A_3

Fan

Motor

(*a*) Open-circuit tunnel

Fan

Motor

Nozzle

Test section

Diffuser

Settling chamber (reservoir)

V_1 p_1 A_1

V_2

p_2, A_2

V_3 p_3 A_3

(*b*) Closed-circuit tunnel

Figure 3.8 (*a*) Open-circuit tunnel. (*b*) Closed-circuit tunnel.

were designed to produce airflows with velocities from 0 to 250 mi/h. Such low-speed wind tunnels are still much in use today, along with a complement of transonic, supersonic, and hypersonic tunnels. The principles developed in this section allow us to examine the basic aspects of low-speed wind tunnels, as follows.

In essence, a low-speed wind tunnel is a large venturi where the airflow is driven by a fan connected to some type of motor drive. The wind-tunnel fan blades are similar to airplane propellers and are designed to draw the airflow through the tunnel circuit. The wind tunnel may be open circuit, where the air is drawn in the front directly from the atmosphere and exhausted out the back, again directly to the atmosphere, as shown in Figure 3.8*a*; or the wind tunnel may be closed circuit, where the air from the exhaust is returned directly to the front of the tunnel via a closed duct forming a loop, as shown in Figure 3.8*b*. In either case, the airflow with pressure p_1 enters the nozzle at a low velocity V_1, where the area is A_1. The nozzle converges to a smaller area A_2 at the test section, where the velocity has increased to V_2 and the pressure has decreased to p_2. After flowing over an aerodynamic model (which may be a model of a complete airplane or part of an airplane such as a wing, tail, engine, or nacelle), the air passes into a diverging duct called a *diffuser,* where the area increases to A_3, the velocity decreases to V_3, and the pressure increases to p_3. From the continuity equation, Equation (3.22), the test-section air velocity is

$$V_2 = \frac{A_1}{A_2} V_1 \tag{3.27}$$

In turn, the velocity at the exit of the diffuser is

$$V_3 = \frac{A_2}{A_3} V_2 \tag{3.28}$$

The pressure at various locations in the wind tunnel is related to the velocity by Bernoulli's equation:

$$p_1 + \tfrac{1}{2}\rho V_1^2 = p_2 + \tfrac{1}{2}\rho V_2^2 = p_3 + \tfrac{1}{2}\rho V_3^2 \tag{3.29}$$

The basic factor that controls the air velocity in the test section of a given low-speed wind tunnel is the pressure difference $p_1 - p_2$. To see this more clearly, rewrite Equation (3.29) as

$$V_2^2 = \frac{2}{\rho}(p_1 - p_2) + V_1^2 \tag{3.30}$$

From Equation (3.27), $V_1 = (A_2/A_1)V_2$. Substituting into the right-hand side of Equation (3.30), we have

$$V_2^2 = \frac{2}{\rho}(p_1 - p_2) + \left(\frac{A_2}{A_1}\right)^2 V_2^2 \tag{3.31}$$

Solving Equation (3.31) for V_2, we obtain

$$\boxed{V_2 = \sqrt{\frac{2(p_1 - p_2)}{\rho[1 - (A_2/A_1)^2]}}} \tag{3.32}$$

The area ratio A_2/A_1 is a fixed quantity for a wind tunnel of given design. Moreover, the density is a known constant for incompressible flow. Therefore, Equation (3.32) demonstrates conclusively that the test-section velocity V_2 is governed by the pressure difference $p_1 - p_2$. The fan driving the wind-tunnel flow creates this pressure difference by doing work on the air. When the wind-tunnel operator turns the "control knob" of the wind tunnel and adjusts the power to the fan, he or she is essentially adjusting the pressure difference $p_1 - p_2$ and, in turn, adjusting the velocity via Equation (3.32).

In low-speed wind tunnels, a method of measuring the pressure difference $p_1 - p_2$, hence of measuring V_2 via Equation (3.32), is by means of a manometer as discussed in Section 1.9. In Equation (1.56), the density is the density of the liquid in the manometer (*not* the density of the air in the tunnel). The product of density and the acceleration of gravity g in Equation (1.56) is the weight per unit volume of the manometer fluid. Denote this weight per unit volume by w. Referring to Equation (1.56), if the side of the manometer associated with p_a is connected to a pressure tap in the settling chamber of the wind tunnel, where the pressure is p_1, and if the other side of the manometer (associated with p_b) is connected to a pressure tap in the test section, where the pressure is p_2, then, from Equation (1.56),

$$p_1 - p_2 = w\Delta h$$

where Δh is the difference in heights of the liquid between the two sides of the manometer. In turn, Equation (3.32) can be expressed as

$$V_2 = \sqrt{\frac{2w\Delta h}{\rho[1 - (A_2/A_1)^2]}}$$

The use of manometers is a straightforward mechanical means to measure pressures. They are time-honored devices that date back to their invention by the Italian mathematician Evangelista Torricelli in 1643. The French engineer Gustave Eiffel used manometers to measure pressures on the surface of wing models mounted in his wind tunnel in Paris in 1909, initiating the use of manometers in wind tunnel work throughout much of the twentieth century. Today, in most wind tunnels manometers have been replaced by an array of electronic pressure-measuring instruments. We discuss manometers here because they are part of the tradition of aerodynamics, and they are a good example of an application in fluid statics.

In many low-speed wind tunnels, the test section is vented to the surrounding atmosphere by means of slots in the wall; in others, the test section is not a duct at all, but rather, an open area between the nozzle exit and the diffuser inlet. In both cases, the pressure in the surrounding atmosphere is impressed on the test-section flow; hence, $p_2 = 1$ atm. (In subsonic flow, a jet that is dumped freely into the surrounding air takes on the same pressure as the surroundings; in contrast, a supersonic free jet may have completely different pressures than the surrounding atmosphere, as we see in Chapter 10.)

Keep in mind that the basic equations used in this section have certain limitations—we are assuming a quasi-one-dimensional inviscid flow. Such equations can sometimes lead to misleading results when the neglected phenomena are in reality important. For example, if $A_3 = A_1$ (inlet area of the tunnel is equal to the exit area), then Equations (3.27) and (3.28) yield $V_3 = V_1$. In turn, from

(*a*)

Figure 3.9 (*a*) Test section of a large subsonic wind tunnel (*NACA/NASA*).

Equation (3.29), $p_3 = p_1$; that is, there is no pressure difference across the entire tunnel circuit. If this were true, the tunnel would run without the application of any power—we would have a perpetual motion machine. In reality, there are losses in the airflow due to friction at the tunnel walls and drag on the aerodynamic model in the test section. Bernoulli's equation, Equation (3.29), does not take such losses into account. (Review the derivation of Bernoulli's equation in Section 3.2; note that viscous effects are neglected.) Thus, in an actual wind tunnel, there is a pressure loss due to viscous and drag effects, and $p_3 < p_1$. The function of the wind-tunnel motor and fan is to add power to the airflow in order to increase the pressure of the flow coming out of the diffuser so that it can be exhausted into the atmosphere (Figure 3.8*a*) or returned to the inlet of the nozzle at the higher pressure p_1 (Figure 3.8*b*). Photographs of a typical subsonic wind tunnel are shown in Figure 3.9*a* and *b*.

(b)

Figure 3.9 (*continued*) (*b*) The power fan drive section (*NASA*).

EXAMPLE 3.3

Consider a venturi with a throat-to-inlet area ratio of 0.8 mounted in a flow at standard sea level conditions. If the pressure difference between the inlet and the throat is 7 lb/ft^2, calculate the velocity of the flow at the inlet.

■ **Solution**

At standard sea level conditions, $\rho = 0.002377$ slug/ft^3. Hence,

$$V_1 = \sqrt{\frac{2(p_1 - p_2)}{\rho[(A_1/A_2)^2 - 1]}} = \sqrt{\frac{2(7)}{(0.002377)[(\frac{1}{0.8})^2 - 1]}} = \boxed{102.3 \text{ ft/s}}$$

EXAMPLE 3.4

Consider a low-speed subsonic wind tunnel with a 12/1 contraction ratio for the nozzle. If the flow in the test section is at standard sea level conditions with a velocity of 50 m/s, calculate the height difference in a U-tube mercury manometer with one side connected to the nozzle inlet and the other to the test section.

■ **Solution**

At standard sea level, $\rho = 1.23$ kg/m^3. From Equation (3.32),

$$p_1 - p_2 = \frac{1}{2}\rho V_2^2 \left[1 - \left(\frac{A_2}{A_1}\right)^2\right] = \frac{1}{2}(50)^2(1.23)\left[1 - \left(\frac{1}{12}\right)^2\right] = 1527 \text{ N/m}^2$$

However, $p_1 - p_2 = w\Delta h$. The density of liquid mercury is 1.36×10^4 kg/m^3. Hence,

$$w = (1.36 \times 10^4 \text{ kg/m}^3)(9.8 \text{ m/s}^2) = 1.33 \times 10^5 \text{ N/m}^2$$

$$\Delta h = \frac{p_1 - p_2}{w} = \frac{1527}{1.33 \times 10^5} = \boxed{0.01148 \text{ m}}$$

EXAMPLE 3.5

Consider a model of an airplane mounted in a subsonic wind tunnel, such as shown in Figure 3.10. The wind-tunnel nozzle has a 12-to-1 contraction ratio. The maximum lift coefficient of the airplane model is 1.3. The wing planform area of the model is 6 ft^2. The lift is measured with a mechanical balance that is rated at a maximum force of 1000 lb; that is, if the lift of the airplane model exceeds 1000 lb, the balance will be damaged. During a given test of this airplane model, the plan is to rotate the model through its whole range

Figure 3.10 Typical model installation in the test section of a large wind tunnel (*Jeff Caplan/NASA*).

of angle of attack, including up to that for maximum C_L. Calculate the maximum pressure difference allowable between the wind-tunnel settling chamber and the test section, assuming standard sea level density in the test section (i.e., $\rho_\infty = 0.002377$ slug/ft^3).

■ **Solution**

Maximum lift occurs when the model is at its maximum lift coefficient. Since the maximum allowable lift force is 1000 lb, the freestream velocity at which this occurs is obtained from

$$L_{max} = \tfrac{1}{2}\rho_\infty V_\infty^2 S C_{L,max}$$

or

$$V_\infty = \sqrt{\frac{2L_{max}}{\rho_\infty S C_{L,max}}} = \sqrt{\frac{(2)(1000)}{(0.002377)(6)(1.3)}} = 328.4 \text{ ft/s}$$

From Equation (3.32),

$$p_1 - p_2 = \frac{1}{2}\rho_\infty V_\infty^2 \left[1 - \left(\frac{A_2}{A_1}\right)^2\right]$$

$$= \tfrac{1}{2}(0.002377)(328.4)^2 \left[1 - \left(\frac{1}{12}\right)^2\right] = \boxed{127.3 \text{ lb/ft}^2}$$

EXAMPLE 3.6

a. The flow velocity in the test section of a low-speed subsonic wind tunnel is 100 mph. The test section is vented to the atmosphere, where atmospheric pressure is 1.01×10^5 N/m^2. The air density in the flow is the standard sea level value of 1.23 kg/m^3. The contraction ratio of the nozzle is 10-to-1. Calculate the reservoir pressure in atmospheres.
b. By how much must the reservoir pressure be increased to achieve 200 mph in the test section of this wind tunnel? Comment on the magnitude of this increase in pressure relative to the increase in test-section velocity.

■ **Solution**

a. Miles per hour is not a consistent unit for velocity. To convert to m/s, we note that 1 mi = 1609 m, and 1 h = 3600 s. Hence,

$$1\frac{\text{mi}}{\text{h}} = \left(1\frac{\text{mi}}{\text{h}}\right)\left(\frac{1609 \text{ m}}{1 \text{ mi}}\right)\left(\frac{1 \text{ h}}{3600 \text{ s}}\right) = 0.447 \text{ m/s}$$

[*Note:* This is the author's iron-clad method for carrying out conversion from one set of units to another. Take the original mile/hour, and multiply it by an equivalent "unity." Since 1609 m is the same distance as 1 mi, then the ratio (1609 m/1 mi) is essentially "unity," and since 1 h is the same time as 3600 s, then the ratio (1 h/3600 s) is essentially "unity." Multiplying (mi/h) by these two equivalent unity ratios, the miles cancel, and the hours cancel, and we are left with the proper number of meters per second.] Therefore,

$$V_2 = 100\,\text{mph} = (100)(0.447) = 44.7\,\text{m/s}$$

From Equation (3.31),

$$p_1 - p_2 = \frac{\rho}{2}V_2^2\left[1 - \left(\frac{A_2}{A_1}\right)^2\right]$$

$$p_1 - p_2 = \frac{1.23}{2}(44.7)^2\left[1 - \left(\frac{1}{10}\right)^2\right] = 0.01217 \times 10^5\text{N/m}^2$$

Thus,

$$p_1 = p_2 + 0.01217 \times 10^5 = 1.01 \times 10^5 + 0.01217 \times 10^5$$

$$p_1 = \boxed{1.022 \times 10^5 \text{ N/m}^2}$$

In atm, $p_1 = 1.022 \times 10^5/1.01 \times 10^5 = \boxed{1.01 \text{ atm}}$

b. $V_2 = 200\,\text{mph} = (200)(0.447) = 89.4$ m/s

$$p_1 - p_2 = \frac{\rho}{2}V_2^2\left[1 - \left(\frac{A_2}{A_1}\right)^2\right] = \frac{1.23}{2}(89.4)^2(0.99) = 0.0487 \times 10^5 \text{ N/m}^2$$

$$p_1 = 1.01 \times 10^5 + 0.0487 \times 10^5 = \boxed{1.059 \times 10^5 \text{ N/m}^2}$$

In atm,

$$p_1 = 1.059 \times 10^5/1.01 \times 10^5 = \boxed{1.048 \text{ atm}}$$

Comparing this result with part (a) above, we observe that to achieve a doubling of the test-section flow velocity from 100 mph to 200 mph, the reservoir pressure needed to be increased by only 0.038 atm (i.e., by 3.8 percent). This reinforces the general trend noted in Example 3.2, namely, that in a low-speed flow, a small pressure change results in a large velocity change.

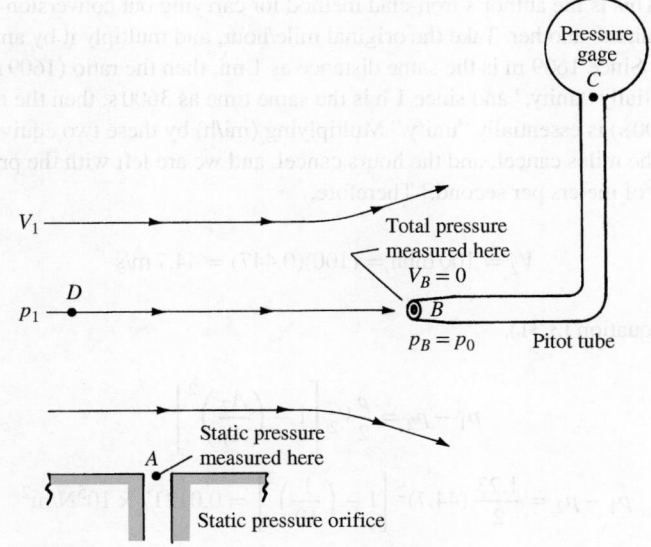

Figure 3.11 Pitot tube and a static pressure orifice.

3.4 PITOT TUBE: MEASUREMENT OF AIRSPEED

In 1732, the French engineer Henri Pitot was busy trying to measure the flow velocity of the Seine River in Paris. One of the instruments he used was his own invention—a strange-looking tube bent into an L shape, as shown in Figure 3.11. Pitot oriented one of the open ends of the tube so that it faced directly into the flow. In turn, he used the pressure inside this tube to measure the water flow velocity. This was the first time in history that a proper measurement of fluid velocity was made, and Pitot's invention has carried through to the present day as the *Pitot tube*—one of the most common and frequently used instruments in any modern aerodynamic laboratory. Moreover, a Pitot tube is the most common device for measuring flight velocities of airplanes. The purpose of this section is to describe the basic principle of the Pitot tube.[5]

Consider a flow with pressure p_1 moving with velocity V_1, as sketched at the left of Figure 3.11. Let us consider the significance of the pressure p_1 more closely. In Section 1.4, the pressure is associated with the time rate of change of momentum of the gas molecules impacting on or crossing a surface; that is, pressure is clearly related to the motion of the molecules. This motion is very random, with molecules moving in all directions with various velocities. Now imagine that you hop on a fluid element of the flow and ride with it at the velocity V_1. The gas molecules, because of their random motion, will still bump

---

[5] See Chapter 4 of Reference 2 for a detailed discussion of the history of the Pitot tube, how Pitot used it to overturn a basic theory in civil engineering, how it created some controversy in engineering, and how it finally found application in aeronautics.

into you, and you will feel the pressure p_1 of the gas. We now give this pressure a specific name: the *static* pressure. Static pressure is a measure of the purely random motion of molecules in the gas; it is the pressure you feel when you ride along with the gas at the local flow velocity. All pressures used in this book so far have been static pressures; the pressure p appearing in all our previous equations has been the static pressure. In engineering, whenever a reference is made to "pressure" without further qualification, that pressure is always interpreted as the *static* pressure. Furthermore, consider a boundary of the flow, such as a wall, where a small hole is drilled perpendicular to the surface. The plane of the hole is parallel to the flow, as shown at point A in Figure 3.11. Because the flow moves over the opening, the pressure felt at point A is due only to the random motion of the molecules; that is, at point A, the static pressure is measured. Such a small hole in the surface is called a *static pressure orifice,* or a *static pressure tap.*

In contrast, consider that a Pitot tube is now inserted into the flow, with an open end facing directly into the flow. That is, the plane of the opening of the tube is perpendicular to the flow, as shown at point B in Figure 3.11. The other end of the Pitot tube is connected to a pressure gage, such as point C in Figure 3.11 (i.e., the Pitot tube is closed at point C). For the first few milliseconds after the Pitot tube is inserted into the flow, the gas will rush into the open end and will fill the tube. However, the tube is closed at point C; there is no place for the gas to go, and hence after a brief period of adjustment, the gas inside the tube will stagnate; that is, the gas velocity inside the tube will go to zero. Indeed, the gas will eventually pile up and stagnate *everywhere* inside the tube, including at the open mouth at point B. As a result, the streamline of the flow that impinges directly at the open face of the tube (streamline DB in Figure 3.11) sees this face as an obstruction to the flow. The fluid elements along streamline DB slow down as they get closer to the Pitot tube and go to zero velocity right at point B. Any point in a flow where $V = 0$ is called a *stagnation point* of the flow; hence, point B at the open face of the Pitot tube is a stagnation point, where $V_B = 0$. In turn, from Bernoulli's equation we know the pressure increases as the velocity decreases. Hence, $p_B > p_1$. The pressure at a stagnation point is called the *stagnation* pressure, or *total* pressure, denoted by p_0. Hence, at point B, $p_B = p_0$.

From the above discussion, we see that two types of pressure can be defined for a given flow: static pressure, which is the pressure you feel by moving with the flow at its local velocity V_1, and total pressure, which is the pressure that the flow achieves when the velocity is reduced to zero. In aerodynamics, the distinction between total and static pressure is important; we have discussed this distinction at some length, and you should make yourself comfortable with the above paragraphs before proceeding further. (Further elaboration on the meaning and significance of total and static pressure will be made in Chapter 7.)

How is the Pitot tube used to measure flow velocity? To answer this question, first note that the total pressure p_0 exerted by the flow at the tube inlet (point B) is impressed throughout the tube (there is no flow inside the tube; hence, the pressure everywhere inside the tube is p_0). Therefore, the pressure gage at point C reads p_0. This measurement, in conjunction with a measurement of the static pressure p_1 at point A, yields the difference between total and static pressure, $p_0 - p_1$,

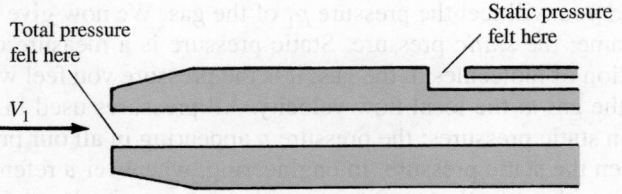

Figure 3.12 Pitot-static probe.

and it is this pressure *difference* that allows the calculation of V_1 via Bernoulli's equation. In particular, apply Bernoulli's equation between point A, where the pressure and velocity are p_1 and V_1, respectively, and point B, where the pressure and velocity are p_0 and $V = 0$, respectively:

$$p_A + \frac{1}{2}\rho V_A^2 = p_B + \frac{1}{2}\rho V_B^2$$

or

$$p_1 + \frac{1}{2}\rho V_1^2 = p_0 + 0 \qquad (3.33)$$

Solving Equation (3.33) for V_1, we have

$$\boxed{V_1 = \sqrt{\frac{2(p_0 - p_1)}{\rho}}} \qquad (3.34)$$

Equation (3.34) allows the calculation of velocity simply from the measured difference between total and static pressure. The total pressure p_0 is obtained from the Pitot tube, and the static pressure p_1 is obtained from a suitably placed static pressure tap.

It is possible to combine the measurement of both total and static pressure in one instrument, a *Pitot-static probe*, as sketched in Figure 3.12. A Pitot-static probe measures p_0 at the nose of the probe and p_1 at a suitably placed static pressure tap on the probe surface downstream of the nose.

In Equation (3.33), the term $\frac{1}{2}\rho V_1^2$ is called the *dynamic pressure* and is denoted by the symbol q_1. The grouping $\frac{1}{2}\rho V^2$ is called the dynamic pressure by definition and is used in all flows, incompressible to hypersonic:

$$q \equiv \frac{1}{2}\rho V^2$$

However, for incompressible flow, the dynamic pressure has special meaning; it is precisely the difference between total and static pressure. Repeating Equation (3.33), we obtain

$$\underset{\substack{\text{static} \\ \text{pressure}}}{p_1} + \underset{\substack{\text{dynamic} \\ \text{pressure}}}{\frac{1}{2}\rho V_1^2} = \underset{\substack{\text{total} \\ \text{pressure}}}{p_0}$$

or

$$p_1 + q_1 = p_0$$

or

$$\boxed{q_1 = p_0 - p_1} \qquad (3.35)$$

It is important to keep in mind that Equation (3.35) comes from Bernoulli's equation, and thus holds for *incompressible flow only*. For compressible flow, where Bernoulli's equation is not valid, the pressure difference $p_0 - p_1$ is *not* equal to q_1. Moreover, Equation (3.34) is valid for incompressible flow only. The velocities of compressible flows, both subsonic and supersonic, can be measured by means of a Pitot tube, but the equations are different from Equation (3.34). (Velocity measurements in subsonic and supersonic compressible flows are discussed in Chapter 8.)

At this stage, it is important to repeat that Bernoulli's equation holds for incompressible flow only, and therefore any result derived from Bernoulli's equation also holds for incompressible flow only, such as Equations (3.26), (3.32), and (3.34). Experience has shown that some students when first introduced to aerodynamics seem to adopt Bernoulli's equation as the standard and tend to use it for all applications, including many cases where it is not valid. Hopefully, the repetitive warnings given above will squelch such tendencies.

EXAMPLE 3.7

An airplane is flying at standard sea level. The measurement obtained from a Pitot tube mounted on the wing tip reads 2190 lb/ft². What is the velocity of the airplane?

■ **Solution**

Standard sea level pressure is 2116 lb/ft². From Equation (3.34), we have

$$V_1 = \sqrt{\frac{2(p_0 - p_1)}{\rho}} = \sqrt{\frac{2(2190 - 2116)}{0.002377}} = \boxed{250 \text{ ft/s}}$$

EXAMPLE 3.8

In the wind-tunnel flow described in Example 3.5, a small Pitot tube is mounted in the flow just upstream of the model. Calculate the pressure measured by the Pitot tube for the same flow conditions as in Example 3.5.

■ **Solution**

From Equation (3.35),

$$p_0 = p_\infty + q_\infty = p_\infty + \frac{1}{2}\rho_\infty V_\infty^2$$
$$= 2116 + \frac{1}{2}(0.002377)(328.4)^2$$
$$= 2116 + 128.2 = \boxed{2244 \text{ lb/ft}^2}$$

Note in this example that the dynamic pressure is $\frac{1}{2}\rho_\infty V_\infty^2 = 128.2$ lb/ft². This is less than 1 percent larger than the pressure difference $(p_1 - p_2)$, calculated in Example 3.5, that is required to produce the test-section velocity in the wind tunnel. Why is $(p_1 - p_2)$ so close to the test-section dynamic pressure? *Answer:* Because the velocity in the settling chamber V_1 is so small that p_1 is close to the total pressure of the flow. Indeed, from Equation (3.22),

$$V_1 = \frac{A_2}{A_1}V_2 = \left(\frac{1}{12}\right)(328.4) = 27.3 \text{ ft/s}$$

Compared to the test-section velocity of 328.4 ft/s, V_1 is seen to be small. In regions of a flow where the velocity is finite but small, the local static pressure is close to the total pressure. (Indeed, in the limiting case of a fluid with zero velocity, the local static pressure is the same as the total pressure; here, the concepts of static pressure and total pressure are redundant. For example, consider the air in the room around you. Assuming the air is motionless, and assuming standard sea level conditions, the pressure is 2116 lb/ft^2, namely, 1 atm. Is this pressure a static pressure or a total pressure? *Answer:* It is *both*. By the definition of total pressure given in the present section, when the local flow velocity is itself zero, then the local static pressure and the local total pressure are exactly the same.)

EXAMPLE 3.9

Consider the P-35 shown in Figure 3.2 cruising at a standard altitude of 4 km. The pressure sensed by the Pitot tube on its right wing (as seen in Figure 3.2) is 6.7×10^4 N/m^2. At what velocity is the P-35 flying?

■ Solution

From Appendix D, at a standard altitude of 4 km, the freestream static pressure and density are 6.166×10^4 N/m^2 and 0.81935 kg/m^3, respectively. The Pitot tube measures the total pressure of 6.7×10^4 N/m^2. From Equation (3.34),

$$V_1 = \sqrt{\frac{2(p_0 - p_1)}{\rho}} = \sqrt{\frac{2(6.7 - 6.166) \times 10^4}{0.81935}} = \boxed{114.2 \text{ m/s}}$$

Note: From the conversion factor between miles per hour and m/s obtained in Example 3.6, we have

$$V_1 = \frac{114.2}{0.447} = 255 \text{ mph}$$

EXAMPLE 3.10

The P-35 in Example 3.9 experiences a certain dynamic pressure at its cruising speed of 114.2 m/s at an altitude of 4 km. Now assume the P-35 is flying at sea level. At what velocity must it fly at sea level to experience the same dynamic pressure?

■ Solution

At $V_1 = 114.2$ m/s and at a standard altitude of 4 km, where $\rho = 0.81935$ kg/m^3,

$$q_1 = \tfrac{1}{2}\rho V_1^2 = \tfrac{1}{2}(0.81935)(114.2)^2 = 5.343 \times 10^3 \text{ N/m}^2$$

For the airplane to experience the same dynamic pressure at sea level where $\rho = 1.23$ kg/m^3, its new velocity, V_e, must satisfy

$$q_1 = \tfrac{1}{2}\rho V_e^2$$

$$5.343 \times 10^3 = \tfrac{1}{2}(1.23)V_e^2$$

or,

$$V_e = \sqrt{\frac{2(5.343 \times 10^3)}{1.23}} = \boxed{93.2 \text{ m/s}}$$

In Example 3.10, the velocity of 93.2 m/s is the *equivalent airspeed*, V_e, of the airplane flying at an altitude of 4 km at a true airspeed of 114.2 m/s. The general definition of equivalent airspeed is as follows. Consider an airplane flying at some true airspeed at some altitude. Its *equivalent airspeed* at this condition is defined as the velocity at which it would have to fly at standard sea level to *experience the same dynamic pressure*. In Example 3.10, we have the P-35 flying at an altitude of 4 km at a true airspeed of 114.2 m/s, and simultaneously at an equivalent airspeed of 93.2 m/s.

DESIGN BOX

The configuration of the Pitot-static probe shown in Figure 3.12 is a schematic only. The design of an actual Pitot-static probe is an example of careful engineering, intended to provide as accurate an instrument as possible. Let us examine some of the overall features of Pitot-static probe design.

Above all, the probe should be a long, streamlined shape such that the surface pressure over a substantial portion of the probe is essentially equal to the freestream static pressure. Such a shape is given in Figure 3.13a. The head of the probe, the nose at which the total pressure is measured, is usually a smooth hemispherical shape in order to encourage smooth, streamlined flow downstream of the nose. The diameter of the tube is denoted by d. A number of static pressure taps are arrayed radially around the

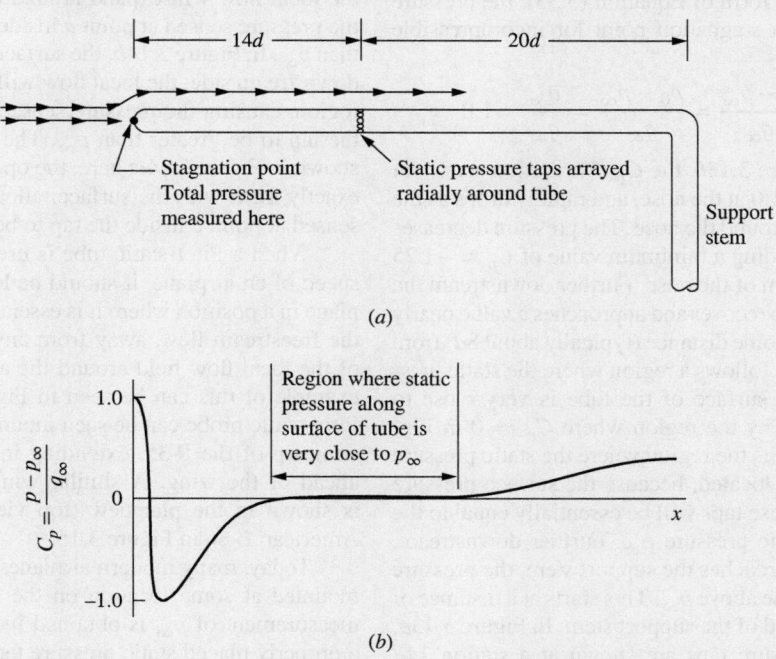

(a)

(b)

Figure 3.13 (a) Pitot-static tube. (b) Schematic of the pressure distribution along the outer surface of the tube.

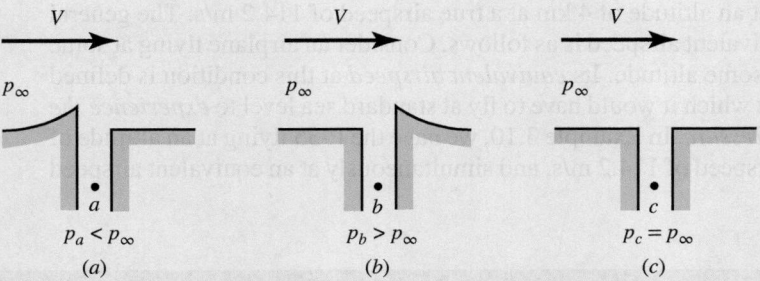

Figure 3.14 Schematic of static pressure taps. (a) and (b) Poor design.
(c) Proper design.

circumference of the tube at a station that should be from 8d to 16d downstream of the nose, and at least 16d ahead of the downstream support stem. The reason for this is shown in Figure 3.13b, which gives the axial distribution of the pressure coefficient along the surface of the tube. From the definition of pressure coefficient given in Section 1.5, and from Bernoulli's equation in the form of Equation (3.35), the pressure coefficient at a stagnation point for incompressible flow is given by

$$C_p = \frac{p - p_\infty}{q_\infty} = \frac{p_0 - p_\infty}{q_\infty} = \frac{q_\infty}{q_\infty} = 1.0$$

Hence, in Figure 3.13b, the C_p distribution starts out at the value of 1.0 at the nose, and rapidly drops as the flow expands around the nose. The pressure decreases below p_∞, yielding a minimum value of $C_p \approx -1.25$ just downstream of the nose. Further downstream the pressure tries to recover and approaches a value nearly equal to p_∞ at some distance (typically about 8d) from the nose. There follows a region where the static pressure along the surface of the tube is very close to p_∞, illustrated by the region where $C_p = 0$ in Figure 3.13b. This is the region where the static pressure taps should be located, because the surface pressure measured at these taps will be essentially equal to the freestream static pressure p_∞. Further downstream, as the flow approaches the support stem, the pressure starts to increase above p_∞. This starts at a distance of about 16d ahead of the support stem. In Figure 3.13a, the static pressure taps are shown at a station 14d downstream of the nose and 20d ahead of the support stem.

The design of the static pressure taps themselves is critical. The surface around the taps should be smooth to insure that the pressure sensed inside the tap is indeed the surface pressure along the tube. Examples of poor design as well as the proper design of the pressure taps are shown in Figure 3.14. In Figure 3.14a, the surface has a burr on the upstream side; the local flow will expand around this burr, causing the pressure sensed at point a inside the tap to be less than p_∞. In Figure 3.14b, the surface has a burr on the downstream side; the local flow will be slowed in this region, causing the pressure sensed at point b inside the tap to be greater than p_∞. The correct design is shown in Figure 3.14c; here, the opening of the tap is exactly flush with the surface, allowing the pressure sensed at point c inside the tap to be equal to p_∞.

When a Pitot-static tube is used to measure the speed of an airplane, it should be located on the airplane in a position where it is essentially immersed in the freestream flow, away from any major influence of the local flow field around the airplane itself. An example of this can be seen in Figure 3.2, where a Pitot-static probe can be seen mounted near the right wing tip of the P-35, extending into the freestream ahead of the wing. A similar wing-mounted probe is shown in the planview (top view) of the North American F-86 in Figure 3.15.

Today, many modern airplanes have a Pitot tube mounted at some location on the fuselage, and the measurement of p_∞ is obtained independently from a properly placed static pressure tap somewhere else on the fuselage. Figure 3.16 illustrates a fuselage-mounted Pitot tube in the nose region of the Boeing

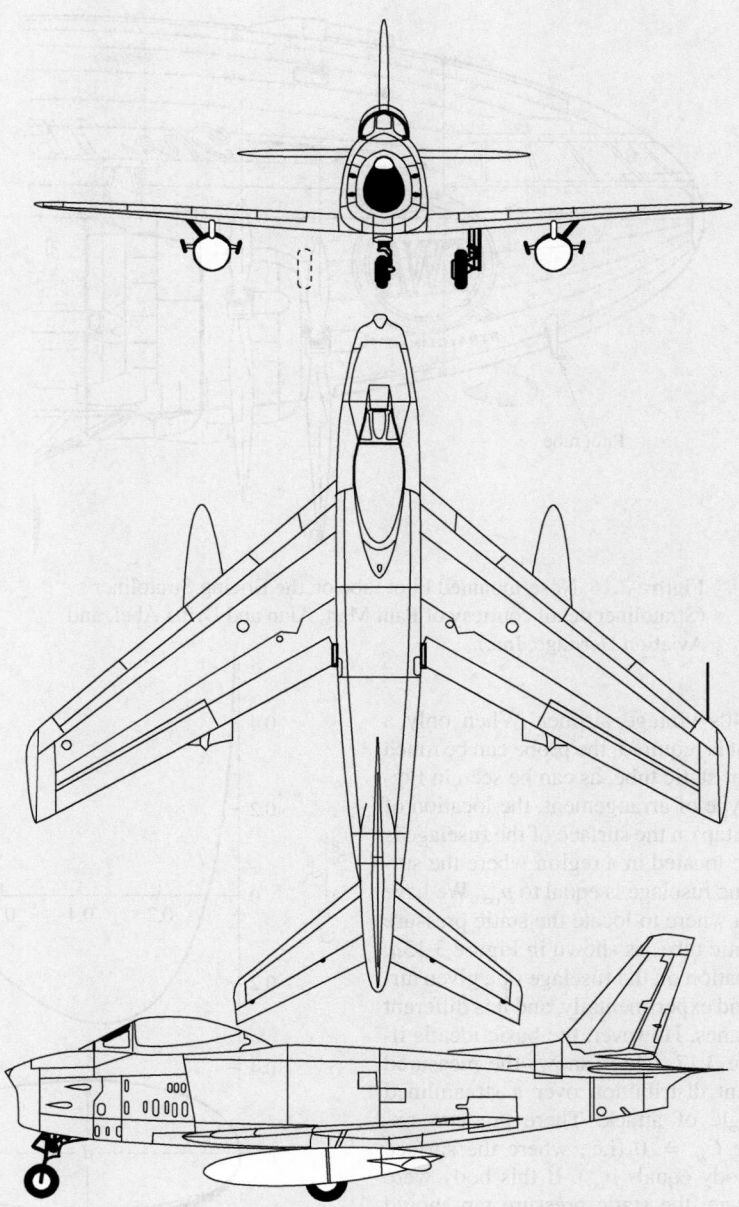

Figure 3.15 Three views of the North American F-86H. Note the wing-mounted Pitot-static tube.

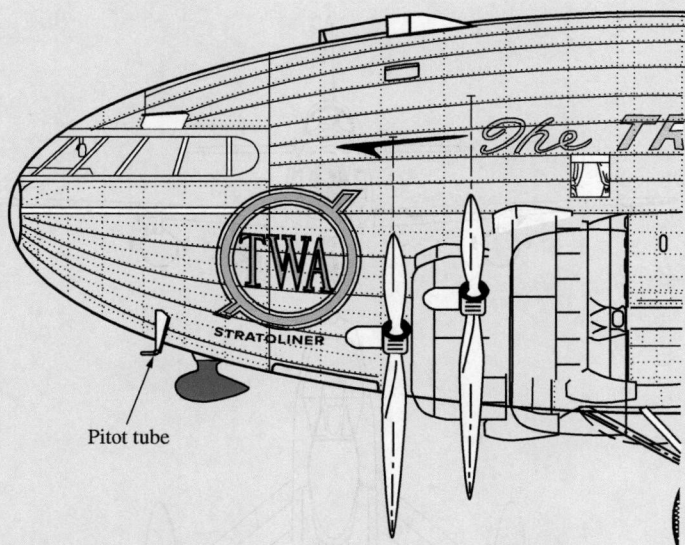

Figure 3.16 Nose-mounted Pitot tube on the Boeing Stratoliner (Stratoliner detail courtesy of Paul Matt, Alan and Drina Abel, and Aviation Heritage, Inc.).

Stratoliner, a 1940s vintage airliner. When only a Pitot measurement is required, the probe can be much shorter than a Pitot-static tube, as can be seen in Figure 3.16. In this type of arrangement, the location of the static pressure tap on the surface of the fuselage is critical; it must be located in a region where the surface pressure on the fuselage is equal to p_∞. We have a pretty good idea where to locate the static pressure taps on a Pitot-static tube, as shown in Figure 3.13*a*. But the proper location on the fuselage of a given airplane must be found experimentally, and it is different for different airplanes. However, the basic idea is illustrated in Figure 3.17, which shows the measured pressure coefficient distribution over a streamlined body at zero angle of attack. There are two axial stations where $C_p = 0$ (i.e., where the surface pressure on the body equals p_∞). If this body were an airplane fuselage, the static pressure tap should be placed at one of these two locations. In practice, the forward location, near the nose, is usually chosen.

Finally, we must be aware that none of these instruments, no matter where they are located, are perfectly accurate. In particular, misalignment of the

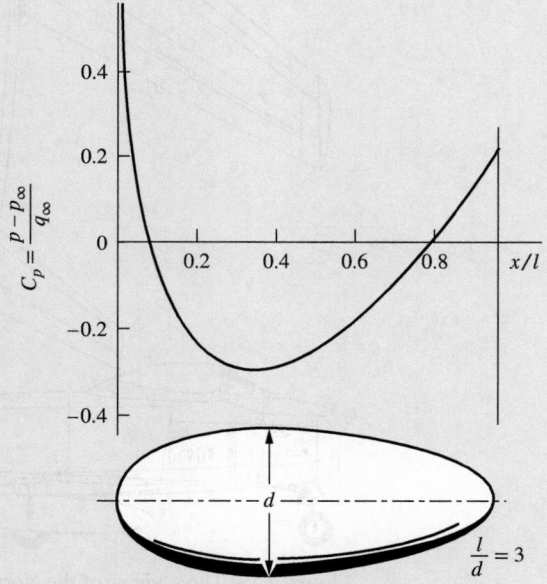

Figure 3.17 Experimentally measured pressure coefficient distribution over a streamlined body with a fineness ratio (length-to-diameter ratio) of 3. Zero angle of attack. Low-speed flow.

probe with respect to the freestream direction causes an error which must be assessed for each particular case. Fortunately, the measurement of the total pressure by means of a Pitot tube is relatively insensitive to misalignment. Pitot tubes with hemispherical noses, such as shown in Figure 3.13a, are insensitive to the mean flow direction up to a few degrees. Pitot tubes with flat faces, such as illustrated in Figure 3.12, are least sensitive. For these tubes, the total pressure measurement varies only 1 percent for misalignment as large as 20°. For more details on this matter, see Reference 61.

3.5 PRESSURE COEFFICIENT

Pressure, by itself, is a dimensional quantity (e.g., pounds per square foot, newtons per square meter). However, in Sections 1.7 and 1.8, we established the usefulness of certain dimensionless parameters such as M, Re, and C_L. It makes sense, therefore, that a dimensionless pressure would also find use in aerodynamics. Such a quantity is the *pressure coefficient* C_p, first introduced in Section 1.5 and defined as

$$C_p \equiv \frac{p - p_\infty}{q_\infty} \tag{3.36}$$

where

$$q_\infty = \tfrac{1}{2}\rho_\infty V_\infty^2$$

The definition given in Equation (3.36) is just that—a definition. It is used throughout aerodynamics, from incompressible to hypersonic flow. In the aerodynamic literature, it is very common to find pressures given in terms of C_p rather than the pressure itself. Indeed, the pressure coefficient is another similarity parameter that can be added to the list started in Sections 1.7 and 1.8.

For *incompressible flow*, C_p can be expressed in terms of velocity only. Consider the flow over an aerodynamic body immersed in a freestream with pressure p_∞ and velocity V_∞. Pick an arbitrary point in the flow where the pressure and velocity are p and V, respectively. From Bernoulli's equation,

$$p_\infty + \tfrac{1}{2}\rho V_\infty^2 = p + \tfrac{1}{2}\rho V^2$$

or

$$p - p_\infty = \tfrac{1}{2}\rho \left(V_\infty^2 - V^2 \right) \tag{3.37}$$

Substituting Equation (3.37) into (3.36), we have

$$C_p = \frac{p - p_\infty}{q_\infty} = \frac{\tfrac{1}{2}\rho \left(V_\infty^2 - V^2 \right)}{\tfrac{1}{2}\rho V_\infty^2}$$

or

$$C_p = 1 - \left(\frac{V}{V_\infty} \right)^2 \tag{3.38}$$

Equation (3.38) is a useful expression for the pressure coefficient; however, note that the form of Equation (3.38) holds for incompressible flow only.

Note from Equation (3.38) that the pressure coefficient at a stagnation point (where $V = 0$) in an incompressible flow is always equal to 1.0. This is the highest allowable value of C_p anywhere in the flow field. (For compressible flows, C_p at a stagnation point is greater than 1.0, as shown in Chapter 14.) Also, keep in mind that in regions of the flow where $V > V_\infty$ or $p < p_\infty$, C_p will be a negative value.

Another interesting property of the pressure coefficient can be seen by rearranging the definition given by Equation (3.36), as follows:

$$p = p_\infty + q_\infty C_p$$

Clearly, the value of C_p tells us how much p differs from p_∞ in multiples of the dynamic pressure. That is, if $C_p = 1$ (the value at a stagnation point in an incompressible flow), then $p = p_\infty + q_\infty$, or the local pressure is "one times" the dynamic pressure above freestream static pressure. If $C_p = -3$, then $p = p_\infty - 3q_\infty$, or the local pressure is three times the dynamic pressure below freestream static pressure.

EXAMPLE 3.11

Consider an airfoil in a flow with a freestream velocity of 150 ft/s. The velocity at a given point on the airfoil is 225 ft/s. Calculate the pressure coefficient at this point.

■ Solution

$$C_p = 1 - \left(\frac{V}{V_\infty} \right)^2 = 1 - \left(\frac{225}{150} \right)^2 = \boxed{-1.25}$$

EXAMPLE 3.12

Consider the airplane model in Example 3.4. When it is at a high angle of attack, slightly less than that when C_L becomes a maximum, the peak (negative) pressure coefficient which occurs at a certain point on the airfoil surface is -5.3. Assuming inviscid, incompressible flow, calculate the velocity at this point when (a) $V_\infty = 80$ ft/s and (b) $V_\infty = 300$ ft/s.

■ Solution
Using Equation (3.38), we have

a. $V = \sqrt{V_\infty^2(1 - C_p)} = \sqrt{(80)^2[1 - (-5.3)]} = \boxed{200.8 \text{ ft/s}}$

b. $V = \sqrt{V_\infty^2(1 - C_p)} = \sqrt{(300)^2[1 - (-5.3)]} = \boxed{753 \text{ ft/s}}$

The above example illustrates two aspects of such a flow, as follows:

1. Consider a given point on the airfoil surface. The C_p is *given* at this point and, from the statement of the problem. C_p is obviously *unchanged* when the velocity is increased from 80 to 300 ft/s. Why? The answer underscores part of our discussion on dimensional analysis in Section 1.7, namely, C_p

should depend only on the Mach number, Reynolds number, shape and orientation of the body, and location on the body. For the low-speed inviscid flow considered here, the Mach number and Reynolds number are not in the picture. For this type of flow, the variation of C_p is a function *only* of location on the surface of the body, and the body shape and orientation. Hence, C_p will *not* change with V_∞ or ρ_∞ as long as the flow can be considered inviscid and incompressible. For such a flow, once the C_p distribution over the body has been determined by some means, the *same* C_p distribution will exist for all freestream values of V_∞ and ρ_∞.

2. In part (*b*) of Example 3.12, the velocity at the point where C_p is a peak (negative) value is a large value, namely, 753 ft/s. Is Equation (3.38) valid for this case? The answer is essentially *no*. Equation (3.38) assumes incompressible flow. The speed of sound at standard sea level is 1117 ft/s; hence, the freestream Mach number is $300/1117 = 0.269$. A flow where the local Mach number is less than 0.3 can be assumed to be essentially incompressible. Hence, the freestream Mach number satisfies this criterion. On the other hand, the flow rapidly expands over the top surface of the airfoil and accelerates to a velocity of 753 ft/s at the point of minimum pressure (the point of peak negative C_p). In the expansion, the speed of sound *decreases*. (We will find out why in Part 3.) Hence, at the point of minimum pressure, the local Mach number is *greater* than $\frac{753}{1117} = 0.674$. That is, the flow has expanded to such a high local Mach number that it is no longer incompressible. Therefore, the answer given in part (*b*) of Example 3.12 is not correct. (We will learn how to calculate the correct value in Part 3.) There is an interesting point to be made here. Just because a model is being tested in a low-speed, subsonic wind tunnel, it does not mean that the assumption of incompressible flow will hold for all aspects of the flow field. As we see here, in some regions of the flow field around a body, the flow can achieve such high local Mach numbers that it must be considered as compressible.

3.6 CONDITION ON VELOCITY FOR INCOMPRESSIBLE FLOW

Consulting our chapter road map in Figure 3.4, we have completed the left branch dealing with Bernoulli's equation. We now begin a more general consideration of incompressible flow, given by the center branch in Figure 3.4. However, before introducing Laplace's equation, it is important to establish a basic condition on velocity in an incompressible flow, as follows.

First, consider the physical definition of incompressible flow, namely, $\rho =$ constant. Since ρ is the mass per unit volume and ρ is constant, then a fluid element of fixed mass moving through an incompressible flow field must also have a fixed, constant volume. Recall Equation (2.32), which shows that $\nabla \cdot \mathbf{V}$ is physically the time rate of change of the volume of a moving fluid element per unit volume. However, for an incompressible flow, we have just stated that the volume of a fluid element is constant [e.g., in Equation (2.32), $D(\delta\mathcal{V})/Dt \equiv 0$]. Therefore, for an incompressible flow,

$$\nabla \cdot \mathbf{V} = 0 \qquad (3.39)$$

The fact that the divergence of velocity is zero for an incompressible flow can also be shown directly from the continuity equation, Equation (2.52):

$$\frac{\partial \rho}{\partial t} + \nabla \cdot \rho\mathbf{V} = 0 \qquad (2.52)$$

For incompressible flow, $\rho = $ constant. Hence, $\partial\rho/\partial t = 0$ and $\nabla \cdot (\rho\mathbf{V}) = \rho\nabla \cdot \mathbf{V}$. Equation (2.52) then becomes

$$0 + \rho\nabla \cdot \mathbf{V} = 0$$

or

$$\nabla \cdot \mathbf{V} = 0$$

which is precisely Equation (3.39).

3.7 GOVERNING EQUATION FOR IRROTATIONAL, INCOMPRESSIBLE FLOW: LAPLACE'S EQUATION

We have seen in Section 3.6 that the principle of mass conservation for an incompressible flow can take the form of Equation (3.39):

$$\nabla \cdot \mathbf{V} = 0 \qquad (3.39)$$

In addition, for an irrotational flow we have seen in Section 2.15 that a velocity potential ϕ can be defined such that [from Equation (2.154)]

$$\mathbf{V} = \nabla\phi \qquad (2.154)$$

Therefore, for a flow that is both incompressible and irrotational, Equations (3.39) and (2.154) can be combined to yield

$$\nabla \cdot (\nabla\phi) = 0$$

or

$$\nabla^2\phi = 0 \qquad (3.40)$$

Equation (3.40) is *Laplace's equation*—one of the most famous and extensively studied equations in mathematical physics. Solutions of Laplace's equation are called *harmonic functions,* for which there is a huge bulk of existing literature. Therefore, it is most fortuitous that incompressible, irrotational flow is described by Laplace's equation, for which numerous solutions exist and are well understood.

For convenience, Laplace's equation is written below in terms of the three common orthogonal coordinate systems employed in Section 2.2:

Cartesian coordinates: $\quad \phi = \phi(x, y, z)$

$$\nabla^2\phi = \frac{\partial^2\phi}{\partial x^2} + \frac{\partial^2\phi}{\partial y^2} + \frac{\partial^2\phi}{\partial z^2} = 0 \qquad (3.41)$$

Cylindrical coordinates: $\quad \phi = \phi(r, \theta, z)$

$$\nabla^2\phi = \frac{1}{r}\frac{\partial}{\partial r}\left(r\frac{\partial\phi}{\partial r}\right) + \frac{1}{r^2}\frac{\partial^2\phi}{\partial\theta^2} + \frac{\partial^2\phi}{\partial z^2} = 0 \qquad (3.42)$$

Spherical coordinates: $\quad \phi = \phi(r, \theta, \Phi)$

$$\nabla^2\phi = \frac{1}{r^2\sin\theta}\left[\frac{\partial}{\partial r}\left(r^2\sin\theta\frac{\partial\phi}{\partial r}\right) + \frac{\partial}{\partial\theta}\left(\sin\theta\frac{\partial\phi}{\partial\theta}\right) + \frac{\partial}{\partial\Phi}\left(\frac{1}{\sin\theta}\frac{\partial\phi}{\partial\Phi}\right)\right] = 0 \qquad (3.43)$$

Recall from Section 2.14 that, for a two-dimensional incompressible flow, a stream function ψ can be defined such that, from Equation (2.150*a* and *b*),

$$u = \frac{\partial\psi}{\partial y} \qquad (2.150a)$$

$$v = -\frac{\partial\psi}{\partial x} \qquad (2.150b)$$

The continuity equation, $\nabla \cdot \mathbf{V} = 0$, expressed in cartesian coordinates, is

$$\nabla \cdot \mathbf{V} = \frac{\partial u}{\partial x} + \frac{\partial v}{\partial y} = 0 \qquad (3.44)$$

Substituting Equation (2.150*a* and *b*) into (3.44), we obtain

$$\frac{\partial}{\partial x}\left(\frac{\partial\psi}{\partial y}\right) + \frac{\partial}{\partial y}\left(-\frac{\partial\psi}{\partial x}\right) = \frac{\partial^2\psi}{\partial x\,\partial y} - \frac{\partial^2\psi}{\partial y\,\partial x} = 0 \qquad (3.45)$$

Since mathematically $\partial^2\psi/\partial x\,\partial y = \partial^2\psi/\partial y\,\partial x$, we see from Equation (3.45) that ψ automatically satisfies the continuity equation. Indeed, the very definition and use of ψ is a statement of the conservation of mass, and therefore Equation (2.150*a* and *b*) can be used in place of the continuity equation itself. If, in addition, the incompressible flow is irrotational, we have, from the irrotationality condition stated in Equation (2.131),

$$\frac{\partial v}{\partial x} - \frac{\partial u}{\partial y} = 0 \qquad (2.131)$$

Substituting Equation (2.150*a* and *b*) into (2.131), we have

$$\frac{\partial}{\partial x}\left(-\frac{\partial\psi}{\partial x}\right) - \frac{\partial}{\partial y}\left(\frac{\partial\psi}{\partial y}\right) = 0$$

or
$$\frac{\partial^2 \psi}{\partial x^2} + \frac{\partial^2 \psi}{\partial y^2} = 0 \qquad\qquad (3.46)$$

which is Laplace's equation. Therefore, the stream function also satisfies Laplace's equation, along with ϕ.

From Equations (3.40) and (3.46), we make the following obvious and important conclusions:

1. Any irrotational, incompressible flow has a velocity potential and stream function (for two-dimensional flow) that both satisfy Laplace's equation.
2. Conversely, any solution of Laplace's equation represents the velocity potential or stream function (two-dimensional) for an irrotational, incompressible flow.

Note that Laplace's equation is a second-order linear partial differential equation. The fact that it is *linear* is particularly important, because the sum of any particular solutions of a linear differential equation is also a solution of the equation. For example, if $\phi_1, \phi_2, \phi_3, \ldots, \phi_n$ represent n separate solutions of Equation (3.40), then the sum

$$\phi = \phi_1 + \phi_2 + \cdots + \phi_n$$

is also a solution of Equation (3.40). Since irrotational, incompressible flow is governed by Laplace's equation and Laplace's equation is linear, we conclude that a complicated flow pattern for an irrotational, incompressible flow can be synthesized by adding together a number of elementary flows that are also irrotational and incompressible. Indeed, this establishes the grand strategy for the remainder of our discussions on inviscid, incompressible flow. We develop flow-field solutions for several different elementary flows, which by themselves may not seem to be practical flows in real life. However, we then proceed to add (i.e., superimpose) these elementary flows in different ways such that the resulting flow fields do pertain to practical problems.

Before proceeding further, consider the irrotational, incompressible flow fields over different aerodynamic shapes, such as a sphere, cone, or airplane wing. Clearly, each flow is going to be distinctly different; the streamlines and pressure distribution over a sphere are quite different from those over a cone. However, these different flows are all governed by the same equation, namely, $\nabla^2 \phi = 0$. How, then, do we obtain different flows for the different bodies? The answer is found in the *boundary conditions*. Although the governing equation for the different flows is the same, the boundary conditions for the equation must conform to the different geometric shapes, and hence yield different flow-field solutions. Boundary conditions are therefore of vital concern in aerodynamic analysis. Let us examine the nature of boundary conditions further.

Consider the external aerodynamic flow over a stationary body, such as the airfoil sketched in Figure 3.18. The flow is bounded by (1) the freestream flow

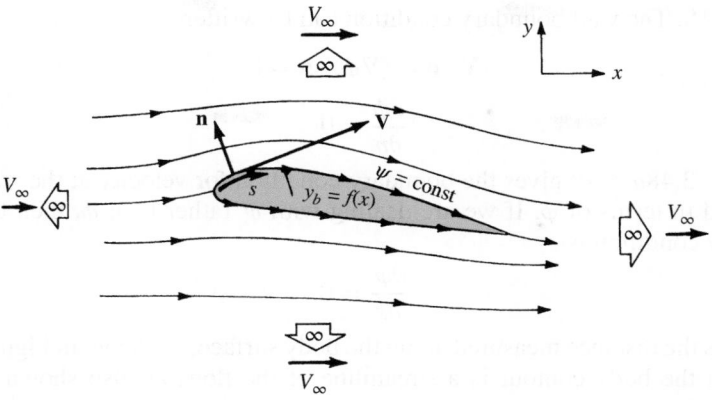

Figure 3.18 Boundary conditions at infinity and on a body; inviscid flow.

that occurs (theoretically) an infinite distance away from the body and (2) the surface of the body itself. Therefore, two sets of boundary conditions apply as follows.

3.7.1 Infinity Boundary Conditions

Far away from the body (toward infinity), in all directions, the flow approaches the uniform freestream conditions. Let V_∞ be aligned with the x direction as shown in Figure 3.18. Hence, at infinity,

$$u = \frac{\partial \phi}{\partial x} = \frac{\partial \psi}{\partial y} = V_\infty \qquad (3.47a)$$

$$v = \frac{\partial \phi}{\partial y} = \frac{\partial \psi}{\partial x} = 0 \qquad (3.47b)$$

Equation (3.47a and b) are the *boundary conditions on velocity at infinity*. They apply at an infinite distance from the body in all directions, above and below, and to the left and right of the body, as indicated in Figure 3.18.

3.7.2 Wall Boundary Conditions

If the body in Figure 3.18 has a solid surface, then it is impossible for the flow to penetrate the surface. Instead, if the flow is viscous, the influence of friction between the fluid and the solid surface creates a zero velocity at the surface. Such viscous flows are discussed in Chapters 15 to 20. In contrast, for inviscid flows the velocity at the surface can be finite, but because the flow cannot penetrate the surface, the velocity vector must be *tangent* to the surface. This "wall tangency" condition is illustrated in Figure 3.18, which shows **V** tangent to the body surface. If the flow is tangent to the surface, then the component of velocity *normal* to the surface must be zero. Let **n** be a unit vector normal to the surface as shown in

Figure 3.18. The wall boundary condition can be written as

$$\mathbf{V} \cdot \mathbf{n} = (\nabla \phi) \cdot \mathbf{n} = 0 \qquad (3.48a)$$

or

$$\frac{\partial \phi}{\partial n} = 0 \qquad (3.48b)$$

Equation (3.48a or b) gives the boundary condition for velocity at the wall; it is expressed in terms of ϕ. If we are dealing with ψ rather than ϕ, then the wall boundary condition is

$$\frac{\partial \psi}{\partial s} = 0 \qquad (3.48c)$$

where s is the distance measured along the body surface, as shown in Figure 3.18. Note that the body contour is a streamline of the flow, as also shown in Figure 3.18. Recall that $\psi =$ constant is the equation of a streamline. Thus, if the shape of the body in Figure 3.18 is given by $y_b = f(x)$, then

$$\psi_{\text{surface}} = \psi_{y=y_b} = \text{const} \qquad (3.48d)$$

is an alternative expression for the boundary condition given in Equation (3.48c).

If we are dealing with neither ϕ nor ψ, but rather with the velocity components u and v themselves, then the wall boundary condition is obtained from the equation of a streamline, Equation (2.118), evaluated at the body surface; that is,

$$\boxed{\frac{dy_b}{dx} = \left(\frac{v}{u}\right)_{\text{surface}}} \qquad (3.48e)$$

Equation (3.48e) states simply that the body surface is a streamline of the flow. The form given in Equation (3.48e) for the flow tangency condition at the body surface is used for all inviscid flows, incompressible to hypersonic, and does not depend on the formulation of the problem in terms of ϕ or ψ (or $\bar{\psi}$).

3.8 INTERIM SUMMARY

Reflecting on our previous discussions, the general approach to the solution of irrotational, incompressible flows can be summarized as follows:

1. Solve Laplace's equation for ϕ [Equation (3.40)] or ψ [Equation (3.46)] along with the proper boundary conditions [such as Equations (3.47) and (3.48)]. These solutions are usually in the form of a sum of elementary solutions (to be discussed in the following sections).
2. Obtain the flow velocity from $\mathbf{V} = \nabla \phi$ or $u = \partial \psi / \partial y$ and $v = -\partial \psi / \partial x$.
3. Obtain the pressure from Bernoulli's equation, $p + \frac{1}{2}\rho V^2 = p_\infty + \frac{1}{2}\rho V_\infty^2$, where p_∞ and V_∞ are known freestream conditions.

Since $\mathbf{V}$ and p are the primary dependent variables for an incompressible flow, steps 1 to 3 are all that we need to solve a given problem as long as the flow is incompressible and irrotational.

3.9 UNIFORM FLOW: OUR FIRST ELEMENTARY FLOW

In this section, we present the first of a series of elementary incompressible flows that later will be superimposed to synthesize more complex incompressible flows. For the remainder of this chapter and in Chapter 4, we deal with two-dimensional steady flows; three-dimensional steady flows are treated in Chapters 5 and 6.

Consider a uniform flow with velocity V_∞ oriented in the positive x direction, as sketched in Figure 3.19. It is easily shown (see Problem 3.8) that a uniform flow is a physically possible incompressible flow (i.e., it satisfies $\nabla \cdot \mathbf{V} = 0$) and that it is irrotational (i.e., it satisfies $\nabla \times \mathbf{V} = 0$). Hence, a velocity potential for uniform flow can be obtained such that $\nabla \phi = \mathbf{V}$. Examining Figure 3.19, and recalling Equation (2.156), we have

$$\frac{\partial \phi}{\partial x} = u = V_\infty \qquad (3.49a)$$

and

$$\frac{\partial \phi}{\partial y} = v = 0 \qquad (3.49b)$$

Integrating Equation (3.49a) with respect to x, we have

$$\phi = V_\infty x + f(y) \qquad (3.50)$$

where $f(y)$ is a function of y only. Integrating Equation (3.49b) with respect to y, we obtain

$$\phi = \text{const} + g(x) \qquad (3.51)$$

where $g(x)$ is a function of x only. In Equations (3.50) and (3.51), ϕ is the same function; hence, by comparing these equations, $g(x)$ must be $V_\infty x$, and $f(y)$ must be constant. Thus,

$$\phi = V_\infty x + \text{const} \qquad (3.52)$$

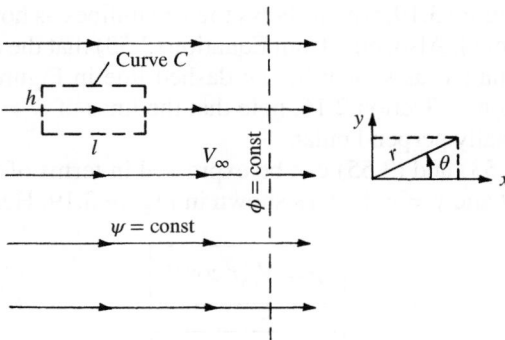

Figure 3.19 Uniform flow.

Note that in a practical aerodynamic problem, the actual value of ϕ is not significant; rather, ϕ is always used to obtain the velocity by differentiation; that is, $\nabla\phi = \mathbf{V}$. Since the derivative of a constant is zero, we can drop the constant from Equation (3.52) without any loss of rigor. Hence, Equation (3.52) can be written as

$$\phi = V_\infty x \tag{3.53}$$

Equation (3.53) is the velocity potential for a uniform flow with velocity V_∞ oriented in the positive x direction. Note that the derivation of Equation (3.53) does not depend on the assumption of incompressibility; it applies to any uniform flow, compressible or incompressible.

Consider the incompressible stream function ψ. From Figure 3.19 and Equation (2.150a and b), we have

$$\frac{\partial \psi}{\partial y} = u = V_\infty \tag{3.54a}$$

and

$$\frac{\partial \psi}{\partial x} = -v = 0 \tag{3.54b}$$

Integrating Equation (3.54a) with respect to y and Equation (3.54b) with respect to x, and comparing the results, we obtain

$$\psi = V_\infty y \tag{3.55}$$

Equation (3.55) is the stream function for an incompressible uniform flow oriented in the positive x direction.

From Section 2.14, the equation of a streamline is given by $\psi = $ constant. Therefore, from Equation (3.55), the streamlines for the uniform flow are given by $\psi = V_\infty y = $ constant. Because V_∞ is itself constant, the streamlines are thus given mathematically as $y = $ constant (i.e., as lines of constant y). This result is consistent with Figure 3.19, which shows the streamlines as horizontal lines (i.e., as lines of constant y). Also, note from Equation (3.53) that the equipotential lines are lines of constant x, as shown by the dashed line in Figure 3.19. Consistent with our discussion in Section 2.16, note that the lines of $\psi = $ constant and $\phi = $ constant are mutually perpendicular.

Equations (3.53) and (3.55) can be expressed in terms of polar coordinates, where $x = r \cos\theta$ and $y = r \sin\theta$, as shown in Figure 3.19. Hence,

$$\phi = V_\infty r \cos\theta \tag{3.56}$$

and

$$\psi = V_\infty r \sin\theta \tag{3.57}$$

Consider the circulation in a uniform flow. The definition of circulation is given by

$$\Gamma \equiv -\oint_C \mathbf{V} \cdot \mathbf{ds} \qquad (2.136)$$

Let the closed curve C in Equation (2.136) be the rectangle shown at the left of Figure 3.19; h and l are the lengths of the vertical and horizontal sides, respectively, of the rectangle. Then

$$\oint_C \mathbf{V} \cdot \mathbf{ds} = -V_\infty l - 0(h) + V_\infty l + 0(h) = 0$$

or

$$\Gamma = 0 \qquad (3.58)$$

Equation (3.58) is true for any arbitrary closed curve in the uniform flow. To show this, note that $\mathbf{V}_\infty$ is constant in both magnitude and direction, and hence

$$\Gamma = -\oint_C \mathbf{V} \cdot \mathbf{ds} = -\mathbf{V}_\infty \cdot \oint_C \mathbf{ds} = \mathbf{V}_\infty \cdot 0 = 0$$

because the line integral of $\mathbf{ds}$ around a closed curve is identically zero. Therefore, from Equation (3.58), we state that *circulation around any closed curve in a uniform flow is zero.*

The above result is consistent with Equation (2.137), which states that

$$\Gamma = -\iint_S (\nabla \times \mathbf{V}) \cdot \mathbf{dS} \qquad (2.137)$$

We stated earlier that a uniform flow is irrotational; that is, $\nabla \times \mathbf{V} = 0$ everywhere. Hence, Equation (2.137) yields $\Gamma = 0$.

Note that Equations (3.53) and (3.55) satisfy Laplace's equation [see Equation (3.41)], which can be easily proved by simple substitution. Therefore, uniform flow is a viable elementary flow for use in building more complex flows.

3.10 SOURCE FLOW: OUR SECOND ELEMENTARY FLOW

Consider a two-dimensional, incompressible flow where all the streamlines are straight lines emanating from a central point O, as shown at the left of Figure 3.20. Moreover, let the velocity along each of the streamlines vary inversely with distance from point O. Such a flow is called a *source flow*. Examining Figure 3.20, we see that the velocity components in the radial and tangential directions are V_r and V_θ, respectively, where $V_\theta = 0$. The coordinate system in Figure 3.20 is a cylindrical coordinate system, with the z axis perpendicular to the page. (Note that polar coordinates are simply the cylindrical coordinates r and θ confined to a single plane given by $z = $ constant.) It is easily shown (see Problem 3.9) that (1) source flow is a physically possible incompressible flow, that is, $\nabla \cdot \mathbf{V} = 0$, at every point *except* the origin, where $\nabla \cdot \mathbf{V}$ becomes infinite, and (2) source flow is *irrotational* at every point.

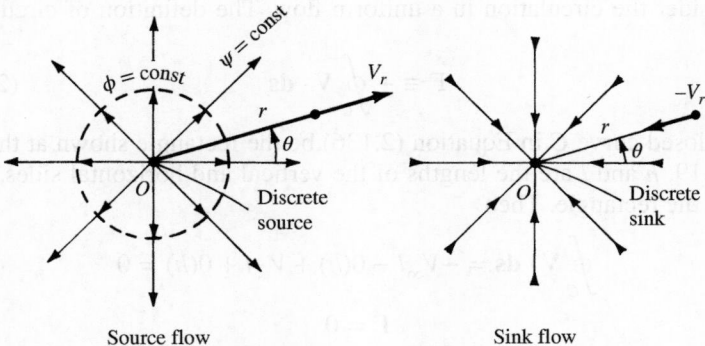

Source flow   Sink flow

Figure 3.20 Source and sink flows.

In a source flow, the streamlines are directed *away from* the origin, as shown at the left of Figure 3.20. The opposite case is that of a *sink flow,* where by definition the streamlines are directed *toward* the origin, as shown at the right of Figure 3.20. For sink flow, the streamlines are still radial lines from a common origin, along which the flow velocity varies inversely with distance from point O. Indeed, a sink flow is simply a negative source flow.

The flows in Figure 3.20 have an alternate, somewhat philosophical interpretation. Consider the origin, point O, as a *discrete* source or sink. Moreover, interpret the radial flow surrounding the origin as simply being *induced* by the presence of the discrete source or sink at the origin (much like a magnetic field is induced in the space surrounding a current-carrying wire). Recall that, for a source flow, $\nabla \cdot \mathbf{V} = 0$ everywhere except at the origin, where it is infinite. Thus, the origin is a *singular point,* and we can interpret this singular point as a discrete source or sink of a given strength, with a corresponding induced flow field about the point. This interpretation is very convenient and is used frequently. Other types of singularities, such as doublets and vortices, are introduced in subsequent sections. Indeed, the irrotational, incompressible flow field about an arbitrary body can be visualized as a flow induced by a proper distribution of such singularities over the surface of the body. This concept is fundamental to many theoretical solutions of incompressible flow over airfoils and other aerodynamic shapes, and it is the very heart of modern numerical techniques for the solution of such flows. You will obtain a greater appreciation for the concept of distributed singularities for the solution of incompressible flow in Chapters 4 through 6. At this stage, however, simply visualize a discrete source (or sink) as a singularity that induces the flows shown in Figure 3.20.

Let us look more closely at the velocity field induced by a source or sink. By definition, the velocity is inversely proportional to the radial distance r. As stated earlier, this velocity variation is a physically possible flow, because it yields $\nabla \cdot \mathbf{V} = 0$. Moreover, it is the *only* such velocity variation for which the relation

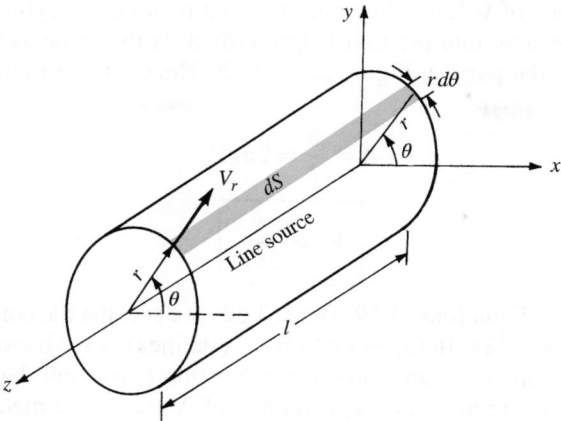

Figure 3.21 Volume flow rate from a line source.

$\nabla \cdot \mathbf{V} = 0$ is satisfied for the radial flows shown in Figure 3.20. Hence,

$$\mathbf{V}_r = \frac{c}{r} \tag{3.59a}$$

and

$$\mathbf{V}_\theta = 0 \tag{3.59b}$$

where c is constant. The value of the constant is related to the volume flow from the source, as follows. In Figure 3.20, consider a depth of length l perpendicular to the page, that is, a length l along the z axis. This is sketched in three-dimensional perspective in Figure 3.21. In Figure 3.21, we can visualize an entire line of sources along the z axis, of which the source O is just part. Therefore, in a two-dimensional flow, the discrete source, sketched in Figure 3.20, is simply a single point on the line source shown in Figure 3.21. The two-dimensional flow shown in Figure 3.20 is the same in any plane perpendicular to the z axis, that is, for any plane given by $z = $ constant. Consider the mass flow across the surface of the cylinder of radius r and height l as shown in Figure 3.21. The elemental mass flow across the surface element $\mathbf{dS}$ shown in Figure 3.21 is $\rho \mathbf{V} \cdot \mathbf{dS} = \rho V_r (r\,d\theta)(l)$. Hence, noting that V_r is the same value at any θ location for the fixed radius r, the total mass flow across the surface of the cylinder is

$$\dot{m} = \int_0^{2\pi} \rho V_r (r\,d\theta) l = \rho r l V_r \int_0^{2\pi} d\theta = 2\pi r l \rho V_r \tag{3.60}$$

Since ρ is defined as the mass per unit volume and $\dot{m}$ is mass per second, then $\dot{m}/\rho$ is the volume flow per second. Denote this rate of volume flow by $\dot{v}$. Thus, from Equation (3.60), we have

$$\dot{v} = \frac{\dot{m}}{\rho} = 2\pi r l V_r \tag{3.61}$$

Moreover, the rate of volume flow per unit length along the cylinder is $\dot{v}/l$. Denote this volume flow rate per unit length (which is the same as per unit depth perpendicular to the page in Figure 3.20) as Λ. Hence, from Equation 3.61, we obtain

$$\Lambda = \frac{\dot{v}}{l} = 2\pi r V_r$$

or

$$\boxed{V_r = \frac{\Lambda}{2\pi r}} \tag{3.62}$$

Hence, comparing Equations (3.59a) and (3.62), we see that the constant in Equation (3.59a) is $c = \Lambda/2\pi$. In Equation (3.62), Λ defines the *source strength*, which is physically the rate of volume flow from the source, per unit depth perpendicular to the page of Figure 3.20. Typical units of Λ are square meters per second or square feet per second. In Equation (3.62), a positive value of Λ represents a source, whereas a negative value represents a sink.

The velocity potential for a source can be obtained as follows. From Equations (2.157), (3.59b), and (3.62),

$$\frac{\partial \phi}{\partial r} = V_r = \frac{\Lambda}{2\pi r} \tag{3.63}$$

and

$$\frac{1}{r}\frac{\partial \phi}{\partial \theta} = V_\theta = 0 \tag{3.64}$$

Integrating Equation (3.63) with respect to r, we have

$$\phi = \frac{\Lambda}{2\pi}\ln r + f(\theta) \tag{3.65}$$

Integrating Equation (3.64) with respect to θ, we have

$$\phi = \text{const} + f(r) \tag{3.66}$$

Comparing Equations (3.65) and (3.66), we see that $f(r) = (\Lambda/2\pi)\ln r$ and $f(\theta) = $ constant. As explained in Section 3.9, the constant can be dropped without loss of rigor, and hence Equation (3.65) yields

$$\boxed{\phi = \frac{\Lambda}{2\pi}\ln r} \tag{3.67}$$

Equation (3.67) is the velocity potential for a two-dimensional source flow.

The stream function can be obtained as follows. From Equations (2.151a and b), (3.59b), and (3.62),

$$\frac{1}{r}\frac{\partial \psi}{\partial \theta} = V_r = \frac{\Lambda}{2\pi r} \tag{3.68}$$

and

$$-\frac{\partial \psi}{\partial r} = V_\theta = 0 \tag{3.69}$$

Integrating Equation (3.68) with respect to θ, we obtain

$$\psi = \frac{\Lambda}{2\pi}\theta + f(r) \tag{3.70}$$

Integrating Equation (3.69) with respect to r, we have

$$\psi = \text{const} + f(\theta) \tag{3.71}$$

Comparing Equations (3.70) and (3.71) and dropping the constant, we obtain

$$\boxed{\psi = \frac{\Lambda}{2\pi}\theta} \tag{3.72}$$

Equation (3.72) is the stream function for a two-dimensional source flow.

The equation of the streamlines can be obtained by setting Equation (3.72) equal to a constant:

$$\psi = \frac{\Lambda}{2\pi}\theta = \text{const} \tag{3.73}$$

From Equation (3.73), we see that $\theta = $ constant, which, in polar coordinates, is the equation of a straight line from the origin. Hence, Equation (3.73) is consistent with the picture of the source flow sketched in Figure 3.20. Moreover, Equation (3.67) gives an equipotential line as $r = $ constant, that is, a circle with its center at the origin, as shown by the dashed line in Figure 3.20. Once again, we see that streamlines and equipotential lines are mutually perpendicular.

To evaluate the circulation for source flow, recall the $\nabla \times \mathbf{V} = 0$ everywhere. In turn, from Equation (2.137),

$$\Gamma = -\iint_S (\nabla \times \mathbf{V}) \cdot \mathbf{dS} = 0$$

for any closed curve C chosen in the flow field. Hence, as in the case of uniform flow discussed in Section 3.9, there is no circulation associated with the source flow.

It is straightforward to show that Equations (3.67) and (3.72) satisfy Laplace's equation, simply by substitution into $\nabla^2\phi = 0$ and $\nabla^2\psi = 0$ written in terms of cylindrical coordinates [see Equation (3.42)]. Therefore, source flow is a viable elementary flow for use in building more complex flows.

3.11 COMBINATION OF A UNIFORM FLOW WITH A SOURCE AND SINK

Consider a polar coordinate system with a source of strength Λ located at the origin. Superimpose on this flow a uniform stream with velocity V_∞ moving from left to right, as sketched in Figure 3.22. The stream function for the resulting flow is the sum of Equations (3.57) and (3.72):

$$\psi = V_\infty r \sin\theta + \frac{\Lambda}{2\pi}\theta \tag{3.74}$$

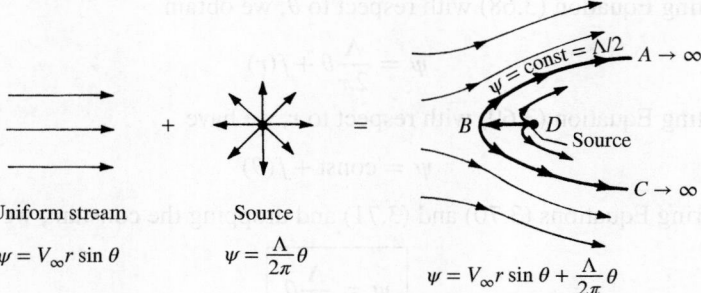

Figure 3.22 Superposition of a uniform flow and a source; flow over a semi-infinite body.

Since both Equations (3.57) and (3.72) are solutions of Laplace's equation, we know that Equation (3.74) also satisfies Laplace's equation; that is, Equation (3.74) describes a viable irrotational, incompressible flow. The streamlines of the combined flow are obtained from Equation (3.74) as

$$\psi = V_\infty r \sin\theta + \frac{\Lambda}{2\pi}\theta = \text{const} \tag{3.75}$$

The resulting streamline shapes from Equation (3.75) are sketched at the right of Figure 3.22. The source is located at point D. The velocity field is obtained by differentiating Equation (3.75):

$$V_r = \frac{1}{r}\frac{\partial\psi}{\partial\theta} = V_\infty \cos\theta + \frac{\Lambda}{2\pi r} \tag{3.76}$$

and

$$V_\theta = -\frac{\partial\psi}{\partial r} = -V_\infty \sin\theta \tag{3.77}$$

Note from Section 3.10 that the radial velocity from a source is $\Lambda/2\pi r$, and from Section 3.9 the component of the freestream velocity in the radial direction is $V_\infty \cos\theta$. Hence, Equation (3.76) is simply the direct sum of the two velocity fields—a result that is consistent with the linear nature of Laplace's equation. Therefore, not only can we add the values of ϕ or ψ to obtain more complex solutions, we can add their derivatives, that is, the velocities, as well.

The stagnation points in the flow can be obtained by setting Equations (3.76) and (3.77) equal to zero:

$$V_\infty \cos\theta + \frac{\Lambda}{2\pi r} = 0 \tag{3.78}$$

and

$$V_\infty \sin\theta = 0 \tag{3.79}$$

Solving for r and θ, we find that one stagnation point exists, located at $(r, \theta) = (\Lambda/2\pi V_\infty, \pi)$, which is labeled as point B in Figure 3.22. That is, the stagnation point is a distance $(\Lambda/2\pi V_\infty)$ directly upstream of the source. From this result, the distance DB clearly grows smaller if V_∞ is increased and larger if Λ is increased—trends that also make sense based on intuition. For example, looking at Figure 3.22, you would expect that as the source strength is increased, keeping V_∞ the same, the stagnation point B will be blown further upstream. Conversely, if V_∞ is increased, keeping the source strength the same, the stagnation point will be blown further downstream.

If the coordinates of the stagnation point at B are substituted into Equation (3.75), we obtain

$$\psi = V_\infty \frac{\Lambda}{2\pi V_\infty} \sin \pi + \frac{\Lambda}{2\pi}\pi = \text{const}$$

$$\psi = \frac{\Lambda}{2} = \text{const}$$

Hence, the streamline that goes through the stagnation point is described by $\psi = \Lambda/2$. This streamline is shown as curve ABC in Figure 3.22.

Examining Figure 3.22, we now come to an important conclusion. Since we are dealing with inviscid flow, where the velocity at the surface of a solid body is tangent to the body, then *any* streamline of the combined flow at the right of Figure 3.22 could be replaced by a solid surface of the same shape. In particular, consider the streamline ABC. Because it contains the stagnation point at B, the streamline ABC is a *dividing* streamline; that is, it separates the fluid coming from the freestream and the fluid emanating from the source. All the fluid outside ABC is from the freestream, and all the fluid inside ABC is from the source. Therefore, as far as the freestream is concerned, the entire region inside ABC could be replaced with a solid body of the same shape, and the external flow, that is, the flow from the freestream, would not feel the difference. The streamline $\psi = \Lambda/2$ extends downstream to infinity, forming a semi-infinite body. Therefore, we are led to the following important interpretation. If we want to construct the flow over a solid semi-infinite body described by the curve ABC as shown in Figure 3.22, then all we need to do is take a uniform stream with velocity V_∞ and add to it a source of strength Λ at point D. The resulting superposition will then represent the flow over the prescribed solid semi-infinite body of shape ABC. This illustrates the practicality of adding elementary flows to obtain a more complex flow over a body of interest.

The superposition illustrated in Figure 3.22 results in the flow over the semi-infinite body ABC. This is a half-body that stretches to infinity in the downstream direction (i.e., the body is not closed). However, if we take a sink of equal strength as the source and add it to the flow downstream of point D, then the resulting body shape will be closed. Let us examine this flow in more detail.

Consider a polar coordinate system with a source and sink placed a distance b to the left and right of the origin, respectively, as sketched in Figure 3.23. The strengths of the source and sink are Λ and $-\Lambda$, respectively (equal and opposite). In addition, superimpose a uniform stream with velocity V_∞, as shown in Figure 3.23. The stream function for the combined flow at any point P with

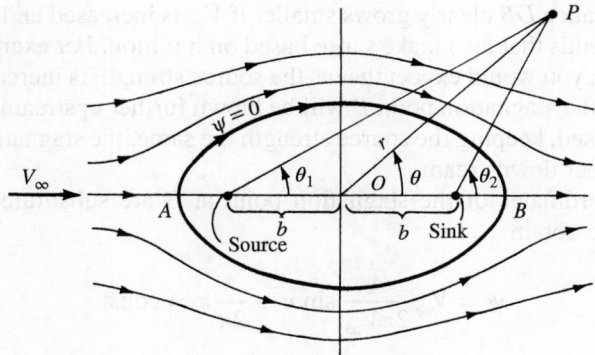

Figure 3.23 Superposition of a uniform flow and a source–sink pair; flow over a Rankine oval.

coordinates (r, θ) is obtained from Equations (3.57) and (3.72):

$$\psi = V_\infty r \sin \theta + \frac{\Lambda}{2\pi}\theta_1 - \frac{\Lambda}{2\pi}\theta_2$$

or

$$\psi = V_\infty r \sin \theta + \frac{\Lambda}{2\pi}(\theta_1 - \theta_2) \qquad (3.80)$$

The velocity field is obtained by differentiating Equation (3.80) according to Equation (2.151a and b). Note from the geometry of Figure 3.23 that θ_1 and θ_2 in Equation (3.80) are functions of r, θ, and b. In turn, by setting $V = 0$, two stagnation points are found, namely, points A and B in Figure 3.23. These stagnation points are located such that (see Problem 3.13)

$$OA = OB = \sqrt{b^2 + \frac{\Lambda b}{\pi V_\infty}} \qquad (3.81)$$

The equation of the streamlines is given by Equation (3.80) as

$$\psi = V_\infty r \sin \theta + \frac{\Lambda}{2\pi}(\theta_1 - \theta_2) = \text{const} \qquad (3.82)$$

The equation of the specific streamline going through the stagnation points is obtained from Equation (3.82) by noting that $\theta = \theta_1 = \theta_2 = \pi$ at point A and $\theta = \theta_1 = \theta_2 = 0$ at point B. Hence, for the stagnation streamline, Equation (3.82) yields a value of zero for the constant. Thus, the stagnation streamline is given by $\psi = 0$, that is,

$$V_\infty r \sin \theta + \frac{\Lambda}{2\pi}(\theta_1 - \theta_2) = 0 \qquad (3.83)$$

the equation of an oval, as sketched in Figure 3.23. Equation (3.83) is also the dividing streamline; all the flow from the source is consumed by the sink and is contained entirely inside the oval, whereas the flow outside the oval has originated with the uniform stream only. Therefore, in Figure 3.23, the region inside the oval can be replaced by a solid body with the shape given by $\psi = 0$, and the region outside the oval can be interpreted as the inviscid, potential (irrotational), incompressible flow over the solid body. This problem was first solved in the nineteenth century by the famous Scottish engineer W. J. M. Rankine; hence, the shape given by Equation (3.83) and sketched in Figure 3.23 is called a *Rankine oval*.

3.12 DOUBLET FLOW: OUR THIRD ELEMENTARY FLOW

There is a special, degenerate case of a source-sink pair that leads to a singularity called a *doublet*. The doublet is frequently used in the theory of incompressible flow; the purpose of this section is to describe its properties.

Consider a source of strength Λ and a sink of equal (but opposite) strength $-\Lambda$ separated by a distance l, as shown in Figure 3.24a. At any point P in the flow, the stream function is

$$\psi = \frac{\Lambda}{2\pi}(\theta_1 - \theta_2) = -\frac{\Lambda}{2\pi}\Delta\theta \tag{3.84}$$

where $\Delta\theta = \theta_2 - \theta_1$ as seen from Figure 3.24a. Equation (3.84) is the stream function for a source-sink pair separated by the distance l.

Now in Figure 3.24a, let the distance l approach zero while the absolute magnitudes of the strengths of the source and sink increase in such a fashion that the product $l\Lambda$ remains constant. This limiting process is shown in Figure 3.24b. In the limit, as $l \to 0$ while $l\Lambda$ remains constant, we obtain a special flow pattern defined as a *doublet*. The *strength* of the doublet is denoted by κ and is defined as $\kappa \equiv l\Lambda$. The stream function for a doublet is obtained from Equation (3.84) as follows:

$$\psi = \lim_{\substack{l\to 0 \\ \kappa=l\Lambda=\text{const}}} \left(-\frac{\Lambda}{2\pi}d\theta\right) \tag{3.85}$$

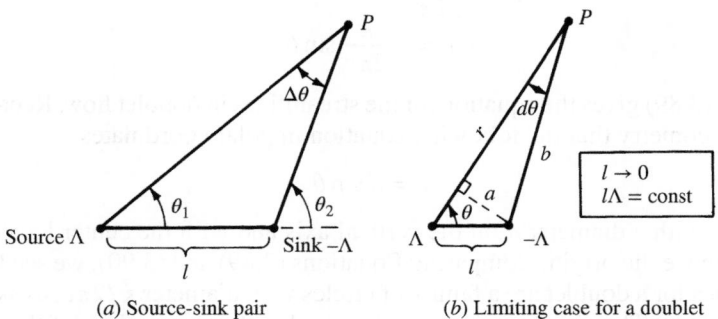

(a) Source-sink pair   (b) Limiting case for a doublet

Figure 3.24 How a source-sink pair approaches a doublet in the limiting case.

where in the limit $\Delta\theta \to d\theta \to 0$. (Note that the source strength Λ approaches an infinite value in the limit.) In Figure 3.24b, let r and b denote the distances to point P from the source and sink, respectively. Draw a line from the sink perpendicular to r, and denote the length along this line by a. For an infinitesimal $d\theta$, the geometry of Figure 3.24b yields

$$a = l\sin\theta$$
$$b = r - l\cos\theta$$
$$d\theta = \frac{a}{b}$$

Hence,

$$d\theta = \frac{a}{b} = \frac{l\sin\theta}{r - l\cos\theta} \tag{3.86}$$

Substituting Equation (3.86) into (3.85), we have

$$\psi = \lim_{\substack{l\to 0 \\ \kappa=\text{const}}} \left(-\frac{\Lambda}{2\pi}\frac{l\sin\theta}{r - l\cos\theta}\right)$$

or

$$\psi = \lim_{\substack{l\to 0 \\ \kappa=\text{const}}} \left(-\frac{\kappa}{2\pi}\frac{\sin\theta}{r - l\cos\theta}\right)$$

or

$$\boxed{\psi = -\frac{\kappa}{2\pi}\frac{\sin\theta}{r}} \tag{3.87}$$

Equation (3.87) is the stream function for a doublet. In a similar fashion, the velocity potential for a doublet is given by (see Problem 3.14)

$$\boxed{\phi = \frac{\kappa}{2\pi}\frac{\cos\theta}{r}} \tag{3.88}$$

The streamlines of a doublet flow are obtained from Equation (3.87):

$$\psi = -\frac{\kappa}{2\pi}\frac{\sin\theta}{r} = \text{const} = c$$

or

$$r = -\frac{\kappa}{2\pi c}\sin\theta \tag{3.89}$$

Equation (3.89) gives the equation for the streamlines in doublet flow. Recall from analytic geometry that the following equation in polar coordinates

$$r = d\sin\theta \tag{3.90}$$

is a circle with a diameter d on the vertical axis and with the center located $d/2$ directly above the origin. Comparing Equations (3.89) and (3.90), we see that the streamlines for a doublet are a family of circles with diameter $\kappa/2\pi c$, as sketched in Figure 3.25. The different circles correspond to different values of the parameter c. Note that in Figure 3.24 we placed the source to the left of the sink; hence, in Figure 3.25 the direction of flow is out of the origin to the left and back into

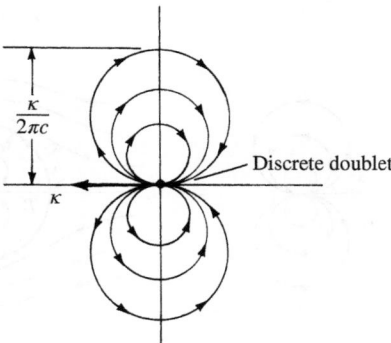

Figure 3.25 Doublet flow with strength κ.

the origin from the right. In Figure 3.24, we could just as well have placed the sink to the left of the source. In such a case, the signs in Equations (3.87) and (3.88) would be reversed, and the flow in Figure 3.25 would be in the opposite direction. Therefore, a doublet has associated with it a sense of direction—the direction with which the flow moves around the circular streamlines. By convention, we designate the direction of the doublet by an arrow drawn from the sink to the source, as shown in Figure 3.25. In Figure 3.25, the arrow points to the left, which is consistent with the form of Equations (3.87) and (3.88). If the arrow would point to the right, the sense of rotation would be reversed, Equation (3.87) would have a positive sign, and Equation (3.88) would have a negative sign.

Returning to Figure 3.24, note that in the limit as $l \to 0$, the source and sink fall on top of each other. However, they do not extinguish each other, because the absolute magnitude of their strengths becomes infinitely large in the limit, and we have a singularity of strength $(\infty - \infty)$; this is an indeterminate form that can have a finite value.

As in the case of a source or sink, it is useful to interpret the doublet flow shown in Figure 3.25 as being *induced* by a discrete doublet of strength κ placed at the origin. Therefore, a doublet is a singularity that induces about it the double-lobed circular flow pattern shown in Figure 3.25.

3.13 NONLIFTING FLOW OVER A CIRCULAR CYLINDER

Consulting our road map given in Figure 3.4, we see that we are well into the third column, having already discussed uniform flow, sources and sinks, and doublets. Along the way, we have seen how the flow over a semi-infinite body can be simulated by the combination of a uniform flow with a source, and the flow over an oval-shaped body can be constructed by superimposing a uniform flow and a source-sink pair. In this section, we demonstrate that the combination of a uniform flow and a doublet produces the flow over a circular cylinder. A circular

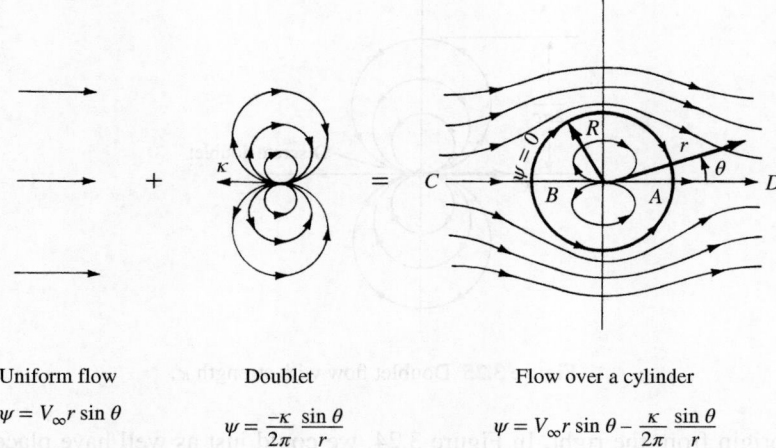

Uniform flow

$\psi = V_\infty r \sin \theta$

Doublet

$\psi = \dfrac{-\kappa}{2\pi} \dfrac{\sin \theta}{r}$

Flow over a cylinder

$\psi = V_\infty r \sin \theta - \dfrac{\kappa}{2\pi} \dfrac{\sin \theta}{r}$

Figure 3.26 Superposition of a uniform flow and a doublet; nonlifting flow over a circular cylinder.

cylinder is one of the most basic geometric shapes available, and the study of the flow around such a cylinder is a classic problem in aerodynamics.

Consider the addition of a uniform flow with velocity V_∞ and a doublet of strength κ, as shown in Figure 3.26. The direction of the doublet is upstream, facing into the uniform flow. From Equations (3.57) and (3.87), the stream function for the combined flow is

$$\psi = V_\infty r \sin \theta - \frac{\kappa}{2\pi} \frac{\sin \theta}{r}$$

or

$$\psi = V_\infty r \sin \theta \left(1 - \frac{\kappa}{2\pi V_\infty r^2} \right) \tag{3.91}$$

Let $R^2 \equiv \kappa/2\pi V_\infty$. Then Equation (3.91) can be written as

$$\boxed{\psi = (V_\infty r \sin \theta) \left(1 - \frac{R^2}{r^2} \right)} \tag{3.92}$$

Equation (3.92) is the stream function for a uniform flow-doublet combination; it is also the stream function for the flow over a circular cylinder of radius R as shown in Figure 3.26 and as demonstrated below.

The velocity field is obtained by differentiating Equation (3.92), as follows:

$$V_r = \frac{1}{r} \frac{\partial \psi}{\partial \theta} = \frac{1}{r} (V_\infty r \cos \theta) \left(1 - \frac{R^2}{r^2} \right)$$

$$V_r = \left(1 - \frac{R^2}{r^2} \right) V_\infty \cos \theta \tag{3.93}$$

$$V_\theta = -\frac{\partial \psi}{\partial r} = -\left[(V_\infty r \sin \theta)\frac{2R^2}{r^3} + \left(1 - \frac{R^2}{r^2}\right)(V_\infty \sin \theta)\right]$$

$$V_\theta = -\left(1 + \frac{R^2}{r^2}\right)V_\infty \sin \theta \tag{3.94}$$

To locate the stagnation points, set Equations (3.93) and (3.94) equal to zero:

$$\left(1 - \frac{R^2}{r^2}\right)V_\infty \cos \theta = 0 \tag{3.95}$$

$$\left(1 + \frac{R^2}{r^2}\right)V_\infty \sin \theta = 0 \tag{3.96}$$

Simultaneously solving Equations (3.95) and (3.96) for r and θ, we find that there are two stagnation points, located at $(r, \theta) = (R, 0)$ and (R, π). These points are denoted as A and B, respectively, in Figure 3.26.

The equation of the streamline that passes through the stagnation point B is obtained by inserting the coordinates of B into Equation (3.92). For $r = R$ and $\theta = \pi$, Equation (3.92) yields $\psi = 0$. Similarly, inserting the coordinates of point A into Equation (3.92), we also find that $\psi = 0$. Hence, the same streamline goes through both stagnation points. Moreover, the equation of this streamline, from Equation (3.92), is

$$\psi = (V_\infty r \sin \theta)\left(1 - \frac{R^2}{r^2}\right) = 0 \tag{3.97}$$

Note that Equation (3.97) is satisfied by $r = R$ for all values of θ. However, recall that $R^2 \equiv \kappa/2\pi V_\infty$, which is a constant. Moreover, in polar coordinates, $r = \text{constant} = R$ is the equation of a circle of radius R with its center at the origin. Therefore, Equation (3.97) describes a circle with radius R, as shown in Figure 3.26. Moreover, Equation (3.97) is satisfied by $\theta = \pi$ and $\theta = 0$ for all values of r; hence, the entire horizontal axis through points A and B, extending infinitely far upstream and downstream, is part of the stagnation streamline.

Note that the $\psi = 0$ streamline, since it goes through the stagnation points, is the dividing streamline. That is, all the flow inside $\psi = 0$ (inside the circle) comes from the doublet, and all the flow outside $\psi = 0$ (outside the circle) comes from the uniform flow. Therefore, we can replace the flow inside the circle by a solid body, and the external flow will not know the difference. Consequently, the inviscid irrotational, incompressible flow over a circular cylinder of radius R can be synthesized by adding a uniform flow with velocity V_∞ and a doublet of strength κ, where R is related to V_∞ and κ through

$$R = \sqrt{\frac{\kappa}{2\pi V_\infty}} \tag{3.98}$$

Note from Equations (3.92) to (3.94) that the entire flow field is symmetrical about both the horizontal and vertical axes through the center of the cylinder, as clearly seen by the streamline pattern sketched in Figure 3.26. Hence, the pressure distribution is also symmetrical about both axes. As a result, the pressure distribution over the top of the cylinder is exactly balanced by the pressure distribution over the bottom of the cylinder (i.e., *there is no net lift*). Similarly, the pressure distribution over the front of the cylinder is exactly balanced by the pressure distribution over the back of the cylinder (i.e., *there is no net drag*). In real life, the result of zero lift is easy to accept, but the result of zero drag makes no sense. We know that any aerodynamic body immersed in a real flow will experience a drag. This paradox between the theoretical result of zero drag, and the knowledge that in real life the drag is finite, was encountered in the year 1744 by the French scientist Jean Le Rond d'Alembert—and it has been known as *d'Alembert's paradox* ever since. For d'Alembert and other fluid dynamic researchers during the eighteenth and nineteenth centuries, this paradox was unexplained and perplexing. Of course, today we know that the drag is due to viscous effects which generate frictional shear stress at the body surface and which cause the flow to separate from the surface on the back of the body, thus creating a large wake downstream of the body and destroying the symmetry of the flow about the vertical axis through the cylinder. These viscous effects are discussed in detail in Chapters 15 through 20. However, such viscous effects are not included in our present analysis of the inviscid flow over the cylinder. As a result, the inviscid theory predicts that the flow closes smoothly and completely behind the body, as sketched in Figure 3.26. It predicts no wake, and no asymmetries, resulting in the theoretical result of zero drag.

Let us quantify the above discussion. The velocity distribution on the surface of the cylinder is given by Equations (3.93) and (3.94) with $r = R$, resulting in

$$V_r = 0 \tag{3.99}$$

and
$$V_\theta = -2V_\infty \sin\theta \tag{3.100}$$

Note that at the surface of the cylinder, V_r is geometrically normal to the surface; hence, Equation (3.99) is consistent with the physical boundary condition that the component of velocity normal to a stationary solid surface must be zero. Equation (3.100) gives the tangential velocity, which is the full magnitude of velocity on the surface of the cylinder, that is, $V = V_\theta = -2V_\infty \sin\theta$ on the surface. The minus sign in Equation (3.100) is consistent with the sign convention in polar coordinates that V_θ is positive in the direction of increasing θ, that is, in the counterclockwise direction as shown in Figure 3.27. However, in Figure 3.26, the surface velocity for $0 \le \theta \le \pi$ is obviously in the opposite direction of increasing θ; hence, the minus sign in Equation (3.100) is proper. For $\pi \le \theta \le 2\pi$, the surface flow is in the same direction as increasing θ, but $\sin\theta$ is itself negative; hence, once again the minus sign in Equation (3.100) is proper. Note from Equation (3.100) that the velocity at the surface reaches a maximum value of $2V_\infty$ at the top and the bottom of the cylinder (where $\theta = \pi/2$ and $3\pi/2$, respectively), as shown in Figure 3.28. Indeed, these are the points of maximum velocity for

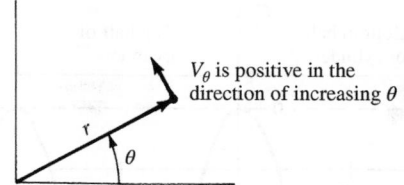

Figure 3.27 Sign convention for V_θ in polar coordinates.

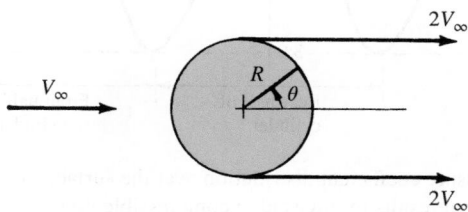

Figure 3.28 Maximum velocity in the flow over a circular cylinder.

the entire flow field around the cylinder, as can be seen from Equations (3.93) and (3.94).

The pressure coefficient is given by Equation (3.38):

$$C_p = 1 - \left(\frac{V}{V_\infty}\right)^2 \tag{3.38}$$

Combining Equations (3.100) and (3.38), we find that the surface pressure coefficient over a circular cylinder is

$$\boxed{C_p = 1 - 4\sin^2\theta} \tag{3.101}$$

Note that C_p varies from 1.0 at the stagnation points to -3.0 at the points of maximum velocity. The pressure coefficient distribution over the surface is sketched in Figure 3.29. The regions corresponding to the top and bottom halves of the cylinder are identified at the top of Figure 3.29. Clearly, the pressure distribution over the top half of the cylinder is equal to the pressure distribution over the bottom half, and hence the lift must be zero, as discussed earlier. Moreover, the regions corresponding to the front and rear halves of the cylinder are identified at the bottom of Figure 3.29. Clearly, the pressure distributions over the front and rear halves are the same, and hence the drag is theoretically zero, as also discussed previously. These results are confirmed by Equations (1.15) and (1.16).

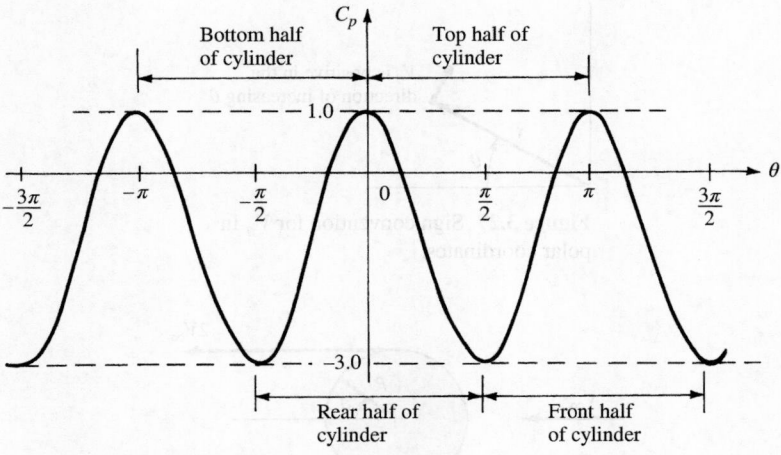

Figure 3.29 Pressure coefficient distribution over the surface of a circular cylinder; theoretical results for inviscid, incompressible flow.

Since $c_f = 0$ (we are dealing with an inviscid flow), Equations (1.15) and (1.16) become, respectively,

$$c_n = \frac{1}{c} \int_0^c (C_{p,l} - C_{p,u}) \, dx \tag{3.102}$$

$$c_a = \frac{1}{c} \int_{LE}^{TE} (C_{p,u} - C_{p,l}) \, dy \tag{3.103}$$

For the circular cylinder, the chord c is the horizontal diameter. From Figure 3.29, $C_{p,l} = C_{p,u}$ for corresponding stations measured along the chord, and hence the integrands in Equations (3.102) and (3.103) are identically zero, yielding $c_n = c_a = 0$. In turn, the lift and drag are zero, thus, again confirming our previous conclusions.

EXAMPLE 3.13

Consider the nonlifting flow over a circular cylinder. Calculate the locations on the surface of the cylinder where the surface pressure equals the freestream pressure.

■ **Solution**

When $p = p_\infty$, then $C_p = 0$. From Equation (3.101),

$$C_p = 0 = 1 - 4 \sin^2 \theta$$

Hence,

$$\sin \theta = \pm \frac{1}{2}$$

$$\boxed{\theta = 30°, 150°, 210°, 330°}$$

These points, as well as the stagnation points and points of minimum pressure, are illustrated in Figure 3.30. Note that at the stagnation point, where $C_p = 1$, the pressure is

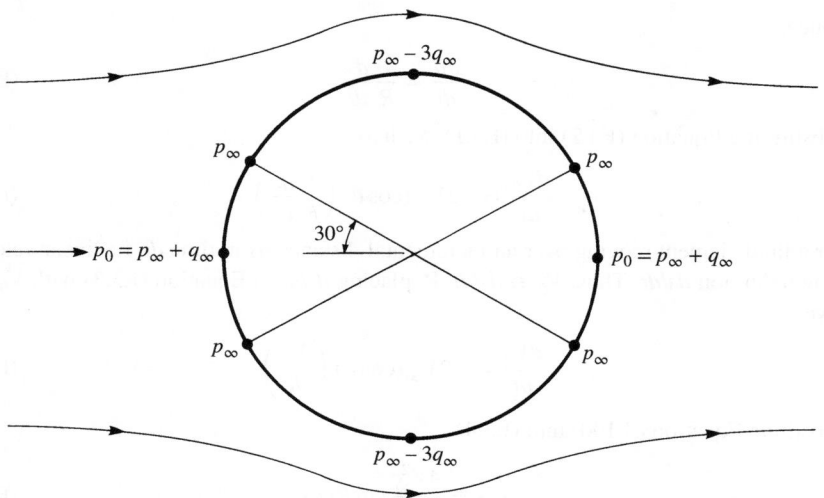

Figure 3.30 Values of pressure at various locations on the surface of a circular cylinder; nonlifting case.

$p_\infty + q_\infty$; the pressure decreases to p_∞ in the first 30° of expansion around the body, and the minimum pressure at the top and bottom of the cylinder, consistent with $C_p = -3$, is $p_\infty - 3q_\infty$.

EXAMPLE 3.14

In the nonlifting flow over a circular cylinder, consider the infinitesimally small fluid elements moving along the surface of the cylinder. Calculate the angular locations over the surface where the acceleration of the fluid elements are a local maximum and minimum. For the case where the radius of the cylinder is 1 m and the freestream flow velocity is 50 m/s, calculate the values of the local maximum and minimum accelerations.

■ **Solution**

From Equation (3.100), the local velocity of the fluid elements on the surface as a function of angular location θ is

$$V_\theta = -2V_\infty \sin \theta \qquad (3.100)$$

The acceleration of the fluid elements is dV_θ/dt. From Equation (3.100),

$$\frac{dV_\theta}{dt} = -2V_\infty (\cos \theta) \frac{d\theta}{dt} \qquad (\text{E3.1})$$

Return to Figure 3.28. Let $d\theta$ be an incremental change in θ. The corresponding incremental distance on the cylinder surface subtended by $d\theta$ is ds, given by

$$ds = R \, d\theta$$

Hence,

$$\frac{d\theta}{dt} = \frac{1}{R}\frac{ds}{dt} \tag{E3.2}$$

Substituting Equation (E3.2) into (E3.1), we have

$$\frac{dV_\theta}{dt} = -2V_\infty(\cos\theta)\left(\frac{1}{R}\frac{ds}{dt}\right) \tag{E3.3}$$

For a fluid element moving over an incremental distance ds in time dt, its linear velocity is by definition ds/dt. Thus, $V_\theta \equiv ds/dt$. Replacing ds/dt in Equation (E3.3) with V_θ, we have

$$\frac{dV_\theta}{dt} = -2V_\infty(\cos\theta)\left(\frac{V_\theta}{R}\right) \tag{E3.4}$$

Substitute Equation (3.100) into (E3.4).

$$\frac{dV_\theta}{dt} = \frac{4V_\infty^2}{R}\sin\theta\cos\theta \tag{E3.5}$$

From the trigonometric identity

$$\sin 2\theta \equiv 2\sin\theta\cos\theta$$

Equation (E3.5) becomes

$$\frac{dV_\theta}{dt} = \frac{2V_\infty^2}{R}\sin 2\theta \tag{E3.6}$$

Equation (E3.6) gives the flow acceleration along the surface, dV_θ/dt, as a function of angular location θ along the surface. To find the θ locations at which the acceleration is a maximum or minimum, differentiate Equation (E3.6) with respect to θ, and set the result equal to zero.

$$\frac{d}{d\theta}\left(\frac{dV_\theta}{dt}\right) = \frac{4V_\infty^2}{R}\cos 2\theta = 0 \tag{E3.7}$$

Solving Equation (E3.7) for θ, we have the location where acceleration is either a local maximum or minimum:

$$\boxed{\theta = 45°, 135°, 225°, 315°} \tag{E3.8}$$

From Equation (E3.6), the values of the local flow acceleration at each one of these locations are respectively

$$\boxed{\frac{2V_\infty^2}{R}, -\frac{2V_\infty^2}{R}, \frac{2V_\infty^2}{R}, -\frac{2V_\infty^2}{R}} \tag{E3.9}$$

Interpretation: Return to Figures 3.27 and 3.28, and note our sign convention that θ is zero at the rearward stagnation point and increases in the *counterclockwise* direction; this is the conventional polar coordinate system where θ increases in the counterclockwise direction, and we have followed this convention throughout the present chapter. That is, as θ sweeps *counterclockwise,* $\theta = 90°$ is the top of the cylinder, $\theta = 180°$ is the location of the forward stagnation point, and $\theta = 270°$ is the bottom of the cylinder. The consistent sign convention for V_θ, as seen in Figure 3.27, is positive in the direction of increasing θ. The direction of V_∞ in Figure 3.28, however, is left to right. Hence the actual left-to-right flow over the *top* of the cylinder has a *negative* velocity because it is running *counter* to the positive direction of V_θ shown in Figure 3.27. The left-to-right flow over the bottom of the cylinder, however, has a positive velocity because it is running in the positive direction of V_θ. This is totally consistent with the result given for V_θ in Equation (3.100).

$$V_\theta = -2V_\infty \sin \theta \qquad (3.100)$$

Over the top of the cylinder, where θ varies from $0°$ to $180°$, Equation (3.100) gives a *negative* value of V_θ, whereas over the bottom of the cylinder, where θ varies from $180°$ to $360°$, Equation (3.100) gives a *positive* value of V_θ. With this in mind, now consider the location and values of the local maximum and minimum acceleration as given by relations (E3.8) and (E3.9), respectively. At $\theta = 45°$, the acceleration is a positive value; i.e., the time rate of change V_θ is positive. However, at $\theta = 45°$, the velocity is a negative value, and with a positive time rate of change, the velocity is becoming less negative; i.e., the absolute value of V_θ is becoming *smaller*. The fluid element at $\theta = 45°$ is *slowing down.* The point $\theta = 45°$ is, therefore, a point of maximum *deceleration* (minimum acceleration). In contrast, at $\theta = 135°$ the velocity is a negative value, but from (E3.9), the local acceleration is also negative. This means that at $\theta = 135°$, with a negative time rate of change of velocity, the velocity itself is becoming more negative, and its absolute value is increasing. At $\theta = 135°$, the fluid element is *speeding up,* and $\theta = 135°$ is a point of maximum *acceleration.*

Over the bottom surface of the cylinder, V_θ is positive. At $\theta = 225°$ (which is on the front face of the cylinder), the acceleration from (E3.9) is also positive; i.e., at $\theta = 225°$ the fluid element is speeding up and hence $\theta = 225°$ is a point of maximum *acceleration.* At $\theta = 315°$ (which is on the back face of the cylinder), the acceleration from (E3.9) is negative; i.e., at $\theta = 315°$ the fluid element is slowing down. Hence, $\theta = 315°$ is a point of maximum deceleration.

In short, again examining Figure 3.28, the flow over the front face of the cylinder starts out at zero velocity at the front stagnation point ($\theta = 180°$), accelerates to a maximum velocity of $2V_\infty$ at the top ($\theta = 90°$) and bottom ($\theta = 270°$) of the cylinder, with a maximum acceleration occurring at $\theta = 135°$ and $225°$. The flow then decelerates over the back face of the cylinder, coming to zero velocity at the rear stagnation point ($\theta = 0$), with maximum deceleration occurring at $\theta = 45°$ and $315°$. It is interesting, and it also makes intuitive sense, that the maximum acceleration and deceleration occur geometrically at the points on the cylinder surface halfway between the stagnation points and the points of maximum velocity at the top and bottom. We might have guessed this right from the beginning of this worked example. But frequently in physical science intuition can betray us, and it is necessary to make an analysis as we have done here to find the correct result.

Note: The interpretation given above is rather long and protracted. It was carried out in order to identify which of the positive or negative values in (E3.9) corresponded to an acceleration or deceleration, and at all times we had to keep our mind focused on the sign convention for the polar coordinate system shown in Figures 3.27 and 3.28. This polar coordinate system with θ increasing in the counterclockwise direction is standard mathematical convention, and we have chosen to use it throughout this chapter.

Finally, let us calculate the actual value of maximum acceleration and deceleration for $R = 1$ m and $V_\infty = 50$ m/s. From (E3.9),

$$\frac{2V_\infty^2}{R} = \frac{2(50)^2}{(1)} = \boxed{5000 \text{ m/s}^2}$$

Since the standard sea level value of the acceleration of gravity on earth is 9.8 ms/s², a fluid element flowing over the surface of the circular cylinder in this example experiences a mind-boggling maximum acceleration and deceleration of

$$\frac{5000}{9.8} = 510.2 \; g$$

So once again, as we saw with Example 2.3, a fluid element in an otherwise seemingly benign flow field can experience tremendously large accelerations.

3.14 VORTEX FLOW: OUR FOURTH ELEMENTARY FLOW

Again, consulting our chapter road map in Figure 3.4, we have discussed three elementary flows—uniform flow, source flow, and doublet flow—and have superimposed these elementary flows to obtain the nonlifting flow over several body shapes, such as ovals and circular cylinders. In this section, we introduce our fourth, and last, elementary flow: vortex flow. In turn, in Sections 3.15 and 3.16, we see how the superposition of flows involving such vortices leads to cases with finite lift.

Consider a flow where all the streamlines are concentric circles about a given point, as sketched in Figure 3.31. Moreover, let the velocity along any given circular streamline be constant, but let it vary from one streamline to another inversely with distance from the common center. Such a flow is called a *vortex flow*. Examine Figure 3.31; the velocity components in the radial and tangential directions are V_r and V_θ, respectively, where $V_r = 0$ and $V_\theta = \text{constant}/r$. It is easily shown (try it yourself) that (1) vortex flow is a physically possible incompressible flow, that is, $\nabla \cdot \mathbf{V} = 0$ at every point, and (2) vortex flow is irrotational, that is, $\nabla \times \mathbf{V} = 0$, at every point except the origin.

From the definition of vortex flow, we have

$$V_\theta = \frac{\text{const}}{r} = \frac{C}{r} \tag{3.104}$$

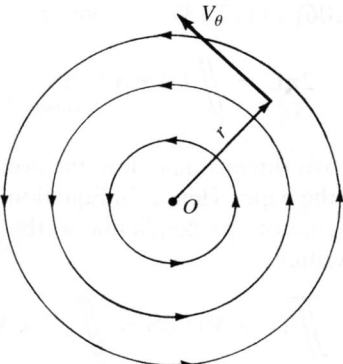

Figure 3.31 Vortex flow.

To evaluate the constant C, take the circulation around a given circular streamline of radius r:

$$\Gamma = -\oint_C \mathbf{V} \cdot \mathbf{ds} = -V_\theta(2\pi r)$$

or

$$\boxed{V_\theta = -\frac{\Gamma}{2\pi r}} \qquad (3.105)$$

Comparing Equations (3.104) and (3.105), we see that

$$C = -\frac{\Gamma}{2\pi} \qquad (3.106)$$

Therefore, for vortex flow, Equation (3.106) demonstrates that the circulation taken about all streamlines is the same value, namely, $\Gamma = -2\pi C$. By convention, Γ is called the *strength* of the vortex flow, and Equation (3.105) gives the velocity field for a vortex flow of strength Γ. Note from Equation (3.105) that V_θ is negative when Γ is positive; that is, a vortex of positive strength rotates in the *clockwise* direction. (This is a consequence of our sign convention on circulation defined in Section 2.13, namely, positive circulation is clockwise.)

We stated earlier that vortex flow is irrotational except at the origin. What happens at $r = 0$? What is the value of $\nabla \times \mathbf{V}$ at $r = 0$? To answer these questions, recall Equation (2.137) relating circulation to vorticity:

$$\Gamma = -\iint_S (\nabla \times \mathbf{V}) \cdot \mathbf{dS} \qquad (2.137)$$

Combining Equations (3.106) and (2.137), we obtain

$$2\pi C = \iint_S (\nabla \times \mathbf{V}) \cdot \mathbf{dS} \tag{3.107}$$

Since we are dealing with two-dimensional flow, the flow sketched in Figure 3.31 takes place in the plane of the paper. Hence, in Equation (3.107), both $\nabla \times \mathbf{V}$ and $\mathbf{dS}$ are in the same direction, both perpendicular to the plane of the flow. Thus, Equation (3.107) can be written as

$$2\pi C = \iint_S (\nabla \times \mathbf{V}) \cdot \mathbf{dS} = \iint_S |\nabla \times \mathbf{V}| \, dS \tag{3.108}$$

In Equation (3.108), the surface integral is taken over the circular area inside the streamline along which the circulation $\Gamma = -2\pi C$ is evaluated. However, Γ is the same for all the circulation streamlines. In particular, choose a circle as close to the origin as we wish (i.e., let $r \to 0$). The circulation will still remain $\Gamma = -2\pi C$. However, the area inside this small circle around the origin will become infinitesimally small, and

$$\iint_S |\nabla \times \mathbf{V}| \, dS \to |\nabla \times \mathbf{V}| \, dS \tag{3.109}$$

Combining Equations (3.108) and (3.109), in the limit as $r \to 0$, we have

$$2\pi C = |\nabla \times \mathbf{V}| \, dS$$

or

$$|\nabla \times \mathbf{V}| = \frac{2\pi C}{dS} \tag{3.110}$$

However, as $r \to 0$, $dS \to 0$. Therefore, in the limit as $r \to 0$, from Equation (3.110), we have

$$|\nabla \times \mathbf{V}| \to \infty$$

Conclusion: Vortex flow is irrotational everywhere except at the point $r = 0$, where the vorticity is infinite. Therefore, the origin, $r = 0$, is a singular point in the flow field. We see that along with sources, sinks, and doublets, the vortex flow contains a singularity. Hence, we can interpret the singularity itself, that is, point O in Figure 3.31, to be a point vortex which induces about it the circular vortex flow shown in Figure 3.31.

The velocity potential for vortex flow can be obtained as follows:

$$\frac{\partial \phi}{\partial r} = V_r = 0 \tag{3.111a}$$

$$\frac{1}{r}\frac{\partial \phi}{\partial \theta} = V_\theta = -\frac{\Gamma}{2\pi r} \tag{3.111b}$$

Integrating Equation (3.111a and b), we find

$$\boxed{\phi = -\frac{\Gamma}{2\pi}\theta} \tag{3.112}$$

Equation (3.112) is the velocity potential for vortex flow.
   The stream function is determined in a similar manner:

$$\frac{1}{r}\frac{\partial \psi}{\partial \theta} = V_r = 0 \tag{3.113a}$$

$$-\frac{\partial \psi}{\partial r} = V_\theta = -\frac{\Gamma}{2\pi r} \tag{3.113b}$$

Integrating Equation (3.113a and b), we have

$$\boxed{\psi = \frac{\Gamma}{2\pi}\ln r} \tag{3.114}$$

Equation (3.114) is the stream function for vortex flow. Note that since $\psi =$ constant is the equation of the streamline, Equation (3.114) states that the stream-lines of vortex flow are given by $r =$ constant (i.e., the streamlines are circles). Thus, Equation (3.114) is consistent with our definition of vortex flow. Also, note from Equation (3.112) that equipotential lines are given by $\theta =$ constant, that is, straight radial lines from the origin. Once again, we see that equipotential lines and streamlines are mutually perpendicular.
   At this stage, we summarize the pertinent results for our four elementary flows in Table 3.1.

Table 3.1

Type of flow	Velocity	ϕ	ψ
Uniform flow in x direction	$u = V_\infty$	$V_\infty x$	$V_\infty y$
Source	$V_r = \dfrac{\Lambda}{2\pi r}$	$\dfrac{\Lambda}{2\pi}\ln r$	$\dfrac{\Lambda}{2\pi}\theta$
Vortex	$V_\theta = -\dfrac{\Gamma}{2\pi r}$	$-\dfrac{\Gamma}{2\pi}\theta$	$\dfrac{\Gamma}{2\pi}\ln r$
Doublet	$V_r = -\dfrac{\kappa}{2\pi}\dfrac{\cos\theta}{r^2}$	$\dfrac{\kappa}{2\pi}\dfrac{\cos\theta}{r}$	$-\dfrac{\kappa}{2\pi}\dfrac{\sin\theta}{r}$
	$V_\theta = -\dfrac{\kappa}{2\pi}\dfrac{\sin\theta}{r^2}$		

EXAMPLE 3.15

Consider the vortex flow discussed in this section. Imagine that you are standing at a location 20 feet from the center of the vortex, and you are feeling a 100 mi/h wind. What is the strength of the vortex?

■ Solution
From Equation (3.105),

$$V_\theta = -\frac{\Gamma}{2\pi r}$$

In this example, the direction of the 100 mi/h wind is not stipulated; for our purpose it is not relevant. We are interested only in the magnitude of the strength of the vortex. Hence, recalling that 88 ft/s = 60 mi/h, we have

$$|\Gamma| = 2\pi r V_\theta = 2\pi (20)(100)\frac{88}{60} = \boxed{1.843 \times 10^4 \text{ ft}^2/\text{s}}$$

Comment: Actual numbers for circulation are not something frequently quoted. Therefore, most aerodynamicists do not have a "feel" for the magnitude of Γ in most applications. In contrast, we do have a feel for more common properties such as velocity; we have a feel for what a 100 mi/h wind is like, especially anybody who has ventured outside in a tropical storm. The purpose of this example is to show us a number for Γ; in this case it is over 18,000 in the English engineering units of feet and seconds. In most cases we really do not care what the value of Γ is because it is immediately used to obtain more practical data, such as aerodynamic lift (as we will see in the next section).

3.15 LIFTING FLOW OVER A CYLINDER

In Section 3.13, we superimposed a uniform flow and a doublet to synthesize the flow over a circular cylinder, as shown in Figure 3.26. In addition, we proved that both the lift and drag were zero for such a flow. However, the streamline pattern shown at the right of Figure 3.26 is not the only flow that is theoretically possible around a circular cylinder. It *is* the only flow that is consistent with zero lift. However, there are other possible flow patterns around a circular cylinder— different flow patterns that result in a nonzero lift on the cylinder. Such lifting flows are discussed in this section.

Now you might be hesitant at this moment, perplexed by the question as to how a lift could possibly be exerted on a circular cylinder. Is not the body perfectly symmetric, and would not this geometry always result in a symmetric flow field with a consequent zero lift, as we have already discussed? You might be so perplexed that you run down to the laboratory, place a stationary cylinder in a low-speed tunnel, and measure the lift. To your satisfaction, you measure no lift, and you walk away muttering that the subject of this section is ridiculous—there

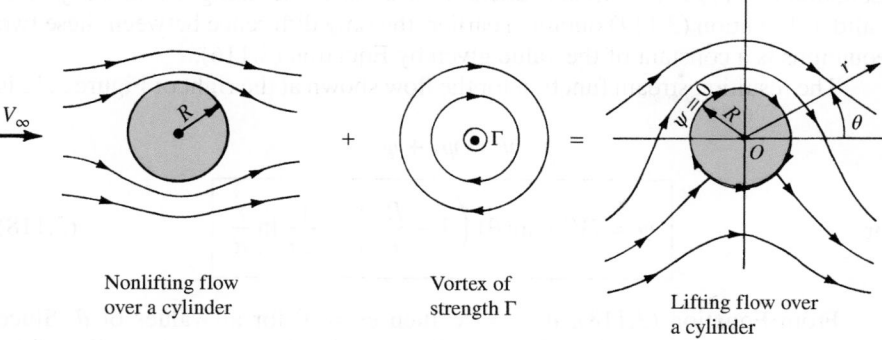

Nonlifting flow over a cylinder

Vortex of strength Γ

Lifting flow over a cylinder

Figure 3.32 The synthesis of lifting flow over a circular cylinder.

is no lift on the cylinder. However, go back to the wind tunnel, and this time run a test with the cylinder *spinning* about its axis at relatively high revolutions per minute. This time you measure a *finite* lift. Also, by this time you might be thinking of other situations: spin on a baseball causes it to curve, and spin on a golfball causes it to hook or slice. Clearly, in real life there are nonsymmetric aerodynamic forces acting on these symmetric, spinning bodies. So, maybe the subject matter of this section is not so ridiculous after all. Indeed, as you will soon appreciate, the concept of lifting flow over a cylinder will start us on a journey which leads directly to the theory of the lift generated by airfoils, as discussed in Chapter 4.

Consider the flow synthesized by the addition of the nonlifting flow over a cylinder and a vortex of strength Γ, as shown in Figure 3.32. The stream function for nonlifting flow over a circular cylinder of radius R is given by Equation (3.92):

$$\psi_1 = (V_\infty r \sin\theta)\left(1 - \frac{R^2}{r^2}\right) \tag{3.92}$$

The stream function for a vortex of strength Γ is given by Equation (3.114). Recall that the stream function is determined within an arbitrary constant; hence, Equation (3.114) can be written as

$$\psi_2 = \frac{\Gamma}{2\pi}\ln r + \text{const} \tag{3.115}$$

Since the value of the constant is arbitrary, let

$$\text{Const} = -\frac{\Gamma}{2\pi}\ln R \tag{3.116}$$

Combining Equations (3.115) and (3.116), we obtain

$$\psi_2 = \frac{\Gamma}{2\pi}\ln\frac{r}{R} \tag{3.117}$$

Equation (3.117) is the stream function for a vortex of strength Γ and is just as valid as Equation (3.114) obtained earlier; the only difference between these two equations is a constant of the value given by Equation (3.116).

The resulting stream function for the flow shown at the right of Figure 3.32 is

$$\psi = \psi_1 + \psi_2$$

or

$$\psi = (V_\infty r \sin \theta)\left(1 - \frac{R^2}{r^2}\right) + \frac{\Gamma}{2\pi} \ln \frac{r}{R} \qquad (3.118)$$

From Equation (3.118), if $r = R$, then $\psi = 0$ for all values of θ. Since $\psi = $ constant is the equation of a streamline, $r = R$ is therefore a streamline of the flow, but $r = R$ is the equation of a circle of radius R. Hence, Equation (3.118) is a valid stream function for the inviscid, incompressible flow over a circular cylinder of radius R, as shown at the right of Figure 3.32. Indeed, our previous result given by Equation (3.92) is simply a special case of Equation (3.118) with $\Gamma = 0$.

The resulting streamline pattern given by Equation (3.118) is sketched at the right of Figure 3.32. Note that the streamlines are no longer symmetrical about the horizontal axis through point O, and you might suspect (correctly) that the cylinder will experience a resulting finite normal force. However, the streamlines are symmetrical about the vertical axis through O, and as a result the drag will be zero, as we prove shortly. Note also that because a vortex of strength Γ has been added to the flow, the circulation about the cylinder is now finite and equal to Γ.

The velocity field can be obtained by differentiating Equation (3.118). An equally direct method of obtaining the velocities is to add the velocity field of a vortex to the velocity field of the nonlifting cylinder. (Recall that because of the linearity of the flow, the velocity components of the superimposed elementary flows add directly.) Hence, from Equations (3.93) and (3.94) for nonlifting flow over a cylinder of radius R, and Equation (3.111a and b) for vortex flow, we have, for the lifting flow over a cylinder of radius R,

$$V_r = \left(1 - \frac{R^2}{r^2}\right) V_\infty \cos \theta \qquad (3.119)$$

$$V_\theta = -\left(1 + \frac{R^2}{r^2}\right) V_\infty \sin \theta - \frac{\Gamma}{2\pi r} \qquad (3.120)$$

To locate the stagnation points in the flow, set $V_r = V_\theta = 0$ in Equations (3.119) and (3.120) and solve for the resulting coordinates (r, θ):

$$V_r = \left(1 - \frac{R^2}{r^2}\right) V_\infty \cos \theta = 0 \qquad (3.121)$$

$$V_\theta = -\left(1 + \frac{R^2}{r^2}\right) V_\infty \sin \theta - \frac{\Gamma}{2\pi r} = 0 \qquad (3.122)$$

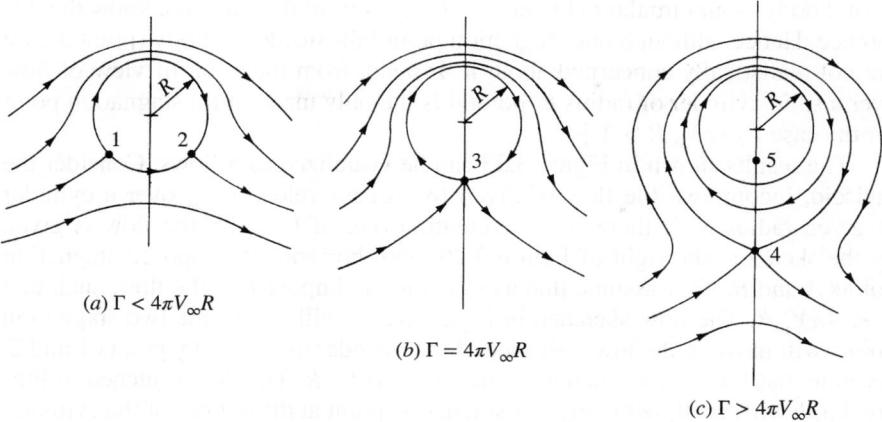

(a) $\Gamma < 4\pi V_\infty R$

(b) $\Gamma = 4\pi V_\infty R$

(c) $\Gamma > 4\pi V_\infty R$

Figure 3.33 Stagnation points for the lifting flow over a circular cylinder.

From Equation (3.121), $r = R$. Substituting this result into Equation (3.122) and solving for θ, we obtain

$$\theta = \arcsin\left(-\frac{\Gamma}{4\pi V_\infty R}\right) \tag{3.123}$$

Since Γ is a positive number, from Equation (3.123) θ must be in the third and fourth quadrants. That is, there can be two stagnation points on the bottom half of the circular cylinder, as shown by points 1 and 2 in Figure 3.33a. These points are located at (R, θ), where θ is given by Equation (3.123). However, this result is valid only when $\Gamma/4\pi V_\infty R < 1$. If $\Gamma/4\pi V_\infty R > 1$, then Equation (3.123) has no meaning. If $\Gamma/4\pi V_\infty R = 1$, there is only one stagnation point on the surface of the cylinder, namely, point $(R, -\pi/2)$ labeled as point 3 in Figure 3.33b. For the case of $\Gamma/4\pi V_\infty R > 1$, return to Equation (3.121). We saw earlier that it is satisfied by $r = R$; however, it is also satisfied by $\theta = \pi/2$ or $-\pi/2$. Substituting $\theta = -\pi/2$ into Equation (3.122), and solving for r, we have

$$r = \frac{\Gamma}{4\pi V_\infty} \pm \sqrt{\left(\frac{\Gamma}{4\pi V_\infty}\right)^2 - R^2} \tag{3.124}$$

Hence, for $\Gamma/4\pi V_\infty R > 1$, there are two stagnation points, one inside and the other outside the cylinder, and both on the vertical axis, as shown by points 4 and 5 in Figure 3.33c. [How does one stagnation point fall *inside* the cylinder? Recall that $r = R$, or $\psi = 0$, is just one of the allowed streamlines of the flow. There is a theoretical flow inside the cylinder—flow that is issuing from the doublet at the origin superimposed with the vortex flow for $r < R$. The circular streamline $r = R$ is the dividing streamline between this flow and the flow from the freestream. Therefore, as before, we can replace the dividing streamline by a solid body—our circular cylinder—and the *external* flow will not know the difference. Hence, although one stagnation point falls inside the body (point 5), we are not realistically concerned about it. Instead, from the point of view of flow over a solid cylinder of radius R, point 4 is the only meaningful stagnation point for the case $\Gamma/4\pi V_\infty R > 1$.]

The results shown in Figure 3.33 can be visualized as follows. Consider the inviscid, incompressible flow of given freestream velocity V_∞ over a cylinder of given radius R. If there is no circulation (i.e., if $\Gamma = 0$), the flow is given by the sketch at the right of Figure 3.26, with horizontally opposed stagnation points A and B. Now assume that a circulation is imposed on the flow, such that $\Gamma < 4\pi V_\infty R$. The flow sketched in Figure 3.33a will result; the two stagnation points will move to the lower surface of the cylinder as shown by points 1 and 2. Assume that Γ is further increased until $\Gamma = 4\pi V_\infty R$. The flow sketched in Figure 3.33b will result, with only one stagnation point at the bottom of the cylinder, as shown by point 3. When Γ is increased still further such that $\Gamma > 4\pi V_\infty R$, the flow sketched in Figure 3.33c will result. The stagnation point will lift from the cylinder's surface and will appear in the flow directly below the cylinder, as shown by point 4.

From the above discussion, Γ is clearly a parameter that can be chosen freely. There is no single value of Γ that "solves" the flow over a circular cylinder; rather, the circulation can be any value. Therefore, for the incompressible flow over a circular cylinder, there are an infinite number of possible potential flow solutions, corresponding to the infinite choices for values of Γ. This statement is not limited to flow over circular cylinders, but rather, it is a general statement that holds for the incompressible potential flow over all smooth two-dimensional bodies. We return to these ideas in subsequent sections.

From the symmetry, or lack of it, in the flows sketched in Figures 3.32 and 3.33, we intuitively concluded earlier that a finite normal force (lift) must exist on the body but that the drag is zero; that is, d'Alembert's paradox still prevails. Let us quantify these statements by calculating expressions for lift and drag, as follows.

The velocity on the surface of the cylinder is given by Equation (3.120) with $r = R$:

$$V = V_\theta = -2V_\infty \sin \theta - \frac{\Gamma}{2\pi R} \tag{3.125}$$

In turn, the pressure coefficient is obtained by substituting Equation (3.125) into Equation (3.38):

$$C_p = 1 - \left(\frac{V}{V_\infty}\right)^2 = 1 - \left(-2\sin\theta - \frac{\Gamma}{2\pi R V_\infty}\right)^2$$

or

$$C_p = 1 - \left[4\sin^2\theta + \frac{2\Gamma\sin\theta}{\pi R V_\infty} + \left(\frac{\Gamma}{2\pi R V_\infty}\right)^2\right] \tag{3.126}$$

In Section 1.5, we discussed in detail how the aerodynamic force coefficients can be obtained by integrating the pressure coefficient and skin friction coefficient over the surface. For inviscid flow, $c_f = 0$. Hence, the drag coefficient c_d is given by Equation (1.16) as

$$c_d = c_a = \frac{1}{c} \int_{\text{LE}}^{\text{TE}} (C_{p,u} - C_{p,l}) \, dy$$

or

$$c_d = \frac{1}{c} \int_{\text{LE}}^{\text{TE}} C_{p,u} \, dy - \frac{1}{c} \int_{\text{LE}}^{\text{TE}} C_{p,l} \, dy \tag{3.127}$$

Converting Equation (3.127) to polar coordinates, we note that

$$y = R \sin \theta \qquad dy = R \cos \theta \, d\theta \tag{3.128}$$

Substituting Equation (3.128) into (3.127), and noting that $c = 2R$, we have

$$c_d = \frac{1}{2} \int_{\pi}^{0} C_{p,u} \cos \theta \, d\theta - \frac{1}{2} \int_{\pi}^{2\pi} C_{p,l} \cos \theta \, d\theta \tag{3.129}$$

The limits of integration in Equation (3.129) are explained as follows. In the first integral, we are integrating from the leading edge (the front point of the cylinder), moving over the *top* surface of the cylinder. Hence, θ is equal to π at the leading edge and, moving over the top surface, *decreases* to 0 at the trailing edge. In the second integral, we are integrating from the leading edge to the trailing edge while moving over the *bottom* surface of the cylinder. Hence, θ is equal to π at the leading edge and, moving over the bottom surface, *increases* to 2π at the trailing edge. In Equation (3.129), both $C_{p,u}$ and $C_{p,l}$ are given by the same analytic expression for C_p, namely, Equation (3.126). Hence, Equation (3.129) can be written as

$$c_d = -\frac{1}{2} \int_{0}^{\pi} C_p \cos \theta \, d\theta - \frac{1}{2} \int_{\pi}^{2\pi} C_p \cos \theta \, d\theta$$

or

$$c_d = -\frac{1}{2} \int_{0}^{2\pi} C_p \cos \theta \, d\theta \tag{3.130}$$

Substituting Equation (3.126) into (3.130), and noting that

$$\int_{0}^{2\pi} \cos \theta \, d\theta = 0 \tag{3.131a}$$

$$\int_{0}^{2\pi} \sin^2 \theta \cos \theta \, d\theta = 0 \tag{3.131b}$$

$$\int_{0}^{2\pi} \sin \theta \cos \theta \, d\theta = 0 \tag{3.131c}$$

we immediately obtain

$$\boxed{c_d = 0} \tag{3.132}$$

Equation (3.132) confirms our intuitive statements made earlier. The drag on a cylinder in an inviscid, incompressible flow is zero, regardless of whether or not the flow has circulation about the cylinder.

The lift on the cylinder can be evaluated in a similar manner as follows. From Equation (1.15) with $c_f = 0$,

$$c_l = c_n = \frac{1}{c} \int_0^c C_{p,l}\, dx - \frac{1}{c} \int_0^c C_{p,u}\, dx \tag{3.133}$$

Converting to polar coordinates, we obtain

$$x = R\cos\theta + R \qquad dx = -R\sin\theta\, d\theta \tag{3.134}$$

Substituting Equation (3.134) into (3.133), we have

$$c_l = -\frac{1}{2} \int_\pi^{2\pi} C_{p,l}\sin\theta\, d\theta + \frac{1}{2} \int_\pi^0 C_{p,u}\sin\theta\, d\theta \tag{3.135}$$

Again, noting that $C_{p,l}$ and $C_{p,u}$ are both given by the same analytic expression, namely, Equation (3.126), Equation (3.135) becomes

$$c_l = -\frac{1}{2} \int_0^{2\pi} C_p \sin\theta\, d\theta \tag{3.136}$$

Substituting Equation (3.126) into (3.136), and noting that

$$\int_0^{2\pi} \sin\theta\, d\theta = 0 \tag{3.137a}$$

$$\int_0^{2\pi} \sin^3\theta\, d\theta = 0 \tag{3.137b}$$

$$\int_0^{2\pi} \sin^2\theta\, d\theta = \pi \tag{3.137c}$$

we immediately obtain

$$c_l = \frac{\Gamma}{RV_\infty} \tag{3.138}$$

From the definition of c_l (see Section 1.5), the lift per unit span L' can be obtained from

$$L' = q_\infty S c_l = \tfrac{1}{2}\rho_\infty V_\infty^2 S c_l \tag{3.139}$$

Here, the planform area $S = 2R(1)$. Therefore, combining Equations (3.138) and (3.139), we have

$$L' = \frac{1}{2}\rho_\infty V_\infty^2 2R \frac{\Gamma}{RV_\infty}$$

or

$$\boxed{L' = \rho_\infty V_\infty \Gamma} \tag{3.140}$$

Equation (3.140) gives the lift per unit span for a circular cylinder with circulation Γ. It is a remarkably simple result, and it states that *the lift per unit span is directly proportional to circulation*. Equation (3.140) is a powerful relation in theoretical aerodynamics. It is called the *Kutta-Joukowski theorem,* named after the German mathematician M. Wilhelm Kutta (1867–1944) and the Russian physicist Nikolai E. Joukowski (1847–1921), who independently obtained it during the first decade of twentieth century. We will have more to say about the Kutta-Joukowski theorem in Section 3.16.

What are the connections between the above theoretical results and real life? As stated earlier, the prediction of zero drag is totally erroneous—viscous effects cause skin friction and flow separation which always produce a finite drag. The inviscid flow treated in this chapter simply does not model the proper physics for drag calculations. On the other hand, the prediction of lift via Equation (3.140) is quite realistic. Let us return to the wind-tunnel experiments mentioned at the beginning of this chapter. If a stationary, nonspinning cylinder is placed in a low-speed wind tunnel, the flow field will appear as shown in Figure 3.34a. The streamlines over the front of the cylinder are similar to theoretical predictions, as sketched at the right of Figure 3.26. However, because of viscous effects, the flow separates over the rear of the cylinder, creating a recirculating flow in the wake downstream of the body. This separated flow greatly contributes to the finite drag measured for the cylinder. On the other hand, Figure 3.34a shows a reasonably symmetric flow about the horizontal axis, and the measurement of lift is essentially zero. Now let us *spin* the cylinder in a clockwise direction about its axis. The resulting flow fields are shown in Figure 3.34b and c. For a moderate amount of spin (Figure 3.34b), the stagnation points move to the lower part of the cylinder, similar to the theoretical flow sketched in Figure 3.33a. If the spin is sufficiently increased (Figure 3.34c), the stagnation point lifts off the surface, similar to the theoretical flow sketched in Figure 3.33c. And what is most important, a *finite lift* is measured for the spinning cylinder in the wind tunnel. What is happening here? Why does spinning the cylinder produce lift? In actuality, the friction between the fluid and the surface of the cylinder tends to drag the fluid near the surface in the same direction as the rotational motion. Superimposed on top of the usual nonspinning flow, this "extra" velocity contribution creates a higher-than-usual velocity at the top of the cylinder and a lower-than-usual velocity at the bottom, as sketched in Figure 3.35. These velocities are assumed to be just outside the viscous boundary layer on the surface. Recall from Bernoulli's equation that as the velocity increases, the pressure decreases. Hence, from Figure 3.35, the pressure

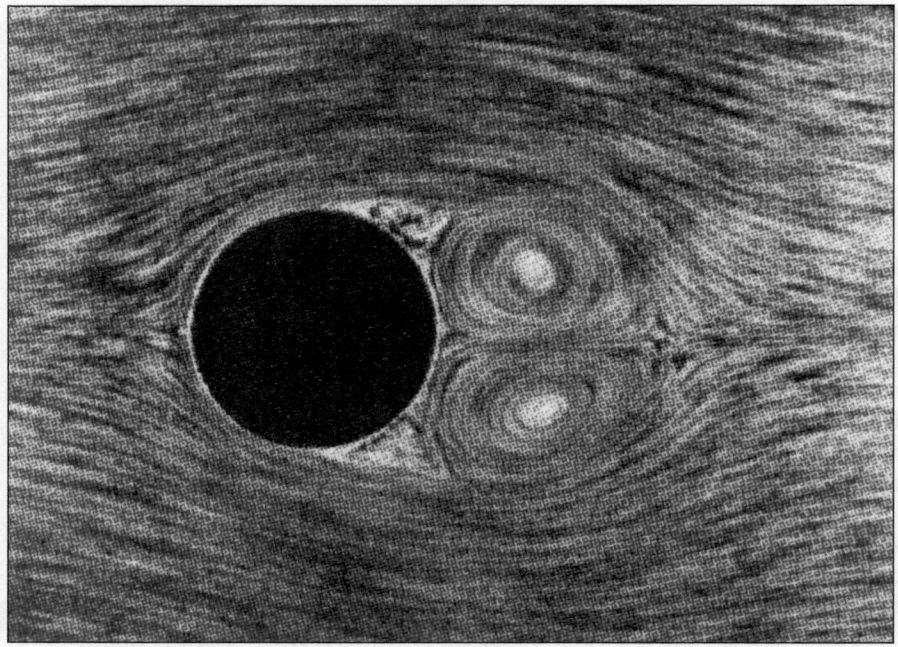

(a)

Figure 3.34 These flow-field pictures were obtained in water, where aluminum filings were scattered on the surface to show the direction of the streamlines. (*a*) Shown above is the case for the nonspinning cylinder (Prandtl, L., and O. G. Tietjens. *Applied Hydro and Aeromechanics Based on Lectures of L. Prandtl*, United Engineering Trustees Inc. New York: McGraw Hill, 1934).

on the top of the cylinder is lower than on the bottom. This pressure imbalance creates a net upward force, that is, a finite lift. Therefore, the theoretical prediction embodied in Equation (3.140), namely that the flow over a circular cylinder can produce a finite lift, is verified by experimental observation.

The general ideas discussed above concerning the generation of lift on a spinning circular cylinder in a wind tunnel also apply to a spinning sphere. This explains why a baseball pitcher can throw a curve and how a golfer can hit a hook or slice—all of which are due to nonsymmetric flows about the spinning bodies, and hence the generation of an aerodynamic force perpendicular to the body's angular velocity vector. This phenomenon is called the *Magnus effect,* named after the German engineer who first observed and explained it in Berlin in 1852.

It is interesting to note that a rapidly spinning cylinder can produce a much higher lift than an airplane wing of the same planform area; however, the drag on the cylinder is also much higher than a well-designed wing. As a result, the Magnus effect is not employed for powered flight. On the other hand, in the 1920s,

(b)

(c)

Figure 3.34 (*continued*) These flow-field pictures were obtained in water, where aluminum filings were scattered on the surface to show the direction of the streamlines. (*b*) Spinning cylinder: peripheral velocity of the surface $= 3V_\infty$. (*c*) Spinning cylinder: peripheral velocity of the surface $= 6V_\infty$. [(b) and (c) Prandtl, L., and O. G. Tietjens. *Applied Hydro and Aeromechanics Based on Lectures of L. Prandtl*, United Engineering Trustees Inc. New York: McGraw Hill, 1934.]

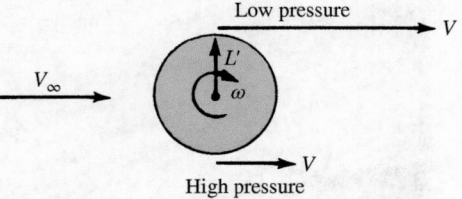

Figure 3.35 Creation of lift on a spinning
cylinder.

the German engineer Anton Flettner replaced the sail on a boat with a rotating
circular cylinder with its axis vertical to the deck. In combination with the wind,
this spinning cylinder provided propulsion for the boat. Moreover, by the action
of two cylinders in tandem and rotating in opposite directions, Flettner was able to
turn the boat around. Flettner's device was a technical success, but an economic
failure because the maintenance on the machinery to spin the cylinders at the
necessary high rotational speeds was too costly. Today, the Magnus effect has an
important influence on the performance of spinning missiles; indeed, a certain
amount of modern high-speed aerodynamic research has focused on the Magnus
forces on spinning bodies for missile applications.

EXAMPLE 3.16

Consider the lifting flow over a circular cylinder. The lift coefficient is 5. Calculate the
peak (negative) pressure coefficient.

■ **Solution**

Examining Figure 3.32, note that the maximum velocity for the *nonlifting* flow over a
cylinder is $2V_\infty$ and that it occurs at the top and bottom points on the cylinder. When the
vortex in Figure 3.32 is added to the flow field, the direction of the vortex velocity is in the
same direction as the flow on the top of the cylinder, but opposes the flow on the bottom
of the cylinder. Hence, the maximum velocity for the lifting case occurs at the *top* of the
cylinder and is equal to the sum of the nonlifting value, $-2V_\infty$, and the vortex, $-\Gamma/2\pi R$.
(*Note:* We are still following the usual sign convention; since the velocity on the top of
the cylinder is in the opposite direction of increasing θ for the polar coordinate system,
the velocity magnitudes here are negative.) Hence,

$$V = -2V_\infty - \frac{\Gamma}{2\pi R} \tag{E.1}$$

The lift coefficient and Γ are related through Equation (3.138):

$$c_l = \frac{\Gamma}{RV_\infty} = 5$$

Hence,

$$\frac{\Gamma}{R} = 5V_\infty \tag{E.2}$$

Substituting Equation (E.2) into (E.1), we have

$$V = -2V_\infty - \frac{5}{2\pi}V_\infty = -2.796V_\infty \tag{E.3}$$

Substituting Equation (E.3) into Equation (3.38), we obtain

$$C_p = 1 - \left(\frac{V}{V_\infty}\right)^2 = 1 - (2.796)^2 = \boxed{-6.82}$$

This example is designed in part to make the following point. Recall that, for an inviscid, incompressible flow, the distribution of C_p over the surface of a body depends only on the shape and orientation of the body—the flow properties such as velocity and density are irrelevant here. Recall Equation (3.101), which gives C_p as a function of θ only, namely, $C_p = 1 - 4\sin^2\theta$. However, for the case of lifting flow, the distribution of C_p over the surface is a function of one additional parameter—namely, the lift coefficient. Clearly, in this example, only the value of c_l is given. However, this is powerful enough to define the flow uniquely; the value of C_p at any point on the surface follows directly from the value of lift coefficient, as demonstrated in the above problem.

EXAMPLE 3.17

For the flow field in Example 3.16, calculate the location of the stagnation points and the points on the cylinder where the pressure equals freestream static pressure.

■ **Solution**
From Equation (3.123), the stagnation points are given by

$$\theta = \arcsin\left(-\frac{\Gamma}{4\pi V_\infty R}\right)$$

From Example 3.16,

$$\frac{\Gamma}{RV_\infty} = 5$$

Thus, $\quad \theta = \arcsin\left(-\frac{5}{4\pi}\right) = \boxed{203.4° \quad \text{and} \quad 336.6°}$

To find the locations where $p = p_\infty$, first construct a formula for the pressure coefficient on the cylinder surface:

$$C_p = 1 - \left(\frac{V}{V_\infty}\right)^2$$

where $\quad V = -2V_\infty \sin\theta - \frac{\Gamma}{2\pi R}$

Thus,

$$C_p = 1 - \left(-2 \sin \theta - \frac{\Gamma}{2\pi R} \right)^2$$

$$= 1 - 4 \sin^2 \theta - \frac{2\Gamma \sin \theta}{\pi R V_\infty} - \left(\frac{\Gamma}{2\pi R V_\infty} \right)^2$$

From Example 3.16, $\Gamma / R V_\infty = 5$. Thus,

$$C_p = 1 - 4 \sin^2 \theta - \frac{10}{\pi} \sin \theta - \left(\frac{5}{2\pi} \right)^2$$

$$= 0.367 - 3.183 \sin \theta - 4 \sin^2 \theta$$

A check on this equation can be obtained by calculating C_p at $\theta = 90°$ and seeing if it agrees with the result obtained in Example 3.16. For $\theta = 90°$, we have

$$C_p = 0.367 - 3.183 - 4 = \boxed{-6.82}$$

This is the same result as in Example 3.16; the equation checks.

To find the values of θ where $p = p_\infty$, set $C_p = 0$:

$$0 = 0.367 - 3.183 \sin \theta - 4 \sin^2 \theta$$

From the quadratic formula,

$$\sin \theta = \frac{3.183 \pm \sqrt{(3.183)^2 + 5.872}}{-8} = \boxed{-0.897 \quad \text{or} \quad 0.102}$$

Hence,

$$\theta = 243.8° \quad \text{and} \quad 296.23°$$

Also,

$$\theta = 5.85° \quad \text{and} \quad 174.1°$$

There are four points on the circular cylinder where $p = p_\infty$. These are sketched in Figure 3.36, along with the stagnation point locations. As shown in Example 3.16, the minimum pressure occurs at the top of the cylinder and is equal to $p_\infty - 6.82 q_\infty$. A *local* minimum pressure occurs at the bottom of the cylinder, where $\theta = 3\pi / 2$. This local minimum is given by

$$C_p = 0.367 - 3.183 \sin \frac{3\pi}{2} - 4 \sin^2 \frac{3\pi}{2}$$

$$= 0.367 + 3.183 - 4 = \boxed{-0.45}$$

Hence, at the bottom of the cylinder, $p = p_\infty - 0.45 q_\infty$.

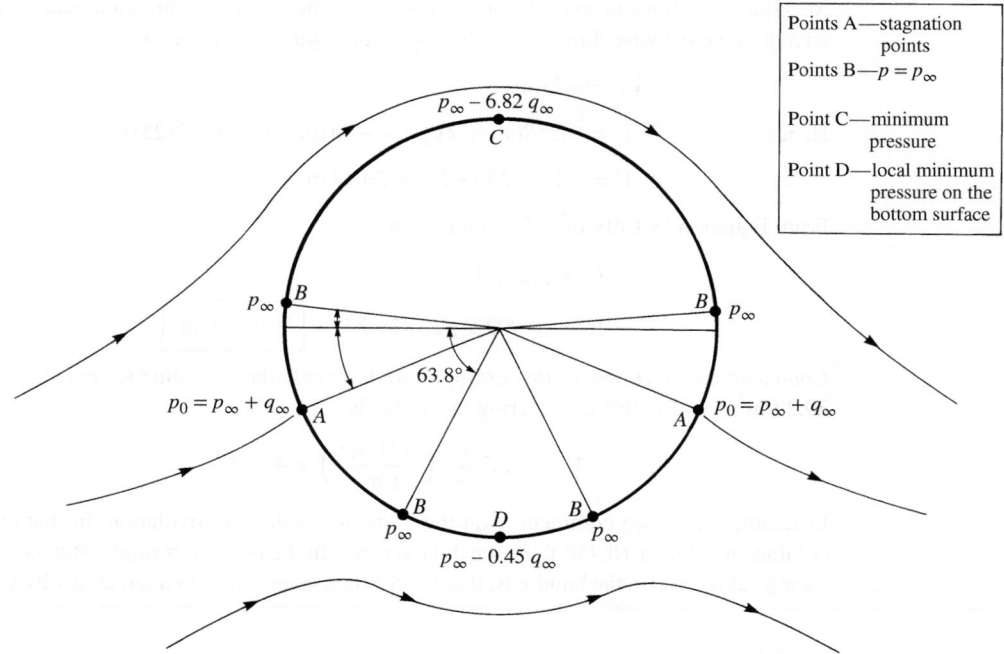

Figure 3.36 Values of pressure at various locations on the surface of a circular cylinder; lifting case with finite circulation. The values of pressure correspond to the case discussed in Example 3.17.

Consider the lifting flow over a circular cylinder with a diameter of 0.5 m. The freestream velocity is 25 m/s, and the maximum velocity on the surface of the cylinder is 75 m/s. The freestream conditions are those for a standard altitude of 3 km. Calculate the lift per unit span on the cylinder.

■ **Solution**
From Appendix D, at an altitude of 3 km, $\rho = 0.90926$ kg/m^3. The maximum velocity occurs at the top of the cylinder, where $\theta = 90°$. From Equation (3.125),

$$V_\theta = -2V_\infty \sin \theta - \frac{\Gamma}{2\pi R}$$

At $\theta = 90°$,

$$V_\theta = -2V_\infty - \frac{\Gamma}{2\pi R}$$

or,

$$\Gamma = -2\pi R(V_\theta + 2V_\infty)$$

Recalling our sign convention that Γ is positive in the clockwise direction, and V_θ is negative in the clockwise direction (reflect again on Figure 3.32), we have

$$V_\theta = -75\,\text{m/s}$$

Hence,

$$\Gamma = -2\pi R(V_\theta + 2V_\infty) = -2\pi(0.25)[-75 + 2(25)]$$

$$\Gamma = -2\pi(0.25)(-25) = 39.27\,\text{m}^2/\text{s}$$

From Equation (3.140), the lift per unit span is

$$L' = \rho_\infty V_\infty \Gamma$$

$$L' = (0.90926)(25)(39.27) = \boxed{892.7\,\text{N/m}}$$

Comment: In the course of this example, we have calculated a value for circulation; $\Gamma = 39.27\,\text{m}^2/\text{s}$. In English engineering units, this is

$$\Gamma = 39.27\frac{\text{m}^2}{\text{s}}\left(\frac{3.28\,\text{ft}^2}{1\,\text{m}}\right) = 422.5\,\frac{\text{ft}^2}{\text{s}}.$$

In Example 3.15 we commented on the numerical value of circulation; in that case we obtained a value of 18,430 ft^2/s, much larger than in the present example. But we are still seeing values of Γ in the hundreds, if not in the thousands—just an interesting observation.

3.16 THE KUTTA-JOUKOWSKI THEOREM AND THE GENERATION OF LIFT

Although the result given by Equation (3.140) was derived for a circular cylinder, it applies in general to cylindrical bodies of arbitrary cross section. For example, consider the incompressible flow over an airfoil section, as sketched in Figure 3.37. Let curve A be any curve in the flow *enclosing* the airfoil. If the airfoil is producing lift, the velocity field around the airfoil will be such that the line integral of velocity around A will be finite, that is, the circulation

$$\Gamma \equiv \oint_A \mathbf{V} \cdot \mathbf{ds}$$

is finite. In turn, the lift per unit span L' on the airfoil will be given by the *Kutta-Joukowski theorem,* as embodied in Equation (3.140):

$$\boxed{L' = \rho_\infty V_\infty \Gamma} \tag{3.140}$$

This result underscores the importance of the concept of circulation, defined in Section 2.13. The Kutta-Joukowski theorem states that lift per unit span on a two-dimensional body is directly proportional to the circulation around the body. Indeed, the concept of circulation is so important at this stage of our discussion that you should reread Section 2.13 before proceeding further.

The general derivation of Equation (3.140) for bodies of arbitrary cross section can be carried out using the method of complex variables. Such mathematics

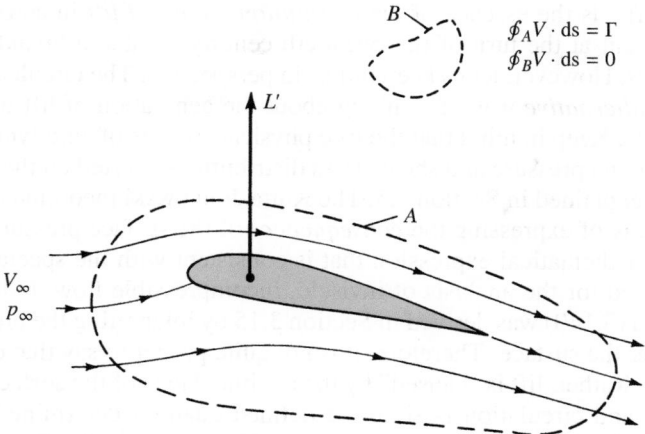

$$\oint_A V \cdot ds = \Gamma$$
$$\oint_B V \cdot ds = 0$$

Figure 3.37 Circulation around a lifting airfoil.

is beyond the scope of this book. (It can be shown that arbitrary functions of complex variables are general solutions of Laplace's equation, which in turn governs incompressible potential flow. Hence, more advanced treatments of such flows utilize the mathematics of complex variables as an important tool. See Reference 9 for a particularly lucid treatment of inviscid, incompressible flow at a more advanced level.)

In Section 3.15, the lifting flow over a circular cylinder was synthesized by superimposing a uniform flow, a doublet, and a vortex. Recall that all three elementary flows are irrotational at all points, except for the vortex, which has infinite vorticity at the origin. Therefore, the lifting flow over a cylinder as shown in Figure 3.33 is irrotational at every point except at the origin. If we take the circulation around any curve *not* enclosing the origin, we obtain from Equation (2.137) the result that $\Gamma = 0$. It is only when we choose a curve that encloses the origin, where $\nabla \times \mathbf{V}$ is infinite, that Equation (2.137) yields a finite Γ, equal to the strength of the vortex. The same can be said about the flow over the airfoil in Figure 3.37. As we show in Chapter 4, the flow *outside* the airfoil is irrotational, and the circulation around any closed curve *not* enclosing the airfoil (such as curve B in Figure 3.37) is consequently zero. On the other hand, we also show in Chapter 4 that the flow over an airfoil is synthesized by distributing vortices either on the surface or inside the airfoil. These vortices have the usual singularities in $\nabla \times \mathbf{V}$, and therefore, if we choose a curve that encloses the airfoil (such as curve A in Figure 3.37), Equation (2.137) yields a finite value of Γ, equal to the *sum* of the vortex strengths distributed on or inside the airfoil. The important point here is that, in the Kutta-Joukowski theorem, the value of Γ used in Equation (3.140) must be evaluated around a closed curve that *encloses the body;* the curve can be otherwise arbitrary, but it must have the body inside it.

At this stage, let us pause and assess our thoughts. The approach we have discussed above—the definition of circulation and the use of Equation (3.140) to obtain the lift—is the essence of the *circulation theory of lift* in aerodynamics. Its development at the turn of the twentieth century created a breakthrough in aerodynamics. However, let us keep things in perspective. The circulation theory of lift is an *alternative* way of thinking about the generation of lift on an aerodynamic body. Keep in mind that the true physical sources of aerodynamic force on a body are the pressure and shear stress distributions exerted on the surface of the body, as explained in Section 1.5. The Kutta-Joukowski theorem is simply an alternative way of expressing the *consequences* of the surface pressure distribution; it is a mathematical expression that is consistent with the special tools we have developed for the analysis of inviscid, incompressible flow. Indeed, recall that Equation (3.140) was derived in Section 3.15 by integrating the pressure distribution over the surface. Therefore, it is not quite proper to say that circulation "causes" lift. Rather, lift is "caused" by the net imbalance of the surface pressure distribution, and circulation is simply a defined quantity determined from the same pressures. The relation between the surface pressure distribution (which produces lift L') and circulation is given by Equation (3.140). However, in the theory of incompressible, potential flow, it is generally much easier to determine the circulation around the body rather than calculate the detailed surface pressure distribution. Therein lies the power of the circulation theory of lift.

Consequently, the theoretical analysis of lift on two-dimensional bodies in incompressible, inviscid flow focuses on the calculation of the circulation about the body. Once Γ is obtained, then the lift per unit span follows directly from the Kutta-Joukowski theorem. As a result, in subsequent sections, we constantly address the question: How can we calculate the circulation for a given body in a given incompressible, inviscid flow?

3.17 NONLIFTING FLOWS OVER ARBITRARY BODIES: THE NUMERICAL SOURCE PANEL METHOD

In this section, we return to the consideration of nonlifting flows. Recall that we have already dealt with the nonlifting flows over a semi-infinite body and a Rankine oval and both the nonlifting and the lifting flows over a circular cylinder. For those cases, we added our elementary flows in certain ways and discovered that the dividing streamlines turned out to fit the shapes of such special bodies. However, this indirect method of starting with a given combination of elementary flows and seeing what body shape comes out of it can hardly be used in a practical sense for bodies of arbitrary shape. For example, consider the airfoil in Figure 3.37. Do we know in advance the correct combination of elementary flows to synthesize the flow over this specified body? The answer is no. Rather, what we want is a direct method; that is, let us *specify* the shape of an arbitrary body and *solve* for the distribution of singularities which, in combination with a uniform stream, produce the flow over the given body. The purpose of this section is to present such a direct method, limited for the present to nonlifting flows. We consider a numerical method appropriate for solution on a high-speed digital

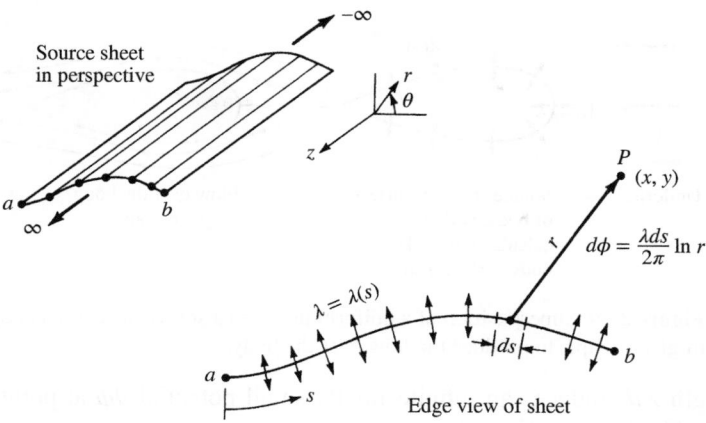

Figure 3.38 Source sheet.

computer. The technique is called the *source panel method,* which, since the late 1960s, has become a standard aerodynamic tool in industry and most research laboratories. In fact, the numerical solution of potential flows by both source and vortex panel techniques has revolutionized the analysis of low-speed flows. We return to various numerical panel techniques in Chapters 4 through 6. As a modern student of aerodynamics, it is necessary for you to become familiar with the fundamentals of such panel methods. The purpose of the present section is to introduce the basic ideas of the source panel method, which is a technique for the numerical solution of nonlifting flows over arbitrary bodies.

First, let us extend the concept of a source or sink introduced in Section 3.10. In that section, we dealt with a single line source, as sketched in Figure 3.21. Now imagine that we have an infinite number of such line sources side by side, where the strength of each line source is infinitesimally small. These side-by-side line sources form a *source sheet,* as shown in perspective in the upper left of Figure 3.38. If we look along the series of line sources (looking along the z axis in Figure 3.38), the source sheet will appear as sketched at the lower right of Figure 3.38. Here, we are looking at an edge view of the sheet; the line sources are all perpendicular to the page. Let s be the distance measured along the source sheet in the edge view. Define $\lambda = \lambda(s)$ to be the *source strength per unit length along s.* [To keep things in perspective, recall from Section 3.10 that the strength of a single line source Λ was defined as the volume flow rate per unit depth, that is, per unit length in the z direction. Typical units for Λ are square meters per second or square feet per second. In turn, the strength of a source sheet $\lambda(s)$ is the volume flow rate per unit depth (in the z direction) and per unit length (in the s direction). Typical units for λ are meters per second or feet per second.] Therefore, the strength of an infinitesimal portion ds of the sheet, as shown in Figure 3.38, is $\lambda\,ds$. This small section of the source sheet can be treated as a distinct source of strength $\lambda\,ds$. Now consider point P in the flow, located a distance r from ds; the cartesian coordinates of P are (x, y). The small section of the source sheet

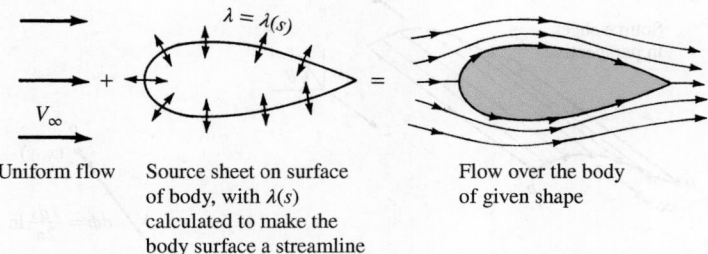

Uniform flow   Source sheet on surface   Flow over the body
              of body, with $\lambda(s)$ of given shape
              calculated to make the
              body surface a streamline

Figure 3.39 Superposition of a uniform flow and a source sheet on a body of given shape, to produce the flow over the body.

of strength $\lambda\, ds$ induces an infinitesimally small potential $d\phi$ at point P. From Equation (3.67), $d\phi$ is given by

$$d\phi = \frac{\lambda\, ds}{2\pi} \ln r \tag{3.141}$$

The complete velocity potential at point P, induced by the entire source sheet from a to b, is obtained by integrating Equation (3.141):

$$\phi(x,y) = \int_a^b \frac{\lambda\, ds}{2\pi} \ln r \tag{3.142}$$

Note that, in general, $\lambda(s)$ can change from positive to negative along the sheet; that is, the "source" sheet is really a combination of line sources and line sinks.

Next, consider a given body of arbitrary shape in a flow with freestream velocity V_∞, as shown in Figure 3.39. Let us cover the surface of the prescribed body with a source sheet, where the strength $\lambda(s)$ varies in such a fashion that the combined action of the uniform flow and the source sheet makes the airfoil surface a streamline of the flow. Our problem now becomes one of finding the appropriate $\lambda(s)$. The solution of this problem is carried out numerically, as follows.

Let us approximate the source sheet by a series of straight panels, as shown in Figure 3.40. Moreover, let the source strength λ per unit length be constant over a given panel, but allow it to vary from one panel to the next. That is, if there are a total of n panels, the source panel strengths per unit length are $\lambda_1, \lambda_2, \ldots, \lambda_j \ldots, \lambda_n$. These panel strengths are unknown; the main thrust of the panel technique is to solve for $\lambda_j, j = 1$ to n, such that the body surface becomes a streamline of the flow. This boundary condition is imposed numerically by defining the midpoint of each panel to be a *control point* and by determining the λ_j's such that the normal component of the flow velocity is zero at each control point. Let us now quantify this strategy.

Let P be a point located at (x, y) in the flow, and let r_{pj} be the distance from any point on the jth panel to P, as shown in Figure 3.40. The velocity potential induced at P due to the jth panel $\Delta\phi_j$ is, from Equation (3.142),

$$\Delta\phi_j = \frac{\lambda_j}{2\pi} \int_j \ln r_{pj}\, ds_j \tag{3.143}$$

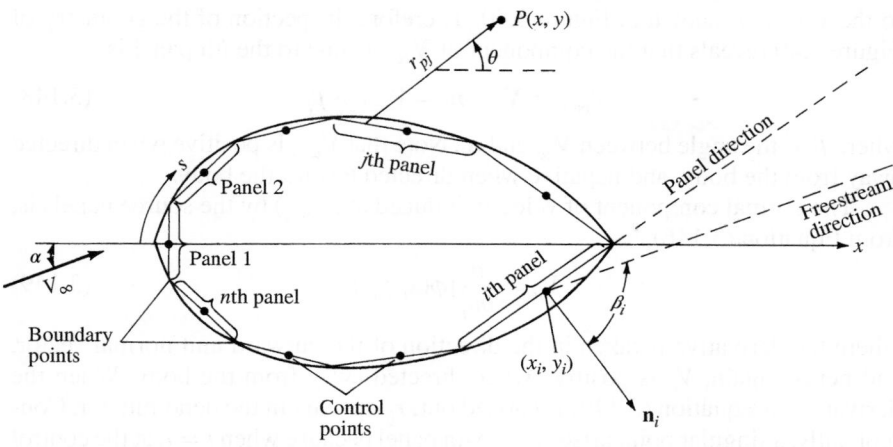

Figure 3.40 Source panel distribution over the surface of a body of arbitrary shape.

In Equation (3.143), λ_j is constant over the jth panel, and the integral is taken over the jth panel only. In turn, the *potential at P* due to *all* the panels is Equation (3.143) summed over all the panels:

$$\phi(P) = \sum_{j=1}^{n} \Delta\phi_j = \sum_{j=1}^{n} \frac{\lambda_j}{2\pi} \int_j \ln r_{pj} \, ds_j \qquad (3.144)$$

In Equation (3.144), the distance r_{pj} is given by

$$r_{pj} = \sqrt{(x - x_j)^2 + (y - y_j)^2} \qquad (3.145)$$

where (x_j, y_j) are coordinates along the surface of the jth panel. Since point P is just an arbitrary point in the flow, let us put P at the control point of the ith panel. Let the coordinates of this control point be given by (x_i, y_i), as shown in Figure 3.40. Then Equations (3.144) and (3.145) become

$$\phi(x_i, y_i) = \sum_{j=1}^{n} \frac{\lambda_j}{2\pi} \int_j \ln r_{ij} \, ds_j \qquad (3.146)$$

and

$$r_{ij} = \sqrt{(x_i - x_j)^2 + (y_i - y_j)^2} \qquad (3.147)$$

Equation (3.146) is physically the contribution of *all* the panels to the potential at the control point of the ith panel.

Recall that the boundary condition is applied at the control points; that is, the normal component of the flow velocity is zero at the control points. To evaluate this component, first consider the component of freestream velocity perpendicular to the panel. Let $\mathbf{n}_i$ be the unit vector normal to the ith panel, directed out of the body, as shown in Figure 3.40. Also, note that the slope of the ith panel is $(dy/dx)_i$. In general, the freestream velocity will be at some incidence angle α

to the x axis, as shown in Figure 3.40. Therefore, inspection of the geometry of Figure 3.40 reveals that the component of $\mathbf{V}_\infty$ normal to the ith panel is

$$V_{\infty,n} = \mathbf{V}_\infty \cdot \mathbf{n}_i = V_\infty \cos \beta_i \qquad (3.148)$$

where β_i is the angle between $\mathbf{V}_\infty$ and $\mathbf{n}_i$. Note that $V_{\infty,n}$ is positive when directed away from the body, and negative when directed toward the body.

The normal component of velocity induced at (x_i, y_i) by the source panels is, from Equation (3.146),

$$V_n = \frac{\partial}{\partial n_i}[\phi(x_i, y_i)] \qquad (3.149)$$

where the derivative is taken in the direction of the outward unit normal vector, and hence, again, V_n is positive when directed away from the body. When the derivative in Equation (3.149) is carried out, r_{ij} appears in the denominator. Consequently, a singular point arises on the ith panel because when $j = i$, at the control point itself $r_{ij} = 0$. It can be shown that when $j = i$, the contribution to the derivative is simply $\lambda_i/2$. Hence, Equation (3.149) combined with Equation (3.146) becomes

$$V_n = \frac{\lambda_i}{2} + \sum_{\substack{j=1 \\ (j \neq i)}}^{n} \frac{\lambda_j}{2\pi} \int_j \frac{\partial}{\partial n_i}(\ln r_{ij})\, ds_j \qquad (3.150)$$

In Equation (3.150), the first term $\lambda_i/2$ is the normal velocity induced at the ith control point by the ith panel itself, and the summation is the normal velocity induced at the ith control point by all the other panels.

The normal component of the flow velocity at the ith control point is the sum of that due to the freestream [Equation (3.148)] and that due to the source panels [Equation (3.150)]. The boundary condition states that this sum must be zero:

$$V_{\infty,n} + V_n = 0 \qquad (3.151)$$

Substituting Equations (3.148) and (3.150) into (3.151), we obtain

$$\frac{\lambda_i}{2} + \sum_{\substack{j=1 \\ (j \neq i)}}^{n} \frac{\lambda_j}{2\pi} \int_j \frac{\partial}{\partial n_i}(\ln r_{ij})\, ds_j + V_\infty \cos \beta_i = 0 \qquad (3.152)$$

Equation (3.152) is the crux of the source panel method. The values of the integrals in Equation (3.152) depend simply on the panel geometry; they are not properties of the flow. Let $I_{i,j}$ be the value of this integral when the control point is on the ith panel and the integral is over the jth panel. Then Equation (3.152) can be written as

$$\frac{\lambda_i}{2} + \sum_{\substack{j=1 \\ (j \neq i)}}^{n} \frac{\lambda_j}{2\pi} I_{i,j} + V_\infty \cos \beta_i = 0 \qquad (3.153)$$

Equation (3.153) is a linear *algebraic* equation with n unknowns $\lambda_1, \lambda_2, \dots, \lambda_n$. It represents the flow boundary condition evaluated at the control point of the ith panel. Now apply the boundary condition to the control points of *all* the panels; that is, in Equation (3.153), let $i = 1, 2, \dots, n$. The results will be a system of n

linear algebraic equations with n unknowns $(\lambda_1, \lambda_2, \ldots, \lambda_n)$, which can be solved simultaneously by conventional numerical methods.

Look what has happened! After solving the system of equations represented by Equation (3.153) with $i = 1, 2, \ldots, n$, we now have the distribution of source panel strengths which, in an appropriate fashion, cause the body surface in Figure 3.40 to be a streamline of the flow. This approximation can be made more accurate by increasing the number of panels, hence more closely representing the source sheet of continuously varying strength $\lambda(s)$ as shown in Figure 3.39. Indeed, the accuracy of the source panel method is amazingly good; a circular cylinder can be accurately represented by as few as 8 panels, and most airfoil shapes, by 50 to 100 panels. (For an airfoil, it is desirable to cover the leading-edge region with a number of small panels to represent accurately the rapid surface curvature and to use larger panels over the relatively flat portions of the body. Note that, in general, all the panels in Figure 3.40 can be of different lengths.)

Once the λ_i's $(i = 1, 2, \ldots, n)$ are obtained, the velocity *tangent* to the surface at each control point can be calculated as follows. Let s be the distance along the body surface, measured positive from front to rear, as shown in Figure 3.40. The component of freestream velocity tangent to the surface is

$$V_{\infty,s} = V_\infty \sin \beta_i \qquad (3.154)$$

The tangential velocity V_s at the control point of the ith panel induced by all the panels is obtained by differentiating Equation (3.146) with respect to s:

$$V_s = \frac{\partial \phi}{\partial s} = \sum_{j=1}^{n} \frac{\lambda_j}{2\pi} \int_j \frac{\partial}{\partial s} (\ln r_{ij}) \, ds_j \qquad (3.155)$$

[The tangential velocity on a flat source panel induced by the panel itself is zero; hence, in Equation (3.155), the term corresponding to $j = i$ is zero. This is easily seen by intuition, because the panel can only emit volume flow from its surface in a direction perpendicular to the panel itself.] The total surface velocity at the ith control point V_i is the sum of the contribution from the freestream [Equation (3.154)] and from the source panels [Equation (3.155)]:

$$V_i = V_{\infty,s} + V_s = V_\infty \sin \beta_i + \sum_{j=1}^{n} \frac{\lambda_j}{2\pi} \int_j \frac{\partial}{\partial s} (\ln r_{ij}) \, ds_j \qquad (3.156)$$

In turn, the pressure coefficient at the ith control point is obtained from Equation (3.38):

$$C_{p,i} = 1 - \left(\frac{V_i}{V_\infty} \right)^2$$

In this fashion, the source panel method gives the pressure distribution over the surface of a nonlifting body of arbitrary shape.

When you carry out a source panel solution as described above, the accuracy of your results can be tested as follows. Let S_j be the length of the jth panel. Recall that λ_j is the strength of the jth panel *per unit length*. Hence, the strength of the jth panel itself is $\lambda_j S_j$. For a closed body, such as in Figure 3.40, the *sum* of all the source and sink strengths must be zero, or else the body itself would be adding or absorbing mass from the flow—an impossible situation for the case we are considering here. Hence, the values of the λ_j's obtained above should obey the relation

$$\sum_{j=1}^{n} \lambda_j S_j = 0 \tag{3.157}$$

Equation (3.157) provides an independent check on the accuracy of the numerical results.

EXAMPLE 3.19

Calculate the pressure coefficient distribution around a circular cylinder using the source panel technique.

■ Solution

We choose to cover the body with eight panels of equal length, as shown in Figure 3.41. This choice is arbitrary; however, experience has shown that, for the case of a circular cylinder, the arrangement shown in Figure 3.41 provides sufficient accuracy. The panels are numbered from 1 to 8, and the control points are shown by the dots in the center of each panel.

Let us evaluate the integrals $I_{i,j}$ which appear in Equation (3.153). Consider Figure 3.42, which illustrates two arbitrary chosen panels. In Figure 3.42, (x_i, y_i) are the coordinates of the control point of the ith panel and (x_j, y_j) are the running coordinates over the entire jth panel. The coordinates of the boundary points for the ith panel are (X_i, Y_i) and (X_{i+1}, Y_{i+1}); similarly, the coordinates of the boundary points for the jth panel are (X_j, Y_j) and (X_{j+1}, Y_{j+1}). In this problem, $\mathbf{V}_\infty$ is in the x direction; hence, the angles between the x axis and the unit vectors $\mathbf{n}_i$ and $\mathbf{n}_j$ are β_i and β_j, respectively. Note that, in

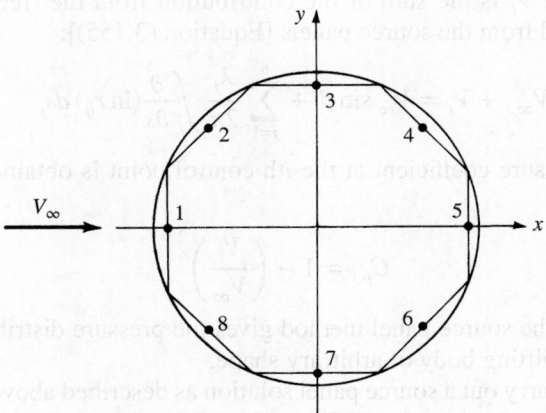

Figure 3.41 Source panel distribution around a circular cylinder.

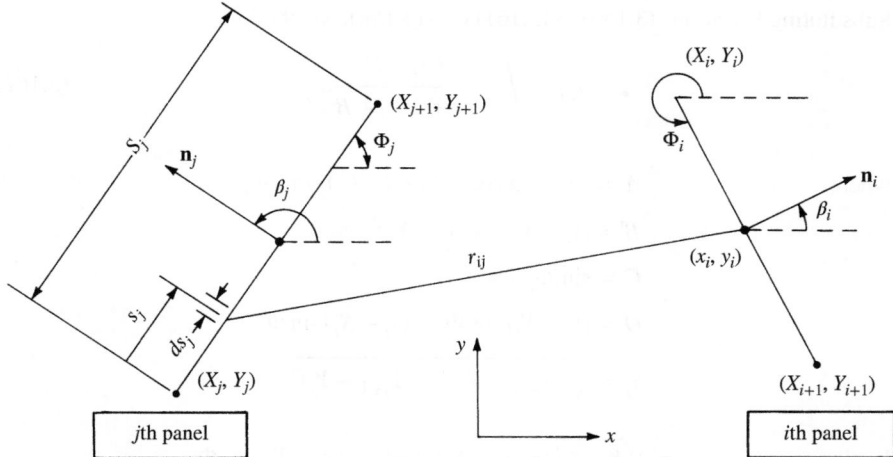

Figure 3.42 Geometry required for the evaluation of I_{ij}.

general, both β_i and β_j vary from 0 to 2π. Recall that the integral $I_{i,j}$ is evaluated at the ith control point and the integral is taken over the complete jth panel:

$$I_{i,j} = \int_j \frac{\partial}{\partial n_i} (\ln r_{ij}) \, ds_j \tag{3.158}$$

Since

$$r_{ij} = \sqrt{(x_i - x_j)^2 + (y_i - y_j)^2}$$

then

$$\frac{\partial}{\partial n_i} (\ln r_{ij}) = \frac{1}{r_{ij}} \frac{\partial r_{ij}}{\partial n_i}$$

$$= \frac{1}{r_{ij}} \frac{1}{2} [(x_i - x_j)^2 + (y_i - y_j)^2]^{-1/2}$$

$$\times \left[2(x_i - x_j) \frac{dx_i}{dn_i} + 2(y_i - y_j) \frac{dy_i}{dn_i} \right]$$

or

$$\frac{\partial}{\partial n_i} (\ln r_{ij}) = \frac{(x_i - x_j) \cos \beta_i + (y_i - y_j) \sin \beta_i}{(x_i - x_j)^2 + (y_i - y_j)^2} \tag{3.159}$$

Note in Figure 3.42 that Φ_i and Φ_j are angles measured in the counterclockwise direction from the x axis to the bottom of each panel. From this geometry,

$$\beta_i = \Phi_i + \frac{\pi}{2}$$

Hence,

$$\sin \beta_i = \cos \Phi_i \tag{3.160a}$$

$$\cos \beta_i = -\sin \Phi_i \tag{3.160b}$$

Also, from the geometry of Figure 3.38, we have

$$x_j = X_j + s_j \cos \Phi_j \tag{3.161a}$$

and

$$y_j = Y_j + s_j \sin \Phi_j \tag{3.161b}$$

Substituting Equations (3.159) to (3.161) into (3.158), we obtain

$$I_{i,j} = \int_0^{S_j} \frac{Cs_j + D}{s_j^2 + 2As_j + B} ds_j \tag{3.162}$$

where

$$A = -(x_i - X_j)\cos\Phi_j - (y_i - Y_j)\sin\Phi_j$$

$$B = (x_i - X_j)^2 + (y_i - Y_j)^2$$

$$C = \sin(\Phi_i - \Phi_j)$$

$$D = (y_i - Y_j)\cos\Phi_i - (x_i - X_j)\sin\Phi_i$$

$$S_j = \sqrt{(X_{j+1} - X_j)^2 + (Y_{j+1} - Y_j)^2}$$

Letting

$$E = \sqrt{B - A^2} = (x_i - X_j)\sin\Phi_j - (y_i - Y_j)\cos\Phi_j$$

we obtain an expression for Equation (3.162) from any standard table of integrals:

$$I_{i,j} = \frac{C}{2}\ln\left(\frac{S_j^2 + 2AS_j + B}{B}\right) \tag{3.163}$$

$$+ \frac{D - AC}{E}\left(\tan^{-1}\frac{S_j + A}{E} - \tan^{-1}\frac{A}{E}\right)$$

Equation (3.163) is a general expression for two arbitrarily oriented panels; it is not restricted to the case of a circular cylinder.

We now apply Equation (3.163) to the circular cylinder shown in Figure 3.41. For purposes of illustration, let us choose panel 4 as the ith panel and panel 2 as the jth panel; that is, let us calculate $I_{4,2}$. From the geometry of Figure 3.41, assuming a unit radius for the cylinder, we see that

$$X_j = -0.9239 \quad X_{j+1} = -0.3827 \quad Y_j = 0.3827$$
$$Y_{j+1} = 0.9239 \qquad \Phi_i = 315° \qquad \Phi_j = 45°$$
$$x_i = 0.6533 \qquad y_i = 0.6533$$

Hence, substituting these numbers into the above formulae, we obtain

$$A = -1.3065 \quad B = 2.5607 \quad C = -1 \quad D = 1.3065$$
$$S_j = 0.7654 \qquad E = 0.9239$$

Inserting the above values into Equation (3.163), we obtain

$$I_{4,2} = 0.4018$$

Return to Figures 3.41 and 3.42. If we now choose panel 1 as the jth panel, keeping panel 4 as the ith panel, we obtain, by means of a similar calculation, $I_{4,1} = 0.4074$. Similarly, $I_{4,3} = 0.3528$, $I_{4,5} = 0.3528$, $I_{4,6} = 0.4018$, $I_{4,7} = 0.4074$, and $I_{4,8} = 0.4084$.

Return to Equation (3.153), which is evaluated for the ith panel in Figures 3.40 and 3.42. Written for panel 4, Equation (3.153) becomes (after multiplying each term by 2 and noting that $\beta_i = 45°$ for panel 4)

$$0.4074\lambda_1 + 0.4018\lambda_2 + 0.3528\lambda_3 + \pi\lambda_4 + 0.3528\lambda_5$$

$$+ 0.4018\lambda_6 + 0.4074\lambda_7 + 0.4084\lambda_8 = -0.7071\, 2\pi V_\infty \qquad (3.164)$$

Equation (3.164) is a linear algebraic equation in terms of the eight unknowns, $\lambda_1, \lambda_2, \ldots,$ λ_8. If we now evaluate Equation (3.153) for each of the seven other panels, we obtain a total of eight equations, including Equation (3.164), which can be solved simultaneously for the eight unknown λ's. The results are

$$\lambda_1/2\pi V_\infty = 0.3765 \qquad \lambda_2/2\pi V_\infty = 0.2662 \qquad \lambda_3/2\pi V_\infty = 0$$
$$\lambda_4/2\pi V_\infty = -0.2662 \qquad \lambda_5/2\pi V_\infty = -0.3765 \qquad \lambda_6/2\pi V_\infty = -0.2662$$
$$\lambda_7/2\pi V_\infty = 0 \qquad \lambda_8/2\pi V_\infty = 0.2662$$

Note the symmetrical distribution of the λ's, which is to be expected for the nonlifting circular cylinder. Also, as a check on the above solution, return to Equation (3.157). Since each panel in Figure 3.41 has the same length, Equation (3.157) can be written simply as

$$\sum_{j=1}^{n} \lambda_j = 0$$

Substituting the values for the λ's obtained into Equation (3.157), we see that the equation is identically satisfied.

The velocity at the control point of the ith panel can be obtained from Equation (3.156). In that equation, the integral over the jth panel is a geometric quantity that is evaluated in a similar manner as before. The result is

$$\int_j \frac{\partial}{\partial s}(\ln r_{ij})ds_j = \frac{D - AC}{2E} \ln \frac{S_j^2 + 2AS_j + B}{B} \qquad (3.165)$$

$$- C\left(\tan^{-1}\frac{S_j + A}{E} - \tan^{-1}\frac{A}{E}\right)$$

With the integrals in Equation (3.156) evaluated by Equation (3.165), and with the values for $\lambda_1, \lambda_2, \ldots, \lambda_8$ obtained above inserted into Equation (3.156), we obtain the velocities $V_1, V_2, \ldots, V_8$. In turn, the pressure coefficients $C_{p,1}, C_{p,2}, \ldots, C_{p,8}$ are obtained directly from

$$C_{p,i} = 1 - \left(\frac{V_i}{V_\infty}\right)^2$$

Results for the pressure coefficients obtained from this calculation are compared with the exact analytical result, Equation (3.101) in Figure 3.43. Amazingly enough, in spite of the relatively crude paneling shown in Figure 3.41, the numerical pressure coefficient results are excellent.

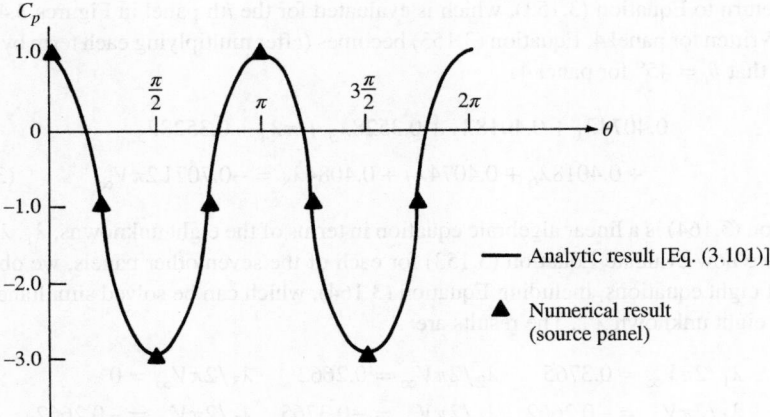

Figure 3.43 Pressure distribution over a circular cylinder; comparison of the source panel results and theory.

3.18 APPLIED AERODYNAMICS: THE FLOW OVER A CIRCULAR CYLINDER—THE REAL CASE

The inviscid, incompressible flow over a circular cylinder was treated in Section 3.13. The resulting theoretical streamlines are sketched in Figure 3.26, characterized by a symmetrical pattern where the flow "closes in" behind the cylinder. As a result, the pressure distribution over the front of the cylinder is the same as that over the rear (see Figure 3.29). This leads to the theoretical result that the pressure drag is zero—d'Alembert's paradox.

The real flow over a circular cylinder is quite different from that studied in Section 3.13, the difference due to the influence of friction. Moreover, the drag coefficient for the real flow over a cylinder is certainly not zero. For a *viscous* incompressible flow, the results of dimensional analysis (Section 1.7) clearly demonstrate that the drag coefficient is a function of the Reynolds number. The variation of $C_D = f(\text{Re})$ for a circular cylinder is shown in Figure 3.44, which is based on a wealth of experimental data. Here, $\text{Re} = (\rho_\infty V_\infty d)/\mu_\infty$, where d is the diameter of the cylinder. Note that C_D is very large for the extremely small values of $\text{Re} < 1$, but decreases monotonically until $\text{Re} \approx 300,000$. At this Reynolds number, there is a precipitous drop of C_D from a value near 1 to about 0.3, then a slight recovery to about 0.6 for $\text{Re} = 10^7$. (*Note:* These results are consistent with the comparison shown in Figure 1.54*d* and *e*, contrasting C_D for a circular cylinder at low and high Re.) What causes this precipitous drop in C_D when the Reynolds number reaches about 300,000? A detailed answer must await our discussion of viscous flow in Chapter 4, and later in Part 4. However, we state now that the phenomenon is caused by a sudden transition of laminar flow within the boundary layer at the lower values of Re to a turbulent boundary layer at the

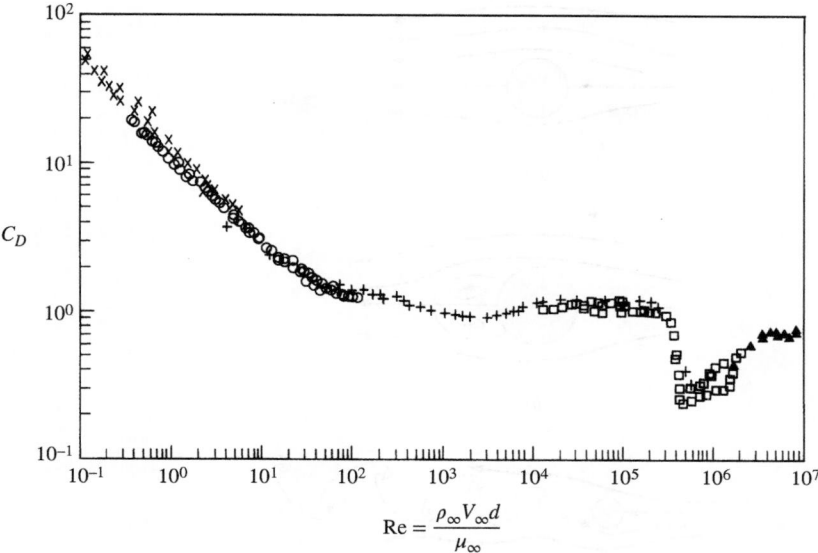

$$\mathrm{Re} = \frac{\rho_\infty V_\infty d}{\mu_\infty}$$

Figure 3.44 Variation of cylinder-drag coefficient with Reynolds number
(*Source:* Experimental data as compiled in Panton, Ronald, *Incompressible Flow,*
Wiley-Interscience, New York, 1984).

higher values of Re. Why does a turbulent boundary layer result in a smaller C_D
for this case? Stay tuned; the answer is given in Chapter 4.

The variation of C_D shown in Figure 3.44 across a range of Re from 10^{-1}
to 10^7 is accompanied by tremendous variations in the qualitative aspects of the
flow field, as itemized, and as sketched in Figure 3.45.

1. For very low values of Re, say, $0 < \mathrm{Re} < 4$, the streamlines are almost
    (but not exactly) symmetrical, and the flow is attached, as sketched in
    Figure 3.45*a*. This regime of viscous flow is called *Stokes flow*; it is
    characterized by a near balance of pressure forces with friction forces
    acting on any given fluid element; the flow velocity is so low that inertia
    effects are very small. A photograph of this type of flow is shown in
    Figure 3.46, which shows the flow of water around a circular cylinder,
    where Re = 1.54. The streamlines are made visible by aluminum powder
    on the surface, along with a time exposure of the film.

2. For $4 < \mathrm{Re} < 40$, the flow becomes separated on the back of the cylinder,
    forming two distinct, stable vortices that remain in the position shown in
    Figure 3.45*b*. A photograph of this type of flow is given in Figure 3.47,
    where Re = 26.

3. As Re is increased above 40, the flow behind the cylinder becomes unstable;
    the vortices which were in a fixed position in Figure 3.45*b* now are alternately
    shed from the body in a regular fashion and flow downstream. This flow is

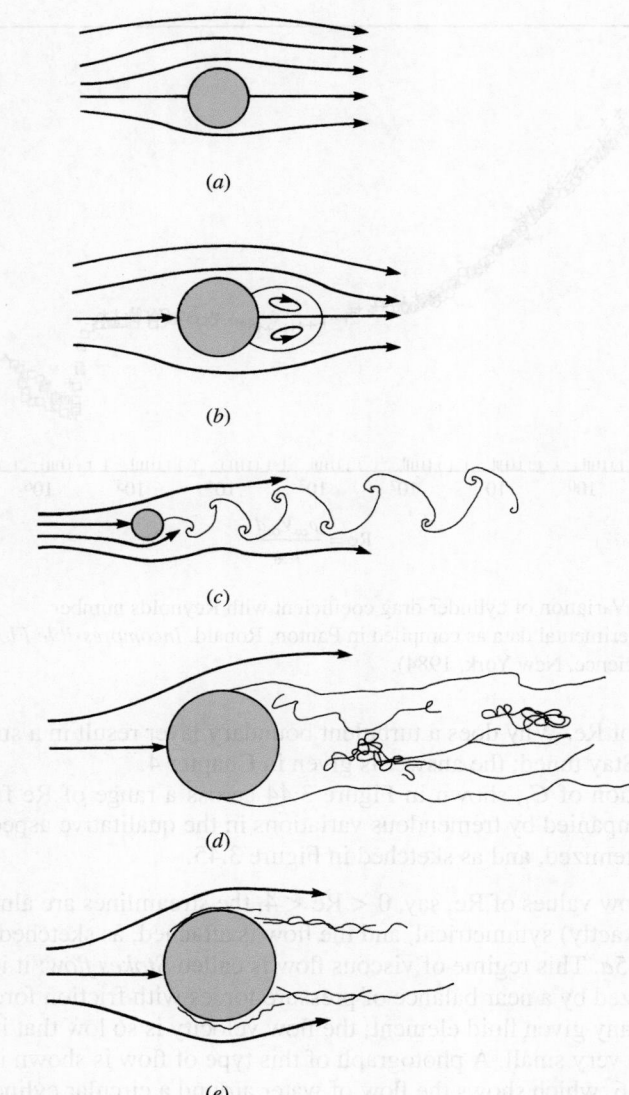

(a)

(b)

(c)

(d)

(e)

Figure 3.45 Various types of flow over a circular cylinder. (*Source:* Panton, Ronald. *Incompressible Flow*. New York: Wiley-Interscience, 1984.)

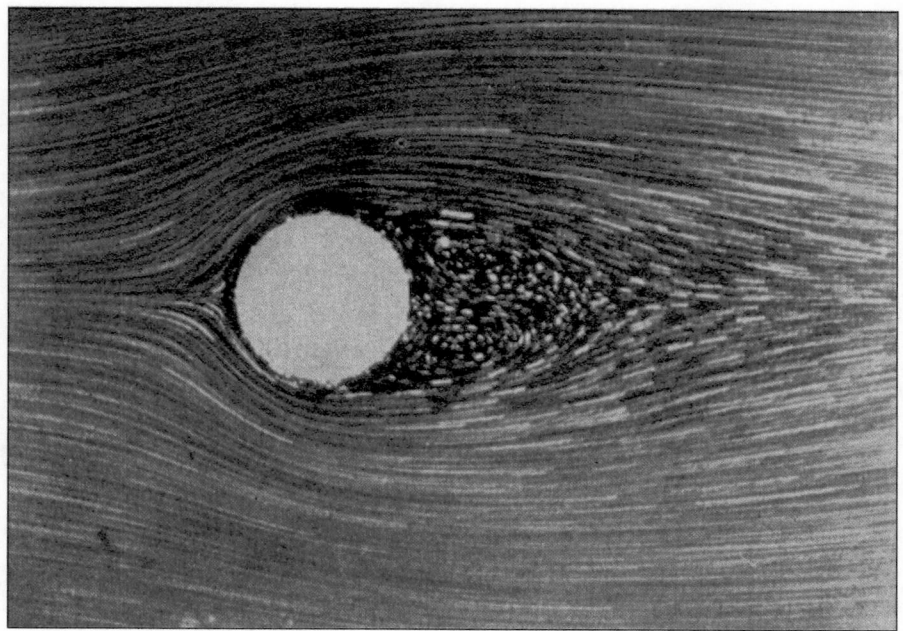

Figure 3.46 Flow over a circular cylinder. Re = 1.54. (Prandtl, L., and O. G. Tietjens. *Applied Hydro and Aeromechanics Based on Lectures of L. Prandtl*, United Engineering Trustees Inc. New York: McGraw Hill, 1934.)

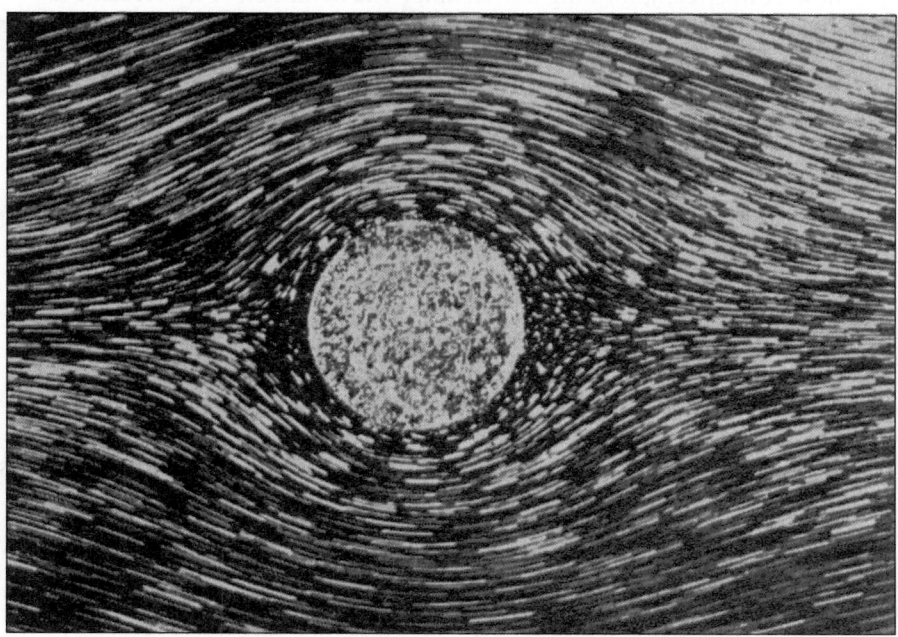

Figure 3.47 Flow over a circular cylinder. Re = 26. (Prandtl, L., and O. G. Tietjens. *Applied Hydro and Aeromechanics Based on Lectures of L. Prandtl*, United Engineering Trustees Inc. New York: McGraw Hill, 1934.)

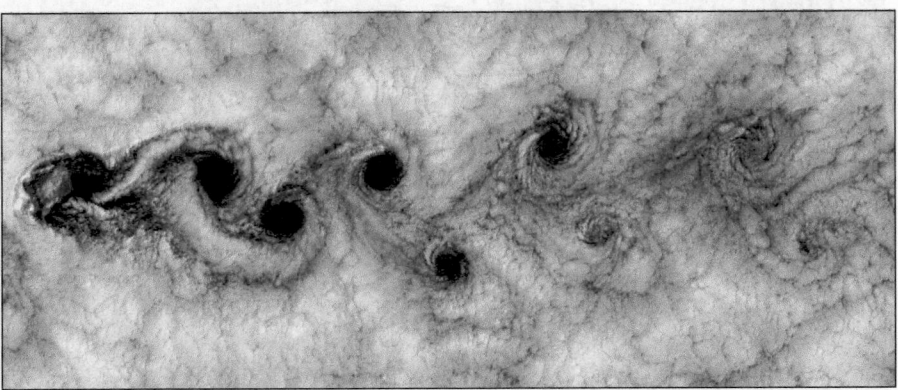

Figure 3.48 Flow over a circular cylinder. Re = 140. A Karman vortex street exists behind the cylinder at this Reynolds number. (*H. S. Photos/Alamy Stock Photo*)

sketched in Figure 3.45*c*. A photograph of this type of flow is shown in Figure 3.48, where Re = 140. This is a water flow where the streaklines are made visible by the electrolytic precipitation method. (In this method, metal plating on the cylinder surface acts as an anode, white particles are precipitated by electrolysis near the anode, and these particles subsequently flow downstream, forming a *streakline*. The definition of a streakline is given in Section 2.11.) The alternately shed vortex pattern shown in Figures 3.45*c* and 3.48 is called a *Karman vortex street,* named after Theodor von Kármán, who began to study and analyze this pattern in 1911 while at Göttingen University in Germany. (von Karman subsequently had a long and very distinguished career in aerodynamics, moving to the California Institute of Technology in 1930, and becoming America's best-known aerodynamicist in the mid-twentieth century. An autobiography of von Karman was published in 1967; see Reference 46. This reference is "must" reading for anyone interested in a riveting perspective on the history of aerodynamics in the twentieth century.)

4. As the Reynolds number is increased to large numbers, the Karman vortex street becomes turbulent and begins to metamorphose into a distinct wake. The laminar boundary layer on the cylinder separates from the surface on the forward face, at a point about 80° from the stagnation point. This is sketched in Figure 3.45*d*. The value of the Reynolds number for this flow is on the order of 10^5. Note, from Figure 3.44, that C_D is a relatively constant value near unity for $10^3 < \text{Re} < 3 \times 10^5$.

5. For $3 \times 10^5 < \text{Re} < 3 \times 10^6$, the separation of the laminar boundary layer still takes place on the forward face of the cylinder. However, in the free shear layer over the top of the separated region, transition to turbulent flow takes place. The flow then reattaches on the back face of the cylinder, but separates again at about 120° around the body measured from the stagnation

point. This flow is sketched in Figure 3.45*e*. This transition to turbulent flow, and the corresponding thinner wake (comparing Figure 3.45*e* with Figure 3.45*d*), reduces the pressure drag on the cylinder and is responsible for the precipitous drop in C_D at Re $= 3 \times 10^5$ shown in Figure 3.44. (More details on this phenomenon are covered in Chapter 4 and Part 4.)

6. For Re $> 3 \times 10^6$, the boundary layer transits directly to turbulent flow at some point on the forward face, and the boundary layer remains totally attached over the surface until it separates at an angular location slightly less than 120° on the back surface. For this regime of flow, C_D actually increases slightly with increasing Re because the separation points on the back surface begin to move closer to the top and bottom of the cylinder, producing a wider wake, and hence larger pressure drag.

In summary, from the photographs and sketches in this section, we see that the real flow over a circular cylinder is dominated by friction effects, namely, the separation of the flow over the rearward face of the cylinder. In turn, a finite pressure drag is created on the cylinder, and d'Alembert's paradox is resolved.

Let us examine the production of drag more closely. The theoretical pressure distribution over the surface of a cylinder in an inviscid, incompressible flow was given in Figure 3.29. In contrast, several real pressure distributions based on experimental measurements for different Reynolds numbers are shown in Figure 3.49, and are compared with the theoretical inviscid flow results obtained in Section 3.13. Note that theory and experiment agree well on the forward face

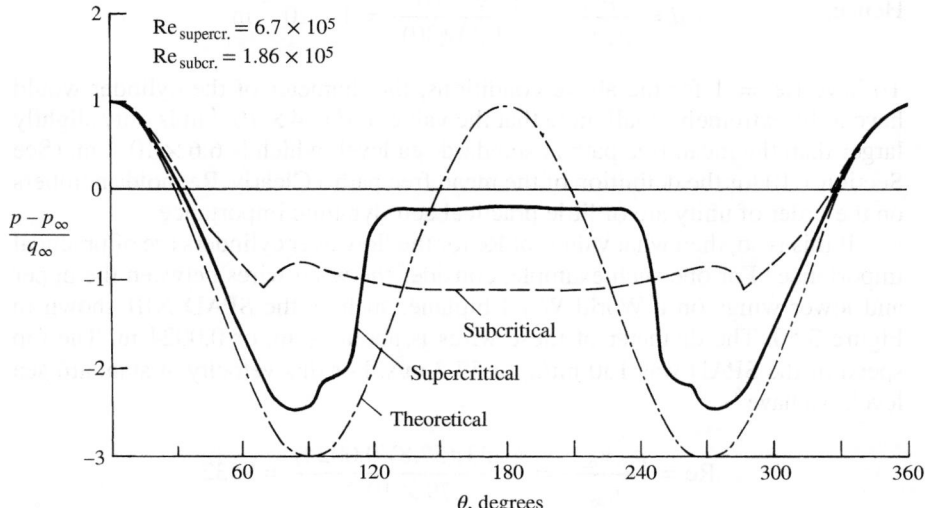

Figure 3.49 Pressure distribution over a circular cylinder in low-speed flow. Comparison of the theoretical pressure distribution with two experimental pressure distributions—one for a subcritical Re and the other for a supercritical Re.

of the cylinder, but that dramatic differences occur over the rearward face. The theoretical results show the pressure decreasing around the forward face from the initial total pressure at the stagnation point, reaching a minimum pressure at the top and bottom of the cylinder ($\theta = 90°$ and $270°$), and then increasing again over the rearward face, recovering to the total pressure at the rear stagnation point. In contrast, in the real case where flow separation occurs, the pressures are relatively constant in the separated region over the rearward face and have values slightly less than freestream pressure. (In regions of separated flow, the pressure frequently exhibits a nearly constant value.) In the separated region over the rearward face, the pressure clearly does not recover to the higher values that exist on the front face. There is a net imbalance of the pressure distribution between the front and back faces, with the pressures on the front being higher than on the back, and this imbalance produces the drag on the cylinder.

Return to Figure 3.44, and examine again the variation of C_D as a function of Re. The regimes associated with the very low Reynolds numbers, such as Stokes flow for Re ≈ 1, are usually of no interest to aeronautical applications. For example, consider a circular cylinder in an airflow of 30 m/s (about 100 ft/s, or 68 mi/h) at standard sea level conditions, where $\rho_\infty = 1.23$ kg/m^3 and $\mu_\infty = 1.79 \times 10^{-5}$ kg/(m · s). The smaller the diameter of the cylinder, the smaller will be the Reynolds number. *Question:* What is the required cylinder diameter in order to have Re = 1? The answer is obtained from

$$\text{Re} = \frac{\rho_\infty V_\infty d}{\mu_\infty} = 1$$

Hence,

$$d = \frac{\mu_\infty}{\rho_\infty V_\infty} = \frac{1.79 \times 10^{-5}}{(1.23)(30)} = 4 \times 10^{-7} \text{ m}$$

To have Re $= 1$ for the above conditions, the diameter of the cylinder would have to be extremely small; note that the value of $d = 4 \times 10^{-7}$ m is only slightly larger than the mean free path at standard sea level, which is 6.6×10^{-8} m. (See Section 1.10 for the definition of the mean free path.) Clearly, Reynolds numbers on the order of unity are of little practical aerodynamic importance.

If this is so, then what values of Re for the flow over cylinders are of practical importance? For one such example, consider the wing wires between the upper and lower wings on a World War I biplane, such as the SPAD XIII shown in Figure 3.50. The diameter of these wires is about $\frac{3}{32}$ in, or 0.0024 m. The top speed of the SPAD was 130 mi/h, or 57.8 m/s. For this velocity at standard sea level, we have

$$\text{Re} = \frac{\rho_\infty V_\infty d}{\mu_\infty} = \frac{(1.23)(57.8)(0.0024)}{1.79 \times 10^{-5}} = 9532$$

With this value of Re, we are beginning to enter the world of practical aerodynamics for the flow over cylinders. It is interesting to note that, from Figure 3.44, $C_D = 1$ for the wires on the SPAD. In terms of airplane aerodynamics, this is a

Figure 3.50 The French SPAD XIII, an example of a strut-and-wire biplane from World War I. Captain Eddie Rickenbacker is shown at the front of the airplane. (*MPI/Archive Photos/Getty Images*)

high drag coefficient for any component of an aircraft. Indeed, the bracing wires used on biplanes of the World War I era were a source of high drag for the aircraft, so much so that early in the war, bracing wire with a symmetric airfoil-like cross section was utilized to help reduce this drag. Such wire was developed by the British at the Royal Aircraft Factory at Farnborough, and was first tested experimentally as early as 1914 on an SE-4 biplane. Interestingly enough, the SPAD used ordinary round wire, and in spite of this was the fastest of all World War I aircraft.

This author was struck by another example of the effect of cylinder drag while traveling in Charleston, South Carolina, shortly after hurricane Hugo devastated the area on September 28, 1989. Traveling north out of Charleston on U.S. Route 17, near the small fishing town of McClellanville, one passes through the Francis Marion National Forest. This forest was virtually destroyed by the hurricane; 60-ft pine trees were snapped off near their base, and approximately 8 out of every 10 trees were down. The sight bore an eerie resemblance to scenes from the battlefields in France during World War I. What type of force can destroy an entire forest in this fashion? To answer this question, we note that the Weather Bureau measured wind gusts as high as 175 mi/h during the hurricane. Let us approximate the wind force on a typical 60-ft pine tree by the aerodynamic drag on a cylinder of a length of 60 ft and a diameter of 5 ft. Since $V = 175$ mi/h $= 256.7$ ft/s,

$\rho_\infty = 0.002377$ slug/ft^3, and $\mu_\infty = 3.7373 \times 10^{-7}$ slug/(ft · s), then the Reynolds number is

$$\text{Re} = \frac{\rho_\infty V_\infty d}{\mu_\infty} = \frac{(0.002377)(256.7)(5)}{3.7373 \times 10^{-7}} = 8.16 \times 10^6$$

Examining Figure 3.44, we see that $C_D = 0.7$. Since C_D is based on the drag per unit length of the cylinder as well as the projected frontal area, we have for the total drag exerted on an entire tree that is 60 ft tall

$$D = q_\infty S C_D = \tfrac{1}{2}\rho_\infty V_\infty^2 (d)(60)C_D$$
$$= \tfrac{1}{2}(0.002377)(256.7)^2(5)(60)(0.7) = 16{,}446 \text{ lb}$$

a 16,000 lb force on the tree—it is no wonder a whole forest was destroyed. (In the above analysis, we neglected the end effects of the flow over the end of the vertical cylinder. Moreover, we did not correct the standard sea level density for the local reduction in barometric pressure experienced inside a hurricane. However, these are relatively small effects in comparison to the overall force on the cylinder.) The aerodynamics of a tree, and especially that of a forest, are more sophisticated than discussed here. Indeed, the aerodynamics of trees have been studied experimentally with trees actually mounted in a wind tunnel.[6]

3.19 HISTORICAL NOTE: BERNOULLI AND EULER—THE ORIGINS OF THEORETICAL FLUID DYNAMICS

Bernoulli's equation, expressed by Equations (3.14) and (3.15), is historically the most famous equation in fluid dynamics. Moreover, we derived Bernoulli's equation from the general momentum equation in partial differential equation form. The momentum equation is just one of the three fundamental equations of fluid dynamics—the others being continuity and energy. These equations are derived and discussed in Chapter 2 and applied to an incompressible flow in this chapter. Where did these equations first originate? How old are they, and who is responsible for them? Considering the fact that all of fluid dynamics in general, and aerodynamics in particular, is built on these fundamental equations, it is important to pause for a moment and examine their historical roots.

As discussed in Section 1.1, Isaac Newton, in his *Principia* of 1687, was the first to establish on a rational basis the relationships between force, momentum, and acceleration. Although he tried, he was unable to apply these concepts properly to a moving fluid. The real foundations of theoretical fluid dynamics were not laid until the next century—developed by a triumvirate consisting of Daniel Bernoulli, Leonhard Euler, and Jean Le Rond d'Alembert.

---

[6] For more details, see the interesting discussion on forest aerodynamics in the book by John E. Allen entitled *Aerodynamics, The Science of Air in Motion,* McGraw Hill, New York, 1982.

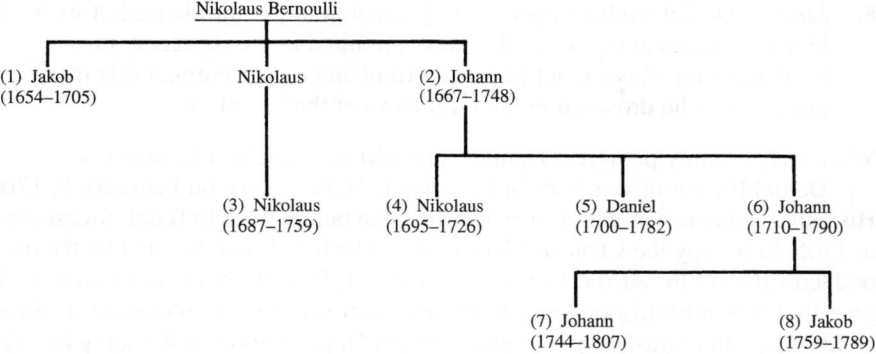

Figure 3.51 A portion of Bernoulli's family tree.

First, consider Bernoulli. Actually, we must consider the whole family of Bernoulli's because Daniel Bernoulli was a member of a prestigious family that dominated European mathematics and physics during the early part of the eighteenth century. Figure 3.51 is a portion of the Bernoulli family tree. It starts with Nikolaus Bernoulli, who was a successful merchant and druggist in Basel, Switzerland, during the seventeenth century. With one eye on this family tree, let us simply list some of the subsequent members of this highly accomplished family:

1. Jakob—Daniel's uncle. Mathematician and physicist, he was professor of mathematics at the University of Basel. He made major contributions to the development of calculus and coined the term "integral."

2. Johann—Daniel's father. He was a professor of mathematics at Groningen, Netherlands, and later at the University of Basel. He taught the famous French mathematician L'Hospital the elements of calculus, and after the death of Newton in 1727 he was considered Europe's leading mathematician at that time.

3. Nikolaus—Daniel's cousin. He studied mathematics under his uncles and held a master's degree in mathematics and a doctor of jurisprudence.

4. Nikolaus—Daniel's brother. He was Johann's favorite son. He held a master of arts degree, and assisted with much of Johann's correspondence to Newton and Liebniz concerning the development of calculus.

5. Daniel himself—to be discussed below.

6. Johann—Daniel's other brother. He succeeded his father in the Chair of Mathematics at Basel and won the prize of the Paris Academy four times for his work.

7. Johann—Daniel's nephew. A gifted child, he earned the master of jurisprudence at the age of 14. When he was 20, he was invited by Frederick II to reorganize the astronomical observatory at the Berlin Academy.

8.   Jakob—Daniel's other nephew. He graduated in jurisprudence but worked in mathematics and physics. He was appointed to the Academy in St. Petersburg, Russia, but he had a promising career prematurely that ended when he drowned in the river Neva at the age of 30.

With such a family pedigree, Daniel Bernoulli was destined for success.

Daniel Bernoulli was born in Groningen, Netherlands, on February 8, 1700. His father, Johann, was a professor at Groningen but returned to Basel, Switzerland, in 1705, to occupy the Chair of Mathematics which had been vacated by the death of Jacob Bernoulli. At the University of Basel, Daniel obtained a master's degree in 1716 in philosophy and logic. He went on to study medicine in Basel, Heidelburg, and Strasbourg, obtaining his Ph.D. in anatomy and botany in 1721. During these studies, he maintained an active interest in mathematics. He followed this interest by moving briefly to Venice, where he published an important work entitled *Exercitationes Mathematicae* in 1724. This earned him much attention and resulted in his winning the prize awarded by the Paris Academy—the first of 10 he was eventually to receive. In 1725, Daniel moved to St. Petersburg, Russia, to join the academy. The St. Petersburg Academy had gained a substantial reputation for scholarship and intellectual accomplishment at that time. During the next 8 years, Bernoulli experienced his most creative period. While at St. Petersburg, he wrote his famous book *Hydrodynamica,* completed in 1734, but not published until 1738. In 1733, Daniel returned to Basel to occupy the Chair of Anatomy and Botany; and in 1750, moved to the Chair of Physics created exclusively for him. He continued to write, give very popular and well-attended lectures in physics, and make contributions to mathematics and physics until his death in Basel on March 17, 1782.

Daniel Bernoulli was famous in his own time. He was a member of virtually all the existing learned societies and academies, such as Bologna, St. Petersburg, Berlin, Paris, London, Bern, Turin, Zurich, and Mannheim. His importance to fluid dynamics is centered on his book *Hydrodynamica* (1738). (With this book, Daniel introduced the term "hydrodynamics" to the literature.) In this book, he ranged over such topics as jet propulsion, manometers, and flow in pipes. Of most importance, he attempted to obtain a relationship between pressure and velocity. Unfortunately, his derivation was somewhat obscure, and Bernoulli's equation, ascribed by history to Daniel via his *Hydrodynamica,* is not to be found in this book, at least not in the form we see it today [such as Equations (3.14) and (3.15)]. The propriety of Equations (3.14) and (3.15) is further complicated by his father, Johann, who also published a book in 1743 entitled *Hydraulica.* It is clear from this latter book that the father understood Bernoulli's theorem better than his son; Daniel thought of pressure strictly in terms of the height of a manometer column, whereas Johann had the more fundamental understanding that pressure was a force acting on the fluid. (It is interesting to note that Johann Bernoulli was a person of some sensitivity and irritability, with an overpowering drive for recognition. He tried to undercut the impact of Daniel's *Hydrodynamica* by predating the publication date of *Hydraulica* to 1728, to make it appear to have been the first of the two. There was little love lost between son and father.)

During Daniel Bernoulli's most productive years, partial differential equations had not yet been introduced into mathematics and physics; hence, he could not approach the derivation of Bernoulli's equation in the same fashion as we have in Section 3.2. The introduction of partial differential equations to mathematical physics was due to d'Alembert in 1747. d'Alembert's role in fluid mechanics is detailed in Section 3.20. Suffice it to say here that his contributions were equally if not more important than Bernoulli's, and d'Alembert represents the second member of the triumvirate which molded the foundations of theoretical fluid dynamics in the eighteenth century.

The third and probably pivotal member of this triumvirate was Leonhard Euler. He was a giant among the eighteenth-century mathematicians and scientists. As a result of his contributions, his name is associated with numerous equations and techniques, for example, the Euler numerical solution of ordinary differential equations, Eulerian angles in geometry, and the momentum equations for inviscid fluid flow [see Equation (3.12)].

Leonhard Euler was born on April 15, 1707, in Basel, Switzerland. His father was a Protestant minister who enjoyed mathematics as a pastime. Therefore, Euler grew up in a family atmosphere that encouraged intellectual activity. At the age of 13, Euler entered the University of Basel which at that time had about 100 students and 19 professors. One of those professors was Johann Bernoulli, who tutored Euler in mathematics. Three years later, Euler received his master's degree in philosophy.

It is interesting that three of the people most responsible for the early development of theoretical fluid dynamics—Johann and Daniel Bernoulli and Euler—lived in the same town of Basel, were associated with the same university, and were contemporaries. Indeed, Euler and the Bernoullis were close and respected friends—so much that, when Daniel Bernoulli moved to teach and study at the St. Petersburg Academy in 1725, he was able to convince the academy to hire Euler as well. At this invitation, Euler left Basel for Russia; he never returned to Switzerland, although he remained a Swiss citizen throughout his life.

Euler's interaction with Daniel Bernoulli in the development of fluid mechanics grew strong during these years at St. Petersburg. It was here that Euler conceived of pressure as a point property that can vary from point to point throughout a fluid and obtained a differential equation relating pressure and velocity, that is, *Euler's equation* given by Equation (3.12). In turn, Euler integrated the differential equation to obtain, for the first time in history, Bernoulli's equation in the form of Equations (3.14) and (3.15). Hence, we see that Bernoulli's equation is really a misnomer; credit for it is legitimately shared by Euler.

When Daniel Bernoulli returned to Basel in 1733, Euler succeeded him at St. Petersburg as a professor of physics. Euler was a dynamic and prolific person; by 1741, he had prepared 90 papers for publication and written the two-volume book *Mechanica*. The atmosphere surrounding St. Petersburg was conducive to such achievement. Euler wrote in 1749: "I and all others who had the good fortune to be for some time with the Russian Imperial Academy cannot but acknowledge that we owe everything which we are and possess to the favorable conditions which we had there."

However, in 1740, political unrest in St. Petersburg caused Euler to leave for the Berlin Society of Sciences, at that time just formed by Frederick the Great. Euler lived in Berlin for the next 25 years, where he transformed the society into a major academy. In Berlin, Euler continued his dynamic mode of working, preparing at least 380 papers for publication. Here, as a competitor with d'Alembert (see Section 3.20), Euler formulated the basis for mathematical physics.

In 1766, after a major disagreement with Frederick the Great over some financial aspects of the academy, Euler moved back to St. Petersburg. This second period of his life in Russia became one of physical suffering. In that same year, he became blind in one eye after a short illness. An operation in 1771 resulted in restoration of his sight, but only for a few days. He did not take proper precautions after the operation, and within a few days, he was completely blind. However, with the help of others, he continued his work. His mind was sharp as ever, and his spirit did not diminish. His literary output even increased—about half of his total papers were written after 1765!

On September 18, 1783, Euler conducted business as usual—giving a mathematics lesson, making calculations of the motion of balloons, and discussing with friends the planet of Uranus, which had recently been discovered. At about 5 P.M., he suffered a brain hemorrhage. His only words before losing consciousness were "I am dying." By 11 P.M., one of the greatest minds in history had ceased to exist.

With the lives of Bernoulli, Euler, and d'Alembert (see Section 3.20) as background, let us now trace the genealogy of the basic equations of fluid dynamics. For example, consider the continuity equation in the form of Equation (2.52). Although Newton had postulated the obvious fact that the mass of a specified object was constant, this principle was not appropriately applied to fluid mechanics until 1749. In this year, d'Alembert gave a paper in Paris, entitled "Essai d'une nouvelle theorie de la resistance des fluides," in which he formulated differential equations for the conservation of mass in special applications to plane and axisymmetric flows. Euler took d'Alembert's results and, 8 years later, generalized them in a series of three basic papers on fluid mechanics. In these papers, Euler published, for the first time in history, the continuity equation in the form of Equation (2.52) and the momentum equations in the form of Equation (2.113a and c), without the viscous terms. Hence, two of the three basic conservation equations used today in modern fluid dynamics were well established long before the American Revolutionary War—such equations were contemporary with the time of George Washington and Thomas Jefferson!

The origin of the energy equation in the form of Equation (2.96) without viscous terms has its roots in the development of thermodynamics in the nineteenth century. Its precise first use is obscure and is buried somewhere in the rapid development of physical science in the nineteenth century.

The purpose of this section has been to give you some feeling for the historical development of the fundamental equations of fluid dynamics. Maybe we can appreciate these equations more when we recognize that they have been with us for quite some time and that they are the product of much thought from some of the greatest minds of the eighteenth century.

3.20 HISTORICAL NOTE: D'ALEMBERT AND HIS PARADOX

You can well imagine the frustration that Jean le Rond d'Alembert felt in 1744 when, in a paper entitled "Traite de l'equilibre et des mouvements de fluids pour servir de suite au traite de dynamique," he obtained the result of zero drag for the inviscid, incompressible flow over a closed two-dimensional body. Using different approaches, d'Alembert encountered this result again in 1752 in his paper entitled "Essai sur la resistance" and again in 1768 in his "Opuscules mathematiques." In this last paper, the quote given at the beginning of Chapter 15 can be found in essence, he had given up trying to explain the cause of this paradox. Even though the prediction of fluid-dynamic drag was a very important problem in d'Alembert's time, and in spite of the number of great minds that addressed it, the fact that viscosity is responsible for drag was not appreciated. Instead, d'Alembert's analyses used momentum principles in a frictionless flow, and quite naturally he found that the flow field closed smoothly around the downstream portion of the bodies, resulting in zero drag. Who was this person, d'Alembert? Considering the role his paradox played in the development of fluid dynamics, it is worth our time to take a closer look at this person.

d'Alembert was born illegitimately in Paris on November 17, 1717. His mother was Madame De Tenun, a famous salon host of that time, and his father was Chevalier Destouches-Canon, a cavalry officer. d'Alembert was immediately abandoned by his mother (she was an ex-nun who was afraid of being forcibly returned to the convent). However, his father quickly arranged for a home for d'Alembert—with a family of modest means named Rousseau. d'Alembert lived with this family for the next 47 years. Under the support of his father, d'Alembert was educated at the College de Quatre-Nations, where he studied law and medicine, and later turned to mathematics. For the remainder of his life, d'Alembert would consider himself a mathematician. By a program of self-study, d'Alembert learned the works of Newton and the Bernoullis. His early mathematics caught the attention of the Paris Academy of Sciences, of which he became a member in 1741. d'Alembert published frequently and sometimes rather hastily, in order to be in print before his competition. However, he made substantial contributions to the science of his time. For example, he was (1) the first to formulate the wave equation of classical physics, (2) the first to express the concept of a partial differential equation, (3) the first to solve a partial differential equation—he used separation of variables, and (4) the first to express the differential equations of fluid dynamics in terms of a field. His contemporary, Leonhard Euler (see Sections 1.1 and 3.18) later expanded greatly on these equations and was responsible for developing them into a truly rational approach for fluid-dynamic analysis.

During the course of his life, d'Alembert became interested in many scientific and mathematical subjects, including vibrations, wave motion, and celestial mechanics. In the 1750s, he had the honored position of science editor for the *Encyclopedia*—a major French intellectual endeavor of the eighteenth century which attempted to compile all existing knowledge into a large series of books. As he grew older, he also wrote papers on nonscientific subjects, mainly musical structure, law, and religion.

In 1765, d'Alembert became very ill. He was helped to recover by the nursing of Mlle. Julie de Lespinasse, who was d'Alembert's only love throughout his life. Although he never married, d'Alembert lived with Julie de Lespinasse until she died in 1776. d'Alembert had always been charming, renowned for his intelligence, gaiety, and considerable conversational ability. However, after Mlle. de Lespinasse's death, he became frustrated and morose—living a life of despair. He died in this condition on October 29, 1783, in Paris.

d'Alembert was one of the great mathematicians and physicists of the eighteenth century. He maintained active communications and dialogue with both Bernoulli and Euler and ranks with them as one of the founders of modern fluid dynamics. This, then, is the person behind the paradox, which has existed as an integral part of fluid dynamics for the past two centuries.

3.21 SUMMARY

Return to the road map given in Figure 3.4. Examine each block of the road map to remind yourself of the route we have taken in this discussion of the fundamentals of inviscid, incompressible flow. Before proceeding further, make certain that you feel comfortable with the detailed material represented by each block, and how each block is related to the overall flow of ideas and concepts.

For your convenience, some of the highlights of this chapter are summarized next:

Bernoulli's equation

$$p + \frac{1}{2}\rho V^2 = \text{const}$$

(a) Applies to inviscid, incompressible flows only.

(b) Holds along a streamline for a rotational flow.

(c) Holds at every point throughout an irrotational flow.

(d) In the form given above, body forces (such as gravity) are neglected, and steady flow is assumed.

Quasi-one-dimensional continuity equation

$$\rho A V = \text{const} \qquad \text{(for compressible flow)}$$

$$A V = \text{const} \qquad \text{(for incompressible flow)}$$

From a measurement of the Pitot pressure p_0 and static pressure p_1, the velocity of an incompressible flow is given by

$$V_1 = \sqrt{\frac{2(p_0 - p_1)}{\rho}} \tag{3.34}$$

Pressure coefficient

Definition:

$$C_p = \frac{p - p_\infty}{q_\infty} \tag{3.36}$$

where dynamic pressure is $q_\infty \equiv \frac{1}{2}\rho_\infty V_\infty^2$.

For incompressible steady flow with no friction:

$$C_p = 1 - \left(\frac{V}{V_\infty}\right)^2 \tag{3.38}$$

Governing equations

$$\nabla \cdot \mathbf{V} = 0 \qquad \text{(condition of incompressibility)} \tag{3.39}$$

$$\nabla^2 \phi = 0 \qquad \text{(Laplace's equation; holds for} \tag{3.40}$$
$$\text{irrotational, incompressible flow)}$$

or $\qquad \nabla^2 \psi = 0$ $\qquad\qquad\qquad\qquad\qquad\qquad\qquad\qquad$ (3.46)

Boundary conditions

$$u = \frac{\partial \phi}{\partial x} = \frac{\partial \psi}{\partial y} = V_\infty$$
$$\qquad\qquad\qquad\qquad\qquad \text{at infinity}$$
$$v = \frac{\partial \phi}{\partial y} = -\frac{\partial \psi}{\partial x} = 0$$

$$\mathbf{V} \cdot \mathbf{n} = 0 \qquad \text{at body (flow tangency condition)}$$

Elementary flows

(a) Uniform flow:

$$\phi = V_\infty x = V_\infty r \cos\theta \tag{3.53}$$

$$\psi = V_\infty y = V_\infty r \sin\theta \tag{3.55}$$

(b) Source flow:

$$\phi = \frac{\Lambda}{2\pi}\ln r \tag{3.67}$$

$$\psi = \frac{\Lambda}{2\pi}\theta \tag{3.72}$$

$$V_r = \frac{\Lambda}{2\pi r} \qquad V_\theta = 0 \tag{3.62}$$

(c) Doublet flow:

$$\phi = \frac{\kappa}{2\pi}\frac{\cos\theta}{r} \tag{3.88}$$

$$\psi = -\frac{\kappa}{2\pi}\frac{\sin\theta}{r} \tag{3.87}$$

(d) Vortex flow:

$$\phi = -\frac{\Gamma}{2\pi}\theta \tag{3.112}$$

$$\psi = \frac{\Gamma}{2\pi}\ln r \tag{3.114}$$

$$V_\theta = -\frac{\Gamma}{2\pi r} \qquad V_r = 0 \tag{3.105}$$

Inviscid flow over a cylinder

(a) Nonlifting (uniform flow and doublet)

$$\psi = (V_\infty r \sin\theta)\left(1 - \frac{R^2}{r^2}\right) \tag{3.92}$$

where R = radius of cylinder = $\kappa/2\pi V_\infty$.

Surface velocity: $$V_\infty = -2V_\infty \sin\theta \tag{3.100}$$

Surface pressure coefficient: $$C_p = 1 - 4\sin^2\theta \tag{3.101}$$

$$L = D = 0$$

(b) Lifting (uniform flow + doublet + vortex)

$$\psi = (V_\infty r \sin\theta)\left(1 - \frac{R^2}{r^2}\right) + \frac{\Gamma}{2\pi}\ln\frac{r}{R} \tag{3.118}$$

(continued)

Surface velocity: $\qquad V_\theta = -2V_\infty \sin\theta - \dfrac{\Gamma}{2\pi R}$ $\qquad\qquad$ (3.125)

$$L' = \rho_\infty V_\infty \Gamma \qquad \text{(lift per unit span)} \qquad (3.140)$$
$$D = 0$$

Kutta-Joukowski theorem
For a closed two-dimensional body of arbitrary shape, the lift per unit span is
$L' = \rho_\infty V_\infty \Gamma$.

Source panel method
This is a numerical method for calculating the nonlifting flow over bodies of arbitrary shape. Governing equations:

$$\frac{\lambda_i}{2} + \sum_{\substack{j=1 \\ (j\neq1)}}^{n} \frac{\lambda_j}{2\pi} \int_j \frac{\partial}{\partial n_i}(\ln r_{ij})\,ds_j + V_\infty \cos\beta_i = 0 \qquad (i=1,2,\dots,n) \quad (3.152)$$

3.22 INTEGRATED WORK CHALLENGE: RELATION BETWEEN AERODYNAMIC DRAG AND THE LOSS OF TOTAL PRESSURE IN THE FLOW FIELD

Concept: The concept of total pressure in a flow is introduced in Section 3.4 in conjunction with the use of a Pitot tube for airspeed measurement. However, total pressure is more significant than just the pressure that exists at a stagnation point on a body. It is, by general definition, the pressure that would exist at any point in a flow if the flow velocity were somehow magically slowed to zero velocity at that point without the influence of the dissipative effects of friction or thermal conduction. A more complete examination of the definition of total pressure is discussed in Section 7.5 dealing with the general case of a compressible flow. For our purposes here, we will continue our treatment of an incompressible flow, where from Equation (3.35), the total pressure, p_0, is given by $p_0 = p_1 + \frac{1}{2}\rho V_1^2$. This relation stems from Bernoulli's equation, and hence applies to an incompressible flow with no friction. Note, however, that the "no friction" criterion applies only to that part of the flow associated with the slowing-down process from the local flow velocity to zero. As discussed in Section 3.4, this slowing-down process can be real, as occurs when the flow comes to a stop at the entrance of a Pitot tube or at a stagnation point on a body. Or, the slowing-down process can be imaginary, where at any given point in a flow we imagine the flow slowed to zero velocity *at that point*, and then the pressure that *would* exist *at that point*, is by definition the *total pressure* at that point. For example, consider two points in a flow, points 1 and 2, where the static pressure and flow velocity are p_1, V_1, p_2, and V_2, respectively. At point 1,

$$p_{0_1} = p_1 + \frac{1}{2}\rho V_1^2$$

and at point 2,

$$p_{0_2} = p_2 + \frac{1}{2}\rho V_2^2$$

In general, at these two points, due to dissipative effects in the flow, the total pressure will not be the same, i.e., $p_{0_1} \neq p_{0_2}$. Only in the case of flows without internal dissipation, i.e., without external friction or thermal conduction, total pressure will be the same. Of course, in this chapter we are dealing with a frictionless flow, so for all the flows considered here, assuming a uniform freestream ahead of the body or entering a wind tunnel, the total pressure is constant throughout the flow.

Total pressure is a measure of the ability of a flow to do some kind of "useful work." When there is the *loss* of total pressure, the flow loses some of its capacity to do useful work. For example, in the flow through a jet engine, a loss of total pressure causes a net loss of engine thrust. In the external flow over an aerodynamic body, the drag on that body causes a net loss of total pressure in the flow. It is precisely this effect that is the subject of the present Integrated Work Challenge.

Challenge: Examine and derive a relation between the aerodynamic drag on a body and the loss of total pressure in the flow field.

Solution: Consider the control volume sketched in Figure 2.20a. Here, the drag per unit span on a body inside the control volume is related to the loss of momentum in the flow, and the relation is given by a combination of Equations (2.78) and (2.79) as

$$D' = -\oiint_S (\rho \mathbf{V} \cdot \mathbf{ds})u = \int_i^a \rho_1 u_1^2 dy - \int_h^b \rho_2 u_2^2 dy \qquad (C3.1)$$

Along station $i - a$,

$$p_{0_1} = p_1 + \frac{1}{2}\rho u_1^2$$

where p_1 and u_1 are constant over the station, and u_1 is the freestream velocity V_∞ ahead of the body. Thus,

$$p_{0_1} = p_1 + \frac{1}{2}\rho V_\infty^2 \qquad (C3.2)$$

Along station $h - b$,

$$p_{0_2} = p_2 + \frac{1}{2}\rho u_2^2 \qquad (C3.3)$$

where p_2 is constant over the station but u_2 is a variable in the downstream wake of the body.

The total pressure integrated over the inflow station is

$$\int_i^a p_{0_1}\, dy = \int_i^a \left(p_1 + \frac{1}{2}\rho V_\infty^2\right) dy$$

and the total pressure integrated over the outflow station is

$$\int_h^b p_{0_1}\, dy = \int_h^b \left(p_2 + \frac{1}{2}\rho u_2^2\right) dy$$

The net integrated *loss* of total pressure between the inflow and outflow stations is the difference between the two integrals. Denoting the net integrated loss of total pressure by *IL*, we have

$$IL = \int_i^a \left(p_1 + \frac{1}{2}\rho V_\infty^2\right) dy - \int_h^b \left(p_2 + \frac{1}{2}\rho u_2^2\right) dy \qquad (C3.4)$$

The static pressure across stations *ai* and *hb* is constant and equal to p_∞. Also, consider the streamlines forming the upper and lower boundaries of the control volume in Figure 2.20a to be far enough apart to be essentially straight, parallel streamlines; thus the distances *ia* and *hb* are the same. Hence, in Equation (3.4), the terms involving the static pressure become

$$\int_i^a p_1\, dy - \int_h^b p_2\, dy = \int_i^a p_\infty\, dy - \int_i^a p_\infty\, dy = 0$$

and Equation (3.4) reduces to

$$IL = \int_i^a \left(\frac{1}{2}\rho V_\infty^2 - \frac{1}{2}\rho u_2^2\right) dy \qquad (C3.5)$$

or, since ρ is constant

$$IL = \frac{1}{2}\rho \int_i^a \left(V_\infty^2 - u_2^2\right) dy \qquad (C3.6)$$

However, from Equation (C3.1),

$$\rho \int_i^a \left(V_\infty^2 - u_2^2\right) dy = D' \qquad (C3.7)$$

Comparing Equations (C3.6) and (C3.7), we have

$$\boxed{IL = \frac{1}{2}D'} \qquad (C3.8)$$

That is, there is a net loss of total pressure per unit depth from the inflow to the outflow boundaries, and this integrated loss of total pressure is equal to one-half of the aerodynamic drag per unit span exerted on the body immersed in the flow.

So we see that there is a connection between drag and the loss of total pressure. Total pressure is a measure of the capacity of the flow to do useful work. Drag is a force opposing the forward motion of a body; drag is a mechanism that creates a kind of "negative work" that must be countered by an increase in the positive work to move the body. In the flow over a body, drag serves to reduce the local flow field velocity, which is reflected in a reduction of total pressure.

3.23 INTEGRATED WORK CHALLENGE: CONCEPTUAL DESIGN OF A SUBSONIC WIND TUNNEL

Concept: The aerodynamic flow in the test section of a low-speed subsonic wind tunnel is discussed in Section 3.3. Two basic configurations for such wind tunnels are sketched in Figure 3.8a and b, an open-circuit tunnel and a closed-circuit tunnel, respectively. These wind tunnels are aerodynamic devices, in the same spirit that an airplane is an aerodynamic device. And like an airplane that is powered by some type of motor, so is a wind tunnel powered by a motor, as shown in Figure 3.8. The conceptual design of an airplane starts with the specification of the requirements to be met by the airplane, and a preliminary calculation of the power required to meet these specifications is an essential part of the conceptual design process. (See Anderson, *Aircraft Performance and Design*, 1999, McGraw Hill, Chapter 7.) Similarly, the conceptual design of a wind tunnel starts with a specification for the aerodynamic flow in the test section, and the power required to drive the wind tunnel flow to meet the specified conditions is an essential part of the conceptual design process of the tunnel. The purpose of this Integrated Work Challenge is to illustrate this design process.

Challenge: Consider that you have been given the job of designing a new subsonic wind tunnel. The specifications call for a maximum test-section flow velocity of 120 m/s and a test section size that can accommodate an airplane model with a wing span of 2 m. The Reynolds number capability in the flow direction must be 25 million. The specifications also require a closed-circuit wind tunnel because it takes less power to run such a tunnel compared with an open circuit for the same test section conditions. Of course, just a glance at Figure 3.8 indicates that a closed-circuit tunnel requires a great deal more laboratory space than an open-circuit tunnel, but for the present design we assume there are no space limitations. Your challenge is to prepare a conceptual design for the new wind tunnel, and in particular to estimate the power required to drive the flow around the tunnel circuit.

Solution: Step one is to determine the *size* of the *test section*. We choose a rectangular shape for the test section, and we need to determine the width and height of the cross section perpendicular to the flow and the length of the test section in the flow direction. For the most part, there is no precise technical calculation for the test section; indeed, there are no "right" dimensions to be had, just like there is no one "right" shape for a new airplane. There are a multitude of viable shapes and sizes. When a designer chooses a size and shape, he or she is guided by existing designs that have been proven to work. For our Integrated Work Challenge, we will examine previous wind tunnel designs and make some decisions based on these designs. An excellent source of such design information is Barlow, Rae, and Pope, *Low-Speed Wind Tunnel Testing,* Third Edition, 1999, John Wiley and Sons, New York. Our specifications call for a test section with a width large enough such that an airplane model with a 2-meter wingspan would comfortably fit. Barlow et al. recommend that the maximum wing span of a model be less than 0.8 of the tunnel width in order to minimize the effects of the tunnel walls on the aerodynamic measurements. Let us design our test section width such that the model wingspan will be about 0.7 of the tunnel width. Thus,

$$\text{Width} = \frac{2 \text{ m}}{0.7} = 2.86 \text{ m}$$

For convenience during the fabrication of the tunnel, we will round the width to the whole number of 3 m. Barlow et al. also suggest that for testing airplane models, a rectangular test section with a width-to-height ratio of about 1.5 will minimize the wall correction factor on the measured data. Thus, we choose the height of the test section to be

$$\text{Height} = \frac{\text{width}}{1.5} = \frac{3 \text{ m}}{1.5} = 2 \text{ m}$$

So our tunnel will have a 2 × 3-m test section. Parenthetically, we note (see Barlow et al., p. 102) that the subsonic wind tunnel established years ago by Boeing at the University of Washington has a rectangular test section that is 8 × 12 ft, the same 2 to 3 ratio chosen for our design.

Our test section must be long enough to provide the maximum specified test Reynolds number of 25 million (25×10^6). To achieve this Reynolds number, recalling that the specified flow velocity in the test section is 120 m/s, noting that the test-section flow temperature and density correspond to standard sea level conditions (see Appendix D) of $T_\infty = 288.16$ K and $\rho_\infty = 1.225$ kg/m^3, and the viscosity coefficient at this temperature is 1.7894×10^{-5} kg/(m)(s) from Section 1.11, the required test section length ℓ is obtained from:

$$\text{Re} = \rho_\infty V_\infty \frac{\ell}{\mu_\infty} = 25 \times 10^6$$

or,

$$\ell = \frac{\mu_\infty \text{Re}}{\rho_\infty V_\infty} = \frac{(1.7894 \times 10^{-5})(25 \times 10^6)}{(1.225)(120)}$$

$$\ell = 3.046 \text{ m}$$

Again we will be conservative and choose the test section length to be slightly longer, namely 3.2 m. So the overall size of our rectangular-shaped text section will be

$$\text{Height:width:length} = 2 \text{ m} \times 3 \text{ m} \times 3.2 \text{ m}$$

Caution: An airplane that is designed larger than it needs to be will have a larger weight and more surface area that generates more drag than necessary, hence requiring engines with more power than they need. Similarly, a wind tunnel with a test section that is larger than it needs to be will require more mass flow of air, and the wetted surface area of the walls being more than it needs to be will generate a larger friction loss than necessary, both requiring more motor power to drive the wind tunnel than necessary. Hence, in our determination of the size of the test section for our wind tunnel, we have been a little conservative to ensure the generation of a proper flow for accurate aerodynamic measurements, but we must be careful not to be too conservative.

Step two of our conceptual design is an *estimate of the power required to drive the wind tunnel*. For this, we first calculate the energy of the flow in the test section. The kinetic energy per unit mass of gas in the test section, where V_t is the flow velocity in the test section, is $\frac{1}{2}V_t^2$. The mass flow of air through the test section with cross-sectional area A_t is $\dot{m}_t = \rho_t A_t V_t$. The jet power in the test section, P_t, is the time rate of flow of energy through the test section, given by

$$P_t = (\text{mass flow})(\text{kinetic energy per unit mass})$$

$$= (\rho_t A_t V_t)\left(\frac{1}{2}V_t^2\right)$$

or,

$$P_t = \frac{1}{2}\rho_t A_t V_t^3 \qquad\qquad (C3.9)$$

The function of the motor in a closed-return wind tunnel is to provide power to first get the air moving through the tunnel, but after the flow is up to speed, then the motor power is required to overcome losses in the flow as it moves around the tunnel circuit. If there were no losses, no motor power would be required, and the tunnel would be a perpetual motion machine. Nature ensures that this does not happen, so our next conceptual design step is to estimate the losses as the air flows around the tunnel circuit.

Referring to our discussion in Section 3.22 of the connection between drag on an aerodynamic body and the loss of total pressure in the flow around that body, in a similar fashion the loss in any section of the wind tunnel is defined as the mean loss of total pressure of the air flow as it passes through that section. The total pressure is

$$p_0 = p + \frac{1}{2}\rho V^2 \qquad\qquad (C3.10)$$

and for the external flow over an aerodynamic body, the loss in p_0 is due to the loss of velocity and hence loss of dynamic pressure $\frac{1}{2}\rho V^2$ in Equation (C3.10). However, for the internal flow through a wind tunnel, the principle of mass conservation applied at any particular section of the tunnel dictates that there can be no loss of velocity due to dissipative effects. From the continuity equation for an incompressible flow, Equation (9.22), we have

$$AV = \text{constant}$$

and since A is fixed by the wind tunnel design, any change in velocity between two locations in the tunnel is simply due to the area change between these two sections. Nevertheless, dissipation occurs in the flow (friction due to the flow along the tunnel walls, etc.), and this loss does show up as a loss of total pressure. However, referring to Equation (C3.10), this loss appears as a decrease in the static pressure, p, compared with what the static pressure would be if there were no dissipative losses. In Equation (C3.10), p_0 and p are frequently referred to as the "total head" and "static head," respectively. As stated by Barlow et al. (p. 73):

> "There will be equal drops in static head and in total head corresponding to the friction loss. Throughout the wind tunnel the losses that occur appear as successive pressure drops to be balanced by the pressure rise through the fan. The total pressure drop will be the pressure rise required by the fan."

As shown in Figure 3.8, the fan is connected to the motor and is the mechanism that supplies energy to the air flow, analogous to the propeller on an airplane that converts engine power to thrust power to propel the airplane.

In order to estimate the power required to drive the wind tunnel, referring to Figure 3.8, we should estimate the total pressure drop in each section of the wind tunnel, such as the nozzle, test section, diffuser, turning vanes in the corners, etc. Let us denote the rate of flow losses in the complete circuit as P_c. Defining the energy ratio, E_R, as

$$E_R = \frac{P_t}{P_c} \tag{C3.11}$$

we could estimate the energy ratio that pertains to each tunnel section. There are engineering methods for making such estimates, as nicely detailed in Barlow et al. For our purposes, however, we will make an estimate of the value of E_R for the complete wind tunnel by looking at values from existing wind tunnels. From a tabulation in Barlow et al. (p. 102), the wind tunnel closest in size to our design is the University of Washington 8×12 foot tunnel with an energy ratio of 8.3. This is the highest value of any of the tunnels listed in the tabulation and reflects a highly efficient design for the university's tunnel. Using this energy ratio for our conceptual design, we have from Equation (C3.11)

$$P_c = \frac{P_t}{E_R} \tag{C3.12}$$

where the jet power in our 2×3-m test section is obtained from Equation (C3.9)

$$P_t = \frac{1}{2} \rho_t A_t V_t^3$$

$$= \frac{1}{2} (1.23)(2)(3)(120)^3 = 6.376 \times 10^6 \text{ W}$$

From Equation (C3.9),

$$P_c = \frac{P_t}{E_R} = \frac{6.376 \times 10^6}{8.3} = 7.682 \times 10^6 \text{ W}$$

Noting that 746 W = 1 hp, the rate of flow losses in the wind tunnel is

$$P_c = \frac{7.682 \times 10^5}{746} = \boxed{1030 \text{ hp}}$$

This is the power that the fan must supply to the flow. Assuming no losses from the motor to the fan and then to the flow (i.e., 100% motor and fan efficiency), the motor must supply 1030 hp to drive the wind tunnel.

To summarize the basic conceptual design of our subsonic wind tunnel:

Tunnel circuit: Closed return, such as in Figure 3.8*b*
Size of test section: 2 m × 3 m × 3.2 m
Flow velocity in the test section: 120 m/s
Motor power required: 1030 hp

3.24 PROBLEMS

Note: All the following problems assume an inviscid, incompressible flow. Also, standard sea level density and pressure are 1.23 kg/m³ (0.002377 slug/ft³) and 1.01×10^5 N/m² (2116 lb/ft²), respectively.

3.1 For an irrotational flow, show that Bernoulli's equation holds between *any* points in the flow, not just along a streamline.

3.2 Consider a venturi with a throat-to-inlet area ratio of 0.8, mounted on the side of an airplane fuselage. The airplane is in flight at standard sea level. If the static pressure at the throat is 2100 lb/ft², calculate the velocity of the airplane.

3.3 Consider a venturi with a small hole drilled in the side of the throat. This hole is connected via a tube to a closed reservoir. The purpose of the venturi is to create a vacuum in the reservoir when the venturi is placed in an airstream. (The *vacuum* is defined as the pressure difference *below* the outside ambient pressure.) The venturi has a throat-to-inlet area ratio of 0.85. Calculate the maximum vacuum obtainable in the reservoir when the venturi is placed in an airstream of 90 m/s at standard sea level conditions.

3.4 Consider a low-speed open-circuit subsonic wind tunnel with an inlet-to-throat area ratio of 12. The tunnel is turned on, and the pressure difference between the inlet (the settling chamber) and the test section is read as a height difference of 10 cm on a U-tube mercury manometer. (The density of liquid mercury is 1.36×10^4 kg/m³.) Calculate the velocity of the air in the test section.

3.5 Assume that a Pitot tube is inserted into the test-section flow of the wind tunnel in Problem 3.4. The tunnel test section is completely sealed from the outside ambient pressure. Calculate the pressure measured by the Pitot tube, assuming the static pressure at the tunnel inlet is atmospheric.

3.6 A Pitot tube on an airplane flying at standard sea level reads 1.07×10^5 N/m^2. What is the velocity of the airplane?

3.7 At a given point on the surface of the wing of the airplane in Problem 3.6, the flow velocity is 130 m/s. Calculate the pressure coefficient at this point.

3.8 Consider a uniform flow with velocity V_∞. Show that this flow is a physically possible incompressible flow and that it is irrotational.

3.9 Show that a source flow is a physically possible incompressible flow everywhere except at the origin. Also show that it is irrotational everywhere.

3.10 Prove that the velocity potential and the stream function for a uniform flow, Equations (3.53) and (3.55), respectively, satisfy Laplace's equation.

3.11 Prove that the velocity potential and the stream function for a source flow, Equations (3.67) and (3.72), respectively, satisfy Laplace's equation.

3.12 Consider the flow over a semi-infinite body as discussed in Section 3.11. If V_∞ is the velocity of the uniform stream, and the stagnation point is 1 ft upstream of the source:

a. Draw the resulting semi-infinite body to scale on graph paper.

b. Plot the pressure coefficient distribution over the body; that is, plot C_p versus distance along the centerline of the body.

3.13 Derive Equation (3.81). *Hint:* Make use of the symmetry of the flow field shown in Figure 3.23; that is, start with the knowledge that the stagnation points must lie on the axis aligned with the direction of V_∞.

3.14 Derive the velocity potential for a doublet; that is, derive Equation (3.88). *Hint:* The easiest method is to start with Equation (3.87) for the stream function and extract the velocity potential.

3.15 Consider the nonlifting flow over a circular cylinder. Derive an expression for the pressure coefficient at an arbitrary point (r, θ) in this flow, and show that it reduces to Equation (3.101) on the surface of the cylinder.

3.16 Consider the nonlifting flow over a circular cylinder of a given radius, where $V_\infty = 20$ ft/s. If V_∞ is doubled, that is, $V_\infty = 40$ ft/s, does the shape of the streamlines change? Explain.

3.17 Consider the lifting flow over a circular cylinder of a given radius and with a given circulation. If V_∞ is doubled, keeping the circulation the same, does the shape of the streamlines change? Explain.

3.18 The lift on a spinning circular cylinder in a freestream with a velocity of 30 m/s and at standard sea level conditions is 6 N/m of span. Calculate the circulation around the cylinder.

3.19 A typical World War I biplane fighter (such as the French SPAD shown in Figure 3.50) has a number of vertical interwing struts and diagonal bracing wires. Assume for a given airplane that the total length for the vertical struts (summed together) is 25 ft, and that the struts are cylindrical with a diameter of 2 in. Assume also that the total length of the bracing wires is 80 ft, with a cylindrical diameter of $\frac{3}{32}$ in. Calculate the drag (in pounds) contributed by these struts and bracing wires when the airplane is flying at 120 mi/h at standard sea level. Compare this component of drag with the total zero-lift drag for the airplane, for which the total wing area is 230 ft^2 and the zero-lift drag coefficient is 0.036.

3.20 The Kutta-Joukowski theorem, Equation (3.140), was derived exactly for the case of the lifting cylinder. In Section 3.16, it is stated without proof that Equation (3.140) also applies in general to a two-dimensional body of arbitrary shape. Although this general result can be proven mathematically, it also can be accepted by making a physical argument as well. Make this physical argument by drawing a closed curve around the body where the closed curve is very far away from the body, so far away that in perspective the body becomes a very small speck in the middle of the domain enclosed by the closed curve.

3.21 Consider the streamlines over a circular cylinder as sketched at the right of Figure 3.26. Single out the first three streamlines flowing over the top of the cylinder. Designate each streamline by its stream function, ψ_1, ψ_2, and ψ_3. The first streamline wets the surface of the cylinder; designate $\psi_1 = 0$. The streamline above that is ψ_2, and the next one above that is ψ_3. Assume the streamlines start out in the freestream equally spaced. Hence, the volume flow rates between the streamlines are the same. The streamline ψ_2 passes through the point $(1.2R, \pi/2)$ directly above the top of the cylinder. Calculate the location of the point directly above the top of the cylinder through which the streamline ψ_3 flows. Comment on the spacing between the streamlines directly above the top.

3.22 Consider the flow field over a circular cylinder mounted perpendicular to the flow in the test section of a low-speed subsonic wind tunnel. At standard sea level conditions, if the flow velocity at some region of the flow field exceeds about 250 mi/h, compressibility begins to have an effect in that region. Calculate the velocity of the flow in the test section of the wind tunnel above which compressibility effects begin to become important, i.e., above which we cannot accurately assume totally incompressible flow over the cylinder for the wind tunnel tests.

3.23 Prove that the flow field specified in Example 2.1 is not incompressible; i.e., it is a compressible flow as stated without proof in Example 2.1.

3.24 Consider the velocity given by:

$$V = (9x^2 + 5y)\hat{i} + (5x)\hat{j} + \hat{k}$$

Does a velocity potential exist and if so, what is it?

3.25 Prove that flows that can be represented using a velocity potential are irrotational.

3.26 Calculate the stream function and angular velocity associated with the following velocity field:

$$V = -y^3\hat{i} + x^4\hat{j}$$

Can this flow have a velocity potential?

CHAPTER 4

Incompressible Flow
over Airfoils

Of the many problems now engaging attention, the following are considered of immediate importance and will be considered by the committee as rapidly as funds can be secured for the purpose.... The evolution of more efficient wing sections of practical form, embodying suitable dimensions for an economical structure, with moderate travel of the center-of-pressure and still affording a large range of angle-of-attack combined with efficient action.

**From the first Annual Report of the
NACA, 1915**

PREVIEW BOX

Imagine that you have just been given an airfoil of a particular shape at a certain angle of attack to a given low-speed flow, and you have been asked to obtain the lift (or more importantly, the lift coefficient) of the airfoil. What do you do (besides panicking)? Your first inclination might be to make a model of the airfoil, put it in a low-speed wind tunnel, and measure the lift coefficient. This is indeed what aerodynamicists have been doing for more than 100 years. The early part of this chapter discusses such experimental measurements of airfoil properties in low-speed wind tunnels. These measurements give you an immediate feel for airfoil lift, drag, and moment coefficients as a function

of angle of attack. The experimental measurements give you a fast track toward obtaining a comfortable and practical understanding as to how airfoils behave. That is what the first three sections of this chapter are all about.

Most of the rest of this chapter, however, deals with our second inclination as to how to obtain the airfoil properties, namely, to *calculate* them. This is a horse of a different color. You will be introduced to the elegant circulation theory of lift—the crowning jewel of inviscid, incompressible flow theory for the calculation of lift. At the turn of the twentieth century, the circulation theory of lift was a breakthrough in the theoretical prediction of lift. In this chapter, we first apply this theory to thin airfoils at small angles of attack; thin airfoil theory was developed in Germany during World War I and is by far the most tractable means of obtaining analytical solutions for lift and moments on an airfoil. But, as we shall see, thin airfoil theory, as its name implies, holds only for thin airfoils at small angles of attack. This is not as restrictive as it seems, however, because many airplanes over the past years have relatively thin airfoils, and cruise at relatively small angles of attack. Thin airfoil theory gives us a lot of practical results, plus the intellectual gratification of carrying through some elegant theoretical thinking—give it a good read and I think you will like it.

Since the 1960s, the advent and development of the high-speed digital computer allowed detailed numerical solutions based on the circulation theory of lift, solutions for the lift on a body of arbitrary shape and thickness at any angle of attack, subject of course to the assumption of inviscid potential flow. These numerical solutions, an extension of the panel solutions discussed in Section 3.17, are discussed toward the end of this chapter—they are the "gold standard" for low-speed, inviscid-flow airfoil calculations, and are used throughout the aeronautical industry and by many aeronautical research and development laboratories. The concept of panel solutions is an inspired numerical application of the circulation theory of lift, and it has opened the door to the analysis of practically any airfoil shape at any angle of attack.

Airfoils come in many different shapes. An historical sequence of airfoil shapes through 1935 is shown in Figure 4.1. Beginning in 1938, the National Advisory Committee for Aeronautics (NACA) developed a revolutionary series of airfoil shapes designed to encourage laminar flow in the boundary layer over the airfoil, hence dramatically reducing skin friction drag on the airfoil; the shape of a representative laminar-flow airfoil is given in Figure 4.2. Although these shapes never produced the desired amount of laminar flow in practice, by a stroke of serendipity they proved to be excellent high-speed airfoils for jet-powered airplanes after 1945. Beginning in 1965, National Aeronautics and Space Administration (NASA)

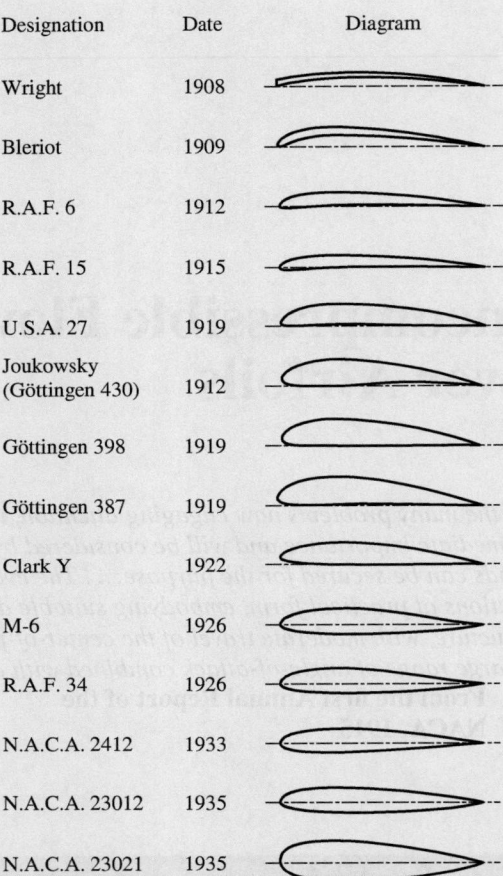

Designation	Date	Diagram
Wright	1908	
Bleriot	1909	
R.A.F. 6	1912	
R.A.F. 15	1915	
U.S.A. 27	1919	
Joukowsky (Göttingen 430)	1912	
Göttingen 398	1919	
Göttingen 387	1919	
Clark Y	1922	
M-6	1926	
R.A.F. 34	1926	
N.A.C.A. 2412	1933	
N.A.C.A. 23012	1935	
N.A.C.A. 23021	1935	

Figure 4.1 Historical sequence of airfoil sections. (*Source: NASA.*)

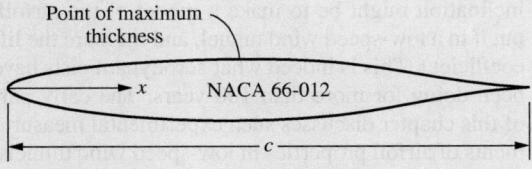

Figure 4.2 Laminar-flow airfoil shape.

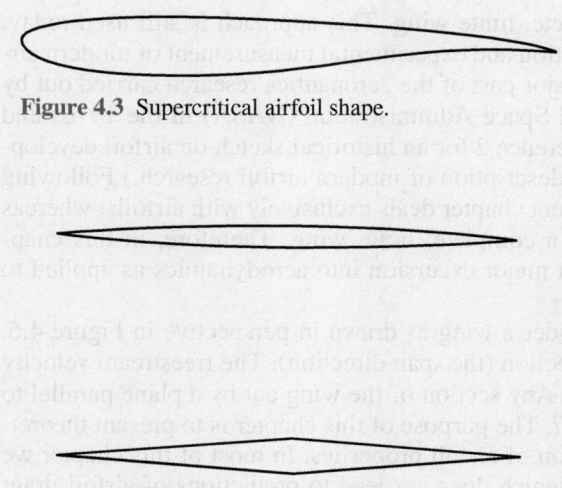

Figure 4.3 Supercritical airfoil shape.

Figure 4.4 Supersonic airfoil shapes.

developed another series of revolutionary airfoil shapes, supercritical airfoils, designed for efficient flight near Mach one; a typical supercritical airfoil shape is shown in Figure 4.3. Classic airfoil shapes for supersonic flow are shown in Figure 4.4; note the very slender profiles with sharp leading edges. All the airfoil shapes shown in Figure 4.1 through 4.4 were designed for specific purposes in their time and have been used on untold numbers of different airplanes. Today, the proper design of new airfoil shapes is more important than ever. Using numerical

techniques, aircraft companies usually custom-design the airfoil shapes for new airplanes, shapes that best fit the design requirements for the specific airplane. This chapter is exclusively devoted to the study of airfoils; it discusses the fundamental aspects of airfoil aerodynamics—aspects that are the heart of airfoil design and performance.

Figure 4.5 shows an airplane in flight, sustained in the air by the aerodynamic action of its wing. Airplane wings are made up of airfoil shapes. The first step in understanding the aerodynamics of wings is to understand the aerodynamics of airfoils. Airfoil aerodynamics is important stuff—it is the stuff of this chapter. Moreover, it is truly interesting. It is fun to visualize the flow over an airfoil and to learn how to calculate the resulting lift on the airfoil. Read on and enjoy.

Figure 4.5 DeHaviland DHC-6 Twin Otter. (*Alan & Sandy Carey/Photodisc/Alamy Stock Photo*)

4.1 INTRODUCTION

With the advent of successful powered flight at the turn of the twentieth century, the importance of aerodynamics ballooned almost overnight. In turn, interest grew in the understanding of the aerodynamic action of such lifting surfaces as fixed wings on airplanes and, later, rotors on helicopters. In the period 1912–1918, the analysis of airplane wings took a giant step forward when Ludwig Prandtl and his colleagues at Göttingen, Germany, showed that the aerodynamic consideration of wings could be split into two parts: (1) the study of the *section* of a wing—an airfoil—and (2) the modification of such airfoil properties to account for the complete, finite wing. This approach is still used today; indeed, the theoretical calculation and experimental measurement of modern airfoil properties have been a major part of the aeronautics research carried out by the National Aeronautics and Space Administration (NASA) in the 1970s and 1980s. (See Chapter 5 of Reference 2 for an historical sketch on airfoil development and Reference 10 for a description of modern airfoil research.) Following Prandtl's philosophy, the present chapter deals exclusively with airfoils, whereas Chapter 5 treats the case of a complete, finite wing. Therefore, in this chapter and Chapter 5, we make a major excursion into aerodynamics as applied to airplanes.

What is an airfoil? Consider a wing as drawn in perspective in Figure 4.6. The wing extends in the y direction (the span direction). The freestream velocity V_∞ is parallel to the xz plane. Any section of the wing cut by a plane parallel to the xz plane is called an *airfoil*. The purpose of this chapter is to present theoretical methods for the calculation of airfoil properties. In most of this chapter we will deal with inviscid flow, which does not lead to predictions of airfoil drag; indeed, d'Alembert's paradox says that the drag on an airfoil is zero—clearly not a realistic answer. We will have to wait until Section 4.12 and Chapter 15 and a discussion of viscous flow before predictions of drag can be made. However, the lift and moments on the airfoil are due mainly to the pressure distribution, which (below the stall) is dictated by inviscid flow. Therefore, this chapter concentrates on the theoretical prediction of airfoil lift and moments.

The road map for this chapter is given in Figure 4.7. After some initial discussion on airfoil nomenclature and characteristics, we present two approaches to low-speed airfoil theory. One is the classical thin airfoil theory developed during the period 1910–1920 (the right-hand branch of Figure 4.7). The other is the modern numerical approach for arbitrary airfoils using vortex panels

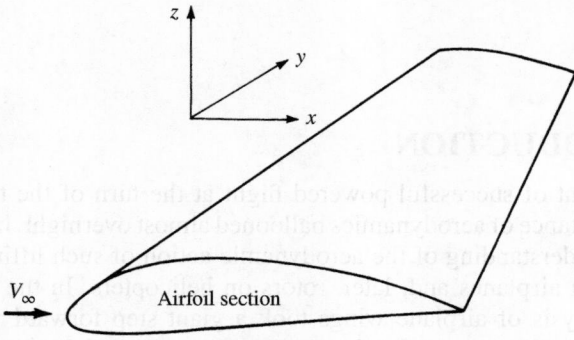

Figure 4.6 Definition of an airfoil.

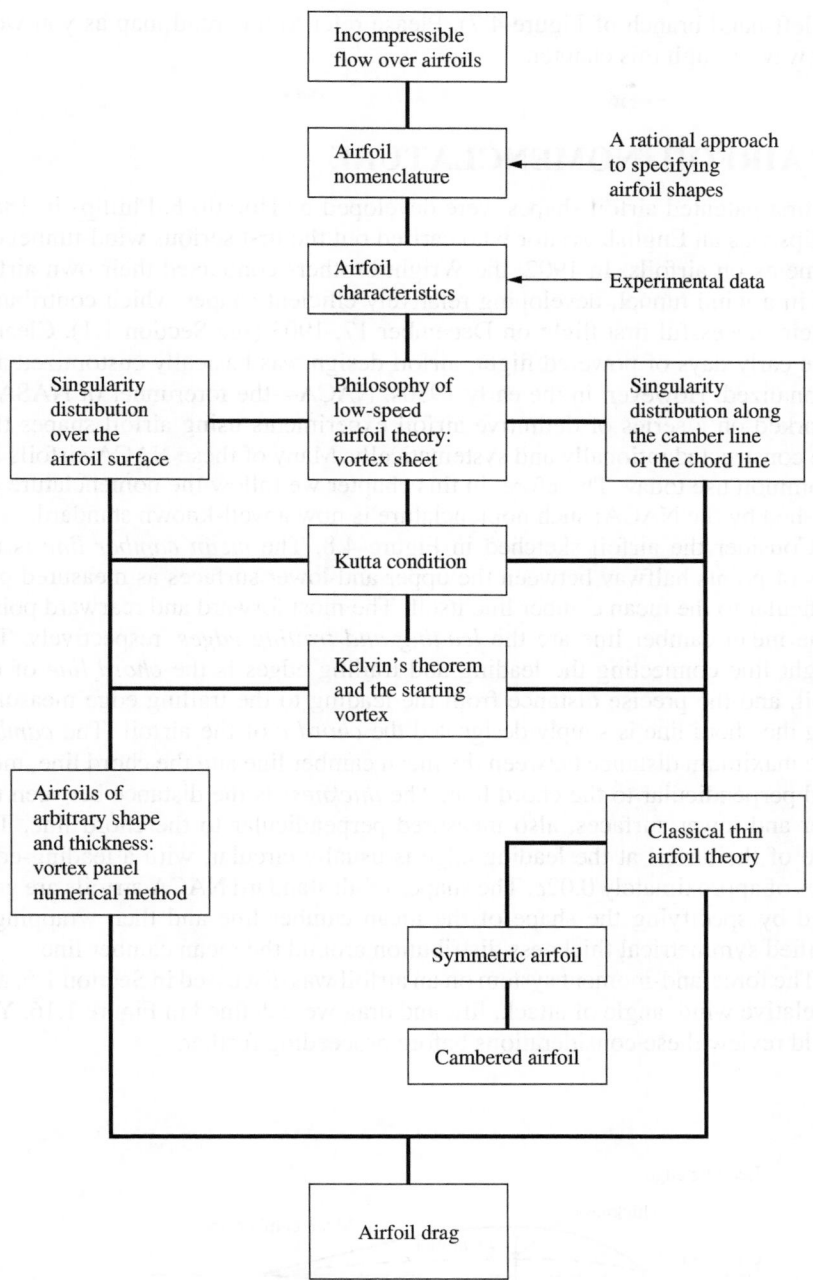

Figure 4.7 Road map for Chapter 4.

(the left-hand branch of Figure 4.7). Please refer to this road map as you work your way through this chapter.

4.2 AIRFOIL NOMENCLATURE

The first patented airfoil shapes were developed by Horatio F. Phillips in 1884. Phillips was an English aviator who carried out the first serious wind-tunnel experiments on airfoils. In 1902, the Wright brothers conducted their own airfoil tests in a wind tunnel, developing relatively efficient shapes which contributed to their successful first flight on December 17, 1903 (see Section 1.1). Clearly, in the early days of powered flight, airfoil design was basically customized and personalized. However, in the early 1930s, NACA—the forerunner of NASA—embarked on a series of definitive airfoil experiments using airfoil shapes that were constructed rationally and systematically. Many of these NACA airfoils are in common use today. Therefore, in this chapter we follow the nomenclature established by the NACA; such nomenclature is now a well-known standard.

Consider the airfoil sketched in Figure 4.8. The *mean camber line* is the locus of points halfway between the upper and lower surfaces as measured perpendicular to the mean camber line itself. The most forward and rearward points of the mean camber line are the *leading and trailing edges,* respectively. The straight line connecting the leading and trailing edges is the *chord line* of the airfoil, and the precise distance from the leading to the trailing edge measured along the chord line is simply designated the *chord c* of the airfoil. The *camber* is the maximum distance between the mean camber line and the chord line, measured perpendicular to the chord line. The *thickness* is the distance between the upper and lower surfaces, also measured perpendicular to the chord line. The shape of the airfoil at the leading edge is usually circular, with a leading-edge radius of approximately $0.02c$. The shapes of all standard NACA airfoils are generated by specifying the shape of the mean camber line and then wrapping a specified symmetrical thickness distribution around the mean camber line.

The force-and-moment system on an airfoil was discussed in Section 1.5, and the relative wind, angle of attack, lift, and drag were defined in Figure 1.16. You should review these considerations before proceeding further.

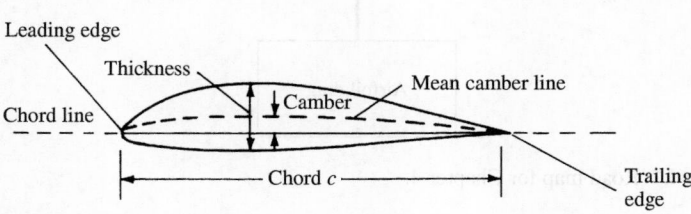

Figure 4.8 Airfoil nomenclature.

The NACA identified different airfoil shapes with a logical numbering system. For example, the first family of NACA airfoils, developed in the 1930s, was the "four-digit" series, such as the NACA 2412 airfoil. Here, the first digit is the maximum camber in hundredths of chord, the second digit is the location of maximum camber along the chord from the leading edge in tenths of chord, and the last two digits give the maximum thickness in hundredths of chord. For the NACA 2412 airfoil, the maximum camber is $0.02c$ located at $0.4c$ from the leading edge, and the maximum thickness is $0.12c$. It is common practice to state these numbers in percent of chord, that is, 2 percent camber at 40 percent chord, with 12 percent thickness. An airfoil with no camber, that is, with the camber line and chord line coincident, is called a *symmetric airfoil*. Clearly, the shape of a symmetric airfoil is the same above and below the chord line. For example, the NACA 0012 airfoil is a symmetric airfoil with a maximum thickness of 12 percent.

The second family of NACA airfoils was the "five-digit" series, such as the NACA 23012 airfoil. Here, the first digit when multiplied by $\frac{3}{2}$ gives the design lift coefficient[1] in tenths, the next two digits when divided by 2 give the location of maximum camber along the chord from the leading edge in hundredths of chord, and the final two digits give the maximum thickness in hundredths of chord. For the NACA 23012 airfoil, the design lift coefficient is 0.3, the location of maximum camber is at $0.15c$, and the airfoil has 12 percent maximum thickness.

One of the most widely used family of NACA airfoils is the "6-series" laminar flow airfoils, developed during World War II. An example is the NACA 65-218. Here, the first digit simply identifies the series, the second gives the location of minimum pressure in tenths of chord from the leading edge (for the basic symmetric thickness distribution at zero lift), the third digit is the design lift coefficient in tenths, and the last two digits give the maximum thickness in hundredths of chord. For the NACA 65-218 airfoil, the 6 is the series designation, the minimum pressure occurs at $0.5c$ for the basic symmetric thickness distribution at zero lift, the design lift coefficient is 0.2, and the airfoil is 18 percent thick.

The complete NACA airfoil numbering system is given in Reference 11. Indeed, Reference 11 is a definitive presentation of the classic NACA airfoil work up to 1949. It contains a discussion of airfoil theory, its application, coordinates for the shape of NACA airfoils, and a huge bulk of experimental data for these airfoils. This author strongly encourages you to read Reference 11 for a thorough presentation of airfoil characteristics.

As a matter of interest, the following is a short partial listing of airplanes currently in service that use standard NACA airfoils.

---

[1] The design lift coefficient is the theoretical lift coefficient for the airfoil when the angle of attack is such that the slope of the mean camber line at the leading edge is parallel to the freestream velocity. In terms of the Kutta condition to be discussed in Section 4.5, this configuration corresponds to the Kutta condition holding at the leading edge as well as the trailing edge, i.e., the vortex sheet strength at the leading edge must be zero because the flow velocity just above the leading edge is the same as the flow velocity just below the leading edge.

Airplane	Airfoil
Beechcraft Sundowner	NACA 63A415
Beechcraft Bonanza	NACA 23016.5 (at root)
	NACA 23012 (at tip)
Cessna 150	NACA 2412
Fairchild A-10	NACA 6716 (at root)
	NACA 6713 (at tip)
Gates Learjet 24D	NACA 64A109
General Dynamics F-16	NACA 64A204
Lockheed C-5 Galaxy	NACA 0012 (modified)

In addition, many of the large aircraft companies today design their own special-purpose airfoils; for example, the Boeing 727, 737, 747, 757, 767, and 777 have specially designed Boeing airfoils. Such capability is made possible by modern airfoil design computer programs utilizing either panel techniques or direct numerical finite-difference solutions of the governing partial differential equations for the flow field. (Such equations are developed in Chapter 2.)

4.3 AIRFOIL CHARACTERISTICS

Before discussing the theoretical calculation of airfoil properties, let us examine some typical results. During the 1930s and 1940s, the NACA carried out numerous measurements of the lift, drag, and moment coefficients on the standard NACA airfoils. These experiments were performed at low speeds in a wind tunnel where the constant-chord wing spanned the entire test section from one sidewall to the other. In this fashion, the flow "sees" a wing without wing tips—a so-called infinite wing, which theoretically stretches to infinity along the span (in the y direction in Figure 4.6). Because the airfoil section is the same at any spanwise location along the infinite wing, the properties of the airfoil and the infinite wing are identical. Hence, airfoil data are frequently called infinite wing data. (In contrast, we see in Chapter 5 that the properties of a finite wing are somewhat different from its airfoil properties.)

The typical variation of lift coefficient with angle of attack for an airfoil is sketched in Figure 4.9. At low-to-moderate angles of attack, c_l varies *linearly* with α; the slope of this straight line is denoted by a_0 and is called the *lift slope*. In this region, the flow moves smoothly over the airfoil and is attached over most of the surface, as shown in the streamline picture at the left of Figure 4.9. However, as α becomes large, the flow tends to separate from the top surface of the airfoil, creating a large wake of relatively "dead air" behind the airfoil as shown at the right of Figure 4.9. Inside this separated region, the flow is recirculating, and part of the flow is actually moving in a direction opposite to the freestream—so-called reversed flow. (Refer also to Figure 1.42.) This separated flow is due to viscous effects and is discussed in Section 4.12 and Chapter 15. The consequence of this separated flow at high α is a precipitous decrease in lift and a large increase in drag; under such conditions the airfoil is said to be *stalled*. The maximum

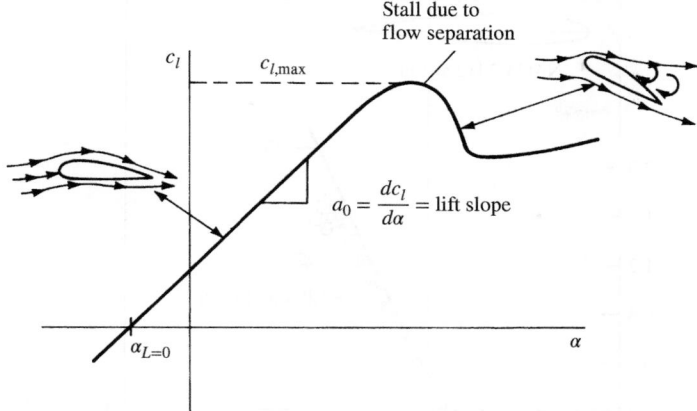

Figure 4.9 Schematic of lift-coefficient variation with angle of attack for an airfoil.

value of c_l, which occurs just prior to the stall, is denoted by $c_{l,\max}$; it is one of the most important aspects of airfoil performance, because it determines the stalling speed of an airplane. The higher is $c_{l,\max}$, the lower is the stalling speed. A great deal of modern airfoil research has been directed toward increasing $c_{l,\max}$. Again examining Figure 4.9, we see that c_l increases linearly with α until flow separation begins to have an effect. Then the curve becomes nonlinear, c_l reaches a maximum value, and finally the airfoil stalls. At the other extreme of the curve, noting Figure 4.9, the lift at $\alpha = 0$ is finite; indeed, the lift goes to zero only when the airfoil is pitched to some negative angle of attack. The value of α when lift equals zero is called the *zero-lift angle of attack* and is denoted by $\alpha_{L=0}$. For a symmetric airfoil, $\alpha_{L=0} = 0$, whereas for all airfoils with positive camber (camber *above* the chord line), $\alpha_{L=0}$ is a negative value, usually on the order of -2 or $-3°$.

The inviscid flow airfoil theory discussed in this chapter allows us to predict the lift slope a_0 and $\alpha_{L=0}$ for a given airfoil. It does not allow us to calculate $c_{l,\max}$, which is a difficult viscous flow problem, to be discussed in Chapters 15 to 20.

Experimental results for lift and moment coefficients for the NACA 2412 airfoil are given in Figure 4.10. Here, the moment coefficient is taken about the quarter-chord point. Recall from Section 1.6 that the force-and-moment system on an airfoil can be transferred to any convenient point; however, the quarter-chord point is commonly used. (Refresh your mind on this concept by reviewing Section 1.6, especially Figure 1.25.) Also shown in Figure 4.10 are theoretical results to be discussed later. Note that the experimental data are given for two different Reynolds numbers. The lift slope a_0 is not influenced by Re; however, $c_{l,\max}$ is dependent upon Re. This makes sense, because $c_{l,\max}$ is governed by viscous effects, and Re is a similarity parameter that governs the strength of inertia forces relative to viscous forces in the flow. [See Section 1.7 and Equation (1.35).]

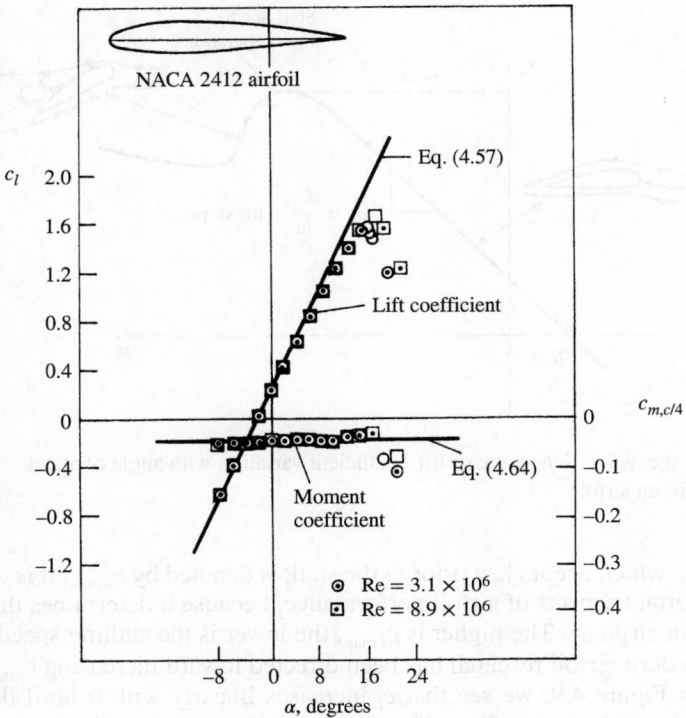

Figure 4.10 Experimental data for lift coefficient and moment coefficient about the quarter-chord point for an NACA 2412 airfoil. (*Source:* Data obtained from Abbott, I. H., and A. E. von Doenhoff: *Theory of Wing Sections*, McGraw Hill Book Company, New York, 1949; also, Dover Publications, Inc., New York, 1959.) Also shown is a comparison with theory described in Section 4.8.

The moment coefficient is also insensitive to Re except at large α. The NACA 2412 airfoil is a commonly used airfoil, and the results given in Figure 4.10 are quite typical of airfoil characteristics. For example, note from Figure 4.10 that $\alpha_{L=0} = -2.1°$, $c_{l,\max} \approx 1.6$, and the stall occurs at $\alpha \approx 16°$.

This chapter deals primarily with airfoil theory for an inviscid, incompressible flow; such theory is incapable of predicting airfoil drag, as noted earlier. However, for the sake of completeness, experimental data for the drag coefficient c_d for the NACA 2412 airfoil are given in Figure 4.11 as a function of the angle of attack.[2] The physical source of this drag coefficient is both skin friction drag and pressure drag due to flow separation (so-called form drag). The sum of these

---

[2] In many references, such as Reference 11, it is common to plot c_d versus c_l, rather than versus α. A plot of c_d versus c_l is called a *drag polar*. For the sake of consistency with Figure 4.10, we choose to plot c_d versus α here.

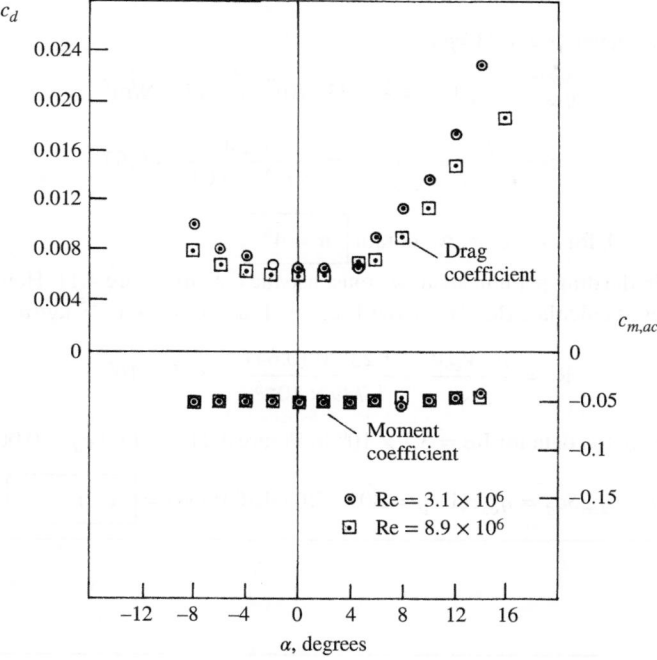

Figure 4.11 Experimental data for profile drag coefficient and moment coefficient about the aerodynamic center for the NACA 2412 airfoil. (*Source:* Data obtained from Abbott, I. H., and A. E. von Doenhoff: *Theory of Wing Sections,* McGraw Hill Book Company, New York, 1949; also, Dover Publications, Inc., New York, 1959.)

two effects yields the *profile* drag coefficient c_d for the airfoil, which is plotted in Figure 4.11. Note that c_d is sensitive to Re, which is to be expected since both skin friction and flow separation are viscous effects. Again, we must wait until Section 4.12 and Chapters 15 to 20 to obtain some tools for theoretically predicting c_d.

Also plotted in Figure 4.11 is the moment coefficient about the aerodynamic center $c_{m,\text{ac}}$. In general, moments on an airfoil are a function of α. However, there is one point on the airfoil about which the moment is independent of angle of attack; such a point is defined as the *aerodynamic center.* Clearly, the data in Figure 4.11 illustrate a constant value for $c_{m,\text{ac}}$ over a wide range of α.

For an elementary but extensive discussion of airfoil and wing properties, see Chapter 5 of Reference 2.

EXAMPLE 4.1

Consider an NACA 2412 airfoil with a chord of 0.64 m in an airstream at standard sea level conditions. The freestream velocity is 70 m/s. The lift per unit span is 1254 N/m. Calculate the angle of attack and the drag per unit span.

■ **Solution**

At standard sea level, $\rho = 1.23$ kg/m^3:

$$q_\infty = \tfrac{1}{2}\rho_\infty V_\infty^2 = \tfrac{1}{2}(1.23)(70)^2 = 3013.5 \text{ N/m}^2$$

$$c_l = \frac{L'}{q_\infty S} = \frac{L'}{q_\infty c(1)} = \frac{1254}{3013.5(0.64)} = 0.65$$

From Figure 4.10, for $c_l = 0.65$, we obtain $\boxed{\alpha = 4°}$.

To obtain the drag per unit span, we must use the data in Figure 4.11. However, since $c_d = f(\text{Re})$, let us calculate Re. At standard sea level, $\mu = 1.789 \times 10^{-5}$ kg/(m · s). Hence,

$$\text{Re} = \frac{\rho_\infty V_\infty c}{\mu_\infty} = \frac{1.23(70)(0.64)}{1.789 \times 10^{-5}} = 3.08 \times 10^6$$

Therefore, using the data for Re = 3.1×10^6 in Figure 4.11, we find $c_d = 0.0068$. Thus,

$$D' = q_\infty S c_d = q_\infty c(1) c_d = 3013.5(0.64)(0.0068) = \boxed{13.1 \text{ N/m}}$$

EXAMPLE 4.2

For the airfoil and flow conditions given in Example 4.1, calculate the moment per unit span about the aerodynamic center.

■ **Solution**

From Figure 4.11, $c_{m,\text{ac}}$, which is independent of angle of attack, is -0.05. The moment per unit span about the aerodynamic center is

$$M'_{\text{ac}} = q_\infty S c c_{m,\text{ac}}$$

$$= (3013.5)(0.64)(0.64)(-0.05) = \boxed{-61.7 \text{ N}}$$

Recall the sign convention for aerodynamic moments introduced in Section 1.5, namely, that a negative moment, as obtained here, is a pitch-down moment, tending to reduce the angle of attack.

EXAMPLE 4.3

For the NACA 2412 airfoil, calculate and compare the lift-to-drag ratios at angles of attack of 0, 4, 8, and 12 degrees. The Reynolds number is 3.1×10^6.

■ **Solution**

The lift-to-drag ratio, L/D, is given by

$$\frac{L}{D} = \frac{q_\infty S c_\ell}{q_\infty S c_d} = \frac{c_\ell}{c_d}$$

From Figures 4.10 and 4.11, we have

α	c_ℓ	c_d	c_ℓ/c_d
0	0.25	0.0065	38.5
4	0.65	0.0070	93
8	1.08	0.0112	96
12	1.44	0.017	85

Note that, as the angle of attack increases, the lift-to-drag ratio first increases, reaches a maximum, and then decreases. The maximum lift-to-drag ratio, $(L/D)_{max}$, is an important parameter in airfoil performance; it is a direct measure of aerodynamic efficiency. The higher the value of $(L/D)_{max}$, the more efficient is the airfoil. The values of L/D for airfoils are quite large numbers in comparison to that for a complete airplane. Due to the extra drag associated with all parts of the airplane, values of $(L/D)_{max}$ for real airplanes are on the order of 10 to 20.

4.4 PHILOSOPHY OF THEORETICAL SOLUTIONS FOR LOW-SPEED FLOW OVER AIRFOILS: THE VORTEX SHEET

In Section 3.14, the concept of vortex flow was introduced; refer to Figure 3.31 for a schematic of the flow induced by a point vortex of strength Γ located at a given point O. (Recall that Figure 3.31, with its counterclockwise flow, corresponds to a negative value of Γ. By convention, a positive Γ induces a clockwise flow.) Let us now expand our concept of a point vortex. Referring to Figure 3.31, imagine a straight line perpendicular to the page, going through point O, and extending to infinity both out of and into the page. This line is a straight *vortex filament* of strength Γ. A straight vortex filament is drawn in perspective in Figure 4.12. (Here, we show a clockwise flow, which corresponds to a positive value of Γ.) The flow induced in any plane perpendicular to the straight vortex filament by the filament itself is identical to that induced by a point vortex of strength Γ; that is, in Figure 4.12, the flows in the planes perpendicular to the vortex filament at O and O' are identical to each other and are identical to the flow induced by a point vortex of strength Γ. Indeed, the point vortex described in Section 3.14 is simply a section of a straight vortex filament.

In Section 3.17, we introduced the concept of a source sheet, which is an infinite number of line sources side by side, with the strength of each line source being infinitesimally small. For vortex flow, consider an analogous situation. Imagine an infinite number of straight vortex filaments side by side, where the strength of each filament is infinitesimally small. These side-by-side vortex filaments form a *vortex sheet,* as shown in perspective in the upper left of Figure 4.13. If we look along the series of vortex filaments (looking along the y axis in Figure 4.13), the vortex sheet will appear as sketched at the lower right of Figure 4.13. Here, we are looking at an edge view of the sheet; the vortex filaments are all perpendicular

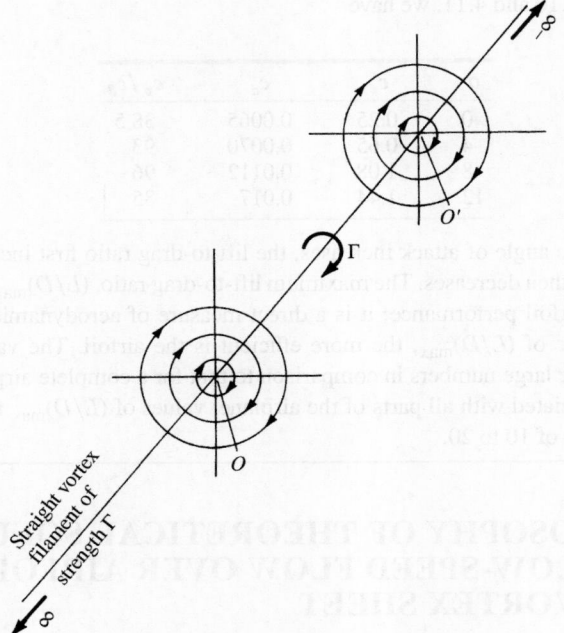

Straight vortex filament of strength Γ

Figure 4.12 Vortex filament.

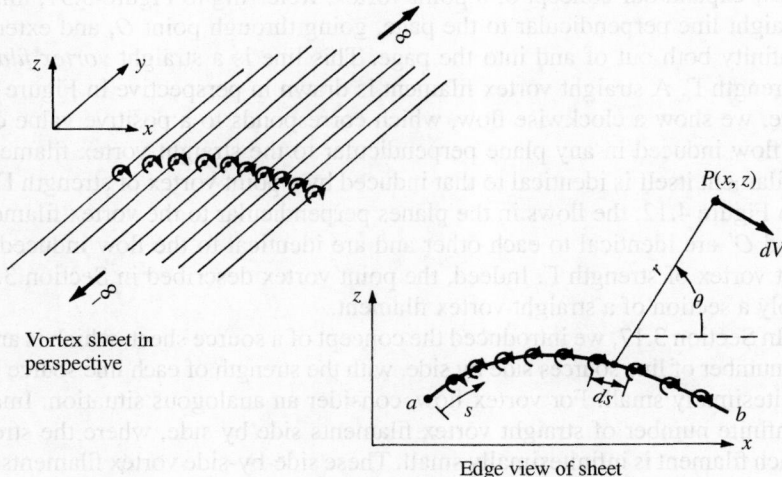

Vortex sheet in perspective

Edge view of sheet

Figure 4.13 Vortex sheet.

to the page. Let s be the distance measured along the vortex sheet in the edge view. Define $\gamma = \gamma(s)$ as the strength of the vortex sheet, per unit length along s. Thus, the strength of an infinitesimal portion ds of the sheet is $\gamma\, ds$. This small section of the vortex sheet can be treated as a distinct vortex of strength $\gamma\, ds$. Now consider point P in the flow, located at a distance r from ds; the cartesian coordinates of P are (x, z). The small section of the vortex sheet of strength $\gamma\, ds$ induces an infinitesimally small velocity dV at point P. From Equation (3.105), dV is given by

$$dV = -\frac{\gamma\, ds}{2\pi r} \tag{4.1}$$

and is in a direction perpendicular to r, as shown in Figure 4.13. The velocity at P induced by the entire vortex sheet is the summation of Equation (4.1) from point a to point b. Note that dV, which is perpendicular to r, changes direction at point P as we sum from a to b; hence, the incremental velocities induced at P by different sections of the vortex sheet must be added vectorially. Because of this, it is sometimes more convenient to deal with the velocity potential. Again referring to Figure 4.13, the increment in velocity potential $d\phi$ induced at point P by the elemental vortex $\gamma\, ds$ is, from Equation (3.112),

$$d\phi = -\frac{\gamma\, ds}{2\pi}\theta \tag{4.2}$$

In turn, the velocity potential at P due to the entire vortex sheet from a to b is

$$\phi(x, z) = -\frac{1}{2\pi} \int_a^b \theta\gamma\, ds \tag{4.3}$$

Equation (4.1) is particularly useful for our discussion of classical thin airfoil theory, whereas Equation (4.3) is important for the numerical vortex panel method.

Recall from Section 3.14 that the circulation Γ around a point vortex is equal to the strength of the vortex. Similarly, the circulation around the vortex sheet in Figure 4.13 is the sum of the strengths of the elemental vortices; that is,

$$\Gamma = \int_a^b \gamma\, ds \tag{4.4}$$

Recall that the source sheet introduced in Section 3.17 has a discontinuous change in the direction of the *normal* component of velocity across the sheet (from Figure 3.38, note that the normal component of velocity changes direction by 180° in crossing the sheet), whereas the tangential component of velocity is the same immediately above and below the source sheet. In contrast, for a vortex sheet, there is a discontinuous change in the tangential component of velocity across the sheet, whereas the normal component of velocity is preserved across the sheet. This change in tangential velocity across the vortex sheet is related to the strength of the sheet as follows. Consider a vortex sheet as sketched in

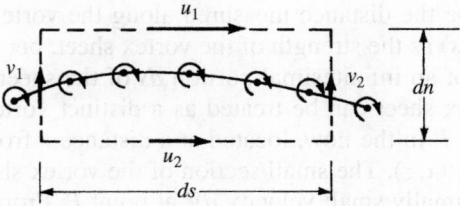

Figure 4.14 Tangential velocity jump across a vortex sheet.

Figure 4.14. Consider the rectangular dashed path enclosing a section of the sheet of length ds. The velocity components tangential to the top and bottom of this rectangular path are u_1 and u_2, respectively, and the velocity components tangential to the left and right sides are v_1 and v_2, respectively. The top and bottom of the path are separated by the distance dn. From the definition of circulation given by Equation (2.136), the circulation around the dashed path is

$$\Gamma = -(v_2\,dn - u_1\,ds - v_1\,dn + u_2\,ds)$$

or

$$\Gamma = (u_1 - u_2)\,ds + (v_1 - v_2)\,dn \tag{4.5}$$

However, since the strength of the vortex sheet contained inside the dashed path is $\gamma\,ds$, we also have

$$\Gamma = \gamma\,ds \tag{4.6}$$

Therefore, from Equations (4.5) and (4.6),

$$\gamma\,ds = (u_1 - u_2)\,ds + (v_1 - v_2)\,dn \tag{4.7}$$

Let the top and bottom of the dashed line approach the vortex sheet; that is, let $dn \to 0$. In the limit, u_1 and u_2 become the velocity components tangential to the vortex sheet immediately above and below the sheet, respectively, and Equation (4.7) becomes

$$\gamma\,ds = (u_1 - u_2)\,ds$$

or

$$\boxed{\gamma = u_1 - u_2} \tag{4.8}$$

Equation (4.8) is important; it states that *the local jump in tangential velocity across the vortex sheet is equal to the local sheet strength.*

We have now defined and discussed the properties of a vortex sheet. The concept of a vortex sheet is instrumental in the analysis of the low-speed characteristics of an airfoil. A philosophy of airfoil theory of inviscid, incompressible flow is as follows. Consider an airfoil of arbitrary shape and thickness in a freestream with velocity V_∞, as sketched in Figure 4.15. Replace the airfoil surface with a vortex sheet of variable strength $\gamma(s)$, as also shown in Figure 4.15. Calculate the variation of γ as a function of s such that the induced velocity field from the vortex sheet when added to the uniform velocity of magnitude V_∞ will make

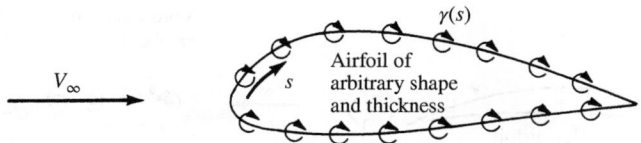

Figure 4.15 Simulation of an arbitrary airfoil by distributing a vortex sheet over the airfoil surface.

the vortex sheet (hence the airfoil surface) a *streamline of the flow*. In turn, the circulation around the airfoil will be given by

$$\Gamma = \int \gamma \, ds$$

where the integral is taken around the complete surface of the airfoil. Finally, the resulting lift is given by the Kutta-Joukowski theorem:

$$L' = \rho_\infty V_\infty \Gamma$$

This philosophy is not new. It was first espoused by Ludwig Prandtl and his colleagues at Göttingen, Germany, during the period 1912–1922. However, no general analytical solution for $\gamma = \gamma(s)$ exists for an airfoil of arbitrary shape and thickness. Rather, the strength of the vortex sheet must be found numerically, and the practical implementation of the above philosophy had to wait until the 1960s with the advent of large digital computers. Today, the above philosophy is the foundation of the modern vortex panel method, to be discussed in Section 4.9.

The concept of replacing the airfoil surface in Figure 4.15 with a vortex sheet is more than just a mathematical device; it also has physical significance. In real life, there is a thin boundary layer on the surface, due to the action of friction between the surface and the airflow (see Figures 1.41 and 1.46). This boundary layer is a highly viscous region in which the large velocity gradients produce substantial vorticity; that is, $\nabla \times \mathbf{V}$ is finite within the boundary layer. (Review Section 2.12 for a discussion of vorticity.) Hence, in real life, there is a distribution of vorticity along the airfoil surface due to viscous effects, and our philosophy of replacing the airfoil surface with a vortex sheet (such as in Figure 4.15) can be construed as a way of modeling this effect in an inviscid flow.[3]

Imagine that the airfoil in Figure 4.15 is made very thin. If you were to stand back and look at such a thin airfoil from a distance, the portions of the vortex sheet

---

[3] It is interesting to note that some recent research by NASA is hinting that even as complex a problem as flow separation, heretofore thought to be a completely viscous-dominated phenomenon, may in reality be an inviscid-dominated flow which requires only a rotational flow. For example, some inviscid flow-field numerical solutions for flow over a circular cylinder, when vorticity is introduced either by means of a nonuniform freestream or a curved shock wave, are accurately predicting the separated flow on the rearward side of the cylinder. However, as exciting as these results may be, they are too preliminary to be emphasized in this book. We continue to talk about flow separation in Chapters 15 to 20 as being a viscous-dominated effect, until definitely proved otherwise. This recent research is mentioned here only as another example of the physical connection between vorticity, vortex sheets, viscosity, and real life.

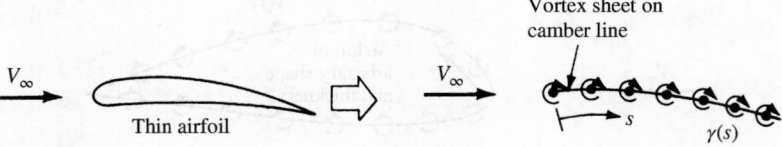

Figure 4.16   Thin airfoil approximation.

on the top and bottom surface of the airfoil would almost coincide. This gives rise to a method of approximating a thin airfoil by replacing it with a single vortex sheet distributed over the camber line of the airfoil, as sketched in Figure 4.16. The strength of this vortex sheet $\gamma(s)$ is calculated such that, in combination with the freestream, the camber line becomes a streamline of the flow. Although the approach shown in Figure 4.16 is approximate in comparison with the case shown in Figure 4.15, it has the advantage of yielding a closed-form analytical solution. This philosophy of thin airfoil theory was first developed by Max Munk, a colleague of Prandtl, in 1922 (see Reference 12). It is discussed in Sections 4.7 and 4.8.

4.5 THE KUTTA CONDITION

The lifting flow over a circular cylinder was discussed in Section 3.15, where we observed that an infinite number of potential flow solutions were possible, corresponding to the infinite choice of Γ. For example, Figure 3.33 illustrates three different flows over the cylinder, corresponding to three different values of Γ. The same situation applies to the potential flow over an airfoil; for a given airfoil at a given angle of attack, there are an infinite number of valid theoretical solutions, corresponding to an infinite choice of Γ. For example, Figure 4.17 illustrates two different flows over the same airfoil at the same angle of attack but with different values of Γ. At first, this may seem to pose a dilemma. We know from experience that a given airfoil at a given angle of attack produces a single value of lift (e.g., see Figure 4.10). So, although there is an infinite number of possible potential flow solutions, nature knows how to pick a particular solution. Clearly, the philosophy discussed in the previous section is not complete—we need an additional condition that *fixes* Γ for a given airfoil at a given α.

To attempt to find this condition, let us examine some experimental results for the development of the flow field around an airfoil which is set into motion from an initial state of rest. Figure 4.18 shows a series of classic photographs of the flow over an airfoil, taken from Prandtl and Tietjens (Reference 8). In Figure 4.18*a*, the flow has just started, and the flow pattern is just beginning to develop around the airfoil. In these early moments of development, the flow tries to curl around the sharp trailing edge from the bottom surface to the top surface, similar to the sketch shown at the left of Figure 4.17. However, more advanced considerations of inviscid, incompressible flow (see, e.g., Reference 9) show the theoretical result that the velocity becomes infinitely large at a sharp corner. Hence, the type

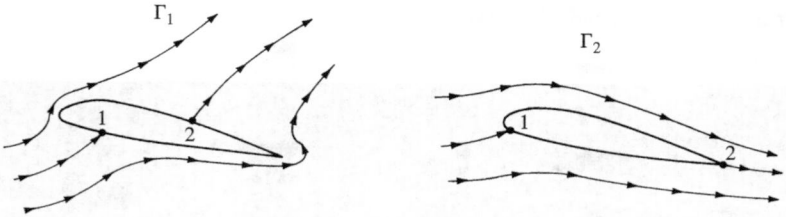

Figure 4.17 Effect of different values of circulation on the potential flow over a given airfoil at a given angle of attack. Points 1 and 2 are stagnation points.

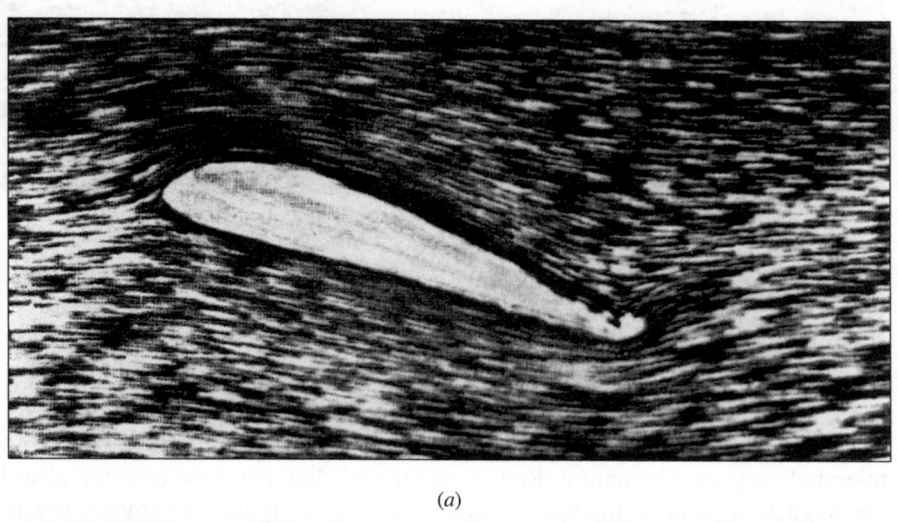

(a)

(b)

Figure 4.18 The development of steady flow over an airfoil; the airfoil is impulsively started from rest and attains a steady velocity through the fluid. (a) A moment just after starting. (b) An intermediate time. (Prandtl, L., and O. G. Tietjens. *Applied Hydro and Aeromechanics Based on Lectures of L. Prandtl.* New York: McGraw Hill, United Engineering Trustees, Inc., 1934.)

(*c*)

Figure 4.18 (*continued*) The development of steady flow over an airfoil; the airfoil is impulsively started from rest and attains a steady velocity through the fluid. (*c*) The final steady flow. (Prandtl, L., and O. G. Tietjens. *Applied Hydro and Aeromechanics Based on Lectures of L. Prandtl.* New York: McGraw Hill, United Engineering Trustees, Inc., 1934.)

of flow sketched at the left of Figure 4.17, and shown in Figure 4.18*a*, is not tolerated very long by nature. Rather, as the real flow develops over the airfoil, the stagnation point on the upper surface (point 2 in Figure 4.17) moves toward the trailing edge. Figure 4.18*b* shows this intermediate stage. Finally, after the initial transient process dies out, the steady flow shown in Figure 4.18*c* is reached. This photograph demonstrates that the flow is *smoothly leaving the top and the bottom surfaces of the airfoil at the trailing edge.* This flow pattern is sketched at the right of Figure 4.17 and represents the type of pattern to be expected for the steady flow over an airfoil.

Reflecting on Figures 4.17 and 4.18, we emphasize again that in establishing the steady flow over a given airfoil at a given angle of attack, nature adopts that particular value of circulation (Γ_2 in Figure 4.17) which results in the flow leaving smoothly at the trailing edge. This observation was first made and used in a theoretical analysis by the German mathematician M. Wilhelm Kutta in 1902. Therefore, it has become known as the *Kutta condition.*

In order to apply the Kutta condition in a theoretical analysis, we need to be more precise about the nature of the flow at the trailing edge. The trailing edge can have a finite angle, as shown in Figures 4.17 and 4.18 and as sketched at the left of Figure 4.19, or it can be cusped, as shown at the right of Figure 4.19. First, consider the trailing edge with a finite angle, as shown at the left of Figure 4.19. Denote the velocities along the top surface and the bottom surface as V_1 and V_2, respectively. V_1 is parallel to the top surface at point *a*, and V_2 is parallel to the bottom surface at point *a*. For the finite-angle trailing edge, if these velocities were

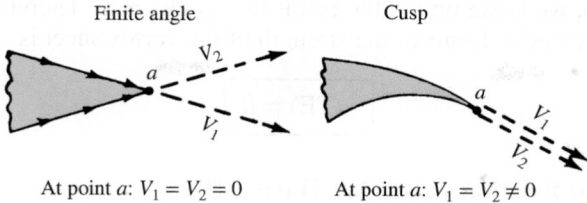

Figure 4.19 Different possible shapes of the trailing edge and their relation to the Kutta condition.

finite at point a, then we would have two velocities in two different directions at the same point, as shown at the left of Figure 4.19. However, this is not physically possible, and the only recourse is for both V_1 and V_2 to be zero at point a. That is, for the finite trailing edge, point a is a stagnation point, where $V_1 = V_2 = 0$. In contrast, for the cusped trailing edge shown at the right of Figure 4.19, V_1 and V_2 are in the same direction at point a, and hence both V_1 and V_2 can be finite. However, the pressure p_2 at point a, is a single, unique value, and Bernoulli's equation applied at both the top and bottom surfaces immediately adjacent to point a yields

$$p_a + \frac{1}{2}\rho V_1^2 = p_a + \frac{1}{2}\rho V_2^2$$

or
$$V_1 = V_2$$

Hence, for the cusped trailing edge, we see that the velocities leaving the top and bottom surfaces of the airfoil at the trailing edge are finite and equal in magnitude and direction.

We can summarize the statement of the Kutta condition as follows:

1. For a given airfoil at a given angle of attack, the value of Γ around the airfoil is such that the flow leaves the trailing edge smoothly.
2. If the trailing-edge angle is finite, then the trailing edge is a stagnation point.
3. If the trailing edge is cusped, then the velocities leaving the top and bottom surfaces at the trailing edge are finite and equal in magnitude and direction.

Consider again the philosophy of simulating the airfoil with vortex sheets placed either on the surface or on the camber line, as discussed in Section 4.4. The strength of such a vortex sheet is variable along the sheet and is denoted by $\gamma(s)$. The statement of the Kutta condition in terms of the vortex sheet is as follows. At the trailing edge (TE), from Equation (4.8), we have

$$\gamma(\text{TE}) = \gamma(a) = V_1 - V_2 \tag{4.9}$$

However, for the finite-angle trailing edge, $V_1 = V_2 = 0$; hence, from Equation (4.9), $\gamma(\text{TE}) = 0$. For the cusped trailing edge, $V_1 = V_2 \neq 0$; hence, from

Equation (4.9), we again obtain the result that $\gamma(\text{TE}) = 0$. Therefore, the Kutta condition expressed in terms of the strength of the vortex sheet is

$$\boxed{\gamma(\text{TE}) = 0} \tag{4.10}$$

4.5.1 Without Friction Could We Have Lift?

In Section 1.5, we emphasized that the resultant aerodynamic force on a body immersed in a flow is due to the net integrated effect of the pressure and shear stress distributions over the body surface. Moreover, in Section 4.1, we noted that lift on an airfoil is primarily due to the surface pressure distribution, and that shear stress has virtually no effect on lift. It is easy to see why. Look at the airfoil shapes in Figures 4.17 and 4.18, for example. Recall that pressure acts *normal* to the surface, and for these airfoils the direction of this normal pressure is essentially in the vertical direction, that is, the lift direction. In contrast, the shear stress acts *tangential* to the surface, and for these airfoils the direction of this tangential shear stress is mainly in the horizontal direction, that is, the drag direction. Hence, pressure is the dominant player in the generation of lift, and shear stress has a negligible effect on lift. It is for this reason that the lift on an airfoil below the stall can be accurately predicted by *inviscid* theories such as that discussed in this chapter.

However, if we lived in a perfectly inviscid world, an airfoil could not produce lift. Indeed, the presence of friction is the very reason why we have lift. These sound like strange, even contradictory statements to our discussion in the preceding paragraph. What is going on here? The answer is that in real life, the *way* that nature insures that the flow will leave smoothly at the trailing edge, that is, the mechanism that nature uses to choose the flow shown in Figure 4.18*c*, is that the viscous boundary layer remains attached to the surface all the way to the trailing edge. *Nature enforces the Kutta condition by means of friction.* If there were no boundary layer (i.e., no friction), there would be no physical mechanism in the real world to achieve the Kutta condition.

So we are led to the most ironic situation that lift, which is created by the surface pressure distribution—an inviscid phenomenon, would not exist in a frictionless (inviscid) world. In this regard, we can say that without friction we could not have lift. However, we say this in the informed manner as discussed above.

4.6 KELVIN'S CIRCULATION THEOREM AND THE STARTING VORTEX

In this section, we put the finishing touch to the overall philosophy of airfoil theory before developing the quantitative aspects of the theory itself in subsequent sections. This section also ties up a loose end introduced by the Kutta condition described in the previous section. Specifically, the Kutta condition states that the circulation around an airfoil is just the right value to ensure that the flow smoothly

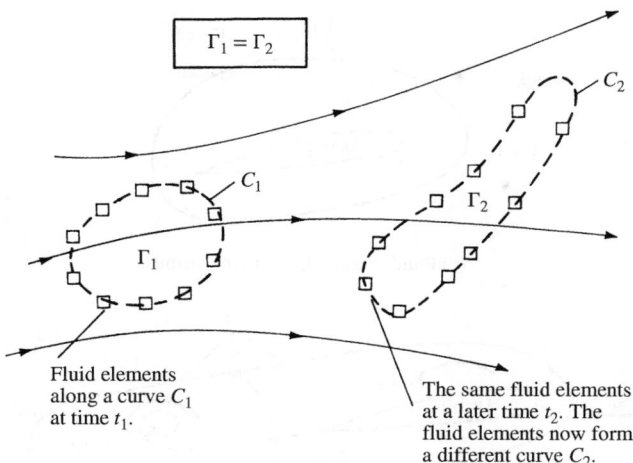

Figure 4.20 Kelvin's theorem.

leaves the trailing edge. *Question:* How does nature generate this circulation? Does it come from nowhere, or is circulation somehow conserved over the whole flow field? Let us examine these matters more closely.

Consider an arbitrary inviscid, incompressible flow as sketched in Figure 4.20. Assume that all body forces $\mathbf{f}$ are zero. Choose an arbitrary curve C_1 and identify the fluid elements that are on this curve at a given instant in time t_1. Also, by definition the circulation around curve C_1 is $\Gamma_1 = -\int_{C_1} \mathbf{V} \cdot \mathbf{ds}$. Now let these specific fluid elements move downstream. At some later time t_2, these *same* fluid elements will form another curve C_2, around which the circulation is $\Gamma_2 = -\int_{C_2} \mathbf{V} \cdot \mathbf{ds}$. For the conditions stated above, we can readily show that $\Gamma_1 = \Gamma_2$. In fact, since we are following a set of specific fluid elements, we can state that circulation around a closed curve formed by a set of contiguous fluid elements remains constant as the fluid elements move throughout the flow. Recall from Section 2.9 that the substantial derivative gives the time rate of change following a given fluid element. Hence, a mathematical statement of the above discussion is simply

$$\frac{D\Gamma}{Dt} = 0 \qquad (4.11)$$

which says that the time rate of change of circulation around a closed curve consisting of the same fluid elements is zero. Equation (4.11) along with its supporting discussion is called *Kelvin's circulation theorem.*[4] Its derivation from first

---

[4] Kelvin's theorem also holds for an inviscid compressible flow in the special case where $\rho = \rho(p)$; that is, the density is some single-valued function of pressure. Such is the case for isentropic flow, to be treated in later chapters.

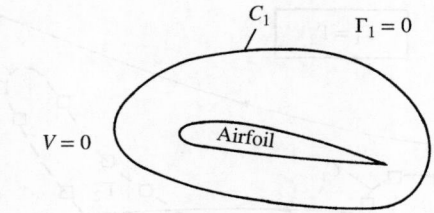

(*a*) Fluid at rest relative to the airfoil

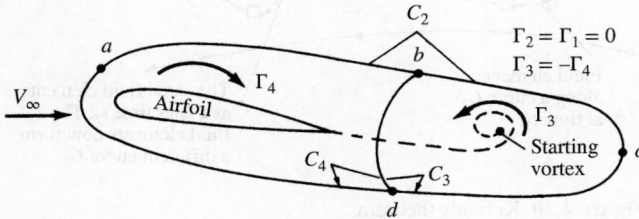

(*b*) Picture some moments after the start of the flow

Figure 4.21 The creation of the starting vortex and the resulting generation of circulation around the airfoil.

principles is left as Problem 4.3. Also, recall our definition and discussion of a vortex sheet in Section 4.4. An interesting consequence of Kelvin's circulation theorem is proof that a stream surface which is a vortex sheet at some instant in time remains a vortex sheet for all times.

Kelvin's theorem helps to explain the generation of circulation around an airfoil, as follows. Consider an airfoil in a fluid at rest, as shown in Figure 4.21a. Because $\mathbf{V} = 0$ everywhere, the circulation around curve C_1 is zero. Now start the flow in motion over the airfoil. Initially, the flow will tend to curl around the trailing edge, as explained in Section 4.5 and illustrated at the left of Figure 4.17. In doing so, the velocity at the trailing edge theoretically becomes infinite. In real life, the velocity tends toward a very large finite number. Consequently, during the very first moments after the flow is started, a thin region of very large velocity gradients (and therefore high vorticity) is formed at the trailing edge. This high-vorticity region is fixed to the same fluid elements, and consequently it is flushed downstream as the fluid elements begin to move downstream from the trailing edge. As it moves downstream, this thin sheet of intense vorticity is unstable, and it tends to roll up and form a picture similar to a point vortex. This vortex is called the *starting vortex* and is sketched in Figure 4.21b. After the flow around the airfoil has come to a steady state where the flow leaves the trailing edge smoothly (the Kutta condition), the high velocity gradients at the trailing edge disappear and vorticity is no longer produced at that point. However, the starting vortex has already been formed during the starting process, and it moves steadily downstream with the flow forever after. Figure 4.21*b* shows the flow field sometime after steady flow has been achieved over the airfoil, with the starting vortex somewhere downstream. The fluid elements that initially made up curve C_1 in Figure 4.21*a* have moved downstream and now make up curve C_2, which is the complete circuit *abcda* shown in Figure 4.21*b*. Thus, from Kelvin's theorem, the circulation Γ_2 around curve C_2 (which encloses both the airfoil and the starting vortex) is the same as that around curve C_1, namely, zero. So,

$$\Gamma_2 = \Gamma_1 = 0.$$

Now let us subdivide C_2 into two loops by making the cut *bd*, thus forming curves C_3 (circuit *bcdb*) and C_4 (circuit *abda*). Curve C_3 encloses the starting vortex, and curve C_4 encloses the airfoil. The circulation Γ_3 around curve C_3 is due to the starting vortex; by inspecting Figure 4.21*b*, we see that Γ_3 is in the counterclockwise direction (i.e., a negative value). The circulation around curve C_4 enclosing the airfoil is Γ_4. Since the cut *bd* is common to both C_3 and C_4, the sum of the circulations around C_3 and C_4 is simply equal to the circulation around C_2:

$$\Gamma_3 + \Gamma_4 = \Gamma_2$$

However, we have already established that $\Gamma_2 = 0$. Hence,

$$\Gamma_4 = -\Gamma_3$$

that is, the circulation around the airfoil is equal and opposite to the circulation around the starting vortex.

This brings us to the summary as well as the crux of this section. As the flow over an airfoil is started, the large velocity gradients at the sharp trailing edge result in the formation of a region of intense vorticity which rolls up downstream of the trailing edge, forming the starting vortex. This starting vortex has associated with it a counterclockwise circulation. Therefore, as an equal-and-opposite reaction, a clockwise circulation around the airfoil is generated. As the starting process continues, vorticity from the trailing edge is constantly fed into the starting vortex, making it stronger with a consequent larger counterclockwise circulation. In turn, the clockwise circulation around the airfoil becomes stronger, making the flow at the trailing edge more closely approach the Kutta condition, thus weakening the vorticity shed from the trailing edge. Finally, the starting vortex builds up to just the right strength such that the equal-and-opposite clockwise circulation around the airfoil leads to smooth flow from the trailing edge (the Kutta condition is exactly satisfied). When this happens, the vorticity shed from the trailing edge becomes zero, the starting vortex no longer grows in strength, and a steady circulation exists around the airfoil.

EXAMPLE 4.4

For the NACA 2412 airfoil at the conditions given in Example 4.1, calculate the strength of the steady-state starting vortex.

■ **Solution**

From the given conditions in Example 4.1,

$$L' = 1254 \text{ N/m}$$

$$V_\infty = 70 \text{ m/s}$$

$$\rho_\infty = 1.23 \text{ kg/m}^3$$

From the Kutta-Joukowski theorem, Equation (3.140),

$$L' = \rho_\infty V_\infty \Gamma$$

the circulation associated with the flow over the airfoil is

$$\Gamma = \frac{L'}{\rho_\infty V_\infty} = \frac{1254}{(1.23)(70)} = 14.56 \text{ m}^2/\text{s}$$

Referring to Figure 4.21, the steady-state starting vortex has strength equal and opposite to the circulation around the airfoil. Hence,

$$\boxed{\text{Strength of starting vortex} = -14.56 \text{ m}^2/\text{s}}$$

Note: For practical calculations in aerodynamics, an actual number for circulation is rarely needed. Circulation is a mathematical quantity defined by Equation (2.136), and it is an essential theoretical concept within the framework of the circulation theory of lift. For example, in Section 4.7, an analytical expression for circulation is derived as Equation (4.30), and then immediately inserted into the Kutta-Joukowski theorem, Equation (4.31), yielding a formula for the lift coefficient, Equation (4.33). Nowhere do we need to calculate an actual number for Γ. In the present example, however, the *strength* of the starting vortex is indeed given by its circulation, and hence to compare the strengths of various starting vortices, calculating numbers for Γ is relevant. Even this can be considered only as an academic exercise. In this author's experience, no practical aerodynamic calculation requires the strength of a starting vortex. The starting vortex is simply a theoretical construct that is consistent with the generation of circulation around a lifting two-dimensional body.

4.7 CLASSICAL THIN AIRFOIL THEORY: THE SYMMETRIC AIRFOIL

Some experimentally observed characteristics of airfoils and a philosophy for the theoretical prediction of these characteristics have been discussed in the preceding sections. Referring to our chapter road map in Figure 4.7, we have now completed the central branch. In this section, we move to the right-hand branch of Figure 4.7, namely, a quantitative development of thin airfoil theory. The basic equations necessary for the calculation of airfoil lift and moments are established in this section, with an application to symmetric airfoils. The case of cambered airfoils will be treated in Section 4.8.

For the time being, we deal with *thin* airfoils; for such a case, the airfoil can be simulated by a vortex sheet placed along the camber line, as discussed in Section 4.4. Our purpose is to calculate the variation of $\gamma(s)$ such that the camber line becomes a streamline of the flow and such that the Kutta condition is satisfied at the trailing edge; that is, $\gamma(\text{TE}) = 0$ [see Equation (4.10)]. Once we have found the particular $\gamma(s)$ that satisfies these conditions, then the total circulation Γ around the airfoil is found by integrating $\gamma(s)$ from the leading edge to the trailing edge. In turn, the lift is calculated from Γ via the Kutta-Joukowski theorem.

Consider a vortex sheet placed on the camber line of an airfoil, as sketched in Figure 4.22a. The freestream velocity is V_∞, and the airfoil is at the angle of attack α. The x axis is oriented along the chord line, and the z axis is perpendicular to the chord. The distance measured along the camber line is denoted by s. The shape of the camber line is given by $z = z(x)$. The chord length is c. In Figure 4.22a, w' is the component of velocity normal to the camber line induced by the vortex sheet; so, $w' = w'(s)$. For a thin airfoil, we rationalized in Section 4.4 that the distribution of a vortex sheet over the surface of the airfoil, when viewed from a distance, looks almost the same as a vortex sheet placed on the camber line. Let us stand back once again and view Figure 4.22a from a distance. If the airfoil is thin, the

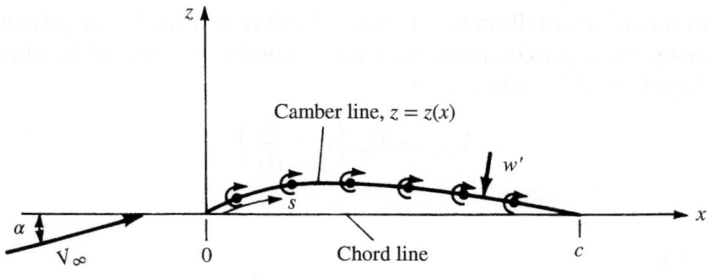

(a) Vortex sheet on the camber line

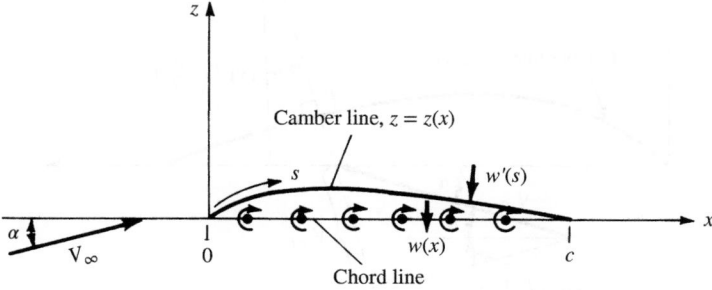

(b) Vortex sheet on the chord line

Figure 4.22 Placement of the vortex sheet for thin airfoil analysis.

camber line is close to the chord line; when viewed from a distance, the vortex sheet appears to fall approximately on the chord line. Therefore, once again, let us reorient our thinking and place the vortex sheet on the chord line, as sketched in Figure 4.22b. Here, $\gamma = \gamma(x)$. We still wish the camber line to be a streamline of the flow, and $\gamma = \gamma(x)$ is calculated to satisfy this condition as well as the Kutta condition $\gamma(c) = 0$. That is, the strength of the vortex sheet on the chord line is determined such that the camber line (not the chord line) is a streamline.

For the camber line to be a streamline, the component of velocity normal to the camber line must be zero at all points along the camber line. The velocity at any point in the flow is the sum of the uniform freestream velocity and the velocity induced by the vortex sheet. Let $V_{\infty,n}$ be the component of the freestream velocity normal to the camber line. Thus, for the camber line to be a streamline,

$$V_{\infty,n} + w'(s) = 0 \tag{4.12}$$

at every point along the camber line.

An expression for $V_{\infty,n}$ in Equation (4.12) is obtained by the inspection of Figure 4.23. At any point P on the camber line, where the slope of the camber line is dz/dx, the geometry of Figure 4.23 yields

$$V_{\infty,n} = V_\infty \sin\left[\alpha + \tan^{-1}\left(-\frac{dz}{dx}\right)\right] \tag{4.13}$$

For a thin airfoil at small angle of attack, both α and $\tan^{-1}(-dz/dx)$ are small values. Using the approximation that $\sin\theta \approx \tan\theta \approx \theta$ for small θ, where θ is in radians, Equation (4.13) reduces to

$$V_{\infty,n} = V_\infty\left(\alpha - \frac{dz}{dx}\right) \tag{4.14}$$

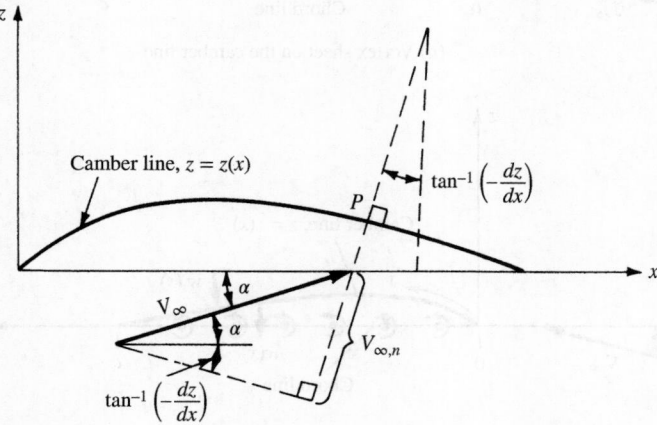

Figure 4.23 Determination of the component of freestream velocity normal to the camber line.

Equation (4.14) gives the expression for $V_{\infty,n}$ to be used in Equation (4.12). Keep in mind that, in Equation (4.14), α is in radians.

Returning to Equation (4.12), let us develop an expression for $w'(s)$ in terms of the strength of the vortex sheet. Refer again to Figure 4.22b. Here, the vortex sheet is along the chord line, and $w'(s)$ is the component of velocity normal to the camber line induced by the vortex sheet. Let $w(x)$ denote the component of velocity normal to the *chord line* induced by the vortex sheet, as also shown in Figure 4.22b. If the airfoil is thin, the camber line is close to the chord line, and it is consistent with thin airfoil theory to make the approximation that

$$w'(s) \approx w(x) \tag{4.15}$$

An expression for $w(x)$ in terms of the strength of the vortex sheet is easily obtained from Equation (4.1), as follows. Consider Figure 4.24, which shows the vortex sheet along the chord line. We wish to calculate the value of $w(x)$ at the location x. Consider an elemental vortex of strength $\gamma\,d\xi$ located at a distance ξ from the origin along the chord line, as shown in Figure 4.24. The strength of the vortex sheet γ varies with the distance along the chord; that is, $\gamma = \gamma(\xi)$. The velocity dw at point x induced by the elemental vortex at point ξ is given by Equation (4.1) as

$$dw = -\frac{\gamma(\xi)\,d\xi}{2\pi(x-\xi)} \tag{4.16}$$

In turn, the velocity $w(x)$ induced at point x by *all* the elemental vortices along the chord line is obtained by integrating Equation (4.16) from the leading edge ($\xi = 0$) to the trailing edge ($\xi = c$):

$$w(x) = -\int_0^c \frac{\gamma(\xi)\,d\xi}{2\pi(x-\xi)} \tag{4.17}$$

Combined with the approximation stated by Equation (4.15), Equation (4.17) gives the expression for $w'(s)$ to be used in Equation (4.12).

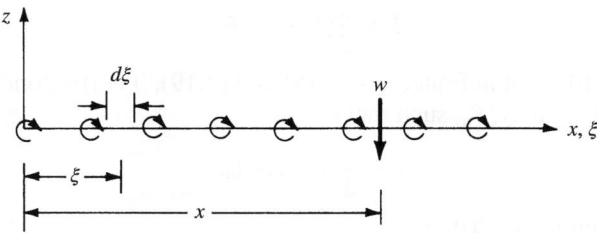

Figure 4.24 Calculation of the induced velocity at the chord line.

Recall that Equation (4.12) is the boundary condition necessary for the camber line to be a streamline. Substituting Equations (4.14), (4.15), and (4.17) into (4.12), we obtain

$$V_\infty \left(\alpha - \frac{dz}{dx} \right) - \int_0^c \frac{\gamma(\xi)\, d\xi}{2\pi(x - \xi)} = 0$$

or

$$\frac{1}{2\pi} \int_0^c \frac{\gamma(\xi)\, d\xi}{x - \xi} = V_\infty \left(\alpha - \frac{dz}{dx} \right) \tag{4.18}$$

the *fundamental equation of thin airfoil theory;* it is simply a statement that the camber line is a streamline of the flow.

Note that Equation (4.18) is written at a given point x on the chord line, and that dz/dx is evaluated at that point x. The variable ξ is simply a dummy variable of integration which varies from 0 to c along the chord line, as shown in Figure 4.24. The vortex strength $\gamma = \gamma(\xi)$ is a variable along the chord line. For a given airfoil at a given angle of attack, both α and dz/dx are known values in Equation (4.18). Indeed, the only unknown in Equation (4.18) is the vortex strength $\gamma(\xi)$. Hence, Equation (4.18) is an integral equation, the solution of which yields the variation of $\gamma(\xi)$ such that the camber line is a streamline of the flow. The central problem of thin airfoil theory is to solve Equation (4.18) for $\gamma(\xi)$, subject to the Kutta condition, namely, $\gamma(c) = 0$.

In this section, we treat the case of a symmetric airfoil. As stated in Section 4.2, a symmetric airfoil has no camber; the camber line is coincident with the chord line. Hence, for this case, $dz/dx = 0$, and Equation (4.18) becomes

$$\frac{1}{2\pi} \int_0^c \frac{\gamma(\xi)\, d\xi}{x - \xi} = V_\infty \alpha \tag{4.19}$$

In essence, within the framework of thin airfoil theory, a symmetric airfoil is treated the same as a flat plate; note that our theoretical development does not account for the airfoil thickness distribution. Equation (4.19) is an *exact* expression for the inviscid, incompressible flow over a flat plate at a small angle of attack.

To help deal with the integral in Equations (4.18) and (4.19), let us transform ξ into θ via the following transformation:

$$\xi = \frac{c}{2}(1 - \cos\theta) \tag{4.20}$$

Since x is a fixed point in Equations (4.18) and (4.19), it corresponds to a particular value of θ, namely, θ_0, such that

$$x = \frac{c}{2}(1 - \cos\theta_0) \tag{4.21}$$

Also, from Equation (4.20),

$$d\xi = \frac{c}{2}\sin\theta\, d\theta \tag{4.22}$$

Substituting Equations (4.20) to (4.22) into (4.19), and noting that the limits of integration become $\theta = 0$ at the leading edge (where $\xi = 0$) and $\theta = \pi$ at the trailing edge (where $\xi = c$), we obtain

$$\frac{1}{2\pi} \int_0^\pi \frac{\gamma(\theta) \sin \theta \, d\theta}{\cos \theta - \cos \theta_0} = V_\infty \alpha \tag{4.23}$$

A rigorous solution of Equation (4.23) for $\gamma(\theta)$ can be obtained from the mathematical theory of integral equations, which is beyond the scope of this book. Instead, we simply state that the solution is

$$\boxed{\gamma(\theta) = 2\alpha V_\infty \frac{(1 + \cos \theta)}{\sin \theta}} \tag{4.24}$$

We can verify this solution by substituting Equation (4.24) into (4.23) yielding

$$\frac{1}{2\pi} \int_0^\pi \frac{\gamma(\theta) \sin \theta \, d\theta}{\cos \theta - \cos \theta_0} = \frac{V_\infty \alpha}{\pi} \int_0^\pi \frac{(1 + \cos \theta) \, d\theta}{\cos \theta - \cos \theta_0} \tag{4.25}$$

The following standard integral appears frequently in airfoil theory and is derived in Appendix E of Reference 9:

$$\int_0^\pi \frac{\cos n\theta \, d\theta}{\cos \theta - \cos \theta_0} = \frac{\pi \sin n\theta_0}{\sin \theta_0} \tag{4.26}$$

Using Equation (4.26) in the right-hand side of Equation (4.25), we find that

$$\frac{V_\infty \alpha}{\pi} \int_0^\pi \frac{(1 + \cos \theta) \, d\theta}{\cos \theta - \cos \theta_0} = \frac{V_\infty \alpha}{\pi} \left(\int_0^\pi \frac{d\theta}{\cos \theta - \cos \theta_0} + \int_0^\pi \frac{\cos \theta \, d\theta}{\cos \theta - \cos \theta_0} \right)$$

$$= \frac{V_\infty \alpha}{\pi} (0 + \pi) = V_\infty \alpha \tag{4.27}$$

Substituting Equation (4.27) into (4.25), we have

$$\frac{1}{2\pi} \int_0^\pi \frac{\gamma(\theta) \sin \theta \, d\theta}{\cos \theta - \cos \theta_0} = V_\infty \alpha$$

which is identical to Equation (4.23). Hence, we have shown that Equation (4.24) is indeed the solution to Equation (4.23). Also, note that at the trailing edge, where $\theta = \pi$, Equation (4.24) yields

$$\gamma(\pi) = 2\alpha V_\infty \frac{0}{0}$$

which is an indeterminant form. However, using L'Hospital's rule on Equation (4.24),

$$\gamma(\pi) = 2\alpha V_\infty \frac{-\sin \pi}{\cos \pi} = 0$$

Thus, Equation (4.24) also satisfies the Kutta condition.

We are now in a position to calculate the lift coefficient for a thin, symmetric airfoil. The total circulation around the airfoil is

$$\Gamma = \int_0^c \gamma(\xi)\, d\xi \tag{4.28}$$

Using Equations (4.20) and (4.22), Equation (4.28) transforms to

$$\Gamma = \frac{c}{2} \int_0^\pi \gamma(\theta) \sin\theta\, d\theta \tag{4.29}$$

Substituting Equation (4.24) into (4.29), we obtain

$$\Gamma = \alpha c V_\infty \int_0^\pi (1 + \cos\theta)\, d\theta = \pi \alpha c V_\infty \tag{4.30}$$

Substituting Equation (4.30) into the Kutta-Joukowski theorem, we find that the lift per unit span is

$$L' = \rho_\infty V_\infty \Gamma = \pi \alpha c \rho_\infty V_\infty^2 \tag{4.31}$$

The lift coefficient is

$$c_l = \frac{L'}{q_\infty S} \tag{4.32}$$

where

$$S = c(1)$$

Substituting Equation (4.31) into (4.32), we have

$$c_l = \frac{\pi \alpha c \rho_\infty V_\infty^2}{\frac{1}{2} \rho_\infty V_\infty^2 c(1)}$$

or

$$\boxed{c_l = 2\pi\alpha} \tag{4.33}$$

and

$$\boxed{\text{Lift slope} = \frac{dc_l}{d\alpha} = 2\pi} \tag{4.34}$$

Equations (4.33) and (4.34) are important results; they state the theoretical result that the lift coefficient is *linearly proportional to angle of attack,* which is supported by the experimental results discussed in Section 4.3. They also state that the theoretical lift slope is equal to 2π rad^{-1}, which is 0.11 degree^{-1}. The experimental lift coefficient data for an NACA 0012 symmetric airfoil are given in Figure 4.25; note that Equation (4.33) accurately predicts c_l over a large range of angle of attack. (The NACA 0012 airfoil section is commonly used on airplane tails and helicopter blades.)

The moment about the leading edge can be calculated as follows. Consider the elemental vortex of strength $\gamma(\xi)\, d\xi$ located at a distance ξ from the leading edge, as sketched in Figure 4.26. The circulation associated with this elemental vortex is $d\Gamma = \gamma(\xi)\, d\xi$. In turn, the increment of lift dL contributed by the elemental vortex is $dL = \rho_\infty V_\infty\, d\Gamma$. This increment of lift creates a moment about

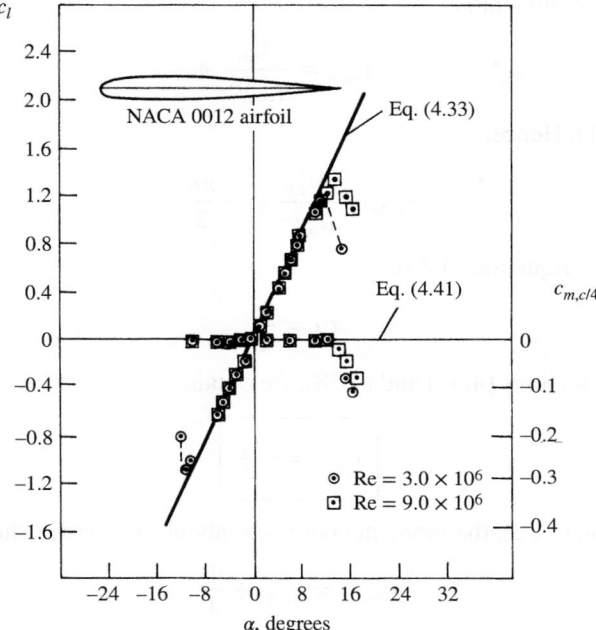

Figure 4.25 Comparison between theory and experiment for
the lift and moment coefficients for an NACA 0012 airfoil.
(*Data Source:* Abbott, I. H., and A. E. von Doenhoff: *Theory
of Wing Sections*, McGraw Hill Book Company, New York,
1949; also, Dover Publications, Inc., New York, 1959.)

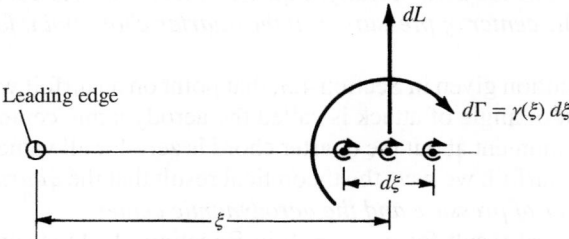

Figure 4.26 Calculation of moments about the leading edge.

the leading edge $dM = -\xi(dL)$. The total moment about the leading edge (LE)
(per unit span) due to the entire vortex sheet is therefore

$$M'_{LE} = -\int_0^c \xi(dL) = -\rho_\infty V_\infty \int_0^c \xi\gamma(\xi)\,d\xi \tag{4.35}$$

Transforming Equation (4.35) via Equations (4.20) and (4.22), and performing
the integration, we obtain (the details are left for Problem 4.4):

$$M'_{LE} = -q_\infty c^2 \frac{\pi\alpha}{2} \tag{4.36}$$

The moment coefficient is

$$c_{m,\text{le}} = \frac{M'_{\text{LE}}}{q_\infty S c}$$

where $S = c(1)$. Hence,

$$c_{m,\text{le}} = \frac{M'_{\text{LE}}}{q_\infty c^2} = -\frac{\pi\alpha}{2} \tag{4.37}$$

However, from Equation (4.33),

$$\pi\alpha = \frac{c_l}{2} \tag{4.38}$$

Combining Equations (4.37) and (4.38), we obtain

$$\boxed{c_{m,\text{le}} = -\frac{c_l}{4}} \tag{4.39}$$

From Equation (1.22), the moment coefficient about the quarter-chord point is

$$c_{m,c/4} = c_{m,\text{le}} + \frac{c_l}{4} \tag{4.40}$$

Combining Equations (4.39) and (4.40), we have

$$\boxed{c_{m,c/4} = 0} \tag{4.41}$$

In Section 1.6, a definition is given for the center of pressure as that point about which the moments are zero. Clearly, Equation (4.41) demonstrates the theoretical result that the *center of pressure is at the quarter-chord point for a symmetric airfoil.*

By the definition given in Section 4.3, that point on an airfoil where moments are independent of angle of attack is called the aerodynamic center. From Equation (4.41), the moment about the quarter chord is zero for all values of α. Hence, for a symmetric airfoil, we have the theoretical result that the *quarter-chord point is both the center of pressure and the aerodynamic center.*

The theoretical result for $c_{m,c/4} = 0$ in Equation (4.41) is supported by the experimental data given in Figure 4.25. Also, note that the experimental value of $c_{m,c/4}$ is constant over a wide range of α, thus demonstrating that the real aerodynamic center is essentially at the quarter chord.

Let us summarize the above results. The essence of thin airfoil theory is to find a distribution of vortex sheet strength along the chord line that will make the camber line a streamline of the flow while satisfying the Kutta condition $\gamma(\text{TE}) = 0$. Such a vortex distribution is obtained by solving Equation (4.18) for $\gamma(\xi)$, or in terms of the transformed independent variable θ, solving Equation (4.23) for $\gamma(\theta)$ [recall that Equation (4.23) is written for a symmetric airfoil]. The resulting vortex distribution $\gamma(\theta)$ for a symmetric airfoil is given by Equation (4.24). In turn, this vortex distribution, when inserted into the Kutta-Joukowski theorem, gives the following important theoretical results for a symmetric airfoil:

1. $c_l = 2\pi\alpha$.
2. Lift slope $= 2\pi$.
3. The center of pressure and the aerodynamic center are both located at the quarter-chord point.

EXAMPLE 4.5

Consider a thin flat plate at angle of attack 5°. Calculate the: (*a*) lift coefficient, (*b*) moment coefficient about the leading edge, (*c*) moment coefficient about the quarter-chord point, and (*d*) moment coefficient about the trailing edge.

■ **Solution**

Recall that the results obtained in Section 4.7, although couched in terms of a thin symmetric airfoil, apply in particular to a flat plate with zero thickness.

(*a*) From Equation (4.33),

$$c_\ell = 2\pi\alpha$$

where α is in radians

$$\alpha = \frac{5}{57.3} = 0.0873 \, \text{rad}$$

$$c_\ell = 2\pi(0.0873) = \boxed{0.5485}$$

(*b*) From Equation (4.39),

$$c_{m,\ell e} = -\frac{c_\ell}{4} = -\frac{0.5485}{4} = \boxed{-0.137}$$

(*c*) From Equation (4.41),

$$c_{m,c/4} = \boxed{0}$$

(*d*) Figure 4.27 is a sketch of the force and moment system on the plate. We place the lift at the quarter-chord point, along with the moment about the quarter-chord point.

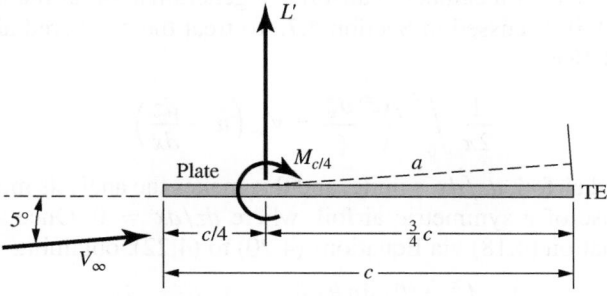

Figure 4.27 Flat plate at angle of attack 5°.

This represents the force and moment system on the plate. Recall from the discussion in Section 1.6 that the force and moment system can be represented by the lift acting through any point on the plate, and giving the moment about that point. Here, for convenience, we place the lift at the quarter-chord point.

The lift acts perpendicular to V_∞. [Part of the statement of the Kutta-Joukowski theorem given by Equation (3.140) is that the direction of the force associated with the circulation Γ is perpendicular to V_∞.] From Figure 4.27, the moment arm from L' to the trailing edge is the length a, where

$$a = \left(\frac{3}{4}c\right)\cos\alpha = \left(\frac{3}{4}c\right)\cos 5°$$

One of the assumptions of thin airfoil theory is that the angle of attack is small, and hence we can assume that $\cos\alpha \approx 1$. Therefore, the moment arm from the point of action of the lift to the trailing edge is reasonably given by $\frac{3}{4}c$. (Note that, in the previous Figure 4.26, the assumption of small α is already implicit because the moment arm is drawn parallel to the plate.)

Examining Figure 4.27, the moment about the trailing edge is

$$M'_{te} = \left(\frac{3}{4}c\right)L' + M'_{c/4}$$

$$c_{m,te} = \frac{M'_{te}}{q_\infty c^2} = \left(\frac{3}{4}c\right)\frac{L'}{q_\infty c^2} + \frac{M'_{c/4}}{q_\infty c^2}$$

$$c_{m,te} = \frac{3}{4}c_\ell + c_{m,c/4}$$

Since $c_{m,c/4} = 0$ we have

$$c_{m,te} = \frac{3}{4}c_\ell$$

$$c_{m,te} = \frac{3}{4}(0.5485) = \boxed{0.411}$$

4.8 THE CAMBERED AIRFOIL

Thin airfoil theory for a cambered airfoil is a generalization of the method for a symmetric airfoil discussed in Section 4.7. To treat the cambered airfoil, return to Equation (4.18):

$$\frac{1}{2\pi}\int_0^c \frac{\gamma(\xi)\,d\xi}{x - \xi} = V_\infty\left(\alpha - \frac{dz}{dx}\right) \qquad (4.18)$$

For a cambered airfoil, dz/dx is finite, and this makes the analysis more elaborate than in the case of a symmetric airfoil, where $dz/dx = 0$. Once again, let us transform Equation (4.18) via Equations (4.20) to (4.22), obtaining

$$\frac{1}{2\pi}\int_0^\pi \frac{\gamma(\theta)\sin\theta\,d\theta}{\cos\theta - \cos\theta_0} = V_\infty\left(\alpha - \frac{dz}{dx}\right) \qquad (4.42)$$

We wish to obtain a solution for $\gamma(\theta)$ from Equation (4.42), subject to the Kutta condition $\gamma(\pi) = 0$. Such a solution for $\gamma(\theta)$ will make the camber line a streamline of the flow. However, as before, a rigorous solution of Equation (4.42) for $\gamma(\theta)$ is beyond the scope of this book. Rather, the result is stated below:

$$\gamma(\theta) = 2V_\infty \left(A_0 \frac{1 + \cos\theta}{\sin\theta} + \sum_{n=1}^{\infty} A_n \sin n\theta \right) \tag{4.43}$$

Note that the above expression for $\gamma(\theta)$ consists of a leading term very similar to Equation (4.24) for a symmetric airfoil, plus a Fourier sine series with coefficients A_n. The values of A_n depend on the shape of the camber line dz/dx, and A_0 depends on both dz/dx and α, as shown below.

The coefficients A_0 and $A_n(n = 1, 2, 3, \ldots)$ in Equation (4.43) must be specific values in order that the camber line be a streamline of the flow. To find these specific values, substitute Equations (4.43) into Equation (4.42):

$$\frac{1}{\pi} \int_0^\pi \frac{A_0(1 + \cos\theta)\,d\theta}{\cos\theta - \cos\theta_0} + \frac{1}{\pi} \sum_{n=1}^{\infty} \int_0^\pi \frac{A_n \sin n\theta \sin\theta\, d\theta}{\cos\theta - \cos\theta_0} = \alpha - \frac{dz}{dx} \tag{4.44}$$

The first integral can be evaluated from the standard form given in Equation (4.26). The remaining integrals can be obtained from another standard form, which is derived in Appendix E of Reference 9, and which is given below:

$$\int_0^\pi \frac{\sin n\theta \sin\theta\, d\theta}{\cos\theta - \cos\theta_0} = -\pi \cos n\theta_0 \tag{4.45}$$

Hence, using Equations (4.26) and (4.45), we can reduce Equation (4.44) to

$$A_0 - \sum_{n=1}^{\infty} A_n \cos n\theta_0 = \alpha - \frac{dz}{dx}$$

or

$$\frac{dz}{dx} = (\alpha - A_0) + \sum_{n=1}^{\infty} A_n \cos n\theta_0 \tag{4.46}$$

Recall that Equation (4.46) was obtained directly from Equation (4.42), which is the transformed version of the fundamental equation of thin airfoil theory, Equation (4.18). Furthermore, recall that Equation (4.18) is evaluated at a given point x along the chord line, as sketched in Figure 4.24. Hence, Equation (4.46) is also evaluated at the given point x; here, dz/dx and θ_0 correspond to the same point x on the chord line. Also, recall that dz/dx is a function of θ_0, where $x = (c/2)(1 - \cos\theta_0)$ from Equation (4.21).

Examine Equation (4.46) closely. It is in the form of a Fourier cosine series expansion for the function of dz/dx. In general, the Fourier cosine series representation of a function $f(\theta)$ over an interval $0 \le \theta \le \pi$ is given by

$$f(\theta) = B_0 + \sum_{n=1}^{\infty} B_n \cos n\theta \tag{4.47}$$

where, from Fourier analysis, the coefficients B_0 and B_n are given by

$$B_0 = \frac{1}{\pi} \int_0^\pi f(\theta)\, d\theta \qquad (4.48)$$

and

$$B_n = \frac{2}{\pi} \int_0^\pi f(\theta) \cos n\theta\, d\theta \qquad (4.49)$$

(See, e.g., page 217 of Reference 6.) In Equation (4.46), the function dz/dx is analogous to $f(\theta)$ in the general form given in Equation (4.47). Thus, from Equations (4.48) and (4.49), the coefficients in Equation (4.46) are given by

$$\alpha - A_0 = \frac{1}{\pi} \int_0^\pi \frac{dz}{dx}\, d\theta_0$$

or

$$A_0 = \alpha - \frac{1}{\pi} \int_0^\pi \frac{dz}{dx}\, d\theta_0 \qquad (4.50)$$

and

$$A_n = \frac{2}{\pi} \int_0^\pi \frac{dz}{dx} \cos n\theta_0\, d\theta_0 \qquad (4.51)$$

Keep in mind that in the above, dz/dx is a function of θ_0. Note from Equation (4.50) that A_0 depends on both α and the shape of the camber line (through dz/dx), whereas from Equation (4.51) the values of A_n depend only on the shape of the camber line.

Pause for a moment and think about what we have done. We are considering the flow over a cambered airfoil of given shape dz/dx at a given angle of attack α. In order to make the camber line a streamline of the flow, the strength of the vortex sheet along the chord line must have the distribution $\gamma(\theta)$ given by Equation (4.43), where the coefficients A_0 and A_n are given by Equations (4.50) and (4.51), respectively. Also, note that Equation (4.43) satisfies the Kutta condition $\gamma(\pi) = 0$. Actual numbers for A_0 and A_n can be obtained for a given shape airfoil at a given angle of attack simply by carrying out the integrations indicated in Equations (4.50) and (4.51). For an example of such calculations applied to an NACA 23012 airfoil, see Example 4.5 at the end of this section. Also, note that when $dz/dx = 0$, Equation (4.43) reduces to Equation (4.24) for a symmetric airfoil. Hence, the symmetric airfoil is a special case of Equation (4.43).

Let us now obtain expressions for the aerodynamic coefficients for a cambered airfoil. The total circulation due to the entire vortex sheet from the leading edge to the trailing edge is

$$\Gamma = \int_0^c \gamma(\xi)\, d\xi = \frac{c}{2} \int_0^\pi \gamma(\theta) \sin \theta\, d\theta \qquad (4.52)$$

Substituting Equation (4.43) for $\gamma(\theta)$ into Equation (4.52), we obtain

$$\Gamma = cV_\infty \left[A_0 \int_0^\pi (1 + \cos \theta)\, d\theta + \sum_{n=1}^\infty A_n \int_0^\pi \sin n\theta \sin \theta\, d\theta \right] \qquad (4.53)$$

From any standard table of integrals,

$$\int_0^\pi (1 + \cos\theta)\, d\theta = \pi$$

and

$$\int_0^\pi \sin n\theta \sin\theta\, d\theta = \begin{cases} \pi/2 & \text{for } n = 1 \\ 0 & \text{for } n \neq 1 \end{cases}$$

Hence, Equation (4.53) becomes

$$\Gamma = cV_\infty \left(\pi A_0 + \frac{\pi}{2} A_1 \right) \tag{4.54}$$

From Equation (4.54), the lift per unit span is

$$L' = \rho_\infty V_\infty \Gamma = \rho_\infty V_\infty^2 c \left(\pi A_0 + \frac{\pi}{2} A_1 \right) \tag{4.55}$$

In turn, Equation (4.55) leads to the lift coefficient in the form

$$c_l = \frac{L'}{\frac{1}{2}\rho_\infty V_\infty^2 c(1)} = \pi(2A_0 + A_1) \tag{4.56}$$

Recall that the coefficients A_0 and A_1 in Equation (4.56) are given by Equations (4.50) and (4.51), respectively. Hence, Equation (4.56) becomes

$$c_l = 2\pi \left[\alpha + \frac{1}{\pi} \int_0^\pi \frac{dz}{dx}(\cos\theta_0 - 1)\, d\theta_0 \right] \tag{4.57}$$

and

$$\text{Lift slope} \equiv \frac{dc_l}{d\alpha} = 2\pi \tag{4.58}$$

Equations (4.57) and (4.58) are important results. Note that, as in the case of the symmetric airfoil, the theoretical lift slope for a cambered airfoil is 2π. It is a general result from thin airfoil theory that $dc_l/d\alpha = 2\pi$ for any shape airfoil. However, the expression for c_l itself differs between a symmetric and a cambered airfoil, the difference being the integral term in Equation (4.57). This integral term has physical significance, as follows. Return to Figure 4.9, which illustrates the lift curve for an airfoil. The angle of zero lift is denoted by $\alpha_{L=0}$ and is a negative value. From the geometry shown in Figure 4.9, clearly

$$c_l = \frac{dc_l}{d\alpha}(\alpha - \alpha_{L=0}) \tag{4.59}$$

Substituting Equation (4.58) into (4.59), we have

$$c_l = 2\pi(\alpha - \alpha_{L=0}) \tag{4.60}$$

Comparing Equations (4.60) and (4.57), we see that the integral term in Equation (4.57) is simply the negative of the zero-lift angle; that is,

$$\alpha_{L=0} = -\frac{1}{\pi}\int_0^{\pi}\frac{dz}{dx}(\cos\theta_0 - 1)\,d\theta_0 \tag{4.61}$$

Hence, from Equation (4.61), thin airfoil theory provides a means to predict the angle of zero lift. Note that Equation (4.61) yields $\alpha_{L=0} = 0$ for a symmetric airfoil, which is consistent with the results shown in Figure 4.25. Also, note that the more highly cambered the airfoil, the larger will be the absolute magnitude of $\alpha_{L=0}$.

Returning to Figure 4.26, the moment about the leading edge can be obtained by substituting $\gamma(\theta)$ from Equation (4.43) into the transformed version of Equation (4.35). The details are left for Problem 4.9. The result for the moment coefficient is

$$c_{m,le} = -\frac{\pi}{2}\left(A_0 + A_1 - \frac{A_2}{2}\right) \tag{4.62}$$

Substituting Equation (4.56) into (4.62), we have

$$c_{m,le} = -\left[\frac{c_l}{4} + \frac{\pi}{4}(A_1 - A_2)\right] \tag{4.63}$$

Note that, for $dz/dx = 0$, $A_1 = A_2 = 0$ and Equation (4.63) reduces to Equation (4.39) for a symmetric airfoil.

The moment coefficient about the quarter chord can be obtained by substituting Equation (4.63) into (4.40), yielding

$$c_{m,c/4} = \frac{\pi}{4}(A_2 - A_1) \tag{4.64}$$

Unlike the symmetric airfoil, where $c_{m,c/4} = 0$, Equation (4.64) demonstrates that $c_{m,c/4}$ is finite for a cambered airfoil. Therefore, the quarter chord is *not* the center of pressure for a cambered airfoil. However, note that A_1 and A_2 depend only on the shape of the camber line and do not involve the angle of attack. Hence, from Equation (4.64), $c_{m,c/4}$ is *independent* of α. Thus, the quarter-chord point is the *theoretical location* of the aerodynamic center for a cambered airfoil.

The location of the center of pressure can be obtained from Equation (1.21):

$$x_{cp} = -\frac{M'_{LE}}{L'} = -\frac{c_{m,le}c}{c_l} \tag{4.65}$$

Substituting Equation (4.63) into (4.65), we obtain

$$x_{cp} = \frac{c}{4}\left[1 + \frac{\pi}{c_l}(A_1 - A_2)\right] \tag{4.66}$$

Equation (4.66) demonstrates that the center of pressure for a cambered airfoil varies with the lift coefficient. Hence, as the angle of attack changes, the center of pressure also changes. Indeed, as the lift approaches zero, x_{cp} moves toward infinity; that is, it leaves the airfoil. For this reason, the center of pressure is not always a convenient point at which to draw the force system on an airfoil. Rather, the force-and-moment system on an airfoil is more conveniently considered at the aerodynamic center. (Return to Figure 1.25 and the discussion at the end of Section 1.6 for the referencing of the force-and-moment system on an airfoil.)

EXAMPLE 4.6

Consider an NACA 23012 airfoil. The mean camber line for this airfoil is given by

$$\frac{z}{c} = 2.6595 \left[\left(\frac{x}{c} \right)^3 - 0.6075 \left(\frac{x}{c} \right)^2 + 0.1147 \left(\frac{x}{c} \right) \right] \qquad \text{for } 0 \leq \frac{x}{c} \leq 0.2025$$

and
$$\frac{z}{c} = 0.02208 \left(1 - \frac{x}{c} \right) \qquad \text{for } 0.2025 \leq \frac{x}{c} \leq 1.0$$

Calculate (a) the angle of attack at zero lift, (b) the lift coefficient when $\alpha = 4°$, (c) the moment coefficient about the quarter chord, and (d) the location of the center of pressure in terms of x_{cp}/c, when $\alpha = 4°$. Compare the results with experimental data.

■ **Solution**

We will need dz/dx. From the given shape of the mean camber line, this is

$$\frac{dz}{dx} = 2.6595 \left[3 \left(\frac{x}{c} \right)^2 - 1.215 \left(\frac{x}{c} \right) + 0.1147 \right] \qquad \text{for } 0 \leq \frac{x}{c} \leq 0.2025$$

and
$$\frac{dz}{dx} = -0.02208 \qquad \text{for } 0.2025 \leq \frac{x}{c} \leq 1.0$$

Transforming from x to θ, where $x = (c/2)(1 - \cos\theta)$, we have

$$\frac{dz}{dx} = 2.6595 \left[\frac{3}{4}(1 - 2\cos\theta + \cos^2\theta) - 0.6075(1 - \cos\theta) + 0.1147 \right]$$

or
$$= 0.6840 - 2.3736\cos\theta + 1.995\cos^2\theta \qquad \text{for } 0 \leq \theta \leq 0.9335 \text{ rad}$$

and
$$= -0.02208 \qquad \text{for } 0.9335 \leq \theta \leq \pi$$

(a) From Equation (4.61),

$$\alpha_{L=0} = -\frac{1}{\pi} \int_0^\pi \frac{dz}{dx}(\cos\theta - 1)\, d\theta$$

[*Note:* For simplicity, we have dropped the subscript zero from θ; in Equation (4.61), θ_0 is the variable of integration—it can just as well be symbolized as θ for the variable of integration.] Substituting the equation for dz/dx into Equation (4.61), we have

$$\alpha_{L=0} = -\frac{1}{\pi} \int_0^{0.9335} (-0.6840 + 3.0576\cos\theta - 4.3686\cos^2\theta + 1.995\cos^3\theta)\, d\theta$$

$$\qquad \qquad \qquad \qquad \qquad \qquad \qquad \qquad \qquad \text{(E.1)}$$

$$-\frac{1}{\pi} \int_{0.9335}^\pi (0.02208 - 0.02208\cos\theta)\, d\theta$$

From a table of integrals, we see that

$$\int \cos \theta \, d\theta = \sin \theta$$

$$\int \cos^2 \theta \, d\theta = \frac{1}{2} \sin \theta \cos \theta + \frac{1}{2} \theta$$

$$\int \cos^3 \theta \, d\theta = \frac{1}{3} \sin \theta (\cos^2 \theta + 2)$$

Hence, Equation (E.1) becomes

$$\alpha_{L=0} = -\frac{1}{\pi} [-2.8683\theta + 3.0576 \sin \theta - 2.1843 \sin \theta \cos \theta$$

$$+ 0.665 \sin \theta (\cos^2 \theta + 2)]_0^{0.9335}$$

$$- \frac{1}{\pi} [0.02208\theta - 0.02208 \sin \theta]_{0.9335}^\pi$$

Hence,

$$\alpha_{L=0} = -\frac{1}{\pi} (-0.0065 + 0.0665) = -0.0191 \text{ rad}$$

or

$$\boxed{\alpha_{L=0} = -1.09°}$$

(b) $\alpha = 4° = 0.0698$ rad
From Equation (4.60),

$$c_l = 2\pi(\alpha - \alpha_{L=0}) = 2\pi(0.0698 + 0.0191) = \boxed{0.559}$$

(c) The value of $c_{m,c/4}$ is obtained from Equation (4.64). For this, we need the two Fourier coefficients A_1 and A_2. From Equation (4.51),

$$A_1 = \frac{2}{\pi} \int_0^\pi \frac{dz}{dx} \cos \theta \, d\theta$$

$$A_1 = \frac{2}{\pi} \int_0^{0.9335} (0.6840 \cos \theta - 2.3736 \cos^2 \theta + 1.995 \cos^3 \theta) \, d\theta$$

$$+ \frac{2}{\pi} \int_{0.9335}^\pi (-0.02208 \cos \theta) \, d\theta$$

$$= \frac{2}{\pi} [0.6840 \sin \theta - 1.1868 \sin \theta \cos \theta - 1.1868\theta + 0.665 \sin \theta (\cos^2 \theta + 2)]_0^{0.9335}$$

$$+ \frac{2}{\pi} [-0.02208 \sin \theta]_{0.09335}^\pi$$

$$= \frac{2}{\pi} (0.1322 + 0.0177) = 0.0954$$

From Equation (4.51),

$$A_2 = \frac{2}{\pi} \int_0^\pi \frac{dz}{dx} \cos 2\theta \, d\theta = \frac{2}{\pi} \int_0^\pi \frac{dz}{dx}(2\cos^2\theta - 1)\, d\theta$$

$$= \frac{2}{\pi} \int_0^{0.9335} (-0.6840 + 2.3736 \cos\theta - 0.627 \cos^2\theta$$

$$- 4.747 \cos^3\theta + 3.99 \cos^4\theta)\, d\theta$$

$$+ \frac{2}{\pi} \int_{0.9335}^\pi (0.02208 - 0.0446 \cos^2\theta)\, d\theta$$

Note:

$$\int \cos^4\theta \, d\theta = \tfrac{1}{4}\cos^3\theta \sin\theta + \tfrac{3}{8}(\sin\theta \cos\theta + \theta)$$

Thus,

$$A_2 = \frac{2}{\pi}\left\{ -0.6840\,\theta + 2.3736 \sin\theta - 0.628 \left(\tfrac{1}{2}\right)(\sin\theta\cos\theta + \theta) \right.$$

$$\left. - 4.747\left(\tfrac{1}{3}\right)\sin\theta(\cos^2\theta + 2) + 3.99\left[\tfrac{1}{4}\cos^3\theta \sin\theta + \tfrac{3}{8}(\sin\theta\cos\theta + \theta)\right]\right\}_0^{0.9335}$$

$$+ \frac{2}{\pi}\left[0.02208\theta - 0.0446\left(\tfrac{1}{2}\right)(\sin\theta\cos\theta + \theta)\right]_{0.9335}^\pi$$

$$= \frac{2}{\pi}(0.11384 + 0.01056) = 0.0792$$

From Equation (4.64),

$$c_{m,c/4} = \frac{\pi}{4}(A_2 - A_1) = \frac{\pi}{4}(0.0792 - 0.0954)$$

$$\boxed{c_{m,c/4} = -0.0127}$$

(*d*) From Equation (4.66),

$$x_{cp} = \frac{c}{4}\left[1 + \frac{\pi}{c_l}(A_1 - A_2)\right]$$

Hence, $$\frac{x_{cp}}{c} = \frac{1}{4}\left[1 + \frac{\pi}{0.559}(0.0954 - 0.0792)\right] = 0.273$$

Comparison with Experimental Data The data for the NACA 23012 airfoil are shown in Figure 4.28. From this, we make the following tabulation:

	Calculated	Experiment
$\alpha_{L=0}$	−1.09°	−1.1°
c_l (at $\alpha = 4°$)	0.559	0.55
$c_{m,c/4}$	−0.0127	−0.01

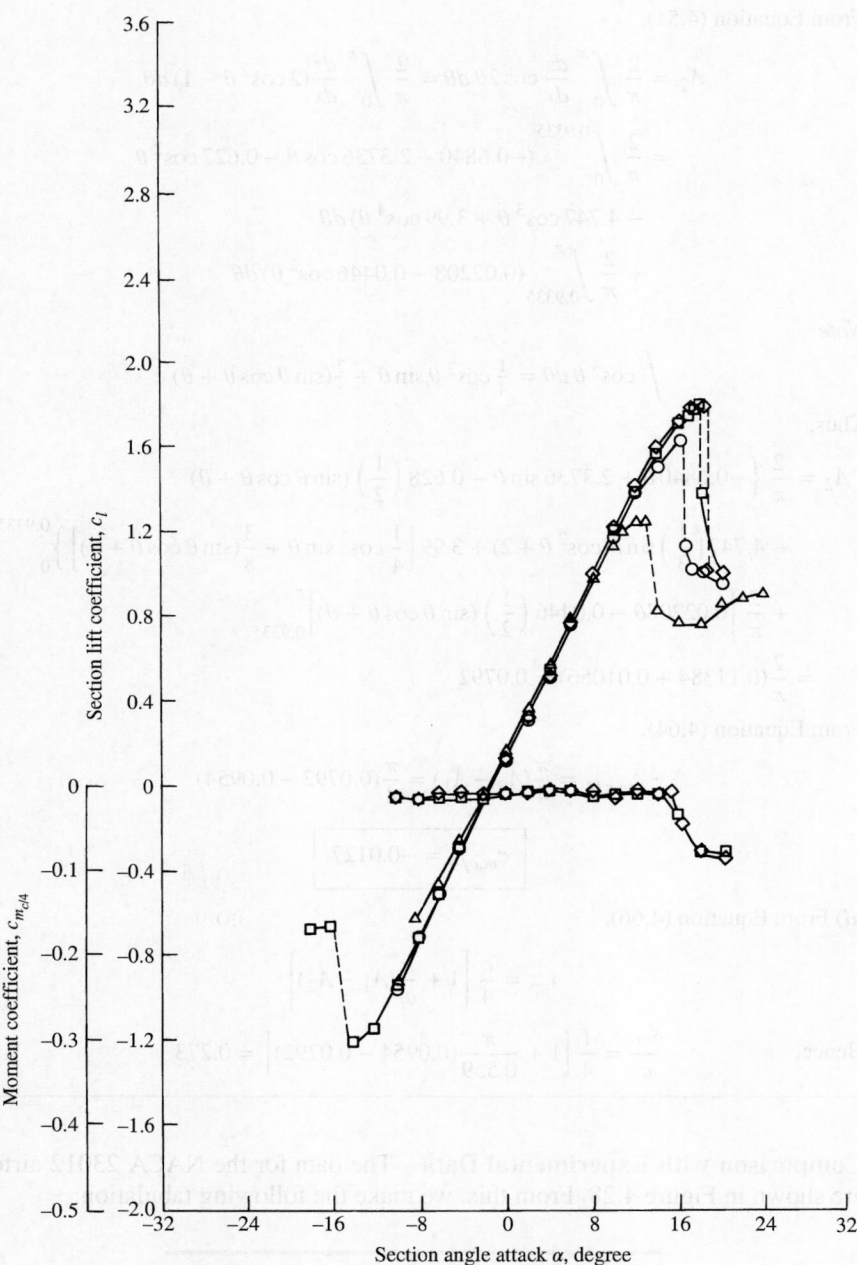

Figure 4.28 Lift- and moment-coefficients data for an NACA 23012 airfoil, for comparison with the theoretical results obtained in Example 4.6.

Note that the results from thin airfoil theory for a cambered airfoil agree very well with the experimental data. Recall that excellent agreement between thin airfoil theory for a symmetric airfoil and experimental data has already been shown in Figure 4.25. Hence, all of the work we have done in this section to develop thin airfoil theory is certainly worth the effort. Moreover, this illustrates that the development of thin airfoil theory in the early 1900s was a crowning achievement in theoretical aerodynamics and validates the mathematical approach of replacing the chord line of the airfoil with a vortex sheet, with the flow tangency condition evaluated along the mean camber line.

This brings to an end our introduction to classical thin airfoil theory. Returning to our road map in Figure 4.7, we have now completed the right-hand branch.

4.9 THE AERODYNAMIC CENTER: ADDITIONAL CONSIDERATIONS

The definition of the aerodynamic center is given in Section 4.3; it is that point on a body about which the aerodynamically generated moment is *independent of angle of attack*. At first thought, it is hard to imagine that such a point could exist. However, the moment coefficient data in Figure 4.11, which are constant with angle of attack, experimentally prove the existence of the aerodynamic center. Moreover, thin airfoil theory as derived in Sections 4.7 and 4.8 clearly shows that, within the assumptions embodied in the theory, not only does the aerodynamic center exist but that it is located at the quarter-chord point on the airfoil. Therefore, to Figure 1.24, which illustrates three different ways of stating the force and moment system on an airfoil, we can now add a fourth way, namely, the specification of the lift and drag acting through the aerodynamic center, and the value of the moment about the aerodynamic center. This is illustrated in Figure 4.29.

For most conventional airfoils, the aerodynamic center is close to, but not necessarily exactly at, the quarter-chord point. Given data for the shape of the lift coefficient curve and the moment coefficient curve taken around an arbitrary

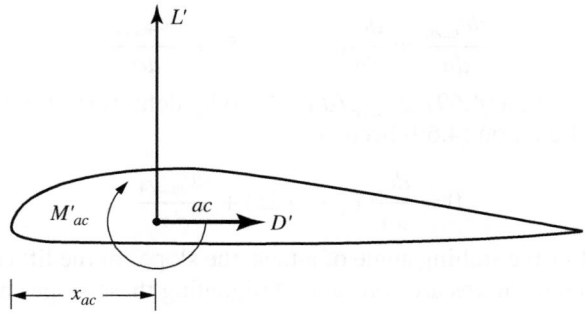

Figure 4.29 Lift, drag, and moments about the aerodynamic center.

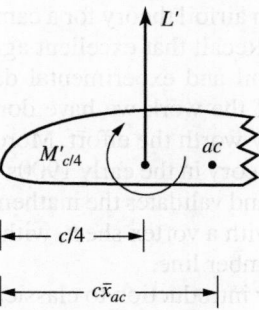

Figure 4.30 Lift and moments about the quarter-chord point, and a sketch useful for locating the aerodynamic center.

point, we can calculate the location of the aerodynamic center as follows. Consider the lift and moment system taken about the quarter-chord point, as shown in Figure 4.30. We designate the location of the aerodynamic center by $c\bar{x}_{ac}$ measured from the leading edge. Here, $\bar{x}_{ac}$ is the location of the aerodynamic center as a fraction of the chord length c. Taking moments about the aerodynamic center designated by ac in Figure 4.30, we have

$$M'_{ac} = L'(c\bar{x}_{ac} - c/4) + M'_{c/4} \tag{4.67}$$

Dividing Equation (4.67) by $q_\infty Sc$, we have

$$\frac{M'_{ac}}{q_\infty Sc} = \frac{L'}{q_\infty S}(\bar{x}_{ac} - 0.25) + \frac{M'_{c/4}}{q_\infty Sc}$$

or

$$c_{m,ac} = c_l(\bar{x}_{ac} - 0.25) + c_{m,c/4} \tag{4.68}$$

Differentiating Equation (4.68) with respect to angle of attack α, we have

$$\frac{dc_{m,ac}}{d\alpha} = \frac{dc_l}{d\alpha}(\bar{x}_{ac} - 0.25) + \frac{dc_{m,c/4}}{d\alpha} \tag{4.69}$$

However, in Equation (4.69), $dc_{m,ac}/d\alpha$ is zero by definition of the aerodynamic center. Hence, Equation (4.69) becomes

$$0 = \frac{dc_l}{d\alpha}(\bar{x}_{ac} - 0.25) + \frac{dc_{m,c/4}}{d\alpha} \tag{4.70}$$

For airfoils below the stalling angle of attack, the slopes of the lift coefficient and moment coefficient curves are constant. Designating these slopes by

$$\frac{dc_l}{d\alpha} \equiv a_0; \quad \frac{dc_{m,c/4}}{d\alpha} \equiv m_0$$

Equation (4.70) becomes

$$0 = a_0(\bar{x}_{ac} - 0.25) + m_0$$

or

$$\boxed{\bar{x}_{ac} = -\frac{m_0}{a_0} + 0.25}$$
(4.71)

Hence, Equation (4.71) proves that, for a body with linear lift and moment curves, that is, where a_0 and m_0 are fixed values, the aerodynamic center exists as a fixed point on the airfoil. Moreover, Equation (4.71) allows the calculation of the location of this point.

EXAMPLE 4.7

Consider the NACA 23012 airfoil studied in Example 4.6. Experimental data for this airfoil is plotted in Figure 4.28, and can be obtained from Reference 11. It shows that, at $\alpha = 4°$, $c_l = 0.55$ and $c_{m,c/4} = -0.005$. The zero-lift angle of attack is $-1.1°$. Also, at $\alpha = -4°$, $c_{m,c/4} = -0.0125$. (Note that the "experimental" value of $c_{m,c/4} = -0.01$ tabulated at the end of Example 4.6 is an average value over a range of angle of attack. Since the calculated value of $c_{m,c/4}$ from thin airfoil theory states that the quarter-chord point is the aerodynamic center, it makes sense in Example 4.6 to compare the calculated $c_{m,c/4}$ with an experimental value averaged over a range of angle of attack. However, in the present example, because $c_{m,c/4}$ in reality varies with angle of attack, we use the actual data at two different angles of attack.) From the given information, calculate the location of the aerodynamic center for the NACA 23012 airfoil.

■ **Solution**
Since $c_l = 0.55$ at $\alpha = 4°$ and $c_l = 0$ at $\alpha = -1.1°$, the lift slope is

$$a_0 = \frac{0.55 - 0}{4 - (-1.1)} = 0.1078 \text{ per degree}$$

The slope of the moment coefficient curve is

$$m_0 = \frac{-0.005 - (-0.0125)}{4 - (-4)} = 9.375 \times 10^{-4} \text{ per degree}$$

From Equation (4.71),

$$\bar{x}_{ac} = -\frac{m_0}{a_0} + 0.25 = -\frac{9.375 \times 10^{-4}}{0.1078} + 0.25 = \boxed{0.241}$$

The result agrees exactly with the measured value quoted on page 183 of Abbott and Von Doenhoff (Reference 11).

DESIGN BOX

The result of Example 4.7 shows that the aerodynamic center for the NACA 23012 airfoil is located *ahead* of, but very close to, the quarter-chord point. For some other families of airfoils, the aerodynamic center is located *behind,* but similarly close to, the quarter-chord point. For a given airfoil family, the location of the aerodynamic center depends on the airfoil thickness, as shown in Figure 4.31. The variation of $\bar{x}_{ac}$ with thickness for the NACA 230XX family is given in Figure 4.31*a*. Here, the aerodynamic center is ahead of the quarter-chord point, and becomes progressively farther ahead as the airfoil thickness is increased. In contrast, the variation of $\bar{x}_{ac}$ with thickness for the NACA 64-2XX family is given in Figure 4.31*b*. Here,

the aerodynamic center is behind the quarter-chord point, and becomes progressively farther behind as the airfoil thickness is increased.

From the point of view of purely aerodynamics, the existence of the aerodynamic center is interesting, but the specification of the force and moment system on the airfoil by placing the lift and drag at the aerodynamic center and giving the value of M'_{ac} as illustrated in Figure 4.29 is not more useful than placing the lift and drag at any other point on the airfoil and giving the value of M' at that point, such as shown in Figure 1.25. However, in flight dynamics, and in particular the consideration of the stability and control of flight vehicles, placing the lift and drag at, and

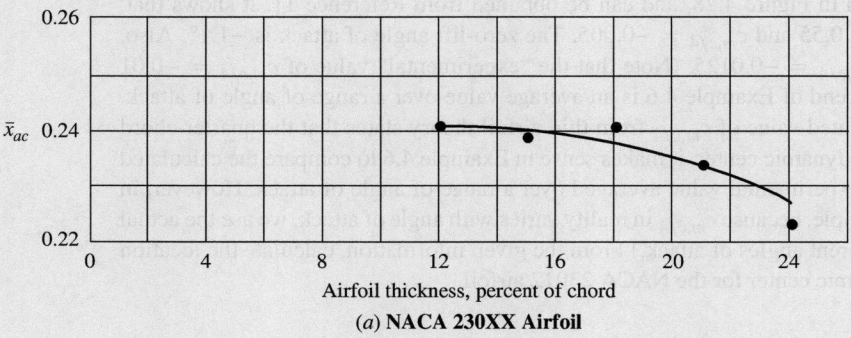
(a) NACA 230XX Airfoil

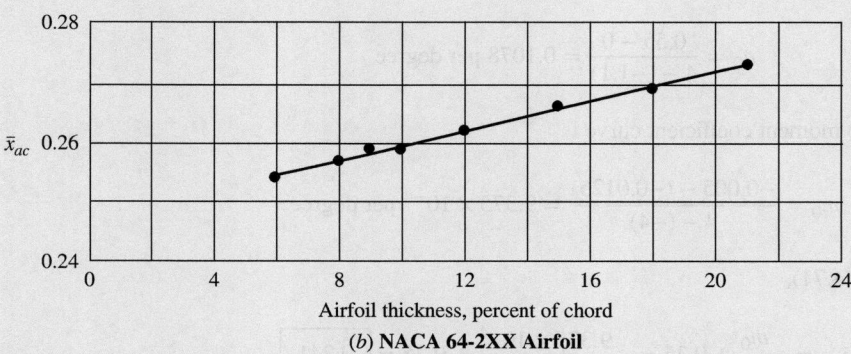
(b) NACA 64-2XX Airfoil

Figure 4.31 Variation of the location of the aerodynamic center with airfoil thickness. (*a*) NACA 230XX airfoil. (*b*) NACA 64-2XX airfoil.

dealing with the moment about, the aerodynamic center, is particularly convenient. The fact that M_{ac} for a flight vehicle is independent of angle of attack simplifies the analysis of the stability and control characteristics, and the use of the aerodynamic center therefore becomes important in airplane design. In the design process, it is important to know where the aerodynamic centers of the various components of the aircraft (wing, tail, fuselage, etc.) are located, and above all the location of the aerodynamic center for the complete flight vehicle. It is for this reason that we have placed extra emphasis on the aerodynamic center in Section 4.9. For an introduction to stability and control, see Chapter 7 of the author's book *Introduction to Flight,* 5th edition, McGraw Hill, Boston, 2005. For more information about the aerodynamic center, and its use in airplane design, see the author's book *Aircraft Performance and Design*, McGraw Hill, Boston, 1999.

4.10 LIFTING FLOWS OVER ARBITRARY BODIES: THE VORTEX PANEL NUMERICAL METHOD

The thin airfoil theory described in Sections 4.7 and 4.8 is just what it says—it applies only to thin airfoils at small angles of attack. (Make certain that you understand exactly where in the development of thin airfoil theory these assumptions are made and the reasons for making them.) The advantage of thin airfoil theory is that closed-form expressions are obtained for the aerodynamic coefficients. Moreover, the results compare favorably with experimental data for airfoils of about 12 percent thickness or less. However, the airfoils on many low-speed airplanes are thicker than 12 percent. Moreover, we are frequently interested in high angles of attack, such as occur during takeoff and landing. Finally, we are sometimes concerned with the generation of aerodynamic lift on other body shapes, such as automobiles or submarines. Hence, thin airfoil theory is quite restrictive when we consider the whole spectrum of aerodynamic applications. We need a method that allows us to calculate the aerodynamic characteristics of bodies of arbitrary shape, thickness, and orientation. Such a method is described in this section. Specifically, we treat the vortex panel method, which is a numerical technique that has come into widespread use since the early 1970s. In reference to our road map in Figure 4.7, we now move to the left-hand branch. Also, since this chapter deals with airfoils, we limit our attention to two-dimensional bodies.

The vortex panel method is directly analogous to the source panel method described in Section 3.17. However, because a source has zero circulation, source panels are useful only for nonlifting cases. In contrast, vortices have circulation, and hence vortex panels can be used for lifting cases. (Because of the similarities between source and vortex panel methods, return to Section 3.17 and review the basic philosophy of the source panel method before proceeding further.)

The philosophy of covering a body *surface* with a vortex sheet of such a strength to make the surface a streamline of the flow was discussed in Section 4.4. We then went on to simplify this idea by placing the vortex sheet on the camber line of the airfoil as shown in Figure 4.16, thus establishing the basis for thin airfoil theory. We now return to the original idea of wrapping the vortex sheet over the complete surface of the body, as shown in Figure 4.15. We wish to find

$\gamma(s)$ such that the body surface becomes a streamline of the flow. There exists no closed-form analytical solution for $\gamma(s)$; rather, the solution must be obtained numerically. This is the purpose of the vortex panel method.

Let us approximate the vortex sheet shown in Figure 4.15 by a series of straight panels, as shown earlier in Figure 3.40. (In Chapter 3, Figure 3.40 was used to discuss source panels; here, we use the same sketch for discussion of vortex panels.) Let the vortex strength $\gamma(s)$ per unit length be constant over a given panel, but allow it to vary from one panel to the next. That is, for the n panels shown in Figure 3.40, the vortex panel strengths per unit length are $\gamma_1, \gamma_2, \ldots, \gamma_j, \ldots, \gamma_n$. These panel strengths are unknowns; the main thrust of the panel technique is to solve for $\gamma_j, j = 1$ to n, such that the body surface becomes a streamline of the flow and such that the Kutta condition is satisfied. As explained in Section 3.17, the midpoint of each panel is a control point at which the boundary condition is applied; that is, at each control point, the normal component of the flow velocity is zero.

Let P be a point located at (x, y) in the flow, and let r_{pj} be the distance from any point on the jth panel to P, as shown in Figure 3.40. The radius r_{pj} makes the angle θ_{pj} with respect to the x axis. The velocity potential induced at P due to the jth panel, $\Delta\phi_j$, is, from Equation (4.3),

$$\Delta\phi_j = -\frac{1}{2\pi} \int_j \theta_{pj} \gamma_j \, ds_j \tag{4.72}$$

In Equation (4.72), γ_j is constant over the jth panel, and the integral is taken over the jth panel only. The angle θ_{pj} is given by

$$\theta_{pj} = \tan^{-1} \frac{y - y_j}{x - x_j} \tag{4.73}$$

In turn, the potential at P due to *all* the panels is Equation (4.72) summed over all the panels:

$$\phi(P) = \sum_{j=1}^{n} \phi_j = -\sum_{j=1}^{n} \frac{\gamma_j}{2\pi} \int_j \theta_{pj} \, ds_j \tag{4.74}$$

Since point P is just an arbitrary point in the flow, let us put P at the control point of the ith panel shown in Figure 3.40. The coordinates of this control point are (x_i, y_i). Then Equations (4.73) and (4.74) become

$$\theta_{ij} = \tan^{-1} \frac{y_i - y_j}{x_i - x_j}$$

and

$$\phi(x_i, y_i) = -\sum_{j=1}^{n} \frac{\gamma_j}{2\pi} \int_j \theta_{ij} \, ds_j \tag{4.75}$$

Equation (4.75) is physically the contribution of *all* the panels to the potential at the control point of the ith panel.

At the control points, the normal component of the velocity is zero; this velocity is the superposition of the uniform flow velocity and the velocity induced by all the vortex panels. The component of V_∞ normal to the ith panel is given by Equation (3.148):

$$V_{\infty,n} = V_\infty \cos \beta_i \qquad (3.148)$$

The normal component of velocity induced at (x_i, y_i) by the vortex panels is

$$V_n = \frac{\partial}{\partial n_i} [\phi(x_i, y_i)] \qquad (4.76)$$

Combining Equations (4.75) and (4.76), we have

$$V_n = -\sum_{j=1}^{n} \frac{\gamma_j}{2\pi} \int_j \frac{\partial \theta_{ij}}{\partial n_i} ds_j \qquad (4.77)$$

where the summation is over all the panels. The normal component of the flow velocity at the ith control point is the sum of that due to the freestream [Equation (3.148)] and that due to the vortex panels [Equation (4.77)]. The boundary condition states that this sum must be zero:

$$V_{\infty,n} + V_n = 0 \qquad (4.78)$$

Substituting Equations (3.148) and (4.77) into (4.78), we obtain

$$V_\infty \cos \beta_i - \sum_{j=1}^{n} \frac{\gamma_j}{2\pi} \int_j \frac{\partial \theta_{ij}}{\partial n_i} ds_j = 0 \qquad (4.79)$$

Equation (4.79) is the crux of the vortex panel method. The values of the integrals in Equation (4.79) depend simply on the panel geometry; they are not properties of the flow. Let $J_{i,j}$ be the value of this integral when the control point is on the ith panel. Then Equation (4.79) can be written as

$$V_\infty \cos \beta_i - \sum_{j=1}^{n} \frac{\gamma_j}{2\pi} J_{i,j} = 0 \qquad (4.80)$$

Equation (4.80) is a linear algebraic equation with n unknowns, $\gamma_1, \gamma_2, \ldots, \gamma_n$. It represents the flow boundary condition evaluated at the control point of the ith panel. If Equation (4.80) is applied to the control points of *all* the panels, we obtain a system of n linear equations with n unknowns.

To this point, we have been deliberately paralleling the discussion of the source panel method given in Section 3.17; however, the similarity stops here. For the source panel method, the n equations for the n unknown source strengths are routinely solved, giving the flow over a nonlifting body. In contrast, for the lifting case with vortex panels, in addition to the n equations given by Equation (4.80) applied at all the panels, we must also satisfy the Kutta condition. This can be done in several ways. For example, consider Figure 4.32, which illustrates a detail of the vortex panel distribution at the trailing edge. Note that

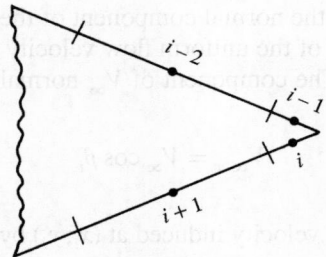

Figure 4.32 Vortex panels at the trailing edge.

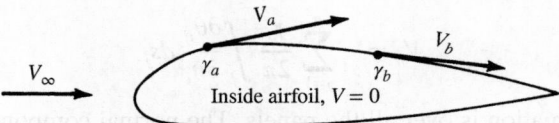

Figure 4.33 Airfoil as a solid body, with zero velocity inside the profile.

the length of each panel can be different; their length and distribution over the body are up to your discretion. Let the two panels at the trailing edge (panels i and $i - 1$ in Figure 4.32) be very small. The Kutta condition is applied *precisely* at the trailing edge and is given by $\gamma(\text{TE}) = 0$. To approximate this numerically, if points i and $i - 1$ are close enough to the trailing edge, we can write

$$\gamma_i = -\gamma_{i-1} \tag{4.81}$$

such that the strengths of the two vortex panels i and $i - 1$ exactly cancel at the point where they touch at the trailing edge. Thus, in order to impose the Kutta condition on the solution of the flow, Equation (4.81) (or an equivalent expression) must be included. Note that Equation (4.80), evaluated at all the panels, and Equation (4.81) constitute an *overdetermined* system of n unknowns with $n + 1$ equations. Therefore, to obtain a determined system, Equation (4.80) is not evaluated at one of the control points on the body. That is, we choose to ignore one of the control points, and we evaluate Equation (4.80) at the other $n - 1$ control points. This, in combination with Equation (4.81), now gives a system of n linear algebraic equations with n unknowns, which can be solved by standard techniques.

At this stage, we have conceptually obtained the values of $\gamma_1, \gamma_2, \dots, \gamma_n$, which make the body surface a streamline of the flow and also satisfy the Kutta condition. In turn, the flow velocity tangent to the surface can be obtained directly from γ. To see this more clearly, consider the airfoil shown in Figure 4.33. We are concerned only with the flow outside the airfoil and on its surface. Therefore, let the velocity be zero at every point *inside* the body, as shown in Figure 4.33. In particular, the velocity just inside the vortex sheet on the surface is zero. This corresponds to $u_2 = 0$ in Equation (4.8). Hence, the velocity just outside the vortex sheet is, from Equation (4.8),

$$\gamma = u_1 - u_2 = u_1 - 0 = u_1$$

In Equation (4.8), u denotes the velocity tangential to the vortex sheet. In terms of the picture shown in Figure 4.33, we obtain $V_a = \gamma_a$ at point a, $V_b = \gamma_b$ at point b, etc. Therefore, *the local velocities tangential to the airfoil surface are equal to the local values of γ*. In turn, the local pressure distribution can be obtained from Bernoulli's equation.

The total circulation and the resulting lift are obtained as follows. Let s_j be the length of the jth panel. Then, the circulation due to the jth panel is $\gamma_j s_j$. In turn, the total circulation due to all the panels is

$$\Gamma = \sum_{j=1}^{n} \gamma_j s_j \tag{4.82}$$

Hence, the lift per unit span is obtained from

$$L' = \rho_\infty V_\infty \sum_{j=1}^{n} \gamma_j s_j \tag{4.83}$$

The presentation in this section is intended to give only the general flavor of the vortex panel method. There are many variations of the method in use today, and you are encouraged to read the modern literature, especially as it appears in the *AIAA Journal* and the *Journal of Aircraft* since 1970. The vortex panel method as described in this section is termed a "first-order" method because it assumes a constant value of γ over a given panel. Although the method may appear to be straightforward, its numerical implementation can sometimes be frustrating. For example, the results for a given body are sensitive to the number of panels used, their various sizes, and the way they are distributed over the body surface (i.e., it is usually advantageous to place a large number of small panels near the leading and trailing edges of an airfoil and a smaller number of larger panels in the middle). The need to ignore one of the control points in order to have a determined system in n equations for n unknowns also introduces some arbitrariness in the numerical solution. Which control point do you ignore? Different choices sometimes yield different numerical answers for the distribution of γ over the surface. Moreover, the resulting numerical distributions for γ are not always smooth, but rather, they have oscillations from one panel to the next as a result of numerical inaccuracies. The problems mentioned above are usually overcome in different ways by different groups who have developed relatively sophisticated panel programs for practical use. For example, what is more common today is to use a *combination* of both source and vortex panels (source panels to basically simulate the airfoil thickness and vortex panels to introduce circulation) in a panel solution. This combination helps to mitigate some of the practical numerical problems just discussed. Again, you are encouraged to consult the literature for more information.

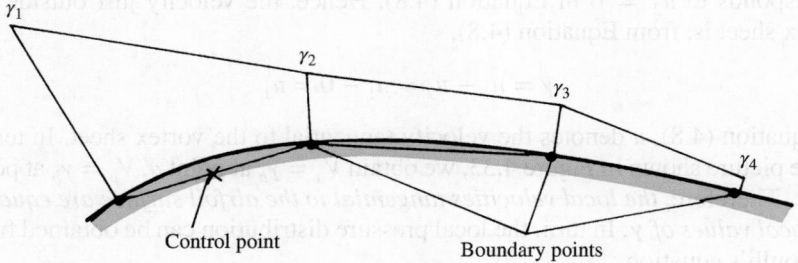

Figure 4.34 Linear distribution of γ over each panel—a second-order panel method.

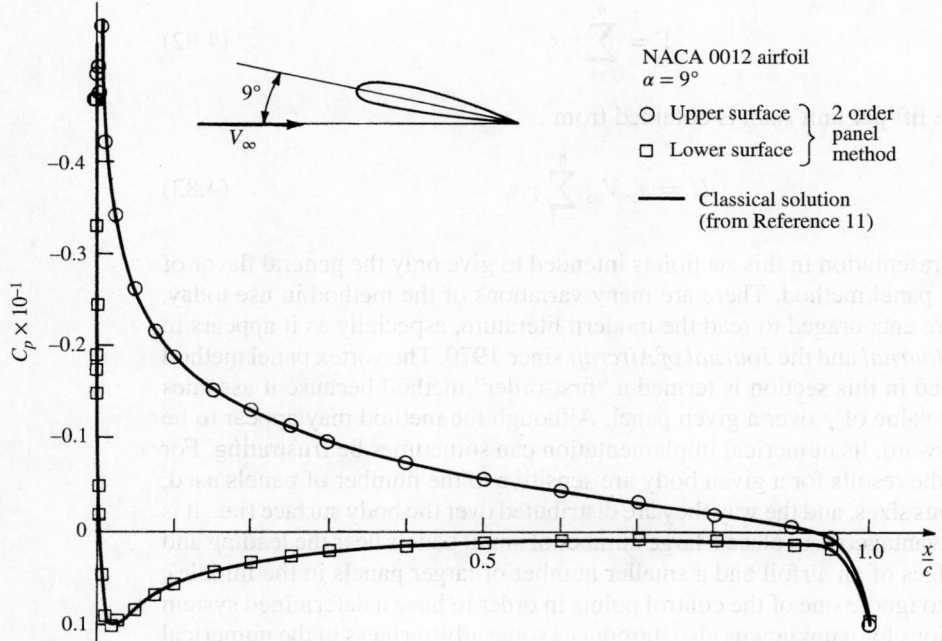

Figure 4.35 Pressure coefficient distribution over an NACA 0012 airfoil; comparison between second-order vortex panel method and NACA theoretical results from Reference 11. The numerical panel results were obtained by one of the author's graduate students, Dr. Tae-Hwan Cho.

Such accuracy problems have also encouraged the development of higher-order panel techniques. For example, a "second-order" panel method assumes a *linear* variation of γ over a given panel, as sketched in Figure 4.34. Here, the value of γ at the edges of each panel is matched to its neighbors, and the values γ_1, γ_2, γ_3, etc., at the *boundary points* become the unknowns to be solved. The flow-tangency boundary condition is still applied at the *control point* of each panel, as before. Some results using a second-order vortex panel technique are given in Figure 4.35, which shows the distribution of pressure coefficients over the upper and lower surfaces of an NACA 0012 airfoil at a 9° angle of attack. The circles and squares are numerical results from a second-order vortex panel technique developed at the University of Maryland, and the solid lines are from NACA results given in Reference 11. Excellent agreement is obtained.

Again, you are encouraged to consult the literature before embarking on any serious panel solutions of your own. For example, Reference 14 is a classic paper on panel methods, and Reference 15 highlights many of the basic concepts of panel methods along with actual computer program statement listings for simple applications. Reference 62 is a modern compilation of papers, several of which deal with current panel techniques. Finally, Katz and Plotkin (Reference 63) give perhaps the most thorough discussion of panel techniques and their foundations to date.

4.11 MODERN LOW-SPEED AIRFOILS

The nomenclature and aerodynamic characteristics of standard NACA airfoils are discussed in Sections 4.2 and 4.3; before progressing further, you should review these sections in order to reinforce your knowledge of airfoil behavior, especially in light of our discussions on airfoil theory. Indeed, the purpose of this section is to provide a modern sequel to the airfoils discussed in Sections 4.2 and 4.3.

During the 1970s, NASA designed a series of low-speed airfoils that have performance superior to the earlier NACA airfoils. The standard NACA airfoils were based almost exclusively on experimental data obtained during the 1930s and 1940s. In contrast, the new NASA airfoils were designed on a computer using a numerical technique similar to the source and vortex panel methods discussed earlier, along with numerical predictions of the viscous flow behavior (skin friction and flow separation). Wind-tunnel tests were then conducted to verify the computer-designed profiles and to obtain the definitive airfoil properties. Out of this work, first came the general aviation, Whitcomb [GA(W)-1] airfoil, which has since been redesignated the LS(1)-0417 airfoil. The shape of this airfoil is given in Figure 4.36, obtained from Reference 16. Note that it has a large leading-edge radius ($0.08c$ in comparison to the standard $0.02c$) in order to flatten the usual peak in pressure coefficient near the nose. Also, note that the bottom surface near the trailing edge is cusped in order to increase the camber and hence the aerodynamic

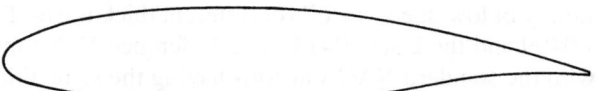

Figure 4.36 Profile for the NASA LS(1)-0417 airfoil. When first introduced, this airfoil was labeled the GA(W)-1 airfoil, a nomenclature which has now been superseded. (*From* McGhee, R. J., and W. D. Beasley: *Low-Speed Aerodynamic Characteristics of a 17-Percent-Thick Airfoil Section Designed for General Aviation Applications*, NASA TN D-7428, December 1973.)

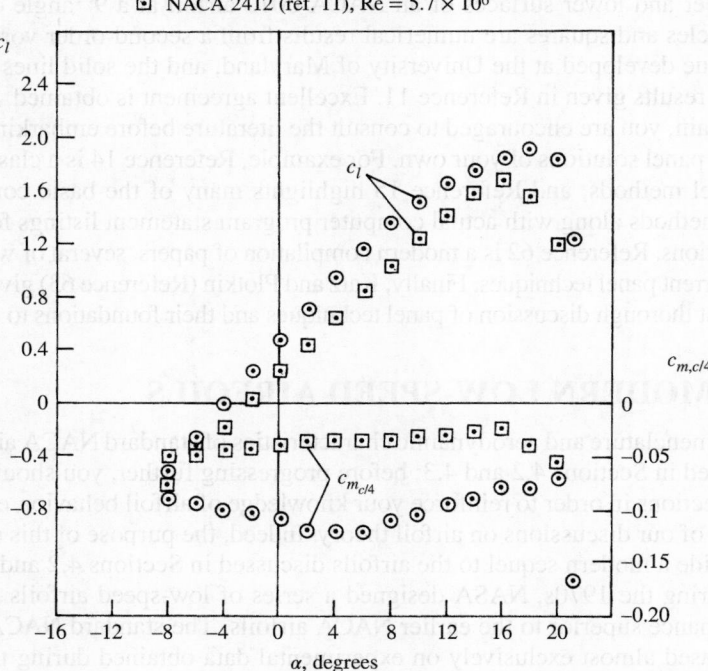

Figure 4.37 Comparison of the modern NASA LS(1)-0417 airfoil with the standard NACA 2412 airfoil.

loading in that region. Both design features tend to discourage flow separation over the top surface at high angle of attack, hence yielding higher values of the maximum lift coefficient. The experimentally measured lift and moment properties (from Reference 16) are given in Figure 4.37, where they are compared with the properties for an NACA 2412 airfoil, obtained from Reference 11. Note that $c_{l,max}$ for the NASA LS(1)-0417 is considerably higher than for the NACA 2412.

The NASA LS(1)-0417 airfoil has a maximum thickness of 17 percent and a design lift coefficient of 0.4. Using the same camber line, NASA has extended this airfoil into a family of low-speed airfoils of different thicknesses, for example, the NASA LS(1)-0409 and the LS(1)-0413. (See Reference 17 for more details.) In comparison with the standard NACA airfoils having the same thicknesses, these new LS(1)-04xx airfoils have:

1. Approximately 30 percent higher $c_{l,max}$.
2. Approximately a 50 percent increase in the ratio of lift to drag (L/D) at a lift coefficient of 1.0. This value of $c_l = 1.0$ is typical of the climb lift coefficient for general aviation aircraft, and a high value of L/D greatly improves the climb performance. (See Reference 2 for a general

introduction to airplane performance and the importance of a high L/D ratio to airplane efficiency.)

It is interesting to note that the shape of the airfoil in Figure 4.36 is very similar to the supercritical airfoils to be discussed in Chapter 11. The development of the supercritical airfoil by NASA aerodynamicist Richard Whitcomb in 1965 resulted in a major improvement in airfoil drag behavior at high subsonic speeds, near Mach 1. The supercritical airfoil was a major breakthrough in high-speed aerodynamics. The LS(1)-0417 low-speed airfoil shown in Figure 4.36, first introduced as the GA(W)-1 airfoil, was a later spin-off from supercritical airfoil research. It is also interesting to note that the first production aircraft to use the NASA LS(1)-0417 airfoil was the Piper PA-38 Tomahawk, introduced in the late 1970s.

In summary, new airfoil development is alive and well in the aeronautics of the late twentieth century. Moreover, in contrast to the purely experimental development of the earlier airfoils, we now enjoy the benefit of powerful computer programs using panel methods and advanced viscous flow solutions for the design of new airfoils. Indeed, in the 1980s, NASA established an official Airfoil Design Center at The Ohio State University, which services the entire general aviation industry with over 30 different computer programs for airfoil design and analysis. For additional information on such new low-speed airfoil development, you are urged to read Reference 16, which is the classic first publication dealing with these airfoils, as well as the concise review given in Reference 17.

DESIGN BOX

This chapter deals with incompressible flow over airfoils. Moreover, the analytical thin airfoil theory and the numerical panel methods discussed here are techniques for calculating the aerodynamic characteristics for a *given airfoil of specified shape*. Such an approach is frequently called the *direct problem*, wherein the shape of the body is given, and the surface pressure distribution (for example) is calculated. For design purposes, it is desirable to turn this process inside-out; it is desirable to specify the surface pressure distribution—a pressure distribution that will achieve enhanced airfoil performance—and calculate the shape of the airfoil that will produce the specified pressure distribution. This approach is called the *inverse problem*. Before the advent of the high-speed digital computer, and the concurrent rise of the discipline of computational fluid dynamics in the 1970s (see Section 2.17.2), the analytical solution of the inverse problem was difficult, and was

not used by the practical airplane designer. Instead, for most of the airplanes designed before and during the twentieth century, the choice of an airfoil shape was based on reasonable experimental data (at best), and guesswork (at worst). This story is told in some detail in Reference 58. The design problem was made more comfortable with the introduction of the various families of NACA airfoils, beginning in the early 1930s. A logical method was used for the geometrical design of these airfoils, and definitive experimental data on the NACA airfoils were made available (such as shown in Figures 4.10, 4.11, and 4.28). For this reason, many airplanes designed during the middle of the twentieth century used standard NACA airfoil sections. Even today, the NACA airfoils are sometimes the most expeditious choice of the airplane designer, as indicated by the tabulation (by no means complete) in Section 4.2 of airplanes using such airfoils.

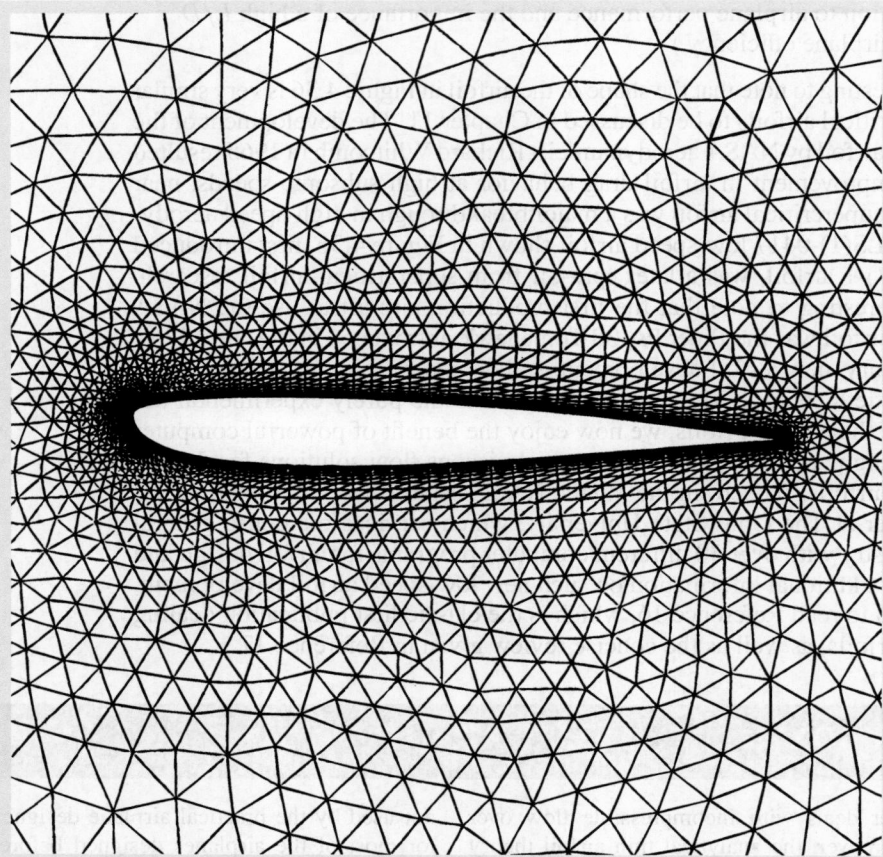

Figure 4.38 Unstructured mesh for the numerical calculation of the flow over an airfoil.
(*Data Source:* Anderson, W. Kyle, and Bonhaus, Daryl L.: "Airfoil Design on Unstructured
Grids for Turbulent Flows," *AIAA J.*, vol. 37, no. 2, Feb. 1999, pp. 185–191.)

However, today the power of computational fluid dynamics (CFD) is revolutionizing airfoil design and analysis. The inverse problem, and indeed the next step—the overall automated procedure that results in a completely optimized airfoil shape for a given design point—are being made tractable by CFD. An example of such work is illustrated in Figures 4.38 and 4.39, taken from the recent work of Kyle Anderson and Daryl Bonhaus (Reference 64). Here, CFD solutions of the continuity, momentum, and energy equations for a compressible, viscous flow (the Navier-Stokes equations, as denoted in Section 2.17.2) are carried out for the purpose of airfoil design. Using a finite

volume CFD technique, and the grid shown in Figure 4.38, the inverse problem is solved. The *specified* pressure distribution over the top and bottom surfaces of the airfoil is given by the circles in Figure 4.39a. The optimization technique is iterative and requires starting with a pressure distribution that is not the desired, specified one; the initial distribution is given by the solid curves in Figure 4.39a, and the airfoil shape corresponding to this initial pressure distribution is shown by the solid curve in Figure 4.39b. (In Figure 4.39b, the airfoil shape appears distorted because an expanded scale is used for the ordinate.) After 10 design cycles, the optimized airfoil shape that supports

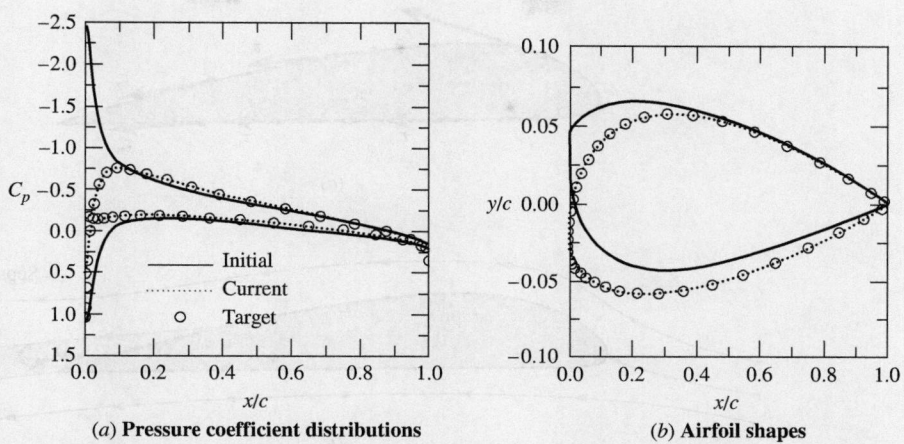

(a) **Pressure coefficient distributions**   (b) **Airfoil shapes**

Figure 4.39 An example of airfoil optimized design using computational fluid dynamics. (*Data Source:* Anderson, W. Kyle, and Bonhaus, Daryl L.: "Airfoil Design on Unstructured Grids for Turbulent Flows," *AIAA J.*, vol. 37, no. 2, Feb. 1999, pp. 185–191.)

the specified pressure distribution is obtained, as given by the circles in Figure 4.39*b*. The initial airfoil shape is also shown in constant scale in Figure 4.38.

The results given in Figures 4.38 and 4.39 are shown here simply to provide the flavor of modern airfoil design and analysis. This is reflective of the wave of future airfoil design procedures, and you are encouraged to read the contemporary literature in order to keep up with this rapidly evolving field. However, keep in mind that the simpler analytical approach of thin airfoil theory discussed in the present chapter, and especially the simple practical results of this theory, will continue to be part of the whole "toolbox" of procedures to be used by the designer in the future. The fundamentals embodied in thin airfoil theory will continue to be part of the fundamentals of aerodynamics and will always be there as a partner with the modern CFD techniques.

4.12 VISCOUS FLOW: AIRFOIL DRAG

This is another "stand-alone" viscous flow section in the same spirit as Section 1.11. It does not break the continuity of our discussions on inviscid flow; rather, it is designed to complement them. Before reading further, you are encouraged to review the introduction to boundary layers given in Section 1.11.

The lift on an airfoil is primarily due to the pressure distribution exerted on its surface; the shear stress distribution acting on the airfoil, when integrated in the lift direction, is usually negligible. The lift, therefore, can be accurately calculated assuming inviscid flow in conjunction with the Kutta condition at the trailing edge. When used to predict drag, however, this same approach yields zero drag, a result that goes against common sense, and is called *d'Alembert's paradox* after Jean le Rond d'Alembert, the eighteenth-century French mathematician and scientist who first made such drag calculations for inviscid flows over two-dimensional bodies (see Sections 3.13 and 3.20).

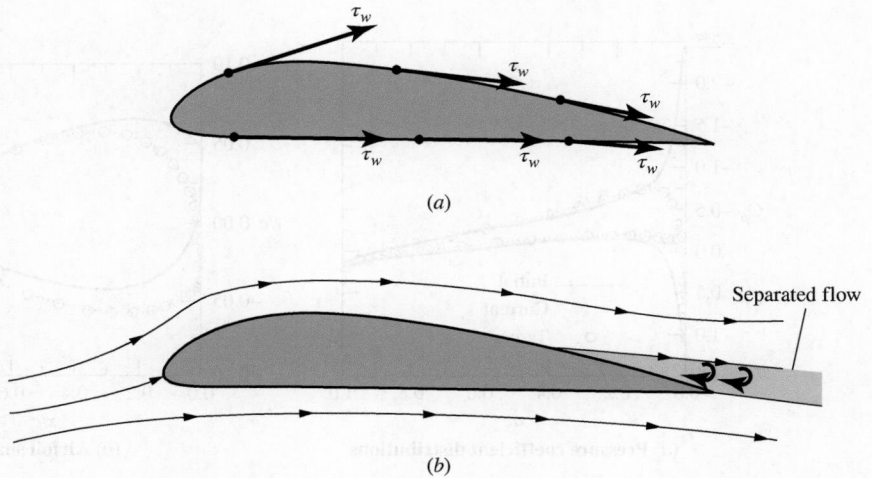

Figure 4.40 Subsonic airfoil drag is due to two components: (*a*) shear stress acting on the surface, and (*b*) pressure drag due to flow separation.

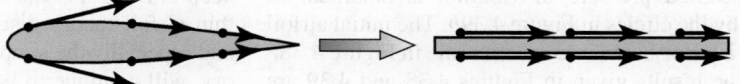

Figure 4.41 Estimation of skin-friction drag on an airfoil from that on a flat plate.

The paradox is immediately removed when viscosity (friction) is included in the flow. Indeed, viscosity in the flow is totally responsible for the aerodynamic drag on an airfoil. It acts through two mechanisms:

1. *Skin-friction drag,* due to the shear stress acting on the surface (Figure 4.40*a*), and

2. *Pressure drag due to flow separation,* sometimes called *form drag* (Figure 4.40*b*).

That shear stress creates drag is self-evident from Figure 4.40*a*. The pressure drag created by flow separation (Figure 4.40*b*) is a more subtle phenomenon and will be discussed toward the end of this section.

4.12.1 Estimating Skin-Friction Drag: Laminar Flow

As a first approximation, we assume that skin-friction drag on an airfoil is essentially the same as the skin-friction drag on a flat plate at zero angle of attack, as illustrated in Figure 4.41. Obviously, this approximation becomes more accurate the thinner the airfoil and the smaller the angle of attack. Consistent with the rest of this chapter, we will continue to deal with low-speed incompressible flow.

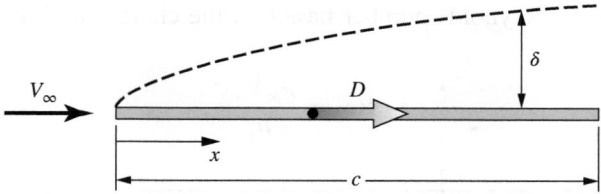

Figure 4.42 Total friction drag on a flat plate.

We first deal with the case of completely laminar flow over the airfoil (and hence the flat plate) in Figure 4.41. There is an exact analytical solution for the laminar boundary-layer flow over a flat plate. The details of this solution are given in Section 18.2, where we present boundary-layer theory in some detail. For the present section, we will use just the results of Section 18.2.

The boundary-layer thickness for incompressible laminar flow over a flat plate at zero angle of attack is given by Equation (18.23), repeated and renumbered below:

$$\delta = \frac{5.0x}{\sqrt{\text{Re}_x}} \tag{4.84}$$

where Re_x is the Reynolds number based on distance x measured from the leading edge (Figure 4.42),

$$\text{Re}_x = \frac{\rho_e V_\infty x}{\mu_\infty}$$

Note from Equation (4.84) that $\delta \propto \sqrt{x}$, that is, the boundary-layer thickness grows parabolically with distance from the leading edge.

The local shear stress, integrated over both the top and bottom surfaces of the flat plate shown in Figure 4.41, yields the net friction drag, D_f, on the plate, illustrated in Figure 4.42. To begin with, however, let us consider just *one* surface of the plate, either the top surface or the bottom surface. The shear stress distribution over the top surface is the same as that over the bottom surface. Let us choose the top surface. The integral of the shear stress over the top surface gives the net friction drag on that surface, $D_{f,\text{top}}$. Clearly, the net friction drag due to the shear stress integrated over the bottom surface, $D_{f,\text{bottom}}$, is the same value, $D_{f,\text{bottom}} = D_{f,\text{top}}$. Hence, the total skin friction drag, D_f, is

$$D_f = 2D_{f,\text{top}} = 2D_{f,\text{bottom}}$$

Define the skin-friction drag coefficient for the flow over *one* surface as

$$C_f \equiv \frac{D_{f,\text{top}}}{q_\infty S} = \frac{D_{f,\text{bottom}}}{q_\infty S} \tag{4.85}$$

The skin-friction drag coefficient is a function of the Reynolds number, and is given by Equation (18.22), repeated and renumbered below,

$$C_f = \frac{1.328}{\sqrt{\text{Re}_c}} \tag{4.86}$$

where Re_c is the Reynolds number based on the chord length c shown in Figure 4.42

$$\mathrm{Re}_c = \frac{\rho_\infty V_\infty c}{\mu_\infty}$$

EXAMPLE 4.8

Consider the NACA 2412 airfoil, data for which is given in Figures 4.10 and 4.11. The data are given for two values of the Reynolds number based on chord length. For the case where $\mathrm{Re}_c = 3.1 \times 10^6$, estimate: (a) the laminar boundary layer thickness at the trailing edge for a chord length of 1.5 m and (b) the net laminar skin-friction drag coefficient for the airfoil.

■ **Solution**

(a) From Equation (4.84) applied at the trailing edge, where $x = c$, we have

$$\delta = \frac{5.0c}{\sqrt{\mathrm{Re}_c}} = \frac{(5.0)(1.5)}{\sqrt{3.1 \times 10^6}} = \boxed{0.00426 \text{ m}}$$

Notice how thin the boundary layer is; at the trailing edge, where its thickness is the largest, the boundary layer is only 0.426 cm thick.

(b) From Equation (4.86),

$$C_f = \frac{1.328}{\sqrt{\mathrm{Re}_c}} = \frac{1.328}{\sqrt{3.1 \times 10^6}} = 7.54 \times 10^{-4}$$

Recall that the above result is for a single surface, either the top or bottom of the plate. Taking both surfaces into account:

$$\text{Net } C_f = 2(7.54 \times 10^{-4}) = \boxed{0.0015}$$

From the data in Figure 4.11, we see that at zero angle of attack for $\mathrm{Re} = 3.1 \times 10^6$, the airfoil drag coefficient is 0.0068. This measured value is about 4.5 times higher than the value of 0.0015 we just calculated. But wait a moment! For the relatively high Reynolds number of 3.1×10^6, the boundary layer over the airfoil will be *turbulent*, not laminar. So our laminar flow calculation is not an appropriate estimate for the boundary layer thickness and the airfoil drag coefficient. Let us take the next step.

4.12.2 Estimating Skin-Friction Drag: Turbulent Flow

In contrast to the situation for laminar flow, there are no exact analytical solutions for turbulent flow. This sad state of affairs is discussed in Chapter 19. The analysis of any turbulent flow requires some amount of empirical data. All analyses of turbulent flow are approximate.

The analysis of the turbulent boundary layer over a flat plate is no exception. From Chapter 19 we lift the following approximate results for the incompressible turbulent flow over a flat plate. From Equation (19.1) repeated and renumbered below:

$$\delta = \frac{0.37x}{\text{Re}_x^{1/5}} \tag{4.87}$$

and from Equation (19.2) repeated and renumbered below:

$$C_f = \frac{0.074}{\text{Re}_c^{1/5}} \tag{4.88}$$

We emphasize again that Equations (4.87) and (4.88) are only approximate results, and they represent only one set of results among a myriad of different turbulent flow analyses for the flat plate boundary layer. Nevertheless, Equatiohs (4.87) and (4.88) give us some reasonable means to estimate the boundary-layer thickness and skin-friction drag coefficient for turbulent flow. Note that, in contrast to the inverse square root variation with Reynolds number for laminar flow, the turbulent flow results show an inverse fifth root variation with Reynolds number.

EXAMPLE 4.9

Repeat Example 4.8 assuming a turbulent boundary layer over the airfoil.

■ **Solution**

Once again we replace the airfoil with a flat plate at zero angle of attack.

(a) The boundary-layer thickness at the trailing edge, where $x = c$ and $\text{Re}_x = \text{Re}_c = 3.1 \times 10^6$, is given by Equation (4.87):

$$\delta = \frac{0.37x}{\text{Re}_x^{1/5}} = \frac{0.37(1.5)}{(3.1 \times 10^6)^{1/5}} = \boxed{0.0279 \text{ m}}$$

The turbulent boundary layer is still thin, 2.79 cm at the trailing edge, but by comparison is much thicker than the laminar boundary layer thickness of 0.426 cm from Example 4.8.

(b) The skin-friction drag coefficient (based on one side of the flat plate) is given by Equation (4.88):

$$C_f = \frac{0.074}{\text{Re}_c^{1/5}} = \frac{0.074}{(3.1 \times 10^6)^{1/5}} = 0.00372$$

The net skin-friction drag coefficient, taking into account both the top and bottom surfaces of the flat plate, is

$$\text{Net } C_f = 2(0.00372) = \boxed{0.00744}$$

This result is a factor of five larger than for the laminar boundary layer, and serves as an illustration of the considerable increase in skin friction caused by a turbulent boundary layer in comparison to that caused by a laminar boundary layer.

The result for the skin friction drag coefficient in Example 4.9 is larger than the measured drag coefficient of the airfoil of 0.0068, which is the *sum* of both skin friction drag and pressure drag due to flow separation. So our result in this example clearly overestimates the skin friction drag coefficient for the airfoil. But wait a minute! In actuality, the boundary layer over a body always starts out as a *laminar* boundary for some distance from the leading edge, and then transists to a turbulent boundary layer at some point downstream of the leading edge. The skin-friction drag is therefore a combination of laminar skin friction over the forward part of the airfoil, and turbulent skin friction over the remaining part. Let us examine this situation.

4.12.3 Transition

In Section 4.12.1, we assumed that the flow over a flat plate was all laminar. Similarly, in Section 4.12.2 we assumed all turbulent flow. In reality, the flow *always* starts out from the leading edge as laminar. Then at some point downstream of the leading edge, the laminar boundary layer becomes unstable and small "bursts" of turbulence begin to grow in the flow. Finally, over a certain region called the *transition region,* the boundary layer becomes completely turbulent. For purposes of analysis, we usually draw the picture shown in Figure 4.43, where a laminar boundary layer starts out from the leading edge of the flat plate and grows parabolically downstream. Then at the *transition point,* it becomes a turbulent boundary layer growing at a faster rate, on the order of $x^{4/5}$ downstream. The value of x where transition is said to take place is the *critical* value x_{cr}. In turn, x_{cr} allows the definition of a *critical Reynolds number* for transition as

$$\text{Re}_{x_{cr}} = \frac{\rho_\infty V_\infty x_{cr}}{\mu_\infty} \tag{4.89}$$

Transition is discussed in more detail in Section 15.2. Volumes of literature have been written on the phenomenon of transition from laminar to turbulent flow. Obviously, because τ_w is different for the two flows—as clearly illustrated by comparing the results of Examples 4.8 and 4.9—knowledge of where on the surface transition occurs is vital to an accurate prediction of skin friction drag. The location of the transition point (in reality, a finite region) depends on many quantities as discussed in Section 15.2. However, if the critical Reynolds number is given to you (usually from experiments for a given type of flow over a given

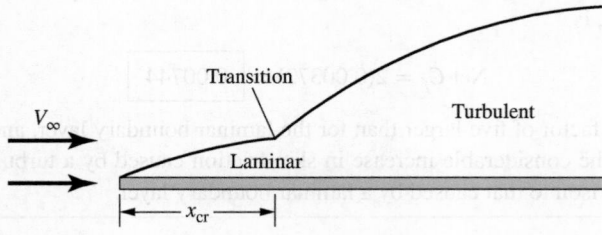

Figure 4.43 Transition from laminar to turbulent flow. The boundary layer thickness is exaggerated for clarity.

body), then the location of transition, x_{cr}, for that type of flow can be obtained directly from the definition, Equation (4.89).

For example, assume that you have an airfoil of given surface roughness (one of the factors that affect the location of transition) in a flow at a freestream velocity of 50 m/s and you wish to predict how far from the leading edge the transition will take place. After searching through the literature for low-speed flows over such surfaces, you may find that the critical Reynolds number determined from experience is approximately $Re_{x_{cr}} = 5 \times 10^5$. Applying this "experience" to your problem, using Equation (4.89), and assuming the thermodynamic conditions correspond to standard sea level, where $\rho_\infty = 1.23 \text{ kg/m}^3$ and (from Section 1.11) $\mu_\infty = 1.789 \times 10^{-5} \text{ kg/(m)(s)}$, you find

$$x_{cr} = \frac{\mu_\infty Re_{x_{cr}}}{\rho_\infty V_\infty} = \frac{(1.789 \times 10^{-5})(5 \times 10^5)}{(1.23)(50)} = 0.145 \text{ m}$$

Note that the region of laminar flow in this example extends from the leading edge to 14.5 cm downstream from the leading edge. If now you double the freestream velocity to 100 m/s, the transition point is still governed by the critical Reynolds number, $Re_{x_{cr}} = 5 \times 10^5$. Thus,

$$x_{cr} = \frac{(1.789 \times 10^{-5})(5 \times 10^5)}{(1.23)(100)} = 0.0727 \text{ m}$$

Hence, when the velocity is doubled, the transition point moves forward one-half the distance to the leading edge.

In summary, once you know the critical Reynolds number, you can find x_{cr} from Equation (4.89). However, an accurate value for $Re_{x_{cr}}$ applicable to your problem must come from somewhere—experiment, free flight, or some semi-empirical theory—and this may be difficult to obtain. This situation provides a little insight into why basic studies of transition and turbulence are needed to advance our understanding of such flows and to allow us to apply more valid reasoning to the prediction of transition in practical problems.

EXAMPLE 4.10

For the NACA 2412 airfoil and the conditions in Example 4.7, calculate the net skin friction drag coefficient assuming that the critical Reynolds number is 500,000.

■ **Solution**

Consider Figure 4.44, which shows a flat plate with a laminar boundary layer extending from the leading edge over the distance x_1 to the transition point (region 1), and a turbulent boundary extending over the distance x_2 from the transition point to the trailing edge (region 2). The critical Reynolds number is

$$Re_{x_{cr}} = \frac{\rho_\infty V_\infty x_{cr}}{\mu_\infty} = 5 \times 10^5$$

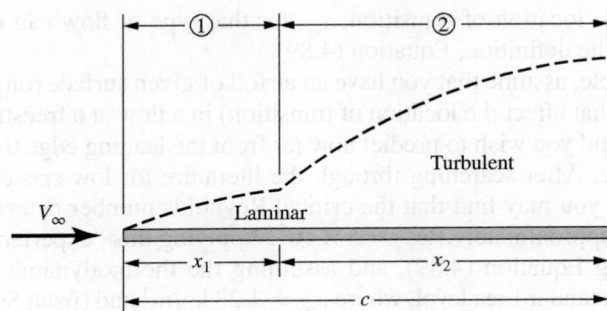

Figure 4.44 Laminar (region 1) and turbulent (region 2) flow over a flat plate.

where in Figure 4.44, $x_{\mathrm{cr}} = x_1$. Hence,

$$\mathrm{Re}_{x_{\mathrm{cr}}} = \frac{\rho_\infty V_\infty x_1}{\mu_\infty} = 5 \times 10^5$$

The Reynolds number based on chord length is given as

$$\mathrm{Re}_c = \frac{\rho_\infty V_\infty c}{\mu_\infty} = 3.1 \times 10^6$$

Thus

$$\frac{\mathrm{Re}_{x_{\mathrm{cr}}}}{\mathrm{Re}_c} = \frac{5 \times 10^5}{3.1 \times 10^6} = 0.1613 = \frac{\left(\rho_\infty V_\infty x_1/\mu_\infty\right)}{\left(\rho_\infty V_\infty c/\mu_\infty\right)} = \frac{x_1}{c}$$

This locates the transition point relative to the chord length, that is, in Figure 4.44, we have

$$\frac{x_1}{c} = 0.1613$$

Because the Reynolds number in the equations for skin friction drag coefficient is always based on length measured *from the leading edge,* we cannot simply calculate the turbulent skin friction drag coefficient for region 2 by using Equation (4.88) with a Reynolds number based on x_2. Rather, we must carry out the following procedure.

Assuming all turbulent flow over the entire length of the plate, the drag (on one side of the plate) is $(D_{f,c})_{\text{turbulent}}$, where

$$(D_{f,c})_{\text{turbulent}} = q_\infty S(C_{f,c})_{\text{turbulent}}$$

As usual, we are dealing with the drag per unit span, hence $S = c(1)$.

$$(D_{f,c})_{\text{turbulent}} = q_\infty c(C_{f,c})_{\text{turbulent}}$$

The turbulent drag on just region 1 is $(D_{f,1})_{\text{turbulent}}$:

$$(D_{f,1})_{\text{turbulent}} = q_\infty S(C_{f,1})_{\text{turbulent}}$$

Here, $S = (x_1)(1)$:

$$(D_{f,1})_{\text{turbulent}} = q_\infty x_1 (C_{f,1})_{\text{turbulent}}$$

Thus, the turbulent drag just on region 2, $(D_{f,2})_{\text{turbulent}}$, is

$$(D_{f,2})_{\text{turbulent}} = (D_{f,c})_{\text{turbulent}} - (D_{f,1})_{\text{turbulent}}$$

$$(D_{f,2})_{\text{turbulent}} = q_\infty c(C_{f,c})_{\text{turbulent}} - q_\infty x_1 (C_{f,1})_{\text{turbulent}}$$

The laminar drag on region 1 is $(D_{f,1})_{\text{laminar}}$

$$(D_{f,1})_{\text{laminar}} = q_\infty S(C_{f,1})_{\text{laminar}} = q_\infty x_1 (C_{f,1})_{\text{laminar}}$$

The total skin-friction drag on the plate, D_f, is then

$$D_f = (D_{f,1})_{\text{laminar}} + (D_{f,2})_{\text{turbulent}}$$

or, $$D_f = q_\infty x_1 (C_{f,1})_{\text{laminar}} + q_\infty c(C_{f,c})_{\text{turbulent}} - q_\infty x_1 (C_{f,1})_{\text{turbulent}} \qquad (4.90)$$

The total skin-friction drag coefficient is

$$C_f = \frac{D_f}{q_\infty S} = \frac{D_f}{q_\infty c} \qquad (4.91)$$

Combining Equations (4.90) and (4.91):

$$C_f = \frac{x_1}{c}(C_{f,1})_{\text{laminar}} + (C_{f,c})_{\text{turbulent}} - \frac{x_1}{c}(C_{f,1})_{\text{turbulent}} \qquad (4.92)$$

Since $x_1/c = 0.1613$, Equation (4.92) becomes

$$C_f = 0.1613(C_{f,1})_{\text{laminar}} + (C_{f,c})_{\text{turbulent}} - 0.1613(C_{f,1})_{\text{turbulent}} \qquad (4.93)$$

The various skin friction drag coefficients in Equation (4.93) are obtained as follows. The Reynolds number for region 1 is

$$\text{Re}_{x_1} = \frac{\rho_\infty V_\infty x_1}{\mu_\infty} = \frac{x_1}{c}\left(\frac{\rho_\infty V_\infty c}{\mu_\infty}\right) = \frac{x_1}{c}\text{Re}_c = 0.1613(3.1\times10^6) = 5\times10^5$$

(Of course, we could have written this down directly because $x = x_1$ is the transition point, *determined* from the critical Reynolds number that is given as 5×10^5.) From Equation (4.86) for laminar flow, with the Reynolds number based on x_1, we have

$$(C_{f,1})_{\text{laminar}} = \frac{1.328}{\sqrt{\text{Re}_{x_1}}} = \frac{1.328}{\sqrt{5\times10^5}} = 0.00188$$

The value of $(C_{f,c})_{\text{turbulent}}$ has already been calculated in Example 4.8, namely,

$$(C_{f,c})_{\text{turbulent}} = 0.00372 \quad \text{(for one side)}$$

From Equation (4.88), with the Reynolds number based on x_1,

$$(C_{f,1})_{\text{turbulent}} = \frac{0.074}{\text{Re}_{x_1}^{1/5}} = \frac{0.074}{(5\times10^5)^{0.2}} = 0.00536$$

Inserting these values into Equation (4.93), we have

$$C_f = 0.1613(0.00188) + 0.00372 - 0.1613(0.00536) = 0.003158$$

Taking into account both sides of the flat plate,

$$\text{Net } C_f = 2(0.003158) = \boxed{0.0063}$$

From the data in Figure 4.11, the measured airfoil drag coefficient is 0.0068, which includes *both* skin friction drag and pressure drag due to flow separation. The result from Example 4.10, therefore, is qualitatively reasonable, giving a skin friction drag coefficient slightly less than the measured total drag coefficient. However, our calculated result of $C_f = 0.0063$ is for a critical Reynolds number of 500,000 for transition from laminar to turbulent flow. We do not know what the critical Reynolds number is for the experiments on which the data in Figure 4.11 are based. In Example 4.10, the assumption of $\text{Re}_{x_{cr}} = 500,000$ is very conservative; more likely the actual value is closer to 1,000,000. If we assume this higher value of $\text{Re}_{x_{cr}}$, what does it do to the calculated result for C_f? Let us take a look.

EXAMPLE 4.11

Repeat Example 4.10, but assuming the critical Reynolds number is 1×10^6.

■ **Solution**

$$\frac{x_1}{c} = \frac{1 \times 10^6}{3.1 \times 10^6} = 0.3226$$

Which, as we could write down immediately, is twice the length from Example 4.10 because the critical Reynolds number is twice as large. Equation (4.93) becomes

$$C_f = 0.3226(C_{f,1})_{\text{laminar}} + (C_{f,c})_{\text{turbulent}} - 0.3226(C_{f,1})_{\text{turbulent}} \tag{4.94}$$

For region 1, we have

$$(C_{f,1})_{\text{laminar}} = \frac{1.328}{\sqrt{\text{Re}_{x_1}}} = \frac{1.328}{\sqrt{1 \times 10^6}} = 0.001328$$

The value of $(C_{f,c})_{\text{turbulent}}$ is the same as before:

$$(C_{f,c})_{\text{turbulent}} = 0.00372$$

Once again, for region 1 assuming turbulent flow, we have

$$(C_{f,1})_{\text{turbulent}} = \frac{0.074}{(\text{Re}_{x_1})^{1/5}} = \frac{0.074}{(1 \times 10^6)^{1/5}} = 0.004669$$

Substituting the above results in Equation (4.94), we have

$$C_f = 0.3226(0.001328) + 0.00372 - 0.3226(0.004669) = 0.002642$$

Since this result is for one side of the plate, the net skin friction drag coefficient is

$$\text{Net } C_f = 2(0.002642) = \boxed{0.00528}$$

Note: Comparing the results from Examples 4.10 and 4.11, we see that an increase in $\text{Re}_{x_{cr}}$ from 500,000 to 1,000,000 resulted in a skin friction drag coefficient that is 16 percent smaller. This difference underscores the importance of knowing where transition takes place on a surface for the calculation of skin friction drag.

Also, comparing the calculated results for skin friction drag coefficient with the measured total drag coefficient of 0.0068, from Example 4.10 the calculated $C_f = 0.0063$ would imply that the pressure drag due to flow separation is about 7.4 percent of the total drag. The result from Example 4.11 of $C_f = 0.00528$ would imply that the pressure drag due to flow separation is about 22 percent of the total drag.

Is this breakdown between skin friction and pressure drag quantitatively reasonable? An answer can be found in the recent results of Lombardi *et al.* given in Reference 88. Here, the authors calculated both the skin friction drag coefficient and the total drag coefficient for an NACA 0012 airfoil using an accurate computational fluid dynamic technique. More details of their calculations are given in Section 20.4. For a Reynolds number based on chord length of 3×10^6, and including a model for transition, they calculated a total drag coefficient of 0.00623 and a skin friction drag coefficient of 0.00534, indicating that the pressure drag due to flow separation is 15 percent of the total drag. For a streamlined body, this drag breakdown is reasonable; the drag on a streamlined two-dimensional shape is mostly skin friction drag, and by comparison the pressure drag is small. For example, it is reasonable to expect 80 percent of the drag to be skin friction drag and 20 percent to be pressure drag due to flow separation.

This is not to say that pressure drag due to flow separation is unimportant; quite the contrary, as the body becomes less streamlined (more like a blunt body), the pressure drag becomes the dominant factor. We need to take a closer look at this phenomenon.

4.12.4 Flow Separation

Pressure drag on an airfoil is caused by the flow separation. For a completely attached flow over an airfoil, the pressure acting on the rear surface gives rise to a force in the forward direction which completely counteracts the pressure acting on the front surface producing a force in the rearward direction, resulting in zero pressure drag. However, if the flow is partially separated over the rear surface, the pressure on the rear surface pushing forward will be smaller than the fully attached case, and the pressure acting on the front surface pushing backwards will not be fully counteracted, giving rise to a net pressure drag on the airfoil—the pressure drag due to flow separation.

What flow conditions are conducive to flow separation? To help answer this question, consider the flow over the NASA LS(1)-0417 airfoil at zero angle of attack, as shown in Figure 4.45. The streamlines move smoothly over the airfoil—there is no flow separation of any consequence. A computational fluid dynamic solution of the variation of pressure coefficient over the upper surface of the airfoil is shown at the bottom of Figure 4.45. Starting at the stagnation point at the leading edge, where for incompressible flow $C_p = 1.0$, the flow rapidly expands around the top surface. The pressure decreases dramatically, dipping to a minimum pressure at a location about 10 percent of the chord length downstream of the leading edge. Then as the flow moves farther downstream, the pressure

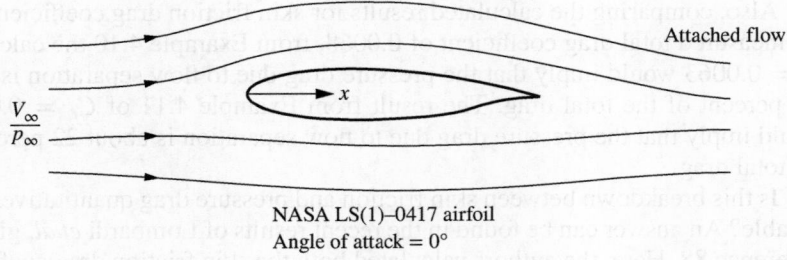

$\dfrac{V_\infty}{p_\infty}$

NASA LS(1)–0417 airfoil
Angle of attack = 0°

Attached flow

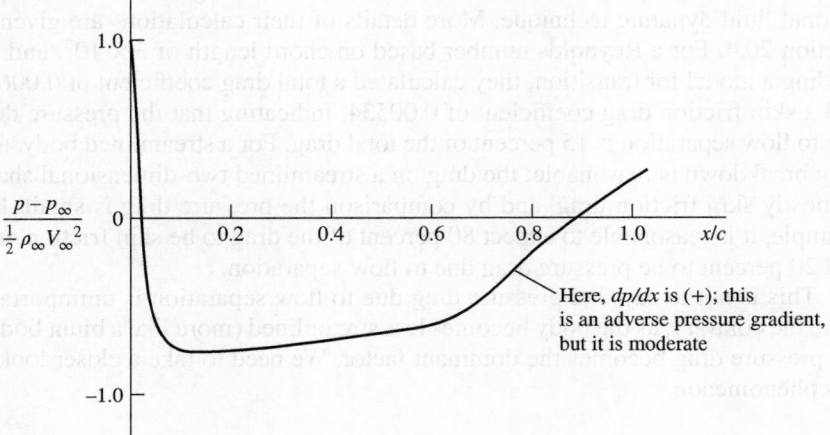

Figure 4.45 Pressure distribution over the top surface for attached flow over an airfoil. (Theoretical data for a modern NASA low-speed airfoil, from NASA Conference Publication 2046, *Advanced Technology Airfoil Research,* vol. II, March 1978, p. 11.)

gradually increases, reaching a value slightly above freestream pressure at the trailing edge. This region of increasing pressure is called a region of *adverse* pressure gradient. By definition, an adverse pressure gradient is a region where the pressure *increases in the flow direction,* that is, in Figure 4.45, the region where *dp/dx* is *positive*. For the conditions shown in Figure 4.45, the adverse pressure gradient is moderate; that is, *dp/dx* is small, and for all practical purposes the flow remains attached to the airfoil surface except for a small region near the trailing edge (not shown in Figure 4.45).

Now consider the same airfoil at the very high angle of attack of 18.4 degrees, as shown in Figure 4.46. First, assume we have a purely inviscid flow with no flow separation—a purely artificial situation. A numerical solution for the inviscid flow gives the results shown by the dashed curve in Figure 4.46. In this artificial situation, the pressure would drop precipitously downstream of the leading edge to a value of C_p almost −9, and then rapidly increase downstream, recovering

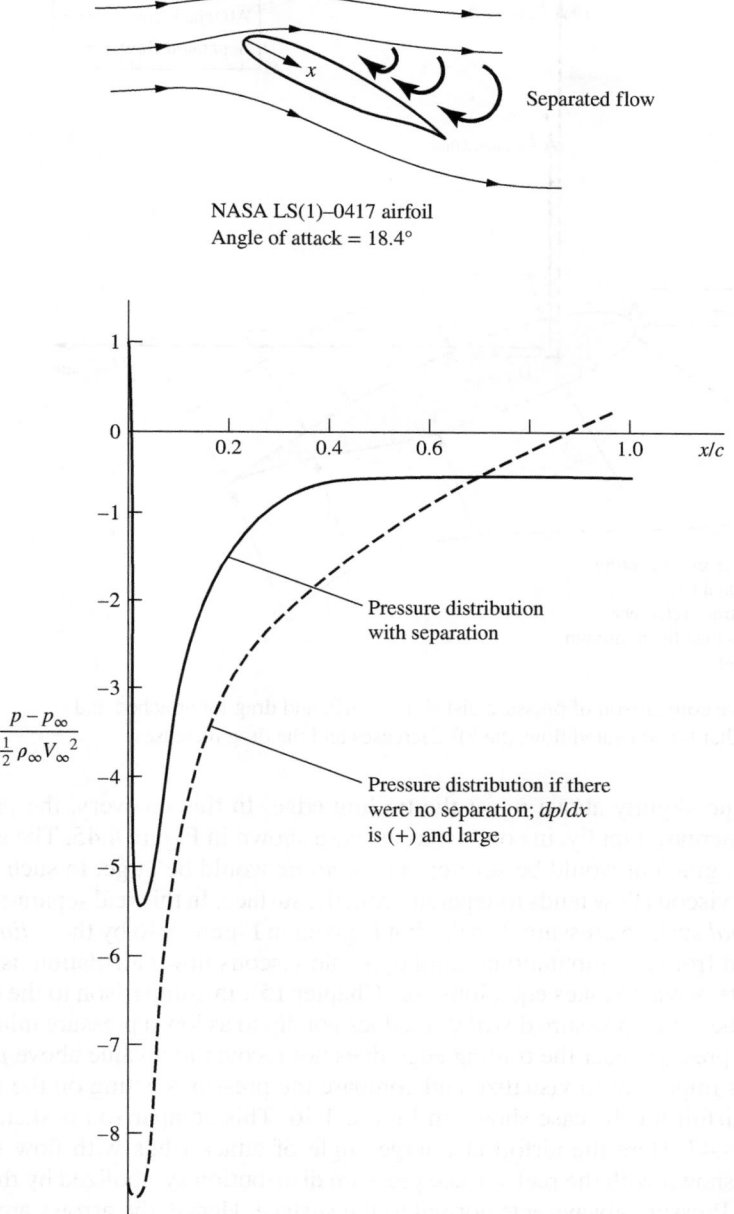

Figure 4.46 Pressure distribution over the top surface for separated flow over an airfoil. (Theoretical data for a modern NASA low-speed airfoil, from NASA Conference Publication 2045, Part 1, *Advanced Technology Airfoil Research,* vol. 1, March 1978, p. 380.)

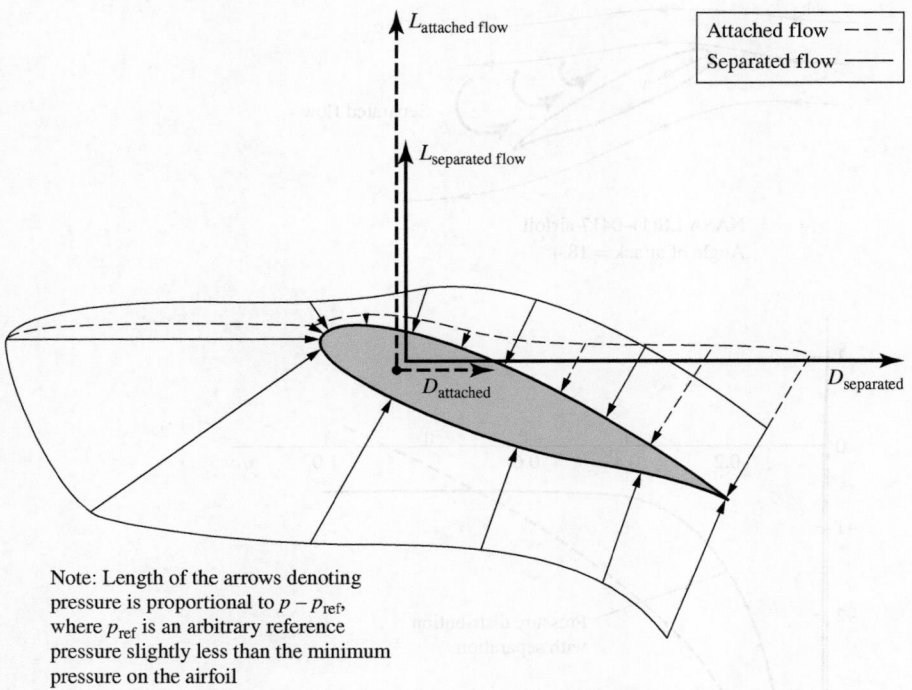

Note: Length of the arrows denoting
pressure is proportional to $p - p_{ref}$,
where p_{ref} is an arbitrary reference
pressure slightly less than the minimum
pressure on the airfoil

Figure 4.47 Qualitative comparison of pressure distribution, lift, and drag for attached and separated flows. Note that for separated flow, the lift decreases and the drag increases.

to a value slightly above p_{∞} at the trailing edge. In this recovery, the pressure would increase rapidly, in contrast to the case shown in Figure 4.45. The adverse pressure gradient would be severe; that is, dp/dx would be large. In such a case, the *real* viscous flow tends to separate from the surface. In this real separated flow, the *actual* surface pressure distribution is given in Figure 4.46 by the *solid* curve, obtained from a computational fluid dynamic viscous flow calculation using the complete Navier-Stokes equations (see Chapter 15). In comparison to the dashed curve, the actual pressure distribution does not dip to as low a pressure minimum, and the pressure near the trailing edge does not recover to a value above p_{∞}.

It is important to visualize and compare the pressures acting on the surface of the airfoil for the case shown in Figure 4.46. This comparison is sketched in Figure 4.47. Here the airfoil at a large angle of attack (thus with flow separation) is shown with the real surface pressure distribution symbolized by the solid arrows. Pressure always acts normal to the surface. Hence, the arrows are all locally perpendicular to the surface. The length of the arrows is representative of the magnitude of the pressure. A solid curve is drawn through the base of the arrows to form an "envelope" to make the pressure distribution easier to visualize. However, if the flow were *not* separated, that is, if the flow were attached, then the pressure distribution would be that shown by the dashed arrows (and the dashed envelope). The solid and dashed arrows in Figure 4.47 qualitatively correspond to the solid and dashed pressure distribution curves, respectively, in Figure 4.46.

The solid and dashed arrows in Figure 4.47 should be compared carefully. They explain the two major consequences of separated flow over the airfoil. The first consequence is a loss of lift. The aerodynamic lift (the vertical force shown in Figure 4.47) is derived from the net component of the pressure distribution in the vertical direction in Figure 4.47 (assuming that the freestream relative wind is horizontal in this figure). High lift is obtained when the pressure on the bottom surface is large and the pressure on the top surface is small. Separation does not affect the bottom surface pressure distribution. However, comparing the solid and dashed arrows on the top surface *just downstream of the leading edge,* we find the solid arrows indicating a higher pressure when the flow is separated. This higher pressure is pushing down, hence reducing the lift. This reduction in lift is also compounded by the geometric effect that the position of the top surface of the airfoil near the leading edge is approximately horizontal in Figure 4.47. When the flow is separated, causing a higher pressure on this part of the airfoil surface, the direction in which the pressure is acting is closely aligned to the vertical, and hence, almost the full effect of the increased pressure is felt by the lift. The combined effect of the increased pressure on the top surface near the leading edge, and the fact that this portion of the surface is approximately horizontal, leads to the rather dramatic loss of lift when the flow separates. Note in Figure 4.47 that the lift for separated flow (the solid vertical vector) is smaller than the lift that would exist if the flow were attached (the dashed vertical vector).

Now let us concentrate on that portion of the top surface *near the trailing edge.* On this portion of the airfoil surface, the pressure for the separated flow is now *smaller* than the pressure that would exist if the flow were attached. Moreover, the top surface near the trailing edge is geometrically inclined more to the horizontal, and, in fact, somewhat faces in the horizontal direction. Recall that the drag is in the horizontal direction in Figure 4.47. Because of the inclination of the top surface near the trailing edge, the pressure exerted on this portion of the surface has a strong component in the horizontal direction. This component acts toward the left, tending to counter the horizontal component of force due to the high pressure acting on the nose of the airfoil pushing toward the right. The net pressure drag on the airfoil is the difference between the force exerted on the front pushing toward the right and the force exerted on the back pushing toward the left. When the flow is separated, the pressure on the back is lower than it would be if the flow were attached. Hence, for the separated flow, there is *less* force on the back pushing toward the left, and the *net* drag acting toward the right is therefore *increased.* Note in Figure 4.47 that the drag for separated flow (the solid horizontal vector) is larger than the drag that would exist if the flow were attached (the dashed horizontal vector).

Therefore, two major consequences of the flow separating over an airfoil are:

1.   A drastic loss of lift (stalling).
2.   A major increase in drag, caused by pressure drag due to flow separation.

Why does a flow separate from a surface? The answer is addressed in detail in Section 15.2. In brief, in a region of adverse pressure gradient the fluid elements moving along a streamline have to work their way "uphill" against an increasing pressure. Consequently, the fluid elements will slow down under the influence of an adverse pressure gradient. For the fluid elements moving outside the boundary layer, where the velocity (and hence kinetic energy) is high, this is not much of a problem. The fluid elements keep moving downstream. However, consider a fluid element deep inside the boundary layer. Its velocity is already small because it is retarded by friction forces. The fluid element still encounters the same adverse pressure gradient because the pressure is transmitted without change normal to the wall, but its velocity is too low to negotiate the increasing pressure. As a result, the element comes to a stop somewhere downstream and reverses its direction. Such reversed flow causes the flow field in general to separate from the surface, as shown at the top of Figure 4.46. This is physically how separated flow develops.

4.12.5 Comment

In this section, we estimated skin friction drag on an airfoil by using the model of a flat plate at zero angle of attack, and calculating the skin friction drag for the airfoil using the formulae for the flat plate such as Equation (4.86) for laminar flow and Equation (4.88) for turbulent flow.[5] How reasonable is this? How close does the flat plate skin friction drag on a flat plate come to that on an airfoil? How close does the local shear stress distribution over the surface of the flat plate resemble that on the airfoil surface?

Some answers can be obtained by comparing the relatively exact computational fluid dynamic calculations of Lombardi et al. (Reference 88) for the viscous flow over an NACA 0012 airfoil at zero angle of attack with that for a flat plate. The variation of the local skin friction coefficient, defined as $c_f = \tau_w/q_\infty$, as a function of distance from the leading edge is given in Figure 4.48 for both the airfoil and the flat plate. They are remarkably close; clearly, for the purpose of the present section the modeling of the airfoil skin friction drag by use of flat plate results is reasonable.

With this, we end our discussion of airfoil drag at low speeds. Although the main thrust of this chapter is the low-speed inviscid potential flow over airfoils with the consequent prediction of lift, the present section provides some balance by exploring the effects of viscous flow on airfoil behavior, and the consequent production of drag. More aspects of the real flow over airfoils are given in Section 4.13.

---

[5] In 1921, Walter Diehl's article in NACA TR111, entitled "The Variation of Aerofoil Lift and Drag Coefficients with Changes in Size and Speed," suggested that the airfoil drag coefficient varies in the same manner as that for a flat plate at zero angle of attack. He did not say that the airfoil drag coefficient equals that for a flat plate, but rather has the same Reynolds number variation. Diehl's suggestion, however, appears to be the first effort to use flat plate data in some fashion to estimate airfoil drag.

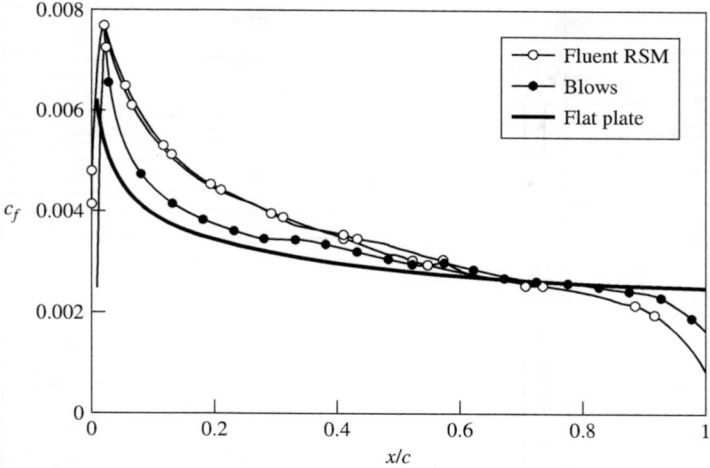

Figure 4.48 Local skin-friction coefficient distributions over an NACA 0012 airfoil, compared with that for a flat plate.

4.13 APPLIED AERODYNAMICS: THE FLOW OVER AN AIRFOIL—THE REAL CASE

In this chapter, we have studied the inviscid, incompressible flow over airfoils. When compared with actual experimental lift and moment data for airfoils in low-speed flows, we have seen that our theoretical results based on the assumption of inviscid flow are quite good—with one glaring exception. In the real case, flow separation occurs over the top surface of the airfoil when the angle of attack exceeds a certain value—the "stalling" angle of attack. As described in Sections 4.3 and 4.12, this is a viscous effect. As shown in Figure 4.9, the lift coefficient reaches a local maximum denoted by $c_{l,max}$, and the angle of attack at which $c_{l,max}$ is achieved is the stalling angle of attack. An increase in α beyond this value usually results in a (sometimes rather precipitous) drop in lift. At angles of attack well below the stalling angle, the experimental data clearly show a *linear* increase in c_l with increasing α—a result that is predicted by the theory presented in this chapter. Indeed, in this linear region, the inviscid flow theory is in excellent agreement with the experiment, as reflected in Figure 4.10 and as demonstrated by Example 4.6. However, the inviscid theory does not predict flow separation, and consequently the prediction of $c_{l,max}$ and the stalling angle of attack must be treated in some fashion by viscous flow theory. Such viscous flow analyses are the purview of Part 4. On the other hand, the purpose of this section is to examine the *physical* features of the real flow over an airfoil, and flow separation is an inherent part of this real flow. Therefore, let us take a more detailed look at how the flow field over an airfoil changes as the angle of attack is increased, and how the lift coefficient is affected by such changes.

The flow fields over an NACA 4412 airfoil at different angles of attack are shown in Figure 4.49. Here, the streamlines are drawn to scale as obtained from

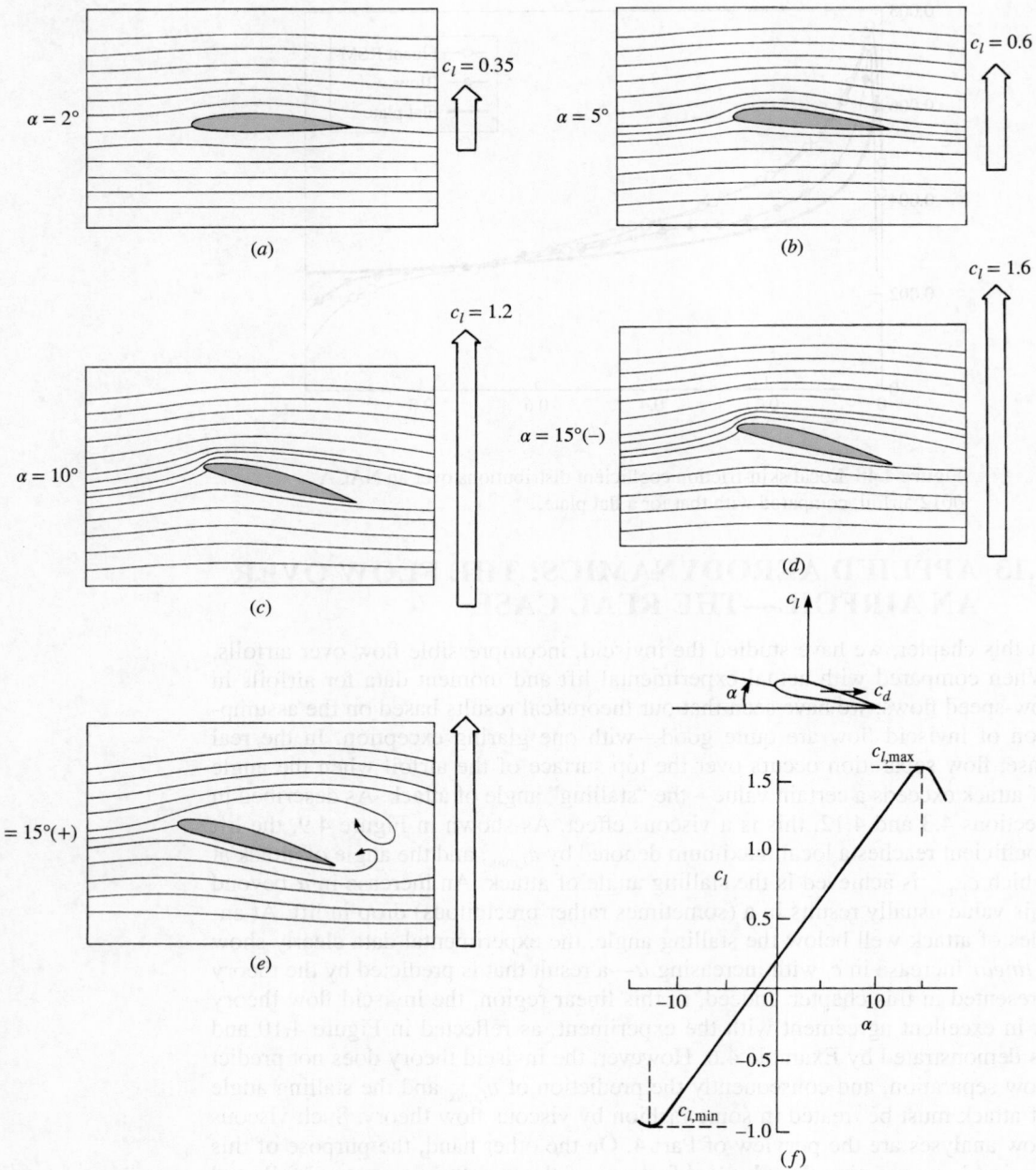

Figure 4.49 Example of leading-edge stall. Streamline patterns for an NACA 4412 airfoil at different angles of attack. [The streamlines are drawn to scale from experimental data given by Nakayama, Y. (ed): *Visualized Flow*, compiled by the Japan Society of Mechanical Engineers, Pergamon Press, New York, 1988.] $Re = 2.1 \times 10^5$ and $V_\infty = 8$ m/s in air. The corresponding experimentally measured lift coefficients are indicated by arrows at the right of each streamline picture, where the length of each arrow indicates the relative magnitude of the lift. The lift coefficient is also shown in part (f).

the experimental results of Hikaru Ito given in Reference 47. The experimental streamline patterns were made visible by a smoke wire technique, wherein metallic wires spread with oil over their surfaces were heated by an electric pulse and the resulting white smoke creates visible streaklines in the flow field. In Figure 4.49, the angle of attack is progressively increased as we scan from Figure 4.49a to e; to the right of each streamline picture is an arrow, the length of which is proportional to the value of the lift coefficient at the given angle of attack. The actual experimentally measured lift curve for the airfoil is given in Figure 4.49f. Note that at low angle of attack, such as $\alpha = 2°$ in Figure 4.49a, the streamlines are relatively undisturbed from their freestream shapes and c_l is small. As α is increased to 5°, as shown in Figure 4.49b, and then to 10°, as shown in Figure 4.49c, the streamlines exhibit a pronounced upward deflection in the region of the leading edge, and a subsequent downward deflection in the region of the trailing edge. Note that the stagnation point progressively moves downstream of the leading edge over the bottom surface of the airfoil as α is increased. Of course, c_l increases as α is increased, and, in this region, the increase is linear, as seen in Figure 4.49f. When α is increased to slightly less than 15°, as shown in Figure 4.49d, the curvature of the streamlines is particularly apparent. In Figure 4.49d, the flow field is still attached over the top surface of the airfoil. However, as α is further increased slightly above 15°, massive flow-field separation occurs over the top surface, as shown in Figure 4.49e. By slightly increasing α from that shown in Figure 4.49d to that in Figure 4.49e, the flow quite suddenly separates from the leading edge and the lift coefficient experiences a precipitous decrease, as seen in Figure 4.49f.

The type of stalling phenomenon shown in Figure 4.49 is called *leading-edge stall;* it is characteristic of relatively thin airfoils with thickness ratios between 10 and 16 percent of the chord length. As seen above, flow separation takes place rather suddenly and abruptly over the entire top surface of the airfoil, with the origin of this separation occurring at the leading edge. Note that the lift curve shown in Figure 4.49f is rather sharp-peaked in the vicinity of $c_{l,max}$ with a rapid decrease in c_l above the stall.

A second category of stall is the *trailing-edge stall*. This behavior is characteristic of thicker airfoils such as the NACA 4421 shown in Figure 4.50. Here, we see a progressive and gradual movement of separation from the trailing edge toward the leading edge as α is increased. The lift curve for this case is shown in Figure 4.51. The solid curve in Figure 4.51 is a repeat of the results for the NACA 4412 airfoil shown earlier in Figure 4.49f—an airfoil with a leading-edge stall. The dot-dashed curve is the lift curve for the NACA 4421 airfoil—an airfoil with a trailing-edge stall. In comparing these two curves, note that:

1. The trailing-edge stall yields a gradual bending-over of the lift curve at maximum lift, in contrast to the sharp, precipitous drop in c_l for the leading-edge stall. The stall is "soft" for the trailing-edge stall.
2. The value of $c_{l,max}$ is not so large for the trailing-edge stall.
3. For both the NACA 4412 and 4421 airfoils, the shape of the mean camber line is the same. From the thin airfoil theory discussed in this chapter, the

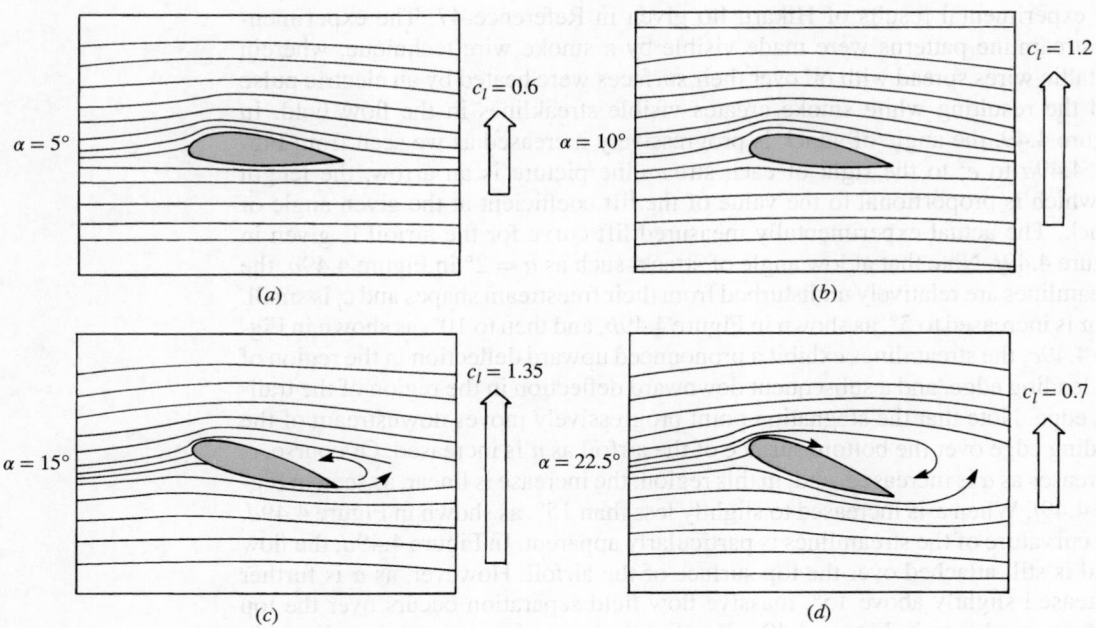

Figure 4.50 Example of trailing-edge stall. Streamline patterns for an NACA 4421 airfoil at different angles of attack. Re = 2.1×10^5 and $V_\infty = 8$ m/s in air. [The streamlines are drawn to scale from the experimental results of Nakayama, Y. (ed): *Visualized Flow*, compiled by the Japan Society of Mechanical Engineers, Pergamon Press, New York, 1988.]

linear lift slope and the zero-lift angle of attack should be the same for both airfoils; this is confirmed by the experimental data in Figure 4.51. The only difference between the two airfoils is that one is thicker than the other. Hence, comparing results shown in Figures 4.49 to 4.51, we conclude that the major effect of thickness of the airfoil is its effect on the value of $c_{l,\max}$, and this effect is mirrored by the leading-edge stall behavior of the thinner airfoil versus the trailing-edge stall behavior of the thicker airfoil.

There is a third type of stall behavior, namely, behavior associated with the extreme thinness of an airfoil. This is sometimes labeled as "thin airfoil stall." An extreme example of a very thin airfoil is a flat plate; the lift curve for a flat plate is shown as the dashed curve in Figure 4.51 labeled "thin airfoil stall." The streamline patterns for the flow over a flat plate at various angles of attack are given in Figure 4.52. The thickness of the flat plate is 2 percent of the chord length. Inviscid, incompressible flow theory shows that the velocity becomes infinitely large at a sharp convex corner; the leading edge of a flat plate at an angle of attack is such a case. In the real flow over the plate as shown in Figure 4.52, nature addresses this singular behavior by having the flow separate at the leading edge, even for very low values of α. Examining Figure 4.52a, where $\alpha = 3°$, we observe a small region of separated flow at the leading edge. This separated

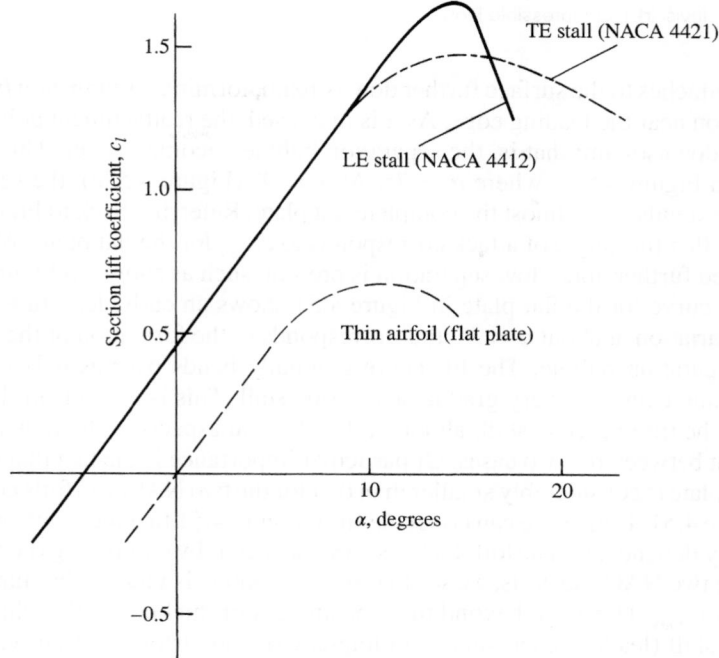

Figure 4.51 Lift-coefficient curves for three airfoils with different aerodynamic behavior: trailing-edge stall (NACA 4421 airfoil), leading-edge stall (NACA 4412 airfoil), thin airfoil stall (flat plate).

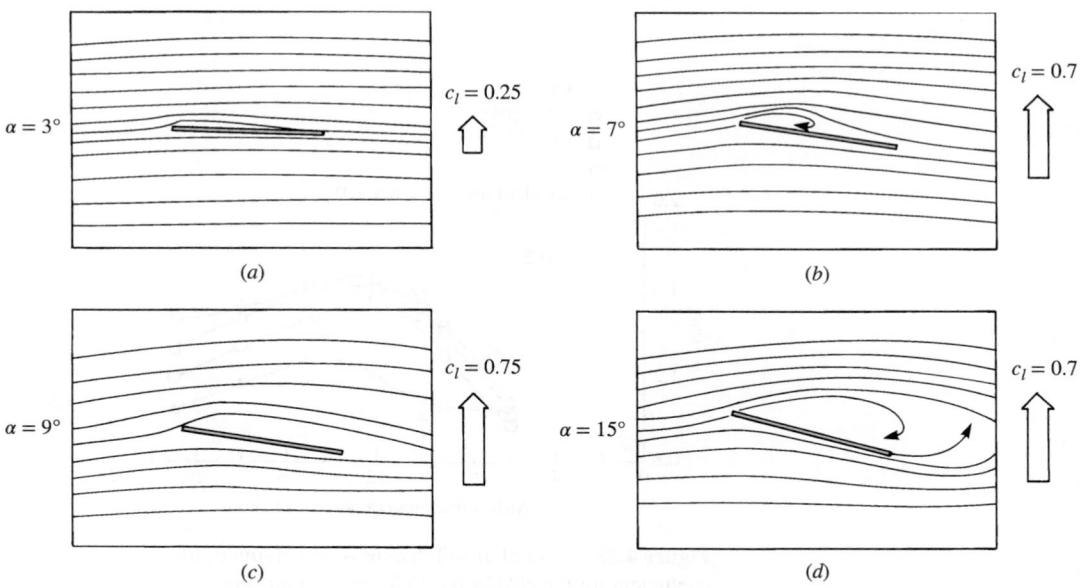

Figure 4.52 Example of thin airfoil stall. Streamline patterns for a flat plate at angle of attack. [The streamlines are drawn to scale from the experimental data of Nakayama, Y. (ed): *Visualized Flow*, compiled by the Japan Society of Mechanical Engineers, Pergamon Press, New York, 1988.]

flow reattaches to the surface further downstream, forming a *separation bubble* in the region near the leading edge. As α is increased, the reattachment point moves further downstream; that is, the separation bubble becomes larger. This is illustrated in Figure 4.52*b*, where $\alpha = 7°$. At $\alpha = 9°$ (Figure 4.52*c*), the separation bubble extends over almost the complete flat plate. Referring back to Figure 4.51, we note that this angle of attack corresponds to $c_{l,\max}$ for the flat plate. When α is increased further, total flow separation is present, such as shown in Figure 4.52*d*. The lift curve for the flat plate in Figure 4.51 shows an early departure from its linear variation at about $\alpha = 3°$; this corresponds to the formation of the leading-edge separation bubble. The lift curve gradually bends over as α is increased further and exhibits a very gradual and "soft" stall. This is a trend similar to the case of the trailing-edge stall, although the physical aspects of the flow are quite different between the two cases. Of particular importance is the fact that $c_{l,\max}$ for the flat plate is considerably smaller than that for the two NACA airfoils compared in Figure 4.51. Hence, we can conclude from Figure 4.51 that the value of $c_{l,\max}$ is critically dependent on airfoil thickness. In particular, by comparing the flat plate with the two NACA airfoils, we see that *some* thickness is vital to obtaining a high value of $c_{l,\max}$. However, beyond that, the amount of thickness will influence the type of stall (leading-edge versus trailing-edge), and airfoils that are very thick tend to exhibit reduced values of $c_{l,\max}$ as the thickness increases. Hence, if we plot $c_{l,\max}$ versus thickness ratio, we expect to see a local maximum. Such is indeed the case, as shown in Figure 4.53. Here, experimental data for $c_{l,\max}$ for the NACA 63-2XX series of airfoils is shown as a function of the thickness ratio. Note that as the thickness ratio increases from a small value, $c_{l,\max}$ first increases,

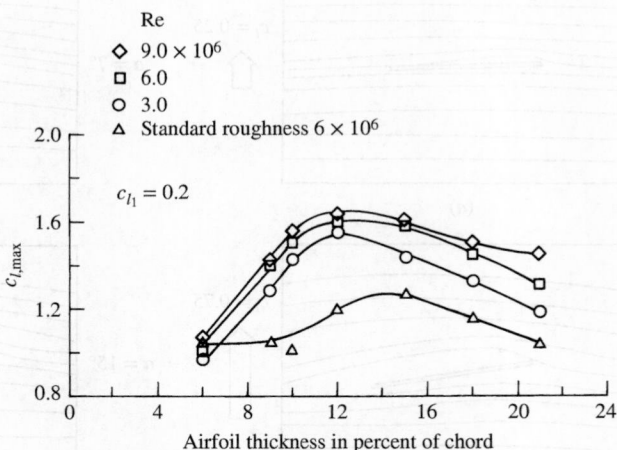

Figure 4.53 Effect of airfoil thickness on maximum lift coefficient for the NACA 63-2XX series of airfoils. (*Data Source:* Abbott, I. H., and A. E. von Doenhoff: *Theory of Wing Sections*, McGraw Hill Book Company, New York, 1949; also, Dover Publications, Inc., New York, 1959.)

reaches a maximum value at a thickness ratio of about 12 percent, and then decreases at larger thickness ratios. The experimental data in Figure 4.53 is plotted with the Reynolds number as a parameter. Note that $c_{l,\max}$ for a given airfoil is clearly a function of Re, with higher values of $c_{l,\max}$ corresponding to higher Reynolds numbers. Since flow separation is responsible for the lift coefficient exhibiting a local maximum, since flow separation is a viscous phenomenon, and since a viscous phenomenon is governed by a Reynolds number, it is no surprise that $c_{l,\max}$ exhibits some sensitivity to Re.

When was the significance of airfoil thickness first understood and appreciated? This question is addressed in the historical note in Section 4.14, where we will see that the aerodynamic properties of thick airfoils even transcended technology during World War I and impacted the politics of the armistice.

Let us examine some other aspects of airfoil aerodynamics—aspects that are not always appreciated in a first study of the subject. The simple generation of lift by an airfoil is not the prime consideration in its design—even a barn door at an angle of attack produces lift. Rather, there are two figures of merit that are primarily used to judge the quality of a given airfoil:

1. *The lift-to-drag ratio L/D.* An efficient airfoil produces lift with a minimum of drag; that is, the ratio of lift-to-drag is a measure of the aerodynamic efficiency of an airfoil. The standard airfoils discussed in this chapter have high *L/D* ratios—much higher than that of a barn door. The *L/D* ratio for a complete flight vehicle has an important impact on its flight performance; for example, the range of the vehicle is directly proportional to the *L/D* ratio. (See Reference 2 for an extensive discussion of the role of *L/D* on flight performance of an airplane.)

2. *The maximum lift coefficient* $c_{l,\max}$. An effective airfoil produces a high value of $c_{l,\max}$—much higher than that produced by a barn door.

The maximum lift coefficient is worth some additional discussion here. For a complete flight vehicle, the maximum lift coefficient $C_{L,\max}$ determines the stalling speed of the aircraft as discussed in the Design Box at the end of Section 1.8. From Equation (1.47), repeated below:

$$V_{\text{stall}} = \sqrt{\frac{2W}{\rho_\infty S C_{L,\max}}} \tag{1.47}$$

Therefore, a tremendous incentive exists to increase the maximum lift coefficient of an airfoil, in order to obtain either lower stalling speeds or higher payload weights at the same speed, as reflected in Equation (1.47). Moreover, the maneuverability of an airplane (i.e., the smallest possible turn radius and the fastest possible turn rate) depends on a large value of $C_{L,\max}$ (see Section 6.17 of Reference 2). On the other hand, for an airfoil at a given Reynolds number, the value of $c_{l,\max}$ is a function primarily of its shape. Once the shape is specified, the value of $c_{l,\max}$ is what nature dictates, as we have already seen. Therefore, to increase $c_{l,\max}$ beyond such a value, we must carry out some special measures. Such special measures include the use of flaps and/or leading-edge slats to increase

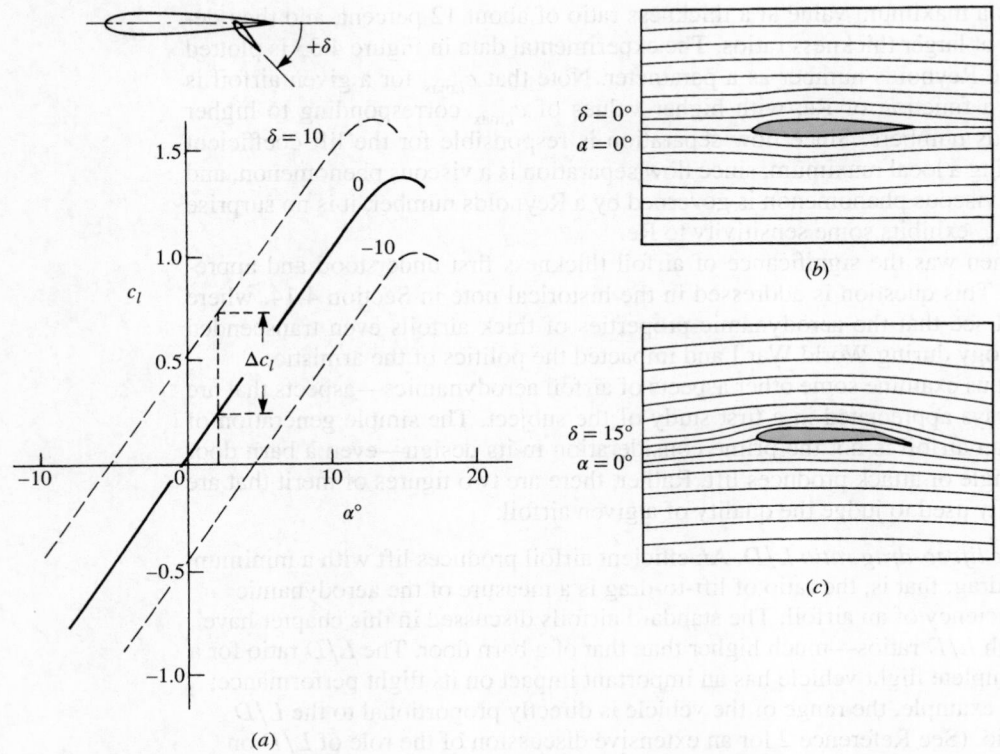

(a)

(b)

(c)

Figure 4.54 Effect of flap deflection on streamline shapes. [The streamlines are drawn to scale from the experimental data of Nakayama, Y. (ed): *Visualized Flow*, compiled by the Japan Society of Mechanical Engineers, Pergamon Press, New York, 1988.] (a) Effect of flap deflection on lift coefficient. (b) Streamline pattern with no flap deflection. (c) Streamline pattern with a 15° flap deflection.

$c_{l,\max}$ above that for the reference airfoil itself. These are called *high-lift devices*, and are discussed in more detail below.

A trailing-edge flap is simply a portion of the trailing-edge section of the airfoil that is hinged and which can be deflected upward or downward, as sketched in the insert in Figure 4.54a. When the flap is deflected downward (a positive angle δ in Figure 4.54a), the lift coefficient is increased, as shown in Figure 4.54a. This increase is due to an effective increase in the camber of the airfoil as the flap is deflected downward. The thin airfoil theory presented in this chapter clearly shows that the zero-lift angle of attack is a function of the amount of camber [see Equation (4.61)], with $\alpha_{L=0}$ becoming more negative as the camber is increased. In terms of the lift curve shown in Figure 4.54a, the original curve for no flap deflection goes through the origin because the airfoil is symmetric; however, as the flap is deflected downward, this lift curve simply translates to the left because $\alpha_{L=0}$ is becoming more negative. In Figure 4.54a, the results are given for flap deflections of $\pm 10°$. Comparing the case for $\delta = 10°$ with the no-deflection

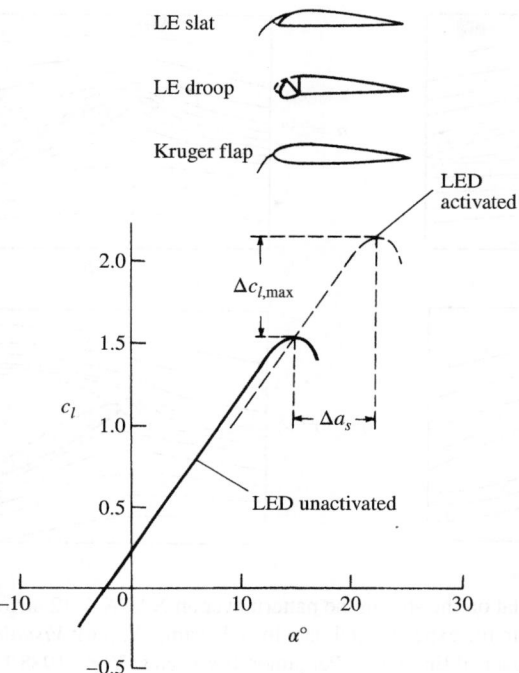

Figure 4.55 Effect of leading-edge flap on lift coefficient.

case, we see that, at a given angle of attack, the lift coefficient is increased by an amount Δc_l due to flap deflection. Moreover, the actual value of $c_{l,\max}$ is increased by flap deflection, although the angle of attack at which $c_{l,\max}$ occurs is slightly decreased. The change in the streamline pattern when the flap is deflected is shown in Figure 4.54b and c. Figure 4.54b is the case for $\alpha = 0$ and $\delta = 0$ (i.e., a symmetric flow). However, when α is held fixed at zero, but the flap is deflected by 15°, as shown in Figure 4.54c, the flow field becomes unsymmetrical and resembles the lifting flows shown (e.g., in Figure 4.49). That is, the streamlines in Figure 4.54c are deflected upward in the vicinity of the leading edge and downward near the trailing edge, and the stagnation point moves to the lower surface of the airfoil—just by deflecting the flap downward.

High-lift devices can also be applied to the leading edge of the airfoil, as shown in the inset in Figure 4.55. These can take the form of a leading-edge slat, leading-edge droop, or a leading-edge flap. Let us concentrate on the leading-edge slat, which is simply a thin, curved surface that is deployed in front of the leading edge. In addition to the primary airflow over the airfoil, there is now a secondary flow that takes place through the gap between the slat and the airfoil leading edge. This secondary flow from the bottom to the top surface modifies

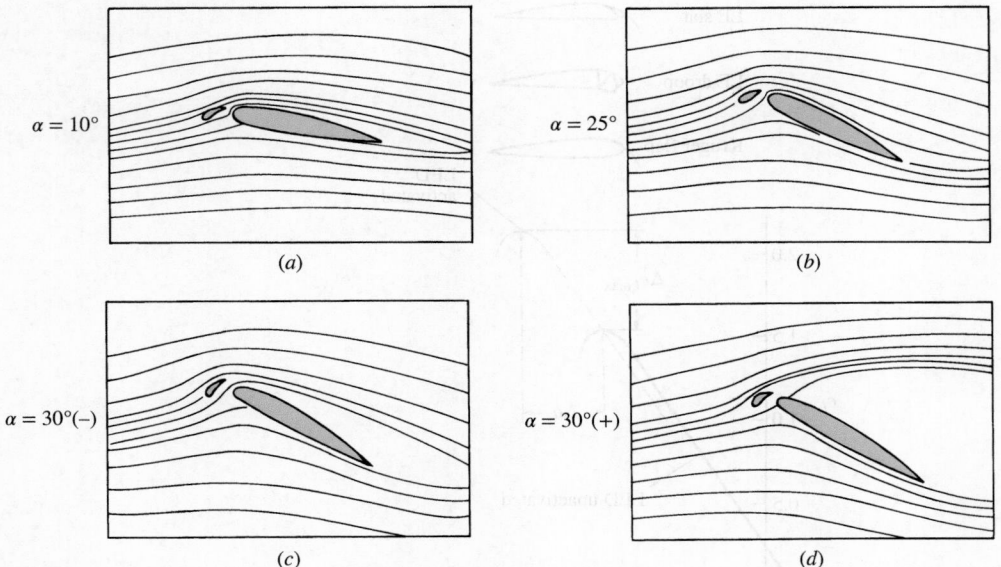

$\alpha = 10°$

(a)

$\alpha = 25°$

(b)

$\alpha = 30°(-)$

(c)

$\alpha = 30°(+)$

(d)

Figure 4.56 Effect of a leading-edge slat on the streamline pattern over an NACA 4412 airfoil. [The streamlines are drawn to scale from the experimental data in Nakayama, Y. (ed): *Visualized Flow*, compiled by the Japan Society of Mechanical Engineers, Pergamon Press, New York, 1988.]

the pressure distribution over the top surface; the adverse pressure gradient which would normally exist over much of the top surface is mitigated somewhat by this secondary flow, hence delaying flow separation over the top surface. Thus, a leading-edge slat increases the stalling angle of attack, and hence yields a higher $c_{l,max}$, as shown by the two lift curves in Figure 4.55, one for the case without a leading-edge device and the other for the slat deployed. Note that the function of a leading-edge slat is inherently different from that of a trailing-edge flap. There is no change in $\alpha_{L=0}$; rather, the lift curve is simply extended to a higher stalling angle of attack, with the attendant increase in $c_{l,max}$. The streamlines of a flow field associated with an extended leading-edge slat are shown in Figure 4.56. The airfoil is in an NACA 4412 section. (*Note:* The flows shown in Figure 4.56 do not correspond exactly with the lift curves shown in Figure 4.55, although the general behavior is the same.) The stalling angle of attack for the NACA 4412 airfoil without slat extension is about 15°, but increases to about 30° when the slat is extended. In Figure 4.56a, the angle of attack is 10°. Note the flow through the gap between the slat and the leading edge. In Figure 4.56b, the angle of attack is 25° and the flow is still attached. This prevails to an angle of attack slightly less than 30°, as shown in Figure 4.56c. At slightly higher than 30°, flow separation suddenly occurs and the airfoil stalls.

The high-lift devices used on modern, high-performance aircraft are usually a combination of leading-edge slats (or flaps) and multielement trailing-edge flaps. Typical airfoil configurations with these devices are sketched in Figure 4.57.

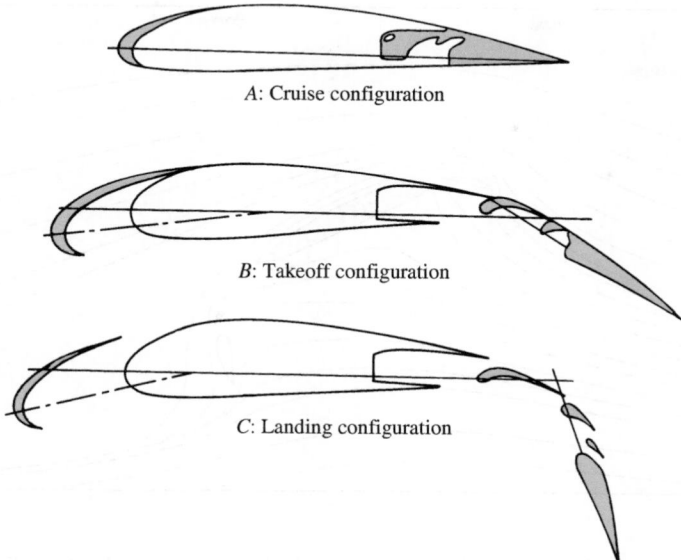

A: Cruise configuration

B: Takeoff configuration

C: Landing configuration

Figure 4.57 Airfoil with leading-edge and trailing-edge high-lift mechanisms. The trailing-edge device is a multielement flap.

Three configurations including the high-lift devices are shown: *A*—the cruise configuration, with no deployment of the high-lift devices; *B*—a typical configuration at takeoff, with both the leading- and trailing-edge devices partially deployed; and *C*—a typical configuration at landing, with all devices fully extended. Note that for configuration *C*, there is a gap between the slat and the leading edge and several gaps between the different elements of the multielement trailing-edge flap. The streamline pattern for the flow over such a configuration is shown in Figure 4.58. Here, the leading-edge slat and the multielement trailing-edge flap are fully extended. The angle of attack is 25°. Although the main flow over the top surface of the airfoil is essentially separated, the local flow through the gaps in the multielement flap is locally attached to the top surface of the flap; because of this locally attached flow, the lift coefficient is still quite high, on the order of 4.5.

With this, we end our discussion of the real flow over airfoils. In retrospect, we can say that the real flow at high angles of attack is dominated by flow separation—a phenomenon that is not properly modeled by the inviscid theories presented in this chapter. On the other hand, at lower angles of attack, such as those associated with the cruise conditions of an airplane, the inviscid theories presented here do an excellent job of predicting both lift and moments on an airfoil. Moreover, in this section, we have clearly seen the importance of airfoil *thickness* in determining the angle of attack at which flow separation will occur, and hence greatly affecting the maximum lift coefficient.

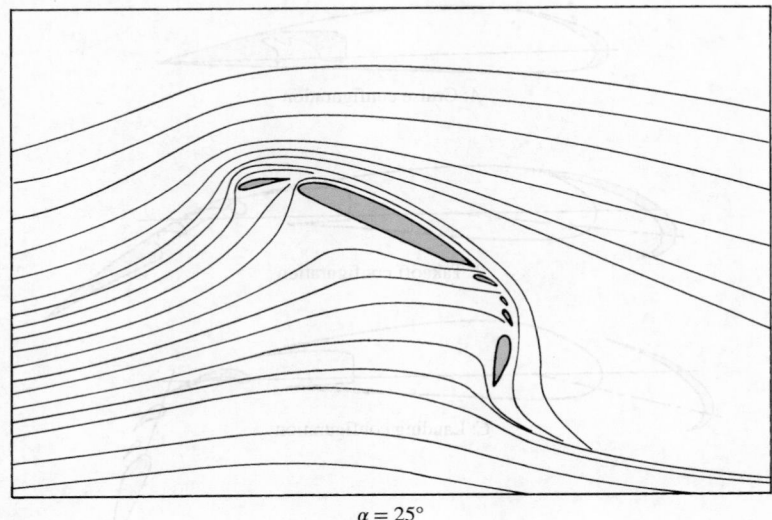

$\alpha = 25°$

Figure 4.58 Effect of leading-edge and multielement flaps on the streamline pattern around an airfoil at angle of attack of 25°. [The streamlines are drawn to scale from the experimental data of Nakayama, Y. (ed): *Visualized Flow*, compiled by the Japan Society of Mechanical Engineers, Pergamon Press, New York, 1988.]

4.14 HISTORICAL NOTE: EARLY AIRPLANE DESIGN AND THE ROLE OF AIRFOIL THICKNESS

In 1804, the first modern configuration aircraft was conceived and built by Sir George Cayley in England—it was an elementary hand-launched glider, about a meter in length, and with a kitelike shape for a wing as shown in Figure 4.59. (For the pivotal role played by George Cayley in the development of the airplane, see the extensive historical discussion in Chapter 1 of Reference 2.) Note that right from the beginning of the modern configuration aircraft, the wing sections were very thin—whatever thickness was present, it was strictly for structural stiffness of the wing. Extremely thin airfoil sections were perpetuated by the work of Horatio Phillips in England. Phillips carried out the first serious wind-tunnel experiments in which the aerodynamic characteristics of a number of different airfoil shapes were measured. (See Section 5.20 of Reference 2 for a presentation of the historical development of airfoils.) Some of Phillips airfoil sections are shown in Figure 4.60—note that they are the epitome of exceptionally thin airfoils. The early pioneers of aviation such as Otto Lilienthal in Germany and Samuel Pierpont Langley in America (see Chapter 1 of Reference 2) continued this thin airfoil tradition. This was especially true of the Wright brothers, who in the period of 1901–1902 tested hundreds of different wing sections and planform shapes in their wind tunnel in Dayton, Ohio (recall our discussion in Section 1.1

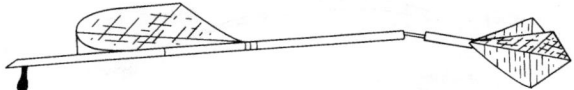

Figure 4.59 The first modern configuration airplane in history: George Cayley's model glider of 1804.

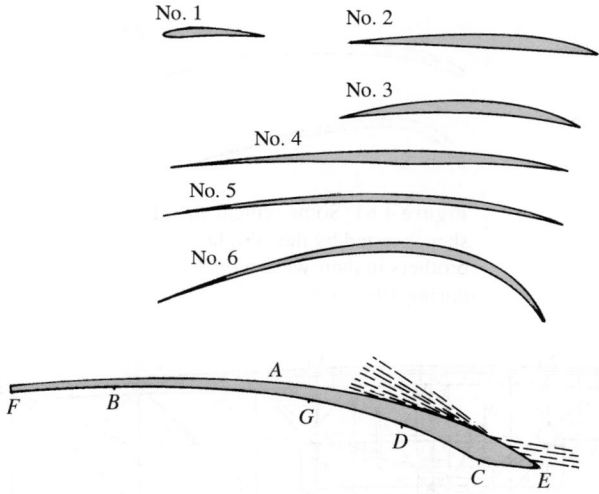

Figure 4.60 Double-surface airfoil sections by Horatio Phillips. The six upper shapes were patented by Phillips in 1884; the lower airfoil was patented in 1891. Note the thin profile shapes.

and the models shown in Figure 1.7). A sketch of some of the Wrights' airfoil sections is given in Figure 4.61—for the most part, very thin sections. Indeed, such a thin airfoil section was used on the 1903 Wright Flyer, as can be readily seen in the side view of the Flyer shown in Figure 4.62. The important point here is that *all* of the early pioneering aircraft, and especially the Wright Flyer, incorporated very thin airfoil sections—airfoil sections that performed essentially like the flat plate results discussed in Section 4.13, and as shown in Figure 4.51 (the dashed curve) and by the streamline pictures in Figure 4.52. Conclusion: These early airfoil sections suffered flow-field separation at small angles of attack and, consequently, had low values of $c_{l,\max}$. By the standards we apply today, these were simply very poor airfoil sections for the production of high lift.

This situation carried into the early part of World War I. In Figure 4.63, we see four airfoil sections that were employed on World War I aircraft. The top three sections had thickness ratios of about 4 to 5 percent and are representative of the type of sections used on all aircraft until 1917. For example, the SPAD XIII (shown in Figure 3.50), the fastest of all World War I fighters, had a thin airfoil section like the Eiffel section shown in Figure 4.63. Why were such thin airfoil

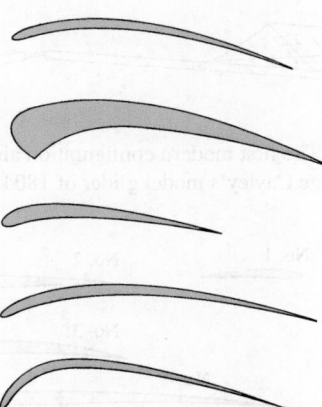

Figure 4.61 Some typical airfoil shapes tested by the Wright brothers in their wind tunnel during 1902–1903.

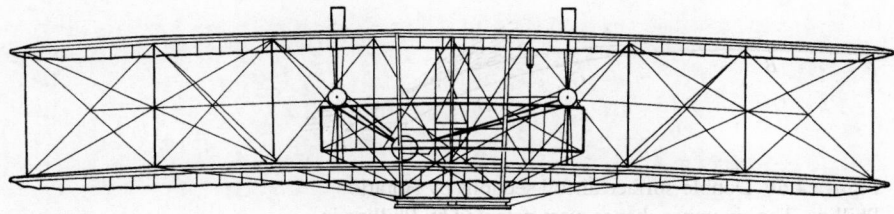

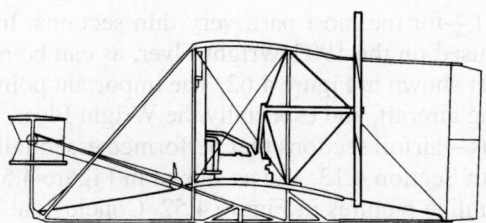

Figure 4.62 Front and side views of the 1903 Wright Flyer. Note the thin airfoil sections. (*Courtesy of the National Air and Space Museum.*)

sections considered to be the best by most designers of World War I aircraft? The historical tradition described above might be part of the answer—a tradition that started with Cayley. Also, there was quite clearly a mistaken notion at that time that *thick* airfoils would produce high drag. Of course, today we know the

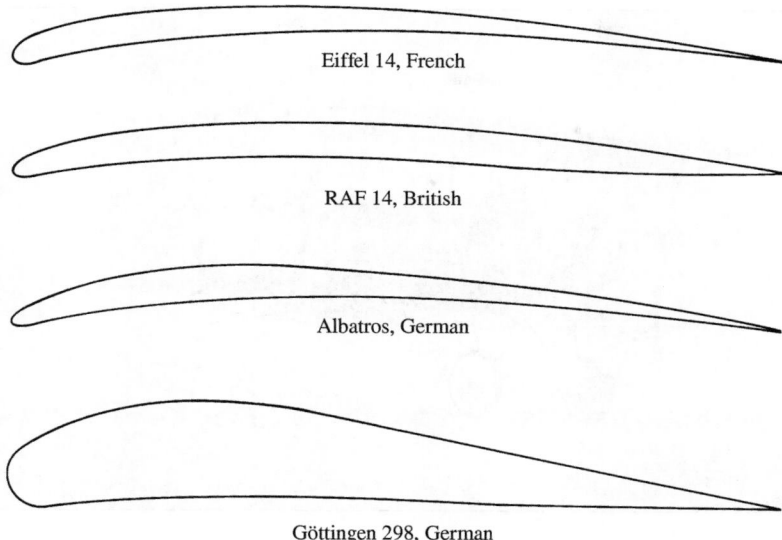

Figure 4.63 Some examples of different airfoil shapes used on World War I aircraft. (*Source:* Loftin, Lawrence K., Jr.: *Quest for Performance: The Evolution of Modern Aircraft*, NASA SP-468, 1985.)

opposite to be true; review our discussion of streamlined shapes in Section 1.12 for this fact. Laurence Loftin in Reference 45 surmises that the mistaken notion might have been fostered by early wind-tunnel tests. By the nature of the early wind tunnels in use—small sizes and very low speeds—the data were obtained at very low Reynolds numbers, less than 100,000 based on the airfoil-chord length. These Reynolds numbers are to be compared with typical values in the millions for actual airplane flight. Modern studies of low Reynolds number flows over conventional thick airfoils (e.g., see Reference 48) clearly show high-drag coefficients, in contrast to the lower values that occur for the high Reynolds number associated with the flight of full-scale aircraft. Also, the reason for the World War I airplane designer's preference for thin airfoils might be as simple as the tendency to follow the example of the wings of birds, which are quite thin. In any event, the design of all English, French, and American World War I aircraft incorporated thin airfoils and, consequently, suffered from poor high-lift performance. The fundamentals of airfoil aerodynamics as we know them today (and as being presented in this book) were simply not sufficiently understood by the designers during World War I. In turn, they never appreciated what they were losing.

This situation changed dramatically in 1917. Work carried out in Germany at the famous Göttingen aerodynamic laboratory of Ludwig Prandtl (see Section 5.8 for a biographical sketch of Prandtl) demonstrated the superiority of a thick airfoil section, such as the Göttingen 298 section shown at the bottom of Figure 4.63. This revolutionary development was immediately picked up by the famous designer Anthony Fokker, who incorporated the 13-percent-thick Göttingen 298 profile in

Figure 4.64 The World War I Fokker Dr-1 triplane, the first fighter aircraft to use a thick airfoil. (*Ed Maker/The Denver Post/Getty Images*)

his new Fokker Dr-1—the famous triplane flown by the "Red Baron," Rittmeister Manfred Freiher von Richthofen. A photograph of the Fokker Dr-1 is shown in Figure 4.64. The major benefits derived from Fokker's use of the thick airfoil were:

1. The wing structure could be completely internal; that is the wings of the Dr-1 were a cantilever design, which removed the need for the conventional wire bracing that used in other aircraft. This, in turn, eliminated the high drag associated with these interwing wires, as discussed at the end of Section 1.11. For this reason, the Dr-1 had a zero-lift drag coefficient of 0.032, among the lowest of World War I airplanes. (By comparison, the zero-lift drag coefficient of the French SPAD XIII was 0.037.)

2. The thick airfoil provided the Fokker Dr-1 with a high maximum lift coefficient. Its performance was analogous to the *upper* curves shown in Figure 4.51. This, in turn, provided the Dr-1 with an exceptionally high rate-of-climb as well as enhanced maneuverability—characteristics that were dominant in dog-fighting combat.

Anthony Fokker continued the use of a thick airfoil in his design of the D-VII, as shown in Figure 4.65. This gave the D-VII a much greater rate-of-climb than its two principal opponents at the end of the war—the English Sopwith Camel and the French SPAD XIII, both of which still used very thin airfoil sections. This rate-of-climb performance, as well as its excellent handling characteristics, singled out the Fokker D-VII as the most effective of all German World War I fighters. The respect given by the Allies to this machine is no more clearly indicated

Figure 4.65 The World War I Fokker D-VII, one of the most effective fighters of the war, due in part to its superior aerodynamic performance allowed by a thick airfoil section. (*Chronicle/Alamy Stock Photo*)

than by a paragraph in article IV of the armistice agreement, which lists war material to be handed over to the Allies by Germany. In this article, the Fokker D-VII is *specifically* listed—the only airplane of any type to be explicitly mentioned in the armistice. To this author's knowledge, this is the one and only time where a breakthrough in airfoil technology is essentially reflected in any major political document, though somewhat implicitly.

4.15 HISTORICAL NOTE: KUTTA, JOUKOWSKI, AND THE CIRCULATION THEORY OF LIFT

Frederick W. Lanchester (1868–1946), an English engineer, automobile manufacturer, and self-styled aerodynamicist, was the first to connect the idea of circulation with lift. His thoughts were originally set forth in a presentation given before the Birmingham Natural History and Philosophical Society in 1894 and later contained in a paper submitted to the Physical Society, which turned it down. Finally, in 1907 and 1908, he published two books, entitled *Aerodynamics* and *Aerodonetics*, where his thoughts on circulation and lift were described in detail. His books were later translated into German in 1909 and French in 1914. Unfortunately, Lanchester's style of writing was difficult to read and understand; this is partly responsible for the general lack of interest shown by British scientists in Lanchester's work. Consequently, little positive benefit was derived from

Lanchester's writings. (See Section 5.7 for a more detailed portrait of Lanchester and his work.)

Quite independently, and with total lack of knowledge of Lanchester's thinking, M. Wilhelm Kutta (1867–1944) developed the idea that lift and circulation are related. Kutta was born in Pitschen, Germany, in 1867, and obtained a Ph.D. in mathematics from the University of Munich in 1902. After serving as professor of mathematics at several German technical schools and universities, he finally settled at the Technische Hochschule in Stuttgart, in 1911, until his retirement in 1935. Kutta's interest in aerodynamics was initiated by the successful glider flights of Otto Lilienthal in Berlin during the period 1890–1896 (see Chapter 1 of Reference 2). Kutta attempted theoretically to calculate the lift on the curved wing surfaces used by Lilienthal. In the process, he surmised from experimental data that the flow left the trailing edge of a sharp-edged body smoothly and that this condition fixed the circulation around the body (the Kutta condition, described in Section 4.5). At the same time, he was convinced that circulation and lift were connected. Kutta was reluctant to publish these ideas, but after the strong insistence of his teacher, S. Finsterwalder, he wrote a paper entitled "Auftriebskrafte in Stromenden Flussigkecten" (Lift in Flowing Fluids). This was actually a short note abstracted from his longer graduation paper in 1902, but it represents the first time in history where the concepts of the Kutta condition as well as the connection of circulation with lift were officially published. Finsterwalder clearly repeated the ideas of his student in a lecture given on September 6, 1909, in which he stated:

> On the upper surface the circulatory motion increases the translatory one, therefore there is high velocity and consequently low pressure, while on the lower surface the two movements are opposite, therefore there is low velocity with high pressure, with the result of a thrust upward.

However, in his 1902 note, Kutta did not give the precise quantitative relation between circulation and lift. This was left to Nikolai E. Joukowski (Zhukouski). Joukowski was born in Orekhovo in central Russia on January 5, 1847. The son of an engineer, he became an excellent student of mathematics and physics, graduating with a Ph.D. in applied mathematics from Moscow University in 1882. He subsequently held a joint appointment as a professor of mechanics at Moscow University and the Moscow Higher Technical School. It was at this latter institution that Joukowski built in 1902 the first wind tunnel in Russia. Joukowski was deeply interested in aeronautics, and he combined a rare gift for both experimental and theoretical work in the field. He expanded his wind tunnel into a major aerodynamics laboratory in Moscow. Indeed, during World War I, his laboratory was used as a school to train military pilots in the principles of aerodynamics and flight. When he died in 1921, Joukowski was by far the most noted aerodynamicist in Russia.

Much of Joukowski's fame was derived from a paper published in 1906, wherein he gives, for the first time in history, the relation $L' = \rho_\infty V_\infty \Gamma$—the Kutta-Joukowski theorem. In Joukowski's own words:

If an irrotational two-dimensional fluid current, having at infinity the velocity V_∞ surrounds any closed contour on which the circulation of velocity is Γ, the force of the aerodynamic pressure acts on this contour in a direction perpendicular to the velocity and has the value

$$L' = \rho_\infty V_\infty \Gamma$$

The direction of this force is found by causing to rotate through a right angle the vector V_∞ around its origin in an inverse direction to that of the circulation.

Joukowski was unaware of Kutta's 1902 note and developed his ideas on circulation and lift independently. However, in recognition of Kutta's contribution, the equation given above has propagated through the twentieth century as the "Kutta-Joukowski theorem."

Hence, by 1906—just 3 years after the first successful flight of the Wright brothers—the circulation theory of lift was in place, ready to aid aerodynamics in the design and understanding of lifting surfaces. In particular, this principle formed the cornerstone of the thin airfoil theory described in Sections 4.7 and 4.8. Thin airfoil theory was developed by Max Munk, a colleague of Prandtl in Germany, during the first few years after World War I. However, the very existence of thin airfoil theory, as well as its amazingly good results, rests upon the foundation laid by Lanchester, Kutta, and Joukowski a decade earlier.

4.16 SUMMARY

Return to the road map given in Figure 4.7. Make certain that you feel comfortable with the material represented by each box on the road map and that you understand the flow of ideas from one box to another. If you are uncertain about one or more aspects, review the pertinent sections before progressing further.

Some important results from this chapter are itemized below:

A vortex sheet can be used to synthesize the inviscid, incompressible flow over an airfoil. If the distance along the sheet is given by s and the strength of the sheet per unit length is $\gamma(s)$, then the velocity potential induced at point (x, y) by a vortex sheet that extends from point a to point b is

$$\phi(x, y) = -\frac{1}{2\pi} \int_a^b \theta \gamma(s)\, ds \tag{4.3}$$

The circulation associated with this vortex sheet is

$$\Gamma = \int_a^b \gamma(s)\, ds \tag{4.4}$$

Across the vortex sheet, there is a tangential velocity discontinuity, where

$$\gamma = u_1 - u_2 \tag{4.8}$$

The Kutta condition is an observation that for a lifting airfoil of given shape at a given angle of attack, nature adopts that particular value of circulation around the airfoil which results in the flow leaving smoothly at the trailing edge. If the trailing-edge angle is finite, then the trailing edge is a stagnation point. If the trailing edge is cusped, then the velocities leaving the top and bottom surfaces at the trailing edge are finite and equal in magnitude and direction. In either case,

$$\gamma(\text{TE}) = 0 \qquad (4.10)$$

Thin airfoil theory is predicated on the replacement of the airfoil by the mean camber line. A vortex sheet is placed along the chord line, and its strength adjusted such that, in conjunction with the uniform freestream, the camber line becomes a streamline of the flow while at the same time satisfying the Kutta condition. The strength of such a vortex sheet is obtained from the fundamental equation of thin airfoil theory:

$$\frac{1}{2\pi} \int_0^c \frac{\gamma(\xi)\, d\xi}{x - \xi} = V_\infty \left(\alpha - \frac{dz}{dx} \right) \qquad (4.18)$$

Results of thin airfoil theory:

Symmetric airfoil

1. $c_l = 2\pi\alpha$.
2.   Lift slope $= dc_l/d\alpha = 2\pi$.
3.   The center of pressure and the aerodynamic center are both at the quarter-chord point.
4. $c_{m,c/4} = c_{m,\text{ac}} = 0$.

Cambered airfoil

1.
$$c_l = 2\pi \left[\alpha + \frac{1}{\pi} \int_0^\pi \frac{dz}{dx} (\cos\theta_0 - 1)\, d\theta_0 \right] \qquad (4.57)$$

2.   Lift slope $= dc_l/d\alpha = 2\pi$.
3.   The aerodynamic center is at the quarter-chord point.
4.   The center of pressure varies with the lift coefficient.

The vortex panel method is an important numerical technique for the solution of the inviscid, incompressible flow over bodies of arbitrary shape, thickness, and angle of attack. For panels of constant strength, the governing equations are

$$V_\infty \cos \beta_i - \sum_{j=1}^{n} \frac{\gamma_j}{2\pi} \int_j \frac{\partial \theta_{ij}}{\partial n_i} \, ds_j = 0 \qquad (i = 1, 2, \ldots, n)$$

and

$$\gamma_i = -\gamma_{i-1}$$

which is one way of expressing the Kutta condition for the panels immediately above and below the trailing edge.

4.17 INTEGRATED WORK CHALLENGE: WALL EFFECTS ON MEASUREMENTS MADE IN SUBSONIC WIND TUNNELS

Concept: Low-speed subsonic wind tunnels were discussed in Chapter 3, including the conceptual design of such tunnels in Section 3.23. When models of flight vehicles are tested in wind tunnels, we want the aerodynamic measurements taken with the models (lift and drag coefficients, pressure coefficients, etc.) to simulate those quantities that prevail on the actual flight vehicle in free flight through the atmosphere. There are, however, many reasons why the wind tunnel measurements may not be quite the same results as in free flight (scale effects, instrumentation errors, etc.), but one fundamental source of error is that the flow-field over a model in a wind tunnel is bounded and constrained by the walls of the tunnel, whereas no such constraints exist in free flight through the atmosphere (except when the vehicle is flying in close proximity to the ground, with consequent "ground effects" on the aerodynamics of the vehicle). In this section, we will address the tunnel wall effects on a model in the tunnel and look at ways to compensate for them on the data.

The corrections to measurements made in wind tunnels due to wall effects have been a concern since the beginning of the serious use of wind tunnels. Indeed, three chapters in the definitive book on tunnel testing by Barlow, Rae, and Pope, *Low-Speed Wind Tunnel Testing*, 3rd Edition, Wiley, 1999, are devoted to wind tunnel wall corrections. It is beyond our scope here to exhaustively look at the subject. However, in Chapters 3 and 4 we have developed some elementary potential flow methods to calculate flows over bodies, and this leads to the following challenge.

Challenge: How can we use our potential flow methods to examine to what extent wall effects might affect the flow over a model mounted in a wind tunnel?

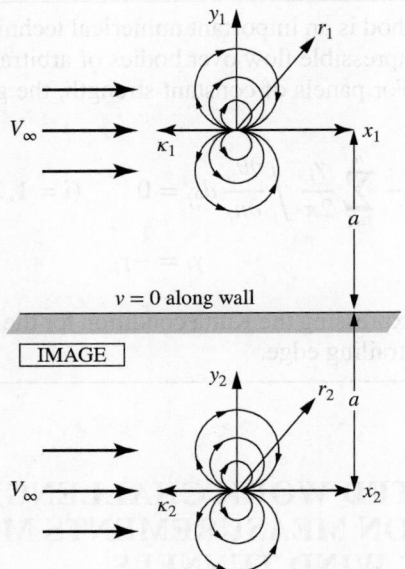

Figure 4.66 Image of a doublet.

Solution: Let us take the classic case of the nonlifting flow over a circular cylinder, as discussed in Section 3.13, where the streamlines are sketched in Figure 3.26. Using potential flow theory, this flow is synthesized by combining a doublet flow with a uniform freestream. The flow field extends to infinity in all directions, and the resulting streamline pattern has straight, uniform streamlines only at infinity. In contrast, imagine a circular cylinder mounted in a wind tunnel flow that is bounded above and below by straight, parallel walls. The flow streamlines along the walls have to be straight and parallel. Simultaneously, another streamline must be a circle in order to simulate the surface streamline of the cylinder. How can we simulate this condition?

One answer is to use the *method of images*, the basic idea of which is as follows. Return to Figure 3.26, which illustrates the combination of a doublet with a uniform freestream. This same picture is sketched in the upper part of Figure 4.66. Here, the doublet is located at distance a above the wall. The streamline along the wall must be straight and parallel to the wall; hence, the vertical component of the flow velocity must be zero at the wall, i.e., $v = 0$. This cannot happen with just the doublet above the wall. However, if we imagine a second doublet of equal strength placed below the wall at an equal distance a below the wall, then by symmetry the vertical components of the two flows along the wall will cancel, yielding $v = 0$ for the combined flow along the wall. This second doublet is called an *image* of the first; it is there simply to provide a straight, parallel streamline that represents the wall.

Let us examine the stream function for the combination of the doublet and its image. From Equation (3.91), for doublet 1, with strength κ_1,

$$\Psi_1 = -\frac{\kappa_1}{2\pi}\frac{\sin\theta_1}{r_1} + V_\infty r_1 \sin\theta_1 \qquad \text{(C4.1)}$$

and for doublet 2 with strength κ_2,

$$\Psi_2 = -\frac{\kappa_2}{2\pi}\frac{\sin\theta_2}{r_2} + V_\infty r_2 \sin\theta_2 \qquad \text{(C4.2)}$$

Using the cartesian coordinates x_1, y_1, x_2, and y_2 shown in Figure 4.66,

$$\sin\theta_1 = \frac{y_1}{\sqrt{(x_1^2 + y_1^2)}}$$

$$\sin\theta_2 = \frac{y_2}{\sqrt{(x_2^2 + y_2^2)}}$$

$$r_1 = \sqrt{x_1^2 + y_1^2}$$

$$r_2 = \sqrt{x_2^2 + y_2^2}$$

$$r_1 \sin\theta_1 = y_1$$

$$r_2 \sin\theta_2 = y_2$$

and therefore Equations (C4.1) and (C4.2) become, respectively,

$$\Psi_1 = -\frac{\kappa_1}{2\pi}\frac{y_1}{(x_1^2 + y_2^2)} + V_\infty r_1 \qquad \text{(C4.3)}$$

$$\Psi_2 = -\frac{\kappa_2}{2\pi}\frac{y_2}{(x_2^2 + y_2^2)} + V_\infty r_2 \qquad \text{(C4.4)}$$

Consider point P in the flow shown in Figure 4.66. This point represents any point in the flow above the wall. The relation between the two sets of cartesian coordinates (x_1, y_1) and (x_2, y_2) is, from Figure 4.66:

$$x_2 = x_1$$

$$y_2 = 2a + y_1$$

Hence, the combined stream function at point P is

$$\psi = \Psi_1 + \Psi_2 = -\frac{\kappa_1}{2\pi}\frac{y_1}{(x_1^2 + y_1^2)} + V_\infty y_1 - \frac{\kappa_2}{2\pi}\frac{(2a + y_1)}{x_2^2 + (2a + y_1)^2} + V_\infty(2a + y_1) \quad \text{(C4.5)}$$

Unfortunately, the streamline through the stagnation points in the flow described by Equation (4.5) is not the shape of a circular cylinder; rather, it is the distorted shape sketched qualitatively in Figure 4.67. There would be a mirror image of

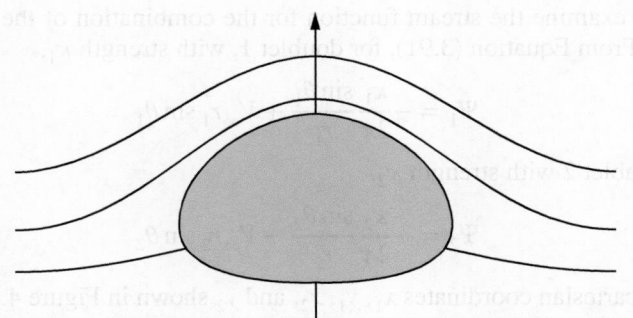

Figure 4.67 Body shape simulated by a doublet and its image in a uniform flow (qualitative sketch).

the same distorted shape below the wall. In order to more accurately represent the flow over the circular cylinder, we need an image of the image to be placed above the cylinder, and then more images of the images arranged in a vertical array above and below. In this fashion, the flow synthesized by all these images will approach closer to the flow over a circular cylinder, while at the same time preserving the straight streamlines at the walls. This is the essence of the "method of images" used to represent the flow over a body in the presence of a wall. Also, note that we have considered only the generation of a flow with a straight stream-line below the body; this would simulate the flow over the body in the presence of a ground plane. In the wind tunnel case, there is also an upper wall, and our "method of images" becomes more complex. Is there a better way of taking into account wall effects in the wind tunnel?

Yes—one technique is to use computational fluid dynamics to calculate the flow in the test section with the test model mounted in it. The panel methods described in Sections 3.17 and 4.10 could be used by covering both the surface of the model and the surface of the tunnel walls with source and vortex panels. Results from such panel solutions for the flow over a circular cylinder are sketched in Figure 4.68*a* and *b*. The streamlines over a nonlifting cylinder in free air are shown in Figure 4.68*a*, and those for a nonlifting cylinder in a wind tunnel test section are shown in Figure 4.68*b*. The streamlines in the wind tunnel case are more constrained than in the free air case, and there would be an effect on the surface pressure distribution on the circular cylinder. The computational results from a panel solution would identify such effects. The use of computational fluid dynamics, such as panel methods, for wind tunnel wall corrections can be expensive and time consuming, so the many empirical methods developed over the years are still in vogue. See Barlow, Rae and Pope, Chapters 9 to 11 for an in-depth discussion. For our purposes here, the "solution" to this Integrated Work Challenge has not been a solution per se, but rather a discussion of the problem. This is because the actual solution is long and detailed and beyond the scope of this book. Our purpose here is to simply open the thought process.

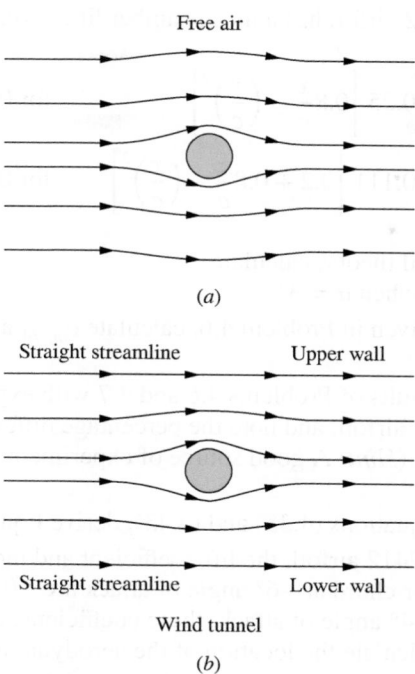

Figure 4.68 Comparison between (a) the streamlines over a cylinder in free air and (b) the streamlines over a cylinder in a wind tunnel.

4.18 PROBLEMS

4.1 Consider the data for the NACA 2412 airfoil given in Figure 4.10. Calculate the lift and moment about the quarter chord (per unit span) for this airfoil when the angle of attack is 4° and the freestream is at standard sea level conditions with a velocity of 50 ft/s. The chord of the airfoil is 2 ft.

4.2 Consider an NACA 2412 airfoil with a 2-m chord in an airstream with a velocity of 50 m/s at standard sea level conditions. If the lift per unit span is 1353 N/m, what is the angle of attack?

4.3 Starting with the definition of circulation, derive Kelvin's circulation theorem, Equation (4.11).

4.4 Starting with Equation (4.35), derive Equation (4.36).

4.5 Consider a thin, symmetric airfoil at 1.5° angle of attack. From the results of thin airfoil theory, calculate the lift coefficient and the moment coefficient about the leading edge.

4.6 The NACA 4412 airfoil has a mean camber line given by

$$
\frac{z}{c} =
\begin{cases}
0.25 \left[0.8\dfrac{x}{c} - \left(\dfrac{x}{c} \right)^2 \right] & \text{for } 0 \le \dfrac{x}{c} \le 0.4 \\[4mm]
0.111 \left[0.2 + 0.8\dfrac{x}{c} - \left(\dfrac{x}{c} \right)^2 \right] & \text{for } 0.4 \le \dfrac{x}{c} \le 1
\end{cases}
$$

Using thin airfoil theory, calculate
(a) $\alpha_{L=0}$ (b) c_l when $\alpha = 3°$

4.7 For the airfoil given in Problem 4.6, calculate $c_{m,c/4}$ and x_{cp}/c when $\alpha = 3°$.

4.8 Compare the results of Problems 4.6 and 4.7 with experimental data for the NACA 4412 airfoil, and note the percentage difference between theory and experiment. (*Hint:* A good source of experimental airfoil data is Reference 11.)

4.9 Starting with Equations (4.35) and (4.43), derive Equation (4.62).

4.10 For the NACA 2412 airfoil, the lift coefficient and moment coefficient about the quarter-chord at $-6°$ angle of attack are -0.39 and -0.045, respectively. At $4°$ angle of attack, these coefficients are 0.65 and -0.037, respectively. Calculate the location of the aerodynamic center.

4.11 Consider again the NACA 2412 airfoil discussed in Problem 4.10. The airfoil is flying at a velocity of 60 m/s at a standard altitude of 3 km (see Appendix D). The chord length of the airfoil is 2 m. Calculate the lift per unit span when the angle of attack is $4°$.

4.12 For the airfoil in Problem 4.11, calculate the value of the circulation around the airfoil.

4.13 In Section 3.15, we studied the case of the lifting flow over a circular cylinder. In real life, a rotating cylinder in a flow will produce lift; such real flow fields are shown in the photographs in Figure 3.34b and c. Here, the viscous shear stress acting between the flow and the surface of the cylinder drags the flow around in the direction of rotation of the cylinder. For a cylinder of radius R rotating with an angular velocity ω in an otherwise stationary fluid, the viscous flow solution for the velocity field obtained from the Navier-Stokes equations (Chapter 15) is

$$
V_\theta = \frac{R^2 \omega}{r}
$$

where V_θ is the tangential velocity along the circular streamlines and r is the radial distance from the center of the cylinder. (See Schlichting, *Boundary-Layer Theory,* 6th ed., McGraw Hill, 1968, page 81.) Note that V_θ varies inversely with r and is of the same form as the inviscid flow velocity for a point vortex given by Equation (3.105). If the rotating cylinder has a radius of 1 m and is flying at the same velocity and altitude as the airfoil in Problem 4.11, what must its angular velocity be to produce the same lift as the airfoil in Problem 4.11? (*Note:* You can check your results with the experimental data for lift on rotating cylinders in Hoerner, *Fluid-Dynamic Lift,* published by the author, 1975, pp. 21–4, Fig. 5.)

4.14 The question is often asked: Can an airfoil fly upside-down? To answer this, make the following calculation. Consider a positively cambered airfoil with a zero-lift angle of −3°. The lift slope is 0.1 per degree. (*a*) Calculate the lift coefficient at an angle of attack of 5°. (*b*) Now imagine the same airfoil turned upside-down, but at the same 5° angle of attack as part (*a*). Calculate its lift coefficient. (*c*) At what angle of attack must the upside-down airfoil be set to generate the same lift as that when it is right-side-up at a 5° angle of attack?

4.15 The airfoil section of the wing of the British Spitfire of World War II fame (see Figure 5.19) is an NACA 2213 at the wing root, tapering to an NACA 2205 at the wing tip. The root chord is 8.33 ft. The measured profile drag coefficient of the NACA 2213 airfoil is 0.006 at a Reynolds number of 9×10^6. Consider the Spitfire cruising at an altitude of 18,000 ft. (a) At what velocity is it flying for the root chord Reynolds number to be 9×10^6? (b) At this velocity and altitude, assuming completely turbulent flow, estimate the skin-friction drag coefficient for the NACA 2213 airfoil, and compare this with the total profile drag coefficient. Calculate the percentage of the profile drag coefficient that is due to pressure drag. *Note:* Assume that μ varies as the square root of temperature, as first discussed in Section 1.8.

4.16 For the conditions given in Problem 4.15, a more reasonable calculation of the skin friction coefficient would be to assume an initially laminar boundary layer starting at the leading edge, and then transitioning to a turbulent boundary layer at some point downstream. Calculate the skin-friction coefficient for the Spitfire's airfoil described in Problem 4.15, but this time assuming a critical Reynolds number of 10^6 for transition.

Incompressible Flow over Finite Wings

The one who has most carefully watched the soaring birds of prey sees man with wings and the faculty of using them.
   James Means, Editor of
   the Aeronautical Annual, 1895

PREVIEW BOX

The Beechcraft Baron 58 twin-engine business aircraft is shown in three-view in Figure 5.1. The wing on this airplane has a 15 percent thick NACA 23015 airfoil at the root, tapering to a 10 percent thickness at the tip. We studied airfoil properties in Chapter 4, ostensibly to be able to predict the lift and drag characteristics of an airplane wing utilizing a given airfoil shape. Wind tunnel data for the NACA 23015 airfoil is given in Figure 5.2 in the standard format used by the NACA as described in Reference 11. The airfoil lift coefficient and the moment coefficient about the quarter-chord point are given as a function of the airfoil section angle of attack in Figure 5.2a. The drag coefficient and the moment coefficient about the aerodynamic center are given as a function of the lift coefficient in Figure 5.2b. The airfoil shape is shown to scale at the top of Figure 5.2b. Results for three different values of the Reynolds number based on chord length are shown, as indicated by the code near the bottom of Figure 5.2b. The data labeled "standard roughness" applies to a special case where the model surface was covered with a sandpaper-like grit; we are not concerned with this special case.

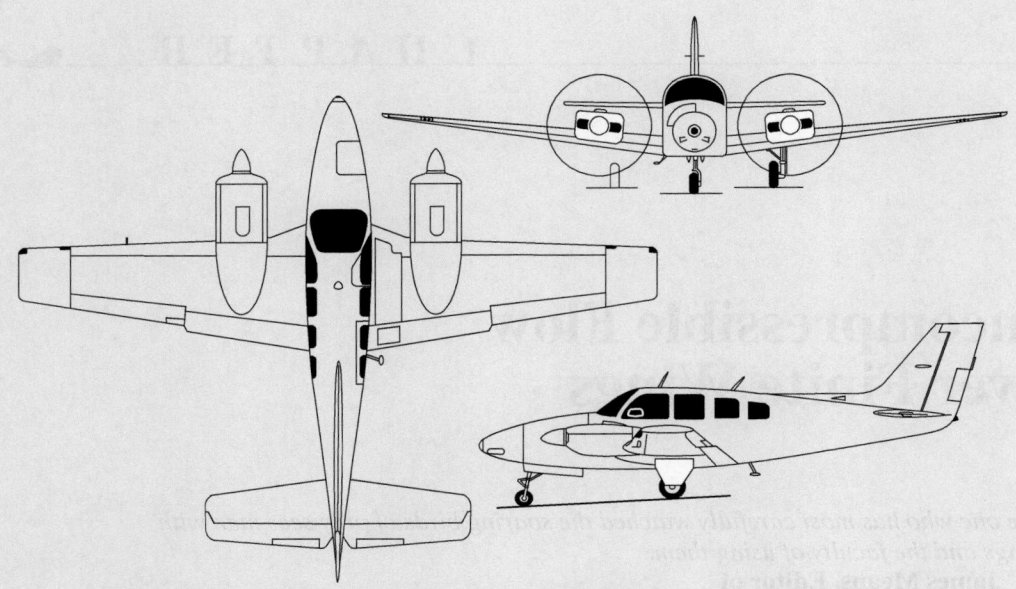

Figure 5.1 Beechcraft Baron 58, four- to six-seat general aviation airplane.

Consider the Beechcraft Baron 58 cruising such that the wing is at a four-degree angle of attack. What are the lift and drag coefficients of the wing? From Figure 5.2a, for the airfoil at $\alpha = 4°$, we have $c_l = 0.54$. Using the drag coefficient data at the highest Reynolds number shown in Figure 5.2b, for $c_l = 0.54$ the corresponding drag coefficient $c_d = 0.0068$. Using capital letters for the aerodynamic coefficients of a wing, in contrast to the lower case letters for an airfoil, we pose the question: Are C_L and C_D for the wing the same as those for the airfoil? That is,

$$C_L = 0.54 \text{ (?)} \qquad C_D = 0.0068 \text{ (?)}$$

The answer is a resounding NO! Not even close! Surprised? How can this be? Why are the aerodynamic coefficients of the wing *not* the same as those for the airfoil shape from which the wing is made? Surely, the aerodynamic properties of the airfoil must have something to do with the aerodynamic properties of the finite wing. Or else our studies of airfoils in Chapter 4 would be a big waste of time.

The single, most important purpose of this chapter is to solve this mystery. This chapter is focused on the aerodynamic properties of real, finite wings. The solution involves some interesting and (yes) some rather intense mathematics combined with some knowledge of the physical properties of the flow over a finite wing. This is very important stuff. This chapter gets at the very core of the aerodynamics of a real airplane flying at speeds appropriate to the assumption of incompressible flow. Read on with vigor, and let us solve this mystery.

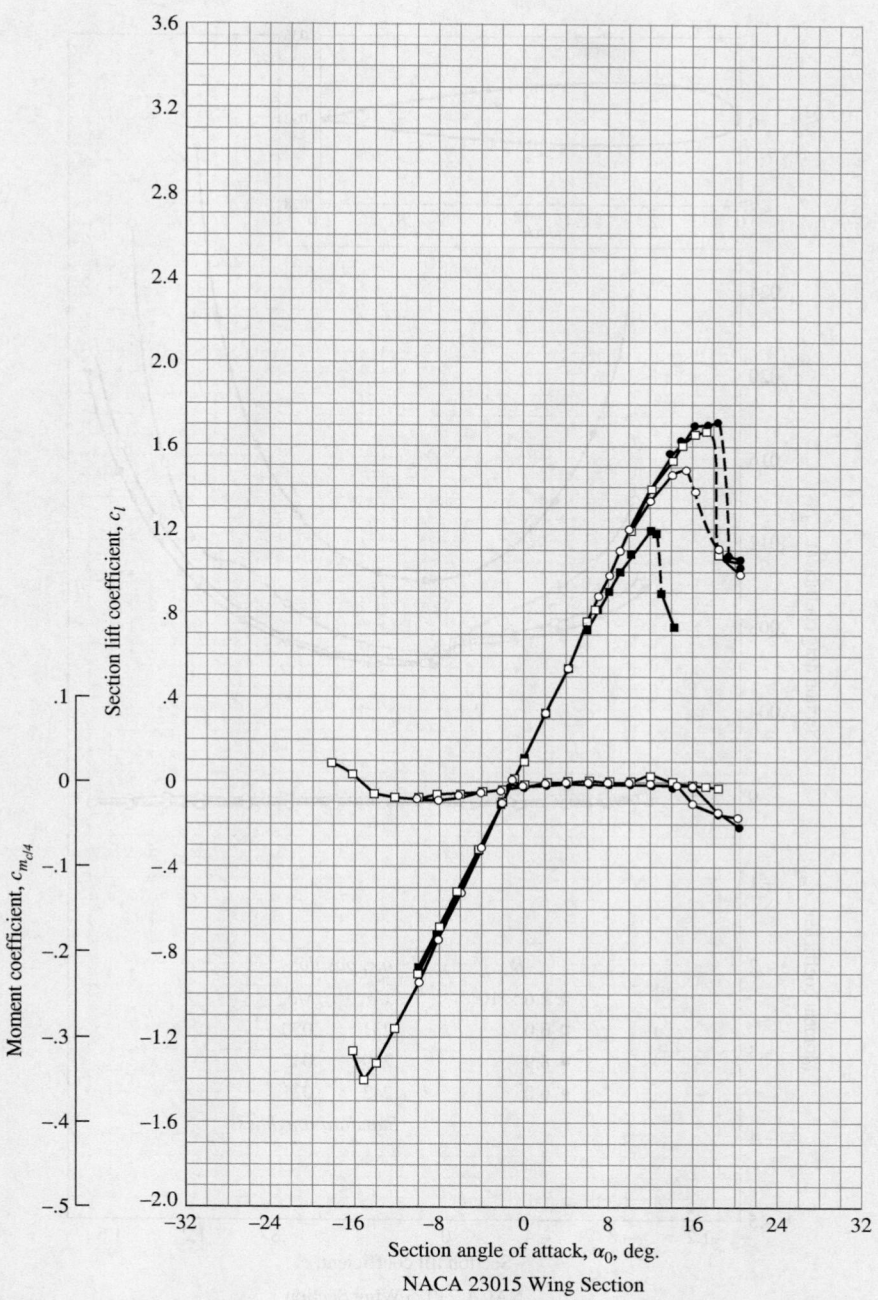

NACA 23015 Wing Section

Figure 5.2a Lift coefficient and moment coefficient about the quarter-chord for the NACA 23015 airfoil.

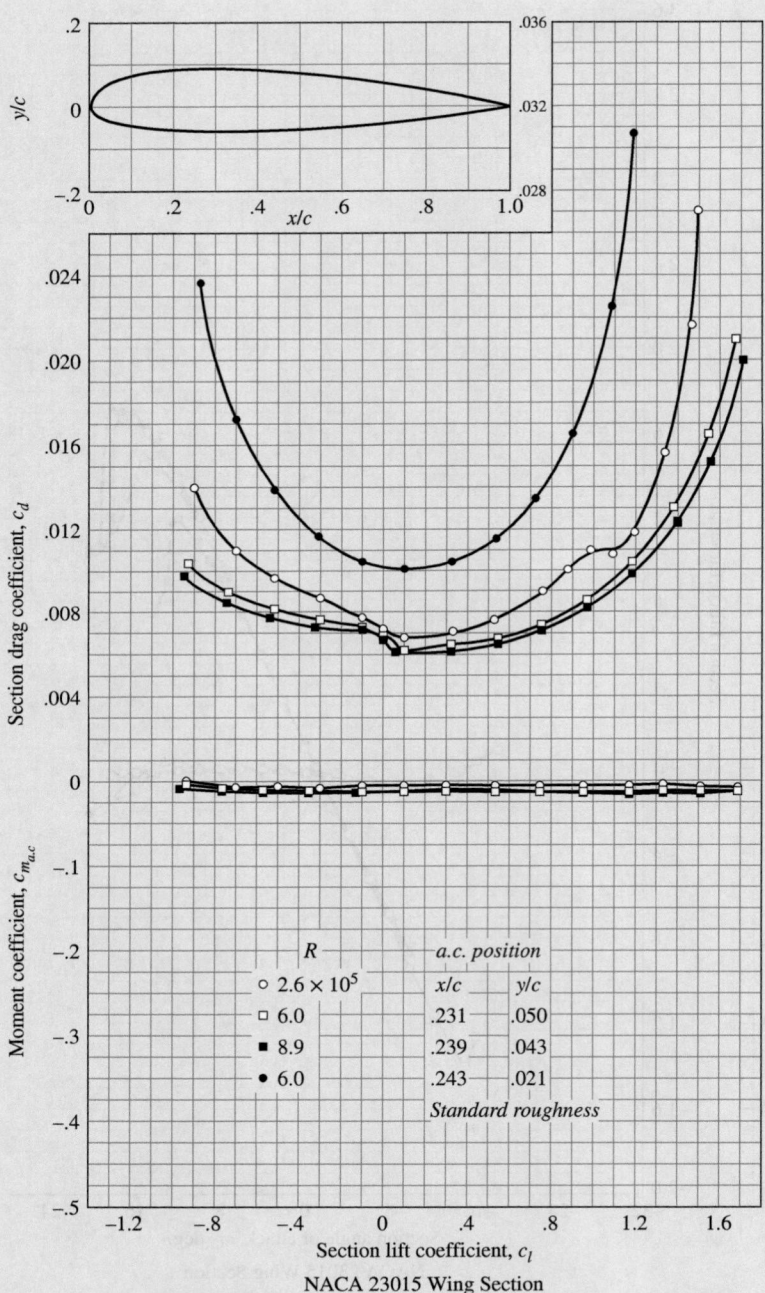

NACA 23015 Wing Section

Figure 5.2*b* (*Continued*) Drag coefficient and moment coefficient about the aerodynamic center for the NACA 23015 airfoil.

5.1 INTRODUCTION: DOWNWASH AND INDUCED DRAG

In Chapter 4 we discussed the properties of airfoils, which are the same as the properties of a wing of infinite span; indeed, airfoil data are frequently denoted as "infinite wing" data. However, all real airplanes have wings of finite span, and the purpose of the present chapter is to apply our knowledge of airfoil properties to the analysis of such finite wings. This is the second step in Prandtl's philosophy of wing theory, as described in Section 4.1. You should review Section 4.1 before proceeding further.

Question: Why are the aerodynamic characteristics of a finite wing any different from the properties of its airfoil sections? Indeed, an airfoil is simply a section of a wing, and at first thought, you might expect the wing to behave exactly the same as the airfoil. However, as studied in Chapter 4, the flow over an airfoil is two-dimensional. In contrast, a finite wing is a three-dimensional body, and consequently the flow over the finite wing is three-dimensional; that is, there is a component of flow in the spanwise direction. To see this more clearly, examine Figure 5.3, which gives the top and front views of a finite wing. The physical mechanism for generating lift on the wing is the existence of high pressure on the bottom surface and low pressure on the top surface. The net imbalance of the pressure distribution creates the lift, as discussed in Section 1.5. However, as

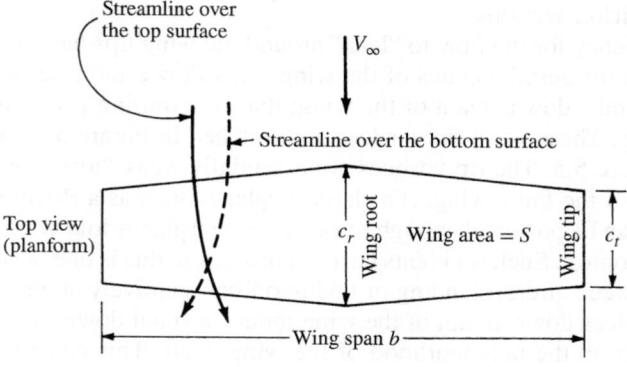

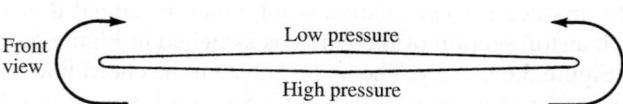

Figure 5.3 Finite wing. In this figure, the curvature of the streamlines over the top and bottom of the wing is exaggerated for clarity.

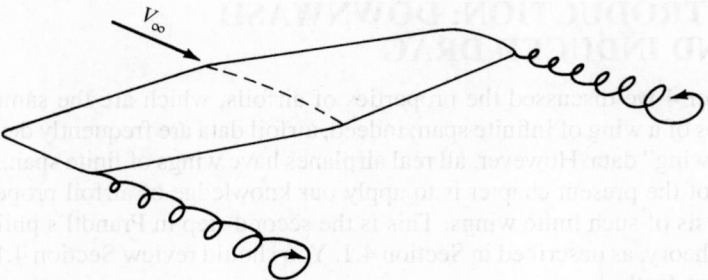

Figure 5.4 Schematic of wing-tip vortices.

a by-product of this pressure imbalance, the flow near the wing tips tends to curl around the tips, being forced from the high-pressure region just underneath the tips to the low-pressure region on top. This flow around the wing tips is shown in the front view of the wing in Figure 5.3. As a result, on the top surface of the wing, there is generally a spanwise component of flow from the tip toward the wing root, causing the streamlines over the top surface to bend toward the root, as sketched on the top view shown in Figure 5.3. Similarly, on the bottom surface of the wing, there is generally a spanwise component of flow from the root toward the tip, causing the streamlines over the bottom surface to bend toward the tip. Clearly, the flow over the finite wing is three-dimensional, and therefore you would expect the overall aerodynamic properties of such a wing to differ from those of its airfoil sections.

The tendency for the flow to "leak" around the wing tips has another important effect on the aerodynamics of the wing. This flow establishes a circulatory motion that trails downstream of the wing; that is, a trailing *vortex* is created at each wing tip. These wing-tip vortices are sketched in Figure 5.4 and are illustrated in Figure 5.5. The tip vortices are essentially weak "tornadoes" that trail downstream of the finite wing. (For large airplanes such as a Boeing 747, these tip vortices can be powerful enough to cause light airplanes following too closely to go out of control. Such accidents have occurred, and this is one reason for large spacings between aircraft landing or taking off consecutively at airports.) These wing-tip vortices downstream of the wing induce a small downward component of air velocity in the neighborhood of the wing itself. This can be seen by inspecting Figure 5.5; the two vortices tend to drag the surrounding air around with them, and this secondary movement induces a small velocity component in the downward direction at the wing. This downward component is called *downwash,* denoted by the symbol *w.* In turn, the downwash combines with the freestream velocity V_∞ to produce a *local* relative wind which is canted downward in the vicinity of each airfoil section of the wing, as sketched in Figure 5.6.

Examine Figure 5.6 closely. The angle between the chord line and the direction of V_∞ is the angle of attack α, as defined in Section 1.5 and as used throughout our discussion of airfoil theory in Chapter 4. We now more precisely define α as the *geometric* angle of attack. In Figure 5.6, the local relative wind is inclined

Figure 5.5 Wing-tip vortices from a finite wing. (*Steve Whiston – Fallen Log Photography/Getty Images*.)

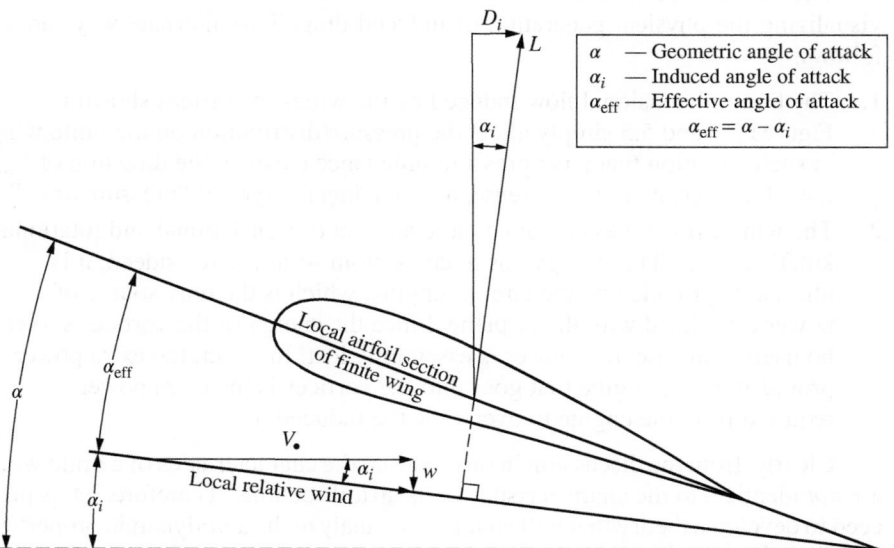

Figure 5.6 Effect of downwash on the local flow over a local airfoil section of a finite wing.

below the direction of V_∞ by the angle α_i, called the *induced* angle of attack. The presence of downwash, and its effect on inclining the local relative wind in the downward direction, has two important effects on the local airfoil section, as follows:

1. The angle of attack actually seen by the local airfoil section is the angle between the chord line and the local relative wind. This angle is given by α_{eff} in Figure 5.4 and is defined as the *effective* angle of attack. Hence, although the wing is at a geometric angle of attack α, the local airfoil section is seeing a smaller angle, namely, the effective angle of attack α_{eff}. From Figure 5.6,

$$\alpha_{\text{eff}} = \alpha - \alpha_i \tag{5.1}$$

2. The local lift vector is aligned perpendicular to the local relative wind, and hence is inclined behind the vertical by the angle α_i, as shown in Figure 5.6. Consequently, there is a component of the local lift vector in the direction of V_∞; that is, there is a *drag* created by the presence of downwash. This drag is defined as *induced drag,* denoted by D_i in Figure 5.6.

Hence, we see that the presence of downwash over a finite wing reduces the angle of attack that each section effectively sees, and moreover, it creates a component of drag—the induced drag D_i. Keep in mind that we are still dealing with an inviscid, incompressible flow, where there is no skin friction or flow separation. For such a flow, there is a *finite* drag—the induced drag—on a finite wing. d'Alembert's paradox does *not* occur for a finite wing.

The tilting backward of the lift vector shown in Figure 5.6 is one way of visualizing the physical generation of induced drag. Two alternate ways are as follows:

1. The three-dimensional flow induced by the wing-tip vortices shown in Figures 5.4 and 5.5 simply alters the pressure distribution on the finite wing in such a fashion that a net pressure imbalance exists in the direction of V_∞ (i.e., drag is created). In this sense, induced drag is a type of "pressure drag."
2. The wing-tip vortices contain a large amount of translational and rotational kinetic energy. This energy has to come from somewhere; indeed, it is ultimately provided by the aircraft engine, which is the only source of power associated with the airplane. Since the energy of the vortices serves no useful purpose, this power is essentially lost. In effect, the extra power provided by the engine that goes into the vortices is the extra power required from the engine to overcome the induced drag.

Clearly, from the discussion in this section, the characteristics of a finite wing are *not* identical to the characteristics of its airfoil sections. Therefore, let us proceed to develop a theory that will enable us to analyze the aerodynamic properties of finite wings. In the process, we follow the road map given in Figure 5.7—keep in touch with this road map as we progress through the present chapter.

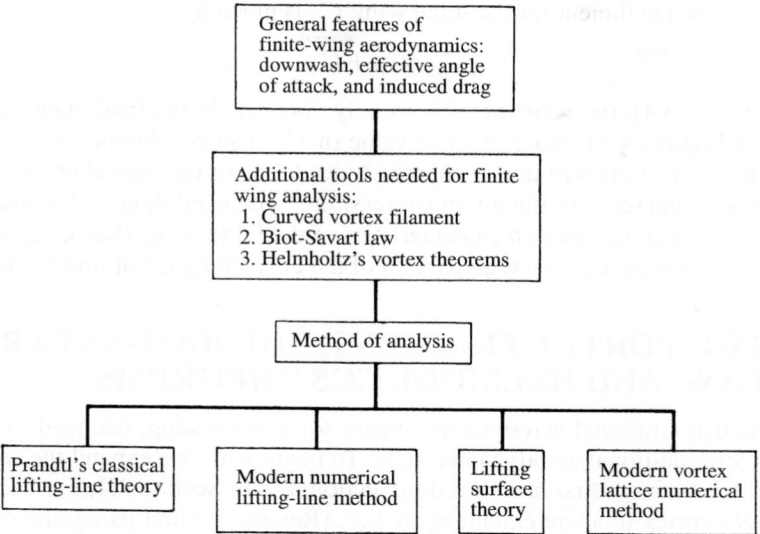

Figure 5.7 Road map for Chapter 5.

In this chapter, we note a difference in nomenclature. For the two-dimensional bodies considered in the previous chapters, the lift, drag, and moments per unit span have been denoted with primes, for example, L', D', and M', and the corresponding lift, drag, and moment coefficients have been denoted by lowercase letters, for example, c_l, c_d, and c_m. In contrast, the lift, drag, and moments on a complete three-dimensional body such as a finite wing are given without primes, for example, L, D, and M, and the corresponding lift, drag, and moment coefficients are given by capital letters, for example, C_L, C_D, and C_M. This distinction has already been mentioned in Section 1.5.

Finally, we note that the total drag on a subsonic finite wing in real life is the sum of the induced drag D_i, the skin friction drag D_f, and the pressure drag D_p due to flow separation. The latter two contributions are due to viscous effects (see Section 4.12 and Chapters 15 to 20). The sum of these two viscous-dominated drag contributions is called profile drag, as discussed in Section 4.3. The profile drag coefficient c_d for an NACA 2412 airfoil was given in Figure 4.11. At moderate angle of attack, the profile drag coefficient for a finite wing is essentially the same as for its airfoil sections. Hence, defining the profile drag coefficient as

$$c_d = \frac{D_f + D_p}{q_\infty S} \tag{5.2}$$

and the induced drag coefficient as

$$C_{D,i} = \frac{D_i}{q_\infty S} \tag{5.3}$$

the total drag coefficient for the finite wing C_D is given by

$$C_D = c_d + C_{D,i} \tag{5.4}$$

In Equation (5.4), the value of c_d is usually obtained from airfoil data, such as given in Figures 4.11 and 5.2*b*. The value of $C_{D,i}$ can be obtained from finite-wing theory as presented in this chapter. Indeed, one of the central objectives of the present chapter is to obtain an expression for induced drag and to study its variation with certain design characteristics of the finite wing. (See Chapter 5 of Reference 2 for an additional discussion of the characteristics of finite wings.)

5.2 THE VORTEX FILAMENT, THE BIOT-SAVART LAW, AND HELMHOLTZ'S THEOREMS

To establish a rational aerodynamic theory for a finite wing, we need to introduce a few additional aerodynamic tools. To begin with, we expand the concept of a vortex filament first introduced in Section 4.4. In Section 4.4, we discussed a *straight* vortex filament extending to $\pm\infty$. (Review the first paragraph of Section 4.4 before proceeding further.)

In general, a vortex filament can be *curved,* as shown in Figure 5.8. Here, only a portion of the filament is illustrated. The filament induces a flow field in the surrounding space. If the circulation is taken about any path enclosing the filament, a constant value Γ is obtained. Hence, the strength of the vortex filament is defined as Γ. Consider a directed segment of the filament **dl**, as shown in Figure 5.8. The radius vector from **dl** to an arbitrary point P in space is **r**. The segment **dl** induces a velocity at P equal to

$$\boxed{d\mathbf{V} = \frac{\Gamma}{4\pi} \frac{d\mathbf{l} \times \mathbf{r}}{|\mathbf{r}|^3}} \tag{5.5}$$

Equation (5.5) is called the *Biot-Savart law* and is one of the most fundamental relations in the theory of inviscid, incompressible flow. Its derivation is given in more advanced books (see, e.g., Reference 9). Here, we must accept it without

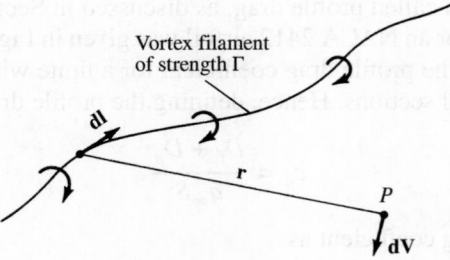

Vortex filament of strength Γ

Figure 5.8 Vortex filament and illustration of the Biot-Savart law.

proof. However, you might feel more comfortable if we draw an analogy with electromagnetic theory. If the vortex filament in Figure 5.8 were instead visualized as a wire carrying an electrical current I, then the magnetic field strength $\mathbf{dB}$ induced at point P by a segment of the wire $\mathbf{dl}$ with the current moving in the direction of $\mathbf{dl}$ is

$$\mathbf{dB} = \frac{\mu I}{4\pi} \frac{\mathbf{dl} \times \mathbf{r}}{|\mathbf{r}|^3} \tag{5.6}$$

where μ is the permeability of the medium surrounding the wire. Equation (5.6) is identical in form to Equation (5.5). Indeed, the Biot-Savart law is a general result of potential theory, and potential theory describes electromagnetic fields as well as inviscid, incompressible flows. In fact, our use of the word "induced" in describing velocities generated by the presence of vortices and sources is a carry over from the study of electromagnet fields induced by electrical currents. When developing their finite-wing theory during the period 1911–1918, Prandtl and his colleagues even carried the electrical terminology over to the generation of drag, hence the term "induced" drag.

Return again to our picture of the vortex filament in Figure 5.8. Keep in mind that this single vortex filament and the associated Biot-Savart law [Equation (5.5)] are simply conceptual aerodynamic tools to be used for synthesizing more complex flows of an inviscid, incompressible fluid. They are, for all practical purposes, a solution of the governing equation for inviscid, incompressible flow— Laplace's equation (see Section 3.7)—and, by themselves, are not of particular value. However, when a number of vortex filaments are used in conjunction with a uniform freestream, it is possible to synthesize a flow which has a practical application. The flow over a finite wing is one such example, as we will soon see.

Let us apply the Biot-Savart law to a straight vortex filament of infinite length, as sketched in Figure 5.9. The strength of the filament is Γ. The velocity induced

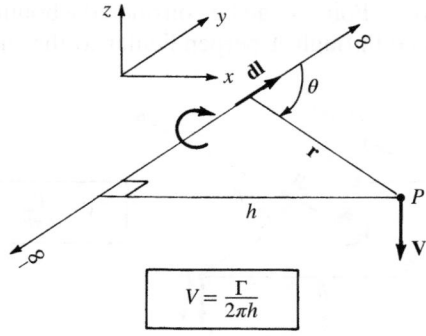

Figure 5.9 Velocity induced at point P by an infinite, straight vortex filament.

at point P by the directed segment of the vortex filament dl is given by Equation (5.5). Hence, the velocity induced at P by the entire vortex filament is

$$\mathbf{V} = \int_{-\infty}^{\infty} \frac{\Gamma}{4\pi} \frac{\mathbf{dl} \times \mathbf{r}}{|\mathbf{r}|^3} \qquad (5.7)$$

From the definition of the vector cross product (see Section 2.2), the direction of $\mathbf{V}$ is downward in Figure 5.9. The magnitude of the velocity, $V = |\mathbf{V}|$, is given by

$$V = \frac{\Gamma}{4\pi} \int_{-\infty}^{\infty} \frac{\sin \theta}{r^2} \, dl \qquad (5.8)$$

In Figure 5.9, let h be the perpendicular distance from point P to the vortex filament. Then, from the geometry shown in Figure 5.9,

$$r = \frac{h}{\sin \theta} \qquad (5.9a)$$

$$l = \frac{h}{\tan \theta} \qquad (5.9b)$$

$$dl = -\frac{h}{\sin^2 \theta} \, d\theta \qquad (5.9c)$$

Substituting Equation (5.9a to c) in Equation (5.8), we have

$$V = \frac{\Gamma}{4\pi} \int_{-\infty}^{\infty} \frac{\sin \theta}{r^2} dl = -\frac{\Gamma}{4\pi h} \int_{\pi}^{0} \sin \theta \, d\theta$$

or

$$V = \frac{\Gamma}{2\pi h} \qquad (5.10)$$

Thus, the velocity induced at a given point P by an infinite, straight vortex filament at a perpendicular distance h from P is simply $\Gamma/2\pi h$, which is precisely the result given by Equation (3.105) for a point vortex in two-dimensional flow. [Note that the minus sign in Equation (3.105) does not appear in Equation (5.10); this is because V in Equation (5.10) is simply the absolute magnitude of V, and hence it is positive by definition.]

Consider the *semi*-infinite vortex filament shown in Figure 5.10. The filament extends from point A to ∞. Point A can be considered a boundary of the flow. Let P be a point in the plane through A perpendicular to the filament. Then, by an

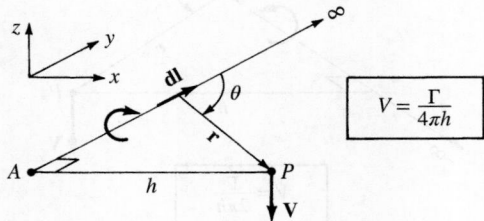

Figure 5.10 Velocity induced at point P by a semi-infinite straight vortex filament.

integration similar to that above (try it yourself), the velocity induced at P by the semi-infinite vortex filament is

$$V = \frac{\Gamma}{4\pi h} \tag{5.11}$$

We use Equation (5.11) in the next section.

The great German mathematician, physicist, and physician Hermann von Helmholtz (1821–1894) was the first to make use of the vortex filament concept in the analysis of inviscid, incompressible flow. In the process, he established several basic principles of vortex behavior which have become known as Helmholtz's vortex theorems:

1. The strength of a vortex filament is constant along its length.
2. A vortex filament cannot end in a fluid; it must extend to the boundaries of the fluid (which can be $\pm\infty$) or form a closed path.

We make use of these theorems in the following sections.

Finally, let us introduce the concept of *lift distribution* along the span of a finite wing. Consider a given spanwise location y_1, where the local chord is c, the local geometric angle of attack is α, and the airfoil section is a given shape. The lift per unit span at this location is $L'(y_1)$. Now consider another location y_2 along the span, where c, α, and the airfoil shape may be different. (Most finite wings have a variable chord, with the exception of a simple rectangular wing. Also, many wings are geometrically twisted so that α is different at different spanwise locations—so-called geometric twist. If the tip is at a lower α than the root, the wing is said to have *washout;* if the tip is at a higher α than the root, the wing has *washin.* In addition, the wings on a number of modern airplanes have different airfoil sections along the span, with different values of $\alpha_{L=0}$; this is called *aerodynamic twist.*) Consequently, the lift per unit span at this different location, $L'(y_2)$, will, in general, be different from $L'(y_1)$. Therefore, there is a distribution of lift per unit span along the wing, that is, $L' = L'(y)$, as sketched in Figure 5.11. In turn, the circulation is also a function of y, $\Gamma(y) = L'(y)/\rho_\infty V_\infty$. Note from Figure 5.11 that the lift distribution goes to zero at the tips; that is because there is a pressure equalization from the bottom to the top of the wing precisely at $y = -b/2$ and $b/2$, and hence no lift is created at these points. The calculation of

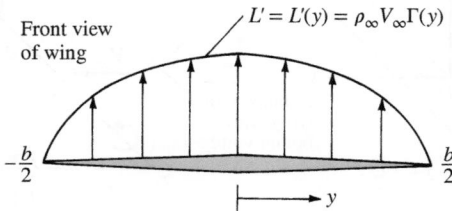

Figure 5.11 Sketch of the lift distribution along the span of a wing.

the lift distribution $L(y)$ [or the circulation distribution $\Gamma(y)$] is one of the central problems of finite-wing theory. It is addressed in the following sections.

In summary, we wish to calculate the induced drag, the total lift, and the lift distribution for a finite wing. This is the purpose of the remainder of this chapter.

5.3 PRANDTL'S CLASSICAL LIFTING-LINE THEORY

The first practical theory for predicting the aerodynamic properties of a finite wing was developed by Ludwig Prandtl and his colleagues at Göttingen, Germany, during the period 1911–1918, spanning World War I. The utility of Prandtl's theory is so great that it is still in use today for preliminary calculations of finite-wing characteristics. The purpose of this section is to describe Prandtl's theory and to lay the groundwork for the modern numerical methods described in subsequent sections.

Prandtl reasoned as follows. A vortex filament of strength Γ that is somehow bound to a fixed location in a flow—a so-called bound vortex—will experience a force $L' = \rho_\infty V_\infty \Gamma$ from the Kutta-Joukowski theorem. This bound vortex is in contrast to a *free vortex,* which moves with the same fluid elements throughout a flow. Therefore, let us replace a finite wing of span b with a bound vortex, extending from $y = -b/2$ to $y = b/2$, as sketched in Figure 5.12. However, due to Helmholtz's theorem, a vortex filament cannot end in the fluid. Therefore, assume the vortex filament continues as two free vortices trailing downstream from the wing tips to infinity, as also shown in Figure 5.12. This vortex (the bound plus the two free) is in the shape of a horseshoe, and therefore is called a *horseshoe vortex.*

A single horseshoe vortex is shown in Figure 5.13. Consider the downwash w induced along the bound vortex from $-b/2$ to $b/2$ by the horseshoe vortex. Examining Figure 5.13, we see that the bound vortex induces no velocity along itself; however, the two trailing vortices both contribute to the induced velocity

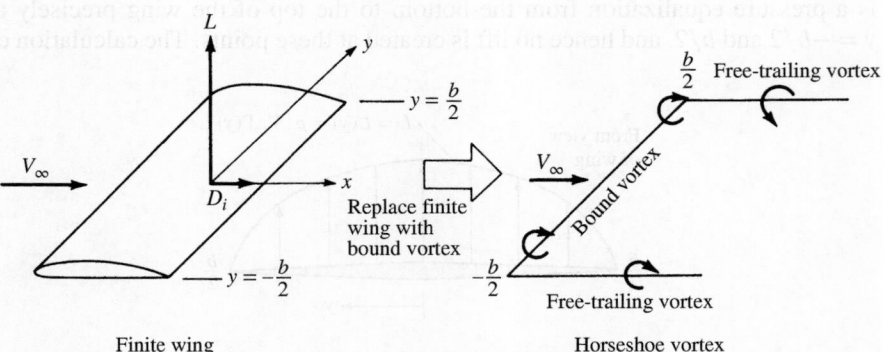

Finite wing   Horseshoe vortex

Figure 5.12 Replacement of the finite wing with a bound vortex.

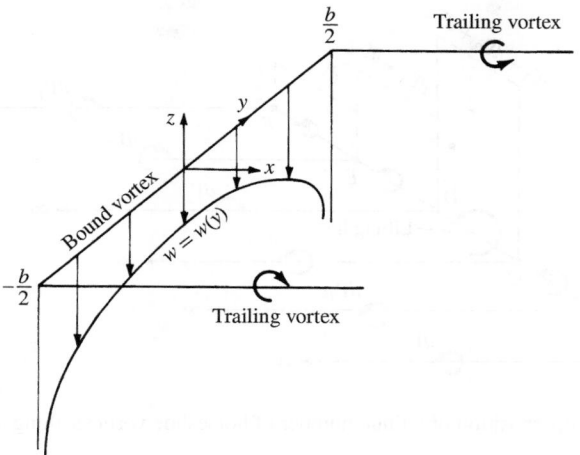

Figure 5.13 Downwash distribution along the y axis for a single horseshoe vortex.

along the bound vortex, and both contributions are in the downward direction. Consistent with the xyz coordinate system in Figure 5.13, such a downward velocity is negative; that is, w (which is in the z direction) is a negative value when directed downward and a positive value when directed upward. If the origin is taken at the center of the bound vortex, then the velocity at any point y along the bound vortex induced by the trailing semi-infinite vortices is, from Equation (5.11),

$$w(y) = -\frac{\Gamma}{4\pi(b/2 + y)} - \frac{\Gamma}{4\pi(b/2 - y)} \qquad (5.12)$$

In Equation (5.12), the first term on the right-hand side is the contribution from the left trailing vortex (trailing from $-b/2$), and the second term is the contribution from the right trailing vortex (trailing from $b/2$). Equation (5.12) reduces to

$$w(y) = -\frac{\Gamma}{4\pi} \frac{b}{(b/2)^2 - y^2} \qquad (5.13)$$

This variation of $w(y)$ is sketched in Figure 5.13. Note that w approaches $-\infty$ as y approaches $-b/2$ or $b/2$.

The downwash distribution due to the single horseshoe vortex shown in Figure 5.13 does not realistically simulate that of a finite wing; the downwash approaching an infinite value at the tips is especially disconcerting. During the early evolution of finite-wing theory, this problem perplexed Prandtl and his colleagues. After several years of effort, a resolution of this problem was obtained which, in hindsight, was simple and straightforward. Instead of representing the wing by a single horseshoe vortex, let us superimpose a large number of horseshoe vortices, each with a different length of the bound vortex, but with all the

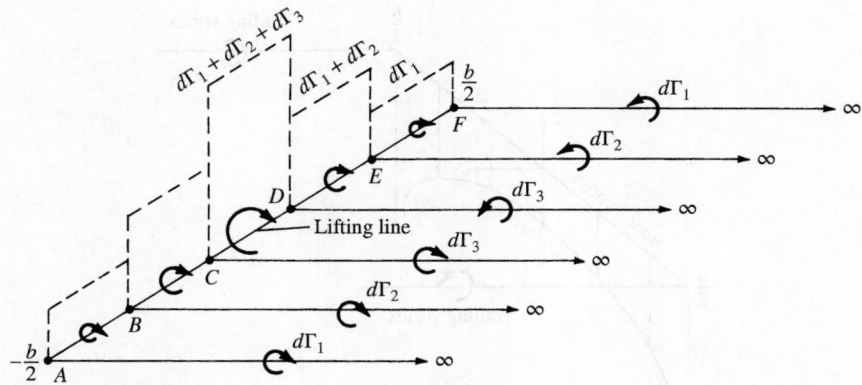

Figure 5.14 Superposition of a finite number of horseshoe vortices along the lifting line.

bound vortices coincident along a single line, called the *lifting line*. This concept is illustrated in Figure 5.14, where only three horseshoe vortices are shown for the sake of clarity. In Figure 5.14, a horseshoe vortex of strength $d\Gamma_1$ is shown, where the bound vortex spans the entire wing from $-b/2$ to $b/2$ (from point A to point F). Superimposed on this is a second horseshoe vortex of strength $d\Gamma_2$, where its bound vortex spans only part of the wing, from point B to point E. Finally, superimposed on this is a third horseshoe vortex of strength $d\Gamma_3$, where its bound vortex spans only the part of the wing from point C to point D. As a result, the circulation varies along the line of bound vortices—the lifting line defined above. Along AB and EF, where only one vortex is present, the circulation is $d\Gamma_1$. However, along BC and DE, where two vortices are superimposed, the circulation is the sum of their strengths $d\Gamma_1 + d\Gamma_2$. Along CD, three vortices are superimposed, and hence the circulation is $d\Gamma_1 + d\Gamma_2 + d\Gamma_3$. This variation of Γ along the lifting line is denoted by the vertical bars in Figure 5.14. Also, note from Figure 5.14 that we now have a series of trailing vortices distributed over the span, rather than just two vortices trailing downstream of the tips as shown in Figure 5.13. The series of trailing vortices in Figure 5.14 represents pairs of vortices, each pair associated with a given horseshoe vortex. Note that the strength of each trailing vortex is equal to the *change in circulation* along the lifting line.

Let us extrapolate Figure 5.14 to the case where an *infinite number* of horseshoe vortices are superimposed along the lifting line, each with a vanishingly small strength $d\Gamma$. This case is illustrated in Figure 5.15. Note that the vertical bars in Figure 5.14 have now become a continuous distribution of $\Gamma(y)$ along the lifting line in Figure 5.15. The value of the circulation at the origin is Γ_0. Also, note that the finite number of trailing vortices in Figure 5.14 have become a *continuous vortex sheet* trailing downstream of the lifting line in Figure 5.15. This vortex sheet is parallel to the direction of V_∞. The total strength of the sheet integrated across the span of the wing is zero, because it consists of pairs of trailing vortices of equal strength but in opposite directions.

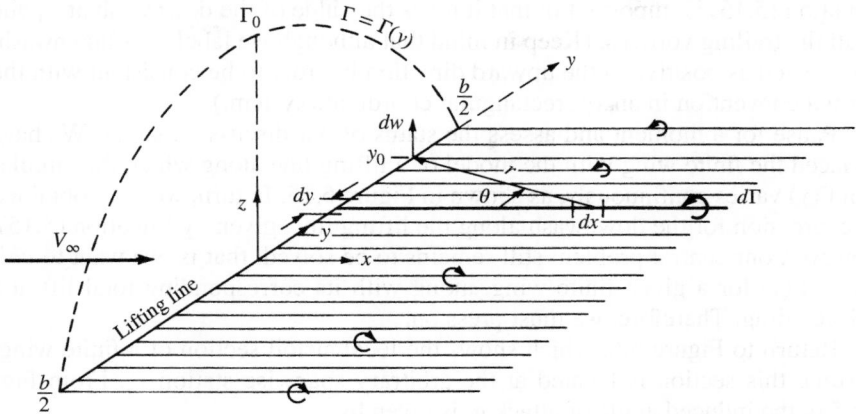

Figure 5.15 Superposition of an infinite number of horseshoe vortices along the lifting line.

Let us single out an infinitesimally small segment of the lifting line dy located at the coordinate y as shown in Figure 5.15. The circulation at y is $\Gamma(y)$, and the change in circulation over the segment dy is $d\Gamma = (d\Gamma/dy)\,dy$. In turn, the strength of the trailing vortex at y must equal the change in circulation $d\Gamma$ along the lifting line; this is simply an extrapolation of our result obtained for the strength of the finite trailing vortices in Figure 5.14. Consider more closely the trailing vortex of strength $d\Gamma$ that intersects the lifting line at coordinate y, as shown in Figure 5.15. Also consider the arbitrary location y_0 along the lifting line. Any segment of the trailing vortex dx will induce a velocity at y_0 with a magnitude and direction given by the Biot-Savart law, Equation (5.5). In turn, the velocity dw at y_0 induced by the entire semi-infinite trailing vortex located at y is given by Equation (5.11), which in terms of the picture given in Figure 5.15 yields

$$dw = -\frac{(d\Gamma/dy)\,dy}{4\pi(y_0 - y)} \tag{5.14}$$

The minus sign in Equation (5.14) is needed for consistency with the picture shown in Figure 5.15; for the trailing vortex shown, the direction of dw at y_0 is upward and hence is a positive value, whereas Γ is decreasing in the y direction, making $d\Gamma/dy$ a negative quantity. The minus sign in Equation (5.14) makes the positive dw consistent with the negative $d\Gamma/dy$.

The total velocity w induced at y_0 by the entire trailing vortex sheet is the summation of Equation (5.14) over all the vortex filaments, that is, the integral of Equation (5.14) from $-b/2$ to $b/2$:

$$w(y_0) = -\frac{1}{4\pi}\int_{-b/2}^{b/2}\frac{(d\Gamma/dy)\,dy}{y_0 - y} \tag{5.15}$$

Equation (5.15) is important in that it gives the value of the downwash at y_0 due to all the trailing vortices. (Keep in mind that although we label w as downwash, w is treated as positive in the upward direction in order to be consistent with the normal convention in an xyz rectangular coordinate system.)

Pause for a moment and assess the status of our discussion so far. We have replaced the finite wing with the model of a lifting line along which the circulation $\Gamma(y)$ varies continuously, as shown in Figure 5.15. In turn, we have obtained an expression for the downwash along the lifting line, given by Equation (5.15). However, our central problem still remains to be solved; that is, we want to *calculate* $\Gamma(y)$ for a given finite wing, along with its corresponding total lift and induced drag. Therefore, we must press on.

Return to Figure 5.6, which shows the local airfoil section of a finite wing. Assume this section is located at the arbitrary spanwise station y_0. From Figure 5.6, the induced angle of attack α_i is given by

$$\alpha_i(y_0) = \tan^{-1}\left(\frac{-w(y_0)}{V_\infty}\right) \tag{5.16}$$

[Note in Figure 5.6 that w is downward, and hence is a negative quantity. Since α_i in Figure 5.6 is positive, the negative sign in Equation (5.16) is necessary for consistency.] Generally, w is much smaller than V_∞, and hence α_i is a small angle, on the order of a few degrees at most. For small angles, Equation (5.16) yields

$$\alpha_i(y_0) = -\frac{w(y_0)}{V_\infty} \tag{5.17}$$

Substituting Equation (5.15) into (5.17), we obtain

$$\boxed{\alpha_i(y_0) = \frac{1}{4\pi V_\infty}\int_{-b/2}^{b/2}\frac{(d\Gamma/dy)\,dy}{y_0 - y}} \tag{5.18}$$

that is, an expression for the induced angle of attack in terms of the circulation distribution $\Gamma(y)$ along the wing.

Consider again the *effective* angle of attack α_{eff}, as shown in Figure 5.6. As explained in Section 5.1, α_{eff} is the angle of attack actually seen by the local airfoil section. Since the downwash varies across the span, then α_{eff} is also variable; $\alpha_{\text{eff}} = \alpha_{\text{eff}}(y_0)$. The lift coefficient for the airfoil section located at $y = y_0$ is

$$c_l = a_0[\alpha_{\text{eff}}(y_0) - \alpha_{L=0}] = 2\pi[\alpha_{\text{eff}}(y_0) - \alpha_{L=0}] \tag{5.19}$$

In Equation (5.19), the local section lift slope a_0 has been replaced by the thin airfoil theoretical value of $2\pi(\text{rad}^{-1})$. Also, for a wing with aerodynamic twist, the angle of zero lift $\alpha_{L=0}$ in Equation (5.19) varies with y_0. If there is no aerodynamic twist, $\alpha_{L=0}$ is constant across the span. In any event, $\alpha_{L=0}$ is a known property of the local airfoil sections. From the definition of lift coefficient and from the Kutta-Joukowski theorem, we have, for the local airfoil section located at y_0,

$$L' = \tfrac{1}{2}\rho_\infty V_\infty^2 c(y_0)c_l = \rho_\infty V_\infty \Gamma(y_0) \tag{5.20}$$

From Equation (5.20), we obtain

$$c_l = \frac{2\Gamma(y_0)}{V_\infty c(y_0)} \tag{5.21}$$

Substituting Equation (5.21) into (5.19) and solving for α_{eff}, we have

$$\alpha_{\text{eff}} = \frac{\Gamma(y_0)}{\pi V_\infty c(y_0)} + \alpha_{L=0} \tag{5.22}$$

The above results come into focus if we refer to Equation (5.1):

$$\alpha_{\text{eff}} = \alpha - \alpha_i \tag{5.1}$$

Substituting Equations (5.18) and (5.22) into (5.1), we obtain

$$\boxed{\alpha(y_0) = \frac{\Gamma(y_0)}{\pi V_\infty c(y_0)} + \alpha_{L=0}(y_0) + \frac{1}{4\pi V_\infty} \int_{-b/2}^{b/2} \frac{(d\Gamma/dy)\,dy}{y_0 - y}} \tag{5.23}$$

the *fundamental equation of Prandtl's lifting-line theory;* it simply states that the geometric angle of attack is equal to the sum of the effective angle plus the induced angle of attack. In Equation (5.23), α_{eff} is expressed in terms of Γ, and α_i is expressed in terms of an integral containing $d\Gamma/dy$. Hence, Equation (5.23) is an integro-differential equation, in which the only unknown is Γ; all the other quantities, α, c, V_∞, and $\alpha_{L=0}$, are known for a finite wing of given design at a given geometric angle of attack in a freestream with given velocity. Thus, a solution of Equation (5.23) yields $\Gamma = \Gamma(y_0)$, where y_0 ranges along the span from $-b/2$ to $b/2$.

The solution $\Gamma = \Gamma(y_0)$ obtained from Equation (5.23) gives us the three main aerodynamic characteristics of a finite wing, as follows:

1. The lift distribution is obtained from the Kutta-Joukowski theorem:

$$L'(y_0) = \rho_\infty V_\infty \Gamma(y_0) \tag{5.24}$$

2. The total lift is obtained by integrating Equation (5.24) over the span:

$$L = \int_{-b/2}^{b/2} L'(y)\,dy$$

or

$$L = \rho_\infty V_\infty \int_{-b/2}^{b/2} \Gamma(y)\,dy \tag{5.25}$$

(Note that we have dropped the subscript on y, for simplicity.) The lift coefficient follows immediately from Equation (5.25):

$$C_L = \frac{L}{q_\infty S} = \frac{2}{V_\infty S} \int_{-b/2}^{b/2} \Gamma(y)\,dy \tag{5.26}$$

3. The induced drag is obtained by inspection of Figure 5.4. The induced drag per unit span is

$$D_i' = L_i' \sin \alpha_i$$

Since α_i is small, this relation becomes

$$D_i' = L_i' \alpha_i \tag{5.27}$$

The total induced drag is obtained by integrating Equation (5.27) over the span:

$$D_i = \int_{-b/2}^{b/2} L'(y) \alpha_i(y) \, dy \tag{5.28}$$

or

$$D_i = \rho_\infty V_\infty \int_{-b/2}^{b/2} \Gamma(y) \alpha_i(y) \, dy \tag{5.29}$$

In turn, the induced drag coefficient is

$$C_{D,i} = \frac{D_i}{q_\infty S} = \frac{2}{V_\infty S} \int_{-b/2}^{b/2} \Gamma(y) \alpha_i(y) \, dy \tag{5.30}$$

In Equations (5.27) to (5.30), $\alpha_i(y)$ is obtained from Equation (5.18). Therefore, in Prandtl's lifting-line theory, the solution of Equation (5.23) for $\Gamma(y)$ is clearly the key to obtaining the aerodynamic characteristics of a finite wing. Before discussing the general solution of this equation, let us consider a special case, as outlined below.

5.3.1 Elliptical Lift Distribution

Consider a circulation distribution given by

$$\Gamma(y) = \Gamma_0 \sqrt{1 - \left(\frac{2y}{b}\right)^2} \tag{5.31}$$

In Equation (5.31), note the following:

1. Γ_0 is the circulation at the origin, as shown in Figure 5.15.
2. The circulation varies elliptically with distance y along the span; hence, it is designated as an *elliptical circulation distribution*. Since $L'(y) = \rho_\infty V_\infty \Gamma(y)$, we also have

$$L'(y) = \rho_\infty V_\infty \Gamma_0 \sqrt{1 - \left(\frac{2y}{b}\right)^2}$$

Hence, we are dealing with an *elliptical lift distribution*.

3. $\Gamma(b/2) = \Gamma(-b/2) = 0$. Thus, the circulation, hence lift, properly goes to zero at the wing tips, as shown in Figure 5.15. We have not obtained Equation (5.31) as a direct solution of Equation (5.23); rather, we are simply stipulating a lift distribution that is elliptic. We now ask the

question, "What are the aerodynamic properties of a finite wing with such an elliptic lift distribution?"

First, let us calculate the downwash. Differentiating Equation (5.31), we obtain

$$\frac{d\Gamma}{dy} = -\frac{4\Gamma_0}{b^2}\frac{y}{(1 - 4y^2/b^2)^{1/2}} \tag{5.32}$$

Substituting Equation (5.32) into (5.15), we have

$$w(y_0) = \frac{\Gamma_0}{\pi b^2}\int_{-b/2}^{b/2}\frac{y}{(1 - 4y^2/b^2)^{1/2}(y_0 - y)}\,dy \tag{5.33}$$

The integral can be evaluated easily by making the substitution

$$y = \frac{b}{2}\cos\theta \quad dy = -\frac{b}{2}\sin\theta\,d\theta$$

Hence, Equation (5.33) becomes

$$w(\theta_0) = -\frac{\Gamma_0}{2\pi b}\int_{\pi}^{0}\frac{\cos\theta}{\cos\theta_0 - \cos\theta}\,d\theta$$

or

$$w(\theta_0) = -\frac{\Gamma_0}{2\pi b}\int_{0}^{\pi}\frac{\cos\theta}{\cos\theta - \cos\theta_0}\,d\theta \tag{5.34}$$

The integral in Equation (5.34) is the standard form given by Equation (4.26) for $n = 1$. Hence, Equation (5.34) becomes

$$\boxed{w(\theta_0) = -\frac{\Gamma_0}{2b}} \tag{5.35}$$

which states the interesting and important result that the *downwash is constant over the span for an elliptical lift distribution*. In turn, from Equation (5.17), we obtain, for the induced angle of attack,

$$\alpha_i = -\frac{w}{V_\infty} = \frac{\Gamma_0}{2bV_\infty} \tag{5.36}$$

For an elliptic lift distribution, the induced angle of attack is also constant along the span. Note from Equations (5.35) and (5.36) that both the downwash and induced angle of attack go to zero as the wing span becomes infinite—which is consistent with our previous discussions on airfoil theory.

A more useful expression for α_i can be obtained as follows. Substituting Equation (5.31) into (5.25), we have

$$L = \rho_\infty V_\infty \Gamma_0 \int_{-b/2}^{b/2}\left(1 - \frac{4y^2}{b^2}\right)^{1/2}\,dy \tag{5.37}$$

Again, using the transformation $y = (b/2)\cos\theta$, Equation (5.37) readily integrates to

$$L = \rho_\infty V_\infty \Gamma_0 \frac{b}{2} \int_0^\pi \sin^2\theta\, d\theta = \rho_\infty V_\infty \Gamma_0 \frac{b}{4}\pi \tag{5.38}$$

Solving Equation (5.38) for Γ_0, we have

$$\Gamma_0 = \frac{4L}{\rho_\infty V_\infty b\pi} \tag{5.39}$$

However, $L = \frac{1}{2}\rho_\infty V_\infty^2 S C_L$. Hence, Equation (5.39) becomes

$$\Gamma_0 = \frac{2V_\infty S C_L}{b\pi} \tag{5.40}$$

Substituting Equation (5.40) into (5.36), we obtain

$$\alpha_i = \frac{2V_\infty S C_L}{b\pi}\frac{1}{2bV_\infty}$$

or

$$\alpha_i = \frac{S C_L}{\pi b^2} \tag{5.41}$$

An important geometric property of a finite wing is the *aspect ratio,* denoted by AR and defined as

$$AR \equiv \frac{b^2}{S}$$

Hence, Equation (5.41) becomes

$$\boxed{\alpha_i = \frac{C_L}{\pi AR}} \tag{5.42a}$$

Equation (5.42a) is a useful expression for the induced angle of attack, as shown below.

The induced drag coefficient is obtained from Equation (5.30), noting that α_i is constant:

$$C_{D,i} = \frac{2\alpha_i}{V_\infty S}\int_{-b/2}^{b/2}\Gamma(y)\,dy = \frac{2\alpha_i\Gamma_0}{V_\infty S}\frac{b}{2}\int_0^\pi \sin^2\theta\, d\theta = \frac{\pi\alpha_i\Gamma_0 b}{2V_\infty S} \tag{5.42b}$$

Substituting Equations (5.40) and (5.42a) into (5.42b), we obtain

$$C_{D,i} = \frac{\pi b}{2V_\infty S}\left(\frac{C_L}{\pi AR}\right)\frac{2V_\infty S C_L}{b\pi}$$

or

$$\boxed{C_{D,i} = \frac{C_L^2}{\pi AR}} \tag{5.43}$$

Equation (5.43) is an important result. It states that the induced drag coefficient is directly proportional to the square of the lift coefficient. The dependence of induced drag on the lift is not surprising for the following reason. In Section 5.1, we saw that induced drag is a consequence of the presence of the wing-tip vortices, which in turn are produced by the difference in pressure between the lower and upper wing surfaces. The lift is produced by this same pressure difference. Hence, induced drag is intimately related to the production of lift on a finite wing; indeed, induced drag is frequently called the *drag due to lift*. Equation (5.43) dramatically illustrates this point. Clearly, an airplane cannot generate lift for free; the induced drag is the price for the generation of lift. The power required from an aircraft engine to overcome the induced drag is simply the power required to generate the lift of the aircraft. Also, note that because $C_{D,i} \propto C_L^2$, the induced drag coefficient increases rapidly as C_L increases and becomes a substantial part of the total drag coefficient when C_L is high (e.g., when the airplane is flying slowly, such as on takeoff or landing). Even at relatively high cruising speeds, induced drag is typically 25 percent of the total drag.

Another important aspect of induced drag is evident in Equation (5.43); that is, $C_{D,i}$ is *inversely proportional to aspect ratio*. Hence, to reduce the induced drag, we want a finite wing with the highest possible aspect ratio. Wings with high and low aspect ratios are sketched in Figure 5.16. Unfortunately, the design of very high aspect ratio wings with sufficient structural strength is difficult. Therefore, the aspect ratio of a conventional aircraft is a compromise between conflicting aerodynamic and structural requirements. It is interesting to note that the aspect ratio of the 1903 Wright Flyer was 6 and that today the aspect ratios of conventional subsonic aircraft range typically from 6 to 8. (Exceptions are the Lockheed U-2 high-altitude reconnaissance aircraft with AR = 14.3 and sailplanes with aspect ratios as high as 51. For example, the Schempp-Hirth

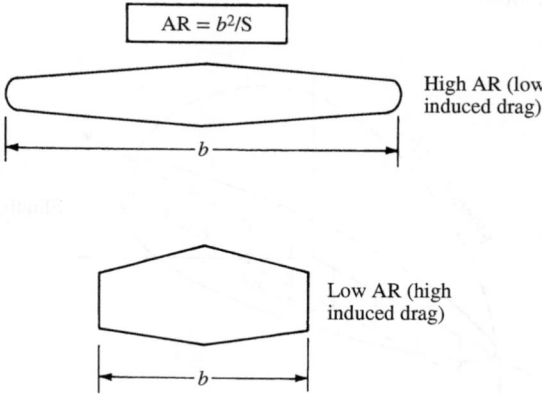

Figure 5.16 Schematic of high- and low-aspect-ratio wings.

Nimbus 4 sailplane, designed in 1994 with over 100 built by 2004, has an aspect ratio of 39. The ETA sailplane, designed in 2000 with 6 built by 2004, has an aspect ratio of 51.3.)

Another property of the elliptical lift distribution is as follows. Consider a wing with no geometric twist (i.e., α is constant along the span) and no aerodynamic twist (i.e., $\alpha_{L=0}$ is constant along the span). From Equation (5.36), we have seen that α_i is constant along the span. Hence, $\alpha_{\text{eff}} = \alpha - \alpha_i$ is also constant along the span. Since the local section lift coefficient c_l is given by

$$c_l = a_0(\alpha_{\text{eff}} - \alpha_{L=0})$$

then assuming that a_0 is the same for each section ($a_0 = 2\pi$ from thin airfoil theory), c_l must be constant along the span. The lift per unit span is given by

$$L'(y) = q_\infty c c_l \tag{5.44}$$

Solving Equation (5.44) for the chord, we have

$$c(y) = \frac{L'(y)}{q_\infty c_l} \tag{5.45}$$

In Equation (5.45), q_∞ and c_l are constant along the span. However, $L'(y)$ varies elliptically along the span. Thus, Equation (5.45) dictates that for such an elliptic lift distribution, the *chord must vary elliptically along the span;* that is, for the conditions given above, the *wing planform is elliptical.*

The related characteristics—the elliptic lift distribution, the elliptic planform, and the constant downwash—are sketched in Figure 5.17. Although an elliptical lift distribution may appear to be a restricted, isolated case, in reality it gives a reasonable approximation for the induced drag coefficient for an arbitrary finite wing. The form of $C_{D,i}$ given by Equation (5.43) is only slightly modified for the general case. Let us now consider the case of a finite wing with a general lift distribution.

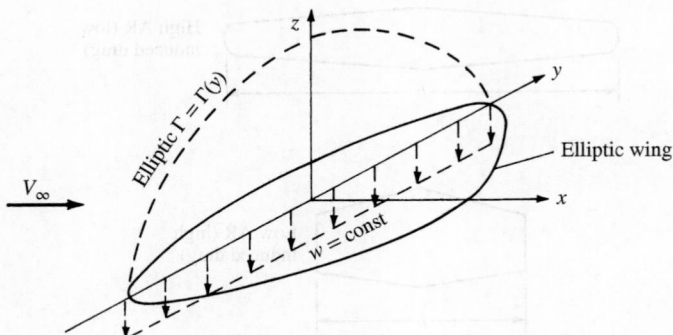

Figure 5.17 Illustration of the related quantities: an elliptic lift distribution, elliptic planform, and constant downwash.

5.3.2 General Lift Distribution

Consider the transformation

$$y = -\frac{b}{2}\cos\theta \tag{5.46}$$

where the coordinate in the spanwise direction is now given by θ, with $0 \le \theta \le \pi$. In terms of θ, the elliptic lift distribution given by Equation (5.31) is written as

$$\Gamma(\theta) = \Gamma_0 \sin\theta \tag{5.47}$$

Equation (5.47) hints that a Fourier sine series would be an appropriate expression for the general circulation distribution along an arbitrary finite wing. Hence, assume for the general case that

$$\Gamma(\theta) = 2bV_\infty \sum_1^N A_n \sin n\theta \tag{5.48}$$

where as many terms N in the series can be taken as we desire for accuracy. The coefficients A_n (where $n = 1, \dots, N$) in Equation (5.48) are unknowns; however, they must satisfy the fundamental equation of Prandtl's lifting-line theory; that is, the A_n's must satisfy Equation (5.23). Differentiating Equation (5.48), we obtain

$$\frac{d\Gamma}{dy} = \frac{d\Gamma}{d\theta}\frac{d\theta}{dy} = 2bV_\infty \sum_1^N nA_n \cos n\theta \frac{d\theta}{dy} \tag{5.49}$$

Substituting Equations (5.48) and (5.49) into (5.23), we obtain

$$\alpha(\theta_0) = \frac{2b}{\pi c(\theta_0)}\sum_1^N A_n \sin n\theta_0 + \alpha_{L=0}(\theta_0) + \frac{1}{\pi}\int_0^\pi \frac{\sum_1^N nA_n \cos n\theta}{\cos\theta - \cos\theta_0}d\theta \tag{5.50}$$

The integral in Equation (5.50) is the standard form given by Equation (4.26). Hence, Equation (5.50) becomes

$$\boxed{\alpha(\theta_0) = \frac{2b}{\pi c(\theta_0)}\sum_1^N A_n \sin n\theta_0 + \alpha_{L=0}(\theta_0) + \sum_1^N nA_n \frac{\sin n\theta_0}{\sin\theta_0}} \tag{5.51}$$

Examine Equation (5.51) closely. It is evaluated at a given spanwise location; hence, θ_0 is specified. In turn, b, $c(\theta_0)$, and $\alpha_{L=0}(\theta_0)$ are known quantities from the geometry and airfoil section of the finite wing. The only unknowns in Equation (5.51) are the A_n's. Hence, written at a given spanwise location (a specified θ_0), Equation (5.51) is one algebraic equation with N unknowns, $A_1, A_2, \dots, A_n$. However, let us choose N different spanwise stations, and let us evaluate Equation (5.51) at each of these N stations. We then obtain a system of N independent algebraic equations with N unknowns, namely, $A_1, A_2, \dots, A_N$. In this fashion, actual numerical values are obtained for the A_n's—numerical values that ensure that the general circulation distribution given by Equation (5.48) satisfies the fundamental equation of finite-wing theory, Equation (5.23).

Now that $\Gamma(\theta)$ is known via Equation (5.48), the lift coefficient for the finite wing follows immediately from the substitution of Equation (5.48) into (5.26):

$$C_L = \frac{2}{V_\infty S} \int_{-b/2}^{b/2} \Gamma(y)\,dy = \frac{2b^2}{S} \sum_{1}^{N} A_n \int_{0}^{\pi} \sin n\theta \sin \theta \, d\theta \qquad (5.52)$$

In Equation (5.52), the integral is

$$\int_{0}^{\pi} \sin n\theta \sin \theta \, d\theta = \begin{cases} \pi/2 & \text{for } n = 1 \\ 0 & \text{for } n \neq 1 \end{cases}$$

Hence, Equation (5.52) becomes

$$C_L = A_1 \pi \frac{b^2}{S} = A_1 \pi \text{AR} \qquad (5.53)$$

Note that C_L depends only on the leading coefficient of the Fourier series expansion. (However, although C_L depends on A_1 only, we must solve for all the A_n's simultaneously in order to obtain A_1.)

The induced drag coefficient is obtained from the substitution of Equation (5.48) into Equation (5.30) as follows:

$$C_{D,i} = \frac{2}{V_\infty S} \int_{-b/2}^{b/2} \Gamma(y)\alpha_i(y)\,dy \qquad (5.54)$$

$$= \frac{2b^2}{S} \int_{0}^{\pi} \left(\sum_{1}^{N} A_n \sin n\theta \right) \alpha_i(\theta) \sin \theta \, d\theta$$

The induced angle of attack $\alpha_i(\theta)$ in Equation (5.54) is obtained from the substitution of Equations (5.46) and (5.49) into (5.18), which yields

$$\alpha_i(y_0) = \frac{1}{4\pi V_\infty} \int_{-b/2}^{b/2} \frac{(d\Gamma/dy)\,dy}{y_0 - y}$$

$$= \frac{1}{\pi} \sum_{1}^{N} n A_n \int_{0}^{\pi} \frac{\cos n\theta}{\cos \theta - \cos \theta_0}\,d\theta \qquad (5.55)$$

The integral in Equation (5.55) is the standard form given by Equation (4.26). Hence, Equation (5.55) becomes

$$\alpha_i(\theta_0) = \sum_{1}^{N} n A_n \frac{\sin n\theta_0}{\sin \theta_0} \qquad (5.56)$$

In Equation (5.56), θ_0 is simply a dummy variable which ranges from 0 to π across the span of the wing; it can therefore be replaced by θ, and Equation (5.56) can be written as

$$\alpha_i(\theta) = \sum_{1}^{N} n A_n \frac{\sin n\theta}{\sin \theta} \qquad (5.57)$$

Substituting Equation (5.57) into (5.54), we have

$$C_{D,i} = \frac{2b^2}{S} \int_0^\pi \left(\sum_1^N A_n \sin n\theta \right) \left(\sum_1^N nA_n \sin n\theta \right) d\theta \qquad (5.58)$$

Examine Equation (5.58) closely; it involves the product of two summations. Also, note that, from the standard integral,

$$\int_0^\pi \sin m\theta \sin k\theta = \begin{cases} 0 & \text{for } m \neq k \\ \pi/2 & \text{for } m = k \end{cases} \qquad (5.59)$$

Hence, in Equation (5.58), the mixed product terms involving unequal subscripts (e.g., $A_1 A_2, A_2 A_4$) are, from Equation (5.59), equal to zero. Hence, Equation (5.58) becomes

$$C_{D,i} = \frac{2b^2}{S} \left(\sum_1^N nA_n^2 \right) \frac{\pi}{2} = \pi\text{AR} \sum_1^N nA_n^2$$

$$= \pi\text{AR} \left(A_1^2 + \sum_2^N nA_n^2 \right)$$

$$= \pi\text{AR}\, A_1^2 \left[1 + \sum_2^N n\left(\frac{A_n}{A_1}\right)^2 \right] \qquad (5.60)$$

Substituting Equation (5.53) for C_L into Equation (5.60), we obtain

$$C_{D,i} = \frac{C_L^2}{\pi\text{AR}}(1+\delta) \qquad (5.61)$$

where $\delta = \sum_2^N n(A_n/A_1)^2$. Note that $\delta \geq 0$; hence, the factor $1 + \delta$ in Equation (5.61) is either greater than 1 or at least equal to 1. Let us define a span efficiency factor, e, as $e = (1+\delta)^{-1}$. Then Equation (5.61) can be written as

$$C_{D,i} = \frac{C_L^2}{\pi e\text{AR}} \qquad (5.62)$$

where $e \leq 1$. Comparing Equations (5.61) and (5.62) for the general lift distribution with Equation (5.43) for the elliptical lift distribution, note that $\delta = 0$ and $e = 1$ for the elliptical lift distribution. Hence, the lift distribution which yields minimum induced drag is the *elliptical lift distribution*. This is why we have a practical interest in the elliptical lift distribution.

Recall that for a wing with no aerodynamic twist and no geometric twist, an elliptical lift distribution is generated by a wing with an elliptical planform,

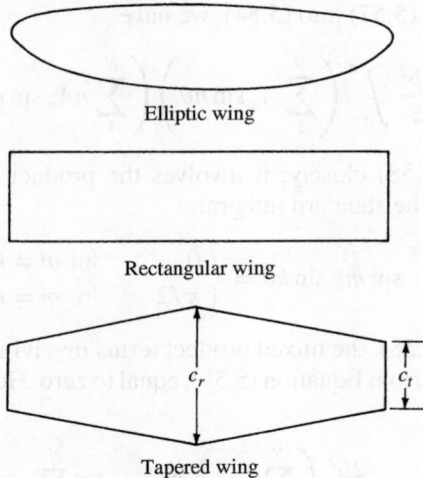

Figure 5.18 Various planforms for straight wings.

as sketched at the top of Figure 5.18. Several aircraft have been designed in the past with elliptical wings; the most famous, perhaps, being the British Spitfire from World War II, shown in Figure 5.19. However, elliptic planforms are more expensive to manufacture than, say, a simple rectangular wing as sketched in the middle of Figure 5.18. On the other hand, a rectangular wing generates a lift distribution far from optimum. A compromise is the tapered wing shown at the bottom of Figure 5.18. The tapered wing can be designed with a taper ratio, that is, tip chord/root chord $\equiv c_t/c_r$, such that the lift distribution closely approximates the elliptic case. The variation of δ as a function of taper ratio for wings of different aspect ratio is illustrated in Figure 5.20. Such calculations of δ were first performed by the famous English aerodynamicist, Hermann Glauert, and published in Reference 18 in the year 1926. Glauert used only four terms in the series expansion given in Equation (5.60). The results shown in Figure 5.20 are based on the recent computer calculations carried out by B.W. McCormick at Penn State University using the equivalent of 50 terms in the series expansion. Note from Figure 5.20 that a tapered wing can be designed with an induced drag coefficient reasonably close to the minimum value. In addition, tapered wings with straight leading and trailing edges are considerably easier to manufacture than elliptic planforms. Therefore, most conventional aircraft employ tapered rather than elliptical wing planforms.

5.3.3 Effect of Aspect Ratio

Returning to Equations (5.61) and (5.62), note that the induced drag coefficient for a finite wing with a general lift distribution is inversely proportional to the aspect ratio, as was discussed earlier in conjunction with the case of the elliptic lift

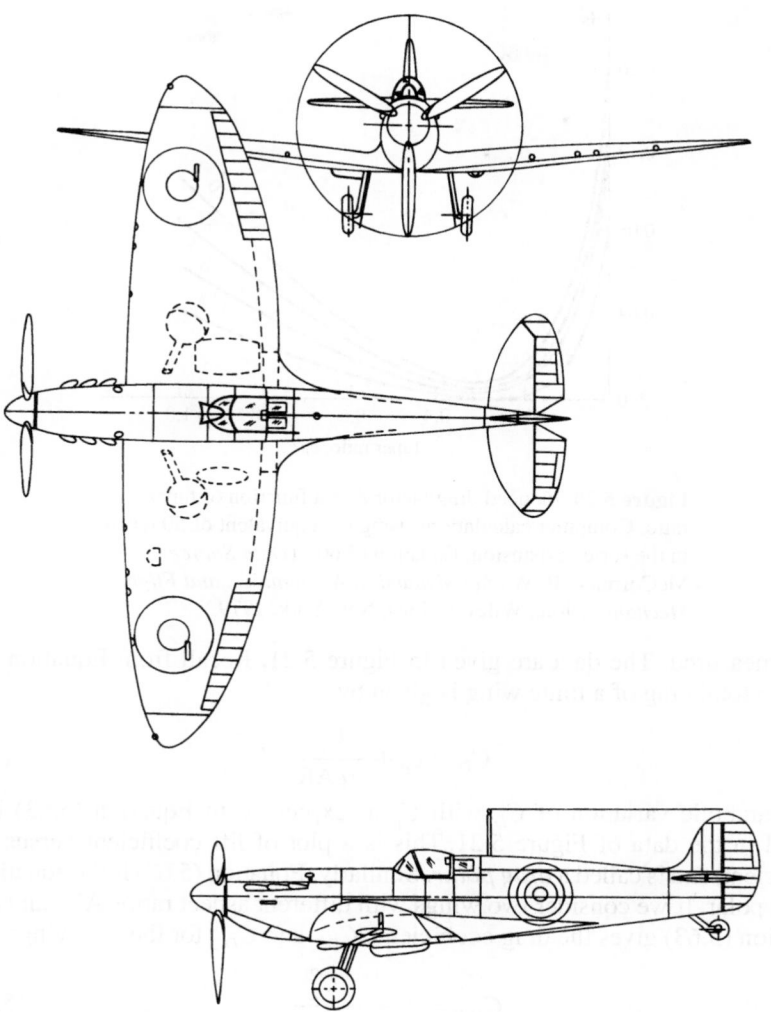

Figure 5.19 Three views of the Supermarine Spitfire, a famous British World War II fighter.

distribution. Note that AR, which typically varies from 6 to 22 for standard subsonic airplanes and sailplanes, has a much stronger effect on $C_{D,i}$ than the value of δ, which from Figure 5.20 varies only by about 10 percent over the practical range of taper ratio. Hence, the primary design factor for minimizing induced drag is not the closeness to an elliptical lift distribution, but rather, the ability to make the aspect ratio as large as possible. The determination that $C_{D,i}$ is inversely proportional to AR was one of the great victories of Prandtl's lifting-line theory. In 1915, Prandtl verified this result with a series of classic experiments wherein the lift and drag of seven rectangular wings with different aspect ratios

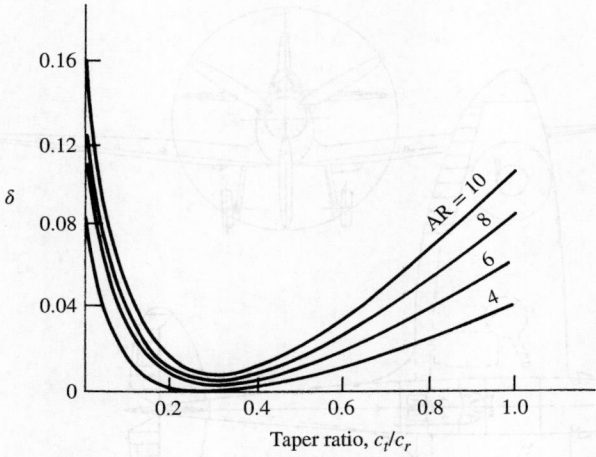

Figure 5.20 Induced drag factor δ as a function of taper
ratio. Computer calculations using the equivalent of 50 terms
in the series expansion, Equation (5.60). (*Data Source:*
McCormick, B. W., *Aerodynamics, Aeronautics, and Flight
Mechanics,* John Wiley & Sons, New York, 1979.)

were measured. The data are given in Figure 5.21. Recall from Equation (5.4)
that the total drag of a finite wing is given by

$$C_D = c_d + \frac{C_L^2}{\pi e AR} \tag{5.63}$$

The parabolic variation of C_D with C_L as expressed in Equation (5.63) is reflected in the data of Figure 5.21. This is a plot of lift coefficient versus drag
coefficient, and is called a *drag polar*. Similarly, Equation (5.63) is the equation of
a drag polar. If we consider two wings with different aspect ratios AR_1 and AR_2,
Equation (5.63) gives the drag coefficients $C_{D,1}$ and $C_{D,2}$ for the two wings as

$$C_{D,1} = c_d + \frac{C_L^2}{\pi e AR_1} \tag{5.64a}$$

and

$$C_{D,2} = c_d + \frac{C_L^2}{\pi e AR_2} \tag{5.64b}$$

Assume that the wings are at the same C_L. Also, since the airfoil section is the
same for both wings, c_d is essentially the same. Moreover, the variation of e between the wings is only a few percent and can be ignored. Hence, subtracting
Equation (5.64b) from (5.64a), we obtain

$$C_{D,1} = C_{D,2} + \frac{C_L^2}{\pi e}\left(\frac{1}{AR_1} - \frac{1}{AR_2}\right) \tag{5.65}$$

Equation (5.65) can be used to scale the data of a wing with aspect ratio AR_2 to
correspond to the case of another aspect ratio AR_1. For example, Prandtl scaled

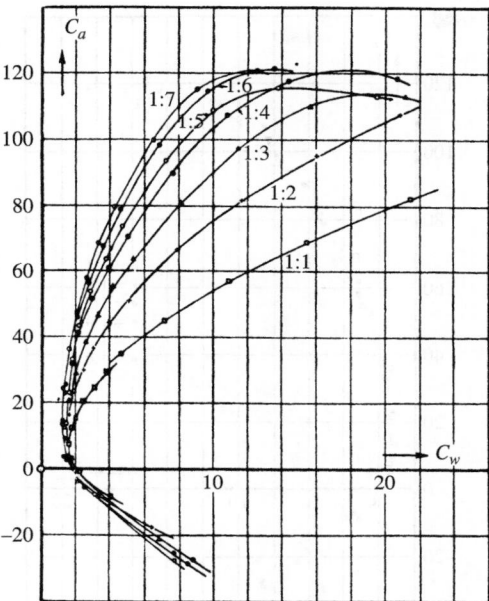

Figure 5.21 Prandtl's classic rectangular wing
data for seven different aspect ratios from 1 to 7;
variation of lift coefficient versus drag coefficient.
For historical interest, we reproduce here Prandtl's
actual graphs. Note that, in his nomenclature,
C_a = lift coefficient and C_w = drag coefficient.
Also, the numbers on both the ordinate and
abscissa are 100 times the actual values of the
coefficients. (*Source:* Prandtl, L., *Applications of
Modern Hydrodynamics to Aeronautics,* NACA
Report No. 116, 1921.)

the data of Figure 5.21 to correspond to a wing with an aspect ratio of 5. For this
case, Equation (5.65) becomes

$$C_{D,1} = C_{D,2} + \frac{C_L^2}{\pi e}\left(\frac{1}{5} - \frac{1}{AR_2}\right) \tag{5.66}$$

Inserting the respective values of $C_{D,2}$ and AR_2 from Figure 5.21 into Equation (5.66), Prandtl found that the resulting data for $C_{D,1}$ versus C_L collapsed
to essentially the same curve, as shown in Figure 5.22. Hence, the inverse dependence of $C_{D,i}$ on AR was substantially verified as early as 1915.

There are two primary differences between airfoil and finite-wing properties.
We have discussed one difference, namely, a finite wing generates induced drag.
However, a second major difference appears in the lift slope. In Figure 4.9, the
lift slope for an airfoil was defined as $a_0 \equiv dc_l/d\alpha$. Let us denote the lift slope

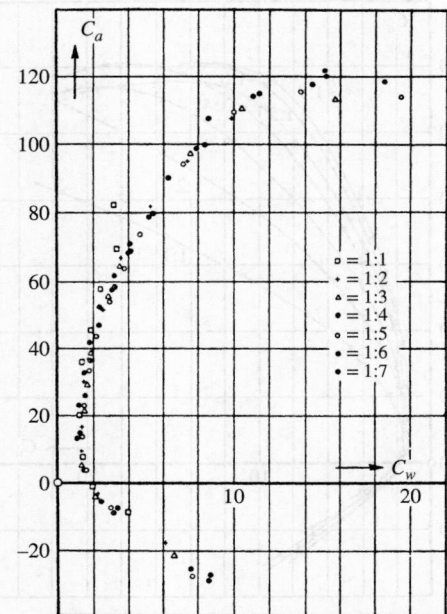

Figure 5.22 Data of Figure 5.21 scaled by Prandtl to an aspect ratio of 5.

for a finite wing as $a \equiv dC_L/d\alpha$. When the lift slope of a finite wing is compared with that of its airfoil section, we find that $a < a_0$. To see this more clearly, return to Figure 5.6, which illustrates the influence of downwash on the flow over a local airfoil section of a finite wing. Note that although the geometric angle of attack of the finite wing is α, the airfoil section effectively senses a smaller angle of attack, namely, $\alpha_{\rm eff}$, where $\alpha_{\rm eff} = \alpha - \alpha_i$. For the time being, consider an elliptic wing with no twist; hence, α_i and $\alpha_{\rm eff}$ are both constants along the span. Moreover, c_l is also constant along the span, and therefore $C_L = c_l$. Assume that we plot C_L for the finite wing versus $\alpha_{\rm eff}$, as shown at the top of Figure 5.23. Because we are using $\alpha_{\rm eff}$ the lift slope corresponds to that for an infinite wing a_0. However, in real life, our naked eyes cannot see $\alpha_{\rm eff}$; instead, what we actually observe is a finite wing with a certain angle between the chord line and the relative wind; that is, in practice, we always observe the geometric angle of attack α. Hence, C_L for a finite wing is generally given as a function of α, as sketched at the bottom of Figure 5.23. Since $\alpha > \alpha_{\rm eff}$, the bottom abscissa is stretched, and hence the bottom lift curve is less inclined; it has a slope equal to a, and Figure 5.23 clearly shows that $a < a_0$. The effect of a finite wing is to *reduce* the lift slope. Also, recall that at zero lift, there are no induced effects; that is, $\alpha_i = C_{D,i} = 0$. Thus, when $C_L = 0$, $\alpha = \alpha_{\rm eff}$. As a result, $\alpha_{L=0}$ is the same for the finite and the infinite wings, as shown in Figure 5.23.

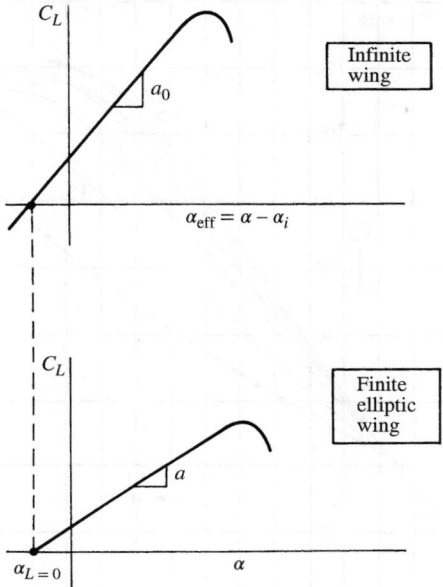

Figure 5.23 Lift curves for an infinite wing versus a finite elliptic wing.

The values of a_0 and a are related as follows. From the top of Figure 5.23,

$$\frac{dC_L}{d(\alpha - \alpha_i)} = a_0$$

Integrating, we find

$$C_L = a_0(\alpha - \alpha_i) + \text{const} \tag{5.67}$$

Substituting Equation (5.42a) into (5.67), we obtain

$$C_L = a_0\left(\alpha - \frac{C_L}{\pi\text{AR}}\right) + \text{const} \tag{5.68}$$

Differentiating Equation (5.68) with respect to α, and solving for $dC_L/d\alpha$, we obtain

$$\frac{dC_L}{d\alpha} = a = \frac{a_0}{1 + a_0/\pi\text{AR}} \tag{5.69}$$

Equation (5.69) gives the desired relation between a_0 and a for an elliptic finite wing. For a finite wing of general planform, Equation (5.69) is slightly modified, as given below:

$$a = \frac{a_0}{1 + (a_0/\pi\text{AR})(1 + \tau)} \tag{5.70}$$

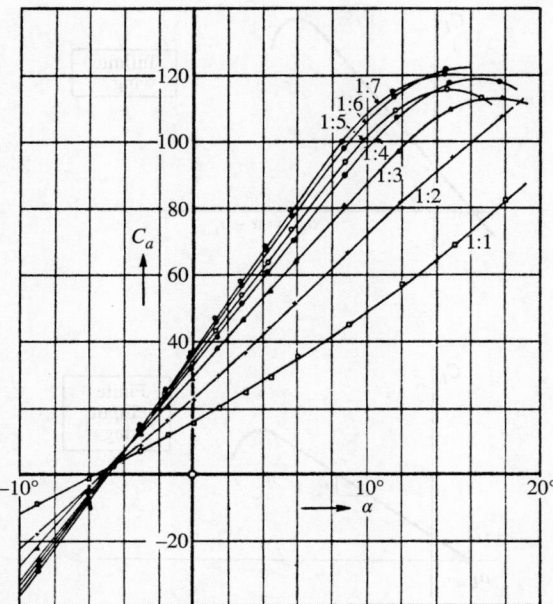

Figure 5.24 Prandtl's classic rectangular wing data.
Variation of lift coefficient with angle of attack for seven
different aspect ratios from 1 to 7. Nomenclature and scale
are the same as given in Figure 5.21.

In Equation (5.70), τ is a function of the Fourier coefficients A_n. Values of τ were first calculated by Glauert in the early 1920s and were published in Reference 18, which should be consulted for more details. Values of τ typically range between 0.05 and 0.25.

Of most importance in Equations (5.69) and (5.70) is the aspect-ratio variation. Note that for low-AR wings, a substantial difference can exist between a_0 and a. However, as AR $\rightarrow \infty$, $a \rightarrow a_0$. The effect of aspect ratio on the lift curve is dramatically shown in Figure 5.24, which gives classic data obtained on rectangular wings by Prandtl in 1915. Note the reduction in $dC_L/d\alpha$ as AR is reduced. Moreover, using the equations obtained above, Prandtl scaled the data in Figure 5.24 to correspond to an aspect ratio of 5; his results collapsed to essentially the same curve, as shown in Figure 5.25. In this manner, the aspect-ratio variation given in Equations (5.69) and (5.70) was confirmed as early as the year 1915.

5.3.4 Physical Significance

Consider again the basic model underlying Prandtl's lifting-line theory. Return to Figure 5.15 and study it carefully. An infinite number of infinitesimally weak horseshoe vortices are superimposed in such a fashion as to generate a lifting line which spans the wing, along with a vortex sheet which trails downstream.

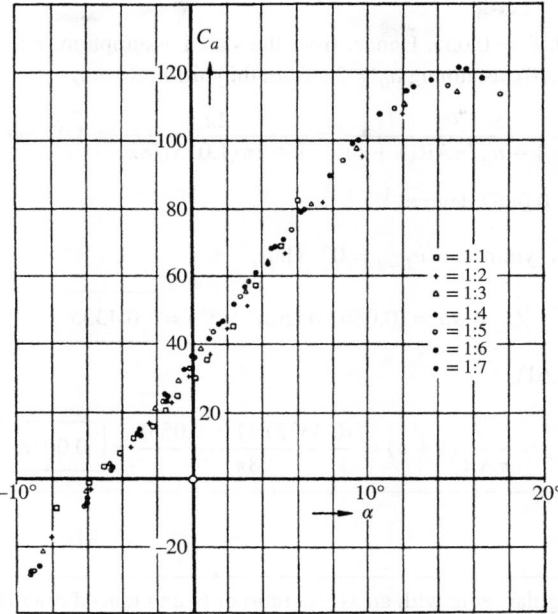

Figure 5.25 Data of Figure 5.24 scaled by Prandtl to an aspect ratio of 5.

This trailing-vortex sheet is the instrument that induces downwash at the lifting line. At first thought, you might consider this model to be somewhat abstract—a mathematical convenience that somehow produces surprisingly useful results. However, the model shown in Figure 5.15 has real physical significance. To see this more clearly, return to Figure 5.3. Note that in the three-dimensional flow over a finite wing, the streamlines leaving the trailing edge from the top and bottom surfaces are in different directions; that is, there is a discontinuity in the tangential velocity at the trailing edge. We know from Chapter 4 that a discontinuous change in tangential velocity is theoretically allowed across a vortex sheet. In real life, such discontinuities do not exist; rather, the different velocities at the trailing edge generate a thin region of large velocity gradients—a thin region of shear flow with very large vorticity. Hence, a sheet of vorticity actually trails downstream from the trailing edge of a finite wing. This sheet tends to roll up at the edges and helps to form the wing-tip vortices sketched in Figure 5.4. Thus, Prandtl's lifting-line model with its trailing-vortex sheet is physically consistent with the actual flow downstream of a finite wing.

EXAMPLE 5.1

Consider a finite wing with an aspect ratio of 8 and a taper ratio of 0.8. The airfoil section is thin and symmetric. Calculate the lift and induced drag coefficients for the wing when it is at an angle of attack of 5°. Assume that $\delta = \tau$.

■ **Solution**

From Figure 5.20, $\delta = 0.055$. Hence, from the stated assumption, τ also equals 0.055. From Equation (5.70), assuming $a_0 = 2\pi$ from thin airfoil theory,

$$a = \frac{a_0}{1 + a_0/\pi AR(1 + \tau)} = \frac{2\pi}{1 + 2\pi(1.055)/8\pi} = 4.97 \text{ rad}^{-1}$$

$$= 0.0867 \text{ degree}^{-1}$$

Since the airfoil is symmetric, $\alpha_{L=0} = 0°$. Thus,

$$C_L = a\alpha = (0.0867 \text{ degree}^{-1}(5°) = \boxed{0.4335}$$

From Equation (5.61),

$$C_{D,i} = \frac{C_L^2}{\pi AR}(1 + \delta) = \frac{(0.4335)^2(1 + 0.055)}{8\pi} = \boxed{0.00789}$$

EXAMPLE 5.2

Consider a rectangular wing with an aspect ratio of 6, an induced drag factor $\delta = 0.055$, and a zero-lift angle of attack of $-2°$. At an angle of attack of 3.4°, the induced drag coefficient for this wing is 0.01. Calculate the induced drag coefficient for a similar wing (a rectangular wing with the same airfoil section) at the same angle of attack, but with an aspect ratio of 10. Assume that the induced factors for drag and the lift slope, δ and τ, respectively, are equal to each other (i.e., $\delta = \tau$). Also, for AR = 10, $\delta = 0.105$.

■ **Solution**

We must recall that although the angle of attack is the same for the two cases compared here (AR = 6 and 10), the value of C_L is different because of the aspect-ratio effect on the lift slope. First, let us calculate C_L for the wing with an aspect ratio of 6. From Equation (5.61),

$$C_L^2 = \frac{\pi AR C_{D,i}}{1 + \delta} = \frac{\pi(6)(0.01)}{1 + 0.055} = 0.1787$$

Hence,

$$C_L = 0.423$$

The lift slope of this wing is therefore

$$\frac{dC_L}{d\alpha} = \frac{0.423}{3.4° - (-2°)} = 0.078/\text{degree} = 4.485/\text{rad}$$

The lift slope for the airfoil (the infinite wing) can be obtained from Equation (5.70):

$$\frac{dC_L}{d\alpha} = a = \frac{a_0}{1 + (a_0/\pi AR)(1 + \tau)}$$

$$4.485 = \frac{a_0}{1 + [(1.055)a_0/\pi(6)]} = \frac{a_0}{1 + 0.056a_0}$$

Solving for a_0, we find that this yields $a_0 = 5.989$/rad. Since the second wing (with AR = 10) has the same airfoil section, then a_0 is the same. The lift slope of the second wing is given by

$$a = \frac{a_0}{1 + (a_0/\pi \text{AR})(1 + \tau)} = \frac{5.989}{1 + [(5.989)(1.105)/\pi(10)]} = 4.95/\text{rad}$$

$$= 0.086/\text{degree}$$

The lift coefficient for the second wing is therefore

$$C_L = a(\alpha - \alpha_{L=0}) = 0.086[3.4° - (-2°)] = 0.464$$

In turn, the induced drag coefficient is

$$C_{D,i} = \frac{C_L^2}{\pi \text{AR}}(1 + \delta) = \frac{(0.464)^2(1.105)}{\pi(10)} = \boxed{0.0076}$$

Note: This problem would have been more straightforward if the lift coefficients had been stipulated to be the same between the two wings rather than the angle of attack. Then Equation (5.61) would have yielded the induced drag coefficient directly. A purpose of this example is to reinforce the rationale behind Equation (5.65), which readily allows the scaling of drag coefficients from one aspect ratio to another, as long as the *lift coefficient is the same*. This allows the scaled drag-coefficient data to be plotted versus C_L (not the angle of attack) as in Figure 5.22. However, in the present example, where the angle of attack is the same in both cases, the effect of aspect ratio on the lift slope must be explicitly considered, as we have done above.

EXAMPLE 5.3

Consider the twin-jet executive transport discussed in Example 1.6. In addition to the information given in Example 1.6, for this airplane the zero-lift angle of attack is $-2°$, the lift slope of the airfoil section is 0.1 per degree, the lift efficiency factor $\tau = 0.04$, and the wing aspect ratio is 7.96. At the cruising condition treated in Example 1.6, calculate the angle of attack of the airplane.

■ **Solution**

The lift slope of the airfoil section in radians is

$$a_0 = 0.1 \text{ per degree} = 0.1(57.3) = 5.73 \text{ rad}$$

From Equation (5.70) repeated below

$$a = \frac{a_0}{1 + (a_0/\pi \text{AR})(1 + \tau)}$$

we have

$$a = \frac{5.73}{1 + \left(\dfrac{5.73}{7.96\pi}\right)(1 + 0.04)} = 4.627 \text{ per rad}$$

or

$$a = \frac{4.627}{57.3} = 0.0808 \text{ per degree}$$

From Example 1.6, the airplane is cruising at a lift coefficient equal to 0.21. Since

$$C_L = a(\alpha - \alpha_{L=0})$$

we have

$$\alpha = \frac{C_L}{a} + \alpha_{L=0} = \frac{0.21}{0.0808} + (-2) = \boxed{0.6°}$$

EXAMPLE 5.4

In the Preview Box for this chapter, we considered the Beechcraft Baron 58 (Figure 5.1) flying such that the wing is at a 4-degree angle of attack. The wing of this airplane has an NACA 23015 airfoil at the root, tapering to a 23010 airfoil at the tip. The data for the NACA 23015 airfoil is given in Figure 5.2. In the Preview Box, we teased you by reading from Figure 5.2 the airfoil lift and drag coefficients at $\alpha = 4°$, namely, $c_l = 0.54$ and $c_d = 0.0068$, and posed the question: Are the lift and drag coefficients of the wing the same values, that is, $C_L = 0.54$ and $C_D = 0.0068$? The answer given in the Preview Box was a resounding NO! We now know why. Moreover, we now know how to calculate C_L and C_D for the wing. Let us proceed to do just that. Consider the wing of the Beechcraft Baron 58 at a 4-degree angle of attack. The wing has an aspect ratio of 7.61 and a taper ratio of 0.45. Calculate C_L and C_D for the wing.

■ **Solution**

From Figure 5.2a, the zero-lift angle of attack of the airfoil, which is the same for the finite wing, is

$$\alpha_{L=0} = -1°$$

The airfoil lift slope is also obtained from Figure 5.2a. Since the lift curve is linear below the stall, we arbitrarily pick two points on this curve: $\alpha = 7°$ where $c_l = 0.9$, and $\alpha = -1°$ where $c_l = 0$. Thus,

$$a_0 = \frac{0.9 - 0}{7 - (-1)} = \frac{0.9}{8} = 0.113 \text{ per degree}$$

The lift slope in radians is:

$$a_0 = 0.113(57.3) = 6.47 \text{ per rad}$$

From Figure 5.20, for AR = 7.61 and taper ratio = 0.45

$$\delta = 0.01$$

Hence,

$$e = \frac{1}{1+\delta} = \frac{1}{1+0.01} = 0.99$$

From Equation (5.70), assuming $\tau = \delta$,

$$a = \frac{a_0}{1 + \left(\dfrac{a_0}{\pi \text{AR}}\right)(1+\tau)} \quad (a \text{ and } a_0 \text{ are per rad})$$

where
$$\frac{a_0}{\pi \text{AR}} = \frac{6.47}{\pi(7.61)} = 0.271$$

$$(1 + \tau) = 1 + 0.01 = 1.01$$

we have
$$a = \frac{6.47}{1 + (0.271)(1.01)} = 5.08 \text{ per rad}$$

Converting back to degrees:
$$a = \frac{5.08}{57.3} = 0.0887 \text{ per degree}$$

For the linear lift curve for the finite wing
$$C_L = a(\alpha - \alpha_{L=0})$$

For $\alpha = 4°$, we have
$$C_L = 0.0887[4 - (-1)] = 0.0887(5)$$

$$C_L = \boxed{0.443}$$

The drag coefficient is given by Equation (5.63);

$$C_D = c_d + \frac{C_L^2}{\pi e \text{AR}}$$

Here, c_d is the section drag coefficient given in Figure 5.2b. Note that in Figure 5.2b, c_d is plotted versus the section lift coefficient c_l. To accurately read c_d from Figure 5.2b, we need to know the value of c_l actually sensed by the airfoil section on the finite wing, that is, the value of the airfoil c_l for the airfoil at its effective angle of attack, α_{eff}. To estimate α_{eff}, we will assume an elliptical lift distribution over the wing. We know this is not quite correct, but with a value of $\delta = 0.01$, it is not very far off. From Equation (5.42a) for an elliptical lift distribution, the induced angle of attack is

$$\alpha_i = \frac{C_L}{\pi \text{AR}} = \frac{(0.443)}{\pi(7.61)} = 0.0185 \text{ rad}$$

In degrees
$$\alpha_i = (0.0185)(57.3) = 1.06°$$

From Figure 5.6,
$$\alpha_{\text{eff}} = \alpha - \alpha_i = 4° - 1.06° = 2.94° \approx 3°$$

The lift coefficient sensed by the airfoil is then
$$c_l = a_0(\alpha_{\text{eff}} - \alpha_{L=0})$$
$$= 0.113[3 - (-1)] = 0.113(4) = 0.452$$

(Note how close this section lift coefficient is to the overall lift coefficient of the wing of 0.443.) From Figure 5.2b, taking the data at the highest Reynolds number shown, for $c_l = 0.452$, we have

$$c_d = 0.0065$$

Returning to Equation (5.63),

$$C_D = c_d + \frac{C_L^2}{\pi e \mathrm{AR}}$$

$$= 0.0065 + \frac{(0.443)^2}{\pi(0.99)(7.61)}$$

$$= 0.0065 + 0.0083 = \boxed{0.0148}$$

So finally, with the results of this worked example, we see why the answer given in the Preview Box was a resounding NO! The lift coefficient for the finite wing is 0.443 compared to the airfoil value of 0.54 given in the Preview Box; the finite wing value is 18 percent lower than the airfoil value—a substantial difference. The drag coefficient for the finite wing is 0.0148 compared to the airfoil value of 0.0068; the finite wing value is more than a factor of two larger—a dramatic difference. These differences are the reason for the studies covered in this chapter.

DESIGN BOX

In airplane design, the aspect ratio is a much more important consideration than wing planform shape from the point of view of reducing the induced drag coefficient. Although the elliptical planform, as sketched at the top of Figure 5.18, leads to the optimum lift distribution to minimize induced drag, the tapered wing, as sketched at the bottom of Figure 5.18, can yield a near-optimum lift distribution, with induced drag coefficients only a few percent higher than the elliptical wing. Because a tapered wing with its straight leading and trailing edges is much cheaper and easier to manufacture, the design choice for wing planform is almost always a tapered wing and hardly ever an elliptical wing.

Why then did the Supermarine Spitfire, shown in Figure 5.19, have such a beautiful elliptical wing planform? The answer has nothing to do with aerodynamics. In 1935, the Supermarine company was responding to the British Air Ministry specification F.37/34 for a new fighter aircraft. Designer Reginald Mitchell had originally laid out on the drawing board an aircraft with tapered wings. However, Mitchell was also coping with the Air Ministry requirement that the airplane be armed with eight 0.303 caliber Browning machine guns. Mounting four of these guns in each wing far enough outboard to be outside of the propeller disk, Mitchell had a problem—the outboard sections of the tapered wing did not have enough chord length to accommodate the guns. His solution was an elliptical planform, which provided sufficient chord length far enough out on the span to allow the guns to fit. The result was the beautiful elliptic planform shown in Figure 5.19. The enhanced aerodynamic efficiency of this wing was only a by-product of a practical design solution. Predictably, the elliptic wings were difficult to produce, and this contributed to production delays in the critical months before the beginning of World War II. A second enhanced aerodynamic by-product was afforded by the large chord lengths along most of the elliptic wing. This allowed Mitchell to choose a thinner airfoil section, with 13 percent thickness ratio at the wing root and 7 percent at the wing tip, and still maintain sufficient absolute thickness for internal structural design. Because of this, the Spitfire had a larger critical Mach number (to be discussed in Section 11.6) and could reach the unusually high freestream Mach number of 0.92 in a dive.

For aerodynamic efficiency at subsonic speeds, the airplane designer would love to have very large aspect ratio wings—wings that would look like a long slat out of a common venetian blind. However, existing airplanes do not fly around with venetian blind slats for wings. The reason is that the structural design of such wings poses a compromise. The larger the aspect ratio, the larger are the bending movements at the wing root caused by the lift distribution reaching farther away from the root. Such wings require heavier internal structure. Hence, as the aspect ratio of a wing increases, so does the structural weight of the wing. As a result of this compromise between aerodynamics and structures, typical aspect ratios for conventional subsonic airplanes are on the order of 6 to 8.

However, examine the three-view of the Lockheed U-2 high-altitude reconnaissance aircraft shown in Figure 5.26. This airplane has the unusually high aspect ratio of 14.3. Why? The answer is keyed to its mission. The U-2 was essentially a *point design;* it was to cruise at the exceptionally high altitude of 70,000 ft or higher in order to not be reached by interceptor aircraft or ground-to-air-missiles during

overflights of the Soviet Union in the 1950s. To achieve this mission, the need for incorporating a very high aspect ratio wing was paramount for the following reason. In steady, level flight, where the airplane lift L must equal its weight W,

$$L = W = q_\infty S C_L = \frac{1}{2}\rho_\infty V_\infty^2 S C_L \qquad (5.71)$$

As the airplane flies higher, ρ_∞ decreases hence, from Equation (5.71) C_L must be increased in order to keep the lift equal to the weight. As its high-altitude cruise design point, the U-2 flies at a high value of C_L, just on the verge of stalling. (This is in stark contrast to the normal cruise conditions of conventional airplanes at conventional altitudes, where the cruise lift coefficient is relatively small.) At the high value of C_L for the U-2 at cruising altitude, its induced drag coefficient [which from Equation (5.62) varies as C_L^2] would be unacceptably high if a conventional aspect ratio were used. Hence, the Lockheed design group (at the Lockheed Skunk Works) had to opt for as high an aspect ratio as possible to keep the induced drag coefficient within reasonable bounds. The wing design shown in Figure 5.26 was the result.

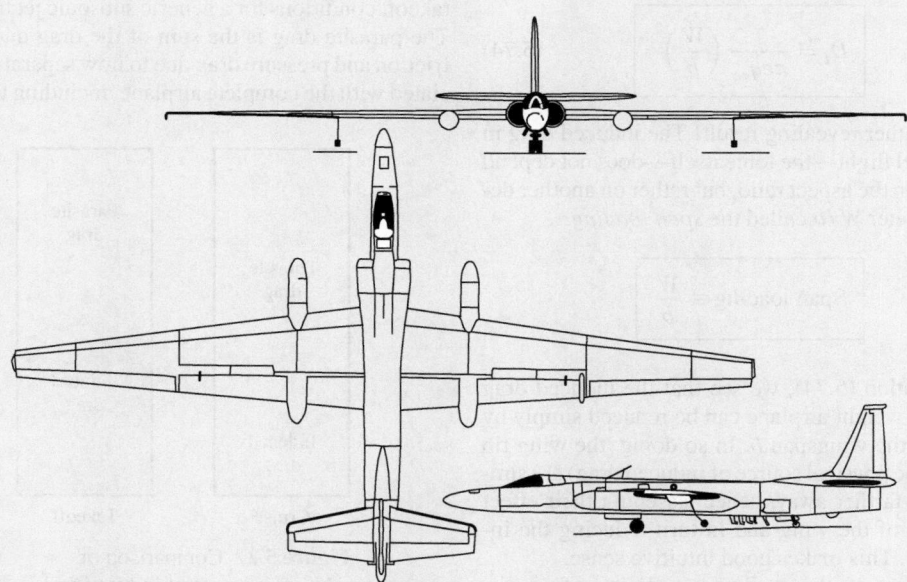

Figure 5.26 Three-view of the Lockheed U-2 high-altitude reconnaissance airplane.

We made an observation about induced drag D_i itself, in contrast to the induced drag coefficient $C_{D,i}$. We have emphasized, based on Equation (5.62), that $C_{D,i}$ can be reduced by increasing the aspect ratio. For an airplane in steady, level flight, however, the induced drag force itself is governed by another design parameter, rather than the aspect ratio per se, as follows. From Equation (5.62), we have

$$D_i = q_\infty S C_{D,i} = q_\infty S \frac{C_L^2}{\pi e \mathrm{AR}} \qquad (5.72)$$

For steady, level flight, from Equation (5.71), we have

$$C_L^2 = \left(\frac{L}{q_\infty S} \right)^2 = \left(\frac{W}{q_\infty S} \right)^2 \qquad (5.73)$$

Substituting Equation (5.73) into (5.72), we have

$$D_i = q_\infty S \left(\frac{W}{q_\infty S} \right)^2 \frac{1}{\pi e \mathrm{AR}}$$

$$= \frac{1}{\pi e} q_\infty S \left(\frac{W}{q_\infty S} \right)^2 \left(\frac{S}{b^2} \right)$$

or

$$\boxed{D_i = \frac{1}{\pi e q_\infty} \left(\frac{W}{b} \right)^2} \qquad (5.74)$$

This is a rather revealing result! The induced drag in steady, level flight—the force itself—does not depend explicitly on the aspect ratio, but rather on another design parameter W/b called the *span loading*:

$$\boxed{\text{Span loading} \equiv \frac{W}{b}}$$

From Equation (5.74), we see that the induced drag for a given weight airplane can be reduced simply by increasing the wingspan b. In so doing, the wing tip vortices (the physical source of induced drag) are simply moved farther away, hence lessening their effect on the rest of the wing and in turn reducing the induced drag. This makes good intuitive sense.

However, in the preliminary design of an airplane, the wing area S is usually dictated by the landing or take-off speed, which is only about 10 to 20 percent above V_{stall}. This is seen by Equation (1.47) repeated below

$$V_{\text{stall}} = \sqrt{\frac{2W}{\rho_\infty S C_{L,\max}}} \qquad (1.47)$$

For a specified V_{stall} at sea level, and given $C_{L,\max}$ for the airplane, Equation (1.47) determines the necessary wing area for a given weight airplane. Therefore, reflecting on Equation (5.74) when we say that D_i can be reduced for a given weight airplane simply by increasing the wing span, since S is usually fixed for the given weight, then clearly the aspect ratio b^2/S increases as we increase b. So when we use Equation (5.74) to say that D_i can be reduced by increasing the span for a given weight airplane, this also has the connotation of increasing the aspect ratio. However, it is instructional to note that D_i depends explicitly on the design parameter, W/b, and not the aspect ratio; this is the message in Equation (5.74).

Question: How much of the total drag of an airplane is induced drag? A generic answer to this question is shown in the bar chart in Figure 5.27. Here, the induced drag (shaded portion) relative to parasite drag (white portion) is shown for typical cruise and takeoff conditions for a generic subsonic jet transport. The parasite drag is the sum of the drag due to skin friction and pressure drag due to flow separation associated with the complete airplane, including the wing.

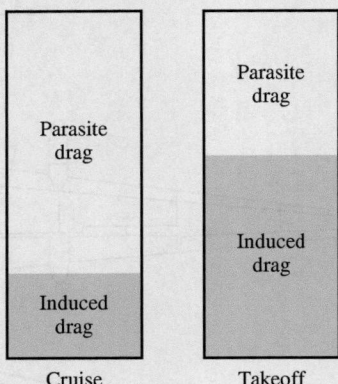

Figure 5.27 Comparison of relative amounts of induced and parasite drag for cruise (high speed) and takeoff (low speed).

The total drag of the airplane is the sum of the parasite drag and the induced drag. Figure 5.27 indicates that induced drag is about 25 percent of the total drag at cruise, but can be 60 percent or more of the total drag at takeoff (where the airplane is flying at high C_L).

For more details on the drag characteristics of airplanes, and their associated impact on airplane design, see the author's book *Aircraft Performance and Design*, McGraw Hill, Boston, 1999.

5.4 A NUMERICAL NONLINEAR LIFTING-LINE METHOD

The classical Prandtl lifting-line theory described in Section 5.3 assumes a linear variation of c_l versus α_{eff}. This is clearly seen in Equation (5.19). However, as the angle of attack approaches and exceeds the stall angle, the lift curve becomes nonlinear, as shown in Figure 4.9. This high-angle-of-attack regime is of interest to modern aerodynamicists. For example, when an airplane is in a spin, the angle of attack can range from 40 to 90°; an understanding of high-angle-of-attack aerodynamics is essential to the prevention of such spins. In addition, modern fighter airplanes achieve optimum maneuverability by pulling high angles of attack at subsonic speeds. Therefore, there are practical reasons for extending Prandtl's classical theory to account for a nonlinear lift curve. One simple extension is described in this section.

The classical theory developed in Section 5.4 is essentially closed form; that is, the results are analytical equations as opposed to a purely numerical solution. Of course, in the end, the Fourier coefficients A_n for a given wing must come from a solution of a system of simultaneous linear algebraic equations. Until the advent of the modern digital computer, these coefficients were calculated by hand. Today, they are readily solved on a computer using standard matrix methods. However, the elements of the lifting-line theory lend themselves to a straightforward purely numerical solution which allows the treatment of nonlinear effects. Moreover, this numerical solution emphasizes the fundamental aspects of lifting-line theory. For these reasons, such a numerical solution is outlined in this section.

Consider the most general case of a finite wing of given planform and geometric twist, with different airfoil sections at different spanwise stations. Assume that we have experimental data for the lift curves of the airfoil sections, including the nonlinear regime (i.e., assume we have the conditions of Figure 4.9 for all the given airfoil sections). A numerical iterative solution for the finite-wing properties can be obtained as follows:

1. Divide the wing into a number of spanwise stations, as shown in Figure 5.28. Here $k + 1$ stations are shown, with n designating any specific station.
2. For the given wing at a given α, *assume* the lift distribution along the span; that is, assume values for Γ at all the stations $\Gamma_1, \Gamma_2, \ldots, \Gamma_n, \ldots, \Gamma_{k+1}$. An elliptical lift distribution is satisfactory for such an assumed distribution.

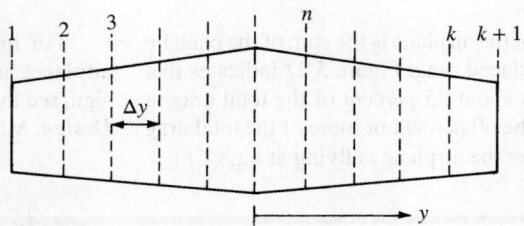

Figure 5.28 Stations along the span for a numerical solution.

3. With this assumed variation of Γ, calculate the induced angle of attack α_i from Equation (5.18) at each of the stations:

$$\alpha_i(y_n) = \frac{1}{4\pi V_\infty} \int_{-b/2}^{b/2} \frac{(d\Gamma/dy)\,dy}{y_n - y} \tag{5.75}$$

The integral is evaluated numerically. If Simpson's rule is used, Equation (5.75) becomes

$$\alpha_i(y_n) = \frac{1}{4\pi V_\infty} \frac{\Delta y}{3} \sum_{j=2,4,6}^{k} \frac{(d\Gamma/dy)_{j-1}}{(y_n - y_{j-1})} + 4\frac{(d\Gamma/dy)_j}{y_n - y_j} + \frac{(d\Gamma/dy)_{j+1}}{y_n - y_{j+1}} \tag{5.76}$$

where Δy is the distance between stations. In Equation (5.76), when $y_n = y_{j-1}, y_j$, or y_{j+1}, a singularity occurs (a denominator goes to zero). When this singularity occurs, it can be avoided by replacing the given term with its average value based on the two adjacent sections.

4. Using α_i from step 3, obtain the effective angle of attack α_{eff} at each station from

$$\alpha_{\text{eff}}(y_n) = \alpha - \alpha_i(y_n)$$

5. With the distribution of α_{eff} calculated from step 4, obtain the section lift coefficient $(c_l)_n$ at each station. These values are read from the known lift curve for the airfoil.

6. From $(c_l)_n$ obtained in step 5, a *new* circulation distribution is calculated from the Kutta-Joukowski theorem and the definition of lift coefficient:

$$L'(y_n) = \rho_\infty V_\infty \Gamma(y_n) = \frac{1}{2}\rho_\infty V_\infty^2 c_n (c_l)_n$$

Hence, $\qquad \Gamma(y_n) = \frac{1}{2} V_\infty c_n (c_l)_n$

where c_n is the local section chord. Keep in mind that in all the above steps, n ranges from 1 to $k + 1$.

7. The new distribution of Γ obtained in step 6 is compared with the values that were initially fed into step 3. If the results from step 6 do not agree with the input to step 3, then a new input is generated. If the previous input to

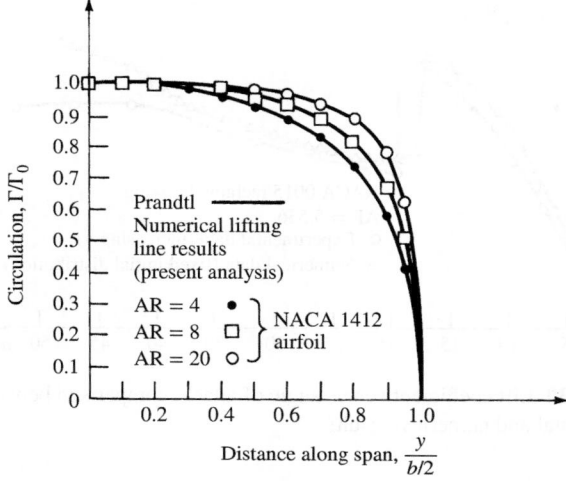

Figure 5.29 Lift distribution for a rectangular wing; comparison between Prandtl's classical theory and the numerical lifting-line method of Reference 20.

step 3 is designated as Γ_{old} and the result of step 6 is designated as Γ_{new}, then the new input to step 3 is determined from

$$\Gamma_{\text{input}} = \Gamma_{\text{old}} + D(\Gamma_{\text{new}} - \Gamma_{\text{old}})$$

where D is a damping factor for the iterations. Experience has found that the iterative procedure requires heavy damping, with typical values of D on the order of 0.05.

8. Steps 3 to 7 are repeated a sufficient number of cycles until Γ_{new} and Γ_{old} agree at each spanwise station to within acceptable accuracy. If this accuracy is stipulated to be within 0.01 percent for a stretch of five previous iterations, then a minimum of 50 and sometimes as many as 150 iterations may be required for convergence.

9. From the converged $\Gamma(y)$, the lift and induced drag coefficients are obtained from Equations (5.26) and (5.30), respectively. The integrations in these equations can again be carried out by Simpson's rule.

The procedure outlined above generally works smoothly and quickly on a high-speed digital computer. Typical results are shown in Figure 5.29, which shows the circulation distributions for rectangular wings with three different aspect ratios. The solid lines are from the classical calculations of Prandtl (Section 5.3), and the symbols are from the numerical method described above. Excellent agreement is obtained, thus verifying the integrity and accuracy of the numerical method. Also, Figure 5.29 should be studied as an example of typical circulation distributions over general finite wings, with Γ reasonably high over the center section of the wing but rapidly dropping to zero at the tips.

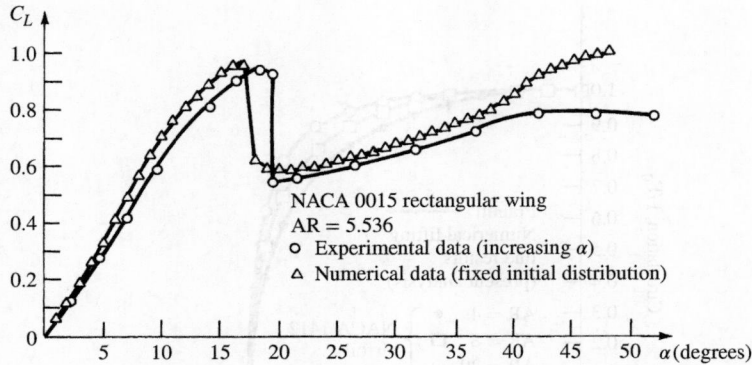

Figure 5.30 Lift coefficient versus angle of attack; comparison between experimental and numerical results.

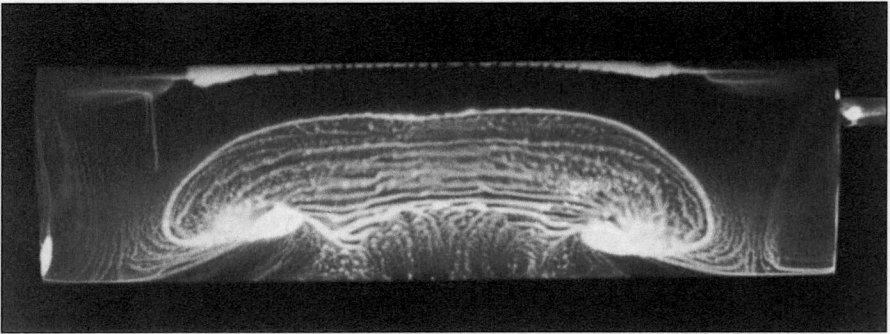

Figure 5.31 Surface oil flow pattern on a stalled, finite rectangular wing with a Clark Y-14 airfoil section. AR = 3.5, $\alpha = 22.8°$, Re = 245,000 (based on chord length). This pattern was established by coating the wing surface with pigmented mineral oil and inserting the model in a low-speed subsonic wind tunnel. In the photograph shown, flow is from top to bottom. Note the highly three-dimensional flow pattern. (©*Allen E. Winkelmann/University of Maryland.*)

An example of the use of the numerical method for the nonlinear regime is shown in Figure 5.30. Here, C_L versus α is given for a rectangular wing up to an angle of attack of 50°—well beyond stall. The numerical results are compared with existing experimental data obtained at the University of Maryland (Reference 19). The numerical lifting-line solution at high angle of attack agrees with the experiment to within 20 percent, and much closer for many cases. Therefore, such solutions give reasonable preliminary engineering results for the high-angle-of-attack post-stall region. However, it is wise not to stretch the applicability of lifting-line theory too far. At high angles of attack, the flow is highly three-dimensional. This is clearly seen in the surface oil pattern on a rectangular wing at high angle of attack shown in Figure 5.31. At high α, there is a strong spanwise flow, in combination with mushroom-shaped flow separation regions. Clearly, the basic assumptions of lifting-line theory, classical or numerical, cannot properly account for such three-dimensional flows.

For more details and results on the numerical lifting-line method, please see Reference 20.

5.5 THE LIFTING-SURFACE THEORY AND THE VORTEX LATTICE NUMERICAL METHOD

Prandtl's classical lifting-line theory (Section 5.3) gives reasonable results for straight wings at moderate to high aspect ratio. However, for low-aspect-ratio straight wings, swept wings, and delta wings, classical lifting-line theory is inappropriate. For such planforms, sketched in Figure 5.32, a more sophisticated model must be used. The purpose of this section is to introduce such a model and to discuss its numerical implementation. However, it is beyond the scope of this book to elaborate on the details of such higher-order models; rather, only the flavor is given here. You are encouraged to pursue this subject by reading the literature and by taking more advanced studies in aerodynamics.

Return to Figure 5.15. Here, a simple lifting line spans the wing, with its associated trailing vortices. The circulation Γ varies with y along the lifting line. Let us extend this model by placing a *series* of lifting lines on the plane of the wing, at different chordwise stations; that is, consider a large number of lifting lines all parallel to the y axis, located at different values of x, as shown in Figure 5.33. In the limit of an infinite number of lines of infinitesimal strength, we obtain a vortex sheet, where the vortex lines run parallel to the y axis. The strength of this sheet (per unit length in the x direction) is denoted by γ, where γ varies in the y direction, analogous to the variation of Γ for the single lifting line in Figure 5.15. Moreover, each lifting line will have, in general, a different overall strength, so that γ varies with x also. Hence, $\gamma = \gamma(x, y)$ as shown in Figure 5.33. In addition, recall that each lifting line has a system of trailing vortices; hence, the series of lifting lines is crossed by a series of superimposed trailing vortices parallel to the x axis. In the limit of an infinite number of infinitesimally weak vortices, these trailing vortices form another vortex sheet of strength δ (per unit length in the

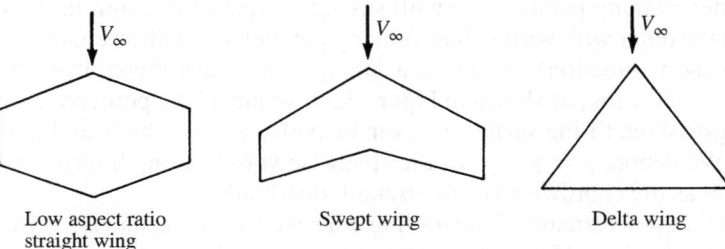

Figure 5.32 Types of wing planforms for which classical lifting-line theory is not appropriate.

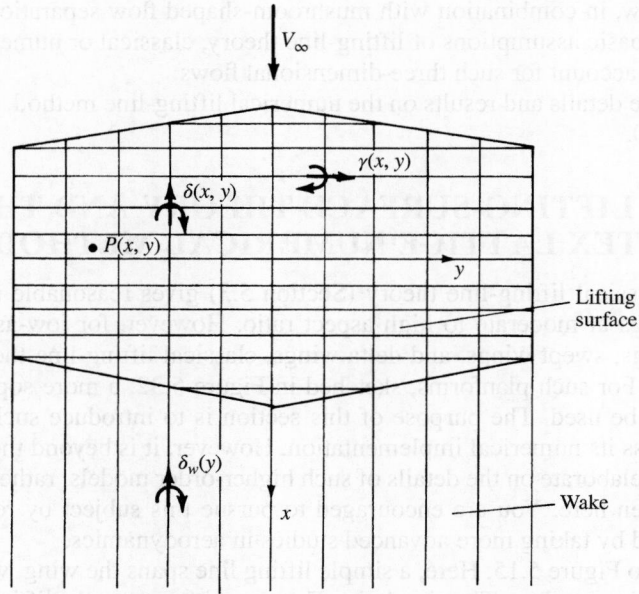

Figure 5.33 Schematic of a lifting surface.

y direction). [Note that this δ is different from the δ used in Equation (5.61); the use of the same symbol in both cases is standard, and there should be no confusion since the meanings and context are completely different.] To see this more clearly, consider a single line parallel to the *x* axis. As we move along this line from the leading edge to the trailing edge, we pick up an additional superimposed trailing vortex each time we cross a lifting line. Hence, δ must vary with *x*. Moreover, the trailing vortices are simply parts of the horseshoe vortex systems, the leading edges of which make up the various lifting lines. Since the circulation about each lifting line varies in the *y* direction, the strengths of different trailing vortices will, in general, be different. Hence, δ also varies in the *y* direction, that is, $\delta = \delta(x, y)$, as shown in Figure 5.33. The two vortex sheets—the one with vortex lines running parallel to *y* with strength γ (per unit length in the *x* direction) and the other with vortex lines running parallel to *x* with strength δ (per unit length in the *y* direction)—result in a *lifting surface* distributed over the entire planform of the wing, as shown in Figure 5.33. At any given point on the surface, the strength of the lifting surface is given by both γ and δ, which are functions of *x* and *y*. We denote $\gamma = \gamma(x, y)$ as the spanwise vortex strength distribution and $\delta = \delta(x, y)$ as the chordwise vortex strength distribution.

Note that downstream of the trailing edge we have no spanwise vortex lines, only trailing vortices. Hence, the wake consists of only chordwise vortices. The strength of this wake vortex sheet is given by δ_w (per unit length in the *y* direction). Since in the wake the trailing vortices do not cross any vortex lines, the strength

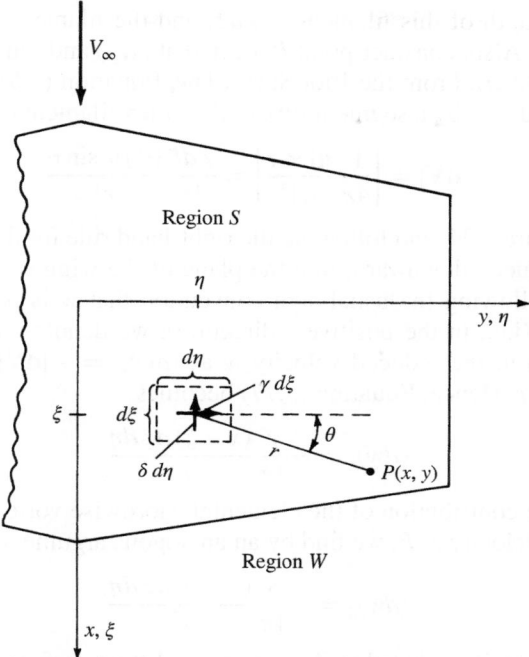

Figure 5.34 Velocity induced at point P by an infinitesimal segment of the lifting surface. The velocity is perpendicular to the plane of the paper.

of any given trailing vortex is constant with x. Hence, δ_w depends only on y and, throughout the wake, $\delta_w(y)$ is equal to its value at the trailing edge.

Now that we have defined the lifting surface, of what use is it? Consider point P located at (x, y) on the wing, as shown in Figure 5.33. The lifting surface and the wake vortex sheet both induce a normal component of velocity at point P. Denote this normal velocity by $w(x, y)$. We want the wing planform to be a stream surface of the flow; that is, we want the sum of the induced $w(x, y)$ and the normal component of the freestream velocity to be zero at point P and for all points on the wing—this is the flow-tangency condition on the wing surface. (Keep in mind that we are treating the wing as a flat surface in this discussion.) The central theme of lifting-surface theory is to find $\gamma(x, y)$ and $\delta(x, y)$ such that the flow-tangency condition is satisfied at all points on the wing. [Recall that in the wake, $\delta_w(y)$ is fixed by the trailing-edge values of $\delta(x, y)$; hence, $\delta_w(y)$ is not, strictly speaking, one of the unknown dependent variables.]

Let us obtain an expression for the induced normal velocity $w(x, y)$ in terms of γ, δ, and δ_w. Consider the sketch given in Figure 5.34, which shows a portion of the planview of a finite wing. Consider the point given by the coordinates (ξ, η). At this point, the spanwise vortex strength is $\gamma(\xi, \eta)$. Consider a thin ribbon, or filament, of the spanwise vortex sheet of incremental length $d\xi$ in the x direction.

Hence, the strength of this filament is $\gamma \, d\xi$, and the filament stretches in the y (or η) direction. Also, consider point P located at (x, y) and removed a distance r from the point (ξ, η). From the Biot-Savart law, Equation (5.5), the incremental velocity induced at P by a segment $d\eta$ of this vortex filament of strength $\gamma \, d\xi$ is

$$|\mathbf{dV}| = \left| \frac{\Gamma}{4\pi} \frac{\mathbf{dl} \times \mathbf{r}}{|\mathbf{r}|^3} \right| = \frac{\gamma \, d\xi}{4\pi} \frac{(d\eta) r \sin \theta}{r^3} \qquad (5.77)$$

Examining Figure 5.34, and following the right-hand rule for the strength γ, note that $|\mathbf{dV}|$ is induced downward, into the plane of the wing (i.e., in the negative z direction). Following the usual sign convention that w is positive in the upward direction (i.e., in the positive z direction), we denote the contribution of Equation (5.77) to the induced velocity w as $(dw)_\gamma = -|\mathbf{dV}|$. Also, note that $\sin \theta = (x - \xi)/r$. Hence, Equation (5.77) becomes

$$(dw)_\gamma = -\frac{\gamma}{4\pi} \frac{(x - \xi) \, d\xi \, d\eta}{r^3} \qquad (5.78)$$

Considering the contribution of the elemental chordwise vortex of strength $\delta \, d\eta$ to the induced velocity at P, we find by an analogous argument that

$$(dw)_\delta = -\frac{\delta}{4\pi} \frac{(y - \eta) \, d\xi \, d\eta}{r^3} \qquad (5.79)$$

To obtain the velocity induced at P by the entire lifting surface, Equations (5.78) and (5.79) must be integrated over the wing planform, designated as region S in Figure 5.34. Moreover, the velocity induced at P by the complete wake is given by an equation analogous to Equation (5.79), but with δ_w instead of δ, and integrated over the wake, designated as region W in Figure 5.34. Noting that

$$r = \sqrt{(x - \xi)^2 + (y - \eta)^2}$$

the normal velocity induced at P by both the lifting surface and the wake is

$$\boxed{\begin{aligned} w(x, y) = &-\frac{1}{4\pi} \iint\limits_S \frac{(x - \xi)\gamma(\xi, \eta) + (y - \eta)\delta(\xi, \eta)}{[(x - \xi)^2 + (y - \eta)^2]^{3/2}} \, d\xi \, d\eta \\ &-\frac{1}{4\pi} \iint\limits_W \frac{(y - \eta)\delta_w(\eta)}{[(x - \xi)^2 + (y - \eta)^2]^{3/2}} \, d\xi \, d\eta \end{aligned}} \qquad (5.80)$$

The central problem of lifting-surface theory is to solve Equation (5.80) for $\gamma(\xi, \eta)$ and $\delta(\xi, \eta)$ such that the sum of $w(x, y)$ and the normal component of the freestream is zero, that is, the flow is tangent to the planform surface S. The details of various lifting-surface solutions are beyond the scope of this book; rather, our purpose here was simply to present the flavor of the basic model.

The advent of the high-speed digital computer has made possible the implementation of numerical solutions based on the lifting-surface concept. These solutions are similar to the panel solutions for two-dimensional flow discussed in Chapters 3 and 4 in that the wing planform is divided into a number of panels, or

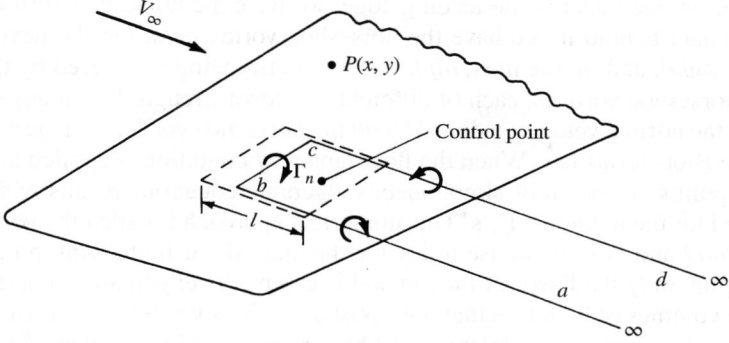

Figure 5.35 Schematic of a single horseshoe vortex, which is part of a vortex system on the wing.

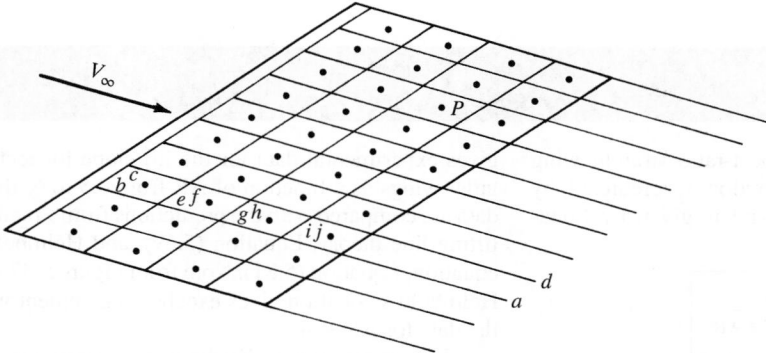

Figure 5.36 Vortex lattice system on a finite wing.

elements. On each panel, either constant or prescribed variations of both γ and δ can be made. Control points on the panels can be chosen, where the net normal flow velocity is zero. The evaluation of equations like Equation (5.80) at these control points results in a system of simultaneous algebraic equations that can be solved for the values of the γ's and δ's on all the panels.

A related but somewhat simpler approach is to superimpose a finite number of horseshoe vortices of different strengths Γ_n on the wing surface. For example, consider Figure 5.35, which shows part of a finite wing. The dashed lines define a panel on the wing planform, where l is the length of the panel in the flow direction. The panel is a trapezoid; it does not have to be a square, or even a rectangle. A horseshoe vortex *abcd* of strength Γ_n is placed on the panel such that the segment *bc* is a distance $l/4$ from the front of the panel. A control point is placed on the centerline of the panel at a distance $\frac{3}{4}l$ from the front. The velocity induced at an arbitrary point P only by the single horseshoe vortex can be calculated from the Biot-Savart law by treating each of the vortex filaments *ab*, *bc*, and *cd* separately. Now consider the entire wing covered by a finite number of panels, as sketched in Figure 5.36. A series of horseshoe vortices is now superimposed. For example, on one panel at the leading edge, we have the horseshoe vortex *abcd*. On the panel behind it, we have the horseshoe vortex *aefd*. On the next panel, we have *aghd*, and on the next, *aijd*, etc. The entire wing is covered by this lattice of horseshoe vortices, each of different unknown strength Γ_n. At any control point P, the normal velocity induced by *all* the horseshoe vortices can be obtained from the Biot-Savart law. When the flow-tangency condition is applied at all the control points, a system of simultaneous algebraic equations results which can be solved for the unknown Γ_n's. This numerical approach is called the *vortex lattice method* and is in wide use today for the analysis of finite-wing properties. Once again, only the flavor of the method is given above; you are encouraged to read the volumes of literature that now exist on various versions of the vortex lattice method. In particular, Reference 13 has an excellent introductory discussion on the vortex lattice method, including a worked example that clearly illustrates the salient points of the technique.

DESIGN BOX

The lift slope for a high-aspect-ratio straight wing with an elliptical lift distribution is predicted by Prandtl's lifting-line theory and is given by Equation (5.69), repeated below:

$$a = \frac{a_0}{1 + a_0/\pi AR}$$

high-aspect-ratio straight wings   (5.69)

This relation, and others like it, is useful for the conceptual design process, where simple formulae, albeit approximate, can lead to fast, back-of-the-envelope calculations. However, Equation (5.69), like all results from simple lifting-line theory, is valid only for high-aspect-ratio straight wings (AR > 4, as a rule of thumb).

The German aerodynamicist H. B. Helmbold in 1942 modified Equation (5.69) to obtain the following form applicable to low-aspect-ratio straight wings:

$$a = \frac{a_0}{\sqrt{1 + (a_0/\pi AR)^2} + a_0/(\pi AR)}$$

low-aspect-ratio straight wing   (5.81)

Equation (5.81) is remarkably accurate for wings with AR < 4. This is demonstrated in Figure 5.37, which

gives experimental data for the lift slope for rectangular wings as a function of AR from 0.5 to 6; these data are compared with the predictions from Prandtl's lifting-line theory, Equation (5.69), and Helmbold's equation, Equation (5.81). Note from Figure 5.37 that Helmbold's equation gives excellent agreement with the data for AR < 4.

For swept wings, Kuchemann (Reference 66) suggests the following modification to Helmbold's equation:

$$a = \frac{a_0 \cos \Lambda}{\sqrt{1 + [(a_0 \cos \Lambda)/(\pi AR)]^2} + [a_0 \cos \Lambda/(\pi AR)]}$$

swept wing   (5.82)

where Λ is the sweep angle of the wing, referenced to the half-chord line, as shown in Figure 5.38.

Keep in mind that Equations (5.69), (5.81), and (5.82) apply to *incompressible flow*. Compressibility corrections to account for high Mach number effects will be discussed in Chapter 10.

Also, the equations above are simple formulae intended to provide quick and easy "back of the envelope" calculations for conceptual design purposes. In contrast, today the power of computational fluid

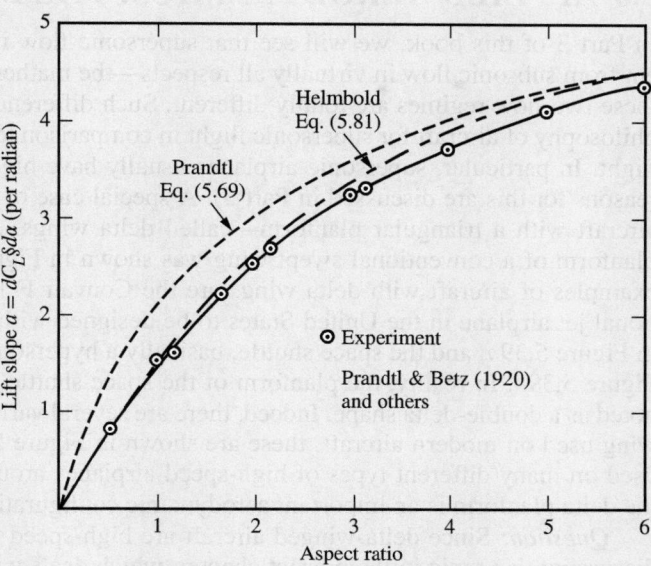

Figure 5.37 Lift slope versus aspect ratio for straight wings in low-speed flow.

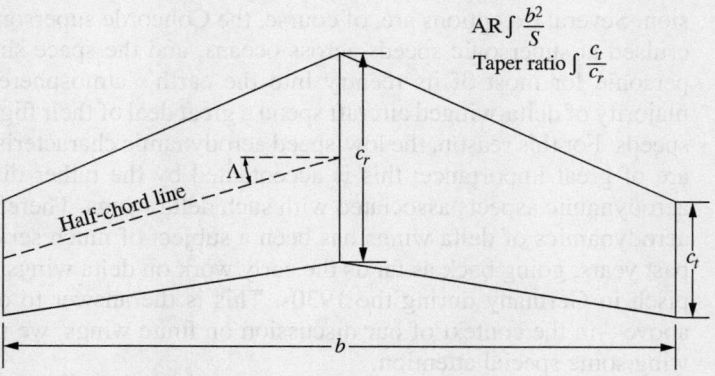

Figure 5.38 Geometry of a swept wing.

dynamics (CFD) allows the detailed calculation of the flow field over a finite wing of any general shape (any aspect ratio, wing sweep, taper ratio, etc.). Moreover, the combination of CFD with modern optimization procedures allows the optimized design of the complete wing, including not only the optimum planform shape, but also the optimum variation of airfoil shape along the span. Because this modern, optimized wing design is usually applied to high-speed airplanes where the flow must be treated as compressible, we defer any discussion of this modern design process until Part 3 of this book.

5.6 APPLIED AERODYNAMICS: THE DELTA WING

In Part 3 of this book, we will see that supersonic flow is dramatically different from subsonic flow in virtually all respects—the mathematics and physics of these two flow regimes are totally different. Such differences impact the design philosophy of aircraft for supersonic flight in comparison to aircraft for subsonic flight. In particular, supersonic airplanes usually have highly swept wings (the reasons for this are discussed in Part 3). A special case of swept wings is those aircraft with a triangular planform—called delta wings. A comparison of the planform of a conventional swept wing was shown in Figure 5.32. Two classic examples of aircraft with delta wings are the Convair F-102A, the first operational jet airplane in the United States to be designed with a delta wing, shown in Figure 5.39a, and the space shuttle, basically a hypersonic airplane, shown in Figure 5.39b. In reality, the planform of the space shuttle is more correctly denoted as a double-delta shape. Indeed, there are several variants of the basic delta wing used on modern aircraft; these are shown in Figure 5.40. Delta wings are used on many different types of high-speed airplanes around the world; hence, the delta planform is an important aerodynamic configuration.

Question: Since delta-winged aircraft are high-speed vehicles, why are we discussing this topic in the present chapter, which deals with the low-speed, incompressible flow over finite wings? The obvious answer is that all high-speed aircraft fly at low speeds for takeoff and landing; moreover, in most cases, these aircraft spend the vast majority of their flight time at subsonic speeds, using their supersonic capability for short "supersonic dashes," depending on their mission. Several exceptions are, of course, the Concorde supersonic transport which cruised at supersonic speeds across oceans, and the space shuttle, which is hypersonic for most of its reentry into the earth's atmosphere. However, a vast majority of delta-winged aircraft spend a great deal of their flight time at subsonic speeds. For this reason, the low-speed aerodynamic characteristics of delta wings are of great importance; this is accentuated by the rather different and unique aerodynamic aspects associated with such delta wings. Therefore, the low-speed aerodynamics of delta wings has been a subject of much serious study over the past years, going back as far as the early work on delta wings by Alexander Lippisch in Germany during the 1930s. This is the answer to our question posed above—in the context of our discussion on finite wings, we must give the delta wing some special attention.

The subsonic flow pattern over the top of a delta wing at angle of attack is sketched in Figure 5.41. The dominant aspect of this flow are the two vortex patterns that occur in the vicinity of the highly swept leading edges. These vortex patterns are created by the following mechanism. The pressure on the bottom surface of the wing at the angle of attack is higher than the pressure on the top surface. Thus, the flow on the bottom surface in the vicinity of the leading edge tries to curl around the leading edge from the bottom to the top. If the leading edge is sharp, the flow will separate along its entire length. (We have already mentioned several times that when low-speed, subsonic flow passes over a sharp

(a)

Figure 5.39 Some delta-winged vehicles. (a) The Convair F-102A. (*NASA.*)

convex corner, inviscid flow theory predicts an infinite velocity at the corner, and that nature copes with this situation by having the flow separate at the corner. The leading edge of a delta wing is such a case.) This separated flow curls into a primary vortex which exists above the wing just inboard of each leading edge, as

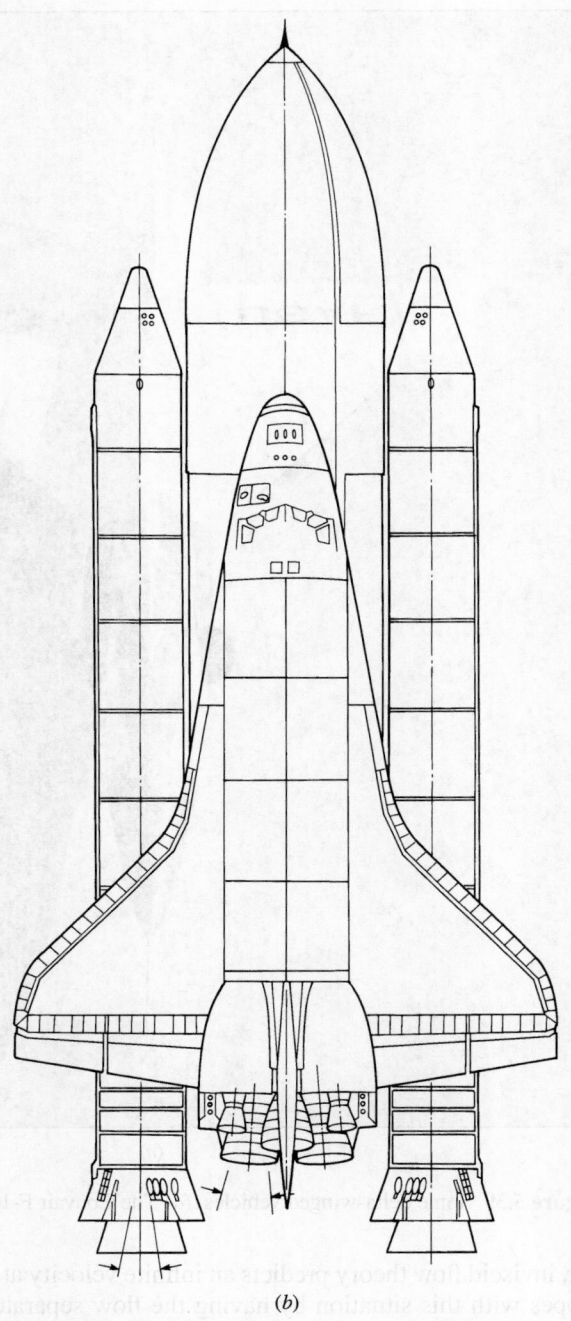

Figure 5.39 (*continued*) Some delta-winged vehicles.
(*b*) The space shuttle. (*NASA*.)

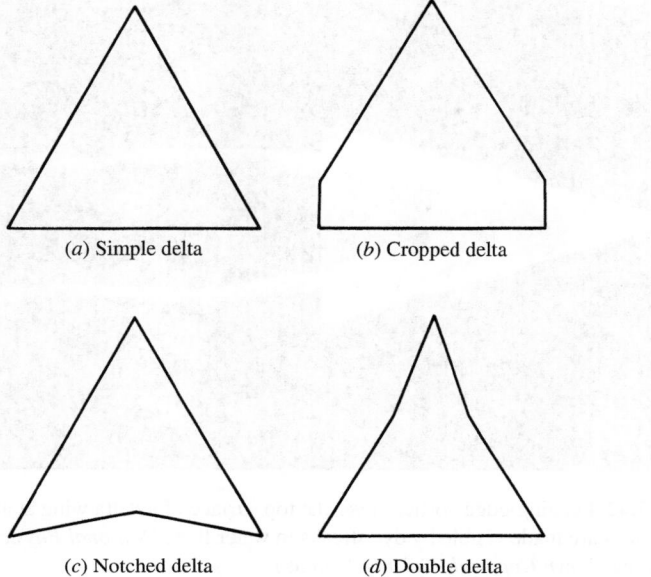

Figure 5.40 Four versions of a delta-wing planform. (From Loftin, Lawrence K., Jr.: *Quest for Performance: The Evolution of Modern Aircraft*, NASA SP-468, 1985.)

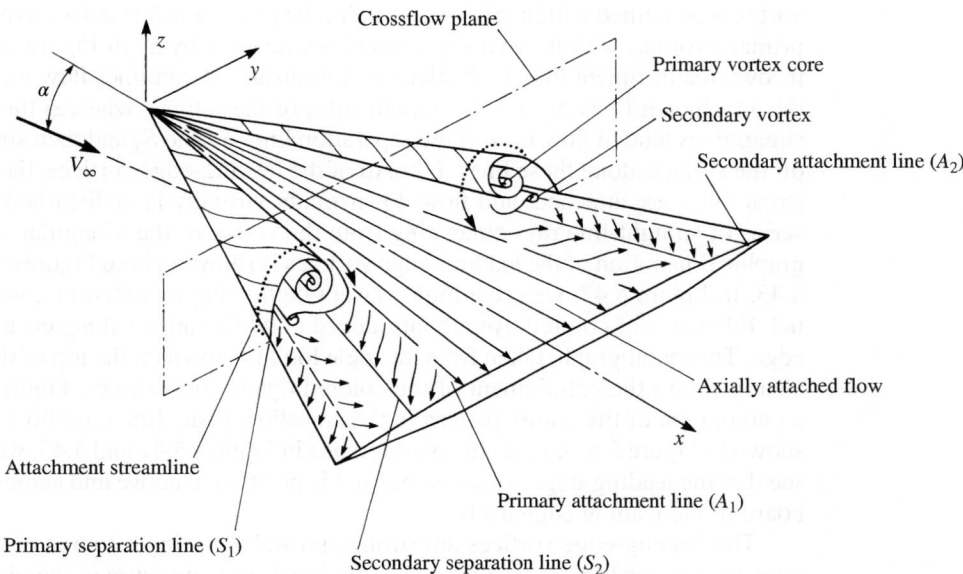

Figure 5.41 Schematic of the subsonic flow field over the top of a delta wing at angle of attack. (*Adapted from John Stollery, Cranfield Institute of Technology, England.*)

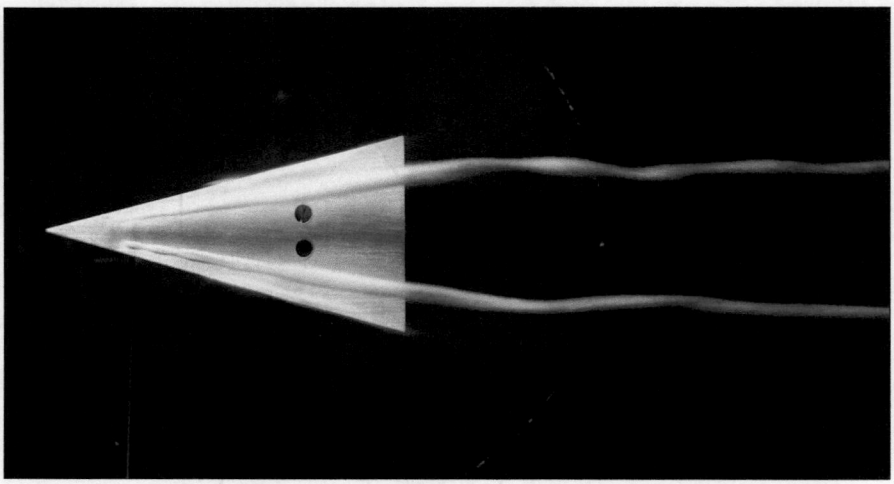

Figure 5.42 Leading-edge vortices over the top surface of a delta wing at angle of attack. The vortices are made visible by dye streaks in water flow. (*National Physical Laboratory/Crown Copyright/Science Source.*)

sketched in Figure 5.41. The stream surface which has separated at the leading edge (the primary separation line S_1 in Figure 5.41) loops above the wing and then reattaches along the primary attachment line (line A_1 in Figure 5.41). The primary vortex is contained within this loop. A secondary vortex is formed underneath the primary vortex, with its own separation line, denoted by S_2 in Figure 5.41, and its own reattachment line A_2. Notice that the surface streamlines flow *away* from the attachment lines A_1 and A_2 on both sides of these lines, whereas the surface streamlines tend to flow *toward* the separation lines S_1 and S_2 and then simply lift off the surface along these lines. Inboard of the leading-edge vortices, the surface streamlines are attached, and flow downstream virtually is undisturbed along a series of straight-line rays emanating from the vertex of the triangular shape. A graphic illustration of the leading-edge vortices is shown in both Figures 5.42 and 5.43. In Figure 5.42, we see a highly swept delta wing mounted in a water tunnel. Filaments of colored dye are introduced at two locations along each leading edge. This photograph, taken from an angle looking down on the top of the wing, clearly shows the entrainment of the colored dye in the vortices. Figure 5.43 is a photograph of the vortex pattern in the crossflow plane (the crossflow plane is shown in Figure 5.41). From the photographs in Figures 5.42 and 5.43, we clearly see that the leading-edge vortex is real and is positioned above and somewhat inboard of the leading edge itself.

The leading-edge vortices are strong and stable. Being a source of high energy and relatively high-vorticity flow, the local static pressure in the vicinity of the vortices is small. Hence, the surface pressure on the top surface of the delta

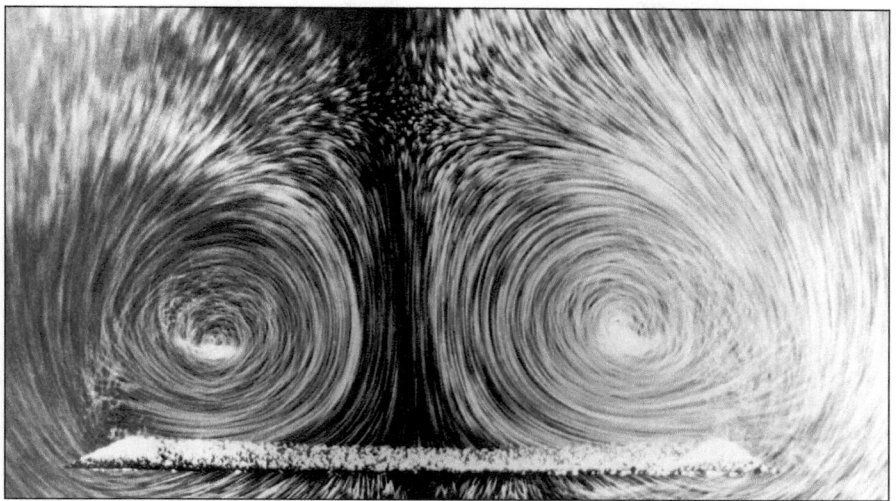

Figure 5.43 The flow field in the crossflow plane above a delta wing at angle of attack, showing the two primary leading-edge vortices. The vortices are made visible by small air bubbles in water. (©*ONERA The French Aerospace Lab*)

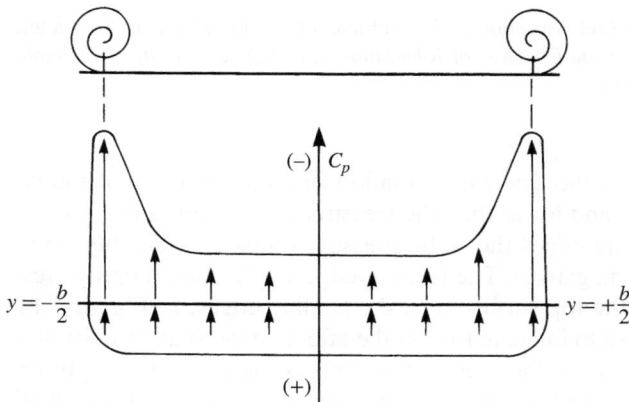

Figure 5.44 Schematic of the spanwise pressure coefficient distribution across a delta wing. (*Data Courtesy of John Stollery, Cranfield Institute of Technology, England.*)

wing is reduced near the leading edge and is higher and reasonably constant over the middle of the wing. The qualitative variation of the pressure coefficient in the spanwise direction (the y direction as shown in Figure 5.41) is sketched in Figure 5.44. The spanwise variation of pressure over the bottom surface is essentially constant and higher than the freestream pressure (a positive C_p). Over

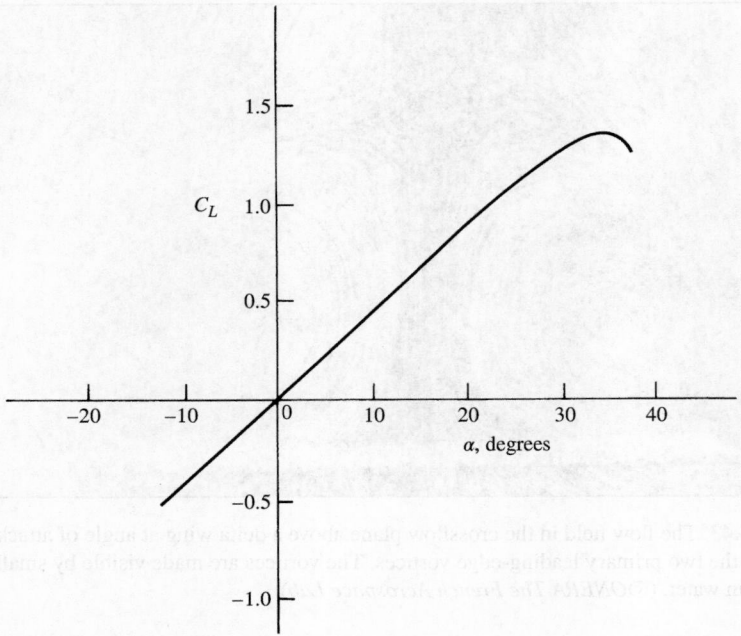

Figure 5.45 Variation of lift coefficient for a flat delta wing with angle of attack. (*Data Courtesy of John Stollery, Cranfield Institute of Technology, England.*)

the top surface, the spanwise variation in the midsection of the wing is essentially constant and lower than the freestream pressure (a negative C_p). However, near the leading edges the static pressure drops considerably (the values of C_p become more negative). The leading-edge vortices are literally creating a strong "suction" on the top surface near the leading edges. In Figure 5.44, vertical arrows are shown to indicate further the effect on the spanwise lift distribution; the upward direction of these arrows as well as their relative length show the local contribution of each section of the wing to the normal force distribution. The suction effect of the leading-edge vortices is clearly shown by these arrows.

The suction effect of the leading-edge vortices enhances the lift; for this reason, the lift coefficient curve for a delta wing exhibits an increase in C_L for values of α at which conventional wing planforms would be stalled. A typical variation of C_L with α for a 60° delta wing is shown in Figure 5.45. Note the following characteristics:

1. The lift slope is small, on the order of 0.05/degree.
2. However, the lift continues to increase to large values of α; in Figure 5.45, the stalling angle of attack is on the order of 35°. The net result is a reasonable value of $C_{L,\max}$, on the order of 1.3.

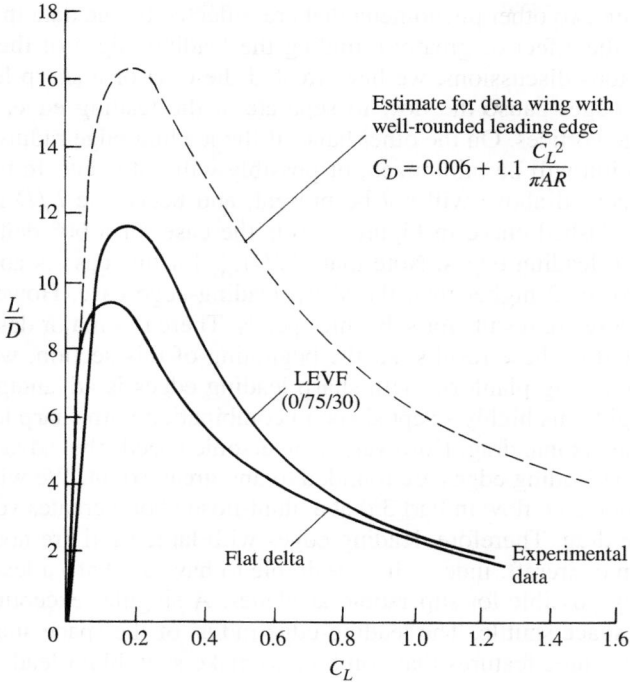

Figure 5.46 The effect of leading-edge shape on the lift-to-drag ratio for a delta wing of aspect ratio 2.31. The two solid curves apply to a sharp leading edge, and the dashed curve applies to a rounded leading edge. LEVF denotes a wing with a leading-edge vortex flap. (*Data Courtesy of John Stollery, Cranfield Institute of Technology, England.*)

The next time you have an opportunity to watch a delta-winged airplane take off or land, say, for example, the televised landing of the space shuttle, note the large angle of attack of the vehicle. Moreover, you will understand why the angle of attack is large—because the lift slope is small, and hence the angle of attack must be large enough to generate the high values of C_L required for low-speed flight.

The suction effect of the leading-edge vortices, in acting to increase the normal force, consequently, increases the drag at the same time it increases the lift. Hence, the aerodynamic effect of these vortices is not necessarily advantageous. In fact, the lift-to-drag ratio L/D for a delta planform is not so high as conventional wings. The typical variation of L/D with C_L for a delta wing is shown in Figure 5.46, the results for the sharp leading edge, 60° delta wing are given by the lower curve. Note that the maximum value of L/D for this case is about 9.3—not a particularly exciting value for a low-speed aircraft.

There are two other phenomena that are reflected by the data in Figure 5.46. The first is the effect of greatly rounding the leading edges of the delta wing. In our previous discussions, we have treated the case of a sharp leading edge; such sharp edges cause the flow to separate at the leading edge, forming the leading-edge vortices. On the other hand, if the leading-edge radius is large, the flow separation will be minimized, or possibly will not occur. In turn, the drag penalty discussed above will not be present, and hence the L/D ratio will increase. The dashed curve in Figure 5.46 is the case for a 60° delta wing with well-rounded leading edges. Note that $(L/D)_{max}$ for this case is about 16.5, almost a factor of 2 higher than the sharp leading-edge case. However, keep in mind that these are results for subsonic speeds. There is a major design compromise reflected in these results. At the beginning of this section, we mentioned that the delta-wing planform with sharp leading edges is advantageous for supersonic flight—its highly swept shape in combination with sharp leading edges has a low supersonic drag. However, at supersonic speeds this advantage will be negated if the leading edges are rounded to any great extent. We will find in our study of supersonic flow in Part 3 that a blunt-nosed body creates very large values of wave drag. Therefore, leading edges with large radii are not appropriate for supersonic aircraft; indeed, it is desirable to have as sharp a leading edge as is practically possible for supersonic airplanes. A singular exception is the design of the space shuttle. The leading-edge radius of the space shuttle is large; this is due to three features that combine to make such blunt leading edges advantageous for the shuttle. First, the shuttle must slow down early during reentry into the earth's atmosphere to avoid massive aerodynamic heating (aspects of aerodynamic heating are discussed in Part 4). Therefore, in order to obtain this deceleration, a high drag is desirable for the space shuttle; indeed, the maximum L/D ratio of the space shuttle during reentry is about 2. A large leading-edge radius, with its attendant high drag, is therefore advantageous. Second, as we will see in Part 4, the rate of aerodynamic heating to the leading edge itself—a region of high heating—is inversely proportional to the square root of the leading-edge radius. Hence, the larger the radius, the smaller will be the heating rate to the leading edge. Third, as already explained above, a highly rounded leading edge is certainly advantageous to the shuttle's subsonic aerodynamic characteristics. Hence, a well-rounded leading edge is an important design feature for the space shuttle on all accounts. However, we must be reminded that this is not the case for more conventional supersonic aircraft, which demand very sharp leading edges. For these aircraft, a delta wing with a sharp leading edge has relatively poor subsonic performance.

This leads to the second of the phenomena reflected in Figure 5.46. The middle curve in Figure 5.46 is labeled LEVF, which denotes the case for a leading-edge vortex flap. This pertains to a mechanical configuration where the leading edges can be deflected downward through a variable angle, analogous to the deflection of a conventional trailing-edge flap. The spanwise pressure-coefficient distribution for this case is sketched in Figure 5.47; note that the direction of the suction due to the leading-edge vortice is now modified in comparison to

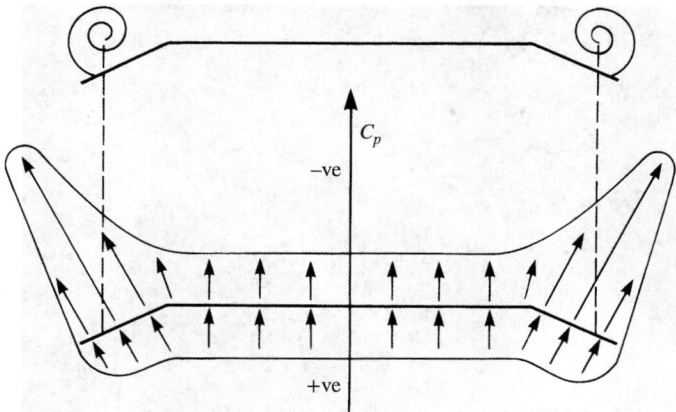

Figure 5.47 A schematic of the spanwise pressure coefficient distribution over the top of a delta wing as modified by leading-edge vortex flaps. (*Data Courtesy of John Stollery, Cranfield Institute of Technology, England.*)

the case with no leading-edge flap shown earlier in Figure 5.44. Also, returning to Figure 5.41, you can visualize what the wing geometry would look like with the leading edge drooped down; a front view of the downward deflected flap would actually show some projected frontal area. Since the pressure is low over this frontal area, the net drag can decrease. This phenomenon is illustrated by the middle curve in Figure 5.46, which shows a generally higher L/D for the leading-edge vortex flap in comparison to the case with no flap (the flat delta wing).

Finally, we note something drastic that occurs in the flow over the top surface of a delta wing when it is at a high enough angle of attack. The primary vortices shown in Figures 5.41 and 5.42 begin to fall apart somewhere along the length of the vortex; this is called *vortex breakdown,* illustrated in Figure 5.48. Compare this photograph with that shown in Figure 5.42 for well-behaved vortices at lower angle of attack. In Figure 5.48, the two leading-edge vortices show vortex breakdown at a location about two-thirds along their length over the top of the wing. This photograph is particularly interesting because it shows two types of vortex breakdown. The vortex at the top of the photograph exhibits a spiral-type of vortex breakdown, where the breakdown occurs progressively along the core and causes the core to twist in various directions. The vortex at the bottom of the photograph exhibits a bubble-type of vortex breakdown, where the vortex suddenly bursts, forming a large bubble of chaotic flow. The spiral type of vortex breakdown is more common. When vortex breakdown occurs, the lift and pitching moment of the delta wing decrease, the flow becomes unsteady, and buffeting of the wing occurs.

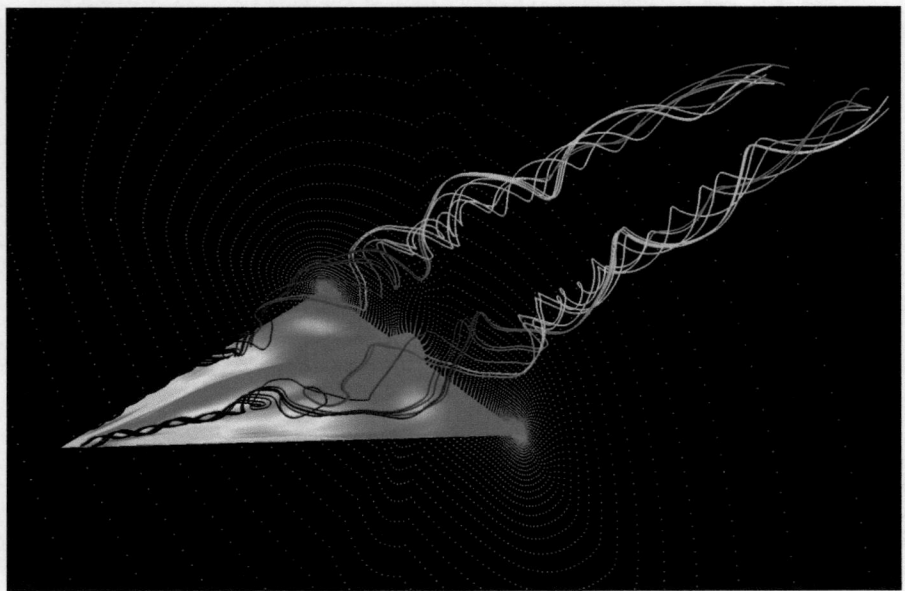

Figure 5.48 Vortex breakdown over a delta wing. (*NASA/Science Source*)

The progressive development of vortex breakdown is shown by the computational fluid dynamic (CFD) results in Figure 5.49 for a delta wing with a 60-degree sweep angle. At $\alpha = 5°$ (Figure 5.49*a*), the vortex core is well-behaved. At $\alpha = 15°$ (Figure 5.49*b*), vortex breakdown is starting. At $\alpha = 40°$ (Figure 5.49*c*), the flow over the top of the delta wing is completely separated, and the wing is stalled. The results in Figure 5.49 are interesting for another reason as well. In a footnote to Section 4.4, we noted that inviscid flow calculations sometimes predict the location and nature of flow separation. Here we see another such case. The vortex breakdown and separated flow shown in Figure 5.49 are calculated from a CFD solution of the Euler equations (i.e., an inviscid flow calculation). It appears that friction does not play a critical role in vortex formation and breakdown.

(For more information on vortex bursting over delta wings, see the recent survey by I. Gursul, "Recent Developments in Delta Wing Aerodynamics," *The Aeronautical Journal,* vol. 108, number 1087, September 2004, pp. 437–452.)

In summary, the delta wing is a common planform for supersonic aircraft. In this section, we have examined the low-speed aerodynamic characteristics of such wings and have found that these characteristics are in some ways quite different from a conventional planform.

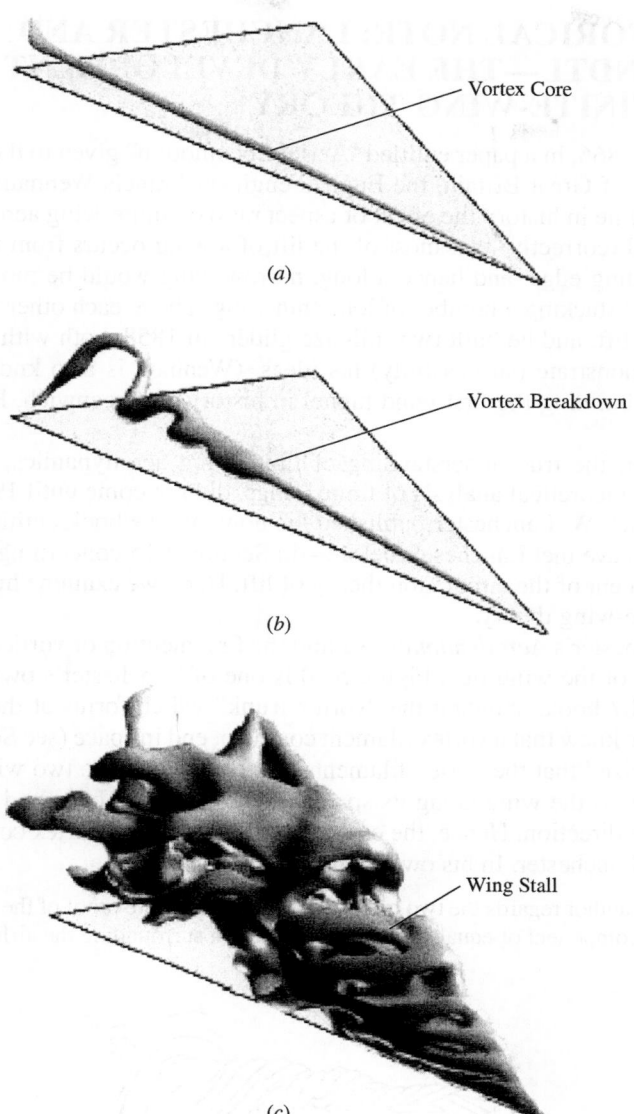

(a)

(b)

(c)

Figure 5.49 Vortex behavior and breakdown progression with increasing angles of attack of (a) 5°, (b) 15°, (c) 40°. (*Source:* R. E. Gordnier and M. R. Visbal, "Computation of the Aeroelastic Response of a Flexible Delta Wing at High Angles-of-Attack," *AIAA Paper 2003-1728*, 2003.)

5.7 HISTORICAL NOTE: LANCHESTER AND PRANDTL—THE EARLY DEVELOPMENT OF FINITE-WING THEORY

On June 27, 1866, in a paper entitled "Aerial Locomotion" given to the Aeronautical Society of Great Britain, the English engineer Francis Wenham expressed for the first time in history the effect of aspect ratio on finite-wing aerodynamics. He theorized (correctly) that most of the lift of a wing occurs from the portion near the leading edge, and hence a long, narrow wing would be most efficient. He suggested stacking a number of long thin wings above each other to generate the required lift, and he built two full-size gliders in 1858, both with five wings each, to demonstrate (successfully) his ideas. (Wenham is also known for designing and building the first wind tunnel in history, at Greenwich, England, in 1871.)

However, the true understanding of finite-wing aerodynamics, as well as ideas for the theoretical analysis of finite wings, did not come until 1907. In that year, Frederick W. Lanchester published his now famous book entitled *Aerodynamics*. We have met Lanchester before—in Section 4.15 concerning his role in the development of the circulation theory of lift. Here, we examine his contributions to finite-wing theory.

In Lanchester's *Aerodynamics,* we find the first mention of vortices that trail downstream of the wing tips. Figure 5.50 is one of Lanchester's own drawings from his 1907 book, showing the "vortex trunk" which forms at the wing tip. Moreover, he knew that a vortex filament could not end in space (see Section 5.2), and he theorized that the vortex filaments that constituted the two wing-tip vortices must cross the wing along its span—the first concept of bound vortices in the spanwise direction. Hence, the essence of the horseshoe vortex concept originated with Lanchester. In his own words:

> Thus the author regards the two trailed vortices as a definite proof of the existence of a cyclic component of equal strength in the motion surrounding the airfoil itself.

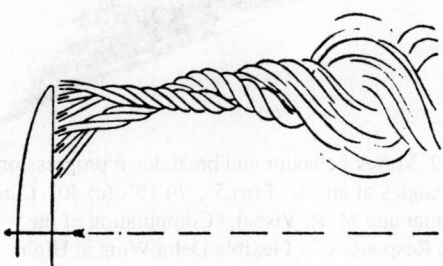

Figure 5.50 A figure from Lanchester's *Aerodynamics,* 1907; this is his own drawing of the wing-tip vortex on a finite wing.

Considering the foresight and originality of Lanchester's thinking, let us pause for a moment and look at the person himself. Lanchester was born on October 23, 1868, in Lewisham, England. The son of an architect, Lanchester became interested in engineering at an early age. (He was told by his family that his mind was made up at the age of 4.) He studied engineering and mining during the years 1886–1889 at the Royal College of Science in South Kensington, London, but never officially graduated. He was a quick-minded and innovative thinker and became a designer at the Forward Gas Engine Company in 1889, specializing in internal combustion engines. He rose to the post of assistant works manager. In the early 1890s, Lanchester became very interested in aeronautics, and along with his development of high-speed engines, he also carried out numerous aerodynamics experiments. It was during this period that he formulated his ideas on both the circulation theory of lift and the finite-wing vortex concepts. A serious paper written by Lanchester first for the Royal Society, and then for the Physical Society, was turned down for publication—something Lanchester never forgot. Finally, his aeronautical concepts were published in his two books *Aerodynamics* and *Aerodonetics* in 1907 and 1908, respectively. To his detriment, Lanchester had a style of writing and a means of explanation that were not easy to follow and his works were not immediately seized upon by other researchers. Lanchester's bitter feelings about the public's receipt of his papers and books are graphically seen in his letter to the Daniel Guggenheim Medal Fund decades later. In a letter dated June 6, 1931, Lanchester writes:

> So far as aeronautical science is concerned, I cannot say that I experienced anything but discouragement; in the early days my theoretical work (backed by a certain amount of experimental verification), mainly concerning the vortex theory of sustentation and the screw propeller, was refused by the two leading scientific societies in this country, and I was seriously warned that my profession as an engineer would suffer if I dabbled in a subject that was merely a dream of madmen! When I published my two volumes in 1907 and 1908 they were well received on the whole, but this was mainly due to the success of the brothers Wright, and the general interest aroused on the subject.

In 1899, he formed the Lanchester Motor Company Limited, and sold automobiles of his own design. Lanchester maintained his interest in automobiles and related mechanical devices until his death on March 8, 1946, at the age of 77.

In 1908, Lanchester visited Göttingen, Germany, and fully discussed his wing theory with Ludwig Prandtl and his student Theodore von Karman. Prandtl spoke no English, Lanchester spoke no German, and in light of Lanchester's unclear way of explaining his ideas, there appeared to be little chance of understanding between the two parties. However, shortly after, Prandtl began to develop his own wing theory, using a bound vortex along the span and assuming that the vortex trails downstream from both wing tips. The first mention of Prandtl's work on finite-wing theory was made in a paper by O. Foppl in 1911, discussing some of Foppl's experimental work on finite wings. Commenting on his results, Foppl says:

> They agree very closely with the theoretical investigation by Professor Prandtl on the current around an airplane with a finite span wing. Already Lanchester in his work, "Aerodynamics" (translated into German by C. and A. Runge), indicated that to the two extremities of an airplane wing are attached two vortex ropes (Wirbelzopfe) which make possible the transition from the flow around the airplane, which occurs nearly according to Kutta's theory, to the flow of the undisturbed fluid at both sides. These two vortex ropes continue the vortex which, according to Kutta's theory, takes place on the lamina.
>
> We are led to admit this owing to the Helmholtz theorem that vortices cannot end in the fluid. At any rate these two vortex ropes have been made visible in the Göttingen Institute by emitting an ammonia cloud into the air. Prandtl's theory is constructed on the consideration of this current in reality existing.

In the same year, Prandtl expressed his own first published words on the subject. In a paper given at a meeting of the Representatives of Aeronautical Science in Göttingen in November 1911, entitled "Results and Purposes of the Model Experimental Institute of Göttingen," Prandtl states:

> Another theoretical research relates to the conditions of the current which is formed by the air behind an airplane. The lift generated by the airplane is, on account of the principle of action and reaction, necessarily connected with a descending current behind the airplane. Now it seemed very useful to investigate this descending current in all its details. It appears that the descending current is formed by a pair of vortices, the vortex filaments of which start from the airplane wing tips. The distance of the two vortices is equal to the span of the airplane, their strength is equal to the circulation of the current around the airplane and the current in the vicinity of the airplane is fully given by the superposition of the uniform current with that of a vortex consisting of three rectilinear sections.

In discussing the results of his theory, Prandtl goes on to state in the same paper:

> The same theory supplies, taking into account the variations of the current on the airplane which came from the lateral vortices, a relationship showing the dependence of the airplane lift on the aspect ratio; in particular it gives the possibility of extrapolating the results thus obtained experimentally to the airplane of infinite span wing. From the maximum aspect ratios measured by us (1:9 to that of 1:∞) the lifts increase further in marked degree—by some 30 or 40 percent. I would add here a remarkable result of this extrapolation, which is, that the results of Kutta's theory of the infinite wing, at least so far as we are dealing with small cambers and small angles of incidence, have been confirmed by these experimental results.
>
> Starting from this line of thought we can attack the problem of calculating the surface of an airplane so that lift is distributed along its span in a determined manner, previously fixed. The experimental trial of these calculations has not yet been made, but it will be in the near future.

It is clear from the above comments that Prandtl was definitely following the model proposed earlier by Lanchester. Moreover, the major concern of the finite-wing theory was first in the calculation of lift—no mention is made of induced drag. It is interesting to note that Prandtl's theory first began with a single horse-shoe vortex, as sketched in Figure 5.13. The results were not entirely satisfactory. During the period 1911–1918, Prandtl and his colleagues expanded and refined his finite-wing theory, which evolved into the concept of a lifting line consisting of an infinite number of horseshoe vortices, as sketched in Figure 5.15. In 1918, the term "induced drag" was coined by Max Munk, a colleague of Prandtl at Göttingen. Much of Prandtl's development of finite-wing theory was classified as secret by the German government during World War I. Finally, his lifting-line theory was released to the outside world, and his ideas were published in English in a special NACA report written by Prandtl and published in 1922, entitled "Applications of Modern Hydrodynamics to Aeronautics" (NACA TR 116). Hence, the theory we have outlined in Section 5.3 was well-established more than 80 years ago.

One of Prandtl's strengths was the ability to base his thinking on sound ideas, and to apply intuition that resulted in relatively straightforward theories that most engineers could understand and appreciate. This is in contrast to the difficult writings of Lanchester. As a result, the lifting theory for finite wings has come down through the years identified as *Prandtl's lifting-line theory,* although we have seen that Lanchester was the first to propose the basic model on which lifting-line theory is built.

In light of Lanchester's 1908 visit with Prandtl and Prandtl's subsequent development of the lifting-line theory, there has been some discussion over the years that Prandtl basically stole Lanchester's ideas. However, this is clearly not the case. We have seen in the above quotes that Prandtl's group at Göttingen was giving full credit to Lanchester as early as 1911. Moreover, Lanchester never gave the world a clear and practical theory with which results could be readily obtained—Prandtl did. Therefore, in this book we have continued the tradition of identifying the lifting-line theory with Prandtl's name. On the other hand, for very good reasons, in England and various places in western Europe, the theory is labeled the *Lanchester-Prandtl theory*.

To help put the propriety in perspective, Lanchester was awarded the Daniel Guggenheim Medal in 1936 (Prandtl had received this award some years earlier). In the medal citation, we find the following words:

> Lanchester was the foremost person to propound the now famous theory of flight based on the Vortex theory, so brilliantly followed up by Prandtl and others. He first put forward his theory in a paper read before the Birmingham Natural History and Philosophical Society on 19th June, 1894. In a second paper in 1897, in his two books published in 1907 and 1908, and in his paper read before the Institution of Automobile Engineers in 1916, he further developed this doctrine.

Perhaps the best final words on Lanchester are contained in this excerpt from his obituary found in the British periodical *Flight* in March 1946:

> And now Lanchester has passed from our ken but not from our thoughts. It is to be hoped that the nation which neglected him during much of his lifetime will at any rate perpetuate his work by a memorial worthy of the "Grand Old Man" of aerodynamics.

5.8 HISTORICAL NOTE: PRANDTL—THE PERSON

The modern science of aerodynamics rests on a strong fundamental foundation, a large percentage of which was established in one place by one person—at the University of Göttingen by Ludwig Prandtl. Prandtl never received a Nobel Prize, although his contributions to aerodynamics and fluid mechanics are felt by many to be of that caliber. Throughout this book, you will encounter his name in conjunction with major advances in aerodynamics: thin airfoil theory in Chapter 4, finite-wing theory in Chapter 5, supersonic shock- and expansion-wave theory in Chapter 9, compressibility corrections in Chapter 11, and what may be his most important contribution, namely, the boundary-layer concept in Chapter 17. Who was this person who has had such a major impact on fluid dynamics? Let us take a closer look.

Ludwig Prandtl was born on February 4, 1874, in Freising, Bavaria. His father was Alexander Prandtl, a professor of surveying and engineering at the agricultural college at Weihenstephan, near Freising. Although three children were born into the Prandtl family, two died at birth, and Ludwig grew up as an only child. His mother, the former Magdalene Ostermann, had a protracted illness, and partly as a result of this, Prandtl became very close to his father. At an early age, Prandtl became interested in his father's books on physics, machinery, and instruments. Much of Prandtl's remarkable ability to go intuitively to the heart of a physical problem can be traced to his environment at home as a child, where his father, a great lover of nature, induced Ludwig to observe natural phenomena and to reflect on them.

In 1894, Prandtl began his formal scientific studies at the Technische Hochschule in Munich, where his principal teacher was the well-known mechanics professor, August Foppl. Six years later, he graduated from the University of Munich with a Ph.D., with Foppl as his advisor. However, by this time Prandtl was alone, his father having died in 1896 and his mother in 1898.

By 1900, Prandtl had not done any work or shown any interest in fluid mechanics. Indeed, his Ph.D. thesis at Munich was in solid mechanics, dealing with unstable elastic equilibrium in which bending and distortion acted together. (It is not generally recognized by people in fluid dynamics that Prandtl continued his interest and research in solid mechanics through most of his life—this work is eclipsed, however, by his major contributions to the study of fluid flow.) However, soon after graduation from Munich, Prandtl had his first major encounter with fluid mechanics. Joining the Nuremburg works of the Maschinenfabrick Augsburg as an engineer, Prandtl worked in an office designing mechanical equipment for the new factory. He was made responsible for redesigning an apparatus for removing machine shavings by suction. Finding no reliable information in the scientific literature about the fluid mechanics of suction, Prandtl arranged his own experiments to answer a few fundamental questions about the flow. The result of this work was his new design for shavings cleaners. The apparatus was modified with pipes of improved shape and size, and carried out satisfactory operation at one-third its original power consumption. Prandtl's contributions to fluid mechanics had begun.

One year later, in 1901, he became Professor of Mechanics in the Mathematical Engineering Department at the Technische Hochschule in Hanover. (Please note that in Germany a "technical high school" is equivalent to a technical university in the United States.) It was at Hanover that Prandtl enhanced and continued his new-found interest in fluid mechanics. He also developed his boundary-layer theory and became interested in supersonic flow through nozzles at Hanover. In 1904, Prandtl delivered his famous paper on the concept of the boundary layer to the Third Congress on Mathematicians at Heidelberg. Entitled "Über Flussigkeitsbewegung bei sehr kleiner Reibung," Prandtl's Heidelberg paper established the basis for most modern calculations of skin friction, heat transfer, and flow separation (see Chapters 15 to 20). From that time on, the star of Prandtl was to rise meteorically. Later that year, he moved to the prestigious University of Göttingen to become Director of the Institute for Technical Physics, later to be renamed Applied Mechanics. Prandtl spent the remainder of his life at Göttingen, building his laboratory into the world's greatest aerodynamic research center of the 1904–1930 time period.

At Göttingen, during 1905–1908, Prandtl carried out numerous experiments on supersonic flow through nozzles and developed oblique shock- and expansion-wave theory (see Chapter 9). He took the first photographs of the supersonic flow through nozzles, using a special schlieren optical system (see Chapter 4 of Reference 21). From 1910 to 1920, he devoted most of his efforts to low-speed aerodynamics, principally airfoil and wing theory, developing the famous lifting-line theory for finite wings (see Section 5.3). Prandtl returned to high-speed flows in the 1920s, during which he contributed to the evolution of the famous Prandtl-Glauert compressibility correction (see Sections 11.4 and 11.11).

By the 1930s, Prandtl was recognized worldwide as the "elder statesman" of fluid dynamics. Although he continued to do research in various areas, including structural mechanics and meteorology, his "Nobel Prize-level" contributions to fluid dynamics had all been made. Prandtl remained at Göttingen throughout the turmoil of World War II, engrossed in his work and seemingly insulated from the intense political and physical disruptions brought about by Nazi Germany. In fact, the German Air Ministry provided Prandtl's laboratory with new equipment and financial support. Prandtl's attitude at the end of the war is reflected in his comments to a U.S. Army interrogation team that swept through Göttingen in 1945; he complained about bomb damage to the roof of his house, and he asked how the Americans planned to support his current and future research. Prandtl was 70 at the time and was still going strong. However, the fate of Prandtl's laboratory at this time is summed up in the words of Irmgard Flugge-Lotz and Wilhelm Flugge, colleagues of Prandtl, who wrote 28 years later in the *Annual Review of Fluid Mechanics* (Vol. 5, 1973):

> World War II swept over all of us. At its end some of the research equipment was dismantled, and most of the research staff was scattered with the winds. Many are now in this country (the United States) and in England, some have returned. The seeds sown by Prandtl have sprouted in many places, and there are now many "second growth" Göttingers who do not even know that they are.

Figure 5.51 Ludwig Prandtl (1875–1953). (*Emilio Segrè Visual Archives/American Institute of Physics/Science Source.*)

What type of person was Prandtl? By all accounts he was gracious, studious, likable, friendly, and totally focused on those things that interested him. He enjoyed music and was an accomplished pianist. Figure 5.51 shows a rather introspective person busily at work. One of Prandtl's most famous students, Theodor von Kármán, wrote in his autobiography *The Wind and Beyond* (Little, Brown and Company, Boston, 1967) that Prandtl bordered on being naive. A favorite story along these lines is that, in 1909, Prandtl decided that he should be married, but he did not know quite what to do. He finally wrote to Mrs. Foppl, the wife of his respected teacher, asking permission to marry one of her two daughters. Prandtl and Foppl's daughters were acquainted, but nothing more than that. Moreover, Prandtl did not stipulate which daughter. The Foppl's made a family decision that Prandtl should marry the elder daughter, Gertrude. This story from von Karman has been disputed in recent years by a Prandtl relative. Nevertheless, the marriage took place, leading to a happy relationship. The Prandtl's had two daughters, born in 1914 and 1917.

Prandtl was considered a tedious lecturer because he could hardly make a statement without qualifying it. However, he attracted excellent students who later went on to distinguish themselves in fluid mechanics—such as Jakob Ackeret in Zurich, Switzerland, Adolf Busemann in Germany, and Theodor von Kármán at Aachen, Germany, and later at Cal Tech in the United States.

Prandtl died in 1953. He was clearly the founder of modern aerodynamics—a monumental figure in fluid dynamics. His impact will be felt for centuries to come.

5.9 SUMMARY

Return to the chapter road map in Figure 5.7, and review the straightforward path we have taken during the development of finite-wing theory. Make certain that you feel comfortable with the flow of ideas before proceeding further.

A brief summary of the important results of this chapter follows:

The wing-tip vortices from a finite wing induce a downwash which reduces the angle of attack effectively seen by a local airfoil section:

$$\alpha_{\text{eff}} = \alpha - \alpha_i \tag{5.1}$$

In turn, the presence of downwash results in a component of drag defined as induced drag D_i.

Vortex sheets and vortex filaments are useful in modeling the aerodynamics of finite wings. The velocity induced by a directed segment $\mathbf{dl}$ of a vortex filament is given by the Biot-Savart law:

$$\mathbf{dV} = \frac{\Gamma}{4\pi} \frac{\mathbf{dl} \times \mathbf{r}}{|\mathbf{r}|^3} \tag{5.5}$$

In Prandtl's classical lifting-line theory, the finite wing is replaced by a single spanwise lifting line along which the circulation $\Gamma(y)$ varies. A system of vortices trails downstream from the lifting line, which induces a downwash at the lifting line. The circulation distribution is determined from the fundamental equation

$$\alpha(y_0) = \frac{\Gamma(y_0)}{\pi V_\infty c(y_0)} + \alpha_{L=0}(y_0) + \frac{1}{4\pi V_\infty} \int_{-b/2}^{b/2} \frac{(d\Gamma/dy)\,dy}{y_0 - y} \tag{5.23}$$

Results from classical lifting-line theory:

Elliptic wing:

Downwash is constant:

$$w = -\frac{\Gamma_0}{2b} \tag{5.35}$$

$$\alpha_i = \frac{C_L}{\pi AR} \tag{5.42a}$$

$$C_{D,i} = \frac{C_L^2}{\pi AR} \tag{5.43}$$

$$a = \frac{a_0}{1 + a_0/\pi AR} \tag{5.69}$$

General wing:

$$C_{D,i} = \frac{C_L^2}{\pi AR}(1 + \delta) = \frac{C_L^2}{\pi e AR} \tag{5.61 and 5.62}$$

$$a = \frac{a_0}{1 + (a_0/\pi AR)(1 + \tau)} \tag{5.70}$$

For low-aspect-ratio wings, swept wings, and delta wings, lifting-surface theory must be used. In modern aerodynamics, such lifting-surface theory is implemented by the vortex panel or the vortex lattice techniques.

5.10 PROBLEMS

5.1 Consider a vortex filament of strength Γ in the shape of a closed circular loop of radius R. Obtain an expression for the velocity induced at the center of the loop in terms of Γ and R.

5.2 Consider the same vortex filament as in Problem 5.1. Consider also a straight line through the center of the loop, perpendicular to the plane of the loop. Let A be the distance along this line, measured from the plane of the loop. Obtain an expression for the velocity at distance A on the line, as induced by the vortex filament.

5.3 The measured lift slope for the NACA 23012 airfoil is 0.1080 degree^{-1}, and $\alpha_{L=0} = -1.3°$. Consider a finite wing using this airfoil, with AR = 8 and taper ratio = 0.8. Assume that $\delta = \tau$. Calculate the lift and induced drag coefficients for this wing at a geometric angle of attack = 7°.

5.4 The Piper Cherokee (a light, single-engine general aviation aircraft) has a wing area of 170 ft^2 and a wing span of 32 ft. Its maximum gross weight is 2450 lb. The wing uses an NACA 65-415 airfoil, which has a lift slope of 0.1033 degree^{-1} and $\alpha_{L=0} = -3°$. Assume $\tau = 0.12$. If the airplane is cruising at 120 mi/h at standard sea level at its maximum gross weight and is in straight-and-level flight, calculate the geometric angle of attack of the wing.

5.5 Consider the airplane and flight conditions given in Problem 5.4. The span efficiency factor e for the complete airplane is generally much less than that for the finite wing alone. Assume $e = 0.64$. Calculate the induced drag for the airplane in Problem 5.4.

5.6 Consider a finite wing with an aspect ratio of 6. Assume an elliptical lift distribution. The lift slope for the airfoil section is 0.1/degree. Calculate and compare the lift slopes for (a) a straight wing, and (b) a swept wing, with a half-chord line sweep of 45 degrees.

5.7 Repeat Problem 5.6, except for a lower aspect ratio of 3. From a comparison of the results from these two problems, draw some conclusions about the effect of wing sweep on the lift slope, and how the magnitude of this effect is affected by aspect ratio.

5.8 In Problem 1.19, we noted that the Wright brothers, in the design of their 1900 and 1901 gliders, used aerodynamic data from the Lilienthal table given in Figure 1.65. They chose a design angle of attack of 3 degrees, corresponding to a design lift coefficient of 0.546. When they tested their gliders at Kill Devil Hills near Kitty Hawk, North Carolina, in 1900 and 1901, however, they measured only one-third the amount of lift they had originally calculated on the basis of the Lilienthal table. This led the Wrights to question the validity of Lilienthal's data, and this cast a pall on the Lilienthal table that has persisted to the present time. However, in Reference 58, this author shows that the Lilienthal data are reasonably valid, and that the Wrights misinterpreted the data in the Lilienthal table in three respects (see pages 209–216 of Reference 58). One of these respects was the difference in aspect ratio. The Wrights' 1900 glider had rectangular wings with an aspect ratio of 3.5, whereas the data in the Lilienthal table were taken with a wing with an ogival planform tapering to a point at the tip and with an aspect ratio of 6.48. The Wrights seemed not to appreciate the aerodynamic importance of aspect ratio at the time, and even if they had, there was no existing theory that would have allowed them to correct the Lilienthal data for their design. (Prandtl's lifting line theory appeared 18 years later.) Given just the difference in aspect ratio between the Wrights' glider and the test model used by Lilienthal, what value of lift coefficient should the Wrights have used instead of the value of 0.546 they took straight from the table? (*Note:* There are two other misinterpretations by the Wrights that resulted in their calculation of lift being too high; see Reference 58 for details.)

5.9 Consider the Supermarine Spitfire shown in Figure 5.19. The first version
of the Spitfire was the Mk I, which first flew in 1936. Its maximum
velocity is 362 mi/h at an altitude of 18,500 ft. Its weight is 5820 lb, wing
area is 242 ft^2, and wing span is 36.1 ft. It is powered by a supercharged
Merlin engine, which produced 1050 horsepower at 18,500 ft. (a) Calculate the induced drag coefficient of the Spitfire at the flight condition of
V_{max} at 18,500 ft. (b) What percentage of the total drag coefficient is the
induced drag coefficient? *Note:* To calculate the total drag, we note that in
steady, level flight of the airplane, $T = D$, where T is the thrust from the
propeller. In turn, the thrust is related to the power by the basic mechanical
relation $TV_\infty = P$, where P is the power supplied by the propeller-engine
combination. Because of aerodynamic losses experienced by the
propeller, P is less than the shaft power provided by the engine by a ratio,
η, defined as the propeller efficiency. That is, if HP is the shaft horsepower
provided by the engine, and since 550 ft · lb/s equals 1 horsepower, then
the power provided by the engine-propeller combination in foot-pounds
per second is $P = 550\eta$HP. See Chapter 6 of Reference 2 for more details.
For this problem, assume the propeller efficiency for the Spitfire is 0.9.

5.10 If the elliptical wing of the Spitfire in Problem 5.9 were replaced by a
tapered wing with a taper ratio of 0.4, everything else remaining the same,
calculate the induced drag coefficient. Compare this value with that
obtained in Problem 5.9. What can you conclude about the relative effect
of planform shape change on the drag of the airplane at high speeds?

5.11 Consider the Spitfire in Problem 5.9 on its landing approach at sea level
with a landing velocity of 70 mi/h. Calculate the induced drag coefficient
for this low-speed case. Compare your result with the high-speed case in
Problem 5.9. From this, what can you conclude about the relative
importance of the induced drag coefficient at low speeds compared
to that at high speeds?

CHAPTER 6

Three-Dimensional Incompressible Flow

Treat nature in terms of the cylinder, the sphere, the cone, all in perspective.
  Paul Cézanne, 1890

PREVIEW BOX

We go three-dimensional in this chapter. For such a huge and complex subject, this chapter is mercifully short. It has only three objectives. This first is to see what happens when the circular cylinder studied in Chapter 3 morphs into a sphere—how do we modify the theory to account for the three-dimensional flow over a sphere, and how are the results changed from those for a circular cylinder? The second is to demonstrate a general phenomenon in aerodynamics known as the *three-dimensional relieving effect*. The third is to briefly examine the aerodynamic flow over a complete three-dimensional flight vehicle. Achieving these objectives is important; they will further open your mind to the wonders of aerodynamics. Read on, and enjoy.

6.1 INTRODUCTION

To this point in our aerodynamic discussions, we have been working mainly in a two-dimensional world; the flows over the bodies treated in Chapter 3 and the airfoils in Chapter 4 involved only two dimensions in a single plane—the so-called planar flows. In Chapter 5, the analyses of a finite wing were carried out in the plane of the wing, in spite of the fact that the detailed flow over a finite wing is truly three-dimensional. The relative simplicity of dealing with two dimensions (i.e., having only two independent variables) is self-evident and is the reason why a large bulk of aerodynamic theory deals with two-dimensional flows.

Fortunately, the two-dimensional analyses go a long way toward understanding many practical flows, but they also have distinct limitations.

The real world of aerodynamic applications is three-dimensional. However, because of the addition of one more independent variable, the analyses generally become more complex. The accurate calculation of three-dimensional flow fields has been, and still is, one of the most active areas of aerodynamic research.

The purpose of this book is to present the fundamentals of aerodynamics. Therefore, it is important to recognize the predominance of three-dimensional flows, although it is beyond our scope to go into detail. Therefore, the purpose of this chapter is to introduce some very basic considerations of three-dimensional incompressible flow. This chapter is short; we do not even need a road map to guide us through it. Its function is simply to open the door to the analysis of three-dimensional flow.

The governing fluid flow equations have already been developed in three dimensions in Chapters 2 and 3. In particular, if the flow is irrotational, Equation (2.154) states that

$$\mathbf{V} = \nabla \phi \qquad (2.154)$$

where, if the flow is also incompressible, the velocity potential is given by Laplace's equation:

$$\nabla^2 \phi = 0 \qquad (3.40)$$

Solutions of Equation (3.40) for flow over a body must satisfy the flow-tangency boundary condition on the body, that is,

$$\mathbf{V} \cdot \mathbf{n} = 0 \qquad (3.48a)$$

where $\mathbf{n}$ is a unit vector normal to the body surface. In all of the above equations, ϕ is, in general, a function of three-dimensional space; for example, in spherical coordinates $\phi = \phi(r, \theta, \Phi)$. Let us use these equations to treat some elementary three-dimensional incompressible flows.

6.2 THREE-DIMENSIONAL SOURCE

Return to Laplace's equation written in spherical coordinates, as given by Equation (3.43). Consider the velocity potential given by

$$\phi = -\frac{C}{r} \qquad (6.1)$$

where C is a constant and r is the radial coordinate from the origin. Equation (6.1) satisfies Equation (3.43), and hence it describes a physically possible incompressible, irrotational three-dimensional flow. Combining Equation (6.1) with the definition of the gradient in spherical coordinates, Equation (2.18), we obtain

$$\mathbf{V} = \nabla \phi = \frac{C}{r^2} \mathbf{e}_r \qquad (6.2)$$

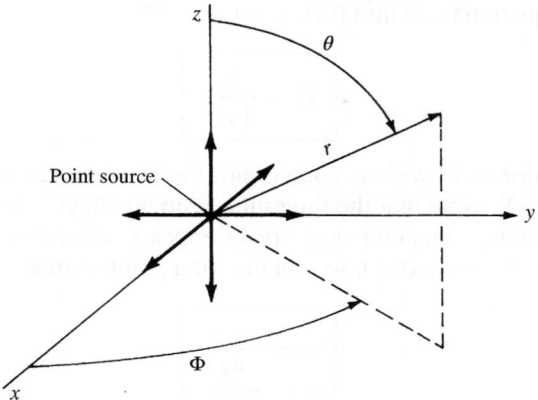

Figure 6.1 Three-dimensional (point) source.

In terms of the velocity components, we have

$$V_r = \frac{C}{r^2} \qquad\qquad (6.3a)$$

$$V_\theta = 0 \qquad\qquad (6.3b)$$

$$V_\Phi = 0 \qquad\qquad (6.3c)$$

Clearly, Equation (6.2), or Equation (6.3a to c), describes a flow with straight streamlines emanating from the origin, as sketched in Figure 6.1. Moreover, from Equation (6.2) or (6.3a), the velocity varies inversely as the square of the distance from the origin. Such a flow is defined as a *three-dimensional source*. Sometimes it is called simply a *point source,* in contrast to the two-dimensional line source discussed in Section 3.10.

To evaluate the constant C in Equation (6.3a), consider a sphere of radius r and surface S centered at the origin. From Equation (2.46), the mass flow across the surface of this sphere is

$$\text{Mass flow} = \oiint_S \rho\mathbf{V}\cdot\mathbf{dS}$$

Hence, the *volume* flow, denoted by λ, is

$$\lambda = \oiint_S \mathbf{V}\cdot\mathbf{dS} \qquad\qquad (6.4)$$

On the surface of the sphere, the velocity is a constant value equal to $V_r = C/r^2$ and is normal to the surface. Hence, Equation (6.4) becomes

$$\lambda = \frac{C}{r^2}4\pi r^2 = 4\pi C$$

Hence, $$C = \frac{\lambda}{4\pi} \qquad\qquad (6.5)$$

Substituting Equation (6.5) into (6.3a), we find

$$\boxed{V_r = \frac{\lambda}{4\pi r^2}} \tag{6.6}$$

Compare Equation (6.6) with its counterpart for a two-dimensional source given by Equation (3.62). Note that the three-dimensional effect is to cause an inverse *r-squared* variation and that the quantity 4π appears rather than 2π. Also, substituting Equation (6.5) into (6.1), we obtain, for a point source,

$$\boxed{\phi = -\frac{\lambda}{4\pi r}} \tag{6.7}$$

In the above equations, λ is defined as the *strength* of the source. When λ is a negative quantity, we have a point sink.

6.3 THREE-DIMENSIONAL DOUBLET

Consider a sink and source of equal but opposite strength located at points O and A, as sketched in Figure 6.2. The distance between the source and sink is l. Consider an arbitrary point P located a distance r from the sink and a distance r_1 from the source. From Equation (6.7), the velocity potential at P is

$$\phi = -\frac{\lambda}{4\pi}\left(\frac{1}{r_1} - \frac{1}{r}\right)$$

or

$$\phi = -\frac{\lambda}{4\pi}\frac{r - r_1}{rr_1} \tag{6.8}$$

Let the source approach the sink as their strengths become infinite; that is, let

$$l \to 0 \qquad \text{as} \qquad \lambda \to \infty$$

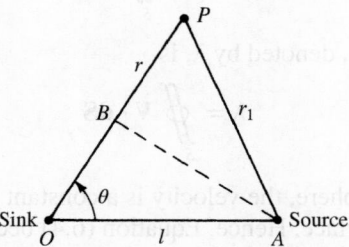

Figure 6.2 Source-sink pair. In the limit as $l \to 0$, a three-dimensional doublet is obtained.

In the limit, as $l \rightarrow 0$, $r - r_1 \rightarrow OB = l \cos \theta$, and $rr_1 \rightarrow r^2$. Thus, in the limit, Equation (6.8) becomes

$$\phi = - \lim_{\substack{l \to 0 \\ \lambda \to \infty}} \frac{\lambda}{4\pi} \frac{r - r_1}{rr_1} = -\frac{\lambda}{4\pi} \frac{l \cos \theta}{r^2}$$

or

$$\boxed{\phi = -\frac{\mu}{4\pi} \frac{\cos \theta}{r^2}} \tag{6.9}$$

where $\mu = \Lambda l$. The flow field produced by Equation (6.9) is a *three-dimensional doublet*; μ is defined as the strength of the doublet. Compare Equation (6.9) with its two-dimensional counterpart given in Equation (3.88). Note that the three-dimensional effects lead to an inverse r-squared variation and introduce a factor 4π, versus 2π for the two-dimensional case.

From Equations (2.18) and (6.9), we find

$$\mathbf{V} = \nabla\phi = \frac{\mu}{2\pi} \frac{\cos \theta}{r^3} \mathbf{e}_r + \frac{\mu}{4\pi} \frac{\sin \theta}{r^3} \mathbf{e}_\theta + 0\mathbf{e}_\Phi \tag{6.10}$$

The streamlines of this velocity field are sketched in Figure 6.3. Shown are the streamlines in the zr plane; they are the same in all the zr planes (i.e., for all values of Φ). Hence, the flow induced by the three-dimensional doublet is a series of stream surfaces generated by revolving the streamlines in Figure 6.3 about the

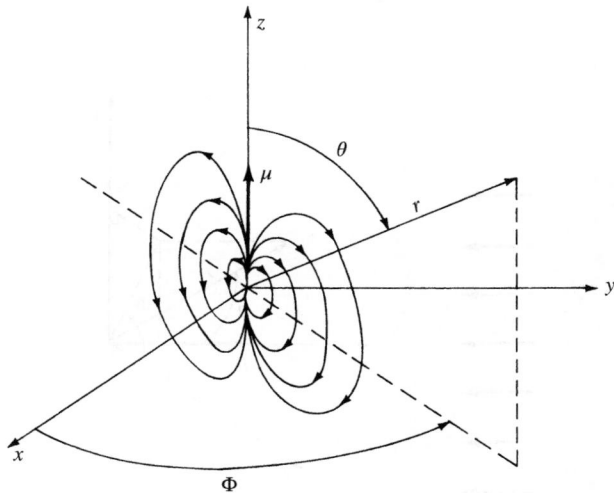

Figure 6.3 Sketch of the streamlines in the zr plane (Φ = constant plane) for a three-dimensional doublet.

z axis. Compare these streamlines with the two-dimensional case illustrated in Figure 3.18; they are qualitatively similar but quantitatively different.

Note that the flow in Figure 6.3 is independent of Φ; indeed, Equation (6.10) clearly shows that the velocity field depends only on *r* and θ. Such a flow is defined as *axisymmetric flow*. Once again, we have a flow with two independent variables. For this reason, axisymmetric flow is sometimes labeled "two-dimensional" flow. However, it is quite different from the two-dimensional planar flows discussed earlier. In reality, axisymmetric flow is a degenerate three-dimensional flow, and it is somewhat misleading to refer to it as "two-dimensional." Mathematically, it has only two independent variables, but it exhibits some of the same physical characteristics as general three-dimensional flows, such as the three-dimensional relieving effect to be discussed later.

6.4 FLOW OVER A SPHERE

Consider again the flow induced by the three-dimensional doublet illustrated in Figure 6.3. Superimpose on this flow a uniform velocity field of magnitude V_∞ in the negative *z* direction. Since we are more comfortable visualizing a freestream that moves horizontally, say, from left to right, let us flip the coordinate system in Figure 6.3 on its side. The picture shown in Figure 6.4 results.

Examining Figure 6.4, the spherical coordinates of the freestream are

$$V_r = -V_\infty \cos \theta \qquad (6.11a)$$

$$V_\theta = V_\infty \sin \theta \qquad (6.11b)$$

$$V_\Phi = 0 \qquad (6.11c)$$

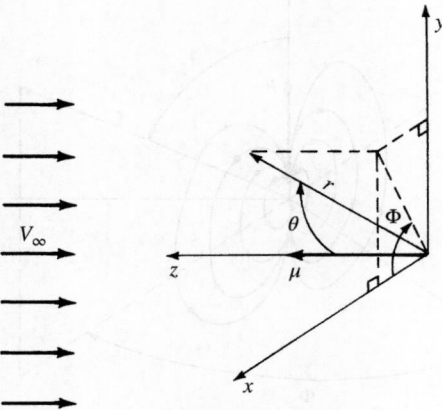

Figure 6.4 The superposition of a uniform flow and a three-dimensional doublet.

Adding V_r, V_θ, and V_Φ for the free stream, Equation (6.11a to c), to the representative components for the doublet given in Equation (6.10), we obtain, for the combined flow,

$$V_r = -V_\infty \cos\theta + \frac{\mu}{2\pi} \frac{\cos\theta}{r^3} = -\left(V_\infty - \frac{\mu}{2\pi r^3}\right)\cos\theta \qquad (6.12)$$

$$V_\theta = V_\infty \sin\theta + \frac{\mu}{4\pi} \frac{\sin\theta}{r^3} = \left(V_\infty + \frac{\mu}{4\pi r^3}\right)\sin\theta \qquad (6.13)$$

$$V_\Phi = 0 \qquad (6.14)$$

To find the stagnation points in the flow, set $V_r = V_\theta = 0$ in Equations (6.12) and (6.13). From Equation (6.13), $V_\theta = 0$ gives $\sin\theta = 0$; hence, the stagnation points are located at $\theta = 0$ and π. From Equation (6.12), with $V_r = 0$, we obtain

$$V_\infty - \frac{\mu}{2\pi R^3} = 0 \qquad (6.15)$$

where $r = R$ is the radial coordinate of the stagnation points. Solving Equation (6.15) for R, we obtain

$$R = \left(\frac{\mu}{2\pi V_\infty}\right)^{1/3} \qquad (6.16)$$

Hence, there are two stagnation points, both on the z axis, with (r, θ) coordinates

$$\left[\left(\frac{\mu}{2\pi V_\infty}\right)^{1/3}, 0\right] \quad \text{and} \quad \left[\left(\frac{\mu}{2\pi V_\infty}\right)^{1/3}, \pi\right]$$

Insert the value of $r = R$ from Equation (6.16) into the expression for V_r given by Equation (6.12). We obtain

$$V_r = -\left(V_\infty - \frac{\mu}{2\pi R^3}\right)\cos\theta = -\left[V_\infty - \frac{\mu}{2\pi}\left(\frac{2\pi V_\infty}{\mu}\right)\right]\cos\theta$$

$$= -(V_\infty - V_\infty)\cos\theta = 0$$

Thus, $V_r = 0$ when $r = R$ for *all values* of θ and Φ. This is precisely the flow-tangency condition for flow over a sphere of radius R. Hence, the velocity field given by Equations (6.12) to (6.14) is the *incompressible flow over a sphere of radius R*. This flow is shown in Figure 6.5; it is qualitatively similar to the flow over the cylinder shown in Figure 3.19, but quantitatively the two flows are different.

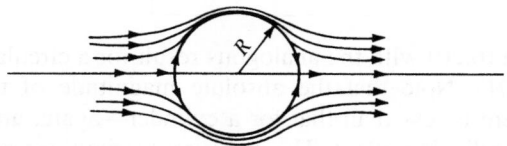

Figure 6.5 Schematic of the incompressible flow over a sphere.

On the surface of the sphere, where $r = R$, the tangential velocity is obtained from Equation (6.13) as follows:

$$V_\theta = \left(V_\infty + \frac{\mu}{4\pi R^3} \right) \sin\theta \tag{6.17}$$

From Equation (6.16),

$$\mu = 2\pi R^3 V_\infty \tag{6.18}$$

Substituting Equation (6.18) into (6.17), we have

$$V_\theta = \left(V_\infty + \frac{1}{4\pi} \frac{2\pi R^3 V_\infty}{R^3} \right) \sin\theta$$

or

$$\boxed{V_\theta = \tfrac{3}{2} V_\infty \sin\theta} \tag{6.19}$$

The maximum velocity occurs at the top and bottom points of the sphere, and its magnitude is $\frac{3}{2} V_\infty$. Compare these results with the two-dimensional circular cylinder case given by Equation (3.100). For the two-dimensional flow, the maximum velocity is $2V_\infty$. Hence, for the same V_∞, the maximum surface velocity on a sphere is *less* than that for a cylinder. The flow over a sphere is somewhat "relieved" in comparison with the flow over a cylinder. The flow over a sphere has an extra dimension in which to move out of the way of the solid body; the flow can move sideways as well as up and down. In contrast, the flow over a cylinder is more constrained; it can only move up and down. Hence, the maximum velocity on a sphere is less than that on a cylinder. This is an example of the *three-dimensional relieving effect,* which is a general phenomenon for all types of three-dimensional flows.

The pressure distribution on the surface of the sphere is given by Equations (3.38) and (6.19) as follows:

$$C_p = 1 - \left(\frac{V}{V_\infty} \right)^2 = 1 - \left(\frac{3}{2} \sin\theta \right)^2$$

or

$$\boxed{C_p = 1 - \tfrac{9}{4} \sin^2\theta} \tag{6.20}$$

Compare Equation (6.20) with the analogous result for a circular cylinder given by Equation (3.101). Note that the absolute magnitude of the pressure coefficient on a sphere is less than that for a cylinder—again, an example of the three-dimensional relieving effect. The pressure distributions over a sphere and a cylinder are compared in Figure 6.6, which dramatically illustrates the three-dimensional relieving effect.

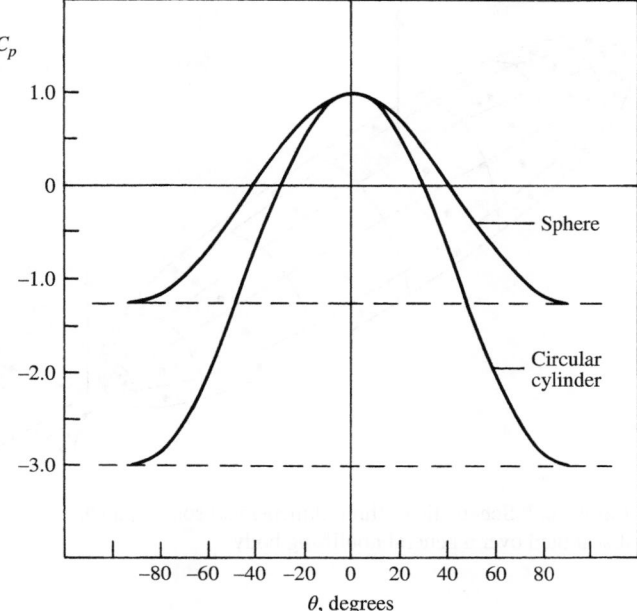

Figure 6.6 The pressure distribution over the surface of a sphere and a cylinder. Illustration of the three-dimensional relieving effect.

6.4.1 Comment on the Three-Dimensional Relieving Effect

There is a good physical reason for the three-dimensional relieving effect. First, visualize the two-dimensional flow over a circular cylinder. In order to move out of the way of the cylinder, the flow has only two ways to go: riding up-and-over and down-and-under the cylinder. In contrast, visualize the three-dimensional flow over a sphere. In addition to moving up-and-over and down-and-under the sphere, the flow can now move sideways, to the left and right over the sphere. This sidewise movement relieves the previous constraint on the flow; the flow does not have to speed up so much to get out of the way of the sphere, and therefore the pressure in the flow does not have to change so much. The flow is "less stressed"; it moves around the sphere in a more relaxed fashion—it is "relieved," and consequently the changes in velocity and pressure are smaller.

6.5 GENERAL THREE-DIMENSIONAL FLOWS: PANEL TECHNIQUES

In modern aerodynamic applications, three-dimensional, inviscid, incompressible flows are almost always calculated by means of numerical panel techniques. The philosophy of the two-dimensional panel methods discussed in previous chapters is readily extended to three dimensions. The details are beyond the scope of

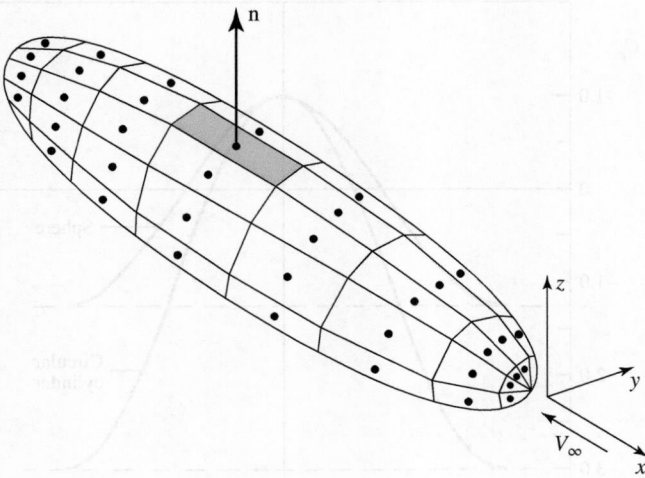

Figure 6.7 Schematic of three-dimensional source panels distributed over a general nonlifting body.

this book—indeed, there are dozens of different variations, and the resulting computer programs are frequently long and sophisticated. However, the general idea behind all such panel programs is to cover the three-dimensional body with panels over which there is an unknown distribution of singularities (such as point sources, doublets, or vortices). Such paneling is illustrated in Figure 6.7. These unknowns are solved through a system of simultaneous linear algebraic equations generated by calculating the induced velocity at control points on the panels and applying the flow-tangency condition. For a nonlifting body such as illustrated in Figure 6.7, a distribution of source panels is sufficient. However, for a lifting body, both source and vortex panels (or their equivalent) are necessary. A striking example of the extent to which panel methods are now used for three-dimensional lifting bodies is shown in Figure 6.8, which illustrates the paneling used for calculations made by the Boeing Company of the potential flow over a Boeing 747–space shuttle piggyback combination. Such applications are very impressive; moreover, they have become an industry standard and are today used routinely as part of the airplane design process by the major aircraft companies.

Examining Figures 6.7 and 6.8, one aspect stands out, namely, the geometric complexity of distributing panels over the three-dimensional bodies. How do you get the computer to "see" the precise shape of the body? How do you distribute the panels over the body; that is, do you put more at the wing leading edges and less on the fuselage, etc.? How many panels do you use? These are all nontrivial questions. It is not unusual for an aerodynamicist to spend weeks or even a few months determining the best geometric distribution of panels over a complex body.

We end this section on the following note. From the time they were introduced in the 1960s, panel techniques have revolutionized the calculation of

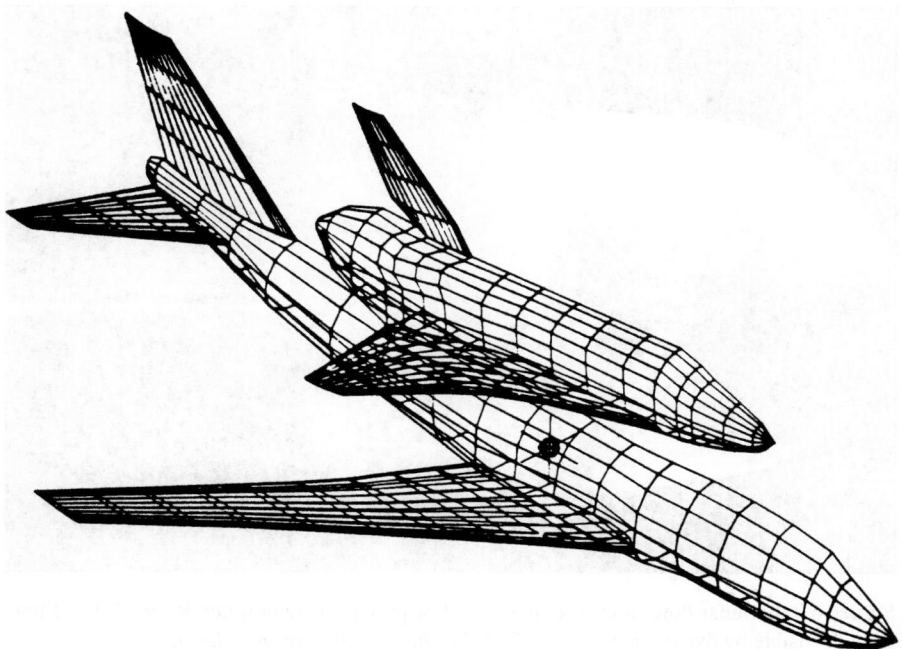

Figure 6.8 Panel distribution for the analysis of the Boeing 747 carrying the space shuttle orbiter.

three-dimensional potential flows. However, no matter how complex the application of these methods may be, the techniques are still based on the fundamentals we have discussed in this and all the preceding chapters. You are encouraged to pursue these matters further by reading the literature, particularly as it appears in such journals as the *Journal of Aircraft* and the *AIAA Journal*.

6.6 APPLIED AERODYNAMICS: THE FLOW OVER A SPHERE—THE REAL CASE

The present section is a complement to Section 3.18, in which the real flow over a circular cylinder was discussed. Since the present chapter deals with three-dimensional flows, it is fitting at this stage to discuss the three-dimensional analog of the circular cylinder, namely, the sphere. The qualitative features of the real flow over a sphere are similar to those discussed for a cylinder in Section 3.18—the phenomenon of flow separation, the variation of drag coefficient with a Reynolds number, the precipitous drop in drag coefficient when the flow transits from laminar to turbulent ahead of the separation point at the critical Reynolds number, and the general structure of the wake. These items are similar for both cases. However, because of the three-dimensional relieving effect, the flow over

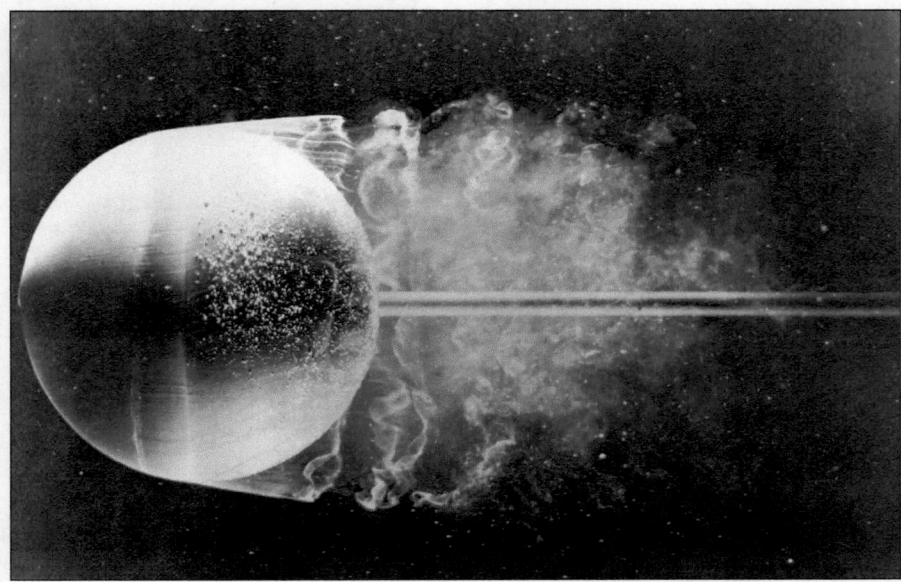

Figure 6.9 Laminar flow case: Instantaneous flow past a sphere in water. Re = 15,000. Flow is made visible by dye in the water. (©*ONERA The French Aerospace Lab*).

a sphere is *quantitatively* different from that for a cylinder. These differences are the subject of the present section.

The laminar flow over a sphere is shown in Figure 6.9. Here, the Reynolds number is 15,000, certainly low enough to maintain laminar flow over the spherical surface. However, in response to the adverse pressure gradient on the back surface of the sphere predicted by inviscid, incompressible flow theory (see Section 6.4 and Figure 6.6), the laminar flow readily separates from the surface. Indeed, in Figure 6.9, separation is clearly seen on the *forward* surface, slightly ahead of the vertical equator of the sphere. Thus, a large, wide wake trails downstream of the sphere, with a consequent large pressure drag on the body (analogous to that discussed in Section 3.18 for a cylinder). In contrast, the turbulent flow case is shown in Figure 6.10. Here, the Reynolds number is 30,000, still a low number normally conducive to laminar flow. However, in this case, turbulent flow is induced artificially by the presence of a wire loop in a vertical plane on the forward face. (Trip wires are frequently used in experimental aerodynamics to induce transition to turbulent flow; this is in order to study such turbulent flows under conditions where they would not naturally exist.) Because the flow is turbulent, separation takes place much farther over the back surface, resulting in a thinner wake, as can be seen by comparing Figures 6.9 and 6.10. Consequently, the pressure drag is less for the turbulent case.

The variation of drag coefficient C_D with the Reynolds number for a sphere is shown in Figure 6.11. Compare this figure with Figure 3.44 for a circular cylinder; the C_D variations are qualitatively similar, both with a precipitous decrease in C_D

Figure 6.10 Turbulent flow case: Instantaneous flow past a sphere in water. Re = 30,000. The turbulent flow is forced by a trip wire hoop ahead of the equator, causing the laminar flow to become turbulent suddenly. The flow is made visible by air bubbles in water (©*ONERA The French Aerospace Lab*).

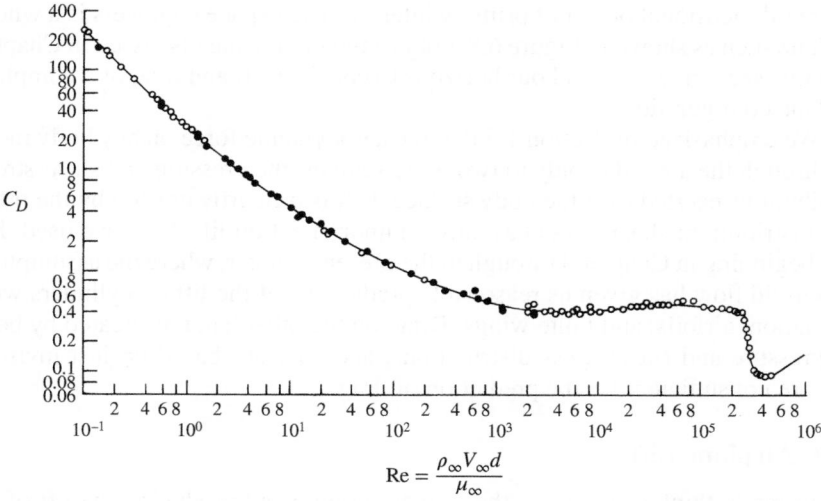

$$\text{Re} = \frac{\rho_\infty V_\infty d}{\mu_\infty}$$

Figure 6.11 Variation of drag coefficient with Reynolds number for a sphere (Data taken from Schlichting, H.: *Boundary Layer Theory*, 7th ed., McGraw Hill Book Company, New York, 1979).

near a critical Reynolds number of 300,000, coinciding with natural transition from laminar to turbulent flow. However, *quantitatively* the two curves are quite different. In the Reynolds number range most appropriate to practical problems, that is, for Re > 1000, the values of C_D for the sphere are considerably smaller than those for a cylinder—a classic example of the three-dimensional relieving effect. Reflecting on Figure 3.44 for the cylinder, note that the value of C_D for Re slightly less than the critical value is about 1 and drops to 0.3 for Re slightly above the critical value. In contrast, for the sphere as shown in Figure 6.11, C_D is about 0.4 in the Reynolds number range below the critical value and drops to about 0.1 for Reynolds numbers above the critical value. These variations in C_D for both the cylinder and sphere are classic results in aerodynamics; you should keep the actual C_D values in mind for future reference and comparisons.

As a final point in regard to both Figures 3.44 and 6.11, the value of the critical Reynolds number at which transition to turbulent flow takes place upstream of the separation point is not a fixed, universal number. Quite the contrary, transition is influenced by many factors, as will be discussed in Part 4. Among these is the amount of turbulence in the freestream; the higher the freestream turbulence, the more readily transition takes place. In turn, the higher the freestream turbulence, the lower is the value of the critical Reynolds number. Because of this trend, calibrated spheres are used in wind-tunnel testing actually to assess the degree of freestream turbulence in the test section, simply by measuring the value of the critical Reynolds number on the sphere.

6.7 APPLIED AERODYNAMICS: AIRPLANE LIFT AND DRAG

A three-dimensional object of primary interest to aerospace engineers is a whole airplane such as shown in Figure 6.8, not just the finite wing discussed in Chapter 5. In this section, we expand our horizons to consider lift and drag of a complete airplane configuration.

We emphasized in Section 1.5 that the aerodynamic force on any body moving through the air is due only to two basic sources, the pressure and shear stress distributions exerted over the body surface. Lift is primarily created by the pressure distribution; shear stress has only a minor effect on lift. We have used this fact, beginning in Chapter 3 through to the present chapter, where the assumption of inviscid flow has given us reasonable predictions of the lift on cylinders with circulation, airfoils, and finite wings. Drag, on the other hand, is created by both the pressure and shear stress distributions, and analyses based on just inviscid flow are not sufficient for the prediction of drag.

6.7.1 Airplane Lift

We normally think of wings as the primary component producing the lift of an airplane in flight, and quite rightly so. However, even a pencil at an angle of attack will generate lift, albeit small. Hence, lift is produced by the fuselage of an airplane as well as the wing. The mating of a wing with a fuselage is called a *wing-body combination*. The lift of a wing-body combination is *not* obtained by simply adding the lift of the wing alone to the lift of the body alone. Rather, as soon as the wing and body are mated, the flow field over the body modifies the flow field over the wing, and vice versa—this is called the *wing-body interaction*.

There is no accurate analytical equation that can predict the lift of a wing-body combination, properly taking into account the nature of the wing-body aerodynamic interaction. Either the configuration must be tested in a wind tunnel, or a computational fluid dynamic calculation must be made. We cannot even say in advance whether the combined lift will be greater or smaller than the sum of the two parts. For subsonic speeds, however, data obtained using different fuselage thicknesses, d, mounted on wings with different spans, b, show that the total lift for a wing-body combination is essentially constant for d/b ranging from 0 (wing only) to 6 (which would be an inordinately wide fuselage, with a short, stubby wing). Hence, the lift of the wing-body combination can be treated as simply the lift on the complete wing by itself, including that portion of the wing that is masked by the fuselage. This is illustrated in Figure 6.12. See Chapter 2 of Reference 65 for more details.

Of course, other components of the airplane such as a horizontal tail, canard surfaces, and wing strakes can contribute to the lift, either in a positive or negative sense. Once again we emphasize that reasonably accurate predictions of lift on a complete airplane can come only from wind tunnel tests, detailed computational fluid dynamic calculations (such as the panel calculations illustrated by Figure 6.8), and, of course, from actual flight tests of the airplane.

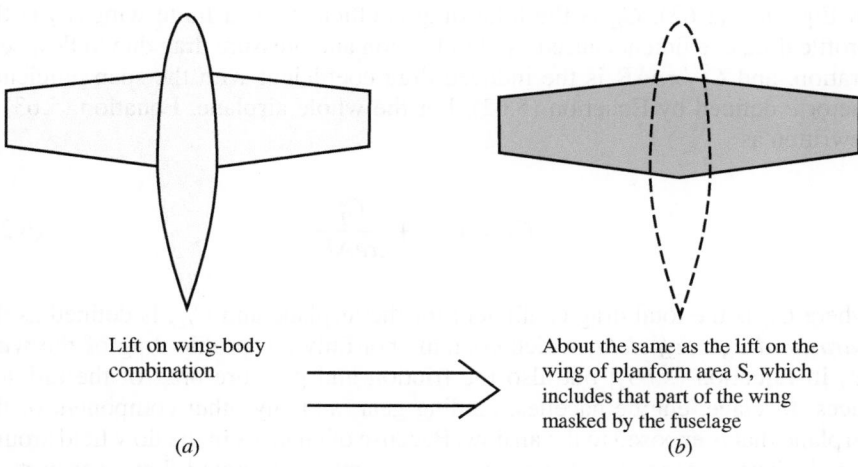

Lift on wing-body combination   →   About the same as the lift on the wing of planform area S, which includes that part of the wing masked by the fuselage

(*a*)   (*b*)

Figure 6.12 Lift on a wing-body combination.

6.7.2 Airplane Drag

When you watch an airplane flying overhead, or when you ride in an airplane, it is almost intuitive that your first aerodynamic thought is about lift. You are witnessing a machine that, in straight and level flight, is producing enough aerodynamic lift to equal the weight of the machine. This keeps it in the air—a vital concern. But this is only part of the role of airplane aerodynamics. It is equally important to produce this lift as *efficiently as possible*, that is, with as little drag as possible. The ratio of lift to drag, *L/D*, is a good measure of aerodynamic efficiency. A barn door will produce lift at angle of attack, but it also produces a lot of drag at the same time—the *L/D* for a barn door is terrible. For such reasons, minimizing drag has been one of the strongest drivers in the historical development of applied aerodynamics. And to minimize drag, we first have to provide methods for its estimation.

As in the case of lift, the drag of an airplane *cannot* be obtained as the simple sum of the drag on each component. For example, for a wing-body combination, the drag is usually higher than the sum of the separate drag forces on the wing and the body, giving rise to an extra drag component called interference drag. For a more detailed discussion of airplane drag prediction, see Reference 65. The subject of drag prediction is so complex that whole books have been written about it; one classic is the book by Hoerner, Reference 112.

In this section, we will limit our discussion to the simple extension of Equation (5.63) for application to the whole airplane. Equation (5.63), copied below, applies to a finite wing.

$$C_D = c_d + \frac{C_L^2}{\pi e \text{AR}} \qquad (5.63)$$

In Equation (5.63), C_D is the total drag coefficient for a finite wing, c_d is the profile drag coefficient caused by skin friction and pressure drag due to flow separation, and $C_L^2/\pi e\text{AR}$ is the induced drag coefficient with the span efficiency factor e defined by Equation (5.62). For the whole airplane, Equation (5.63) is rewritten as

$$C_D = C_{D,e} + \frac{C_L^2}{\pi e \text{AR}} \qquad (6.21)$$

where C_D is the total drag coefficient for the airplane and $C_{D,e}$ is defined as the *parasite drag coefficient*, which contains not only the profile drag of the wing [c_d in Equation (5.63)] but also the friction and pressure drag of the tail surfaces, fuselage, engine nacelles, landing gear, and any other component of the airplane that is exposed to the airflow. Because of changes in the flow field around the airplane—especially changes in the amount of separated flow over parts of the airplane—as the angle of attack is varied, $C_{D,e}$ will change with angle of attack. Because the lift coefficient, C_L, is a specific function of angle of attack, we can consider that $C_{D,e}$ is a function of C_L. A reasonable approximation for this function is

$$C_{D,e} = C_{D,o} + rC_L^2 \tag{6.22}$$

where r is an empirically determined constant. Since at zero lift, $C_L = 0$, then Equation (6.22) defines $C_{D,o}$ as the parasite drag coefficient at zero lift, or more commonly, the *zero-lift drag coefficient*. With Equation (6.22), we can write Equation (6.21) as

$$C_D = C_{D,o} + \left(r + \frac{1}{\pi e AR}\right) C_L^2 \tag{6.23}$$

In Equations (6.21) and (6.23), e is the familiar span efficiency factor, which takes into account the nonelliptical lift distribution on wings of general shape (see Section 5.3.2). Let us now *redefine e* so that it also includes the effect of the variation of parasite drag with lift; that is, let us write Equation (6.23) in the form

$$\boxed{C_D = C_{D,o} + \frac{C_L^2}{\pi e AR}} \tag{6.24}$$

where $C_{D,o}$ is the parasite drag coefficient at *zero lift* (or simply the *zero-lift drag coefficient* for the airplane) and the term $C_L^2/(\pi e AR)$ is the *drag coefficient due to lift* including both induced drag and the contribution to parasite drag due to lift.

In Equation (6.24), the redefined e is called the *Oswald efficiency factor* (named after W. Bailey Oswald, who first established this terminology in NACA Report No. 408 in 1932). The use of the symbol e for the Oswald efficiency factor has become standard in the literature, and that is why we continue this standard here. To avoid confusion, keep in mind that e introduced for a finite wing in Section 5.3.2 and used in Equation (6.21) is the span efficiency factor for a finite wing, and the e used in Equation (6.24) is the Oswald efficiency factor for a complete airplane. These are two different numbers; the Oswald efficiency factor for different airplanes typically varies between 0.7 and 0.85, whereas the span efficiency factor typically varies between 0.9 and at most 1.0 and is a function of wing aspect ratio and taper ratio as demonstrated in Figure 5.20. Daniel Raymer in Reference 113 gives the following empirical expression for the Oswald efficiency factor for straight-wing aircraft, based on data obtained from actual airplanes:

$$e = 1.78\left(1 - 0.045\,AR^{0.68}\right) - 0.64 \tag{6.25}$$

Raymer notes that Equation (6.25) should be used for conventional aspect ratios for normal airplanes, and not for the very large aspect ratios (on the order of 25 or higher) associated with sailplanes.

Equation (6.24) conveys all the information you need to calculate the drag of a complete airplane, but to use it you have to know the zero-lift drag coefficient and the Oswald efficiency factor. Equation (6.24) is called the *drag polar* for the airplane, representing the variation of C_D with C_L. It is the cornerstone for conceptual airplane design and for predictions of the performance of a given aircraft (see Reference 65 for more details).

EXAMPLE 6.1

Return again to the photograph of the Seversky P-35 shown in Figure 3.2. This airplane has a wing planform area of 220 ft^2 and a wingspan of 36 ft. Also, examine again the drag breakdown for the Seversky XP-41 given in Figure 1.58. In Example 1.12 we assumed that the drag breakdown for the XP-41, being an airplane very similar to the P-35, applied to the P-35 as well. We do the same here. Using the data given in Figure 1.58, calculate the zero-lift drag coefficient for the P-35.

■ **Solution**

For the drag breakdown shown in Figure 1.58, condition 18 is that for the complete airplane configuration. For condition 18, the total drag coefficient is given as $C_D = 0.0275$ when the aircraft is at the particular angle of attack where $C_L = 0.15$. That is, we know simultaneous values of C_D and C_L that can be used in Equation (6.24). In that equation,

$$\text{AR} = \frac{b^2}{s} = \frac{(36)^2}{220} = 5.89$$

and the Oswald efficiency factor from Equation (6.25) is

$$e = 1.78\,(1 - 0.045\text{AR}^{0.68}) - 0.64$$

$$= 1.78\,[1 - 0.045(5.89)^{0.68}] - 0.64$$

$$= 1.78\,[1 - 0.045(3.339)] - 0.64$$

$$= 0.873$$

Thus, Equation (6.24) gives for the zero-lift drag coefficient

$$C_{D,o} = C_D - \frac{C_L^2}{\pi e\,\text{AR}}$$

$$= 0.0275 - \frac{(0.15)^2}{\pi(0.873)(5.89)}$$

or,

$$\boxed{C_{D,o} = 0.026}$$

The late Larry Loftin, in his excellent book *Quest for Performance: The Evolution of Modern Aircraft* (Reference 45), tabulated the zero-lift drag coefficient extracted from flight performance data for a large number of historic airplanes from the twentieth century. His tabulated value for the Seversky P-35 is $C_{D,o} = 0.0251$. Note that the value of $C_{D,o} = 0.026$ calculated in this example agrees within 3.6 percent.

Airplane Lift-to-Drag Ratio   The dimensional analysis discussed in Section 1.7 proves that C_L, C_D, and hence the lift-to-drag ratio C_L/C_D, at a given Mach number and Reynolds number depend only on the shape of the body and the angle of attack. This is reinforced by the sketches shown in Figure 6.13. For a

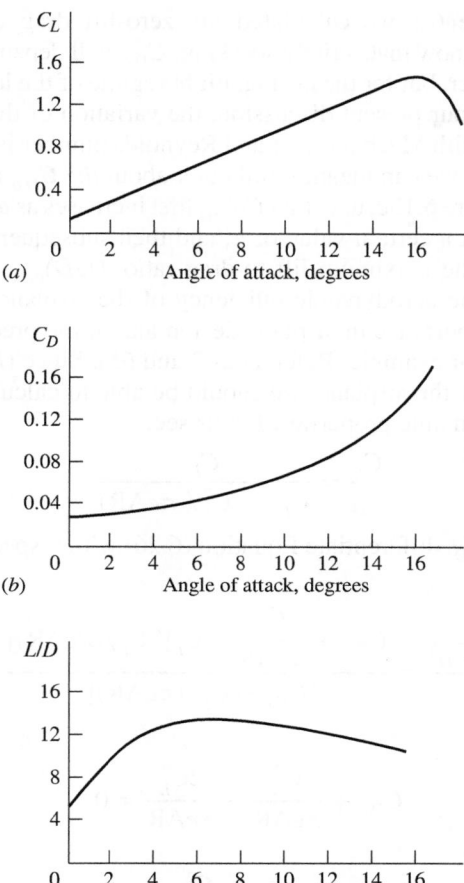

Figure 6.13 Typical variations of lift and drag coefficients and lift-to-drag ratio for a generic small propeller-driven general aviation airplane (based on calculations from Chapter 6 of Anderson, John D., Jr.: *Introduction to Flight*, 6th ed., McGraw Hill Book Company, Boston, 2008).

given airplane shape, Figure 6.13*a* gives the variation of C_L with the airplane angle of attack, α; Figure 6.13*b* gives the variation of C_D with α; and Figure 6.13*c* gives the lift-to-drag ratio C_L/C_D as a function of α. These are *aerodynamic properties* associated with a given airplane. Within reasonable Mach number and Reynolds number ranges, we can simply talk about *the* lift coefficient, drag coefficient, and lift-to-drag ratio as specific values at any specific angle of attack.

Indeed, in Example 6.1 we calculated *the* zero-lift drag coefficient for the Seversky P-35. We know that, strictly speaking, $C_{D,o}$ will depend on Mach number and Reynolds number, but for the normal flight regime of the low-speed subsonic aircraft germane to our present discussion, the variation of the airplane aerodynamic coefficients with Mach number and Reynolds number is considered small. Hence, for example, we can meaningfully talk about *the* $C_{D,o}$ for the airplane.

Examining Figure 6.13c, note that C_L/C_D first increases as α increases, reaches a maximum value at a certain value of α, and then subsequently decreases as α increases further. The maximum lift-to-drag ratio, $(L/D)_{\max} = (C_L/C_D)_{\max}$, is a direct measure of the aerodynamic efficiency of the airplane, and therefore its value is of great importance in airplane design and in the prediction of airplane performance (see, for example, References 2 and 65). Since $(L/D)_{\max}$ is an aerodynamic property of the airplane, we should be able to calculate its value from other known aerodynamic properties. Let us see:

$$\frac{C_L}{C_D} = \frac{C_L}{C_{D,o} + C_L^2/(\pi e \text{AR})} \tag{6.26}$$

For maximum C_L/C_D, differentiate Equation (6.26) with respect to C_L and set the result equal to 0:

$$\frac{d(C_L/C_D)}{dC_L} = \frac{C_{D,o} + \dfrac{C_L^2}{\pi e \text{AR}} - C_L[2C_L/(\pi e \text{AR})]}{[C_{D,o} + C_L^2/(\pi e \text{AR})]^2} = 0$$

Thus,

$$C_{D,o} + \frac{C_L^2}{\pi e \text{AR}} - \frac{2C_L^2}{\pi e \text{AR}} = 0$$

or

$$C_{D,o} = \frac{C_L^2}{\pi e \text{AR}} \tag{6.27}$$

Equation (6.27) is an interesting intermediate result. It states that when the airplane is flying at the specific angle of attack where the lift-to-drag ratio is maximum, the zero-lift drag and the drag due to lift are precisely equal. Solving Equation (6.27) for C_L, we have

$$C_L = \sqrt{\pi e \text{AR} \, C_{D,o}} \tag{6.28}$$

Equation (6.28) gives the value of C_L when the airplane is flying at $(L/D)_{\max}$. Return to Equation (6.25), which gives C_L/C_D as a function of C_L. By substituting the value of C_L from Equation (6.28), which pertains just to the maximum value of L/D, into Equation (6.25), we obtain for the maximum lift-to-drag ratio

$$\left(\frac{C_L}{C_D}\right)_{\max} = \frac{(\pi e \text{AR} \, C_{D,o})^{1/2}}{C_{D,o} + \dfrac{\pi e \text{AR} C_{D,o}}{\pi e \text{AR}}}$$

or,

$$\boxed{\left(\frac{C_L}{C_D}\right)_{max} = \frac{(\pi e \mathrm{AR}\ C_{D,o})^{1/2}}{2C_{D,o}}} \tag{6.29}$$

Equation (6.29) is *powerful*. It tells us that the maximum value of lift-to-drag ratio for a given airplane depends *only* on the zero-lift drag coefficient $C_{D,o}$, the Oswald efficiency factor e, and the wing aspect ratio. So our earlier supposition that $(L/D)_{max}$, being an aerodynamic property of the given airplane, should depend only on other aerodynamic properties, is correct, the other aerodynamic properties being simply $C_{D,o}$ and e.

EXAMPLE 6.2

Using the information obtained in Example 6.1, calculate the maximum lift-to-drag ratio for the Seversky P-35.

■ **Solution**

From Example 6.1, we have

$$C_{D,o} = 0.026$$

$$e = 0.873$$

$$\mathrm{AR} = 5.89$$

From Equation (6.29), we have

$$\left(\frac{C_L}{C_D}\right)_{max} = \frac{(\pi e \mathrm{AR}\ C_{D,o})^{1/2}}{2C_{D,o}}$$

$$= \frac{[\pi(0.873)(5.89)(0.026)]^{1/2}}{2(0.026)}$$

$$\left(\frac{C_L}{C_D}\right)_{max} = \boxed{12.46}$$

The value for $(L/D)_{max}$ for the P-35 as tabulated by Loftin in Reference 45 is $(L/D)_{max} = 11.8$, which is within 5 percent of the value calculated here.

6.7.3 Application of Computational Fluid Dynamics for the Calculation of Lift and Drag

The role of computational fluid dynamics (CFD) for the numerical solution of the continuity, momentum, and energy equations is discussed in Section 2.17.2. Numerical solutions of the purely inviscid flow equations are labeled "Euler solutions"; the CFD results discussed in Chapter 13 are examples of such Euler solutions. Numerical solutions of the general viscous flow equations are labeled "Navier-Stokes solutions"; examples of such Navier-Stokes solutions are given in Chapter 20.

These numerical solutions of the continuity, momentum, and energy equations give the variation of the flow field properties (p, T, V, etc.) as a function of space and time throughout the flow. This includes, of course, the pressure at the body surface. The shear stress at the surface is obtained from Equation (1.59), repeated below

$$\tau_w = \mu \left(\frac{dV}{dy} \right)_{y=0} \tag{1.59}$$

where the velocity gradient at the wall, $(dV/dy)_{y=0}$, is obtained from the CFD solution of the flow velocity at gridpoints adjacent to the wall using one-sided differences (see Section 2.17.2). Finally, by numerically integrating the pressure and shear stress distributions over the surface, the lift and drag of the airplane can be obtained (see Section 1.5). This is how CFD results can be used to give lift and drag on a body.

Some very recent CFD results for the flow field over a complete airplane are described in seven coordinated papers in the *Journal of Aircraft*, Vol. 46, No. 2, March–April 2009. These papers report CFD results obtained by different investigators using different computer programs and algorithms for the flow field over the F-16XL cranked-wing configuration shown in Figure 6.14. As part of

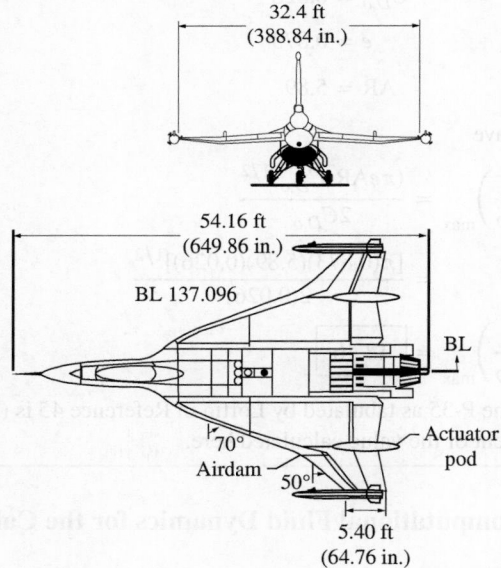

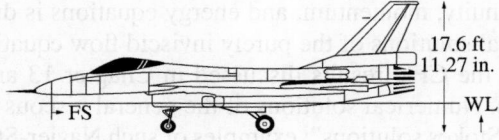

Figure 6.14 Three-view of the F-16XL.

the Cranked-Arrow Wing Aerodynamics Project (CAWAP) organized by NASA and administered through the AIAA Applied Aerodynamics Technical Committee, various investigators were invited to make CFD flow field calculations over the F-16XL at various flight conditions. The purpose is to compare the results in order to assess the state of the art of CFD calculations of flow fields around complete airplane configurations, particularly with the airplane at relatively high angle of attack with large regions of separated flow. A summary of the comparisons and conclusions is given in Reference 114. Although the main thrust of this project was to evaluate and compare calculations of detailed flow field structure and surface pressure distributions, some comparisons of lift and drag coefficients were made. A representative comparison is given in Table 6.1, where C_L and C_D obtained from seven different investigations are tabulated. The results apply to the F-16XL flying at $M_\infty = 0.36$, angle of attack $\alpha = 11.85°$, sideslip angle $= 0.612°$, and Reynolds number $= 46.8 \times 10^6$. In Table 6.1, the different investigators are simply labeled by number; the actual sources and the particular CFD codes are identified in Reference 114.

Note that the discrepancy between the lowest and highest numbers obtained is 26 percent for C_L and 42 percent for C_D. However, if the results from Investigator 3 are not counted, the discrepancies are 6.7 percent for C_L and 21.5 percent for C_D.

The three-dimensional flow field associated with the values of C_L and C_D in Table 6.1 is complex; it contains primary and secondary vortices much like those shown for flow over a delta wing in Figure 5.41. The airplane is at both an angle of attack and angle of sideslip, and the resulting flow field is highly three-dimensional with embedded vortices and large regions of flow separation. This is a severe test for any CFD code, and indeed is the reason why this case has been chosen here to illustrate the use of CFD for the calculation of lift and drag of a complete airplane. The test cases used in the Cranked-Arrow Wing Aerodynamics Project represent perhaps the upper limit of complexity, and therefore the discrepancies between the results of the different CFD codes may represent an upper bound—a kind of worst-case scenario.

With that caveat in mind, note that the discrepancies in the calculation of C_L are remarkably small, but the results for C_D vary considerably. The accurate calculation of drag for most practical aerodynamic vehicles has been a challenge

Table 6.1 Tabulation of the calculated values of lift and drag coefficients for the Cranked-Arrow Wing Aerodynamics Project by various investigators.

Investigator No.	C_L	C_D
1	0.43846	0.13289
2	0.44693	0.13469
3	0.37006	0.11084
4	0.43851	0.15788
5	0.46798	0.13648
6	0.44190	0.16158
7	0.44590	0.14265

and a problem for centuries, going back to the early nineteenth century flying machine inventors (see References 2, 58, and 111, for example). Amazingly, in our modern world of high technology and advanced CFD techniques, accurate drag prediction remains a problem, although improvements are gradually being made. Accurate CFD predictions of drag are compromised by at least the following:

1. The calculation of skin friction drag requires the accurate calculation of shear stress, which requires an accurate calculation of the velocity gradient at the surface [see, for example, Equation (1.59)], which requires a very fine, closely spaced computational grid adjacent to the wall to obtain very accurate values of the flow velocity at the first several gridpoints above the wall. The velocity gradient at the surface is then obtained from these velocities by using one-sided differencing.

2. The boundary layers on any practical-sized vehicle are turbulent, and any CFD calculation of this flow must include this effect. Turbulence remains one of the few unsolved problems in classical physics, so its effect must be *modeled* in aerodynamic calculations. Most CFD calculations of turbulent flows use the Reynolds-averaged Navier-Stokes (RANS) equations, discussed in Part 4 of this book, and must incorporate some type of turbulence model. There are literally dozens of different turbulence models in existence, each one depending, in one way or another, on empirical data. Turbulence models by themselves introduce a great deal of uncertainty in the calculation of drag. All the seven different CFD calculations noted in Table 6.1 used different turbulence models.

3. The calculation of locations on a body where the flow separates is also uncertain. For CFD, the calculation of separated flows can only be made with Navier-Stokes solutions; only in a few (but interesting) instances can a solution of the Euler equations yield a semblance of flow separation. The nature and location of flow separation are different for laminar and turbulent flows (see, for example, the discussion of the real flow over a sphere in Section 6.6). The uncertainty in the calculation of separated flows, which is in part related to the uncertainty in turbulence modeling discussed earlier, is another reason for the discrepancies in drag coefficient as tabulated in Table 6.1. The high angle of attack flows associated with these test cases for the F-16XL have large regions of complex flow separation.

Considering these uncertainties, the discrepancy in the calculations of C_D listed in Table 6.1, all told, are not bad. Further advances in algorithms and modeling will inevitably lead to even better results. Furthermore, because C_L was accurately calculated and the details of the flow field itself were accurately captured by the CFD calculations, the investigators participating in the Cranked-Arrow Wing Aerodynamics Project International (CAWAPI) were moved to conclude (Reference 114):

> Although differences were observed in the comparison of results from 10 different CFD solvers with measurements, these solvers all functioned robustly on an actual aircraft at flight conditions, with sufficient agreement among them to conclude

that the overall objectives of the CAWAPI endeavor have been achieved. In particular, the status of CFD as a tool for understanding flight-test observations has been confirmed.

This is also an appropriate conclusion to end our discussion of airplane lift and drag.

6.8 SUMMARY

For a three-dimensional (point) source,

$$V_r = \frac{\lambda}{4\pi r^2} \tag{6.6}$$

and

$$\phi = -\frac{\lambda}{4\pi r} \tag{6.7}$$

For a three-dimensional doublet,

$$\phi = -\frac{\mu}{4\pi} \frac{\cos\theta}{r^2} \tag{6.9}$$

and

$$\mathbf{V} = \frac{\mu}{2\pi} \frac{\cos\theta}{r^3}\mathbf{e}_r + \frac{\mu}{4\pi} \frac{\sin\theta}{r^3}\mathbf{e}_\theta \tag{6.10}$$

The flow over a sphere is generated by superimposing a three-dimensional doublet and a uniform flow. The resulting surface velocity and pressure distributions are given by

$$V_\theta = \tfrac{3}{2} V_\infty \sin\theta \tag{6.19}$$

and

$$C_p = 1 - \tfrac{9}{4}\sin^2\theta \tag{6.20}$$

In comparison with flow over a cylinder, the surface velocity and magnitude of the pressure coefficient are smaller for the sphere—an example of the three-dimensional relieving effect.

In modern aerodynamic applications, inviscid, incompressible flows over complex three-dimensional bodies are usually computed via three-dimensional panel techniques.

6.9 PROBLEMS

6.1 Prove that three-dimensional source flow is irrotational.

6.2 Prove that three-dimensional source flow is a physically possible incompressible flow.

6.3 A sphere and a circular cylinder (with its axis perpendicular to the flow) are mounted in the same freestream. A pressure tap exists at the top of the sphere, and this is connected via a tube to one side of a manometer. The other side of the manometer is connected to a pressure tap on the surface of the cylinder. This tap is located on the cylindrical surface such that no deflection of the manometer fluid takes place. Calculate the location of this tap.

Inviscid, Compressible Flow

I n Part 3, we deal with high-speed flows—subsonic, supersonic, and hypersonic. In such flows, the density is a variable—this is compressible flow.

■

CHAPTER 7

Compressible Flow: Some Preliminary Aspects

With the realization of aeroplane and missile speeds equal to or even surpassing many times the speed of sound, thermodynamics has entered the scene and will never again leave our considerations.
   Source: Jakob Ackeret, 1962

PREVIEW BOX

With this chapter we move into the wild and exciting world of high-speed flow. To jump into this world, however, we need some preparation. This chapter prepares us to deal with the high-speed subsonic, transonic, supersonic, and hypersonic flows that are discussed in subsequent chapters. In the present chapter, we once again add to our inventory of fundamental principles and relations—those necessary to understand and predict high-speed flow. View this chapter as a continuation of Chapter 2 on some fundamental principles and equations, but a continuation with a flair laced by high energy (figuratively and literally).

Energy? What do we mean? Consider that you are in an automobile traveling at 65 miles per hour along a highway. If you stick your hand out the window (not recommended by the way, for safety reasons) you will sense a certain amount of energy in the airflow, and your hand will be forced back to some

extent. Now imagine that you are traveling at 650 mi/h and you stick your hand out the window (figuratively). You can just imagine what tremendous energy you would sense, and what disaster would happen to your hand. The point is that high-speed flow is *high-energy flow*. The science of energy is *thermodynamics*. So this chapter is, for the most part, a discussion of thermodynamics—but only to the extent necessary for our subsequent applications in high-speed flow.

High-speed flow is also *compressible flow*. With this chapter, we can no longer assume that the density is constant. Rather, the density is a variable, which introduces some neat physics and analytical challenges for our study of compressible flow. Indeed, the world we are entering, beginning with this chapter, is quite different than the incompressible world we dealt with in Chapters 3 to 6. This is like a fresh start. So get on board, and let's start.

7.1 INTRODUCTION

On September 30, 1935, the leading aerodynamicists from all corners of the world converged on Rome, Italy. Some of them arrived in airplanes which, in those days, lumbered along at speeds of 130 mi/h. Ironically, these people were gathering to discuss airplane aerodynamics not at 130 mi/h but rather at the unbelievable speeds of 500 mi/h and faster. By invitation only, such aerodynamic giants as Theodore von Karman and Eastman Jacobs from the United States, Ludwig Prandtl and Adolf Busemann from Germany, Jakob Ackeret from Switzerland, G. I. Taylor from England, Arturo Crocco and Enrico Pistolesi from Italy, and others assembled for the fifth Volta Conference, which had as its topic "High Velocities in Aviation." Although the jet engine had not yet been developed, these men were convinced that the future of aviation was "faster and higher." At that time, some aeronautical engineers felt that airplanes would never fly faster than the speed of sound—the myth of the "sound barrier" was propagating through the ranks of aviation. However, the people who attended the fifth Volta Conference knew better. For six days, inside an impressive Renaissance building that served as the city hall during the Holy Roman Empire, these individuals presented papers that discussed flight at high subsonic, supersonic, and even hypersonic speeds. Among these presentations was the first public revelation of the concept of a swept wing for high-speed flight; Adolf Busemann, who originated the concept, discussed the technical reasons why swept wings would have less drag at high speeds than conventional straight wings. (One year later, the swept-wing concept was classified by the German Luftwaffe as a military secret. The Germans went on to produce a large bulk of swept-wing research during World War II, resulting in the design of the first operational jet airplane—the Me 262—which had a moderate degree of sweep.) Many of the discussions at the Volta Conference centered on the effects of "compressibility" at high subsonic speeds, that is, the effects of variable density, because this was clearly going to be the first problem to be encountered by future high-speed airplanes. For example, Eastman Jacobs presented wind-tunnel test results for compressibility effects on standard NACA four- and five-digit airfoils at high subsonic speeds and noted extraordinarily large increases in drag beyond certain freestream Mach numbers. In regard to supersonic flows, Ludwig Prandtl presented a series of photographs showing shock waves inside nozzles and on various bodies—with some of the photographs dating as far back as 1907, when Prandtl started serious work in supersonic aerodynamics. (Clearly, Ludwig Prandtl was busy with much more than just the development of his incompressible airfoil and finite-wing theory discussed in Chapters 4 and 5.) Jakob Ackeret gave a paper on the design of supersonic wind tunnels, which, under his direction, were being established in Italy, Switzerland, and Germany. There were also presentations on propulsion techniques for high-speed flight, including rockets and ramjets. The atmosphere surrounding the participants in the Volta Conference was exciting and heady; the conference launched the world aerodynamic community into the area of high-speed subsonic and supersonic flight—an area which today is as commonplace as the 130-mi/h flight speeds of 1935. Indeed, the purpose of the next eight chapters of this book is to present the fundamentals of such high-speed flight.

In contrast to the low-speed, incompressible flows discussed in Chapters 3 to 6, the pivotal aspect of high-speed flow is that the density is a variable. Such flows are called *compressible flows* and are the subject of Chapters 7 to 14. Return to Figure 1.45, which gives a block diagram categorizing types of aerodynamic flows. In Chapters 7 to 14, we discuss flows which fall into blocks *D* and *F*; that is, we will deal with *inviscid compressible* flow. In the process, we touch all the flow regimes itemized in blocks *G* through *J*. These flow regimes are illustrated in Figure 1.44; study Figures 1.44 and 1.45 carefully, and review the surrounding discussion in Section 1.10 before proceeding further.

In addition to variable density, another pivotal aspect of high-speed compressible flow is *energy*. A high-speed flow is a high-energy flow. For example, consider the flow of air at standard sea level conditions moving at twice the speed of sound. The internal energy of 1 kg of this air is 2.07×10^5 J, whereas the kinetic energy is larger, namely, 2.31×10^5 J. When the flow velocity is decreased, some of this kinetic energy is lost and reappears as an increase in internal energy, hence increasing the temperature of the gas. Therefore, in a high-speed flow, energy transformations and temperature changes are important considerations. Such considerations come under the science of *thermodynamics*. For this reason, thermodynamics is a vital ingredient in the study of compressible flow. One purpose of the present chapter is to review briefly the particular aspects of thermodynamics which are essential to our subsequent discussions of compressible flow.

The road map for this chapter is given in Figure 7.1. As our discussion proceeds, refer to this road map in order to provide an orientation for our ideas.

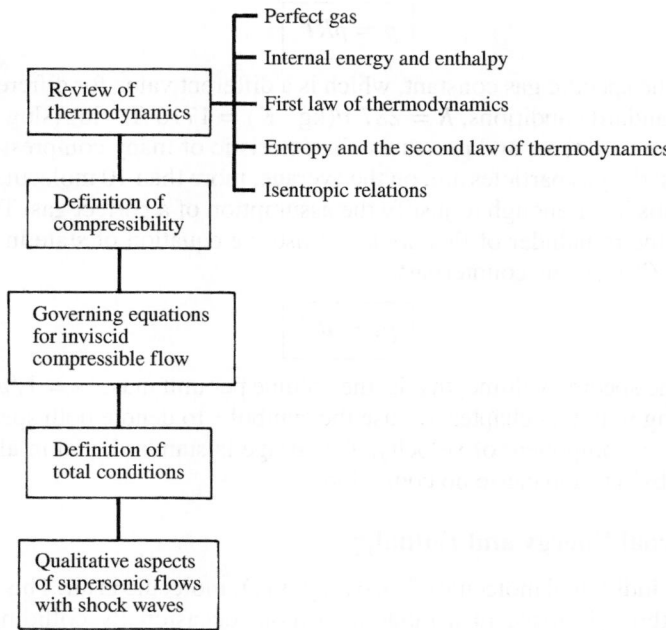

Figure 7.1 Road map for Chapter 7.

7.2 A BRIEF REVIEW OF THERMODYNAMICS

The importance of thermodynamics in the analysis and understanding of compressible flow was underscored in Section 7.1. Hence, the purpose of the present section is to review those aspects of thermodynamics that are important to compressible flows. This is in no way intended to be an exhaustive discussion of thermodynamics; rather, it is a review of only those fundamental ideas and equations that will be of direct use in subsequent chapters. If you have studied thermodynamics, this review should serve as a ready reminder of some important relations. If you are not familiar with thermodynamics, this section is somewhat self-contained so as to give you a feeling for the fundamental ideas and equations that we use frequently in subsequent chapters.

7.2.1 Perfect Gas

As described in Section 1.2, a gas is a collection of particles (molecules, atoms, ions, electrons, etc.) which are in more or less random motion. Due to the electronic structure of these particles, a force field pervades the space around them. The force field due to one particle reaches out and interacts with neighboring particles, and vice versa. Hence, these fields are called *intermolecular forces*. However, if the particles of the gas are far enough apart, the influence of the intermolecular forces is small and can be neglected. A gas in which the intermolecular forces are neglected is defined as a *perfect gas*. For a perfect gas, p, ρ, and T are related through the following *equation of state:*

$$\boxed{p = \rho RT} \tag{7.1}$$

where R is the specific gas constant, which is a different value for different gases. For air at standard conditions, $R = 287$ J/(kg $\cdot$ K) $= 1716$ (ft $\cdot$ lb)/(slug $\cdot$ °R).

At the temperatures and pressures characteristic of many compressible flow applications, the gas particles are, on the average, more than 10 molecular diameters apart; this is far enough to justify the assumption of a perfect gas. Therefore, throughout the remainder of this book, we use the equation of state in the form of Equation (7.1), or its counterpart,

$$\boxed{pv = RT} \tag{7.2}$$

where v is the specific volume, that is, the volume per unit mass; $v = 1/\rho$. (*Please note:* Starting with this chapter, we use the symbol v to denote both specific volume and the y component of velocity. This usage is standard, and in all cases it should be obvious and cause no confusion.)

7.2.2 Internal Energy and Enthalpy

Consider an individual molecule of a gas, say, an O_2 molecule in air. This molecule is moving through space in a random fashion, occasionally colliding with a neighboring molecule. Because of its velocity through space, the molecule has translational kinetic energy. In addition, the molecule is made up of individual atoms which we can visualize as connected to each other along various axes; for example, we can visualize the O_2 molecule as a "dumbbell" shape, with an O atom at each end of a connecting axis. In addition to its translational motion, such a molecule can execute a rotational motion in space; the kinetic energy of this rotation contributes to the net energy of the molecule. Also, the atoms of a given molecule can vibrate back and forth along and across the molecular axis, thus contributing a potential and kinetic energy of vibration to the molecule. Finally, the motion of the electrons around each of the nuclei of the molecule contributes an "electronic" energy to the molecule. Hence, the energy of a given molecule is the sum of its translational, rotational, vibrational, and electronic energies.

Now consider a finite volume of gas consisting of a large number of molecules. The sum of the energies of all the molecules in this volume is defined as the *internal energy* of the gas. The internal energy per unit mass of gas is defined as the specific internal energy, denoted by e. A related quantity is the specific enthalpy, denoted by h and defined as

$$h = e + pv \tag{7.3}$$

For a perfect gas, both e and h are functions of temperature only:

$$e = e(T) \tag{7.4a}$$

$$h = h(T) \tag{7.4b}$$

Let de and dh represent differentials of e and h, respectively. Then, for a perfect gas,

$$de = c_v \, dT \tag{7.5a}$$

$$dh = c_p \, dT \tag{7.5b}$$

where c_v and c_p are the specific heats at constant volume and constant pressure, respectively. In Equation (7.5a and b), c_v and c_p can themselves be functions of T. However, for moderate temperatures (for air, for $T < 1000$ K), the specific heats are reasonably constant. A perfect gas where c_v and c_p are constants is defined as a *calorically perfect gas,* for which Equation (7.5a and b) become

$$\boxed{\begin{aligned} e &= c_v T \\ h &= c_p T \end{aligned}}$$

$$\tag{7.6a}$$
$$\tag{7.6b}$$

For a large number of practical compressible flow problems, the temperatures are moderate. For this reason, in this book we always treat the gas as calorically perfect; that is, we assume that the specific heats are constant. For a discussion of compressible flow problems where the specific heats are not constant (such as the high-temperature chemically reacting flow over a high-speed atmospheric entry vehicle, that is, the space shuttle), see Reference 21.

Note that e and h in Equations (7.3) through (7.6) are thermodynamic state variables—they depend only on the state of the gas and are independent of any process. Although c_v and c_p appear in these equations, there is no restriction to just a constant volume or a constant pressure process. Rather, Equation (7.5a and b) and (7.6a and b) are relations for thermodynamic state variables, namely, e and h as functions of T, and have nothing to do with the process that may be taking place.

For a specific gas, c_p and c_v are related through the equation

$$c_p - c_v = R \tag{7.7}$$

Dividing Equation (7.7) by c_p, we obtain

$$1 - \frac{c_v}{c_p} = \frac{R}{c_p} \tag{7.8}$$

Define $\gamma \equiv c_p/c_v$. For air at standard conditions, $\gamma = 1.4$. Then Equation (7.8) becomes

$$1 - \frac{1}{\gamma} = \frac{R}{c_p}$$

or

$$\boxed{c_p = \frac{\gamma R}{\gamma - 1}} \tag{7.9}$$

Similarly, dividing Equation (7.7) by c_v, we obtain

$$\boxed{c_v = \frac{R}{\gamma - 1}} \tag{7.10}$$

Equations (7.9) and (7.10) are particularly useful in our subsequent discussion of compressible flow.

EXAMPLE 7.1

Consider a room with a rectangular floor that is 5 m by 7 m, and a 3.3 m high ceiling. The air pressure and temperature in the room are 1 atm and 25°C, respectively. Calculate the internal energy and the enthalpy of the air in the room.

■ **Solution**

We first need to calculate the mass of air in the room. From Equation (7.1)

$$\rho = \frac{p}{RT}$$

where each quantity must be expressed in consistent SI units. Since 1 atm = 1.01×10^5 N/m² and 0°C is 273 K, we have

$$p = 1.01 \times 10^5 \text{ N/m}^2 \quad \text{and} \quad T = 273 + 25 = 298 \text{ K}$$

Hence,

$$\rho = \frac{p}{RT} = \frac{1.01 \times 10^5}{(287)(298)} = 1.181 \text{ kg/m}^3$$

The volume of the room is $(5)(7)(3.3) = 115.5 \text{ m}^3$. The mass of air in the room is therefore

$$M = (1.181)(115.5) = 136.4 \text{ kg}$$

From Equation (7.6a), the internal energy per unit mass is

$$e = c_v T$$

where

$$c_v = \frac{R}{\gamma - 1} = \frac{(287)}{1.4 - 1} = \frac{287}{0.4} = 717.5 \text{ joule/(kg} \cdot \text{K)}$$

Thus,
$$e = c_v T = (717.5)(298) = 2.138 \times 10^5 \text{ joule/kg}$$

The internal energy in the room, E, is then

$$E = Me = (136.4)(2.138 \times 10^5) = \boxed{2.92 \times 10^7 \text{ joule}}$$

From Equation (7.6b), the enthalpy per unit mass is

$$h = c_p T$$

where

$$c_p = \frac{\gamma R}{\gamma - 1} = \frac{(1.4)(287)}{0.4} = 1004.5 \text{ joule/(kg} \cdot \text{K)}$$

Thus,
$$h = c_p T = (1004.5)(298) = 2.993 \times 10^5 \text{ joule/kg}$$

The enthalpy in the room, H, is then

$$H = Mh = (136.4)(2.993 \times 10^5) = \boxed{4.08 \times 10^7 \text{ joule}}$$

A check on two answers can be made knowing that

$$\frac{h}{e} = \frac{c_p T}{c_v T} = \frac{c_p}{c_v} = \gamma = 1.4$$

From the answers,

$$\frac{H}{E} = \frac{4.08 \times 10^7}{2.92 \times 10^7} = 1.4 \qquad \text{It checks.}$$

This simple example is intended to reinforce two basic points.

1. Consistent units must be used when making a calculation using basic equations from physics, such as Equations (7.1), (7.6a), and (7.6b). Here we used SI units, because the given information was given in terms of meters and degrees Celsius, which is readily put in terms of the consistent unit of temperature, degrees Kelvin.
2. The internal energy and enthalpy per unit mass were calculated directly from the temperature of the gas, using Equation (7.6a and b). There was no need to consider a "constant volume process" or a "constant pressure process"; there is no process at all to consider here. Internal energy and enthalpy per unit mass, e and h, are simply

state variables, depending only on the thermodynamic state of the system. Even though Equation (7.6a) contains c_v, a constant volume process is not relevant here. Similarly, even though Equation (7.6b) contains c_p, a constant pressure process is not relevant here.

EXAMPLE 7.2

One type of supersonic wind tunnel is a blow-down tunnel, where air is stored in a high-pressure reservoir, and then, upon the opening of a valve, exhausted through the tunnel into a vacuum tank or simply into the open atmosphere at the downstream end of the tunnel. Supersonic wind tunnels are discussed in Chapter 10. For this example, we consider just the high-pressure reservoir as a storage tank that is being charged with air by a high-pressure pump. As air is being pumped into the constant-volume reservoir, the air pressure inside the reservoir increases. The pump continues to charge the reservoir until the desired pressure is achieved.

Consider a reservoir with an internal volume of 30 m³. As air is pumped into the reservoir, the air pressure inside the reservoir continually increases with time. Consider the instant during the charging process when the reservoir pressure is 10 atm. Assume the air temperature inside the reservoir is held constant at 300 K by means of a heat exchanger. Air is pumped into the reservoir at the rate of 1 kg/s. Calculate the time rate of increase of pressure in the reservoir at this instant.

■ Solution

Let M be the total mass of air inside the reservoir at any instant. Since air is being pumped into the reservoir at the rate of 1 kg/s, then the total mass of air is increasing at the rate of $dM/dt = 1$ kg/s. The density of the air at any instant is

$$\rho = \frac{M}{V} \tag{E.7.1}$$

where V is the total volume of the reservoir and is constant; $V = 30$ m³. Since V is constant, from Equation (E.7.1),

$$\frac{d\rho}{dt} = \frac{1}{V} \frac{dM}{dt} = \frac{1 \text{ kg/s}}{30 \text{ m}^3} = 0.0333$$

Differentiating Equation (7.1) with respect to time, and recalling that R and T are constant,

$$\frac{dp}{dt} = RT \frac{d\rho}{dt} \tag{E.7.2}$$

$$\frac{dp}{dt} = (287)(300)(0.0333) = \boxed{2867.13 \frac{\text{N}}{\text{m}^2\text{s}}}$$

EXAMPLE 7.3

In Example 7.2, if the pumping rate of 1 kg/s were maintained constant throughout the charging process, how long will it take to increase the reservoir pressure from 10 to 20 atm?

■ **Solution**

If the pumping rate is held constant at 1 kg/s, then the time rate of change of density, $d\rho/dt = 0.0333$ kg/m^{3}s will remain constant, and consequently the time rate of change of pressure, $dp/dt = 2867.13$ N/m^{2}s will remain constant. Hence, the time required to increase the reservoir pressure from 10 atm to 20 atm is

$$\frac{(20 \text{ atm} - 10 \text{ atm})(1.01 \times 10^5)}{2867.13} = 352.27 \text{ s}$$

or, in minutes,

$$\frac{352.27}{60} = \boxed{5.87 \text{ min}}$$

7.2.3 First Law of Thermodynamics

Consider a fixed mass of gas, which we define as the *system*. (For simplicity, assume a unit mass, for example, 1 kg or 1 slug.) The region outside the system is called the *surroundings*. The interface between the system and its surroundings is called the *boundary,* as shown in Figure 7.2. Assume that the system is stationary. Let δq be an incremental amount of heat added to the system across the boundary, as sketched in Figure 7.2. Examples of the source of δq are radiation from the surroundings that is absorbed by the mass in the system and thermal conduction due to temperature gradients across the boundary. Also, let δw denote the work done on the system by the surroundings (say, by a displacement of the boundary, squeezing the volume of the system to a smaller value). As discussed earlier, due to the molecular motion of the gas, the system has an internal energy e. The heat added and work done on the system cause a change in energy, and since the system is stationary, this change in energy is simply de:

$$\boxed{\delta q + \delta w = de} \tag{7.11}$$

This is the *first law of thermodynamics:* It is an empirical result confirmed by experience. In Equation (7.11), e is a state variable. Hence, de is an exact

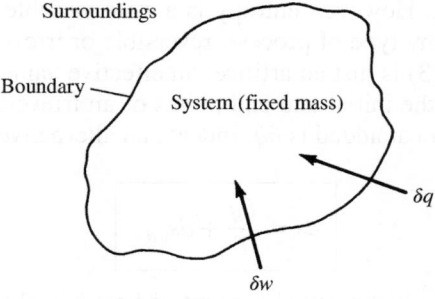

Figure 7.2 Thermodynamic system.

differential, and its value depends only on the initial and final states of the system. In contrast, δq and δw depend on the process in going from the initial to the final states.

For a given de, there are in general an infinite number of different ways (processes) by which heat can be added and work done on the system. We are primarily concerned with three types of processes:

1. *Adiabatic process.* One in which no heat is added to or taken away from the system
2. *Reversible process.* One in which no dissipative phenomena occur, that is, where the effects of viscosity, thermal conductivity, and mass diffusion are absent
3. *Isentropic process.* One that is both adiabatic and reversible

For a reversible process, it can be easily shown that $\delta w = -p\,dv$, where dv is an incremental change in the volume due to a displacement of the boundary of the system. Thus, Equation (7.11) becomes

$$\delta q - p\,dv = de \tag{7.12}$$

7.2.4 Entropy and the Second Law of Thermodynamics

Consider a block of ice in contact with a red-hot plate of steel. Experience tells us that the ice will warm up (and probably melt) and the steel plate will cool down. However, Equation (7.11) does not necessarily say this will happen. Indeed, the first law allows that the ice may get cooler and the steel plate hotter—just as long as energy is conserved during the process. Obviously, in real life this does not happen; instead, nature imposes another condition on the process, a condition that tells us *which direction* a process will take. To ascertain the proper direction of a process, let us define a new state variable, the entropy, as follows:

$$ds = \frac{\delta q_{\text{rev}}}{T} \tag{7.13}$$

where s is the entropy of the system, δq_{rev} is an incremental amount of heat added reversibly to the system, and T is the system temperature. Do not be confused by the above definition. It defines a change in entropy in terms of a reversible addition of heat δq_{rev}. However, entropy is a state variable, and it can be used in conjunction with any type of process, reversible or irreversible. The quantity δq_{rev} in Equation (7.13) is just an artifice; an effective value of δq_{rev} can always be assigned to relate the initial and end points of an irreversible process, where the actual amount of heat added is δq. Indeed, an alternative and probably more lucid relation is

$$\boxed{ds = \frac{\delta q}{T} + ds_{\text{irrev}}} \tag{7.14}$$

In Equation (7.14), δq is the actual amount of heat added to the system during an actual irreversible process, and ds_{irrev} is the generation of entropy due to the irreversible, dissipative phenomena of viscosity, thermal conductivity, and mass diffusion occurring *within* the system. These dissipative phenomena always increase the entropy:

$$ds_{\text{irrev}} \geq 0 \tag{7.15}$$

In Equation (7.15), the equals sign denotes a reversible process, where by definition no dissipative phenomena occur within the system. Combining Equations (7.14) and (7.15), we have

$$ds \geq \frac{\delta q}{T} \tag{7.16}$$

Furthermore, if the process is adiabatic, $\delta q = 0$, and Equation (7.16) becomes

$$ds \geq 0 \tag{7.17}$$

Equations (7.16) and (7.17) are forms of the *second law of thermodynamics*. The second law tells us in what direction a process will take place. A process will proceed in a direction such that the entropy of the system plus that of its surroundings always increases or, at best, stays the same. In our example of the ice in contact with hot steel, consider the system to be both the ice and steel plate combined. The simultaneous heating of the ice and cooling of the plate yield a net increase in entropy for the system. On the other hand, the impossible situation of the ice getting cooler and the plate hotter would yield a net decrease in entropy, a situation forbidden by the second law. In summary, the concept of entropy in combination with the second law allows us to predict the *direction* that nature takes.

The practical calculation of entropy is carried out as follows. In Equation (7.12), assume that heat is added reversibly; then the definition of entropy, Equation (7.13), substituted in Equation (7.12) yields

$$T\,ds - p\,dv = de$$

or

$$T\,ds = de + p\,dv \tag{7.18}$$

From the definition of enthalpy, Equation (7.3), we have

$$dh = de + p\,dv + v\,dp \tag{7.19}$$

Combining Equations (7.18) and (7.19), we obtain

$$T\,ds = dh - v\,dp \tag{7.20}$$

Equations (7.18) and (7.20) are important; they are essentially alternate forms of the first law expressed in terms of entropy. For a perfect gas, recall Equation (7.5*a* and *b*), namely, $de = c_v\,dT$ and $dh = c_p\,dT$. Substituting these relations into Equations (7.18) and (7.20), we obtain

$$ds = c_v\frac{dT}{T} + \frac{p\,dv}{T} \tag{7.21}$$

and

$$ds = c_p\frac{dT}{T} - \frac{v\,dp}{T} \tag{7.22}$$

Working with Equation (7.22), substitute the equation of state $pv = RT$, or $v/T = R/p$, into the last term:

$$ds = c_p\frac{dT}{T} - R\frac{dp}{p} \tag{7.23}$$

Consider a thermodynamic process with initial and end states denoted by 1 and 2, respectively. Equation (7.23), integrated between states 1 and 2, becomes

$$s_2 - s_1 = \int_{T_1}^{T_2} c_p\frac{dT}{T} - \int_{p_1}^{p_2} R\frac{dp}{p} \tag{7.24}$$

For a calorically perfect gas, both R and c_p are constants; hence, Equation (7.24) becomes

$$s_2 - s_1 = c_p\ln\frac{T_2}{T_1} - R\ln\frac{p_2}{p_1} \tag{7.25}$$

In a similar fashion, Equation (7.21) leads to

$$s_2 - s_1 = c_v\ln\frac{T_2}{T_1} + R\ln\frac{v_2}{v_1} \tag{7.26}$$

Equations (7.25) and (7.26) are practical expressions for the calculation of the entropy change of a calorically perfect gas between two states. Note from these equations that s is a function of two thermodynamic variables, for example, $s = s(p, T)$, $s = s(v, T)$.

7.2.5 Isentropic Relations

We have defined an isentropic process as one which is both adiabatic and reversible. Consider Equation (7.14). For an adiabatic process, $\delta q = 0$. Also, for a reversible process, $ds_{\text{irrev}} = 0$. Thus, for an adiabatic, reversible process, Equation (7.14) yields $ds = 0$, or entropy is constant; hence, the word "isentropic." For such an isentropic process, Equation (7.25) is written as

$$0 = c_p\ln\frac{T_2}{T_1} - R\ln\frac{p_2}{p_1}$$

$$\ln\frac{p_2}{p_1} = \frac{c_p}{R}\ln\frac{T_2}{T_1}$$

or

$$\frac{p_2}{p_1} = \left(\frac{T_2}{T_1}\right)^{c_p/R} \tag{7.27}$$

However, from Equation (7.9),

$$\frac{c_p}{R} = \frac{\gamma}{\gamma - 1}$$

and hence Equation (7.27) is written as

$$\frac{p_2}{p_1} = \left(\frac{T_2}{T_1}\right)^{\gamma/(\gamma-1)} \tag{7.28}$$

In a similar fashion, Equation (7.26) written for an isentropic process gives

$$0 = c_v \ln \frac{T_2}{T_1} + R \ln \frac{v_2}{v_1}$$

$$\ln \frac{v_2}{v_1} = -\frac{c_v}{R} \ln \frac{T_2}{T_1}$$

$$\frac{v_2}{v_1} = \left(\frac{T_2}{T_1}\right)^{-c_v/R} \tag{7.29}$$

From Equation (7.10),

$$\frac{c_v}{R} = \frac{1}{\gamma - 1}$$

and hence Equation (7.29) is written as

$$\frac{v_2}{v_1} = \left(\frac{T_2}{T_1}\right)^{-1/(\gamma-1)} \tag{7.30}$$

Since $\rho_2/\rho_1 = v_1/v_2$, Equation (7.30) becomes

$$\frac{\rho_2}{\rho_1} = \left(\frac{T_2}{T_1}\right)^{1/(\gamma-1)} \tag{7.31}$$

Combining Equations (7.28) and (7.31), we can summarize the isentropic relations as

$$\boxed{\frac{p_2}{p_1} = \left(\frac{\rho_2}{\rho_1}\right)^{\gamma} = \left(\frac{T_2}{T_1}\right)^{\gamma/(\gamma-1)}} \tag{7.32}$$

Equation (7.32) is very important; it relates pressure, density, and temperature for an isentropic process. We use this equation frequently, so make certain to brand it on your mind. Also, keep in mind the source of Equation (7.32); it stems from the first law and the definition of entropy. Therefore, Equation (7.32) is basically an energy relation for an isentropic process.

Why is Equation (7.32) so important? Why is it frequently used? Why are we so interested in an isentropic process when it seems so restrictive—requiring both adiabatic and reversible conditions? The answers rest on the fact that a large number of practical compressible flow problems can be assumed to be isentropic—contrary to what you might initially think. For example, consider the flow over an airfoil or through a rocket engine. In the regions adjacent to the airfoil surface and the rocket nozzle walls, a boundary layer is formed wherein the dissipative mechanisms of viscosity, thermal conduction, and diffusion are strong. Hence, the entropy increases within these boundary layers. However, consider the fluid elements moving outside the boundary layer. Here, the dissipative effects of viscosity, etc., are very small and can be neglected. Moreover, no heat is being transferred to or from the fluid element (i.e., we are not heating the fluid element with a Bunsen burner or cooling it in a refrigerator); thus, the flow outside the boundary layer is adiabatic. Consequently, the fluid elements outside the boundary layer are experiencing an adiabatic reversible process—namely, isentropic flow. In the vast majority of practical applications, the viscous boundary layer adjacent to the surface is thin compared with the entire flow field, and hence large regions of the flow can be assumed isentropic. This is why a study of isentropic flow is directly applicable to many types of practical compressible flow problems. In turn, Equation (7.32) is a powerful relation for such flows, valid for a calorically perfect gas.

This ends our brief review of thermodynamics. Its purpose has been to give a quick summary of ideas and equations that will be employed throughout our subsequent discussions of compressible flow. Indeed, this author knows of no practical problem dealing with compressible flow that can be solved without invoking some aspects of thermodynamics—it is that important. For a more thorough discussion of the power and beauty of thermodynamics, see any good thermodynamics text, such as References 22 to 24.

EXAMPLE 7.4

Consider a Boeing 747 flying at a standard altitude of 36,000 ft. The pressure at a point on the wing is 400 lb/ft^2. Assuming isentropic flow over the wing, calculate the temperature at this point.

■ Solution

From Appendix E, at a standard altitude of 36,000 ft, $p_\infty = 476$ lb/ft^2 and $T_\infty = 391$ °R. From Equation (7.32),

$$\frac{p}{p_\infty} = \left(\frac{T}{T_\infty}\right)^{\gamma/(\gamma-1)}$$

or

$$T = T_\infty \left(\frac{p}{p_\infty}\right)^{(\gamma-1)/\gamma} = 391\left(\frac{400}{476}\right)^{0.4/1.4} = \boxed{372\ \text{°R}}$$

EXAMPLE 7.5

Consider the gas in the reservoir of the supersonic wind tunnel discussed in Examples 7.2 and 7.3. The pressure and temperature of the air in the reservoir are 20 atm and 300 K, respectively. The air in the reservoir expands through the wind tunnel duct. At a certain location in the duct, the pressure is 1 atm. Calculate the air temperature at this location if: (a) the expansion is isentropic and (b) the expansion is nonisentropic with an entropy increase through the duct to this location of 320 J/(kg · K).

■ **Solution**

(a) From Equation (7.32),

$$\frac{p_2}{p_1} = \left(\frac{T_2}{T_1}\right)^{\gamma/(\gamma-1)}$$

or,

$$T_2 = T_1 \left(\frac{p_2}{p_1}\right)^{\frac{\gamma-1}{\gamma}} = 300 \left(\frac{1}{20}\right)^{\frac{0.4}{1.4}} = 300\,(0.05)^{0.2857}$$

$$= 300\,(0.4249) = \boxed{127.5 \text{ K}}$$

(b) From Equation (7.25),

$$s_2 - s_1 = c_p \ln \frac{T_2}{T_1} - R \ln \frac{p_2}{p_1}$$

Using Equation (7.9) to obtain the value of c_p,

$$c_p = \frac{\gamma R}{\gamma - 1} = \frac{(1.4)(287)}{0.4} = 1004.5 \; \frac{\text{J}}{\text{kg} \cdot \text{K}}$$

we have from Equation (7.25),

$$320 = 1004.5 \ln \left(\frac{T_2}{300}\right) - (287) \ln \left(\frac{1}{20}\right)$$

$$= 1004.5 \ln \left(\frac{T_2}{300}\right) - (-859.78)$$

Thus,

$$\ln \left(\frac{T_2}{300}\right) = \frac{320 - 859.78}{1004.5} = -0.5374$$

$$\frac{T_2}{300} = e^{-0.5374} = 0.5843$$

$$T_2 = (0.5843)(300) = \boxed{175.3 \text{ K}}$$

Comment: Comparing the results from parts (a) and (b), note that the entropy increase results in a higher temperature at the point in the expansion where $p = 1$ atm compared to that for the isentropic expansion. This makes sense. From Equation (7.25) we see that entropy is a function of both temperature and pressure, increasing with an increase in temperature and decreasing with an increase in pressure. In this example, the final pressure for both cases (a) and (b) is the same, but the entropy for case (b) is higher. Thus, from Equation (7.25), we see that the final temperature for case (b) must be higher than that for case (a). On a more qualitative basis, the physical mechanisms that could produce the change in entropy would be viscous dissipation (friction), the presence of shock waves in the duct, or heat addition from the surroundings through the walls of the duct. Intuitively, all these irreversible mechanisms would result in a higher gas temperature than the isentropic expansion that, by definition, assumes an adiabatic and reversible (no friction) expansion.

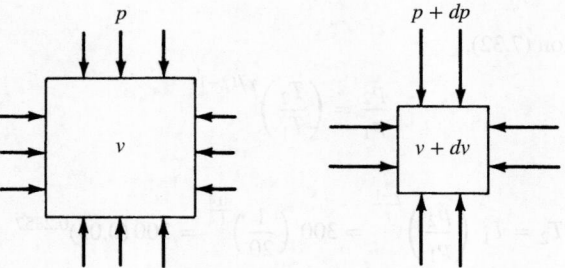

Figure 7.3 Definition of compressibility.

7.3 DEFINITION OF COMPRESSIBILITY

All real substances are compressible to some greater or lesser extent; that is, when you squeeze or press on them, their density will change. This is particularly true of gases, much less so for liquids, and virtually unnoticeable for solids. The amount by which a substance can be compressed is given by a specific property of the substance called the *compressibility,* defined below.

Consider a small element of fluid of volume v, as sketched in Figure 7.3. The pressure exerted on the sides of the element is p. Assume the pressure is now increased by an infinitesimal amount dp. The volume of the element will change by a corresponding amount dv; here, the volume will decrease; hence, dv shown in Figure 7.3 is a negative quantity. By definition, the compressibility τ of the fluid is

$$\tau = -\frac{1}{v}\frac{dv}{dp} \tag{7.33}$$

Physically, the compressibility is the fractional change in volume of the fluid element per unit change in pressure. However, Equation (7.33) is not precise enough. We know from experience that when a gas is compressed (say, in a bicycle pump), its temperature tends to increase, depending on the amount of heat transferred into or out of the gas through the boundaries of the system. If the temperature of the fluid element in Figure 7.3 is held constant (due to some heat transfer mechanism), then τ is identified as the *isothermal compressibility* τ_T, defined from Equation (7.33) as

$$\tau_T = -\frac{1}{v}\left(\frac{\partial v}{\partial p}\right)_T \tag{7.34}$$

On the other hand, if no heat is added to or taken away from the fluid element, and if friction is ignored, the compression of the fluid element takes place isentropically, and τ is identified as the isentropic compressibility τ_s, defined from Equation (7.33) as

$$\tau_s = -\frac{1}{v}\left(\frac{\partial v}{\partial p}\right)_s \tag{7.35}$$

where the subscript s denotes that the partial derivative is taken at constant entropy. Both τ_T and τ_s are precise thermodynamic properties of the fluid; their values for different gases and liquids can be obtained from various handbooks of physical properties. In general, the compressibility of gases is several orders of magnitude larger than that of liquids.

The role of the compressibility τ in determining the properties of a fluid in motion is seen as follows. Define v as the specific volume (i.e., the volume per unit mass). Hence, $v = 1/\rho$. Substituting this definition into Equation (7.33), we obtain

$$\tau = \frac{1}{\rho}\frac{d\rho}{dp} \tag{7.36}$$

Thus, whenever the fluid experiences a change in pressure dp, the corresponding change in density $d\rho$ from Equation (7.36) is

$$d\rho = \rho\tau\, dp \tag{7.37}$$

Consider a fluid flow, say, for example, the flow over an airfoil. If the fluid is a *liquid,* where the compressibility τ is very small, then for a given pressure change dp from one point to another in the flow, Equation (7.37) states that $d\rho$ will be negligibly small. In turn, we can reasonably assume that ρ is constant and that the flow of a liquid is incompressible. On the other hand, if the fluid is a *gas,* where the compressibility τ is large, then for a given pressure change dp from one point to another in the flow, Equation (7.37) states that $d\rho$ can be large. Thus, ρ is *not* constant, and in general, the flow of a gas is a *compressible flow*. The exception to this is the *low-speed flow* of a gas; in such flows, the actual magnitude of the pressure changes throughout the flow field is small compared with the pressure itself. Thus, for a low-speed flow, dp in Equation (7.37) is small, and even though τ is large, the value of $d\rho$ can be dominated by the small dp. In such cases, ρ can be assumed to be constant, hence allowing us to analyze low-speed gas flows as incompressible flows (such as discussed in Chapters 3 to 6).

Later, we demonstrate that the most convenient index to gage whether a gas flow can be considered incompressible, or whether it must be treated as compressible, is the Mach number M, defined in Chapter 1 as the ratio of local flow velocity V to the local speed of sound a:

$$M \equiv \frac{V}{a} \tag{7.38}$$

We show that, when $M > 0.3$, the flow should be considered compressible. Also, we show that the speed of sound in a gas is related to the isentropic compressibility τ_s, given by Equation (7.35).

7.4 GOVERNING EQUATIONS FOR INVISCID, COMPRESSIBLE FLOW

In Chapters 3 to 6, we studied inviscid, incompressible flow; recall that the primary dependent variables for such flows are p and $\mathbf{V}$, and hence we need only two basic equations, namely, the continuity and momentum equations, to solve for these two unknowns. Indeed, the basic equations are combined to obtain Laplace's equation and Bernoulli's equation, which are the primary tools used for the applications discussed in Chapters 3 to 6. Note that both ρ and T are assumed to be constant throughout such inviscid, incompressible flows. As a result, no additional governing equations are required; in particular, there is no need for the energy equation or energy concepts in general. Basically, incompressible flow obeys purely mechanical laws and does not require thermodynamic considerations.

In contrast, for compressible flow, ρ is variable and becomes an unknown. Hence, we need an additional governing equation—the energy equation—which in turn introduces internal energy e as an unknown. Since e is related to temperature, then T also becomes an important variable. Therefore, the primary dependent variables for the study of compressible flow are p, $\mathbf{V}$, ρ, e, and T; to solve for these five variables, we need five governing equations. Let us examine this situation further.

To begin with, the flow of a compressible fluid is governed by the basic equations derived in Chapter 2. At this point in our discussion, it is most important for you to be familiar with these equations as well as their derivation. Therefore, before proceeding further, return to Chapter 2 and carefully review the basic ideas and relations contained therein. This is a serious study tip, and if you follow it, the material in our next seven chapters will flow much easier for you. In particular, review the integral and differential forms of the continuity equation (Section 2.4), the momentum equation (Section 2.5), and the energy equation (Section 2.7); indeed, pay particular attention to the energy equation because this is an important aspect which sets compressible flow apart from incompressible flow.

For convenience, some of the more important forms of the governing equations for an inviscid, compressible flow from Chapter 2 are repeated below:

Continuity: From Equation (2.48),

$$\frac{\partial}{\partial t}\iiint_V \rho\, dV + \oiint_S \rho\mathbf{V}\cdot d\mathbf{S} = 0 \tag{7.39}$$

From Equation (2.52),

$$\frac{\partial\rho}{\partial t} + \nabla\cdot\rho\mathbf{V} = 0 \tag{7.40}$$

Momentum: From Equation (2.64),

$$\frac{\partial}{\partial t}\iiint_V \rho\mathbf{V}\, dV + \oiint_S (\rho\mathbf{V}\cdot d\mathbf{S})\mathbf{V} = -\oiint_S p\, d\mathbf{S} + \iiint_V \rho\mathbf{f}\, dV \tag{7.41}$$

From Equation (2.113a to c),

$$\rho\frac{Du}{Dt} = -\frac{\partial p}{\partial x} + \rho f_x \tag{7.42a}$$

$$\rho\frac{Dv}{Dt} = -\frac{\partial p}{\partial y} + \rho f_y \tag{7.42b}$$

$$\rho\frac{Dw}{Dt} = -\frac{\partial p}{\partial z} + \rho f_z \tag{7.42c}$$

Energy: From Equation (2.95),

$$\frac{\partial}{\partial t} \iiint_{\mathcal{V}} \rho \left(e + \frac{V^2}{2} \right) d\mathcal{V} + \oiint_{S} \rho \left(e + \frac{V^2}{2} \right) \mathbf{V} \cdot \mathbf{dS}$$

$$= \oiint_{\mathcal{V}} \dot{q} \rho \, d\mathcal{V} - \oiint_{S} p\mathbf{V} \cdot \mathbf{dS} + \iiint_{\mathcal{V}} \rho(\mathbf{f} \cdot \mathbf{V}) \, d\mathcal{V} \qquad (7.43)$$

From Equation (2.114),

$$\rho \frac{D(e + V^2/2)}{Dt} = \rho\dot{q} - \nabla \cdot p\mathbf{V} + \rho(\mathbf{f} \cdot \mathbf{V}) \qquad (7.44)$$

The above continuity, momentum, and energy equations are three equations in terms of the five unknowns p, $\mathbf{V}$, ρ, T, and e. Assuming a calorically perfect gas, the additional two equations needed to complete the system are obtained from Section 7.2:

Equation of state: $\qquad\qquad\qquad p = \rho R T \qquad\qquad\qquad (7.1)$

Internal energy: $\qquad\qquad\qquad e = c_v T \qquad\qquad\qquad (7.6a)$

In regard to the basic equations for compressible flow, please note that Bernoulli's equation as derived in Section 3.2 and given by Equation (3.13) does *not* hold for compressible flow; it clearly contains the assumption of constant density, and hence is invalid for compressible flow. This warning is necessary because experience shows that a certain number of students of aerodynamics, apparently attracted by the simplicity of Bernoulli's equation, attempt to use it for all situations, compressible as well as incompressible. Do not do it! Always remember that Bernoulli's equation in the form of Equation (3.13) holds for incompressible flow only, and we must dismiss it from our thinking when dealing with compressible flow.

As a final note, we use both the integral and differential forms of the above equations in our subsequent discussions. Make certain that you feel comfortable with these equations before proceeding further.

7.5 DEFINITION OF TOTAL (STAGNATION) CONDITIONS

At the beginning of Section 3.4, the concept of static pressure p was discussed in some detail. Static pressure is a measure of the purely random motion of molecules in a gas; it is the pressure you feel when you ride along with the gas at the local flow velocity. In contrast, the total (or stagnation) pressure was defined in Section 3.4 as the pressure existing at a point (or points) in the flow where $\mathbf{V} = 0$. Let us now define the concept of total conditions more precisely.

Consider a fluid element passing through a given point in a flow where the local pressure, temperature, density, Mach number, and velocity are p, T, ρ, M, and $\mathbf{V}$, respectively. Here, p, T, and ρ are static quantities (i.e., static pressure, static temperature, and static density, respectively); they are the pressure, temperature, and density you feel when you ride along with the gas at the local flow velocity. Now imagine that you grab hold of the fluid element and *adiabatically* slow it down to zero velocity. Clearly, you would expect (correctly) that the values of p, T, and ρ would change as the fluid element is brought to rest. In particular, the value of the temperature of the fluid element after it has been brought to rest adiabatically is defined as the *total temperature,* denoted by T_0. The corresponding value of enthalpy is defined as the *total enthalpy* h_0, where $h_0 = c_p T_0$ for a calorically perfect gas. Keep in mind that we do not *actually* have to bring the flow to rest in real life in order to talk about the total temperature or total enthalpy; rather, they are *defined quantities* that would exist at a point in a flow *if* (in our imagination) the fluid element passing through that point were brought to rest adiabatically. Therefore, at a given point in a flow, where the static temperature and enthalpy are T and h, respectively, we can also assign a value of total temperature T_0 and a value of total enthalpy h_0 defined as above.

The energy equation, Equation (7.44), provides some important information about total enthalpy and hence total temperature, as follows. Assume that the flow is adiabatic ($\dot{q} = 0$), and that body forces are negligible ($\mathbf{f} = 0$). For such a flow, Equation (7.44) becomes

$$\rho \frac{D(e + V^2/2)}{Dt} = -\nabla \cdot p\mathbf{V} \tag{7.45}$$

Expand the right-hand side of Equation (7.45) using the following vector identity:

$$\nabla \cdot p\mathbf{V} \equiv p\nabla \cdot \mathbf{V} + \mathbf{V} \cdot \nabla p \tag{7.46}$$

Also, note that the substantial derivative defined in Section 2.9 follows the normal laws of differentiation; for example,

$$\rho \frac{D(p/\rho)}{Dt} = \rho \frac{\rho Dp/Dt - pD\rho/Dt}{\rho^2} = \frac{Dp}{Dt} - \frac{p}{\rho}\frac{D\rho}{Dt} \tag{7.47}$$

Recall the form of the continuity equation given by Equation (2.108):

$$\frac{D\rho}{Dt} + \rho\nabla \cdot \mathbf{V} = 0 \tag{2.108}$$

Substituting Equation (2.108) into (7.47), we obtain

$$\rho \frac{D(p/\rho)}{Dt} = \frac{Dp}{Dt} + p\nabla \cdot \mathbf{V} = \frac{\partial p}{\partial t} + \mathbf{V} \cdot \nabla p + p\nabla \cdot \mathbf{V} \tag{7.48}$$

Substituting Equation (7.46) into (7.45), and adding Equation (7.48) to the result, we obtain

$$\rho \frac{D}{Dt}\left(e + \frac{p}{\rho} + \frac{V^2}{2}\right) = -p\nabla \cdot \mathbf{V} - \mathbf{V} \cdot \nabla p + \frac{\partial p}{\partial t} + \mathbf{V} \cdot \nabla p + p\nabla \cdot \mathbf{V} \tag{7.49}$$

Note that

$$e + \frac{p}{\rho} = e + pv \equiv h \tag{7.50}$$

Substituting Equation (7.50) into (7.49), and noting that some of the terms on the right-hand side of Equation (7.49) cancel each other, we have

$$\rho\frac{D(h + V^2/2)}{Dt} = \frac{\partial p}{\partial t} \tag{7.51}$$

If the flow is steady, $\partial p/\partial t = 0$, and Equation (7.51) becomes

$$\rho\frac{D(h + V^2/2)}{Dt} = 0 \tag{7.52}$$

From the definition of the substantial derivative given in Section 2.9, Equation (7.52) states that the time rate of change of $h + V^2/2$ following a moving fluid element is zero; that is,

$$\boxed{h + \frac{V^2}{2} = \text{const}} \tag{7.53}$$

along a streamline. Recall that the assumptions which led to Equation (7.53) are that the flow is steady, adiabatic, and inviscid. In particular, since Equation (7.53) holds for an adiabatic flow, it can be used to elaborate on our previous definition of total enthalpy. Since h_0 is defined as that enthalpy that would exist at a point if the fluid element were brought to rest adiabatically, we find from Equation (7.53) with $V = 0$ and hence $h = h_0$ that the value of the constant in Equation (7.53) is h_0. Hence, Equation (7.53) can be written as

$$\boxed{h + \frac{V^2}{2} = h_0} \tag{7.54}$$

Equation (7.54) is important; it states that at any point in a flow, the total enthalpy is given by the sum of the static enthalpy plus the kinetic energy, all per unit mass. Whenever we have the combination $h + V^2/2$ in any subsequent equations, it can be identically replaced by h_0. For example, Equation (7.52), which was derived for a steady, adiabatic, inviscid flow, states that

$$\rho\frac{Dh_0}{Dt} = 0$$

that is, the total enthalpy is constant along a streamline. Moreover, if all the streamlines of the flow originate from a common uniform freestream (as is usually the case), then h_0 is the same for each streamline. Consequently, we have for such a steady, adiabatic flow that

$$\boxed{h_0 = \text{const}} \tag{7.55}$$

throughout the *entire* flow, and h_0 is equal to its freestream value. Equation (7.55), although simple in form, is a powerful tool. For steady, inviscid, adiabatic flow, Equation (7.55) is a statement of the energy equation, and hence it can be used in *place of* the more complex partial differential equation given by Equation (7.52). This is a great simplification, as we will see in subsequent discussions.

For a calorically perfect gas, $h_0 = c_p T_0$. Thus, the above results also state that the total temperature is constant throughout the steady, inviscid, adiabatic flow of a calorically perfect gas; that is,

$$\boxed{T_0 = \text{const}} \tag{7.56}$$

For such a flow, Equation (7.56) can be used as a form of the governing energy equation.

Keep in mind that the above discussion marbled two trains of thought: On the one hand, we dealt with the general concept of an adiabatic flow field [which led to Equations (7.51) to (7.53)], and on the other hand, we dealt with the definition of total enthalpy [which led to Equation (7.54)]. These two trains of thought are really separate and should not be confused. Consider, for example, a general *nonadiabatic* flow, such as a viscous boundary layer with heat transfer. A generic nonadiabatic flow is sketched in Figure 7.4a. Clearly, Equations (7.51) to (7.53) do not hold for such a flow. However, Equation (7.54) holds locally at each point in the flow, because the assumption of an adiabatic flow contained in Equation (7.54) is made through the *definition* of h_0 and has nothing to do with the general overall flow field. For example, consider two different points, 1 and 2, in the general flow, as shown in Figure 7.4a. At point 1, the local static enthalpy and velocity are h_1 and V_1, respectively. Hence, the local total enthalpy at point 1 is $h_{0,1} = h_1 + V_1^2/2$. At point 2, the local static enthalpy and velocity are h_2 and V_2, respectively. Hence, the local total enthalpy at point 2 is $h_{0,2} = h_2 + V_2^2/2$. If the flow between points 1 and 2 is nonadiabatic, then $h_{0,1} \neq h_{0,2}$. Only for the special case where the flow is adiabatic between the two points would $h_{0,1} = h_{0,2}$. This case is illustrated in Figure 7.4b. Of course, this is the special case treated by Equations (7.55) and (7.56).

Return to the beginning of this section, where we considered a fluid element passing through a point in a flow where the local properties are p, T, ρ, M, and $\mathbf{V}$. Once again, imagine that you grab hold of the fluid element and slow it down to zero velocity, but this time, let us slow it down both adiabatically *and* reversibly. That is, let us slow the fluid element down to zero velocity *isentropically*. When the fluid element is brought to rest isentropically, the resulting pressure and density are defined as the *total pressure* p_0 and *total density* ρ_0. (Since an isentropic process is also adiabatic, the resulting temperature is the same total temperature T_0 as discussed earlier.) As before, keep in mind that we do not have to actually bring the flow to rest in real life in order to talk about total pressure and total density; rather, they are *defined* quantities that would exist at a point in a flow *if* (in our imagination) the fluid element passing through that point were brought to rest isentropically. Therefore, at a given point in a flow, where the static pressure and static density are p and ρ, respectively, we can also assign a value of total pressure p_0, and total density ρ_0 defined as above.

The definition of p_0 and ρ_0 deals with an isentropic compression to zero velocity. Keep in mind that the isentropic assumption is involved with the definition only. The concept of total pressure and density can be applied throughout any general *nonisentropic* flow. For example, consider two different points, 1 and 2, in a

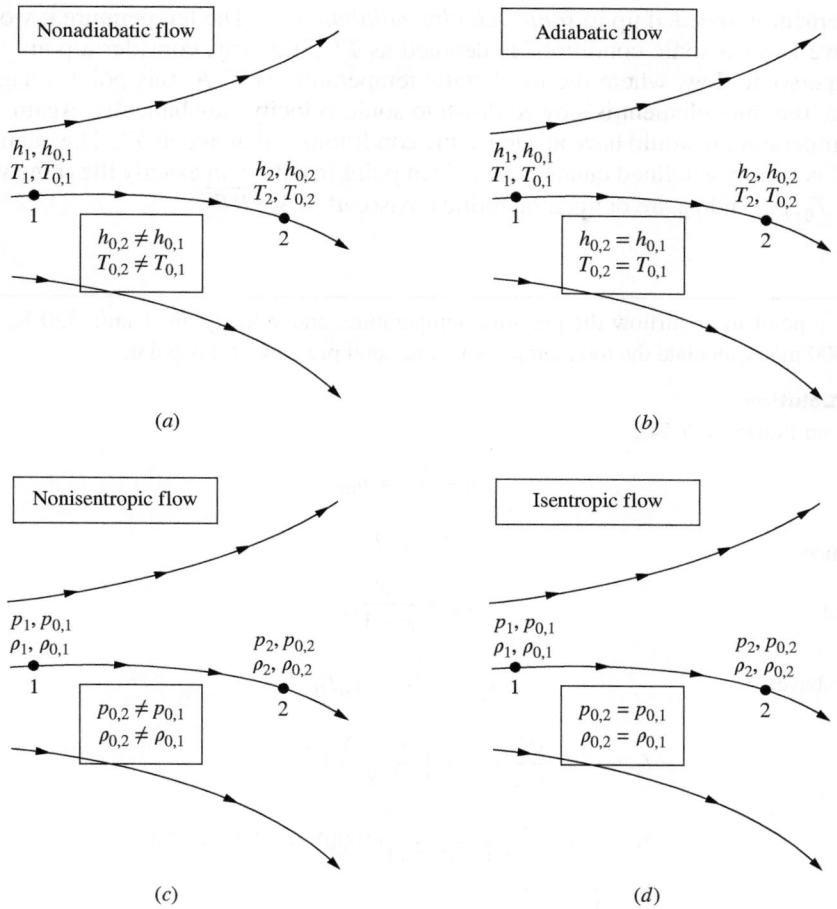

Figure 7.4 Comparisons between (*a*) nonadiabatic, (*b*) adiabatic, (*c*) nonisentropic, and (*d*) isentropic flows.

general flow field, as sketched in Figure 7.4*c*. At point 1, the local static pressure and static density are p_1 and ρ_1, respectively; also the local total pressure and total density are $p_{0,1}$ and $\rho_{0,1}$, respectively, defined as above. Similarly, at point 2, the local static pressure and static density are p_2 and ρ_2, respectively, and the local total pressure and total density are $p_{0,2}$ and $\rho_{0,2}$, respectively. If the flow is nonisentropic between points 1 and 2, then $p_{0,1} \neq p_{0,2}$ and $\rho_{0,1} \neq \rho_{0,2}$, as shown in Figure 7.4*c*. On the other hand, if the flow is isentropic between points 1 and 2, then $p_{0,1} = p_{0,2}$ and $\rho_{0,1} = \rho_{0,2}$, as shown in Figure 7.4*d*. Indeed, if the general flow field is isentropic throughout, then both p_0 and ρ_0 are constant values throughout the flow.

As a corollary to the above considerations, we need another defined temperature, denoted by T^*, and defined as follows. Consider a point in a subsonic flow where the local static temperature is T. At this point, imagine that the fluid element is speeded up to *sonic velocity, adiabatically*. The temperature it would have at such sonic conditions is denoted as T^*. Similarly, consider a point in a supersonic flow, where the local static temperature is T. At this point, imagine that the fluid element is slowed down to sonic velocity, adiabatically. Again, the temperature it would have at such sonic conditions is denoted as T^*. The quantity T^* is simply a defined quantity at a given point in a flow, in exactly the same vein as T_0, p_0, and ρ_0 are defined quantities. Also, $a^* = \sqrt{\gamma R T^*}$.

EXAMPLE 7.6

At a point in an airflow the pressure, temperature, and velocity are 1 atm, 320 K, and 1000 m/s. Calculate the total temperature and total pressure at this point.

■ Solution

From Equation (7.54),

$$h + \frac{V^2}{2} = h_0$$

Since

$$h = c_p T$$

and

$$c_p = \frac{\gamma R}{\gamma - 1},$$

we have

$$c_p T + \frac{V^2}{2} = c_p T_0$$

$$T_0 = T + \frac{V^2}{2c_p} = T + \left(\frac{\gamma - 1}{2\gamma R}\right) V^2$$

$$T_0 = 320 + \left[\frac{0.4}{2(1.4)(287)}\right](1000)^2 = 320 + 497.8$$

$$T_0 = \boxed{817.8 \text{ K}}$$

By definition, the total pressure is the pressure that would exist if the flow at the point were slowed isentropically to zero velocity. Hence, we can use the isentropic relations in Equation (7.32) to relate total to static conditions. That is, from Equation (7.32),

$$\frac{p_0}{p} = \left(\frac{T_0}{T}\right)^{\frac{\gamma}{\gamma - 1}}$$

Hence,

$$p_0 = p\left(\frac{T_0}{T}\right)^{\frac{\gamma}{\gamma - 1}} = (1 \text{ atm})\left(\frac{817.8}{320}\right)^{\frac{1.4}{0.4}}$$

$$P_0 = \boxed{26.7 \text{ atm}}$$

Note: In the above calculation of total pressure, we continued to use the *nonconsistent* unit of atmospheres. This is okay because Equation (7.32) contains a *ratio* of pressures, and therefore it would just be extra work to convert 1 atm to 1.01×10^5 N/m^2, and then carry out the calculation. The result is the same.

EXAMPLE 7.7

An airplane is flying at a standard altitude of 10,000 ft. A Pitot tube mounted at the nose measures a pressure of 2220 lb/ft^2. The airplane is flying at a high subsonic speed, faster than 300 mph. From our comments in Section 3.1, the flow should be considered *compressible*. Calculate the velocity of the airplane.

■ Solution

From our discussion in Section 3.4, the pressure measured by a Pitot tube immersed in an incompressible flow is the total pressure. For the same *physical reasons* discussed in Section 3.4, a Pitot tube also measures the total pressure in a high-speed subsonic compressible flow. (This is further discussed in Section 8.7.1 on the measurement of velocity in a subsonic compressible flow.) *Caution:* Because we are dealing with a compressible flow in this example, we *cannot* use Bernoulli's equation to calculate the velocity.

The flow in front of the Pitot tube is compressed isentropically to zero velocity at the mouth of the tube, hence the pressure at the mouth is the total pressure, p_0. From Equation (7.32), we can write:

$$\frac{p_0}{p_\infty} = \left(\frac{T_0}{T_\infty}\right)^{\gamma/(\gamma-1)} \tag{E.7.3}$$

where p_0 and T_0 are the total pressure and temperature, respectively, at the mouth of the Pitot tube, and p_∞ and T_∞ are the freestream static pressure and static temperature, respectively. Solving Equation (E7.3) above for T_0, we get

$$T_0 = T_\infty \left(\frac{p_0}{p_\infty}\right)^{(\gamma-1)/\gamma} \tag{E.7.4}$$

From Appendix E, the pressure and temperature at a standard altitude of 10,000 ft are 1455.6 lb/ft^2 and 483.04 °R, respectively. These are the values of p_∞ and T_∞ in Equation (E7.4). Thus, from Equation (E7.4),

$$T_0 = (483.04)\left(\frac{2220}{1455.6}\right)^{0.4/1.4} = 544.9 \text{ °R}$$

From the energy equation, Equation (7.54), written in terms of temperature, we have

$$c_p T + \frac{V^2}{2} = c_p T_0 \tag{E.7.5}$$

In Equation (E7.5), both T and V are the freestream values, hence we have

$$c_p T_\infty + \frac{V_\infty^2}{2} = c_p T_0 \tag{E.7.6}$$

Also,

$$c_p = \frac{\gamma R}{\gamma - 1} = \frac{(1.4)(1716)}{0.4} = 6006 \frac{\text{ft} \cdot \text{lb}}{\text{slug} \cdot \text{°R}}$$

Solving Equation (E7.6) for V_∞, we have

$$V_\infty = [2 c_p (T_0 - T_\infty)]^{1/2}$$
$$= [2 (6006)(544.9 - 483.04)]^{1/2}$$
$$= \boxed{862 \text{ ft/s}}$$

Note: From this example, we see that the total pressure measured by a Pitot tube in a subsonic compressible flow is a measure of the flow velocity, but we need also the value of the flow static temperature in order to calculate the velocity. In Section 8.7, we show more fundamentally that the *ratio* of Pitot pressure to flow static pressure in a compressible flow, subsonic or supersonic, is a *direct measure* of the *Mach number*, not the velocity. But more on this later.

7.6 SOME ASPECTS OF SUPERSONIC FLOW: SHOCK WAVES

Return to the different regimes of flow sketched in Figure 1.44. Note that subsonic compressible flow is qualitatively (but not quantitatively) the same as incompressible flow; Figure 1.44*a* shows a subsonic flow with a smoothly varying streamline pattern, where the flow far ahead of the body is forewarned about the presence of the body and begins to adjust accordingly. In contrast, supersonic flow is quite different, as sketched in Figure 1.44*d* and *e*. Here, the flow is dominated by shock waves, and the flow upstream of the body does not know about the presence of the body until it encounters the leading-edge shock wave. In fact, any flow with a supersonic region, such as those sketched in Figure 1.44*b* to *e*, is subject to shock waves. Thus, an essential ingredient of a study of supersonic flow is the calculation of the shape and strength of shock waves. This is the main thrust of Chapters 8 and 9.

A shock wave is an extremely thin region, typically on the order of 10^{-5} cm, across which the flow properties can change drastically. The shock wave is usually at an oblique angle to the flow, such as sketched in Figure 7.5*a*; however, there are many cases where we are interested in a shock wave normal to the flow, as sketched in Figure 7.5*b*. Normal shock waves are discussed at length in Chapter 8, whereas oblique shocks are considered in Chapter 9. In both cases, the shock wave is an almost explosive compression process, where the pressure increases almost discontinuously across the wave. Examine Figure 7.5 closely. In region 1 ahead of the shock, the Mach number, flow velocity, pressure, density, temperature, entropy, total pressure, and total enthalpy are denoted by $M_1, V_1, p_1, \rho_1, T_1, s_1, p_{0,1}$, and $h_{0,1}$, respectively. The analogous quantities in region 2 behind the shock are $M_2, V_2, p_2, \rho_2, T_2, s_2, p_{0,2}$, and $h_{0,2}$, respectively. The qualitative changes across the wave are noted in Figure 7.5. The pressure, density, temperature, and entropy increase across the shock, whereas the total pressure, Mach number, and velocity decrease. Physically, the flow across a shock wave is adiabatic (we are not heating the gas with a laser beam or cooling it in a refrigerator, for example). Therefore, recalling the discussion in Section 7.5, the total enthalpy is constant across the wave. In both the oblique shock and normal shock cases, the flow ahead of the shock wave must be supersonic (i.e., $M_1 > 1$). Behind the oblique shock, the flow usually remains supersonic (i.e., $M_2 > 1$), but at a reduced Mach number (i.e., $M_2 < M_1$). However, as discussed in Chapter 9, there are special cases where the oblique shock is strong enough to decelerate the downstream flow to a subsonic

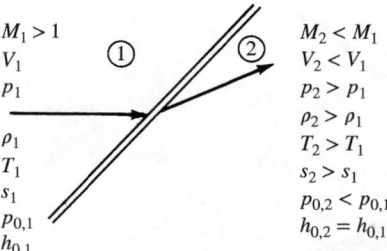

$M_1 > 1$ ①   ② $M_2 < M_1$
V_1 $V_2 < V_1$
p_1 $p_2 > p_1$
                   $\rho_2 > \rho_1$
ρ_1 $T_2 > T_1$
T_1 $s_2 > s_1$
s_1
$p_{0,1}$ $p_{0,2} < p_{0,1}$
$h_{0,1}$ $h_{0,2} = h_{0,1}$

(*a*) Oblique shock wave

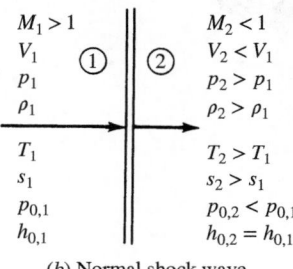

$M_1 > 1$ ①   ② $M_2 < 1$
V_1 $V_2 < V_1$
p_1 $p_2 > p_1$
ρ_1 $\rho_2 > \rho_1$

T_1 $T_2 > T_1$
s_1 $s_2 > s_1$
$p_{0,1}$ $p_{0,2} < p_{0,1}$
$h_{0,1}$ $h_{0,2} = h_{0,1}$

(*b*) Normal shock wave

Figure 7.5 Qualitative pictures of flow through oblique and normal shock waves.

Mach number; hence, $M_2 < 1$ can occur behind an oblique shock. For the normal shock, as sketched in Figure 7.5*b*, the downstream flow is always subsonic (i.e., $M_2 < 1$). Study the qualitative variations illustrated in Figure 7.5 closely. They are important, and you should have them in mind for our subsequent discussions. One of the primary purposes of Chapters 8 and 9 is to develop a shock-wave theory that allows the quantitative evaluation of these variations. We prove that pressure increases across the shock, that the upstream Mach number must be supersonic, etc. Moreover, we obtain equations that allow the direct calculation of changes across the shock.

Several photographs of shock waves are shown in Figure 7.6. Since air is transparent, we cannot usually see shock waves with the naked eye. However, because the density changes across the shock wave, light rays propagating through the flow will be refracted across the shock. Special optical systems, such as shadowgraphs, schlieren, and interferometers, take advantage of this refraction and allow the visual imaging of shock waves on a screen or a photographic negative. For details of the design and characteristics of these optical systems, see References 25 and 26. (Under certain conditions, you can see the refracted light from a shock wave with your naked eye. Recall from Figure 1.43*b* that a shock wave can form in the locally supersonic region on the top surface of an airfoil if the freestream subsonic Mach number is high enough. The next time you are flying in a jet transport, and the sun is directly overhead, look out the window

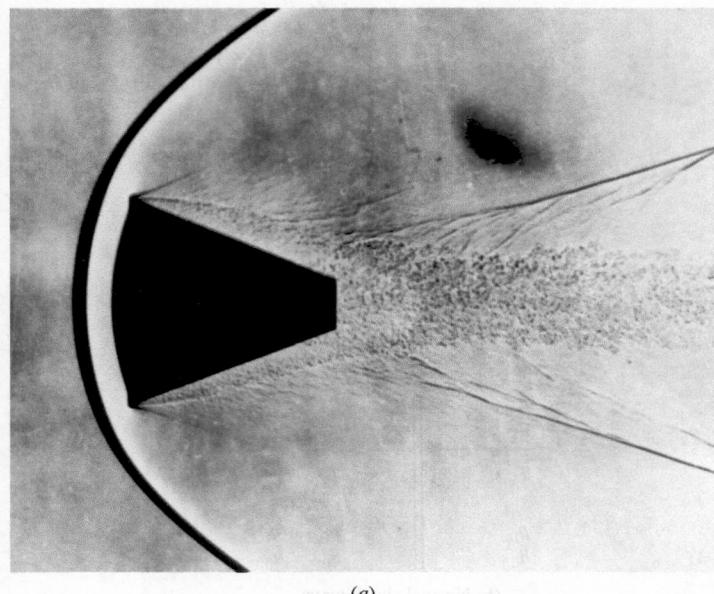

(a)

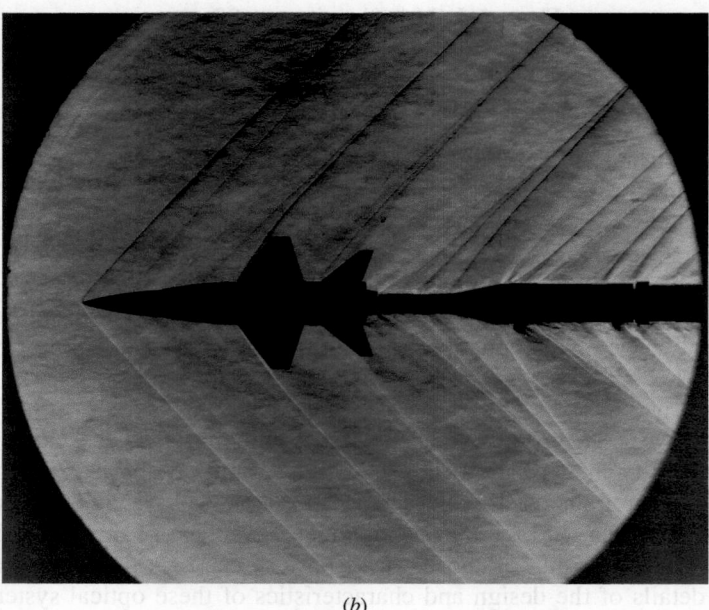

(b)

Figure 7.6 Schlieren photographs illustrating shock waves on various bodies. (*a*) Mercury capsule wind-tunnel model at Mach 8. (*b*) X-15 wind tunnel model at Mach 7. (Both photos: *NASA*)

(c)

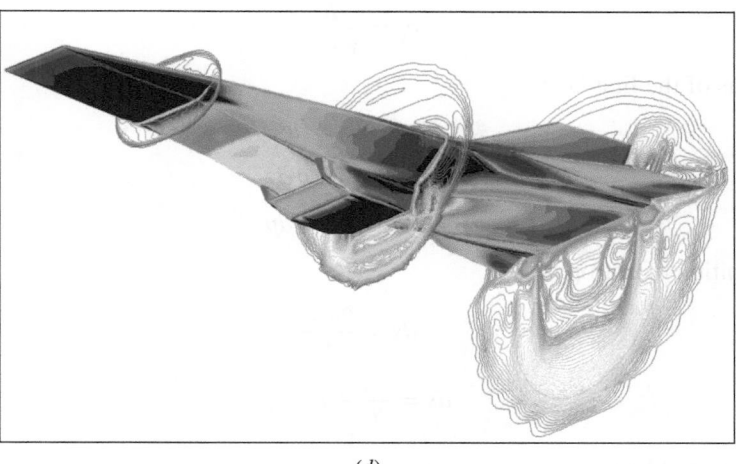

(d)

Figure 7.6 (*continued*) (*c*) Space Shuttle Orbiter model at Mach 6. (*d*) A computational fluid dynamics generated flow (not a Schileren photograph) for the X-43 Hypersonic test vehicle. (Both photos: *NASA*)

along the span of the wing. If you are lucky, you will see the shock wave dancing back and forth over the top of the wing.)

In summary, compressible flows introduce some very exciting physical phenomena into our aerodynamic studies. Moreover, as the flow changes from subsonic to supersonic, the complete nature of the flow changes, not the least of which is the occurrence of shock waves. The purpose of the next seven chapters is to describe and analyze these flows.

7.7 SUMMARY

As usual, examine the road map for this chapter (Figure 7.1), and make certain that you feel comfortable with the material represented by this road map before continuing further.

Some of the highlights of this chapter are summarized below:

Thermodynamic relations:

Equation of state:
$$p = \rho RT \tag{7.1}$$

For a calorically perfect gas,
$$e = c_v T \qquad \text{and} \qquad h = c_p T \tag{7.6a and b}$$

$$c_p = \frac{\gamma R}{\gamma - 1} \tag{7.9}$$

$$c_v = \frac{R}{\gamma - 1} \tag{7.10}$$

Forms of the first law:
$$\delta q + \delta w = de \tag{7.11}$$
$$T\,ds = de + p\,dv \tag{7.18}$$
$$T\,ds = dh - v\,dp \tag{7.20}$$

Definition of entropy:
$$ds = \frac{\delta q_{\text{rev}}}{T} \tag{7.13}$$

Also
$$ds = \frac{\delta q}{T} + ds_{\text{irrev}} \tag{7.14}$$

The second law:
$$ds \geq \frac{\delta q}{T} \tag{7.16}$$

or, for an adiabatic process,
$$ds \geq 0 \tag{7.17}$$

Entropy changes can be calculated from (for a calorically perfect gas)
$$s_2 - s_1 = c_p \ln \frac{T_2}{T_1} - R \ln \frac{p_2}{p_1} \tag{7.25}$$

and
$$s_2 - s_1 = c_v \ln \frac{T_2}{T_1} + R \ln \frac{v_2}{v_1} \tag{7.26}$$

For an isentropic flow,

$$\frac{p_2}{p_1} = \left(\frac{\rho_2}{\rho_1}\right)^{\gamma} = \left(\frac{T_2}{T_1}\right)^{\gamma/(\gamma-1)} \tag{7.32}$$

General definition of compressibility:

$$\tau = -\frac{1}{v}\frac{dv}{dp} \tag{7.33}$$

For an isothermal process,

$$\tau_T = -\frac{1}{v}\left(\frac{\partial v}{\partial p}\right)_T \tag{7.34}$$

For an isentropic process,

$$\tau_s = -\frac{1}{v}\left(\frac{\partial v}{\partial p}\right)_s \tag{7.35}$$

The governing equations for inviscid, compressible flow are
Continuity:

$$\frac{\partial}{\partial t}\iiint_{\mathcal{V}} \rho\, d\mathcal{V} + \oiint_{S} \rho\mathbf{V}\cdot d\mathbf{S} = 0 \tag{7.39}$$

$$\frac{\partial \rho}{\partial t} + \nabla\cdot\rho\mathbf{V} = 0 \tag{7.40}$$

Momentum:

$$\frac{\partial}{\partial t}\iiint_{\mathcal{V}} \rho\mathbf{V}\, d\mathcal{V} + \oiint_{S} (\rho\mathbf{V}\cdot d\mathbf{S})\mathbf{V} = -\oiint_{S} p\, d\mathbf{S} + \iiint_{\mathcal{V}} \rho\mathbf{f}\, d\mathcal{V} \tag{7.41}$$

$$\rho\frac{Du}{Dt} = -\frac{\partial p}{\partial x} + \rho f_x \tag{7.42a}$$

$$\rho\frac{Dv}{Dt} = -\frac{\partial p}{\partial y} + \rho f_y \tag{7.42b}$$

$$\rho\frac{Dw}{Dt} = -\frac{\partial p}{\partial z} + \rho f_z \tag{7.42c}$$

Energy:

$$\frac{\partial}{\partial t}\iiint_{\mathcal{V}} \rho\left(e + \frac{V^2}{2}\right) d\mathcal{V} + \oiint_{S} \rho\left(e + \frac{V^2}{2}\right)\mathbf{V}\cdot d\mathbf{S}$$

(continued)

$$= \oiiint_{\mathcal{V}} \dot{q}\rho \, d\mathcal{V} - \oiint_{S} p\mathbf{V} \cdot d\mathbf{S} + \oiiint_{\mathcal{V}} \rho(\mathbf{f} \cdot \mathbf{V}) \, d\mathcal{V} \qquad (7.43)$$

$$\rho \frac{D(e + V^2/2)}{Dt} = \rho\dot{q} - \nabla \cdot p\mathbf{V} + \rho(\mathbf{f} \cdot \mathbf{V}) \qquad (7.44)$$

If the flow is steady and adiabatic, Equations (7.43) and (7.44) can be replaced with

$$h_0 = h + \frac{V^2}{2} = \text{const}$$

Equation of state (perfect gas):

$$p = \rho RT \qquad (7.1)$$

Internal energy (calorically perfect gas):

$$e = c_v T \qquad (7.6a)$$

Total temperature T_0 and total enthalpy h_0 are defined as the properties that would exist if (in our imagination) we slowed the fluid element at a point in the flow to zero velocity adiabatically. Similarly, total pressure p_0 and total density ρ_0 are defined as the properties that would exist if (in our imagination) we slowed the fluid element at a point in the flow to zero velocity isentropically. If a general flow field is adiabatic, h_0 is constant throughout the flow; in contrast, if the flow field is nonadiabatic, h_0 varies from one point to another. Similarly, if a general flow field is isentropic, p_0 and ρ_0 are constant throughout the flow; in contrast, if the flow field is nonisentropic, p_0 and ρ_0 vary from one point to another.

Shock waves are very thin regions in a supersonic flow across which the pressure, density, temperature, and entropy increase; the Mach number, flow velocity, and total pressure decrease; and the total enthalpy stays the same.

7.8 PROBLEMS

Note: In the following problems, you will deal with both the International System of Units (SI) (N, kg, m, s, K) and the English Engineering System (lb, slug, ft, s, °R). Which system to use will be self-evident in each problem.

All problems deal with calorically perfect air as the gas, unless otherwise noted. Also, recall that 1 atm $= 2116$ lb/ft$^2 = 1.01 \times 10^5$ N/m^2.

7.1 The temperature and pressure at the stagnation point of a high-speed missile are 934 °R and 7.8 atm, respectively. Calculate the density at this point.

7.2 Calculate c_p, c_v, e, and h for

  a. The stagnation point conditions given in Problem 7.1

  b. Air at standard sea level conditions

  (If you do not remember what standard sea level conditions are, find them in an appropriate reference, such as Reference 2.)

7.3 Just upstream of a shock wave, the air temperature and pressure are 288 K and 1 atm, respectively; just downstream of the wave, the air temperature and pressure are 690 K and 8.656 atm, respectively. Calculate the changes in enthalpy, internal energy, and entropy across the wave.

7.4 Consider the isentropic flow over an airfoil. The freestream conditions are $T_\infty = 245$ K and $p_\infty = 4.35 \times 10^4$ N/m^2. At a point on the airfoil, the pressure is 3.6×10^4 N/m^2. Calculate the density at this point.

7.5 Consider the isentropic flow through a supersonic wind-tunnel nozzle. The reservoir properties are $T_0 = 500$ K and $p_0 = 10$ atm. If $p = 1$ atm at the nozzle exit, calculate the exit temperature and density.

7.6 Consider air at a pressure of 0.2 atm. Calculate the values of τ_T and τ_s. Express your answer in SI units.

7.7 Consider a point in a flow where the velocity and temperature are 1300 ft/s and 480 °R, respectively. Calculate the total enthalpy at this point.

7.8 In the reservoir of a supersonic wind tunnel, the velocity is negligible, and the temperature is 1000 K. The temperature at the nozzle exit is 600 K. Assuming adiabatic flow through the nozzle, calculate the velocity at the exit.

7.9 An airfoil is in a freestream where $p_\infty = 0.61$ atm, $\rho_\infty = 0.819$ kg/m^3, and $V_\infty = 300$ m/s. At a point on the airfoil surface, the pressure is 0.5 atm. Assuming isentropic flow, calculate the velocity at that point.

7.10 Calculate the percentage error obtained if Problem 7.9 is solved using (incorrectly) the incompressible Bernoulli equation.

7.11 Repeat Problem 7.9, considering a point on the airfoil surface where the pressure is 0.3 atm.

7.12 Repeat Problem 7.10, considering the flow of Problem 7.11.

7.13 Bernoulli's equation, Equation (3.13), (3.14), or (3.15), was derived in Chapter 3 from Newton's second law; it is fundamentally a statement that force = mass × acceleration. However, the terms in Bernoulli's equation have dimensions of energy per unit volume (check it out), which prompt some argument that Bernoulli's equation is an energy equation for incompressible flow. If this is so, then it should be derivable from the energy equation for compressible flow discussed in the present chapter. Starting with Equation (7.53) for inviscid, adiabatic compressible flow, make the appropriate assumptions for an incompressible flow and see what you need to do to obtain Bernoulli's equation.

7.14 Derive the 1-D energy equation for adiabatic compressible flow staring from the first law of thermodynamics and Euler's equation, that is, derive $c_p T + \frac{1}{2} u^2 = C$ from $de = \delta q + \delta w$ and $u \, du = -\frac{dP}{\rho}$, where C is a constant. *Hint:* Assume you have a simple compressible system with constant specific heats undergoing an internally reversible (i.e., isentropic) process.

7.15 An airplane flies at Mach 0.8 at an altitude of 10 km. Its engine draws in air from the free stream and raises its pressure by a factor of 80 by the time it exits the engine's compressor. Assume that the flow is steady, inviscid, and that air is a calorically perfect gas with $c_p = 1004.5 \, J/kg \, K$ and $\gamma = 1.4$. Also assume that the velocity of the flow exiting the engine's compressor is much smaller than the speed of sound (i.e., $M_2 \approx 0$). Please:

a.   Find the total pressure and temperature of the air exiting the compressor and the work done on the air per unit mass (i.e., J/kg) assuming that the entire compression process is isentropic. Note that the work input to the compression process equals the increase in enthalpy. You will learn why in the next problem (7.16).

b.   Repeat part a assuming that 0.1 kJ/kg of entropy is generated during the compression process.

c.   The efficiency of the compression process is the ratio of the work required by the ideal (i.e., isentropic) process divided by the work required by the real process. Calculate the efficiency of this compression process.

d.   Sketch the ideal and real compression processes on a T-s diagram. Be sure that your diagram includes isobars corresponding to conditions in the free stream, compressor entrance, and compressor exit.

e.   How does the T-s diagram change if the Mach number downstream of the compressor is not negligible? Where will the conditions at the exit of the engine's inlet—which sits between the free stream and the compressor's entrance appear in the T-s diagram? Sketch a new T-s diagram that shows these features and explain.

7.16 Starting with the integral form of energy conservation [Equations (2.95 and 7.43)], derive an expression for energy conservation in a steady, inviscid, nonchemically reacting flow without body forces along a streamline that passes through a compressor or turbine. Use W_{shaft} to represent the work done on the system by the compressor or turbine per unit mass of flow in J/kg. Your result should be analogous to Equation (7.53) or (7.54) except that the total enthalpy will not be constant.

7.17 In a thermally perfect, but not calorically perfect gas, the specific heat at constant pressure can no longer be considered constant. How would this affect the prediction of the stagnation temperature of air flowing at 300 K and 1500 m/s (about Mach 4.3)?

    a. Begin by showing that the differential form of the energy equation for steady, 1-D, constant area, adiabatic flow is:

$$dh + udu = 0 \tag{7.2}$$

    b. Integrate Equation 1 assuming c_p is constant to recover the simple 1-D expression for energy conservation along a streamline.

    c. Now, incorporate the effect of temperature variation by representing the specific heat of air as a power law in temperature (T is the temperature and the units are J/kg K)

$$c_p = 996.96 + (0.0357)\,T + (0.0002)\,T^2 - (8E-8)\,T^3$$

    d. Use your result to determine the total temperature of air having a static temperature of 300 K and a velocity of 1500 m/s assuming variable specific heat. Note that you will have to use an iterative approach to get the temperature.

    e. How does this differ from the result assuming constant specific heat (pick the value corresponding to 300 K)?

    f. Based on what you have seen here, under what conditions would it be important to consider the temperature variation of specific heat when predicting the total temperature?

CHAPTER 8

Normal Shock Waves and Related Topics

Shock wave: A large-amplitude compression wave, such as that produced by an explosion, caused by supersonic motion of a body in a medium.

From the American Heritage Dictionary of the English Language, 1969

PREVIEW BOX

When you have a supersonic flow, chances are that you also have shock waves. Shock waves are exceptionally thin regions—usually much thinner than the thickness of this page—across which the flow properties change drastically. Take pressure for example. The gas pressure in front of the shock wave may be 1 atm, and immediately behind it the pressure may be 20 atm. Imagine that you are a fluid element crossing the shock wave; at one moment you are at 1 atm pressure and a split second later you are at 20 atm. I do not know about you, but if it were me I would be shocked out of my mind. (Perhaps this is why we call them "shock" waves.)

This chapter is all about shock waves. Here we learn how to calculate the change in flow properties across a shock, and we examine the important physical aspects and consequences of shock waves. In this chapter, we focus on shock waves that are perpendicular to the flow—normal shock waves. What we learn here is directly transferable to the study of shock waves that make an oblique angle to the flow—oblique shock waves—discussed in the next chapter. A study of shock-wave phenomena is one of the most important aspects of learning about supersonic flows. So take this chapter very seriously. Besides, shock waves are rather exciting events in nature, so enjoy this excitement.

8.1 INTRODUCTION

The purpose of this chapter and Chapter 9 is to develop shock-wave theory, thus giving us the means to calculate the changes in the flow properties across a wave. These changes were discussed qualitatively in Section 7.6; make certain that you are familiar with these changes before continuing.

The focus of this chapter is on normal shock waves, as sketched in Figure 7.5*b*. At first thought, a shock wave that is normal to the upstream flow may seem to be a very special case—and therefore a case of little practical interest—but nothing could be further from the truth. Normal shocks occur frequently in nature. Two such examples are sketched in Figure 8.1; there are many more. The supersonic flow over a blunt body is shown at the left of Figure 8.1. Here, a strong bow shock wave exists in front of the body. (We study such bow shocks in Chapter 9.) Although this wave is curved, the region of the shock closest to the nose is essentially normal to the flow. Moreover, the streamline that passes through this normal portion of the bow shock later impinges on the nose of the body and controls the values of stagnation (total) pressure and temperature at the nose. Since the nose region of high-speed blunt bodies is of practical interest in the calculation of drag and aerodynamic heating, the properties of the flow behind the normal portion of the shock wave take on some importance. In another example, shown at the right of Figure 8.1, supersonic flow is established inside a nozzle (which can be a supersonic wind tunnel, a rocket engine, etc.) where the back pressure is high enough to cause a normal shock wave to stand inside the nozzle. (We discuss such "overexpanded" nozzle flows in Chapter 10.) The conditions under which this shock wave will occur and the determination of flow properties at the nozzle exit downstream of the normal shock are both important questions to be answered. In summary, for these and many other applications, the study of normal shock waves is important.

Finally, we will find that many of the normal shock relations derived in this chapter carry over directly to the analysis of oblique shock waves, as discussed

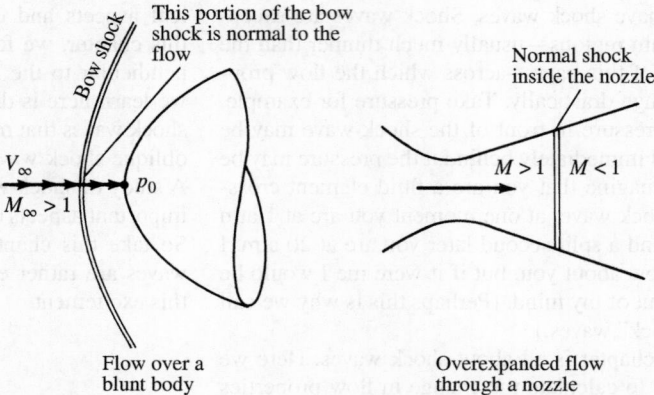

Figure 8.1 Two examples where normal shock waves are of interest.

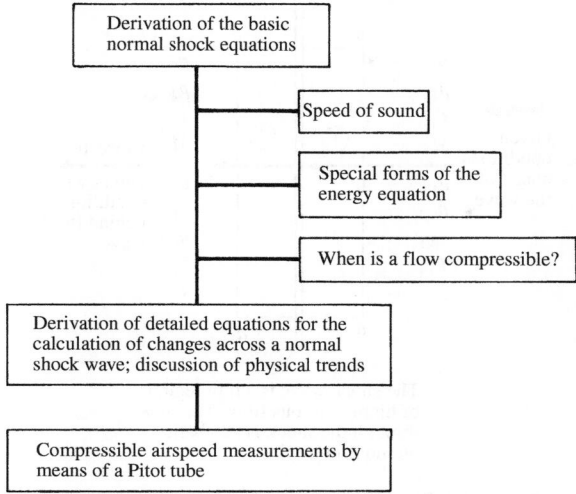

Figure 8.2 Road map for Chapter 8.

in Chapter 9. So once again, time spent on normal shock waves is time well spent.

The road map for this chapter is given in Figure 8.2. As you can see, our objectives are fairly short and straightforward. We start with a derivation of the basic continuity, momentum, and energy equations for normal shock waves, and then we employ these basic relations to obtain detailed equations for the calculation of flow properties across the shock wave. In addition, we emphasize the physical trends indicated by the equations. On the way toward this objective, we take three side streets having to do with (1) the speed of sound, (2) special forms of the energy equation, and (3) a further discussion of the criteria used to judge when a flow must be treated as compressible. Finally, we apply the results of this chapter to the measurement of airspeed in a compressible flow using a Pitot tube. Keep the road map in Figure 8.2 in mind as you progress through the chapter.

8.2 THE BASIC NORMAL SHOCK EQUATIONS

Consider the normal shock wave sketched in Figure 8.3. Region 1 is a uniform flow upstream of the shock, and region 2 is a different uniform flow downstream of the shock. The pressure, density, temperature, Mach number, velocity, total pressure, total enthalpy, total temperature, and entropy in region 1 are p_1, ρ_1, T_1, M_1, u_1, $p_{0,1}$, $h_{0,1}$, $T_{0,1}$, and s_1, respectively. The corresponding variables in region 2 are denoted by p_2, ρ_2, T_2, M_2, u_2, $p_{0,2}$, $h_{0,2}$, $T_{0,2}$, and s_2. (Note that we are denoting the magnitude of the flow velocity by u rather than V; reasons for this will become obvious as we progress.) The problem of the normal shock wave

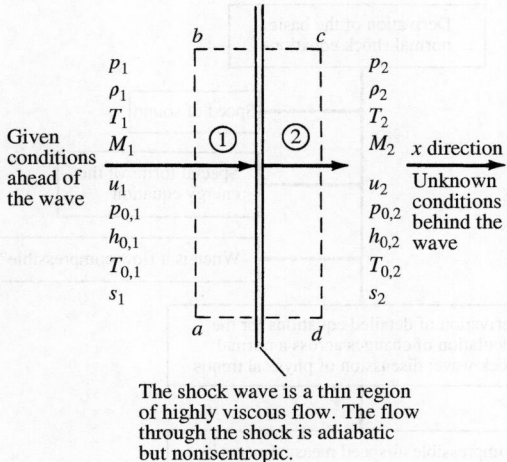

The shock wave is a thin region
of highly viscous flow. The flow
through the shock is adiabatic
but nonisentropic.

Figure 8.3 Sketch of a normal wave.

is simply stated as follows: given the flow properties upstream of the wave (p_1, T_1, M_1, etc.), calculate the flow properties (p_2, T_2, M_2, etc.) downstream of the wave. Let us proceed.

Consider the rectangular control volume *abcd* given by the dashed line in Figure 8.3. The shock wave is inside the control volume, as shown. Side *ab* is the edge view of the left face of the control volume; this left face is perpendicular to the flow, and its area is A. Side *cd* is the edge view of the right face of the control volume; this right face is also perpendicular to the flow, and its area is A. We apply the integral form of conservation equations to this control volume. In the process, we observe three important physical facts about the flow given in Figure 8.3:

1. The flow is steady, that is, $\partial/\partial t = 0$.

2. The flow is adiabatic, that is, $\dot{q} = 0$. We are not adding or taking away heat from the control volume (we are not heating the shock wave with a Bunsen burner, for example). The temperature increases across the shock wave, not because heat is being added, but rather, because kinetic energy is converted to internal energy across the shock wave.

3. There are no viscous effects on the sides of the control volume. The shock wave itself is a thin region of extremely high velocity and temperature gradients; hence, friction and thermal conduction play an important role on the flow structure inside the wave. However, the wave itself is buried inside the control volume, and with the integral form of the conservation equations, we are not concerned about the details of what goes on inside the control volume.

4. There are no body forces; $\mathbf{f} = 0$.

Consider the continuity equation in the form of Equation (7.39). For the conditions described above, Equation (7.39) becomes

$$\oiint_S \rho \mathbf{V} \cdot \mathbf{dS} = 0 \tag{8.1}$$

To evaluate Equation (8.1) over the face ab, note that $\mathbf{V}$ is pointing into the control volume, whereas $\mathbf{dS}$ by definition is pointing out of the control volume, in the opposite direction of $\mathbf{V}$; hence, $\mathbf{V} \cdot \mathbf{dS}$ is negative. Moreover, ρ and $|\mathbf{V}|$ are uniform over the face ab and equal to ρ_1 and u_1, respectively. Hence, the contribution of face ab to the surface integral in Equation (8.1) is simply $-\rho_1 u_1 A$. Over the right face cd both $\mathbf{V}$ and $\mathbf{dS}$ are in the same direction, and hence $\mathbf{V} \cdot \mathbf{dS}$ is positive. Moreover, ρ and $|\mathbf{V}|$ are uniform over the face cd and equal to ρ_2 and u_2, respectively. Thus, the contribution of face cd to the surface integral is $\rho_2 u_2 A$. On sides bc and ad, $\mathbf{V}$ and $\mathbf{dS}$ are always perpendicular; hence, $\mathbf{V} \cdot \mathbf{dS} = 0$, and these sides make no contribution to the surface integral. Hence, for the control volume shown in Figure 8.3, Equation (8.1) becomes

$$-\rho_1 u_1 A + \rho_2 u_2 A = 0$$

or

$$\boxed{\rho_1 u_1 = \rho_2 u_2} \tag{8.2}$$

Equation (8.2) is the continuity equation for normal shock waves.

Consider the momentum equation in the form of Equation (7.41). For the flow we are treating here, Equation (7.41) becomes

$$\oiint_S (\rho \mathbf{V} \cdot \mathbf{dS}) \mathbf{V} = - \oiint_S p \, \mathbf{dS} \tag{8.3}$$

Equation (8.3) is a vector equation. Note that in Figure 8.3, the flow is moving only in one direction (i.e., in the x direction). Hence, we need to consider only the scalar x component of Equation (8.3), which is

$$\oiint_S (\rho \mathbf{V} \cdot \mathbf{dS}) u = - \oiint_S (p \, dS)_x \tag{8.4}$$

In Equation (8.4), $(p \, dS)_x$ is the x component of the vector $(p \, \mathbf{dS})$. Note that over the face ab, $\mathbf{dS}$ points to the left (i.e., in the negative x direction). Hence, $(p \, dS)_x$ is negative over face ab. By similar reasoning, $(p \, dS)_x$ is positive over the face cd. Again noting that all the flow variables are uniform over the faces ab and cd, the surface integrals in Equation (8.4) become

$$\rho_1(-u_1 A)u_1 + \rho_2(u_2 A)u_2 = -(-p_1 A + p_2 A) \tag{8.5}$$

or

$$\boxed{p_1 + \rho_1 u_1^2 = p_2 + \rho_2 u_2^2} \tag{8.6}$$

Equation (8.6) is the momentum equation for normal shock waves.

Consider the energy equation in the form of Equation (7.43). For steady, adiabatic, inviscid flow with no body forces, this equation becomes

$$\oiint\limits_{S} \rho \left(e + \frac{V^2}{2} \right) \mathbf{V} \cdot \mathbf{dS} = -\oiint\limits_{S} p\mathbf{V} \cdot \mathbf{dS} \tag{8.7}$$

Evaluating Equation (8.7) for the control surface shown in Figure 8.3, we have

$$-\rho_1 \left(e_1 + \frac{u_1^2}{2} \right) u_1 A + \rho_2 \left(e_2 + \frac{u_2^2}{2} \right) u_2 A = -(-p_1 u_1 A + p_2 u_2 A)$$

Rearranging, we obtain

$$p_1 u_1 + \rho_1 \left(e_1 + \frac{u_1^2}{2} \right) u_1 = p_2 u_2 + \rho_2 \left(e_2 + \frac{u_2^2}{2} \right) u_2 \tag{8.8}$$

Dividing by Equation (8.2), that is, dividing the left-hand side of Equation (8.8) by $\rho_1 u_1$ and the right-hand side by $\rho_2 u_2$, we have

$$\frac{p_1}{\rho_1} + e_1 + \frac{u_1^2}{2} = \frac{p_2}{\rho_2} + e_2 + \frac{u_2^2}{2} \tag{8.9}$$

From the definition of enthalpy, $h \equiv e + pv = e + p/\rho$. Hence, Equation (8.9) becomes

$$\boxed{h_1 + \frac{u_1^2}{2} = h_2 + \frac{u_2^2}{2}} \tag{8.10}$$

Equation (8.10) is the energy equation for normal shock waves. Equation (8.10) should come as no surprise; the flow through a shock wave is adiabatic, and we derived in Section 7.5 the fact that for a steady, adiabatic flow, $h_0 = h + V^2/2 =$ const. Equation (8.10) simply states that h_0 (hence, for a calorically perfect gas T_0) is constant across the shock wave. Therefore, Equation (8.10) is consistent with the general results obtained in Section 7.5.

Repeating the above results for clarity, the basic normal shock equations are

Continuity: $$\rho_1 u_1 = \rho_2 u_2 \tag{8.2}$$

Momentum: $$p_1 + \rho_1 u_1^2 = p_2 + \rho_2 u_2^2 \tag{8.6}$$

Energy: $$h_1 + \frac{u_1^2}{2} = h_2 + \frac{u_2^2}{2} \tag{8.10}$$

Examine these equations closely. Recall from Figure 8.3 that all conditions upstream of the wave, ρ_1, u_1, p_1, etc., are known. Thus, the above equations are a system of three algebraic equations in four unknowns, ρ_2, u_2, p_2, and h_2. However, if we add the following thermodynamic relations

Enthalpy: $$h_2 = c_p T_2$$

Equation of state: $$p_2 = \rho_2 R T_2$$

we have five equations for five unknowns, namely, ρ_2, u_2, p_2, h_2, and T_2. In Section 8.6, we explicitly solve these equations for the unknown quantities behind the shock. However, rather than going directly to that solution, we first take three side trips as shown in the road map in Figure 8.2. These side trips involve discussions of the speed of sound (Section 8.3), alternate forms of the energy equation (Section 8.4), and compressibility (Section 8.5)—all of which are necessary for a viable discussion of shock-wave properties in Section 8.6.

Finally, we note that Equations (8.2), (8.6), and (8.10) are not limited to normal shock waves; they describe the changes that take place in any steady, adiabatic, inviscid flow where only one direction is involved. That is, in Figure 8.3, the flow is in the x direction only. This type of flow, where the flow-field variables are functions of x only [$p = p(x)$, $u = u(x)$, etc.], is defined as *one-dimensional flow*. Thus, Equations (8.2), (8.6), and (8.10) are governing equations for one-dimensional, steady, adiabatic, and inviscid flow.

8.3 SPEED OF SOUND

Common experience tells us that sound travels through air at some finite velocity. For example, you see a flash of lightning in the distance, but you hear the corresponding thunder at some later moment. What is the physical mechanism of the propagation of sound waves? How can we calculate the speed of sound? What properties of the gas does it depend on? The speed of sound is an extremely important quantity which dominates the physical properties of compressible flow, and hence the answers to the above questions are vital to our subsequent discussions. The purpose of this section is to address these questions.

The physical mechanism of sound propagation in a gas is based on molecular motion. For example, imagine that you are sitting in a room, and suppose that a firecracker goes off in one corner. When the firecracker detonates, chemical energy (basically a form of heat release) is transferred to the air molecules adjacent to the firecracker. These energized molecules are moving about in a random fashion. They eventually collide with some of their neighboring molecules and transfer their high energy to these neighbors. In turn, these neighboring molecules eventually collide with their neighbors and transfer energy in the process. By means of this "domino" effect, the energy released by the firecracker is propagated through the air by molecular collisions. Moreover, because T, p, and ρ for a gas are macroscopic averages of the detailed microscopic molecular motion, the regions of energized molecules are also regions of slight variations in the local temperature, pressure, and density. Hence, as this energy wave from the firecracker passes over our eardrums, we "hear" the slight pressure changes in the wave. This is *sound*, and the propagation of the energy wave is simply the propagation of a *sound wave* through the gas.

Because a sound wave is propagated by molecular collisions, and because the molecules of a gas are moving with an average velocity of $\sqrt{8RT/\pi}$ given by kinetic theory, then we would expect the velocity of propagation of a sound wave

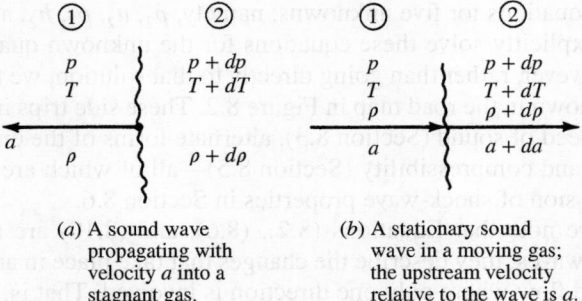

(a) A sound wave
propagating with
velocity a into a
stagnant gas.

(b) A stationary sound
wave in a moving gas;
the upstream velocity
relative to the wave is a.

Figure 8.4 Moving and stationary sound waves; two
analogous pictures, only the perspective is different.

to be approximately the average molecular velocity. Indeed, the speed of sound is
about three-quarters of the average molecular velocity. In turn, because the kinetic
theory expression given above for the average molecular velocity depends only
on the *temperature* of the gas, we might expect the speed of sound to also depend
on temperature only. Let us explore this matter further; indeed, let us now derive
an equation for the speed of sound in a gas. Although the propagation of sound
is due to molecular collisions, we do not use such a microscopic picture for our
derivation. Rather, we take advantage of the fact that the macroscopic properties
p, T, ρ, etc., change across the wave, and we use our macroscopic equations of
continuity, momentum, and energy to analyze these changes.

Consider a sound wave propagating through a gas with velocity a, as sketched
in Figure 8.4a. Here, the sound wave is moving from right to left into a stagnant
gas (region 1), where the local pressure, temperature, and density are p, T, and
ρ, respectively. Behind the sound wave (region 2), the gas properties are slightly
different and are given by $p + dp$, $T + dT$, and $\rho + d\rho$, respectively. Now imagine
that you hop on the wave and ride with it. When you look upstream, into region 1,
you see the gas moving toward you with a relative velocity a, as sketched in Figure 8.4b. When you look downstream, into region 2, you see the gas receding
away from you with a relative velocity $a + da$, as also shown in Figure 8.4. (We
have enough fluid-dynamic intuition by now to realize that because the pressure
changes across the wave by the amount dp, then the relative flow velocity must
also change across the wave by some amount da. Hence, the relative flow velocity behind the wave is $a + da$.) Consequently, in Figure 8.4b, we have a picture
of a stationary sound wave, with the flow ahead of it moving left to right with
velocity a. The pictures in Figure 8.4a and b are analogous; only the perspective
is different. For purposes of analysis, we use Figure 8.4b.

(*Note:* Figure 8.4b is similar to the picture of a normal shock wave shown
in Figure 8.3. In Figure 8.3, the normal shock wave is stationary, and the upstream flow is moving left to right at a velocity u_1. If the upstream flow were to
be suddenly shut off, then the normal shock wave in Figure 8.3 would suddenly
propagate to the left with a wave velocity of u_1, similar to the moving sound wave

shown in Figure 8.4a. The analysis of moving waves is slightly more subtle than the analysis of stationary waves; hence, it is simpler to begin a study of shock waves and sound waves with the pictures of stationary waves as shown in Figures 8.3 and 8.4b. Also, please note that the sound wave in Figure 8.4b is nothing more than an infinitely weak normal shock wave.)

Examine closely the flow through the sound wave sketched in Figure 8.4b. The flow is one-dimensional. Moreover, it is adiabatic, because we have no source of heat transfer into or out of the wave (e.g., we are not "zapping" the wave with a laser beam or heating it with a torch). Finally, the gradients within the wave are very small—the changes dp, dT, $d\rho$, and da are infinitesimal. Therefore, the influence of dissipative phenomena (viscosity and thermal conduction) is negligible. As a result, the flow through the sound wave is both adiabatic and reversible—the flow is *isentropic*. Since we have now established that the flow is one-dimensional and isentropic, let us apply the appropriate governing equations to the picture shown in Figure 8.4b.

Applying the continuity equation, Equation (8.2), to Figure 8.4b, we have

$$\rho a = (\rho + d\rho)(a + da)$$

or
$$\rho a = \rho a + a\,d\rho + \rho\,da + d\rho\,da \tag{8.11}$$

The product of two differentials, $d\rho\,da$, can be neglected in comparison with the other terms in Equation (8.11). Hence, solving Equation (8.11) for a, we obtain

$$a = -\rho\frac{da}{d\rho} \tag{8.12}$$

Now consider the one-dimensional momentum equation, Equation (8.6), applied to Figure 8.4b:

$$p + \rho a^2 = (p + dp) + (\rho + d\rho)(a + da)^2 \tag{8.13}$$

Again ignoring products of differentials, Equation (8.13) becomes

$$dp = -2a\rho\,da - a^2\,d\rho \tag{8.14}$$

Solving Equation (8.14) for da, we have

$$da = \frac{dp + a^2\,d\rho}{-2a\rho} \tag{8.15}$$

Substituting Equation (8.15) into (8.12), we obtain

$$a = -\rho\frac{dp/d\rho + a^2}{-2a\rho} \tag{8.16}$$

Solving Equation (8.16) for a^2, we have

$$a^2 = \frac{dp}{d\rho} \tag{8.17}$$

As discussed above, the flow through a sound wave is isentropic; hence, in Equation (8.17), the rate of change of pressure with respect to density, $dp/d\rho$, is an isentropic change. Hence, we can rewrite Equation (8.17) as

$$a = \sqrt{\left(\frac{\partial p}{\partial \rho}\right)_s} \qquad (8.18)$$

Equation (8.18) is a fundamental expression for the speed of sound in a gas.

Assume that the gas is calorically perfect. For such a case, the isentropic relation given by Equation (7.32) holds, namely,

$$\frac{p_1}{p_2} = \left(\frac{\rho_1}{\rho_2}\right)^{\gamma} \qquad (8.19)$$

From Equation (8.19), we have

$$\frac{p}{\rho^{\gamma}} = \text{const} = c$$

or

$$p = c\rho^{\gamma} \qquad (8.20)$$

Differentiating Equation (8.20) with respect to ρ, we obtain

$$\left(\frac{\partial p}{\partial \rho}\right)_s = c\gamma\rho^{\gamma-1} \qquad (8.21)$$

Substituting Equation (8.20) for the constant c in Equation (8.21), we have

$$\left(\frac{\partial p}{\partial \rho}\right)_s = \left(\frac{p}{\rho^{\gamma}}\right)\gamma\rho^{\gamma-1} = \frac{\gamma p}{\rho} \qquad (8.22)$$

Substituting Equation (8.22) into (8.18), we obtain

$$a = \sqrt{\frac{\gamma p}{\rho}} \qquad (8.23)$$

Equation (8.23) is an expression for the speed of sound in a calorically perfect gas. At first glance, Equation (8.23) seems to imply that the speed of sound would depend on both p and ρ. However, pressure and density are related through the perfect gas equation of state,

$$\frac{p}{\rho} = RT \qquad (8.24)$$

Hence, substituting Equation (8.24) into (8.23), we have

$$a = \sqrt{\gamma RT} \qquad (8.25)$$

which is our final expression for the speed of sound; it clearly states that the *speed of sound in a calorically perfect gas is a function of temperature only.* This is consistent with our earlier discussion of the speed of sound being a molecular phenomenon, and therefore it is related to the average molecular velocity $\sqrt{8RT/\pi}$.

The speed of sound at standard sea level is a useful value to remember; it is

$$a_s = 340.9 \text{ m/s} = 1117 \text{ ft/s}$$

Recall the definition of compressibility given in Section 7.3. In particular, from Equation (7.35) for the isentropic compressibility, repeated below,

$$\tau_s = -\frac{1}{v}\left(\frac{\partial v}{\partial p}\right)_s$$

and recalling that $v = 1/\rho$ (hence, $dv = -d\rho/\rho^2$), we have

$$\tau_s = -\rho\left[-\frac{1}{\rho^2}\left(\frac{\partial \rho}{\partial p}\right)_s\right] = \frac{1}{\rho(\partial p/\partial \rho)_s} \tag{8.26}$$

However, recall from Equation (8.18) that $(\partial p/\partial \rho)_s = a^2$. Hence, Equation (8.26) becomes

$$\tau_s = \frac{1}{\rho a^2}$$

or

$$a = \sqrt{\frac{1}{\rho \tau_s}} \tag{8.27}$$

Equation (8.27) relates the speed of sound to the compressibility of a gas. The lower the compressibility, the higher the speed of sound. Recall that for the limiting case of an incompressible fluid $\tau_s = 0$. Hence, Equation (8.27) states that the speed of sound in a theoretically incompressible fluid is infinite. In turn, for an incompressible flow with finite velocity V, the Mach number, $M = V/a$, is zero. Hence, the incompressible flows treated in Chapters 3 to 6 are theoretically zero-Mach-number flows.

Finally, in regard to additional physical meaning of the Mach number, consider a fluid element moving along a streamline. The kinetic and internal energies per unit mass are $V^2/2$ and e, respectively. Their ratio is [recalling Equations (7.6a), (7.10), and (8.25)]

$$\frac{V^2/2}{e} = \frac{V^2/2}{c_v T} = \frac{V^2/2}{RT/(\gamma-1)} = \frac{(\gamma/2)V^2}{a^2/(\gamma-1)} = \frac{\gamma(\gamma-1)}{2}M^2$$

Hence, we see that the square of the Mach number is proportional to the ratio of kinetic energy to internal energy of a gas flow. In other words, the Mach number is a measure of the directed motion of the gas compared with the random thermal motion of the molecules.

EXAMPLE 8.1

Consider an airplane flying at a velocity of 250 m/s. Calculate its Mach number if it is flying at a standard altitude of (a) sea level, (b) 5 km, (c) 10 km.

■ **Solution**

(a) From Appendix D for the standard atmosphere, at sea level, $T_\infty = 288$ K.

$$a_\infty = \sqrt{\gamma RT} = \sqrt{(1.4)(287)(288)} = 340.2 \text{ m/s}$$

Hence,

$$M_\infty = \frac{V_\infty}{a_\infty} = \frac{250}{340.2} = \boxed{0.735}$$

(b) At 5 km, from Appendix D, $T_\infty = 255.7$.

$$a_\infty = \sqrt{(1.4)(287)(255.7)} = 320.5 \text{ m/s}$$

$$M_\infty = \frac{V_\infty}{a_\infty} = \frac{250}{320.2} = \boxed{0.78}$$

(c) At 10 km, from Appendix D, $T_\infty = 223.3$.

$$a_\infty = \sqrt{(1.4)(287)(223.3)} = 299.5 \text{ m/s}$$

$$M_\infty = \frac{V_\infty}{a_\infty} = \frac{250}{299.5} = \boxed{0.835}$$

Note: (1) The Mach number used here is the freestream Mach number. When we refer to the Mach number of an airplane or any other object in flight, it is the velocity of the object divided by the freestream speed of sound.

(2) The Mach number of the airplane in this example obviously depends on the altitude at which it is flying, because the speed of sound is different at different altitudes. In this example, an airplane velocity of 250 m/s corresponds to a Mach number of 0.735 at sea level, but a higher Mach number of 0.835 at an altitude of 10 km.

EXAMPLE 8.2

Consider the flow properties at the point in the flow described in Example 7.3, where the temperature is 320 K and the velocity is 1000 m/s. Calculate the Mach number at this point.

■ **Solution**

$$a = \sqrt{\gamma RT} = \sqrt{(1.4)(287)(320)} = 358.6 \text{ m/s}$$

$$M = \frac{V}{a} = \frac{1000}{358.6} = \boxed{2.79}$$

Note: This simple calculation is given here to demonstrate that Mach number is a local property of the flow; it varies from point-to-point throughout the flow field. This is in contrast to the freestream Mach number calculated in Example 8.1. A purpose of these two examples is to illustrate the two uses of Mach number.

EXAMPLE 8.3

Calculate the ratio of kinetic energy to internal energy at a point in an airflow where the Mach number is: (a) $M = 2$, and (b) $M = 20$.

■ Solution

(a) $\dfrac{V^2/2}{e} = \dfrac{\gamma(\gamma-1)}{2}M^2 = \dfrac{(1.4)(0.4)}{2}(2)^2 = \boxed{1.12}$

(b) $\dfrac{V^2/2}{e} = \dfrac{\gamma(\gamma-1)}{2}M^2 = \dfrac{(1.4)(0.4)}{2}(20)^2 = \boxed{112}$

Note: Examining these two results, we see that at Mach 2, the kinetic energy and internal energy are about the same, whereas at the large hypersonic Mach number of 20, the kinetic energy is more than a hundred times larger than the internal energy. This is one characteristic of hypersonic flows—high ratios of kinetic to internal energy.

EXAMPLE 8.4

Consider a point in a flow of air where the pressure and density are 0.7 atm and 0.0019 slug/ft^3, respectively. (a) Calculate the corresponding value of the isentropic compressibility. (b) From that value of the isentropic compressibility, calculate the speed of sound at the point in the flow.

■ Solution

(a) The isentropic compressibility, τ_s, is defined by Equation (7.35),

$$\tau_s = -\frac{1}{v}\left(\frac{\partial v}{\partial p}\right)_s \tag{7.35}$$

The relation between p and v for an isentropic process is given by Equation (7.32), which can be written in the form:

$$p = c\,\rho^\gamma = c\left(\frac{1}{v}\right)^\gamma \tag{E8.1}$$

where c is a constant. Solving Equation (E8.1) for v gives

$$v = c_1\,p^{-(1/\gamma)} \tag{E8.2}$$

where $c_1 = c^{(1/\gamma)}$, another constant value. Differentiating Equation (E8.2), we have

$$\left(\frac{\partial v}{\partial p}\right)_s = c_1\left(-\frac{1}{\gamma}\right)p^{-(1/\gamma)-1} \tag{E8.3}$$

From Equation (E8.2),

$$c_1 = vp^{1/\gamma}$$

Substituting this expression for c_1 into Equation (E8.3),

$$\left(\frac{\partial v}{\partial p}\right)_s = -\frac{1}{\gamma}\left(vp^{1/\gamma}\right)p^{-(1/\gamma)-1} = -\frac{v}{\gamma p}$$

Substituting this result into Equation (7.35), we have

$$\tau_s = -\frac{1}{v}\left(-\frac{v}{\gamma p}\right) = \frac{1}{\gamma p} \tag{E8.4}$$

Thus, for air at a pressure of 0.7 atm, Equation (E8.4) yields

$$\tau_s = \frac{1}{(1.4)(0.7)} = \boxed{1.02 \text{ atm}^{-1}}$$

(b) From Equation (8.27)

$$a = \sqrt{\frac{1}{\rho \tau_s}} \qquad (8.27)$$

and using consistent units for τ_s gives

$$\tau_s = 1.02 \text{ atm}^{-1} \left(\frac{1 \text{ atm}}{2116 \text{ lb/ft}^2} \right)$$

$$= 4.82 \times 10^{-4} (\text{lb/ft}^2)^{-1}$$

we have

$$a = \sqrt{\frac{1}{\rho \tau_s}} = \sqrt{\frac{1}{0.0019(4.82 \times 10^{-4})}} = \boxed{1045 \text{ ft/s}}$$

Check:

$$T = \frac{p}{\rho R} = \frac{(0.7)(2116)}{(0.0019)(1716)} = 454.3 \text{ °R}$$

From Equation (8.25),

$$a = \sqrt{\gamma R T} = \sqrt{(1.4)(1716)(454.3)} = 1045 \text{ ft/s}$$

The answer checks!

EXAMPLE 8.5

By the seventeenth century it was understood that sound propagates through air at some finite velocity. By the time Isaac Newton published his *Principia* in 1687, artillery tests had already shown that the sea-level speed of sound was approximately 1140 ft/s. Comparing that result with today's knowledge of the standard speed of sound at sea level, namely 1117 ft/s, shows that these early seventeenth-century measurements were remarkably accurate. Armed with this experimental result for the speed of sound, Isaac Newton in his *Principia* made the first calculation of the speed of sound (see Reference 58). Here Newton correctly theorized that the speed of sound was related to the "elasticity" of the air, which is the reciprocal of the compressibility, τ. However, he incorrectly assumed that the changes of properties in a sound wave took place isothermally. Using Newton's assumption of isothermal changes through the sound wave, calculate the value obtained by Newton for sea-level speed of sound.

■ **Solution**

The isothermal compressibility is defined as

$$\tau_T = -\frac{1}{v} \left(\frac{\partial v}{\partial p} \right)_T \qquad (7.34)$$

From the equation of state, $v = RT/p$. Thus

$$\left(\frac{\partial v}{\partial p}\right)_T = -\frac{RT}{p^2}$$

Substituting this result into Equation (7.34), we have

$$\tau_T = -\frac{1}{v}\left(-\frac{RT}{p^2}\right) = \frac{RT}{(pv)p} = \frac{RT}{(RT)_p} = \frac{1}{p} \qquad\text{(E8.5)}$$

Using this isothermal value for the compressibility in Equation (8.27) rather than the correct isentropic compressibility, we have (incorrectly)

$$a_T = \sqrt{\frac{1}{\rho\tau_T}} \qquad \text{(Newton's \textit{incorrect} result)} \qquad\text{(E8.6)}$$

At standard sea level, $p = 2116$ lb/ft^2 and $\rho = 0.002377$ slug/ft^3. Hence, the calculation of the speed of sound assuming isothermal conditions through the sound wave is, from Equations (E8.5) and (E8.6),

$$a_T = \sqrt{\frac{1}{\rho\tau_T}} = \sqrt{\frac{p}{\rho}} = \sqrt{\frac{2116}{0.002377}} = \boxed{943.5 \text{ ft/s}}$$

Isaac Newton reported in his *Principia* a value of 979 ft/s for the speed of sound, about 4 percent higher than the value calculated in this example. The difference is most likely due to the rather imprecise values for sea-level atmospheric properties known in Newton's time. *Note:* The isentropic compressibility is given by

$$\tau_s = \frac{1}{\gamma p} \qquad\text{(E8.4)}$$

The isothermal compressibility is given by

$$\tau_T = \frac{1}{p} \qquad\text{(E8.5)}$$

The two values differ by the factor γ. The speed of sound calculated from the isothermal compressibility will be smaller than that calculated from the isentropic compressibility by the factor $(\gamma)^{-1/2}$, or by 0.845. The isothermal calculation will yield a speed of sound about 15 percent lower than the correct value. It is interesting that Newton's calculated value of the speed of sound, as reported in his *Principia*, was about 15 percent lower than the measured value at that time of 1140 ft/s. Undaunted, Newton tried to explain away the differences as due to the presence of dust particles and water vapor in the atmosphere. Finally, a century later the French mathematician Laplace corrected Newton's error by correctly assuming that a sound wave was adiabatic, not isothermal. Therefore, by the 1820s, the process and relationship for the propagation of sound in a gas were fully understood.

8.3.1 Comments

In Examples 8.4 and 8.5 we dealt with the role of compressibility, τ, in the determination of the speed of sound. We found that both τ_T and τ_s are functions of pressure. When τ_s is used in Equation (8.27) to obtain the speed of sound, density also appears in the equation, making it seem that $a = a(\rho, p)$. Indeed, for both Examples 8.4 and 8.5, we used values of both p and ρ to calculate the speed of sound. But keep in mind that p and ρ in the equation for speed of sound always appear in the form p/ρ [see Equation (8.23), for example], and from the perfect gas equation of state, $p/\rho = RT$. Hence, we emphasize again that the speed of sound in a perfect gas is a function of *temperature only*. If we have a gas, and we double the pressure of this gas while keeping the temperature constant, the speed of sound remains the same. If we halve the density, keeping the temperature constant, the speed of sound remains the same. Only in the case of an equilibrium chemically reacting gas and/or a gas where intermolecular forces are important (a gas that is *not* a perfect gas, as discussed in Section 7.2.1), does the speed of sound become a function of both temperature and pressure (see, for example, References 21 and 52).

EXAMPLE 8.6

(a) Consider a long tube with a length of 300 m. The tube is filled with air at a temperature of 320 K. A sound wave is generated at one end of the tube. How long will it take for the wave to reach the other end?

(b) If the tube is filled with helium at a temperature of 320 K, and a sound wave is generated at one end of the tube, how long will it take the sound wave to reach the other end? For a monatomic gas such as helium, $\gamma = 1.67$. Also, for helium R = 2078.5 J/(Kg · K).

■ **Solution**

(a)
$$a = \sqrt{\gamma RT} = \sqrt{(1.4)(287)(320)} = 358.6 \text{ m/s}$$

Letting $l =$ length of the tube and $t =$ time for the sound wave to traverse the length l,

$$t = \frac{l}{a} = \frac{300}{358.6} = \boxed{0.837 \text{ s}}$$

(b)
$$a = \sqrt{\gamma RT} = \sqrt{(1.67)(2078.5)(320)} = 1054 \text{ m/s}$$

$$t = \frac{l}{a} = \frac{300}{1054} = \boxed{0.285 \text{ s}}$$

Note: The speed of sound in helium is much faster than in air at the same temperature for two reasons: (1) γ is larger for helium and, more important, (2) helium has a molecular weight $M = 4$, which is much lighter than that for air with $M = 28$. Because $R = \Re/M$ where $\Re$ is the universal gas constant, the same value for all gases, then R for helium is much larger than for air.

8.4 SPECIAL FORMS OF THE ENERGY EQUATION

In this section, we elaborate upon the energy equation for adiabatic flow, as originally given by Equation (7.44). In Section 7.5, we obtained for a steady, adiabatic, inviscid flow the result that

$$h_1 + \frac{V_1^2}{2} = h_2 + \frac{V_2^2}{2} \tag{8.28}$$

where V_1 and V_2 are velocities at any two points along a three-dimensional stream-line. For the sake of consistency in our current discussion of one-dimensional flow, let us use u_1 and u_2 in Equation (8.28):

$$h_1 + \frac{u_1^2}{2} = h_2 + \frac{u_2^2}{2} \tag{8.29}$$

However, keep in mind that all the subsequent results in this section hold in general along a streamline and are by no means limited to just one-dimensional flows.

Specializing Equation (8.29) to a calorically perfect gas, where $h = c_p T$, we have

$$\boxed{c_p T_1 + \frac{u_1^2}{2} = c_p T_2 + \frac{u_2^2}{2}} \tag{8.30}$$

From Equation (7.9), Equation (8.30) becomes

$$\frac{\gamma R T_1}{\gamma - 1} + \frac{u_1^2}{2} = \frac{\gamma R T_2}{\gamma - 1} + \frac{u_2^2}{2} \tag{8.31}$$

Since $a = \sqrt{\gamma R T}$, Equation (8.31) can be written as

$$\boxed{\frac{a_1^2}{\gamma - 1} + \frac{u_1^2}{2} = \frac{a_2^2}{\gamma - 1} + \frac{u_2^2}{2}} \tag{8.32}$$

If we consider point 2 in Equation (8.32) to be a stagnation point, where the stagnation speed of sound is denoted by a_0, then, with $u_2 = 0$, Equation (8.32) yields (dropping the subscript 1)

$$\boxed{\frac{a^2}{\gamma - 1} + \frac{u^2}{2} = \frac{a_0^2}{\gamma - 1}} \tag{8.33}$$

In Equation (8.33), a and u are the speed of sound and flow velocity, respectively, at any given point in the flow, and a_0 is the stagnation (or total) speed of sound *associated* with that same point. Equivalently, if we have any two points along a streamline, Equation (8.33) states that

$$\frac{a_1^2}{\gamma - 1} + \frac{u_1^2}{2} = \frac{a_2^2}{\gamma - 1} + \frac{u_2^2}{2} = \frac{a_0^2}{\gamma - 1} = \text{const} \tag{8.34}$$

Recalling the definition of a^* given at the end of Section 7.5, let point 2 in Equation (8.32) represent sonic flow, where $u = a^*$. Then

$$\frac{a^2}{\gamma - 1} + \frac{u^2}{2} = \frac{a^{*2}}{\gamma - 1} + \frac{a^{*2}}{2}$$

or

$$\boxed{\frac{a^2}{\gamma - 1} + \frac{u^2}{2} = \frac{\gamma + 1}{2(\gamma - 1)} a^{*2}} \qquad (8.35)$$

In Equation (8.35), a and u are the speed of sound and flow velocity, respectively, at any given point in the flow, and a^* is a characteristic value *associated* with that same point. Equivalently, if we have any two points along a streamline, Equation (8.35) states that

$$\frac{a_1^2}{\gamma - 1} + \frac{u_1^2}{2} = \frac{a_2^2}{\gamma - 1} + \frac{u_2^2}{2} = \frac{\gamma + 1}{2(\gamma - 1)} a^{*2} = \text{const} \qquad (8.36)$$

Comparing the right-hand sides of Equations (8.34) and (8.36), the two properties a_0 and a^* associated with the flow are related by

$$\frac{\gamma + 1}{2(\gamma - 1)} a^{*2} = \frac{a_0^2}{\gamma - 1} = \text{const} \qquad (8.37)$$

Clearly, these defined quantities, a_0 and a^*, are both constants along a given streamline in a steady, adiabatic, inviscid flow. If all the streamlines emanate from the same uniform freestream conditions, then a_0 and a^* are constants throughout the entire flow field.

Recall the definition of total temperature T_0, as discussed in Section 7.5. In Equation (8.30), let $u_2 = 0$; hence $T_2 = T_0$. Dropping the subscript 1, we have

$$\boxed{c_p T + \frac{u^2}{2} = c_p T_0} \qquad (8.38)$$

Equation (8.38) provides a formula from which the defined total temperature T_0 can be calculated from the given actual conditions of T and u at any given point in a general flow field. Equivalently, if we have any two points along a streamline in a steady, adiabatic, inviscid flow, Equation (8.38) states that

$$c_p T_1 + \frac{u_1^2}{2} = c_p T_2 + \frac{u_2^2}{2} = c_p T_0 = \text{const} \qquad (8.39)$$

If all the streamlines emanate from the same uniform freestream, then Equation (8.39) holds throughout the entire flow, not just along a streamline.

For a calorically perfect gas, the ratio of total temperature to static temperature T_0/T is a function of Mach number only, as follows. From Equations (8.38) and (7.9), we have

$$\frac{T_0}{T} = 1 + \frac{u^2}{2c_p T} = 1 + \frac{u^2}{2\gamma RT/(\gamma - 1)} = 1 + \frac{u^2}{2a^2/(\gamma - 1)}$$

$$= 1 + \frac{\gamma - 1}{2}\left(\frac{u}{a}\right)^2$$

Hence,

$$\boxed{\frac{T_0}{T} = 1 + \frac{\gamma - 1}{2}M^2} \tag{8.40}$$

Equation (8.40) is very important; it states that only M (and, of course, the value of γ) dictates the ratio of total temperature to static temperature.

Recall the definition of total pressure p_0 and total density ρ_0, as discussed in Section 7.5. These definitions involve an *isentropic* compression of the flow to zero velocity. From Equation (7.32), we have

$$\frac{p_0}{p} = \left(\frac{\rho_0}{\rho}\right)^\gamma = \left(\frac{T_0}{T}\right)^{\gamma/(\gamma-1)} \tag{8.41}$$

Combining Equations (8.40) and (8.41), we obtain

$$\boxed{\begin{aligned} \frac{p_0}{p} &= \left(1 + \frac{\gamma - 1}{2}M^2\right)^{\gamma/(\gamma-1)} \\ \frac{\rho_0}{\rho} &= \left(1 + \frac{\gamma - 1}{2}M^2\right)^{1/(\gamma-1)} \end{aligned}}$$

$$\tag{8.42}$$
$$\tag{8.43}$$

Similar to the case of T_0/T, we see from Equations (8.42) and (8.43) that the total-to-static ratios p_0/p and ρ_0/ρ are determined by M and γ only. Hence, for a given gas (i.e., given γ), the ratios T_0/T, p_0/p, and ρ_0/ρ depend only on Mach number.

Equations (8.40), (8.42), and (8.43) are very important; they should be branded on your mind. They provide formulae from which the defined quantities T_0, p_0, and ρ_0 can be calculated from the actual conditions of M, T, p, and p at a given point in a general flow field (assuming a calorically perfect gas). They are so important that values of T_0/T, p_0/p, and ρ_0/ρ obtained from Equations (8.40), (8.42), and (8.43), respectively, are tabulated as functions of M in Appendix A for $\gamma = 1.4$ (which corresponds to air at standard conditions).

Consider a point in a general flow where the velocity is exactly sonic (i.e., where $M = 1$). Denote the static temperature, pressure, and density at this sonic condition as T^*, p^*, and ρ^*, respectively. Inserting $M = 1$ into Equations (8.40), (8.42), and (8.43), we obtain

$$\frac{T^*}{T_0} = \frac{2}{\gamma + 1} \tag{8.44}$$

$$\frac{p^*}{p_0} = \left(\frac{2}{\gamma + 1}\right)^{\gamma/(\gamma-1)} \tag{8.45}$$

$$\frac{\rho^*}{\rho_0} = \left(\frac{2}{\gamma + 1}\right)^{1/(\gamma-1)} \tag{8.46}$$

For $\gamma = 1.4$, these ratios are

$$\frac{T^*}{T_0} = 0.833 \qquad \frac{p^*}{p_0} = 0.528 \qquad \frac{\rho^*}{\rho_0} = 0.634$$

which are useful numbers to keep in mind for subsequent discussions.

We have one final item of business in this section. In Chapter 1, we defined the Mach number as $M = V/a$ (or, following the one-dimensional notation in this chapter, $M = u/a$). In turn, this allowed us to define several regimes of flow, among them being

$M < 1$ (subsonic flow)

$M = 1$ (sonic flow)

$M > 1$ (supersonic flow)

In the definition of M, a is the local speed of sound, $a = \sqrt{\gamma R T}$. In the theory of supersonic flow, it is sometimes convenient to introduce a "characteristic" Mach number M^* defined as

$$M^* \equiv \frac{u}{a^*}$$

where a^* is the value of the speed of sound at sonic conditions, *not* the actual local value. This is the same a^* introduced at the end of Section 7.5 and used in Equation (8.35). The value of a^* is given by $a^* = \sqrt{\gamma R T^*}$. Let us now obtain a relation between the actual Mach number M and this defined characteristic Mach number M^*. Dividing Equation (8.35) by u^2, we have

$$\frac{(a/u)^2}{\gamma - 1} + \frac{1}{2} = \frac{\gamma + 1}{2(\gamma - 1)}\left(\frac{a^*}{u}\right)^2$$

$$\frac{(1/M)^2}{\gamma - 1} = \frac{\gamma + 1}{2(\gamma - 1)}\left(\frac{1}{M^*}\right)^2 - \frac{1}{2}$$

$$M^2 = \frac{2}{(\gamma + 1)/M^{*2} - (\gamma - 1)} \tag{8.47}$$

Equation (8.47) gives M as a function of M^*. Solving Equation (8.47) for M^{*2}, we have

$$M^{*2} = \frac{(\gamma + 1)M^2}{2 + (\gamma - 1)M^2} \tag{8.48}$$

which gives M^* as a function of M. As can be shown by inserting numbers into Equation (8.48) (try some yourself),

$M^* = 1$ if $M = 1$

$M^* < 1$ if $M < 1$

$M^* > 1$ if $M > 1$

$M^* \to \sqrt{\dfrac{\gamma + 1}{\gamma - 1}}$ if $M \to \infty$

Therefore, M^* acts qualitatively in the same fashion as M except that M^* approaches a finite value when the actual Mach number approaches infinity.

In summary, a number of equations have been derived in this section, all of which stem in one fashion or another from the basic energy equation for steady, inviscid, adiabatic flow. Make certain that you understand these equations and become very familiar with them before progressing further. These equations are pivotal in the analysis of shock waves and in the study of compressible flow in general.

EXAMPLE 8.7

Repeat Example 7.3 using the equations from the present section.

■ **Solution**

In Example 8.2, the local Mach number was calculated to be $M = 2.79$. From Equation (8.40),

$$\frac{T_0}{T} = 1 + \frac{\gamma - 1}{2}M^2 = 1 + \frac{0.4}{2}(2.79)^2 = 2.557$$

From Example 7.3, $T = 320$ K. Thus,

$$T_0 = 2.557T = (2.557)(320) = \boxed{818 \text{ K}}$$

From Equation (8.42),

$$\frac{p_0}{p} = \left[1 + \frac{\gamma - 1}{2}(M)^2\right]^{\frac{\gamma}{\gamma - 1}} = (2.557)^{\frac{1.4}{0.4}} = 26.7$$

From Example 7.3, $p = 1$ atm. Thus,

$$p_0 = 26.7p = 26.7(1) = \boxed{26.7 \text{ atm}}$$

These answers agree with the results obtained in Example 7.3. The technique used here for calculating T_0 and p_0 from the Mach number is, philosophically, more fundamental than that used in Example 7.3. As we proceed with our discussions, you will find that Mach number is the major governing parameter for compressible flow.

Note: In this example, we used analytical equations to obtain the answers. For our subsequent examples we will use the tabulations in Appendix A, which are obtained from the analytical equations. These tabulations are a convenience that saves us from working through the equations each time.

EXAMPLE 8.8

Consider a point in an airflow where the local Mach number, static pressure, and static temperature are 3.5, 0.3 atm, and 180 K, respectively. Calculate the local values of p_0, T_0, T^*, a^*, and M^* at this point.

■ Solution

From Appendix A, for $M = 3.5$, $p_0/p = 76.27$ and $T_0/T = 3.45$. Hence,

$$p_0 = \left(\frac{p_0}{p}\right) p = 76.27(0.3 \text{ atm}) = \boxed{22.9 \text{ atm}}$$

$$T_0 = \frac{T_0}{T} T = 3.45(180) = \boxed{621 \text{ K}}$$

For $M = 1$, $T_0/T^* = 1.2$. Hence,

$$T^* = \frac{T_0}{1.2} = \frac{621}{1.2} = \boxed{517.5 \text{ K}}$$

$$a^* = \sqrt{\gamma R T^*} = \sqrt{1.4(287)(517.5)} = \boxed{456 \text{ m/s}}$$

$$a = \sqrt{\gamma R T} = \sqrt{1.4(287)(180)} = 268.9 \text{ m/s}$$

$$V = Ma = 3.5(268.9) = 941 \text{ m/s}$$

$$M^* = \frac{V}{a^*} = \frac{941}{456} = \boxed{2.06}$$

The above result for M^* can also be obtained directly from Equation (8.48):

$$M^{*2} = \frac{(\gamma + 1)M^2}{2 + (\gamma - 1)M^2} = \frac{2.4(3.5)^2}{2 + 0.4(3.5)^2} = 4.26$$

Hence, $M^* = \sqrt{4.26} = 2.06$, as obtained above.

EXAMPLE 8.9

In Example 3.1, we illustrated for an incompressible flow, the calculation of the velocity at a point on an airfoil when we were given the pressure at that point and the freestream velocity and pressure. (It would be useful to review Example 3.1 before going further.) The solution involved the use of Bernoulli's equation. Let us now examine the compressible flow analog of Example 3.1. Consider an airfoil in a freestream where $M_\infty = 0.6$ and $p_\infty = 1$ atm, as

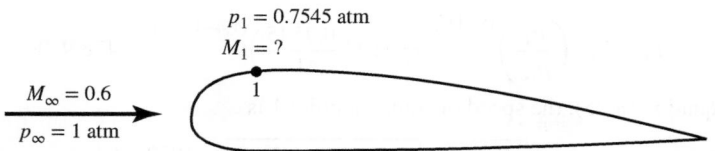

$p_1 = 0.7545$ atm

$M_1 = ?$

$M_\infty = 0.6$

$p_\infty = 1$ atm

1

Figure 8.5 Figure for Example 8.2.

sketched in Figure 8.5. At point 1 on the airfoil, the pressure is $p_1 = 0.7545$ atm. Calculate the local Mach number at point 1. Assume isentropic flow over the airfoil.

■ **Solution**

We cannot use Bernoulli's equation because the freestream Mach number is high enough that the flow should be treated as compressible. The free stream total pressure for $M_\infty = 0.6$ is, from Appendix A

$$p_{0,\infty} = \frac{p_{0,\infty}}{p_\infty}p_\infty = (1.276)(1) = 1.276 \text{ atm}$$

Recall that for an isentropic flow, the total pressure is constant throughout the flow. Hence,

$$p_{0,1} = p_{0,\infty} = 1.276 \text{ atm}$$

or

$$\frac{p_{0,1}}{p_1} = \frac{1.276}{0.7545} = 1.691$$

From Appendix A, for a ratio of total to static pressure equal to 1.69, we have

$$\boxed{M_1 = 0.9}$$

This is the local Mach number at point 1 on the airfoil in Figure 8.5.

EXAMPLE 8.10

Note that flow velocity did not enter the calculations in Example 8.9. For compressible flow, Mach number is a more fundamental variable than velocity; we will see this time-and-time again in the subsequent sections and chapters dealing with compressible flow. However, we can certainly calculate velocities for compressible flow problems, but in such cases we usually need to know something about the temperature level of the flow. For the conditions that prevail in Example 8.9, calculate the velocity at point 1 on the airfoil when the free stream temperature is 59 °F.

■ **Solution**

We will need to deal with consistent units. Since 0 °F is the same as 460 °R,

$$T_\infty = 460 + 59 = 519 \text{ °R}$$

The flow is isentropic, hence, from Equation (7.32)

$$\frac{p_1}{p_\infty} = \left(\frac{T_1}{T_\infty}\right)^{\gamma/(\gamma-1)}$$

or $\qquad T_1 = T_\infty \left(\dfrac{p_1}{p_\infty} \right)^{(\gamma-1)/\gamma} = 519 \left(\dfrac{0.7545}{1} \right)^{(1.4-1)/1.4} = 478.9 \ °\text{R}$

From Equation (8.25), the speed of sound at point 1 is

$$a_1 = \sqrt{\gamma R T_1} = \sqrt{(1.4)(1716)(478.9)} = 1072.6 \ \text{ft/s}$$

Hence, $\qquad\qquad V_1 = M_1 a_1 = (0.9)(1072.6) = \boxed{965.4 \ \text{ft/s}}$

8.5 WHEN IS A FLOW COMPRESSIBLE?

As a corollary to Section 8.4, we are now in a position to examine the question, When does a flow have to be considered compressible, that is, when do we have to use analyses based on Chapters 7 to 14 rather than the incompressible techniques discussed in Chapters 3 to 6? There is no specific answer to this question; for subsonic flows, it is a matter of the degree of accuracy desired whether we treat ρ as a constant or as a variable, whereas for supersonic flow the qualitative aspects of the flow are so different that the density *must* be treated as variable. We have stated several times in the preceding chapters the rule of thumb that a flow can be reasonably assumed to be incompressible when $M < 0.3$, whereas it should be considered compressible when $M > 0.3$. There is nothing magic about the value 0.3, but it is a convenient dividing line. We are now in a position to add substance to this rule of thumb.

Consider a fluid element initially at rest, say, an element of the air around you. The density of this gas at rest is ρ_0. Let us now accelerate this fluid element isentropically to some velocity V and Mach number M, say, by expanding the air through a nozzle. As the velocity of the fluid element increases, the other flow properties will change according to the basic governing equations derived in Chapter 7 and in this chapter. In particular, the density ρ of the fluid element will change according to Equation (8.43):

$$\frac{\rho_0}{\rho} = \left(1 + \frac{\gamma - 1}{2} M^2 \right)^{1/(\gamma-1)} \tag{8.43}$$

For $\gamma = 1.4$, this variation is illustrated in Figure 8.6, where ρ/ρ_0 is plotted as a function of M from zero to sonic flow. Note that at low subsonic Mach numbers, the variation of ρ/ρ_0 is relatively flat. Indeed, for $M < 0.32$, the value of ρ deviates from ρ_0 by less than 5 percent, and for all practical purposes the flow can be treated as incompressible. However, for $M > 0.32$, the variation in ρ is larger than 5 percent, and its change becomes even more pronounced as M increases. As a result, many aerodynamicists have adopted the rule of thumb that the density variation should be accounted for at Mach numbers above 0.3; that is, the flow should be treated as compressible. Of course, keep in mind that all flows, even at the lowest Mach numbers, are, strictly speaking, compressible. Incompressible flow is really a myth. However, as shown in Figure 8.6, the

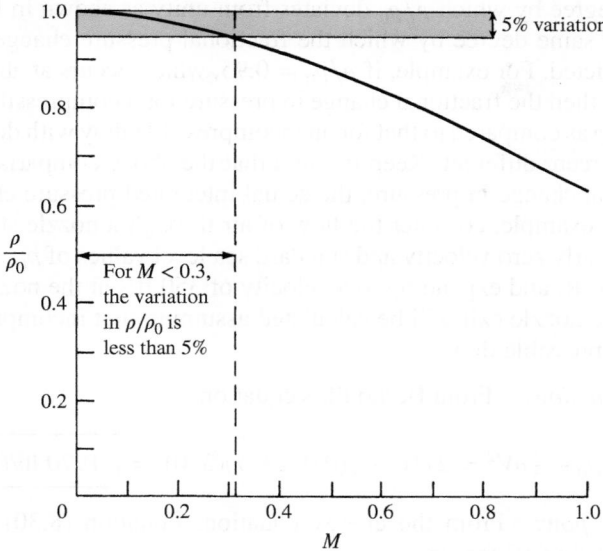

Figure 8.6 Isentropic variation of density with Mach number.

assumption of incompressible flow is very reasonable at low Mach numbers. For this reason, the analyses in Chapters 3 to 6 and the vast bulk of existing literature for incompressible flow are quite practical for many aerodynamic applications.

To obtain additional insight into the significance of Figure 8.6, let us ask how the ratio ρ/ρ_0 affects the change in pressure associated with a given change in velocity. The differential relation between pressure and velocity for a compressible flow is given by Euler's equation, Equation (3.12) repeated below:

$$dp = -\rho V \, dV \tag{3.12}$$

This can be written as

$$\frac{dp}{p} = -\frac{\rho}{p} V^2 \frac{dV}{V}$$

This equation gives the fractional change in pressure for a given fractional change in velocity for a compressible flow with local density ρ. Now, if we *assume* that the density is constant, say, equal to ρ_0 as denoted in Figure 8.6, then Equation (3.12) yields

$$\left(\frac{dp}{p}\right)_0 = -\frac{\rho_0}{p} V^2 \frac{dV}{V}$$

where the subscript zero implies the assumption of constant density. Dividing the last two equations, and assuming the same dV/V and p, we have

$$\frac{dp/p}{(dp/p)_0} = \frac{\rho}{\rho_0}$$

Hence, the degree by which ρ/ρ_0 deviates from unity as shown in Figure 8.6 is related to the same degree by which the fractional pressure change for a given dV/V is predicted. For example, if $\rho/\rho_0 = 0.95$, which occurs at about $M = 0.3$ in Figure 8.5, then the fractional change in pressure for a compressible flow with local density ρ as compared to that for an incompressible flow with density ρ_0 will be about 5 percent different. Keep in mind that the above comparison is for the local fractional change in pressure, the actual integrated pressure change is less sensitive. For example, consider the flow of air through a nozzle starting in the reservoir at nearly zero velocity and standard sea level values of $p_0 = 2116$ lb/ft^2 and $T_0 = 510$ °R, and expanding to a velocity of 350 ft/s at the nozzle exit. The pressure at the nozzle exit will be calculated assuming first incompressible flow and then compressible flow.

Incompressible flow: From Bernoulli's equation,

$$p = p_0 - \tfrac{1}{2}\rho V^2 = 2116 - \tfrac{1}{2}(0.002377)(350)^2 = \boxed{1970 \text{ lb/ft}^2}$$

Compressible flow: From the energy equation, Equation (8.30), with $c_p = 6006$ [(ft) (lb)/slug°R] for air,

$$T = T_0 - \frac{V^2}{2c_p} = 519 - \frac{(350)^2}{2(6006)} = 508.8 \text{ °R}$$

From Equation (7.32),

$$\frac{p}{p_0} = \left(\frac{T}{T_0}\right)^{\gamma/(\gamma-1)} = \left(\frac{508.8}{519}\right)^{3.5} = 0.9329$$

$$p = 0.9329 p_0 = 0.9329(2116) = \boxed{1974 \text{ lb/ft}^2}$$

Note that the two results are almost the same, with the compressible value of pressure only 0.2 percent higher than the incompressible value. Clearly, the assumption of incompressible flow (hence, the use of Bernoulli's equation) is certainly justified in this case. Also, note that the Mach number at the exit is 0.317 (work this out for yourself). Hence, we have shown that for a flow wherein the Mach number ranges from zero to about 0.3, Bernoulli's equation yields a reasonably accurate value for the pressure—another justification for the statement that flows wherein $M < 0.3$ are essentially incompressible flows. On the other hand, if this flow were to continue to expand to a velocity of 900 ft/s, a repeat of the above calculation yields the following results for the static pressure at the end of the expansion:

Incompressible (Bernoulli's equation): $p = 1153$ lb/ft^2

Compressible: $p = 1300$ lb/ft^2

Here, the difference between the two sets of results is considerable—a 13 percent difference. In this case, the Mach number at the end of the expansion is 0.86. Clearly, for such values of Mach number, the flow must be treated as compressible.

In summary, although it may be somewhat conservative, this author suggests on the strength of all the above information, including Figure 8.6, that flows wherein the local Mach number exceeds 0.3 should be treated as compressible. Moreover, when $M < 0.3$, the assumption of incompressible flow is quite justified.

8.6 CALCULATION OF NORMAL SHOCK-WAVE PROPERTIES

Consider again the road map given in Figure 8.2. We have finished our three side trips (Sections 8.3 to 8.5) and are now ready to get back on the main road toward the calculation of changes of flow properties across a normal shock wave. Return again to Section 8.2, and recall the basic normal shock equations given by Equations (8.2), (8.6), and (8.10):

Continuity:
$$\rho_1 u_1 = \rho_2 u_2 \tag{8.2}$$

Momentum:
$$p_1 + \rho_1 u_1^2 = p_2 + \rho_2 u_2^2 \tag{8.6}$$

Energy:
$$h_1 + \frac{u_1^2}{2} = h_2 + \frac{u_2^2}{2} \tag{8.10}$$

In addition, for a calorically perfect gas, we have

$$h_2 = c_p T_2 \tag{8.49}$$

$$p_2 = \rho_2 R T_2 \tag{8.50}$$

Return again to Figure 8.3, and recall the basic normal shock-wave problem: given the conditions in region 1 ahead of the shock, calculate the conditions in region 2 behind the shock. Examining the five equations given above, we see that they involve five unknowns, namely, ρ_2, u_2, p_2, h_2, and T_2. Hence, Equations (8.2), (8.6), (8.10), (8.49), and (8.50) are sufficient for determining the properties behind a normal shock wave in a calorically perfect gas. Let us proceed.

First, dividing Equation (8.6) by (8.2), we obtain

$$\frac{p_1}{\rho_1 u_1} + u_1 = \frac{p_2}{\rho_2 u_2} + u_2$$

$$\frac{p_1}{\rho_1 u_1} - \frac{p_2}{\rho_2 u_2} = u_2 - u_1 \tag{8.51}$$

Recalling from Equation (8.23) that $a = \sqrt{\gamma p / \rho}$, Equation (8.51) becomes

$$\frac{a_1^2}{\gamma u_1} - \frac{a_2^2}{\gamma u_2} = u_2 - u_1 \tag{8.52}$$

Equation (8.52) is a combination of the continuity and momentum equations. The energy equation, Equation (8.10), can be used in one of its alternate forms, namely, Equation (8.35), rearranged below, and applied first in region 1 and then in region 2:

$$a_1^2 = \frac{\gamma + 1}{2} a^{*2} - \frac{\gamma - 1}{2} u_1^2 \tag{8.53}$$

and
$$a_2^2 = \frac{\gamma + 1}{2}a^{*2} - \frac{\gamma - 1}{2}u_2^2 \qquad (8.54)$$

In Equations (8.53) and (8.54), a^* is the same constant value because the flow across the shock wave is adiabatic (see Sections 7.5 and 8.5). Substituting Equations (8.53) and (8.54) into Equation (8.52), we have

$$\frac{\gamma + 1}{2}\frac{a^{*2}}{\gamma u_1} - \frac{\gamma - 1}{2\gamma}u_1 - \frac{\gamma + 1}{2}\frac{a^{*2}}{\gamma u_2} + \frac{\gamma - 1}{2\gamma}u_2 = u_2 - u_1$$

or
$$\frac{\gamma + 1}{2\gamma u_1 u_2}(u_2 - u_1)a^{*2} + \frac{\gamma - 1}{2\gamma}(u_2 - u_1) = u_2 - u_1$$

Dividing by $u_2 - u_1$, we obtain

$$\frac{\gamma + 1}{2\gamma u_1 u_2}a^{*2} + \frac{\gamma - 1}{2\gamma} = 1$$

Solving for a^*, we obtain

$$\boxed{a^{*2} = u_1 u_2} \qquad (8.55)$$

Equation (8.55) is called the *Prandtl relation* and is a useful intermediate relation for normal shock waves. For example, from Equation (8.55),

$$1 = \frac{u_1}{a^*}\frac{u_2}{a^*} \qquad (8.56)$$

Recall the definition of characteristic Mach number, $M^* = u/a^*$, given in Section 8.4. Hence, Equation (8.56) becomes

$$1 = M_1^* M_2^*$$

or
$$M_2^* = \frac{1}{M_1^*} \qquad (8.57)$$

Substituting Equation (8.48) into (8.57), we have

$$\frac{(\gamma + 1)M_2^2}{2 + (\gamma - 1)M_2^2} = \left[\frac{(\gamma + 1)M_1^2}{2 + (\gamma - 1)M_1^2}\right]^{-1} \qquad (8.58)$$

Solving Equation (8.58) for M_2^2, we obtain

$$\boxed{M_2^2 = \frac{1 + [(\gamma - 1)/2]M_1^2}{\gamma M_1^2 - (\gamma - 1)/2}} \qquad (8.59)$$

Equation (8.59) is our first major result for a normal shock wave. Examine Equation (8.59) closely; it states that the Mach number behind the wave M_2 is a function only of the Mach number ahead of the wave M_1. Moreover, if $M_1 = 1$, then $M_2 = 1$. This is the case of an infinitely weak normal shock wave, defined as a *Mach wave*. Furthermore, if $M_1 > 1$, then $M_2 < 1$; that is, the Mach number behind the normal shock wave is *subsonic*. As M_1 increases above 1, the normal shock wave becomes stronger, and M_2 becomes progressively less than 1. However, in the limit as $M_1 \to \infty$, M_2 approaches a finite minimum value, $M_2 \to \sqrt{(\gamma - 1)/2\gamma}$, which for air is 0.378.

Let us now obtain the ratios of the thermodynamic properties ρ_2/ρ_1, p_2/p_1, and T_2/T_1 across a normal shock wave. Rearranging Equation (8.2) and using Equation (8.55), we have

$$\frac{\rho_2}{\rho_1} = \frac{u_1}{u_2} = \frac{u_1^2}{u_2 u_1} = \frac{u^2}{a^{*2}} = M_1^{*2} \tag{8.60}$$

Substituting Equation (8.48) into (8.60), we obtain

$$\boxed{\frac{\rho_2}{\rho_1} = \frac{u_1}{u_2} = \frac{(\gamma + 1)M_1^2}{2 + (\gamma - 1)M_1^2}} \tag{8.61}$$

To obtain the pressure ratio, return to the momentum equation, Equation (8.6), combined with the continuity equation, Equation (8.2):

$$p_2 - p_1 = \rho_1 u_1^2 - \rho_2 u_2^2 = \rho_1 u_1 (u_1 - u_2) = \rho_1 u_1^2 \left(1 - \frac{u_2}{u_1}\right) \tag{8.62}$$

Dividing Equation (8.62) by p_1, and recalling that $a_1^2 = \gamma p_1/\rho_1$, we obtain

$$\frac{p_2 - p_1}{p_1} = \frac{\gamma \rho_1 u_1^2}{\gamma p_1}\left(1 - \frac{u_2}{u_1}\right) = \frac{\gamma u_1^2}{a_1^2}\left(1 - \frac{u_2}{u_1}\right) = \gamma M_1^2\left(1 - \frac{u_2}{u_1}\right) \tag{8.63}$$

For u_2/u_1 in Equation (8.63), substitute Equation (8.61):

$$\frac{p_2 - p_1}{p_1} = \gamma M_1^2\left[1 - \frac{2 + (\gamma - 1)M_1^2}{(\gamma + 1)M_1^2}\right] \tag{8.64}$$

Equation (8.64) simplifies to

$$\boxed{\frac{p_2}{p_1} = 1 + \frac{2\gamma}{\gamma + 1}(M_1^2 - 1)} \tag{8.65}$$

To obtain the temperature ratio, recall the equation of state $p = \rho R T$. Hence,

$$\frac{T_2}{T_1} = \left(\frac{p_2}{p_1}\right)\left(\frac{\rho_1}{\rho_2}\right) \tag{8.66}$$

Substituting Equations (8.61) and (8.65) into (8.66), and recalling that $h = c_p T$, we obtain

$$\boxed{\frac{T_2}{T_1} = \frac{h_2}{h_1} = \left[1 + \frac{2\gamma}{\gamma + 1}(M_1^2 - 1)\right]\frac{2 + (\gamma - 1)M_1^2}{(\gamma + 1)M_1^2}} \tag{8.67}$$

Equations (8.61), (8.65), and (8.67) are important. Examine them closely. Note that ρ_2/ρ_1, p_2/p_1, and T_2/T_1 are *functions of the upstream Mach number M_1 only*. Therefore, in conjunction with Equation (8.59) for M_2, we see that the upstream Mach number M_1 is *the* determining parameter for changes across a normal shock wave in a calorically perfect gas. This is a dramatic example of the power of the Mach number as a governing parameter in compressible flows. In the above equations, if $M_1 = 1$, then $p_2/p_1 = \rho_2/\rho_1 = T_2/T_1 = 1$; that is, we have the case of a normal shock wave of vanishing strength—a Mach wave. As M_1 increases above 1, p_2/p_1, ρ_2/ρ_1, and T_2/T_1 progressively increase above 1. In the limiting case of $M_1 \to \infty$ in Equations (8.59), (8.61), (8.65), and (8.67), we find, for $\gamma = 1.4$,

$$\lim_{M_1 \to \infty} M_2 = \sqrt{\frac{\gamma - 1}{2\gamma}} = 0.378 \qquad \text{(as discussed previously)}$$

$$\lim_{M_1 \to \infty} \frac{\rho_2}{\rho_1} = \frac{\gamma + 1}{\gamma - 1} = 6$$

$$\lim_{M_1 \to \infty} \frac{p_2}{p_1} = \infty \qquad \lim_{M_1 \to \infty} \frac{T_2}{T_1} = \infty$$

Note that, as the upstream Mach number increases toward infinity, the pressure and temperature increase without bound, whereas the density approaches a rather moderate finite limit.

We have stated earlier that shock waves occur in supersonic flows; a stationary normal shock such as shown in Figure 8.3 does not occur in subsonic flow. That is, in Equations (8.59), (8.61), (8.65), and (8.67), the upstream Mach number is supersonic $M_1 \geq 1$. However, on a *mathematical basis,* these equations also allow solutions for $M_1 \leq 1$. These equations embody the continuity, momentum, and energy equations, which in principle do not care whether the value of M_1 is subsonic or supersonic. Here is an ambiguity which can only be resolved by appealing to the second law of thermodynamics (see Section 7.2). Recall that the second law of thermodynamics determines the *direction* which a given process can take. Let us apply the second law to the flow across a normal shock wave, and examine what it tells us about allowable values of M_1.

First, consider the entropy change across the normal shock wave. From Equation (7.25),

$$s_2 - s_1 = c_p \ln \frac{T_2}{T_1} - R \ln \frac{p_2}{p_1} \tag{7.25}$$

with Equations (8.65) and (8.67), we have

$$s_2 - s_1 = c_p \ln \left\{ \left[1 + \frac{2\gamma}{\gamma + 1}(M_1^2 - 1) \right] \frac{2 + (\gamma - 1)M_1^2}{(\gamma + 1)M_1^2} \right\}$$

$$- R \ln \left[1 + \frac{2\gamma}{\gamma + 1}(M_1^2 - 1) \right] \tag{8.68}$$

From Equation (8.68), we see that the entropy change $s_2 - s_1$ across the shock is a function of M_1 only. The second law dictates that

$$s_2 - s_1 \geq 0$$

In Equation (8.68), if $M_1 = 1$, $s_2 = s_1$, and if $M_1 > 1$, then $s_2 - s_1 > 0$, both of which obey the second law. However, if $M_1 < 1$, then Equation (8.68) gives $s_2 - s_1 < 0$, which is *not* allowed by the second law. Consequently, in nature, only cases involving $M_1 \geq 1$ are valid; that is, normal shock waves can occur only in supersonic flow.

Why does the entropy increase across the shock wave? The second law tells us that it must, but what mechanism does nature use to accomplish this increase? To answer these questions, recall that a shock wave is a very thin region (on the order of 10^{-5} cm) across which some large changes occur almost discontinuously. Therefore, within the shock wave itself, large gradients in velocity and temperature occur; that is, the mechanisms of friction and thermal conduction are strong. These are dissipative, irreversible mechanisms that always increase the entropy. Therefore, the precise entropy increase predicted by Equation (8.68) for a given supersonic M_1 is appropriately provided by nature in the form of friction and thermal conduction within the interior of the shock wave itself.

In Section 7.5, we defined the total temperature T_0 and total pressure p_0. What happens to these total conditions across a shock wave? To help answer this question, consider Figure 8.7, which illustrates the definition of total conditions ahead of and behind the shock. In region 1 ahead of the shock, a fluid element has the actual conditions of M_1, p_1, T_1, and s_1. Now imagine that we bring this fluid element to rest isentropically, creating the "imaginary" state $1a$ ahead of the shock. In state $1a$, the fluid element at rest would have a pressure and temperature $p_{0,1}$ and $T_{0,1}$, respectively, that is, the total pressure and total temperature,

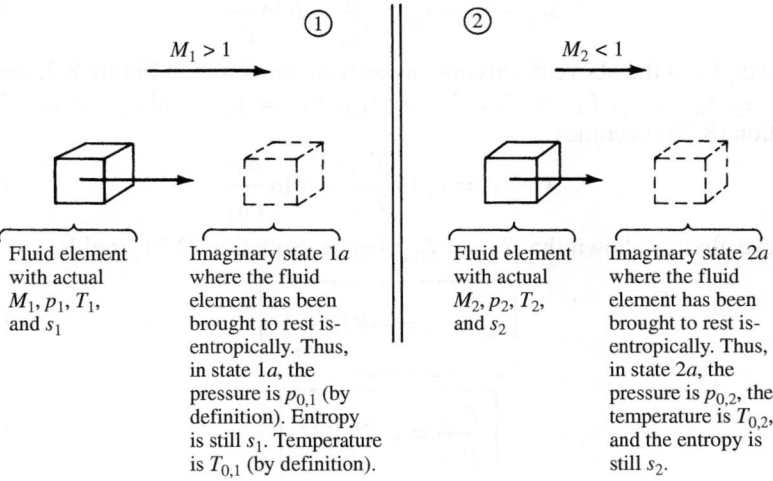

$M_1 > 1$ ①   ② $M_2 < 1$

| Fluid element with actual M_1, p_1, T_1, and s_1 | Imaginary state $1a$ where the fluid element has been brought to rest isentropically. Thus, in state $1a$, the pressure is $p_{0,1}$ (by definition). Entropy is still s_1. Temperature is $T_{0,1}$ (by definition). | Fluid element with actual M_2, p_2, T_2, and s_2 | Imaginary state $2a$ where the fluid element has been brought to rest isentropically. Thus, in state $2a$, the pressure is $p_{0,2}$, the temperature is $T_{0,2}$, and the entropy is still s_2. |

Figure 8.7 Total conditions ahead of and behind a normal shock wave.

respectively, in region 1. The entropy in state $1a$ would still be s_1 because the fluid element is brought to rest isentropically; $s_{1a} = s_1$. Now consider region 2 behind the shock. Again consider a fluid element with the actual conditions of M_2, p_2, T_2, and s_2, as sketched in Figure 8.7. And again let us imagine that we bring this fluid element to rest isentropically, creating the "imaginary" state $2a$ behind the shock. In state $2a$, the fluid element at rest would have pressure and temperature $p_{0,2}$ and $T_{0,2}$, respectively, that is, the total pressure and total temperature, respectively, in region 2. The entropy in state $2a$ would still be s_2 because the fluid element is brought to rest isentropically; $s_{2a} = s_2$. The questions are now asked: How does $T_{0,2}$ compare with $T_{0,1}$, and how does $p_{0,2}$ compare with $p_{0,1}$?

To answer the first of these questions, consider Equation (8.30):

$$c_p T_1 + \frac{u_1^2}{2} = c_p T_2 + \frac{u_2^2}{2} \tag{8.30}$$

From Equation (8.38), the total temperature is given by

$$c_p T_0 = c_p T + \frac{u^2}{2} \tag{8.38}$$

Combining Equations (8.30) and (8.38), we have

$$c_p T_{0,1} = c_p T_{0,2}$$

or

$$\boxed{T_{0,1} = T_{0,2}} \tag{8.69}$$

Equation (8.69) states that *total temperature is constant across a stationary normal shock wave*. This should come as no surprise; the flow across a shock wave is adiabatic, and in Section 7.5 we demonstrated that in a steady, adiabatic, inviscid flow of a calorically perfect gas, the total temperature is constant.

To examine the variation of total pressure across a normal shock wave, write Equation (7.25) between the imaginary states $1a$ and $2a$:

$$s_{2a} - s_{1a} = c_p \ln \frac{T_{2a}}{T_{1a}} - R \ln \frac{p_{2a}}{p_{1a}} \tag{8.70}$$

However, from the above discussion, as well as the sketch in Figure 8.7, we have $s_{2a} = s_2$, $s_{1a} = s_1$, $T_{2a} = T_{0,2}$, $T_{1a} = T_{0,1}$, $p_{2a} = p_{0,2}$, and $p_{1a} = p_{0,1}$. Thus, Equation (8.70) becomes

$$s_2 - s_1 = c_p \ln \frac{T_{0,2}}{T_{0,1}} - R \ln \frac{p_{0,2}}{p_{0,1}} \tag{8.71}$$

We have already shown that $T_{0,2} = T_{0,1}$; hence, Equation (8.71) yields

$$\boxed{s_2 - s_1 = -R \ln \frac{p_{0,2}}{p_{0,1}}} \tag{8.72}$$

or

$$\boxed{\frac{p_{0,2}}{p_{0,1}} = e^{-(s_2 - s_1)/R}} \tag{8.73}$$

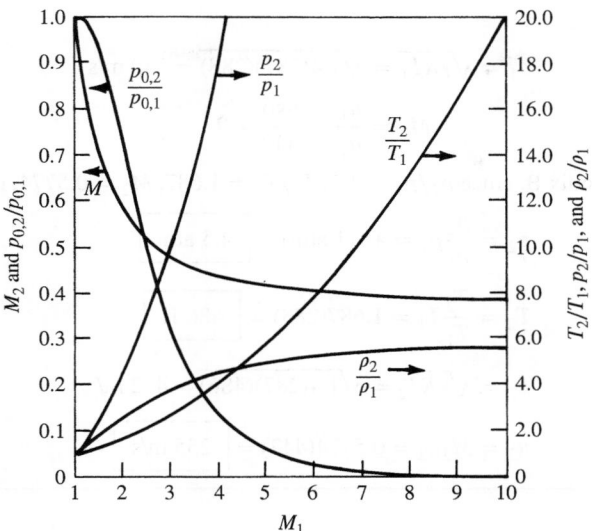

Figure 8.8 The variation of properties across a normal shock wave as a function of upstream Mach number: $\gamma = 1.4$.

From Equation (8.68), we know that $s_2 - s_1 > 0$ for a normal shock wave. Hence, Equation (8.73) states that $p_{0,2} < p_{0,1}$. *The total pressure decreases across a shock wave.* Moreover, since $s_2 - s_1$ is a function of M_1 only [from Equation (8.68)], then Equation (8.73) clearly states that the total pressure ratio $p_{0,2}/p_{0,1}$ across a normal shock wave is a function of M_1 only.

In summary, we have now verified the qualitative changes across a normal shock wave as sketched in Figure 7.4*b* and as originally discussed in Section 7.6. Moreover, we have obtained closed-form analytic expressions for these changes in the case of a calorically perfect gas. We have seen that p_2/p_1, ρ_2/ρ_1, T_2/T_1, M_2, and $p_{0,2}/p_{0,1}$ are functions of the upstream Mach number M_1 only. To help you obtain a stronger physical feeling of normal shock-wave properties, these variables are plotted in Figure 8.8 as a function of M_1. Note that (as stated earlier) these curves show how, as M_1 becomes very large, T_2/T_1 and p_2/p_1 also become very large, whereas ρ_2/ρ_1 and M_2 approach finite limits. Examine Figure 8.8 carefully, and become comfortable with the trends shown.

The results given by Equations (8.59), (8.61), (8.65), (8.67), and (8.73) are so important that they are tabulated as a function of M_1 in Appendix B for $\gamma = 1.4$.

EXAMPLE 8.11

Consider a normal shock wave in air where the upstream flow properties are $u_1 = 680$ m/s, $T_1 = 288$ K, and $p_1 = 1$ atm. Calculate the velocity, temperature, and pressure downstream of the shock.

■ **Solution**

$$a_1 = \sqrt{\gamma R T_1} = \sqrt{1.4(287)(288)} = 340 \text{ m/s}$$

$$M_1 = \frac{u_1}{a_1} = \frac{680}{340} = 2$$

From Appendix B, since $p_2/p_1 = 4.5$, $T_2/T_1 = 1.687$, $M_2 = 0.5774$, then

$$p_2 = \frac{p_2}{p_1} p_1 = 4.5(1 \text{ atm}) = \boxed{4.5 \text{ atm}}$$

$$T_2 = \frac{T_2}{T_1} T_1 = 1.687(288) = \boxed{486 \text{ K}}$$

$$a_2 = \sqrt{\gamma R T_2} = \sqrt{1.4(287)(486)} = 442 \text{ m/s}$$

$$u_2 = M_2 a_2 = 0.5774(442) = \boxed{255 \text{ m/s}}$$

EXAMPLE 8.12

Consider a normal shock wave in a supersonic airstream where the pressure upstream of the shock is 1 atm. Calculate the loss of total pressure across the shock wave when the upstream Mach number is (a) $M_1 = 2$, and (b) $M_1 = 4$. Compare these two results and comment on their implication.

■ **Solution**

(a) The upstream total pressure is obtained from

$$p_{0,1} = \left(\frac{p_{0,1}}{p_1} \right) p_1$$

where from Appendix A for $M_1 = 2$, $p_{0,1}/p_1 = 7.824$. Hence,

$$p_{0,1} = (7.824)(1 \text{ atm}) = 7.824 \text{ atm}$$

The total pressure behind the normal shock is obtained from

$$p_{0,2} = \left(\frac{p_{0,2}}{p_{0,1}} \right) p_{0,1}$$

where from Appendix B, for $M_1 = 2$, $p_{0,2}/p_{0,1} = 0.7209$. Hence,

$$p_{0,2} = (0.7209)(7.824) = 5.64 \text{ atm}$$

The *loss* of total pressure is

$$p_{0,1} - p_{0,2} = 7.824 - 5.64 = \boxed{2.184 \text{ atm}}$$

(b) For $M_1 = 4$, from Appendix A,

$$p_{0,1} = \left(\frac{p_{0,1}}{p_1} \right) p_1 = (151.8)(1 \text{ atm}) = 151.8 \text{ atm}$$

The total pressure behind the normal shock is obtained from Appendix B, for $M_1 = 4$, as

$$p_{0,2} = \left(\frac{p_{0,2}}{p_{0,1}}\right) p_{0,1} = (0.1388)(151.8) = 21.07 \text{ atm}$$

The *loss* of total pressure is

$$p_{0,1} - p_{0,2} = 151.8 - 21.07 = \boxed{130.7 \text{ atm}}$$

Note: In any flow, total pressure is a precious commodity. Any loss of total pressure reduces the flow's ability to do useful work. Losses of total pressure reduce the performance of any flow device, and cost money. We will see this time-and-time-again in subsequent chapters. In this example, we see that for a normal shock at Mach 2, the loss of total pressure was 2.184 atm, whereas simply by doubling the Mach number to 4, the loss of total pressure was a whopping 130.7 atm. The moral of this story is that, if you are going to suffer a normal shock wave in a flow, everything else being equal, you want the normal shock to occur at the lowest possible upstream Mach number.

EXAMPLE 8.13

A ramjet engine is an air-breathing propulsion device with essentially no rotating machinery (no rotating compressor blades, turbine, etc.). The basic generic parts of a conventional ramjet are sketched in Figure 8.9. The flow, moving from left to right, enters the inlet, where it is compressed and slowed down. The compressed air then enters the combustor at very low subsonic speed, where it is mixed and burned with a fuel. The hot gas then expands through a nozzle. The net result is the production of thrust toward the left in Figure 8.9. In this figure the ramjet is shown in a supersonic freestream with a detached shock wave ahead of the inlet. The portion of the shock just to the left of point 1 is a normal shock. (A detached normal shock wave in front of the inlet of a ramjet in a supersonic flow is not the ideal operating condition; rather, it is desirable that the flow pass through one or more *oblique* shock waves before entering the inlet. Oblique shock waves are

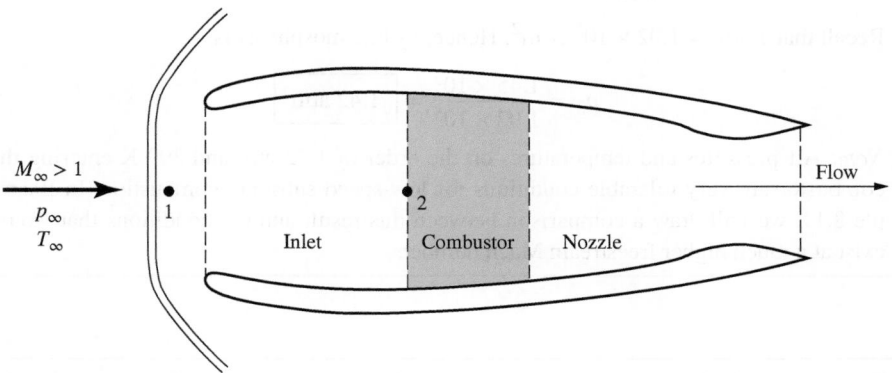

Figure 8.9 Schematic of a conventional subsonic-combustion ramjet engine.

discussed in Chapter 9.) After passing through the shock wave, the flow from point 1 to point 2, located at the entrance to the combustor, is isentropic. The ramjet is flying at Mach 2 at a standard altitude of 10 km, where the air pressure and temperature are 2.65×10^4 N/m^2 and 223.3 K, respectively. Calculate the air temperature and pressure at point 2 when the Mach number at that point is 0.2.

■ **Solution**

The total pressure and total temperature of the freestream at $M_\infty = 2$ can be obtained from Appendix A.

$$p_{0,\infty} = \left(\frac{p_{0,\infty}}{p_\infty} \right) p_\infty = (7.824)(2.65 \times 10^4) = 2.07 \times 10^5 \text{ N/m}^2$$

$$T_{0,\infty} = \left(\frac{T_{0,\infty}}{T_\infty} \right) T_\infty = (1.8)(223.3) = 401.9 \text{ K}$$

At point 1 behind the normal shock, the total pressure is, from Appendix B, for $M_\infty = 2$

$$p_{0,1} = \left(\frac{p_{0,1}}{p_{0,\infty}} \right) p_{0,\infty} = (0.7209)(2.07 \times 10^5) = 1.49 \times 10^5 \text{ N/m}^2$$

The total temperature is constant across the shock, hence

$$T_{0,1} = T_{0,\infty} = 401.9 \text{ K}$$

The flow is isentropic between points 1 and 2, hence p_0 and T_0 are constant between these points. Therefore, $p_{0,2} = 1.49 \times 10^5$ N/m^2 and $T_{0,2} = 401.9$ K. At point 2, where $M_2 = 0.2$, the ratios of the total-to-static pressure and total-to-static temperature, from Appendix A, are $p_{0,2}/p_2 = 1.028$ and $T_{0,2}/T_2 = 1.008$. Hence,

$$p_2 = \left(\frac{p_2}{p_{0,2}} \right) (p_{02}) = \frac{1.49 \times 10^5}{1.028} = \boxed{1.45 \times 10^5 \text{ N/m}^2}$$

$$T_2 = \left(\frac{T_2}{T_{0,2}} \right) (T_{02}) = \frac{401.9}{1.008} = \boxed{399 \text{ K}}$$

Recall that 1 atm $= 1.02 \times 10^5$ N/m^2. Hence, p_2 in atmospheres is

$$p_2 = \frac{1.45 \times 10^5}{1.02 \times 10^5} = \boxed{1.42 \text{ atm}}$$

Note: Air pressures and temperatures on the order of 1.42 atm and 399 K entering the combustor are very tolerable conditions for low-speed subsonic combustion. In Example 8.11, we will draw a comparison between this result and the conditions that would exist at a much higher freestream Mach number.

EXAMPLE 8.14

Repeat Example 8.13, except for a freestream Mach number $M_\infty = 10$. Assume that the ramjet has been redesigned so that the Mach number at point 2 remains equal to 0.2.

■ **Solution**

From Appendix A, for $M = 10$, we have

$$p_{0,\infty} = \left(\frac{p_{0,\infty}}{p_\infty}\right) p_\infty = (0.4244 \times 10^5)(2.65 \times 10^4) = 1.125 \times 10^9 \text{ N/m}^2$$

$$T_{0,\infty} = \left(\frac{T_{0,\infty}}{T_\infty}\right) T_\infty = (21)(223.3) = 4690 \text{ K}$$

At point 1, from Appendix B for $M_\infty = 10$, we have

$$p_{0,1} = \left(\frac{p_{0,1}}{p_{0,\infty}}\right)(p_{0,\infty}) = (0.3045 \times 10^{-2})(1.125 \times 10^9) = 3.43 \times 10^6 \text{ N/m}^2$$

and $$T_{0,1} = T_{0,\infty} = 4690 \text{ K}$$

At point 2, where $M_2 = 0.2$, we have from Example 8.8, $p_{0,2}/p_2 = 1.028$ and $T_{0,2}/T_2 = 1.008$. Also at point 2, since the flow is isentropic between points 1 and 2,

$$p_{0,2} = p_{0,1} = 3.43 \times 10^6 \text{ N/m}^2$$

$$T_{0,2} = T_{0,1} = 4690 \text{ K}$$

Hence,

$$p_2 = \left(\frac{p_2}{p_{0,2}}\right)(p_{0,2}) = \frac{3.43 \times 10^6}{1.028} = \boxed{3.34 \times 10^6 \text{ N/m}^2}$$

$$T_2 = \left(\frac{T_2}{T_{0,2}}\right)(T_{0,2}) = \frac{4690}{1.008} = \boxed{4653 \text{ K}}$$

In atmospheres,

$$p_2 = \frac{3.34 \times 10^6}{1.02 \times 10^5} = \boxed{32.7 \text{ atm}}$$

Compared to the rather benign conditions at point 2 existing for the case treated in Example 8.13, in the present example the air entering the combustor is at a pressure and temperature of 32.7 atm and 4653 K—both extremely severe conditions. The temperature is so hot that the fuel injected into the combustor will decompose rather than burn, with little or no thrust being produced. Moreover, the pressure is so high that the structural design of the combustor would have to be extremely heavy, assuming in the first place that some special heat-resistant material could be found that could handle the high temperature. In short, a conventional ramjet, where the flow is slowed down to a low subsonic Mach number before entering the combustor, *will not work at high, hypersonic Mach numbers*. The solution to this problem is not to slow the flow inside the engine to low subsonic speeds, but rather to slow it only to a lower but still supersonic speed. In this manner, the temperature and pressure increase inside the engine will be smaller and can be made tolerable. In such a ramjet, the entire flowpath through the engine remains at supersonic speed, including inside the combustor. This necessitates the injection and mixing of the fuel in a supersonic stream—a challenging technical problem. This type of ramjet, where the flow is supersonic throughout, is called a supersonic combustion ramjet—SCRAMjet for short. SCRAMjets are a current area of intense research and advanced development. In November 2005, for the first time in history, a SCRAMjet engine successfully powered a hypersonic flight vehicle, the experimental X-43 shown in Figure 9.31, achieving a Mach number of almost 10. SCRAMjet engines are the only viable airbreathing power plants for hypersonic cruise vehicles. Aspects of SCRAMjet engine design will be discussed in Chapter 9.

EXAMPLE 8.15

The pressure ratio across a normal shock wave in air is 4.5. What are the Mach numbers in front of and behind the wave? What are the density and temperature ratios across the wave?

■ **Solution**

From Appendix B, for $p_2/p_1 = 4.5$,

$$M_1 = \boxed{2} \quad \text{and} \quad M_2 = \boxed{0.5774}$$

Also, from the same table,

$$\frac{\rho_2}{\rho_1} = \boxed{2.667} \quad \text{and} \quad \frac{T_2}{T_1} = \boxed{1.687}$$

Note: For a normal shock, the specification of the pressure *ratio* across the wave uniquely determines the Mach number in front of the wave, the Mach number behind the wave, and the ratio of all other thermodynamic properties across the wave.

EXAMPLE 8.16

The temperature ratio across a normal shock wave in air is 5.8. What are the Mach numbers in front of and behind the wave? What are the density and pressure ratios across the wave?

■ **Solution**

From Appendix B, for $T_2/T_1 = 5.8$,

$$M_1 = \boxed{5} \quad \text{and} \quad M_2 = \boxed{0.4152}$$

Also, from the table,

$$\frac{\rho_2}{\rho_1} = \boxed{5} \quad \text{and} \quad \frac{p_2}{p_1} = \boxed{29}$$

Note: For a normal shock wave, the specification of the temperature *ratio* uniquely determines the Mach number in front of the wave, the Mach number behind the wave, and the ratio of all other thermodynamic properties across the wave.

EXAMPLE 8.17

The Mach number behind a normal shock wave is 0.4752. What is the Mach number in front of the wave? What are the density, pressure, and temperature ratios across the shock?

■ **Solution**

From Appendix B, for $M_2 = 0.4752$,

$$M_1 = \boxed{3} \qquad \frac{\rho_2}{\rho_1} = \boxed{3.857} \qquad \frac{p_2}{p_1} = \boxed{10.33} \qquad \frac{T_2}{T_1} = \boxed{2.679}$$

Note: For a normal shock wave, the specification of the Mach number behind the shock uniquely determines the Mach number in front of the wave and the ratios of all thermodynamic properties across the shock.

EXAMPLE 8.18

The velocity and temperature of the flow ahead of a normal shock wave are 1215 m/s and 300 K, respectively. Calculate the velocity of the flow behind the shock.

■ **Solution**

$$a_1 = \sqrt{\gamma R T_1} = \sqrt{(1.4)(287)(300)} = 347.2 \text{ m/s}$$

$$M_1 = \frac{u_1}{a_1} = \frac{1215}{347.2} = 3.5$$

From Appendix B, for $M_1 = 3.5$, $M_2 = 0.4512$ and $T_2/T_1 = 3.315$,

$$T_2 = \left(\frac{T_2}{T_1}\right) T_1 = (3.315)(300) = 994.5 \text{ K}$$

$$a_2 = \sqrt{\gamma R T_2} = \sqrt{(1.4)(287)(994.5)} = 632.1 \text{ m/s}$$

$$u_2 = M_2 \, a_2 = (0.4512)(632.1) = \boxed{285.2 \text{ m/s}}$$

Note: Unlike the previous three examples where only *one dimensionless* quantity (M_1 or M_2, or p_2/p_1, etc.) uniquely specified the shock wave, in this example two quantities are needed to specify the shock wave. This is because we were given actual dimensional quantities such as velocity in meters per second and temperature in kelvins. Just the velocity by itself will not define a specific normal shock, nor will the temperature by itself. We needed both quantities to define the specific shock. Of course, to solve this example, the first thing we did was to calculate the Mach number from the given u_1 and T_1. Emphasis is again made that Mach number, not velocity, is the powerful single quantity that specifies a particular normal shock.

EXAMPLE 8.19

The velocity and temperature behind a normal shock wave are 329 m/s and 1500 K, respectively. Calculate the velocity in front of the shock wave.

■ **Solution**

$$a_2 = \sqrt{\gamma R T_2} = \sqrt{(1.4)(287)(1500)} = 776.3 \text{ m/s}$$

$$M_2 = \frac{u_2}{a_2} = \frac{329}{776.3} = 0.4238$$

Examining Appendix B, we see there is no precise entry for $M_2 = 0.4238$; rather, this number lies between 0.4236 at $M_1 = 4.5$ and 0.4245 at $M_1 = 4.45$. By interpolation,

$$M_1 = 4.45 + \frac{(0.4245 - 0.4238)}{(0.4245 - 0.4236)}(4.5 - 4.45)$$

$$M_1 = 4.45 + 0.0389 = 4.4898$$

From Appendix B we note that $T_2/T_1 = 4.875$ at $M_2 = 0.4236$ and $T_2/T_1 = 4.788$ at $M_2 = 0.4245$. Interpolating to find T_2/T_1 at $M_2 = 0.4238$, we find

$$\frac{T_2}{T_1} = 4.788 + \frac{0.4245 - 0.4238}{0.4245 - 0.4236}(4.875 - 4.788)$$

$$\frac{T_2}{T_1} = 4.788 + 0.068 = 4.856$$

Thus,

$$T_1 = \frac{T_2}{T_2/T_1} = \frac{1500}{4.856} = 308.9$$

$$a_1 = \sqrt{\gamma R T_1} = \sqrt{(1.4)(287)(308.9)} = 352.3 \text{ m/s}$$

$$u_1 = M_1 \, a_1 = (4.489)(352.3) = \boxed{1,581.5 \text{ m/s}}$$

Note: Once again we see that a single velocity does not specify a normal shock wave. However, a velocity in combination with temperature does specify the normal shock. In contrast to Example 8.18, where u_1 and T_1 ahead of the shock specified the shock, we see that u_2 and T_2 behind the shock also are sufficient to specify the shock.

EXAMPLE 8.20

Repeat Example 8.19, but use the "nearest entry" in tables rather than interpolating between entries. Using the nearest entry is a less accurate calculation than interpolation, but it is simpler and quicker. Compare this less accurate result with the more accurate result from Example 8.19.

■ **Solution**
From Example 8.19, $M_2 = 0.4238$. The nearest entry in Appendix B is $M_2 = 0.4236$, which corresponds to $M_1 = 4.5$ and $T_2/T_1 = 4.875$. Using the nearest entry, we have

$$T_1 = \frac{T_2}{T_2/T_1} = \frac{1500}{4.875} = 307.7 \text{ K}$$

$$a_1 = \sqrt{\gamma R T_1} = \sqrt{(1.4)(287)(307.7)} = 351.6 \text{ m/s}$$

$$u_1 = M_1 \, a_1 = 4.5(351.6) = \boxed{1582 \text{ m/s}}$$

Comparing this result with that from Example 8.19, we have:

$$u_1 = 1581 \text{ m/s} \qquad \text{(interpolation)}$$
$$u_1 = 1582 \text{ m/s} \qquad \text{(nearest entry)}$$

We conclude, at least in this case, that using the nearest entry caused only a 0.06 percent error, not enough to worry about in the context of a worked example.

EXAMPLE 8.21

Consider a normal shock with an upstream Mach number of 3.53. Obtain the downstream Mach number by:

(a) Using the nearest entry in the tables.

(b) Interpolating the tabulated values.

(c) Exact analytical calculations.

Compare the accuracy of the results.

■ **Solution**
(a) The nearest entry in Appendix B is for $M_1 = 3.55$. For this entry in the tables,

$$M_2 = \boxed{0.4492}$$

(b) $M_1 = 3.53$ lies between the entries for $M_1 = 3.5$, where $M_2 = 0.4512$, and $M_1 = 3.55$, where $M_2 = 0.4492$. Interpolating to obtain M_2 corresponding to $M_1 = 3.53$, we have

$$M_2 = 0.4492 + \frac{(3.55 - 3.53)}{(3.55 - 3.5)}(0.4512 - 0.4492)$$

$$M_2 = 0.4492 + 0.0008 = \boxed{0.45}$$

(c) From Equation (8.59),

$$M_2^2 = \frac{1 + [\gamma - 1)/2]M_1^2}{\gamma M_1^2 - (\gamma - 1)/2} = \frac{1 + 0.2(3.53)^2}{(1.4)(3.53)^2 - 0.2} = 0.2025$$

Thus,

$$M_2 = (0.2025)^{1/2} = \boxed{0.45}$$

Compare the results:

(a) $M_2 = 0.4492$ (nearest entry)

(b) $M_2 = 0.45$ (interpolation)

(c) $M_2 = 0.45$ (exact)

Conclusion: For all practical purposes, all three approaches yield almost identical results.

8.6.1 Comment on the Use of Tables to Solve Compressible Flow Problems

Appendices A, B, and C provide tables for the convenient calculation of certain problems in compressible flow. Many of the previous worked examples in this chapter illustrate the usefulness of these tables. Even when you are dealing with conditions that do not correspond *exactly* to a direct entry in the tables, and in practice this is usually the case, simple linear interpolation between lines in the table gave quite accurate numbers for answers, as the results of Examples 8.19, 8.20, and 8.21 demonstrated. This accuracy is verified by calculations made with exact analytical formulae as demonstrated in Example 8.21. Various compressible flow tables have been in existence since the 1940s, and their purpose was, as it is now, to provide a quick and convenient tool for the solution of various compressible flow problems. They are particularly convenient when we adopt the method of using the nearest entry in the tables, rather than take the time to interpolate between entries. Examples 8.20 and 8.21 demonstrate that little accuracy is lost by using tables that contain many more closely spaced entries than presented in Appendices A, B, and C in this book (limited in length because of space constraints). A classic example is the compressible flow tables contained in NACA TR-1135 (Reference 115), an important and frequently used reference on the desk of most aerodynamicists working in high-speed flow.

The modern alternative to these tables is, of course, the digital computer into which the analytical equations, such as Equations (8.40), (8.42), (8.43), (8.59), (8.61), and (8.65), can easily be programmed, and the numbers in Appendices A, B, and C can be reproduced on your hand calculator. This is particularly straightforward if you have an explicit calculation, such as calculating M_2 behind a normal shock explicitly from a known value of M_1 ahead of the shock using Equation (8.59). But return to Equation (8.59) for a moment. What if you are given M_2, and you want to find M_1? This is a not-so-convenient implicit calculation in Equation (8.59), whereas, using the tables, you can immediately go to Appendix B, scan down the column for M_2, find the given value of M_2, and then find the corresponding value of M_1 by reading directly across the page.

For our purposes, for the remainder of our discussions on compressible flow in this book, we will frequently use the tables, and we will for simplicity adopt the "nearest entry" method.

8.7 MEASUREMENT OF VELOCITY IN A COMPRESSIBLE FLOW

The use of a Pitot tube for measuring the velocity of a low-speed, incompressible flow was discussed in Section 3.4. Before progressing further, return to Section 3.4, and review the principal aspects of a Pitot tube, as well as the formulae used to obtain the flow velocity from the Pitot pressure, assuming incompressible flow.

For low-speed, incompressible flow, we saw in Section 3.4 that the velocity can be obtained from a knowledge of both the total pressure and the static pressure at a point. The total pressure is measured by a Pitot tube, and the static pressure is obtained from a static pressure orifice or by some independent means. The important aspect of Section 3.4 is that the pressure sensed by a Pitot tube, along with the static pressure, is all that is necessary to extract the flow velocity for an incompressible flow. In the present section, we see that the same is true for a compressible flow, both subsonic and supersonic, if we consider the Mach number rather than the velocity. In both subsonic and supersonic compressible flows, a knowledge of the Pitot pressure and the static pressure is sufficient to calculate Mach number, although the formulae are different for each Mach-number regime. Let us examine this matter further.

8.7.1 Subsonic Compressible Flow

Consider a Pitot tube in a subsonic, compressible flow, as sketched in Figure 8.10a. As usual, the mouth of the Pitot tube (point b) is a stagnation region. Hence, a fluid element moving along streamline ab is brought to rest isentropically at point b. In turn, the pressure sensed at point b is the total pressure of the freestream, $p_{0,1}$. This is the Pitot pressure read at the end of the tube. If, in addition, we know the freestream static pressure p_1, then the Mach number in region 1 can be obtained from Equation (8.42),

$$\frac{p_{0,1}}{p_1} = \left(1 + \frac{\gamma - 1}{2} M_1^2 \right)^{\gamma/(\gamma-1)} \tag{8.42}$$

or solving for M_1^2,

$$\boxed{M_1^2 = \frac{2}{\gamma - 1} \left[\left(\frac{p_{0,1}}{p_1} \right)^{(\gamma-1)/\gamma} - 1 \right]} \tag{8.74}$$

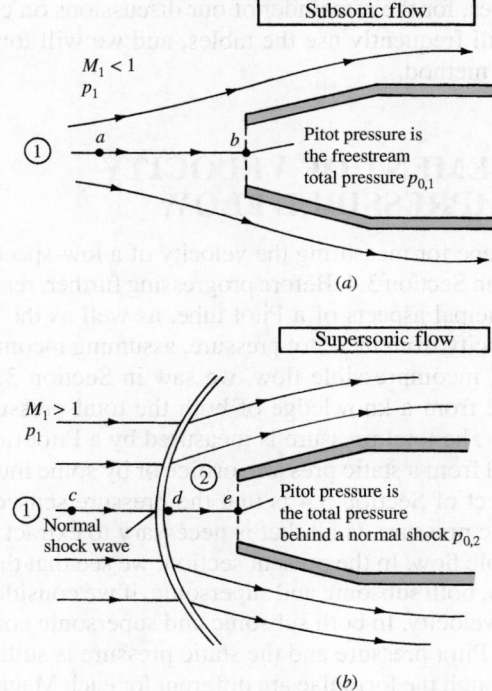

Figure 8.10 A Pitot tube in (*a*) subsonic flow and
(*b*) supersonic flow.

Clearly, from Equation (8.74), the Pitot pressure $p_{0,1}$ and the static pressure p_1 allow the direct calculation of Mach number.

The flow velocity can be obtained from Equation (8.74) by recalling that $M_1 = u_1/a_1$. Hence,

$$u_1^2 = \frac{2a_1^2}{\gamma - 1}\left[\left(\frac{p_{0,1}}{p_1}\right)^{(\gamma-1)/\gamma} - 1\right] \qquad (8.75)$$

From Equation (8.75), we see that, unlike incompressible flow, a knowledge of $p_{0,1}$ and p_1 is not sufficient to obtain u_1; we also need the freestream speed of sound, a_1.

8.7.2 Supersonic Flow

Consider a Pitot tube in a supersonic freestream, as sketched in Figure 8.10*b*. As usual, the mouth of the Pitot tube (point *e*) is a stagnation region. Hence, a fluid element moving along streamline *cde* is brought to rest at point *e*. However, because the freestream is supersonic and the Pitot tube presents an obstruction to the flow, there is a strong bow shock wave in front of the tube, much like the picture shown at the left of Figure 8.1 for supersonic flow over a blunt body. Hence, streamline *cde* crosses the normal portion of the bow shock. A fluid element moving along streamline *cde* will first be decelerated *nonisentropically* to a subsonic velocity at point *d* just behind the shock. Then it is isentropically compressed to zero velocity at point *e*. As a result, the pressure at point *e* is *not* the total pressure of the freestream but rather the total pressure *behind a normal shock wave*, $p_{0,2}$. This is the Pitot pressure read at the end of the tube. Keep in mind that because of the entropy increase across the shock, there is a loss in total pressure across the shock, $p_{0,2} < p_{0,1}$. However, knowing $p_{0,2}$ and the freestream static pressure p_1 is still sufficient to calculate the freestream Mach number M_1, as follows:

$$\frac{p_{0,2}}{p_1} = \frac{p_{0,2}}{p_2}\frac{p_2}{p_1} \tag{8.76}$$

Here, $p_{0,2}/p_2$ is the ratio of total pressure to static pressure in region 2 immediately behind the normal shock, and p_2/p_1 is the static pressure ratio across the shock. From Equation (8.42),

$$\frac{p_{0,2}}{p_2} = \left(1 + \frac{\gamma - 1}{2}M_2^2\right)^{\gamma/(\gamma-1)} \tag{8.77}$$

where, from Equation (8.59),

$$M_2^2 = \frac{1 + [(\gamma - 1)/2]M_1^2}{\gamma M_1^2 - (\gamma - 1)/2} \tag{8.78}$$

Also, from Equation (8.65),

$$\frac{p_2}{p_1} = 1 + \frac{2\gamma}{\gamma + 1}(M_1^2 - 1) \tag{8.79}$$

Substituting Equation (8.78) into (8.77), and substituting the result as well as Equation (8.79) into Equation (8.76), we obtain, after some algebraic simplification (see Problem 8.14),

$$\boxed{\frac{p_{0,2}}{p_1} = \left(\frac{(\gamma + 1)^2 M_1^2}{4\gamma M_1^2 - 2(\gamma - 1)}\right)^{\gamma/(\gamma-1)}\frac{1 - \gamma + 2\gamma M_1^2}{\gamma + 1}} \tag{8.80}$$

Equation (8.80) is called the *Rayleigh Pitot tube formula*. It relates the Pitot pressure $p_{0,2}$ and the freestream static pressure p_1 to the freestream Mach number M_1. Equation (8.80) gives M_1 as an implicit function of $p_{0,2}/p_1$ and allows the calculation of M_1 from a known $p_{0,2}/p_1$. For convenience in making calculations, the ratio $p_{0,2}/p_1$ is tabulated versus M_1 in Appendix B.

EXAMPLE 8.22

A Pitot tube is inserted into an airflow where the static pressure is 1 atm. Calculate the flow Mach number when the Pitot tube measures (*a*) 1.276 atm, (*b*) 2.714 atm, and (*c*) 12.06 atm.

■ **Solution**

First, we must assess whether the flow is subsonic or supersonic. At Mach 1, the Pitot tube would measure $p_0 = p/0.528 = 1.893p$. Hence, when $p_0 < 1.893$ atm, the flow is subsonic, and when $p_0 > 1.893$ atm, the flow is supersonic.

(a) Pitot tube measurement = 1.276 atm. The flow is subsonic. Hence, the Pitot tube is directly sensing the total pressure of the flow. From Appendix A, for $p_0/p = 1.276$,

$$\boxed{M = 0.6}$$

(b) Pitot tube measurement = 2.714 atm. The flow is supersonic. Hence, the Pitot tube is sensing the total pressure behind a normal shock wave. From Appendix B, for $p_{0,2}/p_1 = 2.714$,

$$\boxed{M_1 = 1.3}$$

(c) Pitot tube measurement = 12.06 atm. The flow is supersonic. From Appendix B, for $p_{0,2}/p_1 = 12.06$,

$$\boxed{M_1 = 3.0}$$

EXAMPLE 8.23

Consider a hypersonic missile flying at Mach 8 at an altitude of 20,000 ft, where the pressure is 973.3 lb/ft^2. The nose of the missile is blunt and is shaped like that shown at the left of Figure 8.1. Calculate the pressure at the stagnation point on the nose.

■ **Solution**

Examining the blunt body shown in Figure 8.1, the streamline that impinges at the stagnation point has traversed the normal portion of the bow shock wave. By definition, $V = 0$ at the stagnation point. Since the flow is isentropic between the shock and the body, the pressure at the stagnation point on the body is the total pressure behind a normal shock with an upstream Mach number of 8. Let us denote the pressure at the stagnation point by p_s. Since $p_{0,2}$ is the total pressure behind the normal shock, then $p_s = p_{0,2}$. From Appendix B, for Mach 8, $p_{0,2}/p_1 = 82.87$. Hence,

$$p_s = p_{0,2} = \left(\frac{p_{0,2}}{p_1}\right)(p_1) = 82.87(973.3) = \boxed{8.07 \times 10^4 \text{ lb/ft}^2}$$

Since 1 atm = 2116 lb/ft^2,

$$p_s = \frac{8.07 \times 10^4}{2116} = \boxed{38.1 \text{ atm}}$$

Note that the pressure at the nose of the missile is quite high—38.1 atm. This is typical of hypersonic flight at low altitude.

Check on the calculation This problem can also be solved by first calculating the upstream total pressure from Appendix A, and then using the total pressure ratio across the normal shock from Appendix B. From Appendix A for Mach 8, $p_{0,1}/p_1 = 0.9763 \times 10^4$. Hence,

$$p_{0,1} = \left(\frac{p_{0,1}}{p_1}\right) p_1 = (0.9763 \times 10^4)973.3 = 9.502 \times 10^6$$

From Appendix B for Mach 8, $p_{0,2}/p_{0,1} = 8.8488 \times 10^{-2}$. Hence,

$$p_s = p_{0,2} = \left(\frac{p_{0,2}}{p_{0,1}}\right) p_{0,1} = (0.8488 \times 10^{-2})(9.502 \times 10^6) = \boxed{8.07 \times 10^4 \text{ lb/ft}^2}$$

This is the same result as obtained earlier.

EXAMPLE 8.24

Consider the Lockheed SR-71 Blackbird shown in Figure 8.11 flying at a standard altitude of 25 km. The pressure measured by a Pitot tube on this airplane is 3.88×10^4 N/m². Calculate the velocity of the airplane.

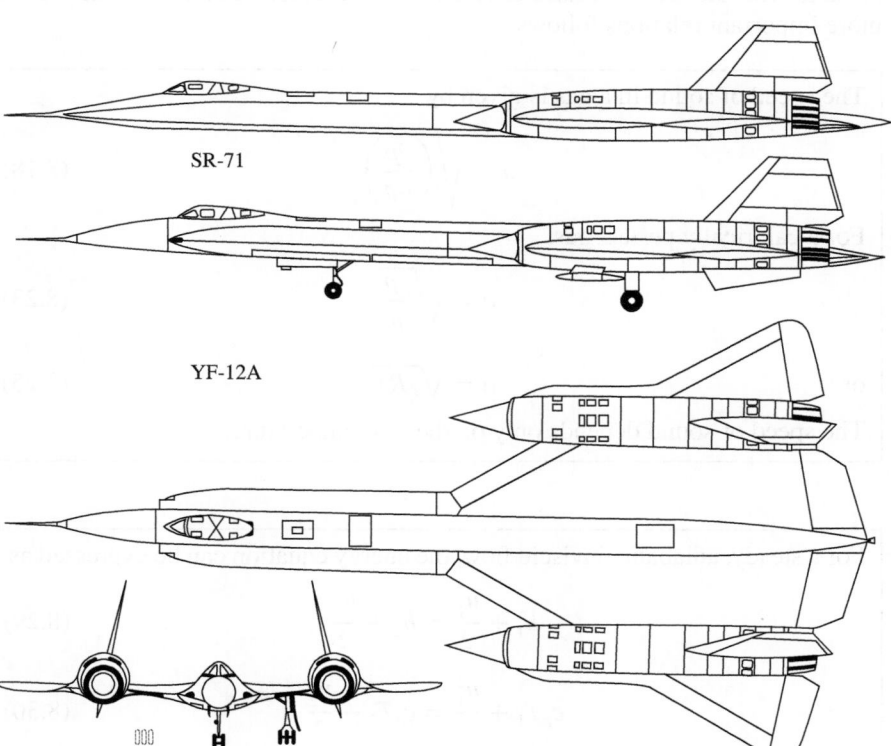

SR-71

YF-12A

Figure 8.11 The Lockheed SR-71/YF-12A Blackbird.

■ **Solution**

From Appendix D, at an altitude of 25 km, $p = 2.5273 \times 10^3$ N/m^2 and $T = 216.66$ K. Hence,

$$\frac{p_{0,2}}{p_1} = \frac{3.88 \times 10^4}{2.5273 \times 10^3} = 15.35$$

From Appendix B, for $p_{0,2}/p_1 = 15.35$, $M_1 = 3.4$:

$$a_1 = \sqrt{\gamma RT} = \sqrt{(1.4)(287)(216.66)} = 295 \text{ m/s}.$$

Thus, the velocity of the airplane is

$$V_1 = M_1 a_1 = (3.4)(295) = \boxed{1003 \text{ m/s}}$$

8.8 SUMMARY

Return to the road map given in Figure 8.2, and make certain that you are comfortable with the areas we have covered in this chapter. A brief summary of the more important relations follows.

The speed of sound in a gas is given by

$$a = \sqrt{\left(\frac{\partial p}{\partial \rho}\right)_s} \tag{8.18}$$

For a calorically perfect gas,

$$a = \sqrt{\frac{\gamma p}{\rho}} \tag{8.23}$$

or

$$a = \sqrt{\gamma RT} \tag{8.25}$$

The speed of sound depends only on the gas temperature.

For a steady, adiabatic, inviscid flow, the energy equation can be expressed as

$$h_1 + \frac{u_1^2}{2} = h_2 + \frac{u_2^2}{2} \tag{8.29}$$

$$c_p T_1 + \frac{u_1^2}{2} = c_p T_2 + \frac{u_2^2}{2} \tag{8.30}$$

(*continued*)

$$\frac{a_1^2}{\gamma - 1} + \frac{u_1^2}{2} = \frac{a_2^2}{\gamma - 1} + \frac{u_2^2}{2} \tag{8.32}$$

$$\frac{a^2}{\gamma - 1} + \frac{u^2}{2} = \frac{a_0^2}{\gamma - 1} \tag{8.33}$$

$$\frac{a^2}{\gamma - 1} + \frac{u^2}{2} = \frac{\gamma + 1}{2(\gamma - 1)} a^{*2} \tag{8.35}$$

Total conditions in a flow are related to static conditions via

$$c_p T + \frac{u^2}{2} = c_p T_0 \tag{8.38}$$

$$\frac{T_0}{T} = 1 + \frac{\gamma - 1}{2} M^2 \tag{8.40}$$

$$\frac{p_0}{p} = \left(1 + \frac{\gamma - 1}{2} M^2\right)^{\gamma/(\gamma-1)} \tag{8.42}$$

$$\frac{\rho_0}{\rho} = \left(1 + \frac{\gamma - 1}{2} M^2\right)^{1/(\gamma-1)} \tag{8.43}$$

Note that the ratios of total to static properties are a function of local Mach number only. These functions are tabulated in Appendix A.

The basic normal shock equations are

Continuity: $\rho_1 u_1 = \rho_2 u_2$ \hfill (8.2)

Momentum: $p_1 + \rho_1 u_1^2 = p_2 + \rho_2 u_2^2$ \hfill (8.6)

Energy: $h_1 + \dfrac{u_1^2}{2} = h_2 + \dfrac{u_2^2}{2}$ \hfill (8.10)

These equations lead to relations for changes across a normal shock as a function of upstream Mach number M_1 only:

$$M_2^2 = \frac{1 + [(\gamma - 1)/2]M_1^2}{\gamma M_1^2 - (\gamma - 1)/2} \tag{8.59}$$

(continued)

$$\frac{\rho_2}{\rho_1} = \frac{u_1}{u_2} = \frac{(\gamma + 1)M_1^2}{2 + (\gamma - 1)M_1^2} \qquad (8.61)$$

$$\frac{p_2}{p_1} = 1 + \frac{2\gamma}{\gamma + 1}(M_1^2 - 1) \qquad (8.65)$$

$$\frac{T_2}{T_1} = \frac{h_2}{h_1} = \left[1 + \frac{2\gamma}{\gamma + 1}(M_1^2 - 1)\right] \frac{2 + (\gamma - 1)M_1^2}{(\gamma + 1)M_1^2} \qquad (8.67)$$

$$s_2 - s_1 = c_p \ln \left\{ \left[1 + \frac{2\gamma}{\gamma + 1}(M_1^2 - 1)\right] \frac{2 + (\gamma - 1)M_1^2}{(\gamma + 1)M_1^2} \right\}$$

$$- R \ln \left[1 + \frac{2\gamma}{\gamma + 1}(M_1^2 - 1)\right] \qquad (8.68)$$

$$\frac{p_{0,2}}{p_{0,1}} = e^{-(s_2 - s_1)/R} \qquad (8.73)$$

The normal shock properties are tabulated versus M_1 in Appendix B.

For a calorically perfect gas, the total temperature is constant across a normal shock wave:

$$T_{0,2} = T_{0,1}$$

However, there is a loss in total pressure across the wave:

$$p_{0,2} < p_{0,1}$$

For subsonic and supersonic compressible flow, the freestream Mach number is determined by the ratio of Pitot pressure to freestream static pressure. However, the equations are different:

Subsonic flow: $\quad M_1^2 = \frac{2}{\gamma - 1} \left[\left(\frac{p_{0,1}}{p_1}\right)^{(\gamma-1)/\gamma} - 1 \right]$

$$(8.74)$$

Supersonic flow: $\quad \frac{p_{0,2}}{p_1} = \left[\frac{(\gamma + 1)^2 M_1^2}{4\gamma M_1^2 - 2(\gamma - 1)} \right]^{\gamma/(\gamma-1)} \frac{1 - \gamma + 2\gamma M_1^2}{\gamma + 1} \qquad (8.80)$

8.9 PROBLEMS

8.1 Consider air at a temperature of 230 K. Calculate the speed of sound.

8.2 The temperature in the reservoir of a supersonic wind tunnel is 519 °R. In the test section, the flow velocity is 1385 ft/s. Calculate the test-section Mach number. Assume the tunnel flow is adiabatic.

8.3 At a given point in a flow, $T = 300$ K, $p = 1.2$ atm, and $V = 250$ m/s. At this point, calculate the corresponding values of p_0, T_0, p^*, T^*, and M^*.

8.4 At a given point in a flow, $T = 700$ °R, $p = 1.6$ atm, and $V = 2983$ ft/s. At this point, calculate the corresponding values of p_0, T_0, p^*, T^*, and M^*.

8.5 Consider the isentropic flow through a supersonic nozzle. If the test-section conditions are given by $p = 1$ atm, $T = 230$ K, and $M = 2$, calculate the reservoir pressure and temperature.

8.6 Consider the isentropic flow over an airfoil. The freestream conditions correspond to a standard altitude of 10,000 ft and $M_\infty = 0.82$. At a given point on the airfoil, $M = 1.0$. Calculate p and T at this point. (*Note:* You will have to use the standard atmosphere table in Appendix E for this problem.)

8.7 The flow just upstream of a normal shock wave is given by $p_1 = 1$ atm, $T_1 = 288$ K, and $M_1 = 2.6$. Calculate the following properties just downstream of the shock: p_2, T_2, ρ_2, M_2, $p_{0,2}$, $T_{0,2}$, and the change in entropy across the shock.

8.8 The pressure upstream of a normal shock wave is 1 atm. The pressure and temperature downstream of the wave are 10.33 atm and 1390 °R, respectively. Calculate the Mach number and temperature upstream of the wave and the total temperature and total pressure downstream of the wave.

8.9 The entropy increase across a normal shock wave is 199.5 J/(kg · K). What is the upstream Mach number?

8.10 The flow just upstream of a normal shock wave is given by $p_1 = 1800$ lb/ft^2, $T_1 = 480$ °R, and $M_1 = 3.1$. Calculate the velocity and M^* behind the shock.

8.11 Consider a flow with a pressure and temperature of 1 atm and 288 K. A Pitot tube is inserted into this flow and measures a pressure of 1.555 atm. What is the velocity of the flow?

8.12 Consider a flow with a pressure and temperature of 2116 lb/ft^2 and 519 °R, respectively. A Pitot tube is inserted into this flow and measures a pressure of 7712.8 lb/ft^2. What is the velocity of this flow?

8.13 Repeat Problems 8.11 and 8.12 using (incorrectly) Bernoulli's equation for incompressible flow. Calculate the percent error induced by using Bernoulli's equation.

8.14 Derive the Rayleigh Pitot tube formula, Equation (8.80).

8.15 On March 16, 1990, an Air Force SR-71 set a new continental speed record, averaging a velocity of 2112 mi/h at an altitude of 80,000 ft.

Calculate the temperature (in degrees Fahrenheit) at a stagnation point on the vehicle.

8.16 In the test section of a supersonic wind tunnel, a Pitot tube in the flow reads a pressure of 1.13 atm. A static pressure measurement (from a pressure tap on the sidewall of the test section) yields 0.1 atm. Calculate the Mach number of the flow in the test section.

8.17 When the Apollo command module returned to earth from the moon, it entered the earth's atmosphere at a Mach number of 36. Using the results from the present chapter for a calorically perfect gas with the ratio of specific heats equal to 1.4, predict the gas temperature at the stagnation point of the Apollo spacecraft at Mach 36 at an altitude where the freestream temperature is 300 K. Comment on the validity of your answer.

8.18 The stagnation temperature on the Apollo vehicle at Mach 36 as it entered the atmosphere was 11,000 K, a much different value than predicted in Problem 8.17 for the case of a calorically perfect gas with a ratio of specific heats equal to 1.4. The difference is due to chemical reactions that occur in air at these high temperatures—dissociation and ionization. The analyses in this book assuming a calorically perfect gas with constant specific heats are not valid for such chemically reacting flows. However, as an engineering approximation, the calorically perfect gas results are sometimes applied with a lower value of the ratio of specific heats, a so-called "effective gamma," in order to try to simulate the effects of high temperature chemically reacting flows. For the condition stated in this problem, calculate the value of the effective gamma necessary to yield a temperature of 11,000 K at the stagnation point. Assume the freestream temperature is 300 K.

8.19 Prove that the total pressure is constant throughout an isentropic flow.

8.20 Plot p_2/p_1, T_2/T_1, p_{02}/p_{01}, T_{02}/T_{01}, and M_2 from $M_1 = 1$ to $M_1 = 10$ assuming that the gas is air ($\gamma = 1.4$) and Krypton ($\gamma = 1.68$). What is the effect of changing the specific heat ratio of the gas? How accurate do you think your results at Mach 10 are?

8.21 Plot p_0/p for a pitot-static system from $M_1 = 0$ to $M_1 = 5$.

Oblique Shock and Expansion Waves

In the case of air (and the same is true for all gases) the shock wave is extremely thin so that calculations based on one-dimensional flow are still applicable for determining the changes in velocity and density on passing through it, even when the rest of the flow system is not limited to one dimension, provided that only the velocity component normal to the wave is considered.

G. I. Taylor and J. W. Maccoll, 1934

PREVIEW BOX

Take a look at Figure 9.1. What you see is the computed wave pattern—both shock and expansion waves—generated by a generic supersonic transport configuration flying at Mach 1.7 at an altitude of 15 km. All these waves are *oblique* to the flow, in contrast to the normal shock waves discussed in Chapter 8. The present chapter is all about *oblique* shock and expansion waves.

The material in this chapter is vital to a fundamental understanding of supersonic flow. Moreover, it is vital to you if you are interested in designing an economically feasible and environmentally acceptable supersonic transport. The shock waves in Figure 9.1 create the major source of drag (wave drag) on the airplane, and the waves, when they propagate to the ground, cause the much discussed "sonic boom." The material in this chapter is also vital to you if you are interested in designing SCRAMjet engines for

hypersonic airplanes. The performance of such engines depends in part on the nature of the oblique wave patterns both upstream of and inside the engine. The material in this chapter is vital to you if you are interested in designing supersonic and hypersonic wind tunnels, where the oblique wave patterns created by models in the tunnel and in the diffuser downstream of the models affect the performance of the tunnel. In fact, the material in this chapter is vital to a whole host of applications in supersonic flow.

By now you get the message—the material in this chapter is simply vital to your study of supersonic flows. It is the bread and butter of such flows. So lay out the bread, spread it with butter, and consume this chapter. And on top of everything else, learning about oblique shock and expansion waves is exciting. I predict that you are going to enjoy this.

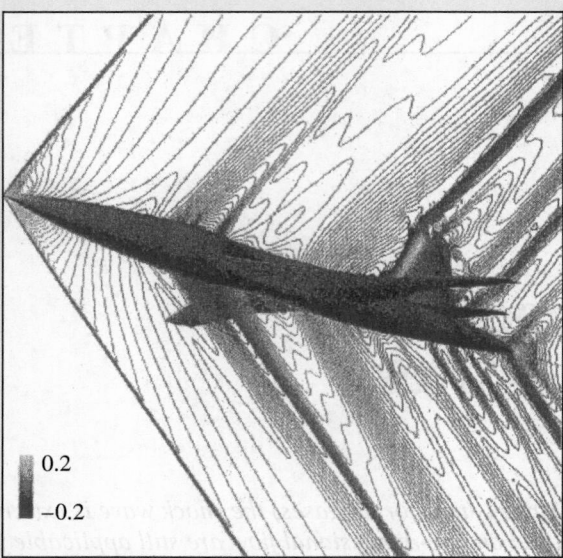

0.2
−0.2

Figure 9.1 Wave pattern on a supersonic transport
configuration (without nacelles) at Mach 1.7.
(*Source:* Computational fluid dynamic calculations by
Y. Makino et al., "Nonaxisymmetrical Fuselage Shape
Modification for Drag Reduction of Low-Sonic-Boom
Airplane," *AIAA Journal,* vol. 41, no. 8, August 2003,
p. 1415.)

9.1 INTRODUCTION

In Chapter 8, we discussed normal shock waves, that is, shock waves that make an
angle of 90° with the upstream flow. The behavior of normal shock waves is important; moreover, the study of normal shock waves provides a relatively straightforward introduction to shock-wave phenomena. However, examining Figure 7.5
and the photographs shown in Figure 7.6, we see that, in general, a shock wave
will make an oblique angle with respect to the upstream flow. These are called
oblique shock waves and are the subject of part of this chapter. A normal shock
wave is simply a special case of the general family of oblique shocks, namely, the
case where the wave angle is 90°.

In addition to oblique shock waves, where the pressure increases discontinuously across the wave, supersonic flows are also characterized by oblique
expansion waves, where the pressure *decreases continuously* across the wave.
Let us examine these two types of waves further. Consider a supersonic flow over
a wall with a corner at point *A*, as sketched in Figure 9.2. In Figure 9.2*a*, the

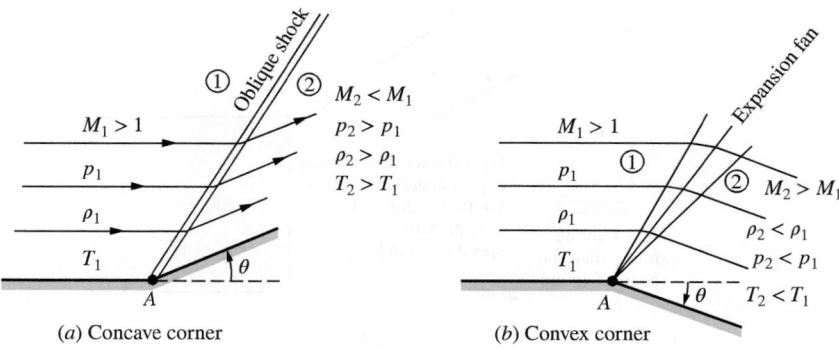

(*a*) Concave corner   (*b*) Convex corner

Figure 9.2 Supersonic flow over a corner.

wall is turned upward at the corner through the deflection angle θ; that is, the corner is concave. The flow at the wall must be tangent to the wall; hence, the streamline at the wall is also deflected upward through the angle θ. The bulk of the gas is above the wall, and in Figure 9.2*a*, the streamlines are turned upward, *into* the main bulk of the flow. Whenever a supersonic flow is "turned into itself" as shown in Figure 9.2*a*, an oblique shock wave will occur. The originally horizontal streamlines ahead of the wave are uniformly deflected in crossing the wave, such that the streamlines behind the wave are parallel to each other and inclined upward at the deflection angle θ. Across the wave, the Mach number discontinuously decreases, and the pressure, density, and temperature discontinuously increase. In contrast, Figure 9.2*b* shows the case where the wall is turned downward at the corner through the deflection angle θ; that is, the corner is convex. Again, the flow at the wall must be tangent to the wall; hence, the streamline at the wall is deflected downward through the angle θ. The bulk of the gas is above the wall, and in Figure 9.2*b*, the streamlines are turned downward, *away from* the main bulk of the flow. Whenever a supersonic flow is "turned away from itself" as shown in Figure 9.2*b*, an expansion wave will occur. This expansion wave is in the shape of a fan centered at the corner. The fan continuously opens in the direction away from the corner, as shown in Figure 9.2*b*. The originally horizontal streamlines ahead of the expansion wave are deflected smoothly and continuously through the expansion fan such that the streamlines behind the wave are parallel to each other and inclined downward at the deflection angle θ. Across the expansion wave, the Mach number increases, and the pressure, temperature, and density decrease. Hence, an expansion wave is the direct antithesis of a shock wave.

Oblique shock and expansion waves are prevalent in two- and three-dimensional supersonic flows. These waves are inherently two-dimensional in nature, in contrast to the one-dimensional normal shock waves discussed in Chapter 8. That is, in Figure 9.2*a* and *b*, the flow-field properties are a function of *x* and *y*. The purpose of the present chapter is to determine and study the properties of these oblique waves.

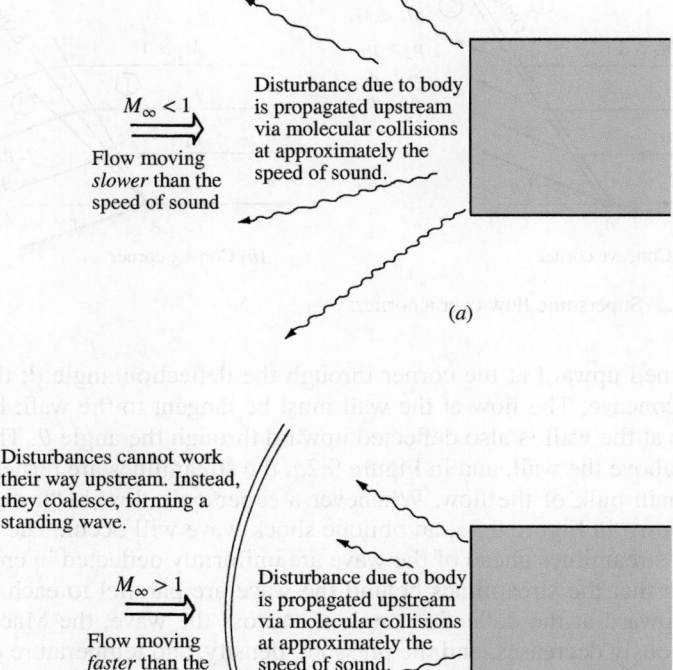

Figure 9.3 Propagation of disturbances. (*a*) Subsonic flow. (*b*) Supersonic flow.

What is the physical mechanism that creates waves in a supersonic flow? To address this question, recall our picture of the propagation of a sound wave via molecular collisions, as portrayed in Section 8.3. If a slight disturbance takes place at some point in a gas, information is transmitted to other points in the gas by sound waves which propagate in all directions away from the source of the disturbance. Now consider a body in a flow, as sketched in Figure 9.3. The gas molecules which impact the body surface experience a change in momentum. In turn, this change is transmitted to neighboring molecules by random molecular collisions. In this fashion, information about the presence of the body attempts to be transmitted to the surrounding flow via molecular collisions; that is, the information is propagated upstream at approximately the local speed of sound. If the upstream flow is subsonic, as shown in Figure 9.3*a*, the disturbances have no

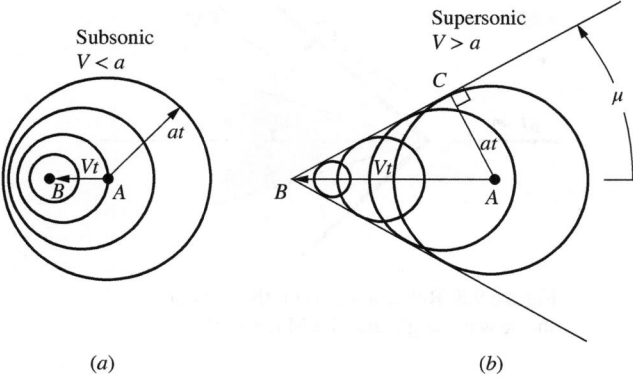

Figure 9.4 Another way of visualizing the propagation of disturbances in (*a*) subsonic and (*b*) supersonic flow.

problem working their way far upstream, thus giving the incoming flow plenty of time to move out of the way of the body. On the other hand, if the upstream flow is supersonic, as shown in Figure 9.3*b*, the disturbances cannot work their way upstream; rather, at some finite distance from the body, the disturbance waves pile up and coalesce, forming a standing wave in front of the body. Hence, the physical generation of waves in a supersonic flow—both shock and expansion waves—is due to the propagation of information via molecular collisions and due to the fact that such propagation cannot work its way into certain regions of the supersonic flow.

Why are most waves oblique rather than normal to the upstream flow? To answer this question, consider a small source of disturbance moving through a stagnant gas. For lack of anything better, let us call this disturbance source a "beeper," which periodically emits sound. First, consider the beeper moving at *subsonic* speed through the gas, as shown in Figure 9.4*a*. The speed of the beeper is V, where $V < a$. At time $t = 0$, the beeper is located at point A; at this point, it emits a sound wave that propagates in all directions at the speed of sound, a. At a later time t, this sound wave has propagated a distance at from point A and is represented by the circle of radius as shown in Figure 9.4*a*. During the same time, the beeper has moved a distance Vt and is now at point B in Figure 9.4*a*. Moreover, during its transit from A to B, the beeper has emitted several other sound waves, which at time t are represented by the smaller circles in Figure 9.4*a*. Note that the beeper always stays *inside* the family of circular sound waves and that the waves continuously move ahead of the beeper. This is because the beeper is traveling at a subsonic speed $V < a$. In contrast, consider the beeper moving at a *supersonic* speed $V > a$ through the gas, as shown in Figure 9.4*b*. At time $t = 0$, the beeper is located at point A, where it emits a sound wave. At a later time t, this sound wave has propagated a distance at from point A and is represented by the circle of radius at shown in Figure 9.4*b*. During the same time, the beeper

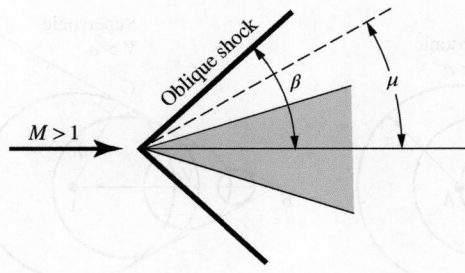

Figure 9.5 Relation between the oblique
shock-wave angle and the Mach angle.

has moved a distance Vt to point B. Moreover, during its transit from A to B, the
beeper has emitted several other sound waves, which at time t are represented by
the smaller circles in Figure 9.4b. However, in contrast to the subsonic case, the
beeper is now constantly *outside* the family of circular sound waves; that is, it is
moving ahead of the wave fronts because $V > a$. Moreover, something new is
happening; these wave fronts form a disturbance envelope given by the straight
line BC, which is tangent to the family of circles. This line of disturbances is defined as a *Mach wave*. In addition, the angle ABC that the Mach wave makes with
respect to the direction of motion of the beeper is defined as the *Mach angle* μ.
From the geometry of Figure 9.4b, we readily find that

$$\sin \mu = \frac{at}{Vt} = \frac{a}{V} = \frac{1}{M}$$

Thus, the Mach angle is simply determined by the local Mach number as

$$\mu = \sin^{-1} \frac{1}{M} \qquad (9.1)$$

Examining Figure 9.4b, the Mach wave, that is, the envelope of disturbances in the
supersonic flow, is clearly *oblique* to the direction of motion. If the disturbances
are stronger than a simple sound wave, then the wave front becomes stronger than
a Mach wave, creating an oblique shock wave at an angle β to the freestream,
where $\beta > \mu$. This comparison is shown in Figure 9.5. However, the physical
mechanism creating the oblique shock is essentially the same as that described
above for the Mach wave. Indeed, a Mach wave is a limiting case for oblique
shock (i.e., it is an infinitely weak oblique shock).

   This finishes our discussion of the physical source of oblique waves in a
supersonic flow. Let us now proceed to develop the equations that allow us to
calculate the change in properties across these oblique waves, first for oblique
shock waves, and then for expansion waves. In the process, we follow the road
map given in Figure 9.6.

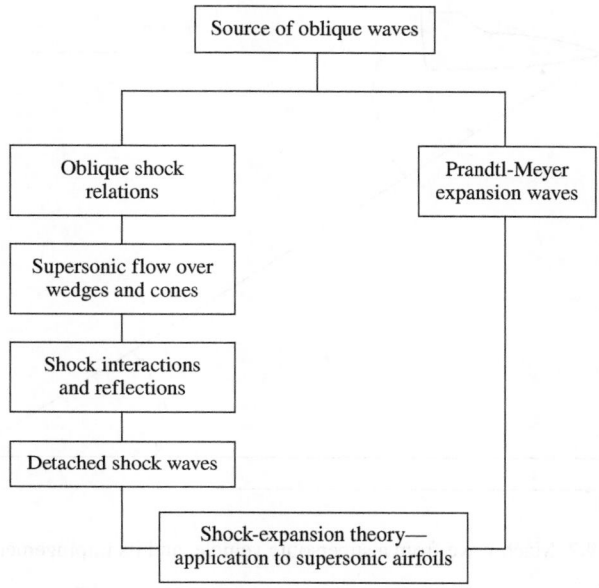

Figure 9.6 Road map for Chapter 9.

A supersonic airplane is flying at Mach 2 at an altitude of 16 km. Assume the shock wave pattern from the airplane (see Figure 9.1) quickly coalesces into a Mach wave that intersects the ground behind the airplane, causing a "sonic boom" to be heard by a bystander on the ground. At the instant the sonic boom is heard, how far ahead of the bystander is the airplane?

■ **Solution**

Examine Figure 9.7, which shows the airplane at an altitude of 16 km with a Mach wave trailing behind it. The Mach wave intersects the ground at a ground distance d from the airplane. From Equation (9.1),

$$\mu = \sin^{-1}\left(\frac{1}{M}\right) = \sin^{-1}\left(\frac{1}{2}\right) = 30°$$

From Figure 9.7,

$$\tan \mu = \frac{16 \text{ km}}{d}$$

or,

$$d = \frac{16 \text{ km}}{\tan \mu} = \frac{16}{0.577} = \boxed{27.7 \text{ km}}$$

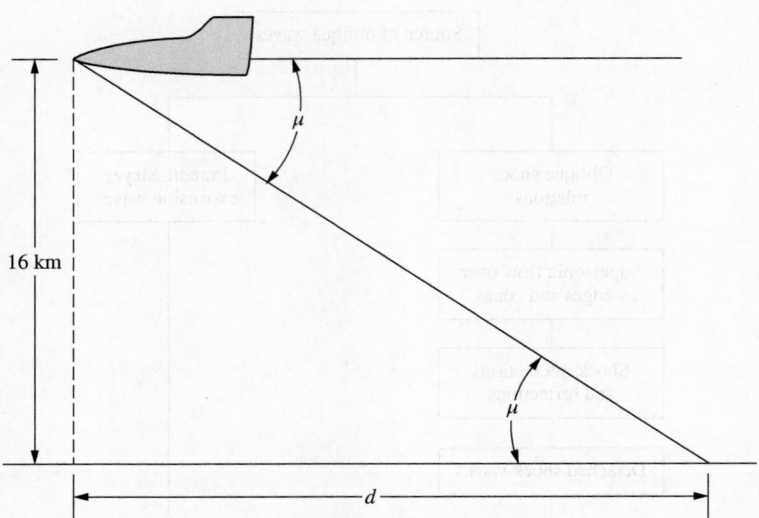

Figure 9.7 Mach wave from a supersonic vehicle, and its impingement on the ground.

9.2 OBLIQUE SHOCK RELATIONS

Consider the oblique shock wave sketched in Figure 9.8. The angle between the shock wave and the upstream flow direction is defined as the *wave angle,* denoted by β. The upstream flow (region 1) is horizontal, with a velocity V_1 and Mach number M_1. The downstream flow (region 2) is inclined upward through the deflection angle θ and has velocity V_2 and Mach number M_2. The upstream velocity V_1 is split into components tangential and normal to the shock wave, w_1 and u_1, respectively, with the associated tangential and normal Mach numbers $M_{t,1}$ and $M_{n,1}$, respectively. Similarly, the downstream velocity is split into tangential and normal components w_2 and u_2, respectively, with the associated Mach numbers $M_{t,2}$ and $M_{n,2}$.

Consider the control volume shown by the dashed lines in the upper part of Figure 9.8. Sides a and d are parallel to the shock wave. Segments b and c follow the upper streamline, and segments e and f follow the lower streamline. Let us apply the integral form of the conservation equations to this control volume, keeping in mind that we are dealing with a steady, inviscid, adiabatic flow with no body forces. For these assumptions, the continuity equation, Equation (2.48), becomes

$$\oiint_S \rho \mathbf{V} \cdot \mathbf{dS} = 0$$

This surface integral evaluated over faces a and d yields $-\rho_1 u_1 A_1 + \rho_2 u_2 A_2$, where $A_1 = A_2 =$ area of faces a and d. The faces b, c, e, and f are parallel to

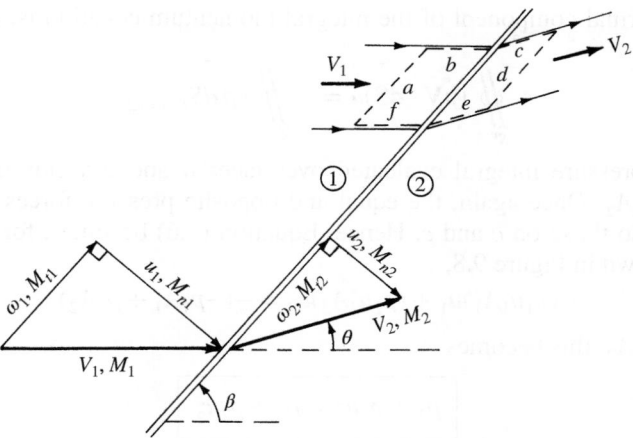

Figure 9.8 Oblique shock geometry.

the velocity, and hence contribute nothing to the surface integral (i.e., $\mathbf{V} \cdot \mathbf{dS} = 0$ for these faces). Thus, the continuity equation for an oblique shock wave is

$$-\rho_1 u_1 A_1 + \rho_2 u_2 A_2 = 0$$

or

$$\boxed{\rho_1 u_1 = \rho_2 u_2} \tag{9.2}$$

Keep in mind that u_1 and u_2 in Equation (9.2) are *normal* to the shock wave.

The integral form of the momentum equation, Equation (2.64), is a vector equation. Hence, it can be resolved into two components, tangential and normal to the shock wave. First, consider the tangential component, keeping in mind the type of flow we are considering:

$$\oiint_S (\rho \mathbf{V} \cdot \mathbf{dS}) w = -\oiint_S (p\, dS)_{\text{tangential}} \tag{9.3}$$

In Equation (9.3), w is the component of velocity tangential to the wave. Since $\mathbf{dS}$ is perpendicular to the control surface, then $(p\, dS)_{\text{tangential}}$ over faces a and d is zero. Also, since the vectors $p\, \mathbf{dS}$ on faces b and f are equal and opposite, the pressure integral in Equation (9.3) involves two tangential forces that cancel each other over faces b and f. The same is true for faces c and e. Hence, Equation (9.3) becomes

$$-(\rho_1 u_1 A_1) w_1 + (\rho_2 u_2 A_2) w_2 = 0 \tag{9.4}$$

Dividing Equation (9.4) by Equation (9.2), we have

$$\boxed{w_1 = w_2} \tag{9.5}$$

Equation (9.5) is an important result; it states that the *tangential component of the flow velocity is constant across an oblique shock.*

The normal component of the integral momentum equation is, from Equation (2.64),

$$\oiint_S (\rho \mathbf{V} \cdot \mathbf{dS})u = -\oiint_S (p\, dS)_{\text{normal}} \qquad (9.6)$$

Here, the pressure integral evaluated over faces a and d yields the net sum $-p_1 A_1 + p_2 A_2$. Once again, the equal and opposite pressure forces on b and f cancel, as do those on c and e. Hence, Equation (9.6) becomes, for the control volume shown in Figure 9.8,

$$-(\rho_1 u_1 A_1)u_1 + (\rho_2 u_2 A_2)u_2 = -(-p_1 A_1 + p_2 A_2)$$

Since $A_1 = A_2$, this becomes

$$\boxed{p_1 + \rho_1 u_1^2 = p_2 + \rho_2 u_2^2} \qquad (9.7)$$

Again, note that the only velocities appearing in Equation (9.7) are the components *normal* to the shock.

Finally, consider the integral form of the energy equation, Equation (2.95). For our present case, this can be written as

$$\oiint_S \rho\left(e + \frac{V^2}{2}\right)\mathbf{V} \cdot \mathbf{dS} = -\oiint_S p\mathbf{V} \cdot \mathbf{dS} \qquad (9.8)$$

Again noting that the flow is tangent to faces b, c, f, and e, and hence $\mathbf{V} \cdot \mathbf{dS} = 0$ on these faces, Equation (9.8) becomes, for the control volume in Figure 9.6,

$$-\rho_1\left(e_1 + \frac{V_1^2}{2}\right)u_1 A_1 + \rho_2\left(e_2 + \frac{V_2^2}{2}\right)u_2 A_2 = -(-p_1 u_1 A_1 + p_2 u_2 A_2) \qquad (9.9)$$

Collecting terms in Equation (9.9), we have

$$-\rho_1 u_1\left(e_1 + \frac{p_1}{\rho_1} + \frac{V_1^2}{2}\right) + \rho_2 u_2\left(e_2 + \frac{p_2}{\rho_2} + \frac{V_2^2}{2}\right) = 0$$

or

$$\rho_1 u_1\left(h_1 + \frac{V_1^2}{2}\right) = \rho_2 u_2\left(h_2 + \frac{V_2^2}{2}\right) \qquad (9.10)$$

Dividing Equation (9.10) by (9.2), we have

$$h_1 + \frac{V_1^2}{2} = h_2 + \frac{V_2^2}{2} \qquad (9.11)$$

Since $h + V^2/2 = h_0$, we have again the familiar result that the *total enthalpy is constant across the shock wave*. Moreover, for a calorically perfect gas, $h_0 = c_p T_0$; hence, the *total temperature is constant across the shock wave*. Carrying Equation (9.11) a bit further, note from Figure 9.8 that $V^2 = u^2 + w^2$. Also, from Equation (9.5), we know that $w_1 = w_2$. Hence,

$$V_1^2 - V_2^2 = \left(u_1^2 + w_1^2\right) - \left(u_2^2 + w_2^2\right) = u_1^2 - u_2^2$$

Thus, Equation (9.11) becomes

$$\boxed{h_1 + \frac{u_1^2}{2} = h_2 + \frac{u_2^2}{2}}$$

(9.12)

Let us now gather our results. Look carefully at Equations (9.2), (9.7), and (9.12). They are the continuity, normal momentum, and energy equations, respectively, for an oblique shock wave. Note that they involve the *normal components only* of velocity u_1 and u_2; the tangential component w does not appear in these equations. Hence, we deduce that *changes across an oblique shock wave are governed only by the component of velocity normal to the wave.*

Again, look hard at Equations (9.2), (9.7), and (9.12). *They are precisely the governing equations for a normal shock wave,* as given by Equations (8.2), (8.6), and (8.10). Hence, precisely the same algebra as applied to the normal shock equations in Section 8.6, when applied to Equations (9.2), (9.7), and (9.12), will lead to identical expressions for changes across an oblique shock in terms of the normal component of the upstream Mach number $M_{n,1}$. Note that

$$M_{n,1} = M_1 \sin \beta$$

(9.13)

Hence, for an oblique shock wave, with $M_{n,1}$ given by Equation (9.13), we have, from Equations (8.59), (8.61), and (8.65),

$$M_{n,2}^2 = \frac{1 + [(\gamma - 1)/2]M_{n,1}^2}{\gamma M_{n,1}^2 - (\gamma - 1)/2}$$

(9.14)

$$\frac{\rho_2}{\rho_1} = \frac{(\gamma + 1)M_{n,1}^2}{2 + (\gamma - 1)M_{n,1}^2}$$

(9.15)

$$\frac{p_2}{p_1} = 1 + \frac{2\gamma}{\gamma + 1}(M_{n,1}^2 - 1)$$

(9.16)

The temperature ratio T_2/T_1 follows from the equation of state:

$$\frac{T_2}{T_1} = \frac{p_2}{p_1}\frac{\rho_1}{\rho_2}$$

(9.17)

Note that $M_{n,2}$ is the *normal* Mach number behind the shock wave. The downstream Mach number itself, M_2, can be found from $M_{n,2}$ and the geometry of Figure 9.8 as

$$M_2 = \frac{M_{n,2}}{\sin(\beta - \theta)}$$

(9.18)

Examine Equations (9.14) to (9.17). They state that oblique shock-wave properties in a calorically perfect gas depend only on the normal component of the upstream Mach number $M_{n,1}$. However, note from Equation (9.13) that $M_{n,1}$ depends on both M_1 and β. Recall from Section 8.6 that changes across a normal shock wave depend on one parameter only—the upstream Mach number M_1.

In contrast, we now see that changes across an oblique shock wave depend on two parameters—say, M_1 and β. However, this distinction is slightly moot because in reality a normal shock wave is a special case of oblique shocks where $\beta = \pi/2$.

Equation (9.18) introduces the deflection angle θ into our oblique shock analysis; we need θ to be able to calculate M_2. However, θ is not an independent, third parameter; rather, θ is a function of M_1 and β, as derived below. From the geometry of Figure 9.8,

$$\tan \beta = \frac{u_1}{w_1} \qquad (9.19)$$

and

$$\tan(\beta - \theta) = \frac{u_2}{w_2} \qquad (9.20)$$

Dividing Equation (9.20) by (9.19), recalling that $w_1 = w_2$, and invoking the continuity equation, Equation (9.2), we obtain

$$\frac{\tan(\beta - \theta)}{\tan \beta} = \frac{u_2}{u_1} = \frac{\rho_1}{\rho_2} \qquad (9.21)$$

Combining Equation (9.21) with Equations (9.13) and (9.15), we obtain

$$\frac{\tan(\beta - \theta)}{\tan \beta} = \frac{2 + (\gamma - 1)M_1^2 \sin^2 \beta}{(\gamma + 1)M_1^2 \sin^2 \beta} \qquad (9.22)$$

which gives θ as an implicit function of M_1 and β. After some trigonometric substitutions and rearrangement, Equation (9.22) can be cast explicitly for θ as

$$\tan \theta = 2 \cot \beta \frac{M_1^2 \sin^2 \beta - 1}{M_1^2(\gamma + \cos 2\beta) + 2} \qquad (9.23)$$

Equation (9.23) is an important equation. It is called the θ-β-M relation, and it specifies θ as a unique function of M_1 and β. This relation is vital to the analysis of oblique shock waves, and results from it are plotted in Figure 9.9 for $\gamma = 1.4$. Examine this figure closely. It is a plot of wave angle versus deflection angle, with the Mach number as a parameter. The results given in Figure 9.9 are plotted in some detail—this is a chart which you will need to use for solving oblique shock problems.

Figure 9.9 illustrates a wealth of physical phenomena associated with oblique shock waves. For example:

1. For any given upstream Mach number M_1, there is a maximum deflection angle θ_{max}. If the physical geometry is such that $\theta > \theta_{max}$, then no solution exists for a *straight* oblique shock wave. Instead, nature establishes a curved shock wave, detached from the corner or the nose of a body. This is illustrated in Figure 9.10. Here, the left side of the figure illustrates flow over a wedge and a concave corner where the deflection angle is less than θ_{max} for the given upstream Mach number. Therefore, we see a straight oblique shock wave attached to the nose of the wedge and to the corner. The right side of Figure 9.10 gives the case where the deflection angle is greater

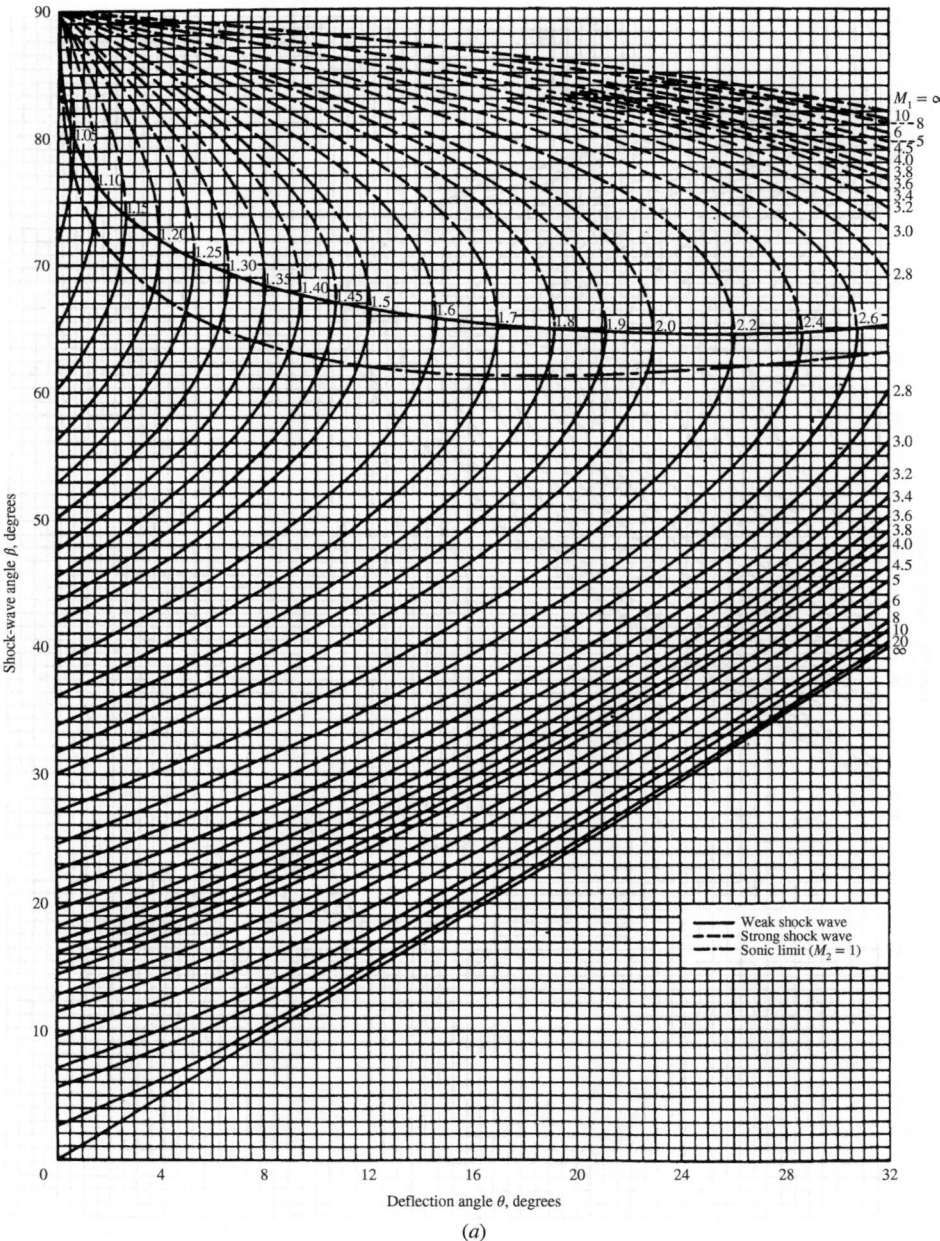

(a)

Figure 9.9 Oblique shock properties: $\gamma = 1.4$. The θ-β-M diagram. (*Source: NACA Report 1135, Ames Research Staff, "Equations, Tables and Charts for Compressible Flow," 1953.*)

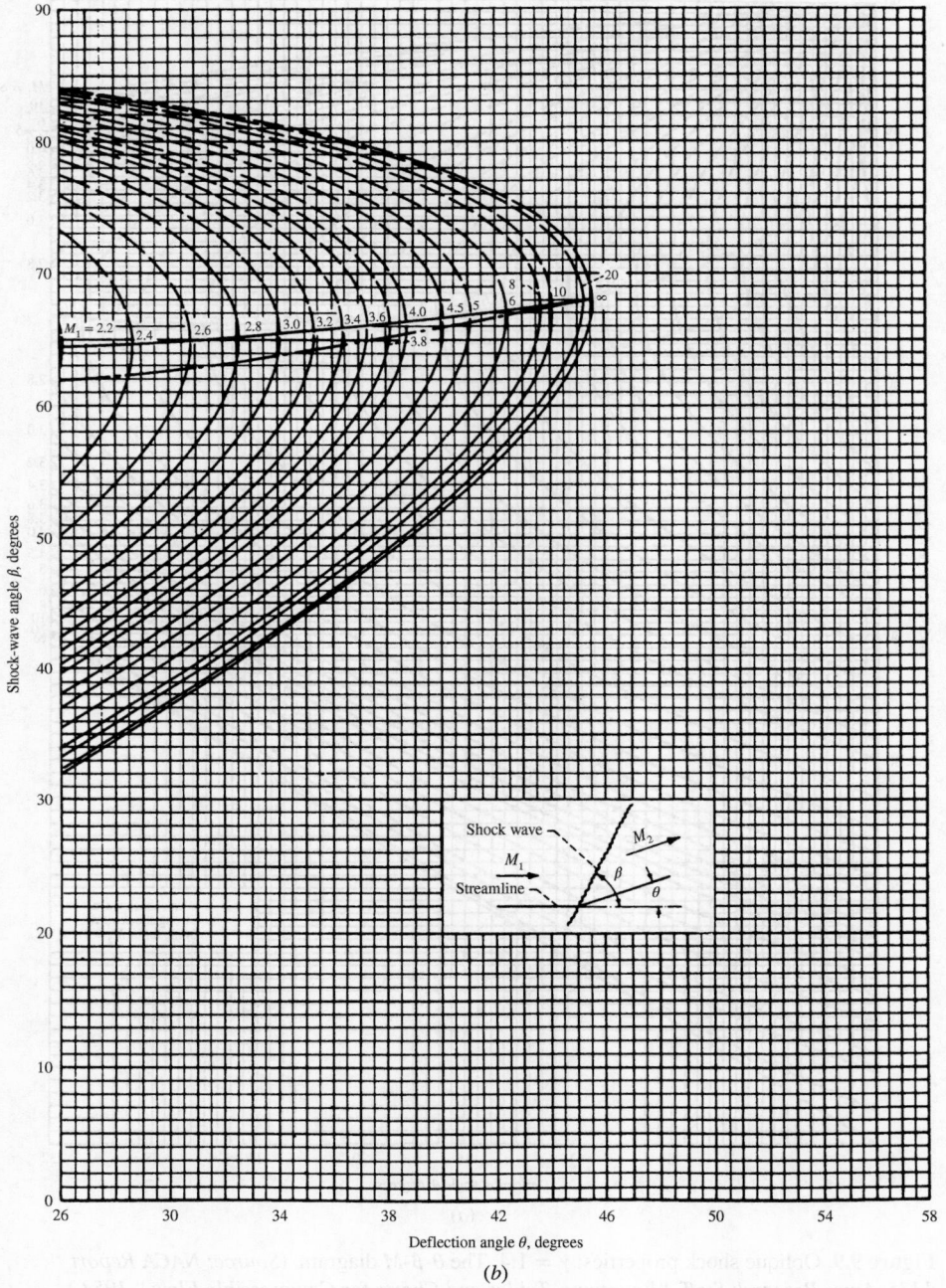

Shock-wave angle β, degrees

Deflection angle θ, degrees

(b)

Figure 9.9 (*continued*) (*Source: NACA Report 1135, Ames Research Staff, "Equations, Tables and Charts for Compressible Flow," 1953.*)

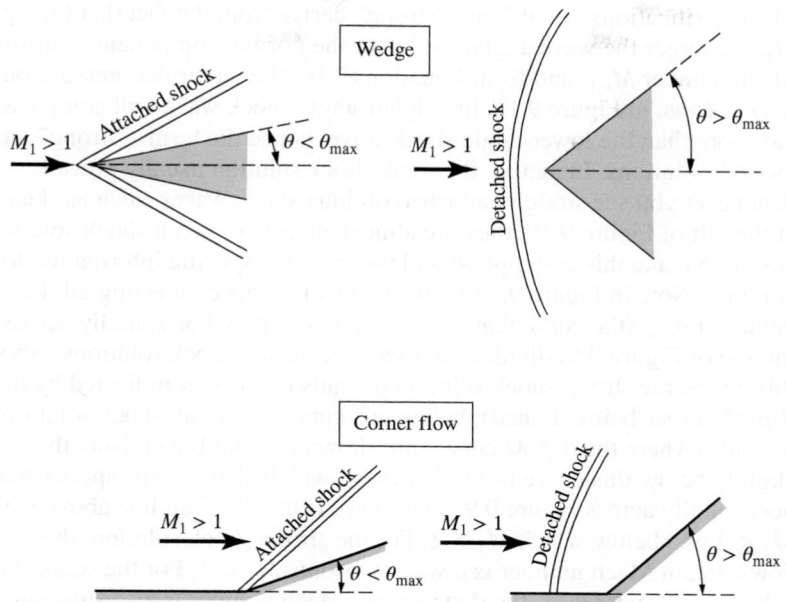

Figure 9.10 Attached and detached shocks.

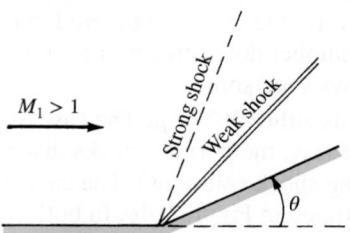

Figure 9.11 The weak and strong shock cases.

than θ_{max}; hence, there is no allowable straight oblique shock solution from the theory developed earlier in this section. Instead, we have a curved shock wave detached from the nose of the wedge or from the corner. Return to Figure 9.9, and note that the value of θ_{max} increases with increasing M_1. Hence, at higher Mach numbers, the straight oblique shock solution can exist at higher deflection angles. However, there is a limit; as M_1 approaches infinity, θ_{max} approaches 45.5° (for $\gamma = 1.4$).

2. For any given θ *less than* θ_{max}, there are *two* straight oblique shock solutions for a given upstream Mach number. For example, if $M_1 = 2.0$ and $\theta = 15°$, then from Figure 9.9, β can equal either 45.3 or 79.8°. The smaller value of β is called the *weak* shock solution, and the larger value of β is the *strong* shock solution. These two cases are illustrated in Figure 9.11.

The classifications "weak" and "strong" derive from the fact that for a given M_1, the larger the wave angle, the larger the normal component of upstream Mach number $M_{n,1}$, and from Equation (9.16) the larger the pressure ratio p_2/p_1. Thus, in Figure 9.11, the higher-angle shock wave will compress the gas more than the lower-angle shock wave, hence the terms "strong" and "weak" solutions. In nature, the weak shock solution usually prevails. Whenever you see straight, attached oblique shock waves, such as sketched at the left of Figure 9.10, they are almost always the weak shock solution. It is safe to make this assumption, unless you have specific information to the contrary. Note in Figure 9.9 that the locus of points connecting all the values of θ_{max} (the curve that sweeps approximately horizontally across the middle of Figure 9.9) divides the weak and strong shock solutions. Above this curve, the strong shock solution prevails (as further indicated by the θ-β-M curves being dashed); below this curve, the weak shock solution prevails (where the θ-β-M curves are shown as solid lines). Note that slightly below this curve is another curve which also sweeps approximately horizontally across Figure 9.9. This curve is the dividing line above which $M_2 < 1$ and below which $M_2 > 1$. For the strong shock solution, the downstream Mach number is always subsonic $M_2 < 1$. For the weak shock solution very near θ_{max}, the downstream Mach number is also subsonic, but barely so. For the vast majority of cases involving the weak shock solution, the downstream Mach number is supersonic $M_2 > 1$. Since the weak shock solution is almost always the case encountered in nature, we can readily state that the Mach number downstream of a straight, attached oblique shock is almost always supersonic.

3. If $\theta = 0$, then β equals either 90° or μ. The case of $\beta = 90°$ corresponds to a normal shock wave (i.e., the normal shocks discussed in Chapter 8 belong to the family of strong shock solutions). The case of $\beta = \mu$ corresponds to the Mach wave illustrated in Figure 9.4*b*. In both cases, the flow streamlines experience no deflection across the wave.

4. (In all of the following discussions, we consider the weak shock solution exclusively, unless otherwise noted.) Consider an experiment where we have supersonic flow over a wedge of given semiangle θ, as sketched in Figure 9.12. Now assume that we increase the freestream Mach number M_1. As M_1 increases, we observe that β decreases. For example, consider $\theta = 20°$ and $M_1 = 2.0$, as shown on the left of Figure 9.12. From Figure 9.9, we find that $\beta = 53.3°$. Now assume M_1 is increased to 5, keeping θ constant at 20°, as sketched on the right of Figure 9.12. Here, we find that $\beta = 29.9°$. Interestingly enough, although this shock is at a lower wave angle, it is a stronger shock than the one on the left. This is because $M_{n,1}$ is larger for the case on the right. Although β is smaller, which decreases $M_{n,1}$, the upstream Mach number M_1 is larger, which increases $M_{n,1}$ by an amount which more than compensates for the decreased β. For example, note the values of $M_{n,1}$ and p_2/p_1 given in Figure 9.12. Clearly,

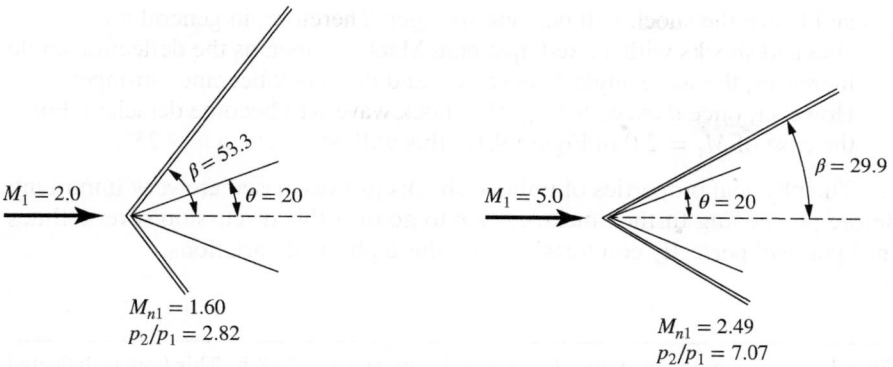

Figure 9.12 Effects of increasing the upstream Mach number.

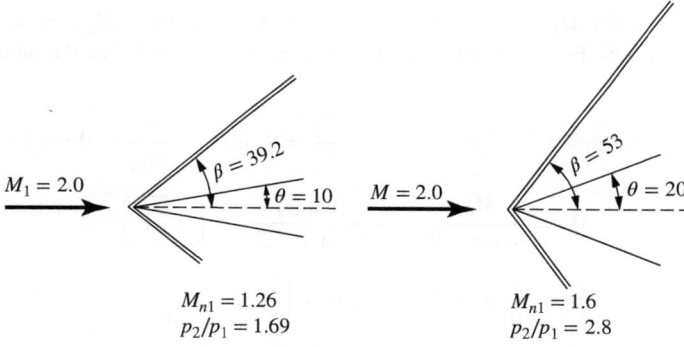

Figure 9.13 Effect of increasing the deflection angle.

the Mach 5 case on the right yields the stronger shock wave. Hence, in general for attached shocks with a fixed deflection angle, as the upstream Mach number M_1 increases, the wave angle β decreases, and the shock wave becomes stronger. Going in the other direction, as M_1 decreases, the wave angle increases, and the shock becomes weaker. Finally, if M_1 is decreased enough, the shock wave will become detached. For the case of $\theta = 20°$ shown in Figure 9.12, the shock will be detached for $M_1 < 1.84$.

5. Consider another experiment. Here, let us keep M_1 fixed and increase the deflection angle. For example, consider the supersonic flow over a wedge shown in Figure 9.13. Assume that we have $M_1 = 2.0$ and $\theta = 10°$, as sketched at the left of Figure 9.13. The wave angle will be 39.2° (from Figure 9.9). Now assume that the wedge is hinged so that we can increase its deflection angle, keeping M_1 constant. In such a case, the wave angle will increase, as shown on the right of Figure 9.13. Also, $M_{n,1}$ will increase,

and hence the shock will become stronger. Therefore, in general for attached shocks with a fixed upstream Mach number, as the deflection angle increases, the wave angle β increases, and the shock becomes stronger. However, once θ exceeds θ_{max}, the shock wave will become detached. For the case of $M_1 = 2.0$ in Figure 9.13, this will occur when $\theta > 23°$.

The physical properties of oblique shocks just discussed are very important. Before proceeding further, make certain to go over this discussion several times until you feel perfectly comfortable with these physical variations.

EXAMPLE 9.2

Consider a supersonic flow with $M = 2$, $p = 1$ atm, and $T = 288$ K. This flow is deflected at a compression corner through $20°$. Calculate M, p, T, p_0, and T_0 behind the resulting oblique shock wave.

■ Solution

From Figure 9.9, for $M_1 = 2$ and $\theta = 20°$, $\beta = 53.4°$. Hence, $M_{n,1} = M_1 \sin \beta = 2 \sin 53.4° = 1.606$. From Appendix B, for $M_{n,1} = 1.60$ (rounded to the nearest table entry),

$$M_{n,2} = 0.6684 \qquad \frac{p_2}{p_1} = 2.82 \qquad \frac{T_2}{T_1} = 1.388 \qquad \frac{p_{0,2}}{p_{0,1}} = 0.8952$$

Hence,

$$M_2 = \frac{M_{n,2}}{\sin(\beta - \theta)} = \frac{0.6684}{\sin(53.4 - 20)} = \boxed{1.21}$$

$$p_2 = \frac{p_2}{p_1} p_1 = 2.82(1 \text{ atm}) = \boxed{2.82 \text{ atm}}$$

$$T_2 = \frac{T_2}{T_1} T_1 = 1.388(288) = \boxed{399.7 \text{ K}}$$

For $M_1 = 2$, from Appendix A, $p_{0,1}/p_1 = 7.824$ and $T_{0,1}/T_1 = 1.8$; thus,

$$p_{0,2} = \frac{p_{0,2}}{p_{0,1}} \frac{p_{0,1}}{p_1} p_1 = 0.8952(7.824)(1 \text{ atm}) = \boxed{7.00 \text{ atm}}$$

The total temperature is constant across the shock. Hence,

$$T_{0,2} = T_{0,1} = \frac{T_{0,1}}{T_1} T_1 = 1.8(288) = \boxed{518.4 \text{ K}}$$

Note: For oblique shocks, the entry for $p_{0,2}/p_1$ in Appendix B *cannot* be used to obtain $p_{0,2}$; this entry in Appendix B is for normal shocks only and is obtained directly from Equation (8.80). In turn, Equation (8.80) is derived using (8.77), where M_2 is the *actual* flow Mach number, not the normal component. Only in the case of a normal shock is this also the Mach number normal to the wave. Hence, Equation (8.80) holds only for normal shocks; it *cannot* be used for oblique shocks with M_1 replaced by $M_{n,1}$. For example, an *incorrect* calculation would be to use

$p_{0,2}/p_1 = 3.805$ for $M_{n,1} = 1.60$. This gives $p_{0,2} = 3.805$ atm, a totally incorrect result compared with the correct value of 7.00 atm obtained above.

EXAMPLE 9.3

Consider an oblique shock wave with a wave angle of 30°. The upstream flow Mach number is 2.4. Calculate the deflection angle of the flow, the pressure and temperature ratios across the shock wave, and the Mach number behind the wave.

■ **Solution**

From Figure 9.9, for $M_1 = 2.4$ and $\beta = 30°$, we have $\boxed{\theta = 6.5°}$. Also,

$$M_{n,1} = M_1 \sin \beta = 2.4 \sin 30° = 1.2$$

From Appendix B,
$$\frac{p_2}{p_1} = \boxed{1.513}$$

$$\frac{T_2}{T_1} = \boxed{1.128}$$

$$M_{n,2} = 0.8422$$

Thus,
$$M_2 = \frac{M_{n,2}}{\sin(\beta - \theta)} = \frac{0.8422}{\sin(30 - 6.5)} = \boxed{2.11}$$

Note: Two aspects are illustrated by this example:

1. This is a fairly weak shock wave—only a 51 percent increase in pressure across the wave. Indeed, examining Figure 9.9, we find that this case is close to that of a Mach wave, where $\mu = \sin^{-1}(1/M) = \sin^{-1}(\frac{1}{2.4}) = 24.6°$. The shock-wave angle of 30° is not much larger than μ; the deflection angle of 6.5° is also small—consistent with the relative weakness of the shock wave.

2. Only two properties need to be specified in order to define uniquely a given oblique shock wave. In this example, M_1 and β were those two properties. In Example 9.2, the specified M_1 and θ were the two properties. Once any two properties about the oblique shock are specified, the shock is uniquely defined. This is analogous to the case of a normal shock wave studied in Chapter 8. There, we proved that all the changes across a normal shock wave were uniquely defined by specifying only *one* property, such as M_1. However, implicit in all of Chapter 8 was an additional property, namely, the wave angle of a normal shock wave is 90°. Of course, a normal shock is simply one example of the whole spectrum of oblique shocks, namely, a shock with $\beta = 90°$. An examination of Figure 9.9 shows that the normal shock belongs to the family of strong shock solutions, as discussed earlier.

EXAMPLE 9.4

Consider an oblique shock wave with $\beta = 35°$ and a pressure ratio $p_2/p_1 = 3$. Calculate the upstream Mach number.

■ **Solution**

From Appendix B, for $p_2/p_1 = 3$, $M_{n,1} = 1.64$ (nearest entry). Since

$$M_{n,1} = M_1 \sin \beta$$

then
$$M_1 = \frac{M_{n,1}}{\sin \beta} = \frac{1.66}{\sin 35°} = \boxed{2.86}$$

Note: Once again, the oblique shock is uniquely defined by two properties, in this case β and p_2/p_1.

EXAMPLE 9.5

Consider a Mach 3 flow. It is desired to slow this flow to a subsonic speed. Consider two separate ways of achieving this: (1) the Mach 3 flow is slowed by passing directly through a normal shock wave; (2) the Mach 3 flow first passes through an oblique shock with a 40° wave angle, and then subsequently through a normal shock. These two cases are sketched in Figure 9.14. Calculate the ratio of the final total pressure values for the two cases, that is, the total pressure behind the normal shock for case 2 divided by the total pressure behind the normal shock for case 1. Comment on the significance of the result.

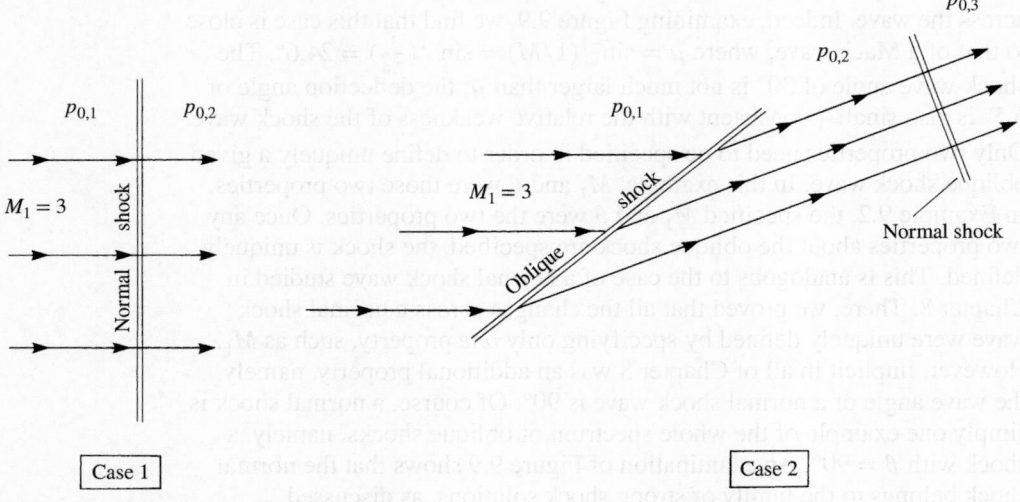

Figure 9.14 Illustration for Example 9.4.

■ **Solution**

For case 1, at $M = 3$, we have, from Appendix B,

$$\left(\frac{p_{0_2}}{p_{0_1}}\right)_{\text{case 1}} = 0.3283$$

For case 2, we have $M_{n,1} = M_1 \sin \beta = 3 \sin 40° = 1.93$. From Appendix B,

$$\frac{p_{0_2}}{p_{0_1}} = 0.7535 \quad \text{and} \quad M_{n,2} = 0.588$$

From Figure 9.9, for $M_1 = 3$ and $\beta = 40°$, we have the deflection angle $\theta = 22°$. Hence,

$$M_2 = \frac{M_{n,2}}{\sin(\beta - \theta)} = \frac{0.588}{\sin(40 - 22)} = 1.90$$

From Appendix B, for a normal shock with an upstream Mach number of 1.9, we have $p_{0_3}/p_{0_2} = 0.7674$. Thus, for case 2,

$$\left(\frac{p_{0_3}}{p_{0_1}}\right)_{\text{case 2}} = \left(\frac{p_{0_2}}{p_{0_1}}\right)\left(\frac{p_{0_3}}{p_{0_2}}\right) = (0.7535)(0.7674) = 0.578$$

Hence, $\quad \left(\frac{p_{0_3}}{p_{0_1}}\right)_{\text{case 2}} \bigg/ \left(\frac{p_{0_2}}{p_{0_1}}\right)_{\text{case 1}} = \frac{0.578}{0.3283} = \boxed{1.76}$

The result of Example 9.5 shows that the final total pressure is 76 percent higher for the case of the multiple shock system (case 2) in comparison to the single normal shock (case 1). In principle, the total pressure is an indicator of how much useful work can be done by the gas; this is described later in Section 10.4. Everything else being equal, the higher the total pressure, the more useful is the flow. Indeed, losses of total pressure are an index of the *efficiency* of a fluid flow— the lower the total pressure loss, the more efficient is the flow process. In this example, case 2 is more efficient in slowing the flow to subsonic speeds than case 1 because the loss in total pressure across the multiple shock system of case 2 is actually less than that for case 1 with a single, strong, normal shock wave. The physical reason for this is straightforward. The loss in total pressure across a normal shock wave becomes particularly severe as the upstream Mach number increases; a glance at the $p_{0,2}/p_{0,1}$ column in Appendix B attests to this. If the Mach number of a flow can be reduced *before* passing through a normal shock, the loss in total pressure is much less because the normal shock is weaker. This is the function of the oblique shock in case 2, namely, to reduce the Mach number of the flow before passing through the normal shock. Although there is a total pressure loss across the oblique shock also, it is much less than across a normal shock at the same upstream Mach number. The net effect of the oblique shock reducing the flow Mach number before passing through the normal shock than makes up for the total pressure loss across the oblique shock, with the beneficial result that the multiple shock system in case 2 produces a *smaller* loss in total pressure than a single normal shock at the same freestream Mach number.

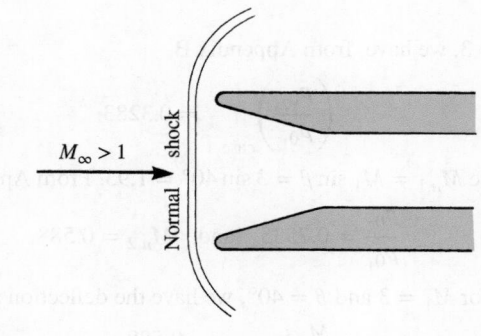

(*a*) Normal shock inlet

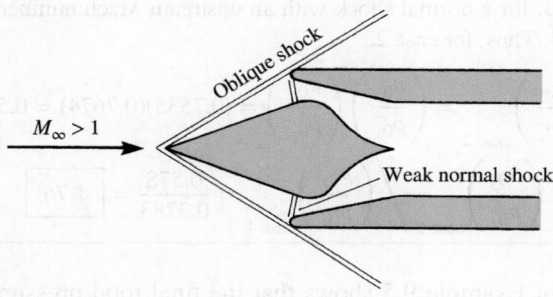

(*b*) Oblique shock inlet

Figure 9.15 Illustration of (*a*) normal shock inlet and (*b*) oblique shock inlet.

A practical application of these results is in the design of supersonic inlets for jet engines. A normal shock inlet is sketched in Figure 9.15*a*. Here, a normal shock forms ahead of the inlet, with an attendant large loss in total pressure. In contrast, an oblique shock inlet is sketched in Figure 9.15*b*. Here, a central cone creates an oblique shock wave, and the flow subsequently passes through a relatively weak normal shock at the lip of the inlet. For the same flight conditions (Mach number and altitude), the total pressure loss for the oblique shock inlet is less than for a normal shock inlet. Hence, everything else being equal, the resulting engine thrust will be higher for the oblique shock inlet. This, of course, is why most modern supersonic aircraft have oblique shock inlets.

9.3 SUPERSONIC FLOW OVER WEDGES AND CONES

For the supersonic flow over wedges, as shown in Figures 9.12 and 9.13, the oblique shock theory developed in Section 9.2 is an *exact* solution of the flow field; no simplifying assumptions have been made. Supersonic flow over a wedge

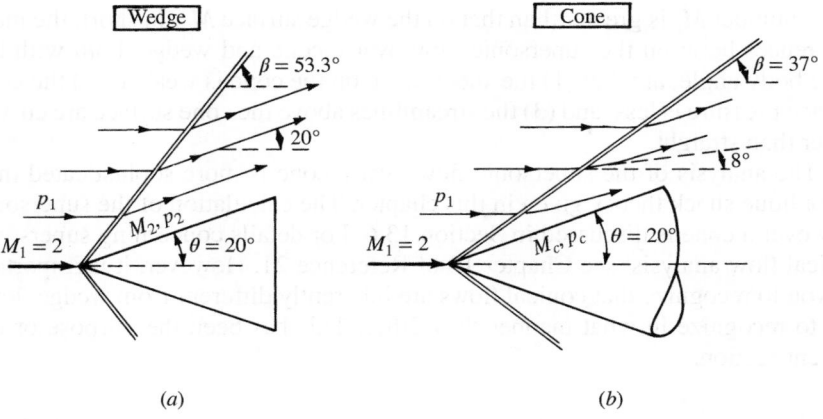

Figure 9.16 Relation between wedge and cone flow; illustration of the three-dimensional relieving effect.

is characterized by an attached, straight oblique shock wave from the nose, a uniform flow downstream of the shock with streamlines parallel to the wedge surface, and a surface pressure equal to the static pressure behind the oblique shock p_2. These properties are summarized in Figure 9.16a. Note that the wedge is a two-dimensional profile; in Figure 9.16a, it is a section of a body that stretches to plus or minus infinity in the direction perpendicular to the page. Hence, wedge flow is, by definition, two-dimensional flow, and our two-dimensional oblique shock theory fits this case nicely.

In contrast, consider the supersonic flow over a cone, as sketched in Figure 9.16b. There is a straight oblique shock which emanates from the tip, just as in the case of a wedge, but the similarity stops there. Recall from Chapter 6 that flow over a three-dimensional body experiences a "three-dimensional relieving effect." That is, in comparing the wedge and cone in Figure 9.16, both with the same 20° angle, the flow over the cone has an extra dimension in which to move, and hence it more easily adjusts to the presence of the conical body in comparison to the two-dimensional wedge. One consequence of this three-dimensional relieving effect is that the shock wave on the cone is weaker than on the wedge; that is, it has a smaller wave angle, as compared in Figure 9.16. Specifically, the wave angles for the wedge and cone are 53.3 and 37°, respectively, for the same body angle of 20° and the same upstream Mach number of 2.0. In the case of the wedge (Figure 9.16a), the streamlines are deflected by exactly 20° through the shock wave, and hence downstream of the shock the flow is exactly parallel to the wedge surface. In contrast, because of the weaker shock on the cone, the streamlines are deflected by only 8° through the shock, as shown in Figure 9.16b. Therefore, between the shock wave and the cone surface, the streamlines must gradually curve upward in order to accommodate the 20° cone. Also, as a consequence of the three-dimensional relieving effect, the pressure on the surface of the cone, p_c, is less than the wedge surface pressure p_2, and the cone surface

Mach number M_c is greater than that on the wedge surface M_2. In short, the main differences between the supersonic flow over a cone and wedge, both with the same body angle, are that (1) the shock wave on the cone is weaker, (2) the cone surface pressure is less, and (3) the streamlines above the cone surface are curved rather than straight.

The analysis of the supersonic flow over a cone is more sophisticated than the oblique shock theory given in this chapter. The calculation of the supersonic flow over a cone is discussed in Section 13.6. For details concerning supersonic conical flow analysis, see Chapter 10 of Reference 21. However, it is important for you to recognize that conical flows are inherently different from wedge flows and to recognize in what manner they differ. This has been the purpose of the present section.

EXAMPLE 9.6

Consider a wedge with a 15° half angle in a Mach 5 flow, as sketched in Figure 9.17. Calculate the drag coefficient for this wedge. (Assume that the pressure over the base is equal to freestream static pressure, as shown in Figure 9.17.)

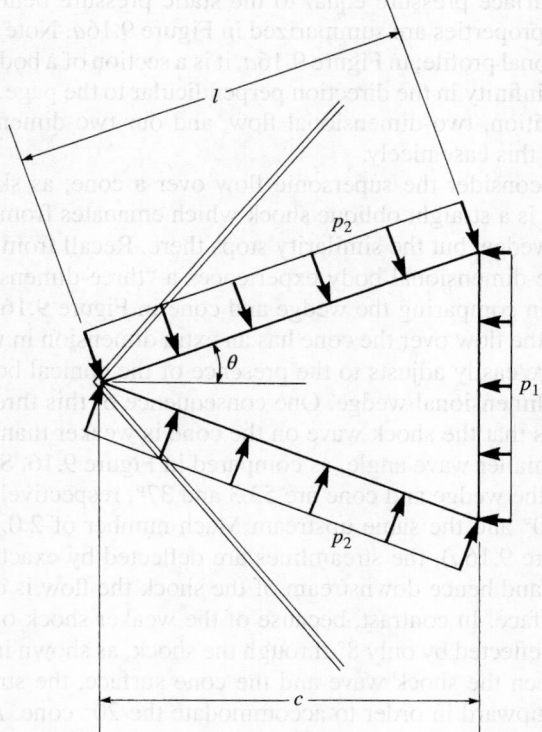

Figure 9.17 Illustration for Example 9.6.

■ **Solution**

Consider the drag on a unit span of the wedge D'. Hence,

$$c_d = \frac{D'}{q_1 S} = \frac{D'}{q_1 c(1)} = \frac{D'}{q_1 c}$$

From Figure 9.17,

$$D' = 2p_2 l \sin\theta - 2p_1 l \sin\theta = (2l \sin\theta)(p_2 - p_1)$$

However,

$$l = \frac{c}{\cos\theta}$$

Thus,

$$D' = (2c \tan\theta)(p_2 - p_1)$$

and

$$c_d = (2\tan\theta)\left(\frac{p_2 - p_1}{q_1}\right)$$

Note that

$$q_1 \equiv \frac{1}{2}\rho_1 V_1^2 = \frac{1}{2}\rho_1 \frac{\gamma p_1}{\gamma p_1} V_1^2 = \frac{\gamma p_1}{2a_1^2} V_1^2 = \frac{\gamma}{2} p_1 M_1^2$$

Thus,

$$c_d = (2\tan\theta)\left(\frac{p_2 - p_1}{(\gamma/2)p_1 M_1^2}\right) = \frac{4\tan\theta}{\gamma M_1^2}\left(\frac{p_2}{p_1} - 1\right)$$

From Figure 9.9, for $M_1 = 5$ and $\theta = 15°$, $\beta = 24.2°$. Hence,

$$M_{n,1} = M_1 \sin\beta = 5\sin(24.2°) = 2.05$$

From Appendix B, for $M_{n,1} = 2.05$, we have

$$\frac{p_2}{p_1} = 4.736$$

Hence,

$$c_d = \frac{4\tan\theta}{\gamma M_1^2}\left(\frac{p_2}{p_1} - 1\right) = \frac{4\tan 15°}{(1.4)(5)^2}(4.736 - 1) = \boxed{0.114}$$

(*Note:* The drag is *finite* for this case. In a supersonic or hypersonic inviscid flow over a two-dimensional body, the drag is always finite. D'Alembert's paradox does *not* hold for freestream Mach numbers such that shock waves appear in the flow. The fundamental reason for the generation of drag here is the presence of shock waves. Shocks are always a dissipative, drag-producing mechanism. For this reason, the drag in this case is called *wave drag*, and c_d is the wave-drag coefficient, more properly denoted as $c_{d,w}$.)

9.3.1 A Comment on Supersonic Lift and Drag Coefficients

The result obtained in Example 9.6 is a stunning verification of the validity of the dimensional analysis discussed in Section 1.7. There we proved that, for a body of a given shape at a given angle of attack, the aerodynamic coefficients are simply a function of Mach number and Reynolds number [see Equations (1.42), (1.43), and (1.44)]. Consider the 15° half-angle wedge at zero angle of attack shown in Figure 9.17. This is a body of a given shape at a given angle of attack. Moreover, in Example 9.6 we are given *only* the freestream Mach number, and are asked to calculate the drag coefficient for the wedge. Since the flow is inviscid, the Reynolds number does not play a role. At first glance, one might intuitively think that we need to be given at least a freestream pressure and velocity to obtain the drag coefficient. After all, the physical source of the drag is the pressure distribution integrated all over the surface of the body, as emphasized in Section 1.5. And the surface pressure distribution is shown schematically in Figure 9.17. Why is it, then, that we are not given some information about the pressure level and the freestream velocity?

The answer, clearly demonstrated by the dimensional analysis in Section 1.7, is that the drag *coefficient* depends just on Mach number. In Example 9.6 the freestream Mach number is given as Mach 5. The solution progresses by treating the surface pressure distribution that is responsible for the drag, but the solution ultimately requires only pressure ratios, not the actual value of the pressure itself. At the end of the calculation in Example 9.6, the drag coefficient is finally obtained, and all we needed for the calculation was the freestream Mach number. What a nice verification of the validity of the dimensional analysis discussed in Section 1.7 and the concept of flow similarity given in Section 1.8! Moreover, we have verified these concepts for a supersonic flow. Of course, the concepts in Sections 1.7 and 1.8 are fundamental; they hold no matter what is the flow regime—subsonic, supersonic, hypersonic, etc.

Finally, if we had been asked for the drag *force* in Example 9.6, additional information would have been required, such as the size of the wedge, and the pressure of the freestream. But one of the beauties of dealing with the aerodynamic *coefficients* rather than the forces or moments themselves is that the *coefficients* for an inviscid flow depend on Mach number, and Mach number only.

9.4 SHOCK INTERACTIONS AND REFLECTIONS

Return to the oblique shock wave illustrated in Figure 9.2a. In this picture, we can imagine the shock wave extending unchanged above the corner to infinity. However, in real life this does not happen. In reality, the oblique shock in Figure 9.2a will impinge somewhere on another solid surface and/or will intersect other waves, either shock or expansion waves. Such wave intersections and interactions are important in the practical design and analysis of supersonic airplanes, missiles, wind tunnels, rocket engines, etc. A perfect historical example of this, as well as the consequences that can be caused by not paying suitable attention to wave interactions, is a ramjet flight-test program conducted in the early 1960s. During this period, a ramjet engine was mounted underneath the X-15 hypersonic airplane for a series of flight tests at high Mach numbers, in the range from 4 to 7. (The X-15, shown in Figure 9.18, was an experimental, rocket-powered airplane designed to probe the lower end of hypersonic manned flight.) During the

Figure 9.18 The X-15 hypersonic research vehicle. Designed and built during the late 1950s, it served as a test vehicle for the U.S. Air Force and NASA. (*NASA*)

first high-speed tests, the shock wave from the engine cowling impinged on the bottom surface of the X-15, and because of locally high aerodynamic heating in the impingement region, a hole was burned in the X-15 fuselage. Although this problem was later fixed, it is a graphic example of what shock-wave interactions can do to a practical configuration.

The purpose of this section is to present a mainly qualitative discussion of shock-wave interactions. For more details, see Chapter 4 of Reference 21.

First, consider an oblique shock wave generated by a concave corner, as shown in Figure 9.19. The deflection angle at the corner is θ, thus generating an oblique shock at point A with a wave angle β_1. Assume that a straight, horizontal wall is present above the corner, as also shown in Figure 9.19. The shock wave generated at point A, called the *incident shock wave,* impinges on the upper wall at point B. *Question:* Does the shock wave simply disappear at point B? If not, what happens to it? To answer this question, we appeal to our knowledge of shock-wave properties. Examining Figure 9.19, we see that the flow in region 2 behind the incident shock is inclined upward at the deflection angle θ. However, the flow must be tangent everywhere along the upper wall; if the flow in region 2 were to continue unchanged, it would run into the wall and have no place to go. Hence, the flow in region 2 must eventually be bent downward through the angle θ in order to maintain a flow tangent to the upper wall. Nature

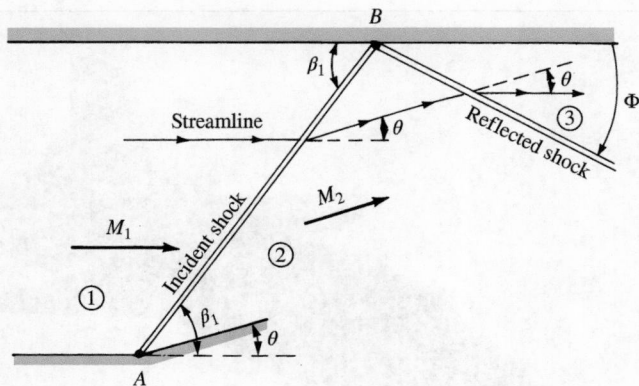

Figure 9.19 Regular reflection of a shock wave from a solid boundary.

accomplishes this downward deflection via a second shock wave originating at the impingement point B in Figure 9.19. This second shock is called the *reflected shock wave*. The purpose of the reflected shock is to deflect the flow in region 2 so that it is parallel to the upper wall in region 3, thus preserving the wall boundary condition.

The strength of the reflected shock wave is weaker than the incident shock. This is because $M_2 < M_1$, and M_2 represents the upstream Mach number for the reflected shock wave. Since the deflection angles are the same, whereas the reflected shock sees a lower upstream Mach number, we know from Section 9.2 that the reflected wave must be weaker. For this reason, the angle the reflected shock makes with the upper wall Φ is not equal to β_1 (i.e., the wave reflection is not specular). The properties of the reflected shock are uniquely defined by M_2 and θ; since M_2 is in turn uniquely defined by M_1 and θ, then the properties in region 3 behind the reflected shock as well as the angle Φ are easily determined from the given conditions of M_1 and θ by using the results of Section 9.2 as follows:

1. Calculate the properties in region 2 from the given M_1 and θ. In particular, this gives us M_2.

2. Calculate the properties in region 3 from the value of M_2 calculated above and the known deflection angle θ.

An interesting situation can arise as follows. Assume that M_1 is only slightly above the minimum Mach number necessary for a straight, attached shock wave at the given deflection angle θ. For this case, the oblique shock theory from Section 9.2 allows a solution for a straight, attached incident shock. However, we know that the Mach number decreases across a shock (i.e., $M_2 < M_1$). This decrease may be enough such that M_2 is *not* above the minimum Mach number

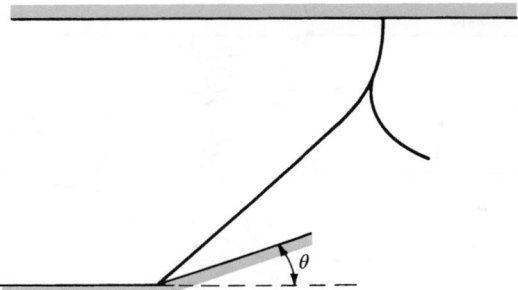

Figure 9.20 Mach reflection.

for the required deflection θ through the reflected shock. In such a case, our oblique shock theory does not allow a solution for a straight reflected shock wave. The regular reflection as shown in Figure 9.19 is not possible. Nature handles this situation by creating the wave pattern shown in Figure 9.20. Here, the originally straight incident shock becomes curved as it nears the upper wall and becomes a normal shock wave at the upper wall. This allows the streamline at the wall to continue parallel to the wall behind the shock intersection. In addition, a curved reflected shock branches from the normal shock and propagates downstream. This wave pattern, shown in Figure 9.20, is called a *Mach reflection*. The calculation of the wave pattern and general properties for a Mach reflection requires numerical techniques such as those to be discussed in Chapter 13.

   Another type of shock interaction is shown in Figure 9.21. Here, a shock wave is generated by the concave corner at point G and propagates upward. Denote this wave as shock A. Shock A is a *left-running wave*, so-called because if you stand on top of the wave and look downstream, you see the shock wave running in front of you toward the left. Another shock wave is generated by the concave corner at point H, and propagates downward. Denote this wave as shock B. Shock B is a *right-running wave,* so-called because if you stand on top of the wave and look downstream, you see the shock running in front of you toward the right. The picture shown in Figure 9.21 is the intersection of right- and left-running shock waves. The intersection occurs at point E. At the intersection, wave A is refracted and continues as wave D. Similarly, wave B is refracted and continues as wave C. The flow behind the refracted shock D is denoted by region 4; the flow behind the refracted shock C is denoted by region $4'$. These two regions are divided by a slip line EF. Across the slip line, the pressures are constant (i.e., $p_4 = p_{4'}$), and the direction (but not necessarily the magnitude) of velocity is the same, namely, parallel to the slip line. All other properties in regions 4 and $4'$ are different, most notably the entropy ($s_4 \neq s_{4'}$). The conditions which must hold across the slip line, along with the known M_1, θ_1,

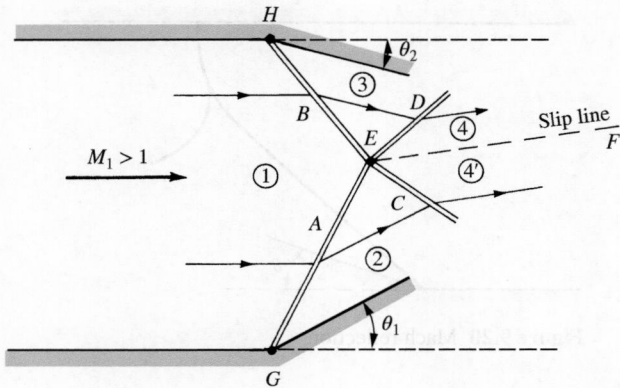

Figure 9.21 Intersection of right- and left-running shock waves.

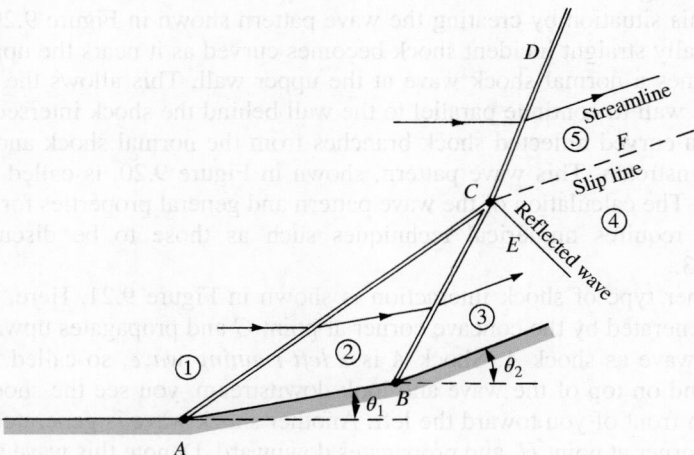

Figure 9.22 Intersection of two left-running shock waves.

and θ_2, uniquely determine the shock-wave interaction shown in Figure 9.21. (See Chapter 4 of Reference 21 for details concerning the calculation of this interaction.)

Figure 9.22 illustrates the intersection of two left-running shocks generated at corners A and B. The intersection occurs at point C, at which the two shocks merge and propagate as the stronger shock CD, usually along with a weak reflected wave CE. This reflected wave is necessary to adjust the flow so that the velocities in regions 4 and 5 are in the same direction. Again, a slip line CF trails downstream of the intersection point.

The above cases are by no means all the possible wave interactions in a supersonic flow. However, they represent some of the more common situations encountered frequently in practice.

EXAMPLE 9.7

Consider an oblique shock wave generated by a compression corner with a 10° deflection angle. The Mach number of the flow ahead of the corner is 3.6; the flow pressure and temperature are standard sea level conditions. The oblique shock wave subsequently impinges on a straight wall opposite the compression corner. The geometry for this flow is given in Figure 9.19. Calculate the angle of the reflected shock wave Φ relative to the straight wall. Also, obtain the pressure, temperature, and Mach number behind the reflected wave.

■ **Solution**

From the θ-β-M diagram, Figure 9.9, for $M_1 = 3.6$ and $\theta = 10°$, $\beta_1 = 24°$. Hence,

$$M_{n,1} = M_1 \sin\beta_1 = 3.6 \sin 24° = 1.464$$

From Appendix B,

$$M_{n,2} = 0.7157 \qquad \frac{p_2}{p_1} = 2.32 \qquad \frac{T_2}{T_1} = 1.294$$

Also,
$$M_2 = \frac{M_{n,2}}{\sin(\beta - \theta)} = \frac{0.7157}{\sin(24 - 10)} = 2.96$$

These are the conditions behind the incident shock wave. They constitute the upstream flow properties for the reflected shock wave. We know that the flow must be deflected again by $\theta = 10°$ in passing through the reflected shock. Thus, from the θ-β-M diagram, for $M_2 = 2.96$ and $\theta = 10°$, we have the wave angle for the reflected shock, $\beta_2 = 27.3°$. Note that β_2 is *not* the angle the reflected shock makes with respect to the upper wall; rather, by definition of the wave angle, β_2 is the angle between the reflected shock and the direction of the flow in region 2. The shock angle relative to the wall is, from the geometry shown in Figure 9.19,

$$\Phi = \beta_2 - \theta = 27.3 - 10 = \boxed{17.3°}$$

Also, the normal component of the upstream Mach number relative to the reflected shock is $M_2 \sin\beta_2 = (2.96)\sin 27.3° = 1.358$. From Appendix B,

$$\frac{p_3}{p_2} = 1.991 \qquad \frac{T_3}{T_2} = 1.229 \qquad M_{n,3} = 0.7572$$

Hence,
$$M_3 = \frac{M_{n,3}}{\sin(\beta_2 - \theta)} = \frac{0.7572}{\sin(27.3 - 10)} = \boxed{2.55}$$

For standard sea level conditions, $p_1 = 2116$ lb/ft^3 and $T_1 = 519°$R. Thus,

$$p_3 = \frac{p_3}{p_2}\frac{p_2}{p_1}p_1 = (1.991)(2.32)(2116) = \boxed{9774 \text{ lb/ft}^3}$$

$$T_3 = \frac{T_3}{T_2}\frac{T_2}{T_1}T_1 = (1.229)(1.294)(519) = \boxed{825°\text{R}}$$

Note that the reflected shock is weaker than the incident shock, as indicated by the smaller pressure ratio for the reflected shock, $p_3/p_2 = 1.991$ as compared to $p_2/p_1 = 2.32$ for the incident shock.

9.5 DETACHED SHOCK WAVE IN FRONT OF A BLUNT BODY

The curved bow shock which stands in front of a blunt body in a supersonic flow is sketched in Figure 8.1. We are now in a position to better understand the properties of this bow shock, as follows.

The flow in Figure 8.1 is sketched in more detail in Figure 9.23. Here, the shock wave stands a distance δ in front of the nose of the blunt body; δ is defined

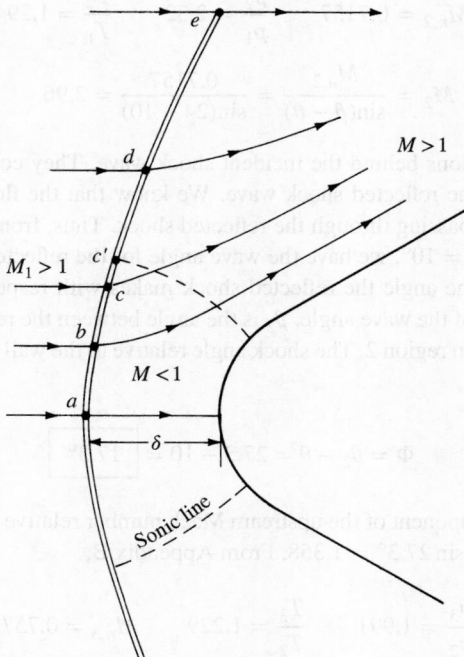

Figure 9.23 Flow over a supersonic blunt body.

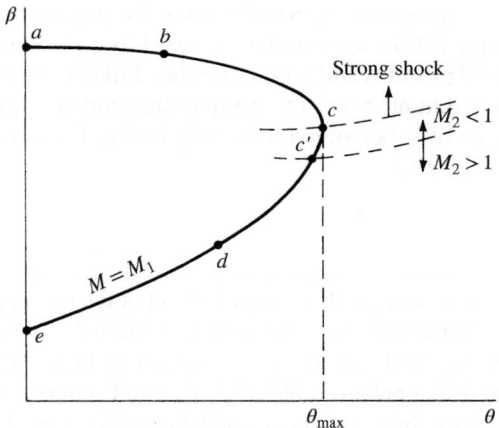

Figure 9.24 θ-β-M diagram for the sketch shown in Figure 9.23.

as the *shock detachment distance*. At point a, the shock wave is normal to the upstream flow; hence, point a corresponds to a normal shock wave. Away from point a, the shock wave gradually becomes curved and weaker, eventually evolving into a Mach wave at large distances from the body (illustrated by point e in Figure 9.23).

A curved bow shock wave is one of the instances in nature when you can observe *all* possible oblique shock solutions at once for a given freestream Mach number M_1. This takes place between points a and e. To see this more clearly, consider the θ-β-M diagram sketched in Figure 9.24 in conjunction with Figure 9.23. In Figure 9.24, point a corresponds to the normal shock, and point e corresponds to the Mach wave. Slightly above the centerline, at point b in Figure 9.23, the shock is oblique but pertains to the strong shock-wave solution in Figure 9.24. The flow is deflected slightly upward behind the shock at point b. As we move further along the shock, the wave angle becomes more oblique, and the flow deflection increases until we encounter point c. Point c on the bow shock corresponds to the maximum deflection angle shown in Figure 9.24. Above point c, from c to e, all points on the shock correspond to the weak shock solution. Slightly above point c, at point c', the flow behind the shock becomes sonic. From a to c', the flow is subsonic behind the bow shock; from c' to e, it is supersonic. Hence, the flow field between the curved bow shock and the blunt body is a mixed region of both subsonic and supersonic flow. The dividing line between the subsonic and supersonic regions is called the *sonic line,* shown as the dashed line in Figure 9.23.

The shape of the detached shock wave, its detachment distance δ, and the complete flow field between the shock and the body depend on M_1 and the size and shape of the body. The solution of this flow field is not trivial. Indeed, the supersonic blunt-body problem was a major focus for supersonic aerodynamicists during the 1950s and 1960s, spurred by the need to understand the high-speed flow over blunt-nosed missiles and reentry bodies. Indeed, it was not until the late 1960s that truly sufficient numerical techniques became available for satisfactory engineering solutions of supersonic blunt-body flows. These modern techniques are discussed in Chapter 13.

EXAMPLE 9.8

Consider the detached curved bow shock wave in front of the two-dimensional parabolic blunt body drawn in Figure 9.25. The freestream is at Mach 8. Consider the two stream-lines passing through the shock at points a and b shown in Figure 9.25. The wave angle at point a is 90°, and that at point b is 60°. Calculate and compare the value of entropy (relative to the freestream) for streamlines a and b in the flow behind the shock.

■ **Solution**

The oblique shock properties studied in this chapter are derived on the basis of a straight oblique shock wave with a uniform flow field behind the shock wave, such as sketched in Figure 9.2a. These solutions do not apply to the nonuniform flow field behind a curved shock wave such as shown in Figure 9.25. Such blunt-body solutions are treated in Section 13.5. The straight oblique shock solutions treated in the present chapter, however, give the shock wave properties at any local point *immediately behind the curved shock* in Figure 9.25 as long as we know the local wave angle at the point. Therefore, immediately behind the shock at point a, because the shock is a normal shock at that point, we have

$$M_{n,1} = 8$$

From Appendix B, for $M_{n,1} = 8$, $p_2/p_1 = 74.5$ and $T_2/T_1 = 13.39$. From Equation (7.25), the entropy increase across the shock is

$$s_2 - s_1 = c_p \ln \frac{T_2}{T_1} - R \ln \frac{p_2}{p_1}$$

Since

$$c_p = \frac{\gamma R}{\gamma - 1} = \frac{(1.4)(287)}{0.4} = 1004.5 \ \frac{J}{Kg \cdot K}$$

Then

$$s_2 - s_1 = (1004.5) \ln 13.39 - (287) \ln 74.5$$

$$s_2 - s_1 = 1370 \ \frac{J}{Kg \cdot K}$$

Downstream of the shock wave, the flow along any given streamline is both adiabatic (no heat transfer) and reversible (no friction, etc.), hence the flow along a given streamline behind the shock wave is *isentropic*. Therefore, along streamline a, the entropy is constant

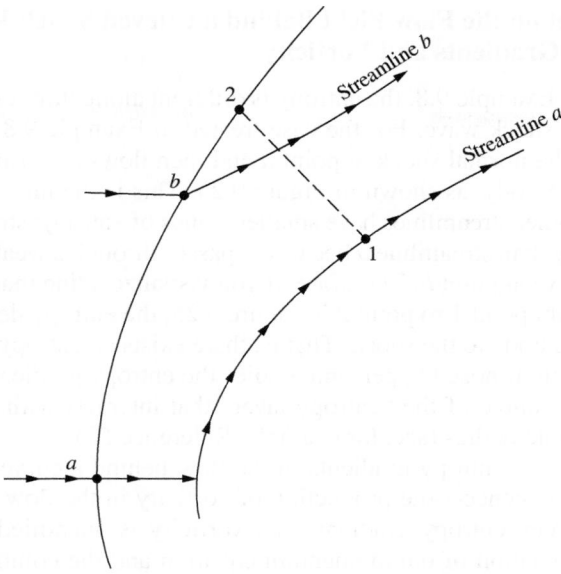

Figure 9.25 Two streamlines crossing a detached bow shockwave in front of a blunt body at Mach 8.

and equal to its value at point a just behind the shock wave. Thus,

$$s_2 - s_1 = \boxed{1370 \, \frac{\text{J}}{\text{Kg} \cdot \text{K}}} \qquad \text{along streamline } a$$

Immediately behind the curved bow shock at point b, where $\beta = 60°$, we have

$$M_{n,1} = M_1 \sin \beta = 8 \sin 60°$$

$$= 8(0.866) = 6.928$$

From Appendix B, using the nearest entry of $M_{n,1} = 6.9$, we have $p_2/p_1 = 55.38$ and $T_2/T_1 = 10.2$. Thus, at point b,

$$s_2 - s_1 = c_p \ln \frac{T_2}{T_1} - R \ln \frac{p_2}{p_1}$$

$$= (1004.5) \ln 10.2 - (287) \ln 55.38$$

$$= \boxed{1180 \, \frac{\text{J}}{\text{Kg} \cdot \text{K}}} \qquad \text{along streamline } b$$

The entropy along streamline b is smaller than that along streamline a because streamline b passes through a weaker part of the bow shock wave.

9.5.1 Comment on the Flow Field Behind a Curved Shock Wave: Entropy Gradients and Vorticity

As illustrated by Example 9.8, the entropy is different along different streamlines behind a curved shock wave. For the case treated in Example 9.8, streamline a passes through the normal shock at point a and then flows downstream, wetting the surface of the body, as shown in Figure 9.25. This is the maximum entropy streamline. All other streamlines have smaller values of entropy; streamline b has a smaller entropy than streamline a because it passes through a weaker part of the curved shock wave at point b. Therefore, if you visualize a line that cuts through the flow field from point 1 to point 2 in Figure 9.25, the entropy decreases along this line from the body to the shock. That is, there exists an entropy gradient, ∇s, in the flow. For blunt-nosed hypersonic bodies the entropy gradient can be quite large, and is the source of the "entropy layer" that interacts with the boundary layer on hypersonic bodies (see, for example, Reference 52).

The presence of entropy gradients in the flow behind a curved shock wave has another consequence—the production of vorticity in the flow. The physical connection between entropy gradients and vorticity is quantified by *Crocco's theorem*, a combination of the momentum equation and the combined first and second laws of thermodynamics:

$$T\nabla s = \nabla h_o - \mathbf{V} \times (\nabla \times \mathbf{V}) \quad \text{Crocco's theorem}$$

In this equation, ∇s is the entropy gradient, ∇h_o is the gradient in the total enthalpy, and $\nabla \times \mathbf{V}$ is the vorticity. For a derivation of Crocco's theorem, see, for example, Section 6.6 of Reference 21. For our discussion, we present Crocco's theorem simply to emphasize an important feature of the flow behind the curved shock shown in Figure 9.25. The flow is adiabatic, hence ∇h_o is zero everywhere in the flow. However, ∇s is finite, and therefore from Crocco's theorem $\nabla \times \mathbf{V}$ must be finite.

Conclusion: The flow field behind a curved shock wave is *rotational*. As a result, a velocity potential with all its analytical advantages discussed earlier in this book cannot be defined for the blunt-body flow field. Consequently, the flow field behind a curved shock is computed by means of numerical solutions of the continuity, momentum, and energy equations. Such computational fluid dynamic solutions are discussed in Section 13.5.

9.6 PRANDTL-MEYER EXPANSION WAVES

Oblique shock waves, as discussed in Sections 9.2 to 9.5, occur when a supersonic flow is turned into itself (see again Figure 9.2a). In contrast, when a supersonic flow is turned away from itself, an expansion wave is formed, as sketched in Figure 9.2b. Examine this figure carefully, and review the surrounding discussion in Section 9.1 before progressing further. The purpose of the present section is to develop a theory which allows us to calculate the changes in flow properties

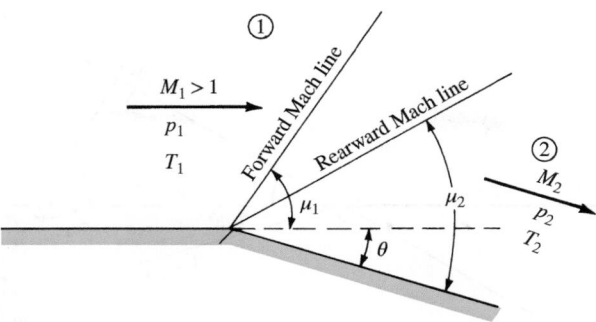

Figure 9.26 Prandtl-Meyer expansion.

across such expansion waves. To this stage in our discussion of oblique waves, we have completed the left-hand branch of the road map in Figure 9.6. In this section, we cover the right-hand branch.

The expansion fan in Figure 9.2b is a *continuous* expansion region that can be visualized as an infinite number of Mach waves, each making the Mach angle μ [see Equation (9.1)] with the local flow direction. As sketched in Figure 9.26, the expansion fan is bounded upstream by a Mach wave which makes the angle μ_1 with respect to the upstream flow, where $\mu_1 = \arcsin(1/M_1)$. The expansion fan is bounded downstream by another Mach wave which makes the angle μ_2 with respect to the downstream flow, where $\mu_2 = \arcsin(1/M_2)$. Since the expansion through the wave takes place across a continuous succession of Mach waves, and since $ds = 0$ for each Mach wave, the expansion is *isentropic*. This is in direct contrast to flow across an oblique shock, which always experiences an entropy increase. The fact that the flow through an expansion wave is isentropic is a greatly simplifying aspect, as we will soon appreciate.

An expansion wave emanating from a sharp convex corner as sketched in Figures 9.2b and 9.26 is called a *centered* expansion wave. Ludwig Prandtl and his student Theodor Meyer first worked out a theory for centered expansion waves in 1907–1908, and hence such waves are commonly denoted as *Prandtl-Meyer expansion waves*.

The problem of an expansion wave is as follows: Referring to Figure 9.26, given the upstream flow (region 1) and the deflection angle θ, calculate the downstream flow (region 2). Let us proceed.

Consider a very weak wave produced by an infinitesimally small flow deflection $d\theta$ as sketched in Figure 9.27. We consider the limit of this picture as $d\theta \rightarrow 0$; hence, the wave is essentially a Mach wave at the angle μ to the upstream flow. The velocity ahead of the wave is V. As the flow is deflected downward through the angle $d\theta$, the velocity is increased by the infinitesimal amount dV, and hence the flow velocity behind the wave is $V + dV$ inclined at the angle $d\theta$. Recall from the treatment of the momentum equation in Section 9.2 that any change in

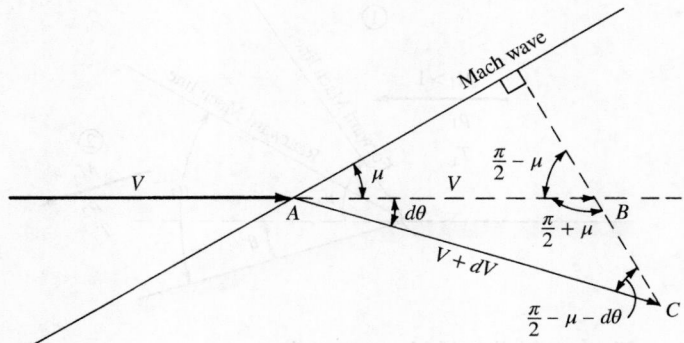

Figure 9.27 Geometrical construction for the infinitesimal changes across an infinitesimally weak wave (in the limit, a Mach wave).

velocity across a wave takes place *normal* to the wave; the tangential component is unchanged across the wave. In Figure 9.27, the horizontal line segment AB with length V is drawn behind the wave. Also, the line segment AC is drawn to represent the new velocity $V + dV$ behind the wave. Then line BC is normal to the wave because it represents the line along which the change in velocity occurs. Examining the geometry in Figure 9.27, from the law of sines applied to triangle ABC, we see that

$$\frac{V + dV}{V} = \frac{\sin(\pi/2 + \mu)}{\sin(\pi/2 - \mu - d\theta)} \tag{9.24}$$

However, from trigonometric identities,

$$\sin\left(\frac{\pi}{2} + \mu\right) = \sin\left(\frac{\pi}{2} - \mu\right) = \cos\mu \tag{9.25}$$

$$\sin\left(\frac{\pi}{2} - \mu - d\theta\right) = \cos(\mu + d\theta) = \cos\mu\cos d\theta - \sin\mu\sin d\theta \tag{9.26}$$

Substituting Equations (9.25) and (9.26) into (9.24), we have

$$1 + \frac{dV}{V} = \frac{\cos\mu}{\cos\mu\cos d\theta - \sin\mu\sin d\theta} \tag{9.27}$$

For small $d\theta$, we can make the small-angle assumptions $\sin d\theta \approx d\theta$ and $\cos d\theta \approx 1$. Then, Equation (9.27) becomes

$$1 + \frac{dV}{V} = \frac{\cos\mu}{\cos\mu - d\theta\sin\mu} = \frac{1}{1 - d\theta\tan\mu} \tag{9.28}$$

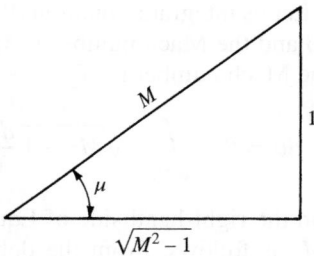

Figure 9.28 Right triangle
associated with the Mach angle.

Note that the function $1/(1 - x)$ can be expanded in a power series (for $x < 1$) as

$$\frac{1}{1 - x} = 1 + x + x^2 + x^3 + \cdots$$

Hence, Equation (9.28) can be expanded as (ignoring terms of second order and higher)

$$1 + \frac{dV}{V} = 1 + d\theta \tan \mu + \cdots \tag{9.29}$$

Thus, from Equation (9.29),

$$d\theta = \frac{dV/V}{\tan \mu} \tag{9.30}$$

From Equation (9.1), we know that $\mu = \arcsin(1/M)$. Hence, the right triangle in Figure 9.28 demonstrates that

$$\tan \mu = \frac{1}{\sqrt{M^2 - 1}} \tag{9.31}$$

Substituting Equation (9.31) into (9.30), we obtain

$$\boxed{d\theta = \sqrt{M^2 - 1}\, \frac{dV}{V}} \tag{9.32}$$

Equation (9.32) relates the infinitesimal change in velocity dV to the infinitesimal deflection $d\theta$ across a wave of vanishing strength. In the precise limit of a Mach wave, of course dV and hence $d\theta$ are zero. In this sense, Equation (9.32) is an approximate equation for a finite $d\theta$, but it becomes a true equality as $d\theta \rightarrow 0$. Since the expansion fan illustrated in Figures 9.2*b* and 9.26 is a region of an infinite number of Mach waves, Equation (9.32) is a differential equation which precisely describes the flow inside the expansion wave.

Return to Figure 9.26. Let us integrate Equation (9.32) from region 1, where the deflection angle is zero and the Mach number is M_1, to region 2, where the deflection angle is θ and the Mach number is M_2:

$$\int_0^\theta d\theta = \theta = \int_{M_1}^{M_2} \sqrt{M^2 - 1} \, \frac{dV}{V} \qquad (9.33)$$

To carry out the integral on the right-hand side of Equation (9.33), dV/V must be obtained in terms of M, as follows. From the definition of Mach number, $M = V/a$, we have $V = Ma$, or

$$\ln V = \ln M + \ln a \qquad (9.34)$$

Differentiating Equation (9.34), we obtain

$$\frac{dV}{V} = \frac{dM}{M} + \frac{da}{a} \qquad (9.35)$$

From Equations (8.25) and (8.40), we have

$$\left(\frac{a_0}{a}\right)^2 = \frac{T_0}{T} = 1 + \frac{\gamma - 1}{2}M^2 \qquad (9.36)$$

Solving Equation (9.36) for a, we obtain

$$a = a_0 \left(1 + \frac{\gamma - 1}{2}M^2\right)^{-1/2} \qquad (9.37)$$

Differentiating Equation (9.37), we obtain

$$\frac{da}{a} = -\left(\frac{\gamma - 1}{2}\right) M \left(1 + \frac{\gamma - 1}{2}M^2\right)^{-1} dM \qquad (9.38)$$

Substituting Equation (9.38) into (9.35), we have

$$\frac{dV}{V} = \frac{1}{1 + [(\gamma - 1)/2]M^2} \frac{dM}{M} \qquad (9.39)$$

Equation (9.39) is a relation for dV/V strictly in terms of M—this is precisely what is desired for the integral in Equation (9.33). Hence, substituting Equation (9.39) into (9.33), we have

$$\theta = \int_{M_1}^{M_2} \frac{\sqrt{M^2 - 1}}{1 + [(\gamma - 1)/2]M^2} \frac{dM}{M} \qquad (9.40)$$

In Equation (9.40), the integral

$$\nu(M) \equiv \int \frac{\sqrt{M^2 - 1}}{1 + [(\gamma - 1)/2]M^2} \frac{dM}{M} \qquad (9.41)$$

is called the *Prandtl-Meyer function,* denoted by ν. Carrying out the integration, Equation (9.41) becomes

$$\nu(M) = \sqrt{\frac{\gamma+1}{\gamma-1}} \tan^{-1} \sqrt{\frac{\gamma-1}{\gamma+1}(M^2-1)} - \tan^{-1}\sqrt{M^2-1} \qquad (9.42)$$

The constant of integration that would ordinarily appear in Equation (9.42) is not important, because it drops out when Equation (9.42) is used for the definite integral in Equation (9.40). For convenience, it is chosen as zero, such that $\nu(M) = 0$ when $M = 1$. Finally, we can now write Equation (9.40), combined with (9.41), as

$$\theta = \nu(M_2) - \nu(M_1) \qquad (9.43)$$

where $\nu(M)$ is given by Equation (9.42) for a calorically perfect gas. The Prandtl-Meyer function ν is very important; it is the key to the calculation of changes across an expansion wave. Because of its importance, ν is tabulated as a function of M in Appendix C. For convenience, values of μ are also tabulated in Appendix C.

How do the above results solve the problem stated in Figure 9.26; that is how can we obtain the properties in region 2 from the known properties in region 1 and the known deflection angle θ? The answer is straightforward:

1. For the given M_1, obtain $\nu(M_1)$ from Appendix C.
2. Calculate $\nu(M_2)$ from Equation (9.43), using the known θ and the value of $\nu(M_1)$ obtained in step 1.
3. Obtain M_2 from Appendix C corresponding to the value of $\nu(M_2)$ from step 2.
4. The expansion wave is isentropic; hence, p_0 and T_0 are constant through the wave. That is, $T_{0,2} = T_{0,1}$ and $p_{0,2} = p_{0,1}$. From Equation (8.40), we have

$$\frac{T_2}{T_1} = \frac{T_2/T_{0,2}}{T_1/T_{0,1}} = \frac{1+[(\gamma-1)/2]M_1^2}{1+[(\gamma-1)/2]M_2^2} \qquad (9.44)$$

From Equation (8.42), we have

$$\frac{p_2}{p_1} = \frac{p_2/p_0}{p_1/p_0} = \left(\frac{1+[(\gamma-1)/2]M_1^2}{1+[(\gamma-1)/2]M_2^2}\right)^{\gamma/(\gamma-1)} \qquad (9.45)$$

Since we know both M_1 and M_2, as well as T_1 and p_1, Equations (9.44) and (9.45) allow the calculation of T_2 and p_2 downstream of the expansion wave.

EXAMPLE 9.9

A supersonic flow with $M_1 = 1.5$, $p_1 = 1$ atm, and $T_1 = 288$ K is expanded around a sharp corner (see Figure 9.26) through a deflection angle of $15°$. Calculate M_2, p_2, T_2, $p_{0,2}$, $T_{0,2}$, and the angles that the forward and rearward Mach lines make with respect to the upstream flow direction.

■ Solution

From Appendix C, for $M_1 = 1.5$, $\nu_1 = 11.91°$. From Equation (9.43), $\nu_2 = \nu_1 + \theta = 11.91 + 15 = 26.91°$. Thus, $\boxed{M_2 = 2.0}$ (rounding to the nearest entry in the table).

From Appendix A, for $M_1 = 1.5$, $p_{0,1}/p_1 = 3.671$ and $T_{0,1}/T_1 = 1.45$, and for $M_2 = 2.0$, $p_{0,2}/p_2 = 7.824$ and $T_{0,2}/T_2 = 1.8$.

Since the flow is isentropic, $T_{0,2} = T_{0,1}$ and $p_{0,2} = p_{0,1}$. Thus,

$$p_2 = \frac{p_2}{p_{0,2}}\frac{p_{0,2}}{p_{0,1}}\frac{p_{0,1}}{p_1}p_1 = \frac{1}{7.824}(1)(3.671)(1\ \text{atm}) = \boxed{0.469\ \text{atm}}$$

$$T_2 = \frac{T_2}{T_{0,2}}\frac{T_{0,2}}{T_{0,1}}\frac{T_{0,1}}{T_1}T_1 = \frac{1}{1.8}(1)(1.45)(288) = \boxed{232\ \text{K}}$$

$$p_{0,2} = p_{0,1} = \frac{p_{0,1}}{p_1}p_1 = 3.671(1\ \text{atm}) = \boxed{3.671\ \text{atm}}$$

$$T_{0,2} = T_{0,1} = \frac{T_{0,1}}{T_1}T_1 = 1.45(288) = \boxed{417.6\ \text{K}}$$

Returning to Figure 9.26, we have

$$\text{Angle of forward Mach line} = \mu_1 = \boxed{41.81°}$$

$$\text{Angle of rearward Mach line} = \mu_2 - \theta = 30 - 15 = \boxed{15°}$$

DESIGN BOX

In Example 8.14 we indicated the reasons why an air-breathing power plant for high Mach number, hypersonic vehicles would have to be a supersonic combustion ramjet engine—a SCRAMjet. The design of such an engine depends heavily on the properties of oblique shock waves and expansion waves—the subjects of this chapter. In this design box, we examine some of the basic design features of SCRAMjet engines. Looking ahead to the future of aerodynamics in the twenty-first century, hypersonic flight is essentially the last frontier of our quest to fly faster and higher. Many of the hypersonic vehicles of the future

will be powered by SCRAMjet engines. So the material in this design box is much like peering through a window into the future.

Two experimental SCRAMjet powered vehicles have successfully flown, the X-51 shown in Figure 9.29a and the X-43 shown in Figure 9.29b (Reference 67). These experimental airplanes are paving the way to the future of hypersonic air-breathing airplanes.

The side view of a generic hypersonic vehicle powered by a SCRAMjet is shown in Figure 9.30. Essentially, the entire bottom surface of the vehicle is an integrated portion of the air-breathing SCRAMjet

(a) SCRAMjet-powered X-51 experimental vehicle.

(b) SCRAMjet-powered X-43 experimental air-breathing hypersonic vehicle.

Figure 9.29 Computer-generated images of possible future SCRAMjet-powered hypersonic vehicles. [(a) *U.S. Air Force Photo;* (b) *NASA*].

engine. The forebody shock wave (1) from the nose of the vehicle is the initial part of the compression process for the engine. Air flowing through this shock wave is compressed, and then enters the SCRAMjet engine module (2) where it is further compressed by reflected shock waves inside the engine duct, mixed with fuel, and then expanded out the back end of the module. The back end of the vehicle is scooped out (3) in order to further enhance the expansion of the exhaust gas. At the design flight condition, the forebody shock wave impinges right at the leading edge of the cowl (4), so that all the flow passing through

the shock will enter the engine, rather than some of the air spilling around the external surface.

It is also possible to further compress the air before it enters the engine module by creating an isentropic compression wave downstream of the shock. This is shown in Figure 9.31, patterned after Reference 68. Here, the bottom surface of the vehicle is contoured just right to form an isentropic compression wave that will focus on the leading edge of the cowl, right where the forebody shock wave is impinging as well. An isentropic compression wave is the opposite of the isentropic expansion wave discussed in Section 9.6, but the calculation of its properties is governed by the same Prandtl-Meyer function given in Equation (9.42), except in this case the local Mach number decreases through the wave, and the pressure increases. To create such an isentropic compression wave in reality is quite difficult; the contour of the body surface must be a specific shape for a specific upstream Mach number, and most efforts over the years to produce isentropic compression waves in various supersonic and hypersonic flow devices have usually resulted in the wave prematurely coalescing into several weak shock waves with associated entropy increases and total pressure loss. SCRAMjet-powered vehicles might incorporate such an isentropic compression surface. Other physical phenomena that influence SCRAMjet engine performance and vehicle aerodynamics are also noted in Figure 9.31. The leading edge must be blunted in order to reduce the aerodynamic heating at the nose (to be discussed in Chapter 14). The viscous boundary layer over the surface of the body creates drag and aerodynamic heating, and when a shock wave impinges on the boundary layer, flow separation and local reattachment may occur, creating local regions of high heat transfer (the shock wave/boundary layer interaction problem). There is always the important question as to where transition from laminar to turbulent boundary layer flow occurs along the body, because turbulent boundary layers result in increased aerodynamic heating and skin friction. Finally, when the forebody shock impinges on the leading edge of the cowl, it will interact with the local shock wave created at the blunt leading edge of the cowl, resulting in a shock–shock interaction

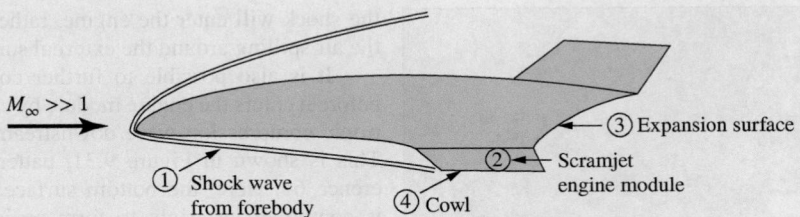

Figure 9.30 Sketch of a generic hypersonic vehicle powered by a SCRAMjet engine.

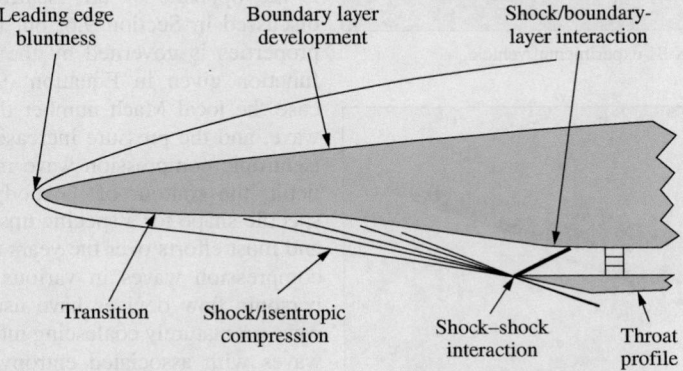

Figure 9.31 Sketch of some of the flow features on the forebody of a SCRAMjet-powered hypersonic vehicle.

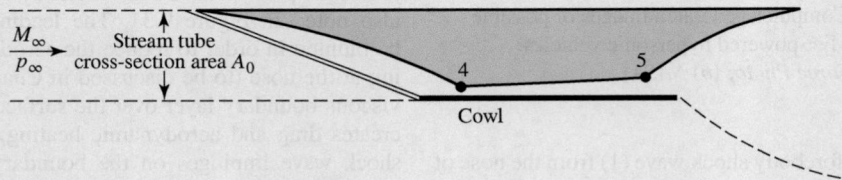

Figure 9.32 Flow path through a SCRAMjet engine.

problem that may create a local region of intense heating at the cowl leading edge. All of these phenomena influence the quality of the flow entering the SCRAMjet module, and they pose challenging problems for the designers of future SCRAMjet engines.

A generic sketch indicating the flow path through the SCRAMjet is shown in Figure 9.32. Here again we see the forebody shock impinging at the leading edge of the cowl, and a contoured compression surface to encourage an isentropic compression behind the shock. The cross-sectional area of the streamtube flowing through the shock wave, as noted in Figure 9.32, is greatly reduced behind the shock and through the compression wave owing to the large increase of air density through these compression processes. Because of this, the flow path through the combustor has a much smaller cross-sectional area. In Figure 9.32, points 4 and 5 denote the entrance

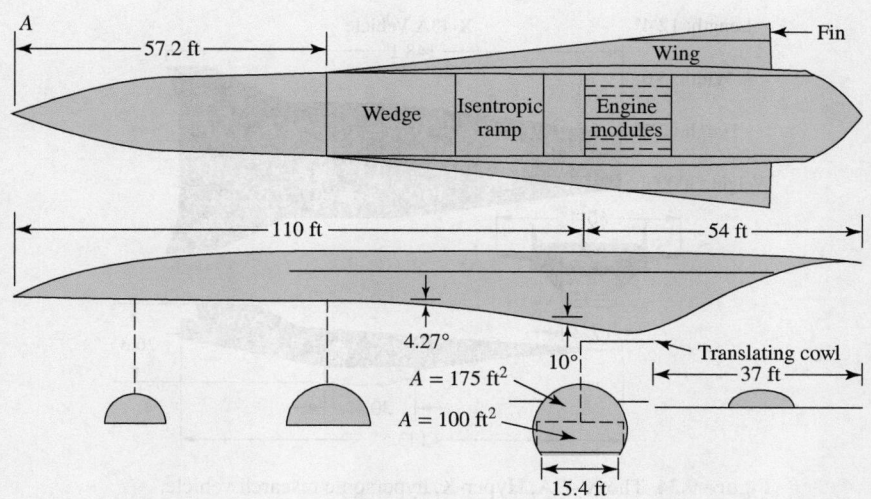

Figure 9.33 Two-sided view of a conceptual vehicle design.
(*Source:* Billig, Frederick S.: "Design and Development of Single-Stage-to-Orbit
Vehicles," *Johns Hopkins Applied Physics Laboratory Technical Digest*, vol. 11,
nos. 3 and 4, July–December 1990, pp. 336–52.)

	Altitude				p_4	T_4	V_4
M_∞	(ft)	M_4	$\dfrac{A_0}{A_4}$	$\dfrac{p_4}{p_\infty}$	(lb/in²)	(°R)	(ft/s)
7	80,077	3.143	10.85	47	19.03	1451	5,757
10	95,500	4.143	16.49	89.6	17.78	1958	8,744
15	114,250	5.502	25.23	185.9	15.94	2880	13,908
20	137,760	6.650	33.11	313.6	10.02	4074	19,648

and exit, respectively, of the combustor. Billig (Reference 69) has calculated typical flow conditions entering the combustor (at point 4) as a function of freestream Mach number M_∞ and flight altitude. Some of his results are tabulated above, where A_0 is the cross-sectional area of the streamtube in the freestream (see Figure 9.32), A_4 is its cross-sectional area at location 4, and M_4, p_4, T_4, and V_4 are the local Mach number, pressure, temperature, and flow velocity, respectively, at location 4.

Note from the tabulation that the local Mach number entering the combustor for the given conditions ranges from about 3 to above 6. The combustion takes place in this high Mach number stream—

the very essence of a SCRAMjet engine. Also note from Figure 9.32 that the cross-sectional area of the combustor increases from point 4 to point 5 along its length; this is to accommodate the local heat addition to the flow due to burning with the fuel, and still keep the flow moving at supersonic speed. (In Reference 21, among others, it is shown that heat addition to a supersonic flow slows the flow, whereas we will prove in Chapter 10 that increasing the cross-sectional area of a supersonic flow increases its speed. Hence, the combustor area in a SCRAMjet must be increased in the flow direction in order to keep the heat addition process from slowing the flow too much.)

A two-sided (bottom and side) view of a conceptual design for a SCRAMjet powered hypersonic vehicle is given in Figure 9.33, patterned after the design discussed by Billig (Reference 69). Note the slender shape for aerodynamic efficiency (high lift-to-drag ratio), the wedge and isentropic ramp to compress the flow before entering the engine modules, and the translating cowl in order to properly position the impinging shock wave on the cowl leading edge as the freestream Mach number changes. We again

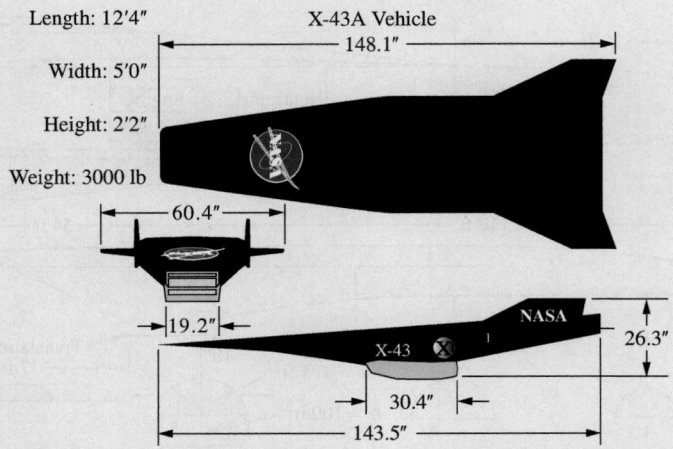

Figure 9.34 The X-43A, Hyper-X, hypersonic research vehicle.
(*NASA*)

make note that the whole undersurface of the vehicle is an integral part of the SCRAMjet engine cycle. For air-breathing hypersonic vehicles, the problem of airframe/propulsion integration is paramount; it is a major driving design feature of such aircraft.

America flew its first SCRAMjet-powered flight vehicle, the X-43, also labeled the Hyper-X, in 2004. A three-view of the X-43 is given in Figure 9.34. This small unpiloted test vehicle was launched from a modified Orbital Sciences Pegasus first stage rocket

booster, which in turn was launched from a B-52 bomber in flight. The primary purpose of the X-43 is to demonstrate the viability of a SCRAMjet engine under actual flight conditions, as opposed to research results in ground test facilities. In particular, in two successful test flights it successively demonstrated performance at $M_\infty = 7$ and 10. The X-43 is a NASA project; it was the first free-flight of an airframe integrated supersonic combustion ramjet engine.

EXAMPLE 9.10

In the preceding discussion on SCRAMjet engines, an isentropic compression wave was mentioned as one of the possible compression mechanisms. Consider the isentropic compression surface sketched in Figure 9.35a. The Mach number and pressure upstream of the wave are $M_1 = 10$ and $p_1 = 1$ atm, respectively. The flow is turned through a total angle of 15°. Calculate the Mach number and pressure in region 2 behind the compression wave.

■ **Solution**

From Appendix C, for $M_1 = 10$, $v_1 = 102.3°$. In region 2,

$$v_2 = v_1 - \theta = 102.3 - 15 = 87.3°$$

From Appendix C for $v_2 = 87.3°$, we have (closest entry)

$$\boxed{M_2 = 6.4}$$

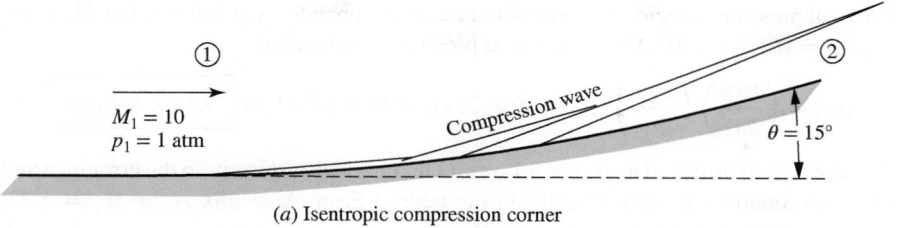

(a) Isentropic compression corner

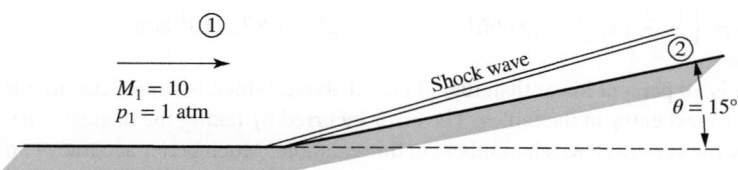

(b) Shock compression corner

Figure 9.35 Figures for (a) Example 9.9 and (b) Example 9.10.

From Appendix A, for $M_1 = 10$, $p_{0,1}/p_1 = 0.4244 \times 10^5$ and for $M_2 = 6.4$, $p_{0,2}/p_2 = 0.2355 \times 10^4$. Since the flow is isentropic, $p_{0,2} = p_{0,1}$, and hence

$$p_2 = \left(\frac{p_2}{p_{0,2}}\right)\left(\frac{p_{0,2}}{p_{0,1}}\right)\left(\frac{p_{0,1}}{p_1}\right)p_1 = \left(\frac{1}{0.2355 \times 10^4}\right)(1)(0.4244 \times 10^5)(1)$$

$$= \boxed{18.02 \text{ atm}}$$

EXAMPLE 9.11

Consider the flow over a compression corner with the same upstream conditions of $M_1 = 10$ and $p_1 = 1$ atm as in Example 9.10, and the same turning angle of 15°, except in this case the corner is sharp and the compression takes place through an oblique shock wave as sketched in Figure 9.35b. Calculate the downstream Mach number, static pressure, and total pressure in region 2. Compare the results with those obtained in Example 9.10, and comment on the significance of the comparison.

■ **Solution**

From Figure 9.9 for $M_1 = 10$ and $\theta = 15°$, the wave angle is $\beta = 20°$. The component of the upstream Mach number perpendicular to the wave is

$$M_{n,1} = M_1 \sin\beta = (10)\sin 20° = 34.2$$

From Appendix B for $M_{n,1} = 3.42$, we have (nearest entry), $p_2/p_1 = 13.32$, $p_{0,2}/p_{0,1} = 0.2322$, and $M_{n,2} = 0.4552$. Hence

$$M_2 = \frac{M_{n,2}}{\sin(\beta - \theta)} = \frac{0.4552}{\sin(20 - 15)} = \boxed{5.22}$$

$$p_2 = (p_2/p_1)p_1 = 13.32(1) = \boxed{13.32 \text{ atm}}$$

The total pressure in region 1 can be obtained from Appendix A as follows. For $M_1 = 10$, $p_{0,1}/p_1 = 0.4244 \times 10^5$. Hence, the total pressure in region 2 is

$$p_{0,2} = \left(\frac{p_{0,2}}{p_{0,1}}\right)\left(\frac{p_{0,1}}{p_1}\right)(p_1) = (0.2322)(0.4244 \times 10^5)(1) = \boxed{9.85 \times 10^3 \text{ atm}}$$

As a check, we can calculate $p_{0,2}$ as follows. (This check also alerts us to the error incurred when we round to the nearest entry in the tables.) From Appendix A for $M_2 = 5.22$, $p_{0,2}/p_2 = 0.6661 \times 10^3$ (nearest entry). Hence,

$$p_{0,2} = \left(\frac{p_{0,2}}{p_2}\right)(p_2) = (0.6661 \times 10^3)(13.32) = 8.87 \times 10^3 \text{ atm}$$

Note this answer is 10 percent lower than that obtained above, which is simply due to our rounding to the nearest entry in the tables. The error incurred by taking the nearest entry is exacerbated by the very high Mach numbers in this example. Much better accuracy can be obtained by properly *interpolating* between table entries.

Comparing the results from this example and Example 9.10, we clearly see that the isentropic compression is a more efficient compression process, yielding a downstream Mach number and pressure that are both considerably higher than in the case of the shock wave. The inefficiency of the shock wave is measured by the loss of total pressure across the shock; total pressure drops by about 77 percent across the shock. This emphasizes why designers of supersonic and hypersonic inlets would love to have the compression process carried out via isentropic compression waves. However, as noted in our discussion on SCRAMjets, it is very difficult to achieve such a compression in real life; the contour of the compression surface must be quite precise, and it is a point design for the given upstream Mach number. At off-design Mach numbers, even the best-designed compression contour will result in shocks.

9.7 SHOCK-EXPANSION THEORY: APPLICATIONS TO SUPERSONIC AIRFOILS

Consider a flat plate of length c at an angle of attack α in a supersonic flow, as sketched in Figure 9.36. On the top surface, the flow is turned away from itself; hence, an expansion wave occurs at the leading edge, and the pressure on the top surface p_2 is less than the freestream pressure $p_2 < p_1$. At the trailing edge, the flow must return to approximately (but not precisely) the freestream direction. Here, the flow is turned back into itself, and consequently a shock wave occurs at the trailing edge. On the bottom surface, the flow is turned into itself; an oblique shock wave occurs at the leading edge, and the pressure on the bottom surface p_3 is greater than the freestream pressure $p_3 > p_1$. At the trailing edge, the flow is turned into approximately (but not precisely) the freestream direction by means of an expansion wave. Examining Figure 9.36, note that the top and bottom surfaces of the flat plate experience uniform pressure distribution of p_2 and p_3, respectively,

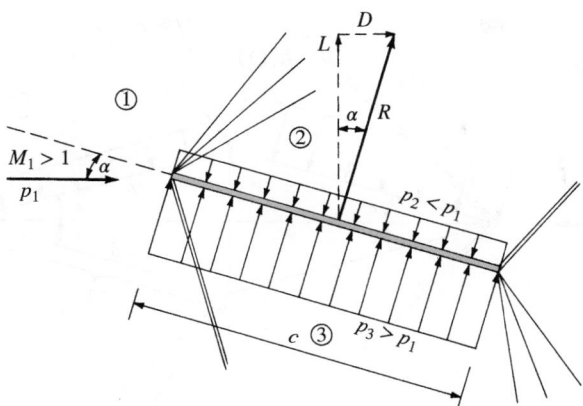

Figure 9.36 Flat plate at an angle of attack in a supersonic flow.

and that $p_3 > p_2$. This creates a net pressure imbalance that generates the resultant aerodynamic force R, shown in Figure 9.36. Indeed, for a unit span, the resultant force and its components, lift and drag, per unit span are

$$R' = (p_3 - p_2)c \tag{9.46}$$
$$L' = (p_3 - p_2)c \cos \alpha \tag{9.47}$$
$$D' = (p_3 - p_2)c \sin \alpha \tag{9.48}$$

In Equations (9.47) and (9.48), p_3 is calculated from oblique shock properties (Section 9.2), and p_2 is calculated from expansion-wave properties (Section 9.6). Moreover, these are *exact* calculations; no approximations have been made. The inviscid, supersonic flow over a flat plate at angle of attack is exactly given by the combination of shock and expansion waves sketched in Figure 9.36.

The flat-plate case given above is the simplest example of a general technique called *shock-expansion theory*. Whenever we have a body made up of straight-line segments and the deflection angles are small enough so that no detached shock waves occur, the flow over the body goes through a series of distinct oblique shock and expansion waves, and the pressure distribution on the surface (hence the lift and drag) can be obtained *exactly* from both the shock- and expansion-wave theories discussed in this chapter.

As another example of the application of shock-expansion theory, consider the diamond-shaped airfoil in Figure 9.37. Assume the airfoil is at 0° angle of attack. The supersonic flow over the airfoil is first compressed and deflected through the angle ε by the oblique shock wave at the leading edge. At midchord, the flow is expanded through an angle 2ε, creating an expansion wave. At the trailing edge, the flow is turned back to the freestream direction through another oblique shock. The pressure distributions on the front and back faces of the airfoil are sketched in Figure 9.37; note that the pressures on faces a and c are uniform and equal

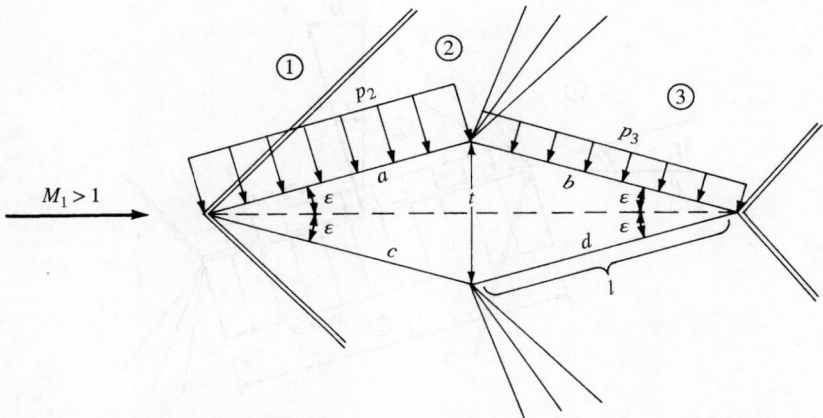

Figure 9.37 Diamond-wedge airfoil at zero angle of attack in a supersonic flow.

to p_2 and that the pressures on faces b and d are also uniform but equal to p_3, where $p_3 < p_2$. In the lift direction, perpendicular to the freestream, the pressure distributions on the top and bottom faces exactly cancel (i.e., $L' = 0$). In contrast, in the drag direction, parallel to the freestream, the pressure on the front faces a and c is larger than on the back faces b and d, and this results in a finite drag. To calculate this drag (per unit span), consider the geometry of the diamond airfoil in Figure 9.37, where l is the length of each face and t is the airfoil thickness. Then,

$$D' = 2(p_2 l \sin \varepsilon - p_3 l \sin \varepsilon) = 2(p_2 - p_3)\frac{t}{2}$$

Hence,
$$D' = (p_2 - p_3)t \tag{9.49}$$

In Equation (9.49), p_2 is calculated from oblique shock theory, and p_3 is obtained from expansion-wave theory. Moreover, these pressures are the *exact* values for supersonic, inviscid flow over the diamond airfoil.

At this stage, it is worthwhile to recall our discussion in Section 1.5 concerning the source of aerodynamic force on a body. In particular, examine Equations (1.1), (1.2), (1.7), and (1.8). These equations give the means to calculate L' and D' from the pressure and shear stress distributions over the surface of a body of general shape. The results of the present section, namely, Equations (9.47) and (9.48) for a flat plate and Equation (9.49) for the diamond airfoil, are simply specialized results from the more general formulae given in Section 1.5. However, rather than formally going through the integration indicated in Equations (1.7) and (1.8), we obtained our results for the simple bodies in Figures 9.36 and 9.37 in a more direct fashion.

The results of this section illustrate a very important aspect of inviscid, supersonic flow. Note that Equation (9.48) for the flat plate and Equation (9.49) for the diamond airfoil predict a *finite drag* for these two-dimensional profiles.

This is in direct contrast to our results for two-dimensional bodies in a low-speed, incompressible flow, as discussed in Chapters 3 and 4, where the drag was theoretically zero. That is, in supersonic flow, d'Alembert's paradox does not occur. In a supersonic, inviscid flow, the drag per unit span on a two-dimensional body is finite. This new source of drag is called *wave drag,* and it represents a serious consideration in the design of all supersonic airfoils. The existence of wave drag is inherently related to the increase in entropy and consequently to the loss of total pressure across the oblique shock waves created by the airfoil.

Finally, the results of this section represent a merger of both the left- and right-hand branches of our road map shown in Figure 9.6. As such, it brings us to a logical conclusion of our discussion of oblique waves in supersonic flows.

EXAMPLE 9.12

Calculate the lift and drag coefficients for a flat plate at a 5° angle of attack in a Mach 3 flow.

■ **Solution**

Refer to Figure 9.36. First, calculate p_2/p_1 on the upper surface. From Equation (9.43),

$$v_2 = v_1 + \theta$$

where $\theta = \alpha$. From Appendix C, for $M_1 = 3$, $v_1 = 49.76°$. Thus,

$$v_2 = 49.76° + 5° = 54.76°$$

From Appendix C,

$$M_2 = 3.27$$

From Appendix A, for $M_1 = 3$, $p_{0_1}/p_1 = 36.73$; for $M_2 = 3.27$, $p_{0_2}/p_2 = 55$. Since $p_{0_1} = p_{0_2}$,

$$\frac{p_2}{p_1} = \frac{p_{0_1}}{p_1} \bigg/ \frac{p_{0_2}}{p_2} = \frac{36.73}{55} = 0.668$$

Next, calculate p_3/p_1 on the bottom surface. From the θ-β-M diagram (Figure 9.9), for $M_1 = 3$ and $\theta = 5°$, $\beta = 23.1°$. Hence,

$$M_{n,1} = M_1 \sin \beta = 3 \sin 23.1° = 1.177$$

From Appendix B, for $M_{n,1} = 1.177$, $p_3/p_1 = 1.458$ (nearest entry).

Returning to Equation (9.47), we have

$$L' = (p_3 - p_2)c \cos \alpha$$

The lift coefficient is obtained from

$$c_l = \frac{L'}{q_1 S} = \frac{L'}{(\gamma/2)p_1 M_1^2 c} = \frac{2}{\gamma M_1^2}\left(\frac{p_3}{p_1} - \frac{p_2}{p_1}\right)\cos \alpha$$

$$= \frac{2}{(1.4)(3)^2}(1.458 - 0.668)\cos 5° = \boxed{0.125}$$

From Equation (9.48),

$$D' = (p_3 - p_2)c \sin \alpha$$

Hence,

$$c_d = \frac{D'}{q_1 S} = \frac{2}{\gamma M_1^2}\left(\frac{p_3}{p_1} - \frac{p_2}{p_1}\right)\sin\alpha$$

$$= \frac{2}{(1.4)(3^2)}(1.458 - 0.668)\sin 5° = \boxed{0.011}$$

A slightly simpler calculation for c_d is to recognize from Equations (9.47) and (9.48), or from the geometry of Figure 9.36, that

$$\frac{c_d}{c_l} = \tan \alpha$$

Hence,

$$c_d = c_l \tan\alpha = 0.125 \tan 5° = 0.011$$

9.8 A COMMENT ON LIFT AND DRAG COEFFICIENTS

Expanding on the comments made in Section 9.3.1, reflect again on the result obtained in Example 9.6, where the drag coefficient was calculated for a 15° half-angle wedge in a Mach 5 flow. Reflect also on the result obtained in Example 9.12, where the lift and drag coefficients were calculated for a flat plate at a 5° angle of attack in a Mach 3 flow. Note that to calculate these coefficients we did not need to know the freestream pressure, density, or velocity. All we needed to know was:

1. The shape of the body
2. The angle of attack
3. The freestream Mach number

These examples are clear-cut illustrations of the results of dimensional analysis discussed in Section 1.7, and are totally consistent with Equations (1.42) and (1.43), which emphasize that lift and drag coefficients for a body of given shape are functions of *only* Reynolds number, Mach number, and angle of attack. For the examples in this chapter, we are dealing with an inviscid flow, so Re is not relevant—only M_∞ and α.

9.9 THE X-15 AND ITS WEDGE TAIL

Examine the photograph of the X-15 hypersonic research vehicle shown in Figure 9.18. The viewpoint in this photograph is looking down at the top of the vehicle. Concentrate on the vertical tail at the rear of the airplane. The cross section of the vertical tail is clearly seen—it is a *wedge* cross section in contrast to the type of thin symmetric airfoil sections usually employed for vertical tails on airplanes. The wedge shape is further emphasized by examining Figure 9.38,

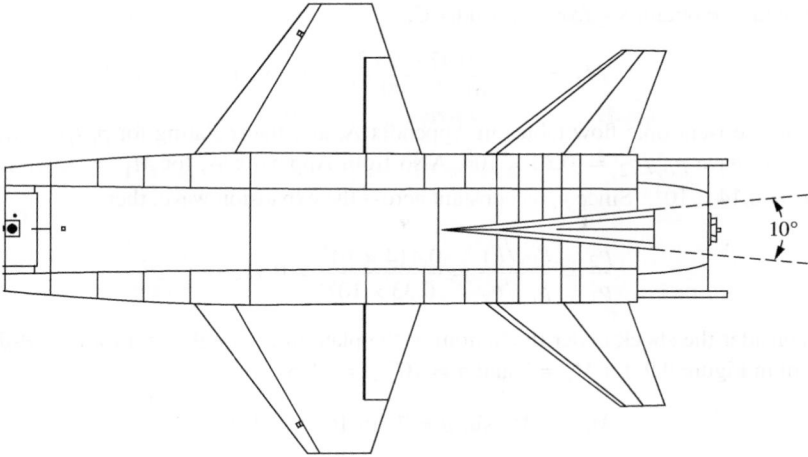

Figure 9.38 Illustration of the wedge-shaped tail on the X-15.

which shows a drawing of the rear top portion of the X-15. The included angle of the wedge cross section is 10°.

The wedge tail is one of the unique design aspects of the X-15. It came out of concern for stability problems at hypersonic speeds. The earlier X-1 and X-2 supersonic research vehicles had encountered such problems at much lower Mach numbers, and one of the major early concerns in the design of the X-15 was to find a solution that would provide stability up to Mach 7. The answer was provided by C. H. McLellan, an NACA engineer at the NACA Langley Memorial Laboratory. McLellan had carried out theoretical calculations of the influence of airfoil shape on normal force at hypersonic speeds. He found that a 10° wedge was more effective than a thin supersonic section. The X-15 designers were aware of McLellan's work, and designed the aircraft with a 10° wedge tail that provided adequate directional stability for the hypersonic vehicle.

Why is a wedge tail more effective than one with a thin section? To help answer this question, consider the following example.

EXAMPLE 9.13

Consider the flat plate shown in Figure 9.39a and the 10° included angle wedge shown in Figure 9.39b, both at an angle of attack of 10° in a Mach 7 airstream. (a) Calculate the lift coefficient of the flat plate. (b) Calculate the lift coefficient of the wedge.

■ **Solution**

(a) First, consider the expansion wave over the top of the plate. From Appendix C, for $M_1 = 7$, $v_1 = 90.97°$. From Equation (9.43)

$$v_2 = v_1 + \alpha = 90.97° + 10° = 100.97°$$

Interpolating to obtain M_2 from Appendix C,

$$M_2 = 9 + \frac{100.97 - 99.32}{102.3 - 99.32} (1) = 9.56$$

Going to the isentropic flow tables in Appendix A, and interpolating for p_o/p between entries, we have $p_{o_2}/p_2 = 0.33 \times 10^5$. Also from Appendix A, for $M_1 = 7$, we have $p_{o_1}/p_1 = 0.14 \times 10^4$. Since p_o is constant across the expansion wave, then

$$\frac{p_2}{p_1} = \frac{p_{o_1}/p_1}{p_{o_2}/p_2} = \frac{0.414 \times 10^4}{0.33 \times 10^5} = 0.1255$$

Now consider the shock under the bottom of the plate in Figure 9.39a. From the θ-β-M diagram in Figure 9.9, for $M_1 = 7$ and $\alpha = 10°, \beta = 16.5°$,

$$M_{n,1} = M_1 \sin \beta = 7 \sin 16.5° = 1.99$$

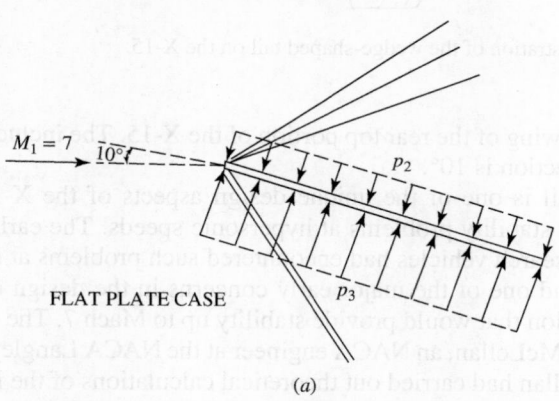

FLAT PLATE CASE

(a)

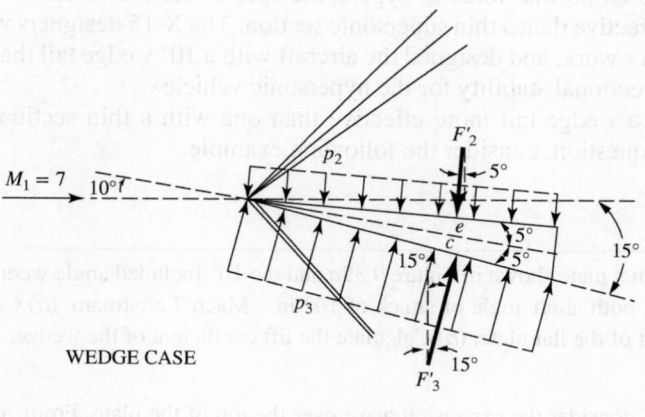

WEDGE CASE

(b)

Note: Angles not to scale.

Figure 9.39 Schematic of hypersonic flow over (a) a flat plate, and (b) a wedge, both at a 10° angle of attack. Not to scale.

From Appendix B, for $M_{n,1} = 1.99$, interpolating, we have

$$\frac{p_3}{p_1} = 4.407 + (0.093)(0.5) = 4.45$$

The lift coefficient for a supersonic or hypersonic flat plate was derived in Example 9.12 as

$$c_\ell = \frac{2}{\gamma M_1^2}\left(\frac{p_3}{p_1} - \frac{p_2}{p_1}\right)\cos\alpha$$

$$= \frac{2}{(1.4)(7)^2}(4.45 - 0.1255) = \boxed{0.126}$$

(*b*) First consider the expansion wave over the top of the wedge.

$$v_2 = v_1 + 5° = 90.97° + 5° = 95.97°$$

From Appendix C, interpolating,

$$M_2 = 8 + \frac{95.97 - 96.62}{99.32 - 95.62}(1) = 8.1$$

From Appendix A, interpolating,

$$\frac{p_{0_2}}{p_2} = 0.9763 \times 10^4 + (0.211 \times 10^5 - 0.9763 \times 10^4)(1) = 1.0897 \times 10^4$$

The relation between the chord length c and the length of the face of the wedge ℓ is

$$\ell = \frac{c}{\cos 5°} = \frac{c}{0.996} = 1.004c$$

The force per unit span, F_2', acting on the top surface of the wedge, is

$$F_2' = p_2 \ell = \left(\frac{p_{0_1}/p_1}{p_{0_2}/p_2}\right)p_1 \ell$$

For $M_1 = 7$, from Appendix A, $p_{0_1}/p_1 = 0.414 \times 10^4$. Thus,

$$F_2' = \left(\frac{0.414 \times 10^4}{1.0897 \times 10^4}\right)p_1\ell = 0.38p_1\ell$$

Considering the shock wave under the bottom of the wedge, we have, from the θ-β-M diagram, for $M_1 = 7$ and $\theta = 15°$, $\beta = 23.5°$. Thus,

$$M_{n,1} = M_1 \sin\beta = 7\sin 23.5° = 2.79$$

From Appendix B, interpolating,

$$\frac{p_3}{p_1} = 8.656 + (8.98 - 8.656)(0.8) = 8.915$$

Thus, the force per unit span, F_3', acting on the bottom surface of the wedge, is

$$F_3' = p_3 \ell = \left(\frac{p_3}{p_1}\right)p_1 \ell = 8.915p_1 \ell$$

The lift per unit span is the combination of the *components* of F_2' and F_3' perpendicular to the freestream. Examining Figure 9.39*b*, we see that

$$L' = F_3' \cos 15° - F_2' \cos 5° = 0.9659F_3' - 0.9962F_2'$$

$$L' = (0.9659)(8.915)p_1 \ell - (0.9962)(0.38)p_1 \ell$$

$$L' = 8.232p_1 \ell$$

However, $\ell = 1.004c$. Thus,

$$L' = 8.232p_1(1.004c) = 8.265p_1 c$$

The lift coefficient is

$$c_\ell = \frac{L'}{q_1 c} = \frac{L'}{(\gamma/2)p_1 M_1^2 c} = \frac{2L'}{\gamma p_1 M_1^2 c}$$

Since $L' = 8.265p_1 c$, we have

$$c_\ell = \frac{2(8.265)p_1 c}{(1.4)p_1(7)^2 c} = \boxed{0.241}$$

From the results of Example 9.13, the lift coefficient for the wedge is double that for the flat plate. For a vertical tail surface, the "lift is the *side force* that creates the restoring yaw moment provided by the vertical tail when the airplane experiences a disturbance or displacement in yaw." Clearly, the wedge provides a much stronger restoring moment than a very thin airfoil shape represented by the flat plate in Figure 9.39*a*.

Physically, the wedge at angle of attack, when compared to a flat plate at the same angle of attack, is taking advantage of the nonlinear nature of supersonic shock waves. When the flat plate in Example 9.13*a* is pitched to a 10° angle of attack, the deflection angle of the flow over the bottom surface is also 10°. In contrast, with the wedge in Example 9.13*b*, the deflection angle of the flow over the bottom surface is already at 5° when the wedge is at zero angle of attack, and then is increased to 15° when the wedge is pitched to an angle of attack of 10°. This gives a shock wave angle for the wedge of 23.5°, larger than the 16.5° wave angle for the flat plate. Examining Equations (9.13) and (9.16), we note that the pressure ratio across an oblique shock varies essentially as the *square* of the wave angle. This is why the pressure ratio on the bottom surface of the wedge from Example 9.13*b* is *twice* as large as that for the flat plate in Example 9.13*a*, $p_3/p_1 = 8.915$ for the wedge as compared to $p_3/p_1 = 4.45$ for the flat plate. The wedge, by already starting with a flow deflection angle at zero angle of attack, gets "more bang for the buck" at angle of attack.

Finally, we note that the wedge tail on the X-15 is a beautiful example of how theoretical aerodynamic research, done to extend the aerodynamic state of the art, was taken off the library shelves later on to solve a show-stopping problem of major importance to the practical design of a pioneering airplane, the X-15. McLellan's research at the NACA helped to make the X-15 possible. (See Reference 116.)

9.10 VISCOUS FLOW: SHOCK-WAVE/ BOUNDARY-LAYER INTERACTION

Shock waves and boundary layers do not mix; bad things can happen when a shock wave impinges on a boundary layer. Unfortunately, shock-wave/boundary-layer interactions frequently occur in practical supersonic flows, and therefore we pay attention to this interaction in the present section. The fluid dynamics of a shock-wave/boundary-layer interaction is complex (and extremely interesting), and a detailed presentation is beyond the scope of this book. Here we give a brief qualitative discussion—just enough to acquaint you with the basic picture.

Consider a supersonic flow over a surface wherein an oblique shock wave impinges on the surface, such as sketched in Figure 9.19. In this figure the flow is assumed to be inviscid, and the incident shock impinges at point B on the upper wall, giving rise to a reflected shock emanating from the same point. There is a discontinuous pressure increase at point B, a combination of the pressure increases across the incident and reflected shocks. Indeed, point B is a singular point where there is an infinitely large adverse pressure gradient.

Imagine that suddenly we have a boundary layer along the wall in Figure 9.19. At point B, the boundary layer would experience an infinitely large adverse pressure gradient. In Section 4.12, we discussed what happens to a boundary layer when it experiences a large adverse pressure gradient—it separates from the surface. These are the basic elements of the shock-wave/boundary-layer interaction. The incident shock wave imposes a strong adverse pressure gradient on the boundary layer, which in turn separates from the surface, and the resulting flow field in the vicinity of the shock wave impingement becomes one of a mutual interaction between the boundary layer and the shock wave.

This mutual interaction is sketched qualitatively in Figure 9.40. Here, for ease of presentation, we show a shock wave impinging on a lower wall rather than on the upper wall as in Figure 9.19. In Figure 9.40, we see a boundary layer growing along a flat plate. Because the external flow is supersonic, the boundary-layer velocity profile is subsonic near the wall and supersonic near the outer edge. At some downstream location, an incident shock impinges on the boundary layer. The large pressure rise across the shock wave acts as a severe adverse pressure gradient imposed on the boundary layer, thus causing the boundary layer to locally separate from the surface. Because the high pressure behind the shock feeds upstream through the subsonic portion of the boundary layer, the separation takes place ahead of the theoretical inviscid flow impingement point of the incident shock wave. In turn, the separated boundary layer deflects the external supersonic flow into itself, thus inducing a second shock wave, identified here as the induced separation shock wave. The separated boundary layer subsequently turns back toward the plate, reattaching to the surface at some downstream location. Here again the supersonic flow is deflected into itself, causing a third shock wave called the *reattachment shock*. Between the separation and reattachment shocks, where the boundary layer is turning back toward the surface, the supersonic flow is turned away from itself, generating expansion waves shown in

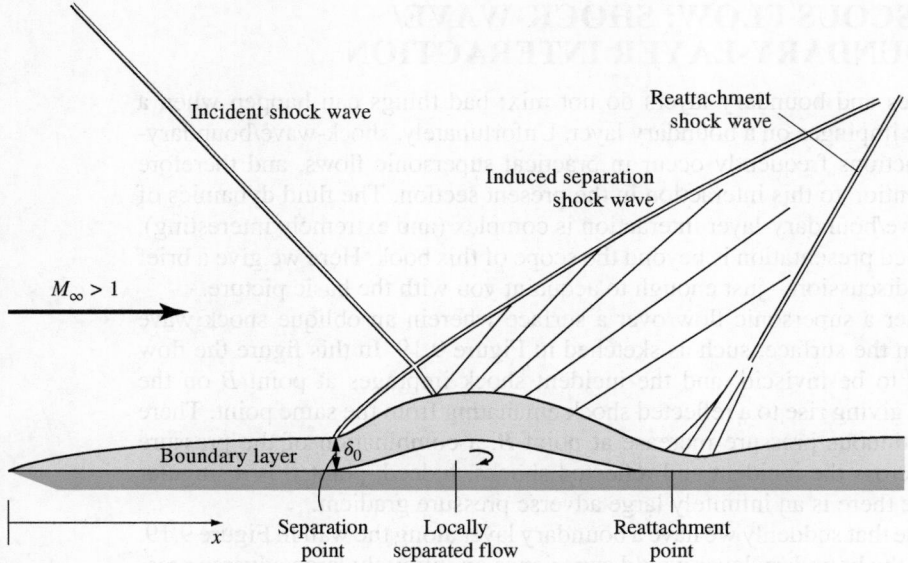

Figure 9.40 Schematic of the shock-wave/boundary-layer interaction.

Figure 9.40. At the point of reattachment, the boundary layer has become relatively thin, the pressure is high, and consequently this becomes a region of high local aerodynamic heating. Further away from the plate, the separation and reattachment shocks merge to form the conventional reflected shock wave that is expected from the inviscid picture, as shown in Figure 9.19. The scale and severity of the interaction shown in Figure 9.40 depends on whether the boundary layer is laminar or turbulent. Since laminar boundary layers separate more readily than turbulent boundary layers (see Section 4.12), the laminar interaction usually takes place more readily with more severe attendant consequences than the turbulent interaction. However, the general qualitative aspects of the interaction shown in Figure 9.40 are the same for both cases.

   The shock-wave/boundary-layer interaction has a major effect on the pressure, shear stress, and heat-transfer distributions along the wall. Of particular consequence is the high local heat-transfer rate at the reattachment point, which at hypersonic speeds can peak to an order of magnitude larger than at neighboring locations. An example of the effect on the wall pressure distribution is shown in Figure 9.41a, patterned after the work of Baldwin and Lomax at the NASA Ames Research Center (B. S. Baldwin and H. Lomax, "Thin Layer Approximation and Algebraic Model for Separated Turbulent Flows," AIAA Paper No. 78-257, January 1978). Here, the pressure distribution along the wall in the interaction region is plotted versus distance along the wall, x, where x_0 is the theoretical point of impingement for the incident shock in the inviscid flow case. The pressure

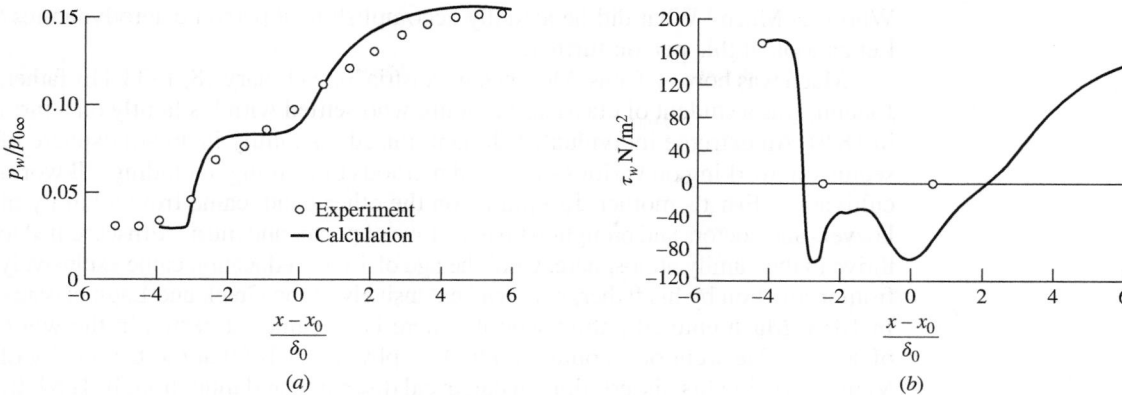

Figure 9.41 Effects of shock-wave/boundary-layer interaction on (*a*) pressure distribution, and (*b*) shear stress for Mach 3 turbulent flow over a flat plate.

distribution shows a steplike increase, with an intermediate plateau; this distribution is typical of the shock-wave/boundary-layer interaction. The external flow is at Mach 3 ahead of the incident shock, and the boundary layer is turbulent. The solid curve is a computational fluid dynamic (CFD) calculation, and the circles are experimental data. Notice that the pressure increase extends a distance equal to about four times the boundary-layer thickness ahead of the theoretical inviscid impingement point. The shear stress distribution is shown in Figure 9.41*b*. In the pocket of separated flow, τ_w becomes small and reverses its direction (negative values of τ_w) due to the low energy recirculating flow.

Because of the consequent creation of separated flow, increased loss of total pressure, and high peak heat-transfer rates, shock-wave/boundary-layer interactions usually should be avoided as much as possible in the design of supersonic aircraft and flow devices. However, this is easier said than done. Shock-wave/boundary-layer interactions are a fact of life in the practical world of supersonic flow, and that is why we have discussed their basic nature in this section. On the other hand, modern creative ideas have led to the beneficial use of the separated flow from a shock-wave/boundary-layer interaction to actually enhance the off-design performance of jet-engine exhaust nozzles and for certain types of flow control. So the picture is not entirely black.

9.11 HISTORICAL NOTE: ERNST MACH—A BIOGRAPHICAL SKETCH

The Mach number is named in honor of Ernst Mach, an Austrian physicist and philosopher who was an illustrious and controversial figure in late nineteenth-century physics. Mach conducted the first meaningful experiments in supersonic flight, and his results triggered a similar interest in Ludwig Prandtl 20 years later.

Who was Mach? What did he actually accomplish in supersonic aerodynamics? Let us look at this person further.

Mach was born at Turas, Moravia, in Austria, on February 18, 1838. His father, Johann, was a student of classical literature who settled with his family on a farm in 1840. An extreme individualist, Johann raised his family in an atmosphere of seclusion, working on various improved methods of farming, including silkworm cultivation. Ernst's mother, Josephine, on the other hand, came from a family of lawyers and doctors and brought with her a love of poetry and music. Ernst seemed to thrive in this family atmosphere. Until the age of 14, his education came exclusively from instruction by his father, who read extensively in the Greek and Latin classics. In 1853, Mach entered public school, where he became interested in the world of science. He went on to obtain a Ph.D. in physics in 1860 at the University of Vienna, writing his dissertation on electrical discharge and induction. In 1864, he became a full professor of mathematics at the University of Graz and was given the title of Professor of Physics in 1866. Mach's work during this period centered on optics—a subject which was to interest him for the rest of his life. The year 1867 was important for Mach—during that year he married, and he also became a professor of experimental physics at the University of Prague, a position he held for the next 28 years. While in Prague, Mach published over 100 technical papers—work that was to constitute the bulk of his technical contributions.

Mach's contribution to supersonic aerodynamics involves a series of experiments covering the period from 1873 to 1893. In collaboration with his son, Ludwig, Mach studied the flow over supersonic projectiles, as well as the propagation of sound waves and shock waves. His work included the flow fields associated with meteorites, explosions, and gas jets. The main experimental data were photographic results. Mach combined his interest in optics and supersonic motion by designing several photographic techniques for making shock waves in air visible. He was the first to use the schlieren system in aerodynamics; this system senses density gradients and allows shock waves to appear on screens or photographic negatives. He also devised an interferometric technique that senses directly the change in density in a flow. A pattern of alternate dark and light bands are set up on a screen by the superposition of light rays passing through regions of different density. Shock waves are visible as a shift in this pattern along the shock. Mach's optical device still perpetuates today in the form of the Mach-Zehnder interferometer, an instrument present in many aerodynamic laboratories. Mach's major contributions in supersonic aerodynamics are contained in a paper given to the Academy of Sciences in Vienna in 1887. Here, for the first time in history, Mach shows a photograph of a weak wave on a slender cone moving at supersonic speed, and he demonstrates that the angle μ between this wave and the direction of flight is given by $\sin \mu = a/V$. This angle was later denoted as the Mach angle by Prandtl and his colleagues after their work on shock and expansion waves in 1907 and 1908. Also, Mach was the first person to point out the discontinuous and marked changes in a flow field as the ratio V/a changes from below 1 to above 1.

It is interesting to note that the ratio V/a was not denoted as Mach number by Mach himself. Rather, the term "Mach number" was coined by the Swiss engineer Jacob Ackeret in his inaugural lecture in 1929 as Privatdozent at the Eidgenossiche Technische Hochschule in Zurich. Hence, the term "Mach number" is of fairly recent usage, not being introduced into the English literature until the mid-1930s.

In 1895, the University of Vienna established the Ernst Mach chair in the philosophy of inductive sciences. Mach moved to Vienna to occupy this chair. In 1897 he suffered a stroke which paralyzed the right side of his body. Although he eventually partially recovered, he officially retired in 1901. From that time until his death on February 19, 1916, near Munich, Mach continued to be an active thinker, lecturer, and writer.

In our time, Mach is most remembered for his early experiments on supersonic flow and, of course, through the Mach number itself. However, Mach's contemporaries, as well as Mach himself, viewed him more as a philosopher and historian of science. Coming at the end of the nineteenth century, when most physicists felt comfortable with Newtonian mechanics, and many believed that virtually all was known about physics, Mach's outlook on science is summarized by the following passage from his book *Die Mechanik:*

> The most important result of our reflections is that precisely the apparently simplest mechanical theorems are of a very complicated nature; that they are founded on incomplete experiences, even on experiences that never can be fully completed; that in view of the tolerable stability of our environment they are, in fact, practically safeguarded to serve as the foundation of mathematical deduction; but that they by no means themselves can be regarded as mathematically established truths, but only as theorems that not only admit of constant control by experience but actually require it.

In other words, Mach was a staunch experimentalist who believed that the established laws of nature were simply theories and that only observations that are apparent to the senses are the fundamental truth. In particular, Mach could not accept the elementary ideas of atomic theory or the basis of relativity, both of which were beginning to surface during Mach's later years and, of course, were to form the basis of twentieth-century modern physics. As a result, Mach's philosophy did not earn him favor with most of the important physicists of his day. Indeed, at the time of his death, Mach was planning to write a book pointing out the flaws of Einstein's theory of relativity.

Although Mach's philosophy was controversial, he was respected for being a thinker. In fact, in spite of Mach's critical outlook on the theory of relativity, Albert Einstein had the following to say in the year of Mach's death: "I even believe that those who consider themselves to be adversaries of Mach scarcely know how much of Mach's outlook they have, so to speak, absorbed with their mother's milk."

Hopefully, this section has given you a new dimension to think about whenever you encounter the term "Mach number." Maybe you will pause now and then to reflect on the namesake and to appreciate that the term "Mach number" is in honor of a person who devoted his life to experimental physics, but who at the same time was bold enough to view the physical world through the eyes of a self-styled philosopher.

9.12 SUMMARY

The road map given in Figure 9.6 illustrates the flow of our discussion on oblique waves in supersonic flow. Review this road map, and make certain that you are familiar with all the ideas and results that are represented in Figure 9.6.

Some of the more important results are summarized as follows:

An infinitesimal disturbance in a multidimensional supersonic flow creates a Mach wave that makes an angle μ with respect to the upstream velocity. This angle is defined as the Mach angle and is given by

$$\mu = \sin^{-1} \frac{1}{M} \tag{9.1}$$

Changes across an oblique shock wave are determined by the normal component of velocity ahead of the wave. For a calorically perfect gas, the normal component of the upstream Mach number is the determining factor. Changes across an oblique shock can be determined from the normal shock relations derived in Chapter 8 by using $M_{n,1}$ in these relations, where

$$M_{n,1} = M_1 \sin \beta \tag{9.13}$$

Changes across an oblique shock depend on two parameters, for example, M_1 and β, or M_1 and θ. The relationship between M_1, β, and θ is given in Figure 9.9, which should be studied closely.

Oblique shock waves incident on a solid surface reflect from that surface in such a fashion to maintain flow tangency on the surface. Oblique shocks also intersect each other, with the results of the intersection depending on the arrangement of the shocks.

The governing factor in the analysis of a centered expansion wave is the Prandtl-Meyer function $\nu(M)$. The key equation which relates the downstream Mach number M_2, the upstream Mach number M_1, and the deflection angle θ is

$$\theta = \nu(M_2) - \nu(M_1) \tag{9.43}$$

The pressure distribution over a supersonic airfoil made up of straight-line segments can usually be calculated exactly from a combination of oblique and expansion waves—that is, from exact shock-expansion theory.

9.13 INTEGRATED WORK CHALLENGE: RELATION BETWEEN SUPERSONIC WAVE DRAG AND ENTROPY INCREASE—IS THERE A RELATION?

Concept: Supersonic wave drag on a body is caused by the high pressures that occur behind shock waves acting on the surface of the body. The simplest illustration of this is shown in Figure 9.17 where the high static pressure behind an oblique shock wave is exerted on the front faces of a wedge, creating a pressure drag; this pressure drag is called *wave drag* because the high pressure on the surface is due to the presence of shock waves. Indeed, the net aerodynamic force on any body in a flow is the net integral of the pressure and shear stress distributions acting over the surface, as explained in Section 1.5. In a supersonic flow, the airflow adjacent to the surface of the body is processed through systems of shock and expansion waves, and the resulting surface pressure is a product of the flow through these waves. The net integral of the resulting surface pressure distribution in the drag direction yields a drag force that is labeled *wave drag*.

This straightforward concept relates surface pressure distributions to a force on a body; it is the most fundamental source of wave drag. However, is there another way of looking at how nature generates wave drag? Is there an alternative explanation? We recall our discussion in Section 3.22 about the relation between drag and the loss of total pressure in the flow. In Chapter 8, we saw a relation between loss of total pressure in flow and a corresponding entropy increase in the flow, as expressed by Equation (8.73). This leads to the question: Is there a relation between the generation of wave drag on a supersonic body and the increase in entropy of the flow over the body?

Challenge: Investigate this question.

Solution: Return to the discussion in Section 2.6 dealing with the application of the momentum equation to the drag of a two-dimensional body and to the control volume sketched in Figure 2.20a. Let us imagine a supersonic flow through the control volume and an aerodynamic body inside the control volume generating a system of shock and expansion waves. These waves will propagate above and below the body in the general downstream direction, as sketched in Figure 9.42. If the upper and lower boundaries of the control volume are drawn far enough away from the body, the waves will exit only through the downstream boundary, as seen in Figure 9.42. The static pressure is constant over the upstream and upper and lower boundaries. Assume that the downstream boundary is taken far enough downstream of the body such that the mutual interaction of the shock and expansion waves yield an *approximately* constant pressure over that boundary (not precisely the case, but reasonable enough for our discussion here). Recall that the fluid dynamic derivation in Section 2.6 was for a general flow. Therefore, Equations (2.78) and (2.79) apply for both incompressible and compressible flows, and the drag on the body in Figure 9.42 is

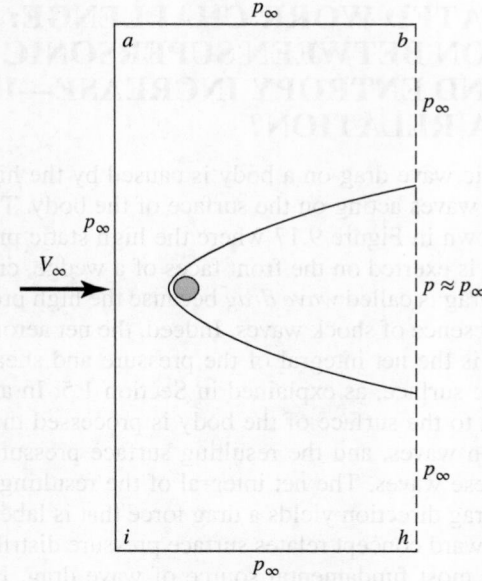

Figure 9.42 Control volume around a body in supersonic flow, with the shock wave exiting the downstream boundary.

given by

$$D' = \int_i^a \rho_1 u_1^2 \, dy - \int_h^b \rho_2 u_2^2 \, dy \tag{C9.1}$$

From Equation (8.42)

$$\frac{p_0}{p} = \left(1 + \frac{\gamma - 1}{2} M^2\right)^{\frac{\gamma}{\gamma - 1}}$$

Hence,

$$1 + \frac{\gamma - 1}{2} M^2 = \left(\frac{p_0}{p}\right)^{\frac{\gamma - 1}{\gamma}}$$

$$M^2 = \frac{2}{\gamma - 1} \left[\left(\frac{p_0}{p}\right)^{\frac{\gamma - 1}{\gamma}} - 1\right] \tag{C9.2}$$

Since

$$M = \frac{u}{a}, \text{Equation (9.2) becomes}$$

$$u^2 = \frac{2a^2}{\gamma - 1} \left[\left(\frac{p_0}{p}\right)^{\frac{\gamma - 1}{\gamma}} - 1\right]$$

Also,

$$a^2 = \gamma R T$$

Thus,

$$u^2 = \frac{2\gamma R T}{\gamma - 1} \left[\left(\frac{p_0}{p} \right)^{\frac{\gamma-1}{\gamma}} - 1 \right]$$

and

$$\rho u^2 = \frac{2\gamma \rho R T}{\gamma - 1} \left[\left(\frac{p_0}{p} \right)^{\frac{\gamma-1}{\gamma}} - 1 \right] \tag{C9.3}$$

Since $p = \rho R T$, Equation (C9.3) becomes

$$\rho u^2 = \frac{2\gamma p}{\gamma - 1} \left[\left(\frac{p_0}{p} \right)^{\frac{\gamma-1}{\gamma}} - 1 \right] \tag{C9.4}$$

Return to Equation (C9.1) and Figure 9.42. The left and right sides of the control volume are of equal length, i.e., $ia = hb$. Thus in Equation (C9.1), we will drop the limits on the integrals. Combining Equations (9.1) and (C9.4), we have

$$D' = \int \left\{ \frac{2\gamma p_1}{\gamma - 1} \left[\left(\frac{p_{0_1}}{p_1} \right)^{\frac{\gamma-1}{\gamma}} - 1 \right] - \frac{2\gamma p_2}{\gamma - 1} \left[\left(\frac{p_{0_2}}{p_2} \right)^{\frac{\gamma-1}{\gamma}} - 1 \right] \tag{C9.5} \right.$$

From Equation (8.72), the relation between change in total pressure and change in entropy between two points in the flow

$$s_2 - s_1 = \Delta s = -R \ell n \frac{p_{0_2}}{p_{0_1}}$$

or,

$$\frac{p_{0_2}}{p_{0_1}} = e^{-\Delta s/R} \tag{C9.6}$$

Also,

$$\frac{p_{0_1}}{p_1} = \frac{p_{0_1}}{p_{0_2}} \frac{p_{0_2}}{p_1} = \left(\frac{1}{e^{-\Delta s/R}} \right) \left(\frac{p_{0_2}}{p_1} \right) = (e^{\Delta s/R}) \frac{p_{0_2}}{p_1} \tag{C9.7}$$

and

$$\frac{p_{0_2}}{p_2} = \frac{p_{0_2}}{p_{0_1}} \frac{p_{0_1}}{p_2} = (e^{-\Delta s/R}) \left(\frac{p_{0_1}}{p_2} \right) \tag{C9.8}$$

Inserting Equations (C9.7) and (C9.8) into (C9.5), we have

$$D' = \int \left\{ \frac{2\gamma p_1}{\gamma - 1} \left[(e^{\Delta s/R})^{\frac{\gamma-1}{\gamma}} \left(\frac{p_{0_2}}{p_1} \right)^{\frac{\gamma-1}{\gamma}} - 1 \right] \right.$$

$$\left. - \frac{2\gamma p_2}{\gamma - 1} \left[(e^{-\Delta s/R})^{\frac{\gamma-1}{\gamma}} \left(\frac{p_{0_1}}{p_2} \right)^{\frac{\gamma-1}{\gamma}} - 1 \right] \right\} dy \tag{C9.9}$$

Equation (C9.9) is rather long, but concentrate just on what it tells us about the effect of increase of entropy, Δs, on the drag. Note that when entropy increases, i.e., Δs is positive, the magnitude of the first term on the right side of Equation (C9.9) increases and the magnitude of the second term decreases, *both of which serve to increase the drag, D'!* Also note that for an isentropic flow, $\Delta s = 0$, $p_{0_2} = p_{0_1}$. And we already have from Equation (9.42) that $p_1 = p_2$; in this case, Equation (9.9) yields $D' = 0$, i.e., for an isentropic flow, there is no wave drag, as we already know from physical considerations.

In summary, Equation (9.9) demonstrates the connection between wave drag and an increase of entropy in the flow. The two go hand in hand; in a supersonic flow over a body the wave drag on the body is related to the entropy increase in the flow.

9.14 INTEGRATED WORK CHALLENGE: THE SONIC BOOM

Concept: In popular culture, the sonic boom is understood to be a loud "boom" that one hears when an airplane flying overhead breaks the speed of sound. In reality, however, one hears a "sonic boom" whenever an airplane, or any flight vehicle, flies overhead at speeds faster than sound, no matter where the vehicle actually first started flying at supersonic speeds; indeed, that could have occurred a thousand miles away. So, just what is the sonic boom?

Challenge: Examine the sonic boom generated from a body in supersonic flight. What is it? How is it created? How can its strength be reduced?

Solution: Return to Figure 9.37 illustrating the wave pattern from a diamond-wedge airfoil at supersonic speeds. Imagine this body flying overheard. The wave pattern from the bottom of the airfoil propagates toward the ground, and when it sweeps past you standing on the ground, your eardrums pick up the pressure changes across the waves, creating a booming sound. What you hear is the "sonic boom." The body is dragging this wave pattern with it as it flies through the atmosphere, and hence the "sonic boom" is sweeping over the ground at the same speed as the body is flying through the air.

Examining Figure 9.37 in more detail, note that two shock waves are produced, one at the nose of the body, and one at the tail of the body. These two shocks propagate downward with the expansion wave contained between them, as illustrated in Figure 9.43. Also shown in Figure 9.43 is the change in pressure, Δp through the wave pattern. At large distances below the body, the variation in Δp shows a jump increase across the bow wave, an almost linear decrease between the bow and tail waves, and another jump increase across the tail wave. The variation of Δp through the wave pattern resembles the capital letter N, and for this reason the sonic boom pressure wave is called an "N-wave." Note that when this N-wave sweeps past you on the ground, you hear two booms separated in time by $\Delta t = \lambda / V_\infty$, where λ is the distance along the ground between the bow

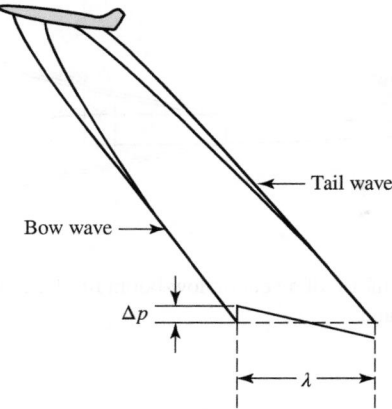

Figure 9.43 Classic N-wave sonic boom generation.

and tail waves, and V_∞ is the flight velocity of the supersonic body. In this case, what you hear as the "sonic boom" is in reality two booms very closely spaced in time, the so-called classic double-boom signature. For example, the Concorde supersonic transport cruised at a Mach number of 2 at 50,000 ft, where the standard atmospheric temperature is 390°R. The speed of sound at this temperature is $a = \sqrt{\gamma R T} = \sqrt{(1.4)(1716)(390)} = 968$ ft/s, hence $V_\infty = (968)(2) = 1936$ ft/s. Assume the distance across the N-wave at ground level is 200 ft. Then the time interval between the two booms would be $\Delta t = \lambda/V_\infty = 200/1936 = 0.103$ s. This is equivalent to the response time sensed by the human ear, and therefore two distinct booms will be heard, the double-boom. However, if Δt is shorter, the average ear cannot distinguish between the two pulses shown in Figure 9.43, and the sonic boom will be heard as only one boom.

The most critical aspect of the sonic boom is the environmental impact caused by the magnitude of Δp felt on the ground. In the early days of supersonic flight, the sonic booms generated by airplanes sometimes exceeded an increase in pressure greater than $\Delta p = 2$ lb/ft^2, greatly affecting human health and causing structural damage (broken windows, cracked walls, etc.). At the time of the design of the Concorde SST during the 1960s, a maximum value of $\Delta p = 2$ lb/ft^2 was considered acceptable. This was an error in judgment because soon after the Concorde went into service with British Airways and Air France, nations all over the world began to ban supersonic flight of the Concorde over land, severely hurting the economic value of the airplane. At the time of writing, such bans on overland supersonic flight are still in effect.

This situation is a critical inhibitor to the design and operation of a second-generation SST, thus promoting extensive research on how to mitigate the strength of the sonic boom. In this situation, aerospace engineers are really fighting the elements; shock waves are a natural consequence of supersonic flight, and it

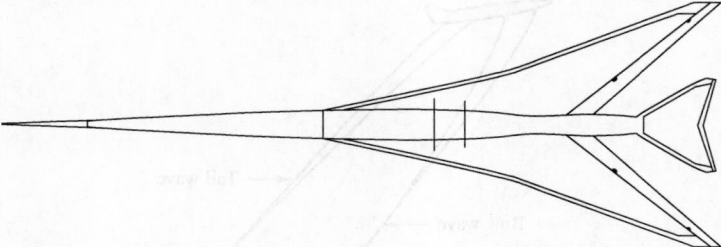

Figure 9.44 Planform of a generic low-boom fuselage, wing, and tail supersonic airplane.

appears that the most direct way of mitigating the strength of the sonic boom is to mitigate the strength of the shock waves causing the sonic boom. As aerospace engineers, we know how to approach this problem. From the wave theory discussed in the present chapter, clearly, sharp-nosed slender bodies will generate weaker shock waves than blunter-nosed thicker bodies. Also, expansion waves generated over portions of the body can interact with, and weaken, shock waves. So aerodynamic *shaping* of the supersonic vehicle is a first line of attack on the sonic boom problem. This leads to the generic low-boom fuselage, wings, tails arrangement sketched in Figure 9.44, with a long, slender fuselage; pointed nose; sharp leading-edge wings and tails; and a necking-down of the fuselage in the region of the wings in order to create favorable expansion waves following a type of supersonic "area rule." The purpose of this shaping is to reduce the magnitude of Δp to an environmentally acceptable level.

There is another, companion approach to mitigating the sonic boom. Referring again to Figure 9.43, an aspect of the severity of the sonic boom is the *sharpness* at which Δp takes place, i.e., the suddenness over which the rise in pressure takes place, which may be as short as 100 microseconds. It is this suddenness in combination with the magnitude of Δp that can be unacceptable. In his excellent book *Shock Waves and Man*, University of Toronto Institute for Aerospace Studies, 1974, author Irvine Glass notes that a change in pressure of 2 lb/ft^2 can readily be sensed by running up or down three flights of stairs, but because the rise time is so gentle our ears are not affected. So another approach to making the sonic boom more acceptable is to smooth out the sharp peak in Δp, and once again this can be achieved by shaping the flight vehicle. Indeed, NASA and the Defense Advanced Research Projects Agency (DARPA) have carried out such experiments using a specially modified Northrop F-5E fighter. The shape of the nose of the airplane was modified to have a flat top and a curved bottom, as generically sketched in Figure 9.45. The expansion wave from the curved bottom surface interacts with the bow shock wave in a fashion that smooths the sharp increase in Δp, as sketched in Figure 9.46, based on experimental measurements of the sonic boom shape generated by the modified F-5E. Note that the peak in Δp from the bow shock has been reduced from 1.2 lb/ft^2 to 0.8 lb/ft^2 and the shape

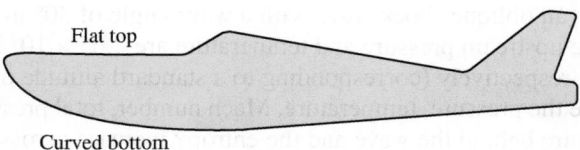

Figure 9.45 Schematic of a supersonic vehicle with a nose shape to produce a shaped sonic boom.

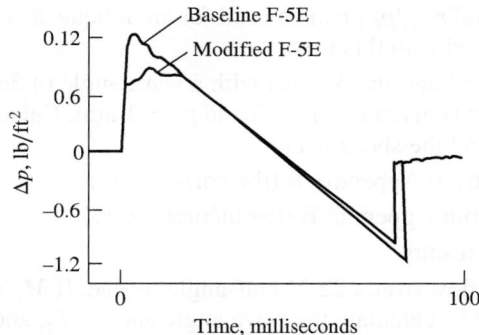

Figure 9.46 Comparison of a classic sonic boom signature (generated by a baseline F-5E aircraft) with a shaped boom signature (generated by a modified F-5E with a curved-bottom nose).

has been smoothed over a time interval of 20 milliseconds, creating what some researchers call a "thump" rather than a boom.

The mitigation of the sonic boom remains one of the most critical aspects of future commercial supersonic flight, and at the time of writing no definitive solution has yet been obtained. In the meantime, wind tunnel experiments and flight tests continue on configurations intended to point the way to a solution. Also, computational fluid dynamic codes are being developed for calculating both the near field and far field structures of the sonic boom. The solution to the sonic boom problem is still a work in progress—stay tuned. Finally, for an extensive survey of the state of the art, see *Sonic Boom: Six Decades of Research*, by D. J. Maglieri, P. J. Bobbitt, K. J. Plotkin, K. P. Shepherd, P. G. Coen, and D. M. Richwine, NASA SP 2014-622.

9.15 PROBLEMS

9.1 A slender missile is flying at Mach 1.5 at low altitude. Assume the wave generated by the nose of the missile is a Mach wave. This wave intersects the ground 559 ft behind the nose. At what altitude is the missile flying?

9.2 Consider an oblique shock wave with a wave angle of 30° in a Mach 4 flow. The upstream pressure and temperature are 2.65×10^4 N/m^2 and 223.3 K, respectively (corresponding to a standard altitude of 10,000 m). Calculate the pressure, temperature, Mach number, total pressure, and total temperature behind the wave and the entropy increase across the wave.

9.3 Equation (8.80) does *not* hold for an oblique shock wave, and hence the column in Appendix B labeled $p_{0,2}/p_1$ *cannot* be used, in conjunction with the normal component of the upstream Mach number, to obtain the total pressure behind an oblique shock wave. On the other hand, the column labeled $p_{0,2}/p_{0,1}$ can be used for an oblique shock wave, using $M_{n,1}$. Explain why all this is so.

9.4 Consider an oblique shock wave with a wave angle of 36.87°. The upstream flow is given by $M_1 = 3$ and $p_1 = 1$ atm. Calculate the total pressure behind the shock using

   a. $p_{0,2}/p_{0,1}$ from Appendix B (the correct way)

   b. $p_{0,2}/p_1$ from Appendix B (the incorrect way)

   Compare the results.

9.5 Consider the flow over a 22.2° half-angle wedge. If $M_1 = 2.5$, $p_1 = 1$ atm, and $T_1 = 300$ K, calculate the wave angle and p_2, T_2, and M_2.

9.6 Consider a flat plate at an angle of attack α to a Mach 2.4 airflow at 1 atm pressure. What is the maximum pressure that can occur on the plate surface and still have an attached shock wave at the leading edge? At what value of α does this occur?

9.7 A 30.2° half-angle wedge is inserted into a freestream with $M_\infty = 3.5$ and $p_\infty = 0.5$ atm. A Pitot tube is located above the wedge surface and behind the shock wave. Calculate the magnitude of the pressure sensed by the Pitot tube.

9.8 Consider a Mach 4 airflow at a pressure of 1 atm. We wish to slow this flow to subsonic speed through a system of shock waves with as small a loss in total pressure as possible. Compare the loss in total pressure for the following three shock systems:

   a. A single normal shock wave

   b. An oblique shock with a deflection angle of 25.3°, followed by a normal shock

   c. An oblique shock with a deflection angle of 25.3°, followed by a second oblique shock of deflection angle of 20°, followed by a normal shock

From the results of (*a*), (*b*), and (*c*), what can you induce about the efficiency of the various shock systems?

9.9 Consider an oblique shock generated at a compression corner with a deflection angle $\theta = 18.2°$. A straight horizontal wall is present above the corner, as shown in Figure 9.19. If the upstream flow has the properties

$M_1 = 3.2$, $p_1 = 1$ atm, and $T_1 = 520°R$, calculate M_3, p_3, and T_3 behind the reflected shock from the upper wall. Also, obtain the angle Φ which the reflected shock makes with the upper wall.

9.10 Consider the supersonic flow over an expansion corner, such as given in Figure 9.25. The deflection angle $\theta = 23.38°$. If the flow upstream of the corner is given by $M_1 = 2$, $p_1 = 0.7$ atm, $T_1 = 630°R$, calculate M_2, p_2, T_2, ρ_2, $p_{0,2}$, and $T_{0,2}$ downstream of the corner. Also, obtain the angles the forward and rearward Mach lines make with respect to the upstream direction.

9.11 A supersonic flow at $M_1 = 1.58$ and $p_1 = 1$ atm expands around a sharp corner. If the pressure downstream of the corner is 0.1306 atm, calculate the deflection angle of the corner.

9.12 A supersonic flow at $M_1 = 3$, $T_1 = 285$ K, and $p_1 = 1$ atm is deflected upward through a compression corner with $\theta = 30.6°$ and then is subsequently expanded around a corner of the same angle such that the flow direction is the same as its original direction. Calculate M_3, p_3, and T_3 downstream of the expansion corner. Since the resulting flow is in the same direction as the original flow, would you expect $M_3 = M_1$, $p_3 = p_1$, and $T_3 = T_1$? Explain.

9.13 Consider an infinitely thin flat plate at an angle of attack α in a Mach 2.6 flow. Calculate the lift and wave-drag coefficients for
(a) $\alpha = 5°$ (b) $\alpha = 15°$ (c) $\alpha = 30°$
(*Note:* Save the results of this problem for use in Chapter 12.)

9.14 Consider a diamond-wedge airfoil such as shown in Figure 9.36, with a half-angle $\varepsilon = 10°$. The airfoil is at an angle of attack $\alpha = 15°$ to a Mach 3 freestream. Calculate the lift and wave-drag coefficients for the airfoil.

9.15 Consider sonic flow. Calculate the maximum deflection angle through which this flow can be expanded via a centered expansion wave.

9.16 Consider a circular cylinder (oriented with its axis perpendicular to the flow) and a symmetric diamond-wedge airfoil with a half-angle of 5° at zero angle of attack; both bodies are in the same Mach 5 freestream. The thickness of the airfoil and the diameter of the cylinder are the same. The drag coefficient (based on projected frontal area) of the cylinder is 4/3. Calculate the *ratio* of the cylinder drag to the diamond airfoil drag. What does this say about the aerodynamic performance of a blunt body compared to a sharp-nosed slender body in supersonic flow?

9.17 Consider the supersonic flow over a flat plate at an angle of attack, as sketched in Figure 9.35. As stated in Section 9.7, the flow direction downstream of the trailing edge of the plate, behind the trailing edge shock and expansion waves, is not precisely in the freestream direction. Why? Outline a method to calculate the strengths of the trailing edge shock and expansion waves, and the direction of the flow downstream of the trailing edge.

9.18 (The purpose of this problem is to calculate a two-dimensional expanding supersonic flow and compare it with the analogous quasi-one-dimensional flow in Problem 10.15.) Consider a two-dimensional duct with a straight horizontal lower wall, and a straight upper wall inclined upward through the angle $\theta = 3°$. The height of the duct entrance is 0.3 m. A uniform horizontal flow at Mach 2 enters the duct and goes through a Prandtl-Mayer expansion wave centered at the top corner of the entrance. The wave propagates to the bottom wall, where the leading edge (the forward Mach line) of the wave intersects the bottom wall at point A located at distance x_A from the duct entrance. Imagine a line drawn perpendicular to the lower wall at point A, and intersecting the upper wall at point B. The local height of the duct at point A is the length of this line AB. Calculate the *average* flow Mach number over AB, assuming that M varies linearly along that portion of AB inside the expansion wave.

9.19 Repeat Problem 9.18, except with $\theta = 30°$. Again, we will use these results to compare with a quasi-one-dimensional calculation in Problem 10.16. The reason for repeating this calculation is to examine the effect of the much more highly two-dimensional flow generated in this case by a much larger expansion angle.

9.20 Consider a Mach 3 flow at 1 atm pressure initially moving over a flat horizontal surface. The flow then encounters a 20 degree expansion corner, followed by a 20 degree compression corner that turns the flow back to the horizontal. Calculate the pressure of the flow downstream of the compression corner. *Note:* You will find that the pressure downstream of the compression corner is different from the pressure upstream of the expansion corner, even though the upstream and downstream flows are in the same direction, namely horizontal. Why?

9.21 Generate a plot of p_2/p_1 across an oblique shock as a function of the turning angle θ for an incoming Mach number of 2. This is called a pressure-deflection diagram and is useful for solving problems involving oblique shock waves.
   a.   Use Excel, MATLAB, or another programming environment of your choice, to solve the θ-β-M relation for β, θ, or M, with the other two parameters fixed. You can also write a subroutine to do this automatically. See Problem 9.33.
   b.   Use your routine to make a plot of p_2/p_1 across an oblique shock for $M = 2$ as a function of turning angle θ.

9.22 The purpose of the inlet in a ramjet engine is to slow the incoming air to subsonic conditions so that it can be mixed and reacted with the fuel. Often, this is accomplished via a system of shock waves that are stabilized in the inlet. As an example, consider the compression process taking place in the inlet of a ramjet engine pictured in Figure 9.47. Two oblique shocks followed by a normal shock are used to decelerate the Mach 3.5 flow. The aircraft is flying at 10 km, where the static pressure and temperature are

26500 N/m^2 and 223.3 K, respectively. Assume that the flow is two-dimensional.

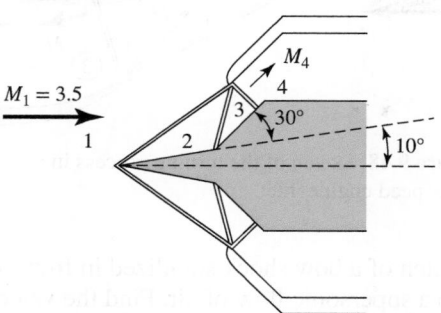

Figure 9.47 Sketch of a planar ramjet inlet showing shock and expansion wave structures.

a. Determine the Mach number, static temperature, static pressure, and total temperature and total pressure at each stage of the <u>compression</u> process (i.e., from stations 1 to 4).

b. Compare the compression process with that due to a single normal shock.

c. Make a T-S diagram (Temperature-Entropy) for the compression process in the inlet.

    i. Compute the entropy change across each shock.

    ii. Plot the static temperature (y axis) against the entropy for each part of the flow field. (Choose the initial value of the entropy equal to zero.)

    iii. Add contours corresponding to the total pressure at each station.

    iv. How does this compression process compare to the "ideal" process?

    v. Compute the total pressure and entropy change if the inlet was replaced with a single normal shock.

    vi. Add the normal shock to your T-S diagram.

    vii. Would you want to use a normal shock to do the compression instead of the shock system?

9.23 Consider an oblique shock wave with $\beta = 29°$, where $T_2 = 2T_1$. Calculate the free stream Mach number upstream of the wave. (*Hint:* Use Appendix B.)

9.24 Find M_3 in the engine inlet illustrated below when $M_\infty = 3$. Assume that the first two shocks are weak and that $\gamma = 1.4$.

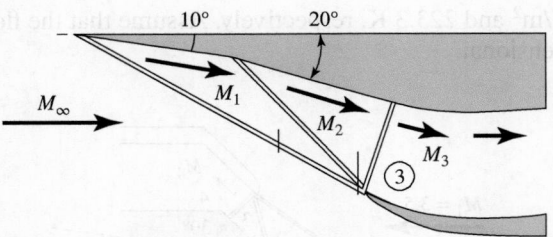

Figure 9.48 Sketch of the turning process in a high-speed engine inlet.

9.25 Below is a sketch of a bow shock stabilized in front of a rounded strut that protrudes into a supersonic flow of air. Find the velocity of the free stream if the angle of the bow shock far away from the strut is 26° with respect to the free stream and the free stream temperature is 300 K.

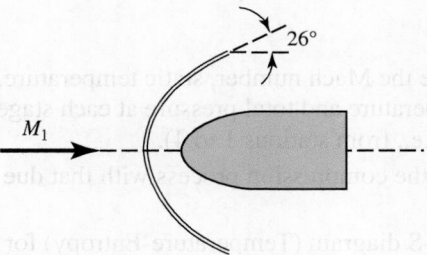

Figure 9.49 Sketch of a 2-D bow shock stabilized in front of a strut.

9.26 Consider supersonic flow of air through the reflected shock system illustrated in Figure 9.50. If the incoming air is at 1 atm and 300 K, please find the angle of the reflected shock (Φ_r) and the conditions (pressure, temperature, and Mach number) in region 3.

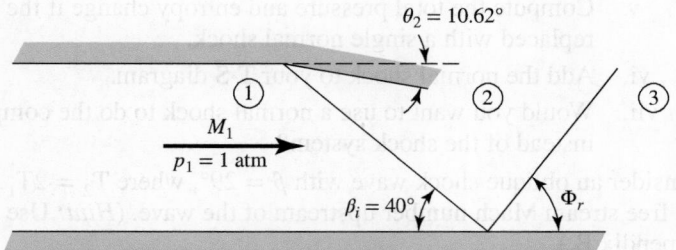

Figure 9.50 Sketch of a reflected shock interacting with the wind tunnel floor.

9.27 Air initially at Mach 3 and 1 atm pressure flows over the series of ramps illustrated below. Referring to Figure 9.22, sketch the structures that form, number the flow regions, and indicate the type of reflected wave that forms. Prove the type of wave by calculating the pressure in regions 3 and 5.

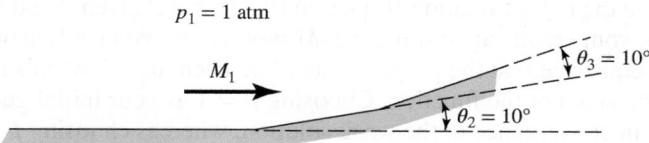

$p_1 = 1$ atm

M_1

$\theta_3 = 10°$

$\theta_2 = 10°$

Figure 9.51 Compression over a series of ramps.

9.28 The static and total pressures of air downstream of a 30° expansion corner are 0.0155 atm and 151.8 atm, respectively. What are the Mach number and static pressure upstream of the expansion corner?

9.29 Consider the smooth expansion corner illustrated below where a Mach 2.55 flow at 1 atm is turned away from itself by 15°. Please sketch the important features of the flow and calculate the Mach number and static pressure downstream of the corner.

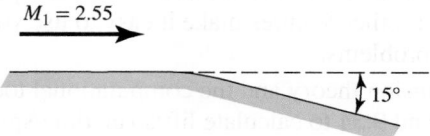

$M_1 = 2.55$

15°

Figure 9.52 A smooth expansion corner.

9.30 The isentropic compression corner illustrated below turns a flow at 1 atm pressure and Mach 4.1 into itself by 10°. Sketch the flow structure being as accurate as possible. Label the forward and rearward Mach lines and calculate their angles. Calculate the pressure and Mach number downstream of the system of compression waves. Do not consider what happens above the point at which the Mach waves coalesce.

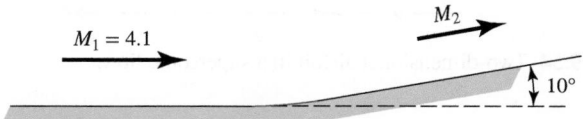

M_2

$M_1 = 4.1$

10°

Figure 9.53 A smooth compression corner.

9.31 Revisit Problem 9.31 recognizing that the Mach waves emanating from the curved wall eventually coalesce into an oblique shock that turns the flow in the far field. What additional structure(s) must form at the coalescence point? Support your reasoning by calculating the Mach number and pressure downstream of the coalescence point. (*Hint:* There will be at least two distinct flow regions downstream of the compression wave-shock system.)

9.32 Write a computer program in the language/environment of your choice to solve the θ-β-M relation [Equation (9.23)] for β given θ and M. Check your results against the θ-β-M plot. Hints: Write a function whose value equals zero at the proper value of β. Then, use Newton's method to find the zeros of the function. Choosing $\beta = 1$ as your initial guess will result in convergence to the weak solution, whereas choosing $\beta = 89$ will result in convergence to the strong solution. MATLAB and MS Excel's Visual Basic environment are good platforms for doing this because their other features make it easy to use your function's output to solve problems.

9.33 Write a computer program in the language/environment of your choice that solves the Prandtl-Meyer function [Equation (9.42)] for M given θ. Check your results against the tabulation of $v(M)$. Hints: Write a function that calculates $v(M)$ and define the "error" as the difference between the desired value of theta and the value associated with the particular choice of Mach number. Then, use Newton's method to find the Mach number that reduces the error to some close approximation of zero. MATLAB and MS Excel's Visual Basic environment are good platforms for doing this because their other features make it easy to use your function's output to solve problems.

9.34 Use shock-expansion theory and the computational tools you built in Problems 9.33 and 9.34 to calculate lift/span, drag/span, and L/D all as functions of angle of attack for the airfoil illustrated below. Sketch the flow field. Assume that $p_1 = 1$ *atm*, $M_1 = 2$, $c = 1m$, $x = 1/3$ and $y = 0.06$. Your plot should have at least five points in the following range: $0° \leq \alpha \leq 10°$. What is the maximum angle of attack on this airfoil that can be investigated using shock-expansion theory?

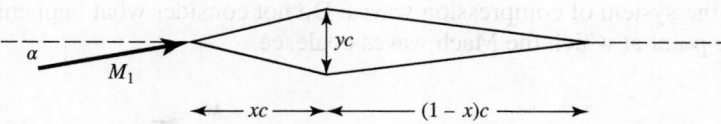

Figure 9.54 Two-dimensional airfoil in a supersonic flow.

9.35 The purpose of this problem is to explain what causes the dramatic white cloud pattern generated in the flow field over the F/A-18C Hornet shown in Figure 9.55. This problem is both a tutorial and a quantitative calculation involving the reader. We first discuss some necessary thermodynamic background, followed by an examination of the physical nature of the flow field.

Figure 9.55 An F/A-18C Hornet Breaking the Sound Barrier. (*U.S. Navy photo by Mass Communication Specialist 2nd Class Ron Reeves*)

Necessary Thermodynamic Background

The white cloud surrounding the F/A-18C Hornet is condensed water vapor in the airflow over the airplane. Inside the cloud, water appears in two phases, vapor and liquid. Both inside and outside the cloud, the medium is a mixture of air and water; outside the cloud the water is completely in vapor form, and inside the cloud there is a mixture of air, water vapor, and condensed liquid water. Both inside and outside the cloud, the gas pressure p is the sum of the partial pressure of air and the much smaller partial pressure of water vapor, p_{air} and p_{H2O}, respectively; i.e., $p = p_{air} + p_{H2O}$. On a graph of p_{H2O} versus temperature for this water-air mixture, we can construct a type of "phase diagram," where over part of this diagram the water is in the vapor phase only, and over another part the water is a liquid-vapor mixture. The curve dividing these two regions is called the saturation curve, which arches upward toward the right (on the saturation curve, p_{H2O} increases with increasing temperature, with the slope of this curve, dp_{H2O}/dT, increasing as T increases). To the right of this curve, the water is completely vapor, and to the left of this curve, the water is a mixture of vapor and liquid. What we see as the white cloud over the top of the F/A-18C Hornet is light reflecting from the condensed droplets of water; i.e., the white cloud is that region of the airflow that is in the liquid-vapor region of the phase diagram, namely that region to the left of the saturation curve.

Every point on the saturation curve corresponds to 100 percent relative humidity in the ambient air. The temperature at each point on the curve is the "dew-point" temperature. In the region to the right of this curve the air is at less than 100 percent humidity, the temperature is greater than the dew point, and there is no condensation of the water vapor. In the region to the left of the curve, the temperature falls below the dew point and condensation starts to occur.

The saturation curve is obtained experimentally. For this problem, we use the saturation curve for an ambient pressure of 1 atm found on page 4-86 of *Marks' Mechanical Engineer's Handbook*, 6th Ed., McGraw Hill, 1958 (we use an older edition because it is on the author's personal book shelf and therefore handy). This is a curve of p_{H2O} versus T for air at 100 percent relative humidity with an ambient pressure of 1 atm.

For this problem, let us choose a particular point on this saturation curve, namely the point where $T = 520$ °R and $p_{H2O} = 38.16$ lb/ft^2. Call this point A. At point A, air has 100 percent humidity and contains an amount of water vapor corresponding to a partial pressure of $p_{H2O} = 38.16$ lb/ft^2. The *slope* of the curve, dp_{H2O}/dT, at point A is important to this problem. The slope is

$$dp_{H2O}/dT = 1.08 \text{ lb}/(\text{ft}^2 \cdot {}^\circ\text{R}) \qquad \text{saturation curve}$$

Nature of the Flow field

The F/A-18C Hornet in Figure 9.55 is flying supersonically at a small angle of attack; note that the trailing wing-tip vortices, also made visible by water condensation, are very thin and tightly wound, characteristic of tip vortices in a supersonic flow. Expansion waves occur over the top surface of the wings and fuselage. The expansion region is terminated by a trailing-edge shock wave. The objective of this problem is to prove that water condensation occurs within the expansion wave, but that the water goes back to the purely vapor phase when it passes through the trailing-edge shock wave. We submit that the white cloud shown around the F/A-18C Hornet is caused by water condensation within the expansion wave, and that the rather sharp termination of the cloud is caused by water evaporation when the flow passes through the trailing-edge shock wave.

For a simplified model of this flow, consider the flat plate at an angle-of-attack in a supersonic flow as sketched in Figure 9.36. The supersonic flow over the top surface of the plate first goes through an expansion wave, then passes through the trailing-edge shock wave downstream. If the freestream upstream of the plate has sufficiently high humidity, a white cloud of condensed water vapor will form through and downstream of the expansion wave. However, the white cloud will disappear when the condensed water vapor is vaporized as the flow passes through the trailing-edge shock wave. So in Figure 9.36, the white cloud will occur in the region bounded upstream by the expansion wave, and downstream by the trailing-edge shock wave. This simulates the nature of the white cloud seen over the top of the F/A-18C Hornet in Figure 9.55.

Statement of the Problem

Consider the flow field sketched in Figure 9.36. Assume the supersonic freestream is at a temperature of 520 °R and a pressure of 1 atm, and that it has 100 percent humidity. Therefore, the thermodynamic state of the freestream corresponds to point A on the saturation curve of the phase diagram discussed earlier.

a. Prove that a fluid element from the freestream, when passing into the expansion wave, will experience water condensation.
b. Prove that a fluid element already within and downstream of the expansion wave and hence with some condensed water, will experience water evaporation when passing through the trailing-edge shock wave.

Comment and Hint

Well, you might say this is obvious. When the flow expands through the expansion wave, the temperature decreases, and water will obviously condense. Hence, the fluid element in the phase diagram will obviously penetrate through point A and enter the liquid-vapor region. But wait a minute! When the fluid element enters the expansion wave, *both* its temperature and pressure will decrease. The temperature decrease, by itself, will move the properties of the fluid element toward the left of point A into the liquid-vapor region, but the pressure decrease by itself will move the properties of the fluid element vertically downward from point A, farther away from the liquid-vapor region. The question is: Which change is more dominant, the temperature decrease or the pressure decease? This is for you to find out. As a hint, calculate the value of the derivative dp_{H2O}/dT for the fluid element as it enters the expansion wave. If this derivative is smaller than the slope of the saturation curve at point A, given earlier as $dp_{H2O}/dT = 1.08$ lb/(ft$^2 \cdot$°R), then the properties of the fluid element as it enters the expansion wave will also penetrate across point A into the liquid-vapor part of the phase diagram, and clearly water condensation will then occur.

Similarly, for a fluid element crossing the shock wave, you might also say that it is obvious that condensed water in the fluid element will evaporate because the temperature increases across the shock, which causes evaporation. But again, wait a minute! Consider that the properties of the fluid element just in front of the shock wave are in the liquid-vapor region just to the left of point A. *Both* the temperature and pressure will increase across the shock wave. The increase in temperature will move the properties of the fluid to the right of point A, away from the liquid-vapor region and into the purely vapor region, but the increase in pressure will move the properties of the fluid element vertically upward from point A, farther into the liquid-vapor region. Again, the question is: Which change is more dominant, the temperature increase or the pressure increase? This is for you to find out. As a hint, calculate the value $dp_{H2O}/dT \approx \Delta p_{H2O}/\Delta T$ across the shock wave. If this derivative is *smaller* than the slope of the saturation curve at point A, given earlier as $dp_{H2O}/dT = 1.08$ lb/$(ft^2 \cdot °R)$, then the properties of the fluid element as it crosses the shock wave will also exit the liquid-vapor phase at point A and enter the purely vapor region, and clearly evaporation will take place. To calculate $\Delta p_{H2O}/\Delta T$, you have to arbitrarily assume the strength of the shock wave. Assume the shock is a normal shock wave with an upstream Mach number of 1.2, and the upstream properties of the flow are those just minutely to the left of point A; i.e., assume ahead of the shock wave $M_1 = 1.2$, $p_1 = 2116$ lb/ft^2, and $p_{H2O} = 38.16$ lb/ft^2.

Answers

a.   When the fluid element enters the expansion wave, $dp_{H2O}/dT = 0.256$ lb/$(ft^2 \cdot °R)$. This is a smaller slope than that of the saturation curve at point A. The properties of the fluid will penetrate into the liquid-vapor region, and condensation will occur.

b.   When the fluid element in front of the shock wave with properties just inside the liquid-vapor region at point A crosses the shock wave, $\Delta p_{H2O}/\Delta T = 0.2923$ lb/$(ft^2 \cdot °R)$. This is a smaller slope than that of the saturation curve at point A. The properties of the fluid element will penetrate into the purely vapor region, and evaporation will occur.

CHAPTER 10

Compressible Flow Through Nozzles, Diffusers, and Wind Tunnels

Having wondered from what source there is so much difficulty in successfully applying the principles of dynamics to fluids than to solids, finally, turning the matter over more carefully in my mind, I found the true origin of the difficulty; I discovered it to consist of the fact that a certain part of the pressing forces important in forming the throat *(so called by me, not considered by others) was neglected, and moreover regarded as if of no importance, for no other reason than the throat is composed of a very small, or even an infinitely small, quantity of fluid, such as occurs whenever fluid passes from a wider place to a narrower, or vice versa, from a narrower to a wider.*

   Johann Bernoulli; from his
   Hydraulics, 1743

PREVIEW BOX

One of the best examples of the harnessing of nature by aerospace engineers is shown in Figure 10.1, which is a photograph of the main rocket engine for the space shuttle. This engine produces over 400,000 lb of thrust. The laws of nature have been used by aerospace engineers to design this engine. But what laws of nature? Take another look at Figure 10.1, and note the large bell-like *divergent* nozzle of the

rocket engine. Why this shape instead of some convergent shape like the low-speed wind tunnel nozzles discussed in Section 3.3? In reality, in Figure 10.1, we are seeing only part of the rocket nozzle; hidden behind all the plumbing to the left of the divergent duct in Figure 10.1 is a combustion chamber that feeds the hot, high-pressure gas into a convergent duct that transitions to the divergent duct seen in

Figure 10.1 Space shuttle main rocket engine. (*NASA*)

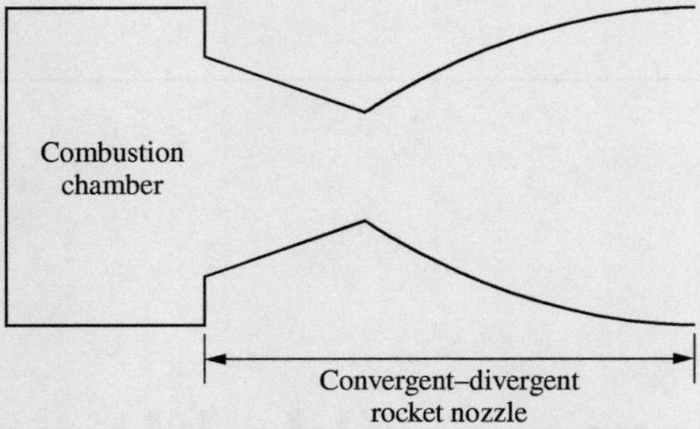

```
Combustion
chamber
                              Convergent–divergent
                                 rocket nozzle
```

Figure 10.2 Schematic of rocket engine nozzle.

Figure 10.1. The whole rocket engine nozzle, from the combustion chamber to the exit, is a convergent–divergent shape such as sketched in Figure 10.2. Why?

The answers to these questions and more are given in the present chapter. Here we deal with the *internal* flow of a high-speed compressible gas through ducts of various shapes. The material in this chapter is essential to the design of rocket engines, jet engines, supersonic and hypersonic wind tunnels, and any flow device that involves the compressible flow of a gas internal to the device. We labeled the study of shock waves in the previous two chapters as the bread and butter of compressible flow. By comparison, the study of the flow through nozzles, diffusers, and wind tunnels in the present chapter is the rolls and jam of compressible flow. Applications of the material in this chapter are made daily in the modern world of aerospace engineering. Moreover, this material is packed with interesting physical phenomena, some of which you likely have not thought about before and which you will find almost amazing in some instances. Take this chapter very seriously. I predict that you will find it enjoyable and rewarding.

10.1 INTRODUCTION

Chapters 8 and 9 treated normal and oblique waves in supersonic flow. These waves are present on any aerodynamic vehicle in supersonic flight. Aeronautical engineers are concerned with observing the characteristics of such vehicles, especially the generation of lift and drag at supersonic speeds, as well as details of the flow field, including the shock- and expansion-wave patterns. To make such observations, we usually have two standard choices: (1) conduct flight tests using the actual vehicle, and (2) run wind-tunnel tests on a small-scale model of the vehicle. Flight tests, although providing the final answers in the full-scale environment, are costly and, not to say the least, dangerous if the vehicle is unproven. Hence, the vast bulk of supersonic aerodynamic data have been obtained in wind tunnels on the ground. What do such supersonic wind tunnels look like? How do

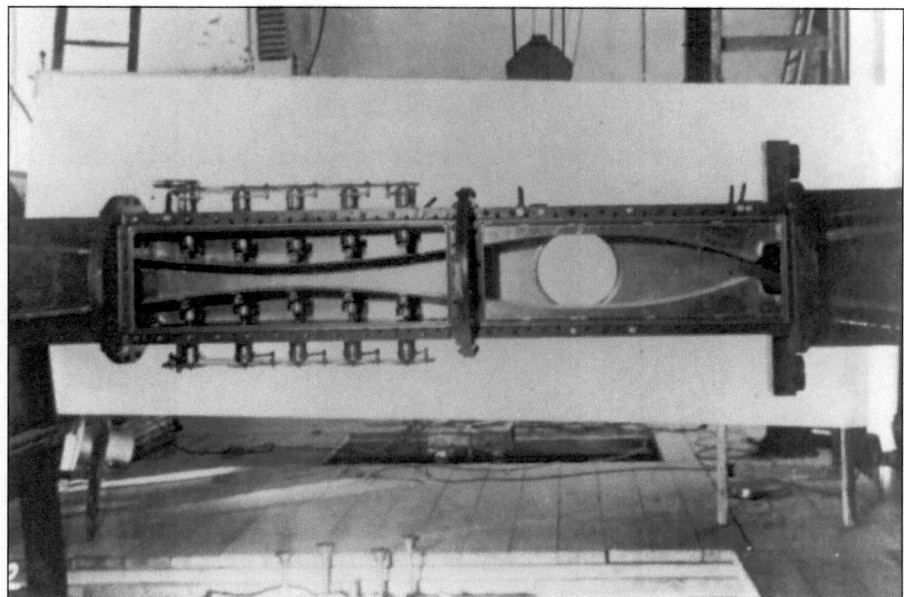

Figure 10.3 The first practical supersonic wind tunnel, built by A. Busemann in Germany in the mid-1930s. (*Courtesy of John Anderson*)

we produce a uniform flow of supersonic gas in a laboratory environment? What are the characteristics of supersonic wind tunnels? The answers to these and other questions are addressed in this chapter.

The first practical supersonic wind tunnel was built and operated by Adolf Busemann in Germany in the mid-1930s, although Prandtl had a small supersonic facility operating as early as 1905 for the study of shock waves. A photograph of Busemann's tunnel is shown in Figure 10.3. Such facilities proliferated quickly during and after World War II. Today, all modern aerodynamic laboratories have one or more supersonic wind tunnels, and many are equipped with hypersonic tunnels as well. Such machines come in all sizes; an example of a moderately large hypersonic tunnel is shown in Figure 10.4.

In this chapter, we discuss the aerodynamic fundamentals of compressible flow through ducts. Such fundamentals are vital to the proper design of high-speed wind tunnels, rocket engines, high-energy gas-dynamic and chemical lasers, and jet engines, to list just a few. Indeed, the material developed in this chapter is used almost daily by practicing aerodynamicists and is indispensable toward a full understanding of compressible flow.

The road map for this chapter is given in Figure 10.5. After deriving the governing equations, we treat the cases of a nozzle and diffuser separately. Then we merge this information to examine the case of supersonic wind tunnels.

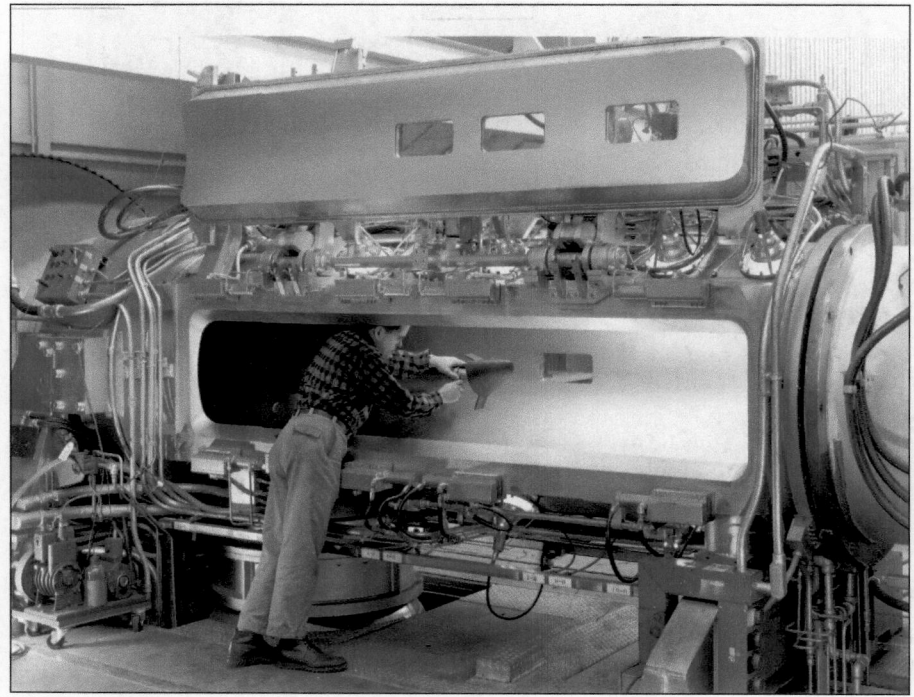

Figure 10.4 A large hypersonic wind tunnel at the U.S. Air Force Wright Aeronautical Laboratory, Dayton, Ohio. (*NASA*)

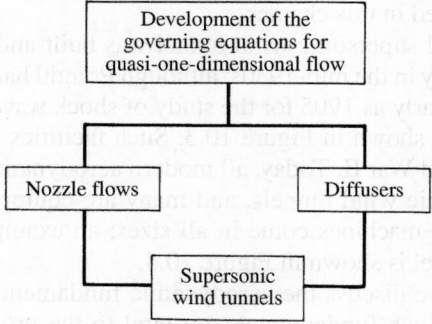

Figure 10.5 Road map for Chapter 10.

10.2 GOVERNING EQUATIONS FOR QUASI-ONE-DIMENSIONAL FLOW

Recall the one-dimensional flow treated in Chapter 8. There, we considered the flow-field variables to be a function of x only, that is, $p = p(x)$, $u = u(x)$, etc. Strictly speaking, a streamtube for such a flow must be of constant area; that is, the

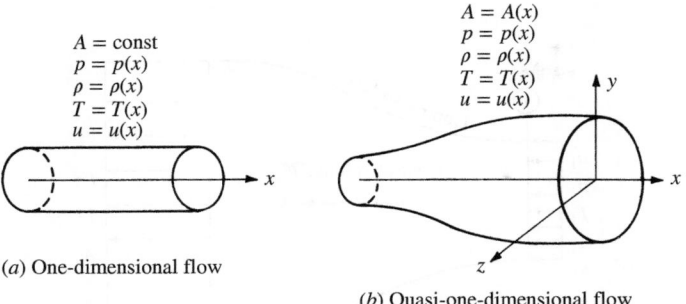

Figure 10.6 One-dimensional and quasi-one-dimensional flows.

one-dimensional flow discussed in Chapter 8 is constant-area flow, as sketched in Figure 10.6a.

In contrast, assume that the area of the streamtube changes as a function of x, that is, $A = A(x)$, as sketched in Figure 10.6b. Strictly speaking, this flow is three-dimensional; the flow-field variables are functions of x, y, and z, as can be seen simply by examining Figure 10.6b. In particular, the velocity at the boundary of the streamtube must be tangent to the boundary, and hence it has components in the y and z directions as well as the axial x direction. However, if the area variation is moderate, the components in the y and z directions are small in comparison with the component in the x direction. In such a case, the flow-field variables can be *assumed* to vary with x only (i.e., the flow can be assumed to be uniform across any cross section at a given x station). Such a flow, where $A = A(x)$, but $p = p(x)$, $\rho = \rho(x)$, $u = u(x)$, etc., is defined as *quasi-one-dimensional flow*, as sketched in Figure 10.6b. Such flow is the subject of this chapter. We have encountered quasi-one-dimensional flow earlier, in our discussion of incompressible flow through a duct in Section 3.3. Return to Section 3.3, and review the concepts presented there before progressing further.

Although the assumption of quasi-one-dimensional flow is an approximation to the actual flow in a variable-area duct, the integral forms of the conservation equations, namely, continuity [Equation (2.48)], momentum [Equation (2.64)], and energy [Equation (2.95)], can be used to obtain governing equations for quasi-one-dimensional flow which are physically consistent, as follows. Consider the control volume given in Figure 10.7. At station 1, the flow across area A_1 is assumed to be uniform with properties p_1, ρ_1, u_1, etc. Similarly, at station 2, the flow across area A_2 is assumed to be uniform with properties p_2, ρ_2, u_2, etc. The application of the integral form of the continuity equation was made to such a variable-area control volume in Section 3.3. The resulting continuity equation for steady, quasi-one-dimensional flow was obtained as Equation (3.21), which in terms of the nomenclature in Figure 10.7 yields

$$\boxed{\rho_1 u_1 A_1 = \rho_2 u_2 A_2} \tag{10.1}$$

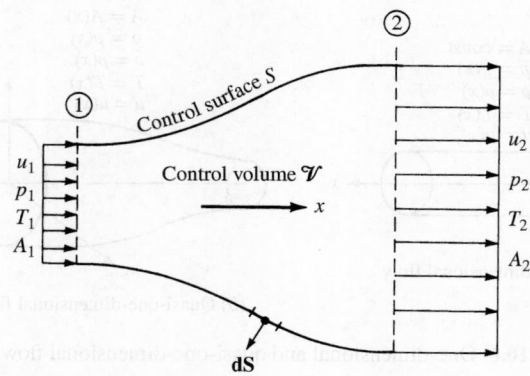

Figure 10.7 Finite control volume for quasi-one-dimensional flow.

Consider the integral form of the momentum equation, Equation (2.64). For a steady, inviscid flow with no body forces, this equation becomes

$$\oiint_S (\rho \mathbf{V} \cdot \mathbf{dS})\mathbf{V} = -\oiint_S p\, \mathbf{dS} \qquad (10.2)$$

Since Equation (10.2) is a vector equation, let us examine its x component, given below:

$$\oiint_S (\rho \mathbf{V} \cdot \mathbf{dS})u = -\oiint_S (p\, dS)_x \qquad (10.3)$$

where $(p\, dS)_x$ denotes the x component of the pressure force. Since Equation (10.3) is a scalar equation, we must be careful about the sign of the x components when evaluating the surface integrals. All components pointing to the right in Figure 10.7 are positive, and those pointing to the left are negative. The upper and lower surfaces of the control volume in Figure 10.7 are streamlines; hence, $\mathbf{V} \cdot \mathbf{dS} = 0$ along these surfaces. Also, recall that across A_1, $\mathbf{V}$ and $\mathbf{dS}$ are in opposite directions; hence, $\mathbf{V} \cdot \mathbf{dS}$ is negative. Therefore, the integral on the left of Equation (10.3) becomes $-\rho_1 u_1^2 A_1 + \rho_2 u_2^2 A_2$. The pressure integral on the right of Equation (10.2), evaluated over the faces A_1 and A_2 of the control volume, becomes $-(-p_1 A_1 + p_2 A_2)$. (The negative sign in front of $p_1 A$ is because $\mathbf{dS}$ over A_1 points to the left, which is the negative direction for the x components.) Evaluated over the upper and lower surface of the control volume, the pressure integral can be expressed as

$$-\int_{A_1}^{A_2} -p\, dA = \int_{A_1}^{A_2} p\, dA \qquad (10.4)$$

where dA is simply the x component of the vector $\mathbf{dS}$, that is, the area dS projected on a plane perpendicular to the x axis. The negative sign *inside* the integral on the left of Equation (10.4) is due to the direction of $\mathbf{dS}$ along the upper and lower surfaces; note that $\mathbf{dS}$ points in the backward direction along these surfaces, as shown in Figure 10.7. Hence, the x component of $p\, \mathbf{dS}$ is to the left, and therefore appears in our equations as a negative component. [Recall from Section 2.5 that the negative sign *outside* the pressure integral, that is, outside the integral on the left of Equation (10.4), is always present to account for the physical fact that the pressure force $p\,\mathbf{dS}$ exerted on a control surface always acts in the opposite direction of $\mathbf{dS}$. If you are unsure about this, review the derivation of the momentum equation in Section 2.5. Also, do not let the signs in the above results confuse you; they are all quite logical if you keep track of the direction of the x components.] With the above results, Equation (10.3) becomes

$$-\rho_1 u_1^2 A_1 + \rho_2 u_2^2 A_2 = -(-p_1 A_1 + p_2 A_2) + \int_{A_1}^{A_2} p\,dA$$

$$p_1 A_1 + \rho_1 u_1^2 A_1 + \int_{A_1}^{A_2} p\,dA = p_2 A_2 + \rho_2 u_2^2 A_2 \qquad (10.5)$$

Equation (10.5) is the momentum equation for steady, quasi-one-dimensional flow.

Consider the energy equation given by Equation (2.95). For inviscid, adiabatic, steady flow with no body forces, this equation becomes

$$\oiint_S \rho\left(e+\frac{V^2}{2}\right)\mathbf{V}\cdot\mathbf{dS} = -\oiint_S p\mathbf{V}\cdot\mathbf{dS} \qquad (10.6)$$

Applied to the control volume in Figure 10.7, Equation (10.6) yields

$$\rho_1\left(e_1+\frac{u_1^2}{2}\right)(-u_1 A_1)+\rho_2\left(e_2+\frac{u_2^2}{2}\right)(u_2 A_2)=-(-p_1 u_1 A_1+p_2 u_2 A_2)$$

or
$$p_1 u_1 A_1+\rho_1 u_1 A_1\left(e_1+\frac{u_1^2}{2}\right)=p_2 u_2 A_2+\rho_2 u_2 A_2\left(e_2+\frac{u_2^2}{2}\right) \qquad (10.7)$$

Dividing Equation (10.7) by Equation (10.1), we have

$$\frac{p_1}{\rho_1}+e_1+\frac{u_1^2}{2}=\frac{p_2}{\rho_2}+e_2+\frac{u_2^2}{2} \qquad (10.8)$$

Recall that $h=e+pv=e+p/\rho$. Hence, Equation (10.8) becomes

$$h_1+\frac{u_1^2}{2}=h_2+\frac{u_2^2}{2} \qquad (10.9)$$

which is the energy equation for steady, adiabatic, inviscid quasi-one-dimensional flow. Examine Equation (10.9) closely; it is a statement that the total enthalpy, $h_0=h+u^2/2$, is a constant throughout the flow. Once again, this should come as no surprise; Equation (10.9) is simply another example of the general result for steady, inviscid, adiabatic flow discussed in Section 7.5. Hence, we can replace Equation (10.9) by

$$h_0 = \text{const} \qquad (10.10)$$

Pause for a moment and examine our results given above. We have applied the integral forms of the conservation equations to the control volume in Figure 10.7. We have obtained, as a result, Equations (10.1), (10.5), and (10.9) or (10.10) as the governing continuity, momentum, and energy equations, respectively, for quasi-one-dimensional flow. Examine these equations—they are *algebraic* equations (with the exception of the single integral term in the momentum equation). In Figure 10.7, assume that the inflow conditions p_1, u_1, p_1, T_1, and h_1 are given and that the area distribution $A = A(x)$ is presented. Also, assume a calorically perfect gas, where

$$p_2 = \rho_2 R T_2 \tag{10.11}$$

and

$$h_2 = c_p T_2 \tag{10.12}$$

Equations (10.1), (10.5), (10.9) or (10.10), (10.11), and (10.12) constitute five equations for the five unknowns ρ_2, u_2, p_2, T_2, and h_2. We could, in principle, solve these equations directly for the unknown flow quantities at station 2 in Figure 10.7. However, such a direct solution would involve substantial algebraic manipulations. Instead, we take a simpler tack, as described in Section 10.3.

Before moving on to a solution of the governing equations, let us examine some physical characteristics of a quasi-one-dimensional flow. To help this examination, we first obtain some *differential* expressions for the governing equations, in contrast to the algebraic equations obtained above. For example, consider Equation (10.1), which states that

$$\rho u A = \text{const} \tag{10.13}$$

through a variable-area duct. Differentiating Equation (10.13), we have

$$\boxed{d(\rho u A) = 0} \tag{10.14}$$

which is the differential form of the continuity equation for quasi-one-dimensional flow.

To obtain a differential form of the momentum equation, apply Equation (10.5) to the infinitesimal control volume sketched in Figure 10.8. The flow going into the volume at station 1, where the area is A, has properties p, u, and ρ. In traversing the length dx, where the area changes by dA, the flow properties change by the corresponding amounts dp, $d\rho$, and du. Hence, the flow leaving at station 2 has the properties $p + dp$, $u + du$, and $\rho + d\rho$, as shown in Figure 10.8. For this case, Equation (10.5) becomes [recognizing that the integral in Equation (10.5) can be replaced by its integrand for the differential volume in Figure 10.8]

$$pA + \rho u^2 A + p\, dA = (p + dp)(A + dA) + (\rho + d\rho)(u + du)^2(A + dA) \tag{10.15}$$

In Equation (10.15), all products of differentials, such as $dp\, dA$, $d\rho(du)^2$, are very small and can be ignored. Hence, Equation (10.15) becomes

$$A\, dp + Au^2\, d\rho + \rho u^2\, dA + 2\rho uA\, du = 0 \tag{10.16}$$

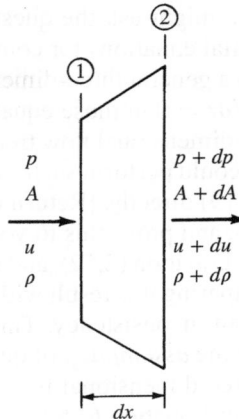

Figure 10.8 Incremental
control volume.

Expanding the continuity equation, Equation (10.14), and multiplying by u, we
have

$$\rho u^2 \, dA + \rho u A \, du + A u^2 \, d\rho = 0 \qquad (10.17)$$

Subtracting Equation (10.17) from (10.16), we obtain

$$\boxed{dp = -\rho u \, du} \qquad (10.18)$$

which is the differential form of the momentum equation for steady, inviscid,
quasi-one-dimensional flow. Equation (10.18) is called *Euler's equation*. We have
seen it before—as Equation (3.12). In Section 3.2, it was derived from the differential form of the general momentum equation in three dimensions. (Make
certain to review that derivation before progressing further.) In Section 3.2, we
demonstrated that Equation (3.12) holds along a streamline in a general three-
dimensional flow. Now we see Euler's equation again, in Equation (10.18), which
was derived from the governing equations for quasi-one-dimensional flow.

A differential form of the energy equation follows directly from Equation (10.9), which states that

$$h + \frac{u^2}{2} = \text{const}$$

Differentiating this equation, we have

$$\boxed{dh + u \, du = 0} \qquad (10.19)$$

In summary, Equations (10.14), (10.18), and (10.19) are differential forms
of the continuity, momentum, and energy equations, respectively, for a steady,
inviscid, adiabatic, quasi-one-dimensional flow. We have obtained them from the
algebraic forms of the equations derived earlier, applied essentially to the picture

0.8. Now you might ask the question, "Since we spent some ... partial differential equations for continuity, momentum, and en- ... hapter 2, applicable to a general three-dimensional flow, why would we ... imply set $\partial/\partial y = 0$ and $\partial/\partial z = 0$ in those equations and obtain differential quations applicable to the one-dimensional flow treated in the present chapter?" The answer is that we certainly could perform such a reduction, and we would obtain Equations (10.18) and (10.19) directly. [Return to the differential equations, Equations (2.113a) and (2.114), and prove this to yourself.] However, if we take the general continuity equation, Equation (2.52), and reduce it to one-dimensional flow, we obtain $d(\rho u) = 0$. Comparing this result with Equation (10.14) for quasi-one-dimensional flow, we see an inconsistency. This is another example of the physical inconsistency between the *assumption* of quasi-one-dimensional flow in a variable-area duct and the three-dimensional flow that actually occurs in such a duct. The result obtained from Equation (2.52), namely, $d(\rho u) = 0$, is a *truly* one-dimensional result, which applies to *constant-area* flows such as considered in Chapter 8. [Recall in Chapter 8 that the continuity equation was used in the form $\rho u = $ constant, which is compatible with Equation (2.52).] However, once we make the quasi-one-dimensional assumption, that is, that uniform properties hold across a given cross section in a variable-area duct, then Equation (10.14) is the only differential form of the continuity equation which insures mass conservation for such an assumed flow.

Let us now use the differential forms of the governing equations, obtained above, to study some physical characteristics of quasi-one-dimensional flow. Such physical information can be obtained from a particular combination of these equations, as follows. From Equation (10.14),

$$\frac{d\rho}{\rho} + \frac{du}{u} + \frac{dA}{A} = 0 \tag{10.20}$$

We wish to obtain an equation that relates the change in velocity du to the change in area dA. Hence, to eliminate $d\rho/\rho$ in Equation (10.20), consider Equation (10.18) written as

$$\frac{dp}{\rho} = \frac{dp}{d\rho}\frac{d\rho}{\rho} = -u\,du \tag{10.21}$$

Keep in mind that we are dealing with inviscid, adiabatic flow. Moreover, for the time being, we are assuming no shock waves in the flow. Hence, the flow is *isentropic*. In particular, any change in density $d\rho$ with respect to a change in pressure dp takes place isentropically; that is,

$$\frac{dp}{d\rho} \equiv \left(\frac{\partial p}{\partial \rho}\right)_s \tag{10.22}$$

From Equation (8.18) for the speed of sound, Equation (10.22) becomes

$$\frac{dp}{d\rho} = a^2 \tag{10.23}$$

Substituting Equation (10.23) into (10.21), we have

$$a^2 \frac{d\rho}{\rho} = -u\,du$$

or

$$\frac{d\rho}{\rho} = -\frac{u\,du}{a^2} = -\frac{u^2}{a^2}\frac{du}{u} = -M^2\frac{du}{u} \tag{10.24}$$

Substituting Equation (10.24) into (10.20), we have

$$-M^2\frac{du}{u} + \frac{du}{u} + \frac{dA}{A} = 0$$

or

$$\boxed{\frac{dA}{A} = (M^2 - 1)\frac{du}{u}} \tag{10.25}$$

Equation (10.25) is the desired equation which relates dA to du; it is called the *area-velocity relation*.

Equation (10.25) is very important; study it closely. In the process, recall the standard convention for differentials; for example, a positive value of du connotes an *increase* in velocity, a negative value of du connotes a *decrease* in velocity, etc. With this in mind, Equation (10.25) tells us the following information:

1. For $0 \leq M < 1$ (subsonic flow), the quantity in parentheses in Equation (10.25) is negative. Hence, an *increase* in velocity (positive du) is associated with a *decrease* in area (negative dA). Likewise, a decrease in velocity (negative du) is associated with an increase in area (positive dA). Clearly, for a subsonic compressible flow, to increase the velocity, we must have a convergent duct, and to decrease the velocity, we must have a divergent duct. These results are illustrated at the top of Figure 10.9. Also, these results are similar to the familiar trends for incompressible flow studied

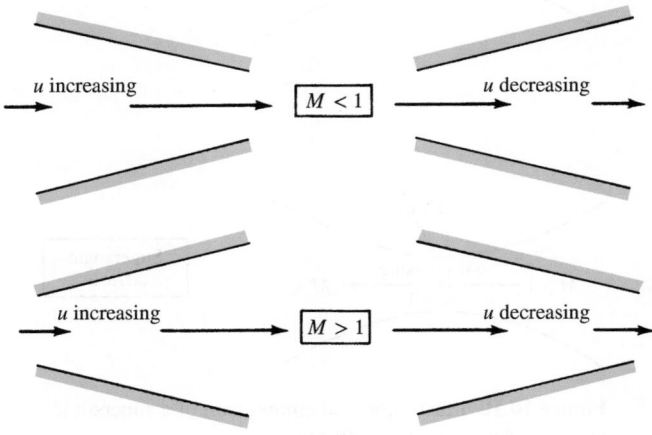

Figure 10.9 Compressible flow in converging and diverging ducts.

in Section 3.3. Once again, we see that subsonic compressible flow is qualitatively (but not quantitatively) similar to incompressible flow.

2. For $M > 1$ (supersonic flow), the quantity in parentheses in Equation (10.25) is positive. Hence, an *increase* in velocity (positive du) is associated with an *increase* in area (positive dA). Likewise, a decrease in velocity (negative du) is associated with a decrease in area (negative dA). For a supersonic flow, to increase the velocity, we must have a divergent duct, and to decrease the velocity, we must have a convergent duct. These results are illustrated at the bottom of Figure 10.9; they are the *direct opposite* of the trends for subsonic flow.

3. For $M = 1$ (sonic flow), Equation (10.25) shows that $dA = 0$ even though a finite du exists. Mathematically, this corresponds to a local maximum or minimum in the area distribution. Physically, it corresponds to a minimum area, as discussed below.

Imagine that we want to take a gas at rest and isentropically expand it to supersonic speeds. The above results show that we must first accelerate the gas subsonically in a convergent duct. However, as soon as sonic conditions are achieved, we must further expand the gas to supersonic speeds by diverging the duct. Hence, a nozzle designed to achieve supersonic flow at its exit is a *convergent–divergent* duct, as sketched at the top of Figure 10.10. The minimum area of the duct is called the *throat*. Whenever an isentropic flow expands from

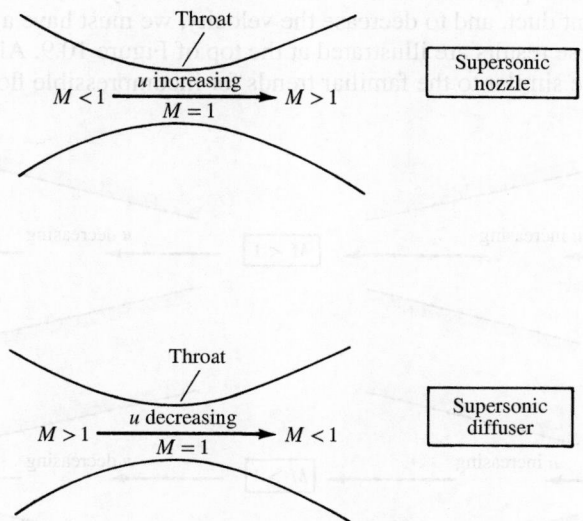

Figure 10.10 Illustration and comparison of a supersonic nozzle and a supersonic diffuser.

subsonic to supersonic speeds, the flow must pass through a throat; moreover, in such a case, $M = 1$ at the throat. The converse is also true; if we wish to take a supersonic flow and slow it down isentropically to subsonic speeds, we must first decelerate the gas in a convergent duct, and then as soon as sonic flow is obtained, we must further decelerate it to subsonic speeds in a divergent duct. Here, the convergent-divergent duct at the bottom of Figure 10.10 is operating as a diffuser. Note that whenever an isentropic flow is slowed from supersonic to subsonic speeds, the flow must pass through a throat; moreover, in such a case, $M = 1$ at the throat.

As a final note on Equation (10.25), consider the case when $M = 0$. Then we have $dA/A = -du/u$, which integrates to $Au = $ constant. This is the familiar continuity equation for incompressible flow in ducts as derived in Section 3.3 and as given by Equation (3.22).

10.3 NOZZLE FLOWS

In this section, we move to the left-hand branch of the road map given in Figure 10.5; that is, we study in detail the compressible flow through nozzles. To expedite this study, we first derive an important equation which relates Mach number to the ratio of duct area to sonic throat area.

Consider the duct shown in Figure 10.11. Assume that sonic flow exists at the throat, where the area is A^*. The Mach number and the velocity at the throat are denoted by M^* and u^*, respectively. Since the flow is sonic at the throat, $M^* = 1$ and $u^* = a^*$. (Note that the use of an asterisk to denote sonic conditions was introduced in Section 7.5; we continue this convention in our present discussion.) At any other section of this duct, the area, the Mach number, and the velocity are denoted by A, M, and u, respectively, as shown in Figure 10.11. Writing Equation (10.1) between A and A^*, we have

$$\rho^* u^* A^* = \rho u A \qquad (10.26)$$

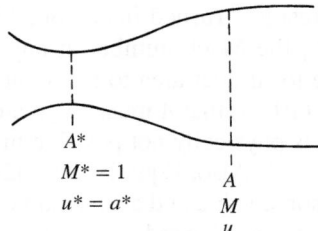

Figure 10.11 Geometry for the derivation of the area–Mach number relation.

Since $u^* = a^*$, Equation (10.26) becomes

$$\frac{A}{A^*} = \frac{\rho^*}{\rho}\frac{a^*}{u} = \frac{\rho^*}{\rho_0}\frac{\rho_0}{\rho}\frac{a^*}{u} \tag{10.27}$$

where ρ_0 is the stagnation density defined in Section 7.5 and is constant throughout an isentropic flow. From Equation (8.46), we have

$$\frac{\rho^*}{\rho_0} = \left(\frac{2}{\gamma + 1}\right)^{1/(\gamma-1)} \tag{10.28}$$

Also, from Equation (8.43), we have

$$\frac{\rho_0}{\rho} = \left(1 + \frac{\gamma - 1}{2}M^2\right)^{1/(\gamma-1)} \tag{10.29}$$

Also, recalling the definition of M^* in Section 8.4, as well as Equation (8.48), we have

$$\left(\frac{u}{a^*}\right)^2 = M^{*2} = \frac{[(\gamma + 1)/2]M^2}{1 + [(\gamma - 1)/2]M^2} \tag{10.30}$$

Squaring Equation (10.27) and substituting Equations (10.28) to (10.30), we obtain

$$\left(\frac{A}{A^*}\right)^2 = \left(\frac{\rho^*}{\rho_0}\right)^2\left(\frac{\rho_0}{\rho}\right)^2\left(\frac{a^*}{u}\right)^2$$

or

$$\left(\frac{A}{A^*}\right)^2 = \left(\frac{2}{\gamma + 1}\right)^{2/(\gamma-1)}\left(1 + \frac{\gamma - 1}{2}M^2\right)^{2/(\gamma-1)}\frac{1 + [(\gamma - 1)/2]M^2}{[(\gamma + 1)/2]M^2} \tag{10.31}$$

Algebraically simplifying Equation (10.31), we have

$$\boxed{\left(\frac{A}{A^*}\right)^2 = \frac{1}{M^2}\left[\frac{2}{\gamma + 1}\left(1 + \frac{\gamma - 1}{2}M^2\right)\right]^{(\gamma+1)/(\gamma-1)}} \tag{10.32}$$

Equation (10.32) is very important; it is called the *area–Mach number relation*, and it contains a striking result. "Turned inside out," Equation (10.32) tells us that $M = f(A/A^*)$; that is, the Mach number at any location in the duct is a function of the ratio of the local duct area to the sonic throat area. Recall from our discussion of Equation (10.25) that A must be greater than or at least equal to A^*; the case where $A < A^*$ is physically not possible in an isentropic flow. Thus, in Equation (10.32), $A/A^* \geq 1$. Also, Equation (10.32) yields *two* solutions for M at a given A/A^*—a subsonic value and a supersonic value. Which value of M that actually holds in a given case depends on the pressures at the inlet and exit of the duct, as explained later. The results for A/A^* as a function of M, obtained from Equation (10.32), are tabulated in Appendix A. Examining Appendix A, we note that for subsonic values of M, as M increases, A/A^* decreases (i.e., the duct converges). At $M = 1$, $A/A^* = 1$ in Appendix A. Finally, for supersonic values

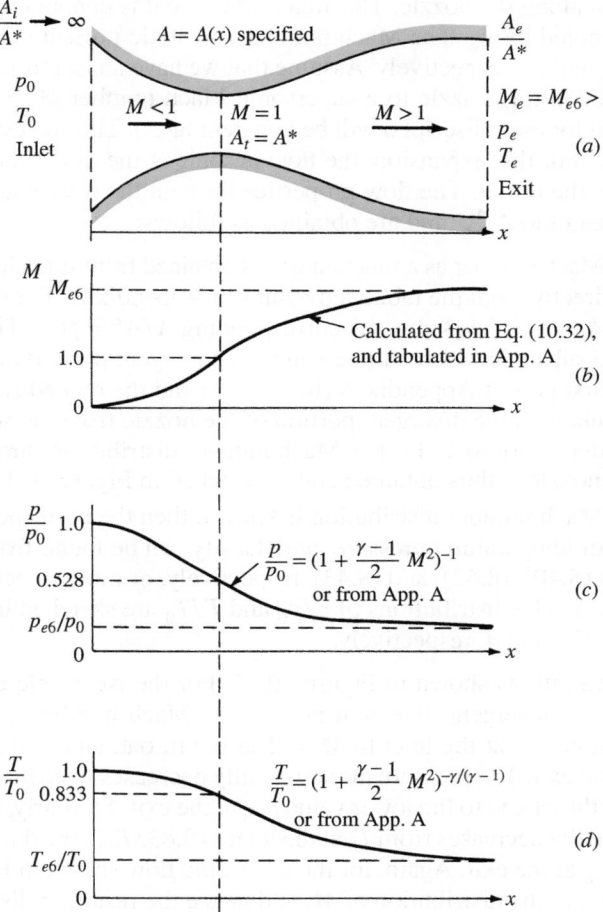

Figure 10.12 Isentropic supersonic nozzle flow.

of M, as M increases, A/A^* increases (i.e., the duct diverges). These trends in Appendix A are consistent with our physical discussion of convergent-divergent ducts at the end of Section 10.2. Moreover, Appendix A shows the double-valued nature of M as a function of A/A^*. For example, for $A/A^* = 2$, we have either $M = 0.31$ or $M = 2.2$.

Consider a given convergent-divergent nozzle, as sketched in Figure 10.12a. Assume that the area ratio at the inlet A_i/A^* is very large and that the flow at the inlet is fed from a large gas reservoir where the gas is essentially stationary. The reservoir pressure and temperature are p_0 and T_0, respectively. Since A_i/A^* is very large, the subsonic Mach number at the inlet is very small, $M \approx 0$. Thus, the pressure and temperature at the inlet are essentially p_0 and T_0, respectively. The area distribution of the nozzle $A = A(x)$ is specified, so that A/A^* is known at every station along the nozzle. The area of the throat is denoted by A_t, and the exit area is denoted by A_e. The Mach number and static pressure at the exit are denoted by M_e and p_e, respectively. Assume that we have an isentropic expansion of the gas through this nozzle to a supersonic Mach number $M_e = M_{e,6}$ at the exit (the reason for the subscript 6 will be apparent later). The corresponding exit pressure is $p_{e,6}$. For this expansion, the flow is sonic at the throat; hence, $M = 1$ and $A_t = A^*$ at the throat. The flow properties through the nozzle are a function of the local area ratio A/A^* and are obtained as follows:

1. The local Mach number as a function of x is obtained from Equation (10.32), or more directly from the tabulated values in Appendix A. For the specified $A = A(x)$, we know the corresponding $A/A^* = f(x)$. Then read the related subsonic Mach numbers in the convergent portion of the nozzle from the first part of Appendix A (for $M < 1$) and the related supersonic Mach numbers in the divergent portion of the nozzle from the second part of Appendix A (for $M > 1$). The Mach number distribution through the complete nozzle is thus obtained and is sketched in Figure 10.12b.

2. Once the Mach number distribution is known, then the corresponding variation of temperature, pressure, and density can be found from Equations (8.40), (8.42), and (8.43), respectively, or more directly from Appendix A. The distributions of p/p_0 and T/T_0 are sketched in Figure 10.12c and d, respectively.

Examine the variations shown in Figure 10.12. For the isentropic expansion of a gas through a convergent-divergent nozzle, the Mach number monotonically increases from near 0 at the inlet to $M = 1$ at the throat, and to the supersonic value $M_{e,6}$ at the exit. The pressure monotonically decreases from p_0 at the inlet to $0.528p_0$ at the throat and to the lower value $p_{e,6}$ at the exit. Similarly, the temperature monotonically decreases from T_0 at the inlet to $0.833T_0$ at the throat and to the lower value $T_{e,6}$ at the exit. Again, for the isentropic flow shown in Figure 10.12, we emphasize that the distribution of M, and hence the resulting distributions of p and T, through the nozzle depends only on the local area ratio A/A^*. This is the key to the analysis of isentropic, supersonic, quasi-one-dimensional nozzle flows.

   Imagine that you take a convergent-divergent nozzle, and simply place it on a table in front of you. What is going to happen? Is the air going to suddenly start flowing through the nozzle of its own accord? The answer is, of course not! Rather, by this stage in your study of aerodynamics, your intuition should tell you that we have to impose a force on the gas in order to produce any acceleration. Indeed, this is the essence of the momentum equation derived in Section 2.5. For the inviscid flows considered here, the only mechanism to produce an accelerating force on a gas is a pressure gradient. Thus, returning to the nozzle on the table, a pressure difference must be created between the inlet and exit; only then will the gas start to flow through the nozzle. The exit pressure must be less than the inlet pressure; that is, $p_e < p_0$. Moreover, if we wish to produce the isentropic supersonic flow sketched in Figure 10.12, the pressure p_e/p_0 must be *precisely*

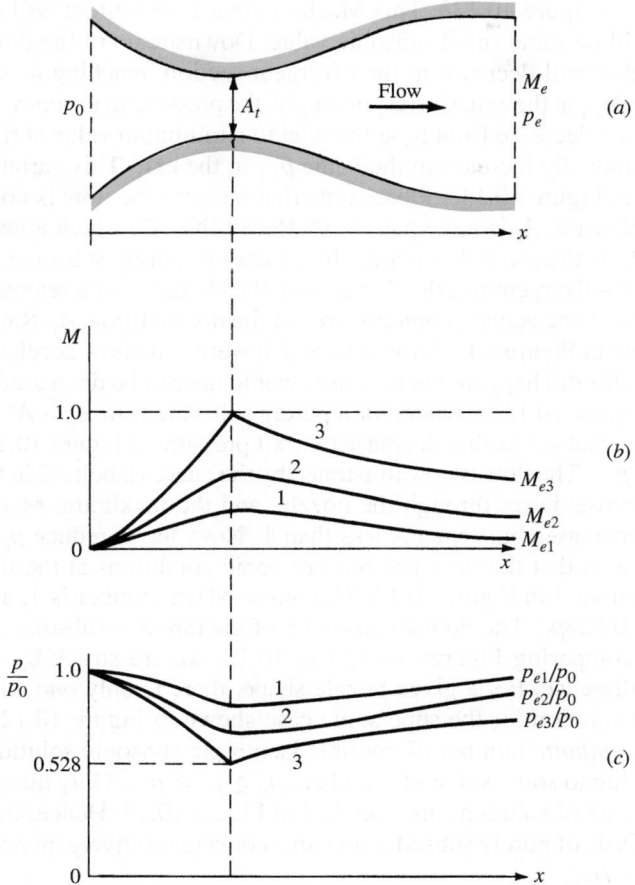

Figure 10.13 Isentropic subsonic nozzle flow.

the value stipulated by Appendix A for the known exit Mach number $M_{e,6}$; that is, $p_e/p_0 = p_{e,6}/p_0$. If the pressure ratio is different from the above isentropic value, the flow either inside or outside the nozzle will be different from that shown in Figure 10.12.

Let us examine the type of nozzle flows that occur when p_e/p_0 is not equal to the precise isentropic value for $M_{e,6}$, that is, when $p_e/p_0 \neq p_{e,6}/p_0$. To begin with, consider the convergent-divergent nozzle sketched in Figure 10.13a. If $p_e = p_0$, no pressure difference exists, and no flow occurs inside the nozzle. Now assume that p_e is minutely reduced below p_0, say, $p_e = 0.999p_0$. This small pressure difference will produce a very low-speed subsonic flow inside the nozzle—essentially a gentle wind. The local Mach number will increase slightly through the convergent portion, reaching a maximum value at the throat, as shown by curve 1 in Figure 10.13b. This Mach number at the throat will *not* be sonic; rather, it will be some small subsonic value. Downstream of the throat, the local Mach number will decrease in the divergent section, reaching a very small but finite value $M_{e,1}$ at the exit. Correspondingly, the pressure in the convergent section will gradually decrease from p_0 at the inlet to a minimum value at the throat, and then will gradually increase to the value $p_{e,1}$ at the exit. This variation is shown as curve 1 in Figure 10.13c. Please note that because the flow is not sonic at the throat in this case, A_t is not equal to A^*. Recall that A^*, which appears in Equation (10.32), is the *sonic* throat area. In the case of purely subsonic flow through a convergent–divergent nozzle, A^* takes on the character of a reference area; it is not the same as the actual geometric area of the nozzle throat A_t. Rather, A^* is the area the flow in Figure 10.13 would have *if* it were somehow accelerated to sonic velocity. If this did happen, the flow area would have to be decreased further than shown in Figure 10.13a. Hence, for a purely subsonic flow $A_t > A^*$.

Assume that we further decrease the exit pressure in Figure 10.13, say, to the value $p_e = p_{e,2}$. The flow is now illustrated by the curves labeled 2 in Figure 10.13. The flow moves faster through the nozzle, and the maximum Mach number at the throat increases but remains less than 1. Now, let us reduce p_e to the value $p_e = p_{e,3}$, such that the flow just reaches sonic conditions at the throat. This is shown by curve 3 in Figure 10.13. The throat Mach number is 1, and the throat pressure is $0.528p_0$. The flow downstream of the throat is subsonic.

Upon comparing Figures 10.12 and 10.13, we are struck by an important physical difference. For a given nozzle shape, there is only *one* allowable isentropic flow solution for the supersonic case shown in Figure 10.12. In contrast, there are an *infinite* number of possible isentropic subsonic solutions, each one corresponding to some value of p_e, where $p_0 \geq p_e \geq p_{e,3}$. Only three solutions of this infinite set of solutions are sketched in Figure 10.13. Hence, the key factors for the analysis of purely subsonic flow in a convergent-divergent nozzle are both A/A^* and p_e/p_0.

Consider the mass flow through the convergent-divergent nozzle in Figure 10.13. As the exit pressure is decreased, the flow velocity in the throat increases; hence, the mass flow increases. The mass flow can be calculated by evaluating Equation (10.1) at the throat; that is, $\dot{m} = \rho_t u_t A_t$. As p_e decreases, u_t increases and ρ_t decreases. However, the percentage increase in u_t is much greater than the decrease in ρ_t. As a result, $\dot{m}$ increases, as sketched in Figure 10.14. When $p_e = p_{e,3}$, sonic flow is achieved at the throat, and $\dot{m} = \rho^* u^* A^* = \rho^* u^* A_t$. Now, if p_e is further reduced below $p_{e,3}$, the conditions at the throat take on a new behavior; they remain *unchanged*. From our discussion in Section 10.2, the Mach number at the throat cannot exceed 1; hence, as p_e is further reduced, M will remain equal to 1 at the throat. Consequently, the mass flow will remain constant as p_e is reduced below $p_{e,3}$, as shown in Figure 10.14. In a sense, the flow at the throat, as well as upstream of the throat, becomes "frozen." Once the flow becomes sonic at the throat, disturbances cannot work their way upstream of the throat. Hence, the flow in the convergent section of the nozzle no longer communicates with the exit pressure and has no way of knowing that the exit pressure

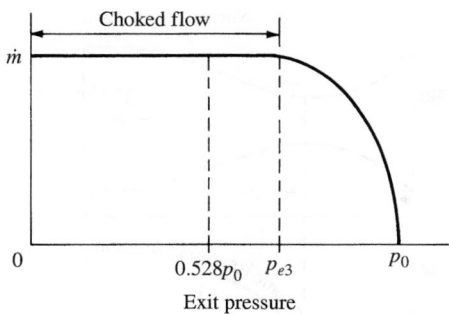

Figure 10.14 Variation of mass flow with
exit pressure; illustration of choked flow.

is continuing to decrease. This situation—when the flow goes sonic at the throat, and the mass flow remains constant no matter how low p_e is reduced—is called *choked flow*. It is a vital aspect of the compressible flow through ducts, and we consider it further in our subsequent discussions.

Return to the subsonic nozzle flows sketched in Figure 10.13. *Question:* What happens in the duct when p_e is reduced below $p_{e,3}$? In the convergent portion, as described above, nothing happens. The flow properties remain fixed at the conditions shown by curve 3 in the convergent section of the duct (the left side of Figure 10.13*b* and *c*). However, a lot happens in the divergent section of the duct. As the exit pressure is reduced below $p_{e,3}$, a region of supersonic flow appears downstream of the throat. However, the exit pressure is too high to allow an isentropic supersonic flow throughout the entire divergent section. Instead, for p_e less than $p_{e,3}$ but substantially higher than the fully isentropic value $p_{e,6}$ (see Figure 10.12*c*), a normal shock wave is formed downstream of the throat. This situation is sketched in Figure 10.15.

In Figure 10.15, the exit pressure has been reduced to $p_{e,4}$, where $p_{e,4} < p_{e,3}$, but where $p_{e,4}$ is also substantially higher than $p_{e,6}$. Here we observe a normal shock wave standing inside the nozzle at a distance d downstream of the throat. Between the throat and the normal shock wave, the flow is given by the supersonic isentropic solution, as shown in Figure 10.15*b* and *c*. Behind the shock wave, the flow is subsonic. This subsonic flow sees the divergent duct and isentropically slows down further as it moves to the exit. Correspondingly, the pressure experiences a discontinuous increase across the shock wave and then is further increased as the flow slows down toward the exit. The flow on both the left and right sides of the shock wave is isentropic; however, the entropy increases across the shock wave. Hence, the flow on the left side of the shock wave is isentropic with one value of entropy s_1, and the flow on the right side of the shock wave is isentropic with another value of entropy s_2, where $s_2 > s_1$. The location of the shock wave inside the nozzle, given by d in Figure 10.15*a*, is determined by the requirement that the increase in static pressure across the wave plus that in the divergent portion of the subsonic flow behind the shock be just right to achieve $p_{e,4}$ at the exit.

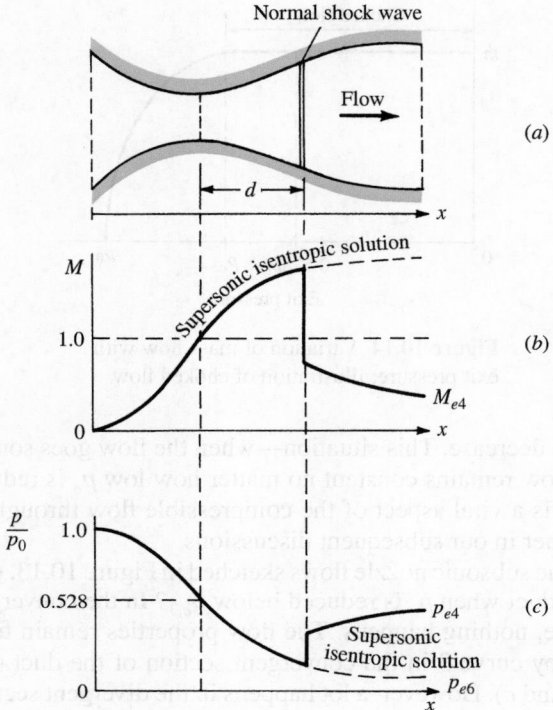

Figure 10.15 Supersonic nozzle flow with a normal shock inside the nozzle.

As p_e is further reduced, the normal shock wave moves downstream, closer to the nozzle exit. At a certain value of exit pressure, $p_e = p_{e,5}$, the normal shock stands precisely at the exit. This is sketched in Figure 10.16a to c. At this stage, when $p_e = p_{e,5}$, the flow through the entire nozzle, except precisely at the exit, is isentropic.

To this stage in our discussion, we have dealt with p_e, which is the pressure right at the nozzle exit. In Figures 10.12, 10.13, 10.15, and 10.16a to c, we have not been concerned with the flow downstream of the nozzle exit. Now imagine that the nozzle in Figure 10.16a exhausts directly into a region of surrounding gas downstream of the exit. These surroundings could be, for example, the atmosphere. In any case, the pressure of the surroundings downstream of the exit is defined as the *back pressure,* denoted by p_B. When the flow at the nozzle exit is subsonic, the exit pressure must equal the back pressure, $p_e = p_B$, because a pressure discontinuity cannot be maintained in a steady subsonic flow. That is, when the exit flow is subsonic, the surrounding back pressure is impressed on the exit flow. Hence, in Figure 10.13, $p_B = p_{e,1}$ for curve 1, $p_B = p_{e,2}$ for curve 2, and $p_B = p_{e,3}$ for curve 3. For the same reason, $p_B = p_{e,4}$ in Figure 10.15, and $p_B = p_{e,5}$ in Figure 10.16. Hence, in discussing these figures, instead of stating

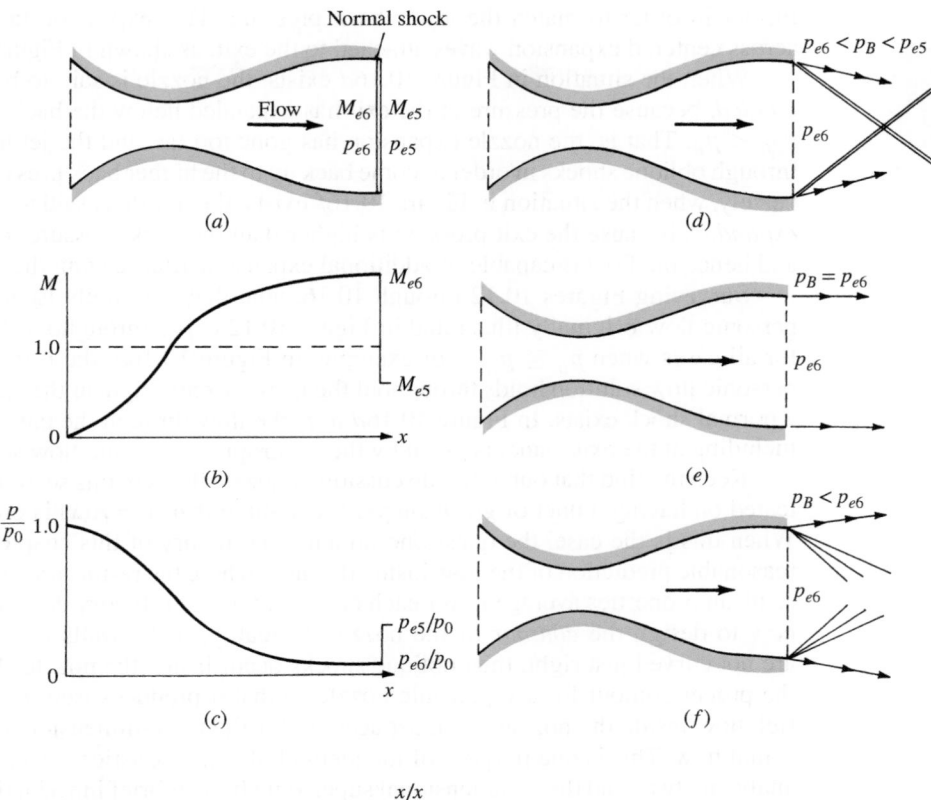

Figure 10.16 Supersonic nozzle flows with waves at the nozzle exit: (*a*), (*b*), and (*c*) pertain to a normal shock at the exit, (*d*) overexpanded nozzle, (*e*) isentropic expansion to the back pressure equal to the exit pressure, and (*f*) underexpanded nozzle.

that we reduced the exit pressure p_e and observed the consequences, we could just as well have stated that we reduced the back pressure p_B. It would have amounted to the same thing.

For the remainder of our discussion in this section, let us now imagine that we have control over p_B and that we are going to continue to decrease p_B. Consider the case when the back pressure is reduced below $p_{e,5}$. When $p_{e,6} < p_B < p_{e,5}$, the back pressure is still above the isentropic pressure at the nozzle exit. Hence, in flowing out to the surroundings, the jet of gas from the nozzle must somehow be compressed such that its pressure is compatible with p_B. This compression takes place across oblique shock waves attached to the exit, as shown in Figure 10.16*d*. When p_B is reduced to the value such that $p_B = p_{e,6}$, there is no mismatch of the exit pressure and the back pressure; the nozzle jet exhausts smoothly into the surroundings without passing through any waves. This is shown in Figure 10.16*e*. Finally, as p_B is reduced below $p_{e,6}$, the jet of gas from the nozzle must expand further in order to match the lower back pressure. This expansion takes place across centered expansion waves attached to the exit, as shown in Figure 10.16f.

When the situation in Figure 10.16d exists, the nozzle is said to be *overexpanded,* because the pressure at the exit has expanded below the back pressure, $p_{e,6} < p_B$. That is, the nozzle expansion has gone too far, and the jet must pass through oblique shocks in order to come back up to the higher back pressure. Conversely, when the situation in Figure 10.16f exists, the nozzle is said to be *underexpanded*, because the exit pressure is higher than the back pressure, $p_{e,6} > p_B$, and hence the flow is capable of additional expansion after leaving the nozzle.

Surveying Figures 10.12 through 10.16, note that the purely isentropic supersonic flow originally illustrated in Figure 10.12 exists throughout the nozzle for all cases when $p_B \leq p_{e,5}$. For example, in Figure 10.16a, the isentropic supersonic flow solution holds throughout the nozzle except right at the exit, where a normal shock exists. In Figure 10.16d to f, the flow through the entire nozzle, including at the exit plane, is given by the isentropic supersonic flow solution.

Keep in mind that our entire discussion of nozzle flows in this section is predicated on having a duct of *given shape.* We assume that $A = A(x)$ is prescribed. When this is the case, the quasi-one-dimensional theory of this chapter gives a reasonable prediction of the flow inside the duct, where the results are interpreted as mean properties averaged over each cross section. This theory does *not* tell us how to design the *contour* of the nozzle. In reality, if the walls of the nozzle are not curved just right, then oblique shocks occur inside the nozzle. To obtain the proper contour for a supersonic nozzle so that it produces isentropic shock-free flow inside the nozzle, we must account for the three-dimensionality of the actual flow. This is one purpose of the method of characteristics, a technique for analyzing two- and three-dimensional supersonic flow. A brief introduction to the method of characteristics is given in Chapter 13.

EXAMPLE 10.1

Consider the isentropic supersonic flow through a convergent-divergent nozzle with an exit-to-throat area ratio of 10.25. The reservoir pressure and temperature are 5 atm and 600°R, respectively. Calculate M, p, and T at the nozzle exit.

■ Solution

From the supersonic portion of Appendix A, for $A_e/A^* = 10.25$,

$$M_e = \boxed{3.95}$$

Also,

$$\frac{p_e}{p_0} = \frac{1}{142} \quad \text{and} \quad \frac{T_e}{T_0} = \frac{1}{4.12}$$

Thus,

$$p_e = 0.007p_0 = 0.007(5) = \boxed{0.035 \text{ atm}}$$

$$T_e = 0.2427T_0 = 0.2427(600) = \boxed{145.6°\text{R}}$$

EXAMPLE 10.2

Consider the isentropic flow through a convergent-divergent nozzle with an exit-to-throat area ratio of 2. The reservoir pressure and temperature are 1 atm and 288 K, respectively. Calculate the Mach number, pressure, and temperature at both the throat and the exit for the cases where (a) the flow is supersonic at the exit and (b) the flow is subsonic throughout the entire nozzle except at the throat, where $M = 1$.

■ **Solution**

(a) At the throat, the flow is sonic. Hence,

$$M_t = \boxed{1.0}$$

$$p_t = p^* = \frac{p^*}{p_0}p_0 = 0.528(1 \text{ atm}) = \boxed{0.528 \text{ atm}}$$

$$T_t = T^* = \frac{T^*}{T_0} = 0.833(288) = \boxed{240 \text{ K}}$$

At the exit, the flow is supersonic. Hence, from the supersonic portion of Appendix A, for $A_e/A^* = 2$,

$$M_e = \boxed{2.2}$$

$$p_e = \frac{p_e}{p_0}p_0 = \frac{1}{10.69}(1 \text{ atm}) = \boxed{0.0935 \text{ atm}}$$

$$T_e = \frac{T_e}{T_0}T_0 = \frac{1}{1.968}(288) = \boxed{146 \text{ K}}$$

(b) At the throat, the flow is still sonic. Hence, from above, $M_t = 1.0$, $p_t = 0.528$ atm, and $T_t = 240$ K. However, at all other locations in the nozzle, the flow is subsonic. At the exit, where $A_e/A^* = 2$, from the subsonic portion of Appendix A,

$$M_e = \boxed{0.3} \qquad \text{(rounded to the nearest entry in Appendix A)}$$

$$p_e = \frac{p_e}{p_0}p_0 = \frac{1}{1.064}(1 \text{ atm}) = \boxed{0.94 \text{ atm}}$$

$$T_e = \frac{T_e}{T_0}T_0 = \frac{1}{1.018}(288) = \boxed{282.9 \text{ K}}$$

EXAMPLE 10.3

For the nozzle in Example 10.2, assume the exit pressure is 0.973 atm. Calculate the Mach numbers at the throat and the exit.

■ **Solution**

In Example 10.2, we saw that if $p_e = 0.94$ atm, the flow is sonic at the throat, but subsonic elsewhere. Hence, $p_e = 0.94$ atm corresponds to $p_{e,3}$ in Figure 10.13. In the present problem, $p_e = 0.973$ atm, which is higher than $p_{e,3}$. Hence, in this case, the flow is subsonic throughout the nozzle, including at the throat. For this case, A^* takes on a reference value, and the actual geometric throat area is denoted by A_t. At the exit,

$$\frac{p_0}{p_e} = \frac{1}{0.973} = 1.028$$

From the subsonic portion of Appendix A, for $p_0/p_e = 1.028$, we have

$$M_e = \boxed{0.2} \quad \text{and} \quad \frac{A_e}{A^*} = 2.964$$

$$\frac{A_t}{A^*} = \frac{A_t}{A_e}\frac{A_e}{A^*} = 0.5(2.964) = 1.482$$

From the subsonic portion of Appendix A, for $A_t/A^* = 1.482$, we have

$$M_t = \boxed{0.44} \qquad \text{(nearest entry)}$$

EXAMPLE 10.4

An equation for the thrust of a jet-propulsion device can be derived by applying the integral form of the momentum equation for a steady inviscid flow [Equation (2.71)] to a control volume wrapped around the jet engine. This derivation is carried out in great detail in Chapter 2 of Reference 21, and in a simpler form in Chapter 9 of Reference 2 where the result is also specialized to a rocket engine. You are encouraged to examine these derivations—they are an excellent example of the use of the control volume concept. The resulting thrust equation for a rocket engine (see Section 9.8 of Reference 2) is

$$T = \dot{m}u_e + (p_e - p_\infty)A_e \tag{E10.1}$$

where T is the thrust, $\dot{m}$ is the mass flow through the engine, u_e is the gas velocity at the nozzle exit, p_e is the gas pressure at the nozzle exit, p_∞ is the surrounding ambient atmospheric pressure, and A_e is the area of the exit.

Consider a rocket engine similar to that shown in Figure 10.1. Liquid hydrogen and oxygen are burned in the combustion chamber producing a combustion gas pressure and temperature of 30 atm and 3500 K, respectively. The area of the rocket nozzle throat is 0.4 m^2. The area of the exit is designed so that the exit pressure exactly equals the ambient atmospheric pressure at a standard altitude of 20 km. Assume an isentropic flow through the rocket engine nozzle with an effective value of the ratio of specific heats $\gamma = 1.22$, and a constant value of the specific gas constant $R = 520$ J/(kg)(K).

(a) Using Equation (E10.1), calculate the thrust of the rocket engine.
(b) Calculate the area of the nozzle exit.

■ **Solution**

(a) Examining Equation (E10.1), we first need to obtain the value of mass flow, $\dot{m}$, and exit velocity, u_e. The mass flow is constant through the nozzle and is equal to $\dot{m} = \rho u A$ evaluated at any location in the nozzle. A convenient location to evaluate $\dot{m}$ is at the throat, where

$$\dot{m} = \rho^* u^* A^*$$

To obtain ρ^*, we need $\rho_0 = p_0/RT_0$. Noting that $(1\ atm) = 1.01 \times 10^5\ N/m^2$,

$$\rho_0 = \frac{(30)(1.01 \times 10^5)}{(520)(3500)} = 1.665\ kg/m^3$$

From Equation (8.46),

$$\frac{\rho^*}{\rho_0} = \left(\frac{2}{\gamma+1}\right)^{\frac{1}{\gamma-1}} = \left(\frac{2}{1.22+1}\right)^{\frac{1}{1.22-1}} = \left(\frac{2}{2.22}\right)^{4.545} = 0.622$$

$$\rho^* = 0.622\rho_0 = 0.622(1.665) = 1.036\ kg/m^3$$

At the throat, the flow velocity is equal to the local speed of sound, $u^* = a^*$. From Equation (8.44),

$$\frac{T^*}{T_0} = \frac{2}{\gamma+1} = \frac{2}{2.22} = 0.901$$

$$T^* = 0.901 T_0 = 0.901(3500) = 3154\ K$$

$$a^* = \sqrt{\gamma R T^*} = \sqrt{(1.22)(520)(3154)} = 1415\ m/s$$

$$\dot{m} = \rho^* u^* A^* = (1.036)(1415)(0.4) = 586.4\ kg/s$$

This is the value of $\dot{m}$ to be used in Equation (E10.1).

Next, we need to obtain the exit velocity u_e. We do this by first obtaining the exit Mach number from Equation (8.42).

$$\frac{p_0}{p_e} = \left(1 + \frac{\gamma-1}{2}M_e^2\right)^{\frac{\gamma}{\gamma-1}}$$

where, from the statement of the problem, p_e is equal to the ambient pressure at a standard altitude of 20 km. From Appendix D, at 20 km, $p_\infty = 5.5293 \times 10^3\ N/m^2$. Hence,

$$p_e = p_\infty = 5529\ N/m^2$$

Thus, from Equation (8.42),

$$1 + \frac{\gamma-1}{2}M_e^2 = \left(\frac{p_0}{p_e}\right)^{\frac{\gamma-1}{\gamma}} = \left[\frac{(30)(1.01 \times 10^5)}{5529}\right]^{\frac{0.22}{1.22}} = (548)^{0.18} = 3.111$$

Note: For future use in this solution, we set aside the value:

$$1 + \frac{\gamma-1}{2}M_e^2 = 3.111 \qquad (E10.2)$$

Thus, from Equation (E10.2),

$$\frac{\gamma-1}{2}M_e^2 = 2.111$$

$$M_e^2 = (2.111)\left(\frac{2}{0.22}\right) = 19.19$$

$$M_e = 4.38$$

To obtain the speed of sound at the exit, from Equations (8.40) and (E10.2),

$$\frac{T_0}{T_e} = 1 + \frac{\gamma - 1}{2}M_e^2 = 3.111$$

$$T_e = \frac{T_0}{3.111} = \frac{3500}{3.111} = 1125 \text{ K}$$

$$a_e = \sqrt{\gamma R T_e} = \sqrt{(1.22)(520)(1125)} = 844.8 \text{ m/s}$$

Thus,
$$u_e = M_e a_e = (4.38)(844.8) = 3700 \text{ m/s}$$

Intermediate check: We can check this value of 3700 m/s for u_e by directly using the energy equation, Equation (8.38)

$$c_p T_0 = c_p T_e + \frac{u_e^2}{2}$$

where, from Equation (7.9),

$$c_p = \frac{\gamma R}{\gamma - 1} = \frac{(1.22)(520)}{0.22} = 2883.6 \frac{\text{J}}{\text{kg} \cdot \text{K}}$$

Thus, from Equation (8.38),

$$u_e^2 = 2c_p(T_0 - T_e) = 2(2883.6)(3500 - 1125) = 1.3697 \times 10^7$$

or,
$$u_e = 3700 \text{ m/s}$$

This checks with the result for u_e obtained earlier.

Finally, we are ready to calculate the thrust from Equation (E10.1). Since the statement of the problem gives $p_e = p_\infty$, the pressure term in Equation (E10.1) drops out, and we have

$$T = \dot{m}u_e = (586.4)(3700) = \boxed{2.17 \times 10^6 \text{ N}}$$

Since
$$1\text{N} = 0.2247 \text{ lb, we have}$$

$$T = (2.17 \times 10^6)(0.2247) = \boxed{487,600 \text{ lb}}$$

(*b*) From Equation (10.32),

$$\left(\frac{A_e}{A^*}\right)^2 = \frac{1}{M_e^2}\left[\frac{2}{\gamma + 1}\left(1 + \frac{\gamma - 1}{2}M_e^2\right)\right]^{\frac{\gamma+1}{\gamma-1}}$$

In this equation, the numerical values of the various terms are

$$\frac{\gamma + 1}{\gamma - 1} = \frac{2.22}{0.22} = 10.1$$

$$\frac{2}{\gamma + 1} = \frac{2}{2.22} = 0.9$$

$$1 + \frac{\gamma - 1}{2}M_e^2 = 3.111 \qquad \text{[from Equation (E10.2)]}$$

and
$$M_e = 4.38$$

Inserting these values into Equation (10.32),

$$\left(\frac{A_e}{A^*}\right)^2 = \frac{1}{(4.38)^2}[(0.9)(3.111)]^{10.1} = 1710.8$$

$$\frac{A_e}{A^*} = 41.36$$

Thus, $A_e = (41.36)A^* = (41.36)(0.4) = \boxed{16.5 \ \text{m}^2}$

EXAMPLE 10.5

Calculate the mass flow through the rocket engine described in Example 10.4 using the closed-form analytical expression given in Problem 10.5 at the end of this chapter. Compare the result with that obtained in Example 10.4.

■ **Solution**

From Problem 10.5, the closed-form expression for the mass flow through a choked nozzle is

$$\dot{m} = \frac{p_0 A^*}{\sqrt{T_0}} \sqrt{\frac{\gamma}{R} \left(\frac{2}{\gamma+1}\right)^{(\gamma+1)/(\gamma-1)}} \qquad \text{(E10.3)}$$

From Example 10.4 we have $p_0 = 30$ atm, $T_0 = 3500$ K, $A^* = 0.4 \ \text{m}^2$, $R = 520$ J/(kg)(K), and $\gamma = 1.22$.

Noting that

$$p_0 = 30 \ \text{atm} = (30)(1.01 \times 10^5) = 3.03 \times 10^6 \ \text{N/m}^2$$

$$\gamma/R = 1.22/510 = 2.346 \times 10^{-3}$$

$$\frac{2}{\gamma+1} = \frac{2}{2.22} = 0.9$$

$$\frac{\gamma+1}{\gamma-1} = \frac{2.22}{0.22} = 10.09$$

from Equation (E10.3) we have

$$\dot{m} = \frac{(3.03 \times 10^6)(0.4)}{\sqrt{3500}} \sqrt{(2.346 \times 10^{-3})(0.9)^{10.09}} = \boxed{583.2 \ \text{kg/s}}$$

This result, obtained from a single equation, compares well with the value of 586.4 kg/s obtained from a sequence of calculations that is subject to a larger cumulative roundoff error (the author is using a hand calculator and usually rounding off to the fourth significant figure). The result obtained here, using Equation (E10.3), should be considered more accurate.

10.3.1 More on Mass Flow

Equation (E10.3) in Example 10.5 has a distinct advantage over the piecemeal calculations of $\dot{m}$ in Example 10.4. Not only does it lead to a straightforward answer in one step, it also shows us exactly on what variables the mass flow depends, and in what manner. From Equation (E10.3), we see that the mass flow depends primarily on p_0, T_0, and A^*, and that it varies directly with reservoir pressure and the area of the throat, and inversely as the square root of the reservoir temperature. If you double the reservoir pressure, you double the mass flow. If you double the throat area, you double the mass flow. If you quadruple the reservoir temperature, you cut the mass flow by a half. These variations are fundamental to the physics of choked flow in a nozzle. Make certain to fix in your mind this proportionality:

$$\dot{m} \propto \frac{p_0 A^*}{\sqrt{T_0}} \tag{10.33}$$

How is this discussion related to the variation of mass flow sketched in Figure 10.14? Recall that Figure 10.14 pertains to a nozzle flow with fixed reservoir conditions, including a fixed value of p_0. The mass flow is plotted versus exit pressure. If the exit pressure equals p_0, there is no pressure difference across the nozzle, hence no flow through the nozzle (i.e., in Figure 10.14 the point for $p_e = p_0$ corresponds to $\dot{m} = 0$). As the exit pressure decreases, $\dot{m}$ first increases, and then reaches a plateau when $p_e \leq p_{e,3}$. For $p_{e,3} \leq p_e \leq p_0$, the nozzle flow is *not* choked, and the value of the mass flow depends not only on p_0, A^*, and T_0, but also on p_e. When p_e falls below $p_{e,3}$, the flow is choked, and the mass flow becomes constant no matter how low p_e is decreased. The horizontal portion of the curve in Figure 10.14 pertains to choked flow, and the magnitude of this choked mass flow depends only on the values of p_0, A^*, and T_0 and *not* on p_e. For the case shown in Figure 10.14, the values of p_0, A^*, and T_0 are fixed, specific values. If, for whatever reason, the value of p_0, or A^*, or T_0 is changed, then the horizontal choked-flow line in Figure 10.14 would be raised or lowered appropriately, governed by the proportionality given by Equation (10.33).

10.4 DIFFUSERS

The role of a diffuser was first introduced in Section 3.3 in the context of a low-speed subsonic wind tunnel. There, a diffuser was a divergent duct downstream of the test section whose role was to slow the higher-velocity air from the test section down to a very low velocity at the diffuser exit (see Figure 3.8). Indeed, in general, we can define a diffuser as any duct designed to slow an incoming gas flow to lower velocity at the exit of the diffuser. The incoming flow can be subsonic, as discussed in Figure 3.8, or it can be supersonic, as discussed in the present section. However, the shape of the diffuser is drastically different, depending on whether the incoming flow is subsonic or supersonic.

Before pursuing this matter further, let us elaborate on the concept of total pressure p_0 as discussed in Section 7.5. In a semiqualitative sense, the total pressure of a flowing gas is a measure of the capacity of the flow to perform useful work. Let us consider two examples:

1. A pressure vessel containing stagnant air at 10 atm
2. A supersonic flow at $M = 2.16$ and $p = 1$ atm

In case 1, the air velocity is zero; hence, $p_0 = p = 10$ atm. Now, imagine that we want to use air to drive a piston in a piston-cylinder arrangement, where useful work is performed by the piston being displaced through a distance. The air is ducted into the cylinder from a large manifold, in the same vein as the reciprocating internal combustion engine in our automobile. In case 1, the pressure vessel can act as the manifold; hence, the pressure on the piston is 10 atm, and a certain amount of useful work is performed, say, W_1. However, in case 2, the supersonic flow must be slowed to a low velocity before we can readily feed it into the manifold. If this slowing process can be achieved without loss of total pressure, then the pressure in the manifold in this case is also 10 atm (assuming $V \approx 0$), and the same amount of useful work W_1 is performed. On the other hand, assume that in slowing down the supersonic stream, a loss of 3 atm takes place in the total pressure. Then the pressure in the manifold is only 7 atm, with the consequent generation of useful work W_2, which is less than in the first case; that is, $W_2 < W_1$. The purpose of this simple example is to indicate that the total pressure of a flowing gas is indeed a measure of its capability to perform useful work. On this basis, a loss of total pressure is always an inefficiency—a loss of the capability to do a certain amount of useful work.

In light of the above, let us expand our definition of a diffuser. A diffuser is a duct designed to slow an incoming gas flow to lower velocity at the exit of the diffuser *with as small a loss in total pressure as possible*. Consequently, an ideal diffuser would be characterized by an *isentropic* compression to lower velocities; this is sketched in Figure 10.17a, where a supersonic flow enters the diffuser at

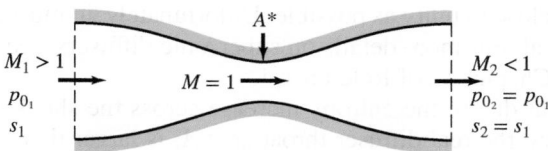

(a) Ideal (isentropic) supersonic diffuser

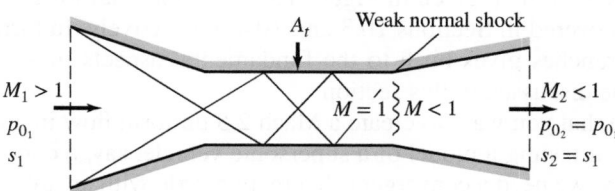

(b) Actual supersonic diffuser

Figure 10.17 The ideal (isentropic) diffuser compared with the actual situation.

M_1, is isentropically compressed in a convergent duct to Mach 1 at the throat, where the area is A^*, and then is further isentropically compressed in a divergent duct to a low subsonic Mach number at the exit. Because the flow is isentropic, $s_2 = s_1$, and from Equation (8.73), $p_{0,2} = p_{0,1}$. Indeed, p_0 is constant throughout the entire diffuser—a characteristic of isentropic flow. However, common sense should tell you that the ideal diffuser in Figure 10.17a can never be achieved. It is extremely difficult to slow a supersonic flow without generating shock waves in the process. For example, examine the convergent portion of the diffuser in Figure 10.17a. Note that the supersonic flow is turned into itself; hence, the converging flow will inherently generate oblique shock waves, which will destroy the isentropic nature of the flow. Moreover, in real life, the flow is viscous; there will be an entropy increase within the boundary layers on the walls of the diffuser. For these reasons, an ideal isentropic diffuser can never be constructed; an ideal diffuser is of the nature of a "perpetual motion machine"—only a utopian wish in the minds of engineers.

An actual supersonic diffuser is sketched in Figure 10.17b. Here, the incoming flow is slowed by a series of reflected oblique shocks, first in a convergent section usually consisting of straight walls, and then in a constant-area throat. Due to the interaction of the shock waves with the viscous flow near the wall, the reflected shock pattern eventually weakens and becomes quite diffuse, sometimes ending in a weak normal shock wave at the end of the constant-area throat. Finally, the subsonic flow downstream of the constant-area throat is further slowed by moving through a divergent section. At the exit, clearly $s_2 > s_1$; hence $p_{0,2} < p_{0,1}$. The art of diffuser design is to obtain as small a total pressure loss as possible, that is, to design the convergent, divergent, and constant-area throat sections so that $p_{0,2}/p_{0,1}$ is as close to unity as possible. Unfortunately, in most cases, we fall far short of that goal. For more details on supersonic diffusers, see Chapter 5 of Reference 21 and Chapter 12 of Reference 1.

Please note that due to the entropy increase across the shock waves and in the boundary layers, the real diffuser throat area A_t is larger than A^*, that is, in Figure 10.17, $A_t > A^*$.

10.5 SUPERSONIC WIND TUNNELS

Return to the road map given in Figure 10.5. The material for the left and right branches is covered in Sections 10.3 and 10.4, respectively. In turn, a mating of these two branches gives birth to the fundamental aspects of supersonic wind tunnels, to be discussed in this section.

Imagine that you want to create a Mach 2.5 uniform flow in a laboratory for the purpose of testing a model of a supersonic vehicle, say, a cone. How do you do it? Clearly, we need a convergent-divergent nozzle with an area ratio $A_e/A^* = 2.637$ (see Appendix A). Moreover, we need to establish a pressure ratio, $p_0/p_e = 17.09$, across the nozzle in order to obtain a shock-free expansion to $M_e = 2.5$ at the exit. Your first thought might be to exhaust the nozzle directly into the laboratory, as sketched in Figure 10.18. Here, the Mach 2.5 flow passes into the

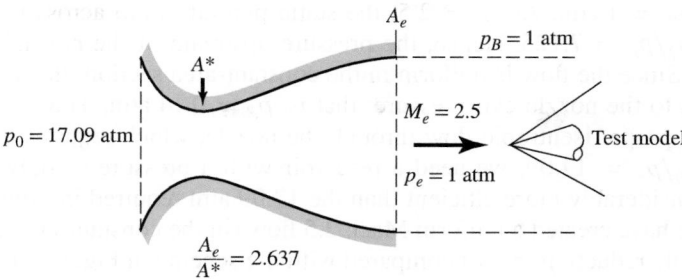

Figure 10.18 Nozzle exhausting directly to the atmosphere.

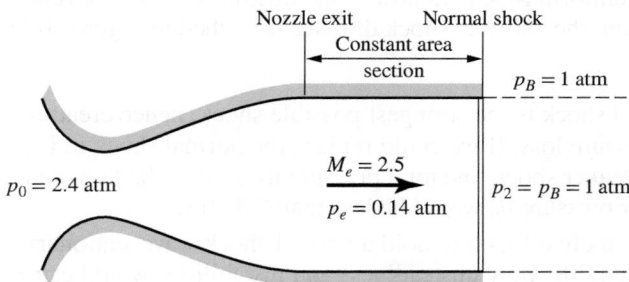

Figure 10.19 Nozzle exhausting into a constant-area duct, where a normal shock stands at the exit of the duct.

surroundings as a "free jet." The test model is placed in the flow downstream of the nozzle exit. In order to make certain that the free jet does not have shock or expansion waves, the nozzle exit pressure p_e must equal the back pressure p_B, as originally sketched in Figure 10.16e. Since the back pressure is simply that of the atmosphere surrounding the free jet, $p_B = p_e = 1$ atm. Consequently, to establish the proper isentropic expansion through the nozzle, you need a high-pressure reservoir with $p_0 = 17.09$ atm at the inlet to the nozzle. In this manner, you would be able to accomplish your objective, namely, to produce a uniform stream of air at Mach 2.5 in order to test a supersonic model, as sketched in Figure 10.18.

In the above example, you may have a problem obtaining the high-pressure air supply at 17.09 atm. You need an air compressor or a bank of high-pressure air bottles—both of which can be expensive. It requires work, hence money, to create reservoirs of high-pressure air—the higher the pressure, the more the cost. So, can you accomplish your objective in a more efficient way, at less cost? The answer is yes, as follows. Instead of the free jet as sketched in Figure 10.18, imagine that you have a long constant-area section downstream of the nozzle exit, with a normal shock wave standing at the end of the constant-area section; this is shown in Figure 10.19. The pressure downstream of the normal shock wave is $p_2 = p_B = 1$ atm. At $M = 2.5$, the static pressure ratio across the normal shock is $p_2/p_e = 7.125$. Hence, the pressure upstream of the normal shock is 0.14 atm. Since the flow is uniform in the constant-area section, this pressure is also equal to the nozzle exit pressure; that is, $p_e = 0.14$ atm. Thus, in order to obtain the proper isentropic flow through the nozzle, which requires a pressure ratio of $p_0/p_e = 17.09$, we need a reservoir with a pressure of only 2.4 atm. This is considerably more efficient than the 17.09 atm required in Figure 10.18. Hence, we have created a uniform Mach 2.5 flow (in the constant-area duct) at a considerable reduction in cost compared with the scheme in Figure 10.18.

In Figure 10.19, the normal shock wave is acting as a diffuser, slowing the air originally at Mach 2.5 to the subsonic value of Mach 0.513 immediately behind the shock. Hence, by the addition of this "diffuser," we can more efficiently produce our uniform Mach 2.5 flow. This illustrates one of the functions of a diffuser. However, the "normal shock diffuser" sketched in Figure 10.19 has several problems:

1. A normal shock is the strongest possible shock, hence creating the largest total pressure loss. If we could replace the normal shock in Figure 10.19 with a weaker shock, the total pressure loss would be less, and the required reservoir pressure p_0 would be less than 2.4 atm.

2. It is extremely difficult to hold a normal shock wave stationary at the duct exit; in real life, flow unsteadiness and instabilities would cause the shock to move somewhere else and to fluctuate constantly in position. Thus, we could never be certain about the quality of the flow in the constant-area duct.

3. As soon as a test model is introduced into the constant-area section, the oblique waves from the model would propagate downstream, causing the flow to become two- or three-dimensional. The normal shock sketched in Figure 10.19 could not exist in such a flow.

Hence, let us replace the normal shock in Figure 10.19 with the oblique shock diffuser shown in Figure 10.17*b*. The resulting duct would appear as sketched in Figure 10.20. Examine this figure closely. We have a convergent-divergent nozzle feeding a uniform supersonic flow into the constant-area duct, which is

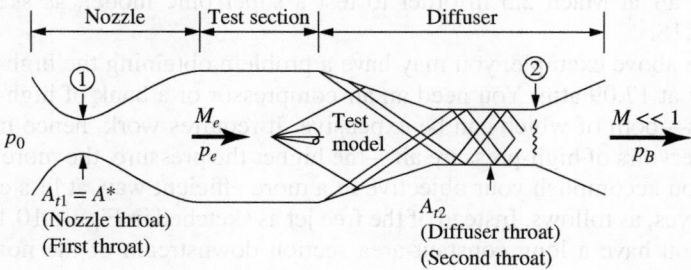

Figure 10.20 Sketch of a supersonic wind tunnel.

called the *test section*. This flow is subsequently slowed to a low subsonic speed by means of a diffuser. This arrangement—namely, a convergent-divergent nozzle, a test section, and a convergent-divergent diffuser—is a *supersonic wind tunnel*. A test model, the cone in Figure 10.20, is placed in the test section, where aerodynamic measurements such as lift, drag, and pressure distribution are made. The wave system from the model propagates downstream and interacts with the multireflected shocks in the diffuser. The pressure ratio required to run the supersonic tunnel is p_0/p_B. This can be obtained by making p_0 large via a high-pressure reservoir at the inlet to the nozzle or by making p_B small via a vacuum source at the exit of the diffuser, or a combination of both.

The main source of total pressure loss in a supersonic wind tunnel is the diffuser. How does the oblique shock diffuser in Figure 10.20 compare with the hypothetical normal shock diffuser in Figure 10.19? Is the total pressure loss across all the reflected oblique shocks in Figure 10.20 greater or less than across the single normal shock wave in Figure 10.19? This is an important question, since the smaller the total pressure loss in the diffuser, the smaller is the pressure ratio p_0/p_B required to run the supersonic tunnel. There is no pat answer to this question. However, it is usually true that progressively reducing the velocity of a supersonic flow through a series of oblique shocks to a low supersonic value, and then further reducing the flow to subsonic speeds across a weak normal shock, results in a *smaller* total pressure loss than simply reducing the flow to subsonic speeds across a single, strong normal shock wave at the initially high supersonic Mach number. This trend is illustrated by Example 9.5. Therefore, the oblique shock diffuser shown in Figures 10.17*b* and 10.20 is usually more efficient than the simple normal shock diffuser shown in Figure 10.19. This is not always true, however, because in an actual real-life oblique shock diffuser, the shock waves interact with the boundary layers on the walls, causing local thickening and even possible separation of the boundary layers. This creates an additional total pressure loss. Moreover, the simple aspect of skin friction exerted on the surface generates a total pressure loss. Hence, actual oblique shock diffusers may have efficiencies greater or less than a hypothetical normal shock diffuser. Nevertheless, virtually all supersonic wind tunnels use oblique shock diffusers qualitatively similar to that shown in Figure 10.20.

Notice that the supersonic wind tunnel shown in Figure 10.20 has two throats: the nozzle throat with area $A_{t,1}$ is called the *first throat,* and the diffuser throat with area $A_{t,2}$ is called the *second throat*. The mass flow through the nozzle can be expressed as $\dot{m} = \rho u A$ evaluated at the first throat. This station is denoted as station 1 in Figure 10.20, and hence the mass flow through the nozzle is $\dot{m}_1 = \rho_1 u_1 A_{t,1} = \rho_1^* a_1^* A_{t,1}$. In turn, the mass flow through the diffuser can be expressed as $\dot{m} = \rho u A$ evaluated at station 2, namely, $\dot{m}_2 = \rho_2 u_2 A_{t,2}$. For steady flow through the wind tunnel, $\dot{m}_1 = \dot{m}_2$. Hence,

$$\rho_1^* a_1^* A_{t,1} = \rho_2 u_2 A_{t,2} \tag{10.34}$$

Since the thermodynamic state of the gas is irreversibly changed in going through the shock waves created by the test model and generated in the diffuser, clearly

ρ_2 and possibly u_2 are different from ρ_1^* and a_1^*, respectively. Hence, from Equation (10.34), *the second throat must have a different area from the first throat;* that is, $A_{t,2} \neq A_{t,1}$.

Question: How does $A_{t,2}$ differ from $A_{t,1}$? Let us assume that sonic flow occurs at *both* stations 1 and 2 in Figure 10.20. Thus, Equation (10.34) can be written as

$$\frac{A_{t,2}}{A_{t,1}} = \frac{\rho_1^* a_1^*}{\rho_2^* a_2^*} \tag{10.35}$$

Recall from Section 8.4 that a^* is constant for an adiabatic flow. Also, recall that the flow across shock waves is adiabatic (but not isentropic). Hence, the flow throughout the wind tunnel sketched in Figure 10.18 is adiabatic, and therefore $a_1^* = a_2^*$. In turn, Equation (10.35) becomes

$$\frac{A_{t,2}}{A_{t,1}} = \frac{\rho_1^*}{\rho_2^*} \tag{10.36}$$

Recall from Section 8.4 that T^* is also constant throughout the adiabatic flow of a calorically perfect gas. Hence, from the equation of state,

$$\frac{\rho_1^*}{\rho_2^*} = \frac{p_1^*/RT_1^*}{p_2^*/RT_2^*} = \frac{p_1^*}{p_2^*} \tag{10.37}$$

Substituting Equation (10.37) into (10.36), we have

$$\frac{A_{t,2}}{A_{t,1}} = \frac{p_1^*}{p_2^*} \tag{10.38}$$

From Equation (8.45), we have

$$p_1^* = p_{0,1} \left(\frac{2}{\gamma + 1} \right)^{\gamma/(\gamma-1)}$$

and

$$p_2^* = p_{0,2} \left(\frac{2}{\gamma + 1} \right)^{\gamma/(\gamma-1)}$$

Substituting the above into Equation (10.38), we obtain

$$\boxed{\frac{A_{t,2}}{A_{t,1}} = \frac{p_{0,1}}{p_{0,2}}} \tag{10.39}$$

Examining Figure 10.20, the total pressure always decreases across shock waves; therefore, $p_{0,2} < p_{0,1}$. In turn, from Equation (10.39), $A_{t,2} > A_{t,1}$. Thus, the second throat must always be *larger* than the first throat. Only in the case of an ideal isentropic diffuser, where $p_0 = $ constant, would $A_{t,2} = A_{t,1}$, and we have already discussed the impossibility of such an ideal diffuser.

Equation (10.39) is a useful relation to size the second throat relative to the first throat *if* we know the total pressure ratio across the tunnel. In the absence of such information, for the preliminary design of supersonic wind tunnels, the total pressure ratio across a normal shock is assumed.

For a given wind tunnel, if $A_{t,2}$ is less than the value given by Equation (10.39), the diffuser will "choke"; that is, the diffuser cannot pass the mass flow coming from the isentropic, supersonic expansion through the nozzle. In this case, nature adjusts the flow through the wind tunnel by creating shock waves in the nozzle, which in turn reduce the Mach number in the test section, producing weaker shocks in the diffuser with an attendant overall reduction in the total pressure loss; that is, nature adjusts the total pressure loss such that $p_{0,1}/p_{0,2} = p_{0,1}/p_B$ satisfies Equation (10.39). Sometimes this adjustment is so severe that a normal shock stands inside the nozzle, and the flow through the test section and diffuser is totally subsonic. Obviously, this choked situation is not desirable because we no longer have uniform flow at the desired Mach number in the test section. In such a case, the supersonic wind tunnel is said to be *unstarted*. The only way to rectify this situation is to make $A_{t,2}/A_{t,1}$ large enough so that the diffuser can pass the mass flow from the isentropic expansion in the nozzle, that is, Equation (10.39) is satisfied along with a shock-free isentropic nozzle expansion.

As a general concluding comment, the basic concepts and relations discussed in this chapter are not limited to nozzles, diffusers, and supersonic wind tunnels. Rather, we have been discussing quasi-one-dimensional flow, which can be applied in many applications involving flow in a duct. For example, inlets on jet engines, which diffuse the flow to lower speeds before entering the engine compressor, obey the same principles. Also, a rocket engine is basically a supersonic nozzle designed to optimize the thrust from the expanded jet. The applications of the ideas presented in this chapter are numerous, and you should make certain that you understand these ideas before progressing further.

In Section 1.2, we subdivided aerodynamics into external and internal flows. You are reminded that the material in this chapter deals exclusively with internal flows.

EXAMPLE 10.6

For the preliminary design of a Mach 2 supersonic wind tunnel, calculate the ratio of the diffuser throat area to the nozzle throat area.

■ **Solution**
Assuming a normal shock wave at the entrance of the diffuser (for starting), from Appendix B, $p_{0,2}/p_{0,1} = 0.7209$ for $M = 2.0$. Hence, from Equation (10.39),

$$\frac{A_{t,2}}{A_{t,1}} = \frac{p_{0,1}}{p_{0,2}} = \frac{1}{0.7209} = \boxed{1.387}$$

10.6 VISCOUS FLOW: SHOCK-WAVE/ BOUNDARY-LAYER INTERACTION INSIDE NOZZLES

Return to Figure 10.15. Here we see the case where the pressure ratio, $p_{e,4}/p_0$ is such that a normal shock wave stands inside the nozzle. This is a classic inviscid flow picture. In reality, there is a boundary layer growing along the nozzle wall, and the shock wave interacts with this boundary layer. One of the possible flow fields resulting from this interaction is sketched in Figure 10.21. The adverse pressure gradient across the shock causes the boundary layer to separate from the nozzle wall. A lambda-type shock pattern occurs at the two feet of the shock near the wall, and the core of the nozzle flow, now separated from the wall, flows downstream at almost constant area.

A series of schlieren photographs showing this type of flow is given in Figure 10.22, obtained from the recent work of Hunter (Craig A. Hunter, "Experimental Investigation of Separated Nozzle Flows," *Journal of Propulsion and Power,* vol. 20, no. 3, May–June 2004, pp. 527–532). For the nozzle shown in Figure 10.22, the exit-to-throat area ratio, A_e/A_t, is 1.797. In Figure 10.22a, the pressure ratio $p_{e,4}/p_0 = 0.5$; the normal shock stands inside the nozzle, and the lambda structure at both ends of the shock is clearly seen. The separated flow is seen trailing downstream from the lambda shock pattern. In Figure 10.22b, c, and d, the shock pattern progressively moves closer to the nozzle exit as the pressure ratio is progressively reduced to 0.417, 0.333, and 0.294, respectively. A detailed schematic of the shock pattern for the pressure ratio of 0.417 is shown in Figure 10.23, corresponding to the flow in Figure 10.22b.

These results are an example of how the realities of a viscous flow can change the ideal picture obtained for an inviscid flow. The recent paper by Craig Hunter, referenced earlier, is an excellent discussion of the real flow in a supersonic nozzle under conditions where shock waves occur inside the nozzle. You are encouraged to study this reference for a revealing discussion of this interesting phenomena.

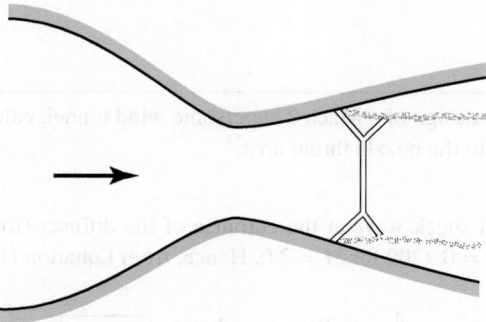

Figure 10.21 Sketch of an overexpanded nozzle flow with flow separation. (*Source: Craig Hunter, NASA*)

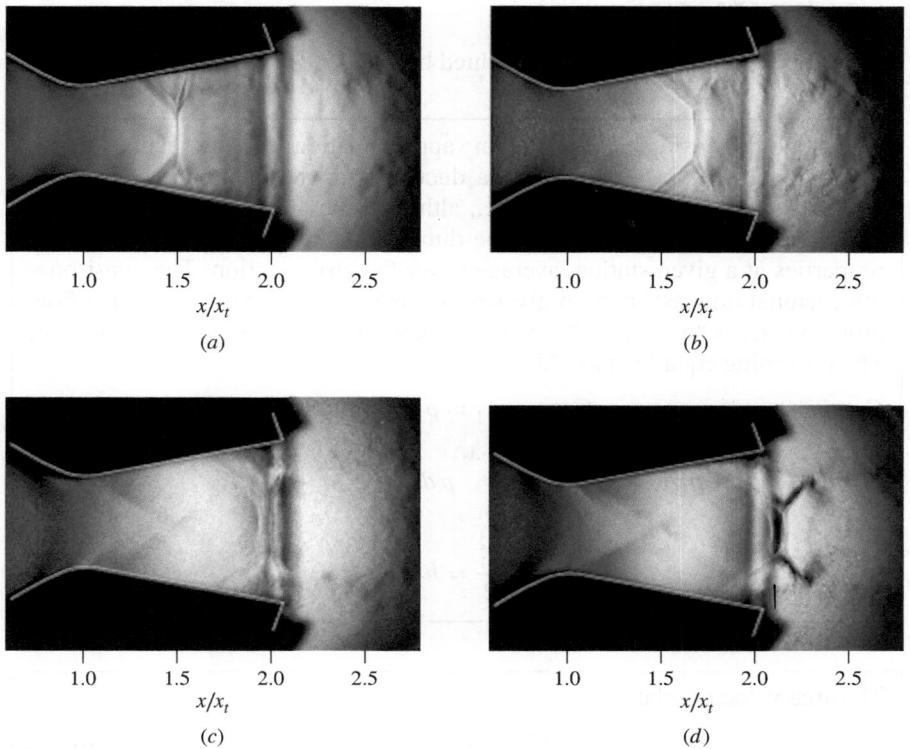

Figure 10.22 Schlieren photographs of the shock-wave/boundary-layer interaction inside an overexpanded nozzle flow. Exit-to-reservoir pressure ratio is (*a*) 0.5, (*b*) 0.417, (*c*) 0.333, and (*d*) 0.294. (*Craig Hunter/NASA*)

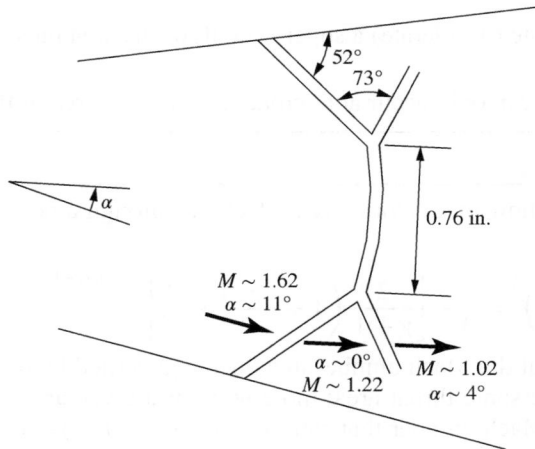

Figure 10.23 Detailed shock schematic for the case with a pressure ratio of 0.417. (*Craig Hunter, NASA*)

10.7 SUMMARY

The results of this chapter are highlighted below:

Quasi-one-dimensional flow is an approximation to the actual three-dimensional flow in a variable-area duct; this approximation assumes that $p = p(x)$, $u = u(x)$, $T = T(x)$, etc., although the area varies as $A = A(x)$. Thus, we can visualize the quasi-one-dimensional results as giving the mean properties at a given station, averaged over the cross section. The quasi-one-dimensional flow assumption gives reasonable results for many internal flow problems; it is a "workhorse" in the everyday application of compressible flow. The governing equations for this are

Continuity:
$$\rho_1 u_1 A_1 = \rho_2 u_2 A_2 \tag{10.1}$$

Momentum:
$$p_1 A_1 + \rho_1 u_1^2 A_1 + \int_{A_1}^{A_2} p\, dA = p_2 A_2 + \rho_2 u_2^2 A_2 \tag{10.5}$$

Energy:
$$h_1 + \frac{u_1^2}{2} = h_2 + \frac{u_2^2}{2} \tag{10.9}$$

The area velocity relation
$$\frac{dA}{A} = (M^2 - 1)\frac{du}{u} \tag{10.25}$$
tells us that

1. To accelerate (decelerate) a subsonic flow, the area must decrease (increase).
2. To accelerate (decelerate) a supersonic flow, the area must increase (decrease).
3. Sonic flow can only occur at a throat or minimum area of the flow.

The isentropic flow of a calorically perfect gas through a nozzle is governed by the relation
$$\left(\frac{A}{A^*}\right)^2 = \frac{1}{M^2}\left[\frac{2}{\gamma + 1}\left(1 + \frac{\gamma - 1}{2}M^2\right)\right]^{(\gamma+1)/(\gamma-1)} \tag{10.32}$$

This tells us that the Mach number in a duct is governed by the ratio of local duct area to the sonic throat area; moreover, for a given area ratio, there are two values of Mach number that satisfy Equation (10.32)—a subsonic value and a supersonic value.

For a given convergent-divergent duct, there is only one possible isentropic flow solution for supersonic flow; in contrast, there are infinite numbers of subsonic isentropic solutions, each one associated with a different pressure ratio across the nozzle, $p_0/p_e = p_0/p_B$.

In a supersonic wind tunnel, the ratio of second throat area to first throat area should be approximately

$$\frac{A_{t,2}}{A_{t,1}} = \frac{p_{0,1}}{p_{0,2}} \tag{10.39}$$

If $A_{t,2}$ is reduced much below this value, the diffuser will choke and the tunnel will unstart.

10.8 INTEGRATED WORK CHALLENGE: CONCEPTUAL DESIGN OF A SUPERSONIC WIND TUNNEL

Concept: A basic sketch of a supersonic wind tunnel is given in Figure 10.20 that illustrates the essential components: nozzle, test section, and diffuser. The pressure ratio p_0/p_B from the inlet to the nozzle to the exit of the diffuser is what makes the tunnel run. Not shown in Figure 10.20 is how this pressure ratio is generated. The answer to this question is an essential first step in the conceptual design of a supersonic wind tunnel.

Four different supersonic wind tunnel configurations for producing the proper pressure ratio across the supersonic nozzle are sketched in Figure 10.24.

(a) **Blowdown Tunnel (Figure 10.24a):** High-pressure air at pressure p_0 is stored in a tank at the entrance to the tunnel. The exit of the tunnel is open to the surrounding atmosphere, where the back pressure p_B is the atmospheric pressure. The pressure ratio across the tunnel is p_0/p_B. Flow is started when a pressure valve at point A is opened. As the air flows out of the storage tank and through the tunnel, the remaining air in the tank expands to fill the tank. During this expansion, both the total pressure p_0 and total temperature T_0 of the remaining air in the tank decrease. These decreases in p_0 and T_0 with time are a disadvantage of the blowdown tunnel, but they can be minimized at the start by having a large enough storage tank such that the mass flow through the tunnel is a small percentage of the mass of air stored in the tank. Eventually, however, during the run of the tunnel there will be some point in time at which p_0 becomes too small and the pressure ratio p_0/p_B required to run the tunnel dips below that required to maintain the proper isentropic flow through the nozzle. This is the effective end of the test time.

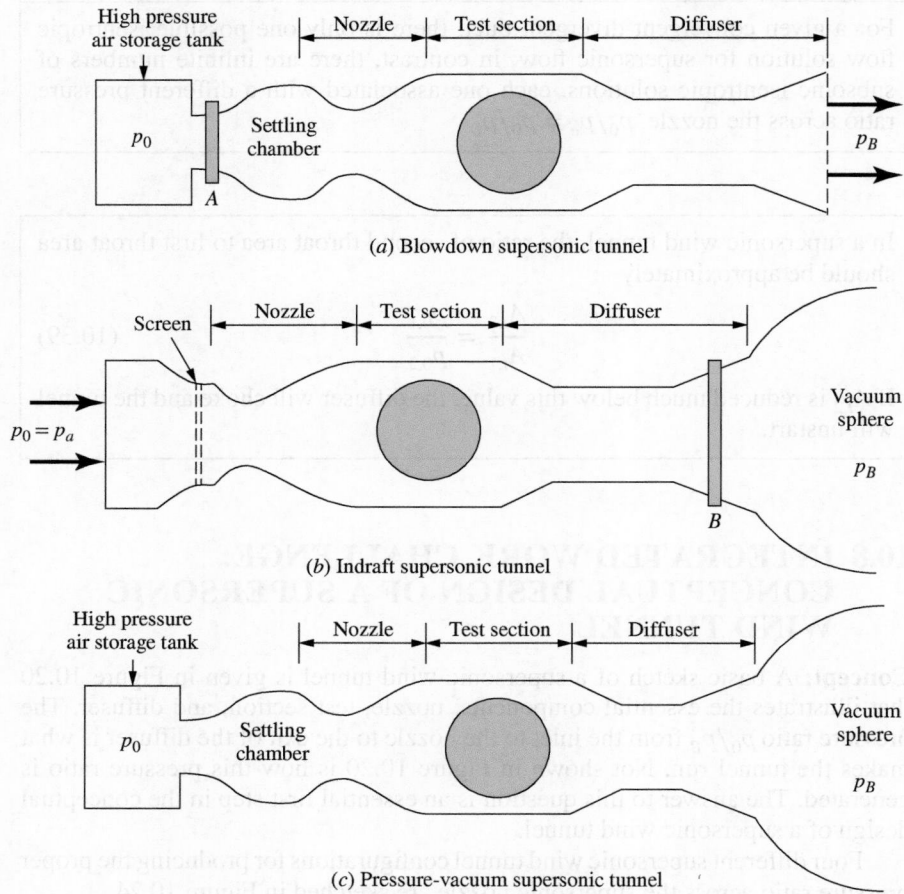

Figure 10.24 Supersonic wind tunnel configurations.

(b) Indraft tunnel (Figure 10.24b): A vacuum tank evacuated to a very low pressure p_B is connected to the exit of the tunnel. The entrance to the tunnel is open to the atmosphere, where the atmospheric pressure is p_a. When valve B in front of the vacuum tank is opened, atmospheric air is sucked in through the tunnel entrance, and flow starts through the tunnel. The total pressure at the entrance to the tunnel is $p_0 = p_a$, and the pressure ratio across the tunnel is p_0/p_B. As the run continues, air fills the vacuum tank and p_B increases. The test run effectively ends when the pressure ratio p_0/p_B becomes smaller than that required to maintain isentropic flow through the nozzle.

(c) Pressure-vacuum Tunnel (Figure 10.24c): This tunnel design is a combination of the two already described earlier—it is a kind of "push-pull"

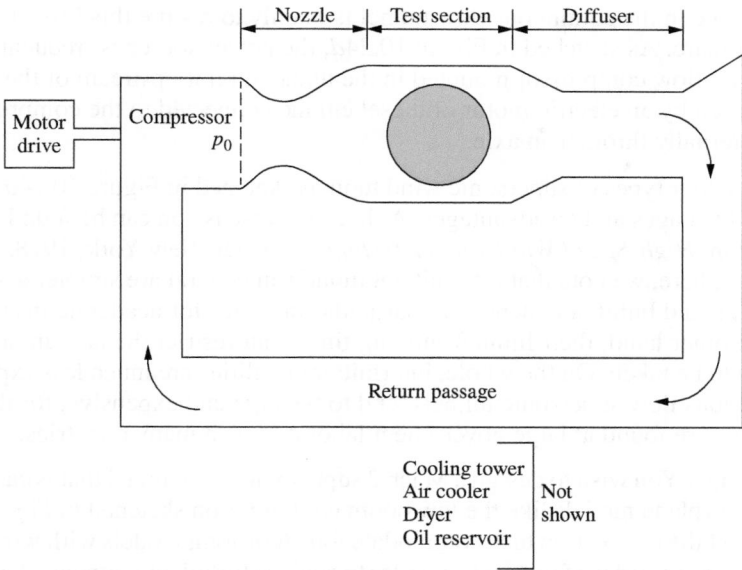

(d) Closed-circuit, continuous flow, supersonic wind tunnel

Figure 10.24 (*Continued*)

arrangement. Here, the inlet to the tunnel is connected to a high-pressure storage tank with pressure p_0, and the exit of the tunnel is connected to an evacuated tank at pressure p_B. The pressure ratio across the tunnel is p_0/p_B. During the running of the tunnel, p_0 decreases and p_B increases, and the run terminates when p_0/p_B becomes too small to support isentropic flow through the supersonic nozzle. This type of arrangement is specifically used for hypersonic wind tunnels where very large pressure ratios are required to produce the requisite test section Mach numbers at Mach 5 and higher.

Note that the tunnels sketched in Figure 10.24*a–c* have run times that are limited by the storage capacity of the high-pressure tank and/or the volume capacity of the vacuum tank. Hence, these tunnels are in a class of *intermittent* tunnels. This leads to a fourth class of wind tunnel, as follows.

(d) Closed-circuit Continuous Flow Tunnel (Figure 10.24*d*): This concept is not unlike the continuous flow closed-circuit subsonic wind tunnels discussed in Sections 3.3 and 3.23, and sketched in Figure 3.8*b*, except that the power source that drives the supersonic tunnel must be much stronger in order to maintain the proper isentropic pressure ratio across the supersonic nozzle. Referring to Figure 10.24*d*, there is a loss of total pressure throughout the circuit due to shock waves occurring on a model in the test section and shock waves in the diffuser, and due to friction losses in the boundary layers formed along the tunnel walls. The function of the power source in the continuous flow tunnel is mainly to restore this loss of total pressure. As sketched in Figure 10.24*d*, the power source is frequently an axial-flow compressor mounted in the tunnel circuit upstream of the nozzle, driven by an electric motor or diesel engine connected to the compressor externally through an axle.

The four types of supersonic wind tunnels sketched in Figure 10.24*a–d* have their advantages and disadvantages. A thorough discussion can be found in Pope and Goin, *High-Speed Wind Tunnel Testing*, Kreieger, New York, 1978. For our purposes here, we note that intermittent tunnels in general are simpler and easier to design and build, and hence are particular favorites for academic institutions. On the other hand, their limited running times can restrict the amount and type of data to be taken. On the whole, intermittent facilities are much less expensive. Continuous flow supersonic tunnels tend to be large and expensive; for the most part they are found at large government laboratories in many countries.

Challenge: You wish to design a Mach 2 supersonic wind tunnel that is capable of testing airplane models like the low boom configuration sketched in Figure 9.43. You want the test section to accommodate four-foot-long models with wing spans that are on the order of two feet. In order to achieve turbulent boundary-layer flow over the test model (to try to simulate the turbulent boundary layers encountered in full-scale flight), you would like to have a Reynolds number of at least 10 million based on the length of the model. Set up the conceptual design of a supersonic wind tunnel to meet these specifications.

Solution: To determine the size of the test section that can properly accommodate the specified model size, we rely once again on previous experience. Pope and Goin in their book *High-Speed Wind Tunnel Testing* recommend a size that will ensure that shock and expansion waves generated by the model will reflect from the walls of the test section far enough downstream that the reflected waves will not impinge back on the model (this is only common sense). Figure 10.25 is a sketch of a four-foot-long slender configuration (such as that sketched in Figure 9.43).

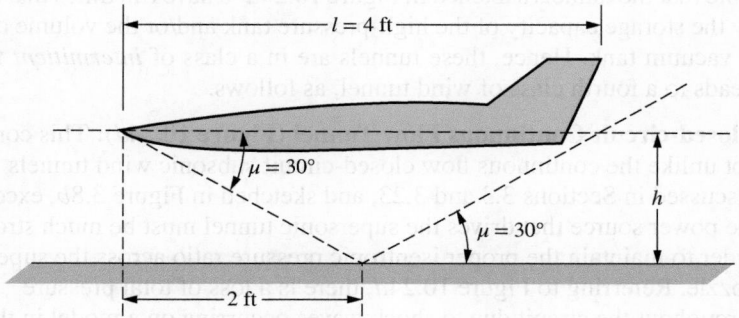

Figure 10.25 Wave generation from a test model, and reflection from the tunnel wall.

For simplicity, assume the bow shock wave is weak and can be simulated by a Mach wave, as sketched in Figure 10.25. The minimum height of the model centerline above the lower wind tunnel wall is denoted by h. At Mach 2,

$$\mu = \text{arc sin} \left(\frac{1}{M} \right) = \text{arc sin} \left(\frac{1}{2} \right) = 30°$$

From the geometry in Figure 10.25,

$$\frac{h}{2} = \tan \mu = \tan 30° = 0.577$$

$$h = 1.155 \text{ ft}$$

The total allowable minimum height of the test section is then $2h = 2.31$ ft.

For the tunnel design, however, we need to be somewhat conservative. The waves from the model are finite shock and expansion waves, not Mach waves, and the wave angles will be larger than 30 degrees. Therefore, h will be larger than 1.155 ft and the total test section height should be made larger than 2.31 ft. A word of caution, however. The overall size of the wind tunnel will be governed by the size of the test section; the larger the size, the more mass flow will pass through the tunnel, with consequent increases in power requirements and costs of operation, not to mention the initial construction cost of the tunnel. Let us apply a conservative "factor of safety" on the height of the test section, i.e., let us design the test section height to be 3.5 ft. Also, since the length of the model is four feet and we want the entire model length to easily fit within the test section, we will design the length of the test section to be five feet. Finally, to accommodate a model wing span of two feet and to minimize side-wall effects on the flow over the wind, we choose a test section width of 3 ft. In summary the conceptual design size of the test section will be a length, width, and height of 5 ft, 3 ft, and 3.5 ft, respectively.

Our next step is the choice of the type of supersonic tunnel: blowdown, indraft, pressure-vacuum, or closed-circuit continuous flow. This choice may be partly determined by the requirement for a Reynolds number of 10^7 based on test model length, i.e., for a length of four feet. Also the choice may be dictated by the laboratory space available to house the tunnel. Other considerations may be the availability of existing equipment in the laboratory such as compressors, high-pressure air storage tanks, and instrumentation—the realities of life in experimental work. So, for our present Integrated Work Challenge, there is no "right choice" of the type of supersonic tunnel to be made—much like the design of an airplane that depends on a number of technical compromises. See Pope and Goin, *High-Speed Wind Tunnel Testing*, for an extensive discussion of such design matters.

In this light, we will proceed as follows. Because of its relative mechanical simplicity and compact size, let us consider a blowdown tunnel, estimate the running conditions, and see if the Reynolds number requirement is satisfied. The flow through the blowdown tunnel is dumped directly to the surrounding atmosphere at the diffuser exit (Figure 10.24a) with a back pressure p_B at the exit that ideally is equal to the atmospheric pressure, p_a. This ideal case, however, corresponds to the flow through the diffuser being slowed to a low subsonic velocity at the exit, and hence nature will impress the surrounding atmospheric pressure p_a directly at the exit, giving an overall pressure ratio across the tunnel of p_0/p_a. If the reservoir pressure in the air storage tank were at a higher value of p_0 required to run the tunnel, the overall pressure levels through the tunnel would be higher, and the pressure at the diffuser exit would be higher than required. At the very least, this will lead to a higher velocity flow at the diffuser exit—a waste of energy and a reduction of operating efficiency. If p_0 in the reservoir is high enough, the flow through the divergent downstream section of the diffuser, after being reduced to Mach 1 in the diffuser throat, might even become supersonic again, with a supersonic exit flow blasting out into the laboratory surroundings. This would create an intolerable noise level in the laboratory, as well as being totally inefficient. Therefore, in this problem we will design a blowdown tunnel with just the minimum overall pressure ratio $p_0/p_B = p_0/p_a$ sufficient to achieve isentropic flow in the nozzle with shock-free flow entering the test section.

In this calculation, we first estimate the *loss* of total pressure through the tunnel. Boundary layers throughout the tunnel increase the entropy level of the flow and result in a loss of total pressure. This loss is small, however, compared with losses across shock waves from the model mounted in the test section, and especially across the shock waves that occur in the diffuser, as sketched in Figure 10.17*b*. For conceptual design purposes, we assume the shock losses are equivalent to the total pressure loss across a normal shock wave at the design test section Mach number; this "normal shock efficiency" is a rule of thumb frequently used for estimating losses in the supersonic diffuser, especially in the design of hypersonic tunnels. We will use this rule of thumb here.

At Mach 2, from Appendix B,

$$\frac{p_{0_2}}{p_{0_1}} = 0.7209$$

Assuming that p_{0_2} is the total pressure at the diffuser exit and the flow velocity is small, we assume $p_{0_2} = p_a = 1$ atm. Thus,

$$p_{0_1} = \frac{p_{0_2}}{0.7209} = \frac{1}{0.7209}$$

$$p_{0_1} = 1.387 \text{ atm}$$

Ignoring all other losses, this implies that we need a reservoir pressure in the air storage tank of 1.387 atm.

There is yet another consideration, namely, *starting* a supersonic tunnel. If we would simply open some valves and impose the minimum pressure ratio of 1.387 across the tunnel, the losses across the waves associated with the transient starting process might be too large, and the flow process struggles to start itself. In this case, the starting pressure ratio across the tunnel needs to be higher than the running pressure ratio. It is difficult to estimate the starting pressure ratio, especially when there is a model mounted in the test section. This is something usually determined empirically. Finally, speaking of the test model, it should not be so large as to cause a "blockage" of the mass flow through the tunnel. The cross-sectional area of the model should be a small fraction of the test section cross-section area. The maximum size model allowed in the test section without causing blockage is a function of model shape (streamlined and slender versus blunt) and test section Mach number. The maximum model size allowable is also something usually determined empirically.

These practical considerations notwithstanding, we might conservatively design for an overall pressure ratio of about 2, which can be throttled down to the minimum design value of 1.387 after the tunnel is properly started.

Question: For an operating value of $p_0 = 1.387$ atm in the reservoir, will the Reynolds number of the flow over the model satisfy our specification of 10^7? To make this estimate, we first calculate the air density, ρ_0, in the reservoir, assuming the air temperature in the reservoir is the standard sea level value of $T_0 = 519 °$R.

$$\rho_0 = \frac{p_0}{RT_0} = \frac{(1.387)(2116)}{(1716)(519)} = 0.00329 \frac{\text{slug}}{\text{ft}^3}$$

At Mach 2 in the test section, from Appendix A,

$$\frac{\rho_0}{\rho} = 4.347 \qquad \text{and} \qquad \frac{T_0}{T} = 1.8$$

Thus, in the test section, the isentropic flow properties are

$$\rho = \frac{\rho_0}{4.347} = \frac{0.00329}{4.347} = 7.568 \times 10^{-4} \frac{\text{slug}}{\text{ft}^3}$$

and

$$T = \frac{T_0}{1.8} = \frac{519}{1.8} = 288°\text{R}$$

The speed of sound in the test section is

$$a = \sqrt{\gamma RT} = \sqrt{(1.4)(1716)(288)} = 831.8 \text{ ft/s}$$

and the flow velocity is

$$V = Ma = 2(831.8) = 1664 \text{ ft/s}$$

The viscosity coefficient as a function of temperature is given in Figure 1.50, but in SI units. Converting T into degrees Kelvin, we have

$$T = \frac{280}{1.8} = 155.6 \text{ K}$$

Extrapolating the linear variation shown in Figure 1.50 to a temperature of 155 K, we have $\mu = 1.05 \times 10^{-5}$ kg/(m)(s). Converting to the Engineering system of units where 1 slug = 14.594 kg, and 1 ft = 0.3048 m, we have

$$\mu = 1.05 \times 10^{-5} \frac{\text{kg}}{\text{(m)(s)}} \left[\frac{1 \text{ slug}}{14.594 \text{ kg}} \right] \frac{0.3048 \text{ m}}{1 \text{ ft}} = 2.19 \times 10^{-7} \frac{\text{slug}}{\text{(ft)(s)}}$$

Therefore, for the reservoir conditions in the high pressure air storage tank, the Reynolds number for the four-footlong model in the test section is:

$$R_e = \frac{\rho V L}{\mu} = \frac{(7.568 \times 10^{-4})(1664)(4)}{2.19 \times 10^{-7}} = 23 \times 10^6$$

This value of a Reynolds number of 23 million exceeds the stipulated requirement of 10 million, so our conceptual design of the blowdown tunnel is on track so far.

Our tunnel is bordering on being rather large, and the next questions are: What is the mass flow through the tunnel? What running times do we want? How large an air storage tank is going to be required? These are all related questions.

First, consider the mass flow. A closed-form expression for the mass flow is given in the end-of-chapter problem 10.5 as

$$\dot{m} = \frac{p_0 A^*}{\sqrt{T_0}} \sqrt{\frac{\gamma}{R} \left(\frac{2}{\gamma + 1} \right)^{(\gamma+1)/(\gamma-1)}} \qquad (C10.1)$$

where A^* is the nozzle throat area. At Mach 2, from Appendix A, $A_e/A^* = 1.687$. Since our tunnel is a "two-dimensional" tunnel with a nozzle and test-section cross-sectional areas being rectangular and with the height of the nozzle exit being $h_e = 3.5$ ft, then the height of the nozzle throat, h^*, is

$$h^* = h_e/1.687 = \frac{3.5}{1.687} = 2.075 \text{ ft}$$

The width of the nozzle is 3 ft. Hence, $A^* = (2.075)(3) = 6.225 \text{ ft}^2$. For $p_0 = 1.387$ atm, where 1 atm $= 2116 \text{ lb/ft}^2$, and hence $p_0 = (1.387)(2116) = 2935 \text{ lb/ft}^2$, and where $T_0 = 519°$R, we have from Equation (C10.1),

$$\dot{m} = \frac{(2935)(6.225)}{\sqrt{519}} \left[\frac{1.4}{1716} \left(\frac{2}{1.4 + 1} \right)^{2.4/0.4} \right]$$

$$= 802(2.73 \times 10^{-4}) = 0.219 \text{ slug/s}$$

Since $32.2 \text{ lb}_m = 1$ slug, the mass flow is

$$\dot{m} = (0.219)(32.2) = 7.05 \text{ lb}_m/\text{s}$$

What about running time? This is a matter of choice. It should be long enough to allow all measurements to be made on the test model—pressure, temperature, force, etc. We choose a running time of one minute for this purpose. Therefore, during that minute, the total mass of air flowing through the tunnel is

$$m = (7.05)(60) = 423 \text{ lb}_m$$

Our high-pressure air storage tank should be large enough to discharge 423 lb_m of air through the tunnel and still have enough air left in the tank to maintain a reservoir pressure of $p_0 = 1.387$ atm throughout the run. For starting the tunnel, we have estimated that initially $p_0 = 2$ atm is required. After the tunnel is started, although the tank pressure initially remains above the stipulated 1.387 atm, the air pressure entering the nozzle can be reduced to the desired 1.387 atm by passing through a throttling valve. What we want during the tunnel run is for the pressure in the tank to remain at or above 1.387 atm. From the equation of state $pV = MRT$, where V is the tank volume and M is the mass of air in the tank we have for fixed V and T,

$$\frac{p_i}{p_f} = \frac{M_i}{M_f} \qquad\qquad (C10.2)$$

where p_i and p_f are the initial and final pressures in the tank, respectively, and M_i and M_f are the initial and final mass in the tank, respectively.

$$\frac{p_i}{p_f} = \frac{2}{1.387} = 1.44$$

Hence,

$$M_i/M_f = 1.44$$

or,

$$M_f = \frac{M_i}{1.44} \qquad\qquad (C10.3)$$

Also, the difference between the initial and final mass in the tank is equal to the mass discharged during the one-minute run,

$$M_i - M_f = 423 \text{ lb} \qquad\qquad (C10.4)$$

Substituting Equation (10.3) into Equation (10.4), we have

$$M_i - \frac{M_i}{1.44} = 423$$

$$1.44 \, M_i - M_i = (423)(1.44)$$

$$0.44 \, M_i = 609.1$$

$$M_i = \frac{609.1}{0.44} = 1384 \text{ lb}_m$$

So the storage tank must contain at least 1384 lb_m of air at $p = 2$ atm at the beginning of the run. Finally, we can now estimate the volume of the storage tank, which must hold 1384 lb_m of air at a pressure of 2 atm initially. The mass in slugs is $M = \frac{1384}{32.2} = 43$ slug, and the pressure in lb/ft^2 is $p = 2(2116) = 4232 \text{ lb/ft}^2$. Thus,

$$V = \frac{MRT}{p} = \frac{(43)(1716)(519)}{4232} = 9049 \text{ ft}^3$$

How large is this tank? Let us assume a cylindrical tank with a 12-foot diameter. Let h be the height of the tank. The volume of the tank is $V = \frac{\pi d^2}{4} h$ or,

$$h = \frac{4V}{\pi d^2} = \frac{(4)(9049)}{\pi(12)^2} = 80 \text{ ft}$$

Wow! This is a tall tank—too tall for normal laboratory space. Here is the importance of a conceptual design process. Instead of storing the air at 2 atm pressure, let us store it at 20 atm and then feed the air to the tunnel through a throttling valve. A 20 atm tank would decrease the tank volume by a factor of 10 compared with the 2 atm tank, yielding a height of 8 ft—a much more reasonable value.

We have barely scratched the surface of the conceptual design of a supersonic tunnel. A more thorough process would be to look at the other types of supersonic tunnels, not just the blowdown tunnel as we have done here. And then we could compare all four types as to which might be the best to satisfy our specifications. But we have done enough here to give you the flavor. Our blowdown tunnel will do the job. It is a big tunnel, driven by the rather large specified size of the test model, and hence requiring a large supersonic test section. Supersonic wind tunnels for testing models of this sort tend to be big—it is just the nature of the beast!

10.9 PROBLEMS

10.1 The reservoir pressure and temperature for a convergent-divergent nozzle are 5 atm and 520°R, respectively. The flow is expanded isentropically to supersonic speed at the nozzle exit. If the exit-to-throat area ratio is 2.193, calculate the following properties at the exit: M_e, p_e, T_e, ρ_e, u_e, $p_{0,e}$, $T_{0,e}$.

10.2 A flow is isentropically expanded to supersonic speeds in a convergent-divergent nozzle. The reservoir and exit pressures are 1 and 0.3143 atm, respectively. What is the value of A_e/A^*?

10.3 A Pitot tube inserted at the exit of a supersonic nozzle reads 8.92×10^4 N/m². If the reservoir pressure is 2.02×10^5 N/m², calculate the area ratio A_e/A^* of the nozzle.

10.4 For the nozzle flow given in Problem 10.1, the throat area is 4 in². Calculate the mass flow through the nozzle.

10.5 A closed-form expression for the mass flow through a choked nozzle is

$$\dot{m} = \frac{p_0 A^*}{\sqrt{T_0}} \sqrt{\frac{\gamma}{R}\left(\frac{2}{\gamma+1}\right)^{(\gamma+1)/(\gamma-1)}}$$

Derive this expression.

10.6 Repeat Problem 10.4, using the formula derived in Problem 10.5, and check your answer from Problem 10.4.

10.7 A convergent-divergent nozzle with an exit-to-throat area ratio of 1.616 has exit and reservoir pressures equal to 0.947 and 1.0 atm, respectively. Assuming isentropic flow through the nozzle, calculate the Mach number and pressure at the throat.

10.8 For the flow in Problem 10.7, calculate the mass flow through the nozzle, assuming that the reservoir temperature is 288 K and the throat area is 0.3 m².

10.9 Consider a convergent-divergent nozzle with an exit-to-throat area ratio of 1.53. The reservoir pressure is 1 atm. Assuming isentropic flow, except for the possibility of a normal shock wave inside the nozzle, calculate the exit Mach number when the exit pressure p_e is
(a) 0.94 atm (b) 0.886 atm (c) 0.75 atm (d) 0.154 atm

10.10 A 20° half-angle wedge is mounted at 0° angle of attack in the test section of a supersonic wind tunnel. When the tunnel is operating, the wave angle from the wedge leading edge is measured to be 41.8°. What is the exit-to-throat area ratio of the tunnel nozzle?

10.11 The nozzle of a supersonic wind tunnel has an exit-to-throat area ratio of 6.79. When the tunnel is running, a Pitot tube mounted in the test section measures 1.448 atm. What is the reservoir pressure for the tunnel?

10.12 We wish to design a supersonic wind tunnel that produces a Mach 2.8 flow at standard sea level conditions in the test section and has a mass flow of air equal to 1 slug/s. Calculate the necessary reservoir pressure and temperature, the nozzle throat and exit areas, and the diffuser throat area.

10.13 Consider a rocket engine burning hydrogen and oxygen. The total mass flow of the propellant plus oxidizer into the combustion chamber is 287.2 kg/s. The combustion chamber temperature is 3600 K. Assume that the combustion chamber is a low-velocity reservoir for the rocket engine. If the area of the rocket nozzle throat is 0.2 m², calculate the combustion chamber (reservoir) pressure. Assume that the gas that flows through the engine has a ratio of specific heats, $\gamma = 1.2$, and a molecular weight of 16.

10.14 For supersonic and hypersonic wind tunnels, a diffuser efficiency, η_D, can be defined as the ratio of the total pressures at the diffuser exit and nozzle reservoir, divided by the total pressure ratio across a normal shock at the test-section Mach number. This is a measure of the efficiency of the diffuser relative to normal shock pressure recovery. Consider a supersonic wind tunnel designed for a test-section Mach number of 3.0 which exhausts directly to the atmosphere. The diffuser efficiency is 1.2. Calculate the minimum reservoir pressure necessary for running the tunnel.

10.15 Return to Problem 9.18, where the average Mach number across the two-dimensional flow in a duct was calculated, and where θ for the upper wall was 3°. Assuming quasi-one-dimensional flow, calculate the Mach number at the location AB in the duct.

10.16 Return to Problem 9.19, where the average Mach number across the two-dimensional flow in a duct was calculated, and where θ for the upper wall was 30°. Assuming quasi-one-dimensional flow, calculate the Mach number at the location AB in the duct.

10.17 A horizontal flow initially at Mach 1 flows over a downward-sloping expansion corner, thus creating a centered Prandtl-Meyer expansion wave. The streamlines that enter the head of the expansion wave curve smoothly and continuously downward through the expansion fan, and emerge parallel to the downward sloping surface downstream of the tail of the wave, as shown in Figure 9.2*b*. Imagine a polar coordinate system r, Φ with its origin at the expansion corner (the vertex of the Prandtl-Meyer expansion wave), with r the usual radial distance along a ray from the origin and Φ the polar angle of r measured from the horizontal. Because the upstream flow is at Mach 1, the head of the expansion fan is a Mach wave perpendicular to the free stream. Consider a given streamline entering the expansion wave at the point $(r, \Phi) = (r^*, \pi/2)$. Construct a method for calculating the shape of this streamline as a function of r and Φ through the expansion fan. *Note:* To solve this problem, material from both Chapters 9 and 10 is required.

10.18 Consider a centered expansion wave where $M_1 = 1.0$ and $M_2 = 1.6$. Using the method developed in Problem 10.17, plot to scale a streamline that passes through the expansion wave.

10.19 Air at 500 kPa and 573°K flows at 0.1 kg/s through a converging-diverging nozzle whose exit is at 100 kPa. Assume $\gamma_{air} = 1.4$, $R_{air} = 288\ J/kgK$, and that the flow is isentropic. Please find:
  a. The exit Mach number
  b. The exit area
  c. The throat area

10.20 Consider the cold gas thruster illustrated below. The propellant is Nitrogen (N_2) which is held at a pressure of 20 atm and a temperature of 300K in a large tank upstream of the nozzle. The nozzle throat area is 0.0005 m² and $A_e/A_t = 5.9$. Assume that the flow through the nozzle is isentropic and that the tank is large enough that the pressure and temperature of the gas in the tank remain approximately constant over the course of a firing. $R_{N_2} = 297\ J/kgK$ and $\gamma_{N_2} = 1.4$. Calculate the net thrust produced on the ground where $p_\infty = 1\ atm$ and in space where $p_\infty \approx 0\ atm$. Why are they different?

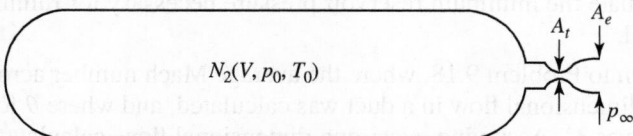

Figure 10.26 Schematic illustration of a cold gas thruster system.

10.21 How does changing A_e/A_t affect the performance of the cold gas thruster discussed in Problem 10.21? Answer this question by calculating the thrust and the specific impulse (I_{sp}) for $1 \leq A_e/A_t \leq 50$. Do this calculation for a thruster firing on the ground where $p_\infty = 1\ atm$ and one firing in space where $p_\infty \approx 0$ *atm*. Explain your results and indicate whether or not you think this is a "good" thruster. Note that the specific impulse is a common parameter used to gauge how effectively a propulsion system uses its fuel to generate thrust. It is defined as follows.

$$I_{sp} = \frac{T}{\dot{m}_p g}$$

where T is the thrust, $\dot{m}_p$ is the mass flow rate of the propellant, and g is the acceleration of gravity. *Hint:* Instead of varying A_e/A_t, vary M_e in a convenient way and then figure out the associated values of A_e/A_t. This will avoid having to interpolate in tables to find M_e for a particular value of A_e/A_t.

10.22 Consider the cold gas thruster described in Problem 10.20 equipped with a fast-acting valve upstream of the nozzle throat. The volume of the tank is 1 m^3. Derive an expression for the pressure in the tank as a function of time that is valid until the nozzle becomes unchoked. Plot the time evolution of the tank pressure ratio p_o/p where p_o is the initial tank pressure. Indicate the range of pressure ratios over which your expression is valid. Assume that the test is being conducted on the ground where $p_\infty = 1$ *atm*.

10.23 Repeat problem 10.22 without assuming that the temperature of the gas inside the tank remains constant. Plot your results and compare to the isothermal case investigated in problem 10.20. How different are your results? How much does the temperature of the N_2 in the tank change as the tank blows down? How does this affect the exit velocity?

10.24 Show that for an *arbitrary* isentropic flow through a nozzle (i.e., one that is not necessarily choked) that the mass flow rate is given by:

$$\dot{m} = A_t \psi \sqrt{2 p_0 \rho_0} \quad \text{where} \quad \psi = \begin{cases} \sqrt{\dfrac{\gamma}{2}\left(\dfrac{2}{\gamma+1}\right)^{\frac{\gamma+1}{\gamma-1}}} & \dfrac{p_0}{p} \geq \left(\dfrac{\gamma+1}{\gamma}\right)^{\frac{\gamma}{\gamma-1}} \\[20pt] \sqrt{\dfrac{\gamma}{\gamma-1}\left[\left(\dfrac{p}{p_0}\right)^{\frac{2}{\gamma}} - \left(\dfrac{p}{p_0}\right)^{\frac{\gamma+1}{\gamma}}\right]} & \dfrac{p_0}{p} < \left(\dfrac{\gamma+1}{\gamma}\right)^{\frac{\gamma}{\gamma-1}} \end{cases}$$

where p is the pressure at the nozzle throat, p_0 and ρ_0 are the total pressure and density upstream of the nozzle, and A_t is the area of the throat. Your task is to find ψ. *Hint:* Use Figure 10.27 to set up your solution.

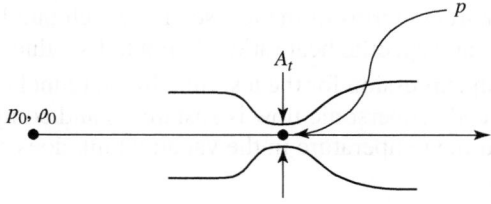

Figure 10.27 Flow along a streamline through a nozzle.

10.25 Figure 10.28 is a schematic diagram of a small supersonic wind tunnel. The downstream side of the test section is connected to a large vacuum chamber. A fast-acting valve located upstream of the throat is initially closed in order to permit the vacuum chamber to be pumped down. An experiment is initiated by opening the valve allowing air to rush into the vacuum chamber via the test section. The pressure in the vacuum chamber immediately before the valve is opened is 5066 N/m² (95% vacuum). The volume of the vacuum chamber is 62.77 m³. The throat and test section cross section are rectangular. The test section height (H) is 0.1651 m and the test section depth is 0.1524 m. The throat height is 0.0826 m. Assume that the temperature of the air in the room and the evacuated tank is 300 K and that the pressure of the air in the room is 101324 N/m² (1 atm). Also assume that the flow is isentropic from the room through the exit of the test section. (For air: $\gamma = 1.4$ and $R = 287$ J/kgK.)

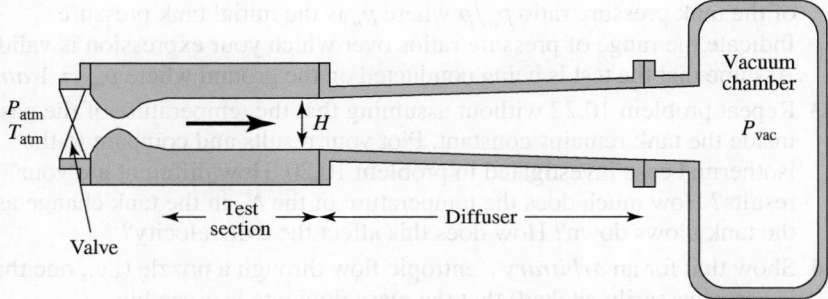

Figure 10.28 Schematic illustration of a small supersonic wind tunnel.

a. What are the total pressure and temperature in the test section?
b. What is the Mach number in the test section?
c. What are the static temperature and pressure in the test section?
d. Describe what happens in the test section and diffuser between the time that the valve is opened and the time that the tunnel "unstarts," that is, when it is no longer possible to sustain steady supersonic flow in the test section.
e. Derive an expression for the vacuum chamber pressure at which the flow unstarts in terms of the test section Mach number, total pressure, and specific heat ratio. Compute its value.
f. Derive an expression for the test time in the tunnel (i.e., the time over which steady supersonic flow is sustained) and compute its value. Note that the temperature in the vacuum tank does not remain constant.

CHAPTER **11**

Subsonic Compressible Flow over Airfoils: Linear Theory

During the war a British engineer named Frank Whittle invented the jet engine, and deHavilland built the first production-type model. He produced a jet plane named Vampire, the first to exceed 500 mph. Then he built the experimental DH 108, and released it to young Geoffrey for test. In the first cautious trials the new plane behaved beautifully; but as Geoffrey stepped up the speed he unsuspectingly drew closer to an invisible wall in the sky then unknown to anyone, later named the sound barrier, which can destroy a plane not designed to pierce it. One evening he hit the speed of sound, and the plane disintegrated. Young Geoffrey's body was not found for ten days.

From the Royal Air Force Flying Review, as condensed in *Reader's Digest*, 1959

PREVIEW BOX

We learned all about airfoils in Chapter 4—but did we really? Chapter 4 dealt with airfoils in low-speed incompressible flow. What happens when we have a high-speed compressible flow over these airfoils? By this stage of our discussions about compressible flow, you might expect compressibility to make some differences, and you would be correct. But what differences, and by how much? This chapter and the next give some answers.

The present chapter deals with airfoils in a high-speed subsonic flow. What happens when an airfoil is flown at high subsonic Mach numbers, near the speed of sound? How does compressibility change the airfoil properties, and by how much? How do we analyze and calculate these compressibility effects? The answers are given in this chapter. In addition, we will see that as Mach 1 is approached, the drag of an airfoil suddenly skyrockets and looks like it is going out of sight (the so-called sound barrier, as sometimes referenced in the popular literature). What is going on here? How do we deal with it?

These are all very important and very practical questions, and they are all addressed in this chapter. Every time you fly in a high-speed jet transport, such as a Boeing 777 cruising at a Mach number of 0.85, you are encountering the phenomena that drive these questions. Without proper answers, we would never be able to design this type of airplane, and we would always be relegated to flying slower—something that the pace of modern life would not easily tolerate. So pay close attention to the material in this chapter.

But be aware. Our first four chapters dealing with the basics of compressible flow (Chapters 7–10) used mathematics at essentially the level of algebra. To go further in our study of compressible flow, especially to answer the questions posed here, we have to return to the world of partial differential equations. But this is no big deal; we have been there before in our discussions of incompressible flow in Part 1 of this book. So jump right into the material of this chapter. I predict that you will experience some increase in technical maturity as you study this material, and that you will enjoy it. After all, here you will be dealing with some of the most important and exciting applications in modern aerodynamics.

11.1 INTRODUCTION

The above quotation refers to an accident that took place on September 27, 1946, when Geoffrey deHavilland, son of the famed British airplane designer Sir Geoffrey deHavilland, took the D. H. 108 Swallow up for an attack on the world's speed record. At that time, no airplane had flown at or beyond the speed of sound. The Swallow was an experimental jet-propelled aircraft with swept wings and no tail. During its first high-speed, low-level run, the Swallow encountered major compressibility problems and broke up in the air. deHavilland was killed instantly. This accident strengthened the opinion of many that Mach 1 stood as a barrier to piloted flight and that no airplane would ever fly faster than the speed of sound. This myth of the "sound barrier" originated in the early 1930s. It was in full force by the time of the Volta Conference in 1935 (see Section 7.1). In light of the opening quotation, the idea of a sound barrier was still being discussed in the popular literature as late as 1959, 12 years after the first successful supersonic flight by Captain Charles Yeager on October 14, 1947.

Of course, we know today that the sound barrier is indeed a myth; the supersonic transport Concorde flew at Mach 2, and some military aircraft are capable of Mach 3 and slightly beyond. The X-15 hypersonic research airplane has flown at Mach 7, and the Apollo lunar return capsule successfully reentered the earth's atmosphere at Mach 36. Supersonic flight is now an everyday occurrence. So, what caused the early concern about a sound barrier? In the present chapter, we develop a theory applicable to high-speed subsonic flight, and we see how the theory predicts a monotonically increasing drag going to infinity as $M_\infty \to 1$. It was this type of result that led some people in the early 1930s to believe that flight beyond the speed of sound was impossible. However, we also show in this chapter that the approximations made in the theory break down near Mach 1 and that in reality, although the drag coefficient at Mach 1 is large, it is still a manageable finite number.

Specifically, the purpose of this chapter is to examine the properties of two-dimensional airfoils at Mach numbers above 0.3, where we can no longer assume incompressible flow, but below Mach 1. That is, this chapter is an extension of the airfoil discussions in Chapter 4 (which applied to incompressible flow) to the high-speed subsonic regime.

In the process, we climb to a new tier in our study of compressible flow. If you survey our discussions so far of compressible flow, you will observe that they treat one-dimensional cases such as normal shock waves and flows in ducts. Even oblique shock waves, which are two- and three-dimensional in nature, depend only on the component of Mach number normal to the wave. Therefore, we have not been explicitly concerned with a multidimensional flow. As a consequence, note that the types of equations which allow an analysis of these flows are *algebraic equations,* and hence are relatively easy to solve in comparison with partial differential equations. In Chapters 8 to 10, we have dealt primarily with such algebraic equations. These algebraic equations were obtained by applying the integral forms of the conservation equations [Equations (2.48), (2.64), and (2.95)] to appropriate control volumes where the flow properties were uniform over the inflow and outflow faces of the control volume. However, for general two- and three-dimensional flows, we are usually not afforded such a luxury. Instead, we must deal directly with the governing equations in their partial differential equation form (see Chapter 2). Such is the nature of the present chapter. Indeed, for the remainder of our aerodynamic discussions in this book, we appeal mainly to the differential forms of the continuity, momentum, and energy equations [such as Equations (2.52), (2.113a to c), and (2.114)].

The road map for this chapter is given in Figure 11.1. We are going to return to the concept of a velocity potential, first introduced in Section 2.15. We are going to combine our governing equations so as to obtain a single equation simply

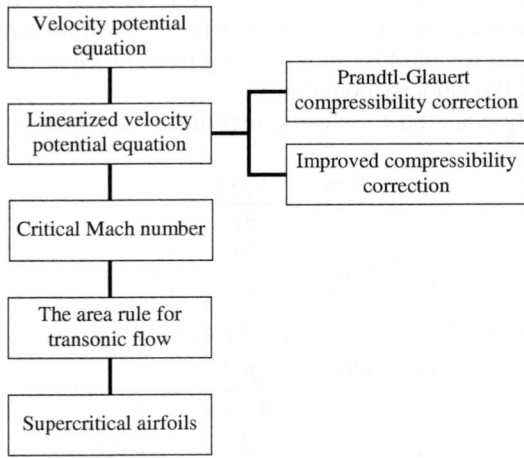

Figure 11.1 Road map for Chapter 11.

in terms of the velocity potential; that is, we are going to obtain for compressible flow an equation analogous to Laplace's equation derived for incompressible flow in Section 3.7 [see Equation (3.40)]. However, unlike Laplace's equation, which is linear, the exact velocity potential equation for compressible flow is nonlinear. By making suitable approximations, we are able to linearize this equation and apply it to thin airfoils at small angles of attack. The results enable us to correct incompressible airfoil data for the effects of compressibility—so-called *compressibility corrections.* Finally, we conclude this chapter by discussing several practical aspects of airfoil and general wing-body aerodynamics at speeds near Mach 1.

11.2 THE VELOCITY POTENTIAL EQUATION

The inviscid, compressible, subsonic flow over a body immersed in a uniform stream is *irrotational;* there is no mechanism in such a flow to start rotating the fluid elements (see Section 2.12). Thus, a velocity potential (see Section 2.15) can be defined. Since we are dealing with irrotational flow and the velocity potential, review Sections 2.12 and 2.15 before progressing further.

Consider two-dimensional, steady, irrotational, isentropic flow. A velocity potential, $\phi = \phi(x, y)$, can be defined such that [from Equation (2.154)]

$$\mathbf{V} = \nabla \phi \tag{11.1}$$

or in terms of the cartesian velocity components,

$$u = \frac{\partial \phi}{\partial x} \tag{11.2a}$$

$$v = \frac{\partial \phi}{\partial y} \tag{11.2b}$$

Let us proceed to obtain an equation for ϕ which represents a combination of the continuity, momentum, and energy equations. Such an equation would be very useful, because it would be simply one governing equation in terms of one unknown, namely the velocity potential ϕ.

The continuity equation for steady, two-dimensional flow is obtained from Equation (2.52) as

$$\frac{\partial(\rho u)}{\partial x} + \frac{\partial(\rho v)}{\partial y} = 0 \tag{11.3}$$

or

$$\rho \frac{\partial u}{\partial x} + u \frac{\partial \rho}{\partial x} + v \frac{\partial \rho}{\partial y} + \rho \frac{\partial v}{\partial y} = 0 \tag{11.4}$$

Substituting Equation (11.2a and b) into (11.4), we have

$$\rho \frac{\partial^2 \phi}{\partial x^2} + \frac{\partial \phi}{\partial x} \frac{\partial \rho}{\partial x} + \frac{\partial \phi}{\partial y} \frac{\partial \rho}{\partial y} + \rho \frac{\partial^2 \phi}{\partial y^2} = 0$$

or

$$\rho \left(\frac{\partial^2 \phi}{\partial x^2} + \frac{\partial^2 \phi}{\partial y^2} \right) + \frac{\partial \phi}{\partial x} \frac{\partial \rho}{\partial x} + \frac{\partial \phi}{\partial y} \frac{\partial \rho}{\partial y} = 0 \tag{11.5}$$

We are attempting to obtain an equation completely in terms of ϕ; hence, we need to eliminate ρ from Equation (11.5). To do this, consider the momentum equation in terms of Euler's equation:

$$dp = -\rho V \, dV \qquad (3.12)$$

This equation holds for a steady, compressible, inviscid flow and relates p and V along a streamline. It can readily be shown that Equation (3.12) holds in *any* direction throughout an irrotational flow, not just along a streamline (try it yourself). Therefore, from Equations (3.12) and (11.2*a* and *b*), we have

$$dp = -\rho V \, dV = -\frac{\rho}{2}d(V^2) = -\frac{\rho}{2}d(u^2 + v^2)$$

or

$$dp = -\frac{\rho}{2}d\left[\left(\frac{\partial \phi}{\partial x}\right)^2 + \left(\frac{\partial \phi}{\partial y}\right)^2\right] \qquad (11.6)$$

Recall that we are also considering the flow to be isentropic. Hence, any change in pressure dp in the flow is automatically accompanied by a corresponding isentropic change in density $d\rho$. Thus, by definition

$$\frac{dp}{d\rho} = \left(\frac{\partial p}{\partial \rho}\right)_s \qquad (11.7)$$

The right-hand side of Equation (11.7) is simply the square of the speed of sound. Thus, Equation (11.7) yields

$$dp = a^2 \, d\rho \qquad (11.8)$$

Substituting Equation (11.8) for the left side of Equation (11.6), we have

$$d\rho = -\frac{\rho}{2a^2}d\left[\left(\frac{\partial \phi}{\partial x}\right)^2 + \left(\frac{\partial \phi}{\partial y}\right)^2\right] \qquad (11.9)$$

Considering changes in the x direction, Equation (11.9) directly yields

$$\frac{\partial \rho}{\partial x} = -\frac{\rho}{2a^2}\frac{\partial}{\partial x}\left[\left(\frac{\partial \phi}{\partial x}\right)^2 + \left(\frac{\partial \phi}{\partial y}\right)^2\right]$$

or

$$\frac{\partial \rho}{\partial x} = -\frac{\rho}{a^2}\left(\frac{\partial \phi}{\partial x}\frac{\partial^2 \phi}{\partial x^2} + \frac{\partial \phi}{\partial y}\frac{\partial^2 \phi}{\partial x \partial y}\right) \qquad (11.10)$$

Similarly, for changes in the y direction, Equation (11.9) gives

$$\frac{\partial \rho}{\partial y} = -\frac{\rho}{a^2}\left(\frac{\partial \phi}{\partial x}\frac{\partial^2 \phi}{\partial x \partial y} + \frac{\partial \phi}{\partial y}\frac{\partial^2 \phi}{\partial y^2}\right) \qquad (11.11)$$

Substituting Equations (11.10) and (11.11) into (11.5), canceling the ρ which appears in each term, and factoring out the second derivatives of ϕ, we obtain

$$\left[1 - \frac{1}{a^2}\left(\frac{\partial \phi}{\partial x}\right)^2\right]\frac{\partial^2 \phi}{\partial x^2} + \left[1 - \frac{1}{a^2}\left(\frac{\partial \phi}{\partial y}\right)^2\right]\frac{\partial^2 \phi}{\partial y^2}$$
$$- \frac{2}{a^2}\left(\frac{\partial \phi}{\partial x}\right)\left(\frac{\partial \phi}{\partial y}\right)\frac{\partial^2 \phi}{\partial x \, \partial y} = 0 \tag{11.12}$$

which is called the *velocity potential equation*. It is almost completely in terms of ϕ; only the speed of sound appears in addition to ϕ. However, a can be readily expressed in terms of ϕ as follows. From Equation (8.33), we have

$$a^2 = a_0^2 - \frac{\gamma-1}{2}V^2 = a_0^2 - \frac{\gamma-1}{2}(u^2 + v^2)$$

$$= a_0^2 - \frac{\gamma-1}{2}\left[\left(\frac{\partial \phi}{\partial x}\right)^2 + \left(\frac{\partial \phi}{\partial y}\right)^2\right] \tag{11.13}$$

Since a_0 is a known constant of the flow, Equation (11.13) gives the speed of sound a as a function of ϕ. Hence, substitution of Equation (11.13) into (11.12) yields a single partial differential equation in terms of the unknown ϕ. This equation represents a combination of the continuity, momentum, and energy equations. In principle, it can be solved to obtain ϕ for the flow field around any two-dimensional shape, subject of course to the usual boundary conditions at infinity and along the body surface. These boundary conditions on ϕ are detailed in Section 3.7, and are given by Equations (3.47a and b) and (3.48b).

Because Equation (11.12) [along with Equation (11.13)] is a single equation in terms of one dependent variable ϕ, the analysis of isentropic, irrotational, steady, compressible flow is greatly simplified—we only have to solve one equation instead of three or more. Once ϕ is known, all the other flow variables are directly obtained as follows:

1. Calculate u and v from Equation (11.2a and b).
2. Calculate a from Equation (11.13).
3. Calculate $M = V/a = \sqrt{u^2 + v^2}/a$.
4. Calculate T, p, and ρ from Equations (8.40), (8.42), and (8.43), respectively. In these equations, the total conditions T_0, p_0, and ρ_0 are known quantities; they are constant throughout the flow field and hence are obtained from the given freestream conditions.

Although Equation (11.12) has the advantage of being one equation with one unknown, it also has the distinct disadvantage of being a *nonlinear* partial differential equation. Such nonlinear equations are very difficult to solve analytically, and in modern aerodynamics, solutions of Equation (11.12) are usually sought by means of sophisticated finite-difference numerical techniques. Indeed, no general analytical solution of Equation (11.12) has been found to this day. Contrast this situation with that for incompressible flow, which is governed by Laplace's equation—a *linear* partial differential equation for which numerous analytical solutions are well known.

Given this situation, aerodynamicists over the years have made assumptions regarding the physical nature of the flow field which are designed to simplify Equation (11.12). These assumptions limit our considerations to the flow over slender bodies at small angles of attack. For subsonic and supersonic flows, these assumptions lead to an *approximate* form of Equation (11.12) which is linear, and hence can be solved analytically. These matters are the subject of the next section.

Keep in mind that, within the framework of steady, irrotational, isentropic flow, Equation (11.12) is *exact* and holds for all Mach numbers, from subsonic to hypersonic, and for all two-dimensional body shapes, thin and thick.

11.3 THE LINEARIZED VELOCITY POTENTIAL EQUATION

Consider the two-dimensional, irrotational, isentropic flow over the body shown in Figure 11.2. The body is immersed in a uniform flow with velocity V_∞ oriented in the x direction. At an arbitrary point P in the flow field, the velocity is $\mathbf{V}$ with the x and y components given by u and v, respectively. Let us now visualize the velocity $\mathbf{V}$ as the sum of the uniform flow velocity plus some extra increments in velocity. For example, the x component of velocity u in Figure 11.2 can be visualized as V_∞ plus an increment in velocity (positive or negative). Similarly, the y component of velocity v can be visualized as a simple increment itself, because the uniform flow has a zero component in the y direction. These increments are called *perturbations,* and

$$u = V_\infty + \hat{u} \qquad v = \hat{v}$$

where $\hat{u}$ and $\hat{v}$ are called the *perturbation velocities*. These perturbation velocities are not necessarily small; indeed, they can be quite large in the stagnation region in front of the blunt nose of the body shown in Figure 11.2. In the same vein, because $\mathbf{V} = \nabla\phi$, we can define a perturbation velocity potential $\hat{\phi}$ such that

$$\phi = V_\infty x + \hat{\phi}$$

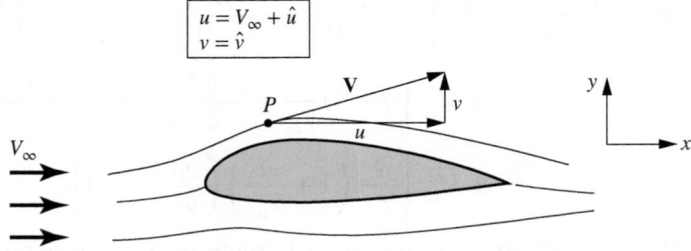

Figure 11.2 Uniform flow and perturbed flow.

where

$$\frac{\partial \hat{\phi}}{\partial x} = \hat{u}$$

$$\frac{\partial \hat{\phi}}{\partial y} = \hat{v}$$

Hence,

$$\frac{\partial \phi}{\partial x} = V_\infty + \frac{\partial \hat{\phi}}{\partial x} \qquad \frac{\partial \phi}{\partial y} = \frac{\partial \hat{\phi}}{\partial y}$$

$$\frac{\partial^2 \phi}{\partial x^2} = \frac{\partial^2 \hat{\phi}}{\partial x^2} \qquad \frac{\partial^2 \phi}{\partial y^2} = \frac{\partial^2 \hat{\phi}}{\partial y^2} \qquad \frac{\partial^2 \phi}{\partial x \, \partial y} = \frac{\partial^2 \hat{\phi}}{\partial x \, \partial y}$$

Substituting the above definitions into Equation (11.12), and multiplying by a^2, we obtain

$$\left[a^2 - \left(V_\infty + \frac{\partial \hat{\phi}}{\partial x} \right)^2 \right] \frac{\partial^2 \hat{\phi}}{\partial x^2} + \left[a^2 - \left(\frac{\partial \hat{\phi}}{\partial y} \right)^2 \right] \frac{\partial^2 \hat{\phi}}{\partial y^2}$$

$$-2 \left(V_\infty + \frac{\partial \hat{\phi}}{\partial x} \right) \left(\frac{\partial \hat{\phi}}{\partial y} \right) \frac{\partial^2 \hat{\phi}}{\partial x \, \partial y} = 0 \tag{11.14}$$

Equation (11.14) is called the *perturbation velocity potential equation*. It is precisely the same equation as Equation (11.12) except that it is expressed in terms of $\hat{\phi}$ instead of ϕ. It is still a nonlinear equation.

To obtain better physical insight in some of our subsequent discussions, let us recast Equation (11.14) in terms of the perturbation velocities. From the definition of $\hat{\phi}$ given earlier, Equation (11.14) can be written as

$$[a^2 - (V_\infty + \hat{u})^2] \frac{\partial \hat{u}}{\partial x} + (a^2 - \hat{v}^2) \frac{\partial \hat{v}}{\partial y} - 2(V_\infty + \hat{u}) \hat{v} \frac{\partial \hat{u}}{\partial y} = 0 \tag{11.14a}$$

From the energy equation in the form of Equation (8.32), we have

$$\frac{a_\infty^2}{\gamma - 1} + \frac{V_\infty^2}{2} = \frac{a^2}{\gamma - 1} + \frac{(V_\infty + \hat{u})^2 + \hat{v}^2}{2} \tag{11.15}$$

Substituting Equation (11.15) into (11.14a), and algebraically rearranging, we obtain

$$\left(1 - M_\infty^2 \right) \frac{\partial \hat{u}}{\partial x} + \frac{\partial \hat{v}}{\partial y} = M_\infty^2 \left[(\gamma + 1) \frac{\hat{u}}{V_\infty} + \frac{\gamma + 1}{2} \frac{\hat{u}^2}{V_\infty^2} + \frac{\gamma - 1}{2} \frac{\hat{v}^2}{V_\infty^2} \right] \frac{\partial \hat{u}}{\partial x}$$

$$+ M_\infty^2 \left[(\gamma - 1) \frac{\hat{u}}{V_\infty} + \frac{\gamma + 1}{2} \frac{\hat{v}^2}{V_\infty^2} + \frac{\gamma - 1}{2} \frac{\hat{u}^2}{V_\infty^2} \right] \frac{\partial \hat{v}}{\partial y}$$

$$+ M_\infty^2 \left[\frac{\hat{v}}{V_\infty} \left(1 + \frac{\hat{u}}{V_\infty} \right) \left(\frac{\partial \hat{u}}{\partial y} + \frac{\partial \hat{v}}{\partial x} \right) \right] \tag{11.16}$$

Equation (11.16) is still exact for irrotational, isentropic flow. Note that the left-hand side of Equation (11.16) is linear but the right-hand side is nonlinear. Also, keep in mind that the size of the perturbations $\hat{u}$ and $\hat{v}$ can be large or small; Equation (11.16) holds for both cases.

Let us now limit our considerations to *small* perturbations; that is, assume that the body in Figure 11.2 is a *slender* body at *small* angle of attack. In such a case, $\hat{u}$ and $\hat{v}$ will be small in comparison with V_∞. Therefore, we have

$$\frac{\hat{u}}{V_\infty}, \frac{\hat{v}}{V_\infty} \ll 1 \qquad \frac{\hat{u}^2}{V_\infty^2}, \frac{\hat{v}^2}{V_\infty^2} \ll 1$$

Keep in mind that products of $\hat{u}$ and $\hat{v}$ with their derivatives are also very small. With this in mind, examine Equation (11.16). Compare like terms (coefficients of like derivatives) on the left- and right-hand sides of Equation (11.16). We find

1. For $0 \leq M_\infty \leq 0.8$ or $M_\infty \geq 1.2$, the magnitude of

$$M_\infty^2 \left[(\gamma + 1)\frac{\hat{u}}{V_\infty} + \cdots \right] \frac{\partial \hat{u}}{\partial x}$$

   is small in comparison with the magnitude of

$$\left(1 - M_\infty^2\right) \frac{\partial \hat{u}}{\partial x}$$

   Thus, *ignore* the former term.

2. For $M_\infty < 5$ (approximately),

$$M_\infty^2 \left[(\gamma - 1)\frac{\hat{u}}{V_\infty} + \cdots \right] \frac{\partial \hat{v}}{\partial y}$$

   is small in comparison with $\partial \hat{v}/\partial y$. So ignore the former term. Also,

$$M_\infty^2 \left[\frac{\hat{v}}{V_\infty} \left(1 + \frac{\hat{u}}{V_\infty}\right) \left(\frac{\partial \hat{u}}{\partial y} + \frac{\partial \hat{v}}{\partial x}\right) \right] \approx 0$$

With the above order-of-magnitude comparisons, Equation (11.16) reduces to

$$\left(1 - M_\infty^2\right)\frac{\partial \hat{u}}{\partial x} + \frac{\partial \hat{v}}{\partial y} = 0 \qquad (11.17)$$

or in terms of the perturbation velocity potential,

$$\boxed{\left(1 - M_\infty^2\right)\frac{\partial^2 \hat{\phi}}{\partial x^2} + \frac{\partial^2 \hat{\phi}}{\partial y^2} = 0} \qquad (11.18)$$

Examine Equation (11.18). It is a *linear* partial differential equation, and therefore is inherently simpler to solve than its parent equation, Equation (11.16). However, we have paid a price for this simplicity. Equation (11.18) is no longer exact. It is only an approximation to the physics of the flow. Due to the assumptions made in obtaining Equation (11.18), it is reasonably valid (but not exact) for the following combined situations:

1. *Small* perturbation, that is, thin bodies at small angles of attack
2. *Subsonic* and *supersonic* Mach numbers

In contrast, Equation (11.18) is not valid for thick bodies and for large angles of attack. Moreover, it cannot be used for transonic flow, where $0.8 < M_\infty < 1.2$, or for hypersonic flow, where $M_\infty > 5$.

We are interested in solving Equation (11.18) in order to obtain the pressure distribution along the surface of a slender body. Since we are now dealing with approximate equations, it is consistent to obtain a linearized expression for the pressure coefficient—an expression that is approximate to the same degree as Equation (11.18), but which is extremely simple and convenient to use. The linearized pressure coefficient can be derived as follows.

First, recall the definition of the pressure coefficient C_p given in Section 1.5:

$$C_p \equiv \frac{p - p_\infty}{q_\infty} \tag{11.19}$$

where $q_\infty = \frac{1}{2}\rho_\infty V_\infty^2 =$ dynamic pressure. The dynamic pressure can be expressed in terms of M_∞ as follows:

$$q_\infty = \frac{1}{2}\rho_\infty V_\infty^2 = \frac{1}{2}\frac{\gamma p_\infty}{\gamma p_\infty}\rho_\infty V_\infty^2 = \frac{\gamma}{2}p_\infty\left(\frac{\rho_\infty}{\gamma p_\infty}\right)V_\infty^2 \tag{11.20}$$

From Equation (8.23), we have $a_\infty^2 = \gamma p_\infty/\rho_\infty$. Hence, Equation (11.20) becomes

$$q_\infty = \frac{\gamma}{2}p_\infty\frac{V_\infty^2}{a_\infty^2} = \frac{\gamma}{2}p_\infty M_\infty^2 \tag{11.21}$$

Substituting Equation (11.21) into (11.19), we have

$$\boxed{C_p = \frac{2}{\gamma M_\infty^2}\left(\frac{p}{p_\infty} - 1\right)} \tag{11.22}$$

Equation (11.22) is simply an alternate form of the pressure coefficient expressed in terms of M_∞. It is still an exact representation of the definition of C_p.

To obtain a linearized form of the pressure coefficient, recall that we are dealing with an adiabatic flow of a calorically perfect gas; hence, from Equation (8.39),

$$T + \frac{V^2}{2c_p} = T_\infty + \frac{V_\infty^2}{2c_p} \tag{11.23}$$

Recalling from Equation (7.9) that $c_p = \gamma R/(\gamma - 1)$, Equation (11.23) can be written as

$$T - T_\infty = \frac{V_\infty^2 - V^2}{2\gamma R/(\gamma - 1)} \tag{11.24}$$

Also, recalling that $a_\infty = \sqrt{\gamma R T_\infty}$, Equation (11.24) becomes

$$\frac{T}{T_\infty} - 1 = \frac{\gamma - 1}{2} \frac{V_\infty^2 - V^2}{\gamma R T_\infty} = \frac{\gamma - 1}{2} \frac{V_\infty^2 - V^2}{a_\infty^2} \qquad (11.25)$$

In terms of the perturbation velocities

$$V^2 = (V_\infty + \hat{u})^2 + \hat{v}^2$$

Equation (11.25) can be written as

$$\frac{T}{T_\infty} = 1 - \frac{\gamma - 1}{2a_\infty^2}(2\hat{u}V_\infty + \hat{u}^2 + \hat{v}^2) \qquad (11.26)$$

Since the flow is isentropic, $p/p_\infty = (T/T_\infty)^{\gamma/(\gamma-1)}$, and Equation (11.26) becomes

$$\frac{p}{p_\infty} = \left[1 - \frac{\gamma - 1}{2a_\infty^2}(2\hat{u}V_\infty + \hat{u}^2 + \hat{v}^2)\right]^{\gamma/(\gamma-1)}$$

or

$$\frac{p}{p_\infty} = \left[1 - \frac{\gamma - 1}{2}M_\infty^2\left(\frac{2\hat{u}}{V_\infty} + \frac{\hat{u}^2 + \hat{v}^2}{V_\infty^2}\right)\right]^{\gamma/(\gamma-1)} \qquad (11.27)$$

Equation (11.27) is still an exact expression. However, let us now make the assumption that the perturbations are small, that is, $\hat{u}/V_\infty \ll 1$, $\hat{u}^2/V_\infty^2 \ll 1$, and $\hat{v}^2/V_\infty^2 \ll 1$. In this case, Equation (11.27) is of the form

$$\frac{p}{p_\infty} = (1 - \varepsilon)^{\gamma/(\gamma-1)} \qquad (11.28)$$

where ε is small. From the binomial expansion, neglecting higher-order terms, Equation (11.28) becomes

$$\frac{p}{p_\infty} = 1 - \frac{\gamma}{\gamma - 1}\varepsilon + \cdots \qquad (11.29)$$

Comparing Equation (11.27) to (11.29), we can express Equation (11.27) as

$$\frac{p}{p_\infty} = 1 - \frac{\gamma}{2}M_\infty^2\left(\frac{2\hat{u}}{V_\infty} + \frac{\hat{u}^2 + \hat{v}^2}{V_\infty^2}\right) + \cdots \qquad (11.30)$$

Substituting Equation (11.30) into the expression for the pressure coefficient, Equation (11.22), we obtain

$$C_p = \frac{2}{\gamma M_\infty^2}\left[1 - \frac{\gamma}{2}M_\infty^2\left(\frac{2\hat{u}}{V_\infty} + \frac{\hat{u}^2 + \hat{v}^2}{V_\infty^2}\right) + \cdots - 1\right]$$

or

$$C_p = -\frac{2\hat{u}}{V_\infty} - \frac{\hat{u}^2 + \hat{v}^2}{V_\infty^2} \qquad (11.31)$$

Since $\hat{u}^2/V_\infty^2$ and $\hat{v}^2/V_\infty^2 \ll 1$, Equation (11.31) becomes

$$\boxed{C_p = -\frac{2\hat{u}}{V_\infty}} \qquad (11.32)$$

Equation (11.32) is the linearized form for the pressure coefficient; it is valid only for *small* perturbations. Equation (11.32) is consistent with the linearized perturbation velocity potential equation, Equation (11.18). Note the simplicity of Equation (11.32); it depends only on the x component of the velocity perturbation, namely, $\hat{u}$.

To round out our discussion on the basics of the linearized equations, we note that any solution to Equation (11.18) must satisfy the usual boundary conditions at infinity and at the body surface. At infinity, clearly $\hat{\phi} = $ constant; that is, $\hat{u} = \hat{v} = 0$. At the body, the flow-tangency condition holds. Let θ be the angle between the tangent to the surface and the freestream. Then, at the surface, the boundary condition is obtained from Equation (3.48e):

$$\tan\theta = \frac{v}{u} = \frac{\hat{v}}{V_\infty + \hat{u}} \tag{11.33}$$

which is an exact expression for the flow-tangency condition at the body surface. A simpler, approximate expression for Equation (11.33), consistent with linearized theory, can be obtained by noting that for small perturbations, $\hat{u} \ll V_\infty$. Hence, Equation (11.33) becomes

$$\hat{v} = V_\infty \tan\theta$$

or

$$\boxed{\frac{\partial\hat{\phi}}{\partial y} = V_\infty \tan\theta} \tag{11.34}$$

Equation (11.34) is an *approximate* expression for the flow-tangency condition at the body surface, with accuracy of the same order as Equations (11.18) and (11.32).

11.4 PRANDTL-GLAUERT COMPRESSIBILITY CORRECTION

The aerodynamic theory for incompressible flow over thin airfoils at small angles of attack was presented in Chapter 4. For aircraft of the period 1903–1940, such theory was adequate for predicting airfoil properties. However, with the rapid evolution of high-power reciprocating engines spurred by World War II, the velocities of military fighter planes began to push close to 450 mi/h. Then, with the advent of the first operational jet-propelled airplanes in 1944 (the German Me 262), flight velocities took a sudden spurt into the 550 mi/h range and faster. As a result, the incompressible flow theory of Chapter 4 was no longer applicable to such aircraft; rather, high-speed airfoil theory had to deal with compressible flow. Because a vast bulk of data and experience had been collected over the years in low-speed aerodynamics, and because there was no desire to totally discard such data, the natural approach to high-speed subsonic aerodynamics was to search for methods that would allow relatively simple *corrections* to existing incompressible flow results which would approximately take into account the effects of

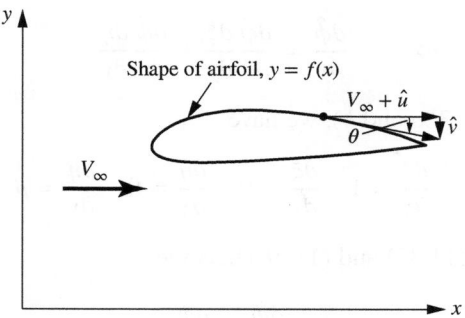

Figure 11.3 Airfoil in physical space.

compressibility. Such methods are called *compressibility corrections*. The first, and most widely known of these corrections is the Prandtl-Glauert compressibility correction, to be derived in this section. The Prandtl-Glauert method is based on the linearized perturbation velocity potential equation given by Equation (11.18). Therefore, it is limited to thin airfoils at small angles of attack. Moreover, it is purely a subsonic theory and begins to give inappropriate results at values of $M_\infty = 0.7$ and above.

Consider the subsonic, compressible, inviscid flow over the airfoil sketched in Figure 11.3. The shape of the airfoil is given by $y = f(x)$. Assume that the airfoil is thin and that the angle of attack is small; in such a case, the flow is reasonably approximated by Equation (11.18). Define

$$\beta^2 \equiv 1 - M_\infty^2$$

so that Equation (11.18) can be written as

$$\beta^2 \frac{\partial^2 \hat{\phi}}{\partial x^2} + \frac{\partial^2 \hat{\phi}}{\partial y^2} = 0 \qquad (11.35)$$

Let us transform the independent variables x and y into a new space, ξ and η, such that

$$\xi = x \qquad (11.36a)$$

$$\eta = \beta y \qquad (11.36b)$$

Moreover, in this transformed space, consider a new velocity potential $\bar{\phi}$ such that

$$\bar{\phi}(\xi, \eta) = \beta \hat{\phi}(x, y) \qquad (11.36c)$$

To recast Equation (11.35) in terms of the transformed variables, recall the chain rule of partial differentiation; that is,

$$\frac{\partial \hat{\phi}}{\partial x} = \frac{\partial \hat{\phi}}{\partial \xi} \frac{\partial \xi}{\partial x} + \frac{\partial \hat{\phi}}{\partial \eta} \frac{\partial \eta}{\partial x} \qquad (11.37)$$

and
$$\frac{\partial \hat{\phi}}{\partial y} = \frac{\partial \hat{\phi}}{\partial \xi}\frac{\partial \xi}{\partial y} + \frac{\partial \hat{\phi}}{\partial \eta}\frac{\partial \eta}{\partial y} \tag{11.38}$$

From Equation (11.36a and b), we have

$$\frac{\partial \xi}{\partial x} = 1 \quad \frac{\partial \xi}{\partial y} = 0 \quad \frac{\partial \eta}{\partial x} = 0 \quad \frac{\partial \eta}{\partial y} = \beta$$

Hence, Equations (11.37) and (11.38) become

$$\frac{\partial \hat{\phi}}{\partial x} = \frac{\partial \hat{\phi}}{\partial \xi} \tag{11.39}$$

$$\frac{\partial \hat{\phi}}{\partial y} = \beta\frac{\partial \hat{\phi}}{\partial \eta} \tag{11.40}$$

Recalling Equation (11.36c), Equations (11.39) and (11.40) become

$$\frac{\partial \hat{\phi}}{\partial x} = \frac{1}{\beta}\frac{\partial \bar{\phi}}{\partial \xi} \tag{11.41}$$

and
$$\frac{\partial \hat{\phi}}{\partial y} = \frac{\partial \bar{\phi}}{\partial \eta} \tag{11.42}$$

Differentiating Equation (11.41) with respect to x (again using the chain rule), we obtain

$$\frac{\partial^2 \hat{\phi}}{\partial x^2} = \frac{1}{\beta}\frac{\partial^2 \bar{\phi}}{\partial \xi^2} \tag{11.43}$$

Differentiating Equation (11.42) with respect to y, we find that the result is

$$\frac{\partial^2 \hat{\phi}}{\partial y^2} = \beta\frac{\partial^2 \bar{\phi}}{\partial \eta^2} \tag{11.44}$$

Substitute Equations (11.43) and (11.44) into (11.35):

$$\beta^2 \frac{1}{\beta}\frac{\partial^2 \bar{\phi}}{\partial \xi^2} + \beta\frac{\partial^2 \bar{\phi}}{\partial \eta^2} = 0$$

or
$$\frac{\partial^2 \bar{\phi}}{\partial \xi^2} + \frac{\partial^2 \bar{\phi}}{\partial \eta^2} = 0 \tag{11.45}$$

Examine Equation (11.45)—it should look familiar. Indeed, Equation (11.45) is Laplace's equation. Recall from Chapter 3 that Laplace's equation is the governing relation for *incompressible flow*. Hence, starting with a subsonic compressible flow in physical (x, y) space where the flow is represented by $\hat{\phi}(x, y)$ obtained from Equation (11.35), we have related this flow to an incompressible flow in transformed (ξ, η) space, where the flow is represented by $\bar{\phi}(\xi, \eta)$ obtained from Equation (11.45). The relation between $\bar{\phi}$ and $\hat{\phi}$ is given by Equation (11.36c).

Consider again the shape of the airfoil given in physical space by $y = f(x)$. The shape of the airfoil in the transformed space is expressed as $\eta = q(\xi)$. Let us compare the two shapes. First, apply the approximate boundary condition, Equation (11.34), in physical space, noting that $df/dx = \tan\theta$. We obtain

$$V_\infty \frac{df}{dx} = \frac{\partial\hat{\phi}}{\partial y} = \frac{1}{\beta}\frac{\partial\bar{\phi}}{\partial y} = \frac{\partial\bar{\phi}}{\partial\eta} \tag{11.46}$$

Similarly, apply the flow-tangency condition in transformed space, which from Equation (11.34) is

$$V_\infty \frac{dq}{d\xi} = \frac{\partial\bar{\phi}}{\partial\eta} \tag{11.47}$$

Examine Equations (11.46) and (11.47) closely. Note that the right-hand sides of these two equations are identical. Thus, from the left-hand sides, we obtain

$$\frac{df}{dx} = \frac{dq}{d\xi} \tag{11.48}$$

Equation (11.48) implies that the shape of the airfoil in the transformed space is the same as in the physical space. Hence, the above transformation relates the compressible flow over an airfoil in (x, y) space to the incompressible flow in (ξ, η) space over the *same* airfoil.

The above theory leads to an immensely practical result, as follows. Recall Equation (11.32) for the linearized pressure coefficient. Inserting the above transformation into Equation (11.32), we obtain

$$C_p = \frac{-2\hat{u}}{V_\infty} = -\frac{2}{V_\infty}\frac{\partial\hat{\phi}}{\partial x} = -\frac{2}{V_\infty}\frac{1}{\beta}\frac{\partial\bar{\phi}}{\partial x} = -\frac{2}{V_\infty}\frac{1}{\beta}\frac{\partial\bar{\phi}}{\partial\xi} \tag{11.49}$$

Question: What is the significance of $\partial\bar{\phi}/\partial\xi$ in Equation (11.49)? Recall that $\bar{\phi}$ is the perturbation velocity potential for an incompressible flow in transformed space. Hence, from the definition of velocity potential, $\partial\bar{\phi}/\partial\xi = \bar{u}$, where $\bar{u}$ is a perturbation velocity for the incompressible flow. Hence, Equation (11.49) can be written as

$$C_p = \frac{1}{\beta}\left(-\frac{2\bar{u}}{V_\infty}\right) \tag{11.50}$$

From Equation (11.32), the expression $(-2\bar{u}/V_\infty)$ is simply the linearized pressure coefficient for the incompressible flow. Denote this incompressible pressure coefficient by $C_{p,0}$. Hence, Equation (11.50) gives

$$C_p = \frac{C_{p,0}}{\beta}$$

or recalling that $\beta \equiv \sqrt{1 - M_\infty^2}$, we have

$$C_p = \frac{C_{p,0}}{\sqrt{1 - M_\infty^2}} \qquad (11.51)$$

Equation (11.51) is called the *Prandtl-Glauert rule;* it states that, if we know the incompressible pressure distribution over an airfoil, then the compressible pressure distribution over the same airfoil can be obtained from Equation (11.51). Therefore, Equation (11.51) is truly a *compressibility correction* to incompressible data.

Consider the lift and moment coefficients for the airfoil. For an inviscid flow, the aerodynamic lift and moment on a body are simply integrals of the pressure distribution over the body, as described in Section 1.5. (If this is somewhat foggy in your mind, review Section 1.5 before progressing further.) In turn, the lift and moment *coefficients* are obtained from the integral of the pressure coefficient via Equations (1.15) to (1.19). Since Equation (11.51) relates the compressible and incompressible pressure coefficients, the same relation must therefore hold for lift and moment coefficients:

$$c_l = \frac{c_{l,0}}{\sqrt{1 - M_\infty^2}} \qquad [11.52]$$

$$c_m = \frac{c_{m,0}}{\sqrt{1 - M_\infty^2}} \qquad [11.53]$$

The Prandtl-Glauert rule, embodied in Equations (11.51) to (11.53), was historically the first compressibility correction to be obtained. As early as 1922, Prandtl was using this result in his lectures at Göttingen, although without written proof. The derivation of Equations (11.51) to (11.53) was first formally published by the British aerodynamicist, Hermann Glauert, in 1928. Hence, the rule is named after both people. The Prandtl-Glauert rule was used exclusively until 1939, when an improved compressibility correction was developed. Because of their simplicity, Equations (11.51) to (11.53) are still used today for initial estimates of compressibility effects.

Recall that the results of Chapters 3 and 4 proved that inviscid, incompressible flow over a closed, two-dimensional body theoretically produces zero drag—the well-known d'Alembert's paradox. Does the same paradox hold for inviscid, subsonic, compressible flow? The answer can be obtained by again noting that the only source of drag is the integral of the pressure distribution. If this integral is zero for an incompressible flow, and since the compressible pressure coefficient differs from the incompressible pressure coefficient by only a constant scale factor, β, then the integral must also be zero for a compressible flow. Hence, d'Alembert's paradox also prevails for inviscid, subsonic, compressible flow. However, as soon as the freestream Mach number is high enough to produce locally supersonic flow on the body surface with attendant shock waves, as shown in Figure 1.43b, then a positive wave drag is produced, and d'Alembert's paradox no longer prevails.

EXAMPLE 11.1

At a given point on the surface of an airfoil, the pressure coefficient is −0.3 at very low speeds. If the freestream Mach number is 0.6, calculate C_p at this point.

■ **Solution**

From Equation (11.51),

$$C_p = \frac{C_{p,0}}{\sqrt{1 - M^2}} = \frac{-0.3}{\sqrt{1 - (0.6)^2}} = \boxed{-0.375}$$

EXAMPLE 11.2

From Chapter 4, the theoretical lift coefficient for a thin, symmetric airfoil in an incompressible flow is $c_l = 2\pi\alpha$. Calculate the lift coefficient for $M_\infty = 0.7$.

■ **Solution**

From Equation (11.52),

$$c_l = \frac{c_{l,0}}{\sqrt{1 - M_\infty^2}} = \frac{2\pi\alpha}{\sqrt{1 - (0.7)^2}} = \boxed{8.8\alpha}$$

Note: The effect of compressibility at Mach 0.7 is to increase the lift slope by the ratio $8.8/2\pi = 1.4$, or by 40 percent.

11.5 IMPROVED COMPRESSIBILITY CORRECTIONS

The importance of accurate compressibility corrections reached new highs during the rapid increase in airplane speeds spurred by World War II. Efforts were made to improve upon the Prandtl-Glauert rule discussed in Section 11.4. Several of the more popular formulae are given below.

The Karman-Tsien rule states

$$C_p = \frac{C_{p,0}}{\sqrt{1 - M_\infty^2} + \left[M_\infty^2 / \left(1 + \sqrt{1 - M_\infty^2} \right) \right] C_{p,0}/2} \tag{11.54}$$

This formula, derived in References 27 and 28, has been widely adopted by the aeronautical industry since World War II.

Laitone's rule states

$$C_p = \frac{C_{p,0}}{\sqrt{1 - M_\infty^2} + \left(M_\infty^2 \left\{ 1 + [(\gamma - 1)/2]M_\infty^2 \right\} / \left(2\sqrt{1 - M_\infty^2} \right) \right) C_{p,0}} \tag{11.55}$$

This formula is more recent than either the Prandtl-Glauert or the Karman-Tsien rule; it is derived in Reference 29.

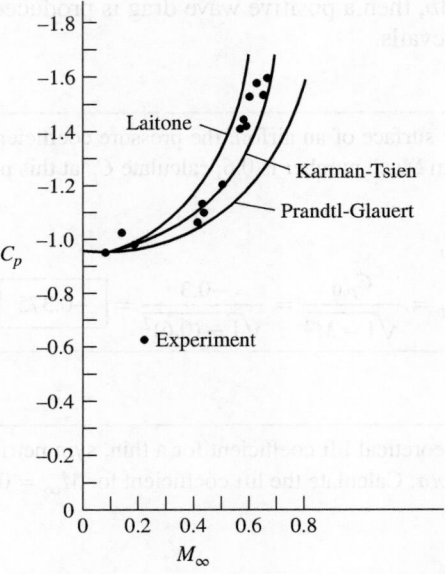

Figure 11.4 Several compressibility corrections compared with experimental results for a NACA 4412 airfoil at an angle of attack $\alpha = 1° \, 53'$. The experimental data are chosen for their historical significance; from Stack, John, W. F. Lindsey, and R. E. Littell: *The Compressibility Burble and the Effect of Compressibility on Pressures and Forces Acting on an Airfoil*, NACA report no. 646, 1938. This was the first major NACA publication to address the compressibility problem in a systematic fashion; it covered work performed in the 2-ft high-speed tunnel at the Langley Aeronautical Laboratory and was carried out during 1935–1936.

These compressibility corrections are compared in Figure 11.4, which also shows experimental data for the C_p variation with M_∞ at the 0.3-chord location on an NACA 4412 airfoil. Note that the Prandtl-Glauert rule, although the simplest to apply, underpredicts the experimental data, whereas the improved compressibility corrections are clearly more accurate. Recall that the Prandtl-Glauert rule is based on linear theory. In contrast, both the Laitone and Karman-Tsien rules attempt to account for some of the nonlinear aspects of the flow.

11.6 CRITICAL MACH NUMBER

Return to the road map given in Figure 11.1. We have now finished our discussion of linearized flow and the associated compressibility corrections. Keep in mind that such linearized theory does *not* apply to the transonic flow regime,

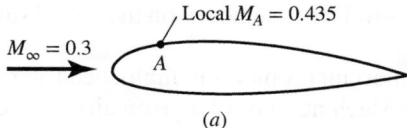

(a)

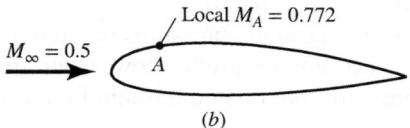

(b)

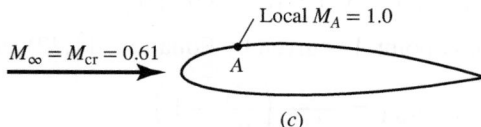

(c)

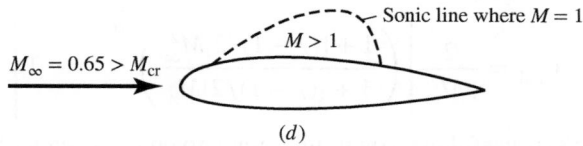

(d)

Figure 11.5 Definition of critical Mach number. Point A is the location of minimum pressure on the top surface of the airfoil. (See end-of-chapter Problem 11.7 for the calculation of the numbers in this figure.)

$0.8 \leq M_\infty \leq 1.2$. Transonic flow is highly nonlinear, and theoretical transonic aerodynamics is a challenging and sophisticated subject. For the remainder of this chapter, we deal with several aspects of transonic flow from a qualitative point of view. The theory of transonic aerodynamics is beyond the scope of this book.

Consider an airfoil in a low-speed flow, say, with $M_\infty = 0.3$, as sketched in Figure 11.5a. In the expansion over the top surface of the airfoil, the local flow Mach number M increases. Let point A represent the location on the airfoil surface where the pressure is a minimum, hence where M is a maximum. In Figure 11.5a, let us say this maximum is $M_A = 0.435$. Now assume that we gradually increase the freestream Mach number. As M_∞ increases, M_A also increases. For example, if M_∞ is increased to $M = 0.5$, the maximum local value of M will be 0.772, as shown in Figure 11.5b. Let us continue to increase M_∞ until we achieve just the right value such that the local Mach number at the minimum pressure point equals 1, that is, such that $M_A = 1.0$, as shown in Figure 11.5c. When this happens, the freestream Mach number M_∞ is called the *critical Mach number,* denoted by M_{cr}. By definition, the critical Mach number is that *freestream* Mach number at which sonic flow is first achieved on the airfoil surface. In Figure 11.5c, $M_{cr} = 0.61$.

One of the most important problems in high-speed aerodynamics is the determination of the critical Mach number of a given airfoil, because at values of M_∞ slightly above M_{cr}, the airfoil experiences a dramatic increase in drag coefficient (discussed in Section 11.7). The purpose of the present section is to give a rather straightforward method for estimating M_{cr}.

Let p_∞ and p_A represent the static pressures in the freestream and at point A, respectively, in Figure 11.5. For isentropic flow, where the total pressure p_0 is constant, these static pressures are related through Equation (8.42) as follows:

$$\frac{p_A}{p_\infty} = \frac{p_A/p_0}{p_\infty/p_0} = \left(\frac{1 + [(\gamma - 1)/2]M_\infty^2}{1 + [(\gamma - 1)/2]M_A^2} \right)^{\gamma/(\gamma-1)} \tag{11.56}$$

The pressure coefficient at point A is given by Equation (11.22) as

$$C_{p,A} = \frac{2}{\gamma M_\infty^2} \left(\frac{p_A}{p_\infty} - 1 \right) \tag{11.57}$$

Combining Equations (11.56) and (11.57), we have

$$C_{p,A} = \frac{2}{\gamma M_\infty^2} \left[\left(\frac{1 + [(\gamma - 1)/2]M_\infty^2}{1 + [(\gamma - 1)/2]M_A^2} \right)^{\gamma/(\gamma-1)} - 1 \right] \tag{11.58}$$

Equation (11.58) is useful in its own right; for a given freestream Mach number, it relates the local value of C_p to the local Mach number. [Note that Equation (11.58) is the compressible flow analogue of Bernoulli's equation, Equation (3.13), which for incompressible flow with a given freestream velocity and pressure relates the local pressure at a point in the flow to the local velocity at that point.] However, for our purposes here, we ask the question, What is the value of the local C_p when the local Mach number is unity? By definition, this value of the pressure coefficient is called the *critical pressure coefficient*, denoted by $C_{p,cr}$. For a given freestream Mach number M_∞, the value of $C_{p,cr}$ can be obtained by inserting $M_A = 1$ into Equation (11.58):

$$C_{p,cr} = \frac{2}{\gamma M_\infty^2} \left[\left(\frac{1 + [(\gamma - 1)/2]M_\infty^2}{1 + (\gamma - 1)/2} \right)^{\gamma/(\gamma-1)} - 1 \right] \tag{11.59}$$

Equation (11.59) allows us to calculate the pressure coefficient at any point in the flow where the local Mach number is 1, for a given freestream Mach number M_∞. For example, if M_∞ is slightly greater than M_{cr}, say, $M_\infty = 0.65$ as shown in Figure 11.5d, then a finite region of supersonic flow will exist above the airfoil; Equation (11.59) allows us to calculate the pressure coefficient at only those points where $M = 1$, that is, at only those points that fall on the sonic line in Figure 11.5d. Now, returning to Figure 11.5c, when the freestream Mach number is precisely equal to the critical Mach number, there is only one point where $M = 1$, namely, point A. The pressure coefficient at point A will be $C_{p,cr}$, which is

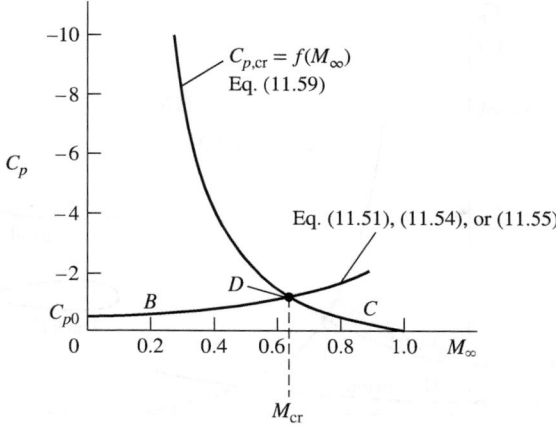

Figure 11.6 Estimation of critical Mach number.

obtained from Equation (11.59). In this case, M_∞ in Equation (11.59) is precisely M_{cr}. Hence,

$$C_{p,cr} = \frac{2}{\gamma M_{cr}^2} \left[\left(\frac{1 + [(\gamma - 1)/2]M_{cr}^2}{1 + (\gamma - 1)/2} \right)^{\gamma/(\gamma-1)} - 1 \right] \quad (11.60)$$

Equation (11.60) shows that $C_{p,cr}$ is a unique function of M_{cr}; this variation is plotted as curve C in Figure 11.6. Note that Equation (11.60) is simply an aerodynamic relation for isentropic flow—it has no connection with the shape of a given airfoil. In this sense, Equation (11.60), and hence curve C in Figure 11.6, is a type of "universal relation" which can be used for all airfoils.

Equation (11.60), in conjunction with any one of the compressibility corrections given by Equation (11.51), (11.54), or (11.55), allows us to estimate the critical Mach number for a given airfoil as follows:

1. By some means, either experimental or theoretical, obtain the low-speed incompressible value of the pressure coefficient $C_{p,0}$ at the minimum pressure point on the given airfoil.

2. Using any of the compressibility corrections, Equation (11.51), (11.54), or (11.55), plot the variation of C_p with M_∞. This is represented by curve B in Figure 11.6.

3. Somewhere on curve B, there will be a single point where the pressure coefficient corresponds to locally sonic flow. Indeed, this point must coincide with Equation (11.60), represented by curve C in Figure 11.6. Hence, the *intersection* of curves B and C represents the point corresponding to sonic flow at the minimum pressure location on the airfoil. In turn, the value of M_∞ at this intersection is, by definition, the critical Mach number, as shown in Figure 11.6.

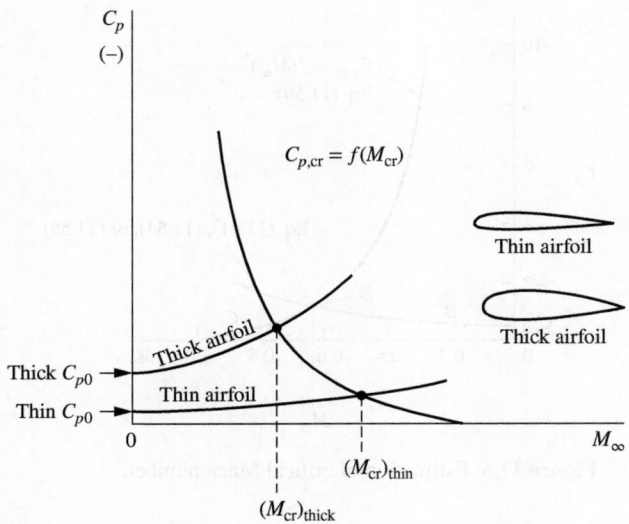

Figure 11.7 Effect of airfoil thickness on critical Mach number.

The graphical construction in Figure 11.6 is not an exact determination of M_{cr}. Although curve C is exact, curve B is approximate because it represents the approximate compressibility correction. Hence, Figure 11.6 gives only an estimation of M_{cr}. However, such an estimation is quite useful for preliminary design, and the results from Figure 11.6 are accurate enough for most applications.

Consider two airfoils, one thin and the other thick, as sketched in Figure 11.7. First consider the low-speed incompressible flow over these airfoils. The flow over the thin airfoil is only slightly perturbed from the freestream. Hence, the expansion over the top surface is mild, and $C_{p,0}$ at the minimum pressure point is a negative number of only small absolute magnitude, as shown in Figure 11.7. [Recall from Equation (11.32) that $C_p \propto \hat{u}$; hence, the smaller the perturbation, the smaller is the absolute magnitude of C_p.] In contrast, the flow over the thick airfoil experiences a large perturbation from the freestream. The expansion over the top surface is strong, and $C_{p,0}$ at the minimum pressure point is a negative number of large magnitude, as shown in Figure 11.7. If we now perform for each airfoil the same construction as given in Figure 11.6, we see that the thick airfoil will have a lower critical Mach number than the thin airfoil. This is clearly illustrated in Figure 11.7. For high-speed airplanes, it is desirable to have M_{cr} as high as possible. Hence, modern high-speed subsonic airplanes are usually designed with relatively thin airfoils. (The development of the supercritical airfoil has somewhat loosened this criterion, as discussed in Section 11.8.) For example, the Gates Lear jet high-speed jet executive transport utilizes a 9 percent thick airfoil; contrast this with the low-speed Piper Aztec, a twin-engine propeller-driven general aviation aircraft designed with a 14 percent thick airfoil.

EXAMPLE 11.3

In this example, we illustrate the estimation of the critical Mach number for an airfoil using (a) the graphical solution discussed in this section, and (b) an analytical solution using a closed-form equation obtained from a combination of Equations (11.51) and (11.60). Consider the NACA 0012 airfoil at zero angle of attack shown at the top of Figure 11.8. The pressure coefficient distribution over this airfoil, measured in a wind tunnel at low speed, is given at the bottom of Figure 11.8. From this information, estimate the critical Mach number of the NACA 0012 airfoil at zero angle of attack.

■ **Solution**

(a) *Graphical Solution.* First, let us accurately plot the curve of $C_{p,\mathrm{cr}}$ versus M_{cr} from Equation (11.60),

$$C_{p,\mathrm{cr}} = \frac{2}{\gamma M_{\mathrm{cr}}^2}\left[\left(\frac{1 + [(\gamma-1)/2]M_{\mathrm{cr}}^2}{1 + (\gamma-1)/2}\right)^{\gamma/(\gamma-1)} - 1\right] \qquad (11.60)$$

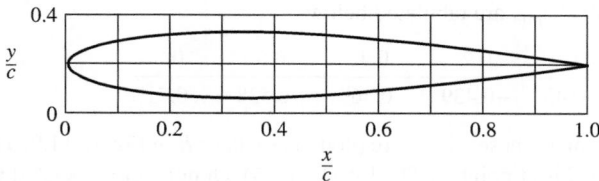

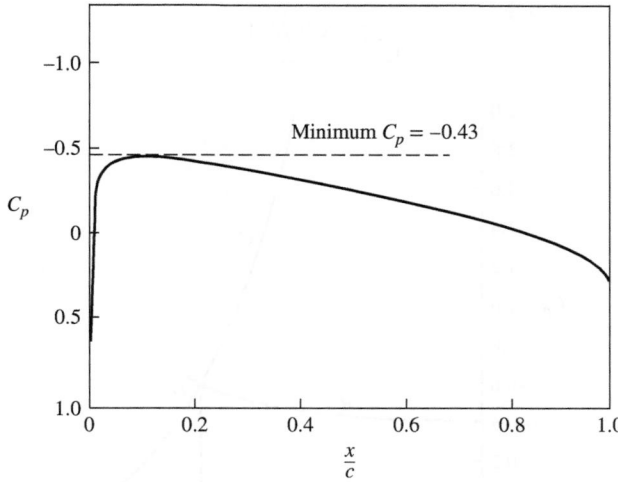

Figure 11.8 Low-speed pressure coefficient distribution over the surface of an NACA 0012 airfoil at zero angle of attack. Re = 3.65×10^6. (*Source:* R. J. Freuler and G. M. Gregorek, "An Evaluation of Four Single Element Airfoil Analytical Methods," in *Advanced Technology Airfoil Research*, NASA CP 2045, 1978, pp. 133–162.)

For $\gamma = 1.4$, from Equation (11.60) we can tabulate

M_∞	0.4	0.5	0.6	0.7	0.8	0.9	1.0
$C_{p,\text{cr}}$	−3.66	−2.13	−1.29	−0.779	−0.435	−0.188	0

These numbers are plotted as curve C in Figure 11.9. Note that $C_{p,\text{cr}} = 0$ when $M_{\text{cr}} = 1.0$. This makes physical sense; if the free stream Mach number is already 1, then no change in the pressure is required to achieve Mach 1 at a local point in the flow, and hence the pressure difference $(p_{\text{cr}} - p_\infty)$ is zero and $C_{p,\text{cr}} = 0$.

Following the three-step procedure mentioned earlier, in step one we obtain the low-speed incompressible value of the minimum pressure coefficient $(C_{p,0})_{\text{min}}$ from the pressure coefficient distribution given in Figure 11.8. The *minimum* value of C_p on the surface is −0.43. Following step two, in Equation (11.51), $(C_{p,0})_{\text{min}} = -0.43$, and we have from Equation (11.51),

$$(C_p)_{\text{min}} = \frac{(C_{p,0})_{\text{min}}}{\sqrt{1 - M_\infty^2}} = \frac{-0.43}{\sqrt{1 - M_\infty^2}} \tag{11.61}$$

Some values of $(C_p)_{\text{min}}$ are tabulated below

M_∞	0	0.2	0.4	0.6	0.8
$(C_p)_{\text{min}}$	−0.43	−0.439	−0.469	−0.538	−0.717

Following step three, these values are plotted as Curve B in Figure 11.9. The intersection of curves B and C is at point D. The freestream Mach number associated with point D is the critical Mach number for the NACA 0012 airfoil. From Figure 11.9, we have

$$\boxed{M_{\text{cr}} = 0.74}$$

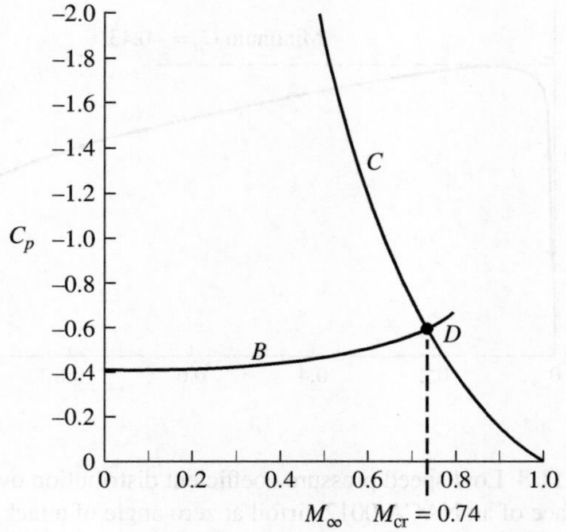

Figure 11.9 Graphical solution for the critical Mach number.

(b) Analytical Solution. In Figure 11.9, curve B is given by Equation (11.61)

$$(C_p)_{\min} = \frac{-0.43}{\sqrt{1 - M_\infty^2}} \tag{11.61}$$

At the intersection point D, $(C_p)_{\min}$ in Equation (11.61) is the critical pressure coefficient and M_∞ is the critical Mach number

$$C_{p,\mathrm{cr}} = \frac{-0.43}{\sqrt{1 - M_{\mathrm{cr}}^2}} \quad \text{(at point } D) \tag{11.62}$$

Also, at point D the value of $C_{p,\mathrm{cr}}$ is given by Equation (11.60). Hence, *at point D we can equate the right-hand sides of Equations (11.62) and (11.60),

$$\frac{-0.43}{\sqrt{1 - M_{\mathrm{cr}}^2}} = \frac{2}{\gamma M_{\mathrm{cr}}^2} \left[\left(\frac{1 + [(\gamma - 1)/2]M_{\mathrm{cr}}^2}{1 + (\gamma - 1)/2} \right)^{\gamma/\gamma - 1} - 1 \right] \tag{11.63}$$

Equation (11.63) is one equation with one unknown, namely, M_{cr}. The solution of Equation (11.63) gives the value of M_{cr} associated with the intersection point D in Figure 11.9, that is, the solution of Equation (11.63) is the critical Mach number for the NACA 0012 airfoil. Since M_{cr} appears in a complicated fashion on both sides of Equation (11.63), we solve the equation by trial-and-error by assuming different values of M_{cr}, calculating the values of both sides of Equation (11.63), and iterating until we find a value of M_{cr} that results in both the right and left sides being the same value.

M_{cr}	$\dfrac{-0.43}{\sqrt{1 - M_{\mathrm{cr}}^2}}$	$\dfrac{2}{\gamma M_{\mathrm{cr}}^2} \left[\left(\dfrac{1 + [(\gamma - 1)/2]M_{\mathrm{cr}}^2}{1 + (\gamma - 1)/2} \right)^{\gamma/\gamma - 1} - 1 \right]$
0.72	−0.6196	−0.6996
0.73	−0.6292	−0.6621
0.74	−0.6393	−0.6260
0.738	−0.6372	−0.6331
0.737	−0.6362	−0.6367
0.7371	−0.6363	−0.6363

To four-place accuracy, when $M_{\mathrm{cr}} = 0.7371$, both the left and right sides of Equation (11.63) have the same value. Therefore, the analytical solution yields

$$\boxed{M_{\mathrm{cr}} = 0.7371}$$

Note: Within the two-place accuracy of the graphical solution in part (a), both the graphical and analytical solutions give the same value of M_{cr}.

Question: How accurate is the estimate of the critical Mach number in this example? To answer this question, we examine some experimental pressure coefficient distributions for the NACA 0012 airfoil obtained at higher freestream Mach numbers. Wind tunnel measurements of the surface pressure distributions for this airfoil at zero angle of attack in a high-speed flow are shown in Figure 11.10; for Figure 11.10a, $M_\infty = 0.575$, and for Figure 11.10b, $M_\infty = 0.725$. In Figure 11.10a, the value of $C_{p,\mathrm{cr}} = -1.465$ at $M_\infty = 0.575$ is shown as the dashed horizontal line. From the definition of critical pressure coefficient, any local value of C_p above this horizontal line corresponds to locally

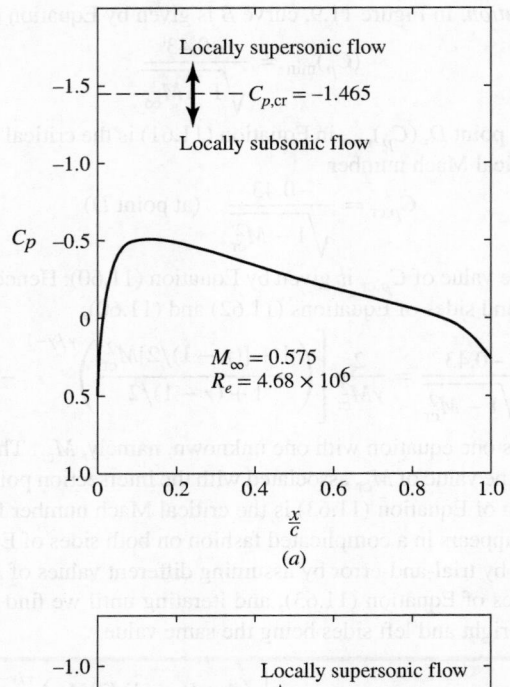

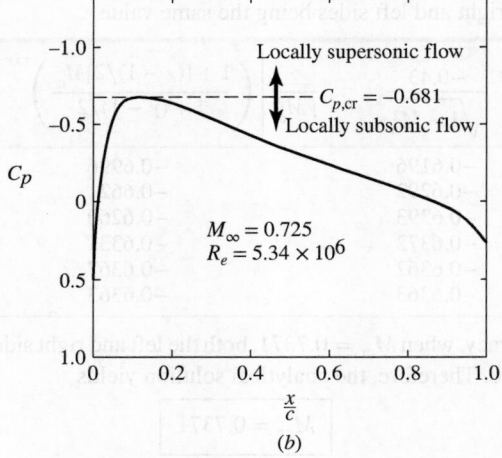

Figure 11.10 Wind tunnel measurements of surface pressure coefficient distribution for the NACA 0012 airfoil at zero angle of attack. Experimental data of Frueler and Gregorek, NASA CP 2045. (a) $M_\infty = 0.575$, (b) $M_\infty = 0.725$.

supersonic flow, and any local value below the horizontal line corresponds to locally subsonic flow. Clearly from the measured surface pressure coefficient distribution at $M_\infty = 0.575$ shown in Figure 11.10a, the flow is locally subsonic at every point on the surface. Hence, $M_\infty = 0.575$ is *below* the critical Mach number. In Figure 11.10b, which is for a higher Mach number, the value of $C_{p,cr} = -0.681$ at $M_\infty = 0.725$ is shown as the dashed horizontal line. Here, the local pressure coefficient on the airfoil is higher than $C_{p,\text{cr}}$ at every point on the surface *except* at the point of minimum pressure, where $(C_p)_{\min}$ is essentially equal to $C_{p,\text{cr}}$. This means that for $M_\infty = 0.725$, the flow is locally subsonic at every point on the surface *except* the point of minimum pressure, where the flow is essentially sonic. Hence, these experimental measurements indicate that the critical Mach number of the NACA 0012 airfoil at zero angle of attack is approximately 0.73. Comparing this experimental result with the calculated value of $M_{\text{cr}} = 0.74$ in this example, we see that our calculations are amazingly accurate, to within about one percent.

11.6.1 A Comment on the Location of Minimum Pressure (Maximum Velocity)

Examining the shape of the NACA 0012 airfoil shown at the top of Figure 11.8, note that the maximum thickness occurs at $x/c = 0.3$. However, examining the pressure coefficient distribution shown at the bottom of Figure 11.8, note that the point of minimum pressure occurs on the surface at $x/c = 0.11$, considerably ahead of the point of maximum thickness. This is a graphic illustration of the general fact that the point of minimum pressure (hence maximum velocity) does *not* correspond to the location of maximum thickness of the airfoil. Intuition might at first suggest that, at least for a symmetric airfoil at zero degrees angle of attack, the location of minimum pressure (maximum velocity) on the surface might be at the maximum thickness of the airfoil, but our intuition would be completely wrong. Nature places the maximum velocity at a point which satisfies the physics of the *whole flow field,* not just what is happening in a local region of the flow. The point of maximum velocity is dictated by the *complete* shape of the airfoil, not just by the shape in a local region.

   We also note that it is implicit in the approximate compressibility corrections discussed in Sections 11.4 and 11.5, and their use for the estimation of the critical Mach number as discussed in Section 11.6, that the point of minimum pressure remains at a fixed location on the body surface as M_∞ is increased from a very low to a high subsonic value. This is indeed approximately the case. Examine the experimental pressure distributions in Figures 11.8 and 11.10, which are for three different Mach numbers ranging from a low, incompressible value (Figure 11.8) to $M_\infty = 0.725$ (Figure 11.10b). Note that in each case the minimum pressure point is at the same approximate location, that is, at $x/c = 0.11$.

11.7 DRAG-DIVERGENCE MACH NUMBER: THE SOUND BARRIER

Imagine that we have a given airfoil at a fixed angle of attack in a wind tunnel, and we wish to measure its drag coefficient c_d as a function of M_∞. To begin with, we measure the drag coefficient at low subsonic speed to be $c_{d,0}$, shown in Figure 11.11. Now, as we gradually increase the freestream Mach number, we observe that c_d remains relatively constant all the way to the critical Mach number, as illustrated in Figure 11.11. The flow fields associated with points a, b, and c in Figure 11.11 are represented by Figure 11.5a, b, and c, respectively. As we very

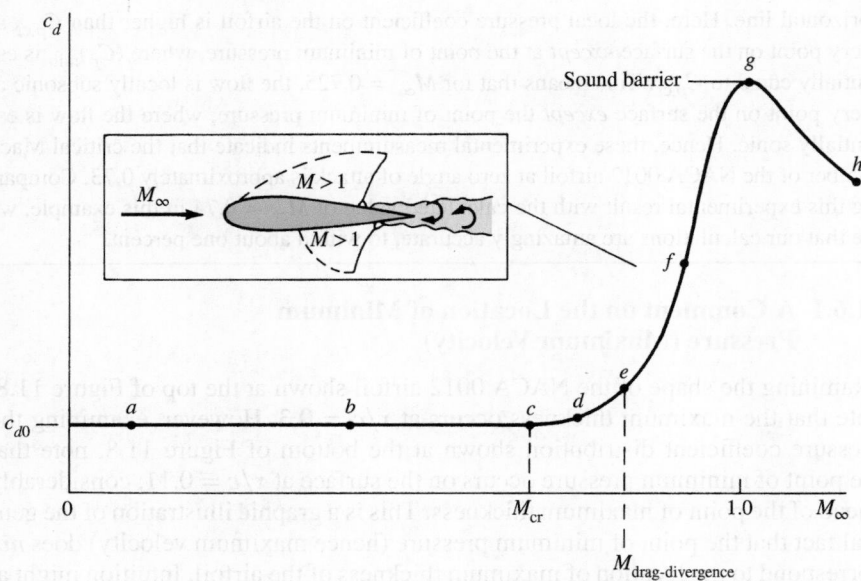

Figure 11.11 Sketch of the variation of profile drag coefficient with freestream Mach number, illustrating the critical and drag-divergence Mach numbers and showing the large drag rise near Mach 1.

carefully increase M_∞ slightly above M_{cr}, say, to point d in Figure 11.11, a finite region of supersonic flow appears on the airfoil, as shown in Figure 11.5d. The Mach number in this bubble of supersonic flow is only slightly above Mach 1, typically 1.02 to 1.05. However, as we continue to nudge M_∞ higher, we encounter a point where the drag coefficient suddenly starts to increase. This is given as point e in Figure 11.11. The value of M_∞ at which this sudden increase in drag starts is defined as the *drag-divergence Mach number*. Beyond the drag-divergence Mach number, the drag coefficient can become very large, typically increasing by a factor of 10 or more. This large increase in drag is associated with an extensive region of supersonic flow over the airfoil, terminating in a shock wave, as sketched in the insert in Figure 11.11. Corresponding to point f on the drag curve, this insert shows that as M_∞ approaches unity, the flow on both the top and bottom surfaces can be supersonic, both terminated by shock waves. For example, consider the case of a reasonably thick airfoil, designed originally for low-speed applications, when M_∞ is beyond drag-divergence; in such a case, the local Mach number can reach 1.2 or higher. As a result, the terminating shock waves can be relatively strong. These shocks generally cause severe flow separation downstream of the shocks, with an attendant large increase in drag.

Now, put yourself in the place of an aeronautical engineer in 1936. You are familiar with the Prandtl-Glauert rule, given by Equation (11.51). You recognize that as $M_\infty \to 1$, this equation shows the absolute magnitude of C_p approaching

Figure 11.12 The Bell XS-1—the first piloted airplane to fly faster than sound, October 14, 1947. (*NASA*)

infinity. This hints at some real problems near Mach 1. Furthermore, you know of some initial high-speed subsonic wind-tunnel tests that have generated drag curves that resemble the portion of Figure 11.11 from points a to f. How far will the drag coefficient increase as we get closer to $M_\infty = 1$? Will c_d go to infinity? At this stage, you might be pessimistic. You might visualize the drag increase to be so large that no airplane with the power plants existing in 1936, or even envisaged for the future, could ever overcome this "barrier." It was this type of thought that led to the popular concept of a sound barrier and that prompted many people to claim that humans would never fly faster than the speed of sound.

Of course, today we know the sound barrier was a myth. We cannot use the Prandtl-Glauert rule to argue that c_d will become infinite at $M_\infty = 1$, because the Prandtl-Glauert rule is invalid at $M_\infty = 1$ (see Sections 11.3 and 11.4). Moreover, early transonic wind-tunnel tests carried out in the late 1940s clearly indicated that c_d peaks at or around Mach 1 and then actually decreases as we enter the supersonic regime, as shown by points g and h in Figure 11.11. All we need is an aircraft with an engine powerful enough to overcome the large drag rise at Mach 1. The myth of the sound barrier was finally put to rest on October 14, 1947, when Captain Charles (Chuck) Yeager became the first human being to fly faster than sound in the sleek, bullet-shaped Bell XS-1. This rocket-propelled research aircraft is shown in Figure 11.12. Of course, today supersonic flight is a common reality; we have jet engines powerful enough to accelerate military fighters through Mach 1 flying straight up! Such airplanes can fly at Mach 3 and beyond. Indeed, we are limited only by aerodynamic heating at high speeds (and the consequent structural problems). Right now, NASA is conducting research on supersonic combustion ramjet engines for flight at Mach 5 and higher (see the Design Box at the end of Section 9.6). Keep in mind, however, that because of the large power requirements for very high-speed flight, the fuel consumption becomes large. In today's energy-conscious world, this constraint can be as much a barrier to high-speed flight as the sound barrier was once envisaged.

Since 1945, research in transonic aerodynamics has focused on reducing the large drag rise shown in Figure 11.11. Instead of living with a factor of 10 increase in drag at Mach 1, can we reduce it to a factor of 2 or 3? This is the subject of the remaining sections of this chapter.

DESIGN BOX

In order to cope with the large drag rise near Mach 1, as seen in Figure 11.11, the designers of high-speed airplanes after World War II utilized two aerodynamic design features to increase the critical Mach number, and hence the drag-divergence Mach number, for such aircraft. These two features are now classic and are discussed here.

The first design ploy was the use of *thin airfoil sections* on the airplane wing. We have already discussed that, everything else being equal, the thinner is the airfoil the higher is the critical Mach number; this is shown in Figure 11.7. This phenomenon was observed as early as 1918 by two research engineers, F. W. Caldwell and E. N. Fales, at the Army's McCook Field in Dayton, Ohio, and was solidly confirmed by various experiments during the 1920s and 1930s. More historical details are given in Section 11.11, and the detailed story of the early research in the twentieth century on compressibility effects is told in Reference 58. Indeed, the Bell X-1 (Figure 11.12) was designed with the full knowledge of the importance of thin airfoils. As a result, the X-1 was designed with two sets of wings, one with a 10 percent thick airfoil for more routine flights and another with an 8 percent thick airfoil for flights intended to penetrate through Mach 1. The airfoil sections for the two wings were NACA 65-110 and NACA 65-108, respectively. These wings were

much thinner than conventional airplanes at that time, which had wing thickness typically of 15 percent or higher. Moreover, the horizontal tail was even thinner, utilizing a 6-percent thick NACA 65-006 airfoil section. This was done to ensure that when the wing encountered major compressibility effects, the horizontal tail and elevator would still be free of such problems and would be functional for stability and control. However, hedging their bets, the Bell engineers also made the tail all-moving, that is, the incidence angle of the complete horizontal tail surface could be changed in flight in case elevator effectiveness was lost, and to help trim the airplane as it flew through Mach 1. Chuck Yeager made good use of this all-moving tail feature during his history-making flights in the X-1.

The use of thin airfoils on high-speed aircraft is almost a standard design practice today. The design trend to thin airfoils is illustrated in Figure 11.13, which shows the variation of airfoil thickness for different airplanes as a function of their design Mach number. Clearly, as M_∞ increases, the trend is toward thinner wings.

The second design ploy was the use of swept wings. The story surrounding the history of the swept wing concept is told in Reference 58; the invention of the concept is shared by two people, the German engineer Adolf Busemann in 1935 and the American aerodynamicist R. T. Jones who independently conceived

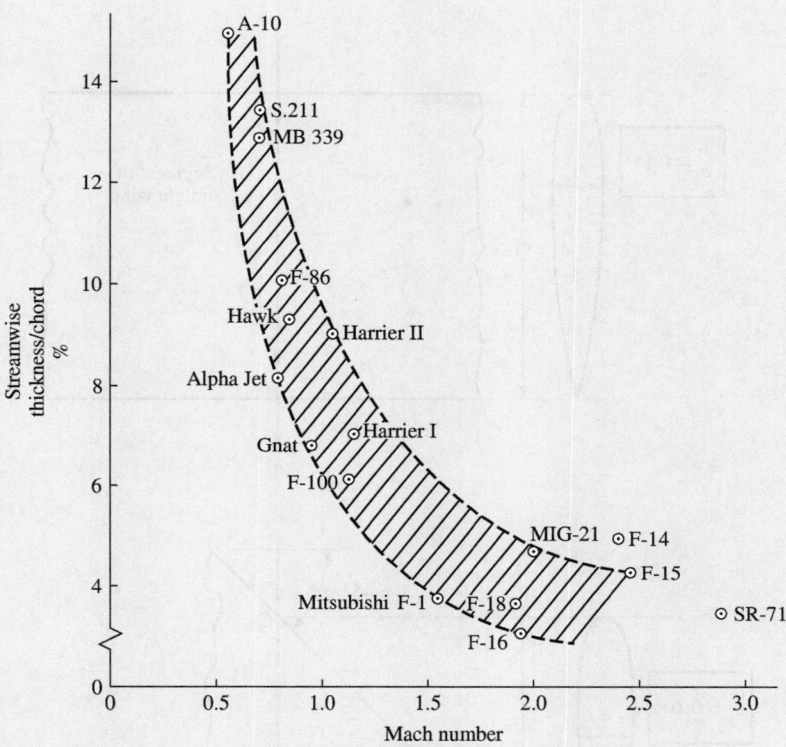

Figure 11.13 Variation of thickness-to-chord ratio with Mach number for a representative sampling of different airplanes. (*Data Source:* Ray Whitford, *Design for Air Combat,* Jane's Information Group, Surry, England, 1989.)

the idea in 1945. There are several ways of explaining how a swept wing works (e.g., see Reference 2). However, for our purposes here we will explain the benefit of a swept wing using the ideas set forth about thin airfoils, as discussed above. Consider a straight wing, a portion of which is sketched in Figure 11.14a. We define a straight wing as one for which the mid-chord line is perpendicular to the freestream; this is certainly the case for the rectangular planform shown in Figure 11.14a. Assume the straight wing has an airfoil section with a thickness-to-chord ratio of 0.15 as shown at the left of Figure 11.14a. Streamline AB flowing over this wing sees the airfoil with $t_1/c_1 = 0.15$. Now consider the same wing swept back through the angle $\Lambda = 45°$, as shown

in Figure 11.14b. Streamline CD that flows over this wing (ignoring any three-dimensional curvature effects) sees an effective airfoil shape with the same thickness as before ($t_2 = t_1$), but the effective chord length c_2 is longer by a factor of 1.41 (i.e., $c_2 = 1.41c_1$). This makes the effective thickness-to-chord ratio seen by streamline CD equal to $t_2/c_2 = 0.106$—thinner by almost one-third compared to the straight-wing case. Hence, by sweeping the wing, the flow behaves as if the airfoil section is thinner, with a consequent increase in the critical Mach number of the wing. Everything else being equal, a swept wing has a larger critical Mach number, hence a larger drag-divergence Mach number than a straightwing. For this reason, most high-speed airplanes designed since the

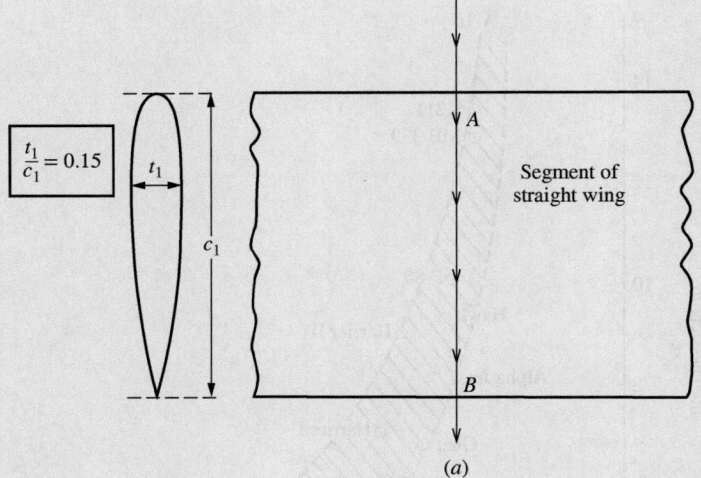

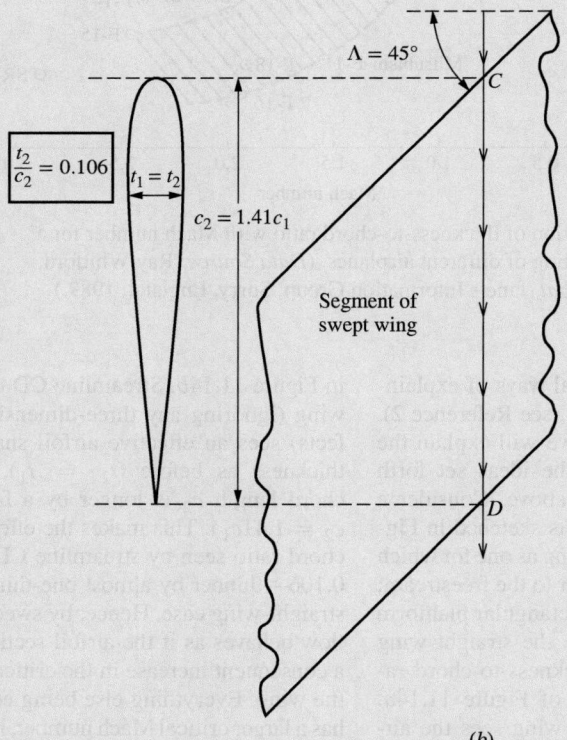

Figure 11.14 By sweeping the wing, a streamline effectively sees a thinner airfoil.

middle 1940s have swept wings. (The only reason why the Bell X-1, shown in Figure 11.12, had straight wings is because its design commenced in 1944 before any knowledge or data about swept wings was available in the United States. Later, when such swept-wing data flooded into the United States from Germany in mid-1945, the Bell designers were conservative, and stuck with the straight wing.) A wonderful example of an early swept-wing fighter is the North American F-86 of Korean War vintage, shown in Figure 11.15.

Finally, we note some compressibility corrections that apply to Equations (5.69), (5.81), and (5.82) given earlier for the estimation of the lift slope for high aspect ratio straight wings, low aspect ratio straight wings, and swept wings respectively. These equations apply to low-speed, incompressible flow. The Prandtl-Glauert rule for an airfoil section expressed in terms of pressure coefficient as given by Equation (11.51) also holds for the lift slope for the airfoil. Letting a_0 be the lift slope for an infinite wing for incompressible flow, and $a_{0,\text{comp}}$ be the lift slope for an infinite wing in a subsonic compressible flow, from the Prandtl-Glauert rule we have

$$a_{0,\text{comp}} = \frac{a_0}{\sqrt{1 - M_\infty^2}} \qquad (11.64)$$

We assume that Equation (5.70) relating the lift slope of a finite wing to that for an infinite wing, as obtained from Prandtl's lifting line theory, holds for subsonic compressible flow as well. Letting the term $(1 + \tau)^{-1}$ in Equation (5.70) be replaced by a span efficiency factor for lift slope denoted by e_1, where $e_1 = (1 + \tau)^{-1}$, and denoting the compressible lift slope for the finite wing by a_{comp}, the compressible analog of Equation (5.70) is

$$a_{\text{comp}} = \frac{a_{0,\text{comp}}}{1 + a_{0,\text{comp}}/(\pi e_1 \text{AR})} \qquad (11.65)$$

Substituting Equation (11.64) into Equation (11.65), and simplifying, we obtain

$$a_{\text{comp}} = \frac{a_0}{\sqrt{1 - M_\infty^2} + a_0/(\pi e_1 \text{AR})} \qquad (11.66)$$

Equation (11.66) is an equation for estimating the lift slope for a *high-aspect-ratio straight wing* in a compressible flow with a subsonic M_∞ from the known lift slope for an infinite wing in an incompressible flow a_0.

Helmbold's equation for low-aspect-ratio straight wings for incompressible flow, Equation (5.81), can be modified for compressibility effects by replacing a_0 in that equation by $a_0/\sqrt{1 - M_\infty^2}$, yielding

$$a_{\text{comp}} = \frac{a_0}{\sqrt{1 - M_\infty^2 + [a_0/(\pi e_1 \text{AR})]^2} + a_0/(\pi \text{AR})} \qquad (11.67)$$

Equation (11.67) is an equation for estimating the lift slope for a *low-aspect-ratio straight wing* in a compressible flow with a subsonic M_∞ from the known lift slope for an infinite wing in an incompressible flow a_0.

Finally, Equation (5.82) for a swept wing can be modified for compressibility effects by replacing a_0 in that equation by $a_0/\sqrt{1 - (M_{\infty,n})^2}$ where $M_{\infty,n}$ is the component of the freestream Mach number perpendicular to the half-chord line of the swept wing. If the half-chord line is swept by the angle Λ, then $M_{\infty,n} = M_\infty \cos \Lambda$. Hence, replace a_0 in Equation (5.82) with $a_0/\sqrt{1 - M_\infty^2 \cos^2 \Lambda}$. The result is in Equation (11.68) below. Equation (11.68) is an equation for estimating the lift slope for a *swept wing* in a compressible flow with a subsonic M_∞ from the known lift slope for an infinite wing in an incompressible flow a_0.

$$a_{\text{comp}} = \frac{a_0 \cos \Lambda}{\sqrt{1 - M_\infty^2 \cos^2 \Lambda + [(a_0 \cos \Lambda)/(\pi \text{AR})]^2} + (a_0 \cos \Lambda)/(\pi \text{AR})} \qquad (11.68)$$

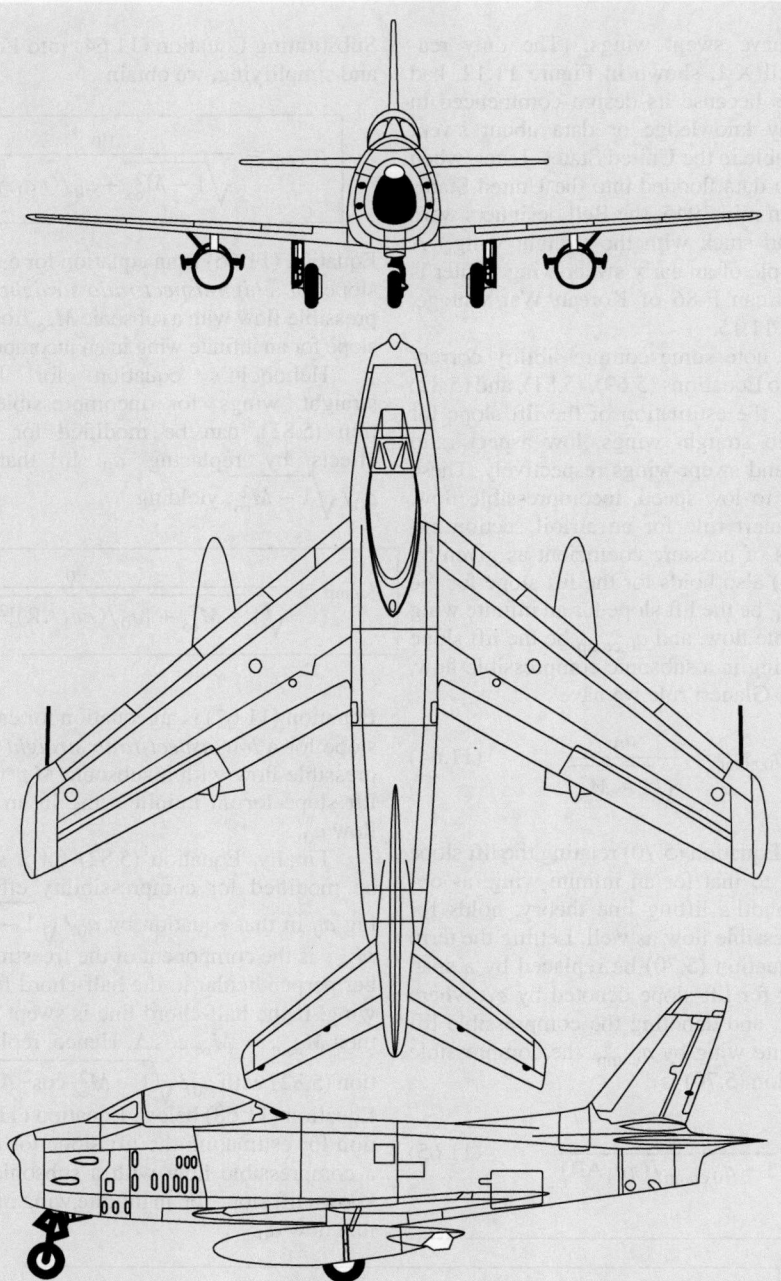

Figure 11.15 A typical example of a swept-wing aircraft. The North American F-86 Sabre of Korean War fame.

11.8 THE AREA RULE

In addition to the classic approaches of using thin airfoils and swept wings to cope with the large drag rise near Mach 1, in recent years two rather revolutionary concepts have helped greatly to break down the "sound barrier" near and beyond the speed of sound. One of these—the area rule—is discussed in this section; the other—the supercritical airfoil—is the subject of Section 11.9.

For a moment, let us expand our discussion from two-dimensional airfoils to a consideration of a complete airplane. In this section, we introduce a design concept that has effectively reduced the drag rise near Mach 1 for a complete airplane.

As stated before, the first practical jet-powered aircraft appeared at the end of World War II in the form of the German Me 262. This was a subsonic fighter plane with a top speed near 550 mi/h. The next decade saw the design and production of many types of jet aircraft—all limited to subsonic flight by the large drag near Mach 1. Even the "century" series of fighter aircraft designed to give the U.S. Air Force supersonic capability in the early 1950s, such as the Convair F-102 delta-wing airplane, ran into difficulty and could not at first readily penetrate the sound barrier in level flight. The thrust of jet engines at that time simply could not overcome the large peak drag near Mach 1.

A planview, cross section, and area distribution (cross-sectional area versus distance along the axis of the airplane) for a typical airplane of that decade are sketched in Figure 11.16. Let A denote the total cross-sectional area at any given station. Note that the cross-sectional area distribution experiences some abrupt

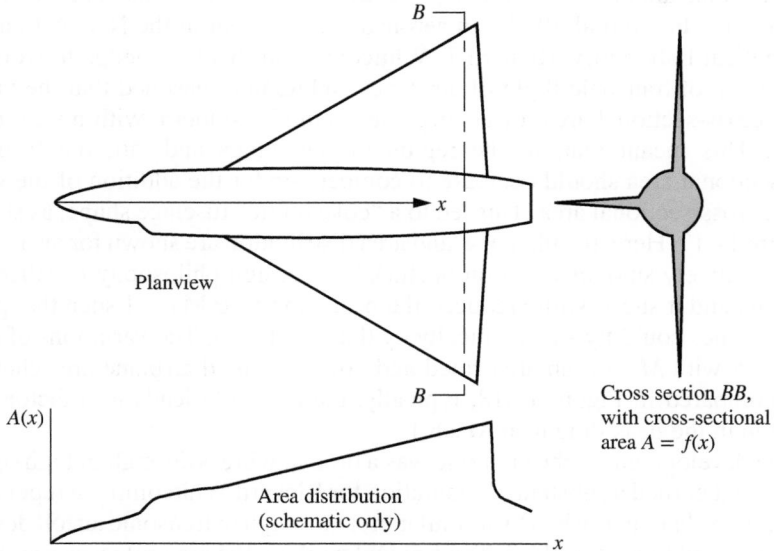

Planview

Cross section *BB*, with cross-sectional area $A = f(x)$

$A(x)$

Area distribution (schematic only)

Figure 11.16 A schematic of a non-area-ruled aircraft.

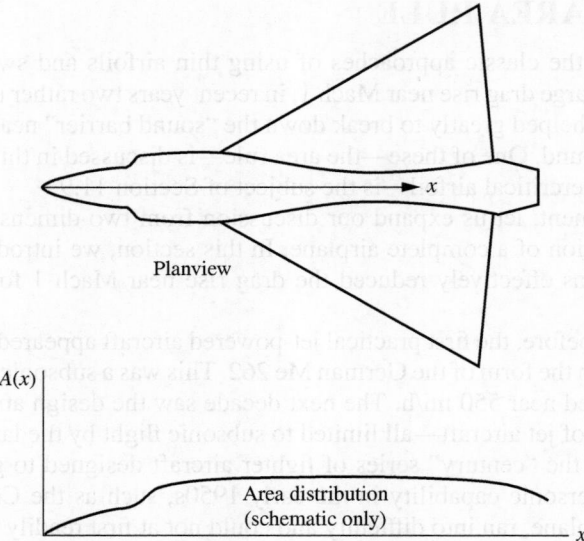

Planview

$A(x)$

Area distribution
(schematic only)

x

Figure 11.17 A schematic of an area-ruled aircraft.

changes along the axis, with discontinuities in both A and dA/dx in the regions
of the wing.

In contrast, for almost a century, it was well known by ballisticians that
the speed of a supersonic bullet or artillery shell with a *smooth* variation of
cross-sectional area was higher than projectiles with abrupt or discontinuous area
distributions. In the mid-1950s, an aeronautical engineer at the NACA Langley
Aeronautical Laboratory, Richard T. Whitcomb, put this knowledge to work on
the problem of transonic flight of airplanes. Whitcomb reasoned that the variation of cross-sectional area for an airplane should be smooth, with no discontinuities. This meant that, in the region of the wings and tail, the fuselage
cross-sectional area should decrease to compensate for the addition of the wing
and tail cross-sectional area. This led to a "coke bottle" fuselage shape, as shown
in Figure 11.17. Here, the planview and area distribution are shown for an aircraft
with a relatively smooth variation of $A(x)$. This design philosophy is called the
area rule, and it successfully reduced the peak drag near Mach 1 such that practical airplanes could fly supersonically by the mid-1950s. The variations of drag
coefficient with M_∞ for an area-ruled and non-area-ruled airplane are schematically compared in Figure 11.18; typically, the area rule leads to a factor-of-2
reduction in the peak drag near Mach 1.

The development of the area rule was a dramatic breakthrough in high-speed
flight, and it earned a substantial reputation for Richard Whitcomb—a reputation
that was to be later garnished by a similar breakthrough in transonic airfoil design,
to be discussed in Section 11.9. The original work on the area rule was presented
by Whitcomb in Reference 30, which should be consulted for more details.

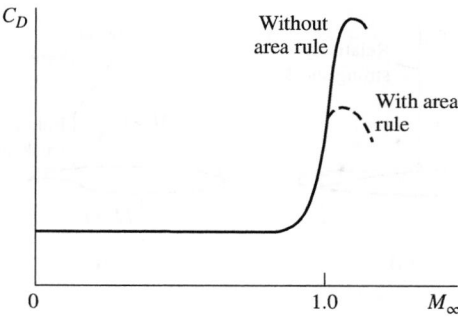

Figure 11.18 The drag-rise properties of area-ruled and non-area-ruled aircraft (schematic only).

11.9 THE SUPERCRITICAL AIRFOIL

Let us return to a consideration of two-dimensional airfoils. A natural conclusion from the material in Section 11.6, and especially from Figure 11.11, is that an airfoil with a high critical Mach number is very desirable, indeed necessary, for high-speed subsonic aircraft. If we can increase M_{cr}, then we can increase $M_{\text{drag-divergence}}$, which follows closely after M_{cr}. This was the philosophy employed in aircraft design from 1945 to approximately 1965. Almost by accident, the NACA 64-series airfoils (see Section 4.2), although originally designed to encourage laminar flow, turned out to have relative high values of M_{cr} in comparison with other NACA shapes. Hence, the NACA 64 series has seen wide application on high-speed airplanes. Also, we know that thinner airfoils have higher values of M_{cr} (see Figure 11.7); hence, aircraft designers have used relatively thin airfoils on high-speed airplanes.

However, there is a limit to how thin a practical airfoil can be. For example, considerations other than aerodynamic influence the airfoil thickness; the airfoil requires a certain thickness for structural strength, and there must be room for the storage of fuel. This prompts the following question: For an airfoil of given thickness, how can we delay the large drag rise to higher Mach numbers? To increase M_{cr} is one obvious tack, as described above, but there is another approach. Rather than increasing M_{cr}, let us strive to increase the Mach number *increment* between M_{cr} and $M_{\text{drag-divergence}}$. That is, referring to Figure 11.11, let us increase the distance between points *e* and *c*. This philosophy has been pursued since 1965, leading to the design of a new family of airfoils called *supercritical airfoils*, which are the subject of this section.

The purpose of a supercritical airfoil is to increase the value of $M_{\text{drag-divergence}}$, although M_{cr} may change very little. The shape of a supercritical airfoil is compared with an NACA 64-series airfoil in Figure 11.19. Here, an NACA 64_2-A215 airfoil is sketched in Figure 11.19a, and a 13-percent thick supercritical airfoil is shown in Figure 11.19c. (Note the similarity between the supercritical profile

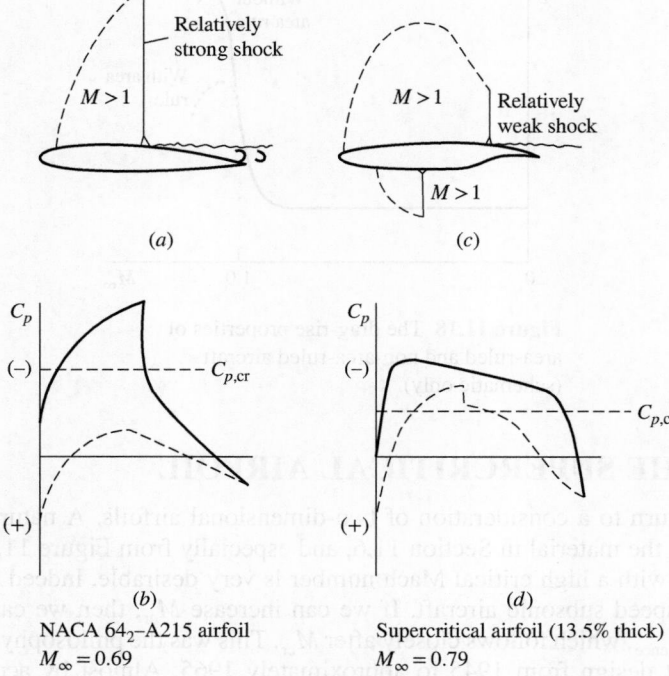

Figure 11.19 Standard NACA 64-series airfoil compared with a supercritical airfoil at cruise lift conditions. (*Source:* Whitcomb, R. T., and L. R. Clark: *An Airfoil Shape for Efficient Flight at Supercritical Mach Numbers*, NASA TMX-1109, July 1965.)

and the modern low-speed airfoils discussed in Section 4.11.) The supercritical airfoil has a relatively flat top, thus encouraging a region of supersonic flow with lower local values of M than the NACA 64 series. In turn, the terminating shock is weaker, thus creating less drag. Similar trends can be seen by comparing the C_p distributions for the NACA 64 series (Figure 11.19b) and the supercritical airfoil (Figure 11.19d). Indeed, Figure 11.19a and b for the NACA 64-series airfoil pertain to a lower freestream Mach number, $M_\infty = 0.69$, than Figure 11.19c and d, which pertain to the supercritical airfoil at a higher freestream Mach number, $M_\infty = 0.79$. In spite of the fact that the 64-series airfoil is at a lower M_∞, the extent of the supersonic flow reaches farther above the airfoil, the local supersonic Mach numbers are higher, and the terminating shock wave is stronger. Clearly, the supercritical airfoil shows more desirable flow field characteristics; namely, the extent of the supersonic flow is closer to the surface, the local supersonic Mach numbers are lower, and the terminating shock wave is weaker. As a result, the value of $M_{\text{drag-divergence}}$ will be higher for the supercritical airfoil. This is verified by the experimental data given in Figure 11.20, taken from Reference 31. Here, the value of $M_{\text{drag-divergence}}$ is 0.79 for the supercritical airfoil in comparison with 0.67 for the NACA 64 series.

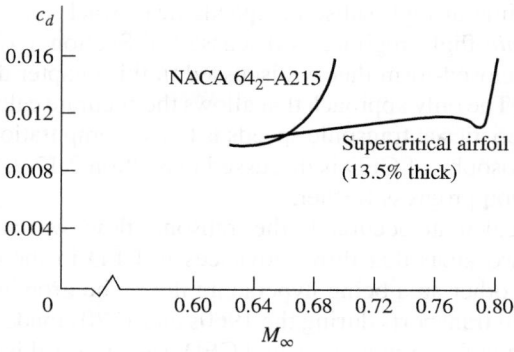

Figure 11.20 The drag-divergence properties of a standard NACA 64-series airfoil and a supercritical airfoil.

Because the top of the supercritical airfoil is relatively flat, the forward 60 percent of the airfoil has negative camber, which lowers the lift. To compensate, the lift is increased by having extreme positive camber on the rearward 30 percent of the airfoil. This is the reason for the cusplike shape of the bottom surface near the trailing edge.

The supercritical airfoil was developed by Richard Whitcomb in 1965 at the NASA Langley Research Center. A detailed description of the rationale and some early experimental data for supercritical airfoils are given by Whitcomb in Reference 31, which should be consulted for more details. The supercritical airfoil and many variations of such are now used by the aircraft industry on modern high-speed airplane designs. Examples are the Boeing 757 and 767, and the latest model Lear jets. The supercritical airfoil is one of two major breakthroughs made in transonic airplane aerodynamics since 1945, the other being the area rule discussed in Section 11.8. It is a testimonial to the person that Richard Whitcomb was mainly responsible for both.

11.10 CFD APPLICATIONS: TRANSONIC AIRFOILS AND WINGS

The analysis of subsonic compressible flow over airfoils discussed in this chapter, resulting in classic compressibility corrections such as the Prandtl-Glauert rule (Section 11.4), fits into the category of "closed-form" theory as discussed in Section 2.17.1. Although this theory is elegant and useful, it is restricted to:

1. Thin airfoils at small angles of attack
2. Subsonic numbers that do not approach too close to one, that is, Mach numbers typically below 0.7
3. Inviscid, irrotational flow

However, modern subsonic transports (Boeing 747, 777, etc.) cruise at freestream Mach numbers on the order of 0.85, and high-performance military combat airplanes spend time at high subsonic speeds near Mach one. These airplanes are in the *transonic* flight regime, as discussed in Section 1.10.4 and noted in Figure 1.44. The closed-form theory discussed in this chapter does not apply in this flight regime. The only approach that allows the accurate calculation of airfoil and wing characteristics at transonic speeds is to use computational fluid dynamics; the basic philosophy of CFD is discussed in Section 2.17.2, which should be reviewed before you progress further.

The need to calculate accurately the transonic flow over airfoils and wings was one of the two areas that drove advances in CFD in the early days of its development, the other area being hypersonic flow. The growing importance of high-speed jet civil transports during the 1960s and 1970s made the accurate calculation of transonic flow imperative, and CFD was (and still is) the only way of making such calculations. In this section we will give only the flavor of such calculations; see Chapter 14 of Reference 21 for more details, as well as the modern aerodynamic literature for the latest developments.

Beginning in the 1960s, transonic CFD calculations historically evolved through four distinct steps, as follows:

1. The earliest calculations numerically solved the nonlinear small-perturbation potential equation for transonic flow, obtained from Equation (11.16) by dropping all terms on the right-hand side except for the leading term, which is not small near $M_\infty = 1$. This yields

$$\left(1 - M_\infty^2\right)\frac{\partial \hat{u}}{\partial x} + \frac{\partial \hat{v}}{\partial y} = M_\infty^2 \left[(\gamma + 1)\frac{\hat{u}}{V_\infty}\right]\frac{\partial \hat{u}}{\partial x}$$

   which in terms of the perturbation velocity potential is

$$\left(1 - M_\infty^2\right)\frac{\partial^2 \hat{\phi}}{\partial x^2} + \frac{\partial^2 \hat{\phi}}{\partial y^2} = M_\infty^2 \left[(\gamma + 1)\frac{\partial \hat{\phi}}{\partial x}\frac{1}{V_\infty}\right]\frac{\partial^2 \hat{\phi}}{\partial x^2} \qquad (11.69)$$

   Equation (11.69) is the transonic small perturbation potential equation; it is non-linear due to the term on the right-hand side, which involves a product of derivatives of the dependent variable $\hat{\phi}$. This necessitated a numerical CFD solution. However, the results were limited to the assumptions embodied in this equation, namely, small perturbations and hence thin airfoils at small angles of attack.

2. The next step was numerical solutions of the full potential equation, Equation (11.12). This allowed applications to airfoils of any shape at any angle of attack. However, the flow was still assumed to be isentropic, and even though shock waves appeared in the results, the properties of these shocks were not always accurately predicted.

3. As CFD algorithms became more sophisticated, numerical solutions of the Euler equations [the full continuity, momentum, and energy equations for inviscid flow, such as Equations (7.40), (7.42), and (7.44)] were obtained. The advantage of these Euler solutions was that shock waves were properly treated. However, none of the approaches discussed in steps 1–3 accounted

for the effects of viscous flow, the importance of which in transonic flows soon became more appreciated because of the interaction of the shock wave with the boundary layer. This interaction, with the attendant flow separation, is dominant in the prediction of drag.

4. This led to the CFD solution of the viscous flow equations [the Navier-Stokes equations, such as Equations (2.43), (2.61), and (2.87) with the viscous terms included] for transonic flow. The Navier-Stokes equations are developed in detail in Chapter 15. Such CFD solutions of the Navier-Stokes equations are currently the state of the art in transonic flow calculations. These solutions contain all of the realistic physics of such flows, with the exception that some type of turbulence model must be included to deal with turbulent boundary layers, and such turbulent models are frequently the Achilles heel of these calculations.

An example of a CFD calculation for the transonic flow over an NACA 0012 airfoil at 2° angle of attack with $M_\infty = 0.8$ is shown in Figure 11.21. The contour lines shown here are lines of constant Mach number, and the bunching of these lines together clearly shows the nearly normal shock wave occurring on the top surface. In reference to our calculation in Example 11.3 showing that the critical Mach number for the NACA 0012 airfoil at zero angle of attack is 0.74, and the experimental confirmation of this shown in Figure 11.10b, clearly the flow over the same airfoil shown in Figure 11.21 is well beyond the critical Mach number. Indeed, the boundary layer downstream of the shock wave in Figure 11.21 is separated, and the airfoil is squarely in the drag-divergence region. The CFD calculations predict this separated flow because a version of the Navier-Stokes equations (called the thin shear layer approximation) is being numerically solved, taking into account the viscous flow effects. The results shown in Figure 11.21 are from the work of Nakahashi and Deiwert at the NASA Ames Research Center (Reference 70); these results are a graphic illustration of the power of CFD applied to transonic flow. For details on these types of CFD calculations, see the definitive books by Hirsch (Reference 71).

Today, CFD is an integral part of modern transonic airfoil and wing design. A recent example of how CFD is combined with modern optimization design techniques for the design of complete wings for transonic aircraft is shown in Figures 11.22 and 11.23, taken from the survey paper by Jameson (Reference 72). On the left side of Figure 11.22a the airfoil shape distribution along the semispan of a baseline, initial wing shape at $M_\infty = 0.83$ is given, with the computed pressure coefficient distributions shown on the right. The abrupt drop in C_p in these distributions is due to a relatively strong shock wave along the wing. After repeated iterations, the optimized design at the same $M_\infty = 0.83$ is shown in Figure 11.22b. Again, the new airfoil shape distribution is shown on the left, and the C_p distribution is given on the right. The new, optimized wing design shown in Figure 11.22b is virtually shock free, as indicated by the smooth C_p distributions, with a consequent reduction in drag of 7.6 percent. The optimization shown in Figure 11.22 was subject to the constraint of keeping the wing thickness the same.

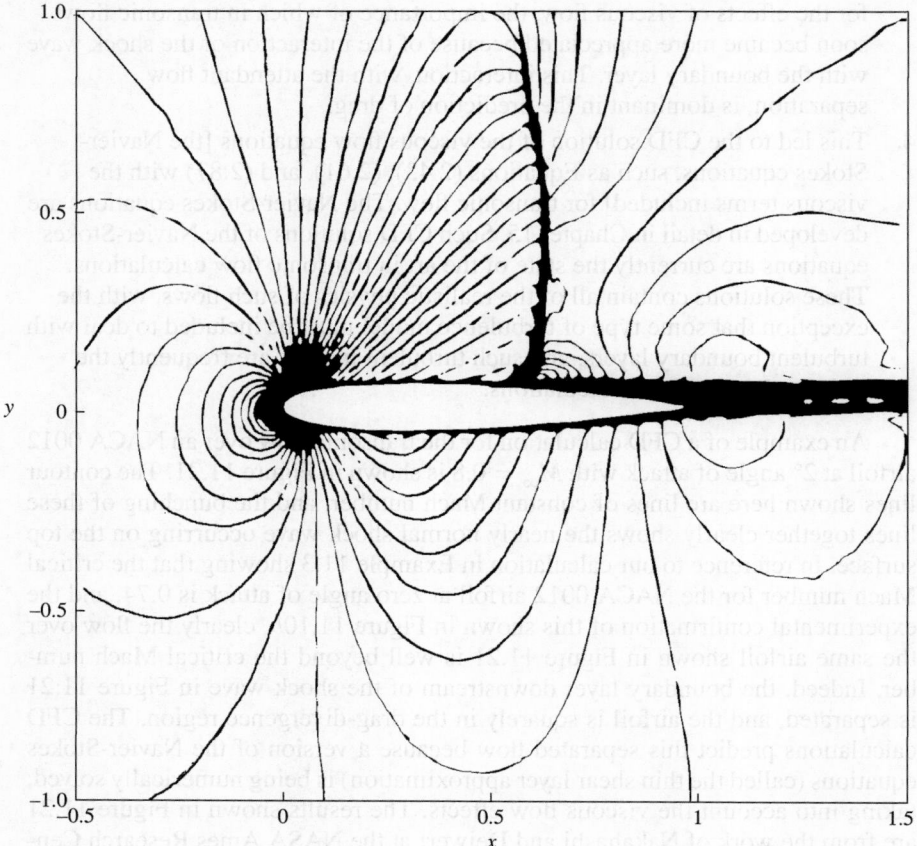

Figure 11.21 Mach number contours in the transonic flow over an NACA 0012 airfoil at $M_\infty = 0.8$ and at 2° angle of attack. (*Source:* Nakahasi, K., and Deiwert, G. S.: "A Self-Adaptive Grid Method with Application to Airfoil Flow," *AIAA Paper 85-1525,* American Institute of Aeronautics and Astronautics, 1985.)

Another but similar case of wing design optimization is shown in Figure 11.23. Here, the final optimized wing planform shape is shown for $M_\infty = 0.86$, with the final computed pressure contour lines shown on the planform. Straddling the wing planform on both the left and right of Figure 11.23 are the pressure coefficient plots at six spanwise stations. The dashed curves show the C_p variations for the initial baseline wing, with the tell-tale oscillations indicating a shock wave, whereas the solid curves are the final C_p variations for the optimized wing, showing smoother variations that are almost shock-free. The results shown in Figures 11.22 and 11.23 are reflective of multidisciplinary design optimization using CFD for transonic wings. For more details on this and other design applications, see the special issue of the *Journal of Aircraft,* vol. 36, no. 1, Jan./Feb. 1999, devoted to aspects of multidisciplinary design optimization.

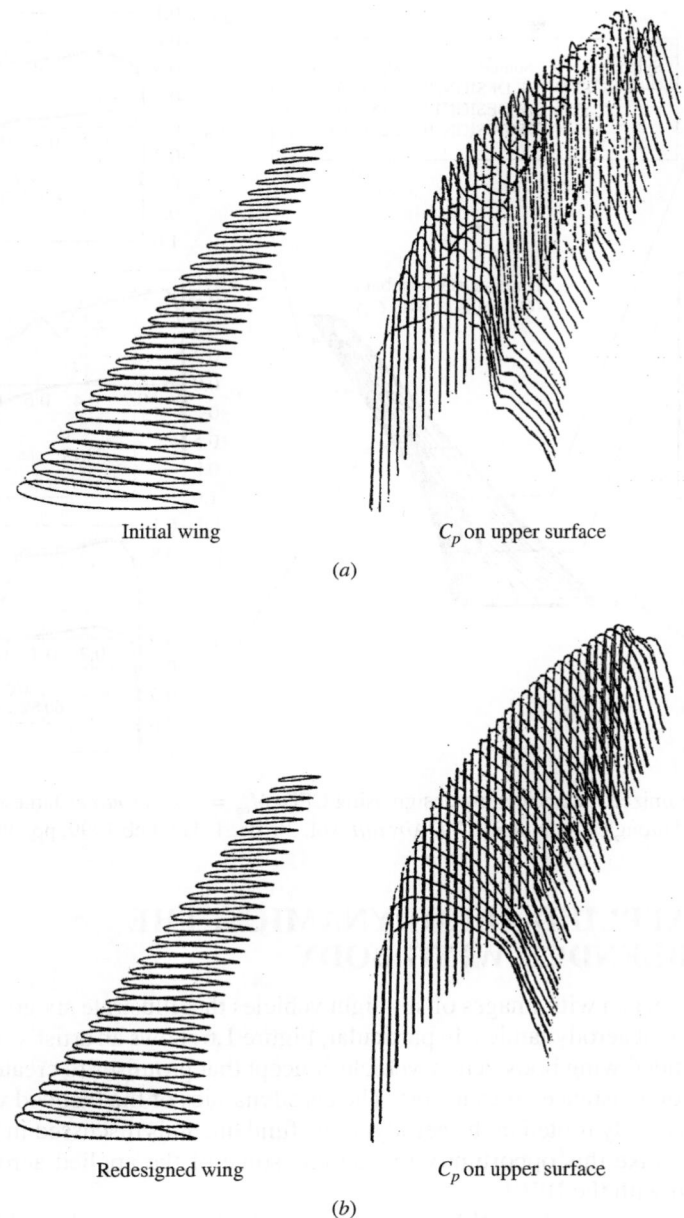

Initial wing C_p on upper surface

(a)

Redesigned wing C_p on upper surface

(b)

Figure 11.22 The use of CFD for optimized transonic wing design. $M_\infty = 0.83$. (*a*) Baseline wing with a shock wave. (*b*) Optimized wing, virtually shock free. (*Source:* Jameson, Antony: "Re-Engineering the Design Process Through Computation," *J. Aircraft*, vol. 36, no. 1, Jan-Feb 1999, pp. 36–50.)

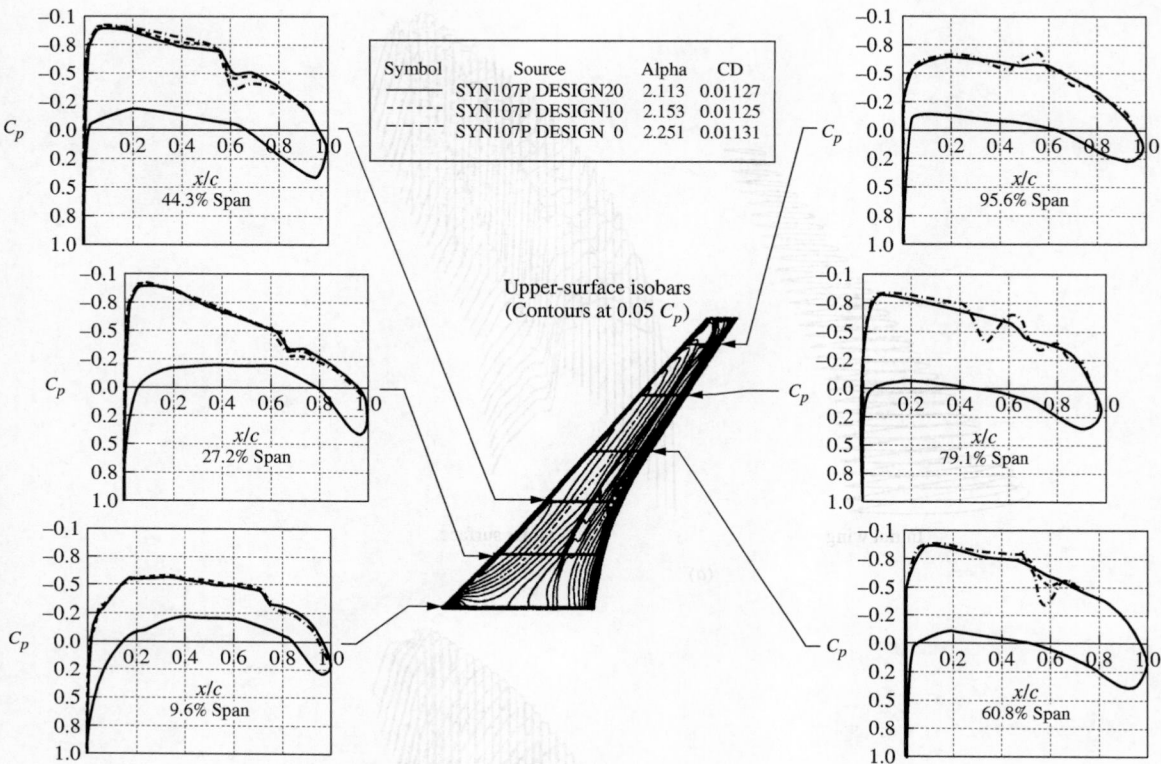

Symbol	Source	Alpha	CD
——	SYN107P DESIGN20	2.113	0.01127
- - - -	SYN107P DESIGN10	2.153	0.01125
—·—·—	SYN107P DESIGN 0	2.251	0.01131

Upper-surface isobars
(Contours at 0.05 C_p)

Figure 11.23 Another example of optimized transonic wing design using CFD. $M_\infty = 0.86$. (*Source:* Jameson, Antony: "Re-Engineering the Design Process Through Computation," *J. Aircraft*, vol. 36. no. 1, Jan–Feb 1999, pp. 36–50.)

11.11 APPLIED AERODYNAMICS: THE BLENDED WING BODY

This book began with images of six flight vehicles that illustrate six good reasons to learn about aerodynamics. In particular, Figure 1.6 shows an artist's conception of the blended wing body, a new vehicle concept that promises to create a renaissance in long distance air transport. The aerodynamics of the blended wing body (BWB) is deeply rooted in the aerodynamic fundamentals discussed in this book, and so we take this opportunity to examine some of the applied aerodynamics associated with the BWB.

As background, the BWB was the outgrowth of a challenge issued by Dennis Bushnell to the aircraft industry in 1988. Bushnell, the chief scientist of the NASA Langley Research Center, asked if new, innovative thinking could result in a commercial jet transport that would provide a quantum leap in efficiency and performance in comparison to the standard tube-fuselage, swept wing airplane with jet engines pod-mounted under the wings, such as pioneered by the historic Boeing 707 (Figure 1.2). After 50 years, this configuration remains the same for virtually all transport aircraft, as reinforced by Boeing's latest design, the 787 Dreamliner shown in Figure 2.2. Responding to this challenge, a small group of aerodynamicists at McDonnell Douglas led by Dr. Robert Liebeck conceived the blended wing body, one version of which is shown in Figure 11.24.

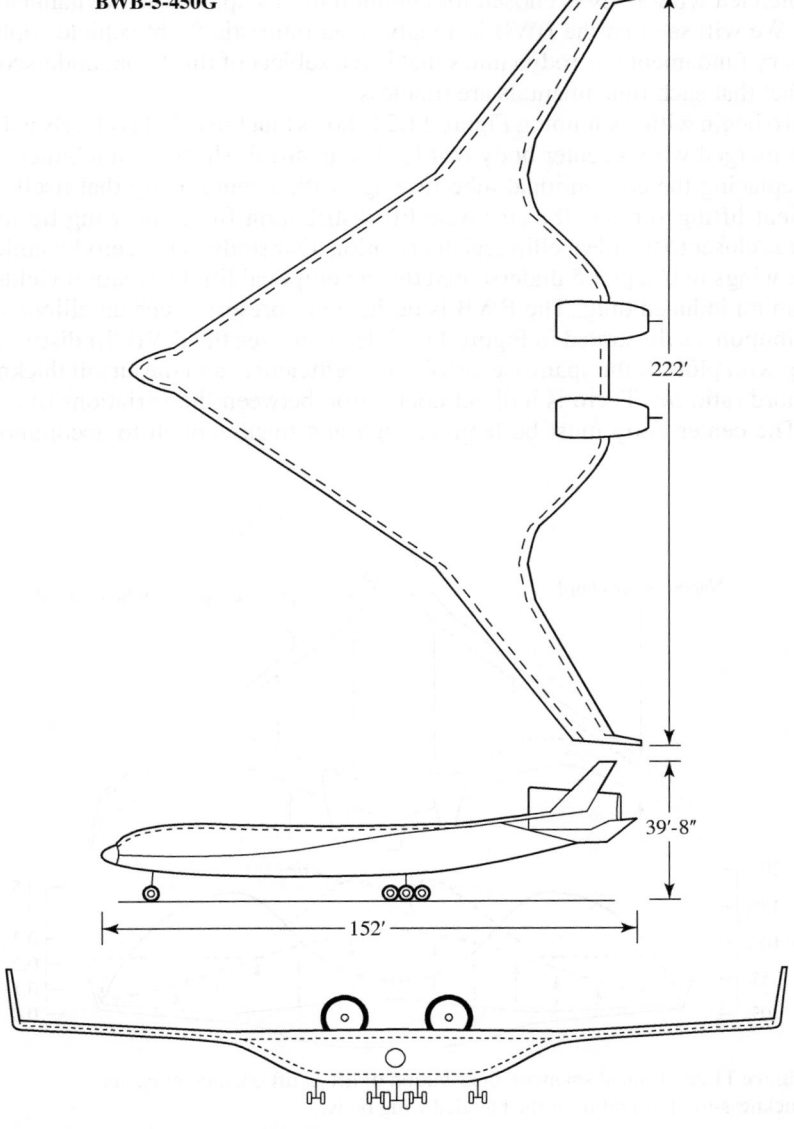

Figure 11.24 Three-view of a blended wing body.

When McDonnell Douglas was absorbed by Boeing, Liebeck continued his work on the BWB funded by NASA. Figure 11.24 shows a configuration obtained from a baseline study of the BWB carried out by Boeing circa 2002. Now a Boeing senior technical fellow, Liebeck continues to spearhead the BWB concept at Boeing.

The aerodynamics of the blended wing body is a graphic illustration of the application of many of the fundamentals highlighted in this book. For this reason, the blended wing body is chosen for attention in this applied aerodynamics section. We will see that the BWB is an advanced futuristic flight vehicle applying the very fundamental aerodynamics that is the subject of this book, underscoring the fact that such fundamentals are timeless.

To begin with, examining Figure 11.24 shows that the BWB is clearly a flying wing merged with a center body that is also an airfoil shape with a bullet nose. By replacing the conventional tube fuselage with a center body that itself is an efficient lifting surface, the spanwise lift distribution from one wing tip to the other is closer to the ideal elliptical distribution. Our study of the aerodynamics of finite wings in Chapter 5 underscored that an elliptical lift distribution yields the minimum induced drag. The BWB is designed to preserve such an elliptical lift distribution, as illustrated in Figure 11.25. Here, we see the BWB lift distribution along with plots of the spanwise airfoil lift coefficient c_l and the airfoil thickness-to-chord ratio t/c. There is a direct connection between the variations of c_l and t/c. The center body must be large enough and thick enough to accommodate

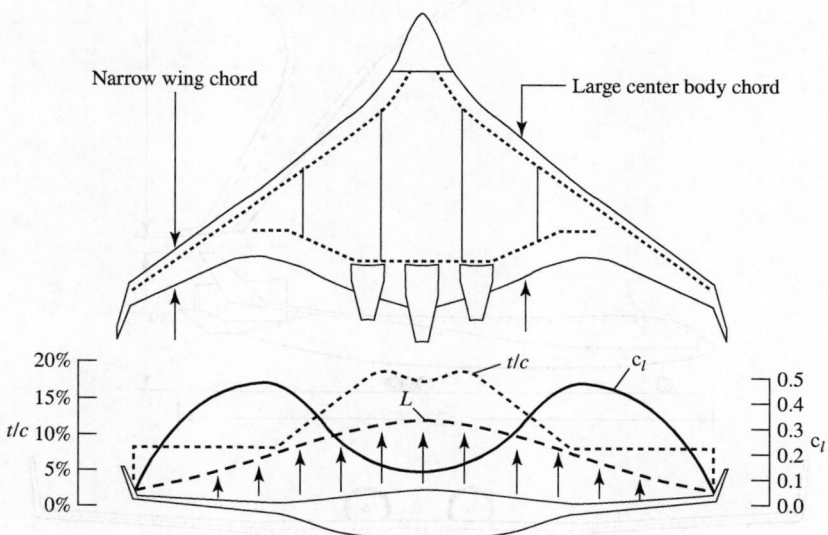

Figure 11.25 Typical spanwise distribution of lift L, lift coefficient c_l, and thickness-to-chord ratio for the blended wing body.

the passenger and cargo load, which drives the increase of both the airfoil chord length and t/c for the center body section. In order to preserve the elliptical lift distribution over the center body, the airfoil section of the center body is different from that used for the outer wing panels. It is chosen to have a lower lift coefficient to counterbalance the larger chord length and thus preserve the smooth spanwise elliptical lift distribution. (Recall that the lift distribution is the variation of the lift *force* per unit span, and this lift force is proportional to both the local value of c_l and the chord length.)

The BWB is a high-speed subsonic airplane intended to fly at the lower end of the transonic flight regime. Hence, major efforts are made to obtain as high a drag-divergence Mach number as possible. Toward that end, the BWB incorporates two design features, both of which deal with aerodynamic fundamentals discussed in this chapter.

1. *Supercritical airfoils.* The function of a supercritical airfoil is discussed in Section 11.9. The outer portions of the BWB wing incorporate a modern supercritical airfoil section with aft camber, similar to that shown in Figure 11.19c. The center body profile is also an airfoil section. In the first generation of the BWB development, the airfoil shape chosen was a Liebeck LW102A airfoil (Reference 89) point designed for $c_l = 0.25$ at a Mach number of 0.7. A side view of the resulting center body profile is shown in Figure 11.26a. The new-generation BWB uses an advanced customized

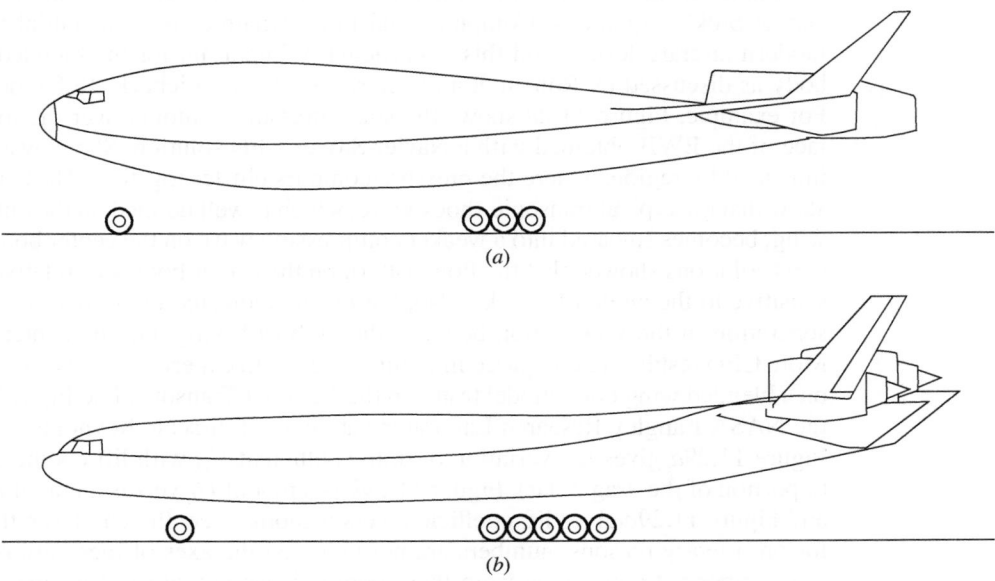

(a)

(b)

Figure 11.26 Center body profiles: (*a*) first generation; (*b*) recent generation.

transonic airfoil design for the center body profile. Taking into account the constraints in cross-sectional area required to effectively hold passengers, baggage, and cargo, the new transonic airfoil design dealt with a careful three-dimensional contouring of the center body smoothly blending into the outer wing panels. The resulting center body profile is shown in Figure 11.26*b*, giving a cleaner, more streamlined appearance than the original profile in Figure 11.26*a*, and providing a higher critical Mach number. Indeed, the new centerbody profile in Figure 11.2*b* increased the BWB lift-to-drag ratio by 4 percent.

2. *Area rule.* The notion of the area rule is discussed in Section 11.8. The blended wing body, with its smooth contours and smoothly varying cross section, is almost naturally area-ruled. Figure 11.27 compares the cross-sectional area distributions as a function of longitudinal coordinate for the BWB (solid curve) and a conventional subsonic transport, the MD-11 (dashed curve). Clearly, the BWB area distribution is much smoother than that of the MD-11, thus exhibiting good area-rule qualities. Liebeck (Reference 90) states that for the BWB "there appears to be no explicit boundary for increasing the cruise Mach number beyond 0.88." Indeed, a set of blended wing bodies have been designed for Mach numbers of 0.85, 0.9, 0.93, and 0.95.

The role of computational fluid dynamics (CFD) in the calculation of transonic inviscid flows is discussed in Section 11.10. This discussion is extended in Chapter 20 to CFD solutions for viscous flows via numerical solutions of the Navier-Stokes equations. Computational fluid dynamics is an essential tool in modern aircraft design, and this is particularly important for the blended wing body as discussed by Roman et al. (Reference 91) and Liebeck (Reference 92). For example, Figure 11.28 shows the static pressure contours over the top surface of the BWB obtained with a Navier-Stokes CFD solution. Shock waves are indicated by regions where the pressure contours cluster together. These results show that the typical transonic shock wave, which is well defined on the outboard wing, becomes smeared into a weaker compression wave on the center body. The CFD solutions showed that the flow pattern on the center body was relatively insensitive to the angle of attack. Also, the results indicated the initiation of flow separation in the kink region between the outboard wing and the center body. More CFD results are compared in Figure 11.29 with experimental data obtained on a blended wing body model tested in the National Transonic Facility (NFT) at the NASA Langley Research Laboratory at almost full-scale Reynolds number. Figure 11.29*a* gives the variation of drag coefficient C_D with lift coefficient C_L (a portion of the drag polar). Figure 11.29*b* is a plot of C_L versus angle of attack, and Figure 11.29*c* gives lift coefficient versus moment coefficient. Even though, for proprietary reasons, numbers are not given on the axes of these graphs, the most important conclusion from these comparisons is that CFD results for the BWB agree within 1% with experimental data. In the words of Robert Liebeck

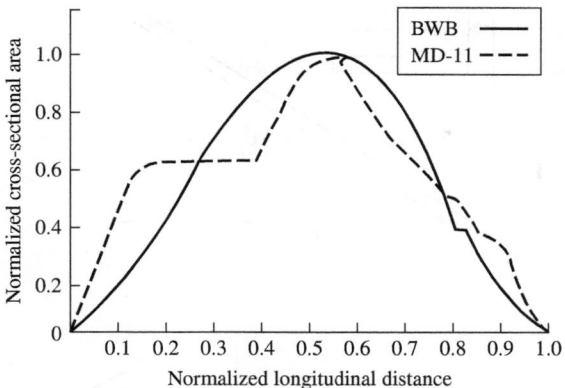

Figure 11.27 Longitudinal distributions of cross-sectional area comparing the blended wing body with a conventional wide-body civil transport, the McDonnell-Douglas MD-11.

(Reference 92), "The remarkable agreement indicated that CFD could be reliably utilized for the aerodynamic design and analysis."

In summary, we offer this applied aerodynamics section as a clear example of how the understanding of the fundamental aerodynamics presented in this book is so essential to the design of the flight vehicles of the future. For more information on the blended wing body, see References 89–92.

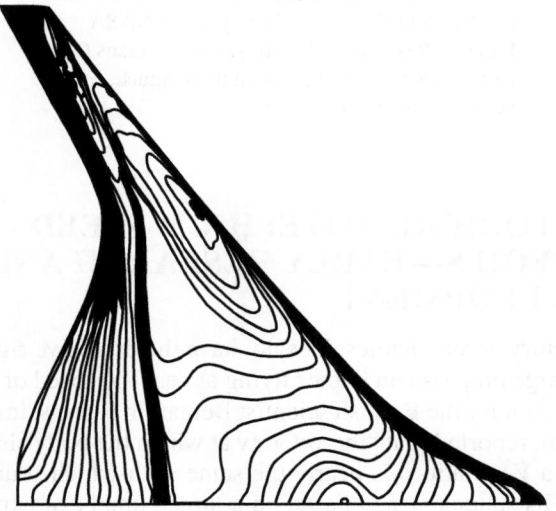

Figure 11.28 Representative CFD calculations of the static pressure contours over the top surface of the BWB at midcruise condition.

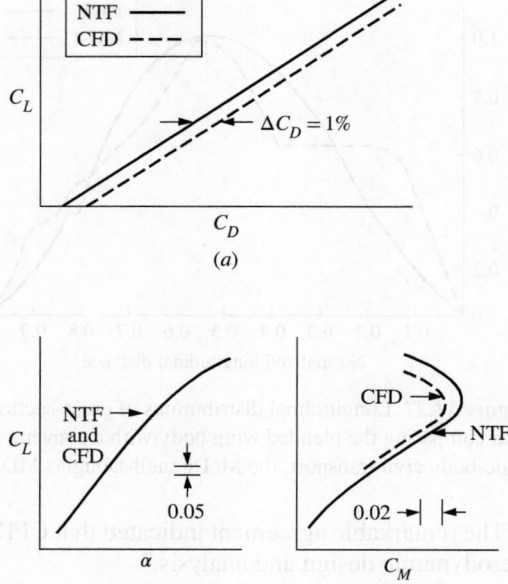

Figure 11.29 Comparison between CFD calculations and experimental data for the blended wing body. The CFD calculations are made with the CFL3D code, and the experimental data are from the National Transonic Facility at the NASA Langley Research Laboratoty. (*a*) C_L versus CD (drag polar); (*b*) C_L versus angle of attack; (*c*) C_L versus moment coefficient.

11.12 HISTORICAL NOTE: HIGH-SPEED AIRFOILS—EARLY RESEARCH AND DEVELOPMENT

Twentieth-century aerodynamics does not have the exclusive rights to the observation of the large drag rise on bodies flying at near the speed of sound; rather, in the eighteenth century the English scientist Benjamin Robins, inventor of the ballistic pendulum, reported that "the velocity at which the body shifts its resistance (from a V^2 to a V^3 relation) is nearly the same with which sound is propagated through air." His statement was based on a large number of experiments during which projectiles were fired into his ballistic pendulum. However, these results had little relevance to the early aerodynamicists of this century, who were struggling to push aircraft speeds to 150 mi/h during and just after World War I. To these people, flight near the speed of sound was just fantasy.

With one exception! World War I airplanes such as the Spad and Nieuport had propeller blades where the tips were moving at near the speed of sound. By 1919, British researchers had already observed the loss in thrust and large increase in blade drag for a propeller with tip speeds up to 1180 ft/s—slightly above the speed of sound. To examine this effect further, F. W. Caldwell and E. N. Fales, both engineers at the U.S. Army's Engineering Division at McCook Field near Dayton, Ohio (the forerunner of the massive Air Force research and development facilities at Wright-Patterson Air Force Base today), conducted a series of high-speed airfoil tests. They designed and built the first high-speed wind tunnel— a facility with a 14-in-diameter test section capable of velocities up to 675 ft/s. In 1918, they conducted the first wind-tunnel tests involving the high-speed flow over a stationary airfoil. Their results showed large decreases in lift coefficient and major increases in drag coefficient for the thicker airfoils at angle of attack. These were the first measured "compressibility effects" on an airfoil in history. Caldwell and Fales noted that such changes occurred at a certain air velocity, which they denoted as the "critical speed"—a term that was to evolve into the critical Mach number at a later date. It is interesting to note that Orville Wright was a consultant to the Army at this time (Wilbur had died prematurely in 1912 of typhoid fever) and observed some of the Caldwell and Fales tests. However, a fundamental understanding and explanation of this critical-speed phenomenon was completely lacking. Nobody at that time had even the remotest idea of what was really happening in this high-speed flow over the airfoil.

Members of the National Advisory Committee for Aeronautics were well aware of the Caldwell-Fales results. Rather than let the matter die, in 1922, the NACA contracted with the National Bureau of Standards (NBS) for a study of high-speed flows over airfoils, with an eye toward improved propeller sections. The work at NBS included the building of a high-speed wind tunnel with a 12-in-diameter test section, capable of producing a Mach number of 0.95. The aerodynamic testing was performed by Lyman J. Briggs (soon to become director of NBS) and Hugh Dryden (soon to become one of the leading aerodynamicists of the twentieth century). In addition to the usual force data, Briggs and Dryden also measured pressure distributions over the airfoil surface. These pressure distributions allowed more insight into the nature of the flow and definitely indicated flow separation on the top surface of the airfoil. We now know that such flow separation is induced by a shock wave, but these early researchers did not at that time know about the presence of such shocks.

During the same period, the only meaningful theoretical work on high-speed airfoil properties was carried out by Ludwig Prandtl in Germany and Hermann Glauert in England—work which led to the Prandtl-Glauert compressibility correction, given by Equation (11.51). As early as 1922, Prandtl is quoted as stating that the lift coefficient increased according to $(1 - M_\infty^2)^{-1/2}$; he mentioned this conclusion in his lectures at Göttingen, but without written proof. This result was mentioned again 6 years later by Jacob Ackeret, a colleague of Prandtl, in the famous German series *Handbuch der Physik,* again without proof. Subsequently, in 1928, the concept was formally established by Hermann Glauert, a British aerodynamicist working for the Royal Aircraft Establishment. (See Chapter 9 of Reference 21 for a biographical sketch of Glauert.) Using only six pages in the *Proceedings of the Royal Society,* vol. 118, p. 113, Glauert presented a derivation based on linearized small-perturbation theory (similar to that described in Section 11.4) which confirmed the $(1 - M_\infty^2)^{-1/2}$ variation. In this paper, entitled "The Effect of Compressibility on the Lift of an Airfoil," Glauert derived the famous Prandtl-Glauert compressibility correction, given here as Equations (11.51) to (11.53). This result was to stand alone, unaltered, for the next 10 years.

Hence, in 1930 the state of the art of high-speed subsonic airfoil research was characterized by experimental proof of the existence of the drag-divergence phenomenon, some idea that it was caused by flow separation, but no fundamental understanding of the basic flow field. In turn, there was virtually no theoretical background outside of the Prandtl-Glauert rule. Also, keep in mind that all the above work was paced by the need to understand propeller performance, because in that day the only component of airplanes to encounter compressibility effects was the propeller tips.

All this changed in the 1930s. In 1928, the NACA had constructed its first rudimentary high-speed subsonic wind tunnel at the Langley Aeronautical Laboratory, utilizing a 1-ft-diameter test section. With Eastman Jacobs as tunnel director and John Stack as the chief researcher, a series of tests was run on various standard airfoil shapes. Frustrated by their continual lack of understanding about the flow field, they turned to optical techniques, following in the footsteps of Ernst Mach (see Section 9.10). In 1933, they assembled a crude schlieren optical system consisting of 3-in-diameter reading-glass-quality lenses and a short-duration-spark light source. In their first test using the schlieren system, dealing with flow over a cylinder, the results were spectacular. Shock waves were seen, along with the resulting flow separation. Visitors flocked to the wind tunnel to observe the results, including Theodore Theodorsen, one of the ranking theoretical aerodynamicists of that period. An indicator of the psychology at that time is given by Theodorsen's comment that since the freestream flow was subsonic, what appeared as shock waves must be an "optical illusion." However, Eastman Jacobs and John Stack knew differently. They proceeded with a major series of airfoil testing, using standard NACA sections. Their schlieren pictures revealed the secrets of flow over the airfoils above the critical Mach number. (See Figure 1.38*b* and its attendant discussion of such supercritical flow.) In 1935, Jacobs traveled to Italy, where he presented results of the NACA high-speed airfoil research at the fifth Volta Conference (see Section 7.1). This is the first time in history that photographs of the transonic flow field over standard-shaped airfoils were presented in a large public forum.

During the course of such work in the 1930s, the incentive for high-speed aerodynamic research shifted from propeller applications to concern about the airframe of the airplane itself. By the mid-1930s, the possibility of the 550 mi/h airplane was more than a dream—reciprocating engines were becoming powerful enough to consider such a speed regime for propeller-driven aircraft. In turn, the entire airplane itself (wings, cowling, tail, etc.) would encounter compressibility effects. This led to the design of a large 8-ft high-speed tunnel at Langley, capable of test-section velocities above 500 mi/h. This tunnel, along with the earlier 1-ft tunnel, established the NACA's dominance in high-speed subsonic research in the late 1930s.

In the decade following 1930, the picture had changed completely. By 1940, the high-speed flow over airfoils was relatively well understood. During this period, Stack and Jacobs had not only highlighted the experimental aspects of such high-speed flow, but they also derived the expression for $C_{p,\mathrm{cr}}$ as a function of M_{cr} given by Equation (11.60), and had shown how to estimate the critical Mach number for a given airfoil as discussed in Section 11.6. Figure 11.30 shows some

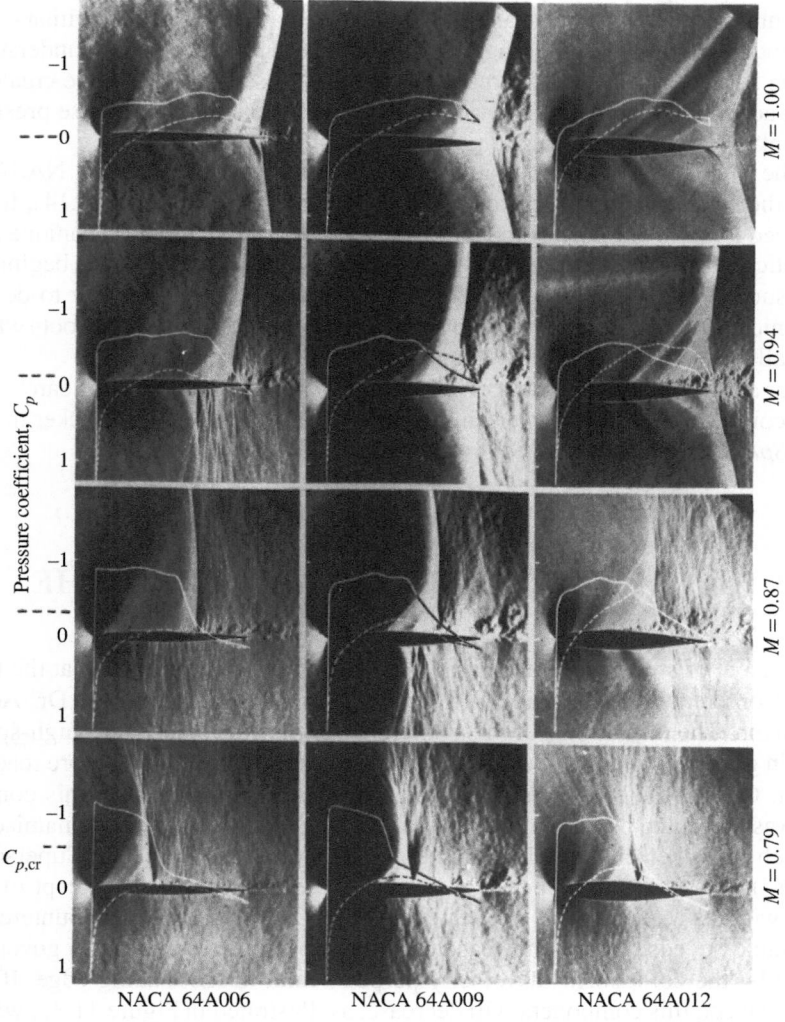

Figure 11.30 Schlieren pictures and pressure distributions for transonic flows over several NACA airfoils. These pictures were taken by the NACA in 1949. (*NASA*)

representative schlieren photographs taken by the NACA of the flow over standard NACA airfoils. Although these photographs were taken in 1949, they are similar to the results obtained by Stack and Jacobs in the 1930s. Superimposed on these photographs are the measured pressure distributions over the top (solid curve) and bottom (dashed curve) surfaces of the airfoil. Study these pictures carefully. Moving from bottom to top, you can see the influence of increasing freestream Mach number, and going from left to right, you can observe the effect of increasing airfoil thickness. Note how the shock wave moves downstream as M_∞ is increased, finally reaching the trailing edge at $M_\infty = 1.0$. For this case, the top row of pictures shows almost completely supersonic flow over the airfoil. Note also the large regions of separated flow downstream of the shock waves for the Mach numbers of 0.79, 0.87, and 0.94—this separated flow is the primary reason for the large increase in drag near Mach 1. By 1940, it was well understood that the almost discontinuous pressure increase across the shock wave creates a strong adverse pressure gradient on the airfoil surface, and this adverse pressure gradient is responsible for separating the flow.

The high-speed airfoil research program continues today within NASA. It led to the supercritical airfoils in the 1960s (see Sections 11.9 and 11.14). It has produced a massive effort in modern times to use computational techniques for theoretically solving the transonic flow over airfoils. Such efforts are beginning to be successful, and in many respects, today we have the capability to design transonic airfoils on the computer. However, such abilities today have roots which reach all the way back to Caldwell and Fales in 1918.

For a more detailed account of the history of high-speed airfoil research, you are encouraged to read the entertaining story portrayed by John V. Becker in *The High-Speed Frontier,* NASA SP-445, 1980.

11.13 HISTORICAL NOTE: THE ORIGIN OF THE SWEPT-WING CONCEPT

The concept of swept wings for high-speed flight was first introduced at the fifth Volta Conference in Rome in 1935 by the German aerodynamicist Dr. Adolf Busemann. The importance of this conference to the advancement of high-speed flight in general is noted in Section 7.1; please review that section before reading further. One of the most farsighted and important papers given at this conference was presented by Busemann (see Figure 11.31). Entitled "Aerodynamischer Auftrieb bei Überschallgeschwindigkert" ("Aerodynamic Forces at Supersonic Speeds"), this paper introduced for the first time in history the concept of the swept wing as a mechanism for reducing the large drag increase encountered at supersonic speeds. Busemann reasoned that the flow over a wing is governed mainly by the component of velocity perpendicular to the leading edge. If the wing is swept, this component will decrease, as illustrated in Figure 11.32, which is taken directly from Busemann's original paper. Consequently, the supersonic wave drag will decrease. If the sweep angle is large enough, the normal component

Figure 11.31 Adolf Busemann (1901–1986). (*Courtesy of John Anderson*)

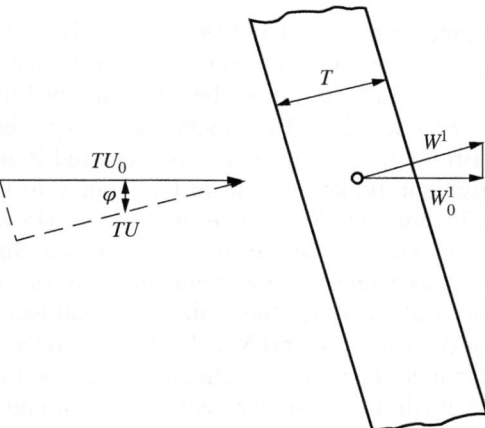

Figure 11.32 The swept-wing concept as it appeared in Busemann's original paper in 1935.

of velocity will be subsonic (the supersonic wing is then said to have a "subsonic leading edge") with a dramatic reduction of wave drag. Figure 11.33, showing a complete swept-wing planform, is also from Busemann's paper.

At the time of the Volta conference, Adolf Busemann was a relatively young (age 34) but accomplished aerodynamicist. Born in Lübeck, Germany, in 1901, he completed high school in his home town and received his engineering diploma

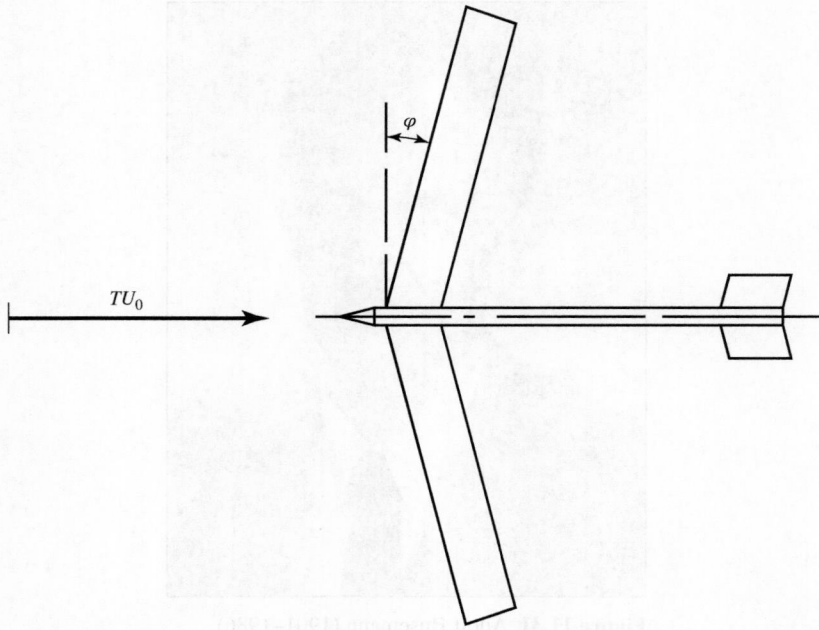

Figure 11.33 A swept-wing airplane planform, from Busemann's original paper in 1935.

and doctorate in engineering in 1924 and 1925, respectively, from the Technische Hochschule in Braunschweig. Busemann was one of the few important German aerodynamicists of that era who did not begin as one of Ludwig Prandtl's students (see Section 5.8), but in 1925 Busemann began his professional career at the Kaiser Wilhelm Institute in Göttingen and soon entered Prandtl's sphere. From 1931 to 1935, Busemann broke away from that sphere to teach in the Engine Laboratory of the Technische Hochschule in Dresden. He was still at Dresden when he gave his seminal paper at the Volta Conference. Shortly thereafter, he went to Braunschweig as chief of the Gas Dynamics Division of the Aeronautical Research Laboratory (DFL). When the Allied technical teams moved into German laboratories at the end of World War II, they not only scooped up masses of technical aerodynamic data but also effectively scooped up Busemann, who first went to the Royal Aircraft Establishment in Farnborough, England, but then accepted an invitation to join the NACA Langley Memorial Laboratory under Operation Paperclip in 1947. Busemann continued his research on high-speed aerodynamics for the NACA after joining Langley. He subsequently became chair of the advanced-study committee at Langley and among other responsibilities supervised the preparation of science lectures used for training the early group of astronauts in the manned space program. In 1963, Busemann became a professor in the Department of Aerospace Engineering Sciences at the University of Colorado in Boulder. After retirement, he remained in Boulder, leading an active life until his death in 1986.

The swept-wing concept in Busemann's 1935 Volta Conference paper was, for everybody outside of Germany, an idea before its time. It is difficult to understand how the premier aerodynamicist in the United States, Theodor von Kármán, and other attendees failed to appreciate the significance of Busemann's idea, even forgetting it entirely, for that very evening Busemann went to dinner with von Kármán, Hugh Dryden (see Section 11.11), and General Arturo Crocco, the organizer of the conference. During dinner, Crocco sketched on the back of a menu card an airplane with swept wings, swept tail, and swept propeller, calling it, facetiously, "Busemann's airplane of the future."

There was no such facetiousness in Germany. The German Luftwaffe recognized the military significance of the swept wing, and classified the concept in 1936—one year after the conference. From that time until the end of World War II, the Germans produced a large bulk of swept-wing research, a secret known only to themselves. Moreover, they expanded the horizons of the swept wing to high-speed subsonic and transonic airplanes, recognizing that the same aerodynamic mechanism described in Busemann's Volta Conference paper would serve to increase the critical Mach number for such aircraft. Experimental results for swept wings at high subsonic Mach numbers were first reported by Hubert Ludwieg in 1939, as shown in Figure 11.34. Here the drag polars (defined in Section 5.3.3 as lift coefficient C_A versus drag coefficient C_W, in the German notation) for a straight wing and a swept wing are compared. These are the original figures from the German report AVA-Bericht 39/H/18, 1939. This was top-secret data obtained in a new high-speed wind tunnel at the Aerodynamische Versuchsanstalt (AVA) in Göttingen, and represented just the tip of the iceberg of German swept-wing research to follow. In his recent article, Peter Hamel ("Birth of Sweepback: Related Research at Luftfahrtforschungsanstalt–Germany," *Journal of Aircraft*, vol. 42, no. 4, July–August 2005, pp. 801–813) labels this data as the first from what was to be "established systematic wind-tunnel tests to generate a world-first database for future transonic aircraft configurations with wing sweep."

Meanwhile, Robert T. Jones (Figure 11.35), a leading aerodynamicist at the NACA Langley Memorial Laboratory, independently discovered the advantages of a swept wing. Jones was a self-made person. Born in Macon, Missouri, in 1910, Jones was totally captivated by aeronautics at an early age. He later wrote in 1977 ("Recollections from an Earlier Period in American Aeronautics," in *Annual Review of Fluid Mechanics*, vol. 9, M. Van Dyke et al. (eds), pp. 1–11):

All during the late twenties the weekly magazine *Aviation* appeared on the local newsstand in my hometown, Macon, Missouri. *Aviation* carried technical articles by eminent aeronautical engineers such as B. V. Korvin-Krovkovsky, Alexander Klemin, and others. Included in both *Aero Digest* and *Aviation* were notices of forthcoming *NACA Technical Reports* and *Notes*. These could be procured from the Government Printing Office usually for ten cents and sometimes even free simply for writing NACA Headquarters in Washington. The contents of these reports seemed much more interesting to me than the regular high school and college curricula, and I suspect that my English teachers may have been quite perplexed by the essays I wrote for them on aeronautical subjects.

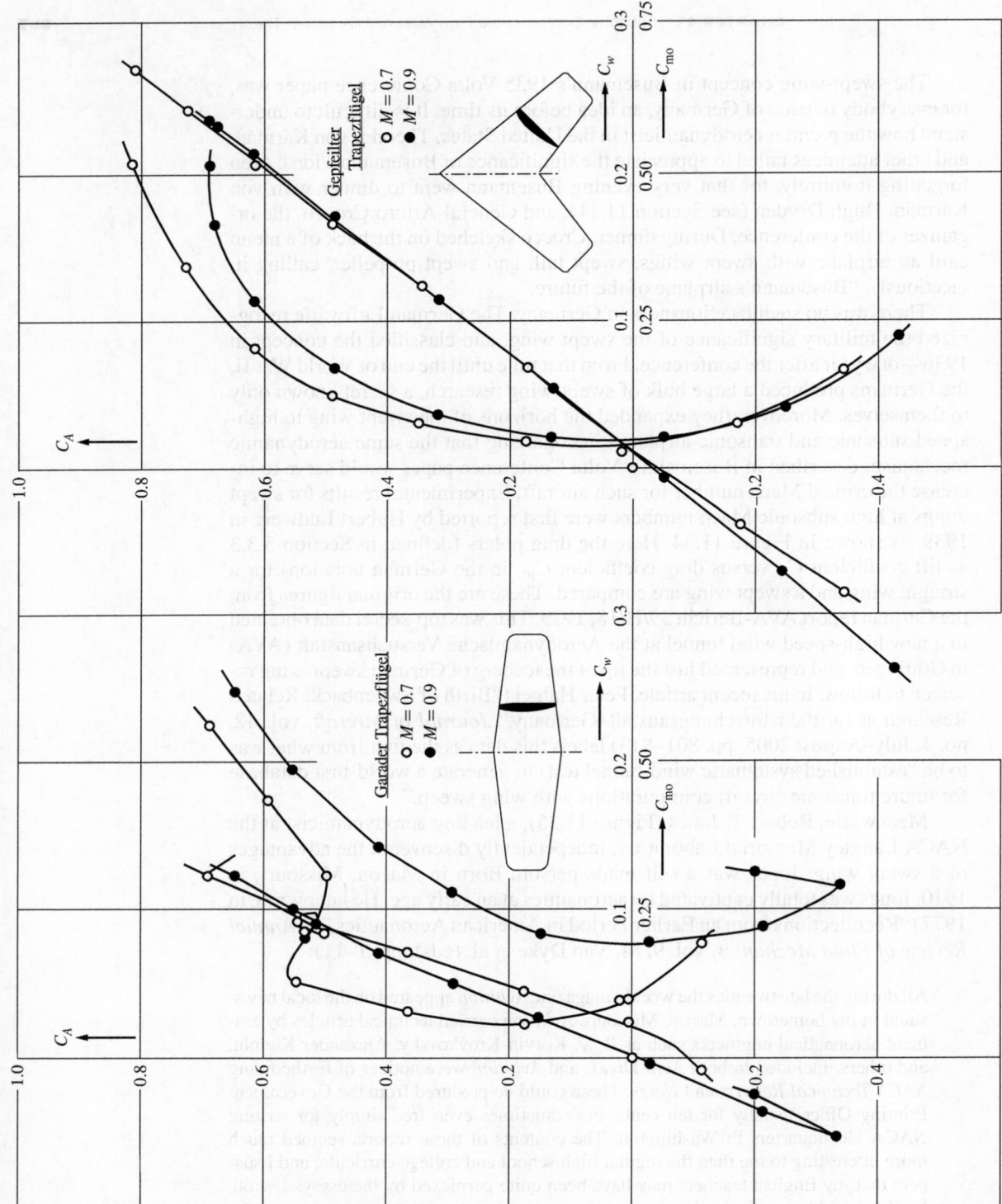

Figure 11.34 Drag polars at Mach 0.7 and 0.9 for a straight wing (left) and swept wing (right), as reported by Ludwieg in 1939.

Figure 11.35 Robert T. Jones (1910–1999). (*NASA*)

Jones attended the University of Missouri for one year, but left to take a series of aeronautics-related jobs, first as a crew member with Marie Meyer Flying Circus, and then with the Nicholas-Beazley Airplane Company in Marshall, Missouri, which was just starting to produce a single-engine, low-wing monoplane designed by the noted British aeronautical engineer Walter H. Barling. At one time, Nicholas-Beazley was producing and selling one of those aircraft each day. However, the company became a victim of the Great Depression, and in 1933 Jones found himself working as an elevator operator in Washington, D.C., and taking night classes in aeronautics at Catholic University, taught by Max Monk. That contact began a lifelong friendship between Jones and Monk. In 1934 the Public Works Administration created a number of temporary scientific positions in the federal government. On the recommendation of Congressman David J. Lewis, from Jones's hometown, Jones received a nine-month appointment at the NACA Langley Memorial Laboratory. That was the beginning of a lifetime career for Jones at the NACA/NASA. Through a passionate interest and self-study in aeronautics, Jones had become exceptionally knowledgeable in aerodynamic theory. His talents were recognized at Langley, and he was kept on at the laboratory through a series of temporary and emergency reappointments for the next two years. Unable to promote him into the lowest professional engineering grade because of civil-service regulations that required a college degree, in 1936 the laboratory management was finally able to hire Jones permanently via a loophole: It hired Jones at the next grade above the lowest, for which the requirement of a college degree was not specifically stated (although presumed).

By 1944, Jones was one of the most respected aerodynamicists in the NACA. At that time he was working on the design of an experimental air-to-air missile for the Army Air Force and he was also studying the aerodynamics of a proposed glide bomb having a low-aspect-ratio delta wing. The Ludington-Griswald Company of Saybrook, Connecticut, had carried out wind-tunnel tests on a dart-shaped missile of their design, and Roger Griswold, president of the company, showed the data to Jones in 1944. Griswold had compared the lift data for the missile's low-aspect-ratio delta wing with calculations made from Prandtl's tried-and-proved lifting-line theory (Section 5.3). Jones realized that Prandtl's lifting-line theory was not valid for low-aspect-ratio wings, and he began to construct a more appropriate theory for the delta-wing planform. Jones obtained rather simple analytical equations for the low-speed, incompressible flow over delta wings, but considered the theory to be "so crude" that "nobody would be interested in it." He placed his analysis in his desk and went on with other matters.

In early 1945, Jones began to look at the mathematical theory of supersonic potential (irrotational) flows. When applied to delta wings, Jones found that he was obtaining equations similar to those he had found for incompressible flow using the crude theory that was now buried in his desk. Searching for an explanation, he recalled the statement by Max Munk in 1924 that the aerodynamic characteristics of a wing were governed mainly by the component of the freestream velocity perpendicular to the leading edge. The answer suddenly was quite simple: For the delta wing, the reason his supersonic findings were the same as his earlier low-speed findings was that the leading edge of the delta wing was swept far enough that the component of the supersonic free stream Mach number perpendicular to the leading edge was subsonic, and hence the supersonic swept wing acted as if it were in a subsonic flow. With that revelation, Jones had independently discovered the high-speed aerodynamic advantage of swept wings, albeit 10 years after Busemann's paper at the Volta conference.

Jones began to discuss his swept-wing theory with colleagues at NACA Langley. In mid-February 1945, he outlined his thoughts to Jean Roche and Ezra Kotcher of the Army Air Force at Wright Field. On March 5, 1945, he sent a memo to Gus Crowley, chief of research at Langley, stated that he had "recently made a theoretical analysis which indicates that a V-shaped wing traveling point foremost would be less affected by compressibility than other planforms. In fact, if the angle of the V is kept small relative to the Mach angle, the lift and center of pressure remain the same at speeds both above and below the speed of sound." In the same memo, Jones asked Crowley to approve experimental work on swept wings. Such work was quickly initiated by the Flight Research Section of Langley, under the direction of Robert Gilruth, beginning with a series of free-flight tests using bodies with swept wings dropped from high altitude.

Jones finished a formal report on his low-aspect-ratio wing theory in late April 1945, including the effects of compressibility and the concept of a swept wing. However, during the in-house editorial review of that report, Theodore Theodorsen raised some serious objections. Theodorsen did not like the heavily intuitive nature of Jones's theory, and he asked Jones to clarify the "hocus-pocus" with some "real mathematics." Furthermore, because supersonic flow was so different physically and mathematically from subsonic flow, Theodorsen could not accept the "subsonic" behavior of Jones's highly swept wings at supersonic speeds. Criticizing Jones's entire swept-wing concept, calling it "a snare and a delusion," Theodorsen insisted that Jones take out the part about swept wings.

Theodorsen's insistence prevailed, and publication of Jones's report was delayed. However, at the end of May 1945, Gilruth's free-flight tests dramatically verified Jones's predictions, showing a factor of 4 reduction in drag due to sweeping the wings. Quickly following those data, wind-tunnel tests carried out in a small supersonic wind tunnel at Langley showed a large reduction in drag on a section of wire in the test section when the wire was placed at a substantial angle of sweep relative to the flow in the test section. With that experimental proof of the validity of the swept-wing concept, Langley forwarded Jones's report to NACA Headquarters in Washington for publication. But Theodorsen would not give up: The transmittal letter to NACA Headquarters contained the statement that "Dr. Theodore Theodorsen (still) does not agree with the arguments presented and the conclusions reached and accordingly declined to participate in editing the paper." Such recalcitrance on the part of Theodorsen is reminiscent of his refusal to believe that the shock waves seen in John Stack's schlieren photographs of the transonic flow over an airfoil 11 years earlier were real (see Section 11.11). Theodorsen certainly made important contributions to airfoil theory in the 1930s, but he was also capable of errors in judgment (i.e., he was human).

On June 21, 1945, the NACA issued Jones's report as a confidential memorandum, chiefly for the Army and Navy. Three weeks later, the report was reissued as an advance confidential report, sent by registered mail to those people in industry with a "need to know." Entitled "Wing Plan Forms for High-Speed Flight," Jones's report quickly spread the idea of the swept wing to selected members of the aeronautical community in the United States, but by that time, information about the German swept-wing research was beginning to reach the same aeronautical community. Jones's work appeared in the open literature about a year later, as NACA TR 863, a technical report only five pages long, but a classic explanation of how a swept wing works aerodynamically.

Credit for the idea of the swept wing for high-speed flight is shared between Busemann and Jones. Separated by a time interval of 10 years, and the closed shops of military security in both Germany and the United States, each independently developed the concept, not knowing of the other's work. The full impact of the swept-wing concept on the aeronautical industry came directly after the end of World War II, with almost simultaneous release of similar information from both sides of the ocean, thus promoting confidence in the validity of the concept.

The speed at which this information was used for airplane design is nothing short of amazing. Boeing's chief aerodynamicist, George Schairer, was a member of one of the Allied technical intelligence teams sent to Germany in April 1945. On the flight over, Schairer, who was aware of Jones's swept-wing concept and that Jones's report was being held up in the editorial process, reported that the concept was the main topic of conversation. Concluding that sweepback was a valid concept, Schairer needed no convincing of its value when the team saw the German swept-wing data at Baunschweig on May 7. On May 10, he sent a seminal and historically important letter to Boeing directing that a new design for a straight-wing jet bomber immediately be changed to a swept-wing configuration, and that other aircraft manufacturers be informed about the German swept-wing research. At Boeing, the result of this letter was the Boeing B-47, shown in Figure 11.36, and at North American the result was the famous F-86 (Figure 11.15). The B-47 sowed the seeds for the Boeing 707 swept-wing jet transport (Figure 1.2) and for all subsequent large commercial jet transports to the present day.

Indeed, by 1948 the swept wing had become an accepted airplane design feature. It had done for the high-speed jet airplane what streamlining had done in the 1930s for the advanced propeller-driven airplane, namely, provided the aerodynamic means for efficient flight in the desired flight regime. Virtually all

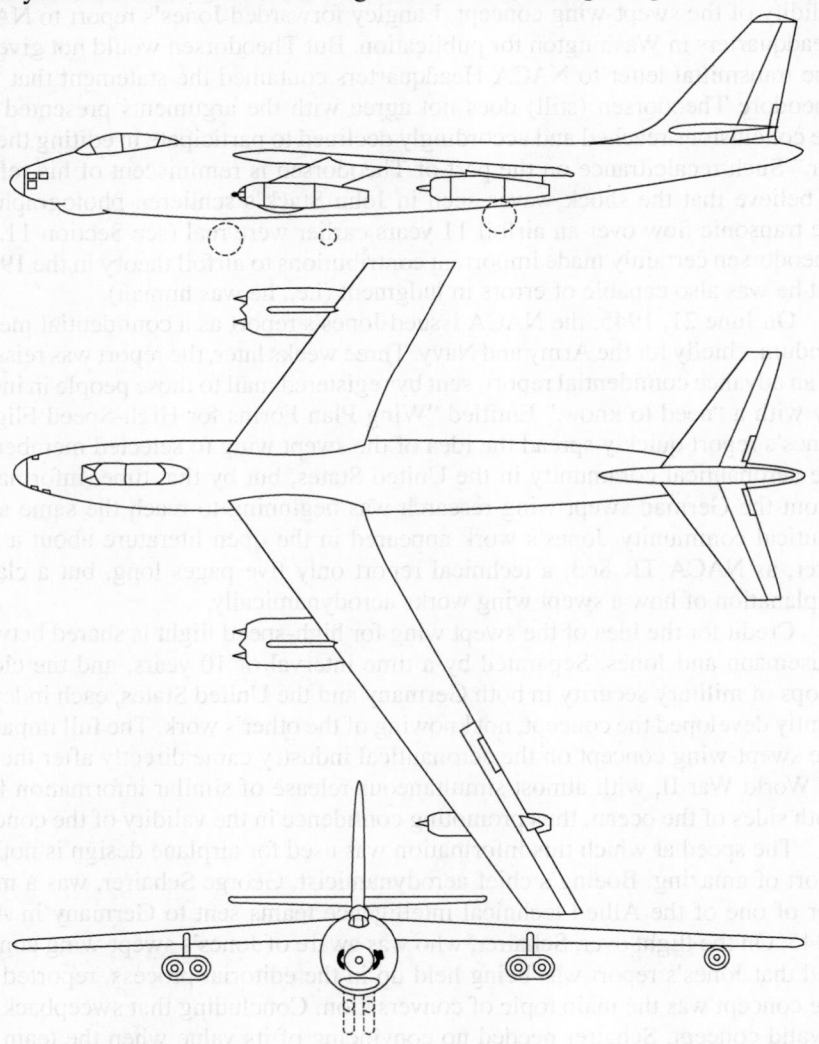

Figure 11.36 Three-view of the Boeing B-47 swept-wing bomber, circa 1948.

high-speed, jet-propelled airplanes today have wings with highly swept leading edges, and these contemporary aircraft can trace their ancestry directly back to the B-47 and the F-86, and to the innovative ideas and genius of Adolf Busemann and Robert Jones.

For a more in-depth discussion of the history of the swept wing, see References 58 and 111.

11.14 HISTORICAL NOTE: RICHARD T. WHITCOMB—ARCHITECT OF THE AREA RULE AND THE SUPERCRITICAL WING

The developments of the area rule (Section 11.8) and the supercritical airfoil (Section 11.9) are two of the most important advancements in aerodynamics since 1950. That both developments were made by the same person—Richard T. Whitcomb—is remarkable. Who is this person? What qualities lead to such accomplishments? Let us pursue these matters further.

Richard Whitcomb was born on February 21, 1921, in Evanston, Illinois. At an early age, he was influenced by his grandfather, who had known Thomas A. Edison. In an interview with *The Washington Post* on August 31, 1969, Whitcomb is quoted as saying: "I used to sit around and hear stories about Edison. He sort of developed into my idol." Whitcomb entered the Worcester Polytechnic Institute in 1939. (This is the same school from which the rocket pioneer, Robert H. Goddard, had graduated 31 years earlier.) Whitcomb distinguished himself in college and graduated with a mechanical engineering degree with honors in 1943. Informed by a *Fortune* magazine article on the research facilities at the NACA Langley Memorial Laboratory, Whitcomb immediately joined the NACA. He became a wind-tunnel engineer, and as an early assignment he worked on design problems associated with the Boeing B-29 Superfortress. He remained with the NACA and later its successor, NASA, until his retirement in 1980—spending his entire career with the wind tunnels at the Langley Research Center. In the process, he rose to become head of the Eight-foot Tunnel Branch at Langley. He died of pneumonia at the age of 88 on October 13, 2009, in Newport News, Virginia.

Whitcomb conceived the idea of the area rule as early as 1951. He tested his idea in the transonic wind tunnel at Langley. The results were so promising that the aeronautical industry changed designs in midstream. For example, the Convair F-102 delta-wing fighter had been designed for supersonic flight, but was having major difficulty even exceeding the speed of sound—the increase in drag near Mach 1 was simply too large. The F-102 was redesigned to incorporate Whitcomb's area rule and afterward was able to achieve its originally intended supersonic Mach number. The area rule was such an important aerodynamic breakthrough that it was classified "secret" from 1952 to 1954, when airplanes incorporating the area rule began to roll off the production line. In 1954, Whitcomb was given the Collier Trophy—an annual award for the "greatest achievement in aviation in America."

In the early 1960s, Whitcomb turned his attention to airfoil design, with the objective again of decreasing the large drag rise near Mach 1. Using the existing knowledge about airfoil properties, a great deal of wind-tunnel testing, and intuition honed by years of experience, Whitcomb produced the supercritical airfoil. Again, this development had a major impact on the aeronautical industry, and today virtually all new commercial transport and executive aircraft designs are incorporating a supercritical wing. Because of his development of the supercritical airfoil, in 1974 NASA gave Whitcomb a cash award of $25,000—the largest cash award ever given by NASA to a single individual.

There are certain parallels between the personalities of the Wright brothers and Richard Whitcomb: (1) they all had powerful intuitive abilities which they brought to bear on the problem of flight, (2) they were totally dedicated to their work, and (3) they did a great deal of their work themselves, trusting only their own results. For example, here is a quote from Whitcomb which appears in the same *Washington Post* interview mentioned above. Concerning the detailed work on the development of the supercritical airfoil, Whitcomb says:

> I modified the shape of the wing myself as we tested it. It's just plain easier this way. In fact my reputation for filing the wing's shape has become so notorious that the people at North American have threatened to provide me with a 10-foot file to work on the real airplane, also.

Perhaps the real ingredient for Whitcomb's success is his personal philosophy, as well as his long hours at work daily. In his own words:

> There's been a continual drive in me ever since I was a teenager to find a better way to do everything. A lot of very intelligent people are willing to adapt, but only to a certain extent. If a human mind can figure out a better way to do something, let's do it. I can't just sit around. I have to think.

Students take note!

11.15 SUMMARY

Review the road map in Figure 11.1, and make certain that you have all the concepts listed on this map well in mind. Some of the highlights of this chapter are as follows:

For two-dimensional, irrotational, isentropic, steady flow of a compressible fluid, the exact velocity potential equation is

$$\left[1 - \frac{1}{a^2}\left(\frac{\partial \phi}{\partial x}\right)^2\right]\frac{\partial^2 \phi}{\partial x^2} + \left[1 - \frac{1}{a^2}\left(\frac{\partial \phi}{\partial y}\right)^2\right]\frac{\partial^2 \phi}{\partial y^2}$$
$$-\frac{2}{a^2}\left(\frac{\partial \phi}{\partial x}\right)\left(\frac{\partial \phi}{\partial y}\right)\frac{\partial^2 \phi}{\partial x \partial y} = 0 \tag{11.12}$$

where

$$a^2 = a_0^2 - \frac{\gamma - 1}{2}\left[\left(\frac{\partial \phi}{\partial x}\right)^2 + \left(\frac{\partial \phi}{\partial y}\right)^2\right] \tag{11.13}$$

This equation is exact, but it is nonlinear and hence difficult to solve. At present, no general analytical solution to this equation exists.

For the case of small perturbations (slender bodies at low angles of attack), the exact velocity potential equation can be approximated by

$$\left(1 - M_\infty^2\right)\frac{\partial^2 \hat{\phi}}{\partial x^2} + \frac{\partial^2 \hat{\phi}}{\partial y^2} = 0 \qquad (11.18)$$

This equation is approximate, but linear, and hence more readily solved. This equation holds for subsonic ($0 \le M_\infty \le 0.8$) and supersonic ($1.2 \le M_\infty \le 5$) flows; it does not hold for transonic ($0.8 \le M_\infty \le 1.2$) or hypersonic ($M_\infty > 5$) flows.

The Prandtl-Glauert rule is a compressibility correction that allows the modification of existing incompressible flow data to take into account compressibility effects:

$$C_p = \frac{C_{p,0}}{\sqrt{1 - M_\infty^2}} \qquad (11.51)$$

Also,

$$c_l = \frac{c_{l,0}}{\sqrt{1 - M_\infty^2}} \qquad (11.52)$$

and

$$c_m = \frac{c_{m,0}}{\sqrt{1 - M_\infty^2}} \qquad (11.53)$$

The critical Mach number is that freestream Mach number at which sonic flow is first obtained at some point on the surface of a body. For thin airfoils, the critical Mach number can be estimated as shown in Figure 11.6.

The drag-divergence Mach number is that freestream Mach number at which a large rise in the drag coefficient occurs, as shown in Figure 11.11.

The area rule for transonic flow states that the cross-sectional area distribution of an airplane, including fuselage, wing, and tail, should have a smooth distribution along the axis of the airplane.

Supercritical airfoils are specially designed profiles to increase the drag-divergence Mach number.

11.16 INTEGRATED WORK CHALLENGE: TRANSONIC TESTING BY THE WING-FLOW METHOD

Concept: At the end of World War II, the speeds of fighter airplanes were approaching Mach 1, and the need for transonic aerodynamic data became paramount. Such data, however, was in short supply. There existed no reliable transonic wind tunnels at that time, and because the governing equations for transonic flow are nonlinear, no accurate analytical solutions were available (see Section 11.2). Faced with this dismal situation, engineers at the NACA Langley Memorial Aeronautical Laboratory (now the NASA Langley Research Center) came up with three innovative techniques for obtaining aerodynamic data in the transonic flight regime. The first was the falling-body method, wherein a heavy aerodynamic model was carried to 30,000 feet by a B-29 bomber and then released. On its way down, the terminal velocity of the model became transonic, and aerodynamic data was radioed to receivers on the ground. A somewhat related second method involved mounting a model on a rocket that was launched from the small NACA facility on Wallops Island, Virginia, and again transonic aerodynamic data was radioed to receivers on the ground using telemetering instrumentation that was originally developed for the falling-body tests. The third method, however, was the most innovative and was totally different from the others. Called the wing-flow method, it is the subject of this Integrated Work Challenge section.

The wing-flow method involved mounting a small wing model vertically on the surface of the wing of a P-51 Mustang fighter airplane at a location inside the bubble of locally supersonic flow existing on the P-51 wing when the airplane exceeded its critical Mach number during a high-speed dive. (This locally supersonic bubble is diagrammed in Figures 11.5*d* and 11.11.) Aerodynamic data such as lift, drag, moments, hinge moments, and pressure distributions on the test model were recorded and stored by instruments housed in an empty ammunition compartment inside the P-51 wing. Both straight- and swept-wing models were tested. Figure 11.37 is a photograph showing a swept-wing model mounted on the P-51 wing.

Integrated Work Challenge: Design the setup for a wing-flow test, i.e., address the following questions:

(a) Where on the P-51 wing is the best location for mounting the test model?
(b) At this location, what are the local flow properties over the model?
(c) How far above the P-51 wing surface can the model reach, i.e., what is the allowable span of the model?
(d) Does the boundary layer on the P-51 wing pose a problem for the model?

Solution: To decide where on the P-51 wing is the best location for the model, we first need to know the distribution of the local flow Mach number over the wing surface. At the time of the original NACA tests in the middle 1940s, this determination was obtained by making static pressure measurements along the

Figure 11.37 Photograph of a test model mounted on the wing of the P-51 used for the NACA Wing-Flow transonic tests, circa 1946. (*NASA*)

surface of the P-51 in flight and by using the isentropic flow relations, obtaining the local Mach number at points along the wing. In essence, we will carry out the same approach here, but instead of using the older pressure data, we will obtain the local Mach numbers from modern computational fluid dynamic calculations of the pressure coefficients over a P-51 wing. These calculations were made by David Lednicer and published among other places in the article "World War II Fighter Aerodynamics," *Sport Aviation*, January 1999, pp. 85–91. Figure 11.38 is a plot of the calculated pressure coefficient over the P-51 wing in cruise, based on Lednicer's calculations. The airfoil used on the P-51 was an NACA laminar flow airfoil; the shape of the airfoil is also shown in Figure 11.38. In conjunction with Figure 4.2, we noted that these airfoils in service in the field never produced the desired amount of laminar flow, but by a stroke of serendipity they had higher critical Mach numbers, prompted by a long favorable pressure gradient with minimum pressure occurring far downstream of the leading edge. For example, in Figure 11.38, the minimum pressure coefficient is −0.575, and it occurs at about midchord along the airfoil.

Returning to our work challenge, we address the question: Where on the wing airfoil should the test model be placed? The answer is straightforward; it should be placed at the location where the local Mach number is maximum, which corresponds to the location of the minimum pressure coefficient. From

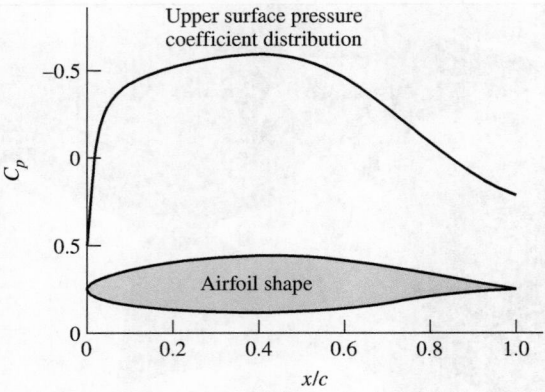

Figure 11.38 Computational fluid dynamic results for the pressure coefficient distribution over the wing of a P-51 at cruise condition, based on the calculations by David Lednicer.

Figure 11.38, this location is in the region between $x/c = 0.4$ and 0.5, where the minimum value of C_p is relatively flat.

To estimate the maximum local Mach number in this region, we need to know the maximum speed of the P-51 during the test. The most definitive paper on the NACA wing-flow tests is by Robert R. Gilruth, chief engineer of the project, entitled "Resume and Analysis of NACA Wing-Flow Tests," Anglo-American Aeronautical Conference, London, 3–5 September 1947, and published by the Royal Aeronautical Society in 1948, pp. 363–383. In this paper, Gilruth notes that the maximum flight Mach number of the P-51 for the tests was 0.76. We will use this value in our analysis here. Also, to obtain the maximum local Mach number on the P-51 wing, we need to know the minimum pressure coefficient on the wing. For this information, we will use the CFD results of Lednicer as plotted in Figure 11.38, suitably modified for compressibility effects as follows.

Lednicer's calculations were made for the P-51 in cruise. Loftin, in his book *Quest for Performance: The Evolution of Modern Aircraft*, NASA SP-468, 1985, gives the cruise velocity for the P-51 as 362 mph. This is a speed low enough that compressibility effects are small, and we will interpret the C_p results in Figure 11.38 to be "low speed results" that need to be corrected for compressibility effects at the maximum Mach number of 0.76. We choose to use the Karman–Tsien compressibility correction rule given in Equation (11.54), which was widely adopted after World War II.

Repeating Equation (11.54)

$$C_p = \frac{C_{p,0}}{\sqrt{1 - M_\infty^2} + \left[\dfrac{M_\infty^2}{1 + \sqrt{1 - M_\infty^2}} \right] \dfrac{C_{p,0}}{2}}$$

and using $C_{p,0} = -0.575$ and $M_\infty = 0.76$, we have

$$C_p = \frac{-0.575}{\sqrt{1 - (0.76)^2} + \left[\dfrac{(0.76)^2}{1 + \sqrt{1 - (0.76)^2}}\right]\left(\dfrac{-0.575}{2}\right)} = -1.0419$$

This is the pressure coefficient, corrected for compressibility effects, at the minimum pressure location on the wing.

To obtain the local Mach number at this location, we use Equation (11.58), repeated next, recognizing that the subscript A now denotes the location of minimum pressure,

$$C_{p,A} = \frac{2}{\gamma M_\infty^2}\left[\left(\frac{1 + \dfrac{\gamma - 1}{2}M_\infty^2}{1 + \dfrac{\gamma - 1}{2}M_A^2}\right)^{\frac{\gamma}{\gamma-1}} - 1\right]$$

and where $C_{p,A} = -1.0419$. Inserting the numbers in this equation, we have

$$-1.0419 = \frac{2}{(1.4)(0.76)^2}\left[\left(\frac{1 + (0.2)(0.76)^2}{1 + (0.2)M_A^2}\right)^{\frac{1.4}{.4}} - 1\right]$$

$$= 2.4733\left[\left(\frac{1.115}{1 + (0.2)M_A^2}\right)^{3.5} - 1\right]$$

Hence,

$$-0.4213 = \left[\frac{1.466}{(1 + 0.2M_A^2)^{3.5}} - 1\right]$$

or,

$$0.5787 = \frac{1.466}{(1 + 0.2M_A^2)^{3.5}}$$

$$\left(1 + 0.2M_A^2\right)^{3.5} = \frac{1.466}{0.5787} = 2.53326$$

$$1 + 0.2M_A^2 = (2.53326)^{\frac{1}{3.5}} = (2.53326)^{0.2857} = 1.30416$$

Thus,

$$0.2M_A^2 = 0.30416$$

$$M_A^2 = \frac{0.30416}{0.2} = 1.5208$$

$$M_A = 1.23$$

Clearly, this result proves that a reasonable region of transonic and low-supersonic flow is available for testing a model mounted on the wing of the P-51.

Gilruth, in his presentation to the Anglo-American Aeronautical Conference in 1947, states that "preliminary tests indicated that the local Mach number increased smoothly from about 0.40 to about 1.15 as the flight Mach number of the airplane was increased from 0.30 to 0.76." Those indications were based on experimental measurements obtained in flight. Our calculated result of a local Mach number of 1.23, based on modern CFD calculations and using a reasonable compressibility correction for pressure coefficient, agrees with Gilruth's data to within 7 percent.

Other local flow field variables such as static pressure and temperature can be calculated for the local Mach number of 1.23 using the isentropic relations discussed in Chapter 8. These results, however, will depend on the altitude of the airplane. The vertical extent of the supersonic bubble above the wing can be obtained by carrying out a detailed computational fluid dynamic calculation of the flow field over the P-51 and tracing the sonic line in the flow. We note that Gilruth carried out measurements of the vertical gradients in Mach number and found that they did not exceed about 1 percent per inch and "were probably somewhat less than this figure," thus presenting a reasonably comfortable vertical extent of the testing region. Finally, the Reynolds numbers associated with the high speed of the P-51 were large, resulting in a thin boundary layer thickness on the wing surface of no real consequence on the model tests.

11.17 PROBLEMS

11.1 Consider a subsonic compressible flow in cartesian coordinates where the velocity potential is given by

$$\phi(x, y) = V_\infty x + \frac{70 \sin(2\pi x)}{\sqrt{1 - M_\infty^2}} e^{-2\pi y \sqrt{1 - M_\infty^2}}$$

If the freestream properties are given by $V_\infty = 700$ ft/s, $p_\infty = 1$ atm, and $T_\infty = 519°$R, calculate the following properties at the location $(x, y) = (0.2$ ft, 0.2 ft$)$: M, p, and T.

11.2 Using the Prandtl-Glauert rule, calculate the lift coefficient for an NACA 2412 airfoil at 5° angle of attack in a Mach 0.6 freestream. (Refer to Figure 4.10 for the original airfoil data.)

11.3 Under low-speed incompressible flow conditions, the pressure coefficient at a given point on an airfoil is −0.54. Calculate C_p at this point when the freestream Mach number is 0.58, using

a.   The Prandtl-Glauert rule

b.   The Karman-Tsien rule

c.   Laitone's rule

11.4 In low-speed incompressible flow, the peak pressure coefficient (at the minimum pressure point) on an airfoil is -0.41. Estimate the critical Mach number for this airfoil, using the Prandtl-Glauert rule.

11.5 For a given airfoil, the critical Mach number is 0.8. Calculate the value of p/p_∞ at the minimum pressure point when $M_\infty = 0.8$.

11.6 Consider an airfoil in a Mach 0.5 freestream. At a given point on the airfoil, the local Mach number is 0.86. Using the compressible flow tables at the back of this book, calculate the pressure coefficient at that point. Check your answer using the appropriate analytical equation from this chapter. [*Note:* This problem is analogous to an incompressible problem where the freestream velocity and the velocity at a point are given, and the pressure coefficient is calculated from Equation (3.38). In an incompressible flow, the pressure coefficient at any point in the flow is a unique function of the local velocity at that point and the freestream velocity. In the present problem, we see that Mach number is the relevant property for a compressible flow—not velocity. The pressure coefficient for an inviscid compressible flow is a unique function of the local Mach number and the freestream Mach number.]

11.7 Figure 11.5 shows four cases for the flow over the same airfoil wherein M_∞ is progressively increased from 0.3 to $M_{cr} = 0.61$. Have you wondered where the numbers on Figure 11.5 came from? Here is your chance to find out. Point A on the airfoil is the point of minimum pressure (hence maximum M) on the airfoil. Assume that the minimum pressure (maximum Mach number) continues to occur at this same point as M_∞ is increased. In part (*a*) of Figure 11.5, for $M_\infty = 0.3$, the local Mach number at point A was arbitrarily chosen as $M_A = 0.435$, this arbitrariness is legitimate because we have not specified the airfoil shape, but rather are stating that, whatever the shape is, a maximum Mach number of 0.435 occurs at point A on the airfoil surface. However, once the numbers are given for part (*a*), then the numbers for parts (*b*), (*c*), and (*d*) are *not* arbitrary. Rather, M_A is a unique function of M_∞ for the remaining pictures. With all this as background information, starting with the data shown in Figure 11.5a, *calculate M_A when $M_\infty = 0.61$.* Obviously, from Figure 11.5d, your result should turn out to be $M_A = 1.0$ because $M_\infty = 0.61$ is said to be the critical Mach number. Said in another way, you are being asked to prove that the critical Mach number for this airfoil is 0.61. *Hint:* For simplicity, assume that the Prandtl-Glauert rule holds for the conditions of this problem.

11.8 Consider the flow over a circular cylinder; the incompressible flow over such a cylinder is discussed in Section 3.13. Consider also the flow over a sphere; the incompressible flow over a sphere is described in Section 6.4. The subsonic compressible flow over both the cylinder and the sphere is qualitatively similar but quantitatively different from their incompressible counterparts. Indeed, because of the "bluntness" of these bodies, their critical Mach numbers are relatively low. In particular:

For a cylinder: $M_{cr} = 0.404$

For a sphere: $M_{cr} = 0.57$

Explain on a physical basis why the sphere has a higher M_{cr} than the cylinder.

11.9 In Problem 11.8, the critical Mach number for a circular cylinder is given as $M_{cr} = 0.404$. This value is based on experimental measurements, and therefore is considered reasonably accurate. Calculate M_{cr} for a circular cylinder using the incompressible result for C_p and the Prandtl-Glauert compressibility correction, and compare your result with the experimental value. *Note:* The Prandtl-Glauert rule is based on linear theory assuming small perturbations, and therefore we would not expect that it would be valid for the case of flow over a circular cylinder. Nevertheless, when you use it to make this calculation of M_{cr}, you will find your calculated result to be within 3.5 percent of the experimental value. Interesting.

11.10 Consider an airfoil flying at zero angle of attack. At low speeds ($M_{\infty} < 0.2$), the minimum pressure coefficient at a particular location on the airfoil is ~0.8. Please find the airfoil's critical Mach number.

11.11 An aircraft is in steady level flight at $M_{\infty} = 0.55$ with $T_{\infty} = 273°K$ and $p_{\infty} = 10,000 \ N/m^2$. Performance data for the aircraft's airfoil measured under incompressible conditions indicate that the lift coefficient is 0.2 at the angle of attack associated with level flight. Estimate the aircraft's weight if the aerodynamic reference area is 15 m^2.

CHAPTER 12

Linearized Supersonic Flow

With the stabilizer setting at 2° the speed was allowed to increase to approximately 0.98 to 0.99 Mach number where elevator and rudder effectiveness were regained and the airplane seemed to smooth out to normal flying characteristics. This development lent added confidence and the airplane was allowed to continue until an indication of 1.02 on the cockpit Mach meter was obtained. At this indication the meter momentarily stopped and then jumped at 1.06, and this hesitation was assumed to be caused by the effect of shock waves on the static source. At this time the power units were cut and the airplane allowed to decelerate back to the subsonic flight condition.

Captain Charles Yeager, describing his flight on October 14, 1947—the first human-operated flight to exceed the speed of sound.

PREVIEW BOX

The calculation of lift and drag for an airfoil at supersonic speeds is about as different from that for lower-speed airfoils as night is from day. The physics of a supersonic flow is completely different from that of a subsonic flow, and therefore virtually nothing discussed in Chapter 4 or Chapter 11 can be used to calculate the properties of an airfoil at supersonic speeds. So what can you do? The answer is that you can read this chapter. It is short and sweet, and it provides some simple results that you can use to estimate the aerodynamic properties of supersonic airfoils.

12.1 INTRODUCTION

The linearized perturbation velocity potential equation derived in Chapter 11, Equation (11.18), is

$$\left(1 - M_\infty^2\right)\frac{\partial^2 \hat{\phi}}{\partial x^2} + \frac{\partial^2 \hat{\phi}}{\partial y^2} = 0 \tag{11.18}$$

and holds for both subsonic and supersonic flow. In Chapter 11, we treated the case of subsonic flow, where $1 - M_\infty^2 > 0$ in Equation (11.18). However, for supersonic flow, $1 - M_\infty^2 < 0$. This seemingly innocent change in sign on the first term of Equation (11.18) is, in reality, a very dramatic change. Mathematically, when $1 - M_\infty^2 > 0$ for subsonic flow, Equation (11.18) is an *elliptic* partial differential equation, whereas when $1 - M_\infty^2 < 0$ for supersonic flow, Equation (11.18) becomes a *hyperbolic* differential equation. The details of this mathematical difference are beyond the scope of this book; however, the important point is that there *is* a difference. Moreover, this portends a fundamental difference in the physical aspects of subsonic and supersonic flow—something we have already demonstrated in previous chapters.

The purpose of this chapter is to obtain a solution of Equation (11.18) for supersonic flow and to apply this solution to the calculation of supersonic airfoil properties. Since our purpose is straightforward, and since this chapter is relatively short, there is no need for a chapter road map to provide guidance on the flow of our ideas.

12.2 DERIVATION OF THE LINEARIZED SUPERSONIC PRESSURE COEFFICIENT FORMULA

For the case of supersonic flow, let us write Equation (11.18) as

$$\lambda^2 \frac{\partial^2 \hat{\phi}}{\partial x^2} - \frac{\partial \hat{\phi}}{\partial y^2} = 0 \tag{12.1}$$

where $\lambda = \sqrt{M_\infty^2 - 1}$. A solution to this equation is the functional relation

$$\hat{\phi} = f(x - \lambda y) \tag{12.2}$$

We can demonstrate this by substituting Equation (12.2) into Equation (12.1) as follows. The partial derivative of Equation (12.2) with respect to x can be written as

$$\frac{\partial \hat{\phi}}{\partial x} = f'(x - \lambda y)\frac{\partial(x - \lambda y)}{\partial x}$$

or

$$\frac{\partial \hat{\phi}}{\partial x} = f' \tag{12.3}$$

In Equation (12.3), the prime denotes differentiation of f with respect to its argument, $x - \lambda y$. Differentiating Equation (12.3) again with respect to x, we obtain

$$\frac{\partial^2 \hat{\phi}}{\partial x^2} = f'' \tag{12.4}$$

Similarly,
$$\frac{\partial \hat{\phi}}{\partial y} = f'(x - \lambda y)\frac{\partial(x - \lambda y)}{\partial y}$$

or
$$\frac{\partial \hat{\phi}}{\partial y} = f'(-\lambda) \tag{12.5}$$

Differentiating Equation (12.5) again with respect to y, we have

$$\frac{\partial^2 \hat{\phi}}{\partial y^2} = \lambda^2 f'' \tag{12.6}$$

Substituting Equations (12.4) and (12.6) into (12.1), we obtain the identity

$$\lambda^2 f'' - \lambda^2 f'' = 0$$

Hence, Equation (12.2) is indeed a solution of Equation (12.1).

Examine Equation (12.2) closely. This solution is not very specific, because f can be *any* function of $x - \lambda y$. However, Equation (12.2) tells us something specific about the flow, namely, that $\hat{\phi}$ is *constant* along lines of $x - \lambda y =$ constant. The slope of these lines is obtained from

$$x - \lambda y = \text{const}$$

Hence,
$$\frac{dy}{dx} = \frac{1}{\lambda} = \frac{1}{\sqrt{M_\infty^2 - 1}} \tag{12.7}$$

From Equation (9.31) and the accompanying Figure 9.25, we know that

$$\tan \mu = \frac{1}{\sqrt{M_\infty^2 - 1}} \tag{12.8}$$

where μ is the Mach angle. Therefore, comparing Equations (12.7) and (12.8), we see that a line along which $\hat{\phi}$ is constant is a *Mach line*. This result is sketched in Figure 12.1, which shows supersonic flow over a surface with a small hump in the middle, where θ is the angle of the surface relative to the horizontal. According to Equations (12.1) to (12.8), all disturbances created at the wall (represented by the perturbation potential $\hat{\phi}$) propagate unchanged away from the wall along Mach waves. All the Mach waves have the same slope, namely, $dy/dx = (M_\infty^2 - 1)^{-1/2}$. Note that the Mach waves slope *downstream* above the wall. Hence, *any disturbance at the wall cannot propagate upstream;* its effect is limited to the region of

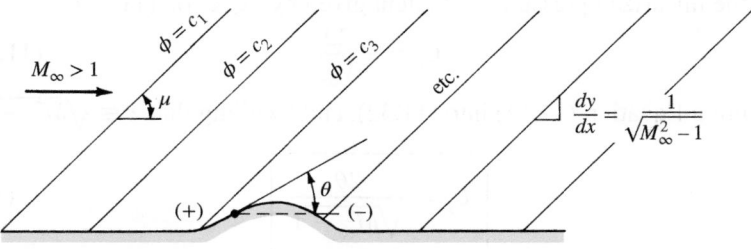

Figure 12.1 Linearized supersonic flow.

the flow downstream of the Mach wave emanating from the point of the disturbance. This is a further substantiation of the major difference between subsonic and supersonic flows mentioned in previous chapters, namely, disturbances propagate *everywhere* throughout a subsonic flow, whereas they cannot propagate upstream in a steady supersonic flow.

Keep in mind that the above results, as well as the picture in Figure 12.1, pertain to *linearized* supersonic flow [because Equation (12.1) is a linear equation]. Hence, these results assume *small perturbations;* that is, the hump in Figure 12.1 is small, and thus θ is small. Of course, we know from Chapter 9 that in reality a shock wave will be induced by the forward part of the hump, and an expansion wave will emanate from the remainder of the hump. These are waves of finite strength and are not a part of linearized theory. Linearized theory is approximate; one of the consequences of this approximation is that waves of finite strength (shock and expansion waves) are not admitted.

The above results allow us to obtain a simple expression for the pressure coefficient in supersonic flow, as follows. From Equation (12.3),

$$\hat{u} = \frac{\partial \hat{\phi}}{\partial x} = f' \tag{12.9}$$

and from Equation (12.5),

$$\hat{v} = \frac{\partial \hat{\phi}}{\partial y} = -\lambda f' \tag{12.10}$$

Eliminating f' from Equations (12.9) and (12.10), we obtain

$$\hat{u} = -\frac{\hat{v}}{\lambda} \tag{12.11}$$

Recall the linearized boundary condition given by Equation (11.34), repeated below:

$$\hat{v} = \frac{\partial \hat{\phi}}{\partial y} = V_\infty \tan \theta \tag{12.12}$$

We can further reduce Equation (12.12) by noting that, for small perturbations, θ is small. Hence, $\tan \theta \approx \theta$, and Equation (12.12) becomes

$$\hat{v} = V_\infty \theta \tag{12.13}$$

Substituting Equation (12.13) into (12.11), we obtain

$$\hat{u} = -\frac{V_\infty \theta}{\lambda} \tag{12.14}$$

Recall the linearized pressure coefficient given by Equation (11.32):

$$C_p = -\frac{2\hat{u}}{V_\infty} \tag{11.32}$$

Substituting Equation (12.14) into (11.32), and recalling that $\lambda \equiv \sqrt{M_\infty^2 - 1}$, we have

$$\boxed{C_p = \frac{2\theta}{\sqrt{M_\infty^2 - 1}}} \tag{12.15}$$

Equation (12.15) is important. It is the linearized supersonic pressure coefficient, and it states that C_p is directly proportional to the local surface inclination with respect to the freestream. It holds for any slender two-dimensional body where θ is small.

Return again to Figure 12.1. Note that θ is positive when measured above the horizontal, and negative when measured below the horizontal. Hence, from Equation (12.15), C_p is positive on the forward portion of the hump, and negative on the rear portion. This is denoted by the (+) and (−) signs in front of and behind the hump shown in Figure 12.1. This is also somewhat consistent with our discussions in Chapter 9; in the real flow over the hump, a compression wave forms on that part of the front portion where the flow is being turned into itself, and hence $p > p_\infty$, whereas an expansion wave occurs over that portion of the hump where the flow is turned away from itself, and the pressure decreases. Think about the picture shown in Figure 12.1; the pressure is higher on the front section of the hump, and lower on the rear section. As a result, a drag force exists on the hump. This drag is called *wave drag* and is a characteristic of supersonic flows. Wave drag was discussed in Section 9.7 in conjunction with shock-expansion theory applied to supersonic airfoils. It is interesting that linearized supersonic theory also predicts a finite wave drag, although shock waves themselves are not treated in such linearized theory.

Examining Equation (12.15), we note that $C_p \propto (M_\infty^2 - 1)^{-1/2}$; hence, for supersonic flow, C_p decreases as M_∞ increases. This is in direct contrast with subsonic flow, where Equation (11.51) shows that $C_p \propto (1 - M_\infty^2)^{-1/2}$; hence, for subsonic flow, C_p increases as M_∞ increases. These trends are illustrated in Figure 12.2. Note that both results predict $C_p \to \infty$ as $M_\infty \to 1$ from either side. However, keep in mind that neither Equation (12.15) nor (11.51) is valid in the transonic range around Mach 1.

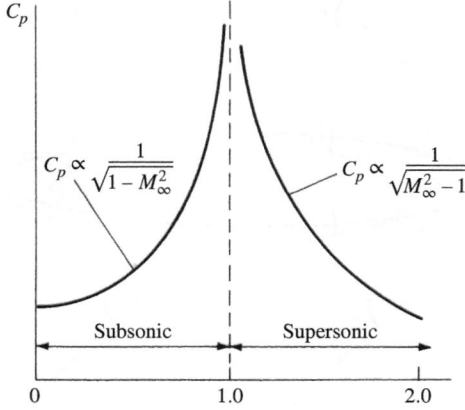

Figure 12.2 Variation of the linearized pressure coefficient with Mach number (schematic).

12.3 APPLICATION TO SUPERSONIC AIRFOILS

Equation (12.15) is very handy for estimating the lift and wave drag for thin super-sonic airfoils, such as sketched in Figure 12.3. When applying Equation (12.15) to any surface, one can follow a formal sign convention for θ, which is different for regions of left-running waves (such as above the airfoil in Figure 12.3) than for regions of right-running waves (such as below the airfoil in Figure 12.3). This sign convention is developed in detail in Reference 21. However, for our purpose here, there is no need to be concerned about the sign associated with θ in Equation (12.15). Rather, keep in mind that when the surface is inclined *into* the freestream direction, linearized theory predicts a positive C_p. For example, points A and B in Figure 12.3 are on surfaces inclined into the freestream, and hence $C_{p,A}$ and $C_{p,B}$ are positive values given by

$$C_{p,A} = \frac{2\theta_A}{\sqrt{M_\infty^2 - 1}} \quad \text{and} \quad C_{p,B} = \frac{2\theta_B}{\sqrt{M_\infty^2 - 1}}$$

In contrast, when the surface is inclined *away from* the freestream direction, lin-earized theory predicts a negative C_p. For example, points C and D in Figure 12.3 are on surfaces inclined away from the freestream, and hence $C_{p,C}$ and $C_{p,D}$ are negative values, given by

$$C_{p,C} = -\frac{2\theta_C}{\sqrt{M_\infty^2 - 1}} \quad \text{and} \quad C_{p,D} = -\frac{2\theta_D}{\sqrt{M_\infty^2 - 1}}$$

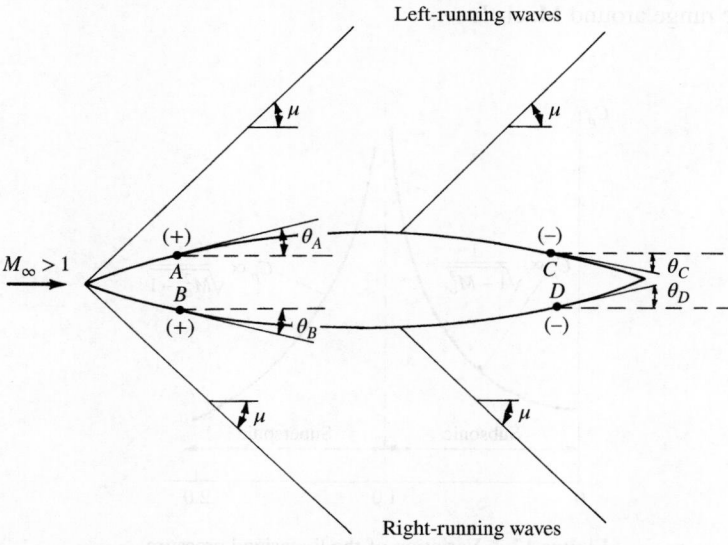

Figure 12.3 Linearized supersonic flow over an airfoil.

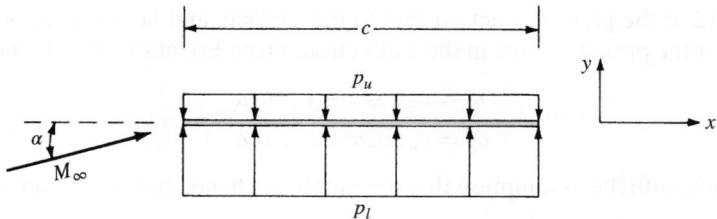

Figure 12.4 A flat plate at angle of attack in a supersonic flow.

In the above expressions, θ is always treated as a positive quantity, and the sign of C_p is determined simply by looking at the body and noting whether the surface is inclined into or away from the freestream.

With the distribution of C_p over the airfoil surface given by Equation (12.15), the lift and drag coefficients, c_l and c_d, respectively, can be obtained from the integrals given by Equations (1.15) to (1.19).

Let us consider the simplest possible airfoil, namely, a flat plate at a small angle of attack α as shown in Figure 12.4. Looking at this picture, the lower surface of the plate is a compression surface inclined at the angle α into the freestream, and from Equation (12.15),

$$C_{p,l} = \frac{2\alpha}{\sqrt{M_\infty^2 - 1}} \tag{12.16}$$

Since the surface inclination angle is constant along the entire lower surface, $C_{p,l}$ is a constant value over the lower surface. Similarly, the top surface is an expansion surface inclined at the angle α away from the freestream, and from Equation (12.15),

$$C_{p,u} = -\frac{2\alpha}{\sqrt{M_\infty^2 - 1}} \tag{12.17}$$

$C_{p,u}$ is constant over the upper surface. The normal force coefficient for the flat plate can be obtained from Equation (1.15):

$$c_n = \frac{1}{c} \int_0^c (C_{p,l} - C_{p,u}) \, dx \tag{12.18}$$

Substituting Equations (12.16) and (12.17) into (12.18), we obtain

$$c_n = \frac{4\alpha}{\sqrt{M_\infty^2 - 1}} \frac{1}{c} \int_0^c dx = \frac{4\alpha}{\sqrt{M_\infty^2 - 1}} \tag{12.19}$$

The axial force coefficient is given by Equation (1.16):

$$c_a = \frac{1}{c} \int_{\text{LE}}^{\text{TE}} (C_{p,u} - C_{p,l}) \, dy \tag{12.20}$$

However, the flat plate has (theoretically) zero thickness. Hence, in Equation (12.20), $dy = 0$, and as a result, $c_a = 0$. This is also clearly seen in

Figure 12.4; the pressures act normal to the surface, and hence there is no component of the pressure force in the x direction. From Equations (1.18) and (1.19),

$$c_l = c_n \cos \alpha - c_a \sin \alpha \tag{1.18}$$

$$c_d = c_n \sin \alpha + c_a \cos \alpha \tag{1.19}$$

and, along with the assumption that α is small and hence $\cos \alpha \approx 1$ and $\sin \alpha \approx \alpha$, we have

$$c_l = c_n - c_a \alpha \tag{12.21}$$

$$c_d = c_n \alpha + c_a \tag{12.22}$$

Substituting Equation (12.19) and the fact that $c_a = 0$ into Equations (12.21) and (12.22), we obtain

$$c_l = \frac{4\alpha}{\sqrt{M_\infty^2 - 1}} \tag{12.23}$$

$$c_d = \frac{4\alpha^2}{\sqrt{M_\infty^2 - 1}} \tag{12.24}$$

Equations (12.23) and (12.24) give the lift and wave-drag coefficients, respectively, for the supersonic flow over a flat plate. Keep in mind that they are results from linearized theory and therefore are valid only for small α.

For a thin airfoil of arbitrary shape at small angle of attack, linearized theory gives an expression for c_l identical to Equation (12.23); that is,

$$c_l = \frac{4\alpha}{\sqrt{M_\infty^2 - 1}}$$

Within the approximation of linearized theory, c_l depends only on α and is independent of the airfoil shape and thickness. However, the same linearized theory gives a wave-drag coefficient in the form of

$$c_d = \frac{4}{\sqrt{M_\infty^2 - 1}}(\alpha^2 + g_c^2 + g_t^2)$$

where g_c and g_t are functions of the airfoil camber and thickness, respectively. For more details, see References 25 and 26.

EXAMPLE 12.1

Using linearized theory, calculate the lift and drag coefficients for a flat plate at a 5° angle of attack in a Mach 3 flow. Compare with the exact results obtained in Example 9.11.

■ **Solution**

$$\alpha = 5° = 0.087 \text{ rad}$$

From Equation (12.23),

$$c_l = \frac{4\alpha}{\sqrt{M_\infty^2 - 1}} = \frac{(4)(0.087)}{\sqrt{(3)^2 - 1}} = \boxed{0.123}$$

From Equation (12.24),

$$c_d = \frac{4\alpha^2}{\sqrt{M_\infty^2 - 1}} = \frac{4(0.087)}{\sqrt{(3)^2 - 1}} = \boxed{0.011}$$

The results calculated in Example 9.11 for the same problem are *exact results,* utilizing the exact oblique shock theory and the exact Prandtl-Meyer expansion-wave analysis. These results were

$$\left.\begin{array}{l} c_l = 0.125 \\ c_d = 0.011 \end{array}\right\} \quad \text{exact results from Example 9.11}$$

Note that, for the relatively small angle of attack of 5°, the linearized theory results are quite accurate—to within 1.6 percent.

EXAMPLE 12.2

The Lockheed F-104 supersonic fighter, shown in Figure 12.5, was the first fighter designed for sustained flight at Mach 2. The F-104 embodies good supersonic aircraft design— long slender fuselage, sharp pointed nose, and a wing with an extremely thin airfoil of 3.4 percent thickness and a razor sharp leading edge (so sharp that a protective covering is placed on the leading edge for ground handling). All these features have one purpose— to reduce supersonic wave drag. The planform area of the wing is 18.21 m^2. Consider the case of the F-104 in steady, level flight at Mach 2 at 11 km altitude. The weight of the airplane is at its combat weight of 9400 kg$_f$. Assume that all the lift of the airplane comes from the lift on the wings (i.e., ignore the lift of the fuselage and tail). Calculate the angle of attack of the wing relative to the freestream.

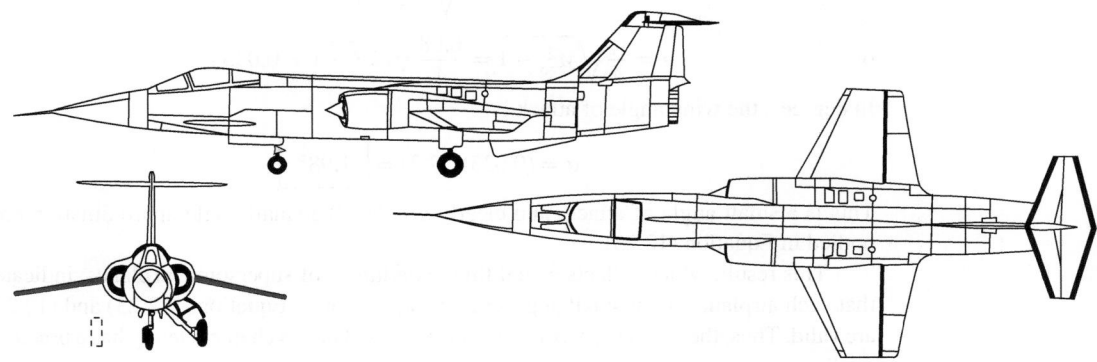

Figure 12.5 Lockheed F-104.

■ **Solution**

We first calculate the lift coefficient associated with the given flight conditions, recognizing that for level flight, the lift of the airplane must equal its weight:

$$C_L = \frac{L}{q_\infty S} = \frac{W}{q_\infty S}$$

At an altitude of 11 km, from Appendix D, $\rho_\infty = 0.3648$ kg/m^3 and $T_\infty = 216.78$ K. Hence,

$$a_\infty = \sqrt{\gamma R T_\infty} = \sqrt{(1.4)(287)(216.78)} = 295 \text{ m/s}$$

$$V_\infty = M_\infty a_\infty = (2)(295) = 590 \text{ m/s}$$

$$q_\infty = \tfrac{1}{2}\rho_\infty V_\infty^2 = \tfrac{1}{2}(0.3648)(590)^2 = 6.35 \times 10^4 \text{ m/s}$$

Also, the weight is given as 9400 kg$_f$. This is an inconsistent unit of force; a newton is the consistent unit of force in the SI system. Since 1 kg$_f$ = 9.8 N, we have

$$W = (9400)(9.8) = 9.212 \times 10^4 \text{ N}$$

$$C_L = \frac{W}{q_\infty S} = \frac{9.212 \times 10^4}{(6.35 \times 10^4)(18.21)} = 0.08$$

We make the assumption that the lift coefficient of the wing, C_L, is the same as the lift coefficient for the airfoil section making up the wing, c_l. For subsonic flight, we know from our discussion in Chapter 5 that such is not the case. For supersonic flight, such is also not the case. For a straight wing in supersonic inviscid flight, however, the wing tip effects are effectively limited to the region inside the Mach cone with its vertex at the tip leading edge. At Mach 2, the semivertex angle of the Mach cone is $\mu = \sin^{-1}\frac{1}{2} = 30°$. Therefore, much of the wing is unaffected by the tip effects, and experiences the two-dimensional flow discussed in this chapter. The airfoil section of the F-104 is thin, and as we will soon see, is at a small angle of attack. We conclude, therefore, that Equation (12.23) should be a good approximation for the lift coefficient of the wing of the F-104. Thus, in Equation (12.23), we use $c_l = C_L = 0.08$. From Equation (12.23),

$$c_l = \frac{4\alpha}{\sqrt{M_\infty^2 - 1}}$$

or,

$$\alpha = \frac{c_l}{4}\sqrt{M_\infty^2 - 1} = \frac{0.08}{4}\sqrt{(2)^2 - 1} = 0.035 \text{ rad}$$

In degrees, the wing angle of attack is

$$\alpha = (0.035)(57.3) = \boxed{1.98°}$$

This is a small angle of attack, and clearly satisfies the small-angle approximation embodied in Equation (12.23).

This result, which reflects actual flight conditions of supersonic airplanes, indicates that such airplanes fly at small angles of attack, for which Equations (12.23) and (12.24) are valid. Thus, the linear supersonic theory discussed in this chapter clearly has a practical application.

DESIGN BOX

The area rule discussed in Section 11.8 can be labeled the *transonic* area rule, because its proper application in an airplane design will reduce the peak drag in the transonic flight regime, allowing the airplane to more readily achieve supersonic flight. The transonic area rule calls for a smooth variation of the cross-sectional area of the airplane measured *normal* to the freestream direction. For example, the transonic area ruling of the F-16 fighter is shown in Figure 12.6; here, the cross-sectional area normal to the freestream is plotted versus distance along the fuselage.

The area rule also applies at supersonic speeds; here, however, the relevant cross-sectional area is not that perpendicular to the freestream relative wind, but rather the area cut by an oblique plane at the freestream Mach angle. For reduced supersonic wave drag, the *supersonic* area rule calls for a smooth distribution of this oblique section area. For example, consider the F-16 flying at $M_\infty = 1.6$, for which the Mach angle is $\mu = \sin^{-1}(1/M_\infty) = \sin^{-1}(1/1.6) = 38.7$ deg. In Figure 12.7a, a side view of the F-16 is shown,

with the oblique cross sections drawn at the angle $\mu = 38.7$ deg. The variation of this oblique cross-sectional area with distance along the fuselage is shown at the top of Figure 12.7b. Note that this area distribution is quite smooth; the F-16 design satisfies the supersonic area rule at $M_\infty = 1.6$. The same holds true for $M_\infty = 1.2$. The oblique cross-sectional area for $\mu = \sin^{-1}(1/1.2) = 56.4$ deg as a function of distance along the fuselage is plotted at the bottom of Figure 12.7b; a smooth distribution holds for this case as well. Figures 12.6 and 12.7 show that the designers of the F-16 incorporated both the transonic and supersonic area rules.

The solid curves in Figure 12.7 show the actual area distributions for the F-16. The adjacent dashed curves illustrate the area distributions proposed in the early stage of the F-16 design process; based on wind tunnel tests, the final configuration was obtained by filling and shaving areas from parts of the airplane (see Figure 12.7a) in order to more faithfully obey the supersonic area rule.

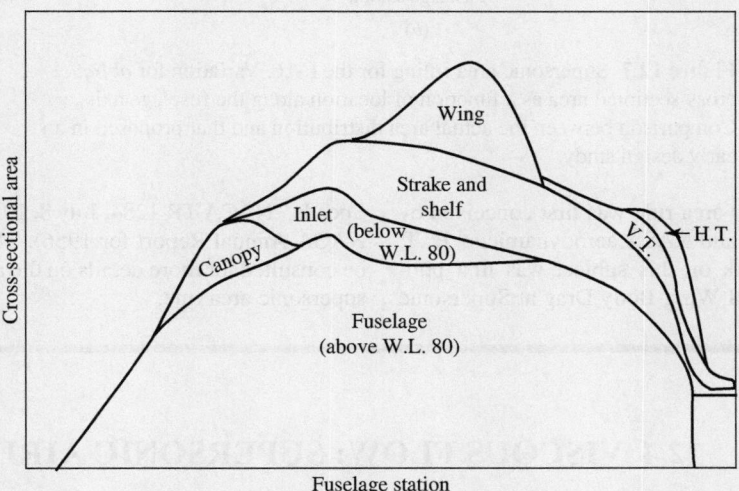

Figure 12.6 Transonic area ruling for the F-16. Variation of normal cross-sectional area as a function of location along the fuselage axis.

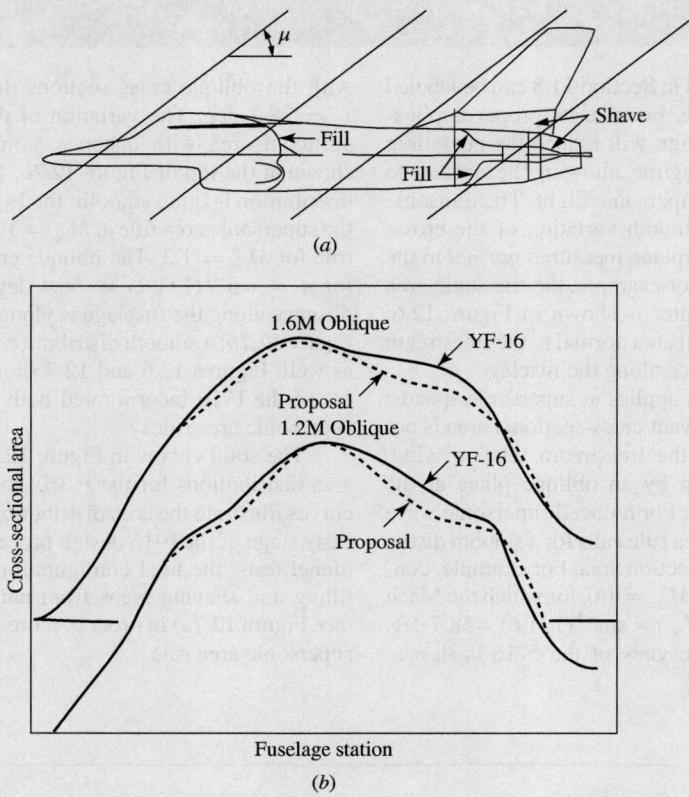

Figure 12.7 Supersonic area ruling for the F-16. Variation for *oblique* cross-sectional area as a function of location along the fuselage axis. Comparison between the actual area distribution and that proposed in an early design study.

The supersonic area rule was first conceived by the famous NACA and NASA aerodynamicist, R. T. Jones, and his work on this subject was first published in "Theory of Wing-Body Drag at Supersonic Speeds," NACA TR 1284, July 8, 1953 (printed in the NACA Annual Report for 1956). This report should be consulted for more details on the application of the supersonic area rule.

12.4 VISCOUS FLOW: SUPERSONIC AIRFOIL DRAG

Drag for an airfoil in low-speed flow was discussed in Section 4.12. Amazingly enough, the influence of compressibility, even supersonic flow, does not change the fundamental approach to the estimation of skin-friction drag as outlined in Section 4.12—only the numbers are different. The details behind the change in numbers are given in Chapters 18 and 19. Basically, the story goes as follows. For laminar incompressible flow the flat plate skin-friction drag coefficient is given by Equation (4.86) as

$$C_f = \frac{1.328}{\sqrt{\text{Re}_c}} \qquad \text{(laminar incompressible)}$$

And for turbulent incompressible flow approximately by Equation (4.88) as

$$C_f = \frac{0.074}{\text{Re}_c^{1/5}} \qquad \text{(turbulent incompressible)}$$

For compressible flow, the numerators of these equations are no longer constants, but rather can be viewed as functions of Mach number, the ratio of wall temperature to the temperature at the outer edge of the boundary layer T_w/T_e, and Prandtl number Pr. The Prandtl number is defined as $\text{Pr} = \mu c_p/k$, where μ, c_p, and k are the viscosity coefficient, specific heat at constant pressure, and thermal conductivity, respectively. The importance of Prandtl number is discussed at length in Chapter 15. In essence, for compressible laminar flow we have

$$C_f = \frac{F(M_e, \text{Pr}, T_w/T_e)}{\sqrt{\text{Re}_c}} \tag{12.25}$$

and for compressible turbulent flow we have approximately

$$C_f = \frac{G(M_e, \text{Pr}, T_w/T_e)}{(\text{Re}_c)^{1/5}} \tag{12.26}$$

For a given set of values of M_e, Pr, and T_w/T_e, the numerical value of the numerators in Equations (12.25) and (12.26) are obtained from numerical solutions of the boundary layer equations as discussed in Chapters 18 and 19. Some classic results are given in Figure 19.1; leap ahead and examine this figure. The results in Figure 19.1 are presented for Pr = 0.75 and for the ratio T_w/T_e corresponding to that for no heat transfer at the wall (an adiabatic wall). *The most important phenomenon to observe in this figure is that C_f decreases as M_∞ increases,* and that the decrease is more dramatic for a turbulent boundary layer than for a laminar boundary layer.

Keep in mind that, in addition to the skin-friction drag, a supersonic airfoil also experiences supersonic wave drag as discussed in Sections 9.7 and 12.3. The source of wave drag is the pressure distribution exerted over the airfoil surface and is a result of the shock-wave and expansion-wave patterns in the flow over the airfoil. The source of skin-friction drag is, of course, the shear stress exerted over the airfoil surface and is the result of friction in the flow. The physical mechanisms of wave drag and skin-friction drag clearly are quite different. How do these two types of drag compare in a practical flow situation? The following example addresses this question.

EXAMPLE 12.3

Consider the same Lockheed F-104 supersonic fighter described in Example 12.2, with the same flight conditions of Mach 2 at an altitude of 11 km. As calculated in Example 12.2, for these conditions the wing angle of attack is $\alpha = 0.035$ rad $= 1.98°$. Assume the chord length of the airfoil is 2.2 m, which is approximately the mean chord length for the wing. Also, assume fully turbulent flow over the airfoil. Calculate: (a) the airfoil skin friction drag coefficient, and (b) the airfoil wave-drag coefficient. Compare the two values of drag.

■ **Solution**

(a) To calculate the skin-friction drag coefficient, C_f, we need the Reynolds number. From Appendix D at 11 km, $\rho_\infty = 0.3648$ kg/m^3 and $T_\infty = 216.78$ K. The speed of sound at 11 km is therefore

$$a_\infty = \sqrt{\gamma R T_\infty} = \sqrt{(1.4)(287)(216.78)} = 295 \text{ m/s}$$

Thus,

$$V_\infty = M_\infty a_\infty = (2)(295) = 590 \text{ m/s}$$

As discussed in Sections 1.11 and 15.3, the viscosity coefficient is a function of temperature. We borrow Equation (15.3), which is Sutherland's law for the temperature variation of viscosity coefficient, given by

$$\frac{\mu}{\mu_0} = \left(\frac{T}{T_0}\right)^{3/2} \frac{T_0 + 110}{T + 110}$$

where T is in Kelvin, and μ_0 is a reference viscosity at a reference temperature, T_0. We choose reference conditions to be the standard sea level values of $\mu_0 = 1.7894 \times 10^{-5}$ kg/(m)(s) and $T_0 = 288.16$ K. Thus, from Sutherland's law, the value of μ at $T = 216.78$ K is

$$\mu = (1.7894 \times 10^{-5}) \left(\frac{216.78}{288.16}\right)^{3/2} \left(\frac{288.16 + 110}{216.78 + 110}\right) = 1.4226 \times 10^{-5} \text{ kg/(m)(s)}$$

The Reynolds number is

$$\text{Re} = \frac{\rho_\infty V_\infty c}{\mu_\infty} = \frac{(0.3648)(590)(2.2)}{1.4226 \times 10^{-5}} = 3.33 \times 10^7$$

Reading Figure 19.1 very carefully, for a turbulent boundary at Mach 2 with Re = 3.33×10^7, we have $C_f = 2.15 \times 10^{-3}$. This is for one side of a flat plate. As we have done before, we assume the skin-friction drag of the thin airfoil (and recall that the airfoil on the F-104 is really thin) to be represented by that on a flat plate. Since the skin-friction drag acts on both the upper and lower surface of the airfoil, we have, for the net skin-friction drag coefficient for the F-104 airfoil,

$$\text{Net } C_f = 2(2.15 \times 10^{-3}) = \boxed{4.3 \times 10^{-3}}$$

(b) The wave-drag coefficient is given by Equation (12.24) as

$$c_d = \frac{4\alpha^2}{\sqrt{M_\infty^2 - 1}}$$

Here, $\alpha = 0.035$ rad, and $M_\infty = 2$. Hence,

$$c_d = \frac{4(0.035)^2}{\sqrt{(2)^2 - 1}} = \boxed{2.83 \times 10^{-3}}$$

From Example 12.3, we see that the total drag coefficient for the supersonic airfoil is $c_d + C_f = 2.83 \times 10^{-3} + 4.3 \times 10^{-3} = 7.13 \times 10^{-3}$. The skin friction is 60 percent of the total drag. At the low angle of attack for the airfoil in Example 12.3, the supersonic wave drag is reasonably small, in this case smaller than the skin friction drag. The wave drag, however, varies as the square of the angle of attack. For the conditions of Example 12.3, the wave drag would equal the skin-friction drag at an angle of attack of 2.47°, and as the angle of attack increases above this value, the wave drag would rapidly become the dominant type of drag.

Let us take a look at the lift-to-drag ratio, L/D, of the airfoil in Example 12.3. First of all, for an inviscid flow (i.e., no friction drag), the results of Examples 12.2 and 12.3 show that

$$\frac{L}{D} = \frac{c_l}{c_d} = \frac{0.08}{2.83 \times 10^{-3}} = 28.3 \quad \text{(inviscid)}$$

By including the skin friction drag, we have

$$\frac{L}{D} = \frac{c_l}{(c_d)_{\text{total}}} = \frac{0.08}{7.13 \times 10^{-3}} = 11.2$$

Clearly, the skin-friction drag greatly diminishes the lift-to-drag ratio of the airfoil. Since the value of L/D is an important measure of aerodynamic efficiency, we see the importance of trying to reduce the skin-friction drag, say by encouraging laminar rather than turbulent boundary layers on the airfoil. Indeed, this is essentially a generic statement that holds true across all parts of the Mach number spectrum, from low-speed subsonic flow to hypersonic flow. In our discussion of supersonic airfoils in this section, we can clearly see the importance of skin-friction drag on aerodynamic performance.

12.5 SUMMARY

In linearized supersonic flow, information is propagated along Mach lines where the Mach angle $\mu = \sin^{-1}(1/M_\infty)$. Since these Mach lines are all based on M_∞, they are straight, parallel lines which propagate away from and downstream of a body. For this reason, disturbances cannot propagate upstream in a steady supersonic flow.

The pressure coefficient, based on linearized theory, on a surface inclined at a small angle θ to the freestream is

$$C_p = \frac{2\theta}{\sqrt{M_\infty^2 - 1}} \qquad (12.15)$$

If the surface is inclined into the freestream, C_p is positive; if the surface is inclined away from the freestream, C_p is negative.

Based on linearized supersonic theory, the lift and wave-drag coefficients for a flat plate at an angle of attack are

$$c_l = \frac{4\alpha}{\sqrt{M_\infty^2 - 1}} \qquad (12.23)$$

and

$$c_d = \frac{4\alpha^2}{\sqrt{M_\infty^2 - 1}} \qquad (12.24)$$

Equation (12.23) also holds for a thin airfoil of arbitrary shape. However, for such an airfoil, the wave-drag coefficient depends on both the shape of the mean camber line and the airfoil thickness.

12.6 PROBLEMS

12.1 Using the results of linearized theory, calculate the lift and wave-drag coefficients for an infinitely thin flat plate in a Mach 2.6 freestream at angles of attack of

a. $\alpha = 5°$
b. $\alpha = 15°$
c. $\alpha = 30°$

Compare these approximate results with those from the exact shock-expansion theory obtained in Problem 9.13. What can you conclude about the accuracy of linearized theory in this case?

12.2 For the conditions of Problem 12.1, calculate the pressures (in the form of p/p_∞) on the top and bottom surfaces of the flat plate, using linearized theory. Compare these approximate results with those obtained from exact shock-expansion theory in Problem 9.13. Make some appropriate conclusions regarding the accuracy of linearized theory for the calculation of pressures.

12.3 Consider a diamond-wedge airfoil such as shown in Figure 9.37, with a half-angle $\varepsilon = 10°$. The airfoil is at an angle of attack $\alpha = 15°$ to a Mach 3 freestream. Using linear theory, calculate the lift and wave-drag coefficients for the airfoil. Compare these approximate results with those from the exact shock-expansion theory obtained in Problem 9.14.

12.4 Equation (12.24), from linear supersonic theory, predicts that c_d for a flat plate decreases as M_∞ increases? Does this mean that the drag force itself decreases as M_∞ increases? To answer this question, derive an equation for drag as a function of M_∞, and evaluate this equation.

12.5 Consider a flat plate at an angle of attack in an inviscid supersonic flow. From linear theory, what is the value of the maximum lift-to-drag ratio, and at what angle of attack does it occur?

12.6 Consider a flat plate at an angle of attack in a viscous supersonic flow; i.e., there is both skin friction drag and wave drag on the plate. Use linear theory for the lift and wave-drag coefficients. Denote the total skin friction drag coefficient by C_f, and assume that it does not change with angle of attack. (a) Derive the expression for the angle of attack at which maximum lift-to-drag ratio occurs as a function of C_f and freestream Mach number. (b) Derive the expression for the maximum lift-to-drag ratio as a function of C_f and freestream Mach number M.

Answers: (a) $\alpha = (C_f)^{1/2}(M^2 - 1)^{1/4}/2$;
(b) $(c_l/c_d)_{max} = (C_f)^{-1/2}(M^2 - 1)^{-1/4}$

12.7 Using the same flight conditions and the same value of the skin-friction coefficient from Example 12.3, and the results of Problem 12.6, calculate the maximum lift-to-drag ratio of the flat plate that is used to simulate the F-104 wing and the angle of attack at which it occurs.

12.8 The result from Problem 12.6 demonstrates that maximum lift-to-drag ratio decreases as the Mach number increases. This is a fact of nature that progressively causes designers of supersonic airplanes grief as they strive toward aerodynamically efficient airplanes at higher supersonic Mach numbers. What physics is nature using against the airplane designer in this case, and how might the designer meet this challenge?

12.9 An F-15 is in straight and level flight at about $36,000\,ft$ where $p_\infty = 22,700\,N/m^2$. Its wing loading (W_S) is about $400\,kg/m^2$. Assume $\gamma = 1.4$ and recall that wing loading is defined as the vehicle weight divided by the total wing area.

a. Use thin airfoil theory to develop an expression for the flight Mach number in terms of α, p_∞, γ and the wing loading (W_S). Calculate the flight Mach number if $\alpha = 1.3°$.

b. The aircraft's weight decreases over the course of the mission. How should the pilot change the angle of attack in order to maintain the same flight Mach number?

12.10 Use linearized theory to calculate the lift/span, drag/span, lift coefficient, drag coefficient, and lift to drag ratio of the symmetric two-dimensional airfoil of Problem 9.35 for $0 \leq \alpha \leq 10°$. Compare to your findings in Problem 9.35 based on shock-expansion theory.

CHAPTER 13

Introduction to Numerical Techniques for Nonlinear Supersonic Flow

Regarding computing as a straightforward routine, some theoreticians still tend to underestimate its intellectual value and challenge, while practitioners often ignore its accuracy and overrate its validity.

C. K. Chu, 1978
Columbia University

PREVIEW BOX

Our discussions of compressible flow have involved algebra (Chapters 7–10) and linear partial differential equations (Chapters 11 and 12)—all tractable mathematics. In the process, we dealt with a number of practical applications. But this is about as far as we can go with our analytical solutions allowed by tractable mathematics. For all other applications, encompassing the vast majority of all other real-world applications, the flows are governed by the more complete *nonlinear* equations of motion, for which there are no closed-form analytical solutions.

Do we throw up our hands and quit? From the title of this chapter, obviously not. Instead, we leave the world of analytical solutions and enter the relatively new world of numerical solutions. Instead of finding nice, neat equations to solve our flows, we are going to crunch numbers for our flows. But we are going to crunch numbers in an intelligent, often elegant fashion. We are going to start with the fundamental nonlinear flow equations of continuity, momentum, and energy obtained in Chapter 2, and solve them numerically for some important practical problems that could not be solved in any other way.

Two such practical problems are addressed in this chapter. The first is the design of the proper contour, the proper *shape*, of a supersonic nozzle. In Chapter 10 we studied the flow properties of nozzle flows, but the shape (i.e., the area-ratio distribution) of the nozzle was always given. Look again at the rocket engine shown in Figure 10.1. *How* was the actual shape of that divergent nozzle actually designed? This is an extremely important question; if the contour of that divergent nozzle is not just right, undesirable shock waves may form inside the nozzle—shock waves that will reduce the performance of the rocket engine. And if you were designing the nozzle for a supersonic wind tunnel, any shock waves formed inside the nozzle are going to ruin the type of high-quality flow you want in the test section. So the design of the proper contour of a supersonic nozzle is extremely important. How do you do it? You will find the answers in this chapter.

The second problem is the solution of the flow around a blunt-nosed body in a supersonic flow. Leap ahead in this book and look at the Space Shuttle shown in Figure 14.17. What you see is a high-speed vehicle that has a blunt nose and a wing with a blunt leading edge. The fundamental reason for these blunt shapes was discussed in Section 1.1 (if you have forgotten, take a quick peek at Section 1.1). In the early days of high-speed flight, the solution of the flow around a blunt-nosed body in supersonic flight was impossible to obtain. And yet, it was, and still is, one of the most important problems in supersonic aerodynamics. We call it the "supersonic blunt body problem." Today, we can readily solve this problem. How? You will find the answer in this chapter.

The material in this chapter is serious business. It reflects modern aspects of compressible flow, and gives some idea of how solutions for complex problems are carried out today. Read on, and see what all this is about.

13.1 INTRODUCTION: PHILOSOPHY OF COMPUTATIONAL FLUID DYNAMICS

The above quotation underscores the phenomenally rapid increase in computer power available to engineers and scientists during 1960 to 1980. This explosion in computer capability is still going on, with no specific limits in sight. As a result, an entirely new discipline in aerodynamics has evolved over the past three decades, namely, computational fluid dynamics (CFD). CFD is a new "third dimension" in aerodynamics, complementing the previous dimensions of both pure experiment and pure theory. It allows us to obtain answers to fluid dynamic problems which heretofore were intractable by classical analytical methods. Consequently, CFD is revolutionizing the airplane design process, and in many ways is modifying the way we conduct modern aeronautical research and development. For these reasons, every modern student of aerodynamics should be aware of the overall philosophy of CFD, because you are bound to be affected by it to some greater or lesser degree in your education and professional life.

The philosophy of computational fluid dynamics was introduced in Section 2.17, where it was compared with the theoretical approach leading to closed-form analytical solutions. Please stop here, return to Section 2.17, and re-read the material presented there; now that you have progressed this far and have seen a number of analytical solutions for both incompressible and compressible flows in the preceding chapters, the philosophy discussed in Section 2.17 will mean much more to you. Do this now, because the present chapter almost exclusively deals with numerical solutions with reference to Section 2.17.2, whereas Chapters 3–12 have dealt almost exclusively with analytical solutions with reference to Section 2.17.1.

In the present chapter we will experience the true essence of computational fluid dynamics for the first time in this book; we will actually see what is meant by the definition of CFD given in Section 2.17.2 as "the art of replacing the integrals or the partial derivatives (as the case may be) in the governing equations of fluid motion with discretized algebraic forms, which in turn are solved to obtain numbers for the flow field values at discrete points in time and/or space." However, because modern CFD is such a sophisticated discipline that is usually the subject of graduate level studies, and which rests squarely on the foundations of applied mathematics, we can only hope to give you an elementary treatment in the present chapter, but a treatment significant enough to represent some of the essence of CFD. For your next step in learning CFD beyond the present book, you are recommended to read Anderson, *Computational Fluid Dynamics: The Basics with Applications* (Reference 60), which the author has written to help undergraduates understand the nature of CFD before going on to more advanced studies of the discipline.

The purpose of this chapter is to provide an introduction to some of the basic ideas of CFD as applied to inviscid supersonic flows. More details are given in Reference 21. Because CFD has developed so rapidly in recent years, we can only scratch the surface here. Indeed, the present chapter is intended to give you only some basic background as well as the incentive to pursue the subject further in the modern literature.

The road map for this chapter is given in Figure 13.1. We begin by introducing the classical method of characteristics—a numerical technique that has been available in aerodynamics since 1929, but which had to wait on the modern computer for practical, everyday implementation. For this reason, the author classifies the method of characteristics under the general heading of numerical

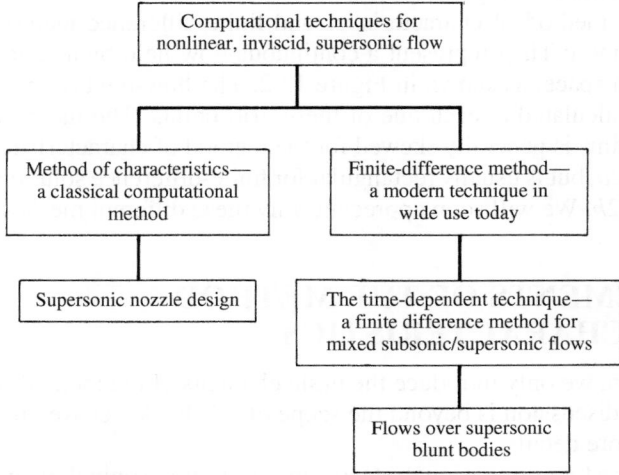

Figure 13.1 Road map for Chapter 13.

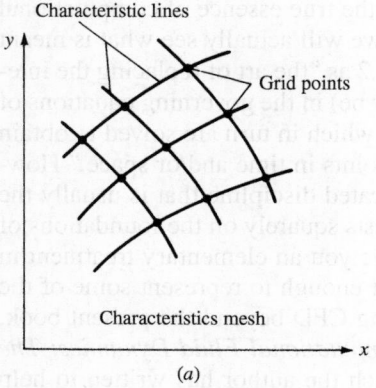

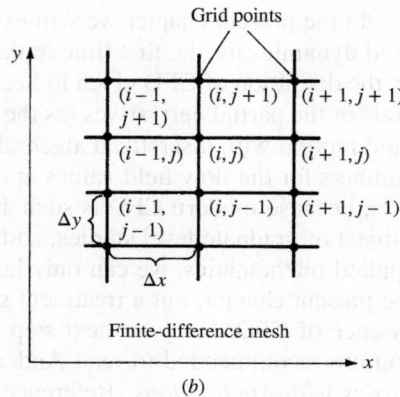

Figure 13.2 Grid points.

techniques, although others may prefer to list it under a more classical heading. We also show how the method of characteristics is applied to design the divergent contour of a supersonic nozzle. Then we move to a discussion of the finite-difference approach, which we will use to illustrate the application of CFD to nozzle flows and the flow over a supersonic blunt body.

In contrast to the linearized solutions discussed in Chapters 11 and 12, CFD represents numerical solutions to the *exact* nonlinear governing equations, that is, the equations without simplifying assumptions such as small perturbations, and which apply to all speed regimes, transonic and hypersonic as well as subsonic and supersonic. Although numerical roundoff and truncation errors are always present in any numerical representation of the governing equations, we still think of CFD solutions as being "exact solutions."

Both the method of characteristics and finite-difference methods have one thing in common: They represent a continuous flow field by a series of distinct grid points in space, as shown in Figure 13.2. The flow-field properties (u, v, p, T, etc.) are calculated at each one of these grid points. The mesh generated by these grid points is generally skewed for the method of characteristics, as shown in Figure 13.2a, but is usually rectangular for finite-difference solutions, as shown in Figure 13.2b. We will soon appreciate why these different meshes occur.

13.2 ELEMENTS OF THE METHOD OF CHARACTERISTICS

In this section, we only introduce the basic elements of the method of characteristics. A full discussion is beyond the scope of this book; see References 21, 25, and 32 for more details.

Consider a two-dimensional, steady, inviscid, supersonic flow in xy space, as given in Figure 13.2a. The flow variables (p, u, T, etc.) are continuous throughout this space. However, there are certain lines in xy space along which the *derivatives* of the flow field variables ($\partial p/\partial x$, $\partial u/\partial y$, etc.) are *indeterminate* and across which may even be discontinuous. Such lines are called *characteristic lines*. This may sound strange at first; however, let us prove that such lines exist, and let us find their precise directions in the xy plane.

In addition to the flow being supersonic, steady, inviscid, and two-dimensional, assume that it is also irrotational. The exact governing equation for such a flow is given by Equation (11.12):

$$\left[1 - \frac{1}{a^2}\left(\frac{\partial \phi}{\partial x}\right)^2\right]\frac{\partial^2 \phi}{\partial x^2} + \left[1 - \frac{1}{a^2}\left(\frac{\partial \phi}{\partial y}\right)^2\right]\frac{\partial^2 \phi}{\partial y^2} - \frac{2}{a^2}\frac{\partial \phi}{\partial x}\frac{\partial \phi}{\partial y}\frac{\partial^2 \phi}{\partial x\, \partial y} = 0$$

$$(11.12)$$

[Keep in mind that we are dealing with the full velocity potential ϕ in Equation (11.12), not the perturbation potential.] Since $\partial \phi/\partial x = u$ and $\partial \phi/\partial y = v$, Equation (11.12) can be written as

$$\left(1 - \frac{u^2}{a^2}\right)\frac{\partial^2 \phi}{\partial x^2} + \left(1 - \frac{v^2}{a^2}\right)\frac{\partial^2 \phi}{\partial y^2} - \frac{2uv}{a^2}\frac{\partial^2 \phi}{\partial x\, \partial y} = 0 \qquad (13.1)$$

The velocity potential and its derivatives are functions of x and y, for example,

$$\frac{\partial \phi}{\partial x} = f(x, y)$$

Hence, from the relation for an exact differential,

$$df = \frac{\partial f}{\partial x}dx + \frac{\partial f}{\partial y}dy$$

we have $\qquad d\left(\frac{\partial \phi}{\partial x}\right) = du = \frac{\partial^2 \phi}{\partial x^2}dx + \frac{\partial^2 \phi}{\partial x\, \partial y}dy \qquad (13.2)$

Similarly, $\qquad d\left(\frac{\partial \phi}{\partial y}\right) = dv = \frac{\partial^2 \phi}{\partial x\, \partial y}dx + \frac{\partial^2 \phi}{\partial y^2}dy \qquad (13.3)$

Examine Equations (13.1) to (13.3) closely. Note that they contain the second derivatives $\partial^2 \phi/\partial x^2$, $\partial^2 \phi/\partial y^2$, and $\partial^2 \phi/\partial x\, \partial y$. If we imagine these derivatives as "unknowns," then Equations (13.1), (13.2), and (13.3) represent three equations with three unknowns. For example, to solve for $\partial^2 \phi/\partial x\, \partial y$, use Cramer's rule as follows:

$$\frac{\partial^2 \phi}{\partial x\, \partial y} = \frac{\begin{vmatrix} 1 - \dfrac{u^2}{a^2} & 0 & 1 - \dfrac{v^2}{a^2} \\ dx & du & 0 \\ 0 & dv & dy \end{vmatrix}}{\begin{vmatrix} 1 - \dfrac{u^2}{a^2} & -\dfrac{2uv}{a^2} & 1 - \dfrac{v^2}{a^2} \\ dx & dy & 0 \\ 0 & dx & dy \end{vmatrix}} = \frac{N}{D} \qquad (13.4)$$

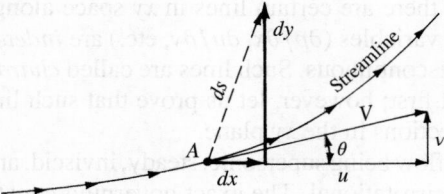

Figure 13.3 An arbitrary direction ds away from point A.

where N and D represent the numerator and denominator determinants, respectively. The physical meaning of Equation (13.4) can be seen by considering point A and its surrounding neighborhood in the flow, as sketched in Figure 13.3. The derivative $\partial^2\phi/\partial x\,\partial y$ has a specific value at point A. Equation (13.4) gives the solution for $\partial^2\phi/\partial x\,\partial y$ for an *arbitrary* choice of dx and dy. The combination of dx and dy defines an arbitrary direction ds away from point A as shown in Figure 13.3. In general, this direction is different from the streamline direction going through point A. In Equation (13.4), the differentials du and dv represent the changes in velocity that take place over the increments dx and dy. Hence, although the choice of dx and dy is arbitrary, the values of du and dv in Equation (13.4) must correspond to this choice. No matter what values of dx and dy are arbitrarily chosen, the corresponding values of du and dv will always ensure obtaining the same value of $\partial^2\phi/\partial x\,\partial y$ at point A from Equation (13.4).

The single exception to the above comments occurs when dx and dy are chosen so that $D = 0$ in Equation (13.4). In this case, $\partial^2\phi/\partial x\,\partial y$ is not defined. This situation will occur for a specific direction ds away from point A in Figure 13.3, defined for that specific combination of dx and dy for which $D = 0$. However, we know that $\partial^2\phi/\partial x\,\partial y$ has a specific defined value at point A. Therefore, the only consistent result associated with $D = 0$ is that $N = 0$, also; that is,

$$\frac{\partial^2\phi}{\partial x\,\partial y} = \frac{N}{D} = \frac{0}{0} \tag{13.5}$$

Here, $\partial^2\phi/\partial x\,\partial y$ is an indeterminate form, which is allowed to be a finite value, that is, that value of $\partial^2\phi/\partial x\,\partial y$ which we know exists at point A. *The important conclusion here is that there is some direction (or directions) through point A along which $\partial^2\phi/\partial x\,\partial y$ is indeterminate.* Since $\partial^2\phi/\partial x\,\partial y = \partial u/\partial y = \partial v/\partial x$, this implies that the derivatives of the flow variables are indeterminate along these lines. Hence, we have proven that lines do exist in the flow field along which derivatives of the flow variables are indeterminate; earlier, we defined such lines as *characteristic lines*.

Consider again point A in Figure 13.3. From our previous discussion, there are one or more characteristic lines through point A. *Question:* How can we calculate the precise direction of these characteristic lines? The answer can be obtained by setting $D = 0$ in Equation (13.4). Expanding the denominator determinant in

Equation (13.4), and setting it equal to zero, we have

$$\left(1 - \frac{u^2}{a^2}\right)(dy)^2 + \frac{2uv}{a^2}dx\,dy + \left(1 - \frac{v^2}{a^2}\right)(dx)^2 = 0$$

or

$$\left(1 - \frac{u^2}{a^2}\right)\left(\frac{dy}{dx}\right)^2_{char} + \frac{2uv}{a^2}\left(\frac{dy}{dx}\right)_{char} + \left(1 - \frac{v^2}{a^2}\right) = 0 \qquad (13.6)$$

In Equation (13.6), dy/dx is the slope of the characteristic lines; hence, the subscript "char" has been added to emphasize this fact. Solving Equation (13.6) for $(dy/dx)_{char}$ by means of the quadratic formula, we obtain

$$\left(\frac{dy}{dx}\right)_{char} = \frac{-2uv/a^2 \pm \sqrt{(2uv/a^2)^2 - 4(1 - u^2/a^2)(1 - v^2/a^2)}}{2(1 - u^2/a^2)}$$

or

$$\left(\frac{dy}{dx}\right)_{char} = \frac{-uv/a^2 \pm \sqrt{(u^2 + v^2)/a^2 - 1}}{1 - u^2/a^2} \qquad (13.7)$$

From Figure 13.3, we see that $u = V\cos\theta$ and $v = V\sin\theta$. Hence, Equation (13.7) becomes

$$\left(\frac{dy}{dx}\right)_{char} = \frac{(-V^2\cos\theta\sin\theta)/a^2 \pm \sqrt{(V^2/a^2)(\cos^2\theta + \sin^2\theta) - 1}}{1 - [(V^2/a^2)\cos^2\theta]} \qquad (13.8)$$

Recall that the local Mach angle μ is given by $\mu = \sin^{-1}(1/M)$, or $\sin\mu = 1/M$. Thus, $V^2/a^2 = M^2 = 1/\sin^2\mu$, and Equation (13.8) becomes

$$\left(\frac{dy}{dx}\right)_{char} = \frac{(-\cos\theta\sin\theta)/\sin^2\mu \pm \sqrt{(\cos^2\theta + \sin^2\theta)/\sin^2\mu - 1}}{1 - (\cos^2\theta)/\sin^2\mu} \qquad (13.9)$$

After considerable algebraic and trigonometric manipulation, Equation (13.9) reduces to

$$\boxed{\left(\frac{dy}{dx}\right)_{char} = \tan(\theta \mp \mu)} \qquad (13.10)$$

Equation (13.10) is an important result; it states that *two* characteristic lines run through point A in Figure 13.3, namely, one line with a slope equal to $\tan(\theta - \mu)$ and the other with a slope equal to $\tan(\theta + \mu)$. The physical significance of this result is illustrated in Figure 13.4. Here, a streamline through point A is inclined at the angle θ with respect to the horizontal. The velocity at point A is V, which also makes the angle θ with respect to the horizontal. Equation (13.10) states that one characteristic line at point A is inclined *below* the streamline direction by the angle μ; this characteristic line is labeled as C_- in Figure 13.4. Equation (13.10) also states that the other characteristic line at point A is inclined *above* the streamline direction by the angle μ; this characteristic line is labeled as C_+ in Figure 13.4. Examining Figure 13.4, we see that the characteristic lines through

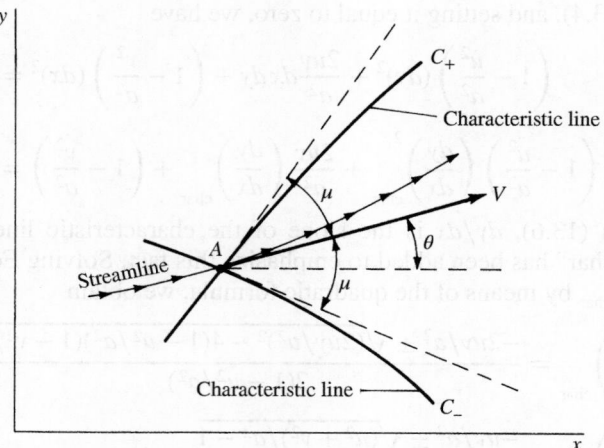

Figure 13.4 Left- and right-running characteristic lines through point A.

point A are simply the left- and right-running *Mach waves* through point A. Hence, *the characteristic lines are Mach lines.* In Figure 13.4, the left-running Mach wave is denoted by C_+, and the right-running Mach wave is denoted by C_-. Hence, returning to Figure 13.2a, the characteristics mesh consists of left- and right-running Mach waves that crisscross the flow field. There are an infinite number of these waves; however, for practical calculations we deal with a finite number of waves, the intersections of which define the grid points shown in Figure 13.2a. Note that the characteristic lines are curved in space because (1) the local Mach angle depends on the local Mach number, which is a function of x and y, and (2) the local streamline direction θ varies throughout the flow.

The characteristic lines in Figure 13.2a are of no use to us by themselves. The practical consequence of these lines is that *the governing partial differential equations that describe the flow reduce to ordinary differential equations along the characteristic lines.* These equations are called the *compatibility equations,* which can be found by setting $N = 0$ in Equation (13.4), as follows. When $N = 0$, the numerator determinant yields

$$\left(1 - \frac{u^2}{a^2}\right) du\, dy + \left(1 - \frac{v^2}{a^2}\right) dx\, dv = 0$$

or

$$\frac{dv}{du} = \frac{-(1 - u^2/a^2)}{1 - v^2/a^2} \frac{dy}{dx} \qquad (13.11)$$

Keep in mind that N is set to zero only when $D = 0$ in order to keep the flow-field derivatives finite, albeit of the indeterminate form $0/0$. When $D = 0$, we are restricted to considering directions *only* along the characteristic lines, as explained earlier. Hence, when $N = 0$, we are held to the same restriction. Therefore,

Equation (13.11) holds only along the characteristic lines. Therefore, in Equation (13.11),

$$\frac{dy}{dx} \equiv \left(\frac{dy}{dx}\right)_{char} \tag{13.12}$$

Substituting Equations (13.12) and (13.7) into (13.11), we obtain

$$\frac{dv}{du} = -\frac{1 - u^2/a^2}{1 - v^2/a^2} \frac{-uv/a^2 \pm \sqrt{(u^2 + v^2)/a^2 - 1}}{1 - u^2/a^2}$$

or

$$\frac{dv}{du} = \frac{uv/a^2 \mp \sqrt{(u^2 + v^2)/a^2 - 1}}{1 - v^2/a^2} \tag{13.13}$$

Recall from Figure 13.3 that $u = V \cos \theta$ and $v = V \sin \theta$. Also, $(u^2 + v^2)/a^2 = V^2/a^2 = M^2$. Hence, Equation (13.13) becomes

$$\frac{d(V \sin \theta)}{d(V \cos \theta)} = \frac{M^2 \cos \theta \sin \theta \mp \sqrt{M^2 - 1}}{1 - M^2 \sin^2 \theta}$$

which, after some algebraic manipulations, reduces to

$$\boxed{d\theta = \mp \sqrt{M^2 - 1}\,\frac{dV}{V}} \tag{13.14}$$

Examine Equation (13.14). It is an *ordinary differential equation* obtained from the original governing partial differential equation, Equation (13.1). However, Equation (13.14) contains the restriction given by Equation (13.12); that is, Equation (13.14) holds *only* along the characteristic lines. Hence, Equation (13.14) gives the *compatibility relations* along the characteristic lines. In particular, comparing Equation (13.14) with Equation (13.10), we see that

$$d\theta = -\sqrt{M^2 - 1}\,\frac{dV}{V} \qquad \text{(applies along the } C_- \text{ characteristic)} \tag{13.15}$$

$$d\theta = \sqrt{M^2 - 1}\,\frac{dV}{V} \qquad \text{(applies along the } C_+ \text{ characteristic)} \tag{13.16}$$

Examine Equation (13.14) further. It should look familiar; indeed, Equation (13.14) is identical to the expression obtained for Prandtl-Meyer flow in Section 9.6, namely, Equation (9.32). Hence, Equation (13.14) can be integrated to obtain a result in terms of the Prandtl-Meyer function, given by Equation (9.42). In particular, the integration of Equations (13.15) and (13.16) yields

$$\theta + v(M) = \text{const} = K_- \qquad \text{(along the } C_- \text{ characteristic)} \tag{13.17}$$

$$\theta - v(M) = \text{const} = K_+ \qquad \text{(along the } C_+ \text{ characteristic)} \tag{13.18}$$

In Equation (13.17), K_- is a constant along a given C_- characteristic; it has different values for different C_- characteristics. In Equation (13.18), K_+ is a constant along a given C_+ characteristic; it has different values for different C_+ characteristics. Note that our compatibility relations are now given by Equations (13.17) and (13.18), which are *algebraic* equations which hold only along the characteristic lines. In a general inviscid, supersonic, steady flow, the compatibility equations are ordinary differential equations; only in the case of two-dimensional irrotational flow do they further reduce to algebraic equations.

What is the advantage of the characteristic lines and their associated compatibility equations discussed above? Simply this—to solve the nonlinear supersonic flow, we need to deal only with ordinary differential equations (or in the present case, algebraic equations) instead of the original partial differential equations. Finding the solution of such ordinary differential equations is usually much simpler than dealing with partial differential equations.

How do we use the above results to solve a practical problem? The purpose of the next section is to give such an example, namely, the calculation of the supersonic flow inside a nozzle and the determination of a proper wall contour so that shock waves do not appear inside the nozzle. To carry out this calculation, we deal with two types of grid points: (1) internal points, away from the wall, and (2) wall points. Characteristics calculations at these two sets of points are carried out as follows.

13.2.1 Internal Points

Consider the internal grid points 1, 2, and 3 as shown in Figure 13.5. Assume that we know the location of points 1 and 2, as well as the flow properties at these points. Define point 3 as the intersection of the C_- characteristic through point 1 and the C_+ characteristic through point 2. From our previous discussion, $(K_-)_1 = (K_-)_3$ because K_- is constant along a given C_- characteristic. The value of $(K_-)_1 = (K_-)_3$ is obtained from Equation (13.17) evaluated at point 1:

$$(K_-)_3 = (K_-)_1 = \theta_1 + \nu_1 \tag{13.19}$$

Similarly, $(K_+)_2 = (K_+)_3$ because K_+ is constant along a given C_+ characteristic. The value of $(K_+)_2 = (K_+)_3$ is obtained from Equation (13.18) evaluated at point 2:

$$(K_+)_3 = (K_+)_2 = \theta_2 - \nu_2 \tag{13.20}$$

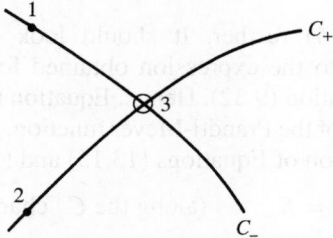

Figure 13.5 Characteristic mesh used for the location of point 3 and the calculation of flow conditions at point 3, knowing the locations and flow properties at points 1 and 2.

Now evaluate Equations (13.17) and (13.18) at point 3:

$$\theta_3 + v_3 = (K_-)_3 \tag{13.21}$$

and

$$\theta_3 - v_3 = (K_+)_3 \tag{13.22}$$

In Equations (13.21) and (13.22), $(K_-)_3$ and $(K_+)_3$ are known values, obtained from Equations (13.19) and (13.20). Hence, Equations (13.21) and (13.22) are two algebraic equations for the two unknowns θ_3 and v_3. Solving these equations, we obtain

$$\theta_3 = \frac{1}{2}[(K_-)_1 + (K_+)_2] \tag{13.23}$$

$$v_3 = \frac{1}{2}[(K_-)_1 - (K_+)_2] \tag{13.24}$$

Knowing θ_3 and v_3, all other flow properties at point 3 can be obtained as follows:

1. From v_3, obtain the associated M_3 from Appendix C.
2. From M_3 and the known p_0 and T_0 for the flow (recall that for inviscid, adiabatic flow, the total pressure and total temperature are constants throughout the flow), find p_3 and T_3 from Appendix A.
3. Knowing T_3, compute $a_3 = \sqrt{\gamma R T_3}$. In turn, $V_3 = M_3 a_3$.

As stated earlier, point 3 is located by the intersection of the C_- and C_+ characteristics through points 1 and 2, respectively. These characteristics are curved lines; however, for purposes of calculation, we assume that the characteristics are straight-line segments between points 1 and 3 and between points 2 and 3. For example, the slope of the C_- characteristic between points 1 and 3 is assumed to be the average value between these two points, that is, $\frac{1}{2}(\theta_1 + \theta_3) - \frac{1}{2}(\mu_1 + \mu_3)$. Similarly, the slope of the C_+ characteristic between points 2 and 3 is approximated by $\frac{1}{2}(\theta_2 + \theta_3) + \frac{1}{2}(\mu_2 + \mu_3)$.

13.2.2 Wall Points

In Figure 13.6, point 4 is an internal flow point near a wall. Assume that we know all the flow properties at point 4. The C_- characteristic through point 4 intersects the wall at point 5. At point 5, the slope of the wall θ_5 is known. The flow properties at the wall point, point 5, can be obtained from the known properties at point 4 as follows. Along the C_- characteristic, K_- is constant. Hence, $(K_-)_4 = (K_-)_5$.

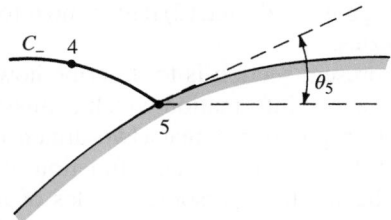

Figure 13.6 Wall point.

Moreover, the value of K_- is known from Equation (13.17) evaluated at point 4:

$$(K_-)_4 = (K_-)_5 = \theta_4 + \nu_4 \qquad (13.25)$$

Evaluating Equation (13.17) at point 5, we have

$$(K_-)_5 = \theta_5 + \nu_5 \qquad (13.26)$$

In Equation (13.26), $(K_-)_5$ and θ_5 are known; thus ν_5 follows directly. In turn, all other flow variables at point 5 can be obtained from ν_5 as explained earlier. The characteristic line between points 4 and 5 is assumed to be a straight-line segment with average slope given by $\frac{1}{2}(\theta_4 + \theta_5) - \frac{1}{2}(\mu_4 + \mu_5)$.

From the above discussion of both internal and wall points, we see that properties at the grid points are calculated from *known* properties at other grid points. Hence, in order to *start* a calculation using the method of characteristics, we have to know the flow properties along some *initial data* line. Then we piece together the characteristics mesh and associated flow properties by "marching downstream" from the initial data line. This is illustrated in the next section.

We emphasize again that the method of characteristics is an exact solution of inviscid, nonlinear supersonic flow. However, in practice, there are numerical errors associated with the finite grid; the approximation of the characteristics mesh by straight-line segments between grid points is one such example. In principle, the method of characteristics is truly exact only in the limit of an infinite number of characteristic lines.

We have discussed the method of characteristics for two-dimensional, irrotational, steady flow. The method of characteristics can also be used for rotational and three-dimensional flows, as well as unsteady flows. See Reference 21 for more details.

13.3 SUPERSONIC NOZZLE DESIGN

In Chapter 10, we demonstrated that a nozzle designed to expand a gas from rest to supersonic speeds must have a convergent-divergent shape. Moreover, the quasi-one-dimensional analysis of Chapter 10 led to the prediction of flow properties as a function of x through a nozzle of specified shape (see, e.g., Figure 10.10). The flow properties at any x station obtained from the quasi-one-dimensional analysis represent an *average* of the flow over the given nozzle cross section. The beauty of the quasi-one-dimensional approach is its simplicity. On the other hand, its disadvantages are (1) it cannot predict the details of the actual three-dimensional flow in a convergent-divergent nozzle and (2) it gives no information on the proper wall contour of such nozzles.

The purpose of the present section is to describe how the method of characteristics can supply the above information which is missing from a quasi-one-dimensional analysis. For simplicity, we treat a two-dimensional flow, as sketched in Figure 13.7. Here, the flow properties are a function of x and y. Such a two-dimensional flow is applicable to supersonic nozzles of rectangular cross section, such as sketched in the insert at the top of Figure 13.7. Two-dimensional

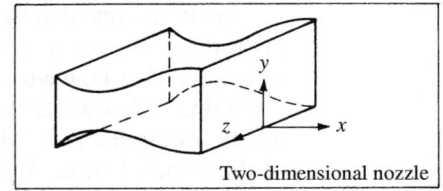

Two-dimensional nozzle

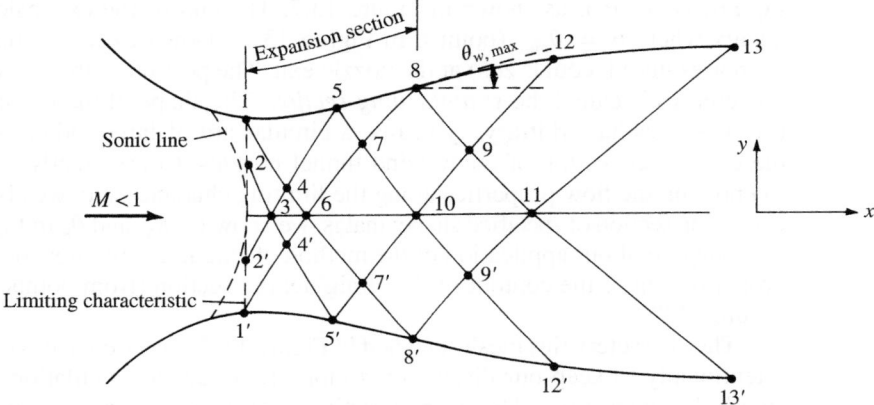

Figure 13.7 Schematic of supersonic nozzle design by the method of characteristics.

(rectangular) nozzles are used in many supersonic wind tunnels. They are also the heart of gas-dynamic lasers (see Reference 1). In addition, there is current discussion of employing rectangular exhaust nozzles on advanced military jet airplanes envisaged for the future.

Consider the following problem. We wish to design a convergent-divergent nozzle to expand a gas from rest to a given supersonic Mach number at the exit M_e. How do we design the proper contour so that we have shock-free, isentropic flow in the nozzle? The answer to this question is discussed in the remainder of this section.

For the convergent, subsonic section, there is no specific contour which is better than any other. There are rules of thumb based on experience and guided by subsonic flow theory; however, we are not concerned with the details here. We simply assume that we have a reasonable contour for the subsonic section.

Due to the two-dimensional nature of the flow in the throat region, the sonic line is generally curved, as sketched in Figure 13.7. A line called the *limiting characteristic* is sketched just downstream of the sonic line. The limiting characteristic is defined such that any characteristic line originating downstream of the limiting characteristic does not intersect the sonic line; in contrast, a characteristic line originating in the small region between the sonic line and the limiting characteristic can intersect the sonic line (for more details on the limiting characteristic, see Reference 21). To begin a method of characteristics solution, we must use an initial data line which is downstream of the limiting characteristic.

Let us assume that by independent calculation of the subsonic-transonic flow in the throat region, we know the flow properties at all points on the limiting characteristic. That is, we use the limiting characteristic as our initial data line. For example, we know the flow properties at points 1 and 2 on the limiting characteristic in Figure 13.7. Moreover, consider the nozzle contour just downstream of the throat. Letting θ denote the angle between a tangent to the wall and the horizontal, the section of the divergent nozzle where θ is increasing is called the *expansion section,* as shown in Figure 13.7. The end of the expansion section occurs where $\theta = \theta_{max}$ (point 8 in Figure 13.7). Downstream of this point, θ decreases until it equals zero at the nozzle exit. The portion of the contour where θ decreases is called the *straightening section.* The shape of the expansion section is somewhat arbitrary; typically, a circular arc of large radius is used for the expansion section of many wind-tunnel nozzles. Consequently, in addition to knowing the flow properties along the limiting characteristic, we also have an expansion section of specified shape; that is, we know θ_1, θ_5, and θ_8 in Figure 13.7. The purpose of our application of the method of characteristics now becomes the proper design of the contour of the straightening section (from points 8 to 13 in Figure 13.7).

The characteristics mesh sketched in Figure 13.7 is very coarse—this is done intentionally to keep our discussion simple. In an actual calculation, the mesh should be much finer. The characteristics mesh and the flow properties at the associated grid points are calculated as follows:

1. Draw a C_- characteristic from point 2, intersecting the centerline at point 3. Evaluating Equation (13.17) at point 3, we have

$$\theta_3 + \nu_3 = (K_-)_3$$

In the above equation, $\theta_3 = 0$ (the flow is horizontal along the centerline). Also, $(K_-)_3$ is known because $(K_-)_3 = (K_-)_2$. Hence, the above equation can be solved for ν_3.

2. Point 4 is located by the intersection of the C_- characteristic from point 1 and the C_+ characteristic from point 3. In turn, the flow properties at the internal point 4 are determined as discussed in the last part of Section 13.2.

3. Point 5 is located by the intersection of the C_+ characteristic from point 4 with the wall. Since θ_5 is known, the flow properties at point 5 are determined as discussed in Section 13.2 for wall points.

4. Points 6 through 11 are located in a manner similar to the above, and the flow properties at these points are determined as discussed before, using the internal point or wall point method as appropriate.

5. Point 12 is a wall point on the straightening section of the contour. The purpose of the straightening section is to cancel the expansion waves generated by the expansion section. Hence, there are no waves which are reflected from the straightening section. In turn, no right-running waves cross the characteristic line between points 9 and 12. As a result, the

characteristic line between points 9 and 12 is a straight line, along which θ is constant, that is, $\theta_{12} = \theta_9$. The section of the wall contour between points 8 and 12 is approximated by a straight line with an average slope of $\frac{1}{2}(\theta_8 + \theta_{12})$.

6. Along the centerline, the Mach number continuously increases. Let us assume that at point 11, the design exit Mach number M_e is reached. The characteristic line from points 11 to 13 is the last line of the calculation. Again, $\theta_{13} = \theta_{11}$, and the contour from point 12 to point 13 is approximated by a straight-line segment with an average slope of $\frac{1}{2}(\theta_{12} + \theta_{13})$.

The above description is intended to give you a "feel" for the application of the method of characteristics. If you wish to carry out an actual nozzle design, and/or if you are interested in more details, read the more complete treatments in References 21 and 32.

Note in Figure 13.7 that the nozzle flow is symmetrical about the centerline. Hence, the points below the centerline ($1'$, $2'$, $3'$, etc.) are simply mirror images of the corresponding points above the centerline. In making a calculation of the flow through the nozzle, we need to concern ourselves only with those points in the upper half of Figure 13.7, above and on the centerline.

13.4 ELEMENTS OF FINITE-DIFFERENCE METHODS

The method of characteristics described in the previous section legitimately can be considered a part of computational fluid dynamics because it uses discrete algebraic forms of the governing equations [such as Equations (13.17) and (13.18)] which are solved at discrete points in the flow (the characteristic mesh illustrated in Figure 13.5). However, most authors consider that CFD is represented by mainly finite difference and finite volume techniques, such as are discussed in Reference 60, and the method of characteristics is usually not included in the study of CFD. The purpose of this section is to give you the flavor of finite-difference techniques by describing one particular method that is readily applicable to a number of compressible flow problems. The method discussed here is representative of mainstream CFD, but it is just the tip of the iceberg of CFD. The intensive research in CFD since 1960 has produced a multitude of different algorithms and philosophies, and it is far beyond the scope of this book to go into the details of such work. See Reference 60 for an in-depth presentation of CFD at the introductory level. In addition, you are strongly encouraged to read the current literature in this regard, in particular the *AIAA Journal, Computers and Fluids,* and the *Journal of Computational Physics.* Finally, in this chapter we are dealing with numerical solutions of *inviscid supersonic* flows. See Reference 21 for an expanded discussion of finite difference methods applied to supersonic flows.

First, recall the discrete finite difference representations for partial derivatives that were derived in Section 2.17.2 using Taylors series. In particular, we recall

Equations (2.168), (2.171), and (2.174), repeated and renumbered, respectively, below for convenience:

$$\left(\frac{\partial u}{\partial x}\right)_{i,j} = \frac{u_{i+1,j} - u_{i,j}}{\Delta x} \qquad \text{(forward difference)} \tag{13.27}$$

$$\left(\frac{\partial u}{\partial x}\right)_{i,j} = \frac{u_{i,j} - u_{i-1,j}}{\Delta x} \qquad \text{(rearward difference)} \tag{13.28}$$

$$\left(\frac{\partial u}{\partial x}\right)_{i,j} = \frac{u_{i+1,j} - u_{i-1,j}}{2\Delta x} \qquad \text{(central difference)} \tag{13.29}$$

Analogous expressions for the derivatives in the y direction are as follows:

$$\left(\frac{\partial u}{\partial y}\right)_{i,j} = \begin{cases} \dfrac{u_{i,j+1} - u_{i,j}}{\Delta y} & \text{(forward difference)} \\[2ex] \dfrac{u_{i,j} - u_{i,j-1}}{\Delta y} & \text{(rearward difference)} \\[2ex] \dfrac{u_{i,j+1} - u_{i,j-1}}{2\Delta y} & \text{(central difference)} \end{cases}$$

How do we use the finite differences obtained here? Imagine that a flow in xy space is covered by the mesh shown in Figure 13.2b. Assume there are N grid points. At each one of these grid points, evaluate the continuity, momentum, and energy equations with their partial derivatives replaced by the finite-difference expressions derived above. For example, replacing the derivatives in Equations (7.40), (7.42a and b), and (7.44) with finite differences, along with Equations (7.1) and (7.6a), we obtain a system (over all N grid points) of $6N$ simultaneous nonlinear algebraic equations in terms of the $6N$ unknowns, namely, ρ, u, v, p, T, and e, at each of the N grid points. In principle, we could solve this system for the unknown flow variables at all the grid points. In practice, this is easier said than done. There are severe problems in solving such a large number of simultaneous nonlinear equations. Moreover, we have to deal with problems associated with numerical instabilities that sometimes cause such attempted solutions to "blow up" on the computer. Finally, and most importantly, we must properly account for the boundary conditions. These considerations make all finite-difference solutions a nontrivial endeavor. As a result, a number of specialized finite-difference techniques have evolved, directed at solving different types of flow problems and attempting to increase computational efficiency and accuracy. It is beyond the scope of this book to describe these difference techniques in detail. However, one technique in particular was widely used during the 1970s and 1980s. This is an approach developed in 1969 by Robert MacCormack at the NASA Ames Research Center. Because of its widespread use and acceptance at the time, as well as its relative simplicity, we will describe MacCormack's technique in enough detail to give you a reasonable understanding of the method. This description will be carried out in the context of the following example.

Consider the two-dimensional supersonic flow through the divergent duct shown in Figure 13.8a. Assume the flow is supersonic at the inlet, and that all

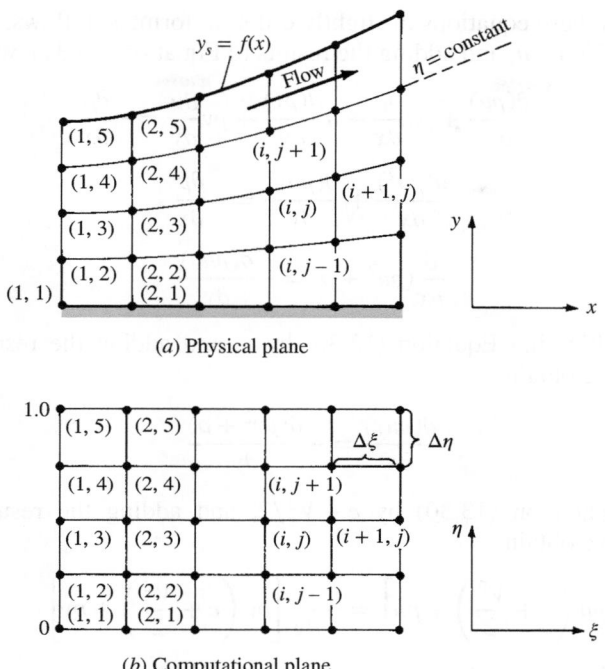

Figure 13.8 Finite-difference meshes in both the physical and computational planes.

properties are known at the inlet. That is, the flow-field variables at grid points $(1, 1), (1, 2), (1, 3), (1, 4)$, and $(1, 5)$ are known. The duct is formed by a flat surface at the bottom and a specified contour, $y_s = f(x)$, at the top. In addition, assume that the flow is inviscid, adiabatic, and steady, and with no body forces. It can be rotational or irrotational—the method of solution is the same. The governing equations are obtained from Equations (7.40), (7.42a and b), (7.44), (7.1), and (7.6a), which yield

$$\frac{\partial(\rho u)}{\partial x} + \frac{\partial(\rho v)}{\partial y} = 0 \tag{13.30}$$

$$\rho u \frac{\partial u}{\partial x} + \rho v \frac{\partial u}{\partial y} = -\frac{\partial p}{\partial x} \tag{13.31}$$

$$\rho u \frac{\partial v}{\partial x} + \rho v \frac{\partial v}{\partial y} = -\frac{\partial p}{\partial y} \tag{13.32}$$

$$\rho u \frac{\partial(e + V^2/2)}{\partial x} + \rho v \frac{\partial(e + V^2/2)}{\partial y} = -\frac{\partial(pu)}{\partial x} - \frac{\partial(pv)}{\partial y} \tag{13.33}$$

$$p = \rho RT \tag{13.34}$$

$$e = c_v T \tag{13.35}$$

Let us express these equations in slightly different form, as follows. Multiplying Equation (13.30) by u, and adding the results to Equation (13.31), we have

$$u\frac{\partial(\rho u)}{\partial x} + \rho u\frac{\partial u}{\partial x} + u\frac{\partial(\rho v)}{\partial y} + \rho v\frac{\partial u}{\partial y} = -\frac{\partial p}{\partial x}$$

or

$$\frac{\partial(\rho u^2)}{\partial x} + \frac{\partial(\rho uv)}{\partial y} = -\frac{\partial p}{\partial x}$$

or

$$\frac{\partial}{\partial x}(\rho u^2 + p) = -\frac{\partial(\rho uv)}{\partial y} \tag{13.36}$$

Similarly, multiplying Equation (13.30) by v, and adding the result to Equation (13.32), we obtain

$$\frac{\partial(\rho uv)}{\partial x} = -\frac{\partial(\rho v^2 + p)}{\partial y} \tag{13.37}$$

Multiplying Equation (13.30) by $e + V^2/2$, and adding the result to Equation (13.33), we obtain

$$\frac{\partial}{\partial x}\left[\rho u\left(e + \frac{V^2}{2}\right) + pu\right] = -\frac{\partial}{\partial y}\left[\rho v\left(e + \frac{V^2}{2}\right) + pv\right] \tag{13.38}$$

Define the following symbols:

$$F = \rho u \tag{13.39a}$$

$$G = \rho u^2 + p \tag{13.39b}$$

$$H = \rho uv \tag{13.39c}$$

$$K = \rho u\left(e + \frac{V^2}{2}\right) + pu \tag{13.39d}$$

Then, Equations (13.30) and (13.36) to (13.38) become

$$\frac{\partial F}{\partial x} = -\frac{\partial(\rho v)}{\partial y} \tag{13.40}$$

$$\frac{\partial G}{\partial x} = -\frac{\partial(\rho uv)}{\partial y} \tag{13.41}$$

$$\frac{\partial H}{\partial x} = -\frac{\partial(\rho v^2 + p)}{\partial y} \tag{13.42}$$

$$\frac{\partial K}{\partial x} = -\frac{\partial}{\partial y}\left[\rho v\left(e + \frac{V^2}{2}\right) + pv\right] \tag{13.43}$$

Equations (13.40) to (13.43) are the continuity, x and y momentum, and energy equations, respectively—but in a slightly different form from those we are used to seeing. The above form of these equations is frequently called the *conservation form.* Let us now treat F, G, H, and K as our primary dependent variables; these quantities are called *flux variables,* in contrast to the usual p, ρ, T, u, v, e, etc., which are called *primitive variables.* It is important to note that once the values of F, G, H, and K are known at a given grid point, the primitive variables at that point can be found from Equations (13.39a to d) and

$$p = \rho RT \qquad (13.44)$$

$$e = c_v T \qquad (13.45)$$

$$V^2 = u^2 + v^2 \qquad (13.46)$$

That is, Equations (13.39a to d) and (13.44) to (13.46) constitute seven algebraic equations for the seven primitive variables, ρ, u, v, p, e, T, and V.

Let us return to the physical problem given in Figure 13.8a. Because the duct diverges, it is difficult to deal with an orthogonal, rectangular mesh; rather, a mesh which conforms to the boundary of the system will be curved, as shown in Figure 13.8a. On the other hand, to use our finite-difference quotients as given in Equation (13.27), (13.28), or (13.29), we desire a rectangular computational mesh. Therefore, we must *transform* the curved mesh shown in Figure 13.8a, known as the *physical plane,* to a rectangular mesh shown in Figure 13.8b, known as the *computational plane.* This transformation can be carried out as follows. Define

$$\xi = x \qquad (13.47a)$$

$$\eta = \frac{y}{y_s}$$

where
$$y_s = f(x) \qquad (13.47b)$$

In the above transformation, η ranges from 0 at the bottom wall to 1.0 at the top wall. In the computational plane (Figure 13.8b), $\eta = $ constant is a straight horizontal line, whereas in the physical plane, $\eta = $ constant corresponds to the curved line shown in Figure 13.8. Because we wish to apply our finite differences in the computational plane, we need the governing equations in terms of ξ and η rather than x and y. To accomplish this transformation, apply the chain rule of differentiation, using Equation (13.47a and b) as follows:

$$\frac{\partial}{\partial x} = \frac{\partial}{\partial \xi}\frac{\partial \xi}{\partial x} + \frac{\partial}{\partial \eta}\frac{\partial \eta}{\partial x} = \frac{\partial}{\partial \xi} - \frac{y}{y_s^2}\frac{dy_s}{dx}\frac{\partial}{\partial \eta}$$

or
$$\frac{\partial}{\partial x} = \frac{\partial}{\partial \xi} - \left(\frac{\eta}{y_s}\frac{dy_s}{dx}\right)\frac{\partial}{\partial \eta} \qquad (13.48)$$

and
$$\frac{\partial}{\partial y} = \frac{\partial}{\partial \xi}\frac{\partial \xi}{\partial y} + \frac{\partial}{\partial \eta}\frac{\partial \eta}{\partial y} = \frac{1}{y_s}\frac{\partial}{\partial \eta} \qquad (13.49)$$

Using Equations (13.48) and (13.49), we see that Equations (13.40) to (13.43) become

$$\frac{\partial F}{\partial \xi} = \left(\frac{\eta}{y_s}\frac{dy_s}{dx}\right)\left(\frac{\partial F}{\partial \eta}\right) - \frac{1}{y_s}\frac{\partial(\rho v)}{\partial \eta} \tag{13.50}$$

$$\frac{\partial G}{\partial \xi} = \left(\frac{\eta}{y_s}\frac{dy_s}{dx}\right)\frac{\partial G}{\partial \eta} - \frac{1}{y_s}\frac{\partial(\rho uv)}{\partial \eta} \tag{13.51}$$

$$\frac{\partial H}{\partial \xi} = \left(\frac{\eta}{y_s}\frac{dy_s}{dx}\right)\frac{\partial H}{\partial \eta} - \frac{1}{y_s}\frac{\partial(\rho v^2 + p)}{\partial \eta} \tag{13.52}$$

$$\frac{\partial K}{\partial \xi} = \left(\frac{\eta}{y_s}\frac{dy_s}{dx}\right)\frac{\partial K}{\partial \eta} - \frac{1}{y_s}\frac{\partial}{\partial \eta}\left[\rho v\left(e + \frac{V^2}{2}\right) + pv\right] \tag{13.53}$$

Note in the above equations that the ξ derivatives are on the left and the η derivatives are all grouped on the right.

Let us now concentrate on obtaining a numerical, finite-difference solution of the problem shown in Figure 13.8. We will deal exclusively with the computational plane, Figure 13.8b, where the governing continuity, x and y momentum, and energy equations are given by Equations (13.50) to (13.53), respectively. Grid points $(1, 1)$, $(2, 1)$, $(1, 2)$, $(2, 2)$, etc., in the computational plane are the same as grid points $(1, 1)$, $(2, 1)$, $(1, 2)$, $(2, 2)$, etc., in the physical plane. All the flow variables are known at the inlet, including F, G, H, and K. The solution for the flow variables downstream of the inlet can be found by using MacCormack's method, which is based on Taylor's series expansions for F, G, H, and K as follows:

$$F_{i+1,j} = F_{i,j} + \left(\frac{\partial F}{\partial \xi}\right)_{\text{ave}} \Delta\xi \tag{13.54a}$$

$$G_{i+1,j} = G_{i,j} + \left(\frac{\partial G}{\partial \xi}\right)_{\text{ave}} \Delta\xi \tag{13.54b}$$

$$H_{i+1,j} = H_{i,j} + \left(\frac{\partial H}{\partial \xi}\right)_{\text{ave}} \Delta\xi \tag{13.54c}$$

$$K_{i+1,j} = K_{i,j} + \left(\frac{\partial K}{\partial \xi}\right)_{\text{ave}} \Delta\xi \tag{13.54d}$$

In Equation (13.54a to d), F, G, H, and K at point (i,j) are considered known, and these equations are used to find F, G, H, and K at point $(i + 1, j)$ assuming that we can calculate the values of $(\partial F/\partial \xi)_{\text{ave}}$, $(\partial G/\partial \xi)_{\text{ave}}$, etc. The main thrust of MacCormack's method is the calculation of these average derivatives. Examining Equation (13.54a to d), we find that this finite-difference method is clearly a "down-stream marching" method; given the flow at point (i,j) we use Equation (13.54a to d) to find the flow at point $(i + 1, j)$. Then the process is repeated to find the flow at point $(i + 2, j)$, etc. This downstream marching is similar to that performed with the method of characteristics.

The average derivatives in Equation (13.54a to d) are found by means of a straightforward "predictor-corrector" approach, outlined below. In carrying out this approach, we assume that the flow properties are known at grid point (i,j), as well as at all points directly above and below (i,j), namely, at $(i,j+1)$, $(i,j+2)$, $(i,j-1)$, $(i,j-2)$, etc.

13.4.1 Predictor Step

First, predict the value of $F_{i+1,j}$ by using a Taylor series where $\partial F/\partial \xi$ is evaluated at point (i,j). Denote this predicted value by $\bar{F}_{i+1,j}$:

$$\bar{F}_{i+1,j} = F_{i,j} + \left(\frac{\partial F}{\partial \xi}\right)_{i,j} \Delta \xi \tag{13.55}$$

In Equation (13.55), $(\partial F/\partial \xi)_{i,j}$ is obtained from the continuity equation, Equation (13.50), using *forward differences* for the η derivatives; that is,

$$\left(\frac{\partial F}{\partial \xi}\right)_{i,j} = \left(\frac{\eta}{y_s}\frac{dy_s}{dx}\right)_{i,j}\left(\frac{F_{i,j+1} - F_{i,j}}{\Delta \eta}\right) - \frac{1}{y_s}\left[\frac{(\rho v)_{i,j+1} - (\rho v)_{i,j}}{\Delta \eta}\right] \tag{13.56}$$

In Equation (13.56), all quantities on the right-hand side are known and allow the calculation of $(\partial F/\partial \xi)_{i,j}$ which is, in turn, inserted into Equation (13.55). A similar procedure is used to find predicted values of G, H, and K, namely, $\bar{G}_{i+1,j}$, $\bar{H}_{i+1,j}$, and $\bar{K}_{i+1,j}$, using forward differences in Equations (13.51) to (13.53). In turn, predicted values of the primitive variables, $\bar{p}_{i+1,j}$, $\bar{\rho}_{i+1,j}$, etc., can be obtained from Equations (13.39a to d) and (13.44) to (13.46).

13.4.2 Corrector Step

The predicted values obtained above are used to obtain predicted values of the derivative $\overline{(\partial F/\partial \xi)}_{i+1,j}$, using *rearward differences* in Equation (13.50):

$$\left(\overline{\frac{\partial F}{\partial \xi}}\right)_{i+1,j} = \left(\frac{\eta}{y_s}\frac{dy_s}{dx}\right)_{i+1,j}\frac{\bar{F}_{i+1,j} - \bar{F}_{i+1,j-1}}{\Delta \eta} - \frac{1}{y_s}\frac{\overline{(\rho v)}_{i+1,j} - \overline{(\rho v)}_{i+1,j-1}}{\Delta \eta} \tag{13.57}$$

In turn, the results from Equations (13.56) and (13.57) allow the calculation of the average derivative

$$\left(\frac{\partial F}{\partial \xi}\right)_{ave} = \frac{1}{2}\left[\left(\frac{\partial F}{\partial \xi}\right)_{i,j} + \left(\overline{\frac{\partial F}{\partial \xi}}\right)_{i+1,j}\right] \tag{13.58}$$

Finally, this average derivative is used in Equation (13.54a) to obtain the corrected value of $F_{i+1,j}$. The same process is followed to find the corrected values of $G_{i+1,j}$, $H_{i+1,j}$, and $K_{i+1,j}$ using rearward differences in Equations (13.51) to (13.53) and calculating the average derivatives $(\partial G/\partial \xi)_{ave}$, etc., in the same manner as Equation (13.58).

The above finite-difference procedure allows the step-by-step calculation of the flow field, marching downstream from some initial data line. In the flow given in Figure 13.8, the initial data line is the inlet, where properties are considered known. Although all the calculations are carried out in the transformed, computational plane, the flow-field results obtained at points $(2, 1)$, $(2, 2)$, etc., in the computational plane are the same values at points $(2, 1)$, $(2, 2)$, etc., in the physical plane.

There are other aspects of the finite-difference solution which have not been described above. For example, what values of $\Delta \eta$ and $\Delta \xi$ in Equations (13.54*a* to *d*), (13.55), (13.56), and (13.57) are allowed in order to maintain numerical stability? How is the flow-tangency condition at the walls imposed on the finite-difference calculations? These are important matters, but we do not take the additional space to discuss them here. See Chapter 11 of Reference 21 for details on these questions. Our purpose here has been to give you only a feeling for the nature of the finite-difference method.

13.5 THE TIME-DEPENDENT TECHNIQUE: APPLICATION TO SUPERSONIC BLUNT BODIES

The method of characteristics described in Section 13.2 is applicable only to supersonic flows; the characteristic lines are not defined in a practical fashion for steady, subsonic flow. Also, the particular finite-difference method outlined in Section 13.4 applies only to supersonic flows; if it were to be used in a locally subsonic region, the calculation would blow up. The reason for both of the above comments is that the method of characteristics and the steady flow, forward-marching finite-difference technique depend on the governing equations being mathematically "hyperbolic." In contrast, the equations for steady subsonic flow are "elliptic." (See Reference 21 for a description of these mathematical classifications.) The fact that the governing equations change their mathematical nature in going from locally supersonic to locally subsonic flow has historically caused theoretical aerodynamicists much grief. One problem in particular, namely, the mixed subsonic-supersonic flow over a supersonic blunt body as described in Section 9.5, was a major research area until a breakthrough was made in the late 1960s for its proper numerical solution. The purpose of this section is to describe a numerical finite-difference solution which readily allows the calculation of mixed subsonic-supersonic flows—the *time-dependent method*—and to show how it is used to solve supersonic blunt-body flows. Time-dependent techniques are very common in modern computational fluid dynamics, and as a student of aerodynamics, you should be familiar with their philosophy. These techniques are also called *time-marching techniques* because the solutions are obtained by marching in steps of time.

Consider a blunt body in a supersonic stream, as sketched in Figure 13.9*a*. The shape of the body is known and is given by $b = b(y)$. For a given freestream Mach number M_∞, we wish to calculate the shape and location of the detached shock wave, as well as the flow-field properties between the shock and the body.

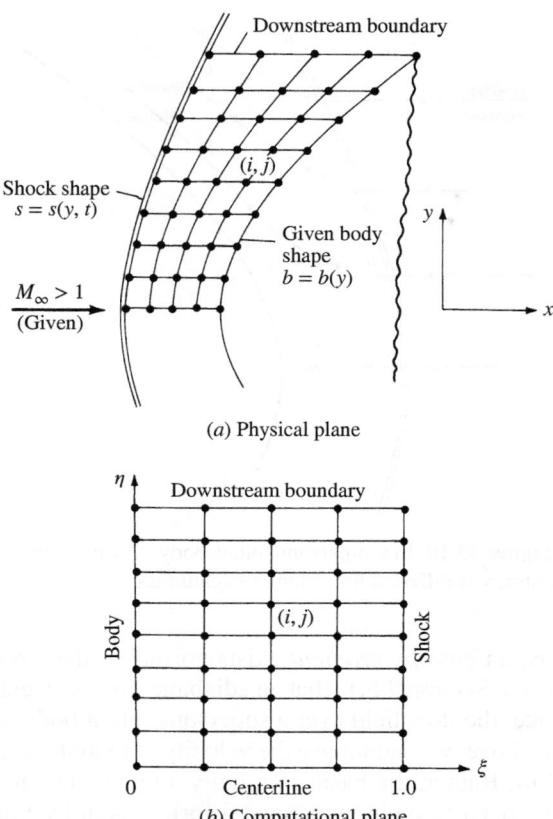

Figure 13.9 Blunt-body flow field in both the physical and computational planes.

The physical aspects of this flow field were described in Section 9.5, which you should review before progressing further.

The flow around a blunt body in a supersonic stream is rotational. Why? Examine Figure 13.10, which illustrates several streamlines around the blunt body. The flow is inviscid and adiabatic. In the uniform freestream ahead of the shock wave, the entropy is the same for each streamline. However, in crossing the shock wave, each streamline traverses a different part of the wave, and hence experiences a different increase in entropy. That is, the streamline at point a in Figure 13.10 crosses a normal shock, and hence experiences a large increase in entropy, whereas the streamline at point b crosses a weaker, oblique shock, and therefore experiences a smaller increase in entropy, $s_b < s_a$. The streamline at point c experiences an even weaker portion of the shock, and hence $s_c < s_b < s_a$. The net result is that in the flow between the shock and the body, the entropy *along* a given streamline is constant, whereas the entropy changes from one streamline

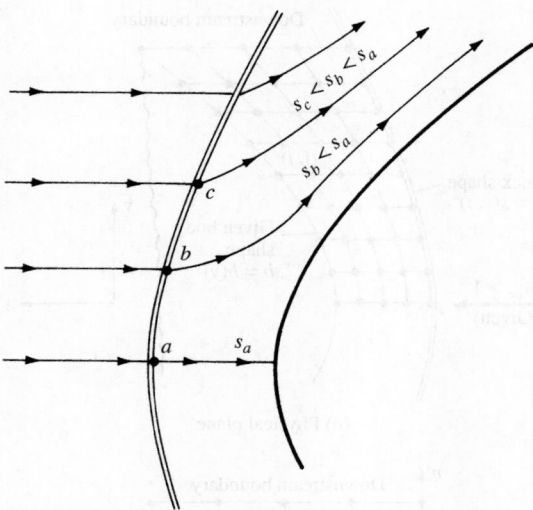

Figure 13.10 In a supersonic blunt-body flow field, the entropy is different for different streamlines.

to the next; that is, an *entropy gradient* exists normal to the streamlines. It can readily be shown (see Section 9.5.1) that an adiabatic flow with entropy gradients is *rotational*. Hence, the flow field over a supersonic blunt body is rotational.

In light of the above, we cannot use the velocity potential equation to analyze the blunt-body flow. Rather, the basic continuity, momentum, and energy equations must be employed in their fundamental form, given by Equations (7.40), (7.42a and b), and (7.44). With no body forces, these equations are

Continuity:
$$\frac{\partial \rho}{\partial t} = -\left(\frac{\partial(\rho u)}{\partial x} + \frac{\partial(\rho v)}{\partial y} \right) \tag{13.59}$$

x momentum:
$$\frac{\partial u}{\partial t} = -\left(u\frac{\partial u}{\partial x} + v\frac{\partial u}{\partial y} + \frac{1}{\rho}\frac{\partial p}{\partial x} \right) \tag{13.60}$$

y momentum:
$$\frac{\partial v}{\partial t} = -\left(u\frac{\partial v}{\partial x} + v\frac{\partial v}{\partial y} + \frac{1}{\rho}\frac{\partial p}{\partial y} \right) \tag{13.61}$$

Energy:
$$\frac{\partial(e + V^2/2)}{\partial t} = -\left(u\frac{\partial(e + V^2/2)}{\partial x} + v\frac{\partial(e + V^2/2)}{\partial y} \right.$$
$$\left. + \frac{1}{\rho}\frac{\partial(pu)}{\partial x} + \frac{1}{\rho}\frac{\partial(pv)}{\partial y} \right) \tag{13.62}$$

Notice the form of the above equations; the time derivatives are on the left, and all spatial derivatives are on the right. These equations are in the form necessary for a time-dependent finite-difference solution, as described below.

Return to Figure 13.9a. Recall that the body shape and freestream conditions are given, and we wish to calculate the shape and location of the shock wave as

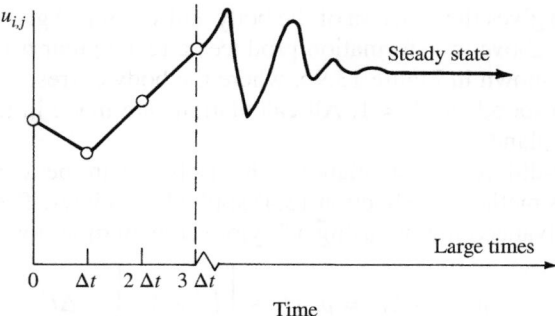

Figure 13.11 Schematic of the time variation of a typical
flow variable—the time-dependent method.

well as the flow field between the shock and body. We are interested in the *steady
flow* over the blunt body; however, we use a time-dependent method to obtain the
steady flow. The basic philosophy of this method is as follows. First, *assume* a
shock-wave shape and location. Also, cover the flow field between the shock and
body with a series of grid points, as sketched in Figure 13.9a. At each of these grid
points, *assume* values of all the flow variables, ρ, u, v, etc. These assumed values
are identified as *initial conditions* at time $t = 0$. With these assumed values, the
spatial derivatives on the right sides of Equations (13.59) to (13.62) are known
values (obtained from finite differences). Hence, Equations (13.59) to (13.62)
allow the calculation of the time derivatives $\partial\rho/\partial t$, $\partial u/\partial t$, etc. In turn, these time
derivatives allow us to calculate the flow properties at each grid point at a later
instant in time, say, Δt. The flow properties at time $t = \Delta t$ are different from at
$t = 0$. A repetition of this cycle gives the flow-field variables at all grid points
at time $t = 2\Delta t$. As this cycle is repeated many hundreds of times, the flow-field
properties at each grid point are calculated as a function of time. For example,
the time variation of $u_{i,j}$ is sketched in Figure 13.11. At each time step, the value
of $u_{i,j}$ is different; however, at large times the changes in $u_{i,j}$ from one time step
to another become small, and $u_{i,j}$ approaches a steady-state value, as shown in
Figure 13.11. *It is this steady-state value that we want; the time-dependent
approach is simply a means to that end.* Moreover, the shock-wave shape and
location will change with time; the new shock location and shape at each time
step are calculated so as to satisfy the shock relations across the wave at each
of the grid points immediately behind the wave. At large times, as the flow-field
variables approach a steady state, the shock shape and location also approach a
steady state. Because of the time-dependent motion of the shock wave, the wave
shape is a function of both t and y as shown in Figure 13.9a, $s = s(y, t)$.

Given this philosophy, let us examine a few details of the method. First, note
that the finite-difference grid in Figure 13.9a is curved. We would like to apply
our finite differences in a rectangular grid; hence, in Equations (13.59) to (13.62)
the independent variables can be transformed as

$$\xi = \frac{x - b}{s - b} \quad \text{and} \quad \eta = y$$

where $b = b(y)$ gives the abscissa of the body and $s = s(y, t)$ gives the abscissa of the shock. The above transformation produces a rectangular grid in the computational plane, shown in Figure 13.9b, where the body corresponds to $\xi = 0$ and the shock corresponds to $\xi = 1$. All calculations are made in this transformed, computational plane.

The finite-difference calculations themselves can be carried out using MacCormack's method (see Section 13.4) applied as follows. The flow-field variables can be advanced in time using a Taylor series in time; for example,

$$\rho_{i,j}(t + \Delta t) = \rho_{i,j}(t) + \left[\left(\frac{\partial \rho}{\partial t} \right)_{i,j} \right]_{\text{ave}} \Delta t \tag{13.63}$$

In Equation (13.63), we know the density at grid point (i,j) at time t; that is, we know $\rho_{i,j}(t)$. Then Equation (13.63) allows us to calculate the density at the same grid point at time $t + \Delta t$, that is, $\rho_{i,j}(t + \Delta t)$, *if* we know a value of the average time derivative $[(\partial \rho / \partial t)_{i,j}]_{\text{ave}}$. This time derivative is an average between times t and $t + \Delta t$ and is obtained from a predictor-corrector process as follows.

13.5.1 Predictor Step

All the flow variables are known at time t at all the grid points. This allows us to replace the spatial derivatives on the right of Equations (13.59) to (13.62) (suitably transformed into $\xi\eta$ space) with known *forward differences*. These equations then give values of the time derivatives at time t, which are used to obtain *predicted* values of the flow-field variables at time $t + \Delta t$; for example,

$$\bar{\rho}_{i,j}(t + \Delta t) = \rho_{i,j}(t) + \left[\left(\frac{\partial \rho}{\partial t} \right)_{i,j} \right]_t \Delta t$$

where $\rho_{i,j}(t)$ is known, $[(\partial \rho / \partial t)_{i,j}]_t$ is obtained from the governing equation, Equation (13.59) (suitably transformed), using *forward differences* for the spatial derivatives, and $\bar{\rho}_{i,j}(t + \Delta t)$ is the predicted density at time $t + \Delta t$. Predicted values of all other flow variables $\bar{u}_{i,j}(t + \Delta t)$, etc., are obtained at all the grid points in a likewise fashion.

13.5.2 Corrector Step

Inserting the flow variables obtained above into the governing equations, Equations (13.59) to (13.62), using *rearward differences* for the spatial derivatives, predicted values of the time derivatives at $t + \Delta t$ are obtained, for example, $[(\partial \rho / \partial t)_{i,j}]_{(t+\Delta t)}$. In turn, these are averaged with the time derivatives from the predictor step to obtain; for example,

$$\left[\left(\frac{\partial \rho}{\partial t} \right)_{i,j} \right]_{\text{ave}} = \frac{1}{2} \left\{ \left[\left(\frac{\partial \rho}{\partial t} \right)_{i,j} \right]_t + \left[\left(\overline{\frac{\partial \rho}{\partial t}} \right)_{i,j} \right]_{(t+\Delta t)} \right\} \tag{13.64}$$

Finally, the average time derivative obtained from Equation (13.64) is inserted into Equation (13.63) to yield the corrected value of density at time $t + \Delta t$. The same procedure is used for all the dependent variables, u, v, etc.

Starting from the assumed initial conditions at $t = 0$, the repeated application of Equation (13.63) along with the above predictor-corrector algorithm at each time step allows the calculation of the flow-field variables and shock shape and location as a function of time. As stated above, after a large number of time steps, the calculated flow-field variables approach a steady state, where $[(\partial\rho/\partial t)_{i,j}]_{ave} \to 0$ in Equation (13.63). Once again, we emphasize that we are interested in the steady-state answer, and the time-dependent technique is simply a means to that end.

Note that the applications of MacCormack's technique to both the steady flow calculations described in Section 13.4 and the time-dependent calculations described in the present section are analogous; in the former, we march forward in the spatial coordinate x, starting with known values along with a constant y line, whereas in the latter, we march forward in time starting with a known flow field at $t = 0$.

Why do we bother with a time-dependent solution? Is it not an added complication to deal with an extra independent variable t in addition to the spatial variables x and y? The answers to these questions are as follows. The governing unsteady flow equations given by Equations (13.59) to (13.62) are hyperbolic with respect to time, independent of whether the flow is locally subsonic or supersonic. In Figure 13.9a, some of the grid points are in the subsonic region and others are in the supersonic region. However, the time-dependent solution progresses in the same manner at all these points, independent of the local Mach number. Hence, the time-dependent technique is the only approach known today which allows the uniform calculation of a mixed subsonic-supersonic flow field of arbitrary extent. For this reason, the application of the time-dependent technique, although it adds one additional independent variable, allows the straightforward solution of a flow field which is extremely difficult to solve by a purely steady-state approach.

A much more detailed description of the time-dependent technique is given in Chapter 12 of Reference 21, and especially in Reference 7, which you should study before attempting to apply this technique to a specific problem. The intent of our description here has been to give you simply a "feeling" for the philosophy and general approach of the technique.

Some typical results for supersonic blunt-body flow fields are given in Figures 13.12 to 13.15. These results were obtained with a time-dependent solution described in Reference 33. Figures 13.12 and 13.13 illustrate the behavior of a time-dependent solution during its approach to the steady state. In Figure 13.12, the time-dependent motion of the shock wave is shown for a parabolic cylinder in a Mach 4 freestream. The shock labeled $0\,\Delta t$ is the initially assumed shock wave at $t = 0$. At early times, the shock wave rapidly moves away from the body; however, after about 300 time steps, it has slowed considerably, and between 300 and 500 time steps, the shock wave is virtually motionless—it has reached its steady-state shape and location. The time variation of the stagnation point pressure is given in

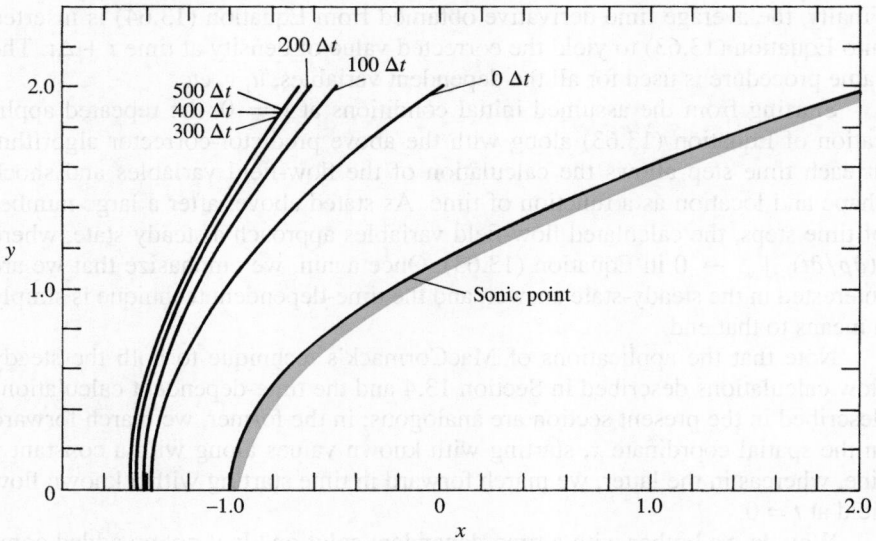

Figure 13.12 Time-dependent shock-wave motion, parabolic cylinder, $M_\infty = 4$.

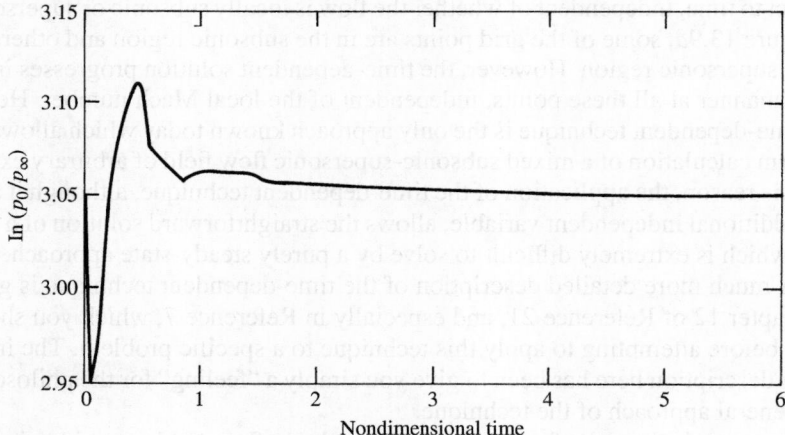

Figure 13.13 Time variation of stagnation point pressure, parabolic cylinder, $M_\infty = 4$.

Figure 13.13. Note that the pressure shows strong timewise oscillations at early times, but then it asymptotically approaches a steady value at large times. Again, it is this asymptotic steady state that we want, and the intermediate transient results are just a means to that end. Concentrating on just the steady-state results, Figure 13.14 gives the pressure distribution (nondimensionalized by stagnation

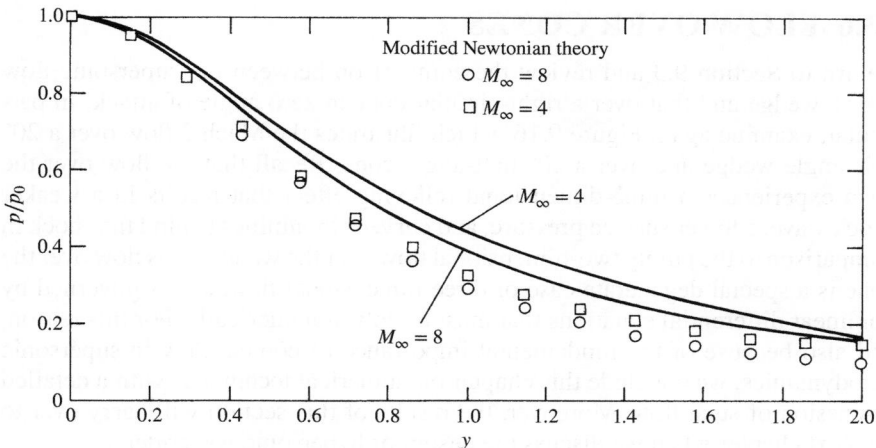

Figure 13.14 Surface pressure distributions, parabolic cylinder.

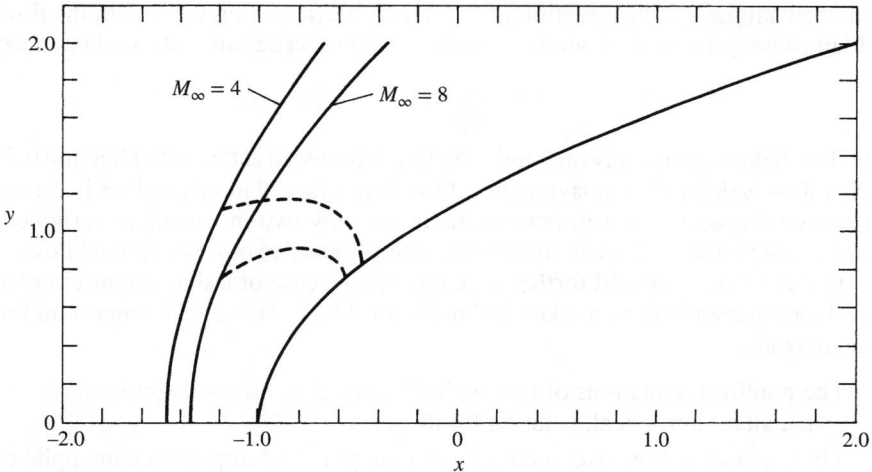

Figure 13.15 Shock shapes and sonic lines, parabolic cylinder.

point pressure) over the body surface for the cases of both $M_\infty = 4$ and 8. The time-dependent numerical results are shown as the solid curves, whereas the open symbols are from Newtonian theory, to be discussed in Chapter 14. Note that the pressure is a maximum at the stagnation point and decreases as a function of distance away from the stagnation point—a variation that we most certainly would expect based on our previous aerodynamic experience. The steady shock shapes and sonic lines are shown in Figure 13.15 for the cases of $M_\infty = 4$ and 8. Note that as the Mach number increases, the shock wave moves closer to the body.

13.6 FLOW OVER CONES

Return to Section 9.3 and review the comparison between the supersonic flow over a wedge and that over a right-circular cone at zero angle of attack. In particular, examine again Figure 9.16, which illustrates the Mach 2 flow over a 20° half-angle wedge and over a 20° half-angle cone. Recall that the flow over the cone experiences a three-dimensional relieving effect that results in a weaker shock wave, a lower surface pressure, and curved streamlines behind the shock in comparison to the purely two-dimensional flow over the wedge. This flow over the cone is a special degenerate case of three-dimensional flow, and is governed by nonlinear differential equations that must be solved numerically. For this reason, and also because of the fundamental importance of conical flow in supersonic aerodynamics, we conclude this chapter on numerical techniques with a detailed discussion of such flow. Moreover, the results of this section will carry over to the next chapter when we discuss the design of hypersonic waveriders.

Consider a body of revolution (a body generated by rotating a given planar curve about a fixed axis) at zero angle of attack as shown in Figure 13.16. A cylindrical coordinate system (r, Φ, z) is drawn, with the z axis as the axis of symmetry aligned in the direction of V_∞. By inspection of Figure 13.16, the flow field must be symmetric about the z axis; i.e., all properties are independent of Φ:

$$\frac{\partial}{\partial \phi} \equiv 0$$

The flow field depends only on r and z. As first introduced at the end of Section 6.3, such a flow is defined as axisymmetric flow. It is a flow that takes place in three-dimensional space; however, because there are only two independent variables, r and z, axisymmetric flow is sometimes called "quasi-two-dimensional" flow.

In this section, we will further specialize to the case of a sharp right-circular cone in a supersonic flow, as sketched in Figure 13.17. This case is important for three reasons.

1. The nonlinear equations of motion lend themselves to a straightforward exact, albeit numerical, solution for this case.
2. The supersonic flow over a cone is of great practical importance in applied aerodynamics; the nose cones of many high-speed missiles and projectiles are approximately conical, as are the nose regions of the fuselages of most supersonic airplanes.

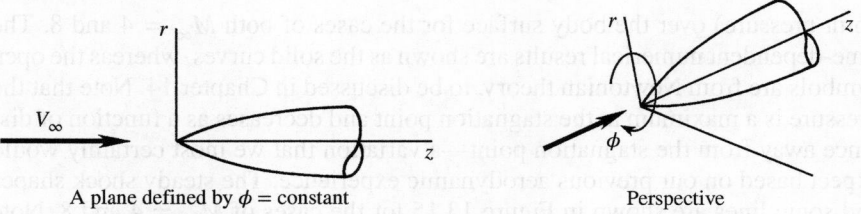

Figure 13.16 Cylindrical coordinate system for an axisymmetric body.

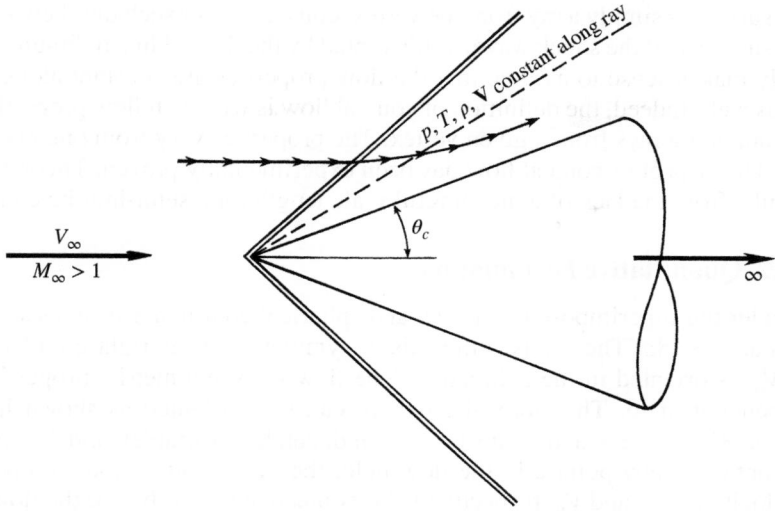

Figure 13.17 Supersonic flow over a cone.

3. The first solution for the supersonic flow over a cone was obtained by A. Busemann in 1929, long before supersonic flow became fashionable (see Reference 93). This solution was essentially graphical, and illustrated some of the important physical phenomena. A few years later, in 1933, G. I. Taylor and J. W. Macoll (see Reference 94) presented a numerical solution that is a hallmark in the evolution of compressible flow. Therefore, the study of conical flow is of historical significance. Moreover, the Taylor-Maccoll solution for the supersonic flow over a cone is a classic case in inviscid supersonic aerodynamics.

13.6.1 Physical Aspects of Conical Flow

Consider a sharp cone of semivertex angle θ_c, sketched in Figure 13.17. Assume this cone extends to infinity in the downstream direction (a semi-infinite cone). The cone is in a supersonic flow, and hence an oblique shock wave is attached at the vertex. The shape of this shock wave is also conical. A streamline from the supersonic freestream discontinuously deflects as it traverses the shock, and then curves continuously downstream of the shock, becoming parallel to the cone surface asymptotically at infinity. Contrast this flow with that over a two-dimensional wedge (Figure 9.16a) where all streamlines behind the shock are immediately parallel to the wedge surface.

Because the cone extends to infinity, distance along the cone becomes meaningless: If the pressure were different at the 1- and 10-m stations along the surface of the cone, then what would it become at infinity? This presents a dilemma that can be reconciled only by assuming that the pressure is constant along the surface of the cone, as well as that all other flow properties are also constant. Since the cone surface is simply a ray from the vertex, consider other such rays between the cone surface and the shock wave, as illustrated by the dashed line in Figure 13.17. It only makes sense to assume that the flow properties are constant along these rays as well. Indeed, the definition of conical flow is where all flow properties are constant along rays from a given vertex. The properties vary from one ray to the next. This aspect of conical flow has been experimentally proven. Theoretically, it results from the lack of a meaningful scale length for a semi-infinite cone.

13.6.2 Quantitative Formulation

Consider the superimposed cartesian and spherical coordinate systems sketched in Figure 13.18a. The z axis is the axis of symmetry for the right-circular cone, and V_∞ is oriented in the z direction. The flow is axisymmetric; properties are independent of Φ. Therefore, the picture can be reoriented as shown in Figure 13.18b, where r and θ are the two independent variables and V_∞ is now horizontal. At any point e in the flow field, the radial and normal components of velocity are V_r and V_θ, respectively. Our objective is to solve for the flow field between the body and the shock wave. Recall that for axisymmetric conical flow

$$\frac{\partial}{\partial \Phi} = 0 \qquad \text{(axisymmetric flow)}$$

$$\frac{\partial}{\partial r} \equiv 0 \qquad \text{(flow properties are constant along a ray from the vertex)}$$

The continuity equation for steady flow is Equation (2.54):

$$\nabla \cdot (\rho \mathbf{V}) = 0$$

The divergence of a vector in spherical coordinates is given by Equation 2.21. Thus, in terms of spherical coordinates, Equation (2.54) becomes

$$\nabla \cdot (\rho \mathbf{V}) = \frac{1}{r^2} \frac{\partial}{\partial r}(r^2 \rho V_r) + \frac{1}{r \sin \theta} \frac{\partial}{\partial \theta}(\rho V_\theta \sin \theta) + \frac{1}{r \sin \theta} \frac{\partial(\rho V_\phi)}{\partial \phi} = 0 \quad (13.65)$$

Evaluating the derivatives, and applying the above conditions for axisymmetric conical flow, we see that Equation (13.65) becomes

$$\frac{1}{r^2} \left[r^2 \frac{\partial(\rho V_r)}{\partial r} + \rho V_r(2r) \right] + \frac{1}{r \sin \theta} \left[\rho V_\theta \cos \theta + \sin \theta \frac{\partial(\rho V_\theta)}{\partial \theta} \right] + \frac{1}{r \sin \theta} \frac{\partial(\rho V_\theta)}{\partial \phi} = 0$$

$$\frac{2\rho V_r}{r} + \frac{\rho V_\theta}{r} \cot \theta + \frac{1}{r} \left(\rho \frac{\partial V_\theta}{\partial \theta} + V_\theta \frac{\partial \rho}{\partial \theta} \right) = 0$$

$$2\rho V_r + \rho V_\theta \cot \theta + \rho \frac{\partial V_\theta}{\partial \theta} + V_\theta \frac{\partial \rho}{\partial \theta} = 0$$

$$(13.66)$$

Equation (13.66) is the continuity equation for axisymmetric conical flow.

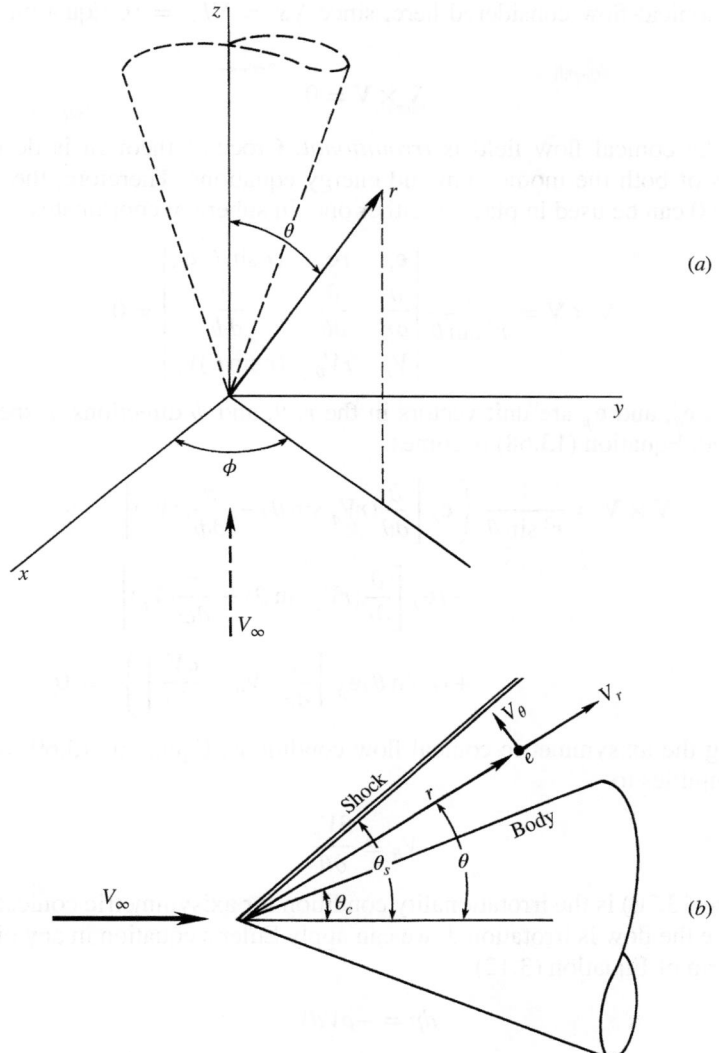

Figure 13.18 Spherical coordinate system for a cone.

Return to the conical flow field sketched in Figures 13.17 and 13.18. The shock wave is straight, and hence the increase in entropy across the shock is the same for all streamlines. Consequently, throughout the conical flow field, $\nabla s = 0$. Moreover, the flow is adiabatic and steady, and hence Equation 7.55 dictates that $\Delta h_o = 0$. In an inviscid compressible flow, entropy gradients ∇s, gradients in total enthalpy Δh_o, and vorticity $\nabla \times \mathbf{V}$, are related through *Crocco's theorem*, derived in Reference 21, and given here without proof:

$$T\nabla s = \nabla h_o - \mathbf{V} \times (\nabla \times \mathbf{V}) \qquad (13.67)$$

For the conical flow considered here, since $\nabla s = \nabla h_o = 0$, Equation (13.67) gives

$$\nabla \times \mathbf{V} = 0$$

Hence, the conical flow field is *irrotational*. Crocco's theorem is derived on the basis of both the momentum and energy equations. Therefore, the relation $\nabla \times \mathbf{V} = 0$ can be used in place of either one. In spherical coordinates,

$$\nabla \times \mathbf{V} = \frac{1}{r^2 \sin\theta} \begin{vmatrix} \mathbf{e}_r & r\mathbf{e}_\theta & (r\sin\theta)\mathbf{e}_\phi \\ \dfrac{\partial}{\partial r} & \dfrac{\partial}{\partial \theta} & \dfrac{\partial}{\partial \phi} \\ V_r & rV_\theta & (r\sin\theta)V_\phi \end{vmatrix} = 0 \qquad (13.68)$$

where $\mathbf{e}_r$, $\mathbf{e}_\theta$, and $\mathbf{e}_\phi$ are unit vectors in the r, θ, and ϕ directions, respectively. Expanded, Equation (13.68) becomes

$$\nabla \times \mathbf{V} = \frac{1}{r^2 \sin\theta} \left\{ \mathbf{e}_r \left[\frac{\partial}{\partial \theta}(rV_\phi \sin\theta) - \frac{\partial}{\partial \phi}(rV_\theta) \right] \right.$$

$$- r\mathbf{e}_\theta \left[\frac{\partial}{\partial r}(rV_\phi \sin\theta) - \frac{\partial}{\partial \phi}(V_r) \right]$$

$$\left. + (r\sin\theta)\mathbf{e}_\phi \left[\frac{\partial}{\partial r}(rV_\theta - \frac{\partial V_r}{\partial \theta} \right] \right\} = 0 \qquad (13.69)$$

Applying the axisymmetric conical flow conditions, Equation (13.69) dramatically simplifies to

$$V_\theta \equiv \frac{\partial V_r}{\partial \theta} \qquad (13.70)$$

Equation (13.70) is the irrotationality condition for axisymmetric conical flow.

Since the flow is irrotational, we can apply Euler's equation in any direction in the form of Equation (3.12):

$$dp = -\rho V dV$$

where

$$V^2 = V_r^2 + V_\theta^2$$

Hence, Equation (3.12) becomes

$$dp = -\rho(V_r \, dV_r + V_\theta \, dV_\theta) \qquad (13.71)$$

Recall that, for isentropic flow,

$$\frac{dp}{d\rho} \equiv \left(\frac{\partial p}{\partial \rho} \right)_s = a^2$$

Thus, Equation (13.71) becomes

$$\frac{d\rho}{\rho} = -\frac{1}{a^2}(V_r\,dV_r + V_\theta\,dV_\theta) \tag{13.72}$$

From Equation (7.55), and defining a new reference velocity V_{max} as the maximum theoretical velocity obtainable from a fixed reservoir condition (when $V = V_{max}$, the flow has expanded theoretically to zero temperature, hence $h = 0$), we have

$$h_o = \text{const} = h + \frac{V^2}{2} = \frac{V_{max}^2}{2}$$

Note that V_{max} is a constant for the flow and is equal to $\sqrt{2h_o}$. For a calorically perfect gas, the above becomes

$$\frac{a^2}{\gamma-1} + \frac{V^2}{2} = \frac{V_{max}^2}{2}$$

or

$$a^2 = \frac{\gamma-1}{2}\left(V_{max}^2 - V^2\right) = \frac{\gamma-1}{2}\left(V_{max}^2 - V_r^2 - V_\theta^2\right) \tag{13.73}$$

Substituting Equation (13.73) into (13.72),

$$\frac{d\rho}{\rho} = -\frac{2}{\gamma-1}\left(\frac{V_r\,dV_r + V_\theta\,dV_\theta}{V_{max}^2 - V_r^2 - V_\theta^2}\right) \tag{13.74}$$

Equation (13.74) is essentially Euler's equation in a form useful for studying conical flow.

Equations (13.66), (13.70), and (13.74) are three equations with three dependent variables: ρ, V_r, and V_θ. Due to the axisymmetric conical flow conditions, there is only one independent variable, namely θ. Hence, the partial derivatives in Equations (13.66) and (13.70) are more properly written as ordinary derivatives. From Equation (13.66),

$$2V_r + V_\theta \cot\theta + \frac{dV_\theta}{d\theta} + \frac{V_\theta}{\rho}\frac{d\rho}{d\theta} = 0 \tag{13.75}$$

From Equation (13.74),

$$\frac{d\rho}{d\theta} = -\frac{2\rho}{\gamma-1}\left(\frac{V_r\dfrac{dV_r}{d\theta} + V_\theta\dfrac{dV_\theta}{d\theta}}{V_{max}^2 - V_r^2 - V_\theta^2}\right) \tag{13.76}$$

Substituting Equation (13.76) into Equation (13.75),

$$2V_r + V_\theta \cot\theta + \frac{dV_\theta}{d\theta} - \frac{2V_\theta}{\gamma-1}\left(\frac{V_r\dfrac{dV_r}{d\theta} + V_\theta\dfrac{dV_\theta}{d\theta}}{V_{max}^2 - V_r^2 - V_\theta^2}\right) = 0$$

or

$$\frac{\gamma - 1}{2}\left(V_{max}^2 - V_r^2 - V_\theta^2\right)\left(2V_r + V_\theta \cot\theta + \frac{dV_\theta}{d\theta}\right) - V_\theta\left(V_r\frac{dV_r}{d\theta} + V_\theta\frac{dV_\theta}{d\theta}\right) = 0$$

(13.77)

Recall from Equation (13.70)

$$V_\theta = \frac{dV_r}{d\theta}$$

Hence,

$$\frac{dV_\theta}{d\theta} = \frac{d^2V_r}{d\theta^2}$$

Substituting this result into Equation (13.77), we have

$$\frac{\gamma - 1}{2}\left[V_{max}^2 - V_r^2 - \left(\frac{dV_r}{d\theta}\right)^2\right]\left[2V_r + \frac{dV_r}{d\theta}\cot\theta + \frac{d^2V_r}{d\theta^2}\right]$$
$$- \frac{dV_r}{d\theta}\left[V_r\frac{dV_r}{d\theta} + \frac{dV_r}{d\theta}\left(\frac{d^2V_r}{d\theta^2}\right)\right] = 0$$

(13.78)

Equation (13.78) is the Taylor-Maccoll equation for the solution of conical flows. Note that it is an ordinary differential equation, with only one dependent variable, V_r. Its solution gives $V_r = f(\theta)$; V_θ follows from Equation (13.70), namely,

$$V_\theta = \frac{dV_r}{d\theta}$$

(13.79)

There is no closed-form solution to Equation (13.78); it must be solved *numerically*. To expedite the numerical solution, define the nondimensional velocity V' as

$$V' \equiv \frac{V}{V_{max}}$$

Then, Equation (13.78) becomes

$$\frac{\gamma - 1}{2}\left[1 - V_r'^2 - \left(\frac{dV_r'}{d\theta}\right)^2\right]\left[2V_r' + \frac{dV_r'}{d\theta}\cot\theta + \frac{d^2V_r'}{d\theta^2}\right]$$
$$- \frac{dV_r'}{d\theta}\left[V_r'\frac{dV_r'}{d\theta} + \frac{dV_r'}{d\theta}\left(\frac{d^2V_r'}{d\theta^2}\right)\right] = 0$$

(13.80)

The nondimensional velocity V' is a function of Mach number only. To see this more clearly recall that

$$h + \frac{V^2}{2} = \frac{V_{max}^2}{2}$$

$$\frac{a^2}{\gamma - 1} + \frac{V^2}{2} = \frac{V_{max}^2}{2}$$

$$\frac{1}{\gamma - 1} \left(\frac{a}{V} \right)^2 + \frac{1}{2} = \frac{1}{2} \left(\frac{V_{max}}{V} \right)^2$$

$$\frac{2}{\gamma - 1} + \left(\frac{1}{M} \right)^2 + 1 = \left(\frac{V_{max}}{V} \right)^2$$

$$\frac{V}{V_{max}} \equiv V' = \left[\frac{2}{(\gamma - 1)M^2} + 1 \right]^{-1/2} \tag{13.81}$$

Clearly, from Equation (13.81), $V' = f(M)$; given M, we can always find V', or vice versa.

13.6.3 Numerical Procedure

For the numerical solution of the supersonic flow over a right-circular cone, we will employ an inverse approach. By this, we mean that a given shock wave will be assumed, and the particular cone that supports the given shock will be calculated. This is in contrast to the direct approach, where the cone is given and the flow field and shock wave are calculated. The numerical procedure is as follows:

1. Assume a shock wave angle θ_s and a freestream Mach number M_∞, as sketched in Figure 13.19. From this, the Mach number and flow deflection angle, M_2 and δ, respectively, immediately behind the shock can be found from the oblique shock relations. Note that, contrary to our previous practice, the flow deflection angle is here denoted by δ so as not to confuse it with the polar coordinate θ.

2. From M_2 and δ, the radial and normal components of flow velocity, V'_r and V'_θ, respectively, directly behind the shock can be found from the geometry of Figure 13.19. Note that V' is obtained by inserting M_2 into Equation (13.81).

3. Using the above value of V'_r directly behind the shock as a boundary value, solve Equation (13.80) for V'_r numerically in steps of θ, marching away from the shock. Here, the flow field is divided into incremental angles $\Delta\theta$, as sketched in Figure 13.19. The ordinary differential equation [Equation (13.80)] can be solved at each $\Delta\theta$ using any standard numerical solution technique, such as the Runge-Kutta method.

4. At each increment in θ, the value of V'_θ is calculated from Equation (13.79). At some value of θ, namely $\theta = \theta_c$, we will find $V'_\theta = 0$. The normal component of velocity at an impermeable surface is zero. Hence, when

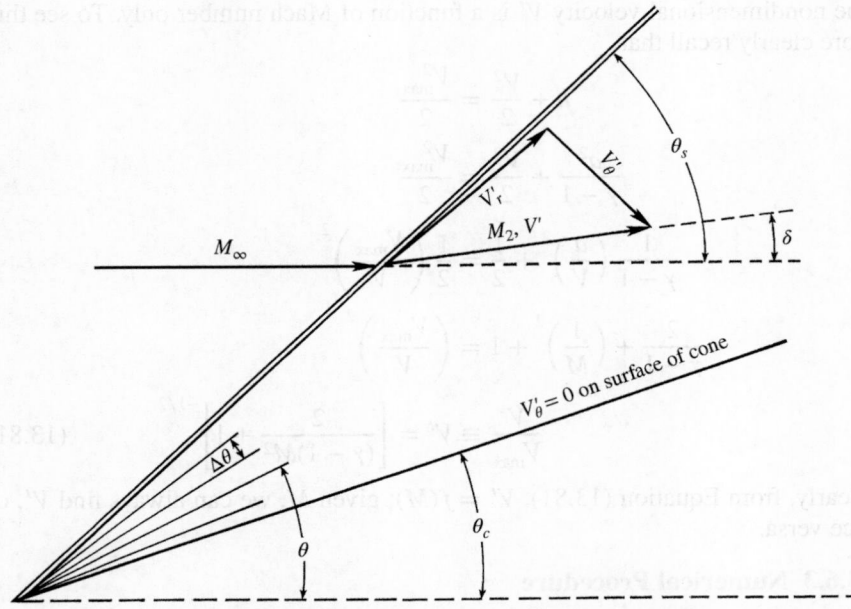

Figure 13.19 Geometry for the numerical solution of flow over a cone.

$V'_\theta = 0$ at $\theta = \theta_c$ then θ_c must represent the surface of the particular cone that supports the shock wave of given wave angle θ_s at the given Mach number M_∞ as assumed in step 1. That is, the cone angle compatible with M_∞ and θ_s is θ_c. The value of V'_r at θ_c gives the Mach number along the cone surface via Equation (13.81).

5. In the process of steps 1 through 4 here, the complete velocity flow field between the shock and the body has been obtained. Note that, at each point (or ray), $V' = \sqrt{(V'_r)^2 + (V'_\theta)^2}$ and M follow from Equation (13.81). The pressure, density, and temperature along each ray can then be obtained from the isentropic relations, Equations (8.42), (8.43), and (8.40).

If a different value of M_∞ and/or θ_s is assumed in step 1, a different flow field and cone angle θ_c will be obtained from steps 1 through 5. By a repeated series of these calculations, tables or graphs of supersonic cone properties can be generated. Such tables exist in the literature, the most common being those of Kopal (Reference 95) and Sims (Reference 96).

13.6.4 Physical Aspects of Supersonic Flow over Cones

Some typical numerical results obtained from the solution in Section 13.6.3 are illustrated in Figure 13.20, which gives the shock wave angle θ_s as a function of cone angle θ_c with M_∞ as a parameter. Figure 13.20 for cones is analogous to Figure 9.9 for two-dimensional wedges; the two figures are qualitatively similar, but the numbers are different.

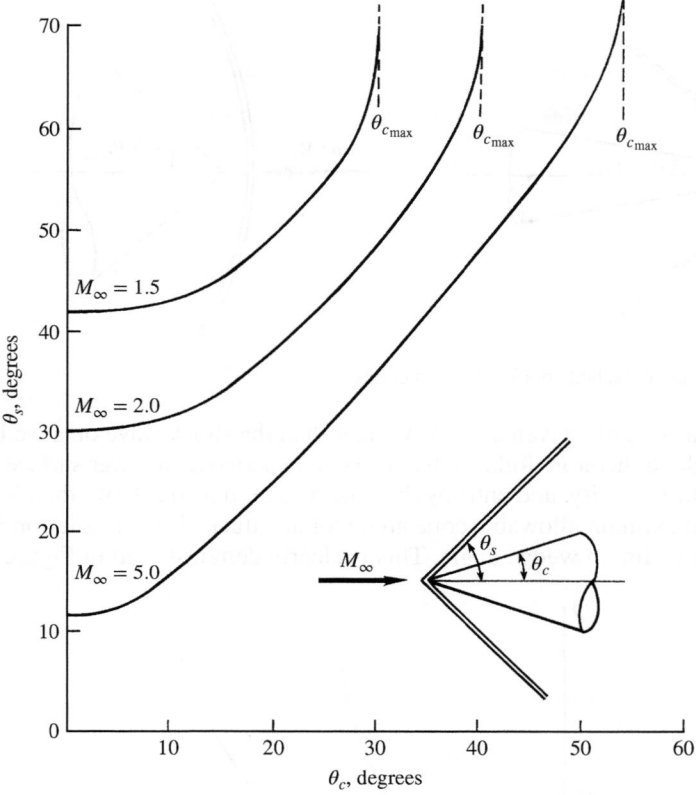

Figure 13.20 θ_c-θ_s-M diagram for cones in supersonic flow. (The top portion of the curves curls back for the strong-shock solution, which is not shown here.)

Examine Figure 13.20 closely. Note that, for a given cone angle θ_c and given M_∞, there are two possible oblique shock waves—the strong- and weak-shock solutions. This is directly analogous to the two-dimensional case discussed in Chapter 9. The weak solution is almost always observed in practice on real finite cones; however, it is possible to force the strong-shock solution by independently increasing the back pressure near the base of the cone.

Also note from Figure 13.20 that, for a given M_∞, there is a maximum cone angle $\theta_{c_{\text{max}}}$, beyond which the shock becomes detached. This is illustrated in Figure 13.21. When $\theta_c > \theta_{c_{\text{max}}}$, there exists no Taylor-Maccoll solution as given here; instead, the flow field with a detached shock must be solved by techniques such as those discussed in Section 13.5.

In comparison to the two-dimensional flow over a wedge, the three-dimensional flow over a cone has an extra dimension in which to expand. This "three-dimensional relieving effect" was discussed in Section 9.3, which should now be reviewed by the reader. In particular, recall from Figure 9.16 that the shock

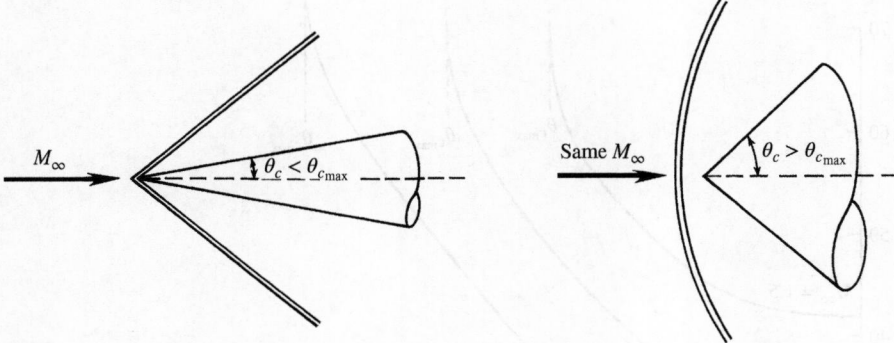

Figure 13.21 Attached and detached shock waves on cones.

wave on a cone of a given angle is weaker than the shock wave on a wedge of the same angle. It therefore follows that the cone experiences a lower surface pressure, temperature, density, and entropy than the wedge. It also follows that, for a given M_∞, the maximum allowable cone angle for an attached shock solution is greater than the maximum wedge angle. This is clearly demonstrated in Figure 13.22.

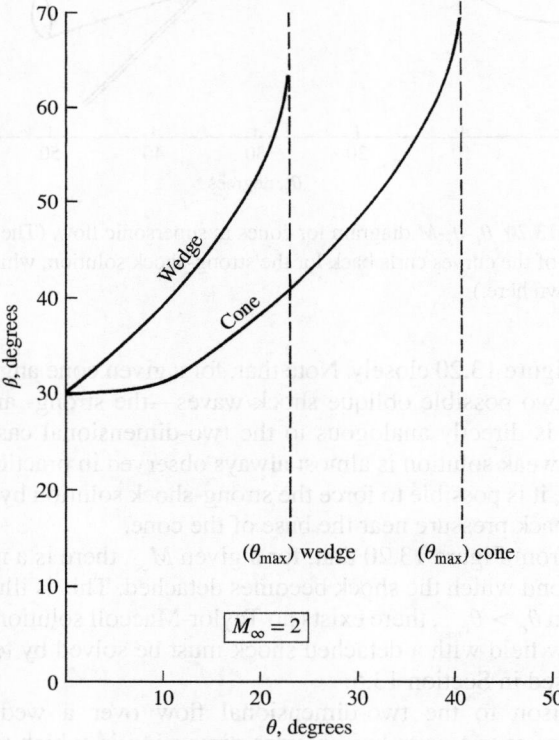

Figure 13.22 Comparison of shock wave angles for wedges and cones at Mach 2.

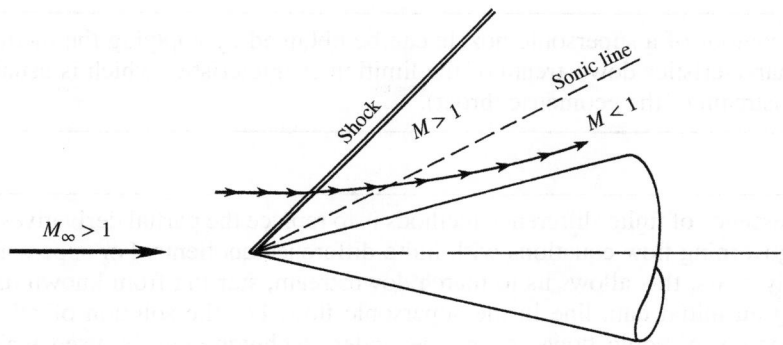

Figure 13.23 Some supersonic conical flow fields are characterized by an isentropic compression to subsonic velocities near the cone surface.

Finally, the numerical results show that any given streamline between the shock wave and cone surface is curved, as sketched in Figure 13.23, and asymptotically becomes parallel to the cone surface at infinity. Also, for most cases, the complete flow field between the shock and the cone is supersonic. However, if the cone angle is large enough, but still less than $\theta_{c_{\max}}$, there are some cases where the flow becomes subsonic near the surface. This case is illustrated in Figure 13.23, where one of the rays in the flow field becomes a sonic line. In this case, we see one of the few instances in nature where a supersonic flow field is actually isentropically compressed from supersonic to subsonic velocities. A transition from supersonic to subsonic flow is almost invariably accompanied by shock waves, as discussed in Chapter 8. However, flow over a cone can be an exception to this observation.

13.7 SUMMARY

We have now completed both branches of our road map shown in Figure 13.1. Make certain that you feel comfortable with all the material represented by this road map. A short summary of the highlights is given below:

For a steady, two-dimensional, irrotational, supersonic flow, the characteristic lines are Mach lines, and the compatibility equations which hold along these characteristic lines are

$$\theta + v = K_- \text{ (along a } C_- \text{ characteristic)}$$

and
$$\theta - v = K_+ \text{ (along a } C_+ \text{ characteristic)}$$

The numerical solution of such a flow can be carried out by solving the compatibility equation along the characteristic lines in a step-by-step fashion, starting from an appropriate initial data line.

> The contour of a supersonic nozzle can be obtained by applying the method of characteristics downstream of the limiting characteristic (which is usually downstream of the geometric throat).

> The essence of finite-difference methods is to replace the partial derivatives in the governing flow equations with finite-difference quotients. For supersonic steady flows, this allows us to march downstream, starting from known data along an initial data line in the supersonic flow. For the solution of mixed subsonic-supersonic flows, a time-dependent technique can be used which allows us to march forward in time, starting with assumed initial conditions at time $t = 0$ and achieving a steady-state result in the limit of large times.

> A popular technique for carrying out finite-difference solutions, whether for supersonic steady flow or for a time-dependent solution of mixed subsonic and supersonic flow, is the straightforward predictor-corrector technique by MacCormack.

13.8 PROBLEM

Note: The purpose of the following problem is to provide an exercise in carrying out a unit process for the method of characteristics. A more extensive application to a complete flow field is left to your specific desires. Also, an extensive practical problem utilizing the finite-difference method requires a large number of arithmetic operations and is practical only on a digital computer. You are encouraged to set up such a problem at your leisure. The main purpose of the present chapter is to present the essence of several numerical methods, not to burden the reader with a lot of calculations or the requirement to write an extensive computer program.

13.1 Consider two points in a supersonic flow. These points are located in a Cartesian coordinate system at $(x_1, y_1) = (0, 0.0684)$ and $(x_2, y_2) = (0.0121, 0)$, where the units are meters. At point (x_1, y_1): $u_1 = 639$ m/s, $v_1 = 232.6$ m/s, $p_1 = 1$ atm, $T_1 = 288$ K. At point (x_2, y_2): $u_2 = 680$ m/s, $v_2 = 0$, $p_2 = 1$ atm, $T_2 = 288$ K. Consider point 3 downstream of points 1 and 2 located by the intersection of the C_+ characteristic through point 2 and the C_- characteristic through point 1. At point 3, calculate: u_3, v_3, p_3, and T_3. Also, calculate the location of point 3, assuming the characteristics between these points are straight lines.

CHAPTER 14

Elements of Hypersonic Flow

Almost everyone has their own definition of the term hypersonic. If we were to conduct something like a public opinion poll among those present, and asked everyone to name a Mach number above which the flow of a gas should properly be described as hypersonic there would be a majority of answers round about 5 or 6, but it would be quite possible for someone to advocate, and defend, numbers as small as 3, or as high as 12.

P. L. Roe,
comment made in a lecture
at the von Karman Institute, Belgium
January 1970

PREVIEW BOX

Airbreathing hypersonic flight is held by many (including this author) to be the last frontier of air-vehicle design. Some progress has been made, but much needs to be done. The practical design of hypersonic vehicles for sustained hypersonic flight in the atmosphere will be a major challenge to the next generation of aerospace engineers.

Aeronautical history was made in March 2004 when the X-43 Hyper-X test vehicle, shown in Figure 14.1, achieved sustained flight for 11 s at Mach 6.9 powered by a supersonic combustion ramjet engine (SCRAMjet). In November, another X-43 achieved sustained flight at nearly Mach 10, making it the fastest airplane in history to date. At these speeds,

aerodynamic heating becomes a major problem, and the vehicle must be fabricated from special high-temperature materials. The X-43 thermal protection design is shown in Figure 14.2. The two successful flights of the X-43 in 2004 made aerospace engineering history; for the first time a SCRAMjet engine operated successfully for a sustained period in atmospheric flight. After more than 40 years of research and technical development, enough progress had been made to allow a successful test of the design methodology. But so much more needs to be done to achieve truly practical devices and flight vehicles. That is where some of you come into the picture; perhaps you will be the ones to tackle this last frontier.

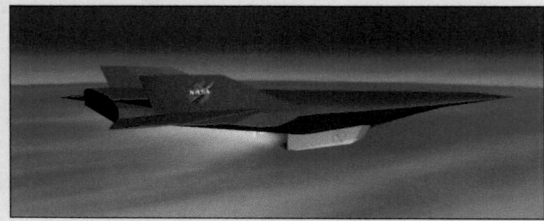

Figure 14.1 X-43 hypersonic test vehicle (*NASA*).

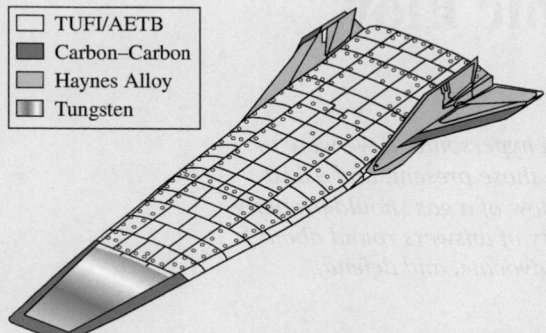

Figure 14.2 X-43 thermal protection system.

Hypersonic flight of other human-made vehicles has been a reality since 1949, when a WAC Corporal rocket, boosted to high altitudes on top of a captured German V-2 rocket, powered itself back into the atmosphere at over 5000 mph at White Sands Proving Ground in New Mexico (see Reference 52). Since then, a whole host of space vehicles, such as the space shuttle and the Apollo return module have returned to Earth after entering and flying through the atmosphere at large hypersonic Mach numbers, from Mach 26

to Mach 36. So hypersonic aerodynamics has been around for a relatively long time, but for all practical purposes it is still a young, developing discipline.

Hypersonic aerodynamics is the subject of this chapter. What is it about hypersonic flows that make them any different from supersonic flows? Why is it that they justify a separate chapter in this book? After all, hypersonic flow is flow with velocities greater than the speed of sound, which is the definition of supersonic flow, except that hypersonic flows are moving at velocities generally a lot larger than the speed of sound. Indeed, an old rule of thumb defines hypersonic flow as flow at Mach 5 or greater. However, there is nothing magic about Mach 5. If you were flying at Mach 4.99, and you accelerate to Mach 5.01, nothing new is going to happen—the flow will not change from green to red, and there will be no clap of thunder. (In contrast, if you were flying at Mach 0.99 and accelerated to Mach 1.01, the flow would "change from green to red," i.e., the physics of the flow would change drastically as we have already seen, and there would be a "clap of thunder," the sudden occurrence of shock waves.) So why is a distinction made between hypersonic and supersonic flows? The answer is that certain physical phenomena that are not so important at supersonic speeds become dominant at hypersonic speeds. These phenomena are described in Section 14.2, which basically constitutes a four-page definition of hypersonic flow. Read on, and find out for yourself what this is all about.

Hypersonic aerodynamics is state-of-the-art aerodynamics. It is exciting, and it is fun. This chapter introduces you to some of the special aspects and analyses of hypersonic flow. It is just a beginning for you, but I predict that it will be an enjoyable beginning.

14.1 INTRODUCTION

The history of aviation has always been driven by the philosophy of "faster and higher," starting with the Wright brothers' sea level flights at 35 mi/h in 1903, and progressing exponentially to the human-operated space flight missions of the 1960s and 1970s. The current altitude and speed records for human-operated flight are the moon and 36,000 ft/s—more than 36 times the speed of sound—set by the Apollo lunar capsule in 1969. Although most of the flights of the Apollo took place in space, outside the earth's atmosphere, one of its most critical aspects was reentry into the atmosphere after completion of the lunar mission. The aerodynamic phenomena associated with very high-speed flight, such as those encountered during atmospheric reentry, are classified as *hypersonic aerodynamics*—the subject of this chapter. In addition to reentry vehicles, both human-operated and pilotless, there are other hypersonic applications on the horizon, such as ramjet-powered hypersonic missiles now under consideration by the military and the concept of a hypersonic transport, the basic technology of which is now being studied by NASA. Therefore, although hypersonic aerodynamics is at one extreme end of the whole flight spectrum (see Section 1.10), it is important enough to justify one small chapter in our presentation of the fundamentals of aerodynamics.

This chapter is short; its purpose is simply to introduce some basic considerations of hypersonic flow. Therefore, we have no need for a chapter road map or a summary at the end. Also, before progressing further, return to Chapter 1 and review the short discussion on hypersonic flow given in Section 1.10. For an in-depth study of hypersonic flow, see the author's book listed as Reference 52.

14.2 QUALITATIVE ASPECTS OF HYPERSONIC FLOW

Consider a 15° half-angle wedge flying at $M_\infty = 36$. From Figure 9.9, we see that the wave angle of the oblique shock is only 18°; that is, the oblique shock wave is very close to the surface of the body. This situation is sketched in Figure 14.3. Clearly, the shock layer between the shock wave and the body is very thin. Such thin shock layers are one characteristic of hypersonic flow. A practical consequence of a thin shock layer is that a major interaction frequently occurs between the inviscid flow behind the shock and the viscous boundary layer on the surface. Indeed, hypersonic vehicles generally fly at high altitudes where the density, hence Reynolds number, is low, and therefore the boundary layers are thick. Moreover, at hypersonic speeds, the boundary-layer thickness on slender bodies is approximately proportional to M_∞^2; hence, the high Mach

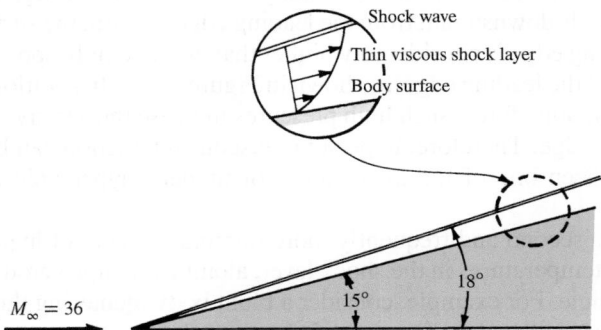

Figure 14.3 For hypersonic flow, the shock layers are thin and viscous.

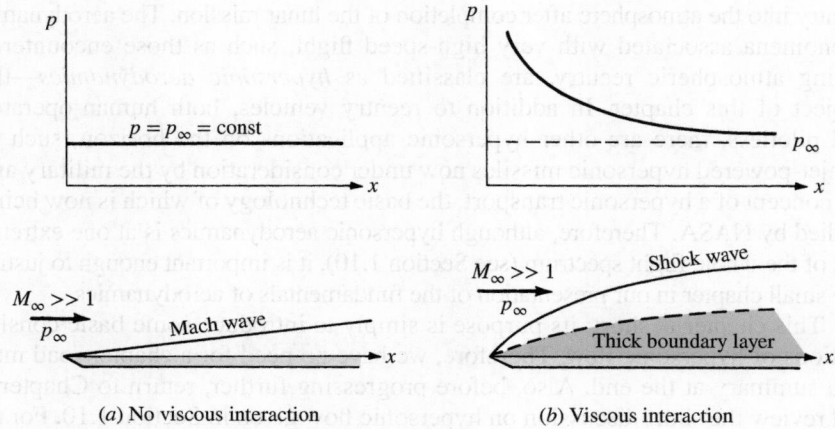

Figure 14.4 The viscous interaction on a flat plate at hypersonic speeds.

numbers further contribute to a thickening of the boundary layer. In many cases, the boundary-layer thickness is of the same magnitude as the shock-layer thickness, such as sketched in the insert at the top of Figure 14.3. Here, the shock layer is fully viscous, and the shock-wave shape and surface pressure distribution are affected by such viscous effects. These phenomena are called *viscous interaction phenomena*—where the viscous flow greatly affects the external inviscid flow, and, of course, the external inviscid flow affects the boundary layer. A graphic example of such viscous interaction occurs on a flat plate at hypersonic speeds, as sketched in Figure 14.4. If the flow were completely inviscid, then we would have the case shown in Figure 14.4a, where a Mach wave trails downstream from the leading edge. Since there is no deflection of the flow, the pressure distribution over the surface of the plate is constant and equal to p_∞. In contrast, in real life there is a boundary layer over the flat plate, and at hypersonic conditions this boundary layer can be thick, as sketched in Figure 14.4b. The thick boundary layer deflects the external, inviscid flow, creating a comparably strong, curved shock wave which trails downstream from the leading edge. In turn, the surface pressure from the leading edge is considerably higher than p_∞, and only approaches p_∞ far downstream of the leading edge, as shown in Figure 14.4b. In addition to influencing the aerodynamic force, such high pressures increase the aerodynamic heating at the leading edge. Therefore, hypersonic viscous interaction can be important, and this has been one of the major areas of modern hypersonic aerodynamic research.

There is a second and frequently more dominant aspect of hypersonic flow, namely, high temperatures in the shock layer, along with large aerodynamic heating of the vehicle. For example, consider a blunt body reentering the atmosphere at Mach 36, as sketched in Figure 14.5. Let us calculate the temperature in the shock layer immediately behind the normal portion of the bow shock wave. From Appendix B, we find that the static temperature ratio across a normal shock wave

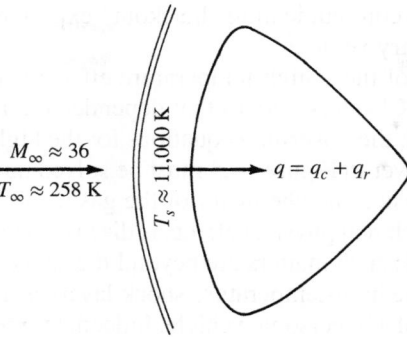

Figure 14.5 High-temperature shock layer.

with $M_\infty = 36$ is 252.9; this is denoted by T_s/T_∞ in Figure 14.5. Moreover, at a standard altitude of 59 km, $T_\infty = 258$ K. Hence, we obtain $T_s = 65,248$ K—an incredibly high temperature, which is more than six times hotter than the surface of the sun! This is, in reality, an incorrect value, because we have used Appendix B which is good only for a calorically perfect gas with $\gamma = 1.4$. However, at high temperatures, the gas will become chemically reacting; γ will no longer equal 1.4 and will no longer be constant. Nevertheless, we get the impression from this calculation that the temperature in the shock layer will be very high, albeit something less than 65,248 K. Indeed, if a proper calculation of T_s is made taking into account the chemically reacting gas, we would find that $T_s \approx 11,000$ K—still a very high value. Clearly, high-temperature effects are very important in hypersonic flow.

Let us examine these high-temperature effects in more detail. If we consider air at $p = 1$ atm and $T = 288$ K (standard sea level), the chemical composition is essentially 20 percent O_2 and 80 percent N_2 by volume. The temperature is too low for any significant chemical reaction to take place. However, if we were to increase T to 2000 K, we would observe that the O_2 begins to dissociate; that is,

$$O_2 \rightarrow 2O \qquad 2000 \text{ K} < T < 4000 \text{ K}$$

If the temperature were increased to 4000 K, most of the O_2 would be dissociated, and N_2 dissociation would commence:

$$N_2 \rightarrow 2N \qquad 4000 \text{ K} < T < 9000 \text{ K}$$

If the temperature were increased to 9000 K, most of the N_2 would be dissociated, and ionization would commence:

$$\begin{aligned} N &\rightarrow N^+ + e^- \\ O &\rightarrow O^+ + e^- \end{aligned} \qquad T > 9000 \text{ K}$$

Hence, returning to Figure 14.5, the shock layer in the nose region of the body is a partially ionized plasma, consisting of the atoms N and O, the ions N^+ and O^+, and electrons, e^-. Indeed, the presence of these free electrons in the shock layer is responsible for the "communications blackout" experienced over portions of the trajectory of a reentry vehicle.

One consequence of these high-temperature effects is that all our equations and tables obtained in Chapters 7 to 13 that depended on a constant $\gamma = 1.4$ are no longer valid. Indeed, the governing equations for the high-temperature, chemically reacting shock layer in Figure 14.5 must be solved numerically, taking into account the proper physics and chemistry of the gas itself. The analysis of aerodynamic flows with such real physical effects is discussed in detail in Chapters 16 and 17 of Reference 21; such matters are beyond the scope of this book.

Associated with the high-temperature shock layers is a large amount of heat transfer to the surface of a hypersonic vehicle. Indeed, for reentry velocities, aerodynamic heating dominates the design of the vehicle, as explained at the end of Section 1.1. (Recall that the third historical example discussed in Section 1.1 was the evolution of the blunt-body concept to reduce aerodynamic heating; review this material before progressing further.) The usual mode of aerodynamic heating is the transfer of energy from the hot shock layer to the surface by means of thermal conduction at the surface; that is, if $\partial T/\partial n$ represents the temperature gradient in the gas normal to the surface, then $q_c = -k(\partial T/\partial n)$ is the heat transfer into the surface. Because $\partial T/\partial n$ is a flow-field property generated by the flow of the gas over the body, q_c is called *convective heating*. For reentry velocities associated with ICBMs (about 28,000 ft/s), this is the only meaningful mode of heat transfer to the body. However, at higher velocities, the shock-layer temperature becomes even hotter. From experience, we know that all bodies emit thermal radiation, and from physics you know that blackbody radiation varies as T^4; hence, radiation becomes a dominant mode of heat transfer at high temperatures. (For example, the heat you feel by standing beside a fire in a fireplace is radiative heating from the flames and the hot walls.) When the shock layer reaches temperatures on the order of 11,000 K, as for the case given in Figure 14.5, thermal radiation from the hot gas becomes a substantial portion of the total heat transfer to the body surface. Denoting radiative heating by q_r, we can express the total aerodynamic heating q as the sum of convective and radiative heating; $q = q_c + q_r$. For Apollo reentry, $q_r/q \approx 0.3$, and hence radiative heating was an important consideration in the design of the Apollo heat shield. For the entry of a space probe into the atmosphere of Jupiter, the velocities will be so high and the shock-layer temperatures so large that the convective heating is negligible, and in this case $q \approx q_r$. For such a vehicle, radiative heating becomes the dominant aspect in its design. Figure 14.6 illustrates the relative importance of q_c and q_r for a typical human-operated reentry vehicle in the earth's atmosphere; note how rapidly q_r dominates the aerodynamic heating of the body as velocities increase above 36,000 ft/s. The details of shock-layer radiative heating are interesting and important; however, they are beyond the scope of this book. For a thorough survey of the engineering aspects of shock-layer radiative heat transfer, see Reference 34.

In summary, the aspects of thin shock-layer viscous interaction and high-temperature, chemically reacting and radiative effects distinguish hypersonic flow from the more moderate supersonic regime. Hypersonic flow has been the subject

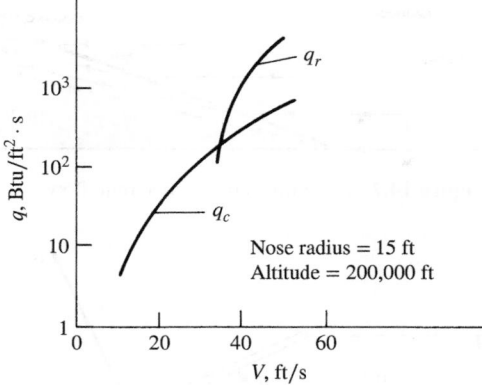

Figure 14.6 Convective and radiative heating rates of a blunt reentry vehicle as a function of flight velocity. (*Source:* Anderson, J. D., Jr.: "An Engineering Survey of Radiating Shock Layers," *AIAA J.*, vol. 7, no. 9, September 1969, pp. 1665–1675.)

of several complete books; see, for example, References 35 to 39. In particular, see Reference 52 for a modern textbook on the subject.

14.3 NEWTONIAN THEORY

Return to Figure 14.3; note how close the shock wave lies to the body surface. This figure is redrawn in Figure 14.7 with the streamlines added to the sketch. When viewed from afar, the straight, horizontal streamlines in the freestream appear to almost impact the body, and then move tangentially along the body. Return to Figure 1.6, which illustrates Isaac Newton's model for fluid flow, and compare it with the hypersonic flow field shown in Figure 14.7; they have certain distinct similarities. (Also, review the discussion surrounding Figure 1.6 before progressing further.) Indeed, the thin shock layers around hypersonic bodies are the closest example in fluid mechanics to Newton's model. Therefore, we might expect that results based on Newton's model would have some applicability in hypersonic flows. This is indeed the case; Newtonian theory is used frequently to estimate the pressure distribution over the surface of a hypersonic body. The purpose of this section is to derive the famous Newtonian sine-squared law first mentioned in Section 1.1 and to show how it is applied to hypersonic flows.

Consider a surface inclined at the angle θ to the freestream, as sketched in Figure 14.8. According to the newtonian model, the flow consists of a large number of individual particles which impact the surface and then move tangentially to the surface. During collision with the surface, the particles lose their component of momentum normal to the surface, but the tangential component is preserved. The time rate of change of the normal component of momentum equals the force exerted on the surface by the particle impacts. To quantify this model, examine

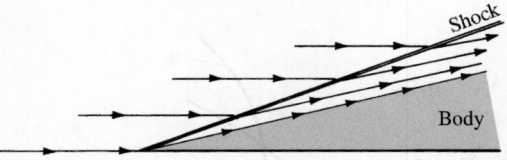

Figure 14.7 Streamlines in a hypersonic flow.

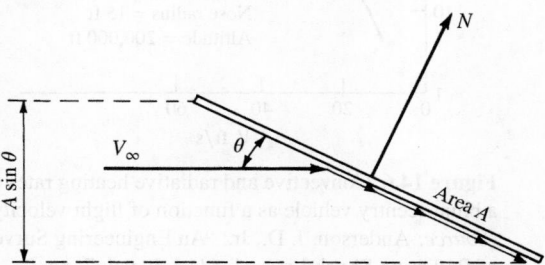

Figure 14.8 Schematic for newtonian impact theory.

Figure 14.8. The component of the freestream velocity normal to the surface is $V_\infty \sin \theta$. If the area of the surface is A, the mass flow incident on the surface is $\rho_\infty (A \sin \theta) V_\infty$. Hence, the time rate of change of momentum is

Mass flow $\times$ change in normal component of velocity

or $(\rho_\infty V_\infty A \sin \theta)(V_\infty \sin \theta) = \rho_\infty V_\infty^2 A \sin^2 \theta$

In turn, from Newton's second law, the force on the surface is

$$N = \rho_\infty V_\infty^2 A \sin^2 \theta \qquad (14.1)$$

This force acts along the same line as the time rate of change of momentum (i.e., normal to the surface), as sketched in Figure 14.8. From Equation (14.1), the normal force per unit area is

$$\frac{N}{A} = \rho_\infty V_\infty^2 \sin^2 \theta \qquad (14.2)$$

Let us now interpret the physical meaning of the normal force per unit area in Equation (14.2), N/A, in terms of our modern knowledge of aerodynamics. Newton's model assumes a stream of individual particles all moving in straight, parallel paths toward the surface; that is, the particles have a completely directed, rectilinear motion. There is no random motion of the particles—it is simply a stream of particles such as pellets from a shotgun. In terms of our modern concepts, we know that a moving gas has molecular motion that is a composite of random motion of the molecules as well as a directed motion. Moreover, we know that the freestream static pressure p_∞ is simply a measure of the purely *random motion* of the molecules. Therefore, when the purely *directed motion* of the particles in Newton's model results in the normal force per unit area, N/A in

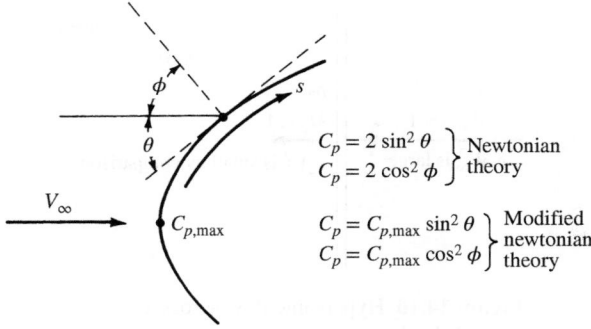

Figure 14.9 Definition of angles for newtonian theory.

Equation (14.2), this normal force per unit area must be construed as the *pressure difference* above p_∞, namely, $p - p_\infty$ on the surface. Hence, Equation (14.2) becomes

$$p - p_\infty = \rho_\infty V_\infty^2 \sin^2 \theta \qquad (14.3)$$

Equation (14.3) can be written in terms of the pressure coefficient $C_p = (p - p_\infty)/\frac{1}{2}\rho_\infty V_\infty^2$, as follows

$$\frac{p - p_\infty}{\frac{1}{2}\rho_\infty V_\infty^2} = 2\sin^2 \theta$$

or

$$\boxed{C_p = 2\sin^2 \theta} \qquad (14.4)$$

Equation (14.4) is Newton's sine-squared law; it states that the pressure coefficient is proportional to the sine square of the angle between a tangent to the surface and the direction of the freestream. This angle θ is illustrated in Figure 14.9. Frequently, the results of newtonian theory are expressed in terms of the angle between a normal to the surface and the freestream direction, denoted by ϕ as shown in Figure 14.9. In terms of ϕ, Equation (14.4) becomes

$$C_p = 2\cos^2 \phi \qquad (14.5)$$

which is an equally valid expression of newtonian theory.

Consider the blunt body sketched in Figure 14.9. Clearly, the maximum pressure, hence the maximum value of C_p, occurs at the stagnation point, where $\theta = \pi/2$ and $\phi = 0$. Equation (14.4) predicts $C_p = 2$ at the stagnation point. Contrast this hypersonic result with the result obtained for incompressible flow theory in Chapter 3, where $C_p = 1$ at a stagnation point. Indeed, the stagnation pressure coefficient increases continuously from 1.0 at $M_\infty = 0$ to 1.28 at $M_\infty = 1.0$ to 1.86 for $\gamma = 1.4$ as $M_\infty \to \infty$. (Prove this to yourself.)

The result that the maximum pressure coefficient approaches 2 at $M_\infty \to \infty$ can be obtained independently from the one-dimensional momentum equation,

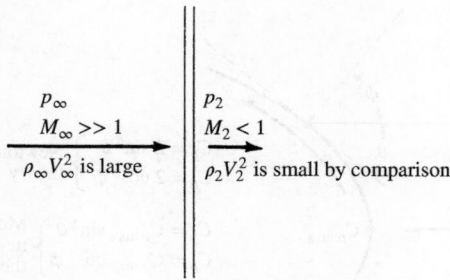

Figure 14.10 Hypersonic flow across a normal shock wave.

namely, Equation (8.6). Consider a normal shock wave at hypersonic speeds, as sketched in Figure 14.10. For this flow, Equation (8.6) gives

$$p_\infty + \rho_\infty V_\infty^2 = p_2 + \rho_2 V_2^2 \tag{14.6}$$

Recall that across a normal shock wave the flow velocity decreases, $V_2 < V_\infty$; indeed, the flow behind the normal shock is subsonic. This change becomes more severe as M_∞ increases. Hence, at hypersonic speeds, we can assume that $(\rho_\infty V_\infty^2) \gg (\rho_2 V_2^2)$, and we can neglect the latter term in Equation (14.6). As a result, Equation (14.6) becomes, at hypersonic speeds in the limiting case as $M_\infty \to \infty$,

$$p_2 - p_\infty = \rho_\infty V_\infty^2$$

or

$$C_p = \frac{p_2 - p_\infty}{\frac{1}{2}\rho_\infty V_\infty^2} = 2$$

thus confirming the newtonian results from Equation (14.4).

As stated above, the result that $C_p = 2$ at a stagnation point is a limiting value as $M_\infty \to \infty$. For large but finite Mach numbers, the value of C_p at a stagnation point is less than 2. Return again to the blunt body shown in Figure 14.9. Considering the distribution of C_p as a function of distance s along the surface, the largest value of C_p will occur at the stagnation point. Denote the stagnation point value of C_p by $C_{p,\max}$, as shown in Figure 14.9. $C_{p,\max}$ for a given M_∞ can be readily calculated from normal shock-wave theory. [If $\gamma = 1.4$, then $C_{p,\max}$ can be obtained from $p_{0,2}/p_1 = p_{0,2}/p_\infty$, tabulated in Appendix B. Recall from Equation (11.22) that $C_{p,\max} = (2/\gamma M_\infty^2)(p_{0,2}/p_\infty - 1)$.] Downstream of the stagnation point, C_p can be assumed to follow the sine-squared variation predicted by newtonian theory; that is,

$$\boxed{C_p = C_{p,\max} \sin^2 \theta} \tag{14.7}$$

Equation (14.7) is called the *modified* newtonian law. For the calculation of the C_p distribution around blunt bodies, Equation (14.7) is more accurate than Equation (14.4).

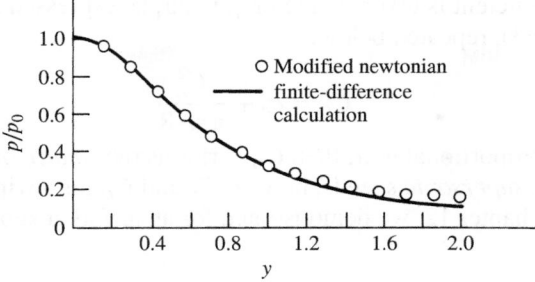

Figure 14.11 Surface pressure distribution, paraboloid, $M_\infty = 4$. Comparison of modified newtonian theory and time-dependent finite-difference calculations.

Return to Figure 13.14, which gives the numerical results for the pressure distributions around a blunt, parabolic cylinder at $M_\infty = 4$ and 8. The open symbols in this figure represent the results of modified newtonian theory, namely, Equation (14.7). For this two-dimensional body, modified newtonian theory is reasonably accurate only in the nose region, although the comparison improves at the higher Mach numbers. It is generally true that newtonian theory is more accurate at larger values of both M_∞ and θ. The case for an axisymmetric body, a paraboloid at $M_\infty = 4$, is given in Figure 14.11. Here, although M_∞ is relatively low, the agreement between the time-dependent numerical solution (see Chapter 13) and newtonian theory is much better. It is generally true that newtonian theory works better for three-dimensional bodies. In general, the modified newtonian law, Equation (14.7), is sufficiently accurate that it is used very frequently in the preliminary design of hypersonic vehicles. Indeed, extensive computer codes have been developed to apply Equation (14.7) to three-dimensional hypersonic bodies of general shape. Therefore, we can be thankful to Isaac Newton for supplying us with a law which holds reasonably well at hypersonic speeds, although such an application most likely never crossed his mind. Nevertheless, it is fitting that three centuries later, Newton's fluid mechanics has finally found a reasonable application.

14.4 THE LIFT AND DRAG OF WINGS AT HYPERSONIC SPEEDS: NEWTONIAN RESULTS FOR A FLAT PLATE AT ANGLE OF ATTACK

Question: At *subsonic* speeds, how do the lift coefficient C_L and drag coefficient C_D for a wing vary with angle of attack α?

Answer: As shown in Chapter 5, we know that:

1. The lift coefficient varies *linearly* with angle of attack, at least up to the stall; see, for example, Figure 5.24.

2. The drag coefficient is given by the drag polar, as expressed in Equation (5.63), repeated below:

$$C_D = c_d + \frac{C_L^2}{\pi e \text{AR}} \qquad (5.63)$$

Since C_L is proportional to α, then C_D varies as the *square* of α.

Question: At *supersonic* speeds, how do C_L and C_D for a wing vary with α?

Answer: In Chapter 12, we demonstrated for an airfoil at supersonic speeds that:

1. Lift coefficient varies *linearly* with α, as seen from Equation (12.23), repeated below:

$$c_l = \frac{4\alpha}{\sqrt{M_\infty^2 - 1}} \qquad (12.23)$$

2. Drag coefficient varies as the *square* of α, as seen from Equation (12.24) for the flat plate, repeated below:

$$c_d = \frac{4\alpha^2}{\sqrt{M_\infty^2 - 1}} \qquad (12.24)$$

The characteristics of a finite wing at supersonic speeds follow essentially the same functional variation with the angle of attack, namely, C_L is proportional to α and C_D is proportional to α^2.

Question: At *hypersonic* speeds, how do C_L and C_D for a wing vary with α? We have shown that C_L is proportional to α for both subsonic and supersonic speeds—does the same proportionality hold for hypersonic speeds? We have shown that C_D is proportional to α^2 for both subsonic and supersonic speeds—does the same proportionality hold for hypersonic speeds? The purpose of the present section is to address these questions.

In an approximate fashion, the lift and drag characteristics of a wing in hypersonic flow can be modeled by a flat plate at an angle of attack, as sketched in Figure 14.12. The exact flow field over the flat plate involves a series of expansion

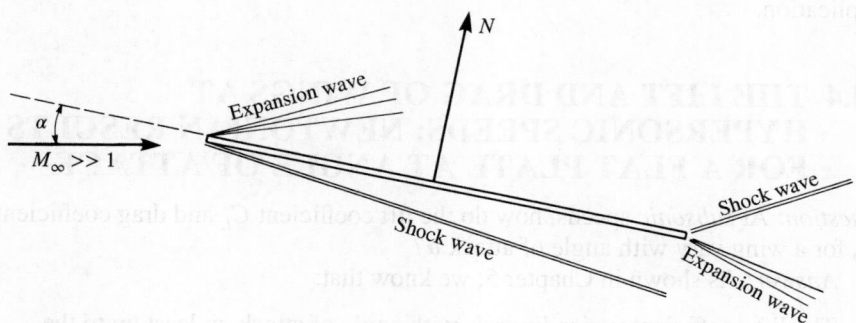

Figure 14.12 Wave system on a flat plate in hypersonic flow.

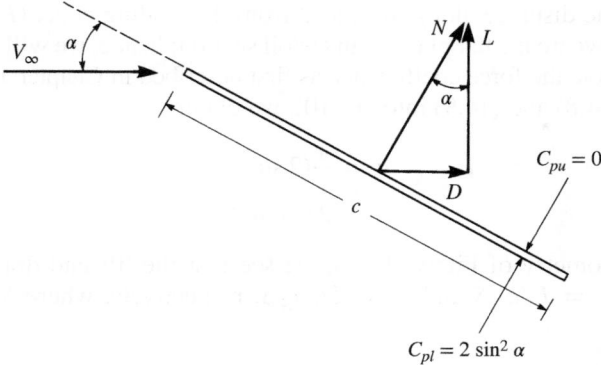

Figure 14.13 Flat plate at angle of attack. Illustration of aerodynamic forces.

and shock waves as shown in Figure 14.12; the exact lift- and wave-drag coefficients can be obtained from the shock-expansion method as described in Section 9.7. However, for hypersonic speeds, the lift- and wave-drag coefficients can be further approximated by the use of newtonian theory, as described in this equation.

Consider Figure 14.13. Here, a two-dimensional flat plate with chord length c is at an angle of attack α to the freestream. Since we are not including friction, and because surface pressure always acts normal to the surface, the resultant aerodynamic force is perpendicular to the plate; that is, in this case, the normal force N is the resultant aerodynamic force. (For an infinitely thin flat plate, this is a general result that is not limited to newtonian theory, or even to hypersonic flow.) In turn, N is resolved into lift and drag, denoted by L and D, respectively, as shown in Figure 14.13. According to newtonian theory, the pressure coefficient on the lower surface is

$$C_{p,l} = 2\sin^2\alpha \qquad (14.8)$$

The upper surface of the flat plate shown in Figure 14.13, in the spirit of newtonian theory, receives no direct "impact" of the freestream particles; the upper surface is said to be in the "shadow" of the flow. Hence, consistent with the basic model of newtonian flow, only freestream pressure acts on the upper surface, and we have

$$C_{p,u} = 0 \qquad (14.9)$$

Returning to the discussion of aerodynamic force coefficients in Section 1.5, we note that the normal force coefficient is given by Equation (1.15). Neglecting friction, this becomes

$$c_n = \frac{1}{c}\int_0^c (C_{p,l} - C_{p,u})\,dx \qquad (14.10)$$

where x is the distance along the chord from the leading edge. (*Please note:* In this section, we treat a flat plate as an airfoil section; hence, we will use lowercase letters to denote the force coefficients, as first described in Chapter 1.) Substituting Equations (14.8) and (14.9) into (14.10), we obtain

$$c_n = \frac{1}{c}(2\sin^2\alpha)c$$

or
$$= 2\sin^2\alpha \tag{14.11}$$

From the geometry of Figure 14.13, we see that the lift and drag coefficients, defined as $c_l = L/q_\infty S$ and $c_d = D/q_\infty S$, respectively, where $S = (c)(l)$, are given by

$$c_l = c_n\cos\alpha \tag{14.12}$$

and
$$c_d = c_n\sin\alpha \tag{14.13}$$

Substituting Equation (14.11) into Equations (14.12) and (14.13), we obtain

$$c_l = 2\sin^2\alpha\cos\alpha \tag{14.14}$$
$$c_d = 2\sin^3\alpha \tag{14.15}$$

Finally, from the geometry of Figure 14.13, the lift-to-drag ratio is given by

$$\frac{L}{D} = \cot\alpha \tag{14.16}$$

[*Note:* Equation (14.16) is a general result for inviscid supersonic or hypersonic flow over a flat plate. For such flows, the resultant aerodynamic force is the normal force N. From the geometry shown in Figure 14.13, the resultant aerodynamic force makes the angle α with respect to lift, and clearly, from the right triangle between L, D, and N, we have $L/D = \cot\alpha$. Hence, Equation (14.16) is not limited to newtonian theory.]

The aerodynamic characteristics of a flat plate based on newtonian theory are shown in Figure 14.14. Although an infinitely thin flat plate, by itself, is not a practical aerodynamic configuration, its aerodynamic behavior at hypersonic speeds is consistent with some of the basic characteristics of other hypersonic shapes. For example, consider the variation of c_l shown in Figure 14.14. First, note that, at a small angle of attack, say, in the range of α from 0 to 15°, c_l varies in a *nonlinear* fashion; that is, the slope of the lift curve is *not* constant. This is in direct contrast to the subsonic case we studied in Chapters 4 and 5, where the lift coefficient for an airfoil or a finite wing was shown to vary linearly with α at small angles of attack, up to the stalling angle. This is also in contrast with the results from linearized supersonic theory as itemized in Section 12.3, leading to Equation (12.23) where a linear variation of c_l with α for a flat plate is indicated. However, the nonlinear lift curve shown in Figure 14.14 is *totally consistent* with the results discussed in Section 11.3, where hypersonic flow was shown to be governed by the *nonlinear* velocity potential equation, *not* by the linear equation expressed by Equation (11.18). In that section, we noted that both transonic and

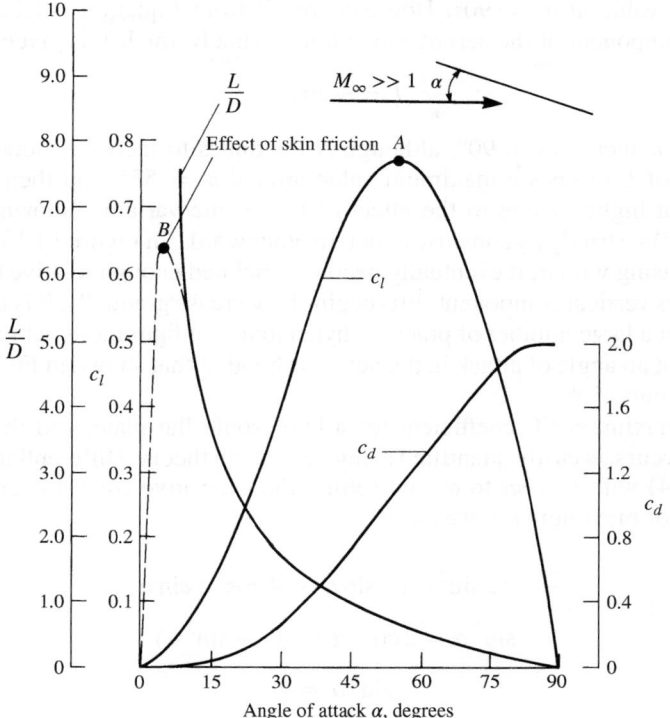

Figure 14.14 Aerodynamic properties of a flat plate based on newtonian theory.

hypersonic flow cannot be described by a linear theory—both these flows are inherently nonlinear regimes, even for low angles of attack. Once again, the flat-plate lift curve shown in Figure 14.14 certainly demonstrates the nonlinearity of hypersonic flow.

Also, note from the lift curve in Figure 14.14 that c_l first increases as α increases, reaches a maximum value at an angle of attack of about 55° (54.7° to be exact), and then decreases, reaching zero at $\alpha = 90°$. However, the attainment of $c_{l,\max}$ (point A) in Figure 14.14 is *not* due to any viscous, separated flow phenomenon analogous to that which occurs in subsonic flow. Rather, in Figure 14.14, the attainment of a maximum c_l is purely a geometric effect. To understand this better, return to Figure 14.13. Note that, as α increases, C_p continues to increase via the newtonian expression

$$C_p = 2 \sin^2 \alpha$$

That is, C_p reaches a maximum value at $\alpha = 90°$. In turn, the *normal force N* shown in Figure 14.13 continues to increase as α increases, also reaching a maximum value at $\alpha = 90°$. However, recall from Equation (14.12) that the vertical component of the aerodynamic force, namely, the lift, is given by

$$L = N \cos \alpha \qquad (14.17)$$

Hence, as α increases to 90°, although N continues to increase monotonically, the value of L reaches a maximum value around $\alpha = 55°$, and then begins to decrease at higher α due to the effect of the cosine variation shown in Equation (14.17)—strictly a geometric effect. In other words, in Figure 14.13, although N is increasing with α, it eventually becomes inclined enough relative to the vertical that its vertical component (lift) begins to decrease gradually. It is interesting to note that a large number of practical hypersonic configurations achieve a maximum C_L at an angle of attack in the neighborhood of that shown in Figure 14.14, namely, around 55°.

The maximum lift coefficient for a hypersonic flat plate, and the angle at which it occurs, is easily quantified using newtonian theory. Differentiating Equation (14.14) with respect to α, and setting the derivative equal to zero (for the condition of maximum c_l), we have

$$\frac{dc_l}{d\alpha} = (2 \sin^2 \alpha)(- \sin \alpha) + 4 \cos^2 \alpha \sin \alpha = 0$$

or

$$\sin^2 \alpha = 2 \cos^2 \alpha = 2(1 - \sin^2 \alpha)$$

or

$$\sin^2 \alpha = \tfrac{2}{3}$$

Hence,

$$\alpha = 54.7°$$

This is the angle of attack at which c_l is a maximum. The maximum value of c_l is obtained by substituting the above result for α into Equation (14.14):

$$c_{l,\max} = 2 \sin^2(54.7°) \cos(54.7°) = 0.77$$

Note, although c_l increases over a wide latitude in the angle of attack (c_l increases in the range from $\alpha = 0$ to $\alpha = 54.7°$), its rate of increase is small (that is, the effective lift slope is small). In turn, the resulting value for the maximum lift coefficient is relatively small—at least in comparison to the much higher $c_{l,\max}$ values associated with low-speed flows (see Figures 4.25 and 4.28). Returning to Figure 14.14, we now note the *precise* values associated with the peak of the lift curve (point A), namely, the peak value of c_l is 0.77, and it occurs at an angle of attack of 54.7°.

Examining the variation of drag coefficient c_d in Figure 14.14, we note that it monotonically increases from zero at $\alpha = 0$ to a maximum of 2 at $\alpha = 90°$. The newtonian result for drag is essentially *wave drag* at hypersonic speeds because we are dealing with an inviscid flow, hence no friction drag. The variation of c_d with α for the low angle of attack in Figure 14.14 is essentially a *cubic* variation, in contrast to the result from linearized supersonic flow, namely, Equation (12.24), which shows that c_d varies as the square angle of attack. The hypersonic result that c_d varies as α^3 is easily obtained from Equation (14.15), which for small α becomes

$$c_d = 2\alpha^3 \tag{14.18}$$

The variation of the lift-to-drag ratio as predicted by newtonian theory is also shown in Figure 14.14. The solid curve is the pure newtonian result; it shows that L/D is infinitely large at $\alpha = 0$ and monotonically decreases to zero at $\alpha = 90°$. The infinite value of L/D at $\alpha = 0$ is purely fictional—it is due to the neglect of skin friction. When skin friction is added to the picture, denoted by the dashed curve in Figure 14.14, L/D reaches a maximum value at a small angle of attack (point B in Figure 14.14) and is equal to zero at $\alpha = 0$. (At $\alpha = 0$, no lift is produced, but there is a finite drag due to friction; hence, $L/D = 0$ at $\alpha = 0$.)

Let us examine the conditions associated with $(L/D)_{max}$ more closely. The value of $(L/D)_{max}$ and the angle of attack at which it occurs (i.e., the coordinates of point B in Figure 14.14) are strictly a function of the zero-lift drag coefficient, denoted by $c_{d,0}$. The zero-lift drag coefficient is simply due to the integrated effect of skin friction over the plate surface at zero angle of attack. At small angles of attack, the skin friction exerted on the plate should be essentially that at zero angle of attack; hence, we can write the total drag coefficient [referring to Equation (14.15)] as

$$c_d = 2\sin^3\alpha + c_{d,0} \tag{14.19}$$

Furthermore, when α is small, we can write Equations (14.14) and (14.19) as

$$c_l = 2\alpha^2 \tag{14.20}$$

and

$$c_d = 2\alpha^3 + c_{d,0} \tag{14.21}$$

Dividing Equation (14.20) by (14.21), we have

$$\frac{c_l}{c_d} = \frac{2\alpha^2}{2\alpha^3 + c_{d,0}} \tag{14.22}$$

The conditions associated with maximum lift-to-drag ratio can be found by differentiating Equation (14.22) and setting the result equal to zero:

$$\frac{d(c_l/c_d)}{d\alpha} = \frac{(2\alpha^3 + c_{d,0})4\alpha - 2\alpha^2(6\alpha^2)}{(2\alpha^3 + c_{d,0})} = 0$$

or

$$8\alpha^4 + 4\alpha c_{d,0} - 12\alpha^4 = 0$$

$$4\alpha^3 = 4c_{d,0}$$

Hence,

$$\boxed{\alpha = (c_{d,0})^{1/3}} \tag{14.23}$$

Substituting Equation (14.23) into Equation (14.21), we obtain

$$\left(\frac{c_l}{c_d}\right)_{max} = \frac{2(c_{d,0})^{2/3}}{2c_{d,0} + c_{d,0}} = \frac{2/3}{(c_{d,0})^{1/3}}$$

$$\text{or} \qquad \left(\frac{L}{D}\right)_{max} = \left(\frac{c_l}{c_d}\right)_{max} = \boxed{0.67/(c_{d,0})^{1/3}} \qquad (14.24)$$

Equations (14.23) and (14.24) are important results. They clearly state that the co-ordinates of the maximum L/D point in Figure 14.14, when friction is included (point B in Figure 14.14), are strictly a function of $c_{d,0}$. In particular, note the expected trend that $(L/D)_{max}$ decreases as $c_{d,0}$ increases—the higher the friction drag, the lower is L/D. Also, the angle of attack at which maximum L/D occurs increases as $c_{d,0}$ increases. There is yet another interesting aerodynamic condition that holds at $(L/D)_{max}$, derived as follows. Substituting Equation (14.23) into (14.21), we have

$$c_d = 2c_{d,0} + c_{d,0} = 3c_{d,0} \qquad (14.25)$$

Since the total drag coefficient is the sum of the wave-drag coefficient $c_{d,w}$ and the friction-drag coefficient $c_{d,0}$, we can write

$$c_d = c_{d,w} + c_{d,0} \qquad (14.26)$$

However, at the point of maximum L/D (point B in Figure 14.14), we know from Equation (14.25) that $c_d = 3c_{d,0}$. Substituting this result into Equation (14.26), we obtain

$$3c_{d,0} = c_{d,w} + c_{d,0}$$

$$\text{or} \qquad \boxed{c_{d,w} = 2c_{d,0}} \qquad (14.27)$$

This clearly shows that, for the hypersonic flat plate using newtonian theory, at the flight condition associated with maximum lift-to-drag ratio, wave drag is twice the friction drag.

This brings to an end our short discussion of the lift and drag of wings at hypersonic speeds as modeled by the newtonian flat-plate problem. The quantitative and qualitative results presented here are reasonable representations of the hypersonic aerodynamic characteristics of a number of practical hypersonic vehicles; the flat-plate problem is simply a straightforward way of demonstrating these characteristics.

14.4.1 Accuracy Considerations

How accurate is newtonian theory in the prediction of pressure distributions over hypersonic bodies? The comparison shown in Figure 14.11 indicates that Equation (14.7) leads to a reasonably accurate pressure distribution over the surface of a *blunt body*. Indeed, for "back-of-the-envelope" estimates of the pressure distributions over blunt bodies at hypersonic speeds, modified newtonian is quite satisfactory. However, what about relatively thin bodies at small angles of attack? We can provide an answer by using the newtonian flat-plate relations derived in the present section, and compare these results with exact shock-expansion theory (Section 9.7), for flat plates at small angles of attack. This is the purpose of the following worked example.

EXAMPLE 14.1

Consider an infinitely thin flat plate at an angle of attack of 15° in a Mach 8 flow. Calculate the pressure coefficients on the top and bottom surface, the lift and drag coefficients, and the lift-to-drag ratio using (*a*) exact shock-expansion theory, and (*b*) newtonian theory. Compare the results.

■ **Solution**

(*a*) Using the diagram in Figure 9.35 showing a flat plate at angle of attack, and following the shock-expansion technique given in Example 9.11, we have for the upper surface, for $M_1 = 8$ and $v_1 = 95.62°$,

$$v_2 = v_1 + \theta = 95.62 + 15 = 110.62°$$

From Appendix C, interpolating between entries,

$$M_2 = 14.32$$

From Appendix A, for $M_1 = 8$, $p_{0_1}/p_1 = 0.9763 \times 10^4$, and for $M_2 = 14.32$, $p_{0_2}/p_2 = 0.4808 \times 10^6$. Since $p_{0_1} = p_{0_2}$,

$$\frac{p_2}{p_1} = \frac{p_{0_1}}{p_1} \bigg/ \frac{p_{0_2}}{p_2} = \frac{0.9763 \times 10^4}{0.4808 \times 10^6} = 0.0203$$

The pressure coefficient is given by Equation (11.22), and the freestream static pressure in Figure 9.28 is denoted by p_1. Hence,

$$C_{p_2} = \frac{2}{\gamma M_1^2}\left(\frac{p_2}{p_1} - 1\right) = \frac{2}{(1.4)(8)^2}(0.0203 - 1) = \boxed{-0.0219}$$

To obtain the pressure coefficient on the bottom surface from the oblique shock theory, we have from the θ-β-M for $M_1 = 8$ and $\theta = 15°$, $\beta = 21°$:

$$M_{n,1} = M_1 \sin\beta = 8\sin 21° = 2.87$$

Interpolating from Appendix B, for $M_{n,1} = 2.87$, $p_3/p_1 = 9.443$. Hence, the pressure coefficient on the bottom surface is

$$C_{p_3} = \frac{2}{\gamma M_1^2}\left(\frac{p_3}{p_2} - 1\right) = \frac{2}{(1.4)(8)^2}(9.443 - 1) = \boxed{0.1885}$$

The lift coefficient can be obtained from the pressure coefficients via Equations (1.15), (1.16), and (1.18).

$$c_n = \frac{1}{c}\int_0^c (C_{p,\ell} - C_{p,u})\,dx = C_{p_3} - C_{p_2} = 0.1885 - (-0.0219) = 0.2104$$

The axial force on the plate is zero, because the pressure acts only perpendicular to the plate. On a formal basis, dy/dx in Equation (1.16) is zero for a flat plate. Hence, from Equation (1.18),

$$c_\ell = c_n \cos\alpha = 0.2104\cos 15° = \boxed{0.2032}$$

From Equation (1.19),

$$c_d = c_n \sin\alpha = 0.2104 \sin 15° = \boxed{0.0545}$$

Hence,

$$\frac{L}{D} = \frac{c_\ell}{c_d} = \frac{0.2032}{0.0545} = \boxed{3.73}$$

(b) From newtonian theory, the pressure coefficient is given by Equation (14.4), where $\theta \equiv \alpha$. This is the pressure coefficient on the lower surface, hence

$$C_{p_3} = 2\sin^2\alpha = 2\sin^2 15° = \boxed{0.134}$$

From Equation (14.9), we have for the upper surface

$$C_{p_2} = \boxed{0}$$

Hence,

$$c_\ell = (C_{p_3} - C_{p_2})\cos\alpha = 0.134\cos 15° = \boxed{0.1294}$$

and

$$c_d = (C_{p_3} - C_{p_2})\sin\alpha = 0.134\sin 15° = \boxed{0.03468}$$

and

$$\frac{L}{D} = \frac{c_\ell}{c_d} = \frac{0.1294}{0.3468} = \boxed{3.73}$$

Discussion. From the above worked example, we see that newtonian theory *underpredicts* the pressure coefficient on the bottom surface by 29 percent, and of course predicts a value of zero for the pressure coefficient on the upper surface in comparison to −0.0219 from exact theory—an error of 100 percent. Also, newtonian theory underpredicts c_ℓ and c_d by 36.6 percent. However, the value of L/D from newtonian theory is *exactly correct*. This is no surprise, for two reasons. First, the newtonian values of c_ℓ and c_d are both underpredicted by the same amount, hence their ratio is not affected. Second, the value of L/D for supersonic or hypersonic inviscid flow over a flat plate, no matter what theory is used to obtain the pressures on the top and bottom surfaces, is simply a matter of *geometry*. Because the pressure acts normal to the surface, the resultant aerodynamic force is perpendicular to the plate (i.e., the resultant force *is* the normal force N). Examining Figure 1.16, we see that when this is the case, the vectors **R** and **N** are the same vectors, and L/D is geometrically given by

$$\frac{L}{D} = \cot\alpha$$

For the above worked example, where $\alpha = 15°$, we have

$$\frac{L}{D} = \cot 15° = 3.73$$

which agrees with the above calculations where c_ℓ and c_d were first obtained, and L/D is found from the ratio, $L/D = c_\ell/c_d$. So, Equation (14.16), derived in our discussion of newtonian theory applied to a flat plate, is not unique to newtonian theory; it is a general result when the resultant aerodynamic force is perpendicular to the plate.

We induce from Example 14.1 the general fact that the newtonian sine-squared law, Equation (14.4), does not accurately predict the hypersonic pressure distribution on the surface of two-dimensional bodies with local tangent lines that are at small or moderate angles to the flow, such as the bi-convex airfoil shape shown in Figure 12.3. On the other hand, it generally turns out that the newtonian prediction of the lift-to-drag ratio for slender shapes at small to moderate angles of attack is reasonably accurate. These statements apply to a gas with the ratio of specific heats substantially greater than one, such as the case of air with $\gamma = 1.4$ treated in Example 14.1. In the next section, we will see that newtonian theory becomes more accurate as $M_\infty \to \infty$ and $\gamma \to 1$. For more information on the accuracy of newtonian theory applied to two-dimensional slender shapes, see Reference 73 which is a study of this specific matter.

Finally, we note that newtonian theory does a better job of predicting the pressure on axisymmetric slender bodies, such as the 15° half-angle cone shown in Figure 14.15.

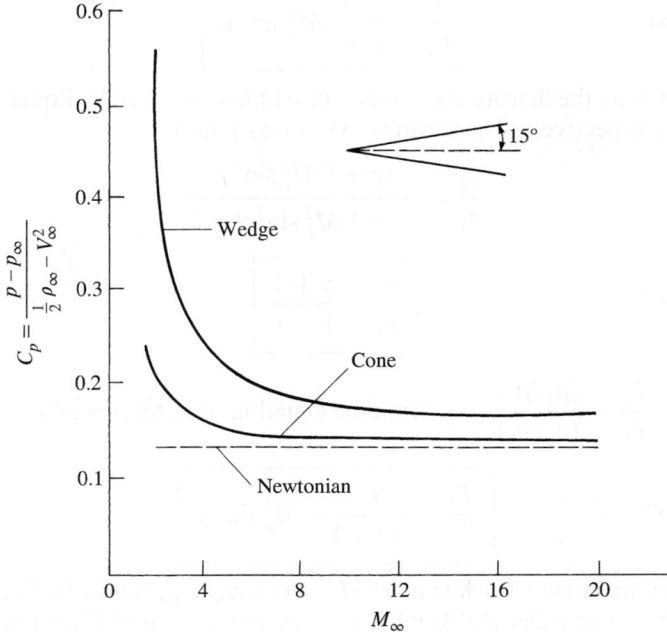

Figure 14.15 Comparison between newtonian and exact results for the pressure coefficient on a sharp wedge and a sharp cone. Also, an illustration of Mach number independence at high Mach numbers.

14.5 HYPERSONIC SHOCK-WAVE RELATIONS AND ANOTHER LOOK AT NEWTONIAN THEORY

The basic oblique shock relations are derived and discussed in Chapter 9. These are exact shock relations and hold for all Mach numbers greater than unity, supersonic or hypersonic (assuming a calorically perfect gas). However, some interesting approximate and simplified forms of these shock relations are obtained in the limit of a high Mach number. These limiting forms are called the hypersonic shock relations; they are obtained below.

Consider the flow through a straight oblique shock wave. (See, e.g., Figure 9.2.) Upstream and downstream conditions are denoted by subscripts 1 and 2, respectively. For a calorically perfect gas, the classical results for changes across the shock are given in Chapter 9. To begin with, the exact oblique shock relation for pressure ratio across the wave is given by Equation (9.16). Since $M_{n,1} = M_1 \sin \beta$, this equation becomes

Exact:
$$\frac{p_2}{p_1} = 1 + \frac{2\gamma}{\gamma + 1}\left(M_1^2 \sin^2 \beta - 1\right) \qquad (14.28)$$

where β is the wave angle. In the limit as M_1 goes to infinity, the term $M_1^2 \sin^2 \beta \gg 1$, and hence Equation (14.28) becomes

as $M_1 \to \infty$:
$$\boxed{\frac{p_2}{p_1} = \frac{2\gamma}{\gamma + 1}M_1^2 \sin^2 \beta} \qquad (14.29)$$

In a similar vein, the density and temperature ratios are given by Equations (9.15) and (9.17), respectively. These can be written as follows:

Exact:
$$\frac{\rho_2}{\rho_1} = \frac{(\gamma + 1)M_1^2 \sin^2 \beta}{(\gamma - 1)M_1^2 \sin^2 \beta + 2} \qquad (14.30)$$

as $M_1 \to \infty$:
$$\boxed{\frac{\rho_2}{\rho_1} = \frac{\gamma + 1}{\gamma - 1}} \qquad (14.31)$$

$$\frac{T_2}{T_1} = \frac{(p_2/p_1)}{(\rho_2/\rho_1)} \qquad \text{(from the equation of state: } p = \rho RT)$$

as $M_1 \to \infty$:
$$\boxed{\frac{T_2}{T_1} = \frac{2\gamma(\gamma - 1)}{(\gamma + 1)^2}M_1^2 \sin^2 \beta} \qquad (14.32)$$

The relationship among Mach number M_1, shock angle β, and deflection angle θ is expressed by the so-called θ-β-M relation given by Equation (9.23), repeated below:

Exact:
$$\tan \theta = 2 \cot \beta \left[\frac{M_1^2 \sin^2 \beta - 1}{M_1^2(\gamma + \cos 2\beta) + 2}\right] \qquad (9.23)$$

This relation is plotted in Figure 9.9, which is a standard plot of the wave angle versus the deflection angle, with the Mach number as a parameter. Returning to Figure 9.9, we note that, in the hypersonic limit, where θ is small, β is also small. Hence, in this limit, we can insert the usual small-angle approximation into Equation (9.23):

$$\sin \beta \approx \beta$$
$$\cos 2\beta \approx 1$$
$$\tan \theta \approx \sin \theta \approx \theta$$

resulting in

$$\theta = \frac{2}{\beta}\left[\frac{M_1^2\beta^2 - 1}{M_1^2(\gamma + 1) + 2}\right] \tag{14.33}$$

Applying the high Mach number limit to Equation (14.33), we have

$$\theta = \frac{2}{\beta}\left[\frac{M_1^2\beta^2}{M_1^2(\gamma + 1)}\right] \tag{14.34}$$

In Equation (14.34), M_1 cancels, and we finally obtain in both the small-angle and hypersonic limits,

as $M_1 \rightarrow \infty$ *and* θ, *hence* β *is small:*
$$\boxed{\frac{\beta}{\theta} = \frac{\gamma + 1}{2}} \tag{14.35}$$

Note that, for $\gamma = 1.4$,

$$\boxed{\beta = 1.2\theta} \tag{14.36}$$

It is interesting to observe that, in the hypersonic limit for a slender wedge, the wave angle is only 20 percent larger than the wedge angle—a graphic demonstration of a thin shock layer in hypersonic flow.

In aerodynamics, pressure distributions are usually quoted in terms of the nondimensional pressure coefficient C_p, rather than the pressure itself. The pressure coefficient is defined as

$$C_p = \frac{p_2 - p_1}{q_1} \tag{14.37}$$

where p_1 and q_1 are the upstream (freestream) static pressure and dynamic pressure, respectively. Recall from Section 11.3 that Equation (14.37) can also be written as Equation (11.22), repeated below:

$$C_p = \frac{2}{\gamma M_1^2}\left(\frac{p_2}{p_1} - 1\right) \tag{11.22}$$

Combining Equations (11.22) and (14.28), we obtain an exact relation for C_p behind an oblique shock wave as follows:

Exact:
$$C_p = \frac{4}{\gamma + 1}\left(\sin^2 \beta - \frac{1}{M_1^2}\right) \tag{14.38}$$

In the hypersonic limit,

$$as\ M_1 \to \infty: \qquad \boxed{C_p = \left(\frac{4}{\gamma + 1}\right) \sin^2 \beta} \qquad (14.39)$$

Pause for a moment, and review our results. We have obtained limiting forms of the oblique shock equations, valid for the case when the upstream Mach number becomes very large. These limiting forms, called the hypersonic shock-wave relations, are given by Equations (14.29), (14.31), and (14.32), which yield the pressure ratio, density ratio, and temperature ratio across the shock when $M_1 \to \infty$. Furthermore, in the limit of both $M_1 \to \infty$ and small θ (such as the hypersonic flow over a slender airfoil shape), the limiting relation for the wave angle as a function of the deflection angle is given by Equation (14.35). Finally, the form of the pressure coefficient behind an oblique shock is given in the limit of hypersonic Mach numbers by Equation (14.39). Note that the limiting forms of the equations are always simpler than their corresponding exact counterparts.

In terms of actual *quantitative* results, it is always recommended that the *exact* oblique shock equations be used, even for hypersonic flow. This is particularly convenient because the exact results are tabulated in Appendix B. The value of the relations obtained in the hypersonic limit (as described above) is more for theoretical analysis rather than for the calculation of actual numbers. For example, in this section, we use the hypersonic shock relations to shed additional understanding of the significance of newtonian theory. In the next section, we will examine the same hypersonic shock relations to demonstrate the principle of Mach number independence.

Newtonian theory was discussed at length in Sections 14.3 and 14.4. For our purposes here, temporarily discard any thoughts of newtonian theory, and simply recall the exact oblique shock relation for C_p as given by Equation (14.38), repeated below (with freestream conditions now denoted by a subscript ∞ rather than a subscript 1, as used earlier):

$$C_p = \frac{4}{\gamma + 1}\left[\sin^2 \beta - \frac{1}{M_\infty^2}\right] \qquad (14.38)$$

Equation (14.39) gave the limiting value of C_p as $M_\infty \to \infty$, repeated below:

$$as\ M_\infty \to \infty: \qquad C_p \to \frac{4}{\gamma + 1} \sin^2 \beta \qquad (14.39)$$

Now take the additional limit of $\gamma \to 1.0$. From Equation (14.39), in both limits as $M_\infty \to \infty$ and $\gamma \to 1.0$, we have

$$C_p \to 2\sin^2 \beta \qquad (14.40)$$

Equation (14.40) is a result from exact oblique shock theory; it has nothing to do with newtonian theory (as yet). Keep in mind that β in Equation (14.40) is the wave angle, not the deflection angle.

Let us go further. Consider the exact oblique shock relation for the density ratio, ρ/ρ_∞, given by Equation (14.30), repeated below (again with a subscript ∞ replacing the subscript 1):

$$\frac{\rho_2}{\rho_\infty} = \frac{(\gamma + 1)M_\infty^2 \sin^2 \beta}{(\gamma - 1)M_\infty^2 \sin^2 \beta + 2} \tag{14.41}$$

Equation (14.31) was obtained as the limit where $M_\infty \to \infty$, namely,

$$as\ M_\infty \to \infty: \qquad \frac{\rho_2}{\rho_\infty} \to \frac{\gamma + 1}{\gamma - 1} \tag{14.42}$$

In the additional limit as $\gamma \to 1$, we find

$$as\ \gamma \to 1\ and\ M_\infty \to \infty: \qquad \boxed{\frac{\rho_2}{\rho_\infty} \to \infty} \tag{14.43}$$

that is, the density behind the shock is infinitely large. In turn, mass flow considerations then dictate that the shock wave is coincident with the body surface. This is further substantiated by Equation (14.35), which is good for $M_\infty \to \infty$ and small deflection angles:

$$\frac{\beta}{\theta} \to \frac{\gamma + 1}{2} \tag{14.35}$$

In the additional limit as $\gamma \to 1$, we have

$$as\ \gamma \to 1\ and\ M_\infty \to \infty\ and\ \theta\ and\ \beta\ are\ small: \qquad \boxed{\beta = \theta}$$

that is, the shock wave lies on the body. In light of this result, Equation (14.40) is written as

$$\boxed{C_p = 2 \sin^2 \theta} \tag{14.44}$$

Examine Equation (14.44). It is a result from exact oblique shock theory, taken in the combined limit of $M_\infty \to \infty$ and $\gamma \to 1$. However, it is also precisely the newtonian results given by Equation (14.4). Therefore, we make the following conclusion. The closer the actual hypersonic flow problem is to the limits $M_\infty \to \infty$ and $\gamma \to 1$, the closer it should be physically described by newtonian flow. In this regard, we gain a better appreciation of the true significance of newtonian theory. We can also state that the application of newtonian theory to practical hypersonic flow problems, where γ is always greater than unity, is theoretically not proper, and the agreement that is frequently obtained with experimental data has to be viewed as somewhat fortuitous. Nevertheless, the simplicity of newtonian theory along with its (sometimes) reasonable results (no matter how fortuitous) has made it a widely used and popular engineering method for the estimation of surface pressure distributions, hence lift- and wave-drag coefficients, for hypersonic bodies.

14.6 MACH NUMBER INDEPENDENCE

Examine again the hypersonic shock-wave relation for pressure ratio as given by Equation (14.29); note that, as the freestream Mach number approaches infinity, the pressure ratio itself also becomes infinitely large. On the other hand, the *pressure coefficient* behind the shock, given in the hypersonic limit by Equation (14.39), *is a constant value at high values of the Mach number*. This hints strongly of a situation where certain aspects of a hypersonic flow do not depend on Mach number, as long as the Mach number is sufficiently high. This is a type of "independence" from the Mach number, formally called the *hypersonic Mach number independence principle*. From the above argument, C_p clearly demonstrates Mach number independence. In turn, recall that the lift- and wave-drag coefficients for a body shape are obtained by integrating the local C_p, as shown by Equations (1.15), (1.16), (1.18), and (1.19). These equations demonstrate that, since C_p is independent of the Mach number at high values of M_∞, the lift and drag coefficients are also Mach number independent. Keep in mind that these conclusions are theoretical, based on the limiting form of the hypersonic shock relations.

Let us examine an example that clearly illustrates the Mach number independence principle. In Figure 14.15, the pressure coefficients for a 15° half-angle wedge and a 15° half-angle cone are plotted versus freestream Mach number for $\gamma = 1.4$. The exact wedge results are obtained from Equation (14.38), and the exact cone results are obtained from the solution of the classical Taylor-Maccoll equation. (See Reference 21 for a detailed discussion of the solution of the supersonic flow over a cone. There, you will find that the governing continuity, momentum, and energy equations for a conical flow cascade into a single differential equation called the Taylor-Maccoll equation. In turn, this equation allows the *exact* solution of this conical flow field.) Both sets of results are compared with newtonian theory, $C_p = 2 \sin^2 \theta$, shown as the dashed line in Figure 14.15. This comparison demonstrates two general aspects of newtonian results:

1. The accuracy of the newtonian results improves as M_∞ increases. This is to be expected from our discussion in Section 14.5. Note from Figure 14.15 that below $M_\infty = 5$ the newtonian results are not even close, but the comparison becomes much closer as M_∞ increases above 5.
2. Newtonian theory is usually more accurate for three-dimensional bodies (e.g., the cone) than for two-dimensional bodies (e.g., the wedge). This is clearly evident in Figure 14.15, where the newtonian result is much closer to the cone results than to the wedge results.

However, more to the point of Mach number independence, Figure 14.15 also shows the following trends. For both the wedge and the cone, the exact results show that, at low supersonic Mach numbers, C_p decreases rapidly as M_∞ is increased. However, at hypersonic speeds, the rate of decrease diminishes considerably, and C_p appears to reach a plateau as M_∞ becomes large; that is, C_p becomes relatively independent of M_∞ at high values of the Mach number. This is the essence of the Mach number independence principle; at high Mach numbers,

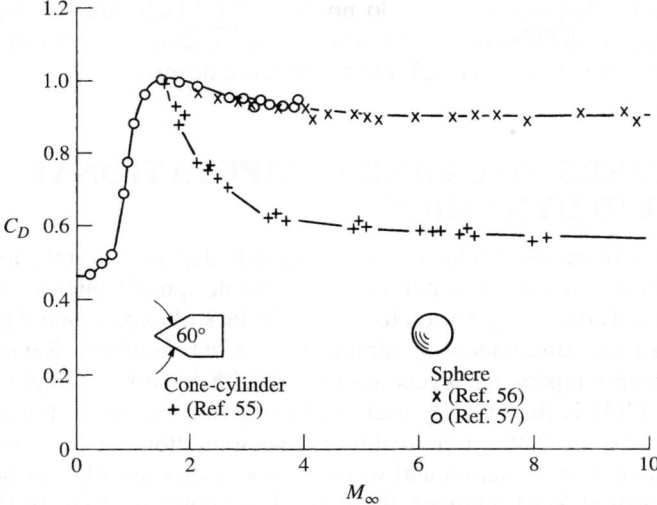

Figure 14.16 Drag coefficient for a sphere and a cone cylinder from ballistic range measurements; an example of Mach number independence at hypersonic speeds. (*Data Source:* Cox, R. N., and L. F. Crabtree: *Elements of Hypersonic Aerodynamics*, Academic Press, New York, 1965.)

certain aerodynamic quantities such as pressure coefficient, lift- and wave-drag coefficients, and flow-field structure (such as shock-wave shapes and Mach wave patterns) become essentially independent of the Mach number. Indeed, newtonian theory gives results that are totally independent of the Mach number, as clearly demonstrated by Equation (14.4).

Another example of Mach number independence is shown in Figure 14.16. Here, the measured drag coefficients for spheres and for a large-angle cone cylinder are plotted versus the Mach number, cutting across the subsonic, supersonic, and hypersonic regimes. Note the large drag rise in the subsonic regime associated with the drag-divergence phenomenon near Mach 1 and the decrease in C_D in the supersonic regime beyond Mach 1. Both of these variations are expected and well understood. For our purposes in the present section, note, in particular, the variation of C_D in the hypersonic regime; for both the sphere and cone cylinder, C_D approaches a plateau and becomes relatively independent of the Mach number as M_∞ becomes large. Note also that the sphere data appear to achieve "Mach number independence" at lower Mach numbers than the cone cylinder.

Keep in mind from the above analysis that it is the nondimensional variables that become Mach number independent. Some of the dimensional variables, such as p, are not Mach number independent; indeed, $p \to \infty$ and $M_\infty \to \infty$.

Finally, the Mach number independence principle is well grounded mathematically. The governing inviscid flow equations (the Euler equations) expressed in terms of suitable nondimensional quantities, along with the boundary conditions for the limiting hypersonic case, do not have the Mach number appearing in them—hence, by definition, the solution to these equations is independent of the Mach number. See References 21 and 52 for more details.

14.7 HYPERSONICS AND COMPUTATIONAL FLUID DYNAMICS

The design of hypersonic vehicles today is greatly dependent on the use of computational fluid dynamics, much more so than the design of vehicles for any other flight regime. The primary reason for this is the lack of experimental ground test facilities that can simultaneously simulate the Mach numbers, Reynolds numbers, and high-temperature levels associated with hypersonic flight. For such simulation, CFD is the primary tool. Reflecting once again on the philosophy illustrated in Figure 2.46, in the realm of hypersonic flow the three partners are not quite equal. Pure experimental work in hypersonics usually involves tests at either the desired Mach number, the desired Reynolds number, or the desired temperature level, but not all at the same time nor in the same test facilities. As a result, experimental data for the design of hypersonic vehicles is a patchwork of different data taken in different facilities under different conditions. Moreover, the data are usually incomplete, especially for the high-temperature effects, which are difficult to simulate in a wind tunnel. The designer must then do his or her best to piece together the information for the specified design conditions. The next partner shown in Figure 2.46, pure theory, is greatly hampered by the nonlinear nature of hypersonic flow, hence making mathematical solutions intractable. In addition, the proper inclusion of high-temperature chemically reacting flows in any pure theory is extremely difficult. For these reasons, the third partner shown in Figure 2.46, computational fluid dynamics, takes on a dominant role. The numerical calculation of both inviscid and viscous hypersonic flows, including all the high-temperature effects discussed in Section 14.2, has been a major thrust of CFD research and design application since the 1960s. Indeed, hypersonics has paced the development of CFD since its beginning.

As an example of CFD applied to a hypersonic flight vehicle appropriate to this chapter, consider the space shuttle shown in Figure 14.17. A numerical solution of the three-dimensional inviscid flow field around the shuttle was carried out by Maus et al. in Reference 74. They made two sets of calculations, one for a perfect gas with $\gamma = 1.4$, and one assuming chemically reacting air in local chemical equilibrium. The freestream Mach number was 23 in both cases. The CFD technique used for these calculations involved a time-dependent solution of the flow in the blunt nose region, patterned after our discussion in Section 13.5, and starting beyond the sonic line a downstream marching approach patterned after our discussion in Section 13.4. The calculated surface pressure distributions along the windward centerline of the space shuttle for both the perfect gas case (the circles) and the chemically reacting case (the triangles) are shown in Figure 14.18. The expansion around the nose, the pressure plateau over the relatively flat bottom

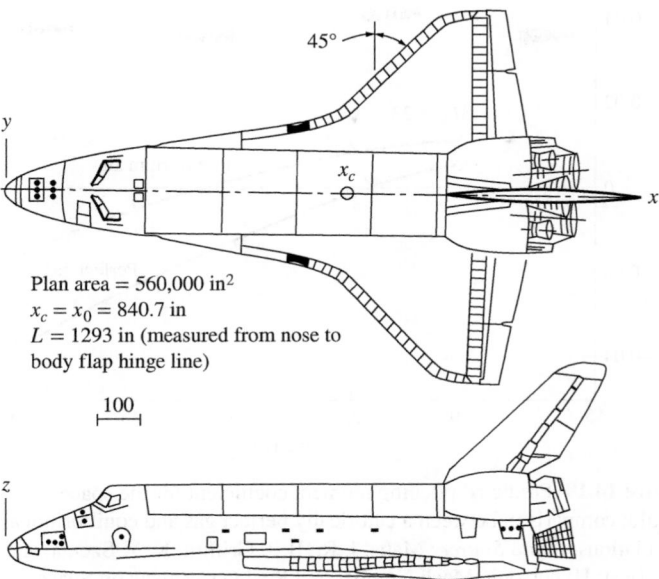

Plan area = 560,000 in^2
$x_c = x_0 = 840.7$ in
$L = 1293$ in (measured from nose to body flap hinge line)

Figure 14.17 Space shuttle geometry.

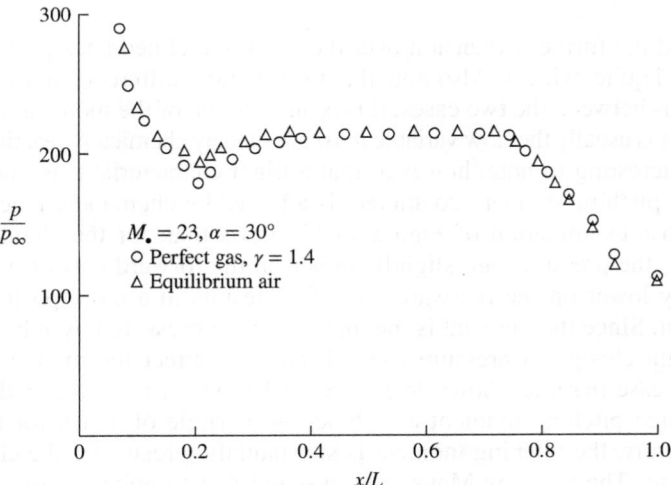

Figure 14.18 Pressure distribution along the windward centerline of the space shuttle; comparison between a calorically perfect gas and chemically reacting equilibrium air calculations. (*Data Source:* Maus, J. R., B. J. Griffith, K. Y. Szema, and J. T. Best: "Hypersonic Mach Number and Real Gas Effects on Space Shuttle Orbiter Aerodynamics," *J. Spacecraft and Rockets*, vol. 21, no. 2, March–April 1984, pp. 136–141.)

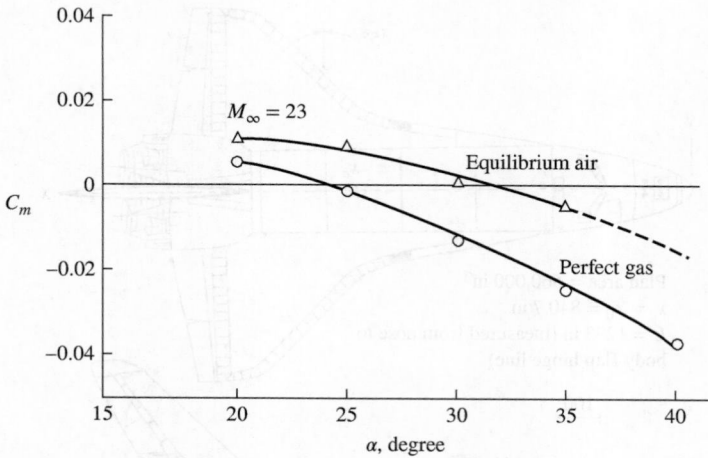

Figure 14.19 Predicted pitching moment coefficient for the space shuttle; comparison between a calorically perfect gas and equilibrium air calculations. (*Data Source:* Maus, J. R., B. J. Griffith, K. Y. Szema, and J. T. Best: Hypersonic Mach Number and Real Gas Effects on Space Shuttle Orbiter Aerodynamics," *J. Spacecraft and Rockets*, vol. 21, no. 2, March–April 1984, pp. 136–141.)

surface, and the further expansion over the slightly inclined back portion of the body, are all quite evident. Also note that there is little difference in the pressure distributions between the two cases; this is an example of the more general result that pressure is usually the flow variable *least* affected by chemically reacting effects.

It is interesting to note, however, that a flight characteristic as mundane as the vehicle pitching moment coefficient is affected by chemically reacting flow effects. Close examination of Figure 14.18 shows that, for the chemically reacting flow, the pressures are slightly higher on the forward part of the shuttle, and slightly lower on the rearward part. This results in a more positive pitching moment. Since the moment is the integral of the pressure through a moment arm, a slight change in pressure can substantially affect the moment. This is indeed the case here, as shown in Figure 14.19, which is a plot of the resulting calculated pitching moment as a function of angle of attack for the space shuttle. Clearly, the pitching moment is substantially greater for the chemically reacting case. The work by Maus et al. was the first to point out this effect on pitching moment, and it serves to reinforce the importance of high-temperature flows on hypersonic aerodynamics. It also serves to reinforce the importance of CFD in the analysis of hypersonic flows. The predicted pitching moment used for the space shuttle design came from "cold-flow" wind tunnel tests which did not simulate the high-temperature effects, that is, the designers used data for a perfect gas with $\gamma = 1.4$ obtained in the wind tunnel. This is represented by the lower curve in Figure 14.19. The early flight experience with the shuttle indicated a much higher pitching moment at hypersonic speeds than predicted, which required that the body flap deflection for trim to be more than twice that predicted—an alarming situation. The reason for this is now known; the actual flight environment encountered by the shuttle at high Mach numbers was that of a high-temperature chemically reacting flow—the situation reflected in the upper curve in Figure 14.19. The difference in the pitching moment between the two curves in Figure 14.19 is enough to account for the unexpected extra body flap deflection required to trim the shuttle. Although these CFD results were obtained well after the design of the shuttle, they serve to underscore the importance of CFD to present and future hypersonic vehicle designs.

14.8 HYPERSONIC VISCOUS FLOW: AERODYNAMIC HEATING

Aerodynamic heating can become so severe at hypersonic speeds that it is the dominant design consideration for hypersonic vehicles. Indeed, for the reason discussed at the end of Section 1.1 and sketched in Figures 1.8 and 1.9, the nose and wing leading edges of hypersonic vehicles *must be blunt* rather than sharp, or else the vehicle will be destroyed by aerodynamic heating. In the history of flight, the most unfortunate example of such destruction occurred on February 1, 2003, when the space shuttle *Columbia* disintegrated over Texas during entry into the earth's atmosphere. Several of the thermal protection tiles near the leading edge of the left wing had been damaged by debris during launch. This allowed hot gases to penetrate the surface and destroy the internal wing structure.

The physical mechanisms that create atmospheric heating, both thermal conduction and radiation, are briefly discussed at the end of Section 14.2. In the present section, we will present some engineering methods for predicting aerodynamic heating, and apply them to some hypersonic flow examples. Aerodynamic heating is a major subject in its own right, and is well beyond the scope of this book. (See Reference 52 for an in-depth discussion of aerodynamic heating applied to hypersonic flows.) However, its importance to the design of hypersonic vehicles demands that we examine a few aspects in the present chapter.

14.8.1 Aerodynamic Heating and Hypersonic Flow—The Connection

What is it about hypersonic flight that makes aerodynamic heating so severe? We address this question by reaching ahead to Chapter 16 and Equation (16.55), which introduces a dimensionless heat transfer coefficient called the Stanton number C_H, defined as

$$C_H \equiv \frac{\dot{q}_w}{\rho_e u_e (h_{aw} - h_w)} \tag{14.45}$$

In Equation (14.45), $\dot{q}_w$ is the heat transfer rate per unit area at a given point on the body surface. In the English engineering system, the unit of $\dot{q}_w$ is ft-lb/(s · ft²); in the international system, the unit is W/m². Also in Equation (14.45), ρ_e is the local density at the edge of the boundary layer at the given point, u_e is the local velocity at the edge of the boundary layer, h_w is the enthalpy of the gas at the wall, and h_{aw} is the adiabatic wall enthalpy defined as the enthalpy of the gas at the wall when the wall temperature is the adiabatic wall temperature—the wall temperature when the wall becomes so hot that no more energy is conducted into the wall from the gas adjacent to the wall.

Consider the hypersonic flow over a flat plate at zero angle of attack, where $\rho_e = \rho_\infty$ and $u_e = V_\infty$ (ignoring any viscous interaction effect as described in Section 14.2). For high Mach number laminar flow over a flat plate, T_{aw} is about 12 percent less than the total temperature in the freestream. For our purposes here, we make the approximation that $T_{aw} \approx T_o$, and hence in Equation (14.45)

$$h_{aw} \approx h_o \tag{14.46}$$

where h_o is the total enthalpy of the freestream. From Equation (7.54), we can write

$$h_o = h_\infty + \frac{V_\infty^2}{2} \tag{14.47}$$

At hypersonic speeds, V_∞ is very large. Also the ambient air far ahead of the vehicle is relatively cool; hence, $h_\infty = c_p T_\infty$ is relatively small. Thus, at high speeds, from Equation (14.47),

$$h_o \approx \frac{V_\infty^2}{2} \tag{14.48}$$

The surface temperature of the plate, although it may be hot by normal standards, still must be maintained at below the melting or decomposition temperature of the surface, which is usually much smaller than the total temperature at high Mach numbers. Thus, we can easily make the assumption that

$$h_o \gg h_w \tag{14.49}$$

It follows from Equations (14.46), (14.48), and (14.49) that

$$h_{aw} - h_w \approx h_o - h_w \approx h_o \approx \frac{V_\infty^2}{2} \tag{14.50}$$

Equation (14.45), written for a flat plate, is

$$C_H = \frac{\dot{q}_w}{\rho_\infty V_\infty (h_{aw} - h_w)}$$

Invoking the approximation given by Equation (14.50), we have

$$C_H \approx \frac{\dot{q}_w}{\rho_\infty V_\infty (V_\infty^2 / 2)}$$

or

$$\dot{q}_w \approx \frac{1}{2} \rho_\infty V_\infty^3 C_H \tag{14.51}$$

Equation (14.51) states that *the aerodynamic heating rate varies as the cube of the velocity*. This is in contrast to aerodynamic drag, which varies only as the square of the velocity. For this reason, at very high velocities, aerodynamic heating becomes a dominant aspect of hypersonic vehicle design. This is the connection between aerodynamic heating and hypersonic flow.

14.8.2 Blunt Versus Slender Bodies in Hypersonic Flow

We have made the claim that the nose and wing leading edges of hypersonic vehicles must be blunt rather than sharp in order to reduce the aerodynamic heating in those regions. In this section, we will begin to demonstrate this fact quantitatively.

In Section 14.8.1, we focused on the local heat transfer rate per unit area at a point on the surface of the vehicle, $\dot{q}_w$. Here we expand our view to the total heat transferred to the vehicle per unit time, dQ/dt, which is equal to the local heat transfer rate integrated over the whole surface area of the vehicle. We can define an integrated overall Stanton number $\overline{C}_H$ by an equation similar to Equation (14.45),

$$\overline{C}_H \equiv \frac{dQ/dt}{\rho_\infty V_\infty (h_o - h_w) S} \tag{14.52}$$

where S is a reference area (planform area of a wing, cross-sectional area of a spherical entry vehicle, or the like) in the same spirit as in the definition of the lift or drag coefficients for a vehicle. Using the approximations made in Section 14.8.1, Equation (14.52) can be approximated by an expression similar to Equation (14.51); i.e.,

$$\frac{dQ}{dt} = \frac{1}{2}\rho_\infty V_\infty^3 S \overline{C}_H \tag{14.53}$$

Again we borrow a result from Chapters 16 and 18, namely that there exists an analogy between skin friction and aerodynamic heating, called Reynolds' analogy, expressed for a laminar flow by Equation (18.50), repeated below:

$$\frac{C_H}{C_f} = \frac{1}{2}\Pr^{-2/3} \tag{18.50}$$

where C_f is the local skin friction coefficient as first defined in Section 1.5, and Pr is the Prandtl number defined in Section 15.6. For our analysis here, it is safe to assume that $\Pr = 1$. Also, Reynolds analogy expressed by Eq. (18.50) can be written in terms of the integrated heat transfer and skin friction coefficients, $\overline{C}_H$ and C_f, respectively. Hence, we have

$$\frac{\overline{C}_H}{C_f} = \frac{1}{2} \tag{14.54}$$

Inserting Equation (14.54) into (14.53) gives

$$\frac{dQ}{dt} = \frac{1}{4}\rho_\infty V_\infty^3 S C_f \tag{14.55}$$

Let us consider the case of a hypersonic vehicle entering the atmosphere at very high Mach number from a mission in space. The force that slows this vehicle during entry is aerodynamic drag. From Newton's second law, we have

$$F = D = -m \frac{dV_\infty}{dt} \tag{14.56}$$

where m is the mass of the vehicle and the minus sign is necessary because dV_∞/dt is negative; i.e., the vehicle is decelerating. From Equation (14.56),

$$\frac{dV_\infty}{dt} = -\frac{D}{m} = -\frac{1}{2m} \rho_\infty V_\infty^2 S C_D \tag{14.57}$$

where C_D is the drag coefficient of the vehicle. Mathematically, we can write dQ/dt as $(dQ/dV_\infty)(dV_\infty/dt)$, where dV_∞/dt is given by Equation (14.57).

$$\frac{dQ}{dt} = \frac{dQ}{dV_\infty} \left(-\frac{1}{2m} \rho_\infty V_\infty^2 S C_D \right) \tag{14.58}$$

Equating Equations (14.55) and (14.58),

$$\frac{dQ}{dV_\infty} \left(-\frac{1}{2m} \rho_\infty V_\infty^2 S C_D \right) = \frac{1}{4} \rho_\infty V_\infty^3 S C_f$$

or

$$\frac{dQ}{dV_\infty} = -\frac{1}{2} m V_\infty \frac{C_f}{C_D}$$

or

$$dQ = -\frac{1}{2} m \frac{C_f}{C_D} \frac{dV_\infty^2}{2} \tag{14.59}$$

Integrate Equation (14.59) from the beginning of entry to the atmosphere, where $Q = 0$ and $V_\infty = V_E$, to the end of entry where $Q = Q_{\text{total}}$ and $V_\infty = 0$:

$$\int_0^{Q_{\text{total}}} dQ = -\frac{1}{2} \frac{C_f}{C_D} \int_{V_E}^0 d \left(m \frac{V_\infty^2}{2} \right)$$

or

$$\boxed{Q_{\text{total}} = \frac{1}{2} \frac{C_f}{C_D} \left(\frac{1}{2} m V_E^2 \right)} \tag{14.60}$$

Equation (14.60) gives the total heat input Q_{total} to the vehicle. It reflects two vital conclusions:

1. The quantity $\frac{1}{2} m V_E^2$ is the initial kinetic energy of the vehicle as it first enters the atmosphere. Equation (14.60) says that the total heat input is directly proportional to this initial kinetic energy.

2. Total heat input is directly proportional to the ratio of skin friction drag to the total drag, C_f/C_D.

The second of these conclusions is of direct relevance to our discussion. Recall from Section 1.5 that the aerodynamic drag on a vehicle is the sum of drag due to the pressure distribution exerted over its surface, called pressure drag D_p, and the drag due to shear stress exerted over its surface, called skin friction drag D_f. In terms of the pressure drag coefficient C_{D_p} and the skin friction drag coefficient C_f, we have:

$$C_D = C_{D_p} + C_f$$

From Equation (14.60), to minimize the total aerodynamic heating, we need to minimize the ratio

$$\frac{C_f}{C_{D_p} + C_f}$$

Now consider two extremes of aerodynamic configurations: a sharp-nosed *slender body* such as the cone shown in Figure 14.20a and the *blunt body* shown in Figure 14.20b. For a slender body, the skin friction drag is large in comparison to the pressure drag, hence $C_D \approx C_f$ and

$$\frac{C_f}{C_D} \approx 1 \qquad \text{slender body}$$

On the other hand, for a blunt body the pressure drag is large in comparison to the skin friction drag, hence $C_D \approx C_{D_p}$ and

$$\frac{C_f}{C_D} \ll 1 \qquad \text{blunt body}$$

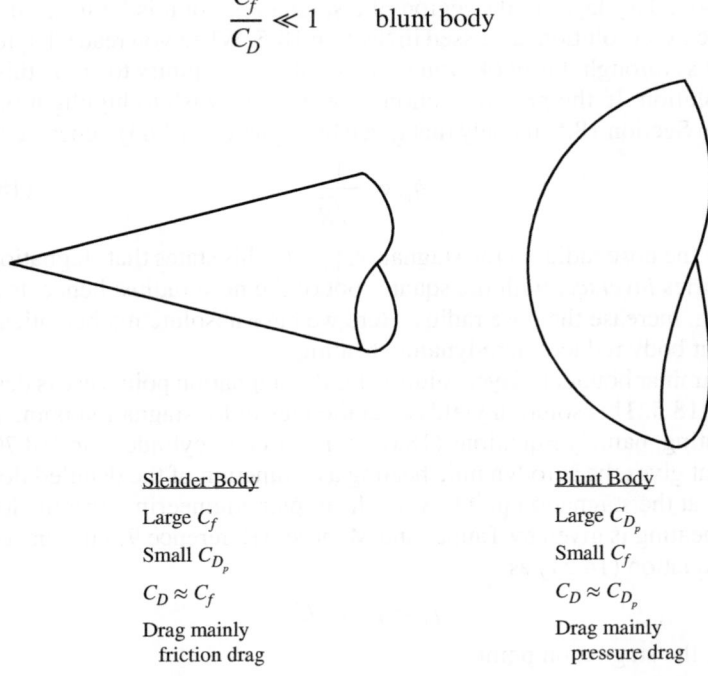

Slender Body

Large C_f

Small C_{D_p}

$C_D \approx C_f$

Drag mainly
    friction drag

(a)

Blunt Body

Large C_{D_p}

Small C_f

$C_D \approx C_{D_p}$

Drag mainly
    pressure drag

(b)

Figure 14.20 Comparison of blunt and slender bodies.

In light of Equation (14.60), this leads to the following vital conclusion:

To minimize aerodynamic heating, the vehicle must be a
blunt body, i.e., have a blunt nose.

For this reason, all successful entry vehicles in practice, from intercontinental ballistic missiles (ICBMs) to the Apollo lunar return capsule and to the space shuttle, have utilized rounded noses and rounded leading edges. Even vehicles designed for sustained hypersonic atmospheric flight such as the X-43 shown in Figures 14.1 and 14.2 have rounded noses and leading edges, although the radii of curvatures are small because the minimization of drag, hence the maximization of lift-to-drag ratio, also becomes important in the design of such vehicles.

14.8.3 Aerodynamic Heating to a Blunt Body

The concept that a blunt body would reduce aerodynamic heating in comparison to a slender body was first advanced by H. Julian "Harvey" Allen in 1951, as discussed in Section 1.1. From that time on, the calculation of blunt body aerodynamic heating has been of paramount importance in the design of hypersonic vehicles. In this section, we examine the calculation of aerodynamic heating to the stagnation point of a blunt body because the stagnation point is frequently (but not always) the point of maximum heat transfer rate to a hypersonic vehicle.

The boundary layer in the region of a stagnation point is laminar, and lends itself to the exact solution discussed in Section 18.5. When you read Chapter 18 as you progress through this book, you will have the opportunity to enjoy this rather elegant solution. In the present section, however, we wish to highlight only one result from Section 18.5, namely that given by Equation (18.83), repeated below.

$$\dot{q}_w \propto \frac{1}{\sqrt{R}} \qquad (18.83)$$

where R is the nose radius at the stagnation point. This states that stagnation point heating varies *inversely* with the square root of the nose radius; hence, to reduce the heating, increase the nose radius. Here we have absolute mathematical proof that a blunt body reduces aerodynamic heating.

The laminar boundary layer solution for the stagnation point case is described in Section 18.5. This solution yields a detailed result for stagnation point aerodynamic heating, namely Equations (18.65) for a circular cylinder and (18.70) for a sphere, that gives the aerodynamic heating as a function of the detailed flow field properties at the stagnation point. A much simpler engineering formula for aerodynamic heating is given by Tauber and Meneses (Reference 97) in a generalized form of Equation (14.51) as

$$\dot{q}_w = \rho_\infty^N V_\infty^M C \qquad (14.61)$$

where, for the stagnation point,

$$M = 3, N = 0.5, C = 1.83 \times 10^{-8}\, R^{-1/2}\left(1 - \frac{h_w}{h_o}\right)$$

when the units for $\dot{q}_w$, V_∞, ρ_∞, and R are respectively, W/cm^2, m/s, kg/m^3, and m. Thus, for the stagnation point, using Equation (14.61), we have

$$\dot{q}_w = \rho_\infty^{0.5} V_\infty^3 (1.83 \times 10^{-8} R^{-0.5}) \left(1 - \frac{h_w}{h_o} \right) \tag{14.62}$$

In Equation (14.62) we see the now familiar result that aerodynamic heating varies with the cube of the velocity, and that the stagnation point heating varies inversely with the square root of the nose radius. Equation (14.62) also shows that $\dot{q}_w$ varies with the square root of the density, which at first glance appears not to be consistent with Equation (14.51), repeated below

$$\dot{q}_w = \frac{1}{2} \rho_\infty V_\infty^3 C_H \tag{14.51}$$

This relation appears to show that $\dot{q}_w$ is proportional to density to the first power. However, once again drawing on results from Chapter 18, and specifically from Equation (18.54), we see that the Stanton number itself for laminar flow is inversely proportional to the square root of the Reynolds number. Since the Reynolds number by definition is in turn proportional to ρ_∞, we can state that

$$C_H \propto \frac{1}{\sqrt{\text{Re}}} \propto \frac{1}{\sqrt{\rho_\infty}}$$

and then from Equation (14.51),

$$\dot{q}_w \propto \sqrt{\rho_\infty}$$

This is consistent with Equation (14.62).

EXAMPLE 14.2

During the entry of the space shuttle into the earth's atmosphere, maximum stagnation point heating occurs at the trajectory point corresponding to an altitude of 68.9 km, where $\rho_\infty = 1.075 \times 10^{-4}$ kg/m^3, and a flight velocity of 6.61 km/s. At this point on its entry trajectory, the shuttle is at a 40.2 degree angle of attack, which presents an effective nose radius at the stagnation point of 1.29 m. If the wall temperature is $T_w = 1110$ K, calculate the stagnation point heating rate.

■ **Solution**

Equation (14.62), repeated here, is

$$\dot{q}_w = \rho_\infty^{0.5} V_\infty^3 (1.83 \times 10^{-8} R^{-0.5}) \left(1 - \frac{h_w}{h_o} \right)$$

To evaluate the ratio h_w/h_o, we have from Equation (14.48)

$$h_o \approx \frac{V_\infty^2}{2} = \frac{(6610)^2}{2} = 2.185 \times 10^7 \text{ J/(kg} \cdot \text{K)}$$

For h_w, when the wall temperature is 1110 K, we can reasonably assume a calorically perfect gas. As calculated in Example 7.1, for calorically perfect air, $C_p = 1004.5$ J/(kg · K).

Thus,

$$h_w = c_p T_w = (1004.5)(1110) = 1.115 \times 10^6 \text{ J/(kg} \cdot \text{K)}$$

Thus,

$$\frac{h_w}{h_o} = \frac{1.115 \times 10^6}{2.185 \times 10^7} = 0.051$$

The stagnation point heat transfer is, from Equation (14.62),

$$\dot{q}_w = \rho_\infty^{0.5} V_\infty^3 (1.83 \times 10^{-8} R^{-0.5}) \left(1 - \frac{h_w}{h_0}\right)$$

$$= (1.075 \times 10^{-4})^{0.5}(6610)^3(1.83 \times 10^{-8}) \times (1.29)^{-0.5}(1 - 0.051)$$

$$= \boxed{45.78 \text{ W/cm}^2}$$

Zoby (Reference 98) quotes a maximum stagnation point heating of 45 W/cm^2 based on experimental data obtained for the space shuttle at the given altitude and velocity on the entry trajectory. Our calculated result from Equation (14.62) agrees very well with the experimental data.

In Example 14.2, we calculated the heat transfer rate to the stagnation point. This is the point on the body of maximum aerodynamic heating rate. Along the windward centerline (on the bottom surface), the heating rate rapidly decreases with distance downstream from the stagnation point. Figure 14.21 gives experimental data for the local aerodynamic heating rate as reported in Reference 98. Note the qualitative similarity between the variation of $\dot{q}_w$ with distance as shown in Figure 14.21 and the variation of pressure with distance as shown in Figure 14.18. Although the results shown in these two figures are for slightly different angles of attack, this comparison illustrates a qualitative trend that is frequently seen in hypersonic aerodynamics, namely that the distribution of the aerodynamic heating rate over a surface tends to qualitatively follow the distribution of pressure over the surface.

14.9 APPLIED HYPERSONIC AERODYNAMICS: HYPERSONIC WAVERIDERS

The maximum lift-to-drag ratio $(L/D)_{\max}$ for a flight vehicle is a measure of its aerodynamic efficiency. Unfortunately, for supersonic and hypersonic flight vehicles, as the freestream Mach number increases, $(L/D)_{\max}$ decreases rather dramatically. This is just a fact of nature, brought about by the rapidly increasing shock-wave strength as Mach number increases, with consequent large increases in wave drag. Return to the variation of L/D versus angle of attack for a flat plate shown in Figure 14.14. The solid curve is from the newtonian analysis discussed in Section 14.4. This is an inviscid flow result, and shows that L/D theoretically approaches infinity as the angle of attack α approaches zero. In reality, the viscous shear stress acting on the plate surface causes L/D to peak at a low value

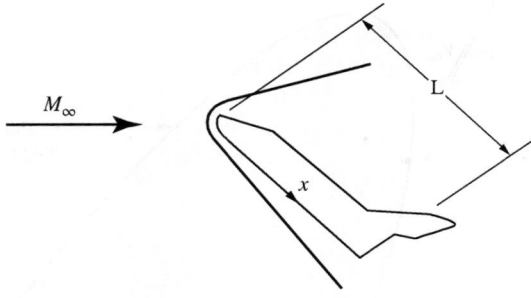

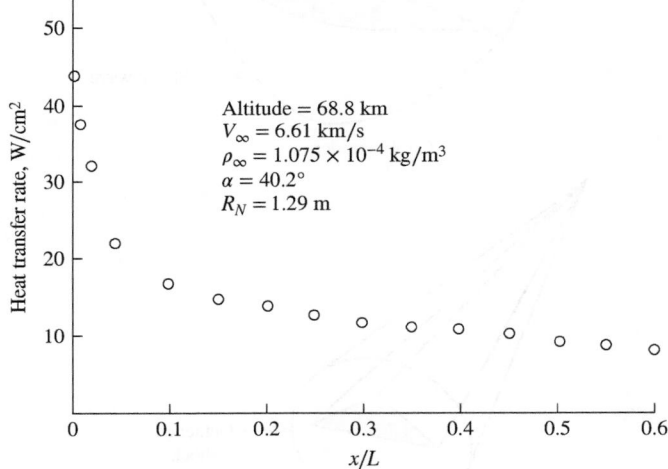

Figure 14.21 Experimental data for the local aerodynamic heating rate along the windward centerline of the space shuttle. (Data from Zoby, E. V., "Approximate Heating Analysis for the Windward Symmetry Plane of Shuttlelike Bodies at Angle of Attack," in *Thermodynamics of Atmospheric Entry*, T. E. Horton (ed.), Vol. 82, *Progress in Astronautics and Aeronautics*, American Institute of Aeronautics and Astronautics, 1982, pp. 229–247.)

of α and to go to zero as $\alpha \rightarrow 0$. This is illustrated by the dashed curve in Figure 14.14, which shows the variation of L/D modified by skin friction as predicted by the reference temperature method discussed in Section 18.4. The skin-friction calculation is for laminar flow at Mach 10 and a Reynolds number of 3×10^6. Note that $(L/D)_{max}$ for the flat plate is about 6.5. By comparison, $(L/D)_{max}$ for a Boeing 747 at normal cruising conditions near Mach 1 is about 20. So the $(L/D)_{max}$ for a hypersonic flat plate, as shown in Figure 14.14, is a low value, reflecting the characteristically low lift-to-drag ratios generated by hypersonic vehicles. And the infinitely thin flat plate is the most efficient lifting surface aerodynamically compared to other hypersonic shapes with finite thickness. *Conclusion:* The L/D value of vehicles at hypersonic Mach numbers is low. This is particularly

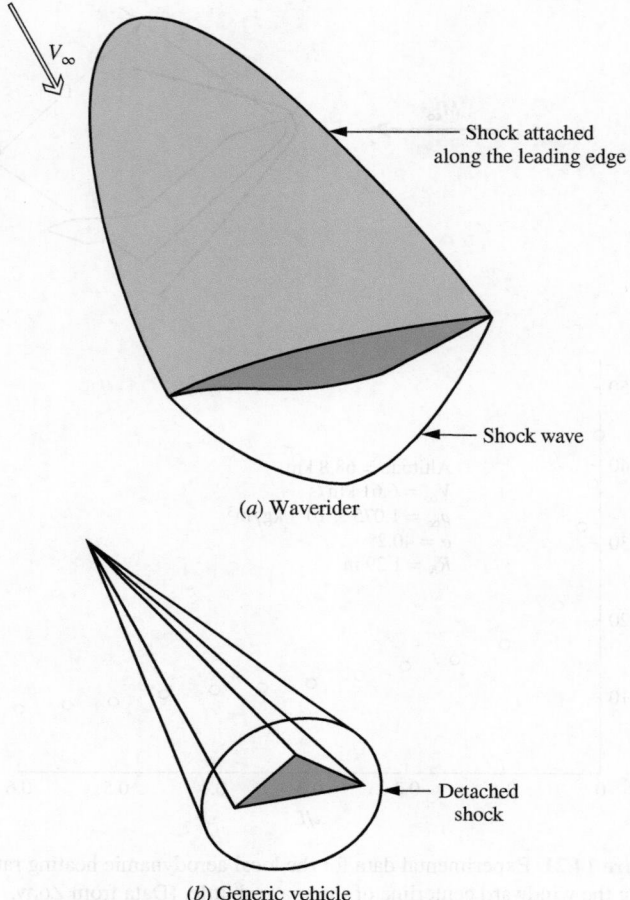

V_∞

Shock attached
along the leading edge

Shock wave

(*a*) Waverider

Detached
shock

(*b*) Generic vehicle

Figure 14.22 Comparison of waverider and generic hypersonic
configurations.

bothersome for future hypersonic vehicles designed for sustained flight in the
atmosphere. Current design practice for such vehicles is illustrated by the X-43
shown in Figures 14.1 and 14.2.

There is a class of hypersonic vehicle shapes, however, that generates higher
value of L/D than other shapes—waveriders. A waverider is a supersonic or hypersonic vehicle that has an attached shock wave all along its leading edge, as
sketched in Figure 14.22*a*. Because of this, the vehicle appears to be riding on
top of its shock wave, hence the term "waverider." This is in contrast to a more
conventional hypersonic vehicle, where the shock wave is usually detached from
the leading edge, as sketched in Figure 14.22*b*. The aerodynamic advantage of the
waverider in Figure 14.22*a* is that the high pressure behind the shock wave under
the vehicle does not "leak" around the leading edge to the top surface; the flow
field over the bottom surface is contained, and the high pressure is preserved, and

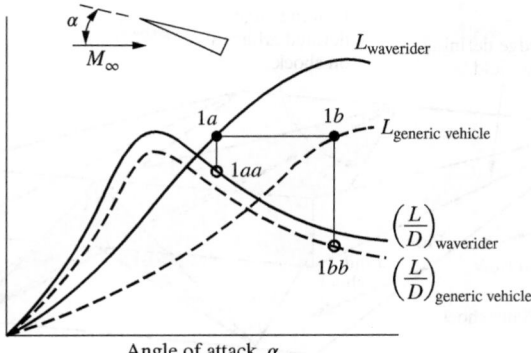

Figure 14.23 Comparison of lift and L/D curves between waverider and a generic hypersonic vehicle.

therefore more lift is generated on the vehicle. In contrast, for the vehicle shown in Figure 14.22b, there is communication between the flows over the bottom and top surfaces; the pressure tends to leak around the leading edge, and the general integrated pressure level on the bottom surface is reduced, resulting in less lift. Because of this, the generic vehicle in Figure 14.22b must fly at a larger angle of attack α to produce the same lift as the waverider in Figure 14.22a. This is illustrated in Figure 14.23, where the lift curves (L versus α) are sketched for the two vehicles shown in Figure 14.22. Note that the lift curve for the waverider is considerably higher because of the pressure containment compared to that for the generic vehicle. At the same lift, points 1a and 1b in Figure 14.23 represent the waverider and generic vehicles, respectively. Also shown in Figure 14.23 are typical variations of L/D versus α, which for slender hypersonic vehicles are not too different for the shapes in Figure 14.22a and b. (Although the lift of the waverider at a given angle of attack is increased by the pressure containment on the bottom surface, so is the wave drag; hence, the L/D ratio at a given angle of attack for the waverider is better, but not greatly so, than that for the generic vehicle.) However, note that because the waverider generates the same lift at a smaller α (point 1a in Figure 14.23) than does the generic vehicle, which must fly at a large α (point 1b in Figure 14.23), the L/D for the waverider is considerably higher (point 1aa) than that for the generic shape point (1bb). Therefore, for sustained hypersonic cruising flight in the atmosphere the waverider configuration has a definite advantage.

Question: How do you design a vehicle shape such that the shock wave is attached all along its leading edge; that is, how do you design a waverider?

One answer is as follows. Consider the simple flow field generated by a wedge in a supersonic or hypersonic freestream as discussed in Section 9.3. Imagine that the top surface of the wedge is parallel to the freestream, and hence the only wave in the flow is the planar shock wave propagating below the wedge, as sketched at the top of Figure 14.24. Now imagine two straight lines arbitrarily traced on the surface of the shock wave, coming to a point at the front of the shock. Consider all of the streamlines of the flow behind the shock that emanate from these

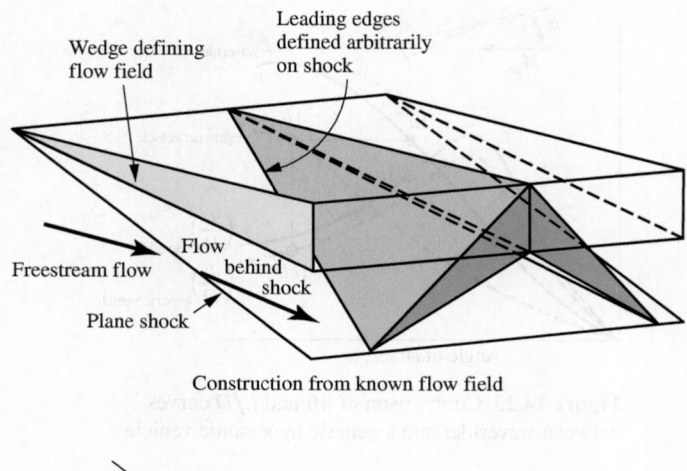

Construction from known flow field

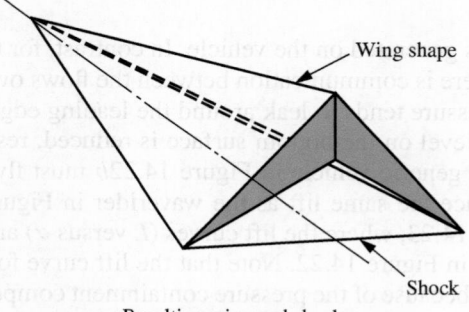

Figure 14.24 Nonweiler or "caret" wing.

arbitrarily traced lines. Taken together, these streamlines form a stream surface that can be considered the surface of a vehicle with its leading edges defined by the two arbitrarily traced lines on the shock wave. Because the flow field behind a planar shock wave is uniform with parallel streamlines, these stream surfaces are flat surfaces that trace out a vehicle shape with a caret cross section as shown in Figure 14.24, named after the caret symbol ∧. If you now mentally strip away the imaginary generating flow field shown at the top of Figure 14.24, you have left the caret-shaped vehicle shown at the bottom of Figure 14.24. Concentrating on the vehicle shape at the bottom of Figure 14.24, the planar surfaces on the bottom of the vehicle are stream surfaces that exist behind a planar oblique shock wave—stream surfaces that are generated by streamlines that begin on the shock surface itself. Hence, the shock wave is, by definition, attached to the leading edge of the vehicle; this planar attached shock is shown stretching between the two straight leading edges of the vehicle sketched at the bottom of Figure 14.24. By definition, therefore, this vehicle is a waverider. *Caution:* The waverider is in principle a point-designed vehicle. The generating oblique shock sketched at the top of Figure 14.24 pertains to a given freestream Mach number M_∞ and a given flow deflection angle of the imaginary wedge that generates the oblique shock. Nevertheless, if you construct the vehicle shape shown at the bottom of Figure 14.24 and put it in a freestream at the given M_∞ and at an angle of attack such that the flow deflection angle of the vehicle's bottom surface is the same as that of the imaginary generating wedge, then nature will make certain that the shock wave is attached all along the vehicle's leading edge; that is, the vehicle will be a waverider. Note that in Figure 14.24 we have oriented the imaginary generating wedge such that its top surface is parallel to the freestream; hence, there is no wave over the top surface of the wedge. Consequently, the top surfaces of the resulting caret waverider shown at the bottom of Figure 14.24 are aligned with the freestream, and there is no wave above the waverider.

In principle, any shape can be used for the imaginary body producing the flow field from which a waverider shape is carved. The simplest case is to use a wedge for the imaginary body as just described. This has the advantage that a wedge produces a simple known flow field that is easily calculated, as treated in Chapter 9. You do not need a CFD solution for this flow. The flow over a cone at zero angle of attack in a supersonic or hypersonic flow is similarly a known flow field that can be used to generate waverider shapes. Because this conical flow field is quasi-three-dimensional, it provides more flexibility in the generation of waverider shapes. The idea is the same. Consider the supersonic or hypersonic conical flow field over a right-circular cone at zero angle of attack as sketched at the top of Figure 14.25. The exact numerical solution of this flow field is discussed in Section 13.6. The flow field is obtained from a solution of the Taylor-Maccoll equation, Equation (13.78), and tabulated results given in References 95 and 96. In short, this is a known flow field. At the top of Figure 14.25, we see a conical shock wave attached at the vertex of the right-circular cone. This cone is simply the imaginary body generating the flow field.

Consider the dashed curve drawn on the bottom surface of the conical shock wave as sketched at the top of Figure 14.25. All of the streamlines flowing through this dashed curve constitute a stream surface. In turn this stream surface defines the bottom surface of a waverider with a leading edge traced out by the dashed curve, as sketched at the bottom of Figure 14.25. Any curve can be traced on the conical shock; hence, any stream surface of the conical flow field downstream of the shock can be used as the surface of a waverider. When this is done, the shock wave will be attached all along the leading edge of the waverider, as shown in Figure 14.25. Moreover, the attached shock wave on this resulting waverider will, of course, be a segment of the conical shock wave shown at the top of Figure 14.25.

The waverider concept was first introduced by Nonweiler (Reference 99) in 1959, who generated caret-shaped waveriders from the two-dimensional flow field behind a planar oblique shock wave generated by a wedge, as described earlier. Nonweiler was interested in such waveriders as lifting atmospheric entry bodies. The first extension of Nonweiler's concept to the use of a conical flow as generating flow field by Jones (Reference 100) in 1963, and further extensions to other axisymmetric generating flows are discussed by Jones et al. in Reference 101. An excellent and authoritative survey of waverider research up to

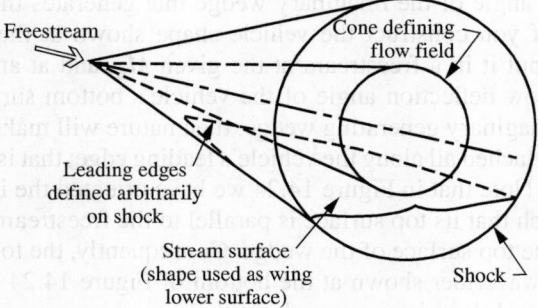

Freestream

Cone defining
flow field

Leading edges
defined arbitrarily
on shock

Stream surface
(shape used as wing
lower surface)

Shock

Construction from known flow field

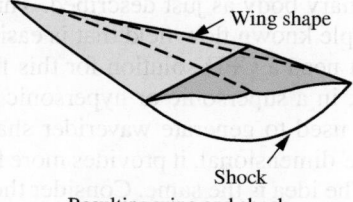

Wing shape

Shock

Resulting wing and shock

Figure 14.25 Cone flow wing.

1979 is given by Townend in Reference 102. In the early 1980s, Rasmussen and his colleagues at the University of Oklahoma (for example, see References 103–105) utilized hypersonic small-disturbance theory to design waveriders from flow fields over right-circular cones as well as elliptic cones. In a manner consistent with his use of analytical solutions of the waverider flows, Rasmussen was also able to use the classic calculus of variations to optimize the waverider shapes utilizing the inviscid properties of the flow.

14.9.1 Viscous-Optimized Waveriders

In the work just described, the waverider configurations were designed (and sometimes optimized) on the basis of inviscid flow fields, not including the effect of skin-friction drag. In turn, the drag predicted by such inviscid analyses was simply wave drag, and the resulting values of the inviscid L/D looked promising. However, waveriders tend to have large wetted surface areas, and the skin-friction drag, always added to the waverider aerodynamics after the fact, tended to greatly decrease the predicted inviscid lift-to-drag ratio. This made the waverider a less interesting prospect and led to a temporary lack of interest, indeed outright skepticism by researchers and vehicle designers in the waverider as a viable hypersonic configuration. Beginning in 1987, the author and his students at the University of Maryland took a different tack. New families of waveriders were generated wherein the skin-friction drag was included within an optimization routine to calculate waveriders with maximum L/D. In this fashion, the trade-offs between wave drag and friction drag were accounted for during the optimization process, and the resulting family of waveriders had a shape and wetted surface area so as to optimize L/D. This family of waveriders is called viscous-optimization hypersonic waveriders, and subsequent CFD calculations and wind-tunnel tests have proven their viability, thus greatly enhancing modern interest in the waverider concept.

The design process for viscous-optimized waveriders was first published in References 106 and 107. This work, beginning in the late 1980s, led to a new class of waveriders where the optimization process is trying to reduce the wetted surface area, hence reducing skin-friction drag, while maximizing L/D. Because detailed viscous effects cannot be couched in simple analytical forms, the formal optimization methods based on the calculus of variations cannot be used. Instead, a numerical optimization technique was used based on the simplex method of Nelder and Mead (Reference 108). By using a numerical optimization technique, other real configuration aspects could be included in the analysis in addition to viscous effects, such as blunted leading edges and an expansion upper surface (in contrast to the standard assumption of a freestream upper surface, i.e., an upper surface with all generators parallel to the freestream direction). The results of the study by Bowcutt et al. led to a new class of waveriders, namely, viscous-optimized waveriders. Moreover, these waveriders produced relatively high values of L/D, as will be discussed later.

For the viscous-optimized waverider configurations, the following philosophy was followed:

1. The lower (compression) surface was generated by a stream surface behind a conical shock wave. The inviscid conical flow field was obtained from the numerical solution of the Taylor-Maccoll equation, derived in Section 13.6.
2. The upper surface was treated as an expansion surface, generated in a manner similar to that for the inviscid flow about a tapered, axisymmetric cylinder at zero angle of attack, and calculated by means of the axisymmetric method of characteristics.
3. The viscous effects were calculated by means of an integral boundary-layer analysis following surface streamlines, including transition from laminar to turbulent flow.
4. Blunt leading edges were included to the extent of determining the maximum leading-edge radius required to yield acceptable leading-edge surface temperatures, and then the leading-edge drag was estimated by modified newtonian theory.
5. The final waverider configuration, optimized for maximum L/D at a given Mach number and Reynolds number with body fineness ratio as a constraint, was obtained from the numerical simplex method taking into account all of the effects itemized in steps 1–4 within the optimization process itself.

The following discussion provides some insight into the optimization process. First, assume a given conical shock wave in a flow at a given Mach number, say, a conical shock wave angle of $\theta_s = 11$ deg. at Mach 6. As discussed previously, now trace a curve on the surface of the shock wave. The stream surface generated from this curve is a bottom surface of a waverider, and the curve itself forms the leading edge of the waverider. An infinite number of such curves can be traced on the conical shock wave, generating an infinite number of waverider shapes using the conical shock with $\theta_s = 11$ deg. at $M_\infty = 6$. Indeed, some of these leading-edge curves are shown in Figure 14.26. The optimization procedure progresses through a series of these leading-edge shapes, each one generating a new waverider with a certain lift-to-drag ratio, and finally settling on that particular leading-edge shape that yields the maximum value of L/D. This is the optimum waverider for the given generating conical shock wave angle of $\theta_s = 11$ deg. This resulting $(L/D)_{max}$ is then plotted as a point in Figure 14.27 for the conical shock wave angle $\theta_s = 11$ deg. Figure 14.27 also gives the corresponding value of lift coefficient C_L and volumetric efficiency $\eta = V^{2/3}/S_p$, where V is the vehicle volume and S_p is the planform area. Now choose another conical shock angle for generating the flow field, say, $\theta_s = 12$ deg., and repeat the preceding procedure, finding that leading-edge shape that yields the waverider shape that produces the highest L/D. This result is now plotted in Figure 14.27 for $\theta_s = 12$ deg. Then another conical shock wave angle, say $\theta_s = 13$ deg., is chosen, and the process is repeated, finding that particular waverider shape that produces the highest L/D. This point is now plotted in Figure 14.27 for $\theta_s = 13$ deg. And so forth. The front views of these optimized waverider shapes are shown in Figure 14.28, each one labeled according to its generating conical shock-wave angle. These same optimized waveriders are shown in perspective in Figure 14.29. Returning to Figure 14.27, note that the curve of L/D versus θ_s itself has a maximum value of L/D, occurring in this case for $\theta_s = 12$ deg. This yields an "optimum of

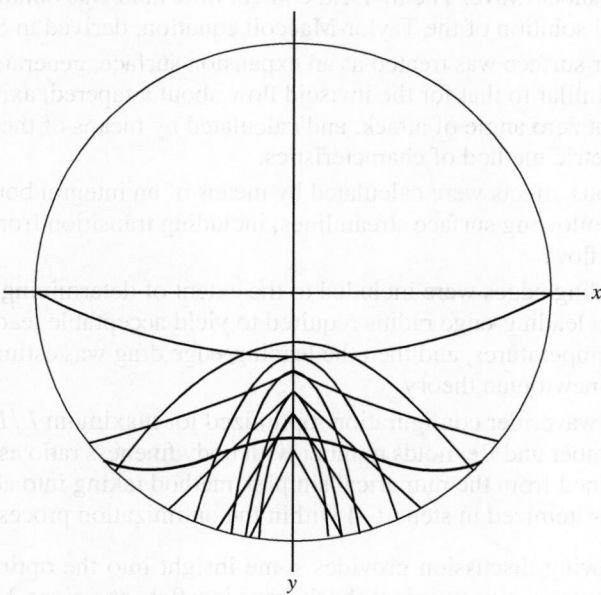

Figure 14.26 Examples of initial and optimized waverider leading-edge shapes.

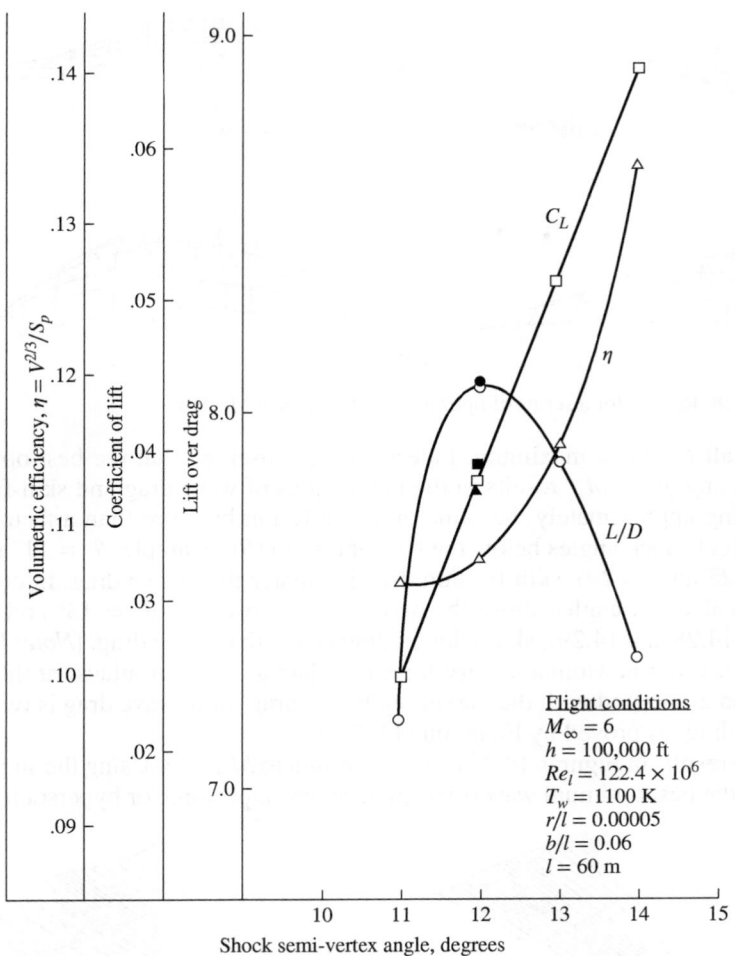

Figure 14.27 Results for a series of optimized waveriders at Mach 6; $l =$ length of waverider, $b/l =$ body fineness ratio, and $r =$ leading edge radius.

the optimums" and defines the final viscous optimized waverider at $M_\infty = 6$ for the flight conditions shown in Figure 14.27. Noted on Figure 14.27 is the body fineness ratio, b/l, where b is the wing span and l is the length of the waverider. Recall that fineness ratio is taken as a constraint in the optimization process. For this case, $b/l = 0.06$. Finally, a summary three-view of the best optimum (the optimum of the optimums) waverider, which here corresponds to $\theta_s = 12$ deg., is given in Figure 14.30. Also, in Figures 14.28 to 14.30 the lines on the upper and lower surfaces of the waveriders are inviscid streamlines. Note in these figures that the shape of the optimum waverider changes considerably with θ_s. Moreover, examining (for example) Figure 14.30, note the rather complex curvature of the leading edge in both the planform and front views; the optimization program is shaping the waverider to adjust both wave drag and skin-friction drag so that

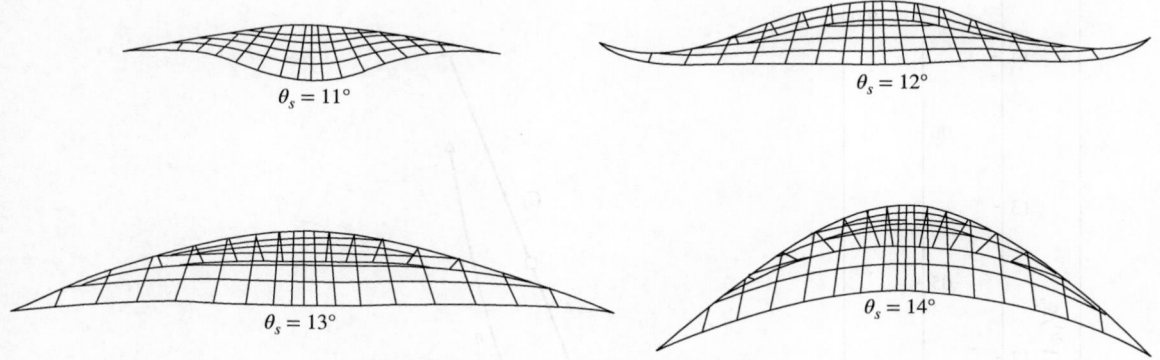

$\theta_s = 11°$ $\theta_s = 12°$

$\theta_s = 13°$ $\theta_s = 14°$

Figure 14.28 Results for a series of optimized waveriders at Mach 6.

the overall L/D is a maximum. Indeed, it was observed that the best optimum shape at any given M_∞ results in the magnitudes of wave drag and skin-friction drag being approximately the same, never differing by more than a factor of 2. For conical shock angles below the best optimum (for example, $\theta_s = 11°$ in Figures 14.28 and 14.29), skin-friction drag is greater than wave drag; in contrast, for conical shock angles above the optimum (for example, $\theta_s = 13°$ and $14°$ in Figures 14.28 and 14.29), skin-friction drag is less than wave drag. [*Note:* In Section 14.4, using newtonian theory for a flat plate at angle of attack, at the flight condition associated with the maximum lift-to-drag ratio, wave drag is twice the friction drag, as proved by Equation (14.27).]

The results in Figures 14.27 to 14.30 pertain to $M_\infty = 6$. Using the same procedure, the best optimum waverider shape at any supersonic or hypersonic Mach

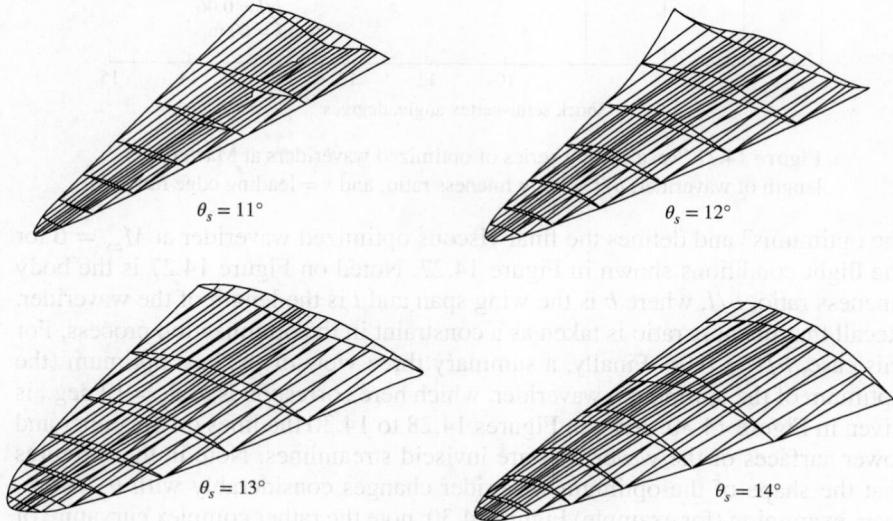

$\theta_s = 11°$ $\theta_s = 12°$

$\theta_s = 13°$ $\theta_s = 14°$

Figure 14.29 Perspective views of a series of optimized waveriders at Mach 6.

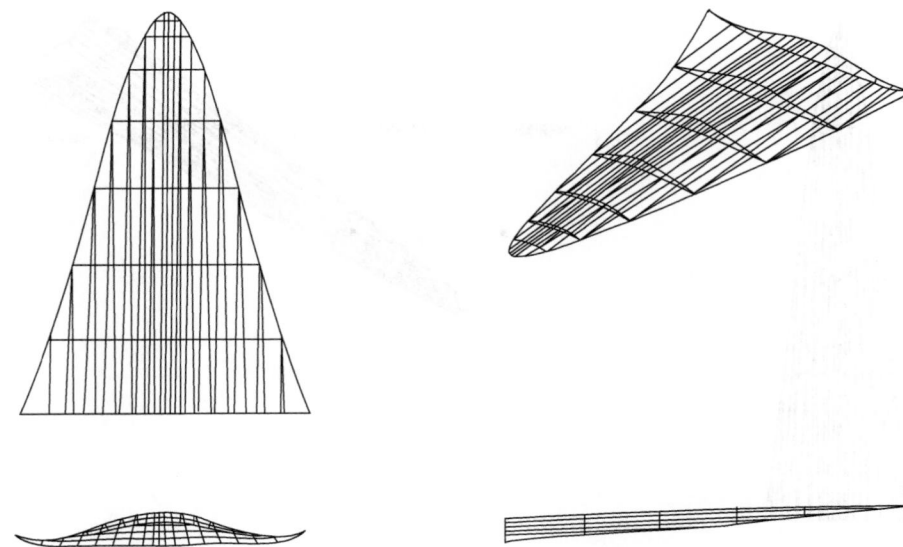

Figure 14.30 Three-view and perspective of the best optimized waverider at Mach 6.

number can be obtained. For example, the shape of the best optimized waverider for $M_\infty = 25$ is given in Figure 14.31. Comparing the optimum configuration of $M_\infty = 6$ (Figure 14.30), note that the Mach 25 shape has more wing sweep. This pertains to a conical flow field with a smaller wave angle, both of which are intuitively expected at higher Mach number. However, the body slenderness ratio at $M_\infty = 6$ is constrained to be $b/l = 0.06$ (analogous to a supersonic transport such as the Concorde), but that $b/l = 0.09$ is the constraint chosen at $M_\infty = 25$ (analogous to a hydrogen-fueled hypersonic airplane). The two different slenderness ratios are chosen on the basis of reality for two different aircraft with two different missions at either extreme of the hypersonic flight spectrum.

For supersonic and hypersonic vehicles, L/D markedly decreases as M_∞ increases. Indeed, Kuchemann (Reference 66) gives the following general empirical correlation for $(L/D)_{max}$ based on actual flight-vehicle experience:

$$(L/D)_{max} = \frac{4(M_\infty + 3)}{M_\infty}$$

This variation is shown as the solid curve in Figure 14.32. This figure is important to our present discussion; it brings home the importance of viscous optimized waveriders. The Kuchemann curve (the solid curve) in Figure 14.32 represents a type of "L/D barrier" for conventional vehicles, which is difficult to break. The open circles in Figure 14.32, which form an almost shotgun scatter of points, are data for a variety of conventional vehicles representing various wind-tunnel and flight tests. (Precise identification of the sources for these points is given in Reference 109.) The solid symbols pertain to the various optimized hypersonic waveriders discussed here. The solid squares are results for the waveriders based

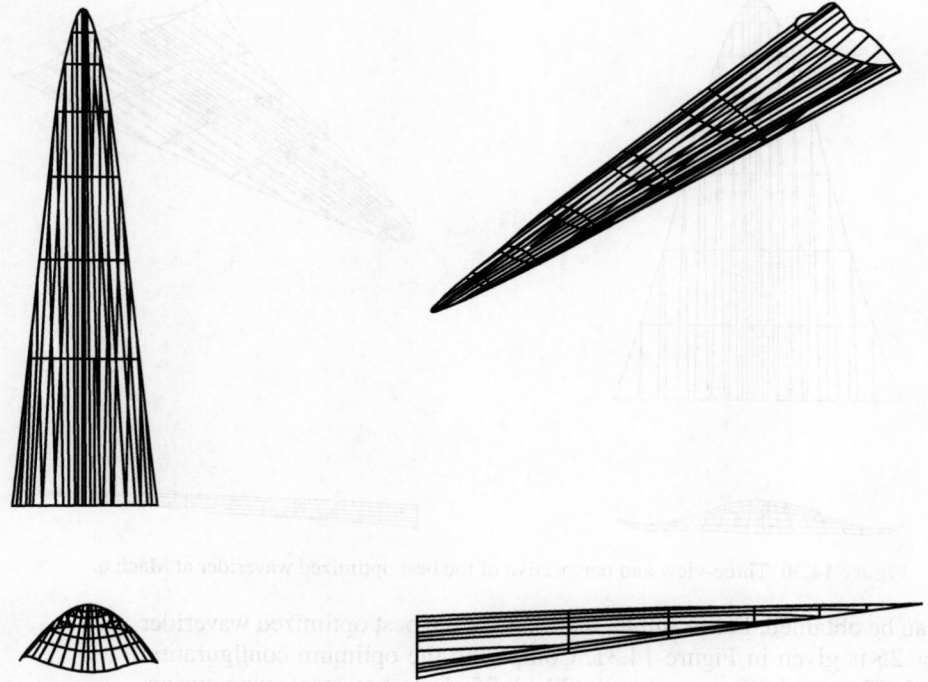

Figure 14.31 Three-view and perspective of the best optimized waverider at Mach 25.

number can be obtained. The three-view and perspective of the best optimized waverider
for $M_\infty = 25$ is given in Figure 14.31. Compare this optimum configuration with that
of $M_\infty = 6$ (Figure 14.30); note that the Mach 25 shape has more wing sweep.
This pertains to the higher Mach number; a more highly swept wing is intuitively expected at a higher Mach number. However, the body slenderness ratio
of $M_\infty = 6$ is constrained to be $b/\ell = 0.06$ (analagous to a supersonic transport
such as the Concorde), but that $b/\ell = 0.09$ is the constraint chosen at $M_\infty = 25$
(analagous to a hydrogen-fueled hypersonic airplane). The two different slenderness ratios are chosen on the basis of realizing, for two different aircraft with two
different missions—a supersonic transport and a hypersonic airplane.

For supersonic and hypersonic vehicles, $(L/D)_{max}$ markedly decreases as M_∞ increases. Indeed, Kuchemann, in 1978, gives the following general empirical
correlation for $(L/D)_{max}$ based on actual flight vehicle experience:

$$(L/D)_{max} = \frac{4(M+3)}{M}$$

This variation is plotted as the solid curve in Figure 14.32. This figure is important
to our present discussion because it illustrates the importance of recent, optimized
waveriders. The Kuchemann curve (the solid curve in Figure 14.32) represents a
type of "L/D barrier" for conventional vehicles, which is difficult to break. The
open circles in Figure 14.32 (which represent various hypersonic configurations) are data from a variety of configurations representing various wind-tunnel
and flight tests. (Specific identification of the sources for these points is given in
Reference 104.) These conventional configurations include some hypersonic
waveriders discussed here. The solid circles and squares are for waveriders based

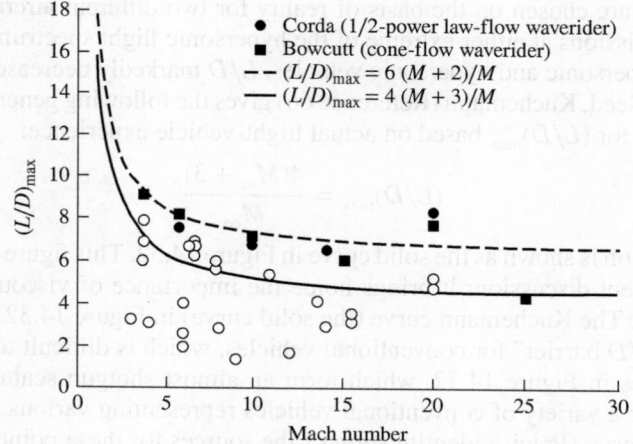

Figure 14.32 Maximum lift-to-drag ratio comparison for
various hypersonic configurations.

on conical generating flows as described below. The solid circles are results for another family of viscous optimized waveriders based on the shock wave and downstream streamsurfaces generated by a one-half power law ogive-shaped body, obtained by Corda and Anderson (Reference 110). From Figure 14.32 we see that the viscous optimized waveriders break the L/D barrier, that is, they give $(L/D)_{\max}$ values that lie above the Kuchemann curve. Indeed, the L/D variation of the viscous optimized waveriders is more closely given by

$$(L/D)_{\max} = \frac{6(M_\infty + 2)}{M_\infty}$$

This variation is shown as the dotted curve in Figure 14.32. The importance of the viscous optimized waveriders is established by the results shown in Figure 14.32. These results have been confirmed by various wind-tunnel tests. They are the reason for renewed interest in the waverider configuration as a hypersonic vehicle, particularly for sustained cruising in the atmosphere.

The physical aspects that define the hypersonic flow regime were discussed in Section 14.2. The influence of viscous interaction effects, high-temperature flows, and aerodynamic heating on waverider design is discussed at length in the second edition of Reference 52 (see pages 361–374, 409–413, and 644–646 of the second edition of Reference 52). Also, hypersonic vehicle design is sensitive to the location of the transition from laminar to turbulent flow, and the design of hypersonic waveriders is no exception. Numerical experiments carried out at $M_\infty = 10$ wherein the transition location was varied over a wide latitude, ranging from all-laminar flow on one hand, to almost all-turbulent flow on the other hand, with various cases in between, are discussed in References 52, 107, and 109. Although these physical phenomena have an effect on the optimized shape of viscous optimized hypersonic waveriders, the resulting values of $(L/D)_{\max}$ are not greatly changed.

Even with these real physical phenomena included in the optimization process, the viscous optimized hypersonic waverider remains a viable configuration for future hypersonic vehicle design. Indeed, the Air Force–sponsored and Boeing-designed X-51, shown in Figure 14.33, is a viscous-optimized waverider. SCRAMjet-powered, and designed for flight at Mach 5 to 6, the X-51 will provide the technology for future atmospheric cruise missiles. On May 1, 2013, the X-51 achieved a flight of over six minutes and reached speeds of over Mach 5 for 210 seconds. At the time of writing, this is the longest duration SCRAMjet-powered hypersonic flight. It is only the second SCRAMjet-powered hypersonic vehicle of any type to achieve sustained atmospheric flight, the first being the X-43 in 2004 (Figures 14.1 and 14.2).

14.10 SUMMARY

Only a few of the basic elements of hypersonic flow are presented here, with special emphasis on newtonian flow results. Useful information on hypersonic flows can be extracted from such results. We have derived the basic newtonian sine-squared law:

$$C_p = 2\sin^2\theta \tag{14.4}$$

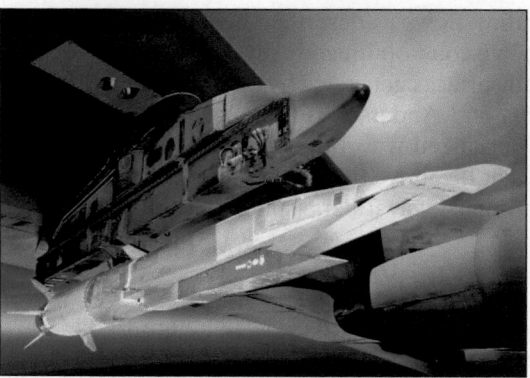

Figure 14.33 X-51 waverider (*U.S. Air Force Photo*).

and used this result to treat the case of a hypersonic flat plate in Section 14.4. We also obtained the limiting form of the oblique shock relations as $M_\infty \to \infty$, that is, the hypersonic shock relations. From these relations, we were able to examine the significance of newtonian theory more thoroughly, namely, Equation (14.4) becomes an exact relation for a hypersonic flow in the combined limit of $M_\infty \to \infty$ and $\gamma \to 1$. Finally, these hypersonic shock relations illustrate the existence of the Mach number independence principle.

14.11 PROBLEMS

14.1 Repeat Problem 9.13 using

    a.   Newtonian theory

    b.   Modified newtonian theory

    Compare these results with those obtained from exact shock-expansion theory (Problem 9.13). From this comparison, what comments can you make about the accuracy of newtonian and modified newtonian theories at low supersonic Mach numbers?

14.2 Consider a flat plate at $\alpha = 20°$ in a Mach 20 freestream. Using straight newtonian theory, calculate the lift- and wave-drag coefficients. Compare these results with exact shock-expansion theory.

14.3 Consider a hypersonic vehicle with a spherical nose flying at Mach 20 at a standard altitude of 150,000 ft, where the ambient temperature and pressure are 500°R and 3.06 lb/ft^2, respectively. At the point on the surface of the nose located 20° away from the stagnation point, estimate the: (*a*) pressure, (*b*) temperature, (*c*) Mach number, and (*d*) velocity of the flow.

Viscous Flow

I n Part 4, we deal with flows that are dominated by viscosity and thermal conduction—viscous flows. We will treat both incompressible and compressible viscous flows. ■

CHAPTER 15

Introduction to the Fundamental Principles and Equations of Viscous Flow

I do not see then, I admit, how one can explain the resistance of fluids by the theory in a satisfactory manner. It seems to me on the contrary that this theory, dealt with and studied with profound attention gives, at least in most cases, resistance absolutely zero: a singular paradox which I leave to geometricians to explain.

Jean LeRond d'Alembert, 1768

PREVIEW BOX

The real life of aerodynamics—that is what the present and remaining chapters are all about. Except for the few earlier sections on viscous flow, most of our previous discussions in this book have dealt with inviscid flows. Do not get the wrong impression; a large number of practical aerodynamic applications are appropriately treated by assuming inviscid flow, as we have already seen. Thank goodness for this, because inviscid flows are usually easier to analyze than viscous flows. But some aspects of aerodynamics are inherently viscous in nature, such as skin-friction drag, aerodynamic heating, and flow separation. To deal with these important aspects, we have to undertake the study of viscous flow, which is the subject of the remainder of this book.

The present chapter is all about the fundamental aspects of viscous flow. Here you will find new definitions, new concepts, and new equations, including the derivation of the Navier-Stokes equations, which are nothing more than the continuity, momentum, and energy equations for a viscous flow. Although some of the basic concepts of viscous flow were first introduced in Chapter 1, the present chapter goes far beyond that discussed in Chapter 1. As you read through this chapter, as well as the subsequent chapters, you will find a small amount of repetition of a few thoughts introduced in earlier chapters. This is intentional. I want to make Part 4 of this book a stand-alone, almost self-contained presentation of viscous flow. Besides, a little bit of repetition never hurt anybody, and in the educational process it can help a lot. So jump into this chapter and immerse yourself in the ideas and thoughts of viscous flow. Come into the real world.

15.1 INTRODUCTION

In the above quotation, the "theory" referred to by d'Alembert is inviscid, incompressible flow theory; we have seen in Chapter 3 that such theory leads to a prediction of zero drag on a closed two-dimensional body—this is d'Alembert's paradox. In reality, there is always a finite drag on any body immersed in a moving fluid. Our earlier predictions of zero drag are a result of the inadequacy of the theory rather than some fluke of nature. With the exception of induced drag and supersonic wave drag, which can be obtained from inviscid theory, the calculation of all other forms of drag must explicitly take into account the presence of viscosity, which has not been included in our previous inviscid analyses. The purpose of the remaining chapters in this book is to discuss the basic aspects of viscous flows, thus "rounding out" our overall presentation of the fundamentals of aerodynamics. In so doing, we address the predictions of aerodynamic drag and aerodynamic heating. To help put our current discussion in perspective, return to the block diagram of flow categories given in Figure 1.45. All of our previous discussions have focused on blocks D, E, and F—inviscid, incompressible, and compressible flows. Now, for the remaining six chapters, we move to the left branch in Figure 1.45, and deal with blocks C, E, and F—*viscous*, incompressible, and compressible flows.

Our treatment of viscous flows will be intentionally brief—our purpose is to present enough of the fundamental concepts and equations to give you the flavor of viscous flows. A thorough presentation of viscous flow theory would double the size of this book (at the very least) and is clearly beyond our scope. A study of viscous flow is an essential part of any serious study of aerodynamics. Many books have been exclusively devoted to the presentation of viscous flows; References 40 and 41 are two good examples. You are encouraged to examine these references closely.

There have been sections on viscous flow topics earlier in this book, namely, Sections 1.11, 4.12, 9.10, 10.6, and 12.4. These are stand-alone sections dealing with viscous flow aspects pertinent to the chapters in which they appear. You, the reader, have had two choices: (1) to read these viscous flow sections in order to discover how friction has an impact on some of the ideal inviscid flows discussed in the main body of the chapter, or (2) to by-pass these sections in order to preserve the intellectual continuity of a study of inviscid flows that is, after all, the main thrust of Parts 2 and 3 of this book. Now, we are at Part 4 dealing exclusively with viscous flow. Readers who took the first choice will find some slight

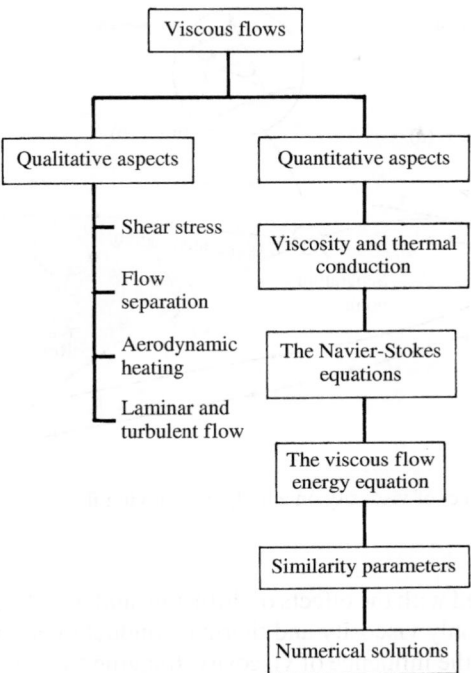

Figure 15.1 Road map for Chapter 15.

repetition in Part 4, but repetition is a good thing in learning a new subject. Readers who took the second choice will find Part 4 a totally self-contained discussion of viscous flow that does not depend on reading the earlier viscous flow sections; however, at appropriate stages of our discussions you will find direct references to specific earlier sections that fit nicely into the continuity of Part 4.

The road map for the present chapter is given in Figure 15.1. Our course is to first examine some qualitative aspects of viscous flows as shown on the left branch of Figure 15.1. Then we quantify some of these aspects as given on the right branch. In the process, we obtain the governing equations for a general viscous flow—in particular, the Navier-Stokes equations (the momentum equations) and the viscous flow energy equation. Finally, we examine a numerical solution to these equations.

15.2 QUALITATIVE ASPECTS OF VISCOUS FLOW

What is a viscous flow? *Answer:* A flow where the effects of viscosity, thermal conduction, and mass diffusion are important. The phenomenon of mass diffusion is important in a gas with gradients in its chemical species, for example, the flow of air over a surface through which helium is being injected or the chemically reacting flow through a jet engine or over a high-speed reentry body. In this book,

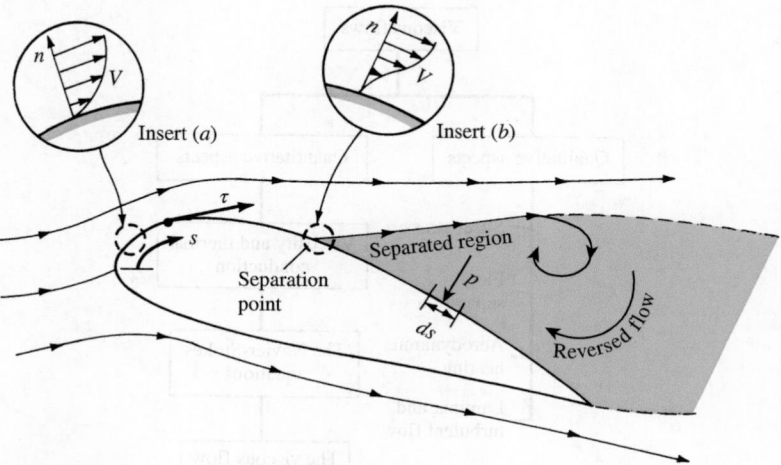

Figure 15.2 Effect of viscosity on a body in a moving fluid: shear stress and separated flow.

we are not concerned with the effects of diffusion, and therefore we treat a viscous flow as one where only viscosity and thermal conduction are important.

First, consider the influence of viscosity. Imagine two solid surfaces slipping over each other, such as this book being pushed across a table. Clearly, there will be a frictional force between these objects which will retard their relative motion. The same is true for the flow of a fluid over a solid surface; the influence of friction between the surface and the fluid adjacent to the surface acts to create a frictional force which retards the relative motion. This has an effect on both the surface and the fluid. The surface feels a "tugging" force in the direction of the flow, tangential to the surface. This tangential force per unit area is defined as the *shear stress* τ, first introduced in Section 1.5 and illustrated in Figure 15.2. As an equal and opposite reaction, the fluid adjacent to the surface feels a retarding force which decreases its local flow velocity, as shown in insert *a* of Figure 15.2. Indeed, the influence of friction is to create $V = 0$ right at the body surface— this is called the *no-slip* condition which dominates viscous flow. In any real continuum fluid flow over a solid surface, the flow velocity is zero at the surface. Just above the surface, the flow velocity is finite, but retarded, as shown in insert *a*. If n represents the coordinate normal to the surface, then in the region near the surface, $V = V(n)$, where $V = 0$ at $n = 0$, and V increases as n increases. The plot of V versus n as shown in insert *a* is called a *velocity profile*. Clearly, the region of flow near the surface has velocity gradients, $\partial V/\partial n$, which are due to the frictional force between the surface and the fluid.

In addition to the generation of shear stress, friction also plays another (but related) role in dictating the flow over the body in Figure 15.2. Consider a fluid element moving in the viscous flow near a surface, as sketched in Figure 15.3. Assume that the flow is in its earliest moments of being started. At the station s_1,

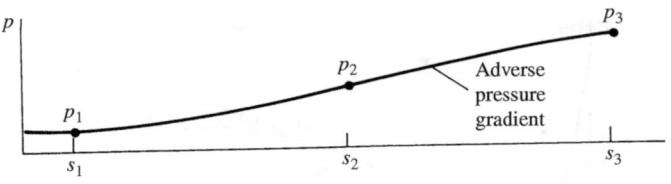

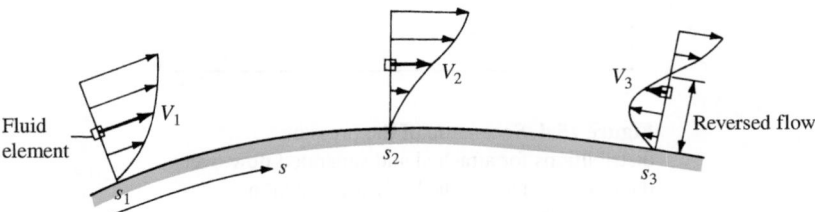

Figure 15.3 Separated flow induced by an adverse pressure gradient. This picture corresponds to the early evolution of the flow; once the flow separates from the surface between points 2 and 3, the fluid element shown at s_3 is in reality different from that shown at s_1 and s_2 because the primary flow moves away from the surface, as shown in Figure 15.2.

the velocity of the fluid element is V_1. Assume that the flow over the surface produces an increasing pressure distribution in the flow direction (i.e., assume $p_3 > p_2 > p_1$). Such a region of increasing pressure is called an *adverse pressure gradient*. Now follow the fluid element as it moves downstream. The motion of the element is already retarded by the effect of friction; in addition, it must work its way along the flow against an increasing pressure, which tends to further reduce its velocity. Consequently, at station 2 along the surface, its velocity V_2 is less than V_1. As the fluid element continues to move downstream, it may completely "run out of steam," come to a stop, and then, under the action of the adverse pressure gradient, actually reverse its direction and start moving back upstream. This "reversed flow" is illustrated at station s_3 in Figure 15.3, where the fluid element is now moving upstream at the velocity V_3. The picture shown in Figure 15.3 is meant to show the flow details very near the surface at the very initiation of the flow. In the bigger picture of this flow at later times shown in Figure 15.2, the consequence of such reversed-flow phenomena is to cause the flow to *separate from the surface* and create a large wake of recirculating flow downstream of the surface. The point of separation on the surface in Figure 15.2 occurs where $\partial V / \partial n = 0$ at the surface, as sketched in insert b of Figure 15.2. Beyond this point, reversed flow occurs. Therefore, in addition to the generation of shear stress, the influence of friction can cause the flow over a body to separate from the surface. When such separated flow occurs, the pressure distribution over the surface is greatly altered. The primary flow over the body in Figure 15.2 no longer sees the complete body shape; rather, it sees the body shape upstream

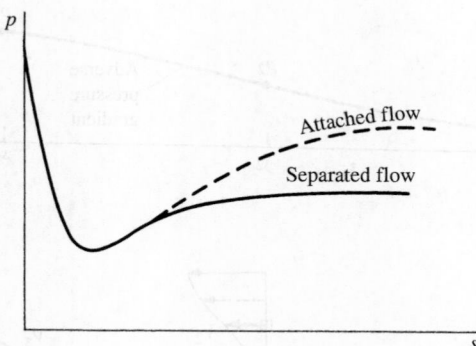

Figure 15.4 Schematic of the pressure distributions for attached and separated flow over the upper surface of the body illustrated in Figure 15.2.

of the separation point, but downstream of the separation point it sees a greatly deformed "effective body" due to the large separated region. The net effect is to create a pressure distribution over the actual body surface which results in an integrated force in the flow direction, that is, a drag. To see this more clearly, consider the pressure distribution over the upper surface of the body as sketched in Figure 15.4. If the flow were attached, the pressure over the downstream portion of the body would be given by the dashed curve. However, for separated flow, the pressure over the downstream portion of the body is smaller, as given by the solid curve in Figure 15.4. Now return to Figure 15.2. Note that the pressure over the upper rearward surface contributes a force in the negative drag direction; that is, p acting over the element of surface ds shown in Figure 15.2 has a horizontal component in the upstream direction. If the flow were inviscid, subsonic, and attached and the body were two-dimensional, the forward-acting components of the pressure distribution shown in Figure 15.2 would exactly cancel the rearward-acting components due to the pressure distribution over other parts of the body such that the net, integrated pressure distribution would give zero drag. This would be d'Alembert's paradox discussed in Chapter 3. However, for the viscous, separated flow, we see that p is reduced in the separated region; hence, it can no longer fully cancel the pressure distribution over the remainder of the body. The net result is the production of drag; this is called the *pressure drag due to flow separation* and is denoted by D_p.

In summary, we see that the effects of viscosity are to produce two types of drag as follows:

D_f is the skin friction drag, that is, the component in the drag direction of the integral of the shear stress τ over the body.

D_p is the pressure drag due to separation, that is, the component in the drag direction of the integral of the pressure distribution over the body.

D_p is sometimes called *form drag*. The sum $D_f + D_p$ is called the *profile drag* of a two-dimensional body. For a three-dimensional body such as a complete airplane, the sum $D_f + D_p$ is frequently called *parasite drag*. (See Reference 2 for a more extensive discussion of the classification of different drag contributions.)

The occurrence of separated flow over an aerodynamic body not only increases the drag but also results in a substantial loss of lift. Such separated flow is the cause of airfoil stall as discussed in Section 4.3. For these reasons, the study, understanding, and prediction of separated flow is an important aspect of viscous flow.

Let us turn our attention to the influence of thermal conduction—another overall physical characteristic of viscous flow in addition to friction. Again, let us draw an analogy from two solid bodies slipping over each other, such as the motion of this book over the top of a table. If we would press hard on the book, and vigorously rub it back and forth over the table, the cover of the book as well as the table top would soon become warm. Some of the energy we expend in pushing the book over the table will be dissipated by friction, and this shows up as a form of heating of the bodies. The same phenomenon occurs in the flow of a fluid over a body. The moving fluid has a certain amount of kinetic energy; in the process of flowing over a surface, the flow velocity is decreased by the influence of friction, as discussed earlier, and hence the kinetic energy is decreased. This lost kinetic energy reappears in the form of internal energy of the fluid, hence causing the temperature to rise. This phenomenon is called *viscous dissipation* within the fluid. In turn, when the fluid temperature increases, there is an overall temperature difference between the warmer fluid and the cooler body. We know from experience that heat is transferred from a warmer body to a cooler body; therefore, heat will be transferred from the warmer fluid to the cooler surface. This is the mechanism of *aerodynamic heating* of a body. Aerodynamic heating becomes more severe as the flow velocity increases, because more kinetic energy is dissipated by friction, and hence the overall temperature difference between the warm fluid and the cool surface increases. As discussed in Chapter 14, at hypersonic speeds, aerodynamic heating becomes a dominant aspect of the flow.

All the aspects discussed above—shear stress, flow separation, aerodynamic heating, etc.—are dominated by a single major question in viscous flow, namely, Is the flow laminar or turbulent? Consider the viscous flow over a surface as sketched in Figure 15.5. If the path lines of various fluid elements are smooth

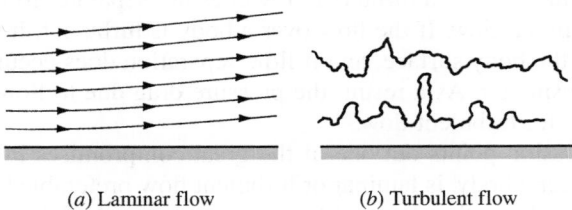

(*a*) Laminar flow   (*b*) Turbulent flow

Figure 15.5 Path lines for laminar and turbulent flows.

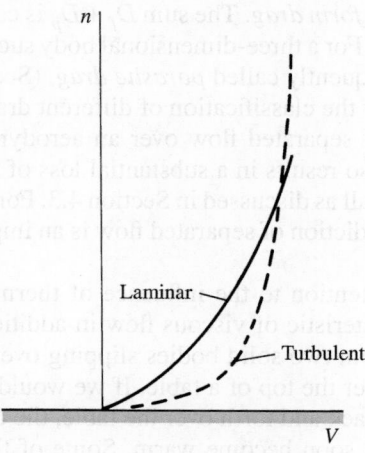

Figure 15.6 Schematic of velocity profiles for laminar and turbulent flows.

and regular, as shown in Figure 15.5*a*, the flow is called *laminar flow*. In contrast, if the motion of a fluid element is very irregular and tortuous, as shown in Figure 15.5*b*, the flow is called *turbulent flow*. Because of the agitated motion in a turbulent flow, the higher-energy fluid elements from the outer regions of the flow are pumped close to the surface. Hence, the average flow velocity near a solid surface is larger for a turbulent flow in comparison with laminar flow. This comparison is shown in Figure 15.6, which gives velocity profiles for laminar and turbulent flow. Note that immediately above the surface, the turbulent flow velocities are much larger than the laminar values. If $(\partial V/\partial n)_{n=0}$ denotes the velocity gradient at the surface, we have

$$\left[\left(\frac{\partial V}{\partial n} \right)_{n=0} \right]_{\text{turbulent}} > \left[\left(\frac{\partial V}{\partial n} \right)_{n=0} \right]_{\text{laminar}}$$

Because of this difference, the frictional effects are more severe for a turbulent flow; both the shear stress and aerodynamic heating are larger for the turbulent flow in comparison with laminar flow. However, turbulent flow has a major redeeming value; because the energy of the fluid elements close to the surface is larger in a turbulent flow, a turbulent flow does not separate from the surface as readily as a laminar flow. If the flow over a body is turbulent, it is less likely to separate from the body surface, and if flow separation does occur, the separated region will be smaller. As a result, the pressure drag due to flow separation D_p will be smaller for turbulent flow.

This discussion points out one of the great compromises in aerodynamics. For the flow over a body, is laminar or turbulent flow preferable? There is no pat answer; it depends on the shape of the body. In general, if the body is slender, as sketched in Figure 15.7*a*, the friction drag D_f is much greater than D_p. For

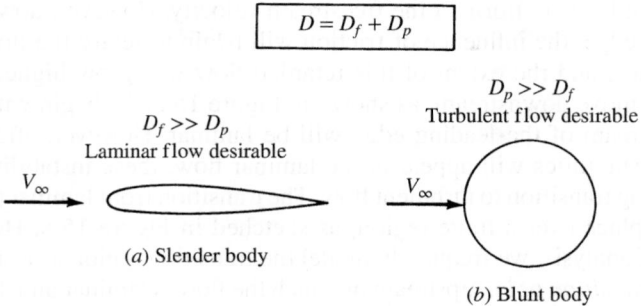

Figure 15.7 Drag on slender and blunt bodies.

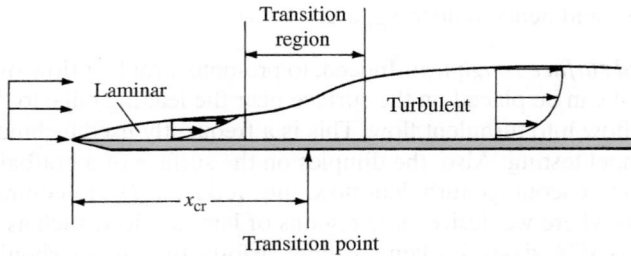

Figure 15.8 Transition from laminar to turbulent flow.

this case, because D_f is smaller for laminar than for turbulent flow, laminar flow is desirable for slender bodies. In contrast, if the body is blunt, as sketched in Figure 15.7b, D_p is much greater than D_f. For this case, because D_p is smaller for turbulent than for laminar flow, turbulent flow is desirable for blunt bodies. The above comments are not all-inclusive; they simply state general trends, and for any given body, the aerodynamic virtues of laminar versus turbulent flow must always be assessed.

Although, from the above discussion, laminar flow is preferable for some cases, and turbulent flow for other cases, in reality we have little control over what actually happens. Nature makes the ultimate decision as to whether a flow will be laminar or turbulent. There is a general principle in nature that a system, when left to itself, will always move toward its state of maximum disorder. To bring order to the system, we generally have to exert some work on the system or expend energy in some manner. (This analogy can be carried over to daily life; a room will soon become cluttered and disordered unless we exert some effort to keep it clean.) Since turbulent flow is much more "disordered" than laminar flow, nature will always favor the occurrence of turbulent flow. Indeed, in the vast majority of practical aerodynamic problems, turbulent flow is usually present.

Let us examine this phenomenon in more detail. Consider the viscous flow over a flat plate, as sketched in Figure 15.8. The flow immediately upstream of the leading edge is uniform at the freestream velocity. However, downstream of the leading edge, the influence of friction will begin to retard the flow adjacent to the surface, and the extent of this retarded flow will grow higher above the plate as we move downstream, as shown in Figure 15.8. To begin with, the flow just downstream of the leading edge will be laminar. However, after a certain distance, instabilities will appear in the laminar flow; these instabilities rapidly grow, causing transition to turbulent flow. The transition from laminar to turbulent flow takes place over a finite region, as sketched in Figure 15.8. However, for purposes of analysis, we frequently model the transition region as a single point, called the *transition point,* upstream of which the flow is laminar and downstream of which the flow is turbulent. The distance from the leading edge to the transition point is denoted by x_{cr}. The value of x_{cr} depends on a whole host of phenomena. For example, some characteristics which encourage transition from laminar to turbulent flow, and hence reduce x_{cr}, are:

1. *Increased surface roughness.* Indeed, to promote turbulent flow over a body, rough grit can be placed on the surface near the leading edge to "trip" the laminar flow into turbulent flow. This is a frequently used technique in wind-tunnel testing. Also, the dimples on the surface of a golfball are designed to encourage turbulent flow, thus reducing D_p. In contrast, in situations where we desire large regions of laminar flow, such as the flow over the NACA six-series laminar-flow airfoils, the surface should be as smooth as possible. The main reason why such airfoils do not produce in actual flight the large regions of laminar flow observed in the laboratory is that manufacturing irregularities and bug spots (believe it or not) roughen the surface and promote early transition to turbulent flow.

2. *Increased turbulence in the freestream.* This is particularly a problem in wind-tunnel testing; if two wind tunnels have different levels of freestream turbulence, then data generated in one tunnel are not repeatable in the other.

3. *Adverse pressure gradients.* In addition to causing flow-field separation as discussed earlier, an adverse pressure gradient strongly favors transition to turbulent flow. In contrast, strong favorable pressure gradients (where p decreases in the downstream direction) tend to preserve initially laminar flow.

4. *Heating of the fluid by the surface.* If the surface temperature is warmer than the adjacent fluid, such that heat is transferred to the fluid from the surface, the instabilities in the laminar flow will be amplified, thus favoring early transition. In contrast, a cold wall will tend to encourage laminar flow.

There are many other parameters which influence transition; see Reference 40 for a more extensive discussion. Among these are the similarity parameters of the flow, principally Mach number and Reynolds number. High values of M_∞ and low values of Re tend to encourage laminar flow; hence, for high-altitude hypersonic flight, laminar flow can be quite extensive. The Reynolds number itself is a dominant factor in transition to turbulent flow. Referring to Figure 15.8, we define a *critical* Reynolds number, Re_{cr}, as

$$\text{Re}_{\text{cr}} \equiv \frac{\rho_\infty V_\infty x_{\text{cr}}}{\mu_\infty}$$

The value of Re_{cr} for a given body under specified conditions is difficult to predict; indeed, the analysis of transition is still a very active area of modern aerodynamic research. As a rule of thumb in practical applications, we frequently take $\text{Re}_{\text{cr}} \approx$ 500,000; if the flow at a given x station is such that $\text{Re} = \rho_\infty V_\infty x / \mu_\infty$ is considerably below 500,000, then the flow at that station is most likely laminar, and if the value of Re is much larger than 500,000, then the flow is most likely turbulent.

To obtain a better feeling for Re_{cr}, let us imagine that the flat plate in Figure 15.8 is a wind-tunnel model. Assume that we carry out an experiment under standard sea level conditions [$\rho_\infty = 1.23$ kg/m^3 and $\mu_\infty = 1.79 \times 10^{-5}$ kg/(m $\cdot$ s)] and *measure* x_{cr} for a certain freestream velocity; for example, say that $x_{\text{cr}} = 0.05$ m when $V_\infty = 120$ m/s. In turn, this measured value of x_{cr} determines the measured Re_{cr} as

$$\text{Re}_{\text{cr}} = \frac{\rho_\infty V_\infty x_{\text{cr}}}{\mu_\infty} = \frac{1.23(120)(0.05)}{1.79 \times 10^{-5}} = 412{,}000$$

Hence, for the given flow conditions and the surface characteristics of the flat plate, transition will occur whenever the local Re exceeds 412,000. For example, if we double V_∞, that is, $V_\infty = 240$ m/s, then we will observe transition to occur at $x_{\text{cr}} = 0.05/2 = 0.025$ m, such that Re_{cr} remains the same value of 412,000.

This brings to an end our introductory qualitative discussion of viscous flow. The physical principles and trends discussed in this section are very important, and you should study them carefully and feel comfortable with them before progressing further.

15.3 VISCOSITY AND THERMAL CONDUCTION

The basic physical phenomena of viscosity and thermal conduction in a fluid are due to the transport of momentum and energy via random molecular motion. Each molecule in a fluid has momentum and energy, which it carries with it when it moves from one location to another in space before colliding with another molecule. The transport of molecular momentum gives rise to the macroscopic effect we call viscosity, and the transport of molecular energy gives rise to the macroscopic effect we call thermal conduction. This is why viscosity and thermal conduction are labeled as *transport phenomena*. A study of these transport phenomena at the molecular level is part of kinetic theory, which is beyond the scope of this book. Instead, in this section we simply state the macroscopic results of such molecular motion.

Consider the flow sketched in Figure 15.9. For simplicity, we consider a one-dimensional shear flow, that is, a flow with horizontal streamlines in the x direction but with gradients in the y direction of velocity, $\partial u / \partial y$, and temperature,

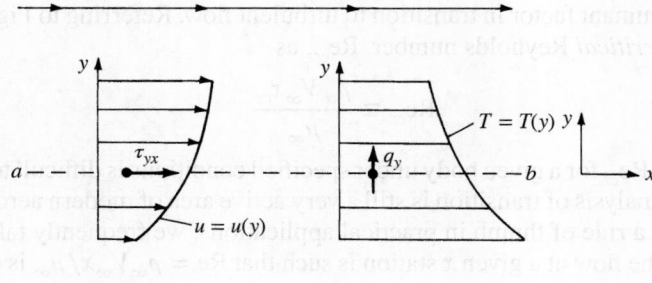

Figure 15.9 Relationship of shear stress and thermal conduction to velocity and temperature gradients, respectively.

$\partial T/\partial y$. Consider a plane ab perpendicular to the y axis, as shown in Figure 15.9. The shear stress exerted on plane ab by the flow is denoted by τ_{yx} and is proportional to the velocity gradient in the y direction, $\tau_{yx} \propto \partial u/\partial y$. The constant of proportionality is defined as the *viscosity coefficient* μ. Hence,

$$\tau_{yx} = \mu \frac{\partial u}{\partial y} \tag{15.1}$$

The subscripts on τ_{yx} denote that the shear stress is acting in the x direction and is being exerted on a plane perpendicular to the y axis. The velocity gradient $\partial u/\partial y$ is also taken perpendicular to this plane (i.e., in the y direction). The dimensions of μ are mass/length × time, as originally stated in Section 1.7 and as can be seen from Equation (15.1). In addition, the time rate of heat conducted per unit area across plane ab in Figure 15.9 is denoted by $\dot{q}_y$ and is proportional to the temperature gradient in the y direction, $\dot{q}_y \propto \partial T/\partial y$. The constant of proportionality is defined as the *thermal conductivity* k. Hence,

$$\dot{q}_y = -k \frac{\partial T}{\partial y} \tag{15.2}$$

where the minus sign accounts for the fact that the heat is transferred from a region of high temperature to a region of lower temperature; that is, $\dot{q}_y$ is in the opposite direction of the temperature gradient. The dimensions of k are mass × length/(s² · K), which can be obtained from Equation (15.2) keeping in mind that $\dot{q}_y$ is energy per second per unit area.

Both μ and k are physical properties of the fluid and, for most normal situations, are functions of temperature only. A conventional relation for the temperature variation of μ for air is given by Sutherland's law,

$$\frac{\mu}{\mu_0} = \left(\frac{T}{T_0}\right)^{3/2} \frac{T_0 + 110}{T + 110} \tag{15.3}$$

where T is in kelvin and μ_0 is a reference viscosity at a reference temperature, T_0. For example, if we choose reference conditions to be standard sea level values, then $\mu_0 = 1.7894 \times 10^{-5}$ kg/(m · s) and $T_0 = 288.16$ K. The temperature variation of k is analogous to Equation (15.3) because the results of elementary kinetic theory show that $k \propto \mu c_p$; for air at standard conditions,

$$k = 1.45 \mu c_p \tag{15.4}$$

where $c_p = 1000$ J/(kg · K).

Equations (15.3) and (15.4) are only approximate and do not hold at high temperatures. They are given here as representative expressions which are handy to use. For any detailed viscous flow calculation, you should consult the published literature for more precise values of μ and k.

In order to simplify our introduction of the relation between shear stress and viscosity, we considered the case of a one-dimensional shear flow in Figure 15.9. In this picture, the y and z components of velocity, v and w, respectively, are zero. However, in a general three-dimensional flow, u, v, and w are finite, and this requires a generalization of our treatment of stress in the fluid. Consider the fluid element sketched in Figure 15.10. In a three-dimensional flow, each face of the fluid element experiences both tangential and normal stresses. For example, on face *abcd*, τ_{xy} and τ_{xz} are the tangential stresses, and τ_{xx} is the normal stress. As before, the nomenclature τ_{ij} denotes a stress in the j direction exerted on a plane perpendicular to the i axis. Similarly, on face *abfe*, we have the tangential stresses τ_{yx} and τ_{yz}, and the normal stress τ_{yy}. On face *adge*, we have the tangential stresses τ_{zx} and τ_{zy}, and the normal stress τ_{zz}. Now recall the discussion in the last part of Section 2.12 concerning the *strain* of a fluid element, that is, the change in the angle κ shown in Figure 2.33. What is the force which causes this deformation shown in Figure 2.33? Returning to Figure 15.10, we have to say that the strain

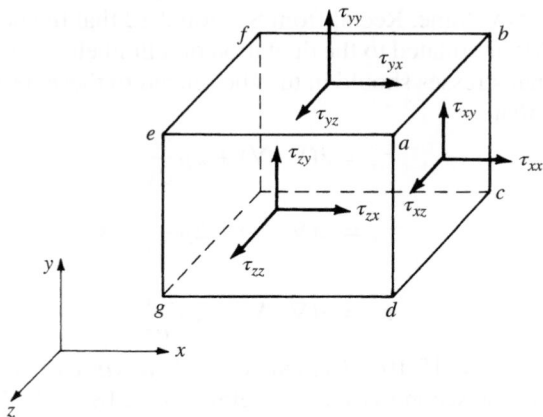

Figure 15.10 Shear and normal stresses caused by viscous action on a fluid element.

is caused by the tangential shear stress. However, in contrast to solid mechanics where stress is proportional to strain, in fluid mechanics the stress is proportional to the *time rate of strain*. The time rate of strain in the xy plane was given in Section 2.12 as Equation (2.135a):

$$\varepsilon_{xy} = \frac{\partial v}{\partial x} + \frac{\partial u}{\partial y} \qquad (2.135a)$$

Examining Figure 15.10, the strain in the xy plane must be carried out by τ_{xy} and τ_{yx}. Moreover, we assume that moments on the fluid element in Figure 15.10 are zero; hence, $\tau_{xy} = \tau_{yx}$. Finally, from the above, we know that $\tau_{xy} = \tau_{yx} \propto \varepsilon_{xy}$. The proportionality constant is the viscosity coefficient μ. Hence, from Equation (2.135a), we have

$$\tau_{xy} = \tau_{yx} = \mu \left(\frac{\partial v}{\partial x} + \frac{\partial u}{\partial y} \right) \qquad (15.5)$$

which is a generalization of Equation (15.1), extended to the case of multidimensional flow. For the shear stresses in the other planes, Equation (2.135b and c) yield

$$\tau_{yz} = \tau_{zy} = \mu \left(\frac{\partial w}{\partial y} + \frac{\partial v}{\partial z} \right) \qquad (15.6)$$

and

$$\tau_{zx} = \tau_{xz} = \mu \left(\frac{\partial u}{\partial z} + \frac{\partial w}{\partial x} \right) \qquad (15.7)$$

The normal stresses τ_{xx}, τ_{yy}, and τ_{zz} shown in Figure 15.10 may at first seem strange. In our previous treatments of inviscid flow, the only force normal to a surface in a fluid is the pressure force. However, if the gradients in velocity $\partial u / \partial x$, $\partial v / \partial y$, and $\partial w / \partial z$ are *extremely large* on the faces of the fluid element, there can be a meaningful viscous-induced normal force on each face which acts *in addition to* the pressure. These normal stresses act to compress or expand the fluid element, hence changing its volume. Recall from Section 2.12 that the derivatives $\partial u / \partial x$, $\partial v / \partial y$, and $\partial w / \partial z$ are related to the dilatation of a fluid element, that is, to $\nabla \cdot \mathbf{V}$. Hence, the normal stresses should in turn be related to these derivatives. Indeed, it can be shown that

$$\tau_{xx} = \lambda (\nabla \cdot \mathbf{V}) + 2\mu \frac{\partial u}{\partial x} \qquad (15.8)$$

$$\tau_{yy} = \lambda (\nabla \cdot \mathbf{V}) + 2\mu \frac{\partial v}{\partial y} \qquad (15.9)$$

$$\tau_{zz} = \lambda (\nabla \cdot \mathbf{V}) + 2\mu \frac{\partial w}{\partial z} \qquad (15.10)$$

In Equations (15.8) to (15.10), λ is called the *bulk viscosity coefficient,* sometimes identified as the second viscosity coefficient. In 1845, the English physicist George Stokes hypothesized that

$$\lambda = -\frac{2}{3}\mu \qquad (15.11)$$

To this day, the correct expression for the bulk viscosity is still somewhat controversial, and so we continue to use the above expression given by Stokes. Once again, the normal stresses are important only where the derivatives $\partial u/\partial x$, $\partial v/\partial y$, and $\partial w/\partial z$ are very large. For most practical flow problems, τ_{xx}, τ_{yy}, and τ_{zz} are small, and hence the uncertainty regarding λ is essentially an academic question. An example where the normal stress is important is inside the internal structure of a shock wave. Recall that, in real life, shock waves have a finite but small thickness. If we consider a normal shock wave across which large changes in velocity occur over a small distance (typically 10^{-5} cm), then clearly $\partial u/\partial x$ will be very large, and τ_{xx} becomes important inside the shock wave.

To this point in our discussion, the transport coefficients μ and k have been considered molecular phenomena, involving the transport of momentum and energy by random molecular motion. This molecular picture prevails in a laminar flow. The values of μ and k are physical properties of the fluid; that is, their values for different gases can be found in standard reference sources, such as the *Handbook of Chemistry and Physics* (The Chemical Rubber Co.). In contrast, for a turbulent flow the transport of momentum and energy can also take place by random motion of large turbulent eddies, or globs of fluid. This turbulent transport gives rise to effective values of viscosity and thermal conductivity defined as *eddy viscosity* ε and *eddy thermal conductivity* κ, respectively. (Please do not confuse this use of the symbols ε and κ with the time rate of strain and strain itself, as used earlier.) These turbulent transport coefficients ε and κ can be much larger (typically 10 to 100 times larger) than the respective molecular values μ and k. Moreover, ε and κ predominantly depend on characteristics of the flow field, such as velocity gradients; they are not just a molecular property of the fluid, such as μ and k. The proper calculation of ε and κ for a given flow has remained a state-of-the-art research question for the past 80 years; indeed, the attempt to model the complexities of turbulence by defining an eddy viscosity and thermal conductivity is even questionable. The details and basic understanding of turbulence remain one of the greatest unsolved problems in physics today. For our purpose here, we simply adopt the ideas of eddy viscosity and eddy thermal conductivity, and for the transport of momentum and energy in a turbulent flow, we replace μ and k in Equations (15.1) to (15.10) by the combination $\mu + \varepsilon$ and $k + \kappa$; that is,

$$\tau_{yx} = (\mu + \varepsilon)\left(\frac{\partial v}{\partial x} + \frac{\partial u}{\partial y}\right)$$

$$\dot{q}_y = -(k + \kappa)\frac{\partial T}{\partial y}$$

An example of the calculation of ε and κ is as follows. In 1925, Prandtl suggested that

$$\varepsilon = \rho l^2 \left|\frac{\partial u}{\partial y}\right| \tag{15.12}$$

for a flow where the dominant velocity gradient is in the y direction. In Equation (15.12), l is called the *mixing length,* which is different for different applications; it is an empirical constant which must be obtained from experiment. Indeed, *all* turbulence models require the input of empirical data; no self-contained purely theoretical turbulence model exists today. Prandtl's mixing length theory, embodied in Equation (15.12), is a simple relation which appears to be adequate for a number of engineering problems. For these reasons, the mixing length model for ε has been used extensively since 1925. In regard to κ, a relation similar to Equation (15.4) can be assumed (using 1.0 for the constant); that is,

$$\kappa = \varepsilon c_p \qquad (15.13)$$

The comments on eddy viscosity and thermal conductivity are purely introductory. The modern aerodynamicist has a whole stable of turbulence models to choose from, and before tackling the analysis of a turbulent flow, you should be familiar with the modern approaches described in such books as References 40 to 43.

15.4 THE NAVIER-STOKES EQUATIONS

In Chapter 2, Newton's second law was applied to obtain the fluid-flow momentum equation in both integral and differential forms. In particular, recall Equation (2.13a to c), where the influence of viscous forces was expressed simply by the generic terms $(\mathcal{F}_x)_{\text{viscous}}$, $(\mathcal{F}_y)_{\text{viscous}}$, and $(\mathcal{F}_z)_{\text{viscous}}$. The purpose of this section is to obtain the analogous forms of Equation (2.13a to c) where the viscous forces are expressed explicitly in terms of the appropriate flow-field variables. The resulting equations are called the *Navier-Stokes equations*—probably the most pivotal equations in all of theoretical fluid dynamics.

In Section 2.3, we discussed the philosophy behind the derivation of the governing equations, namely, certain physical principles are applied to a suitable *model* of the fluid flow. Moreover, we saw that such a model could be either a finite control volume (moving or fixed in space) or an infinitesimally small element (moving or fixed in space). In Chapter 2, we chose the fixed, finite control volume for our model and obtained integral forms of the continuity, momentum, and energy equations directly from this model. Then, indirectly, we went on to extract partial differential equations from the integral forms. Before progressing further, it would be wise for you to review these matters from Chapter 2.

For the sake of variety, let us not use the fixed, finite control volume employed in Chapter 2; rather, in this section, let us adopt an infinitesimally small moving fluid element of fixed mass as our model of the flow, as sketched in Figure 15.11. To this model, let us apply Newton's second law in the form $\mathbf{F} = m\mathbf{a}$. Moreover, for the time being consider only the x component of Newton's second law:

$$F_x = ma_x \qquad (15.14)$$

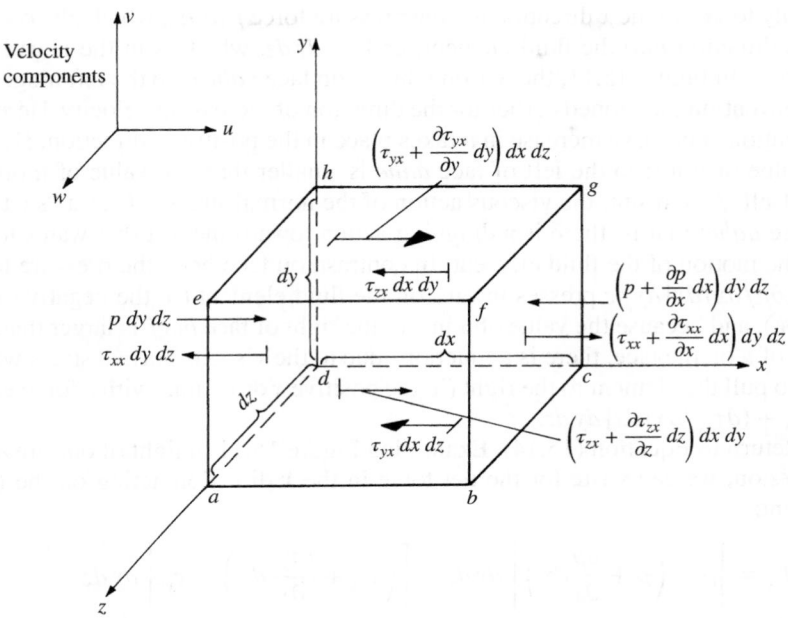

Figure 15.11 Infinitesimally small, moving fluid element. Only the forces in the
x direction are shown.

In Equation (15.14), F_x is the sum of all the body and surface forces acting on
the fluid element in the x direction. Let us ignore body forces; hence, the net force
acting on the element in Figure 15.11 is simply due to the pressure and viscous
stress distributions over the surface of the element. For example, on face *abcd*,
the only force in the x direction is that due to shear stress, $\tau_{yx}\,dx\,dz$. Face *efgh* is a
distance dy above face *abcd*; hence, the shear force in the x direction on face *efgh*
is $[\tau_{yx} + (\partial\tau_{yx}/\partial y)\,dy]\,dx\,dz$. Note the directions of the shear stress on faces *abcd*
and *efgh*; on the bottom face, τ_{yx} is to the left (the negative x direction), whereas
on the top face, $\tau_{yx} + (\partial\tau_{yx}/\partial y)\,dy$ is to the right (the positive x direction). These
directions are due to the convention that positive increases in all three compo-
nents of velocity, u, v, and w, occur in the positive directions of the axes. For
example, in Figure 15.11, u increases in the positive y direction. Therefore, concentrating on face *efgh*, u is higher just above the face than on the face; this causes
a "tugging" action which tries to pull the fluid element in the positive x direction
(to the right) as shown in Figure 15.11. In turn, concentrating on face *abcd*, u is
lower just beneath the face than on the face; this causes a retarding or dragging
action on the fluid element, which acts in the negative x direction (to the left), as
shown in Figure 15.11. The directions of all the other viscous stresses shown in
Figure 15.11, including τ_{xx}, can be justified in a like fashion. Specifically, on face
dcgh, τ_{zx} acts in the negative x direction, whereas on face *abfe*, $\tau_{zx} + (\partial\tau_{zx}/\partial z)\,dz$
acts in the positive x direction. On face *adhe*, which is perpendicular to the x axis,

the only forces in the x direction are the pressure force $p\,dy\,dz$, which always acts in the direction *into* the fluid element, and $\tau_{xx}\,dy\,dz$, which is in the negative x direction. In Figure 15.11, the reason why τ_{xx} on face *adhe* is to the left hinges on the convention mentioned earlier for the direction of increasing velocity. Here, by convention, a positive increase in u takes place in the positive x direction. Hence, the value of u just to the left of face *adhe* is smaller than the value of u on the face itself. As a result, the viscous action of the normal stress acts as a "suction" on face *adhe*; that is, there is a dragging action toward the left that wants to retard the motion of the fluid element. In contrast, on face *bcgf*, the pressure force $[p + (\partial p/\partial x)\,dx]\,dy\,dz$ presses inward on the fluid element (in the negative x direction), and because the value of u just to the right of face *bcgf* is larger than the value of u on the face, there is a "suction" due to the viscous normal stress which tries to pull the element to the right (in the positive x direction) with a force equal to $[\tau_{xx} + (\partial \tau_{xx}/\partial x)\,dx]\,dy\,dz$.

Return to Equation (15.14). Examining Figure 15.11 in light of our previous discussion, we can write for the net force in the x direction acting on the fluid element:

$$F_x = \left[p - \left(p + \frac{\partial p}{\partial x}dx\right)\right] dy\,dz + \left[\left(\tau_{xx} + \frac{\partial \tau_{xx}}{\partial x}dx\right) - \tau_{xx}\right] dy\,dz$$

$$+ \left[\left(\tau_{yx} + \frac{\partial \tau_{yx}}{\partial y}dy\right) - \tau_{yx}\right] dx\,dz + \left[\left(\tau_{zx} + \frac{\partial \tau_{zx}}{\partial z}dz\right) - \tau_{zx}\right] dx\,dy$$

or
$$F_x = \left(-\frac{\partial p}{\partial x} + \frac{\partial \tau_{xx}}{\partial x} + \frac{\partial \tau_{yx}}{\partial y} + \frac{\partial \tau_{zx}}{\partial z}\right) dx\,dy\,dz \qquad (15.15)$$

Equation (15.15) represents the left-hand side of Equation (15.14). Considering the right-hand side of Equation (15.14), recall that the mass of the fluid element is fixed and is equal to

$$m = \rho\,dx\,dy\,dz \qquad (15.16)$$

Also, recall that the acceleration of the fluid element is the time rate of change of its velocity. Hence, the component of acceleration in the x direction, denoted by a_x, is simply the time rate of change of u; since we are following a moving fluid element, this time rate of change is given by the *substantial derivative* (see Section 2.9 for a review of the meaning of substantial derivative). Thus,

$$a_x = \frac{Du}{Dt} \qquad (15.17)$$

Combining Equations (15.14) to (15.17), we obtain

$$\rho\frac{Du}{Dt} = -\frac{\partial p}{\partial x} + \frac{\partial \tau_{xx}}{\partial x} + \frac{\partial \tau_{yx}}{\partial y} + \frac{\partial \tau_{zx}}{\partial z} \qquad (15.18a)$$

which is the x component of the momentum equation for a viscous flow. In a similar fashion, the y and z components can be obtained as

$$\rho\frac{Dv}{Dt} = -\frac{\partial p}{\partial y} + \frac{\partial \tau_{xy}}{\partial x} + \frac{\partial \tau_{yy}}{\partial y} + \frac{\partial \tau_{zy}}{\partial z} \tag{15.18b}$$

$$\rho\frac{Dw}{Dt} = -\frac{\partial p}{\partial z} + \frac{\partial \tau_{xz}}{\partial x} + \frac{\partial \tau_{yz}}{\partial y} + \frac{\partial \tau_{zz}}{\partial z} \tag{15.18c}$$

Equation (15.18a to c) are the momentum equations in the x, y, and z directions, respectively. They are scalar equations and are called the *Navier-Stokes* equations in honor of two people—French engineer M. Navier and the English physicist G. Stokes—who independently obtained the equations in the first half of the nineteenth century.

With the expressions for $\tau_{xy} = \tau_{yx}$, $\tau_{yz} = \tau_{zy}$, $\tau_{zx} = \tau_{xz}$, τ_{xx}, τ_{yy}, and τ_{zz} from Equations (15.5) to (15.10), the Navier-Stokes equations, Equation (15.18a to c) can be written as

$$\rho\frac{\partial u}{\partial t} + \rho u\frac{\partial u}{\partial x} + \rho v\frac{\partial u}{\partial y} + \rho w\frac{\partial u}{\partial z} = -\frac{\partial p}{\partial x} + \frac{\partial}{\partial x}\left(\lambda\nabla\cdot\mathbf{V} + 2\mu\frac{\partial u}{\partial x}\right)$$
$$+ \frac{\partial}{\partial y}\left[\mu\left(\frac{\partial v}{\partial x} + \frac{\partial u}{\partial y}\right)\right] + \frac{\partial}{\partial z}\left[\mu\left(\frac{\partial u}{\partial z} + \frac{\partial w}{\partial x}\right)\right] \tag{15.19a}$$

$$\rho\frac{\partial v}{\partial t} + \rho u\frac{\partial v}{\partial x} + \rho v\frac{\partial v}{\partial y} + \rho w\frac{\partial v}{\partial z} = -\frac{\partial p}{\partial y} + \frac{\partial}{\partial x}\left[\mu\left(\frac{\partial v}{\partial x} + \frac{\partial u}{\partial y}\right)\right]$$
$$+ \frac{\partial}{\partial y}\left(\lambda\nabla\cdot\mathbf{V} + 2\mu\frac{\partial v}{\partial y}\right) + \frac{\partial}{\partial z}\left[\mu\left(\frac{\partial w}{\partial y} + \frac{\partial v}{\partial z}\right)\right] \tag{15.19b}$$

$$\rho\frac{\partial w}{\partial t} + \rho u\frac{\partial w}{\partial x} + \rho v\frac{\partial w}{\partial y} + \rho w\frac{\partial w}{\partial z} = -\frac{\partial p}{\partial z} + \frac{\partial}{\partial x}\left[\mu\left(\frac{\partial u}{\partial z} + \frac{\partial w}{\partial x}\right)\right]$$
$$+ \frac{\partial}{\partial y}\left[\mu\left(\frac{\partial w}{\partial y} + \frac{\partial v}{\partial z}\right)\right] + \frac{\partial}{\partial z}\left(\lambda\nabla\cdot\mathbf{V} + 2\mu\frac{\partial w}{\partial z}\right) \tag{15.19c}$$

Equation (15.19a to c) represent the complete Navier-Stokes equations for an unsteady, compressible, three-dimensional viscous flow. To analyze incompressible viscous flow, Equation (15.19a to c) and the continuity equation [say, Equation (2.52)] are sufficient. However, for a compressible flow, we need an additional equation, namely, the energy equation to be discussed in the next section.

In the above form, the Navier-Stokes equations are suitable for the analysis of laminar flow. For a turbulent flow, the flow variables in Equation (15.19a to c) can be assumed to be time-mean values over the turbulent fluctuations, and μ can be replaced by $\mu + \varepsilon$, as discussed in Section 15.3. For more details, see References 40 and 41.

15.5 THE VISCOUS FLOW ENERGY EQUATION

The energy equation was derived in Section 2.7, where the first law of thermodynamics was applied to a finite control volume fixed in space. The resulting integral form of the energy equation was given by Equation (2.95), and differential forms were obtained in Equations (2.96) and (2.114). In these equations, the influence of viscous effects was expressed generically by such terms as $\dot{Q}'_{\text{viscous}}$ and $\dot{W}'_{\text{viscous}}$. It is recommended that you review Section 2.7 before progressing further.

In the present section, we derive the energy equation for a viscous flow using as our model an infinitesimal moving fluid element. This will be in keeping with our derivation of the Navier-Stokes equation in Section 15.4, where the infinitesimal element was shown in Figure 15.11. In the process, we obtain explicit expressions for $\dot{Q}'_{\text{viscous}}$ and $\dot{W}'_{\text{viscous}}$ in terms of the flow-field variables. That is, we once again derive Equation (2.114), except the viscous terms are now displayed in detail.

Consider again the moving fluid element shown in Figure 15.11. To this element, apply the first law of thermodynamics, which states

Rate of change	net flux of	rate of work
of energy inside =	heat into	+ done on element
fluid element	element	due to pressure and
		stress forces on surface

or A $=$ B $+$ C (15.20)

where A, B, and C denote the respective terms above.

Let us first evaluate C; that is, let us obtain an expression for the rate of work done on the moving fluid element due to the pressure and stress forces on the surface of the element. (Note that we are neglecting body forces in this derivation.) These surface forces are illustrated in Figure 15.11, which for simplicity shows only the forces in the x direction. Recall from Section 2.7 that the rate of doing work by a force exerted on a moving body is equal to the product of the force and the component of velocity in the direction of the force. Hence, the rate of work done on the moving fluid element by the forces in the x direction shown in Figure 15.11 is simply the x component of velocity u multiplied by the forces; for example, on face $abcd$ the rate of work done by $\tau_{yx}\,dx\,dz$ is $u\tau_{yx}\,dx\,dz$, with similar expressions for the other faces. To emphasize these energy considerations, the moving fluid element is redrawn in Figure 15.12, where the rate of work done on each face by forces in the x direction is shown explicitly. Study this

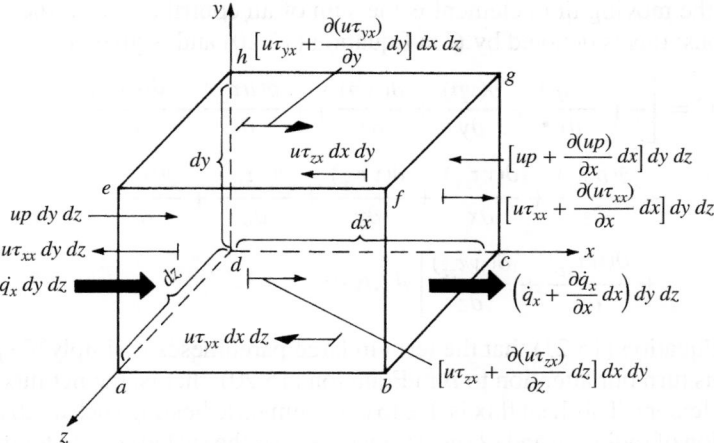

Figure 15.12 Energy fluxes associated with an infinitesimally small, moving fluid element. For simplicity, only the fluxes in the *x* direction are shown.

figure carefully, referring frequently to its companion in Figure 15.11, until you feel comfortable with the work terms given in each face. To obtain the *net* rate of work done on the fluid element by the forces in the *x* direction, note that forces in the positive *x* direction do positive work and that forces in the negative *x* direction do negative work. Hence, comparing the pressure forces on faces *adhe* and *bcgf* in Figure 15.12, the net rate of work done by pressure in the *x* direction is

$$\left[up - \left(up + \frac{\partial(up)}{\partial x}dx \right) \right] dy\, dz = -\frac{\partial(up)}{\partial x}dx\, dy\, dz$$

Similarly, the net rate of work done by the shear stresses in the *x* direction on faces *abcd* and *efgh* is

$$\left[\left(u\tau_{yx} + \frac{\partial(u\tau_{yx})}{\partial y}dy \right) - u\tau_{yx} \right] dx\, dz = \frac{\partial(u\tau_{yx})}{\partial y}dx\, dy\, dz$$

Considering all the forces shown in Figure 15.12, the net rate of work done on the moving fluid element is simply

$$\left[-\frac{\partial(up)}{\partial x} + \frac{\partial(u\tau_{xx})}{\partial x} + \frac{\partial(u\tau_{yx})}{\partial y} + \frac{\partial(u\tau_{zx})}{\partial z} \right] dx\, dy\, dz$$

The above expression considers only forces in the *x* direction. When the forces in the *y* and *z* directions are also included, similar expressions are obtained (draw some pictures and obtain these expressions yourself). In total, the net rate of work done on the moving fluid element is the sum of all contributions in the x, y, and z directions; this is denoted by C in Equation (15.20) and is given by

$$C = \left[-\left(\frac{\partial(up)}{\partial x} + \frac{\partial(vp)}{\partial y} + \frac{\partial(wp)}{\partial z} \right) + \frac{\partial(u\tau_{xx})}{\partial x} + \frac{\partial(u\tau_{yx})}{\partial y} \right. \tag{15.21}$$

$$+ \frac{\partial(u\tau_{zx})}{\partial z} + \frac{\partial(v\tau_{xy})}{\partial x} + \frac{\partial(v\tau_{yy})}{\partial y} + \frac{\partial(v\tau_{zy})}{\partial z} + \frac{\partial(w\tau_{xz})}{\partial x}$$

$$\left. + \frac{\partial(w\tau_{yz})}{\partial y} + \frac{\partial(w\tau_{zz})}{\partial z} \right] dx\, dy\, dz$$

Note in Equation (15.21) that the term in large parentheses is simply $\nabla \cdot p\mathbf{V}$.

Let us turn our attention to B in Equation (15.20), that is, the net flux of heat into the element. This heat flux is due to (1) volumetric heating such as absorption or emission of radiation and (2) heat transfer across the surface due to temperature gradients (i.e., thermal conduction). Let us treat the volumetric heating the same as was done in Section 2.7; that is, define $\dot{q}$ as the rate of volumetric heat addition per unit mass. Noting that the mass of the moving fluid element in Figure 15.12 is $\rho\, dx\, dy\, dz$, we obtain

$$\text{Volumetric heating of element} = \rho\dot{q}\, dx\, dy\, dz \tag{15.22}$$

Thermal conduction was discussed in Section 15.3. In Figure 15.12, the heat transferred by thermal conduction into the moving fluid element across face *adhe* is $\dot{q}_x\, dy\, dz$, and the heat transferred out of the element across face *bcgf* is $[\dot{q}_x + (\partial\dot{q}_x/\partial x)\, dx]\, dy\, dz$. Thus, the net heat transferred in the x direction into the fluid element by thermal conduction is

$$\left[\dot{q}_x - \left(\dot{q}_x + \frac{\partial\dot{q}_x}{\partial x} dx \right) \right] dy\, dz = -\frac{\partial\dot{q}_x}{\partial x} dx\, dy\, dz$$

Taking into account heat transfer in the y and z directions across the other faces in Figure 15.12, we obtain

$$\begin{array}{l} \text{Heating of fluid element} \\ \text{by thermal conduction} \end{array} = -\left(\frac{\partial\dot{q}_x}{\partial x} + \frac{\partial\dot{q}_y}{\partial y} + \frac{\partial\dot{q}_z}{\partial z} \right) dx\, dy\, dz \tag{15.23}$$

The term B in Equation (15.20) is the sum of Equations (15.22) and (15.23). Also, recalling that thermal conduction is proportional to temperature gradient, as exemplified by Equation (15.2), we have

$$B = \left[\rho\dot{q} + \frac{\partial}{\partial x}\left(k\frac{\partial T}{\partial x} \right) + \frac{\partial}{\partial y}\left(k\frac{\partial T}{\partial y} \right) + \frac{\partial}{\partial z}\left(k\frac{\partial T}{\partial z} \right) \right] dx\, dy\, dz \tag{15.24}$$

Finally, the term A in Equation (15.20) denotes the time rate of change of energy of the fluid element. In Section 2.7, we stated that the energy of a moving fluid per unit mass is the sum of the internal and kinetic energies, for example, $e + V^2/2$. Since we are following a moving fluid element, the time rate of change of energy per unit mass is given by the substantial derivative (see Section 2.9). Since the mass of the fluid element is $\rho\,dx\,dy\,dz$, we have

$$A = \rho\frac{D}{Dt}\left(e + \frac{V^2}{2}\right)dx\,dy\,dz \qquad (15.25)$$

The final form of the energy equation for a viscous flow is obtained by substituting Equations (15.21), (15.24), and (15.25) into Equation (15.20), obtaining

$$\rho\frac{D(e + V^2/2)}{Dt} = \rho\dot{q} + \frac{\partial}{\partial x}\left(k\frac{\partial T}{\partial x}\right) + \frac{\partial}{\partial y}\left(k\frac{\partial T}{\partial y}\right)$$
$$+ \frac{\partial}{\partial z}\left(k\frac{\partial T}{\partial z}\right) - \nabla\cdot p\mathbf{V} + \frac{\partial(u\tau_{xx})}{\partial x} + \frac{\partial(u\tau_{yx})}{\partial y}$$
$$+ \frac{\partial(u\tau_{zx})}{\partial z} + \frac{\partial(v\tau_{xy})}{\partial x} + \frac{\partial(v\tau_{yy})}{\partial y} + \frac{\partial(v\tau_{zy})}{\partial z}$$
$$+ \frac{\partial(w\tau_{xz})}{\partial x} + \frac{\partial(w\tau_{yz})}{\partial y} + \frac{\partial(w\tau_{zz})}{\partial z} \qquad (15.26)$$

Equation (15.26) is the general energy equation for unsteady, compressible, three-dimensional, viscous flow. Compare Equation (15.26) with Equation (2.105); the viscous terms are now explicitly spelled out in Equation (15.26). [Note that the body force term in Equation (15.26) has been neglected.] Moreover, the normal and shear stresses that appear in Equation (15.26) can be expressed in terms of the velocity field via Equations (15.5) to (15.10). This substitution will not be made here because the resulting equation would simply occupy too much space.

Reflect on the viscous flow equations obtained in this chapter—the Navier-Stokes equations given by Equation (15.19a to c) and the energy equation given by Equation (15.26). These equations are obviously more complex than the inviscid flow equations dealt with in previous chapters. This underscores the fact that viscous flows are inherently more difficult to analyze than inviscid flows. This is why, in the study of aerodynamics, the student is first introduced to the concepts associated with inviscid flow. Moreover, this is why we attempt to model a number of practical aerodynamic problems in real life as inviscid flows—simply to allow a reasonable analysis of such flows. However, there are many aerodynamic problems, especially those involving the prediction of drag and flow separation, which must take into account viscous effects. For the analysis of such problems, the basic equations derived in this chapter form a starting point.

Question: What is the form of the continuity equation for a viscous flow? To answer this question, review the derivation of the continuity equation in Section 2.4. You will note that the consideration of the viscous or inviscid nature of the flow never enters the derivation—the continuity equation is simply a statement that mass is conserved, which is independent of whether the flow is viscous or inviscid. Hence, Equation (2.52) holds in general.

15.6 SIMILARITY PARAMETERS

In Section 1.7, we introduced the concept of dimensional analysis, from which sprung the similarity parameters necessary to ensure the dynamic similarity between two or more different flows (see Section 1.8). In the present section, we revisit the governing similarity parameters, but cast them in a slightly different light.

Consider a steady, two-dimensional, viscous, compressible flow. The x-momentum equation for such a flow is given by Equation (15.19a), which for the present case reduces to

$$\rho u \frac{\partial u}{\partial x} + \rho v \frac{\partial u}{\partial y} = -\frac{\partial p}{\partial x} + \frac{\partial}{\partial y}\left[\mu\left(\frac{\partial v}{\partial x} + \frac{\partial u}{\partial y}\right)\right] \tag{15.27}$$

In Equation (15.27), ρ, u, p, etc., are the actual dimensional variables, say, $[\rho] =$ kg/m^3, etc. Let us introduce the following dimensionless variables:

$$\rho' = \frac{\rho}{\rho_\infty} \quad u' = \frac{u}{V_\infty} \quad v' = \frac{v}{V_\infty} \quad p' = \frac{p}{p_\infty}$$

$$\mu' = \frac{\mu}{\mu_\infty} \quad x' = \frac{x}{c} \quad y' = \frac{y}{c}$$

where $\rho_\infty, V_\infty, p_\infty$, and μ_∞ are reference values (say, e.g., freestream values) and c is a reference length (say, the chord of an airfoil). In terms of these dimensionless variables, Equation (15.27) becomes

$$\rho' u' \frac{\partial u'}{\partial x'} + \rho' v' \frac{\partial u'}{\partial y'} = -\left(\frac{p_\infty}{\rho_\infty V_\infty^2}\right)\frac{\partial p'}{\partial x'} + \left(\frac{\mu_\infty}{\rho_\infty V_\infty c}\right)\frac{\partial}{\partial y'}\left[\mu'\left(\frac{\partial v'}{\partial x'} + \frac{\partial u'}{\partial y'}\right)\right]$$

$$\tag{15.28}$$

Noting that

$$\frac{p_\infty}{\rho_\infty V_\infty^2} = \frac{\gamma p_\infty}{\gamma \rho_\infty V_\infty^2} = \frac{a_\infty^2}{\gamma V_\infty^2} = \frac{1}{\gamma M_\infty^2}$$

and

$$\frac{\mu_\infty}{\rho_\infty V_\infty c} = \frac{1}{\text{Re}_\infty}$$

where M_∞ and Re_∞ are the freestream Mach and Reynolds numbers, respectively, Equation (15.28) becomes

$$\rho' u' \frac{\partial u'}{\partial x'} + \rho' v' \frac{\partial u'}{\partial y'} = -\frac{1}{\gamma M_\infty^2}\frac{\partial p'}{\partial x'} + \frac{1}{\text{Re}_\infty}\frac{\partial}{\partial y'}\left[\mu'\left(\frac{\partial v'}{\partial x'} + \frac{\partial u'}{\partial y'}\right)\right] \tag{15.29}$$

Equation (15.29) tells us something important. Consider two different flows over two bodies of different shapes. In one flow, the ratio of specific heats, Mach number, and Reynolds number are γ_1, $M_{\infty 1}$, and $\text{Re}_{\infty 1}$, respectively; in the other flow, these parameters have different values, γ_2, $M_{\infty 2}$, and $\text{Re}_{\infty 2}$. Equation (15.29) is valid for both flows. It can, in principle, be solved to obtain u' as a function of x' and y'. However, since γ, M_∞, and Re_∞ are different for the two cases, the coefficients of the derivatives in Equation (15.29) will be different. This will ensure, if

$$u' = f_1(x', y')$$

represents the solution for one flow and

$$u' = f_2(x', y')$$

represents the solution for the other flow, that

$$f_1 \neq f_2$$

However, consider now the case where the two different flows have the *same* values of γ, M_∞, and Re_∞. Now the coefficients of the derivatives in Equation (15.29) will be the same for both flows; that is, Equation (15.29) is *numerically identical* for the two flows. In addition, assume the two bodies are geometrically similar, so that the boundary conditions in terms of the nondimensional variables are the same. Then, the solutions of Equation (15.29) for the two flows in terms of $u' = f_1(x', y')$ and $u' = f_2(x', y')$ must be identical; that is,

$$f_1(x', y') \equiv f_2(x', y') \tag{15.30}$$

Recall the definition of dynamically similar flows given in Section 1.8. There, we stated in part that two flows are dynamically similar if the distributions of V/V_∞, p/p_∞, etc., are the same throughout the flow field when plotted against common nondimensional coordinates. This is precisely what Equation (15.30) is saying—that u' as a function of x' and y' is the same for the two flows. That is, the variation of the *nondimensional* velocity as a function of the *nondimensional* coordinates is the same for the two flows. How did we obtain Equation (15.30)? Simply by saying that γ, M_∞, and Re_∞ are the same for the two flows and that the two bodies are geometrically similar. *These are precisely the criteria for two flows to be dynamically similar,* as originally stated in Section 1.8.

What we have seen in the above derivation is a formal mechanism to identify governing similarity parameters for a flow. By couching the governing flow equations in terms of nondimensional variables, we find that the coefficients of the derivatives in these equations are dimensionless similarity parameters or combinations thereof.

To see this more clearly, and to extend our analysis further, consider the energy equation for a steady, two-dimensional, viscous, compressible flow, which from Equation (15.26) can be written as (assuming no volumetric heating and neglecting the normal stresses)

$$\rho u \frac{\partial(e + V^2/2)}{\partial x} + \rho v \frac{\partial(e + V^2/2)}{\partial y} = \frac{\partial}{\partial x}\left(k\frac{\partial T}{\partial x}\right) + \frac{\partial}{\partial y}\left(k\frac{\partial T}{\partial y}\right) - \frac{\partial(up)}{\partial x}$$

$$- \frac{\partial(vp)}{\partial y} + \frac{\partial(v\tau_{xy})}{\partial x} + \frac{\partial(u\tau_{yx})}{\partial y} \tag{15.31}$$

Substituting Equation (15.5) into (15.31), we have

$$\rho u \frac{\partial(e + V^2/2)}{\partial x} + \rho v \frac{\partial(e + V^2/2)}{\partial y} = \frac{\partial}{\partial x}\left(k\frac{\partial T}{\partial x}\right) + \frac{\partial}{\partial y}\left(k\frac{\partial T}{\partial y}\right) \qquad (15.32)$$

$$- \frac{\partial(up)}{\partial x} - \frac{\partial(vp)}{\partial y}$$

$$+ \frac{\partial}{\partial x}\left[\mu v\left(\frac{\partial v}{\partial x} + \frac{\partial u}{\partial y}\right)\right]$$

$$+ \frac{\partial}{\partial y}\left[\mu u\left(\frac{\partial v}{\partial x} + \frac{\partial u}{\partial y}\right)\right]$$

Using the same nondimensional variables as before, and introducing

$$e' = \frac{e}{c_v T_\infty} \quad k' = \frac{k}{k_\infty} \quad V'^2 = \frac{V^2}{V_\infty^2} = \frac{u^2 + v^2}{V_\infty^2} = (u')^2 + (v')^2$$

Equation (15.32) can be written as

$$\frac{\rho_\infty V_\infty c_v T_\infty}{c}\left(\rho' u'\frac{\partial e'}{\partial x'} + \rho' v'\frac{\partial e'}{\partial y'}\right)$$

$$= -\frac{\rho_\infty V_\infty^3}{2c}\left[\rho' u'\frac{\partial}{\partial x'}(u'^2 + v'^2) + \rho' v'\frac{\partial}{\partial y'}(u'^2 + v'^2)\right]$$

$$+ \frac{k_\infty T_\infty}{c^2}\left[\frac{\partial}{\partial x'}\left(k'\frac{\partial T'}{\partial x'}\right) + \frac{\partial}{\partial y'}\left(k'\frac{\partial T'}{\partial y'}\right)\right] - \frac{V_\infty p_\infty}{c}\left(\frac{\partial(u'p')}{\partial x'} + \frac{\partial(v'p')}{\partial y'}\right)$$

$$+ \frac{\mu_\infty V_\infty^2}{c^2}\left\{\frac{\partial}{\partial x'}\left[\mu' v'\left(\frac{\partial v'}{\partial x'} + \frac{\partial u'}{\partial y'}\right)\right] + \frac{\partial}{\partial y'}\left[\mu' u'\left(\frac{\partial v'}{\partial x'} + \frac{\partial u'}{\partial y'}\right)\right]\right\}$$

or

$$\rho' u'\frac{\partial e'}{\partial x'} + \rho' v'\frac{\partial e'}{\partial y'} = \frac{V_\infty^2}{2c_v T_\infty}\left[\rho' u'\frac{\partial}{\partial x'}(u'^2 + v'^2)\right. \qquad (15.32a)$$

$$\left. + \rho' v'\frac{\partial}{\partial y'}(u'^2 + v'^2)\right]$$

$$+ \frac{k_\infty}{c\rho_\infty V_\infty c_v}\left[\frac{\partial}{\partial x'}\left(k'\frac{\partial T'}{\partial x'}\right) + \frac{\partial}{\partial y'}\left(k'\frac{\partial T'}{\partial y'}\right)\right]$$

$$- \frac{p_\infty}{\rho_\infty c_v T_\infty}\left(\frac{\partial(u'p')}{\partial x'} + \frac{\partial(v'p')}{\partial y'}\right)$$

$$+ \frac{\mu_\infty V_\infty}{c\rho_\infty c_v T_\infty}\left\{\frac{\partial}{\partial x'}\left[\mu' v'\left(\frac{\partial v'}{\partial x'} + \frac{\partial u'}{\partial y'}\right)\right]\right.$$

$$\left. + \frac{\partial}{\partial y'}\left[\mu' u'\left(\frac{\partial v'}{\partial x'} + \frac{\partial u'}{\partial y'}\right)\right]\right\}$$

Examining the coefficients of each term on the right-hand side of Equation (15.32a), we find, consecutively,

$$\frac{V_\infty^2}{c_v T_\infty} = \frac{(\gamma - 1)V_\infty^2}{RT_\infty} = \frac{\gamma(\gamma - 1)V_\infty^2}{\gamma RT_\infty} = \frac{\gamma(\gamma - 1)V_\infty^2}{a_\infty^2} = \gamma(\gamma - 1)M_\infty^2$$

$$\frac{k_\infty}{c\rho_\infty V_\infty c_v} = \frac{k_\infty \gamma \mu_\infty}{c\rho_\infty V_\infty c_p \mu_\infty} = \frac{\gamma}{\text{Pr}_\infty \text{Re}_\infty}$$

Note: In the above, we have introduced a new dimensionless parameter, the *Prandtl number,* $\text{Pr}_\infty \equiv \mu_\infty c_p / k_\infty$, the significance of which will be discussed later:

$$\frac{p_\infty}{\rho_\infty c_v T_\infty} = \frac{(\gamma - 1)p_\infty}{\rho_\infty RT_\infty} = \frac{(\gamma - 1)p_\infty}{p_\infty} = \gamma - 1$$

$$\frac{\mu_\infty V_\infty}{c\rho_\infty c_v T_\infty} = \frac{\mu_\infty}{\rho_\infty V_\infty c}\left(\frac{V_\infty^2}{c_v T_\infty}\right) = \frac{1}{\text{Re}_\infty}(\gamma - 1)\frac{V_\infty^2}{RT_\infty} = \gamma(\gamma - 1)\frac{M_\infty^2}{\text{Re}_\infty}$$

Hence, Equation (15.32) can be written as

$$\rho' u' \frac{\partial e'}{\partial x'} + \rho' v' \frac{\partial e'}{\partial y'} \tag{15.33}$$

$$= \frac{\gamma(\gamma - 1)}{2} M_\infty^2 \left[\rho' u' \frac{\partial}{\partial x'}(u'^2 + v'^2) + \rho' v' \frac{\partial}{\partial y'}(u'^2 + v'^2)\right]$$

$$+ \frac{\gamma}{\text{Pr}_\infty \text{Re}_\infty} \left[\frac{\partial}{\partial x'}\left(k' \frac{\partial T'}{\partial x'}\right) + \frac{\partial}{\partial y'}\left(k' \frac{\partial T'}{\partial y'}\right)\right]$$

$$- (\gamma - 1)\left(\frac{\partial(u'p')}{\partial x'} + \frac{\partial(v'p')}{\partial y'}\right)$$

$$+ \gamma(\gamma - 1)\frac{M_\infty^2}{\text{Re}_\infty}\left\{\frac{\partial}{\partial x'}\left[\mu' v'\left(\frac{\partial v'}{\partial x'} + \frac{\partial u'}{\partial y'}\right)\right]\right.$$

$$\left. + \frac{\partial}{\partial y'}\left[\mu' u'\left(\frac{\partial v'}{\partial x'} + \frac{\partial u'}{\partial y'}\right)\right]\right\}$$

Examine Equation (15.33). It is a nondimensional equation which, in principle, can be solved for $e' = f(x', y')$. If we have two different flows, but with the same values γ, M_∞, Re_∞, and Pr_∞, Equation (15.33) will be numerically identical for the two flows, and if we are considering geometrically similar bodies, then the solution $e' = f(x', y')$ will be identical for the two flows.

Reflecting upon Equations (15.29) and (15.33), which are the nondimensional x-momentum and energy equations, respectively, we clearly see that the governing similarity parameters for a viscous, compressible flow are γ, M_∞, Re_∞, and Pr_∞. If the above parameters are the same for two different flows with geometrically similar bodies, then the flows are dynamically similar. We obtained these results by considering the x-momentum equation and the energy equation, both in two dimensions. The same results would have occurred if we had considered three-dimensional flow and the y- and z-momentum equations.

Note that the similarity parameters γ, M_∞, and Re_∞ were obtained from the momentum equation. When the energy equation is considered, an additional similarity parameter is introduced, namely, the Prandtl number. On a physical basis, the Prandtl number is an index which is proportional to the ratio of energy dissipated by friction to the energy transported by thermal conduction; that is,

$$\text{Pr} = \frac{\mu_\infty c_p}{k} \propto \frac{\text{frictional dissipation}}{\text{thermal conduction}}$$

In the study of compressible, viscous flow, Prandtl number is just as important as γ, Re_∞, or M_∞. For air at standard conditions, $\text{Pr}_\infty = 0.71$. Note that Pr_∞ is a property of the gas. For different gases, Pr_∞ is different. Also, like μ and k, Pr_∞ is, in general, a function of temperature; however, for air over a reasonable temperature range (up to $T_\infty = 600$ K), it is safe to assume $\text{Pr}_\infty = \text{constant} = 0.71$.

15.7 SOLUTIONS OF VISCOUS FLOWS: A PRELIMINARY DISCUSSION

The governing continuity, momentum, and energy equations for a general unsteady, compressible, viscous, three-dimensional flow are given by Equations (2.52), (15.19a to c), and (15.26), respectively. Examine these equations closely. They are nonlinear, coupled, partial differential equations. Moreover, they have additional terms—namely, the viscous terms—in comparison to the analogous equations for an inviscid flow treated in Part 3. Since we have already seen that the nonlinear inviscid flow equations do not lend themselves to a general analytical solution, we can certainly expect the viscous flow equations also not to have any general solutions (at least, at the time of this writing, no general analytical solutions have been found). This leads to the following question: How, then, can we make use of the viscous flow equations in order to obtain some practical results? The answer is much like our approach to the solution of inviscid flows. We have the following options:

1. There are a few viscous flow problems which, by their physical and geometrical nature, allow many terms in the Navier-Stokes solutions to be precisely zero, with the resulting equations being simple enough to solve, either analytically or by simple numerical methods. Sometimes this class of solutions is called "exact solutions" of the Navier-Stokes equations, because no simplifying *approximations* are made to reduce the equations—just *precise* conditions are applied to reduce the equations. Chapter 16 is devoted to this class of solutions; an example is Couette flow (to be defined later).

2. We can simplify the equations by treating certain classes of physical problems for which some terms in the viscous flow equations are small and can be neglected. This is an approximation, not a precise condition. The boundary-layer equations developed and discussed in Chapter 17 are a case

in point. However, as we will see, the boundary-layer equations may be simpler than the full viscous flow equations, but they are still nonlinear.

3.   We can tackle the solution of the full viscous flow equations by modern numerical techniques. For example, some of the computational fluid dynamic algorithms discussed in Chapter 13 in conjunction with "exact" solutions for the inviscid flow equations carry over to exact solutions for the viscous flow equations. These matters will be discussed in Chapter 20.

There are some inherent very important differences between the analysis of viscous flows and the study of inviscid flows that were presented in Parts 2 and 3. The remainder of this section highlights these differences.

First, we have already demonstrated in Example 2.5 that viscous flows are *rotational flows*. Therefore, a velocity potential cannot be defined for a viscous flow, thus losing the attendant advantages that were discussed in Sections 2.15 and 11.2. On the other hand, a stream function can be defined, because the stream function satisfies the continuity equation and has nothing to do with the flow being rotational or irrotational (see Section 2.14).

Second, the boundary condition at a solid surface for a viscous flow is the *no-slip* condition. Due to the presence of friction between the surface material and the adjacent layer of fluid, the fluid velocity right at the surface is zero. This no-slip condition was discussed in Section 15.2. For example, if the surface is located at $y = 0$ in a cartesian coordinate system, then the no-slip boundary condition on velocity is

At y = 0: $u = 0$ $v = 0$ $w = 0$

This is in contrast to the analogous boundary condition for an inviscid flow, namely, the flow-tangency condition at a surface as discussed in Section 3.7, where only the component of the velocity normal to the surface is zero. Also, recall that for an inviscid flow, there is no boundary condition on the temperature; the temperature of the gas adjacent to a solid surface in an inviscid flow is governed by the physics of the flow field and has no connection whatsoever with the actual wall temperature. However, for a viscous flow, the mechanism of thermal conduction ensures that the temperature of the fluid immediately adjacent to the surface is the same as the temperature of the material surface. In this respect, the no-slip condition is more general than that applied to the velocity; in addition to $u = v = 0$ at the wall, we also have $T = T_w$ at the wall, where T is the gas temperature immediately adjacent to the wall and T_w is the temperature of the surface material. Thus,

At y = 0: $T = T_w$ (15.34)

In many problems, T_w is specified and held constant; this boundary condition is easily applied. However, consider the following, more general case. Imagine a viscous flow over a surface where heat is being transferred from the gas to the surface, or vice versa. Also, assume that the surface is at a certain temperature, T_w, when the flow first starts, but that T_w changes as a function of time as the surface is either heated or cooled by the flow [i.e., $T_w = T_w(t)$]. Because this timewise variation is dictated in part by the flow which is being calculated, T_w becomes an unknown in the problem and must be calculated along with the solution of the viscous flow. For this general case, the boundary condition at the surface is obtained from Equation (15.2) applied at the wall; that is,

$$At\ y = 0: \qquad \dot{q}_w = -\left(k\frac{\partial T}{\partial y} \right)_w \qquad (15.35)$$

Here, the surface material is responding to the heat transfer to the wall $\dot{q}_w$, hence changing T_w, which in turn affects $\dot{q}_w$. This general, unsteady heat transfer problem must be solved by treating the viscous flow and the thermal response of the material simultaneously. This problem is beyond the scope of the present book.

Finally, let us imagine the above, unsteady case carried out to the limit of large times. That is, imagine a wind-tunnel model which is at room temperature suddenly inserted in a supersonic or hypersonic stream. At early times, say, for the first few seconds, the surface temperature remains relatively cool, and the assumption of constant wall temperature T_w is reasonable [Equation (15.34)]. However, due to the heat transfer to the model [Equation (15.35)], the surface temperature soon starts to increase and becomes a function of time, as discussed in the previous paragraph. However, as T_w increases, the heating rate decreases. Finally, at large times, T_w increases to a high enough value that the net heat transfer rate to the surface becomes zero, that is, from Equation (15.35),

$$\dot{q}_w = -\left(k\frac{\partial T}{\partial y} \right)_w = 0$$

or

$$\left(\frac{\partial T}{\partial y} \right)_w = 0 \qquad (15.36)$$

When the situation of zero heat transfer is achieved, a state of equilibrium exists; the wall temperature at which this occurs is, by definition, the equilibrium wall temperature, or, as it is more commonly denoted, the *adiabatic wall temperature*, T_{aw}. Hence, for the case of an *adiabatic wall* (no heat transfer), the wall boundary condition is given by Equation (15.36).

In summary, for the wall boundary condition associated with the solution of the energy equation [Equation (15.26)], we have three possible cases:

1. *Constant temperature wall, where T_w is a specified constant* [Equation (15.34)]. For this given wall temperature, the temperature gradient at the wall $(\partial T/\partial y)_w$ is obtained as part of the flow-field solution and allows the direct calculation of the aerodynamic heating to the wall via Equation (15.35).

2. *The general, unsteady case, where the heat transfer to the wall $\dot{q}_w$ causes the wall temperature T_w to change, which in turn causes $\dot{q}_w$ to change.* Here, both T_w and $(\partial T/\partial y)_w$ change as a function of time, and the problem must be solved by treating jointly the viscous flow as well as the thermal response of the wall material (which usually implies a separate thermal conduction heat transfer numerical analysis).

3. *The adiabatic wall case (zero heat transfer), where* $(\partial T/\partial y)_w = 0$
    [Equation (15.36)]. Here, the boundary condition is applied to the
    temperature *gradient* at the wall, not to the wall temperature itself. Indeed,
    the wall temperature for this case is defined as the adiabatic wall
    temperature T_{aw} and is obtained as part of the flow-field solution.

Finally, we emphasize again that, from the point of view of applied aerodynamics, the practical results obtained from a viscous flow analysis are the
skin friction and heat transfer at the surface. However, to obtain these quantities, we usually need a complete solution of the viscous flow field; among the
data obtained from such a solution are the velocity and temperature gradients at
the wall. These, in turn, allow the direct calculation of τ_w and $\dot{q}_w$ from

$$\tau_w = \mu \left(\frac{\partial u}{\partial y} \right)_w$$

and

$$\dot{q}_w = -k \left(\frac{\partial T}{\partial y} \right)_w$$

Another practical result provided by a viscous flow analysis is the prediction and
calculation of flow separation; we have discussed numerous cases in the preceding chapters where the pressure field around an aerodynamic body can be
greatly changed by flow separation; the flows over cylinders and spheres (see
Sections 3.18 and 6.6) are cases in point.

Clearly, the study of viscous flow is important within the entire scope of aerodynamics. The purpose of the following chapters is to provide an introduction to
such flows. We will organize our study following the three options itemized at the
beginning of this section; that is, we will treat, in turn, certain specialized "exact"
solutions of the Navier-Stokes equations, boundary-layer solutions, and then "exact" numerical solutions of Navier-Stokes equations. In so doing, we hope that the
reader will gain an overall, introductory picture of the whole area of viscous flow.
Entire books have been written on this subject, see, for example, References 40
and 41. We cannot possibly present such detail here; rather, our objective is simply
to provide a "feel" for and a basic understanding of the material. Let us proceed.

15.8 SUMMARY

We have now completed the road map given in Figure 15.1. The main results of
this chapter are summarized below:

> Shear stress and flow separation are two major ramifications of viscous flow.
> Shear stress is the cause of skin friction drag D_f, and flow separation is the
> source of pressure drag D_p, sometimes called form drag. Transition from
> laminar to turbulent flow causes D_f to increase and D_p to decrease.

Shear stress in a flow is due to velocity gradients: for example, $\tau_{yx} = \mu \, \partial u / \partial y$ for a flow with gradients in the y direction. Similarly, heat conduction is due to temperature gradients; for example, $\dot{q}_y = -k \, \partial T / \partial y$, etc. Both μ and k are physical properties of the gas and are functions of temperature.

The general equations of viscous flow are

x momentum:
$$\rho \frac{Du}{Dt} = -\frac{\partial p}{\partial x} + \frac{\partial \tau_{xx}}{\partial x} + \frac{\partial \tau_{yx}}{\partial y} + \frac{\partial \tau_{zx}}{\partial z} \qquad (15.18a)$$

y momentum:
$$\rho \frac{Dv}{Dt} = -\frac{\partial p}{\partial y} + \frac{\partial \tau_{xy}}{\partial x} + \frac{\partial \tau_{yy}}{\partial y} + \frac{\partial \tau_{zy}}{\partial z} \qquad (15.18b)$$

z momentum:
$$\rho \frac{Dw}{Dt} = -\frac{\partial p}{\partial z} + \frac{\partial \tau_{xz}}{\partial x} + \frac{\partial \tau_{yz}}{\partial y} + \frac{\partial \tau_{zz}}{\partial z} \qquad (15.18c)$$

Energy:
$$\rho \frac{D(e + V^2/2)}{Dt} = \rho \dot{q} + \frac{\partial}{\partial x}\left(k\frac{\partial T}{\partial x}\right) + \frac{\partial}{\partial y}\left(k\frac{\partial T}{\partial y}\right) \qquad (15.26)$$
$$+ \frac{\partial}{\partial z}\left(k\frac{\partial T}{\partial z}\right) - \nabla \cdot p\mathbf{V} + \frac{\partial(u\tau_{xx})}{\partial x} + \frac{\partial(u\tau_{yx})}{\partial y}$$
$$+ \frac{\partial(u\tau_{zx})}{\partial z} + \frac{\partial(v\tau_{xy})}{\partial x} + \frac{\partial(v\tau_{yy})}{\partial y} + \frac{\partial(v\tau_{zy})}{\partial z}$$
$$+ \frac{\partial(w\tau_{xz})}{\partial x} + \frac{\partial(w\tau_{yz})}{\partial y} + \frac{\partial(w\tau_{zz})}{\partial z}$$

where
$$\tau_{xy} = \tau_{yx} = \mu\left(\frac{\partial v}{\partial x} + \frac{\partial u}{\partial y}\right)$$
$$\tau_{yz} = \tau_{zy} = \mu\left(\frac{\partial w}{\partial y} + \frac{\partial v}{\partial z}\right)$$
$$\tau_{zx} = \tau_{xz} = \mu\left(\frac{\partial u}{\partial z} + \frac{\partial w}{\partial x}\right)$$
$$\tau_{xx} = \lambda(\nabla \cdot \mathbf{V}) + 2\mu\frac{\partial u}{\partial x}$$
$$\tau_{yy} = \lambda(\nabla \cdot \mathbf{V}) + 2\mu\frac{\partial v}{\partial y}$$
$$\tau_{zz} = \lambda(\nabla \cdot \mathbf{V}) + 2\mu\frac{\partial w}{\partial z}$$

> The similarity parameters for a flow can be obtained by nondimensionalizing the governing equations; the coefficients in front of the nondimensionalized derivatives give the similarity parameters or combinations thereof. For a viscous, compressible flow, the main similarity parameters are γ, M_∞, Re_∞, and Pr_∞.

> Exact analytical solutions of the complete Navier-Stokes equations exist for only a few very specialized cases. Instead, the equations are frequently simplified by making appropriate approximations about the flow. In modern times, exact solutions of the complete Navier-Stokes equations for many practical problems can be obtained numerically, using various techniques of computational fluid dynamics.

15.9 PROBLEMS

15.1 Consider the incompressible viscous flow of air between two infinitely long parallel plates separated by a distance h. The bottom plate is stationary, and the top plate is moving at the constant velocity u_e in the direction of the plate. Assume that no pressure gradient exists in the flow direction.

    a.   Obtain an expression for the variation of velocity between the plates.

    b.   If $T = \text{constant} = 320$ K, $u_e = 30$ m/s, and $h = 0.01$ m, calculate the shear stress on the top and bottom plates.

15.2 Assume that the two parallel plates in Problem 15.1 are both stationary but that a constant pressure gradient exists in the flow direction (i.e., $dp/dx = \text{constant}$).

    a.   Obtain an expression for the variation of velocity between the plates.

    b.   Obtain an expression for the shear stress on the plates in terms of dp/dx.

CHAPTER 16

A Special Case: Couette Flow

The resistance arising from the want of lubricity in the parts of a fluid is, other things being equal, proportional to the velocity with which the parts of the fluid are separated from one another.

**Isaac Newton, 1687,
from Section IX of Book II
of his Principia**

PREVIEW BOX

An old but wise expression states that you must learn to walk before you can run. General applications of viscous flow are frequently complex and challenging—you are constantly "running" in order to obtain their solutions. This chapter teaches you to walk first. Here we treat a special viscous flow problem that lends itself to rather straightforward solutions (all we have to do is "walk" to obtain them). Yet these solutions illustrate some of the most important aspects of viscous flows in general, obtaining and highlighting the parameters that dictate skin friction and aerodynamic heating. Here we will learn some new ideas with strange-sounding words such as "recovery factor" and "Reynolds analogy." We will be able to see some of the basic physics of viscous flow, stripped of the extra geometrical complexities that go along with more complex flow applications. Although this chapter involves a peculiar-sounding flow, *Couette flow*, the results are far from peculiar. Indeed, in this chapter we are going to take a walk through some of the most important ideas surrounding the basic analyses of viscous flow. Get going, and enjoy your walk.

16.1 INTRODUCTION

The general equations of viscous flow were derived and discussed in Chapter 15. In particular, the viscous flow momentum equations were treated in Section 15.4 and are given in partial differential equation form by Equation (15.19a to c)—the Navier-Stokes equations. These, along with the viscous flow energy equation,

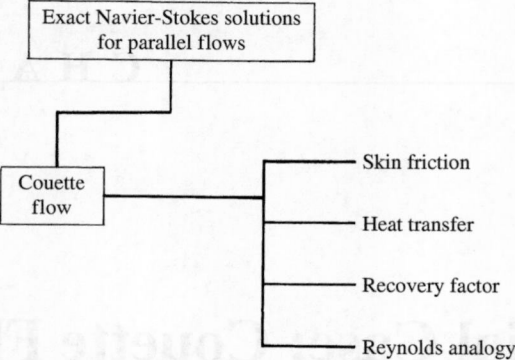

Figure 16.1 Road map for Chapter 16.

Equation (15.26), derived in Section 15.5, are the theoretical tools for the study of viscous flows. However, examine these equations closely; as discussed in Section 15.7, they are a system of coupled, nonlinear partial differential equations—equations which contain more terms and which are inherently more elaborate than the inviscid flow equations treated in Parts 2 and 3 of this book. Three classes of solutions of these equations were itemized in Section 15.5. The first itemized class was that of "exact" solutions of the Navier-Stokes equations for a few specific physical problems which, by their physical and geometrical nature, allow many terms in the governing equations to be precisely zero, resulting in a system of equations simple enough to solve, either analytically or by simple numerical methods. Such exact problems are the subject of this chapter.

The road map for this chapter is given in Figure 16.1. The type of flow considered here is generally labeled as *parallel flow* because the streamlines are straight and parallel to each other. We will consider one of these flows, Couette flow, which will be defined in due course. In addition to representing exact solutions of the Navier-Stokes equations, this flow illustrates some of the important practical facets of any viscous flow, as itemized on the right side of the road map. In a clear, uncomplicated fashion, we will be able to calculate and study the surface skin friction and heat transfer. We will also use the results to define the recovery factor and Reynolds analogy—two practical engineering tools that are frequently used in the analysis of skin friction and heat transfer.

16.2 COUETTE FLOW: GENERAL DISCUSSION

Consider the flow model shown in Figure 16.2. Here we see a viscous fluid contained between two parallel plates separated by a distance D. The upper plate is moving to the right at velocity u_e. Due to the no-slip condition, there can be no relative motion between the plate and the fluid; hence, at $y = D$ the flow velocity is $u = u_e$ and is directed toward the right. Similarly, the flow velocity at $y = 0$, which is the surface of the stationary lower plate, is $u = 0$. In addition, the two plates may be at different temperatures; the upper plate is at temperature T_e and

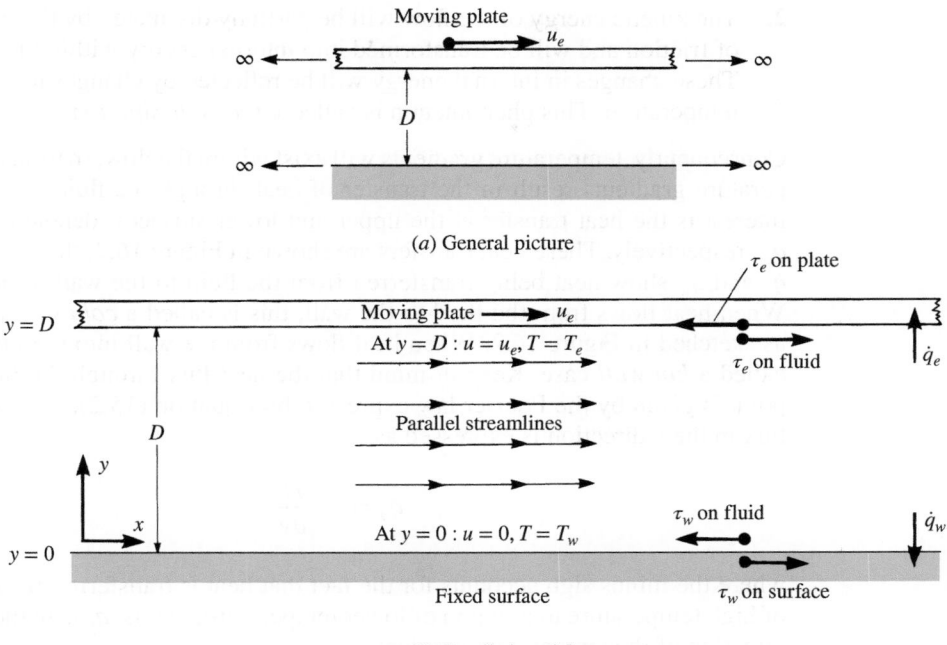

Figure 16.2 Model for Couette flow.

the lower plate is at temperature T_w. Again, due to the no-slip condition as discussed in Section 15.7, the fluid temperature at $y = D$ is $T = T_e$ and that at $y = 0$ is $T = T_w$.

Clearly, there is a flow field between the two plates; the driving force for this flow is the motion of the upper plate, dragging the flow along with it through the mechanism of friction. The upper plate is exerting a shear stress, τ_e, acting toward the right on the fluid at $y = D$, thus causing the fluid to move toward the right. By an equal and opposite reaction, the fluid is exerting a shear stress τ_e on the upper plate acting toward the left, tending to retard its motion. We assume that the upper plate is being driven by some external force that is sufficient to overcome the retarding shear stress and to allow the plate to move at the constant velocity u_e. Similarly, the lower plate is exerting a shear stress τ_w acting toward the left on the fluid at $y = 0$. By an equal and opposite reaction, the fluid is exerting a shear stress τ_w acting toward the right on the lower plate. (In all subsequent diagrams dealing with viscous flow, the only shear stresses shown will be those due to the fluid acting on the surface, unless otherwise noted.)

In addition to the velocity field induced by the relative motion of the two plates, there will also be a temperature field induced by the following two mechanisms:

1. The plates in general will be at different temperatures, thus causing temperature gradients in the flow.

2.   The kinetic energy of the flow will be partially dissipated by the influence of friction and will be transformed into internal energy within the fluid. These changes in internal energy will be reflected by changes in temperature. This phenomenon is called *viscous dissipation*.

Consequently, temperature gradients will exist within the flow; in turn, these temperature gradients result in the transfer of heat through the fluid. Of particular interest is the heat transfer at the upper and lower surfaces, denoted by $\dot{q}_e$ and $\dot{q}_w$, respectively. These heat transfers are shown in Figure 16.2; the directions for $\dot{q}_e$ and $\dot{q}_w$ show heat being transferred from the fluid to the wall in both cases. When heat flows from the fluid to the wall, this is called a *cold wall* case, such as sketched in Figure 16.2. When heat flows from the wall into the fluid, this is called a *hot wall* case. Keep in mind that the heat flux through the fluid at any point is given by the Fourier law expressed by Equation (15.2); that is, the heat flux in the y direction is expressed as

$$\dot{q}_y = -k\frac{\partial T}{\partial y} \qquad (15.2)$$

where the minus sign accounts for the fact that heat is transferred from a region of high temperature to a region of lower temperature; that is, $\dot{q}_y$ is in the opposite direction of the temperature gradient.

Let us examine the geometry of Couette flow as illustrated in Figure 16.2. An x-y cartesian coordinate system is oriented with the x axis in the direction of the flow and the y axis perpendicular to the flow. Since the two plates are parallel, the only possible flow pattern consistent with this picture is that of straight, parallel streamlines. Moreover, since the plates are infinitely long (i.e., stretching to plus and minus infinity in the x direction), then the flow properties cannot change with x. (If the properties did change with x, then the flow-field properties would become infinitely large or infinitesimally small at large values of x—a physical inconsistency.) Thus, all partial derivatives with respect to x are zero. The only changes in the flow-field variables take place in the y direction. Moreover, the flow is steady, so that all time derivatives are zero. With this geometry in mind, return to the governing Navier-Stokes equations given by Equations (15.19a to c) and (15.26). In these equations, for Couette flow,

$$v = w = 0 \qquad \frac{\partial u}{\partial x} = \frac{\partial T}{\partial x} = \frac{\partial p}{\partial x} = 0$$

Hence, from Equations (15.19a to c) and (15.26), we have

x-momentum equation: $\dfrac{\partial}{\partial y}\left(\mu\dfrac{\partial u}{\partial y}\right) = 0 \qquad (16.1)$

y-momentum equation: $\dfrac{\partial p}{\partial y} = 0 \qquad (16.2)$

Energy equation: $$\frac{\partial}{\partial y}\left(k\frac{\partial T}{\partial y}\right) + \frac{\partial}{\partial y}\left(\mu u\frac{\partial u}{\partial y}\right) = 0 \qquad (16.3)$$

Equations (16.1) to (16.3) are the governing equations for Couette flow. Note that these equations are exact forms of the Navier-Stokes equations applied to the geometry of Couette flow—no approximations have been made. Also, note from Equation (16.2) that the variation of pressure in the y direction is zero; this in combination with the earlier result that $\partial p/\partial x = 0$ implies that the *pressure is constant* throughout the entire flow field. Couette flow is a constant pressure flow. It is interesting to note that all the previous flow problems discussed in Parts 2 and 3, being *inviscid* flows, were established and maintained by the existence of *pressure gradients* in the flow. In these problems, the pressure gradient was nature's mechanism of grabbing hold of the flow and making it move. However, in the problem we are discussing now—being a viscous flow—shear stress is another mechanism by which nature can exert a force on a flow. For Couette flow, the shear stress exerted by the moving plate on the fluid is the exclusive driving mechanism that maintains the flow; clearly, no pressure gradient is present, nor is it needed.

This section has presented the general nature of Couette flow. Note that we have made no distinction between incompressible and compressible flow; all aspects discussed here apply to both cases. Also, we note that, although Couette flow appears to be a rather academic problem, the following sections illustrate, in a simple fashion, many of the important characteristics of practical viscous flows in real engineering applications.

The next two sections will treat the separate cases of incompressible and compressible Couette flow. Incompressible flow will be discussed first because of its relative simplicity; this is the subject of Section 16.3. Then compressible Couette flow, and how it differs from the incompressible case, will be examined in Section 16.4.

As a final note in this section, it is obvious from our general discussion of Couette flow that the flow-field properties vary only in the y direction; all derivatives in the x direction are zero. Therefore, as a matter of mathematical preciseness, all the partial derivatives in Equations (16.1) to (16.3) can be written as ordinary derivatives. For example, Equation (16.1) can be written as

$$\frac{d}{dy}\left(\mu\frac{du}{dy}\right) = 0$$

However, our discussion of Couette flow is intended to serve as a straightforward example of a viscous flow problem, "breaking the ice" so-to-speak for the more practical but more complex problems to come—problems which involve changes in both the x and y directions, and which are described by *partial* differential equations. Therefore, on pedagogical grounds, we choose to continue the partial differential notation here, simply to make the reader feel more comfortable when we extend these concepts to the boundary layer and full Navier-Stokes solutions in Chapters 17 and 20, respectively.

16.3 INCOMPRESSIBLE (CONSTANT PROPERTY) COUETTE FLOW

In the study of viscous flows, a flow field in which ρ, μ, and k are treated as constants is sometimes labeled as "constant property" flow. This assumption is made in the present section. On a physical basis, this means that we are dealing with an incompressible flow, where ρ is constant. Also, since μ and k are functions of temperature (see Section 15.3), constant property flow implies that T is constant also. (We will relax this assumption slightly at the end of this section.)

The governing equations for Couette flow were derived in Section 16.2. In particular, the y-momentum equation, Equation (16.2), along with the geometrical property that $\partial p/\partial x = 0$, states that the pressure is constant throughout the flow. Consequently, all the information about the velocity field comes from the x-momentum equation, Equation (16.1), repeated below:

$$\frac{\partial}{\partial y}\left(\mu\frac{\partial u}{\partial y}\right) = 0 \tag{16.1}$$

For constant μ, this becomes

$$\frac{\partial^2 u}{\partial y^2} = 0 \tag{16.4}$$

Integrating with respect to y twice, we obtain

$$u = ay + b \tag{16.5}$$

where a and b are constants of integration. These constants can be obtained from the boundary conditions illustrated in Figure 16.2 as follows:

At $y = 0$, $u = 0$; hence, $b = 0$.

At $y = D$, $u = u_e$; hence, $a = u_e/D$.

Thus, the variation of velocity for incompressible Couette flow is given by Equation (16.5) as

$$\boxed{u = u_e\left(\frac{y}{D}\right)} \tag{16.6}$$

Note the important result that the velocity varies *linearly* across the flow. This result is sketched in Figure 16.3.

Once the velocity profile is obtained, we can obtain the shear stress at any point in the flow from Equation (15.1), repeated below (the subscript yx is dropped here because we know the only shear stress acting in this problem is that in the x direction):

$$\tau = \mu\frac{\partial u}{\partial y} \tag{16.7}$$

From Equation (16.6),

$$\frac{\partial u}{\partial y} = \frac{u_e}{D} \tag{16.8}$$

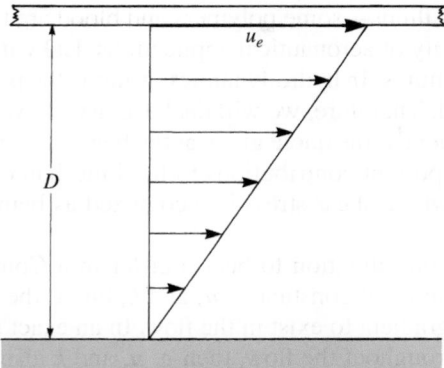

Figure 16.3 Velocity profile for incompressible Couette flow.

Hence, from Equations (16.7) and (16.8), we have

$$\tau = \mu \left(\frac{u_e}{D} \right)$$

(16.9)

Note that the shear stress is *constant* throughout the flow. Moreover, the straightforward result given by Equation (16.9) illustrates two important physical trends—trends that we will find to be almost universally present in all viscous flows:

1. As u_e increases, the shear stress increases. From Equation (16.9), τ increases linearly with u_e; this is a specific result germane to Couette flow. For other problems, the increase is not necessarily linear.

2. As D increases, the shear stress decreases; that is, as the thickness of the viscous shear layer increases, all other things being equal, the shear stress becomes smaller. From Equation (16.9), τ is inversely proportional to D—again a result germane to Couette flow. For other problems, the decrease in τ is not necessarily in *direct* inverse proportion to the shear-layer thickness.

With the above results in mind, reflect for a moment on the quotation from Isaac Newton's *Principia* given at the beginning of this chapter. Here, the "want of lubricity" is, in modern terms, interpreted as the shear stress. This want of lubricity is, according to Newton, "proportional to the velocity with which the parts of the fluid are separated from one another," that is, in the context of the present problem proportional to u_e/D. This is precisely the statement contained in Equation (16.9). In more recent times, Newton's statement is generalized to the form given by Equation (16.7), and even more generalized by Equation (15.1). For this reason, Equations (15.1) and (16.7) are frequently called the newtonian shear stress law, and fluids which obey this law are called *newtonian fluids*. [There are some specialized fluids which do not obey Equation (15.1) or (16.7); they are called non-newtonian fluids—some polymers and blood are two such examples.] By far, the vast majority of aeronautical applications deal with air or other gases, which are newtonian fluids. In hydrodynamics, water is the primary medium, and it is a newtonian fluid. Therefore, we will deal exclusively with newtonian fluids in this book. Consequently, the quote given at the beginning of this chapter is one of Newton's most important contributions to fluid mechanics—it represents the first time in history where shear stress is recognized as being proportional to a velocity gradient.

Let us now turn our attention to heat transfer in a Couette flow. Here, we continue our assumptions of constant ρ, μ, and k, but at the same time, we will allow a temperature gradient to exist in the flow. In an exact sense, this is inconsistent; if T varies throughout the flow, then ρ, μ, and k also vary. However, for this application, we assume that the temperature variations are small—indeed, small enough such that ρ, μ, and k are *approximately* constant—and treat them so in the equations. On the other hand, the temperature changes, although small on an absolute basis, are sufficient to result in meaningful heat flux through the fluid. The results obtained will reflect some of the important trends in aerodynamic heating associated with high-speed flows, to be discussed in subsequent chapters.

For Couette flow with heat transfer, return to Figure 16.2. Here, the temperature of the upper plate is T_e and that of the lower plate is T_w. Hence, we have as boundary conditions for the temperature of the fluid:

At $y = 0$: $\qquad\qquad\qquad\qquad\qquad T = T_w$

At $y = D$: $\qquad\qquad\qquad\qquad\qquad T = T_e$

The temperature profile in the flow is governed by the energy equation, Equation (16.3). For constant μ and k, this equation is written as

$$\frac{k}{\mu}\left(\frac{\partial^2 T}{\partial y^2}\right) + \frac{\partial}{\partial y}\left(u\frac{\partial u}{\partial y}\right) = 0 \qquad (16.10)$$

Also, since μ is assumed to be constant, Equations (16.10) and (16.1) are *totally uncoupled*. That is, for the constant property flow considered here, the solution of the momentum equation [Equation (16.1)] is totally separate from the solution of the energy equation [Equation (16.10)]. Therefore, in this problem, although the temperature is allowed to vary, the velocity field is still given by Equation (16.6), as sketched in Figure 16.3.

In dealing with flows where energy concepts are important, the enthalpy h is frequently a more fundamental variable than temperature; we have seen much evidence of this in Part 3, where energy changes were a vital consideration. In the present problem, where the temperature changes are small enough to justify the assumptions of constant ρ, μ, and k, this is not quite the same situation. However, because we will need to solve Equation (16.10), which is an energy equation for a flow (no matter how small the energy changes), and because we are using Couette flow as an example to set the stage for more complex problems, it is instructional (but by no means necessary) to couch Equation (16.10) in terms of enthalpy. Assuming constant specific heat, we have

$$h = c_p T \tag{16.11}$$

Equation (16.11) is valid for the Couette flow of any fluid with constant heat capacity; here, the germane specific heat is that at constant pressure c_p because the entire flow field is at constant pressure. In this sense, Equation (16.11) is a result of applying the first law of thermodynamics to a constant pressure process and recalling the fundamental definition of heat capacity as the heat added per unit change in temperature, $\delta q/dT$. Of course, if the fluid is a calorically perfect gas, then Equation (16.11) is a basic thermodynamic property of such a gas quite independent of what the process may be [see Section 7.2 and Equation (7.6b)]. Inserting Equation (16.11) into Equation (16.10), we have

$$\frac{k}{\mu c_p} \frac{\partial^2 h}{\partial y^2} + \frac{\partial}{\partial y}\left(u\frac{\partial u}{\partial y}\right) = 0 \tag{16.12}$$

Recall the definition of the Prandtl number from Section 15.6, namely,

$$\text{Pr} = \frac{\mu c_p}{k}$$

Equation (16.12) can be written in terms of the Prandtl number as

$$\frac{1}{\text{Pr}} \frac{\partial^2 h}{\partial y^2} + \frac{\partial}{\partial y}\left(u\frac{\partial u}{\partial y}\right) = 0$$

or

$$\frac{\partial^2 h}{\partial y^2} + \frac{\text{Pr}}{2}\frac{\partial}{\partial y}\left(\frac{\partial u^2}{\partial y}\right) = 0 \tag{16.13}$$

Integrating twice in the y direction, we find that Equation (16.13) yields

$$h + \left(\frac{\text{Pr}}{2}\right)u^2 = ay + b \tag{16.14}$$

where a and b are constants of integration [different from the a and b in Equation (16.5)]. Expressions for a and b are found by applying Equation (16.14) at the boundaries, as follows:

At $y = 0$: $h = h_w$ and $u = 0$

At $y = D$: $h = h_e$ and $u = u_e$

Hence, from Equation (16.14) at the boundaries,

$$b = h_w$$

and

$$a = \frac{h_e - h_w + (\text{Pr}/2)u_e^2}{D}$$

Inserting these values into Equation (16.14) and rearranging, we have

$$h = h_w + \left[h_e - h_w + \left(\frac{\text{Pr}}{2}\right)u_e^2\right]\frac{y}{D} - \left(\frac{\text{Pr}}{2}\right)u^2 \tag{16.15}$$

Inserting Equation (16.6) for the velocity profile in Equation (16.15) yields

$$h = h_w + \left[h_e - h_w + \left(\frac{\text{Pr}}{2}\right)u_e^2\right]\frac{y}{D} - \left(\frac{\text{Pr}}{2}\right)u_e^2\left(\frac{y}{D}\right)^2 \qquad (16.16)$$

Note that h varies *parabolically* with y/D across the flow. Since $T = h/c_p$, then the temperature profile across the flow is also parabolic. The precise shape of the parabolic curve depends on h_w (or T_w), h_e (or T_e), and Pr. Also note that, as expected from our discussion of the viscous flow similarity parameters in Section 15.6, the Prandtl number is clearly a strong player in the results; Equation (16.16) is one such example.

Once the enthalpy (or temperature) profile is obtained, we can obtain the heat flux at any point in the flow from Equation (15.2), repeated below (the subscript y is dropped here because we know the only direction of heat transfer is in the y direction for this problem):

$$\dot{q} = -k\frac{\partial T}{\partial y} \qquad (16.17)$$

Equation (16.17) can be written as

$$\dot{q} = -\frac{k}{c_p}\frac{\partial h}{\partial y} \qquad (16.18)$$

In Equation (16.18), the enthalpy gradient is obtained by differentiating Equation (16.16) as follows:

$$\frac{\partial h}{\partial y} = \left[h_e - h_w + \left(\frac{\text{Pr}}{2}\right)u_e^2\right]\frac{1}{D} - \text{Pr}\,u_e^2\frac{y}{D^2} \qquad (16.19)$$

Inserting Equation (16.19) into Equation (16.18), and writing k/c_p as μ/Pr, we have

$$\dot{q} = -\mu\left(\frac{h_e - h_w}{\text{Pr}} + \frac{u_e^2}{2}\right)\frac{1}{D} + \mu u_e^2\frac{y}{D^2} \qquad (16.20)$$

From Equation (16.20), note that $\dot{q}$ is *not* constant across the flow, unlike the shear stress discussed earlier. Rather, $\dot{q}$ varies linearly with y. The physical reason for the variation of $\dot{q}$ is *viscous dissipation* which takes place within the flow, and which is associated with the shear stress in the flow. Indeed, the last term in Equation (16.20), in light of Equations (16.6) and (16.9), can be written as

$$\mu u_e^2\frac{y}{D^2} = \tau u_e\left(\frac{y}{D}\right) = \tau u$$

Hence, Equation (16.20) becomes

$$\dot{q} = -\mu\left(\frac{h_e - h_w}{\text{Pr}} + \frac{u_e^2}{2}\right)\frac{1}{D} + \tau u \qquad (16.21)$$

The variation of $\dot{q}$ across the flow is due to the last term in Equation (16.21), and this term involves shear stress multiplied by velocity. The term τu is *viscous dissipation;* it is the time rate of heat generated at a point in the flow by one streamline at a given velocity "rubbing" against an adjacent streamline at a slightly different velocity—analogous to the heat you feel when rubbing your hands together vigorously. Note that, if u_e is negligibly small, then the viscous dissipation is small and can be neglected; that is, in Equation (16.20) the last term can be neglected (u_e is small), and in Equation (16.21) the last term can be neglected (τ is small if u_e is small). In this case, the heat flux becomes constant across the flow, simply equal to

$$\dot{q} \approx -\frac{\mu}{\text{Pr}} \left(\frac{h_e - h_w}{D} \right) \tag{16.22}$$

In this case, the "driving potential" for heat transfer across the flow is simply the enthalpy difference $(h_e - h_w)$ or, in other words, the temperature difference $(T_e - T_w)$ across the flow. However, as we have emphasized, if u_e is not negligible, then viscous dissipation becomes another factor that drives the heat transfer across the flow.

Of particular practical interest is the heat flux at the walls—the *aerodynamic heating* as we label it here. We denote the heat transfer at a wall as $\dot{q}_w$. Moreover, it is conventional to quote aerodynamic heating at a wall without any sign convention. For example, if the heat transfer from the fluid to the wall is 10 W/cm², or, if in reverse the heat transfer from the wall to the fluid is 10 W/cm², it is simply quoted as such; in both cases, $\dot{q}_w$ is given as 10 W/cm² without any sign convention. In this sense, we write Equation (16.18) as

$$\dot{q}_w = \frac{k}{c_p} \left| \frac{\partial h}{\partial y} \right|_w = \frac{\mu}{\text{Pr}} \left| \frac{\partial h}{\partial y} \right|_w \tag{16.23}$$

where the subscript w implies conditions at the wall. The direction of the net heat transfer at the wall, whether it is from the fluid to the wall or from the wall to the fluid, is easily seen from the temperature gradient at the wall; if the wall is cooler than the adjacent fluid, heat is transferred into the wall, and if the wall is hotter than the adjacent fluid, heat is transferred into the fluid. Another criterion is to compare the wall temperature with the adiabatic wall temperature, to be defined shortly.

Return to the picture of Couette flow in Figure 16.2. To calculate the heat transfer at the lower wall, use Equation (16.23) with the enthalpy gradient given by Equation (16.19) evaluated at $y = 0$:

$$\text{At } y = 0: \quad \dot{q}_w = \frac{\mu}{\text{Pr}} \left| \frac{h_e - h_w + \frac{1}{2} \text{Pr } u_e^2}{D} \right| \tag{16.24}$$

To calculate the heat transfer at the upper wall, use Equation (16.23) with the enthalpy gradient given by Equation (16.19) evaluated at $y = D$. In this case, Equation (16.19) yields

$$\frac{\partial h}{\partial y} = \frac{h_e - h_w + \frac{1}{2} \text{Pr } u_e^2}{D} - \frac{\text{Pr } u_e^2}{D} = \frac{h_e - h_w - \frac{1}{2} \text{Pr } u_e^2}{D}$$

In turn, from Equation (16.23)

$$At \; y = D: \qquad \dot{q}_w = \frac{\mu}{\text{Pr}} \left| \frac{h_e - h_w - \frac{1}{2}\text{Pr}\,u_e^2}{D} \right| \qquad (16.25)$$

Let us examine the above results for three different scenarios, namely, (1) *negligible viscous dissipation,* (2) *equal wall temperature,* and (3) *adiabatic wall conditions* (no heat transfer to the wall). In the process, we define three important concepts in the analysis of aerodynamic heating: (1) *adiabatic wall temperature,* (2) *recovery factor,* and (3) *Reynolds analogy.*

16.3.1 Negligible Viscous Dissipation

To some extent, we have already discussed this case in regard to the local heat flux at any point within the flow. If u_e is very small, hence τ is very small, then the amount of viscous dissipation is negligibly small, and Equation (16.21) becomes

$$\dot{q} = -\frac{\mu}{\text{Pr}} \left(\frac{h_e - h_w}{D} \right) \qquad (16.26)$$

Clearly, for this case, the heat flux is constant across the flow. Moreover, the enthalpy profile given by Equation (16.16) becomes

$$h = h_w + (h_e - h_w)\frac{y}{D} \qquad (16.27)$$

Since $h = c_p T$, the temperature profile is identical to the enthalpy profile:

$$T = T_w + (T_e - T_w)\frac{y}{D} \qquad (16.28)$$

Note that the temperature varies *linearly* across the flow, as sketched in Figure 16.4. The case shown here is for the upper wall at a higher temperature than the lower

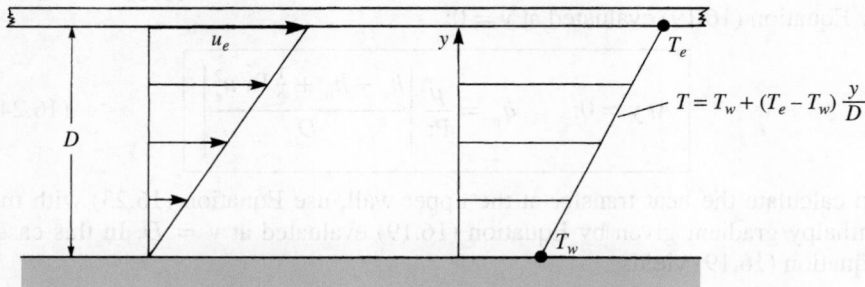

Figure 16.4 Couette flow temperature profile for negligible viscous dissipation.

wall. The heat transfer at the lower wall is obtained from Equation (16.24) with a negligible u_e:

$$At\ y = 0: \qquad \dot{q}_w = \frac{\mu}{\text{Pr}}\left|\frac{h_e - h_w}{D}\right| \qquad (16.29)$$

The heat transfer at the upper wall is similarly obtained as

$$At\ y = D: \qquad \dot{q}_w = \frac{\mu}{\text{Pr}}\left|\frac{h_e - h_w}{D}\right| \qquad (16.30)$$

Equations (16.29) and (16.30) are identical; this is no surprise, since we have already shown that the heat flux is constant across the flow, as shown by Equation (16.26), and therefore the heat transfer at both walls should be the same. Equations (16.29) and (16.30) can also be written in terms of temperature as

$$\dot{q}_w = k\left|\frac{T_e - T_w}{D}\right| \qquad (16.31)$$

Examining Equations (16.29) to (16.31), we can make some conclusions which can be generalized to most viscous flow problems, as follows:

1. Everything else being equal, the larger the temperature difference across the viscous layer, the greater the heat transfer at the wall. The temperature difference $(T_e - T_w)$ or the enthalpy difference $(h_e - h_w)$ takes on the role of a "driving potential" for heat transfer. For the special case treated here, the heat transfer at the wall is directly proportional to this driving potential.
2. Everything else being equal, the thicker the viscous layer (the larger D is), the smaller the heat transfer is at the wall. For the special case treated here, $\dot{q}_w$ is inversely proportional to D.
3. Heat flows from a region of high temperature to low temperature. For negligible viscous dissipation, if the temperature at the top of the viscous layer is higher than that at the bottom, heat flows from the top to the bottom. In the case sketched in Figure 16.4, heat is transferred from the upper plate into the fluid, and then is transferred from the fluid to the lower plate.

16.3.2 Equal Wall Temperatures

Here we assume that $T_e = T_w$; that is, $h_e = h_w$. The enthalpy profile for this case, from Equation (16.16), is

$$or \qquad h = h_w + \frac{1}{2}\text{Pr}\,u_e^2\left(\frac{y}{D}\right) - \frac{1}{2}\text{Pr}\,u_e^2\left(\frac{y}{D}\right)^2 \qquad (16.32)$$

$$= h_w + \frac{1}{2}\text{Pr}\,u_e^2\left[\frac{y}{D} - \left(\frac{y}{D}\right)^2\right]$$

In terms of temperature, this becomes

$$T = T_w + \frac{\text{Pr}\,u_e^2}{2c_p}\left[\frac{y}{D} - \left(\frac{y}{D}\right)^2\right] \qquad (16.33)$$

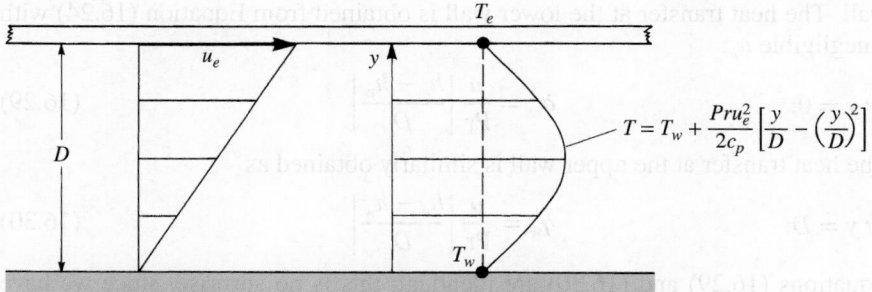

Figure 16.5 Couette flow temperature profile for equal wall temperature with viscous dissipation.

Note that the temperature varies parabolically with y, as sketched in Figure 16.5. The maximum value of temperature occurs at the midpoint, $y = D/2$. This maximum value is obtained by evaluating Equation (16.33) at $y = D/2$.

$$T_{max} = T_w + \frac{\Pr u_e^2}{8c_p} \tag{16.34}$$

The heat transfer at the walls is obtained from Equations (16.24) and (16.25) as

At y = 0:
$$\dot{q}_w = \frac{\mu}{2} \frac{u_e^2}{D} \tag{16.35}$$

At y = D:
$$\dot{q}_w = \frac{\mu}{2} \frac{u_e^2}{D} \tag{16.36}$$

Equations (16.35) and (16.36) are identical; the heat transfers at the upper and lower walls are equal. In this case, as can be seen by inspecting the temperature distribution shown in Figure 16.5, the upper and lower walls are both cooler than the adjacent fluid. Hence, at both the upper and lower walls, heat is transferred from the fluid to the wall.

Question: Since the walls are at equal temperature, where is the heat transfer coming from? *Answer: Viscous dissipation.* The local temperature increase in the flow as sketched in Figure 16.5 is due *solely* to viscous dissipation within the fluid. In turn, both walls experience an aerodynamic heating effect due to this viscous dissipation. This is clearly evident in Equations (16.35) and (16.36), where $\dot{q}_w$ depends on the velocity u_e. Indeed, $\dot{q}_w$ is directly proportional to the *square* of u_e. In light of Equation (16.9), Equations (16.35) and (16.36) can be written as

$$\dot{q}_w = \tau \left(\frac{u_e}{2} \right) \tag{16.37}$$

which further emphasizes that $\dot{q}_w$ is due entirely to the action of shear stress in the flow. From Equations (16.35) to (16.37), we can make the following conclusions that reflect general properties of most viscous flows:

1. Everything else being equal, aerodynamic heating increases as the flow velocity increases. This is why aerodynamic heating becomes an important

design factor in high-speed aerodynamics. Indeed, for most hypersonic vehicles, you can begin to appreciate that viscous dissipation generates extreme temperatures within the boundary layer adjacent to the vehicle surface and frequently makes aerodynamic heating the dominant design factor. In the Couette flow case shown here—a far cry from hypersonic flow—we see that $\dot{q}_w$ varies directly as u_e^2.

2. Everything else being equal, aerodynamic heating decreases as the thickness of the viscous layer increases. For the case considered here, $\dot{q}_w$ is inversely proportional to D. This conclusion is the same as that made for the above case of negligible viscous dissipation but with unequal wall temperature.

16.3.3 Adiabatic Wall Conditions (Adiabatic Wall Temperature)

Let us imagine the following situation. Assume that the flow illustrated in Figure 16.5 is established. We have the parabolic temperature profile established as shown, and we have heat transfer into the walls as just discussed. However, both wall temperatures are considered *fixed,* and both are equal to the same constant value. *Question:* How can the wall temperature remain fixed at the same time that heat is transferred into the wall? *Answer:* There must be some independent mechanism that conducts heat away from the wall at the same rate that the aerodynamic heating is pumping heat into the wall. This is the only way for the wall temperature to remain fixed at some cooler temperature than the adjacent fluid. For example, the wall can be some vast heat sink that can absorb heat without any appreciable change in temperature, or possibly there are cooling coils within the plate that can carry away the heat, much like the water coils that keep the engine of your automobile cool. In any event, to have the picture shown in Figure 16.5 with a constant wall temperature independent of time, some exterior mechanism must carry away the heat that is transferred from the fluid to the walls. Now imagine that, at the lower wall, this exterior mechanism is suddenly shut off. The lower wall will now begin to grow hotter in response to $\dot{q}_w$, and T_w will begin to increase with time. At any given instant during this transient process, the heat transfer to the lower wall is given by Equation (16.24), repeated below:

$$\dot{q}_w = \frac{\mu}{\text{Pr}}\left| \frac{h_e - h_w + \frac{1}{2}\,\text{Pr}\,u_e^2}{D} \right| \qquad (16.24)$$

At time $t = 0$, when the exterior cooling mechanism is just shut off, $h_w = h_e$, and $\dot{q}_w$ is given by Equation (16.35), namely,

At time $t = 0$:
$$\dot{q}_w = \frac{\mu}{2}\frac{u_e^2}{D}$$

However, as time now progresses, T_w (and therefore h_w) increases. From Equation (16.24), as h_w increases, the numerator decreases in magnitude, and hence $\dot{q}_w$ decreases. That is,

At $t > 0$:
$$\dot{q}_w < \frac{\mu}{2}\frac{u_e^2}{D}$$

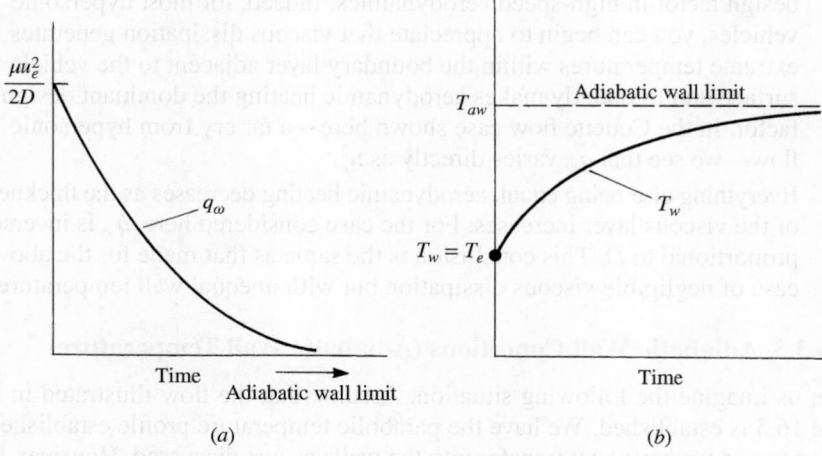

Figure 16.6 Illustration for the definition of an adiabatic wall and the adiabatic wall temperature.

Hence, as time progresses from when the exterior cooling mechanism was first cut off at the lower wall, the wall temperature increases, and the aerodynamic heating to the wall decreases. This in turn slows the rate of increase of T_w as time progresses. The transient variations of both $\dot{q}_w$ and T_w are sketched in Figure 16.6. In Figure 16.6a, we see that, as time increases to large values, the heat transfer to the wall approaches zero—this is defined as the *equilibrium, or the adiabatic wall condition*. For an *adiabatic* wall, the heat transfer is, by definition, equal to *zero*. Simultaneously, the wall temperature T_w approaches asymptotically a limiting value defined as the *adiabatic wall temperature* T_{aw}, and the corresponding enthalpy is defined as the *adiabatic wall enthalpy* h_{aw}.

The purpose of this discussion is to define an adiabatic wall condition; the example involving a timewise approach to this condition was just for convenience and edification. Let us now assume that the lower wall in our Couette flow is an adiabatic wall. For this case, we already know the value of heat transfer to the wall—by definition, it is zero. The question now becomes, What is the value of the adiabatic wall enthalpy h_{aw}, and in turn the adiabatic wall temperature T_{aw}? The answer is given by Equation (16.23), where $\dot{q}_w = 0$ for an adiabatic wall.

Adiabatic wall:
$$\dot{q}_w = 0 \rightarrow \left(\frac{\partial h}{\partial y}\right)_w = \left(\frac{\partial T}{\partial y}\right)_w = 0 \qquad (16.38)$$

Therefore, from Equation (16.19), with $\partial h/\partial y = 0$, $y = 0$, and $h_w = h_{aw}$, by definition

$$h_e - h_{aw} + \frac{1}{2}\operatorname{Pr} u_e^2 = 0$$

or
$$\boxed{h_{aw} = h_e + \operatorname{Pr}\frac{u_e^2}{2}} \qquad (16.39)$$

In turn, the adiabatic wall temperature is given by

$$T_{aw} = T_e + \Pr \frac{u_e^2}{2c_p}$$
(16.40)

Clearly, the higher the value of u_e, the higher is the adiabatic wall temperature.

The enthalpy profile across the flow for this case is given by a combination of Equations (16.16) and (16.40), as follows. Setting $h_w = h_{aw}$ in Equation (16.16), we obtain

$$h = h_{aw} + \left(h_e + h_{aw} + \Pr \frac{u_e^2}{2} \right) \frac{y}{D} - \frac{\Pr}{2} u_e^2 \left(\frac{y}{D} \right)^2$$
(16.41)

From Equation (16.39),

$$h_e - h_{aw} = -\Pr \frac{u_e^2}{2}$$
(16.42)

Inserting Equation (16.42) into (16.41), we have

$$h = h_{aw} - \Pr \frac{u_e^2}{2} \left(\frac{y}{D} \right)^2$$
(16.43)

Equation (16.43) gives the enthalpy profile across the flow. The temperature profile follows from Equation (16.43) as

$$T = T_{aw} - \Pr \frac{u_e^2}{2c_p} \left(\frac{y}{D} \right)^2$$
(16.44)

This variation of T is sketched in Figure 16.7. Note that T_{aw} is the maximum temperature in the flow. Moreover, the temperature curve is perpendicular at the plate for $y = 0$; that is, the temperature gradient at the lower plate is zero, as expected for an adiabatic wall. This result is also obtained by differentiating Equation (16.44):

$$\frac{\partial T}{\partial y} = -\Pr \frac{u_e^2}{c_p D} \left(\frac{y}{D} \right)$$

which gives $\partial T / \partial y = 0$ at $y = 0$.

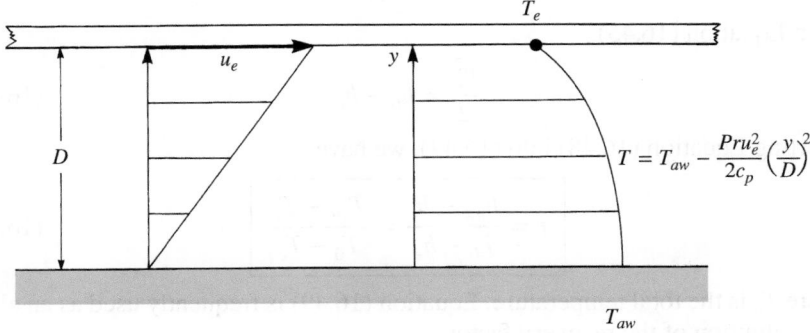

Figure 16.7 Couette flow temperature profile for an adiabatic lower wall.

16.3.4 Recovery Factor

As a corollary to the above case for the adiabatic wall, we take this opportunity to define the recovery factor—a useful engineering parameter in the analysis of aerodynamic heating. The total enthalpy of the flow at the upper plate (which represents the upper boundary of a viscous shear layer) is, by definition,

$$h_0 = h_e + \frac{u_e^2}{2} \tag{16.45}$$

(The significance and definition of total enthalpy are discussed in Section 7.5.) Compare Equation (16.45), which is a general definition, with Equation (16.39), repeated below, which is for the special case of Couette flow:

$$h_{aw} = h_e + \mathrm{Pr}\, \frac{u_e^2}{2} \tag{16.39}$$

Note that h_{aw} is different from h_0, the difference provided by the value of Pr as it appears in Equation (16.39). We now *generalize* Equation (16.39) to a form which holds for any viscous flow, as follows:

$$\boxed{h_{aw} = h_e + r\frac{u_e^2}{2}} \tag{16.46a}$$

Similarly, Equation (16.40) can be generalized to

$$\boxed{T_{aw} = T_e + r\frac{u_e^2}{2c_p}} \tag{16.46b}$$

In Equation (16.46a and b), r is defined as the *recovery factor*. It is the factor that tells us how close the adiabatic wall enthalpy is to the total enthalpy at the upper boundary of the viscous flow. If $r = 1$, then $h_{aw} = h_0$. An alternate expression for the recovery factor can be obtained by combining Equations (16.46) and (16.45) as follows. From Equation (16.46),

$$r = \frac{h_{aw} - h_e}{u_e^2/2} \tag{16.47}$$

From Equation (16.45),

$$\frac{u_e^2}{2} = h_0 - h_e \tag{16.48}$$

Inserting Equation (16.48) into (16.47), we have

$$\boxed{r = \frac{h_{aw} - h_e}{h_0 - h_e} = \frac{T_{aw} - T_e}{T_0 - T_e}} \tag{16.49}$$

where T_0 is the total temperature. Equation (16.49) is frequently used as an alternate definition of the recovery factor.

In the special case of Couette flow, by comparing Equation (16.39) or (16.40) with Equation (16.46a) or (16.46b), we find that

$$r = \text{Pr} \qquad (16.50)$$

For Couette flow, the recovery factor is simply the Prandtl number. Note that, if $\text{Pr} < 1$, then $h_{aw} < h_0$; conversely, if $\text{Pr} > 1$, then $h_{aw} > h_0$.

In more general viscous flow cases, the recovery factor is not simply the Prandtl number; however, in general, for incompressible viscous flows, we will find that the recovery *factor* is some *function* of Pr. Hence, the Prandtl number is playing its role as an important viscous flow parameter. As expected from Section 15.6, for a compressible viscous flow, the recovery factor is a function of Pr along with the Mach number and the ratio of specific heats.

16.3.5 Reynolds Analogy

Another useful engineering relation for the analysis of aerodynamic heating is Reynolds analogy, which can easily be introduced within the context of our discussion of Couette flow. Reynolds analogy is a relation between the skin friction coefficient and the heat transfer coefficient. The skin friction coefficient c_f was first introduced in Section 1.5. In our context here, we define the skin friction coefficient as

$$c_f = \frac{\tau_w}{\frac{1}{2}\rho_e u_e^2} \qquad (16.51)$$

From Equation (16.9), we have, for Couette flow,

$$\tau_w = \mu \left(\frac{u_e}{D} \right) \qquad (16.52)$$

Combining Equations (16.51) and (16.52), we have

$$c_f = \frac{\mu(u_e/D)}{\frac{1}{2}\rho_e u_e^2} = \frac{2\mu}{\rho_e u_e D} \qquad (16.53)$$

Let us define the Reynolds number for Couette flow as

$$\text{Re} = \frac{\rho_e u_e D}{\mu}$$

Then, Equation (16.53) becomes

$$c_f = \frac{2}{\text{Re}} \qquad (16.54)$$

Equation (16.54) is interesting in its own right. It demonstrates that the skin friction coefficient is a function of just the Reynolds number—a result which applies in general for other incompressible viscous flows [although the function is not necessarily the same as given in Equation (16.54)].

Now let us define a *heat transfer coefficient* as

$$C_H = \frac{\dot{q}_w}{\rho_e u_e (h_{aw} - h_w)} \tag{16.55}$$

In Equation (16.55), C_H is called the *Stanton number;* it is one of several different types of heat transfer coefficient that is used in the analysis of aerodynamic heating. It is a dimensionless quantity, in the same vein as the skin-friction coefficient. For Couette flow, from Equation (16.24), and dropping the absolute value signs for convenience, we have

$$\dot{q}_w = \frac{\mu}{Pr} \left(\frac{h_e - h_w + \frac{1}{2} Pr\, u_e^2}{D} \right) \tag{16.56}$$

Inserting Equation (16.39) into (16.56), we have for Couette flow

$$\dot{q}_w = \frac{\mu}{Pr} \left(\frac{h_{aw} - h_w}{D} \right) \tag{16.57}$$

Inserting Equation (16.57) into (16.55), we obtain

$$C_H = \frac{(\mu/Pr)[(h_{aw} - h_w)/D]}{\rho_e u_e (h_{aw} - h_w)} = \frac{\mu/Pr}{\rho_e u_e D} = \frac{1}{Re\ Pr} \tag{16.58}$$

Equation (16.58) is interesting in its own right. It demonstrates that the Stanton number is a function of the Reynolds number and Prandtl number—a result that applies generally for other incompressible viscous flows [although the function is not necessarily the same as given in Equation (16.58)].

We now combine the results for c_f and C_H obtained above. From Equations (16.54) and (16.58), we have

$$\frac{C_H}{c_f} = \left(\frac{1}{Re\ Pr} \right) \frac{Re}{2}$$

or

$$\boxed{\frac{C_H}{c_f} = \frac{1}{2} Pr^{-1}} \tag{16.59}$$

Equation (16.59) is Reynolds analogy as applied to Couette flow. Reynolds analogy is, in general, a relation between the heat transfer coefficient and the skin friction coefficient. For Couette flow, this relation is given by Equation (16.59). Note that the ratio C_H/c_f is simply a function of the Prandtl number—a result that applies usually for other incompressible viscous flows, although not necessarily the same function as given in Equation (16.59).

16.3.6 Interim Summary

In this section, we have studied incompressible Couette flow. Although it is a somewhat academic flow, it has all the trappings of many practical viscous flow problems, with the added advantage of lending itself to a simple, straightforward solution. We have taken this advantage, and have discussed incompressible

Couette flow in great detail. Our major purpose in this discussion is to make the reader familiar with many concepts used in general in the analysis of viscous flows without clouding the picture with more fluid dynamic complexities. In the context of our study of Couette flow, we have one additional question to address, namely, What is the effect of compressibility? This question is addressed in the next section.

EXAMPLE 16.1

Consider the geometry sketched in Figure 16.2. The velocity of the upper plate is 200 ft/s, and the two plates are separated by a distance of 0.01 in. The fluid between the plates is air. Assume incompressible flow. The temperature of both plates is the standard sea level value of 519°R.

(*a*) Calculate the velocity in the middle of the flow.

(*b*) Calculate the shear stress.

(*c*) Calculate the maximum temperature in the flow.

(*d*) Calculate the heat transfer to either wall.

(*e*) If the lower wall is suddenly made adiabatic, calculate its temperature.

■ Solution

Assume that μ is constant throughout the flow, and that it is equal to its value of 3.7373×10^{-7} slug/ft/s at the standard sea level temperature of 519°R.

(*a*) From Equation (16.6),

$$u = u_e \left(\frac{y}{D} \right)$$

$$u = (200)\left(\tfrac{1}{2} \right) = \boxed{100 \text{ ft/s}}$$

(*b*) From Equation (16.9),

$$\tau_w = \mu \frac{u_e}{D} \qquad \text{where } D = 0.01 \text{ in} = 8.33 \times 10^{-4} \text{ ft}$$

$$\tau_w = \frac{(3.7373 \times 10^{-7})(200)}{8.33 \times 10^{-4}} = \boxed{0.09 \text{ lb/ft}^2}$$

Note that the shear stress is relatively small—less than a tenth of a pound acting over a square foot.

(*c*) From Equation (16.34), for equal wall temperatures, the maximum temperature, which occurs at $y/D = 0.5$, is

$$T = T_w + \frac{\text{Pr}}{c_p} \frac{u_e^2}{2} \left[\frac{y}{D} - \left(\frac{y}{D} \right)^2 \right] = T_w + \frac{\text{Pr} \, u_e^2}{8 c_p}$$

For air at standard conditions, Pr = 0.71 and $c_p = 6006$ (ft · lb)/(slug · °R). Hence,

$$T = 519 + \frac{(0.71)(200)^2}{8(6006)} = 519 + 0.6 = \boxed{519.6°\text{R}}$$

Notice that the maximum temperature in the flow is only six-tenths of a degree above the wall temperature—viscous dissipation for this relatively low-speed case is very small.

This certainly justifies our assumption of constant ρ, μ, and k in this section, and gives us a feeling for the energy changes associated with an essentially incompressible flow—they are very small.

(d) From Equation (16.35),

$$\dot{q}_w = \frac{\mu}{2}\left(\frac{u_e^2}{D}\right) = \frac{(3.7373 \times 10^{-7})(200)^2}{(2)(8.33 \times 10^{-4})} = \boxed{8.97 \text{ (ft} \cdot \text{lb)/(ft}^2/\text{s)}}$$

Since there are 778 ft · lb to a Btu (British thermal unit), then

$$\dot{q}_w = 8.97 \text{ (ft} \cdot \text{lb)/(ft}^2/\text{s)} = 0.0115 \text{ Btu/(ft}^2/\text{s)}$$

(e) From Equation (16.40),

$$T_{aw} = T_e + \frac{\text{Pr}}{c_p}\left(\frac{u_e^2}{2}\right) = 519 + \frac{(0.71)(200)^2}{(2)(6006)}$$

$$= 519 + 2.36 = \boxed{521.36°\text{R}}$$

Note in the above example that the adiabatic wall temperature is higher than the maximum flow temperature calculated in part (c) for the cold wall case. In general, for cold wall cases, the viscous dissipation in the flow is not sufficient to heat the gas anywhere in the flow to a temperature as high as the adiabatic wall temperature. Also, we again note the comparatively low temperature increase—T_{aw} is only 2.36° higher than the upper wall temperature. In contrast, for the compressible flow to be treated in the next section, the temperature increases can be substantial—this is one of the major aspects that distinguishes compressible viscous flow from incompressible viscous flow. Note that, in the present problem, the Mach number of the upper plate is

$$M_e = \frac{u_e}{a_e} = \frac{u_e}{\sqrt{\gamma R T_e}} = \frac{200}{\sqrt{(1.4)(1716)(519)}} = 0.18$$

Again, this certainly justifies our assumption of incompressible flow for this problem.

16.4 COMPRESSIBLE COUETTE FLOW

Return to Figure 16.2, which is our general model of Couette flow. We now assume that u_e is large enough; hence, the changes in temperature within the flow are substantial enough, so that we must treat ρ, μ, and k as variables—this is compressible Couette flow. Since $T = T(y)$, then $\mu = \mu(y)$ and $k = k(y)$. Also, since $\partial p/\partial x = 0$ from the geometry and $\partial p/\partial y = 0$ from Equation (16.2), then the pressure is constant throughout the compressible Couette flow, just as in the incompressible case discussed in Section 16.3. From the equation of state, we have $\rho = p/RT$; because $T = T(y)$, ρ is also a function of y, varying inversely with temperature.

The governing equations for compressible Couette flow are Equations (16.1) to (16.3), with μ and k as variables. Let us arrange these equations in a form convenient for solution. From Equation (16.1),

$$\frac{\partial}{\partial y}\left(\mu \frac{\partial u}{\partial y}\right) = \frac{\partial \tau}{\partial y} = 0 \qquad (16.60)$$

or
$$\tau = \text{const} \qquad (16.61)$$

Hence, just as in the incompressible case, the shear stress is constant across the flow. However, keep in mind that $\mu = \mu(y)$, and, from $\tau = \mu(\partial u/\partial y)$, clearly the velocity gradient, $\partial u/\partial y$, is *not* constant across the flow—this is an essential difference between compressible and incompressible flows. With all this in mind, Equation (16.3), repeated below

$$\frac{\partial}{\partial y}\left(k \frac{\partial T}{\partial y}\right) + \frac{\partial}{\partial y}\left(\mu u \frac{\partial u}{\partial y}\right) = 0 \qquad (16.3)$$

can be written as

$$\frac{\partial}{\partial y}\left(k \frac{\partial T}{\partial y}\right) + \tau \frac{\partial u}{\partial y} = 0 \qquad (16.62)$$

The temperature variation of μ is accurately given by Sutherland's law, Equation (15.3), for the temperature range of interest in this book. Hence, from Equation (15.3) and recalling that it is written in the International System of Units, we have

$$\tau = \mu \frac{\partial u}{\partial y} = \mu_0 \left(\frac{T}{T_0}\right)^{3/2} \frac{T_0 + 110}{T + 110} \left(\frac{\partial u}{\partial y}\right) \qquad (16.63)$$

The solution for compressible Couette flow requires a numerical solution of Equation (16.62). Note that, with μ and k as variables, Equation (16.62) is a *nonlinear* differential equation, and for the conditions stated, it does not have a neat, closed-form, analytic solution. Recognizing the need for a numerical solution, let us write Equation (16.62) in terms of the ordinary differential equation that it really is. (Recall that we have been using the partial differential notation only as a carry-over from the Navier-Stokes equations and to make the equations for our study of Couette flow look more familiar when treating the two-dimensional and three-dimensional viscous flows discussed in Chapters 17 to 20—just a pedagogical ploy on our part):

$$\frac{d}{dy}\left(k \frac{\partial T}{\partial y}\right) + \tau \frac{du}{dy} = 0 \qquad (16.64)$$

Equation (16.64) must be solved between $y = 0$, where $T = T_w$, and $y = D$, where $T = T_e$. Note that the boundary conditions must be satisfied at two different locations in the flow, namely, at $y = 0$ and $y = D$; this is called a *two-point boundary value problem*. We present two approaches to the numerical solution of this problem. Both approaches are used for the solutions of more complex viscous flows to be discussed in Chapters 17 through 20, and that is why they are presented here in the context of Couette flow—simply to "break the ice" for our subsequent discussions.

16.4.1 Shooting Method

This method is a classic method for the solution of the boundary-layer equations to be discussed in Chapter 17. For the solution of compressible Couette flow, the same philosophy follows as that to be applied to boundary-layer solutions, and that is why we discuss it now. The method involves a double iteration, that is, two minor iterations nested within a major iteration. The scheme is as follows:

1. *Assume* a value for τ in Equation (16.64). A reasonable assumption to start with is the incompressible value, $\tau = \mu(u_e/D)$. Also, assume that the variation of $u(y)$ is given by the incompressible result from Equation (16.6).

2. Starting at $y = 0$ with the known boundary condition $T = T_w$, integrate Equation (16.64) across the flow until $y = D$. Use any standard numerical technique for ordinary differential equations, such as the well-known Runge-Kutta method (see, e.g., Reference 49). However, to start this numerical integration, because Equation (16.64) is second order, *two* boundary conditions must be specified at $y = 0$. We only have one physical condition, namely, $T = T_w$. Therefore, we have to *assume* a second condition; let us assume a value for the temperature gradient at the wall, that is, assume a value for $(dT/dy)_w$. A value based on the incompressible flow solution discussed in Section 16.3 would be a reasonable assumption. With the *assumed* $(dT/dy)_w$ and the *known* T_w at $y = 0$, then Equation (16.64) is integrated numerically away from the wall, starting at $y = 0$ and moving in small increments, Δy in the direction of increasing y. Values of T at each increment in y are produced by the numerical algorithm.

3. Stop the numerical integration when $y = D$ is reached. Check to see if the numerical value of T at $y = D$ equals the specified boundary condition, $T = T_e$. Most likely, it will not because we have had to assume a value for $(dT/dy)_w$ in step 2. Hence, return to step 2, assume another value of $(dT/dy)_w$, and repeat the integration. Continue to repeat steps 2 and 3 until convergence is obtained, that is, until a value of $(dT/dy)_w$ is found such that, after the numerical integration, $T = T_e$ at $y = D$. From the converged temperature profile obtained by repetition of steps 2 and 3, we now have numerical values for T as a function of y that satisfy both boundary conditions; that is, $T = T_w$ at the lower wall and $T = T_e$ at the upper wall. However, do not forget that this converged solution was obtained for the *assumed* value of τ and the assumed velocity profile $u(y)$ in step 1. Therefore, the converged profile for T is not necessarily the correct profile. We must continue further; this time to find the correct value for τ.

4. From the converged temperature profile obtained by the repetitive iteration in steps 2 and 3, we can obtain $\mu = \mu(y)$ from Equation (15.3).

5. From the definition of shear stress,

$$\tau = \mu \frac{du}{dy}$$

we have
$$\frac{du}{dy} = \frac{\tau}{\mu} \tag{16.65}$$

Recall from the solution of the momentum equation, Equation (16.60), that τ is a constant. Using the assumed value of τ from step 1, and the values of $\mu = \mu(y)$ from step 4, numerically integrate Equation (16.65) starting at $y = 0$ and using the known boundary condition $u = 0$ at $y = 0$. Since Equation (16.65) is first order, this single boundary condition is sufficient to initiate the numerical integration. Values of u at each increment in y, Δy, are produced by the numerical algorithm.

6. Stop the numerical integration when $y = D$ is reached. Check to see if the numerical value of u at $y = D$ equals the specified boundary condition, $u = u_e$. Most likely, it will not, because we have had to *assume* a value of τ and $u(y)$ all the way back in step 1, which has carried through to this point in our iterative solution. Hence, return to step 5, assume another value for τ, and repeat the integration of Equation (16.65). Continue to repeat steps 5 and 6 [using the same values of $\mu = \mu(y)$ from step 4] until convergence is obtained, that is, until a value of τ is found that, after the numerical integration of Equation (16.65), $u = u_e$ at $y = D$. From the converged velocity profile obtained by repetition of steps 5 and 6, we now have numerical values for u as a function of y that satisfy both boundary conditions; that is, $u = 0$ at $y = 0$ and $u = u_e$ at $y = D$. However, do not forget that this converged solution was obtained using $\mu = \mu(y)$ from step 4, which was obtained using the initially assumed τ and $u(y)$ from step 1. Therefore, the converged profile for u obtained here is not necessarily the correct profile. We must continue one big step further.

7. Return to step 2, using the new value of τ and the new $u(y)$ obtained from step 6. Repeat steps 2 through 7 until total convergence is obtained. When this double iteration is completed, then the profile for $T = T(y)$ obtained at the last cycle of step 3, the profile for $u = u(y)$ obtained at the last cycle of step 6, and the value of τ obtained at the last cycle of step 7 are all the correct values for the given boundary conditions. The problem is solved!

Looking over the shooting method as described above, we see two minor iterations nested within a major iteration. Steps 2 and 3 constitute the first minor iteration and provide ultimately the temperature profile. Steps 5 and 6 are the second minor iteration and provide ultimately the velocity profile. Steps 2 to 7 constitute the major iteration and ultimately result in the proper value of τ.

The shooting method described above for the solution of compressible Couette flow is carried over almost directly for the solution of the boundary-layer equations to be described in Chapter 18. In the same vein, there is another completely different approach to the solution of compressible Couette flow which carries over directly for the solution of the Navier-Stokes equations to be described in Chapter 20. This is the time-dependent, finite-difference method, first discussed in Chapter 13 and applied to the inviscid flow over a supersonic blunt body in Section 13.5. In order to prepare ourselves for Chapter 20, we briefly discuss the application of this method to the solution of compressible Couette flow.

16.4.2 Time-Dependent Finite-Difference Method

Return to the picture of Couette flow in Figure 16.2. Imagine, for a moment, that the space between the upper and lower plates is filled with a flow field which is *not* a Couette flow; for example, imagine some arbitrary flow field with gradients in both the *x* and *y* directions, including gradients in pressure. We can imagine such a flow existing at some instant during the start-up process just after the upper plate is set into motion. This would be a transient flow field, where u, T, ρ, etc., would be functions of time t as well as of x and y. Finally, after enough time elapses, the flow will approach a steady state, and this steady state will be the Couette flow solution discussed above. Let us track this picture numerically. That is, starting from an assumed initial flow field at time $t = 0$, let us solve the unsteady Navier-Stokes equations in steps of time until a steady flow is obtained at large times. As discussed in Section 13.5, the time-asymptotic steady flow is the desired result; the time-dependent approach is just a means to that end. At this stage in our discussion, it would be well for you to review the philosophy (not the details) presented in Section 13.5 before progressing further.

The Navier-Stokes equations are given by Equations (15.18*a* to *c*) and (15.26). For an unsteady, two-dimensional flow, they are

Continuity:
$$\frac{\partial p}{\partial t} = -\frac{\partial(\rho u)}{\partial x} - \frac{\partial(\rho v)}{\partial y} \tag{16.66}$$

x momentum:
$$\frac{\partial u}{\partial t} = -u\frac{\partial u}{\partial x} - v\frac{\partial u}{\partial y} - \frac{1}{\rho}\left[\frac{\partial p}{\partial x} - \frac{\partial \tau_{xx}}{\partial x} - \frac{\partial \tau_{yz}}{\partial y}\right] \tag{16.67}$$

y momentum:
$$\frac{\partial v}{\partial t} = -u\frac{\partial v}{\partial x} - v\frac{\partial v}{\partial y} - \frac{1}{\rho}\left[\frac{\partial p}{\partial y} - \frac{\partial \tau_{xy}}{\partial x} - \frac{\partial \tau_{yy}}{\partial y}\right] \tag{16.68}$$

Energy:
$$\frac{\partial(e + V^2/2)}{\partial t} = -u\frac{\partial(e + V^2/2)}{\partial x} - v\frac{\partial(e + V^2/2)}{\partial y} \tag{16.69}$$

$$+ \frac{1}{\rho}\left\{\frac{\partial}{\partial x}\left(k\frac{\partial T}{\partial x}\right) + \frac{\partial}{\partial y}\left(k\frac{\partial T}{\partial y}\right) - \frac{\partial(pu)}{\partial x} - \frac{\partial(pv)}{\partial y}\right.$$

$$\left. + \frac{\partial(u\tau_{xx})}{\partial x} + \frac{\partial(u\tau_{yx})}{\partial y} + \frac{\partial(v\tau_{xy})}{\partial x} + \frac{\partial(v\tau_{yy})}{\partial y}\right\}$$

Note that Equations (16.66) to (16.69) are written with the time derivatives on the left-hand side and spatial derivatives on the right-hand side. These are analogous to the form of the Euler equations given by Equations (13.59) to (13.62). In Equations (16.67) to (16.69), τ_{xy}, τ_{xx}, and τ_{yy} are given by Equations (15.5), (15.8), and (15.9), respectively.

The above equations can be solved by means of MacCormack's method as described in Chapter 13. This is a predictor-corrector approach, and its arrangement for the time-dependent method is described in Section 13.5. The application to compressible Couette flow is outlined as follows:

1. Divide the space between the two plates into a finite-difference grid, as sketched in Figure 16.8a. The length L of the grid is somewhat arbitrary, but it must be longer than a certain minimum, to be described shortly.

2. At $x = 0$ (the inflow boundary), specify some inflow conditions for u, v, ρ, and T (hence, e, since $e = c_v T$). The incompressible solution for Couette flow makes reasonable inflow boundary conditions.

3. At all the remaining grid points, arbitrarily assign values for all the flow-field variables, u, v, ρ, and T. This arbitrary flow field, which constitutes the initial conditions at $t = 0$, can have finite values of v, and can include pressure gradients.

4. Starting with the initial flow field established in step 3, solve Equations (16.66) to (16.69) in steps of time. For example, consider the x-momentum equation in the form of Equation (16.67). MacCormack's predictor-corrector method, applied to this equation, is as follows.

Predictor: Assume that we know the complete flow field at time t, and we wish to advance the flow-field variables to time $t + \Delta t$. Replace the spatial derivatives with forward differences:

$$\left(\frac{\partial u}{\partial t}\right)_{i,j} = -u_{i,j}\left(\frac{u_{i+1,j} - u_{i,j}}{\Delta x}\right) - v_{i,j}\left(\frac{u_{i,j+1} - u_{i,j}}{\Delta y}\right) \qquad (16.70)$$

$$- \frac{1}{\rho_{i,j}}\left\{\frac{p_{i+1,j} - p_{i,j}}{\Delta x} - \left[\frac{(\tau_{xx})_{i+1,j} - (\tau_{xx})_{i,j}}{\Delta x}\right]\right.$$

$$\left. - \left[\frac{(\tau_{yx})_{i,j+1} - (\tau_{yx})_{i,j}}{\Delta y}\right]\right\}$$

All the quantities on the right-hand side are known at time t; we want to advance the flow-field values to the next time, $t + \Delta t$. That is, the right-hand side of Equation (16.70) is a known number at time t. Form the *predicted* value of $u_{i,j}$ at time $t + \Delta t$, denoted by $\bar{u}_{i,j}$ from the first two terms of a Taylor's series as

$$\bar{u}_{i,j} = \underbrace{u_{i,j}}_{\substack{\text{Known at} \\ \text{time } t}} + \underbrace{\left(\frac{\partial u}{\partial t}\right)_{i,j}}_{\substack{\text{Calculated} \\ \text{from} \\ \text{Equation (16.70)}}} \Delta t \qquad (16.71)$$

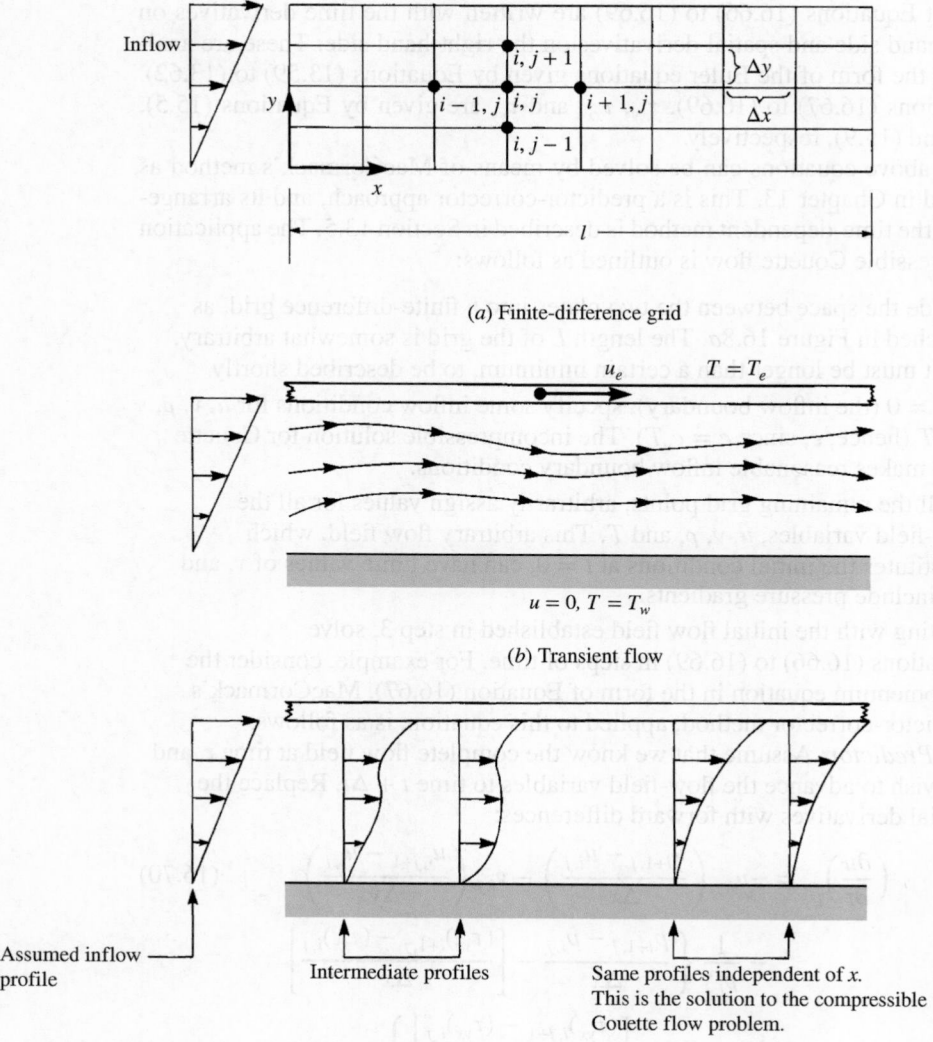

(a) Finite-difference grid

(b) Transient flow

(c) Steady-state flow

Figure 16.8 Illustration of the finite-difference grid, and characteristics of the flow during its transient approach to the steady state.

Calculate predicted values for ρ, v, and e, namely, $\bar{\rho}_{i,j}$, $\bar{v}_{i,j}$, and $\bar{e}_{i,j}$, by the same approach applied to Equations (16.66), (16.68), and (16.69), respectively. Do this for all the grid points in Figure 16.8a.

Corrector: Return to Equation (16.67), and replace the spatial derivatives with rearward differences using the predicted (barred) quantities obtained from the predictor step:

$$\left(\overline{\frac{\partial u}{\partial t}}\right)_{i,j} = -\bar{u}_{i,j}\left(\frac{\bar{u}_{i,j} - \bar{u}_{i-1,j}}{\Delta x}\right) - \bar{v}_{i,j}\left(\frac{\bar{u}_{i,j} - \bar{u}_{i,j-1}}{\Delta y}\right) \qquad (16.72)$$

$$- \frac{1}{\bar{\rho}_{i,j}}\left\{\frac{\bar{p}_{i,j} - \bar{p}_{i-1,j}}{\Delta x} - \left[\frac{(\bar{\tau}_{xx})_{i,j} - (\bar{\tau}_{xx})_{i-1,j}}{\Delta x}\right]\right.$$

$$\left. - \left[\frac{(\bar{\tau}_{yx})_{i,j} - (\bar{\tau}_{yx})_{i,j-1}}{\Delta y}\right]\right\}$$

Finally, calculate the corrected value of $u_{i,j}$ at time $t + \Delta t$, denoted by $u_{i,j}^{t+\Delta t}$, from the first two terms of a Taylor's series using an average time derivative obtained from Equations (16.70) and (16.72). That is,

$$u_{i,j}^{t+\Delta t} = u_{i,j}^t + \frac{1}{2}\left[\underbrace{\left(\frac{\partial u}{\partial t}\right)_{i,j}}_{\substack{\text{From}\\\text{Equation (16.70)}}} + \underbrace{\left(\overline{\frac{\partial u}{\partial t}}\right)_{i,j}}_{\substack{\text{From}\\\text{Equation (16.72)}}}\right]\Delta t \qquad (16.73)$$

Carry out the same process using Equations (16.66), (16.68), and (16.69) to obtain $\rho_{i,j}^{t+\Delta t}$, $v_{i,j}^{t+\Delta t}$, and $e_{i,j}^{t+\Delta t}$. The complete flow field at time $t + \Delta t$ is now obtained.

5. Repeat step 4, except starting with the newly calculated flow-field variables at the previous time. The flow-field variables will change from one time step to the next. This transient flow field will not even have parallel streamlines (i.e., there will be finite values of v throughout the flow). This is sketched in Figure 16.8b. Make the calculations for a large number of time steps; as we go out to large times, the changes in the flow-field variables from one time step to another will become smaller. Finally, if we go out to a large enough time (hundreds, sometimes even thousands, of time steps in some problems), the flow-field variables will not change anymore—a steady flow will be achieved, as sketched in Figure 16.8c. Moving from left to right in Figure 16.8c, we see a developing flow near the entrance, influenced by the assumed inflow profile. However, at the right of Figure 16.8c, the history of the inflow has died out, and the flow-field profiles become independent of distance. Indeed, we have chosen L to be a sufficient length for this to occur. The flow field near the exit is the desired solution to the compressible Couette flow problem.

The value of Δt in Equations (16.71) and (16.73) is not arbitrary. The steps outlined above constitute an *explicit* finite-difference method, and hence there is a stability bound on Δt. The value of Δt must be less than some prescribed maximum, or else the numerical solution will become unstable and "blow up" on the computer. A useful expression for Δt is the Courant-Friedrichs-Lewy (CFL) criterion, which states that Δt should be the minimum of Δt_x and Δt_y, where

$$\Delta t_x = \frac{\Delta x}{u + a} \qquad \Delta t_y = \frac{\Delta y}{v + a} \qquad (16.74)$$

In Equation (16.74), a is the local speed of sound. Equation (16.74) is evaluated at every grid point, and the minimum value is used to advance the whole flow field.

The time-dependent technique described above is a common approach to the solution of the compressible Navier-Stokes equations, and for that reason, it has been outlined here. Our purpose has been not so much to outline the solution of Couette flow by means of this technique, but rather to present the technique as a precursor to our later discussions on Navier-Stokes solutions.

16.4.3 Results for Compressible Couette Flow

Some typical results for compressible Couette flow are shown in Figure 16.9 for a cold wall case, and in Figure 16.10 for an adiabatic lower wall case. These results are obtained from White (Reference 41); they assume a viscosity-temperature relation of $\mu / \mu_{\max} = (T/T_{\mathrm{ref}})^{2/3}$, which is not quite as accurate for a gas as is Sutherland's law [Equation (15.3)]. Recall from Section 15.6 that a compressible viscous flow is governed by the following similarity parameters: the Mach number, the Prandtl number, and the ratio of specific heats, γ. Therefore, we expect the results for compressible Couette flow to be governed by the same parameters. Such is the case, as illustrated in Figures 16.9 and 16.10. Here we see

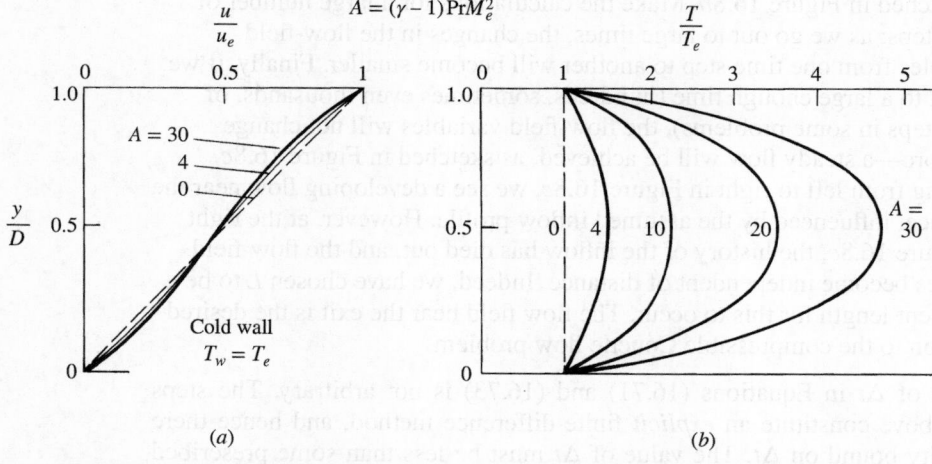

$$A = (\gamma - 1)\mathrm{Pr}M_e^2$$

Figure 16.9 Velocity and temperature profiles for compressible Couette flow. Cold wall cases. (*Data Source:* White, F. M.: *Viscous Fluid Flow*, McGraw Hill Book Company, New York, 1974.)

$$A = (\gamma - 1)\mathrm{Pr}M_e^2$$

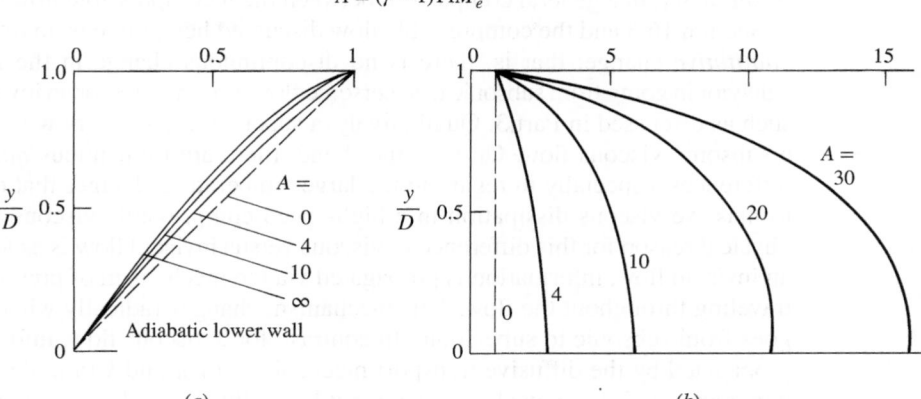

Figure 16.10 Velocity and temperature profiles for compressible Couette flow. Adiabatic lower wall. (*Data Source:* White, F. M.: *Viscous Fluid Flow*, McGraw Hill Book Company, New York, 1974.)

the different flow-field profiles for different values of the combined parameter $A = (\gamma - 1)\mathrm{Pr}\,M_e^2$. In particular, examining Figure 16.9 for the equal temperature, cold wall case, we note that:

1. From Figure 16.9a, the velocity profiles are not greatly affected by compressibility. The profile labeled $A = 0$ is the familiar linear incompressible case, and that labeled $A = 30$ corresponds to M_e approximately 10. Clearly, the velocity profile (in terms of u/u_e versus y/D) does not change greatly over such a large range of Mach number.

2. In contrast, from Figure 16.9b, there are huge temperature changes in the flow; these are due exclusively to viscous dissipation, which is a major effect at high Mach numbers. For example, for $A = 30$ ($M_e \approx 10$), the temperature in the middle of the flow is almost five times the wall temperature. Contrast this with the very small temperature increase calculated in Example 16.1 for an incompressible flow. This is why, on the scale in Figure 16.9b, the incompressible case ($A = 0$) is seen as essentially a vertical line.

For the adiabatic wall case shown in Figure 16.10, we note the following:

1. From Figure 16.10a, the velocity profiles show a pronounced curvature due to compressibility.

2. From Figure 16.10b, the temperature increases are larger than for the cold wall case. Note that, for $A = 30$ ($M_e \approx 10$), the maximum temperature is over 15 times that of the upper wall. Also, note the results, familiar from our discussion in Section 16.3, that the temperature is the largest at the adiabatic wall; that is, T_{aw} is the maximum temperature. As expected, Figure 16.10b shows that T_{aw} increases markedly as M_e increases.

In summary, in a general comparison between the incompressible flow discussed in Section 16.3 and the compressible flow discussed here, there is no tremendous *qualitative* change; that is, there is no discontinuous change in the flow-field behavior in going from subsonic to supersonic flow as is the case for an inviscid flow, such as discussed in Part 3. Qualitatively, a supersonic viscous flow is similar to a subsonic viscous flow. On the other hand, there are tremendous *quantitative* differences, especially in regard to the large temperature changes that occur due to massive viscous dissipation in a high-speed compressible viscous flow. The physical reason for this difference in viscous versus inviscid flow is as follows. In an inviscid flow, information is propagated via the mechanism of pressure waves traveling throughout the flow. This mechanism changes radically when the flow goes from subsonic to supersonic. In contrast, for a viscous flow, information is propagated by the diffusive transport mechanisms of μ and k (a molecular phenomenon), and these mechanisms are not basically changed when the flow goes from subsonic to supersonic. These statements hold in general for any viscous flow, not just for the Couette flow case treated here.

16.4.4 Some Analytical Considerations

For air temperatures up to 1000 K, the specific heats are essentially constant, thus justifying the assumption of a calorically perfect gas for this range. Moreover, the temperature variations of μ and k over this range are virtually identical. As a result, the Prandtl number, $\mu c_p/k$, is essentially *constant* up to temperatures on the order of 1000 K. This is shown in Figure 16.11 (Reference 50). Note that $\text{Pr} \approx 0.71$ for air; this is the value that was used in Example 16.1.

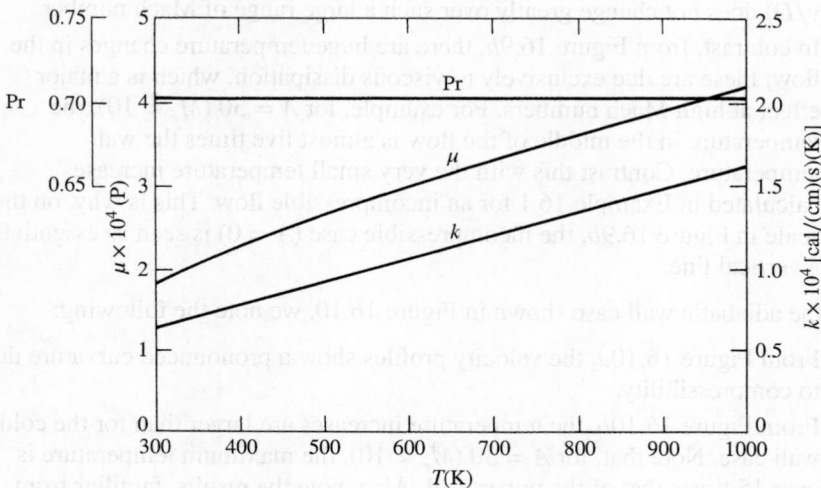

Figure 16.11 Variation of viscosity coefficient, thermal conductivity, and Prandtl number for air as a function of temperature. (*Data Source:* Schetz, Joseph A.: *Foundations of Boundary Layer Theory for Momentum, Heat, and Mass Transfer, Prentice-Hall, Inc.*, Englewood Cliffs, NJ, 1984.)

Question: How high a Mach number can exist before we would expect to encounter temperatures in the flow above 1000 K? *Answer:* An approximate answer is to calculate that Mach number at which the total temperature is 1000 K. Assuming a static temperature $T = 288$ K, from Equation (8.40),

$$M = \sqrt{\frac{2}{\gamma - 1}\left(\frac{T_0}{T} - 1\right)} = \sqrt{\frac{2}{0.4}\left(\frac{1000}{288} - 1\right)} = 3.5$$

Hence, for most aeronautical applications involving flight at a Mach number of 3.5 or less, the temperature within the viscous portions of the flow field will not exceed 1000 K. A Mach number of 3.5 or less encompasses virtually all operational aircraft today, with the exception of a few hypersonic test vehicles.

In light of the above, many viscous flow solutions are carried out making the justifiable assumption of a *constant Prandtl number*. For the case of compressible Couette flow, the assumption of Pr = constant allows the following analysis. Consider the energy equation, Equation (16.3), repeated below:

$$\frac{\partial}{\partial y}\left(k\frac{\partial T}{\partial y}\right) + \frac{\partial}{\partial y}\left(\mu u\frac{\partial u}{\partial y}\right) = 0 \tag{16.3}$$

Since $T = h/c_p$ and Pr $= \mu c_p/k$, Equation (16.3) can be written as

$$\frac{\partial}{\partial y}\left(\frac{\mu}{\text{Pr}}\frac{\partial h}{\partial y}\right) + \frac{\partial}{\partial y}\left(\mu u\frac{\partial u}{\partial y}\right) = 0 \tag{16.75}$$

or

$$\frac{\partial}{\partial y}\left[\mu\left(\frac{1}{\text{Pr}}\frac{\partial h}{\partial y} + u\frac{\partial u}{\partial y}\right)\right] = 0 \tag{16.76}$$

Integrating Equation (16.76) with respect to y, we have

$$\frac{1}{\text{Pr}}\frac{\partial h}{\partial y} + u\frac{\partial u}{\partial y} = \frac{a}{\mu} \tag{16.77}$$

Since $\tau = \mu(\partial u/\partial y)$, we have $\mu = \tau(\partial u/\partial y)^{-1}$. Also, recalling from Equation (16.61) that τ is constant, we can write the right-hand side of Equation (16.77) as

$$\frac{a}{\mu} = \frac{a}{\tau}\frac{\partial u}{\partial y} = b\frac{\partial u}{\partial y}$$

where b is a constant. With this, Equation (16.77) becomes

$$\frac{1}{\text{Pr}}\frac{\partial h}{\partial y} + \frac{\partial(u^2/2)}{\partial y} - b\frac{\partial u}{\partial y} = 0 \tag{16.78}$$

Integrating Equation (16.78) with respect to y, remembering that Pr = constant, we have

$$\frac{h}{\text{Pr}} + \frac{u^2}{2} - bu = c \tag{16.79}$$

where c is another constant of integration. Expressions for b and c can be obtained by evaluating Equation (16.79) at $y = 0$ and $y = D$. At $y = 0$, $h = h_w$ and $u = 0$; hence,

$$c = \frac{h_w}{\text{Pr}}$$

At $y = D$, $h = h_e$ and $u = u_e$; hence,

$$b = \frac{1}{u_e}\left(\frac{h_e - h_w}{\text{Pr}}\right) + \frac{u_e}{2}$$

Inserting b and c into Equation (16.79) and simplifying, we obtain

$$h + \text{Pr}\,\frac{u^2}{2} = h_w + \frac{u}{u_e}(h_e - h_w) + \frac{\text{Pr}}{2}(uu_e) \tag{16.80}$$

Assume the lower wall is adiabatic; that is, $(\partial h/\partial y)_w = 0$. Differentiating Equation (16.80) with respect to y, we have

$$\frac{\partial h}{\partial y} = -\text{Pr}\,u\frac{\partial u}{\partial y} + \left(\frac{h_e + h_w}{u_e}\right)\frac{\partial u}{\partial y} + \frac{u_e\,\text{Pr}}{2}\frac{\partial u}{\partial y}$$

or

$$\frac{\partial h}{\partial y} = \left(-u\,\text{Pr} + \frac{h_e - h_w}{u_e} + \frac{u_e\,\text{Pr}}{2}\right)\frac{\partial u}{\partial y} \tag{16.81}$$

Recall that the condition for an adiabatic wall is that $(\partial h/\partial y)_w = 0$. Applying Equation (16.81) at $y = 0$ for an adiabatic wall, where $u = 0$ and by definition $h_w = h_{aw}$, we have

$$\left(\frac{\partial h}{\partial y}\right)_w = \left(\frac{h_e - h_{aw}}{u_e} + \frac{u_e\,\text{Pr}}{2}\right)\left(\frac{\partial u}{\partial y}\right)_w = 0$$

Since $(\partial u/\partial y)_w$ is finite, then

$$\frac{h_e - h_{aw}}{u_e} + \frac{u_e\,\text{Pr}}{2} = 0$$

or

$$\boxed{h_{aw} = h_e + \text{Pr}\,\frac{u_e^2}{2}} \tag{16.82}$$

This is identical to Equation (16.39) obtained for incompressible flow. Hence, we have just proven that the recovery factor for compressible Couette flow, assuming constant Prandtl number, is also

$$r = \text{Pr} \tag{16.83}$$

   Since the recovery factors for the incompressible and compressible cases are the same (as long as $\text{Pr} = $ constant), what can we say about Reynolds analogy? Does Equation (16.59) hold for the compressible case? Let us examine this question. Return to Equation (16.3), repeated below:

$$\frac{\partial}{\partial y}\left(k\frac{\partial T}{\partial y}\right) + \frac{\partial}{\partial y}\left(\mu u\frac{\partial u}{\partial y}\right) = 0 \tag{16.3}$$

Recalling that, from the definitions,

$$\dot{q} = k\frac{\partial T}{\partial y} \tag{16.84}$$

and

$$\tau = \mu\frac{\partial u}{\partial y} \tag{16.85}$$

then Equation (16.3) can be written as

$$\frac{\partial \dot{q}}{\partial y} + \frac{\partial(\tau u)}{\partial y} = 0 \tag{16.86}$$

Integrating Equation (16.86) with respect to y, we have

$$\dot{q} + \tau u = a \tag{16.87}$$

where a is a constant of integration. Evaluating Equation (16.87) at $y = 0$, where $u = 0$ and $\dot{q} = q_w$, we find that

$$a = \dot{q}_w$$

Hence, Equation (16.87) is

$$\dot{q} + \tau u = \dot{q}_w \tag{16.88}$$

Inserting Equations (16.84) and (16.85) into (16.88), we have

$$\dot{q}_w = k\frac{\partial T}{\partial y} + \mu u\frac{\partial u}{\partial y} \tag{16.89}$$

or

$$\frac{\dot{q}_w}{\mu} = \frac{k}{\mu}\frac{\partial T}{\partial y} + u\frac{\partial u}{\partial y} \tag{16.90}$$

Recall that the shear stress is constant throughout the flow; hence,

$$\tau = \mu\frac{\partial u}{\partial y} = \tau_w$$

or

$$\mu = \frac{\tau_w}{\partial u/\partial y} \tag{16.91}$$

Also,

$$\frac{k}{\mu} = \frac{c_p}{\text{Pr}} \tag{16.92}$$

Inserting Equation (16.91) into the left-hand side of Equation (16.90), and Equation (16.92) into the right-hand side of Equation (16.90), we have

$$\frac{\dot{q}_w}{\tau_w}\frac{\partial u}{\partial y} = \frac{c_p}{\text{Pr}}\frac{\partial T}{\partial y} + \frac{\partial(u^2/2)}{\partial y} \tag{16.93}$$

Integrate Equation (16.93) between the two plates, keeping in mind that $\dot{q}_w$, τ_w, c_p, and Pr are all fixed values:

$$\frac{\dot{q}_w}{\tau_w}\int_0^D \frac{\partial u}{\partial y}dy = \frac{c_p}{\text{Pr}}\int_0^D \frac{\partial T}{\partial y}dy + \int_0^D \frac{\partial(u^2/2)}{\partial y}dy$$

or

$$\frac{\dot{q}_w}{\tau_w} \int_0^{u_e} du = \frac{c_p}{\text{Pr}} \int_{T_w}^{T_e} dT + \int_0^{u_e} d\left(\frac{u^2}{2}\right)$$

which yields

$$\frac{\dot{q}_w}{\tau_w} u_e = \frac{c_p}{\text{Pr}}(T_e - T_w) + \frac{u_e^2}{2} \tag{16.94}$$

Rearranging Equation (16.94), and recalling that $h = c_p T$, we have

$$\dot{q}_w = \frac{\tau_w}{u_e \text{Pr}}\left(h_e - h_w + \text{Pr}\frac{u_e^2}{2}\right) \tag{16.95}$$

Inserting Equation (16.82) into (16.95), we have

$$\dot{q}_w = \frac{\tau_w}{u_e \text{Pr}}(h_{aw} - h_w) \tag{16.96}$$

The skin-friction coefficient and Stanton number are defined by Equations (16.51) and (16.55), respectively. Thus, their ratio is

$$\frac{C_H}{c_f} = \frac{\dot{q}_w/[\rho_e u_e(h_{aw} - h_w)]}{\tau_w/\left(\frac{1}{2}\rho_e u_e^2\right)} = \frac{\dot{q}_w}{\tau_w}\left[\frac{u_e}{2(h_{aw} - h_w)}\right] \tag{16.97}$$

Inserting Equation (16.96) into (16.97), we have

$$\frac{C_H}{c_f} = \frac{(h_{aw} - h_w)}{u_e \text{Pr}}\left[\frac{u_e}{2(h_{aw} - h_w)}\right]$$

or

$$\boxed{\frac{C_H}{c_f} = \frac{1}{2}\text{Pr}^{-1}} \tag{16.98}$$

Equation (16.98) is *Reynolds analogy*—a relation between heat transfer and skin friction coefficients. Moreover, it is precisely the same result as obtained in Equation (16.59) for incompressible flow. Hence, for a constant Prandtl number, we have shown that Reynolds analogy is precisely the same form for incompressible and compressible flow.

EXAMPLE 16.2

Consider the geometry given in Figure 16.2. The two plates are separated by a distance of 0.01 in (the same as in Example 16.1). The temperature of the two plates is equal, at a value of 288 K (standard sea level temperature). The air pressure is constant throughout the flow and equal to 1 atm. The upper plate is moving at Mach 3. The shear stress at the lower wall is 72 N/m². (This is about 1.5 lb/ft²—a much larger value than that associated with the low-speed case treated in Example 16.1.) Calculate the heat transfer to either plate. (Since the shear stress is constant throughout the flow, and the plates are at equal temperature, the heat transfer to the upper and lower plates is the same.)

■ **Solution**

The velocity of the upper plate is

$$u_e = M_e a_e = M_e \sqrt{\gamma R T_e} = 3\sqrt{(1.4)(288)(287)} = 1020 \text{ m/s}$$

The air density at both plates is (noting that 1 atm = 1.01×10^5 N/m^2)

$$\rho_e = \frac{p_e}{RT_e} = \frac{1.01 \times 10^5}{(287)(288)} = 1.22 \text{ kg/m}^3$$

Hence, the skin-friction coefficient is

$$c_f = \frac{\tau_w}{\frac{1}{2}\rho_e u_e^2} = \frac{72}{(0.5)(1.22)(1020)^2} = 1.13 \times 10^{-4}$$

From Reynolds analogy, Equation (16.92), we have

$$C_H = \frac{c_f}{2\,\text{Pr}} = \frac{1.13 \times 10^{-4}}{2(0.71)} = 8 \times 10^{-5}$$

The adiabatic wall enthalpy, from Equation (16.82), is

$$h_{aw} = h_e + \text{Pr}\,\frac{u_e^2}{2} = c_p T_e + \text{Pr}\,\frac{u_e^2}{2}$$

For air, $c_p = 1004.5$ J/kg $\cdot$ K. Thus,

$$h_{aw} = (1004.5)(288) + (0.71)\frac{(1020)^2}{2} = 6.59 \times 10^5 \text{ J/kg}$$

[*Note:* This gives $T_{aw} = h_{aw}/c_p = (6.59 \times 10^5)/1004.5 = 656$ K. In the adiabatic case, the wall would be quite warm.] Hence, from the definition of the Stanton number [Equation (16.55)], and noting that $h_w = c_p T_w = (1004.6)(288) = 2.89 \times 10^5$ J/kg,

$$\dot{q}_w = \rho_e u_e (h_{aw} - h_w) C_H = (1.22)(1020)[(6.59 - 2.89) \times 10^5](8 \times 10^{-5})$$

$$= \boxed{3.68 \times 10^4 \text{ W/m}^2}$$

16.5 SUMMARY

The parallel flow discussed in this chapter illustrates features common to many more complex viscous flows, with the added advantage of lending itself to a relatively straightforward solution. The purpose of this discussion has been to introduce many of the basic concepts of viscous flows in a fashion unencumbered by fluid dynamic complexities. In particular, we have studied Couette flow and found the following.

1. The driving force is the shear stress between the moving wall and the fluid. Shear stress is constant across the flow for both incompressible and compressible cases.

2. For incompressible Couette flow,

$$u = u_e \left(\frac{y}{D} \right) \qquad (16.6)$$

$$\tau = \mu \left(\frac{u_e}{D} \right) \qquad (16.9)$$

3. The heat transfer depends on the wall temperatures and the amount of viscous dissipation. For an adiabatic wall, the wall enthalpy is

$$h_{aw} = h_e + r \frac{u_e^2}{2} \qquad (16.46a)$$

For incompressible and compressible Couette flow with a constant Prandtl number, the recovery factor is

$$r = \text{Pr}$$

and Reynolds analogy holds in both cases;

$$\frac{C_H}{c_f} = \frac{1}{2} \text{Pr}^{-1} \qquad (16.59)$$

CHAPTER 17

Introduction to Boundary Layers

A very satisfactory explanation of the physical process in the boundary layer between a fluid and a solid body could be obtained by the hypothesis of an adhesion of the fluid to the walls, that is, by the hypothesis of a zero relative velocity between fluid and wall. If the viscosity was very small and the fluid path along the wall not too long, the fluid velocity ought to resume its normal value at a very short distance from the wall. In the thin transition layer however, the sharp changes of velocity, even with small coefficient of friction, produce marked results.

   Ludwig Prandtl, 1904

PREVIEW BOX

The revelations in this chapter are simply amazing. Although the influence of friction is present at every point throughout every flow, on a practical basis it is usually of no consequence *except* in a thin region adjacent to the surface of a body immersed in the flow, or in the boundary region between two flows of widely different velocities. For the former, the thin region is called a *boundary layer,* and for the latter it is called a *shear layer*. After centuries of trying to solve flows taking into account friction, most to no avail, the introduction of the boundary-layer concept revolutionized fluid dynamics and made the analysis of flows with friction tractable. It is amazing that what goes on inside the thin boundary layer is the physical mechanism for skin friction and aerodynamic heating to the surface, and for flow separation from the surface. It is truly amazing that to calculate skin friction and aerodynamic heating at the surface, we have only to account for friction and thermal conduction within the thin boundary layer, and *not* in the large region of flow outside the boundary layer.

This chapter is all about boundary layers. Read on, and be prepared to be amazed.

17.1 INTRODUCTION

The above quotation is taken from an historic paper given by Ludwig Prandtl at the third Congress of Mathematicians at Heidelberg, Germany, in 1904. In this paper, the concept of the boundary layer was first introduced—a concept which eventually revolutionized the analysis of viscous flows in the twentieth century and which allowed the practical calculation of drag and flow separation over aerodynamic bodies. Before Prandtl's 1904 paper, the Navier-Stokes equations discussed in Chapter 15 were well known, but fluid dynamicists were frustrated in their attempts to solve these equations for practical engineering problems. After 1904, the picture changed completely. Using Prandtl's concept of a boundary layer adjacent to an aerodynamic surface, the Navier-Stokes equations can be reduced to a more tractable form called the *boundary-layer equations*. In turn, these boundary-layer equations can be solved to obtain the distributions of shear stress and aerodynamic heat transfer to the surface. Prandtl's boundary-layer concept was an advancement in the science of fluid mechanics of the caliber of a Nobel prize, although he never received that honor. The purpose of this chapter is to present the general concept of the boundary layer and to give a few representative samples of its application. Our purpose here is to provide only an introduction to boundary-layer theory; consult Reference 40 for a rigorous and thorough discussion of boundary-layer analysis and applications. Also, a general introduction to boundary layers was given in Section 1.11. If you have not read that section, now would be a good time to pause and return to Section 1.11 before progressing further.

What is a boundary layer? We have used this term in several places in our previous chapters, first introducing the idea in Section 1.10 and illustrating the concept in Figure 1.42. The boundary layer is the thin region of flow adjacent to a surface, where the flow is retarded by the influence of friction between a solid surface and the fluid. For example, a photograph of the flow over a supersonic body is shown in Figure 17.1, where the boundary layer (along with shock and expansion waves and the wake) is made visible by a special optical technique called a *shadow-graph* (see References 25 and 26 for discussions of the shadowgraph method). Note how thin the boundary layer is in comparison with the size of the body; however, although the boundary layer occupies geometrically only a small portion of the flow field, its influence on the drag and heat transfer to the body is immense—in Prandtl's own words as quoted above, it produces "marked results."

The purpose of the remaining chapters is to examine these "marked results." The road map for the present chapter is given in Figure 17.2. In the next section, we discuss some fundamental properties of boundary layers. This is followed by a development of the boundary-layer equations, which are the continuity, momentum, and energy equations written in a special form applicable to the flow in the thin viscous region adjacent to a surface. The boundary-layer equations are partial differential equations that apply *inside* the boundary layer.

Finally, we note that this chapter represents the second of the three options discussed in Section 15.7 for the solution of the viscous flow equations, namely,

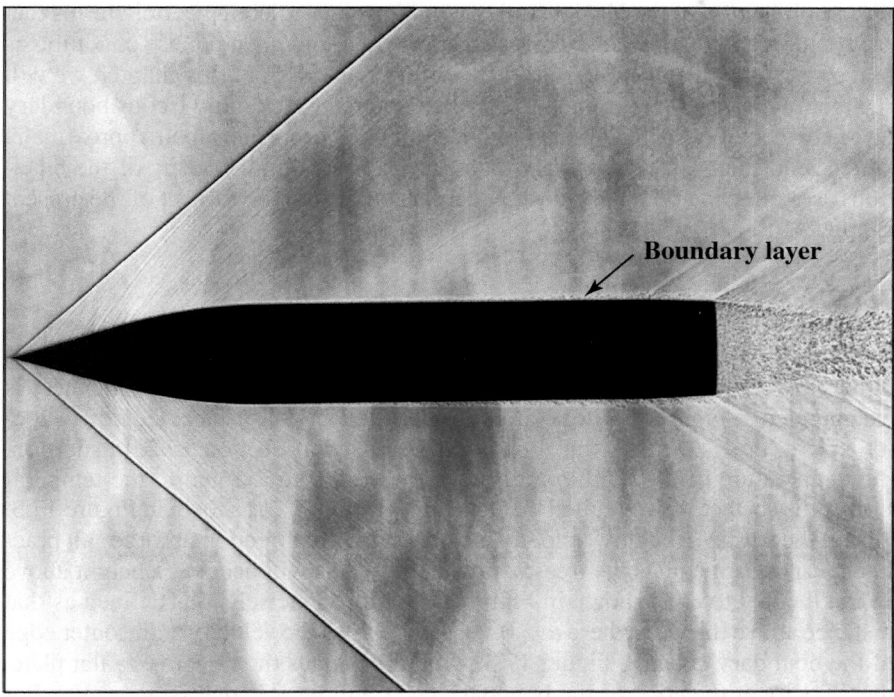

Figure 17.1 The boundary layer on an aerodynamic body (*H.S. Photos/Alamy Stock Photo*).

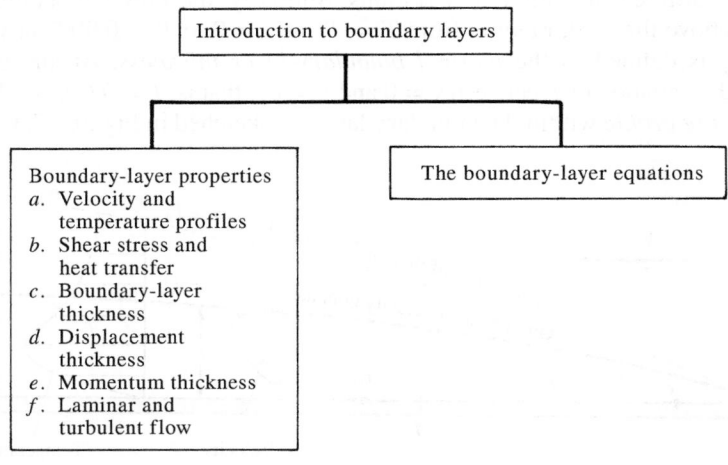

Figure 17.2 Road map for Chapter 17.

the simplification of the Navier-Stokes equations by neglecting certain terms that are smaller than other terms. This is an approximation, not a precise condition as in the case of Couette and Poiseuille flows in Chapter 16. In this chapter, we will see that the Navier-Stokes equations, when applied to the thin viscous boundary layer adjacent to a surface, can be reduced to simpler forms, albeit approximate, which lend themselves to simpler solutions. These simpler forms of the equations are called the boundary-layer equations—they are the subject of the present chapter.

17.2 BOUNDARY-LAYER PROPERTIES

Consider the viscous flow over a flat plate as sketched in Figure 17.3. The viscous effects are contained within a thin layer adjacent to the surface; the thickness is exaggerated in Figure 17.3 for clarity. Immediately at the surface, the flow velocity is zero; this is the "no-slip" condition discussed in Section 15.2. In addition, the temperature of the fluid immediately at the surface is equal to the temperature of the surface; this is called the *wall temperature* T_w, as shown in Figure 17.3. Above the surface, the flow velocity increases in the y direction until, for all practical purposes, it equals the freestream velocity. This will occur at a height above the wall equal to δ, as shown in Figure 17.3. More precisely, δ is defined as that distance above the wall where $u = 0.99u_e$; here, u_e is the velocity at the outer edge of the boundary layer. In Figure 17.3, which illustrates the flow over a flat plate, the velocity at the edge of the boundary layer will be V_∞; that is, $u_e = V_\infty$. For a body of general shape, u_e is the velocity obtained from an inviscid flow solution evaluated at the body surface (or at the "effective body" surface, as discussed later). The quantity δ is called the *velocity boundary-layer thickness*. At any given x station, the variation of u between $y = 0$ and $y = \delta$, that is, $u = u(y)$, is defined as the *velocity profile* within the boundary layer, as sketched in Figure 17.3. This profile is different for different x stations. Similarly, the flow temperature will change above the wall, ranging from $T = T_w$ at $y = 0$ to $T = 0.99T_e$ at $y = \delta_T$. Here, δ_T is defined as the *thermal boundary-layer thickness*. At any given x station, the variation of T between $y = 0$ and $y = \delta_T$, that is, $T = T(y)$, is called the *temperature profile* within the boundary layer, as sketched in Figure 17.3. (In the

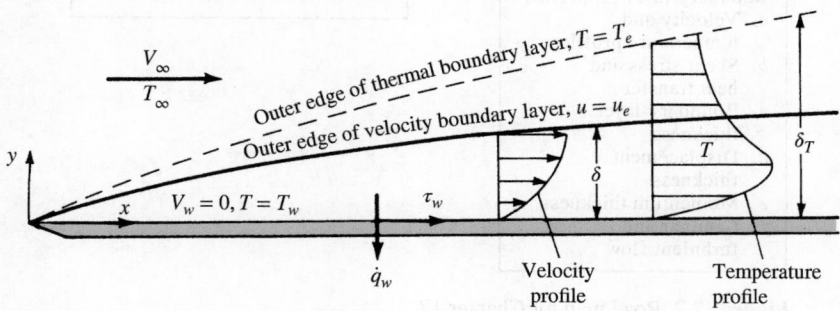

Figure 17.3 Boundary-layer properties.

above, T_e is the temperature at the edge of the thermal boundary layer. For the flow over a flat plate, as sketched in Figure 17.3, $T_e = T_\infty$. For a general body, T_e is obtained from an inviscid flow solution evaluated at the body surface, or at the "effective body" surface, to be discussed later.) Hence, two boundary layers can be defined: a velocity boundary layer with thickness δ and a temperature boundary layer with thickness δ_T. In general, $\delta_T \neq \delta$. The relative thicknesses depend on the Prandtl number: it can be shown that if $\text{Pr} = 1$, then $\delta = \delta_T$; if $\text{Pr} > 1$, then $\delta_T < \delta$; if $\text{Pr} < 1$, then $\delta_T > \delta$. For air at standard conditions, $\text{Pr} = 0.71$; hence, the thermal boundary layer is thicker than the velocity boundary layer, as shown in Figure 17.3. Note that both boundary-layer thicknesses increase with distance from the leading edge; that is, $\delta = \delta(x)$ and $\delta_T = \delta_T(x)$.

The consequence of the velocity gradient at the wall is the generation of shear stress at the wall,

$$\tau_w = \mu \left(\frac{\partial u}{\partial y} \right)_w \tag{17.1}$$

where $(\partial u / \partial y)_w$ is the velocity gradient evaluated at $y = 0$ (i.e., at the wall). Similarly, the temperature gradient at the wall generates heat transfer at the wall,

$$\dot{q}_w = -k \left(\frac{\partial T}{\partial y} \right)_w \tag{17.2}$$

where $(\partial T / \partial y)_w$ is the temperature gradient evaluated at $y = 0$ (i.e., at the wall). In general, both τ_w and $\dot{q}_w$ are functions of distance from the leading edge; that is, $\tau_w = \tau_w(x)$ and $\dot{q}_w = \dot{q}_w(x)$. One of the central purposes of boundary-layer theory is to compute τ_w and $\dot{q}_w$.

A frequently used boundary-layer property is the *displacement thickness* δ^*, defined as

$$\delta^* \equiv \int_0^{y_1} \left(1 - \frac{\rho u}{\rho_e u_e} \right) dy \quad \delta \leq y_1 \to \infty \tag{17.3}$$

The displacement thickness has two physical interpretations:

1. δ^* is an index proportional to the "missing mass flow" due to the presence of the boundary layer. Let us explain. Consider point y_1 above the boundary layer, as shown in Figure 17.4. Consider also the mass flow (per unit depth perpendicular to the page) across the vertical line connecting $y = 0$ and $y = y_1$. Then

$$A = \text{actual mass flow between 0 and } y_1 = \int_0^{y_1} \rho u \, dy$$

$$B = \begin{array}{l} \text{hypothetical mass flow} \\ \text{between 0 and } y_1 \text{ if boundary} \\ \text{layer were not present} \end{array} = \int_0^{y_1} \rho_e u_e \, dy$$

$$B - A = \begin{array}{l} \text{decrement in mass flow due to} \\ \text{presence of boundary layer, that is,} \\ \text{missing mass flow} \end{array} = \int_0^{y_1} (\rho_e u_e - \rho u) \, dy \tag{17.4}$$

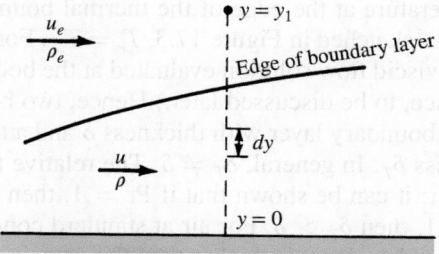

Figure 17.4 Construction for the discussion of displacement thickness.

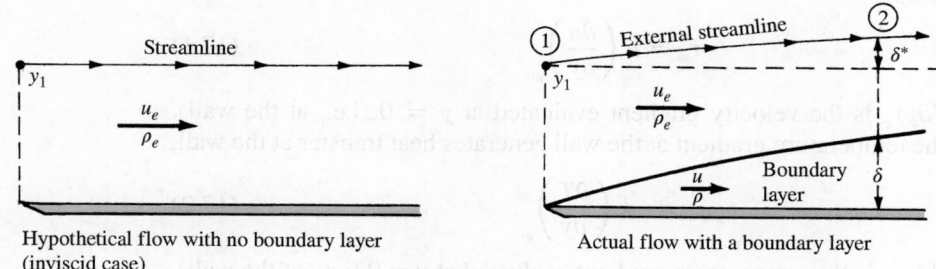

Hypothetical flow with no boundary layer (inviscid case)

Actual flow with a boundary layer

Figure 17.5 Displacement thickness is the distance by which an external flow streamline is displaced by the presence of the boundary layer.

Express this missing mass flow as the product of $\rho_e u_e$ and a height δ^*; that is,

$$\text{Missing mass flow} = \rho_e u_e \delta^* \tag{17.5}$$

Equating Equations (17.4) and (17.5), we have

$$\rho_e u_e \delta^* = \int_0^{y_1} (\rho_e u_e - \rho u)\, dy$$

or

$$\delta^* = \int_0^{y_1} \left(1 - \frac{\rho u}{\rho_e u_e}\right) dy \tag{17.6}$$

Equation (17.6) is identical to the definition of δ^* given in Equation (17.3). Hence, clearly δ^* is a height proportional to the missing mass flow. If this missing mass flow was crammed into a streamtube where the flow properties were constant at ρ_e and u_e, then Equation (17.5) says that δ^* is the height of this hypothetical streamtube.

2. The second physical interpretation of δ^* is more practical than the one discussed above. Consider the flow over a flat surface as sketched in Figure 17.5. At the left is a picture of the hypothetical inviscid flow over the surface; a streamline through point y_1 is straight and parallel to the surface.

The actual viscous flow is shown at the right of Figure 17.5; here, the retarded flow inside the boundary layer acts as a partial obstruction to the freestream flow. As a result, the streamline external to the boundary layer passing through point y_1 is deflected upward through a distance δ^*. We now prove that this δ^* is precisely the displacement thickness defined by Equation (17.3). At station 1 in Figure 17.5, the mass flow (per unit depth perpendicular to the page) between the surface and the external streamline is

$$\dot{m} = \int_0^{y_1} \rho_e u_e \, dy \qquad (17.7)$$

At station 2, the mass flow between the surface and the external streamline is

$$\dot{m} = \int_0^{y_1} \rho u \, dy + \rho_e u_e \, \delta^* \qquad (17.8)$$

Since the surface and the external streamline form the boundaries of a streamtube, the mass flows across stations 1 and 2 are equal. Hence, equating Equations (17.7) and (17.8), we have

$$\int_0^{y_1} \rho_e u_e \, dy = \int_0^{y_1} \rho u \, dy + \rho_e u_e \, \delta^*$$

or

$$\delta^* = \int_0^{y_1} \left(1 - \frac{\rho u}{\rho_e u_e} \right) dy \qquad (17.9)$$

Hence, the height by which the streamline in Figure 17.5 is displaced upward by the presence of the boundary layer, namely, δ^*, is given by Equation (17.9). However, Equation (17.9) is precisely the definition of the displacement thickness given by Equation (17.3). Thus, the displacement thickness, first defined by Equation (17.3), is physically the distance through which the external inviscid flow is displaced by the presence of the boundary layer.

This second interpretation of δ^* gives rise to the concept of an *effective body*. Consider the aerodynamic shape sketched in Figure 17.6. The actual contour of

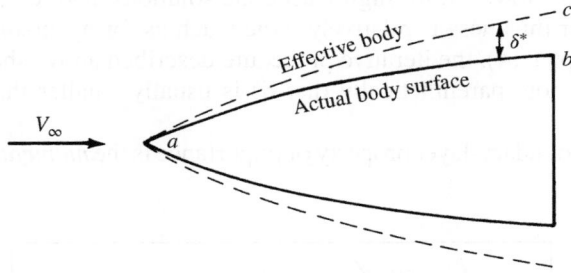

Figure 17.6 The "effective body," equal to the actual body shape plus the displacement thickness distribution.

the body is given by curve *ab*. However, due to the displacement effect of the boundary layer, the shape of the body effectively seen by the freestream is not given by curve *ab*; rather, the freestream sees an effective body given by curve *ac*. In order to obtain the conditions ρ_e, u_e, T_e, etc., at the outer edge of the boundary layer on the actual body *ab*, an inviscid flow solution should be carried out for the effective body, and ρ_e, u_e, T_e, etc., are obtained from this inviscid solution evaluated along curve *ac*.

Note that in order to solve for δ^* from Equation (17.3), we need the profiles of u and ρ from a solution of the boundary-layer flow. In turn, to solve the boundary-layer flow, we need ρ_e, u_e, T_e, etc. However, ρ_e, u_e, T_e, etc., depend on δ^*. This leads to an iterative solution. To calculate accurately the boundary-layer properties as well as the pressure distribution over the surface of the body in Figure 17.6, we proceed as follows:

1. Carry out an inviscid solution for the given body shape *ab*. Evaluate ρ_e, u_e, T_e, etc., along curve *ab*.

2. Using these values of ρ_e, u_e, T_e, etc., solve the boundary-layer equations (discussed in Sections 17.3 to 17.6) for $u = u(y)$, $\rho = \rho(y)$, etc., at various stations along the body.

3. Obtain δ^* at these stations from Equation (17.3). This will not be an accurate δ^* because ρ_e, u_e, T_e, etc., were evaluated on curve *ab*, not the proper effective body. Using this intermediate δ^*, calculate an effective body, given by a curve *ac'* (not shown in Figure 17.6).

4. Carry out an inviscid solution for the flow over the intermediate effective body *ac'*, and evaluate new values of ρ_e, u_e, T_e, etc., along *ac'*.

5. Repeat steps 2 to 4 above until the solution at one iteration essentially does not deviate from the solution at the previous iteration. At this stage, a converged solution will be obtained, and the final results will pertain to the flow over the proper effective body *ac* shown in Figure 17.6.

In some cases, the boundary layers are so thin that the effective body can be ignored, and a boundary-layer solution can proceed directly from ρ_e, u_e, T_e, etc., obtained from an inviscid solution evaluated on the actual body surface (*ab* in Figure 17.6). However, for highly accurate solutions, and for cases where the boundary-layer thickness is relatively large (such as for hypersonic flow as discussed in Chapter 14), the iterative procedure described above should be carried out. Also, we note parenthetically that δ^* is usually smaller than δ; typically, $\delta^* \approx 0.3\,\delta$.

Another boundary-layer property of importance is the *momentum thickness* θ, defined as

$$\theta \equiv \int_0^{y_1} \frac{\rho u}{\rho_e u_e}\left(1 - \frac{u}{u_e}\right) dy \qquad \delta \leq y_1 \to \infty \qquad (17.10)$$

To understand the physical interpretation of θ, return again to Figure 17.4. Consider the mass flow across a segment dy, given by $dm = \rho u\, dy$. Then

$$A = \text{momentum flow across } dy = dm\, u = \rho u^2\, dy$$

If this same elemental mass flow were associated with the freestream, where the velocity is u_e, then

$$B = \begin{cases} \text{momentum flow at freestream} \\ \text{velocity associated with mass } dm = dm\, u_e = (\rho u\, dy)u_e \end{cases}$$

Hence,

$$B - A = \begin{cases} \text{decrement in momentum flow} \\ \text{(missing momentum flow) associated} = \rho u(u_e - u)\, dy \\ \text{with mass } dm \end{cases} \quad (17.11)$$

The *total* decrement in momentum flow across the vertical line from $y = 0$ to $y = y_1$ in Figure 17.4 is the integral of Equation (17.11),

$$\left.\begin{array}{l}\text{Total decrement in momentum} \\ \text{flow, or missing momentum flow}\end{array}\right\} = \int_0^{y_1} \rho u(u_e - u)\, dy \quad (17.12)$$

Assume that the missing momentum flow is the product of $\rho_e u_e^2$ and a height θ. Then,

$$\text{Missing momentum flow} = \rho_e u_e^2 \theta \quad (17.13)$$

Equating Equations (17.12) and (17.13), we obtain

$$\rho_e u_e^2 \theta = \int_0^{y_1} \rho u(u_e - u)\, dy$$

$$\theta = \int_0^{y_1} \frac{\rho u}{\rho_e u_e}\left(1 - \frac{u}{u_e}\right) dy \quad (17.14)$$

Equation (17.14) is precisely the definition of the momentum thickness given by Equation (17.10). Therefore, θ is an index that is proportional to the decrement in momentum flow due to the presence of the boundary layer. It is the height of a hypothetical streamtube, which is carrying the missing momentum flow at freestream conditions.

Note that $\theta = \theta(x)$. In more detailed discussions of boundary-layer theory, it can be shown that θ evaluated at a given station $x = x_1$ is proportional to the integrated friction drag coefficient from the leading edge to x_1; that is,

$$\theta(x_1) \propto \frac{1}{x_1}\int_0^{x_1} c_f\, dx = C_f$$

where c_f is the local skin-friction coefficient defined in Section 1.5 and C_f is the total skin-friction drag coefficient for the length of surface from $x = 0$ to $x = x_1$. Hence, the concept of momentum thickness is useful in the prediction of drag coefficient.

All the boundary-layer properties discussed above are general concepts; they apply to compressible as well as incompressible flows, and to turbulent as well as laminar flows. The differences between turbulent and laminar flows were introduced in Section 15.2. Here, we extend that discussion by noting that the increased momentum and energy exchange that occur within a turbulent flow cause a turbulent boundary layer to be thicker than a laminar boundary layer. That is, for the same edge conditions, ρ_e, u_e, T_e, etc., we have $\delta_{\text{turbulent}} > \delta_{\text{laminar}}$ and $(\delta_T)_{\text{turbulent}} > (\delta_T)_{\text{laminar}}$. When a boundary layer changes from laminar to turbulent flow, as sketched in Figure 15.8, the boundary-layer thickness markedly increases. Similarly, δ^* and θ are larger for turbulent flows.

17.3 THE BOUNDARY-LAYER EQUATIONS

For the remainder of this chapter, we consider two-dimensional, steady flow. The nondimensionalized form of the x-momentum equation (one of the Navier-Stokes equations) was developed in Section 15.6 and was given by Equation (15.29):

$$\rho' u' \frac{\partial u'}{\partial x'} + \rho' v' \frac{\partial u'}{\partial y'} = -\frac{1}{\gamma M_\infty^2} \frac{\partial p'}{\partial x'} + \frac{1}{\text{Re}_\infty} \frac{\partial}{\partial y'} \left[\mu' \left(\frac{\partial v'}{\partial x'} + \frac{\partial u'}{\partial y'} \right) \right] \quad (15.29)$$

Let us now reduce Equation (15.29) to an approximate form which holds reasonably well within a boundary layer.

Consider the boundary layer along a flat plate of length c as sketched in Figure 17.7. The basic assumption of boundary-layer theory is that a boundary layer is very thin in comparison with the scale of the body; that is,

$$\boxed{\delta \ll c} \quad (17.15)$$

Consider the continuity equation for a steady, two-dimensional flow,

$$\frac{\partial(\rho u)}{\partial x} + \frac{\partial(\rho v)}{\partial y} = 0 \quad (17.16)$$

In terms of the nondimensional variables defined in Section 15.6, Equation (17.16) becomes

$$\frac{\partial(\rho' u')}{\partial x'} + \frac{\partial(\rho' v')}{\partial y'} = 0 \quad (17.17)$$

Because u' varies from 0 at the wall to 1 at the edge of the boundary layer, let us say that u' is of the *order of magnitude* equal to 1, symbolized by $O(1)$. Similarly,

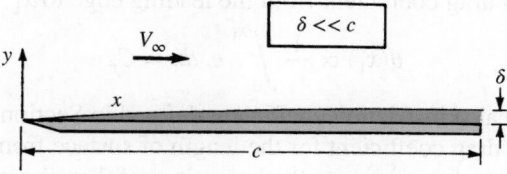

Figure 17.7 The basic assumption of boundary-layer theory: A boundary layer is very thin in comparison with the scale of the body.

$\rho' = O(1)$. Also, since x varies from 0 to c, $x' = O(1)$. However, since y varies from 0 to δ, where $\delta \ll c$, then y' is of the *smaller* order of magnitude, denoted by $y' = O(\delta/c)$. Without loss of generality, we can assume that c is a unit length. Therefore, $y' = O(\delta)$. Putting these orders of magnitude in Equation (17.17), we have

$$\frac{[O(1)][O(1)]}{O(1)} + \frac{[O(1)][v']}{O(\delta)} = 0 \qquad (17.18)$$

Hence, from Equation (17.18), clearly v' must be of an order of magnitude equal to δ; that is, $v' = O(\delta)$. Now examine the order of magnitude of the terms in Equation (15.29). We have

$$\rho'u'\frac{\partial u'}{\partial x'} = O(1) \qquad \rho'v'\frac{\partial u'}{\partial y'} = O(1) \qquad \frac{\partial p'}{\partial x'} = O(1)$$

$$\frac{\partial}{\partial y'}\left(\mu'\frac{\partial v'}{\partial x'}\right) = O(1) \qquad \frac{\partial}{\partial y'}\left(\mu'\frac{\partial u'}{\partial y'}\right) = O\left(\frac{1}{\delta^2}\right)$$

Hence, the order-of-magnitude equation for Equation (15.29) can be written as

$$O(1) + O(1) = -\frac{1}{\gamma M_\infty^2}O(1) + \frac{1}{\text{Re}_\infty}\left[O(1) + O\left(\frac{1}{\delta^2}\right)\right] \qquad (17.19)$$

Let us now introduce another assumption of boundary-layer theory, namely, the *Reynolds number is large,* indeed, large enough such that

$$\boxed{\frac{1}{\text{Re}_\infty} = O(\delta^2)} \qquad (17.20)$$

Then, Equation (17.19) becomes

$$O(1) + O(1) = -\frac{1}{\gamma M_\infty^2}O(1) + O(\delta^2)\left[O(1) + O\left(\frac{1}{\delta^2}\right)\right] \qquad (17.21)$$

In Equation (17.21), there is one term with an order of magnitude that is much smaller than the rest, namely, the product $O(\delta^2)[O(1)] = O(\delta^2)$. This term corresponds to $(1/\text{Re}_\infty)\partial/\partial y'(\mu'\partial v'/\partial x')$ in Equation (15.29). Hence, *neglect* this term in comparison to the remaining terms in Equation (15.29). We obtain

$$\rho'u'\frac{\partial u'}{\partial x'} + \rho'v'\frac{\partial u'}{\partial y'} = -\frac{1}{\gamma M_\infty^2}\frac{\partial p'}{\partial x'} + \frac{1}{\text{Re}_\infty}\frac{\partial}{\partial y'}\left(\mu'\frac{\partial u'}{\partial y'}\right) \qquad (17.22)$$

In terms of dimensional variables, Equation (17.22) is

$$\rho u\frac{\partial u}{\partial x} + \rho v\frac{\partial u}{\partial y} = -\frac{\partial p}{\partial x} + \frac{\partial}{\partial y}\left(\mu\frac{\partial u}{\partial y}\right) \qquad (17.23)$$

Equation (17.23) is the approximate x-momentum equation which holds for flow in a thin boundary layer at a high Reynolds number.

Consider the y-momentum equation for two-dimensional, steady flow, obtained from Equation (15.19b) as (neglecting the normal stress τ_{yy})

$$\rho u \frac{\partial v}{\partial x} + \rho v \frac{\partial v}{\partial y} = -\frac{\partial p}{\partial y} + \frac{\partial}{\partial x}\left[\mu\left(\frac{\partial v}{\partial x} + \frac{\partial u}{\partial y}\right)\right] \tag{17.24}$$

In terms of the nondimensional variables, Equation (17.24) becomes

$$\rho' u' \frac{\partial v'}{\partial x'} + \rho' v' \frac{\partial v'}{\partial y'} = -\frac{1}{\gamma M_\infty^2}\frac{\partial p'}{\partial y'} + \frac{1}{\text{Re}_\infty}\frac{\partial}{\partial x'}\left[\mu'\left(\frac{\partial v'}{\partial x'} + \frac{\partial u'}{\partial y'}\right)\right] \tag{17.25}$$

The order-of-magnitude equation for Equation (17.25) is

$$O(\delta) + O(\delta) = -\frac{1}{\gamma M_\infty^2}\frac{\partial p'}{\partial y'} + O(\delta^2)\left[O(\delta) + O\left(\frac{1}{\delta}\right)\right] \tag{17.26}$$

From Equation (17.26), we see that $\partial p'/\partial y' = O(\delta)$ or smaller, assuming that $\gamma M_\infty^2 = O(1)$. Since δ is very small, this implies that $\partial p'/\partial y'$ is very small. Therefore, from the y-momentum equation specialized to a boundary layer, we have

$$\frac{\partial p}{\partial y} = 0 \tag{17.26a}$$

Equation (17.26a) is important; it states that at a given x station, *the pressure is constant through the boundary layer in a direction normal to the surface*. This implies that the pressure distribution at the outer edge of the boundary layer is impressed directly to the surface without change. Hence, throughout the boundary layer, $p = p(x) = p_e(x)$.

It is interesting to note that if M_∞^2 is very large, as in the case of large hypersonic Mach numbers, then from Equation (17.26) $\partial p'/\partial y'$ does not have to be small. For example, if M_∞ were large enough such that $1/\gamma M_\infty^2 = O(\delta)$, then $\partial p'/\partial y'$ could be as large as $O(1)$, and Equation (17.26) would still be satisfied. Thus, for very large hypersonic Mach numbers, the assumption that p is constant in the normal direction through a boundary layer is not always valid.

Consider the general energy equation given by Equation (15.26). The nondimensional form of this equation for two-dimensional, steady flow is given in Equation (15.33). Inserting $e = h - p/\rho$ into this equation, subtracting the momentum equation multiplied by velocity, and performing an order-of-magnitude analysis similar to those above, we can obtain the boundary-layer energy equation as

$$\rho u \frac{\partial h}{\partial x} + \rho v \frac{\partial h}{\partial y} = \frac{\partial}{\partial y}\left(k\frac{\partial T}{\partial y}\right) + u\frac{\partial p}{\partial x} + \mu\left(\frac{\partial u}{\partial y}\right)^2 \tag{17.27}$$

The details are left to you.

In summary, by making the combined assumptions of $\delta \ll c$ and $\text{Re} \geq 1/\delta^2$, the complete Navier-Stokes equations derived in Chapter 15 can be reduced to simpler forms which apply to a boundary layer. These boundary-layer equations are

$$\text{Continuity:} \qquad \frac{\partial(\rho u)}{\partial x} + \frac{\partial(\rho v)}{\partial y} = 0 \tag{17.28}$$

$$x\ momentum: \qquad \rho u \frac{\partial u}{\partial x} + \rho v \frac{\partial u}{\partial y} = -\frac{dp_e}{dx} + \frac{\partial}{\partial y}\left(\mu \frac{\partial u}{\partial y}\right) \tag{17.29}$$

$$y\ momentum: \qquad \frac{\partial p}{\partial y} = 0 \tag{17.30}$$

$$Energy: \qquad \rho u \frac{\partial h}{\partial x} + \rho v \frac{\partial h}{\partial y} = \frac{\partial}{\partial y}\left(k \frac{\partial T}{\partial y}\right) + u \frac{dp_e}{dx} + \mu \left(\frac{\partial u}{\partial y}\right)^2 \tag{17.31}$$

Note that, as in the case of the Navier-Stokes equations, the boundary-layer equations are nonlinear. However, the boundary-layer equations are simpler, and therefore are more readily solved. Also, since $p = p_e(x)$, the pressure gradient expressed as $\partial p/\partial x$ in Equations (17.23) and (17.27) is reexpressed as dp_e/dx in Equations (17.29) and (17.31). In the above equations, the unknowns are u, v, ρ, and h; p is known from $p = p_e(x)$, and μ and k are properties of the fluid which vary with temperature. To complete the system, we have

$$p = \rho RT \tag{17.32}$$

and

$$h = c_p T \tag{17.33}$$

Hence, Equations (17.28), (17.29), and (17.31) to (17.33) are five equations for the five unknowns, u, v, ρ, T, and h.

The boundary conditions for the above equations are as follows:

At the wall: $\qquad\qquad y = 0 \quad u = 0 \quad v = 0 \quad T = T_w$

At the boundary-layer edge: $\qquad y \to \infty \quad u \to u_e \quad T \to T_e$

Note that since the boundary-layer thickness is not known a priori, the boundary condition at the edge of the boundary layer is given at large y, essentially y approaching infinity.

17.4 HOW DO WE SOLVE THE BOUNDARY-LAYER EQUATIONS?

Examine again the boundary-layer equations given by Equations (17.28) to (17.31). With these equations, are we still in the same "soup" as we are with the complete Navier-Stokes equations, in that the equations are a coupled system of nonlinear partial differential equations for which no general analytical solution has been obtained to date? The answer is partly yes, but with a difference. Because the boundary-layer equations are simpler than the Navier-Stokes equations, especially the boundary-layer y-momentum equation, Equation (17.30), which states that at any axial location along the surface the pressure is constant in the direction normal to the surface, there is more hope of obtaining meaningful solutions for the flow inside a boundary layer. For almost one hundred years, engineers and scientists have "nudged" the boundary-layer equations in many different ways, and have come up with reasonable solutions for a number of practical applications. The most complete and authoritative book on such solutions is by Hermann Schlichting (Reference 40).

In Chapters 18 and 19, we will discuss some of these solutions—their technique and some practical results. Solutions of the boundary-layer equations can be classified into two groups: (1) classical solutions, some of which date back to 1908, and (2) numerical solutions obtained by modern computational fluid dynamic techniques. In the subsequent chapters, we will show examples from both groups. In addition to this subdivision based on the solution technique, boundary-layer solutions also subdivide on a *physical basis* into laminar boundary layers and turbulent boundary layers. This subdivision is natural for the reasons discussed in Section 15.2—the nature of turbulent flow is quite different than that of laminar flow. Indeed, for certain types of flow problems, *exact* solutions have been obtained for laminar boundary layers. To date, *no* exact solution has been obtained for turbulent boundary layers, because we still do not have a complete understanding of turbulence, and hence all turbulent boundary-layer solutions to date have depended on the use of some type of approximate model of turbulence. These contrasts will become more clear when you read Chapter 18, which deals with laminar boundary layers, and Chapter 19, which deals with turbulent boundary layers. You will see that laminar boundary-layer theory is well in hand, but turbulent boundary-layer theory is not. In some sense, what a pity that nature always tends toward turbulent flow, and hence the vast majority of practical boundary-layer problems deal with turbulent boundary layers—if it were the other way around, the engineering calculation of skin friction and aerodynamic heating at the surface would be much simpler and more reliable.

Finally, we note what is meant by a "boundary-layer solution." The solution of Equations (17.28) to (17.31) yields the velocity and temperature profiles throughout the boundary layer. However, the *practical* information we want is the solution for τ_w and $\dot{q}_w$, the *surface* shear stress and heat transfer, respectively. These are given by

$$\tau_w = \mu_w \left(\frac{\partial u}{\partial y} \right)_w \tag{17.34}$$

and

$$\dot{q}_w = k_w \left(\frac{\partial T}{\partial y} \right)_w \tag{17.35}$$

where the subscript w denotes conditions at the wall. *Question:* Where do values of $(\partial u/\partial y)_w$ and $(\partial T/\partial y)_w$ come from? *Answer:* From the velocity and temperature profiles obtained from a solution of the boundary-layer equations, where the profiles evaluated at the wall provide both $(\partial u/\partial y)_w$ and $(\partial T/\partial y)_w$. Hence in order to obtain the values for τ_w and $\dot{q}_w$ along the wall, which are usually considered the engineering results of most importance, we first have to solve the boundary-layer equations for the velocity and temperature profiles throughout the boundary layer, which by themselves usually are of lesser practical interest.

17.5 SUMMARY

Return to the road map given in Figure 17.2, and make certain that you feel at home with the material represented by each box. The highlights of our discussion of boundary layers are summarized as follows:

The basic quantities of interest from boundary-layer theory are the velocity and thermal boundary-layer thicknesses, δ and δ_T, respectively; the shear stress at the wall, τ_w; and heat transfer to the surface, $\dot{q}_w$. In the process, we can define two additional thicknesses: the displacement thickness

$$\delta^* \equiv \int_0^{y_1} \left(1 - \frac{\rho u}{\rho_e u_e}\right) dy \qquad \delta \leq y_1 \to \infty \qquad (17.3)$$

and the momentum thickness

$$\theta \equiv \int_0^{y_1} \frac{\rho u}{\rho_e u_e} \left(1 - \frac{u}{u_e}\right) dy \qquad \delta \leq y_1 \to \infty \qquad (17.10)$$

Both δ^* and θ are related to decrements in the flow due to the presence of the boundary layer; δ^* is proportional to the decrement in mass flow, and θ is proportional to the decrement in momentum flow. Moreover, δ^* is the distance away from the body surface through which the outer inviscid flow is displaced due to the boundary layer. The body shape plus δ^* defines a new effective body seen by the inviscid flow.

By an order-of-magnitude analysis, the complete Navier-Stokes equations for two-dimensional flow reduce to the following boundary-layer equations:

Continuity:
$$\frac{\partial(\rho u)}{\partial x} + \frac{\partial(\rho v)}{\partial y} = 0 \qquad (17.28)$$

x momentum:
$$\rho u \frac{\partial u}{\partial x} + \rho v \frac{\partial u}{\partial y} = -\frac{dp_e}{dx} + \frac{\partial}{\partial y}\left(\mu \frac{\partial u}{\partial y}\right) \qquad (17.29)$$

y momentum:
$$\frac{\partial p}{\partial y} = 0 \qquad (17.30)$$

(continued)

Energy:
$$\rho u \frac{\partial h}{\partial x} + \rho v \frac{\partial h}{\partial y} = \frac{\partial}{\partial y}\left(k\frac{\partial T}{\partial y}\right) + u\frac{dp_e}{dx} + \mu\left(\frac{\partial u}{\partial y}\right)^2 \qquad (17.31)$$

These equations are subject to the boundary conditions:

At the wall: $y = 0$ $u = 0$ $v = 0$ $h = h_w$

At the boundary-layer edge: $y \to \infty$ $u \to u_e$ $h \to h_e$

Inherent in the above boundary-layer equations are the assumptions that $\delta \ll c$, Re is large, and M_∞ is not inordinately large.

CHAPTER 18

Laminar Boundary Layers

Lamina—A thin scale or sheet. A layer or coat lying over another.
   Funk and Wagnalls Standard Desk
   Dictionary, 1964

PREVIEW BOX

Here we get down to the brass tacks of actually solving the boundary-layer equations and obtaining practical formulae for the calculation of skin friction and aerodynamic heating to the surface. The solution, however, depends on whether the flow in the boundary layer is laminar or turbulent. And what a big difference there is! The difference is so important that we have separate chapters for each case—the present chapter dealing with laminar boundary layers, and Chapter 19 dealing with turbulent boundary layers.

Laminar boundary layers are easier to deal with, and that is why we treat them first. The analysis of a laminar boundary layer is more theoretically "pure" than that for a turbulent boundary layer. You will feel intellectually comfortable with this chapter, so sit back and enjoy this intellectual experience.

18.1 INTRODUCTION

Within the panoply of boundary-layer analyses, the solution of *laminar* boundary layers is well in hand compared to the status for turbulent boundary layers. This chapter is exclusively devoted to laminar boundary layers; turbulent boundary layers is the subject of Chapter 19. The basic definitions of laminar and turbulent flows are discussed in Section 15.1, and some characteristics of these flows are illustrated in Figures 15.5 and 15.6; it is recommended that you review that material before progressing further.

   The road map for this chapter is given in Figure 18.1. We will first deal with some well-established classical solutions that come under the heading of

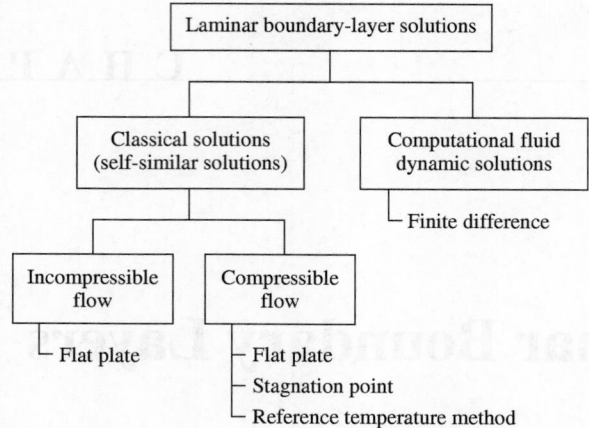

Figure 18.1 Road map for Chapter 18.

self-similar solutions, a term that is defined in Section 18.2. In this regard, we will deal with both incompressible and compressible flows over a flat plate, as noted on the left side of our road map in Figure 18.1. We will also discuss the boundary-layer solution in the region surrounding the stagnation point on a blunt-nosed body, because this solution gives us important information on aerodynamic heating at the stagnation point—vital information for high-speed flight vehicles. As part of the classical solution of compressible boundary layers, we will discuss the reference temperature method—a very useful engineering calculation that makes use of classical incompressible boundary-layer results to predict skin friction and aerodynamic heating for a compressible flow. Then we will move to the right side of the road map in Figure 18.1 and discuss some more modern computational fluid dynamic solutions to laminar boundary layers. Unlike the classical self-similar solutions, which are limited to a few (albeit important) applications such as flat plates and the stagnation region, these CFD numerical solutions deal with the laminar boundary layer over bodies of arbitrary shapes.

Note: As we progress through this chapter, we will encounter ideas and results that are already familiar to us from our discussion of Couette flow in Chapter 16. Indeed, this is one of the primary reasons for Chapter 16—to introduce these concepts within the context of a relatively straightforward flow problem before dealing with the more intricate boundary-layer solutions.

18.2 INCOMPRESSIBLE FLOW OVER A FLAT PLATE: THE BLASIUS SOLUTION

Consider the incompressible, two-dimensional flow over a flat plate at 0° angle of attack, such as sketched in Figure 17.7. For such a flow, $\rho = $ constant, $\mu = $ constant, and $dp_e/dx = 0$ (because the inviscid flow over a flat plate at $\alpha = 0$ yields a constant pressure over the surface). Moreover, recall that the energy equation is not needed to calculate the velocity field for an incompressible flow. Hence, the boundary-layer equations, Equations (17.28) to (17.31), reduce to

$$\frac{\partial u}{\partial x} + \frac{\partial v}{\partial y} = 0 \tag{18.1}$$

$$u\frac{\partial u}{\partial x} + v\frac{\partial u}{\partial y} = v\frac{\partial^2 u}{\partial y^2} \tag{18.2}$$

$$\frac{\partial p}{\partial y} = 0 \tag{18.3}$$

where v is the *kinematic viscosity*, defined as $v \equiv \mu/\rho$.

We now embark on a procedure that is common to many boundary-layer solutions. Let us transform the independent variables (x, y) to (ξ, η), where

$$\xi = x \quad \text{and} \quad \eta = y\sqrt{\frac{V_\infty}{vx}} \tag{18.4}$$

Using the chain rule, we obtain the derivatives

$$\frac{\partial}{\partial x} = \frac{\partial}{\partial \xi}\frac{\partial \xi}{\partial x} + \frac{\partial}{\partial \eta}\frac{\partial \eta}{\partial x} \tag{18.5}$$

$$\frac{\partial}{\partial y} = \frac{\partial}{\partial \xi}\frac{\partial \xi}{\partial y} + \frac{\partial}{\partial \eta}\frac{\partial \eta}{\partial y} \tag{18.6}$$

However, from Equation (18.4) we have

$$\frac{\partial \xi}{\partial x} = 1 \qquad \frac{\partial \xi}{\partial y} = 0 \qquad \frac{\partial \eta}{\partial y} = \sqrt{\frac{V_\infty}{vx}} \tag{18.7}$$

(We do not have to explicitly obtain $\partial\eta/\partial x$ because these terms will eventually cancel from our equations.) Substituting Equations (18.7) into (18.5) and (18.6), we have

$$\frac{\partial}{\partial x} = \frac{\partial}{\partial \xi} + \frac{\partial \eta}{\partial x}\frac{\partial}{\partial \eta} \tag{18.8}$$

$$\frac{\partial}{\partial y} = \sqrt{\frac{V_\infty}{vx}}\frac{\partial}{\partial \eta} \tag{18.9}$$

$$\frac{\partial^2}{\partial y^2} = \frac{V_\infty}{vx}\frac{\partial^2}{\partial \eta^2} \tag{18.10}$$

Also, let us define a stream function ψ such that

$$\psi = \sqrt{vxV_\infty}f(\eta) \tag{18.11}$$

where $f(\eta)$ is strictly a function of η only. This expression for ψ identically satisfies the continuity equation, Equation (18.1); therefore, it is a physically possible stream function. [Show yourself that ψ satisfies Equation (18.1); to do this, you will have to carry out many of the same manipulations described below.]

From the definition of the stream function, and using Equations (18.8), (18.9), and (18.11), we have

$$u = \frac{\partial \psi}{\partial y} = \sqrt{\frac{V_\infty}{\nu x}} \frac{\partial \psi}{\partial \eta} = V_\infty f'(\eta) \tag{18.12}$$

$$v = -\frac{\partial \psi}{\partial x} = -\left(\frac{\partial \psi}{\partial \xi} + \frac{\partial \eta}{\partial x} \frac{\partial \psi}{\partial \eta} \right) = -\frac{1}{2}\sqrt{\frac{\nu V_\infty}{x}} f - \sqrt{\nu x V_\infty} \frac{\partial \eta}{\partial x} f' \tag{18.13}$$

Equation (18.12) is of particular note. The function $f(\eta)$ defined in Equation (18.11) has the property that its derivative f' gives the x component of velocity as

$$f'(\eta) = \frac{u}{V_\infty}$$

Substitute Equations (18.8) to (18.10), (18.12), and (18.13) into the momentum equation, Equation (18.2). Writing each term explicitly so that you can see what is happening, we have

$$V_\infty f' \left(V_\infty \frac{\partial \eta}{\partial x} f'' \right) - \left(\frac{1}{2}\sqrt{\frac{\nu V_\infty}{x}} f + \sqrt{\nu x V_\infty} \frac{\partial \eta}{\partial x} f' \right) V_\infty \sqrt{\frac{V_\infty}{\nu x}} f'' = \nu V_\infty \frac{V_\infty}{\nu x} f'''$$

Simplifying, we obtain

$$V_\infty^2 \frac{\partial \eta}{\partial x} f' f'' - \frac{1}{2} \frac{V_\infty^2}{x} f f'' - V_\infty^2 \left(\frac{\partial \eta}{\partial x} \right) f' f'' = \frac{V_\infty^2}{x} f''' \tag{18.14}$$

The first and third terms cancel, and Equation (18.14) becomes

$$\boxed{2f''' + f f'' = 0} \tag{18.15}$$

Equation (18.15) is important; it is called *Blasius' equation,* after H. Blasius, who obtained it in his Ph.D. dissertation in 1908. Blasius was a student of Prandtl, and his flat-plate solution using Equation (18.15) was the first practical application of Prandtl's boundary-layer hypothesis since its announcement in 1904. Examine Equation (18.15) closely. Amazingly enough, it is an *ordinary differential equation.* Look what has happened! Starting with the partial differential equations for a flat-plate boundary layer given by Equations (18.1) to (18.3), and transforming *both* the independent and dependent variables through Equations (18.4) and (18.11), we obtain an ordinary differential equation for $f(\eta)$. In the same breath, we can say that Equation (18.15) is also an equation for the velocity u because $u = V_\infty f'(\eta)$. Because Equation (18.15) is a single ordinary differential equation, it is simpler to solve than the original boundary-layer equations. However, it is still a nonlinear equation and must be solved numerically, subject to the transformed boundary conditions,

At $\eta = 0$: $\qquad\qquad\qquad\qquad f = 0, f' = 0$

At $\eta \rightarrow \infty$: $\qquad\qquad\qquad\qquad f' = 1$

[Note that at the wall where $\eta = 0$, $f' = 0$ because $u = 0$, and therefore $f = 0$ from Equation (18.13) evaluated at the wall.]

Equation (18.15) is a third-order, nonlinear, ordinary differential equation; it can be solved numerically by means of standard techniques, such as the Runge-Kutta method (such as that described in Reference 49). The integration begins at the wall and is carried out in small increments Δy in the direction of increasing y away from the wall. However, since Equation (18.15) is third order, three boundary conditions must be known at $\eta = 0$; from the above, only two are specified. A third boundary condition, namely, some value for $f''(0)$, must be *assumed;* Equation (18.15) is then integrated across the boundary layer to a large value of η. The value of f' at large eta is then examined. Does it match the boundary condition at the edge of the boundary layer, namely, is $f' = 1$ satisfied at the edge of the boundary layer? If not, assume a different value of $f''(0)$ and integrate again. Repeat this process until convergence is obtained. This numerical approach is called the "shooting technique"; it is a classical approach, and its basic philosophy and details are discussed at great length in Section 16.4. Its application to Equation (18.15) is more straightforward than the discussion in Section 16.4, because here we are dealing with an incompressible flow and only one equation, namely, the momentum equation as embodied in Equation (18.15).

The solution of Equation (18.15) is plotted in Figure 18.2 in the form of $f'(\eta) = u/V_\infty$ as a function of η. Note that this curve is the *velocity profile* and that it is a function of η only. Think about this for a moment. Consider two different x stations along the plate, as shown in Figure 18.3. In general, $u = u(x, y)$, and the velocity profiles in terms of $u = u(y)$ at given x stations will be *different.* Clearly, the variation of u normal to the wall will change as the flow progresses downstream. However, when plotted versus η, we see that the profile, $u = u(\eta)$,

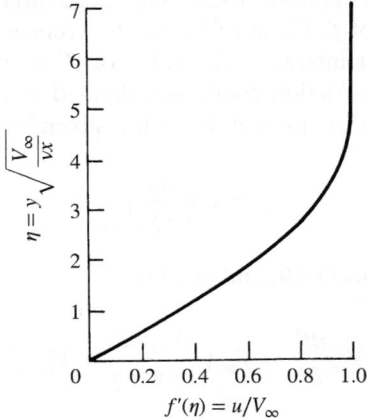

Figure 18.2 Incompressible velocity profile for a flat plate; solution of the Blasius equation.

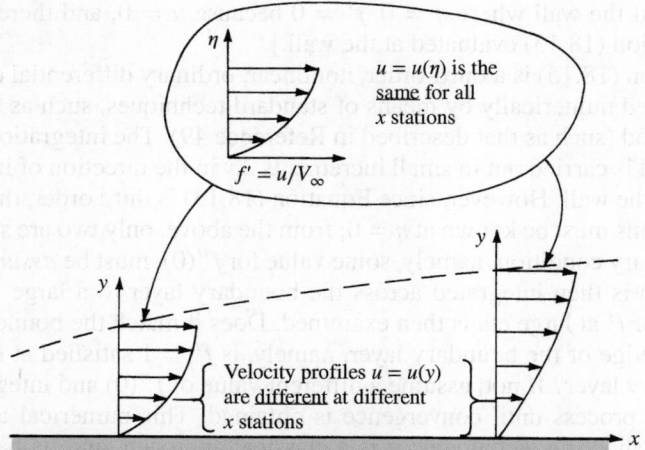

Figure 18.3 Velocity profiles in physical and transformed space, demonstrating the meaning of self-similar solutions.

is the same for *all* x stations, as illustrated in Figure 18.3. This result is an example of a *self-similar solution*—solutions where the boundary-layer profiles, when plotted versus a similarity variable η are the same for all x stations. For such self-similar solutions, the governing boundary-layer equations reduce to one or more ordinary differential equations in terms of a transformed independent variable. Self-similar solutions occur only for certain special types of flows—the flow over a flat plate is one such example. In general, for the flow over an arbitrary body, the boundary-layer solutions are nonsimilar; the governing partial differential equations cannot be reduced to ordinary differential equations.

Numerical values of f, f', and f'' tabulated versus η can be found in Reference 40. Of particular interest is the value of f'' at the wall; $f''(0) = 0.332$. Consider the local skin-friction coefficient defined as $c_f = \tau_w / \frac{1}{2} \rho_\infty V_\infty^2$. From Equation (17.34), the shear stress at the wall is given by

$$\tau_w = \mu \left(\frac{\partial u}{\partial y} \right)_{y=0} \tag{18.16}$$

However, from Equations (18.9) and (18.11),

$$\frac{\partial u}{\partial y} = V_\infty \frac{\partial f'}{\partial y} = V_\infty \sqrt{\frac{V_\infty}{vx}} \frac{\partial f'}{\partial \eta} = V_\infty \sqrt{\frac{V_\infty}{vx}} f'' \tag{18.17}$$

Evaluating Equation (18.17) at the wall, where $y = \eta = 0$, we obtain

$$\left(\frac{\partial u}{\partial y} \right)_{y=0} = V_\infty \sqrt{\frac{V_\infty}{vx}} f''(0) \tag{18.18}$$

Combining Equations (18.16) and (18.18), we have

$$c_f = \frac{\tau_w}{\frac{1}{2}\rho_\infty V_\infty^2} = \frac{2\mu}{\rho_\infty V_\infty^2} V_\infty \sqrt{\frac{V_\infty}{\nu x}} f''(0) \qquad (18.19)$$

$$= 2\sqrt{\frac{\mu}{\rho_\infty V_\infty x}} f''(0) = \frac{2f''(0)}{\sqrt{Re_x}}$$

where Re_x is the local Reynolds number. Since $f''(0) = 0.332$ from the numerical solution of Equation (18.15), then Equation (18.19) yields

$$c_f = \frac{0.664}{\sqrt{Re_x}} \qquad (18.20)$$

which is a classic expression for the local skin-friction coefficient for the incompressible laminar flow over a flat plate—a result that stems directly from boundary-layer theory. Its validity has been amply verified by experiment. Note that $c_f \propto Re_x^{-1/2} \propto x^{-1/2}$; that is, c_f decreases inversely proportional to the square root of distance from the leading edge. Examining the flat plate sketched in Figure 17.7, the total drag on the top surface of the entire plate is the integrated contribution of $\tau_w(x)$ from $x = 0$ to $x = c$. Letting C_f denote the skin friction drag coefficient, we obtain from Equation (1.16)

$$C_f = \frac{1}{c}\int_0^c c_f \, dx \qquad (18.21)$$

Substituting Equation (18.20) into (18.21), we obtain

$$C_f = \frac{1}{c}(0.664)\sqrt{\frac{\mu}{\rho_\infty V_\infty}}\int_0^c x^{-1/2}\,dx = \frac{1.328}{c}\sqrt{\frac{\mu c}{\rho_\infty V_\infty}}$$

or

$$C_f = \frac{1.328}{\sqrt{Re_c}} \qquad (18.22)$$

where Re_c is the Reynolds number based on the total plate length c.

An examination of Figure 18.2 shows that $f' = 0.99$ at approximately $\eta = 5.0$. Hence, the boundary-layer thickness, which was defined earlier as that distance above the surface where $u = 0.99u_e$, is

$$\eta = y\sqrt{\frac{V_\infty}{\nu x}} = \delta\sqrt{\frac{V_\infty}{\nu x}} = 5.0$$

or

$$\delta = \frac{5.0x}{\sqrt{Re_x}} \qquad (18.23)$$

Note that the boundary-layer thickness is inversely proportional to the square root of the Reynolds number (based on the local distance x). Also, $\delta \propto x^{1/2}$;

the laminar boundary layer over a flat plate grows parabolically with distance from the leading edge.

The displacement thickness δ^*, defined by Equation (17.3), becomes for an incompressible flow

$$\delta^* = \int_0^{y_1} \left(1 - \frac{u}{u_e}\right) dy \tag{18.24}$$

In terms of the transformed variables f' and η given by Equations (18.4) and (18.12), the integral in Equation (18.24) can be written as

$$\delta^* = \sqrt{\frac{vx}{V_\infty}} \int_0^{\eta_1} [1 - f'(\eta)] \, d\eta = \sqrt{\frac{vx}{V_\infty}} [\eta_1 - f(\eta_1)] \tag{18.25}$$

where η_1 is an arbitrary point above the boundary layer. The numerical solution for $f(\eta)$ obtained from Equation (18.15) shows that, amazingly enough, $\eta_1 - f(\eta) = 1.72$ for all values of η above 5.0. Therefore, from Equation (18.25), we have

$$\delta^* = 1.72 \sqrt{\frac{vx}{V_\infty}}$$

or

$$\boxed{\delta^* = \frac{1.72x}{\sqrt{\mathrm{Re}_x}}} \tag{18.26}$$

Note that, as in the case of the boundary-layer thickness itself, δ^* varies inversely with the square root of the Reynolds number, and $\delta^* \propto x^{1/2}$. Also, comparing Equations (18.23) and (18.26), we see that $\delta^* = 0.34\delta$; the displacement thickness is smaller than the boundary-layer thickness, confirming our earlier statement in Section 17.2.

The momentum thickness for an incompressible flow is, from Equation (17.10),

$$\theta = \int_0^{y_1} \frac{u}{u_e} \left(1 - \frac{u}{u_e}\right) dy$$

or in terms of our transformed variables,

$$\theta = \sqrt{\frac{vx}{V_\infty}} \int_0^{\eta_1} f'(1 - f') \, d\eta \tag{18.27}$$

Equation (18.27) can be integrated numerically from $\eta = 0$ to any arbitrary point $\eta_1 > 5.0$. The result gives

$$\theta = \sqrt{\frac{\eta x}{V_\infty}} (0.664)$$

or

$$\boxed{\theta = \frac{0.664x}{\sqrt{\mathrm{Re}_x}}} \tag{18.28}$$

Note that, as in the case of our previous thicknesses, θ varies inversely with the square root of the Reynolds number and that $\theta \propto x^{1/2}$. Also, $\theta = 0.39\delta^*$ and $\theta = 0.13\delta$. Another property of momentum thickness can be demonstrated by evaluating θ at the trailing edge of the flat plate sketched in Figure 17.7. In this case, $x = c$, and from Equation (18.28), we obtain

$$\theta_{x=c} = \frac{0.664c}{\sqrt{\mathrm{Re}_c}} \tag{18.29}$$

Comparing Equations (18.22) and (18.29), we have

$$C_f = \frac{2\theta_{x=c}}{c} \tag{18.30}$$

Equation (18.30) demonstrates that the integrated skin-friction coefficient for the flat plate is directly proportional to the value of θ evaluated at the trailing edge.

18.3 COMPRESSIBLE FLOW OVER A FLAT PLATE

The properties of the incompressible, laminar, flat-plate boundary layer were developed in Section 18.2. These results hold at low Mach numbers where the density is essentially constant through the boundary layer. However, what happens to these properties at high Mach numbers where the density becomes a variable; that is, what are the compressibility effects? The purpose of the present section is to outline briefly the effects of compressibility on both the derivations and the final results for laminar flow over a flat plate. We do not intend to present much detail; rather, we examine some of the salient aspects which distinguish compressible from incompressible boundary layers.

The compressible boundary-layer equations were derived in Section 17.3, and were presented as Equations (17.28) to (17.31). For flow over a flat plate, where $dp_e/dx = 0$, these equations become

$$\frac{\partial(\rho u)}{\partial x} + \frac{\partial(\rho v)}{\partial y} = 0 \tag{18.31}$$

$$\rho u \frac{\partial u}{\partial x} + \rho v \frac{\partial u}{\partial y} = \frac{\partial}{\partial y}\left(\mu \frac{\partial u}{\partial y}\right) \tag{18.32}$$

$$\frac{\partial p}{\partial y} = 0 \tag{18.33}$$

$$\rho u \frac{\partial h}{\partial x} + \rho v \frac{\partial h}{\partial y} = \frac{\partial}{\partial y}\left(k \frac{\partial T}{\partial y}\right) + \mu \left(\frac{\partial u}{\partial y}\right)^2 \tag{18.34}$$

Compare these equations with those for the incompressible case given by Equations (18.1) to (18.3). Note that, for a compressible boundary layer, (1) the energy equation must be included, (2) the density is treated as a variable, and (3) in general, μ and k are functions of temperature and hence also must be treated as variables. As a result, the system of equations for the compressible case,

Equations (18.31) to (18.34), is more complex than for the incompressible case, Equations (18.1) to (18.3).

It is sometimes convenient to deal with total enthalpy, $h_0 = h + V^2/2$, as the dependent variable in the energy equation, rather than the static enthalpy as given in Equation (18.34). Note that, consistent with the boundary-layer approximation, where v is small, $h_0 = h + V^2/2 = h + (u + v^2)/2 \approx h + u^2/2$. To obtain the energy equation in terms of h_0, multiply Equation (18.32) by u, and add to Equation (18.34), as follows. From Equation (18.32) multiplied by u,

$$\rho u \frac{\partial (u^2/2)}{\partial x} + \rho v \frac{\partial (u^2/2)}{\partial y} = u \frac{\partial}{\partial y} \left(\mu \frac{\partial u}{\partial y} \right) \tag{18.35}$$

Adding Equation (18.35) to (18.34), we obtain

$$\rho u \frac{\partial (h + u^2/2)}{\partial x} + \rho v \frac{\partial (h + u^2/2)}{\partial y} = \frac{\partial}{\partial y} \left(k \frac{\partial T}{\partial y} \right) + \mu \left(\frac{\partial u}{\partial y} \right)^2 + u \frac{\partial}{\partial y} \left(\mu \frac{\partial u}{\partial y} \right) \tag{18.36}$$

Recall that for a calorically perfect gas, $dh = c_p \, dT$; hence,

$$\frac{\partial T}{\partial y} = \frac{1}{c_p} \frac{\partial h}{\partial y} = \frac{1}{c_p} \frac{\partial}{\partial y} \left(h_0 - \frac{u^2}{2} \right) \tag{18.37}$$

Substituting Equation (18.37) into (18.36), we obtain

$$\rho u \frac{\partial h_0}{\partial x} + \rho v \frac{\partial h_0}{\partial y} = \frac{\partial}{\partial y} \left[\frac{k}{c_p} \frac{\partial}{\partial y} \left(h_0 - \frac{u^2}{2} \right) \right] + \mu \left(\frac{\partial u}{\partial y} \right)^2 + u \frac{\partial}{\partial y} \left(\mu \frac{\partial u}{\partial y} \right) \tag{18.38}$$

Note that

$$\frac{k}{c_p} \frac{\partial}{\partial y} \left(h_0 - \frac{u^2}{2} \right) = \frac{\mu k}{\mu c_p} \frac{\partial}{\partial y} \left(h_0 - \frac{u^2}{2} \right) = \frac{\mu}{\text{Pr}} \left(\frac{\partial h_0}{\partial y} - u \frac{\partial u}{\partial y} \right) \tag{18.39}$$

and

$$\mu \left(\frac{\partial u}{\partial y} \right)^2 + u \frac{\partial}{\partial y} \left(\mu \frac{\partial u}{\partial y} \right) = \frac{\partial}{\partial y} \left(\mu u \frac{\partial u}{\partial y} \right) \tag{18.40}$$

Substituting Equations (18.39) and (18.40) into (18.38), we obtain

$$\rho u \frac{\partial h_0}{\partial x} + \rho v \frac{\partial h_0}{\partial y} = \frac{\partial}{\partial y} \left[\frac{\mu}{\text{Pr}} \frac{\partial h_0}{\partial y} + \left(1 - \frac{1}{\text{Pr}} \right) \mu u \frac{\partial u}{\partial y} \right] \tag{18.41}$$

which is an alternate form of the boundary-layer energy equation. In this equation, Pr is the local Prandtl number, which, in general, is a function of T and hence varies throughout the boundary layer.

For the laminar, compressible flow over a flat plate, the system of governing equations can now be considered to be Equations (18.31) to (18.33) and (18.41). These are nonlinear partial differential equations. As in the incompressible case, let us seek a self-similar solution; however, the transformed independent variables must be defined differently:

$$\xi = \rho_e \mu_e u_e x \qquad \xi = \xi(x)$$

$$\eta = \frac{u_e}{\sqrt{2\xi}} \int_0^y \rho \, dy \qquad \eta = \eta(x, y)$$

The dependent variables are transformed as follows:

$$f' = \frac{u}{u_e} \qquad \text{(which is consistent with defining stream function } \psi = \sqrt{2\xi} f)$$

$$g = \frac{h_0}{(h_0)_e}$$

The mechanics of the transformation using the chain rule are similar to that described in Section 18.2. Hence, without detailing the precise steps (which are left for your entertainment), Equations (18.32) and (18.41) transform into

$$\left(\frac{\rho\mu}{\rho_e \mu_e} f'' \right)' + f f' = 0 \tag{18.42}$$

and

$$\left(\frac{\rho\mu}{\rho_e \mu_e} \frac{1}{\text{Pr}} g' \right)' + f g' + \frac{u_e^2}{(h_0)_e} \left[\left(1 - \frac{1}{\text{Pr}} \right) \frac{\rho\mu}{\rho_e \mu_e} f' f'' \right]' = 0 \tag{18.43}$$

Examine Equations (18.42) and (18.43) closely. They are ordinary differential equations—recall that the primes denote differentiation with respect to η. Therefore, the compressible, laminar flow over a flat plate does lend itself to a self-similar solution, where $f' = f'(\eta)$ and $g = g(\eta)$. That is, the velocity and total enthalpy profiles plotted versus η are the same at any station. Furthermore, the product $\rho\mu$ is a variable and depends in part on temperature. Hence, Equation (18.42) is coupled to the energy equation, Equation (18.43), via $\rho\mu$. Of course, the energy equation is strongly coupled to Equation (18.42) via the appearance of f, f', and f'' in Equation (18.43). Hence, we are dealing with a *system* of coupled ordinary differential equations which must be solved simultaneously. The boundary conditions for these equations are

At $\eta = 0$: $\qquad\qquad\qquad f = f' = 0 \qquad g = g_w$

At $\eta \to \infty$: $\qquad\qquad\qquad\qquad f' = 1 \qquad g = 1$

Note that the coefficient $u_e^2/(h_0)_e$ appearing in Equation (18.43) is simply a function of the Mach number:

$$\frac{u_e^2}{(h_0)_e} = \frac{u_e^2}{h_e + u_e^2/2} = \frac{1}{h_e/u_e^2 + \frac{1}{2}} = \frac{1}{c_p T_e/u_e^2 + \frac{1}{2}} = \frac{1}{RT_e/(\gamma - 1)u_e^2 + \frac{1}{2}}$$

$$= \frac{1}{1/(\gamma - 1)M_e^2 + \frac{1}{2}} = \frac{2(\gamma - 1)M_e^2}{2 + (\gamma - 1)M_e^2}$$

Therefore, Equation (18.43) involves as a parameter the Mach number of the flow at the outer edge of the boundary layer, that is, for the flat-plate case, the

---

freestream Mach number. Hence, we can explicitly see that the compressible boundary-layer solutions will depend on the Mach number. Moreover, because of the appearance of the local Pr in Equation (18.43), the solutions will also depend on the freestream Prandtl number as a parameter. Finally, note from the boundary conditions that the value of g at the wall g_w is a given quantity. Note that at the wall where $u = 0$, $g_w = h_w/(h_0)_e = c_p T_w/(h_0)_e$. Hence, instead of referring to a given enthalpy at the wall g_w, we usually deal with a given wall temperature T_w. An alternative to a given value of T_w is the assumption of an *adiabatic wall,* that is, a case where there is no heat transfer to the wall. If $\dot{q}_w = k(\partial T/\partial y)_w = 0$, then $(\partial T/\partial y)_w = 0$. Hence, for an adiabatic wall, the boundary condition at the wall becomes simply $(\partial T/\partial y)_w = 0$.

In short, we see from the above discussion that a numerical self-similar solution can be obtained for the compressible, laminar flow over a flat plate. However, this solution depends on the Mach number, the Prandtl number, and the condition of the wall (whether it is adiabatic or a constant temperature wall with T_w given). Such numerical solutions have been carried out; see Reference 41 for details. A classic solution to Equations (18.42) and (18.43) is the shooting technique described in Section 16.4. The approach here is directly analogous to that used for the solution of compressible Couette flow discussed in Section 16.4. Since Equation (18.42) is third order, we need three boundary conditions at $\eta = 0$. We have only two, namely, $f = f' = 0$. Therefore, *assume* a value for $f''(0)$, and iterate until the boundary condition at the boundary-layer edge, $f' = 1$, is matched. Similarly, Equation (18.43) is a second-order equation. It requires two boundary conditions at the wall in order to integrate numerically across the boundary layer; we have only one, namely, $g(0) = g_w$. Thus, *assume* $g'(0)$, and integrate Equation (18.43). Iterate until the outer boundary condition is satisfied; that is, $g = 1$. Since Equation (18.42) is coupled to Equation (18.43), that is, since $\rho\mu$ in Equation (18.42) requires a knowledge of the enthalpy (or temperature) profile across the boundary layer, the entire process must be repeated. This is directly analogous to the two minor iterations nested within the major iteration that was described in the discussion of the shooting method in Section 16.4. The approach here is virtually the same philosophy as described in Section 16.4, which should be reviewed at this stage. Therefore, no further details will be given here.

Typical solutions of Equations (18.42) and (18.43) for the velocity and temperature profiles through a compressible boundary layer on a flat plate are shown in Figures 18.4–18.7 (Reference 75). Figures 18.4 and 18.5 contain results for an insulated flat plate (zero-heat transfer) using Sutherland's law for μ, and assuming a constant Pr = 0.75. The velocity profiles are shown in Figure 18.4 for different Mach numbers ranging from 0 (incompressible flow) to the large hypersonic value of 20. Note that at a given x station at a given Re_x, the boundary-layer thickness increases markedly as M_e is increased to hypersonic values. This clearly demonstrates one of the most important aspects of compressible boundary layers, namely, that the boundary-layer thickness becomes large at large Mach numbers. Figure 18.5 illustrates the temperature profiles for the same case as Figure 18.4. Note the obvious physical trend that, as M_e increases to large hypersonic values, the temperatures increase markedly. Also note in Figure 18.5 that at the wall

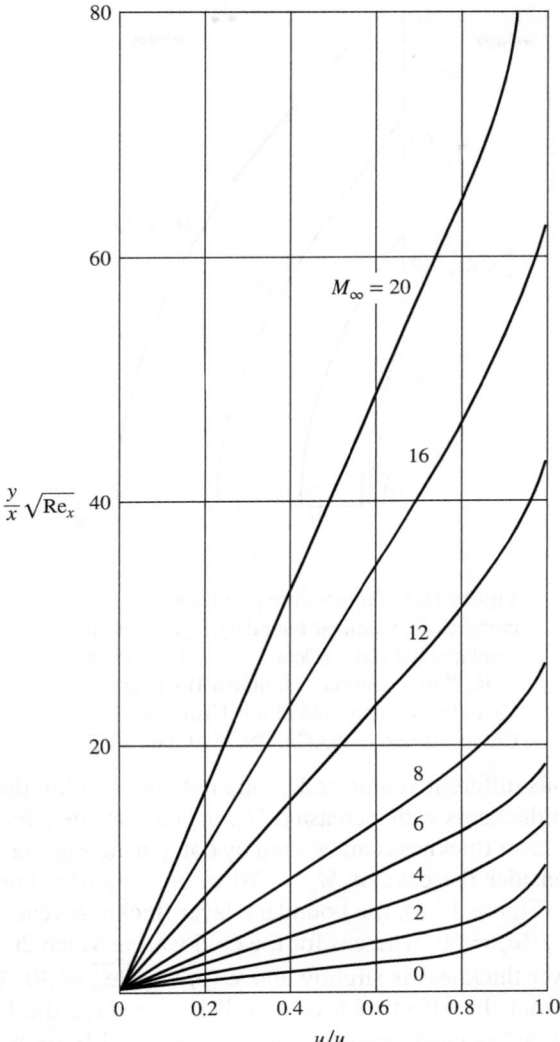

Figure 18.4 Velocity profiles in a compressible laminar
boundary layer over an insulated flat plate. (*Data Source:*
Van Driest, E. R.: "Investigation of Laminar Boundary
Layer in Compressible Fluids Using the Crocco Method,"
NACA TN 2579, Jan. 1952.)

$(y = 0)$, $(\partial T/\partial y)_w = 0$, as it should be for an insulated surface $(q_w = 0)$. Figures 18.6 and 18.7 also contain results by van Driest, but now for the case of heat transfer to the wall. Such a case is called a "cold wall" case, because $T_w < T_{aw}$. (The opposite case would be a "hot wall," where heat is transferred from the wall into the flow; in this case, $T_w > T_{aw}$.) For the results shown in Figures 18.6 and 18.7, $T_w/T_e = 0.25$ and $\mathrm{Pr} = 0.75 = $ constant. Figure 18.6 shows velocity

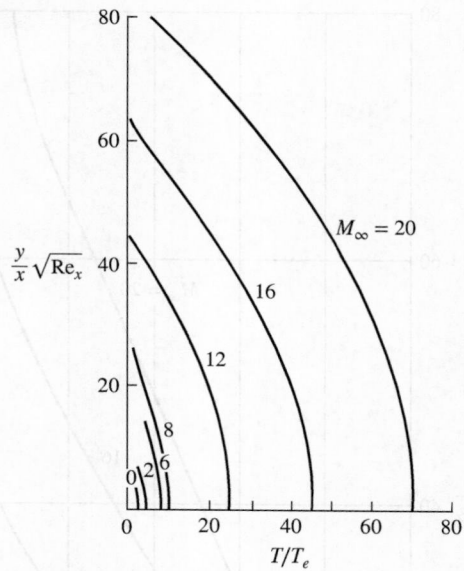

Figure 18.5 Temperature profiles in a compressible laminar boundary layer over an insulated flat plate. (*Data Source:* Van Driest, E. R.: "Investigation of Laminar Boundary Layer in Compressible Fluids Using the Crocco Method," NACA TN 2579, Jan. 1952.)

profiles for various different values of M_e, again demonstrating the rapid growth in boundary-layer thickness with increasing M_e. In addition, the effect of a cold wall on the boundary-layer thickness can be seen by comparing Figures 18.4 and 18.6. For example, consider the case of $M_e = 20$ in both figures. For the insulated wall at Mach 20 (Figure 18.4), the boundary-layer thickness reaches out beyond a value of $(y/x)\sqrt{\mathrm{Re}_x} = 60$, whereas for the cold wall at Mach 20 (Figure 18.6), the boundary-layer thickness is slightly above $(y/x)\sqrt{\mathrm{Re}_w} = 30$. This illustrates the general fact that the effect of a cold wall is to reduce the boundary-layer thickness. This trend is easily explainable on a physical basis when we examine Figure 18.7, which illustrates the temperature profiles through the boundary layer for the cold-wall case. Comparing Figures 18.5 and 18.7, we note that, as expected, the temperature levels in the cold-wall case are considerably lower than in the insulated case. In turn, because the pressure is the same in both cases, we have from the equation of state $p = \rho RT$, that the density in the cold-wall case is much higher. If the density is higher, the mass flow within the boundary layer can be accommodated within a smaller boundary-layer thickness; hence, the effect of a cold wall is to thin the boundary layer. Also note in Figure 18.7 that, starting at the outer edge of the boundary layer and going toward the wall, the temperature first increases, reaches a peak somewhere within the boundary layer, and then decreases to its prescribed cold-wall value of T_w. The peak temperature inside the boundary layer is an indication of the amount of viscous dissipation occurring

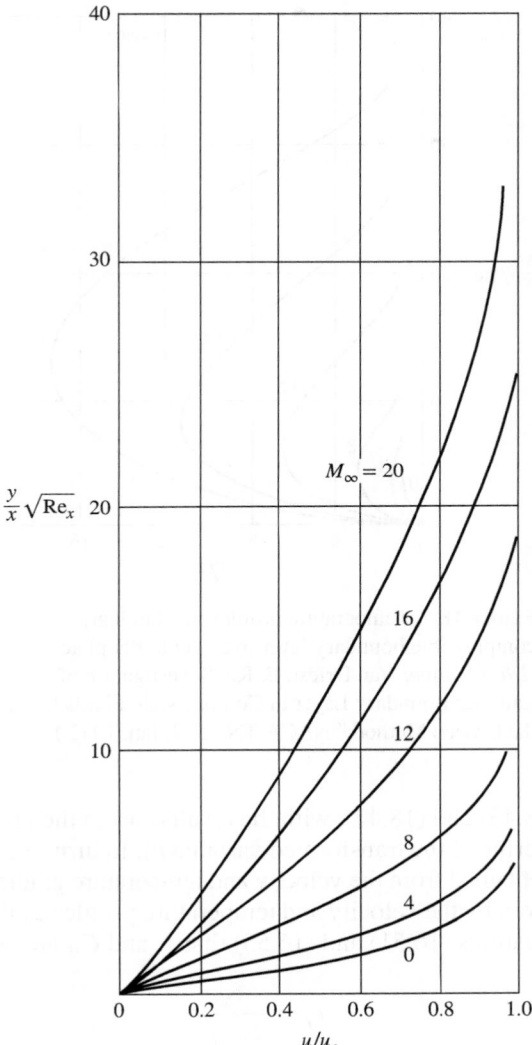

Figure 18.6 Velocity profiles in a laminar, compressible boundary layer over a cold flat plate. (*Data Source:* Van Driest, E. R.: "Investigation of Laminar Boundary Layer in Compressible Fluids Using the Crocco Method," NACA TN 2579, Jan. 1952.)

within the boundary layer. Figure 18.7 clearly demonstrates the rapidly growing effect of this viscous dissipation as M_e increases—yet another basic aspect of compressible boundary layers.

Carefully study the boundary-layer profiles shown in Figures 18.4–18.7. They are an example of the detailed results which emerge from a solution of Equations (18.42) and (18.43); indeed, these figures are graphical representations

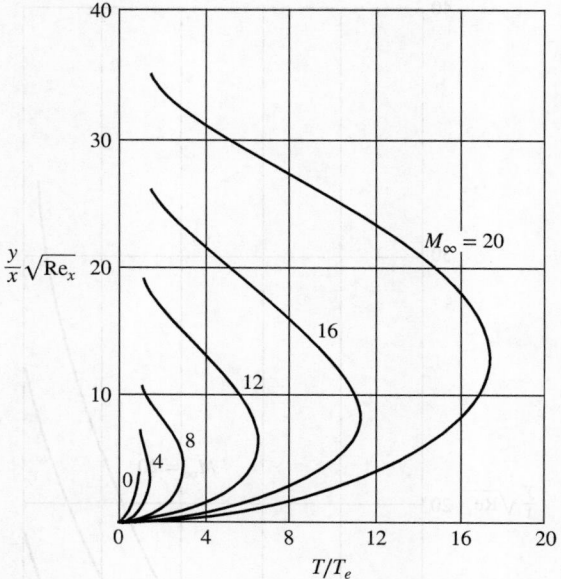

Figure 18.7 Temperature profiles in a laminar, compressible boundary layer over a cold flat plate. (*Data Source:* Van Driest, E. R.: "Investigation of Laminar Boundary Layer in Compressible Fluids Using the Crocco Method," NACA TN 2579, Jan. 1952.)

of Equations (18.43) and (18.42), with the results cast in the physical (x, y) space (rather than in terms of the transformed variable η). In turn, the surface values c_f and C_H can be obtained from the velocity and temperature gradients, respectively, at the wall as given by the velocity and temperature profiles evaluated at the wall. Recall from Equations (16.51) and (16.55) that c_f and C_H are defined as

$$c_f = \frac{\tau_w}{\frac{1}{2}\rho_e u_e^2} \tag{16.51}$$

and

$$C_H = \frac{\dot{q}_w}{\rho_e u_e (h_{aw} - h_e)} \tag{16.54}$$

where

$$\tau_w = \mu \left(\frac{\partial u}{\partial y}\right)_w \quad \text{and} \quad \dot{q}_w = -k \left(\frac{\partial T}{\partial y}\right)_w$$

and where $(\partial u/\partial y)_w$ and $(\partial T/\partial y)_w$ are the values obtained from the velocity and temperature profiles, respectively, evaluated at the wall. In turn, the overall flat plate skin friction drag coefficient C_f can be obtained by integrating c_f over the plate via Equation (18.21).

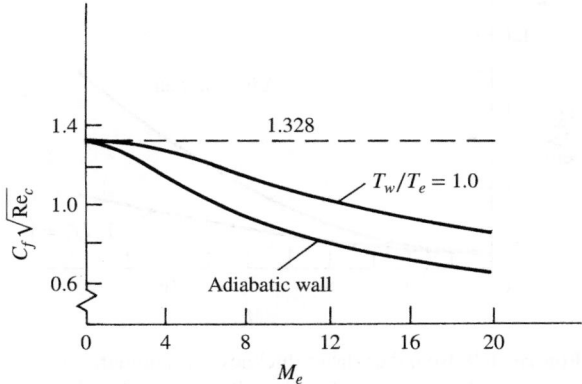

Figure 18.8 Friction drag coefficient for laminar, compressible flow over a flat plate, illustrating the effect of Mach number and wall temperature. Pr = 0.75. (*Calculations by E. R. van Driest, NACA Tech. Note 2597.*)

Return to Equation (18.22) for the friction drag coefficient for incompressible flow. The analogous compressible result can be written as

$$C_f = \frac{1.328}{\sqrt{\text{Re}_c}} F\left(M_e, \text{Pr}, \frac{T_w}{T_e}\right) \qquad (18.44)$$

In Equation (18.44), the function F is determined from the numerical solution. Sample results are given in Figure 18.8, which shows that the product $C_f \sqrt{\text{Re}_c}$ decreases as M_e increases. Moreover, the adiabatic wall is warmer than the wall in the case of $T_w/T_e = 1.0$. Hence, Figure 18.8 demonstrates that a hot wall also reduces $C_f \sqrt{\text{Re}_c}$.

Return to Equation (18.23) for the thickness of the incompressible flat-plate boundary layer. The analogous result for compressible flow is

$$\delta = \frac{5.0x}{\sqrt{\text{Re}_x}} G\left(M_e, \text{Pr}, \frac{T_w}{T_e}\right) \qquad (18.45)$$

In Equation (18.45), the function G is obtained from the numerical solution. Sample results are given in Figure 18.9, which shows that the product $(\delta \sqrt{\text{Re}_x}/x)$ increases as M_e increases. Everything else being equal, boundary layers are thicker at higher Mach numbers. This fact was stated earlier, as shown in Figures 18.4 and 18.6. Note also from Figure 18.9 that a hot wall thickens the boundary layer, as discussed earlier.

Recall our discussion of Couette flow in Chapter 16. There, we introduced the concept of the recovery factor r where

$$h_{aw} = h_e + r\frac{u_e^2}{2} \qquad (18.46)$$

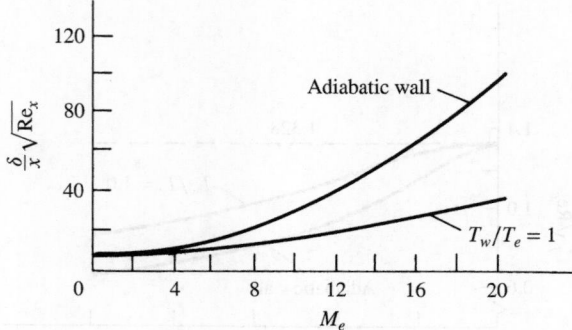

Figure 18.9 Boundary-layer thickness for laminar, compressible flow over a flat plate, illustrating the effect of Mach number and wall temperature. Pr = 0.75. (*Calculations by E. R. van Driest, NACA Tech. Note 2597.*)

This is a general concept, and can be applied to the boundary-layer solutions here. If we assume a constant Prandtl number for the compressible flat-plate flow, the numerical solution shows that

$$r = \sqrt{\text{Pr}} \qquad (18.47)$$

for the flat plate. Note that Equation (18.47) is analogous to the result given for Couette flow in that the recovery factor is a function of the Prandtl number only. However, for the flat plate, $r = \sqrt{\text{Pr}}$, whereas for Couette flow, $r = \text{Pr}$.

Aerodynamic heating for the flat plate can be treated via Reynolds analogy. The Stanton number and skin friction coefficients are defined respectively as

$$C_H = \frac{\dot{q}_w}{\rho_e u_e (h_{aw} - h_w)} \qquad (18.48)$$

and

$$c_f = \frac{\tau_w}{\frac{1}{2}\rho_e u_e^2} \qquad (18.49)$$

(See our discussion of these coefficients in Chapter 16.) Our results for Couette flow proved that a relation existed between C_H and c_f—namely, Reynolds analogy, given by Equation (16.59) for Couette flow. Moreover, in this relation, the ratio C_H/c_f was a function of the Prandtl number only. A directly analogous result holds for the compressible flat-plate flow. If we assume that the Prandtl number is constant, then for a flat plate, Reynolds analogy is, from the numerical solution,

$$\boxed{\frac{C_H}{c_f} = \frac{1}{2}\text{Pr}^{-2/3}} \qquad (18.50)$$

In Equation (18.50), the local skin friction coefficient c_f which is given by Equation (18.20) for the incompressible flat-plate case, becomes the following form for the compressible flat-plate flow:

$$c_f = \frac{0.664}{\sqrt{Re}} F\left(M_e, Pr, \frac{T_w}{T_e}\right) \tag{18.51}$$

In Equation (18.51), F is the same function as appears in Equation (18.44), and its variation with M_e and T_w/T_e is the same as shown in Figure 18.8.

EXAMPLE 18.1

Consider a flat plate at zero angle of attack in an airflow at standard sea level conditions ($p_\infty = 1.01 \times 10^5$ N/m^2 and $T_\infty = 288$ K). The chord length of the plate (distance from the leading edge to the trailing edge) is 2 m. The planform area of the plate is 40 m^2. At standard sea level conditions, $\mu_\infty = 1.7894 \times 10^{-5}$ kg/(m)(s). Assume the wall temperature is the adiabatic wall temperature T_{aw}. Calculate the friction drag on the plate when the freestream velocity is (a) 100 m/s, (b) 1000 m/s.

■ **Solution**

(a) The freestream density is

$$\rho_\infty = \frac{p_\infty}{RT_\infty} = \frac{1.01 \times 10^5}{(287)(288)} = 1.22 \text{ kg/m}^3$$

The speed of sound is

$$a_\infty = \sqrt{\gamma R T_\infty} = \sqrt{(1.4)(287)(288)} = 340.2 \text{ m/s}$$

The Mach number is $M_\infty = 100/340.2 = 0.29$. Hence, M_∞ is low enough to assume incompressible flow, and we can use Equation (18.22),

$$C_f = \frac{1.328}{\sqrt{Re_c}}$$

Please note that for the flow over a flat plate at zero angle of attack, the freestream velocity and density, V_∞ and ρ_∞, are the same as the velocity and density at the outer edge of the boundary layer, u_e and ρ_e. Hence, these quantities can be used interchangeably. Thus,

$$Re_c = \frac{\rho_\infty V_\infty c}{\mu_\infty} = \frac{(1.22)(100)(2)}{1.7894 \times 10^{-5}} = 1.36 \times 10^7$$

Hence, $$C_f = \frac{1.328}{\sqrt{Re_c}} = \frac{1.328}{\sqrt{1.36 \times 10^7}} = 3.60 \times 10^{-4}$$

The friction drag on one surface of the plate is given by

$$D_f = \tfrac{1}{2}\rho_\infty V_\infty^2 S C_f = \tfrac{1}{2}(1.22)(100)^2(40)(3.6 \times 10^{-4}) = 87.8 \text{ N}$$

The *total* drag due to friction is generated by the shear stress acting on *both* the top and bottom of the plate. Since D_f above is the friction drag on only one surface, we have

$$\text{Total friction drag} = D = 2D_f = 2(87.8) = \boxed{175.6 \text{ N}}$$

(b) For $V_\infty = 1000$ m/s, we have

$$M_\infty = \frac{V_\infty}{a_\infty} = \frac{1000}{340.2} = 2.94$$

Clearly, the flow is compressible, and we have to use Equation (18.44), or more directly, Figure 18.8. From Figure 18.8, we have for $M_\infty = M_e = 2.94$ and an adiabatic wall,

$$C_f \sqrt{\text{Re}_c} = 1.2$$

or
$$C_f = \frac{1.2}{\sqrt{\text{Re}_c}}$$

$$\text{Re}_c = \frac{\rho_\infty V_\infty c}{\mu_\infty} = \frac{(1.22)(1000)(2)}{1.7894 \times 10^{-5}} = 1.36 \times 10^8$$

Thus,
$$C_f = \frac{1.2}{\sqrt{1.36 \times 10^8}} = 1.03 \times 10^{-4}$$

The friction drag on one surface is

$$D_f = \tfrac{1}{2}\rho_\infty V_\infty^2 S C_f = \tfrac{1}{2}(1.22)(1000)^2(40)(1.03 \times 10^{-4}) = 2513 \text{ N}$$

Taking into account both the top and bottom surfaces,

$$\boxed{\text{Total friction drag} = D = 2D_f = 2(2513) = 5026 \text{ N}}$$

18.3.1 A Comment on Drag Variation with Velocity

Beginning with Chapter 1, indeed beginning with the most elementary studies of fluid dynamics, the point is usually made that the aerodynamic force on a body immersed in a flowing fluid is proportional to the square of the flow velocity. For example, from Section 1.5,

$$L = \tfrac{1}{2}\rho_\infty V_\infty^2 S C_L$$

and
$$D = \tfrac{1}{2}\rho_\infty V_\infty^2 S C_D$$

As long as C_L and C_D are *independent* of velocity, then clearly $L \propto V_\infty^2$ and $D \propto V_\infty^2$. This is the case for an inviscid, incompressible flow, where C_L and C_D depend only on the shape and angle of attack of the body. However, from the dimensional analysis in Section 1.7, we also discovered that C_L and C_D in general are functions of both Reynolds number and Mach number,

$$C_L = f_1(\text{Re}, M_\infty)$$
$$C_D = f_2(\text{Re}, M_\infty)$$

Of course, for an inviscid, incompressible flow, Re and M_∞ are not players (indeed, for inviscid flow, Re $\to \infty$ and for incompressible flow, $M_\infty \to 0$). However for all other types of flow, Re and M_∞ *are* players, and the values of C_L and C_D depend not only on the shape and angle of attack of the body, but also on Re and M_∞. For this reason, in general, the aerodynamic force is *not* exactly proportional to the square of the velocity. For example, examine the results from Example 18.1. In part (a), we calculated a value for drag to the 175.6 N when $V_\infty = 100$ m/s. If the drag were proportional to V_∞^2, then in part (b) where $V_\infty = 1000$ m/s, a factor of 10 larger, the drag would have been one hundred times larger, or 17,560 N. In contrast, our calculations in part (b) showed the drag to be considerably smaller, namely 5026 N. In other words, when V_∞ was increased by a factor of 10, the drag increased by only a factor of 28.6, *not* by a factor of 100. The reason is obvious. The value of C_f *decreases* when the velocity is increased because: (1) the Reynolds number increases, which from Equation (18.22) causes C_f to decrease, and (2) the Mach number increases, which from Figure 18.8 causes C_f to decrease.

So be careful about thinking that aerodynamic force varies with the square of the velocity. For cases other than inviscid, incompressible flow, this is not true.

18.4 THE REFERENCE TEMPERATURE METHOD

In this section we discuss an approximate engineering method for predicting skin friction and heat transfer for laminar compressible flow. It is based on the simple idea of utilizing the formulae obtained from incompressible flow theory, wherein the thermodynamic and transport properties in these formulae are evaluated at some reference temperature indicative of the temperature somewhere inside the boundary layer. This idea was first advanced by Rubesin and Johnson in Reference 76 and was modified by Eckert (Reference 77) to include a reference enthalpy. In this fashion, in some sense the classical incompressible formulae were "corrected" for compressibility effects. Reference temperature (or reference enthalpy) methods have enjoyed frequent application in engineering-oriented analyses, because of their simplicity. For this reason, we briefly describe the approach here.

Consider the incompressible laminar flow over a flat plate, as discussed in Section 18.2. The local skin friction coefficient is given by Equation (18.20), repeated below:

$$c_f = \frac{0.664}{\sqrt{\mathrm{Re}_x}} \tag{18.20}$$

For the *compressible* laminar flow over a flat plate, we write the analogous expression

$$\boxed{c_f^* = \frac{0.664}{\sqrt{\mathrm{Re}_x^*}}} \tag{18.52}$$

except here c_f^* and Re_x^* are evaluated at a reference temperature T^*. That is,

$$\mathrm{Re}_x^* = \frac{\rho^* u_e x}{\mu^*}$$

and

$$c_f^* = \frac{\tau_w}{\frac{1}{2}\rho^* u_e^2}$$

From Section 18.3, we know that for a compressible boundary layer c_f is a function of both M_e and T_w/T_e. Hence, the reference temperature T^* must be a function of M_e and T_w/T_e. This function is

$$\frac{T^*}{T_e} = 1 + 0.032M_e^2 + 0.58\left(\frac{T_w}{T_e} - 1\right) \tag{18.53}$$

For heat transfer for the compressible boundary layer, we write Reynolds analogy from Equation (18.50) repeated below.

$$\frac{C_H}{c_f} = \frac{1}{2}\,\mathrm{Pr}^{-2/3} \tag{18.50}$$

Inserting the incompressible formula for c_f, Equation (18.20), into Equation (18.50), we have

$$C_H = \frac{0.332}{\sqrt{\mathrm{Re}_x}}\,\mathrm{Pr}^{-2/3} \tag{18.54}$$

Evaluating Equation (18.54) at the reference temperature, we have

$$\boxed{C_H^* = \frac{0.332}{\sqrt{\mathrm{Re}_x^*}}(\mathrm{Pr}^*)^{-2/3}} \tag{18.55}$$

where

$$C_H^* = \frac{\dot{q}_w}{\rho^* u_e^*(h_{aw} - h_w)}$$

EXAMPLE 18.2

Use the reference temperature method to calculate the friction drag on the same flat plate at the same flow conditions as described in Example 18.1b. Compare the reference temperature results with that obtained in Example 8.1b, which reflected the "exact" laminar boundary-layer theory.

■ **Solution**

The reference temperature is calculated from Equation (18.53), where we need the ratio T_w/T_e. For the present case, the flat plate is at the adiabatic wall temperature, hence we need the ratio T_{aw}/T_e. To obtain this, we use the recovery factor, which for a flat plate laminar boundary layer is given by Equation (18.47):

$$r = \sqrt{\mathrm{Pr}} \qquad \text{where} \qquad \mathrm{Pr} = \frac{\mu c_p}{k}$$

For air at standard conditions, $\mathrm{Pr} = 0.71$, and it is relatively constant with temperature up to about 800 K. Hence, we assume $\mathrm{Pr} = \mathrm{Pr}^* = 0.71$, and

$$r = \sqrt{\mathrm{Pr}} = \sqrt{0.71} = 0.843$$

The recovery factor is defined by Equation (16.49) as

$$r = \frac{T_{aw} - T_e}{T_0 - T_e} \tag{16.49}$$

or
$$T_{aw} = T_e + r(T_0 - T_e)$$

or
$$\frac{T_{aw}}{T_e} = 1 + r\left(\frac{T_0}{T_e} - 1\right) \tag{18.56}$$

From Appendix A, for $M_e = 2.94$, $\frac{T_0}{T_e} = 2.74$. Hence, Equation (18.56) yields

$$\frac{T_{aw}}{T_e} = 1 + r\left(\frac{T_0}{T_e} - 1\right) = 1 + 0.843(2.74 - 1) = 2.467$$

From Equation (18.53),

$$\frac{T^*}{T_e} = 1 + 0.032M_e^2 + 0.58\left(\frac{T_{aw}}{T_e} - 1\right)$$

$$= 1 + 0.032(2.94)^2 + 0.58(2.467 - 1) = 2.1275$$

Thus, $\qquad T^* = 2.1275 T_e = 2.1275(288) = 612.7$ K

From the equation of state, the value of ρ^* that corresponds to T^* is

$$\rho^* = \frac{p^*}{RT^*} = \frac{1.01 \times 10^5}{(287)(612.7)} = 0.574 \text{ kg/m}^3$$

Also, the value of μ^* that corresponds to T^* is obtained from Sutherland's law, given by Equation (15.3)

$$\frac{\mu}{\mu_0} = \left(\frac{T}{T_0}\right)^{3/2} \frac{T_0 + 110}{T + 110} \tag{15.3}$$

Recall: In Equation (15.3), μ_0 is the reference viscosity coefficient at the reference temperature T_0. In Equation (15.3), T_0 denotes the *reference temperature,* not the total temperature. Here we have a case of the same notation for two different quantities, but the meaning of T_0 in Equation (15.3) is clear from its context. We will use the standard sea level conditions for the values of T_0 and μ_0, that is,

$$\mu_0 = 1.7894 \times 10^{-5} \text{ kg/(m)(s)} \quad \text{and} \quad T_0 = 288 \text{ K}$$

Hence, from Equation (15.3)

$$\frac{\mu^*}{\mu_0} = \left(\frac{T^*}{T_0}\right)^{3/2} \frac{T_0 + 110}{T^* + 110} = \left(\frac{612.7}{288}\right)^{3/2} \frac{288 + 110}{612.7 + 110} = 1.709$$

or $\qquad \mu^* = 1.709\mu_0 = (1.709)(1.7894 \times 10^{-5}) = 3.058 \times 10^{-5}$ kg/(m)(s)

Thus, $\qquad \text{Re}_c^* = \frac{\rho^* u_e c}{\mu^*} = \frac{(0.574)(1000)(2)}{3.058 \times 10^{-5}} = 3.754 \times 10^7$

From Equation (18.52) integrated over the entire chord of the plate, we have the same form as Equation (18.22), namely,

$$C_f^* = \frac{1.328}{\sqrt{\text{Re}_c^*}} \tag{18.57}$$

Thus,
$$C_f^* = \frac{1.328}{\sqrt{\mathrm{Re}_c^*}} = \frac{1.328}{\sqrt{3.754 \times 10^7}} = 2.167 \times 10^{-4}$$

Hence, the friction drag on one side of the plate is
$$D_f = \tfrac{1}{2}\rho^* V_\infty^2 S C_f^* = \tfrac{1}{2}(0.574)(1000)^2(40)(2.167 \times 10^{-4}) = 2488 \text{ N}$$

The total friction drag taking into account both the top and bottom surfaces of the plate is
$$D = 2(2488) = \boxed{4976 \text{ N}}$$

The result obtained from classical compressible boundary-layer theory in Example 18.1*b* is $D = 5026$ N. The result from the reference temperature method used here is within one percent of the "exact" value found in Example 18.1*b*, a stunning example of the accuracy of the reference temperature method, at least for the case treated here.

18.4.1 Recent Advances: The Meador-Smart Reference Temperature Method

The reference temperature method discussed in Section 18.4 is a concept that dates back to the late 1940s, but it is still a work in progress. Very recently, Meador and Smart (William E. Meador and Michael K. Smart, "Reference Enthalpy Method Developed from Solutions of the Boundary-Layer Equations," *AIAA Journal*, vol. 43, no. 1, January 2005, pp. 135–139) published improved formulae for the calculation of the reference temperature, one for laminar flow and another for turbulent flow. This result for a laminar flow is
$$\frac{T^*}{T_e} = 0.45 + 0.55\frac{T_w}{T_e} + 0.16\, r\left(\frac{\gamma-1}{2}\right)M_e^2$$

where r is the recovery factor for laminar flow, $r = \sqrt{\mathrm{Pr}^*}$.

EXAMPLE 18.3

Repeat Example 18.2, using the Meador-Smart equation for the reference temperature.

■ **Solution**

Assuming the Prandtl number is reasonably constant,
$$r = \sqrt{\mathrm{Pr}^*} = \sqrt{\mathrm{Pr}} = \sqrt{0.71} = 0.843$$

Also, because the flat plate is at the adiabatic wall temperature, from Example 8.2,
$$\frac{T_w}{T_e} = \frac{T_{aw}}{T_e} = 2.467$$

So the Meador-Smart equation becomes
$$\frac{T^*}{T_e} = 0.45 + 0.55(2.467) + 0.16(0.843)(0.2)M_e^2$$

or
$$\frac{T^*}{T_e} = 1.807 + 0.027 M_e^2$$

For $M_e = 2.94$, we have

$$\frac{T^*}{T_e} = 1.807 + 0.027(2.94)^2 = 2.04$$

$$T^* = 2.04\,T_e = 2.04\,(288) = 587.5 \text{ K}$$

$$\rho^* = \frac{p^*}{RT^*} = \frac{1.01 \times 10^5}{(287)(587.5)} = 0.599 \text{ kg/m}^3$$

$$\frac{\mu^*}{\mu_0} = \left(\frac{T^*}{T_0}\right)^{3/2} \frac{T_0 + 110}{T^* + 110} = \left(\frac{587.5}{288}\right)^{3/2} \frac{288 + 110}{587.5 + 110} = 1.664$$

$$\mu^* = 1.664\,\mu_0 = 1.664(1.7894 \times 10^{-5}) = 2.978 \times 10^{-5}$$

$$\text{Re}_c^* = \frac{\rho^* u_e c}{\mu^*} = \frac{(0.500)(1000)(2)}{2.978 \times 10^{-5}} = 4.02 \times 10^7$$

$$C_f^* = \frac{1.328}{\sqrt{\text{Re}_c^*}} = \frac{1.328}{\sqrt{4.02 \times 10^7}} = 2.09 \times 10^4$$

$$D_f = \tfrac{1}{2}\rho^* V_\infty^2 S C_f^* = \tfrac{1}{2}(0.599)(1000)^2(40)(2.09 \times 10^{-4}) = 2504 \text{ N}$$

$$D = 2(2504) = \boxed{5008 \text{ N}}$$

The result from Example 18.2 is $D = 4976$ N, and the exact result from Example 18.1 is 5026 N. The Meador-Smith method is more accurate than Equation (18.53); it agrees to within 0.4 percent of the exact amount.

18.5 STAGNATION POINT AERODYNAMIC HEATING

Contrary to what you might think, even though the flow velocity is zero at a stagnation point, the boundary layer at the stagnation point can be defined and has a finite thickness. The flow conditions at the edge of the stagnation point boundary layer are given by the inviscid solution for a stagnation point; in particular, at the boundary-layer edge, the velocity is zero and the temperature is the total temperature, that is, $u_e = 0$ and $T_e = T_0$. This is shown in Figure 18.10. Moreover, along the vertical line in the η-direction shown in Figure 18.10, $u = 0$ at every point inside the boundary layer. However, the ratio $(u/u_e) = (0/0)$ is an indeterminant form that has a finite value at each point in the boundary layer. As in the case of the flat plate solutions discussed in Sections 18.2 and 18.3, we define a function $f(\eta)$ such that $(u/u_e) = f'(\eta)$, and f' has a definite profile through the boundary layer. Indeed, we can define the edge of the boundary layer as the point where $(u/u_e) = f'(\eta) = 0.99$. Finally, we note that the shear stress at the wall at the stagnation point (point A in Figure 18.10) is zero. This not only comes out of the solution of the boundary-layer equations, but it is obvious by inspection. Along the wall above point A the shear stress acts upward, and below point A it acts downward. Hence, right at point A the shear stress must go through zero.

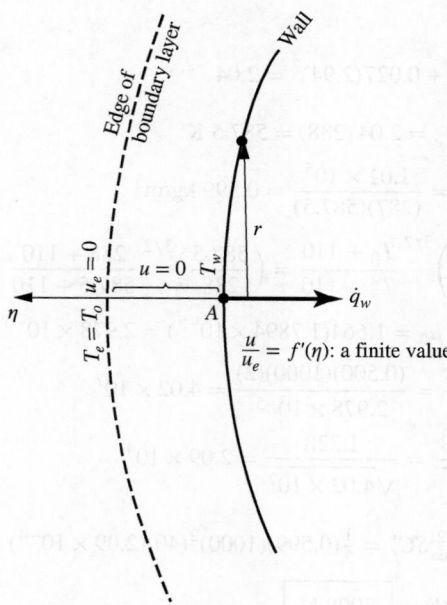

Figure 18.10 Schematic of the stagnation region boundary layer.

If the above discussion sounds rather theoretical, the temperature profile through the stagnation point boundary layer is easier to visualize. The temperature at the outer edge is the total temperature T_0. The temperature at the wall at $\eta = 0$ is T_w. Hence, there is a temperature profile that exists in the normal direction through the stagnation point boundary layer. The heat transfer at the stagnation point is given by the temperature gradient at point A, namely,

$$\dot{q}_w = \left[-k \left(\frac{\partial T}{\partial y} \right)_w \right]_A \tag{18.58}$$

The practical purpose of a stagnation point boundary-layer solution is to calculate the heat transfer, $\dot{q}_w$.

The boundary-layer equations, Equations (17.28)–(17.31), applied at the stagnation point region are transformed using a version of the transformation described in Section 18.3, namely,

$$\xi = \int_0^x \rho_e \mu_e u_r \, dx \tag{18.59}$$

$$\eta = \frac{u_e}{\sqrt{2\xi}} \int_0^y \rho \, dy \tag{18.60}$$

$$f'(\eta) = \frac{u}{u_e} \tag{18.61}$$

$$g(\eta) = \frac{h}{h_e} \tag{18.62}$$

where h is the static enthalpy (since $u = 0$, the static and total enthalpies are the same). This leads to the stagnation point boundary-layer equations given below. For a detailed derivation of these equations, see, for example, Chapter 6 of Reference 52.

$$(Cf'')' + ff'' = (f')^2 - g \tag{18.63}$$

$$\left(\frac{C}{Pr}g'\right)' + fg' = 0 \tag{18.64}$$

where $C = (\rho\mu/\rho_e\mu_e)$. Equations (18.63) and (18.64) are the governing equations for a compressible, stagnation-point boundary layer. Examining these equations, we see no ξ-dependency. Hence, the stagnation point boundary layer is a self-similar case.

Numerical solutions to Equations (18.63) and (18.64) can be obtained by the "shooting technique" as described earlier in the flat-plate case. There is nothing to be gained in going through the details at this stage of our discussion. Instead, we simply state the result of solving Equations (18.63) and (18.64), correlated in the following expression obtained from Reference 78:

$$\text{Cylinder:} \qquad \dot{q}_w = 0.57 \, Pr^{-0.6} (\rho_e\mu_e)^{1/2} \sqrt{\frac{du_e}{dx}} (h_{aw} - h_w) \tag{18.65}$$

If we had considered an axisymmetric body, the original transformation given by Equations (18.59) and (18.60) would have been slightly modified as follows:

$$\xi = \int_0^x \rho_e u_e \mu_e r^2 \, dx \tag{18.66}$$

and

$$\eta = \frac{u_e r}{\sqrt{2\xi}} \int_0^y \rho \, dy \tag{18.67}$$

where r is the vertical coordinate measured from the centerline, as shown in Figure 18.10. Equations (18.66) and (18.67) lead to equations for the axisymmetric stagnation point almost identical to Equations (18.63) and (18.64), namely,

$$(Cf'')' + ff'' = \tfrac{1}{2}[(f')^2 - g] \tag{18.68}$$

and

$$\left(\frac{C}{Pr}g'\right)' + fg' = 0 \tag{18.69}$$

where $C = (\rho\mu/\rho_e\mu_e)$. In turn, the resulting heat transfer expression is (Reference 78):

$$\text{Sphere:} \qquad \dot{q}_w = 0.763 \, Pr^{-0.65} (\rho_e\mu_e) \sqrt{\frac{du_e}{dx}} (h_{aw} - h_w) \tag{18.70}$$

Compare Equation (18.65) for the two-dimensional cylinder with Equation (18.70) for the axisymmetric sphere. The equations are the same except for the leading coefficient, which is higher for the sphere. Everything else being the same, this demonstrates that the stagnation point heating to a sphere is larger than to a two-dimensional cylinder. Why? The answer lies in a basic difference between two- and three-dimensional flows. In a two-dimensional flow, the gas has only two directions to move when it encounters a body—up or down. In contrast, in an axisymmetric flow, the gas has three directions to move—up, down, and sideways—and hence the flow is somewhat "relieved," that is, in comparing two- and three-dimensional flows over bodies with the same longitudinal section (such as a cylinder and a sphere), there is a well-known three-dimensional relieving effect for the three-dimensional flow. As a consequence of this relieving effect, the boundary-layer thickness δ at the stagnation point is smaller for the sphere than for the cylinder. In turn, the temperature gradient at the wall, $(\partial T/\partial y)_w$, which is of the order of (T_e/δ), is larger for the sphere. Since $\dot{q}_w = k(\partial T/\partial y)_w$, then $\dot{q}_w$ is larger for the sphere. This confirms the comparison between Equations (18.65) and (18.70).

The above results for aerodynamic heating to a stagnation point have a stunning impact on hypersonic vehicle design, namely, they impose the requirement for the vehicle to have a blunt, rather than a sharp, nose. To see this, consider the velocity gradient, du_e/dx, which appears in Equations (18.65) and (18.70). From Euler's equation applied at the edge of the boundary layer

$$dp_e = -\rho_e u_e\, du_e \tag{18.71}$$

we have

$$\frac{du_e}{dx} = -\frac{1}{\rho_e u_e}\frac{dp_e}{dx} \tag{18.72}$$

Assuming a Newtonian pressure distribution over the surface, we have from Equation (14.4)

$$C_p = 2\sin^2\theta$$

where θ is defined as the angle between a tangent to the surface and the freestream direction. If we define ϕ as the angle between the normal to the surface and the freestream, then Equation (14.4) can be written as

$$C_p = 2\cos^2\phi \tag{18.73}$$

From the definition of C_p, Equation (18.73) becomes

$$\frac{p_e - p_\infty}{q_\infty} = 2\cos^2\phi$$

or

$$p_e = 2q_\infty\cos^2\phi + p_\infty \tag{18.74}$$

Differentiating Equation (18.74), we obtain

$$\frac{dp_e}{dx} = -4q_\infty\cos\phi\sin\phi\frac{d\phi}{dx} \tag{18.75}$$

Combining Equations (18.72) and (18.75), we have

$$\frac{du_e}{dx} = \frac{4q_\infty}{\rho_e u_e}\cos\phi\sin\phi\frac{d\phi}{dx} \tag{18.76}$$

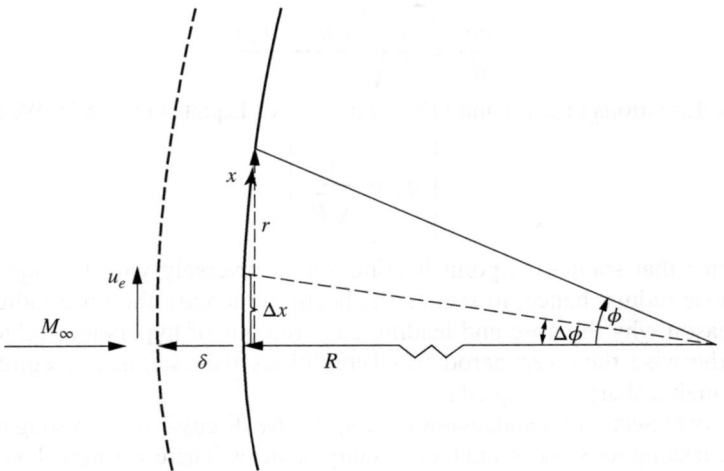

Figure 18.11 Geometry of the stagnation region.

Equation (18.76) is a general result which applies at all points along the body. Now consider the stagnation-point region, as sketched in Figure 18.11. In this region, let Δx be a small increment of surface distance above the stagnation point, corresponding to the small change in ϕ, $\Delta\phi$. The inviscid velocity variation in the stagnation region can be shown to be

$$u_e = \left(\frac{du_e}{dx}\right)_s \Delta x \tag{18.77}$$

Also, in the stagnation region ϕ is small, hence, from Figure 18.11,

$$\cos\phi \approx 1 \tag{18.78}$$

$$\sin\phi \approx \phi \approx \Delta\phi \approx \frac{\Delta x}{R} \tag{18.79}$$

$$\frac{d\phi}{dx} = \frac{1}{R} \tag{18.80}$$

where R is the local radius of curvature of the body at the stagnation point. Finally, at the stagnation point, Equation (18.73) becomes

$$C_p = 2 = \frac{p_e - p_\infty}{q_\infty}$$

or

$$q_\infty = \tfrac{1}{2}(p_e - p_\infty) \tag{18.81}$$

Substituting Equations (18.77)–(18.81) into (18.76), we have

$$\left(\frac{du_e}{dx}\right)^2 = \frac{2(p_e - p_\infty)}{\rho_e \, \Delta x}\left(\frac{\Delta x}{R}\right)\left(\frac{1}{R}\right)$$

or

$$\frac{du_e}{dx} = \frac{1}{R}\sqrt{\frac{2(p_e - p_\infty)}{\rho_e}} \tag{18.82}$$

Examine Equations (18.65) and (18.70) in light of Equation (18.82). We see that

$$\boxed{\dot{q}_w \propto \frac{1}{\sqrt{R}}} \tag{18.83}$$

This states that stagnation-point heating varies inversely with the square root of the nose radius; hence, to reduce the heating, increase the nose radius. This is the reason why the nose and leading edge regions of hypersonic vehicles are blunt; otherwise, the severe aerothermal conditions in the stagnation region would quickly melt a sharp leading edge.

Return to Section 1.1 and review our *qualitative* discussion contrasting the aerodynamic heating for slender and blunt reentry vehicles. There we argued on a qualitative basis that to minimize aerodynamic heating a blunt nose must be used. We have now quantitatively proven this fact with the derivation of Equation (18.83).

The fact that $\dot{q}_w$ is inversely proportional to $\sqrt{R}$ is experimentally verified in Figure 18.12. Here, various sets of experimental data for C_H at the stagnation point are plotted versus Reynolds number based on nose diameter; the abscissa is essentially proportional to R. This is a log-log plot, and the data exhibit a slope of -0.5, hence verifying that $\dot{q}_w \propto 1/\sqrt{R}$.

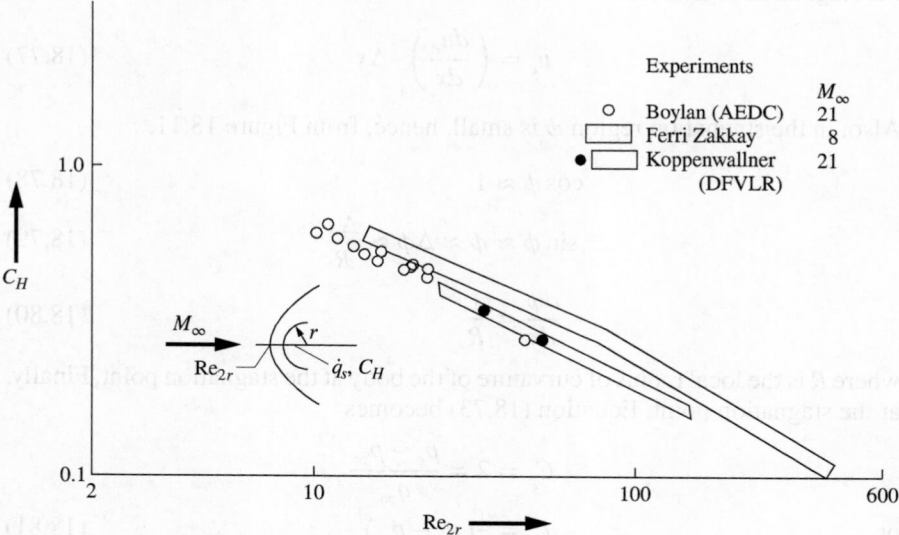

Figure 18.12 Stagnation point Stanton number versus Re based on nose radius. (*Source:* Koppenwallner, G.: "Fundamentals of Hypersonics: Aerodynamics and Heat Transfer," in the *Short Course Notes entitled Hypersonic Aerothermodynamics*, presented at the Von Karman Institute for Fluid Dynamics, Rhode Saint Genese, Belgium, Feb. 1984.)

18.6 BOUNDARY LAYERS OVER ARBITRARY BODIES: FINITE-DIFFERENCE SOLUTION

"Exact" solutions of the boundary-layer equations, Equations (17.28)–(17.31), for the flow over bodies of arbitrary shape did not occur until the advent of the high-speed digital computer and ultimately not until the beginnings of computational fluid dynamics. In this section, we discuss a finite-difference technique for solving the general boundary-layer equations; such finite-difference solutions represent the current state of the art in the analysis of boundary layers.

Let us set the perspective for our discussion. Equations (17.28)–(17.31) are the general boundary-layer equations. For the special case of the flat plate, these equations reduced to Equations (18.42) and (18.43), and for the stagnation region they reduced to Equations (18.63) and (18.64). In both special cases, these equations in terms of the transformed dependent and independent variables led to self-similar solutions (flow variations only in the transformed η direction). For the general case of an arbitrary body, it is still useful to transform the full boundary-layer equations, Equations (17.28)–(17.31), via the transformation given by Equations (18.59)–(18.62). For a detailed derivation of these transformed equations, see Chapter 6 of Reference 52. The resulting form of the equation is:

x momentum:

$$(Cf'')' + ff'' = \frac{2\xi}{u_e}\left[\left((f')^2 - \frac{\rho_e}{\rho}\right)\frac{du_e}{d\xi} + 2\xi\left(f'\frac{\partial f'}{\partial\xi} - \frac{\partial f}{\partial\xi}f''\right)\right] \qquad (18.84)$$

y momentum:

$$\frac{\partial p}{\partial\eta} = 0 \qquad (18.85)$$

Energy:

$$\left(\frac{C}{\text{Pr}}g'\right)' + fg' = 2\xi\left[f'\frac{\partial g}{\partial\xi} + \frac{f'g}{h_e}\frac{\partial h_e}{\partial\xi} - g'\frac{\partial f}{\partial\xi} + \frac{\rho_e u_e}{\rho h_e}f'\frac{du_e}{d\xi}\right] - C\frac{u_e^2}{h_e}(f'')^2$$

$$(18.86)$$

where, as before, $C = \rho\mu/\rho_e\mu_e, f' = u/u_e$, and $g = h/h_e$. In Equations (18.84)–(18.86), the prime denotes the *partial* derivative with respect to η, that is, $f' \equiv \partial f/\partial\eta$. Equations (18.84)–(18.86) are simply the transformed versions of Equations (17.28)–(17.31), with no loss of authority.

Examine Equations (18.84)–(18.86); they are the transformed compressible boundary-layer equations. They are still *partial* differential equations, where both f and g are functions of ξ and η. They contain no further approximations or assumptions beyond those associated with the original boundary-layer equations. However, they are certainly in a less recognizable, somewhat more complicated-looking form than the original equations. However, do not be disturbed by this; in reality, Equations (18.84)–(18.86) are in a form that proves to be practical and useful.

The above transformed boundary-layer equations must be solved subject to the following boundary conditions. The physical boundary conditions were given immediately following Equations (17.28)–(17.31); the corresponding transformed boundary conditions are:

At the wall: $\eta = 0$ $f = f' = 0$ $g = g_w$ (fixed wall temperature)

or $g' = 0$ (adiabatic wall)

At the boundary-layer edge: $\eta \to \infty$ $f' = 1$ $g = 1$

In general, solutions of Equations (18.84), (18.85), and (18.86) along with the appropriate boundary conditions yield variations of velocity and enthalpy throughout the boundary layer, via $u = u_e f'(\xi, \eta)$ and $h = h_e g(\xi, \eta)$. The pressure throughout the boundary layer is known, because the known pressure distribution (or equivalently the known velocity distribution) at the edge of the boundary is given by $p_e = p_e(\xi)$, and this pressure is impressed without change through the boundary layer in the locally normal direction via Equation (18.85), which says that $p = $ constant in the normal direction at any ξ location. Finally, knowing h and p throughout the boundary layer, equilibrium thermodynamics provides the remaining variables through the appropriate equations of state, for example, $T = T(h, p)$, $\rho = \rho(h, p)$, etc.

18.6.1 Finite-Difference Method

Return for a moment to Section 2.17.2 where we introduced some ideas from computation fluid dynamics, and especially review the finite-difference expressions derived there. Recall that we can simulate the partial derivatives with forward, rearward, or central differences. We will use these concepts in the following discussion.

Also consider Figure 18.13, which shows a schematic of a finite-difference grid inside the boundary layer. The grid is shown in the physical *x-y* space, where it is curvilinear and unequally spaced. However, in the ξ-η space, where the calculations are made, the grid takes the form of a rectangular grid with uniform spacing $\Delta\xi$ and $\Delta\eta$. In Figure 18.13, the portion of the grid at four different ξ (or *x*) stations is shown, namely, at $(i-2)$, $(i-1)$, i, and $(i+1)$.

Consider again the general, transformed boundary-layer equations given by Equations (18.84) and (18.86). Assume that we wish to calculate the boundary layer at station $(i+1)$ in Figure 18.13. As discussed in Section 2.17.2, the general philosophy of finite-difference approaches is to evaluate the governing partial differential equations at a given grid point by replacing the derivatives by finite-difference quotients at that point. Consider, for example, the grid point (i, j)

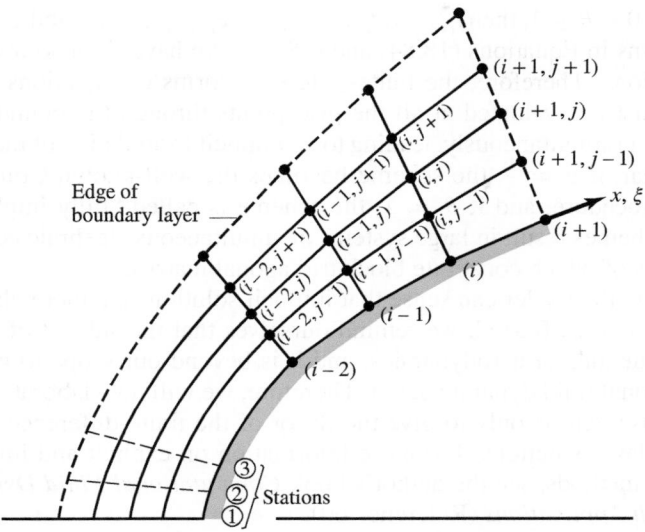

Figure 18.13 Schematic for finite-difference solution of the boundary layer.

in Figure 18.13. At this point, replace the derivatives in Equations (18.84) and (18.86) by finite-difference expressions of the form:

$$\frac{\partial f}{\partial \xi} = \frac{f_{i+1,j} - f_{i,j}}{\Delta \xi} \tag{18.87}$$

$$\frac{\partial f}{\partial \eta} = \frac{\theta(f_{i+1,j+1} - f_{i+1,j-1})}{2\,\Delta \eta} + \frac{(1-\theta)(f_{i,j+1} - f_{i,j-1})}{2\,\Delta \eta} \tag{18.88}$$

$$\frac{\partial^2 f}{\partial \eta^2} = \frac{\theta(f_{i+1,j+1} - 2f_{i+1,j} + f_{i+1,j-1})}{(\Delta \eta)^2} + \frac{(1-\theta)(f_{i,j+1} - 2f_{i,j} + f_{i,j-1})}{(\Delta \eta)^2} \tag{18.89}$$

$$f = \theta f_{i+1,j} + (1-\theta)f_{i,j} \tag{18.90}$$

where θ is a parameter which adjusts Equations (18.87)–(18.90) to various finite-difference approaches (to be discussed below). Similar relations for the derivatives of g are employed. When Equations (18.87)–(18.90) are inserted into Equations (18.84) and (18.86), along with the analogous expressions for g, two algebraic equations are obtained. If $\theta = 0$, the only unknowns that appear are $f_{i+1,j}$ and $g_{i+1,j}$, which can be obtained directly from the two algebraic equations. This is an explicit approach. Using this approach, the boundary-layer properties at grid point $(i + 1,j)$ are solved explicitly in terms of the known properties at points $(i,j+1)$, (i,j), and $(i,j-1)$. The boundary-layer solution is a downstream marching procedure; we are calculating the boundary-layer profiles at station $(i + 1)$ only after the flow at the previous station (i) has been obtained.

When $0 < \theta \le 1$, then $f_{i+1,j+1}, f_{i+1,j}, f_{i+1,j-1}, g_{i+1,j+1}, g_{i+1,j}$, and $g_{i+1,j-1}$ appear as unknowns in Equations (18.84) and (18.86). We have six unknowns and only two equations. Therefore, the finite-difference forms of Equations (18.84) and (18.86) must be evaluated at all the grid points through the boundary layer at station $(i+1)$ simultaneously, leading to an implicit formulation of the unknowns. In particular, if $\theta = \frac{1}{2}$, the scheme becomes the well-known Crank-Nicolson implicit procedure, and if $\theta = 1$, the scheme is called "fully implicit." These implicit schemes result in large systems of simultaneous algebraic equations, the coefficients of which constitute block tridiagonal matrices.

Already the reader can sense that implicit solutions are more elaborate than explicit solutions. Indeed, we remind ourselves that the subject of this book is the fundamentals of aerodynamics, and it is beyond our scope to go into great computational fluid dynamic detail. Therefore, we will not elaborate any further. Our purpose here is only to give the flavor of the finite-difference approach to boundary-layer solutions. For more information on explicit and implicit finite-difference methods, see the author's book *Computational Fluid Dynamics: The Basics with Applications* (Reference 60).

In summary, a finite-difference solution of a general, nonsimilar boundary-layer proceeds as follows:

1. The solution must be started from a given solution at the leading edge, or at a stagnation point (say station 1 in Figure 18.13). This can be obtained from appropriate self-similar solutions.

2. At station 2, the next downstream station, the finite-difference procedure reflected by Equations (18.87)–(18.90) yields a solution of the flow-field variables across the boundary layer.

3. Once the boundary-layer profiles of u and T are obtained, the skin friction and heat transfer at the wall are determined from

$$\tau = \left[\mu \left(\frac{\partial u}{\partial y} \right) \right]_w$$

and

$$\dot{q} = \left(k \frac{\partial T}{\partial y} \right)_w$$

Here, the velocity gradients can be obtained from the known profiles of u and T by using one-sided differences (see Reference 60), such as

$$\left(\frac{\partial u}{\partial y} \right)_w = \frac{-3u_1 + 4u_2 - u_3}{2\,\Delta y} \tag{18.91}$$

$$\left(\frac{\partial T}{\partial y} \right)_w = \frac{-3T_1 + 4T_2 - T_3}{2\,\Delta y} \tag{18.92}$$

In Equations (18.91) and (18.92), the subscripts 1, 2, and 3 denote the wall point and the next two adjacent grid points above the wall. Of course, due to the specified boundary conditions of no velocity slip and a fixed wall temperature, $u_1 = 0$ and $T_1 = T_w$ in Equations (18.91) and (18.92).

4. The above steps are repeated for the next downstream location, say station 3 in Figure 18.13. In this fashion, by repeating applications of these steps, the complete boundary layer is computed, marching downstream from a given initial solution.

An example of results obtained from such finite-difference boundary-layer solutions is given in Figures 18.14 and 18.15 obtained by Blottner (Reference 80). These are calculated for flow over an axisymmetric hyperboloid flying at

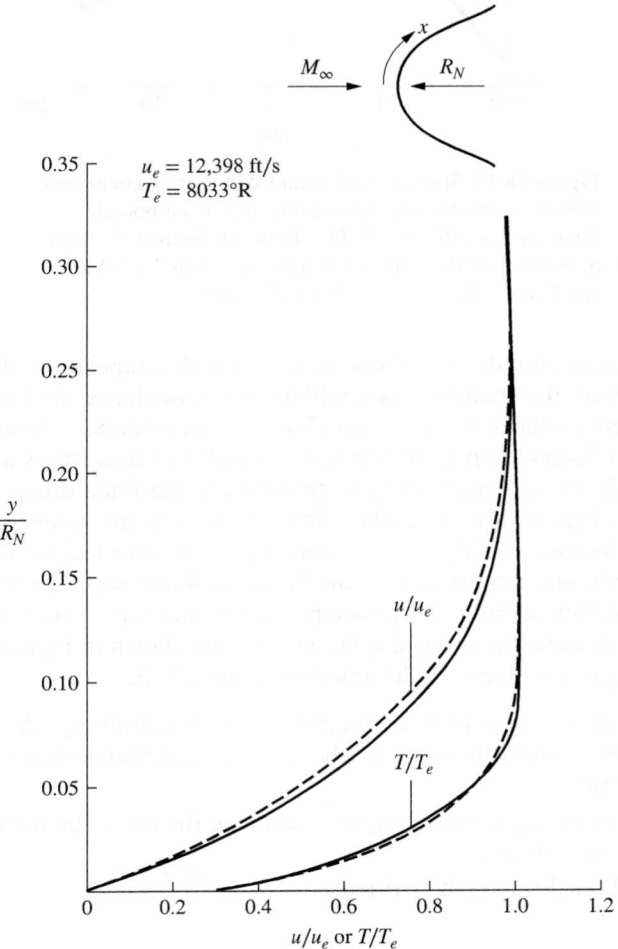

Figure 18.14 Velocity and temperature profiles across the boundary layer at $x/R_N = 50$ on an axisymmetric hyperboloid. (*Data Source:* Blottner, F. G.: "Finite Difference Methods of Solution of the Boundary-Layer Equations," *AIAA J.*, vol. 8, no. 2, February 1970, pp. 193–205.)

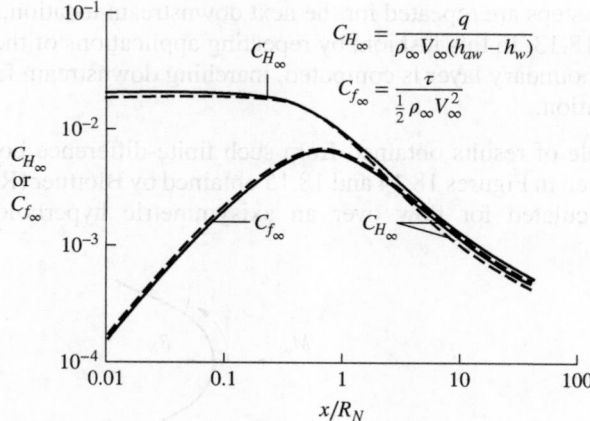

Figure 18.15 Stanton number and skin friction coefficient (based on freestream properties) along a hyperboloid. (*Data Source:* Blottner, F. G.: "Finite Difference Methods of Solution of the Boundary-Layer Equations," *AIAA J.*, vol. 8, no. 2, February 1970, pp. 193–205.)

20,000 ft/s at an altitude of 100,000 ft, with a wall temperature of 1000 K. At these conditions, the boundary layer will involve dissociation, and such chemical reactions were included in the calculations of Reference 80. Chemically reacting boundary layers are not the purview of this book; however, some results of Reference 80 are presented here just to illustrate the finite-difference method. For example, Figure 18.14 gives the calculated velocity and temperature profiles as a station located at $x/R_N = 50$, where R_N is the nose radius. The local values of velocity and temperature at the boundary-layer edge are also quoted in Figure 18.14. Considering the surface properties, the variations of C_H and c_f as functions of distance from the stagnation point are shown in Figure 18.15. Note the following physical trends illustrated in Figure 18.15.

1. The shear stress is zero at the stagnation point (as is always the case), then it increases around the nose, reaches a maximum, and decreases further downstream.

2. The values of C_H are relatively constant near the nose, and then decrease further downstream.

3. Reynolds analogy can be written as

$$C_H = \frac{c_f}{2s} \tag{18.93}$$

where s is called the "Reynolds analogy factor." For the flat-plate case, we see from Equation (18.50) that $s = \text{Pr}^{2/3}$. However, clearly from the results of Figure 18.15 we see that s is a variable in the nose region because C_H is relatively constant while c_f is rapidly increasing. In contrast, for the downstream region, c_f and C_H are essentially equal, and we can state that Reynolds analogy becomes approximately $C_H/c_f = 1$. The point here is that Reynolds analogy is greatly affected by strong pressure gradients in the flow, and hence loses its usefulness as an engineering tool in such cases, at least when C_H and c_f are based on freestream quantities as shown in Figure 18.15.

18.7 SUMMARY

This brings to an end our discussion of laminar boundary layers. Return to the road map in Figure 18.1 and remind yourself of the territory we have covered. Some of the important results are summarized below.

For incompressible laminar flow over a flat plate, the boundary-layer equations reduce to the Blasius equation

$$2f''' + ff'' = 0 \tag{18.15}$$

where $f' = u/u_e$. This produces a self-similar solution where $f' = f'(\eta)$, independent of any particular x station along the surface. A numerical solution of Equation (17.48) yields numbers which lead to the following results.

Local skin friction coefficient: $\quad c_f = \dfrac{\tau_w}{\frac{1}{2}\rho_\infty V_\infty^2} = \dfrac{0.664}{\sqrt{\text{Re}_x}} \tag{18.20}$

Integrated friction drag coefficient: $\quad C_f = \dfrac{1.328}{\sqrt{\text{Re}_c}} \tag{18.22}$

Boundary-layer thickness: $\quad \delta = \dfrac{5.0x}{\sqrt{\text{Re}_x}} \tag{18.23}$

Displacement thickness: $\quad \delta^* = \dfrac{1.72x}{\sqrt{\text{Re}_x}} \tag{18.26}$

Momentum thickness: $\quad \theta = \dfrac{0.664x}{\sqrt{\text{Re}_x}} \tag{18.28}$

Compressibility effects are such as to make boundary-layer solutions a function of Mach number, Prandtl number, and wall-to-freestream temperature ratio. Typical compressibility effects are shown in Figure 18.8. Generally, compressibility reduces C_f and increases δ.

The reference temperature method is an easy engineering calculation of skin friction and heat transfer to a flat plate taking into account compressibility effects, but using the incompressible equations for c_f and C_H. The reference temperature T^* is given by

$$\frac{T^*}{T_e} = 1 + 0.032 M_e^2 + 0.58 \left(\frac{T_w}{T_e} - 1 \right) \tag{18.53}$$

The local skin-friction coefficient is given by

$$c_f^* = \frac{0.644}{\sqrt{\mathrm{Re}_x^*}} \tag{18.52}$$

where

$$c_f^* = \frac{\tau_w}{\frac{1}{2}\rho^* u_e^2}$$

and

$$\mathrm{Re}_x^* = \frac{\rho^* u_e x}{\mu^*}$$

The Stanton number is given by

$$C_H^* = \frac{0.332}{\sqrt{\mathrm{Re}_x^*}} (\mathrm{Pr}^*)^{-2/3} \tag{18.55}$$

where

$$C_H^* = \frac{\dot{q}_w}{\rho^* u_e (h_{aw} - h_w)}$$

18.8 PROBLEMS

Note: The homework problems for this chapter are deferred until the end of Chapter 19 so that both laminar and turbulent boundary layers can be dealt with together.

CHAPTER 19

Turbulent Boundary Layers

The one uncontroversial fact about turbulence is that it is the most complicated kind of fluid motion.

Peter Bradshaw
Imperial College of Science and Technology,
London 1978

Turbulence was, and still is, one of the great unsolved mysteries of science, and it intrigued some of the best scientific minds of the day. Arnold Sommerfeld, the noted German theoretical physicist of the 1920s, once told me, for instance, that before he died he would like to understand two phenomena—quantum mechanics and turbulence. Sommerfeld died in 1924. I believe he was somewhat nearer to an understanding of the quantum, the discovery that led to modern physics, but no closer to the meaning of turbulence.

Theodore von Karman, 1967

PREVIEW BOX

Nature, when left to herself, always goes to the state of maximum disorder. This is particularly true for boundary-layer flows under real conditions. For most practical applications in aerodynamics, the flow in boundary layers is predominantly turbulent—nature going to the state of maximum disorder. Turbulent boundary layers can be bad news; for the same external flow conditions, the turbulent skin friction and aerodynamic heating is larger, frequently *much* larger, than laminar skin friction and aerodynamic heating. But turbulent boundary layers can also be good news

because they remain attached to a surface for much larger distances downstream than a laminar boundary layer under the same external flow conditions. Hence, pressure drag due to flow separation is usually smaller for bodies with turbulent boundary layers compared to those with laminar boundary layers.

Turbulent boundary layers are the subject of this chapter. There are no stand-alone theoretical results for turbulent boundary layers—any analysis must incorporate empirical data in some form. So we will not beat around the bush. We go directly to empirically based formulae that allow the estimation (and it is truly only an estimation) of turbulent boundary-layer thickness and skin friction. There are many different approaches to the calculation of turbulent flows, all requiring some input from experimental data. Whole books have been written on the subject of turbulent flow. The objective of this very short chapter is simply to give you some ability to compute turbulent boundary-layer results, albeit imprecisely. Hang on, and enjoy this short ride.

19.1 INTRODUCTION

The subject of turbulent flow is deep, extensively studied, but at the time of writing still imprecise. The basic nature of turbulence, and therefore our ability to predict its characteristics, is still an unsolved problem in classical physics. Many books have been written on turbulent flows, and many people have spent their professional lives working on the subject. As a result, it is presumptuous for us to try to carry out a thorough discussion of turbulent boundary layers in this chapter. Instead, the purpose of this chapter is simply to provide a contrast with our study of laminar boundary layers in Chapter 18. Here, we will only be able to provide a flavor of turbulent boundary layers, but this is all that is necessary for the present book. Turbulence is a subject that we leave for you to study more extensively as a subject on its own.

Before proceeding further, return to Section 15.2 and review the basic discussion of the nature of turbulence that is given there. In the present chapter, we will pick up where Section 15.2 leaves off.

Also, we note that no *pure theory* of turbulent flow exists. Every analysis of turbulent flow requires some type of *empirical data* in order to obtain a practical answer. As we examine the calculation of turbulent boundary layers in the following sections, the impact of this statement will become blatantly obvious. Finally, because this chapter is short, there is no need for a road map to act as a guide.

19.2 RESULTS FOR TURBULENT BOUNDARY LAYERS ON A FLAT PLATE

In this section, we discuss a few results for the turbulent boundary layer on a flat plate, both incompressible and compressible, simply to provide a basis of comparison with the laminar results described in the previous section. For considerably more detail on the subject of turbulent boundary layers, consult References 40 to 42.

For incompressible flow over a flat plate, the boundary-layer thickness is given approximately by

$$\delta = \frac{0.37x}{\text{Re}_x^{1/5}} \tag{19.1}$$

Note from Equation (19.1) that the turbulent boundary-layer thickness varies approximately as $Re_x^{-1/5}$ in contrast to $Re_x^{-1/2}$ for a laminar boundary layer. Also, turbulent values of δ grow more rapidly with distance along the surface; $\delta \propto x^{4/5}$ for a turbulent flow in contrast to $\delta \propto x^{1/2}$ for a laminar flow. With regard to skin-friction drag, for incompressible turbulent flow over a flat plate, we have

$$C_f = \frac{0.074}{Re_c^{1/5}} \qquad (19.2)$$

Note that for turbulent flow, C_f varies as $Re_c^{-1/5}$ in comparison with the $Re_c^{-1/2}$ variation for laminar flow. Hence, Equation (19.2) yields larger friction drag coefficients for turbulent flow in comparison with Equation (18.22) for laminar flow.

The effects of compressibility on Equation (19.2) are shown in Figure 19.1, where C_f is plotted versus Re_∞ with M_∞ as a parameter. The turbulent flow results are shown toward the right of Figure 19.1, at the higher values of Reynolds numbers where turbulent conditions are expected to occur, and laminar flow results are shown toward the left of the figure, at lower values of Reynolds numbers. This type of figure—friction drag coefficient for both laminar and turbulent flow as a function of Re on a log-log plot—is a classic picture, and it allows a ready contrast of the two types of flow. From this figure, we can see that, for the same Re_∞,

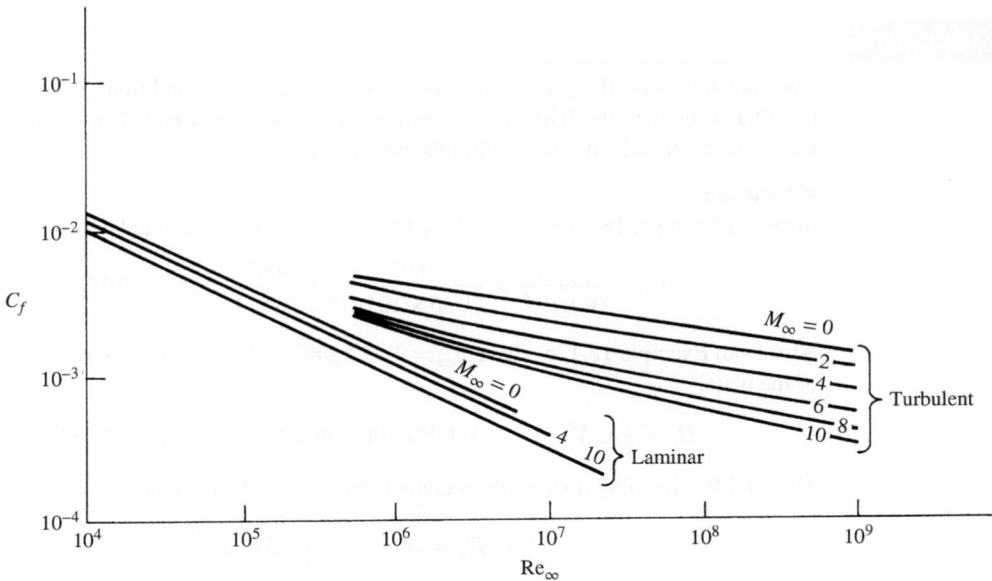

Figure 19.1 Turbulent friction drag coefficient for a flat plate as a function of Reynolds and Mach numbers. Adiabatic wall Pr = 0.75. For contrast, some laminar results are shown. (Data are from the calculations of Van Driest, E. R.: "Turbulent Boundary Layer in Compressible Fluids," *J. Aeronaut. Sci.*, vol. 18, no. 3, March 1951, p. 145.)

turbulent skin friction is higher than laminar; also, the slopes of the turbulent curves are smaller than the slopes of the laminar curves—a graphic comparison of the $\mathrm{Re}^{-1/5}$ variation in contrast to the laminar $\mathrm{Re}^{-1/2}$ variation. Note that the effect of increasing M_∞ is to reduce C_f at constant Re and that this effect is stronger on the turbulent flow results. Indeed, C_f for the turbulent results decreases by almost an order of magnitude (at the higher values of Re_∞) when M_∞ is increased from 0 to 10. For the laminar flow, the decrease in C_f as M_∞ is increased though the same Mach number range is far less pronounced.

19.2.1 Reference Temperature Method for Turbulent Flow

The reference temperature method discussed in Section 18.4 for laminar boundary layers can be applied to turbulent boundary layers as well. With the reference temperature T^* given by Equation (18.53), the incompressible turbulent flat plate result for C_f given by Equation (19.2) can be modified for compressible turbulent flow as

$$C_f^* = \frac{0.074}{(\mathrm{Re}_c^*)^{1/5}} \tag{19.3}$$

where

$$C_f^* = \frac{D_f}{\frac{1}{2}\rho^* u_e^2 S} \tag{19.4}$$

EXAMPLE 19.1

Consider the same flat plate under the same external flow conditions given in Example 18.1. Calculate the friction drag on the plate assuming a *turbulent* boundary layer for a freestream velocity of (*a*) 100 m/s, and (*b*) 1000 m/s.

■ **Solution**

(*a*) From Example 18.1*a*, $\mathrm{Re}_c = 1.36 \times 10^7$. Hence, from Equation (19.2)

$$C_f = \frac{0.074}{(\mathrm{Re}_c)^{1/5}} = \frac{0.074}{(1.36 \times 10^7)^{1/5}} = \frac{0.074}{26.71} = 2.77 \times 10^{-3}$$

Also from Example 18.1, we have $\rho_\infty = 1.22\ \mathrm{kg/m^3}$ and $S = 40\ \mathrm{m^2}$. Hence, for one side of the plate,

$$D_f = \frac{1}{2}\rho_\infty V_\infty^2 S C_f = \frac{1}{2}(1.22)(100)^2(40)(2.77 \times 10^{-3}) = 675.9\ \mathrm{N}$$

The total friction drag taking into account both sides of the plate is

$$D = 2D_f = 2(675.9) = \boxed{1352\ \mathrm{N}}$$

Comparing this result for turbulent flow with the laminar result in Example 1.81*a*, we have

$$\frac{D_{\text{turbulent}}}{D_{\text{laminar}}} = \frac{1352}{175.6} = 7.7$$

Turbulent flow causes a factor of 7.7 *increase* in friction drag compared to the laminar flow. You can easily see why the understanding of, and prediction of, turbulent flow, especially the prediction of when the flow will transist from laminar to turbulent flow, is so important.

(b) For $V_\infty = 1000$ m/s, from Example 18.1b, $Re_c = 1.36 \times 10^8$, and $M_\infty = 2.94$. From Figure 19.1, we have

$$C_f = 1.34 \times 10^{-3}$$

Hence, $D_f = \frac{1}{2}\rho_\infty V_\infty^2 S C_f = \frac{1}{2}(1.22)(1000)^2(40)(1.34 \times 10^{-3}) = 32{,}700$ N.

The total friction drag is

$$D = 2(32{,}700) = \boxed{65{,}400 \text{ N}}$$

Again, comparing this result with that from Example 18.1b, we have

$$\frac{D_{\text{turbulent}}}{D_{\text{laminar}}} = \frac{65{,}400}{5026} = 13$$

Note that, at the higher Mach number of 2.92, turbulence increased the drag by a factor of 13, whereas for the incompressible case, the increase was 7.7, a smaller amount. The difference between the drag for laminar and turbulent flow is more pronounced at higher speeds.

EXAMPLE 19.2

Repeat Example 19.1b, except using the reference temperature method. Assume the plate has an adiabatic wall.

■ **Solution**

We draw on the results calculated in Example 18.2. The recovery factor for a turbulent flow is slightly different than that for a laminar flow. However, we will not account for that difference, and we will assume that the reference temperature for this case is the same as given in Example 18.2. Hence, from Example 18.2, we have

$$Re_c^* = 3.754 \times 10^7 \quad \text{and} \quad \rho^* = 0.574 \text{ kg/m}^3$$

From Equation (19.3), we have

$$C_f^* = \frac{0.074}{(Re_c^*)^{1/5}} = \frac{0.074}{(3.754 \times 10^7)^{1/5}} = 2.26 \times 10^{-3}$$

From Equation (19.4),

$$D_f = \frac{1}{2}\rho^* u_e^2 S C_f^* = \frac{1}{2}(0.574)(1000)^2(40)(2.26 \times 10^{-3}) = 25{,}945 \text{ N}$$

Hence, $\boxed{D = 2(25{,}945) = 51{,}890 \text{ N}}$

Comparing this answer with that obtained in Example 19.1b, we find a 20 percent discrepancy between the two methods of calculations. This is not surprising. It simply points out the great uncertainty in making calculations of turbulent skin friction.

19.2.2 The Meador-Smart Reference Temperature Method for Turbulent Flow

The method developed recently by Meador and Smart, discussed in Section 18.4.1, gives a reference temperature equation for turbulent flow slightly different than that for laminar flow. For a turbulent flow, their equation is

$$\frac{T^*}{T_e} = 0.5 \left(1 + \frac{T_w}{T_e} \right) + 0.16\, r \left(\frac{\gamma - 1}{2} \right) M_e^2$$

They also give a *local* turbulent skin-friction coefficient for incompressible flow as

$$c_f = \frac{\tau_w}{\frac{1}{2}\rho_e u_e^2} = \frac{0.02296}{(\mathrm{Re}_x)^{0.139}}$$

When integrated over the entire plate of length c, this gives for the net skin-friction drag coefficient (prove it to yourself)

$$C_f = \frac{D_f}{\frac{1}{2}\rho_\infty V_\infty^2 S} = \frac{0.02667}{(\mathrm{Re}_c)^{0.139}}$$

EXAMPLE 19.3

Repeat Example 9.2 using the Meador-Smart reference temperature method.

■ **Solution**
From the above equation,

$$\frac{T^*}{T_e} = 0.5 \left(1 + \frac{T_w}{T_e} \right) + 0.16r \left(\frac{\gamma - 1}{2} \right) M_e^2$$

For turbulent flow, the recovery factor is approximately

$$r = \mathrm{Pr}^{1/3} = (0.71)^{1/3} = 0.892$$

$$T_{aw} - T_e = r(T_0 - T_e)$$

or

$$\frac{T_{aw}}{T_e} = 1 + r \left(\frac{T_0}{T_e} - 1 \right)$$

For $M_e = 2.94$,

$$\frac{T_0}{T_e} = 2.74$$

$$\frac{T_{aw}}{T_r} = 1 + 0.892(1.74) = 2.55$$

Since the flat plate has an adiabatic wall, $T_w = T_{aw}$. The Meador-Smith equation then becomes

$$\frac{T^*}{T_e} = 0.5 \left(1 + \frac{T_w}{T_e} \right) + 0.16(0.892)(0.2)(2.94)^2 = 0.5(1 + 2.55) + 0.2467 = 2.02$$

$$T^* = 2.02\,T_e = 2.02\,(288) = 581.8\text{ K}$$

$$\rho^* = \frac{p}{RT^*} = \frac{1.01 \times 10^5}{(287)(581.8)} = 0.605 \text{ kg/m}^3$$

From Sutherland's law (*note that T_0 in Sutherland's law is a reference temperature, not the total temperature*)

$$\frac{\mu^*}{\mu_0} = \left(\frac{T^*}{T_0}\right)^{3/2} \frac{T_0 + 110}{T^* + 110} = \left(\frac{581.8}{288}\right)^{3/2} \frac{398}{691.8} = 1.651$$

$$\mu^* = 1.651\,\mu_0 = 1.651(1.7894 \times 10^{-5}) = 2.05 \times 10^{-5} \text{ kg/m} \cdot \text{s}$$

$$\text{Re}_c^* = \frac{\rho^* u_e c}{\mu^*} = \frac{(0.605)(1000)(2)}{2.95 \times 10^{-5}} = 4.1 \times 10^7$$

From the Meador-Smith choice of the turbulent skin-friction coefficient equation,

$$C_f^* = \frac{0.02667}{(\text{Re}_c^*)^{0.139}} = \frac{0.02667}{(4.1 \times 10^7)^{0.139}} = 2.32 \times 10^{-3}$$

$$D_f = \tfrac{1}{2}\rho^* V_\infty^2 S C_f^* = \tfrac{1}{2}(0.605)(1000)^2(40)(2.32 \times 10^{-3}) = 28{,}070 \text{ N}$$

$$\text{Total drag} = D = 2D_f = 2(28{,}070) = \boxed{56{,}140 \text{ N}}$$

Note: This result is more accurate than that obtained in Example 9.2; it shows a 14 percent discrepancy compared with the result obtained in Example 19.1*b*.

19.2.3 Prediction of Airfoil Drag

The flat-plate results obtained in Chapter 18 for laminar flow, and in the present chapter for turbulent flow, can be used for engineering prediction of skin-friction drag on thin airfoils. Using results from Chapters 18 and 19, airfoil drag in low-speed incompressible flow is treated in Section 4.12, and supersonic airfoil drag is discussed in Section 12.4. If you have not read Sections 4.12 and 12.4, do so now. They give an important practical application of the boundary-layer results we have just covered. Indeed, Sections 4.12 and 12.4 provide a vital continuation of our discussion of viscous flow, and for all practical purposes they can be considered integral sections of Part 4 of this book, although they were inserted in the earlier chapters to provide some viscous flow reality to our otherwise inviscid flow presentations. The prediction of airfoil drag is one of the most important aspects of aerodynamics. Take it seriously, and make certain that you read, or have read, Sections 4.12 and 12.4.

19.3 TURBULENCE MODELING

The simple equations given in Section 19.2 for boundary-layer thickness and skin-friction coefficient for a turbulent flow over a flat plate are simplified results that are heavily empirically based. Modern calculations of turbulent flows over arbitrarily shaped bodies involve the solution of the continuity, momentum, and energy equations along with some *model* of the turbulence. The calculations are carried out by means of computational fluid dynamic techniques. Here we will discuss only one model of turbulence, the Baldwin-Lomax turbulence model, which has become popular over the past two decades. *We emphasize that the following discussion is intended only to give you the flavor of what is meant by a turbulence model.*

19.3.1 The Baldwin-Lomax Model

In order to include the effects of turbulence in any analysis or computation, it is first necessary to have a model for the turbulence itself. Turbulence modeling is a state-of-the-art subject, and a recent survey of such modeling as applied to computations is given in Reference 81. Again, it is beyond the scope of the present book to give a detailed presentation of various turbulence models; the reader is referred to the literature for such matters. Instead, we choose to discuss only one such model here, because: (a) it is a typical example of an engineering-oriented turbulence model; (b) it is the model used in the majority of modern applications in turbulent, subsonic, supersonic, and hypersonic flows; and (c) we will discuss in the next chapter several applications which use this model. The model is called the Baldwin-Lomax turbulence model, first proposed in Reference 82. It is in the class of what is called an "eddy viscosity" model, where the effects of turbulence in the governing viscous flow equations (such as the boundary-layer equations or the Navier-Stokes equations) are included simply by adding an additional term to the transport coefficients. For example, in all our previous viscous flow equations, μ is replaced by $(\mu + \mu_T)$ and k by $(k + k_T)$, where μ_T and k_T are the eddy viscosity and eddy thermal conductivity, respectively—both due to turbulence. In these expressions, μ and k are denoted as the "molecular" viscosity and thermal conductivity, respectively. For example, the x momentum boundary-layer equation for turbulent flow is written as

$$\rho u \frac{\partial u}{\partial x} + \rho v \frac{\partial u}{\partial y} = -\frac{\partial p}{\partial x} + \frac{\partial}{\partial y}\left[(\mu + \mu_T)\frac{\partial u}{\partial y}\right] \tag{19.5}$$

Moreover, the Baldwin-Lomax model is also in the class of "algebraic" or "zero-equation" models meaning that the formulation of the turbulence model utilizes just algebraic relations involving the flow properties. This is in contrast to one- and two-equation models which involve partial differential equations for the convection, creation, and dissipation of the turbulent kinetic energy and (frequently) the local vorticity. (See Reference 83 for a concise description of such one- and two-equation turbulence models.)

The Baldwin-Lomax turbulence model is described below. We give just a "cookbook" prescription for the model; the motivation and justification for the model are described at length in Reference 82. This, like all other turbulence models, is highly empirical. The final justification for its use is that it yields reasonable results across a wide range of Mach numbers, from subsonic to hypersonic. The model assumes that the turbulent-boundary layer is split into two layers, an inner and an outer layer, with different expressions for μ_T in each layer:

$$\mu_T = \begin{cases} (\mu_T)_{\text{inner}} & y \leq y_{\text{crossover}} \\ (\mu_T)_{\text{outer}} & y \geq y_{\text{crossover}} \end{cases} \tag{19.6}$$

where y is the local normal distance from the wall, and the crossover point from the inner to the outer layer is denoted by $y_{\text{crossover}}$. By definition, $y_{\text{crossover}}$ is that point in the turbulent boundary where $(\mu_T)_{\text{outer}}$ becomes less than $(\mu_T)_{\text{inner}}$. For the inner region:

$$(\mu_T)_{\text{inner}} = \rho l^2 |\omega| \tag{19.7}$$

where

$$l = ky \left[1 - \exp\left(\frac{-y^+}{A^+} \right) \right] \tag{19.8}$$

$$y^+ = \frac{\sqrt{\rho_w \tau_w} y}{\mu_w} \tag{19.9}$$

and k and A^+ are two dimensionless constants, specified later. In Equation (19.7), ω is the local vorticity, defined for a two dimensional flow as

$$\omega = \frac{\partial u}{\partial y} - \frac{\partial v}{\partial x} \tag{19.10}$$

For the outer region:

$$(\mu_T)_{\text{outer}} = \rho K C_{cp} F_{\text{wake}} F_{\text{Kleb}} \tag{19.11}$$

where K and C_{cp} are two additional constants, and F_{wake} and F_{Kleb} are related to the function

$$f(y) = y|\omega| \left[1 - \exp\left(\frac{-y^+}{A^+} \right) \right] \tag{19.12}$$

Equation (19.12) will have a maximum value along a given normal distance y; this maximum value and the location where it occurs are denoted by F_{max} and y_{max}, respectively. In Equation (19.11), F_{wake} is taken to be either $y_{\text{max}} F_{\text{max}}$ or $C_{\text{wk}} y_{\text{max}} U_{\text{dif}}^2 / F_{\text{max}}$, whichever is smaller, where C_{wk} is constant, and

$$U_{\text{dif}} = \sqrt{u^2 + v^2} \tag{19.13}$$

Also, in Equation (19.11), F_{Kleb} is the Klebanoff intermittency factor, given by

$$F_{\text{Kleb}}(y) = \left[1 + 5.5 \left(C_{\text{Kleb}} \frac{y}{y_{\text{max}}} \right)^6 \right]^{-1} \tag{19.14}$$

The six dimensionless constants that appear in the above equations are: $A^+ = 26.0$, $C_{cp} = 1.6$, $C_{\text{Kleb}} = 0.3$, $C_{\text{wk}} = 0.25$, $k = 0.4$, and $K = 0.0168$. These constants are taken directly from Reference 82 with the understanding that, while they are not precisely the correct constants for most flows in general, they have been used successfully for a number of different applications. Note that, unlike many algebraic eddy viscosity models that are based on a characteristic length, the Baldwin-Lomax model is based on the local vorticity ω. This is a distinct advantage for the analysis of flows without an obvious mixing length, such as separated flows. Note that, like all eddy-viscosity turbulent models, the value of μ_T obtained above is dependent on the flow-field properties themselves (for example, ω and ρ); this is in contrast to the molecular viscosity μ, which is solely a property of the gas itself.

The molecular values of viscosity coefficient and thermal conductivity are related through the Prandtl number

$$k = \frac{\mu c_p}{\text{Pr}} \qquad (19.15)$$

In lieu of developing a detailed turbulence model for the turbulent thermal conductivity k_T, the usual procedure is to define a "turbulent" Prandtl number as $\text{Pr}_T = \mu_T c_p / k_T$. Thus, analogous to Equation (19.15), we have

$$k_T = \frac{\mu_T c_p}{\text{Pr}_T} \qquad (19.16)$$

where the usual assumption is that $\text{Pr}_T = 1$. Therefore, μ_T is obtained from a given eddy-viscosity model (such as the Baldwin-Lomax model), and the corresponding k_T is obtained from Equation (19.16).

Turbulence itself is a flow field; it is *not* a simple property of the gas. This is why, as mentioned above, in an algebraic eddy viscosity model the values of μ_T and k_T *depend* on the solution of the flow field—they are not pure properties of the gas as are μ and k. This is clearly seen in the Baldwin-Lomax model via Equation (19.7), where μ_T is a function of the local vorticity in the flow, ω—a flow-field variable which comes out as part of the solution for the particular case at hand.

19.4 FINAL COMMENTS

This chapter and the previous two have dealt with boundary layers, especially those on a flat plate. We end with the presentation of an artist's rendering a photograph in Figure 19.2 showing the development of velocity profiles in the boundary layer over a flat plate. The fluid is water, which flows from left to right. The profiles in the original photograph are made visible by the hydrogen bubble technique, the same used for Figure 16.13. The Reynolds number is low (the freestream velocity is only 0.6 m/s); hence, the boundary-layer thickness is large. However, the thickness of the plate is only 0.5 mm, which means that the boundary layer shown here is on the order of 1 mm thick—still small on an absolute scale. In any event, if you need any further proof of the existence of boundary layers, Figure 19.2 is it.

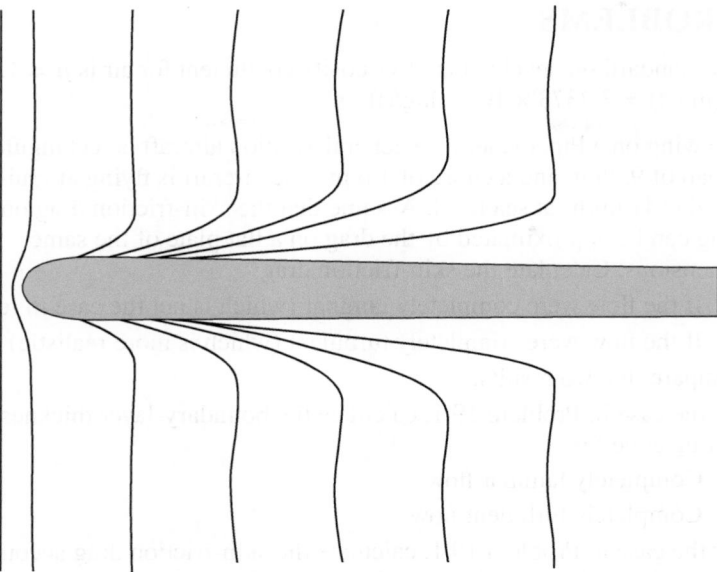

Figure 19.2 Rendering of velocity profiles for the laminar flow over a flat plate. Flow is from left to right.

19.5 SUMMARY

Approximations for the turbulent, incompressible flow over a flat plate are

$$\delta = \frac{0.37x}{\text{Re}_x^{1/5}} \tag{19.1}$$

$$C_f = \frac{0.074}{\text{Re}_c^{1/5}} \tag{19.2}$$

To account for compressibility effects, the data shown in Figure 19.1 can be used, or alternatively the reference temperature method can be employed.

When the continuity, momentum, and energy equations are used to solve a turbulent flow, some type of turbulence model must be used. In the eddy viscosity concept, the viscosity coefficient and thermal conductivity in these equations must be the *sum* of the molecular and turbulent values.

19.6 PROBLEMS

Note: The standard sea level value of viscosity coefficient for air is $\mu = 1.7894 \times 10^{-5}$ kg/(m · s) $= 3.7373 \times 10^{-7}$ slug/(ft · s).

19.1 The wing on a Piper Cherokee general aviation aircraft is rectangular, with a span of 9.75 m and a chord of 1.6 m. The aircraft is flying at cruising speed (141 mi/h) at sea level. Assume that the skin-friction drag on the wing can be approximated by the drag on a flat plate of the same dimensions. Calculate the skin-friction drag:

a. If the flow were completely laminar (which is not the case in real life)

b. If the flow were completely turbulent (which is more realistic)

Compare the two results.

19.2 For the case in Problem 19.1, calculate the boundary-layer thickness at the trailing edge for

a. Completely laminar flow

b. Completely turbulent flow

19.3 For the case in Problem 19.1, calculate the skin-friction drag accounting for transition. Assume the transition Reynolds number $= 5 \times 10^5$.

19.4 Consider Mach 4 flow at standard sea level conditions over a flat plate of chord 5 in. Assuming all laminar flow and adiabatic wall conditions, calculate the skin-friction drag on the plate per unit span.

19.5 Repeat Problem 19.4 for the case of all turbulent flow.

19.6 Consider a compressible, laminar boundary layer over a flat plate. Assuming Pr = 1 and a calorically perfect gas, show that the profile of total temperature through the boundary layer is a function of the velocity profile via

$$T_0 = T_w + (T_{0,e} - T_w)\frac{u}{u_e}$$

where $T_w =$ wall temperature and $T_{0,e}$ and u_e are the total temperature and velocity, respectively, at the outer edge of the boundary layer. [*Hint:* Compare Equations (18.32) and (18.41).]

19.7 Consider a high-speed vehicle flying at a standard altitude of 35 km, where the ambient pressure and temperature are 583.59 N/m² and 246.1 K, respectively. The radius of the spherical nose of the vehicle is 2.54 cm. Assume the Prandtl number for air at these conditions is 0.72, that c_p is 1008 joules/(kg K), and that the viscosity coefficient is given by Sutherland's law. The wall temperature at the nose is 400 K. Assume the recovery factor at the nose is 1.0. Calculate the aerodynamic heat transfer to the stagnation point for flight velocities of (*a*) 1500 m/s, and (*b*) 4500 m/s. From these results, make a comment about how the heat transfer varies with flight velocity.

CHAPTER 20

Navier-Stokes Solutions: Some Examples

A numerical simulation of the flow over an airfoil using the Reynolds averaged Navier-Stokes equations can be conducted on today's supercomputers in less than a half hour for less than $1000 cost in computer time. If just one such simulation had been attempted 20 years ago on computers of that time (e.g., the IBM 704 class) and with algorithms then known, the cost in computer time would have amounted to roughly $10 million, and the results for that single flow would not be available until 10 years from now, since the computation would have taken about 30 years to complete.

Dean R. Chapman, NASA, 1977

PREVIEW BOX

This is a short chapter about a very extensive subject—the numerical solution of general viscous flows. This is the ultimate method for solving general flow fields including friction and thermal conduction throughout the whole flow field. This chapter is in the purview of computational fluid dynamics—a whole subject in itself. The purpose of this chapter is simply to round out our discussion of viscous flow and to bring some closure to the subject. So read on, and allow yourself to be rounded out.

In so doing, pay close attention to the section dealing with the accuracy of the prediction of skin-friction drag. You will see that no matter how hard we try, there is still room for improvement. This is a good thing. Aerodynamics in general is still an evolving intellectual subject, and there is plenty of room for you to make your own personal contributions to its improvement. I hope that by reading this book you are inspired to do so.

As we say in aerodynamics—onward and upward.

20.1 INTRODUCTION

This chapter is short. Its purpose is to discuss the third option for the solution of viscous flows as discussed in Section 15.7, namely, the exact numerical solution of the complete Navier-Stokes equations. This option is the purview of modern computational fluid dynamics—it is a state-of-the-art research activity which is currently in a rapid state of development. This subject now occupies volumes of modern literature; for a basic treatment, see the definitive text on computational fluid dynamics listed as Reference 51. We will only list a few sample calculations here.

20.2 THE APPROACH

Return to the complete Navier-Stokes equations, as derived in Chapter 15, and repeated and renumbered below for convenience:

Continuity:

$$\frac{\partial \rho}{\partial t} = -\left[\frac{\partial(\rho u)}{\partial x} + \frac{\partial(\rho v)}{\partial y} + \frac{\partial(\rho w)}{\partial z} \right] \tag{20.1}$$

x momentum:

$$\frac{\partial u}{\partial t} = -u\frac{\partial u}{\partial x} - v\frac{\partial u}{\partial y} - w\frac{\partial u}{\partial z} + \frac{1}{\rho}\left[-\frac{\partial p}{\partial x} + \frac{\partial \tau_{xx}}{\partial x} + \frac{\partial \tau_{yx}}{\partial y} + \frac{\partial \tau_{zx}}{\partial z} \right] \tag{20.2}$$

y momentum:

$$\frac{\partial v}{\partial t} = -u\frac{\partial v}{\partial x} - v\frac{\partial v}{\partial y} - w\frac{\partial v}{\partial z} + \frac{1}{\rho}\left[-\frac{\partial p}{\partial y} + \frac{\partial \tau_{xy}}{\partial x} + \frac{\partial \tau_{yy}}{\partial y} + \frac{\partial \tau_{zy}}{\partial z} \right] \tag{20.3}$$

z momentum:

$$\frac{\partial w}{\partial t} = -u\frac{\partial w}{\partial x} - v\frac{\partial w}{\partial y} - w\frac{\partial w}{\partial z} + \frac{1}{\rho}\left[-\frac{\partial p}{\partial z} + \frac{\partial \tau_{xz}}{\partial x} + \frac{\partial \tau_{yz}}{\partial y} + \frac{\partial \tau_{zz}}{\partial z} \right] \tag{20.4}$$

Energy:

$$\frac{\partial(e + V^2/2)}{\partial t} = -u\frac{\partial(e + V^2/2)}{\partial x} - v\frac{\partial(e + V^2/2)}{\partial y} - w\frac{\partial(e + V^2/2)}{\partial z} + \dot{q}$$

$$+ \frac{1}{\rho}\left[\frac{\partial}{\partial x}\left(k\frac{\partial T}{\partial x} \right) + \frac{\partial}{\partial y}\left(k\frac{\partial T}{\partial y} \right) + \frac{\partial}{\partial z}\left(k\frac{\partial T}{\partial z} \right) \right.$$

$$- \frac{\partial(pu)}{\partial x} - \frac{\partial(pv)}{\partial y} - \frac{\partial(pw)}{\partial z} + \frac{\partial(u\tau_{xx})}{\partial x} \tag{20.5}$$

$$+ \frac{\partial(u\tau_{yx})}{\partial y} + \frac{\partial(u\tau_{zx})}{\partial z} + \frac{\partial(v\tau_{xy})}{\partial x} + \frac{\partial(v\tau_{yy})}{\partial y} + \frac{\partial(v\tau_{zy})}{\partial z}$$

$$\left. + \frac{\partial(w\tau_{xz})}{\partial x} + \frac{\partial(w\tau_{yz})}{\partial y} + \frac{\partial(w\tau_{zz})}{\partial z} \right]$$

These equations have been written with the time derivatives on the left-hand side and all spatial derivatives on the right-hand side. This is the form suitable to a time-dependent solution of the equations, as discussed in Chapters 13 and 16.

Indeed, Equations (20.1) to (20.5) are partial differential equations that have a mathematically "elliptic" behavior; that is, on a physical basis they treat flow-field information and flow disturbances that can travel throughout the flow field, in both the upstream and downstream directions. The time-dependent technique is particularly suited to such a problem.

The time-dependent solution of Equations (20.1) to (20.5) can be carried out in direct parallel to the discussion in Section 16.4. It is important for you to return to that section and review our discussion of the time-dependent solution of compressible Couette flow using MacCormack's technique. We suggest doing this before reading further. The approach to the solution of Equations (20.1) to (20.5) for other problems is exactly the same. Therefore, we will not elaborate further here.

20.3 EXAMPLES OF SOME SOLUTIONS

In this section, we present samples of a few numerical solutions of the complete Navier-Stokes equations. Most of these solutions have the following in common:

1. They were obtained by means of a time-dependent solution using MacCormack's technique as described in Section 16.4.
2. They utilize the Baldwin-Lomax turbulence model (see Section 19.3.1 for a discussion of this model). Hence, turbulent flow is modeled in these calculations.
3. They require anywhere from thousands to close to a million grid points for their solution. Therefore, these are problems that must be solved on large-scale digital computers.

20.3.1 Flow over a Rearward-Facing Step

The supersonic viscous flow over a rearward-facing step was examined in Reference 44. Some results are shown in Figures 20.1 and 20.2. The flow is moving from left to right. In the velocity vector diagram in Figure 20.1, note the separated, recirculating flow region just downstream of the step. The calculation of such separated flows is the forte of solutions of the complete Navier-Stokes equations. In contrast, the boundary-layer equations discussed in Chapter 17 are not suited for the analysis of separated flows; boundary-layer calculations usually "blow up" in regions of separated flow. Figure 20.2 shows the temperature contours (lines of constant temperature) for the same flow in Figure 20.1.

20.3.2 Flow over an Airfoil

The viscous compressible flow over an airfoil was studied in Reference 53. For the treatment of this problem, a nonrectangular finite-difference grid is wrapped around the airfoil, as shown in Figure 20.3. Equations (20.1) to (20.5) have to be transformed into the new curvilinear coordinate system in Figure 20.3. The details are beyond the scope of this book; see Reference 53 for a complete discussion. Some results for the streamline patterns are shown in Figure 20.4*a* and *b*. Here, the flow over a Wortmann airfoil at zero angle of attack is shown. The freestream

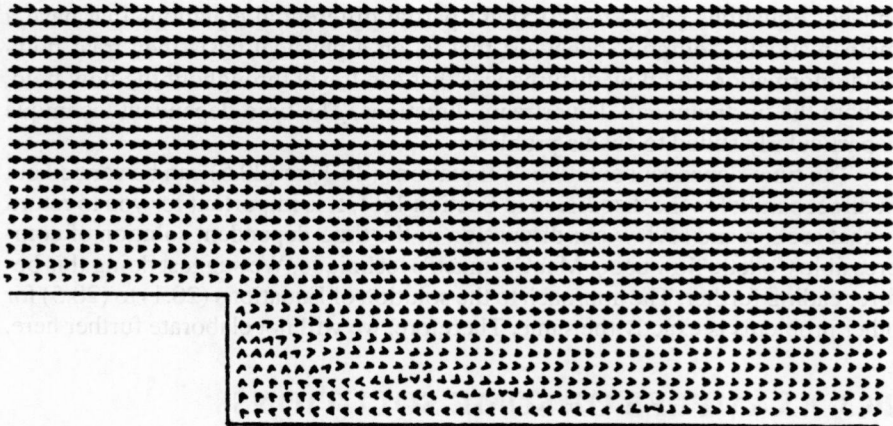

Figure 20.1 Velocity vector diagram for the flow over a rearward-facing step. $M = 2.19$, $T = 1005$ K, Re = 70,000 (based on step height). (Berman, H. A., J. D. Anderson, Jr., and J. P. Drummond: "Supersonic Flow over a Rearward Facing Step with Transverse Nonreacting Hydrogen Injection," *AIAA J.*, vol. 21, no. 12, December 1983, pp. 1707–1713.) Note the recirculating flow region downstream of the step.

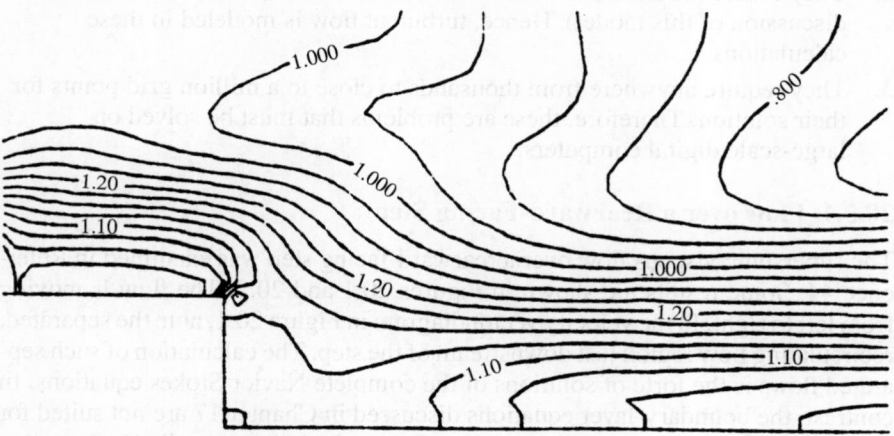

Figure 20.2 Temperature contours for the flow shown in Figure 20.1. The separated region just downstream of the step is a reasonably constant pressure, constant temperature region.

Mach number is 0.5, and the Reynolds number based on chord is relatively low, Re = 100,000. The completely laminar flow over this airfoil is shown in Figure 20.4*a*. Because of the peculiar aerodynamic properties of some low Reynolds number flows over airfoils (see References 48 and 53), we note that the laminar flow separated over both the top and bottom surfaces of the airfoil. However, in Figure 20.4*b*, the turbulence model is turned on for the calculation; note that the flow is now completely attached. The differences in Figure 20.4*a* and *b* vividly demonstrate the basic trend that turbulent flow resists flow separation much more strongly than laminar flow.

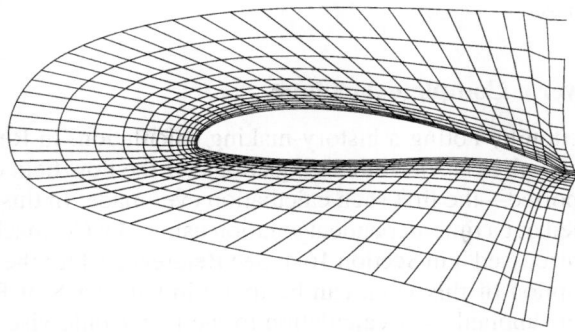

Figure 20.3 Curvilinear, boundary-fitted finite-difference grid for the solution of the flow over an airfoil. (*Data Source:* Kothari, A. P., and J. D. Anderson: "Flow over Low Reynolds Number Airfoils-Compressible Navier-Stokes Solutions," AIAA paper no. 85-0107, January 1985.)

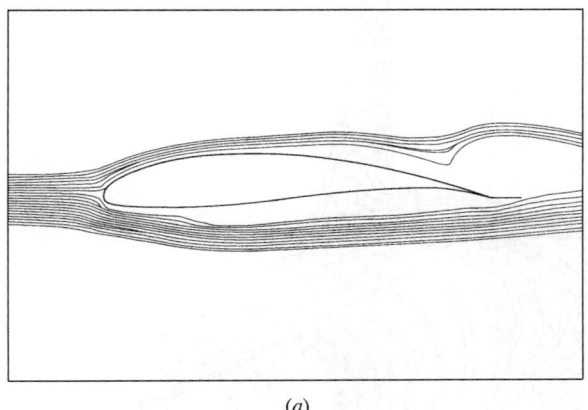

(*a*)

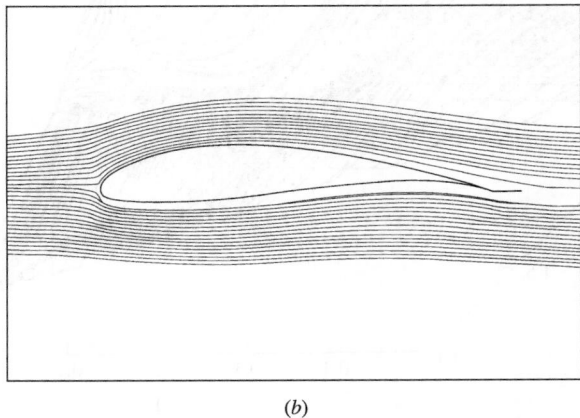

(*b*)

Figure 20.4 Streamlines for the low Reynolds flow over a Wortmann airfoil. Re = 100,000. (*a*) Laminar flow. (*b*) Turbulent flow. (*Data Source:* Kothari, A. P., and J. D. Anderson: "Flow over Low Reynolds Number Airfoils-Compressible Navier-Stokes Solutions," AIAA paper no. 85-0107, January 1985.)

20.3.3 Flow over a Complete Airplane

We end this section by noting a history-making calculation. In Reference 54, a solution of the complete Navier-Stokes equations for the flow field over an entire airplane was reported—the first such calculation ever made. In this work, Shang and Scherr carried out a time-dependent solution using MacCormack's method—just as we have discussed it in Section 16.4. See Reference 54 for the details. Also, a lengthy description of this work can be found in Chapter 8 of Reference 52. Shang and Scherr applied their calculation to the hypersonic viscous flow over the X-24C hypersonic test vehicle. To illustrate the results, the surface streamline pattern is shown in Figure 20.5. In reality, since the flow velocity is zero at the

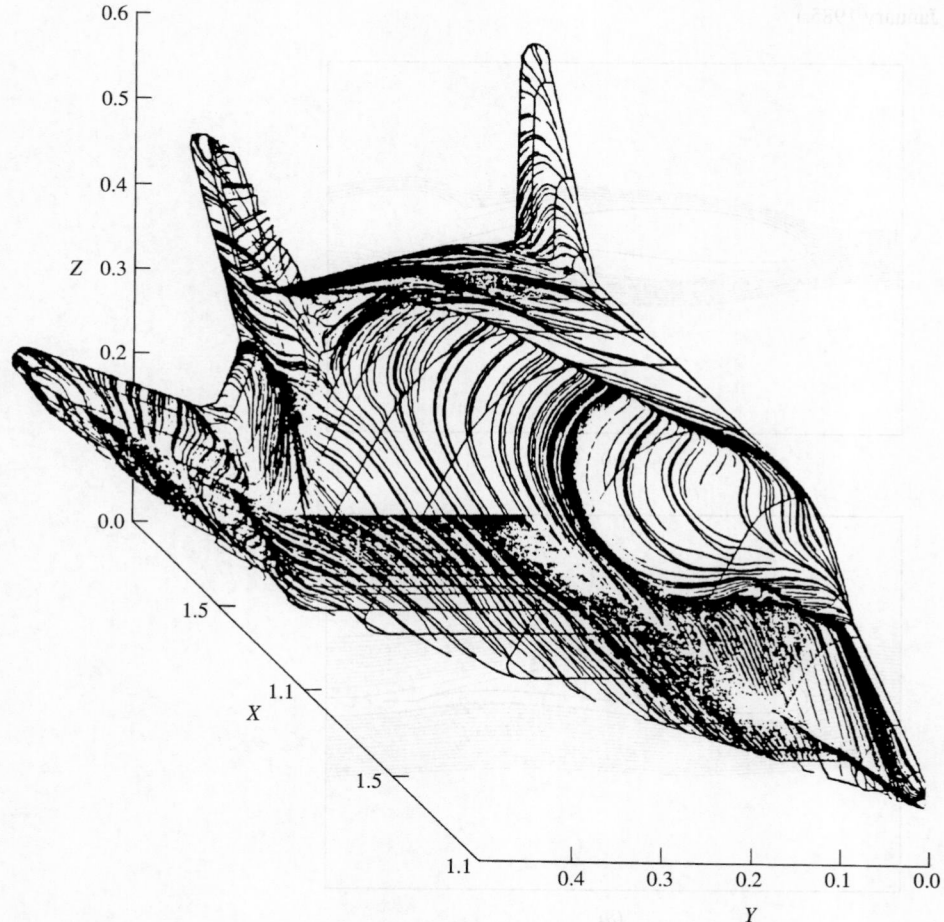

Figure 20.5 Surface shear stress lines on the X-24C. (*Data Source:* Shang, J. S., and S. J. Scherr: "Navier-Stokes Solution for a Complete Re-Entry Configuration," *J. Aircraft*, vol. 23, no. 12, December 1986, pp. 881–888.)

surface in a viscous flow (the no-slip condition), the lines shown in Figure 20.5 are the surface shear stress directions.

20.3.4 Shock-Wave/Boundary-Layer Interaction

The flow field that results when a shock wave impinges on a boundary layer can only be calculated in detail by means of a numerical solution of the complete Navier-Stokes equations. The qualitative physical aspects of a two-dimensional shock-wave/boundary-layer interaction are sketched in Figure 20.6. Here we see a boundary layer growing along a flat plate, where at some downstream location an incident shock wave impinges on the boundary layer. The large pressure rise across the shock wave acts as a severe adverse pressure gradient imposed on the boundary layer, thus causing the boundary layer to locally separate from the surface. Because the high pressure behind the shock feeds upstream through the subsonic portion of the boundary layer, the separation takes place ahead of the impingement point of the incident shock wave. In turn, the separated boundary layer induces a shock wave, identified here as the induced separation shock. The separated boundary layer subsequently turns back toward the plate, reattaching to the surface at the reattachment shock. Between the separation and reattachment shocks, expansion waves are generated where the boundary layer is turning back toward the surface. At the point of reattachment, the boundary layer has become relatively thin, the pressure is high, and consequently this becomes a region of high local aerodynamic heating. Further away from the plate, the separation and reattachment shocks merge to form the conventional "reflected shock wave" that is expected from the classical inviscid picture (see, for example, Figure 9.17). The scale and severity of the interaction picture shown in Figure 20.6 depends on whether the boundary layer is laminar or turbulent. Since laminar boundary layers

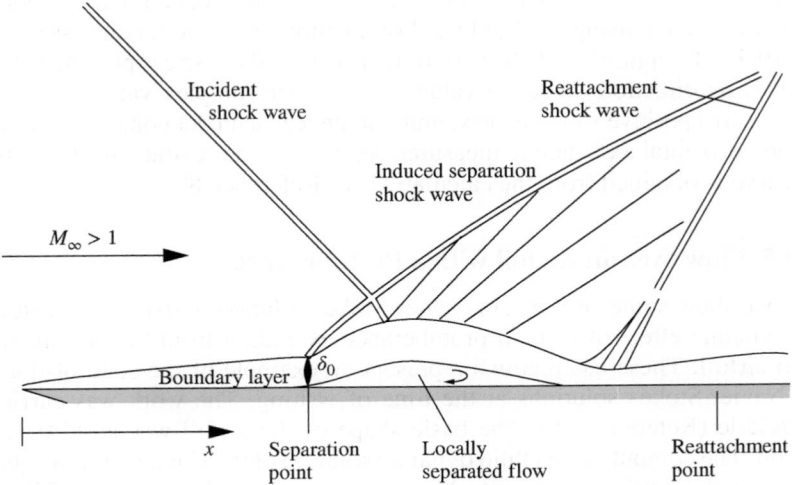

Figure 20.6 Schematic of the shock-wave/boundary-layer interaction.

separate more readily than turbulent boundary layers, the laminar interaction usually takes place more readily with more severe attendant consequences than the turbulent interaction. However, the general qualitative aspects of the interaction as sketched in Figure 20.6 are the same.

The first discussion in this book of the physical aspects of shock-wave/boundary-layer interactions can be found in Section 9.9, with a specific application to shock waves inside nozzles in Section 10.6. If you have not read Sections 9.9 and 10.6, now is the time to pause and read these two sections.

The fluid dynamic and mathematical details of the interaction region sketched in Figure 20.6 are complex, and the full prediction of this flow is still a state-of-the-art research problem. However, great strides have been made in recent years with the application of computational fluid dynamics to this problem, and solutions of the full Navier-Stokes equations for the flow sketched in Figure 20.6 have been obtained. For example, experimental and computational data for the two-dimensional interaction of a shock wave impinging on a turbulent flat plate boundary layer are given in Figure 20.7. In Figure 20.7a, the ratio of surface pressure to freestream total pressure is plotted versus distance along the surface (nondimensionalized by δ_0, the boundary-layer thickness ahead of the interaction). Here, x_0 is taken as the theoretical inviscid flow impingement point for the incident shock wave. The freestream Mach number is 3. The Reynolds number based on δ_0 is about 10^6. Note in Figure 20.7a that the surface pressure first increases at the front of the interaction region (ahead of the theoretical incident shock impingement point), reaches a plateau through the center of the separated region, and then increases again as the reattachment point is approached. The pressure variation shown in Figure 20.7a is typical of that for a two-dimensional shock-wave/boundary-layer interaction. The open circles correspond to experimental measurements of Reda and Murphy (Reference 84). The curve is obtained from a numerical solution of the Navier-Stokes equations as reported in Reference 82 and using the Baldwin-Lomax turbulence model discussed in Section 19.3.1. In Figure 20.7b the variation of surface shear stress plummets to zero, reverses its direction (negative values) in a rather complex variation, and then recovers to a positive value in the vicinity of the reattachment point. The two circles on the horizontal axis denote measured separation and reattachment points, and the curve is obtained from the calculations of Reference 82.

20.3.5 Flow over an Airfoil with a Protuberance

Here we show some very recent Navier-Stokes solutions carried out to study the aerodynamic effect of a small protuberance extending from the bottom surface of an airfoil. These calculations represent an example of the state-of-the-art of full Navier-Stokes solutions at the time of writing. The work was carried out by Beierle (Reference 85). The basic shape of the airfoil was an NACA 0015 section. The computational fluid dynamic solution of the Navier-Stokes equations was carried out using a time-marching finite volume code labeled OVERFLOW, developed by NASA (Reference 86). The flow was low speed, with a freestream

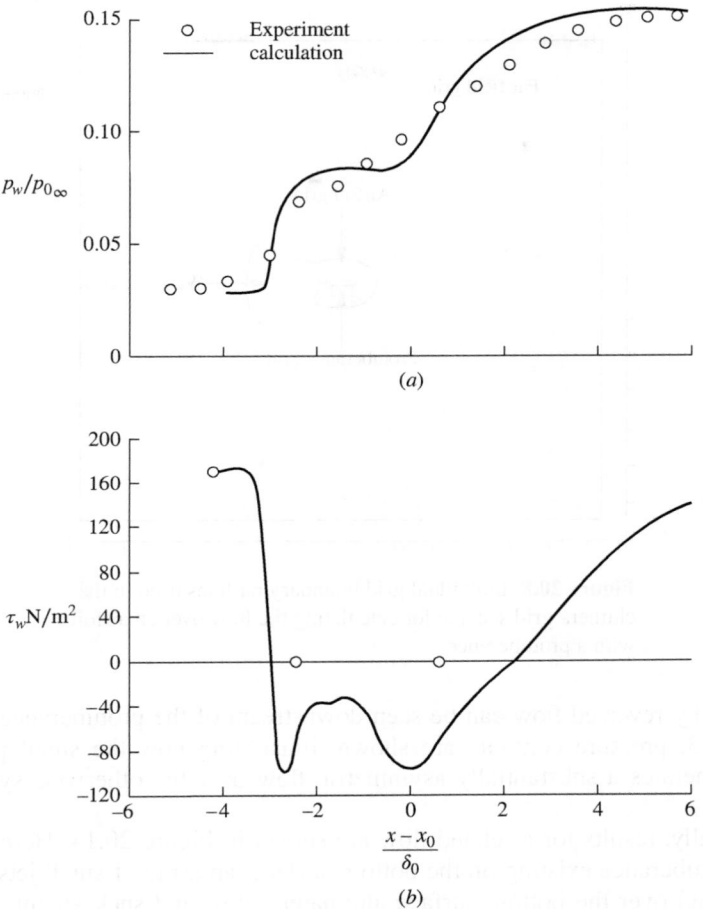

Figure 20.7 Effects of shock-wave/boundary-layer interaction on
(*a*) pressure distribution, and (*b*) shear stress for Mach 3 turbulent flow
over a flat plate.

Mach number of 0.15 and Reynolds number of 1.5×10^6. The fully turbulent flow
field was simulated using a one-equation turbulence model.

Using a proper grid is vital to the integrity of any Navier-Stokes CFD solution.
For the present case, Figures 20.8–20.11 show the grid used, progressing from
the big picture of the whole grid (Figure 20.8) to the detail of the grid around the
small protuberance on the bottom surface of the airfoil (Figure 20.11). The grid is
an example of a chimera grid, a series of independent but overlapping grids that
are generated about individual parts of the body and for specific flow regions.

Some results for the computed flow field are shown in Figures 20.12 and
20.13. In Figure 20.12, the local velocity vector field is shown; the flow separation

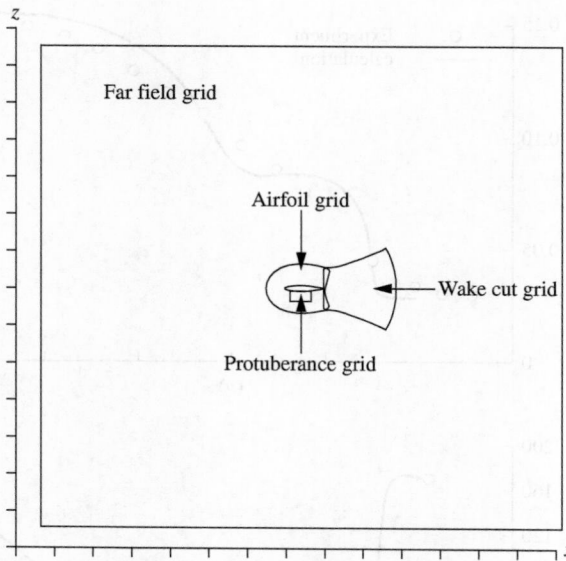

Figure 20.8 Individual grid boundary outlines used in the chimera grid scheme for calculating the flow over an airfoil with a protuberance.

and locally reversed flow can be seen downstream of the protuberance. In Figure 20.13, pressure contours are shown, illustrating how the small protuberance generates a substantially asymmetric flow over the otherwise symmetric airfoil.

Finally, results for a related flow are shown in Figure 20.14. Here, instead of a protuberance existing on the bottom surface, an array of small jets that are distributed over the bottom surface alternately blow and suck air into and out of the flow in such a manner that the net mass flow added is zero, so-called "zero-mass synthetic jets." The resulting series of large-scale vortices is shown in Figure 20.14—another example of a flow field that can only be solved in detail by means of a full Navier-Stokes solution. (See Hassan and JanakiRam, Reference 87, for details.)

20.4 THE ISSUE OF ACCURACY FOR THE PREDICTION OF SKIN FRICTION DRAG

The aerodynamic drag on a body is the sum of pressure drag and skin friction drag. For attached flows, the prediction of pressure drag is obtained from inviscid flow analyses such as those presented in Parts 2 and 3 of this book. For separated flows, various approximate theories for pressure drag have been advanced over the last century, but today the only viable and general method of the analysis of pressure drag for such flows is a complete numerical Navier-Stokes solution.

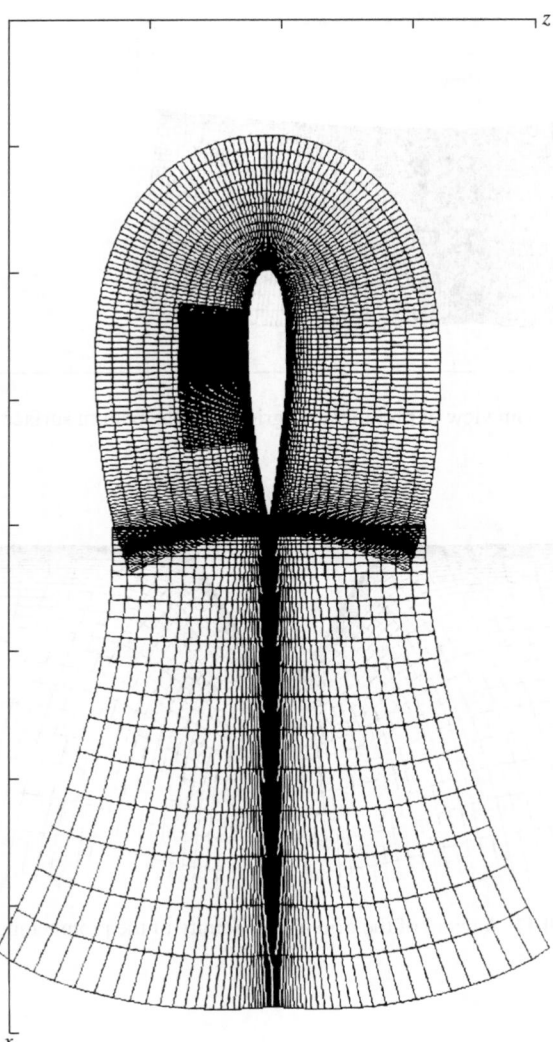

Figure 20.9 Zoom view of the airfoil, wake cut, and protuberance grids.

The prediction of skin friction on the surface of a body in an attached flow is nicely accomplished by means of a boundary-layer solution coupled with an inviscid flow analysis to define the flow conditions at the edge of the boundary layer. Such an approach is well-developed, and the calculations can be rapidly carried out on local computer workstations. Therefore, the use of boundary-layer solutions for skin friction and aerodynamic heating is the preferred engineering approach. However, as mentioned above, if regions of flow separation are present,

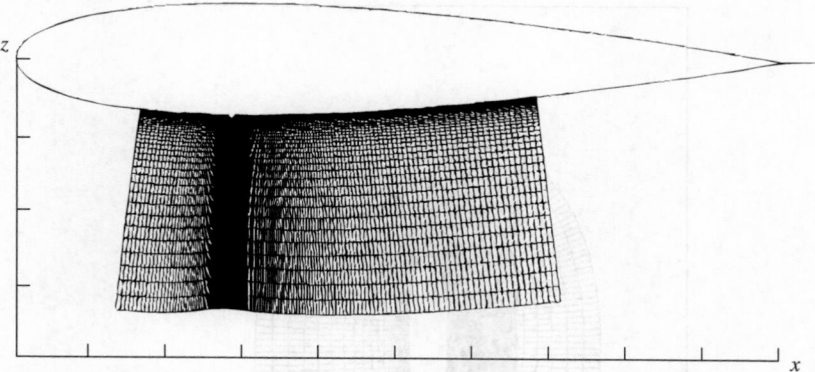

Figure 20.10 Zoom view of protuberance grid along the bottom surface of the airfoil.

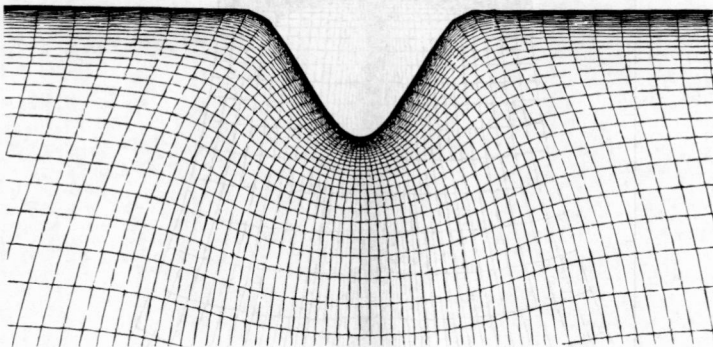

Figure 20.11 A detail of the grid in the vicinity of the protuberance.

Figure 20.12 Computed velocity vector field around and downstream of the protuberance.

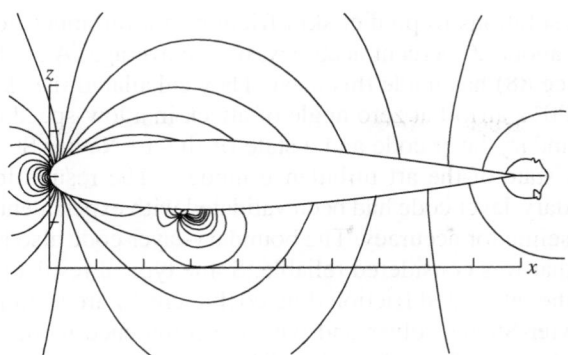

Figure 20.13 Computed pressure contours around the
NACA 0012 airfoil with a protuberance.

Figure 20.14 Streamline pattern over the bottom of an NACA 0012 airfoil, over which
a series of jets is distributed which are alternatively blowing or sucking mass into or out
of the flow, with the overall net mass injected into the flow being zero—a zero-mass jet
array. (*Data Source:* Hassan, A. A., and R. D., JanakiRam: "Effects of Zero-Mass
'Synthetic' Jets on the Aerodynamics of the NACA-0012 Airfoil," AIAA Paper
No. 97-2326, 1997.)

this approach cannot be used. In its place, a full Navier-Stokes solution can be
used to obtain local skin friction and heat transfer, but these Navier-Stokes
solutions are still not in the category of "quick engineering calculations."

This leads us to the question of the accuracy of CFD Navier-Stokes solutions
for skin friction drag and heat transfer. There are three aspects that tend to diminish the accuracy of such solutions for the prediction of τ_w and $\dot{q}_w$ (or alternately,
c_f and C_H):

1. The need to have a *very closely spaced grid in the vicinity of the wall* in
   order to obtain an accurate numerical value of $(\partial u/\partial y)_w$ and $(\partial T/\partial y)_w$,
   from which τ_w and $\dot{q}_w$ are obtained.

2. The uncertainty in the accuracy of turbulence models when a turbulent flow
   is being calculated.

3. The lack of ability of most turbulent models to predict transition from
   laminar to turbulent flow.

In spite of all the advances made in CFD to the present, and all the work
that has gone into turbulence modeling, at the time of writing the ability of

Navier-Stokes solutions to predict skin friction in a turbulent flow seems to be no better than about 20 percent accuracy, on the average. A study by Lombardi et al. (Reference 88) has made this clear. They calculated the skin friction drag on an NACA 0012 airfoil at zero angle of attack in a low-speed flow using both a standard boundary-layer code and a state-of-the-art Navier-Stokes solver with three different state-of-the-art turbulence models. The results for friction drag from the boundary-layer code had been validated with experiment, and were considered the baseline for accuracy. The boundary-layer code also had a prediction for transition that was considered reliable. Some typical results reported in Reference 88 for the integrated friction drag coefficient C_f are as follows, where NS represents Navier-Stokes solver and with the turbulence model in parenthesis. The calculations were all for Re $= 3 \times 10^6$.

	$C_f \times 10^3$
NS (Standard $k - \varepsilon$)	7.486
NS (RNG $k - \varepsilon$)	6.272
NS (Reynolds stress)	6.792
Boundary layer solution	5.340

Clearly, the accuracy of the various Navier-Stokes calculations ranged from 18 percent to 40 percent.

More insight can be gained from the spatial distribution of the local skin-friction coefficient c_f along the surface of the airfoil, as shown in Figure 20.15. Again the three different Navier-Stokes calculations are compared with the results

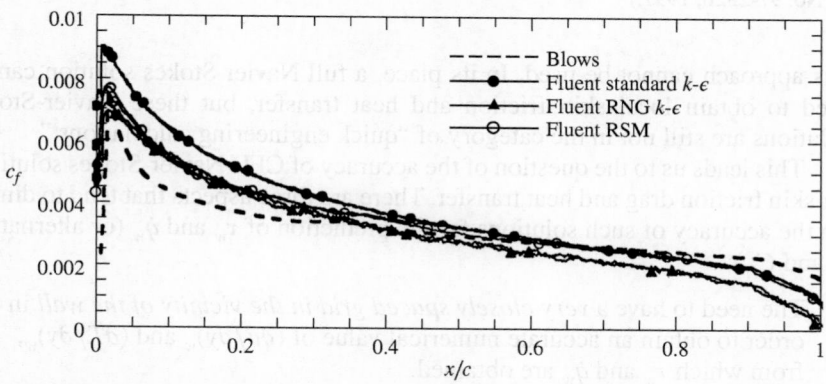

Figure 20.15 Distribution of the skin-friction coefficient over the surface of an NACA 0012 airfoil at zero angle of attack in low-speed flow. Comparison of three Navier-Stokes calculations using different turbulent models, and results obtained from a boundary-layer calculation. The boundary-layer results are given by the dashed curve labeled "Blows." (*Data Source:* Lombardi, G., M. V. Salvetti, and D. Pinelli: "Numerical Evaluation of Airfoil Friction Drag," *J. Aircraft*, vol. 37, no. 2, March-April 2000, pp. 354–356.)

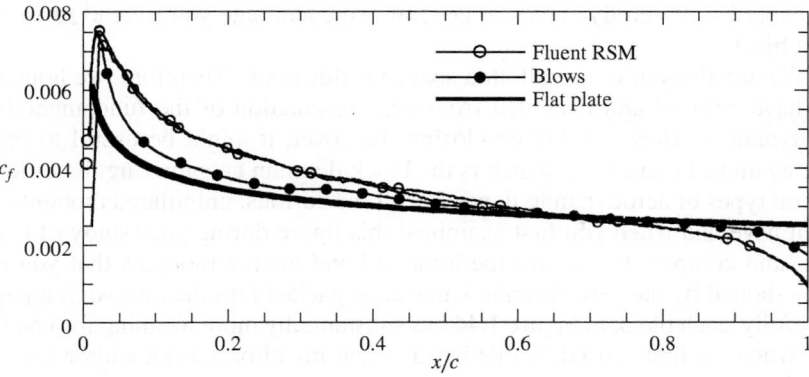

Figure 20.16 Computed skin-friction coefficient distributions over an NACA 0012 airfoil, comparing results from a Navier-Stokes solution and a boundary-layer solution. The heavy curve is for a flat plate, allowing a comparison with the skin friction distribution over a curved airfoil surface. (*Data Source:* Lombardi, G., M. V. Salvetti, and D. Pinelli: "Numerical Evaluation of Airfoil Friction Drag," *J. Aircraft*, vol. 37, no. 2, March-April 2000, pp. 354–356.)

from the boundary layer code. All the Navier-Stokes calculations greatly overestimated the peak in c_f just downstream of the leading edge, and slightly underestimated c_f near the trailing edge.

For a completely different reason not having to do with our discussion of accuracy, but for purposes of showing and contrasting the physically different distribution of c_f along a flat plate compared with that along the surface of the airfoil, we show Figure 20.16. Here the heavy curve is the variation of c_f with distance from the leading edge for a flat plate; the monotonic decrease is expected from our previous discussions of flat plate boundary layers. In contrast, for the airfoil c_f rapidly increases from a value of zero at the stagnation point to a peak value shortly downstream of the leading edge. This rapid increase is due to the rapidly increasing velocity as the flow external to the boundary layer rapidly expands around the leading edge. Beyond the peak, c_f then monotonically decreases in the same qualitative manner as for a flat plate. It is simply interesting to note these different variations for c_f over an airfoil compared to that for a flat plate, especially since we devoted so much attention to flat plates in the previous chapters.

20.5 SUMMARY

With this, we end our discussion of viscous flow. The purpose of all of Part 4 has been to introduce you to the basic aspects of viscous flow. The subject is so vast that it demands a book in itself—many of which have been written (see, e.g., References 39 through 43). Here, we have presented only enough material to give you a flavor for some of the basic ideas and results. This is a subject of great importance in aerodynamics, and if you wish to expand your knowledge and expertise of aerodynamics in general, we encourage you to read further on the subject.

We are also out of our allotted space for this book. Therefore, we hope that you have enjoyed and benefited from our presentation of the fundamentals of aerodynamics. However, before closing the cover, it might be useful to return once again to Figure 1.44, which is the block diagram categorizing the different general types of aerodynamic flows. Recall the curious, uninitiated thoughts you might have had when you first examined this figure during your study of Chapter 1, and compare these with the informed and mature thoughts that you now have—honed by the aerodynamic knowledge packed into the intervening pages. Hopefully, each block in Figure 1.44 has substantially more meaning for you now than when we first started. If this is true, then my efforts as an author have not gone in vain.

Isentropic Flow Properties

M	$\dfrac{p_0}{p}$	$\dfrac{\rho_0}{\rho}$	$\dfrac{T_0}{T}$	$\dfrac{A}{A^*}$
0.2000 − 01	0.1000 + 01	0.1000 + 01	0.1000 + 01	0.2894 + 02
0.4000 − 01	0.1001 + 01	0.1001 + 01	0.1000 + 01	0.1448 + 02
0.6000 − 01	0.1003 + 01	0.1002 + 01	0.1001 + 01	0.9666 + 01
0.8000 − 01	0.1004 + 01	0.1003 + 01	0.1001 + 01	0.7262 + 01
0.1000 + 00	0.1007 + 01	0.1005 + 01	0.1002 + 01	0.5822 + 01
0.1200 + 00	0.1010 + 01	0.1007 + 01	0.1003 + 01	0.4864 + 01
0.1400 + 00	0.1014 + 01	0.1010 + 01	0.1004 + 01	0.4182 + 01
0.1600 + 00	0.1018 + 01	0.1013 + 01	0.1005 + 01	0.3673 + 01
0.1800 + 00	0.1023 + 01	0.1016 + 01	0.1006 + 01	0.3278 + 01
0.2000 + 00	0.1028 + 01	0.1020 + 01	0.1008 + 01	0.2964 + 01
0.2200 + 00	0.1034 + 01	0.1024 + 01	0.1010 + 01	0.2708 + 01
0.2400 + 00	0.1041 + 01	0.1029 + 01	0.1012 + 01	0.2496 + 01
0.2600 + 00	0.1048 + 01	0.1034 + 01	0.1014 + 01	0.2317 + 01
0.2800 + 00	0.1056 + 01	0.1040 + 01	0.1016 + 01	0.2166 + 01
0.3000 + 00	0.1064 + 01	0.1046 + 01	0.1018 + 01	0.2035 + 01
0.3200 + 00	0.1074 + 01	0.1052 + 01	0.1020 + 01	0.1922 + 01
0.3400 + 00	0.1083 + 01	0.1059 + 01	0.1023 + 01	0.1823 + 01
0.3600 + 00	0.1094 + 01	0.1066 + 01	0.1026 + 01	0.1736 + 01
0.3800 + 00	0.1105 + 01	0.1074 + 01	0.1029 + 01	0.1659 + 01
0.4000 + 00	0.1117 + 01	0.1082 + 01	0.1032 + 01	0.1590 + 01
0.4200 + 00	0.1129 + 01	0.1091 + 01	0.1035 + 01	0.1529 + 01
0.4400 + 00	0.1142 + 01	0.1100 + 01	0.1039 + 01	0.1474 + 01
0.4600 + 00	0.1156 + 01	0.1109 + 01	0.1042 + 01	0.1425 + 01
0.4800 + 00	0.1171 + 01	0.1119 + 01	0.1046 + 01	0.1380 + 01
0.5000 + 00	0.1186 + 01	0.1130 + 01	0.1050 + 01	0.1340 + 01
0.5200 + 00	0.1202 + 01	0.1141 + 01	0.1054 + 01	0.1303 + 01
0.5400 + 00	0.1219 + 01	0.1152 + 01	0.1058 + 01	0.1270 + 01
0.5600 + 00	0.1237 + 01	0.1164 + 01	0.1063 + 01	0.1240 + 01
0.5800 + 00	0.1256 + 01	0.1177 + 01	0.1067 + 01	0.1213 + 01
0.6000 + 00	0.1276 + 01	0.1190 + 01	0.1072 + 01	0.1188 + 01

M	$\dfrac{p_0}{p}$	$\dfrac{\rho_0}{\rho}$	$\dfrac{T_0}{T}$	$\dfrac{A}{A^*}$
0.6200 + 00	0.1296 + 01	0.1203 + 01	0.1077 + 01	0.1166 + 01
0.6400 + 00	0.1317 + 01	0.1218 + 01	0.1082 + 01	0.1145 + 01
0.6600 + 00	0.1340 + 01	0.1232 + 01	0.1087 + 01	0.1127 + 01
0.6800 + 00	0.1363 + 01	0.1247 + 01	0.1092 + 01	0.1110 + 01
0.7000 + 00	0.1387 + 01	0.1263 + 01	0.1098 + 01	0.1094 + 01
0.7200 + 00	0.1412 + 01	0.1280 + 01	0.1104 + 01	0.1081 + 01
0.7400 + 00	0.1439 + 01	0.1297 + 01	0.1110 + 01	0.1068 + 01
0.7600 + 00	0.1466 + 01	0.1314 + 01	0.1116 + 01	0.1057 + 01
0.7800 + 00	0.1495 + 01	0.1333 + 01	0.1122 + 01	0.1047 + 01
0.8000 + 00	0.1524 + 01	0.1351 + 01	0.1128 + 01	0.1038 + 01
0.8200 + 00	0.1555 + 01	0.1371 + 01	0.1134 + 01	0.1030 + 01
0.8400 + 00	0.1587 + 01	0.1391 + 01	0.1141 + 01	0.1024 + 01
0.8600 + 00	0.1621 + 01	0.1412 + 01	0.1148 + 01	0.1018 + 01
0.8800 + 00	0.1655 + 01	0.1433 + 01	0.1155 + 01	0.1013 + 01
0.9000 + 00	0.1691 + 01	0.1456 + 01	0.1162 + 01	0.1009 + 01
0.9200 + 00	0.1729 + 01	0.1478 + 01	0.1169 + 01	0.1006 + 01
0.9400 + 00	0.1767 + 01	0.1502 + 01	0.1177 + 01	0.1003 + 01
0.9600 + 00	0.1808 + 01	0.1526 + 01	0.1184 + 01	0.1001 + 01
0.9800 + 00	0.1850 + 01	0.1552 + 01	0.1192 + 01	0.1000 + 01
0.1000 + 01	0.1893 + 01	0.1577 + 01	0.1200 + 01	0.1000 + 01
0.1020 + 01	0.1938 + 01	0.1604 + 01	0.1208 + 01	0.1000 + 01
0.1040 + 01	0.1985 + 01	0.1632 + 01	0.1216 + 01	0.1001 + 01
0.1060 + 01	0.2033 + 01	0.1660 + 01	0.1225 + 01	0.1003 + 01
0.1080 + 01	0.2083 + 01	0.1689 + 01	0.1233 + 01	0.1005 + 01
0.1100 + 01	0.2135 + 01	0.1719 + 01	0.1242 + 01	0.1008 + 01
0.1120 + 01	0.2189 + 01	0.1750 + 01	0.1251 + 01	0.1011 + 01
0.1140 + 01	0.2245 + 01	0.1782 + 01	0.1260 + 01	0.1015 + 01
0.1160 + 01	0.2303 + 01	0.1814 + 01	0.1269 + 01	0.1020 + 01
0.1180 + 01	0.2363 + 01	0.1848 + 01	0.1278 + 01	0.1025 + 01
0.1200 + 01	0.2425 + 01	0.1883 + 01	0.1288 + 01	0.1030 + 01
0.1220 + 01	0.2489 + 01	0.1918 + 01	0.1298 + 01	0.1037 + 01
0.1240 + 01	0.2556 + 01	0.1955 + 01	0.1308 + 01	0.1043 + 01
0.1260 + 01	0.2625 + 01	0.1992 + 01	0.1318 + 01	0.1050 + 01
0.1280 + 01	0.2697 + 01	0.2031 + 01	0.1328 + 01	0.1058 + 01
0.1300 + 01	0.2771 + 01	0.2071 + 01	0.1338 + 01	0.1066 + 01
0.1320 + 01	0.2847 + 01	0.2112 + 01	0.1348 + 01	0.1075 + 01
0.1340 + 01	0.2927 + 01	0.2153 + 01	0.1359 + 01	0.1084 + 01
0.1360 + 01	0.3009 + 01	0.2197 + 01	0.1370 + 01	0.1094 + 01
0.1380 + 01	0.3094 + 01	0.2241 + 01	0.1381 + 01	0.1104 + 01
0.1400 + 01	0.3182 + 01	0.2286 + 01	0.1392 + 01	0.1115 + 01
0.1420 + 01	0.3273 + 01	0.2333 + 01	0.1403 + 01	0.1126 + 01
0.1440 + 01	0.3368 + 01	0.2381 + 01	0.1415 + 01	0.1138 + 01
0.1460 + 01	0.3465 + 01	0.2430 + 01	0.1426 + 01	0.1150 + 01
0.1480 + 01	0.3566 + 01	0.2480 + 01	0.1438 + 01	0.1163 + 01
0.1500 + 01	0.3671 + 01	0.2532 + 01	0.1450 + 01	0.1176 + 01
0.1520 + 01	0.3779 + 01	0.2585 + 01	0.1462 + 01	0.1190 + 01
0.1540 + 01	0.3891 + 01	0.2639 + 01	0.1474 + 01	0.1204 + 01
0.1560 + 01	0.4007 + 01	0.2695 + 01	0.1487 + 01	0.1219 + 01
0.1580 + 01	0.4127 + 01	0.2752 + 01	0.1499 + 01	0.1234 + 01
0.1600 + 01	0.4250 + 01	0.2811 + 01	0.1512 + 01	0.1250 + 01

M	$\dfrac{p_0}{p}$	$\dfrac{\rho_0}{\rho}$	$\dfrac{T_0}{T}$	$\dfrac{A}{A^*}$
0.1620 + 01	0.4378 + 01	0.2871 + 01	0.1525 + 01	0.1267 + 01
0.1640 + 01	0.4511 + 01	0.2933 + 01	0.1538 + 01	0.1284 + 01
0.1660 + 01	0.4648 + 01	0.2996 + 01	0.1551 + 01	0.1301 + 01
0.1680 + 01	0.4790 + 01	0.3061 + 01	0.1564 + 01	0.1319 + 01
0.1700 + 01	0.4936 + 01	0.3128 + 01	0.1578 + 01	0.1338 + 01
0.1720 + 01	0.5087 + 01	0.3196 + 01	0.1592 + 01	0.1357 + 01
0.1740 + 01	0.5244 + 01	0.3266 + 01	0.1606 + 01	0.1376 + 01
0.1760 + 01	0.5406 + 01	0.3338 + 01	0.1620 + 01	0.1397 + 01
0.1780 + 01	0.5573 + 01	0.3411 + 01	0.1634 + 01	0.1418 + 01
0.1800 + 01	0.5746 + 01	0.3487 + 01	0.1648 + 01	0.1439 + 01
0.1820 + 01	0.5924 + 01	0.3564 + 01	0.1662 + 01	0.1461 + 01
0.1840 + 01	0.6109 + 01	0.3643 + 01	0.1677 + 01	0.1484 + 01
0.1860 + 01	0.6300 + 01	0.3723 + 01	0.1692 + 01	0.1507 + 01
0.1880 + 01	0.6497 + 01	0.3806 + 01	0.1707 + 01	0.1531 + 01
0.1900 + 01	0.6701 + 01	0.3891 + 01	0.1722 + 01	0.1555 + 01
0.1920 + 01	0.6911 + 01	0.3978 + 01	0.1737 + 01	0.1580 + 01
0.1940 + 01	0.7128 + 01	0.4067 + 01	0.1753 + 01	0.1606 + 01
0.1960 + 01	0.7353 + 01	0.4158 + 01	0.1768 + 01	0.1633 + 01
0.1980 + 01	0.7585 + 01	0.4251 + 01	0.1784 + 01	0.1660 + 01
0.2000 + 01	0.7824 + 01	0.4347 + 01	0.1800 + 01	0.1687 + 01
0.2050 + 01	0.8458 + 01	0.4596 + 01	0.1840 + 01	0.1760 + 01
0.2100 + 01	0.9145 + 01	0.4859 + 01	0.1882 + 01	0.1837 + 01
0.2150 + 01	0.9888 + 01	0.5138 + 01	0.1924 + 01	0.1919 + 01
0.2200 + 01	0.1069 + 02	0.5433 + 01	0.1968 + 01	0.2005 + 01
0.2250 + 01	0.1156 + 02	0.5746 + 01	0.2012 + 01	0.2096 + 01
0.2300 + 01	0.1250 + 02	0.6076 + 01	0.2058 + 01	0.2193 + 01
0.2350 + 01	0.1352 + 02	0.6425 + 01	0.2104 + 01	0.2295 + 01
0.2400 + 01	0.1462 + 02	0.6794 + 01	0.2152 + 01	0.2403 + 01
0.2450 + 01	0.1581 + 02	0.7183 + 01	0.2200 + 01	0.2517 + 01
0.2500 + 01	0.1709 + 02	0.7594 + 01	0.2250 + 01	0.2637 + 01
0.2550 + 01	0.1847 + 02	0.8027 + 01	0.2300 + 01	0.2763 + 01
0.2600 + 01	0.1995 + 02	0.8484 + 01	0.2352 + 01	0.2896 + 01
0.2650 + 01	0.2156 + 02	0.8965 + 01	0.2404 + 01	0.3036 + 01
0.2700 + 01	0.2328 + 02	0.9472 + 01	0.2458 + 01	0.3183 + 01
0.2750 + 01	0.2514 + 02	0.1001 + 02	0.2512 + 01	0.3338 + 01
0.2800 + 01	0.2714 + 02	0.1057 + 02	0.2568 + 01	0.3500 + 01
0.2850 + 01	0.2929 + 02	0.1116 + 02	0.2624 + 01	0.3671 + 01
0.2900 + 01	0.3159 + 02	0.1178 + 02	0.2682 + 01	0.3850 + 01
0.2950 + 01	0.3407 + 02	0.1243 + 02	0.2740 + 01	0.4038 + 01
0.3000 + 01	0.3673 + 02	0.1312 + 02	0.2800 + 01	0.4235 + 01
0.3050 + 01	0.3959 + 02	0.1384 + 02	0.2860 + 01	0.4441 + 01
0.3100 + 01	0.4265 + 02	0.1459 + 02	0.2922 + 01	0.4657 + 01
0.3150 + 01	0.4593 + 02	0.1539 + 02	0.2984 + 01	0.4884 + 01
0.3200 + 01	0.4944 + 02	0.1622 + 02	0.3048 + 01	0.5121 + 01
0.3250 + 01	0.5320 + 02	0.1709 + 02	0.3112 + 01	0.5369 + 01
0.3300 + 01	0.5722 + 02	0.1800 + 02	0.3178 + 01	0.5629 + 01
0.3350 + 01	0.6152 + 02	0.1896 + 02	0.3244 + 01	0.5900 + 01
0.3400 + 01	0.6612 + 02	0.1996 + 02	0.3312 + 01	0.6184 + 01
0.3450 + 01	0.7103 + 02	0.2101 + 02	0.3380 + 01	0.6480 + 01
0.3500 + 01	0.7627 + 02	0.2211 + 02	0.3450 + 01	0.6790 + 01

M	$\dfrac{p_0}{p}$	$\dfrac{\rho_0}{\rho}$	$\dfrac{T_0}{T}$	$\dfrac{A}{A^*}$
0.3550 + 01	0.8187 + 02	0.2325 + 02	0.3520 + 01	0.7113 + 01
0.3600 + 01	0.8784 + 02	0.2445 + 02	0.3592 + 01	0.7450 + 01
0.3650 + 01	0.9420 + 02	0.2571 + 02	0.3664 + 01	0.7802 + 01
0.3700 + 01	0.1010 + 03	0.2701 + 02	0.3738 + 01	0.8169 + 01
0.3750 + 01	0.1082 + 03	0.2838 + 02	0.3812 + 01	0.8552 + 01
0.3800 + 01	0.1159 + 03	0.2981 + 02	0.3888 + 01	0.8951 + 01
0.3850 + 01	0.1241 + 03	0.3129 + 02	0.3964 + 01	0.9366 + 01
0.3900 + 01	0.1328 + 03	0.3285 + 02	0.4042 + 01	0.9799 + 01
0.3950 + 01	0.1420 + 03	0.3446 + 02	0.4120 + 01	0.1025 + 02
0.4000 + 01	0.1518 + 03	0.3615 + 02	0.4200 + 01	0.1072 + 02
0.4050 + 01	0.1623 + 03	0.3791 + 02	0.4280 + 01	0.1121 + 02
0.4100 + 01	0.1733 + 03	0.3974 + 02	0.4362 + 01	0.1171 + 02
0.4150 + 01	0.1851 + 03	0.4164 + 02	0.4444 + 01	0.1224 + 02
0.4200 + 01	0.1975 + 03	0.4363 + 02	0.4528 + 01	0.1279 + 02
0.4250 + 01	0.2108 + 03	0.4569 + 02	0.4612 + 01	0.1336 + 02
0.4300 + 01	0.2247 + 03	0.4784 + 02	0.4698 + 01	0.1395 + 02
0.4350 + 01	0.2396 + 03	0.5007 + 02	0.4784 + 01	0.1457 + 02
0.4400 + 01	0.2553 + 03	0.5239 + 02	0.4872 + 01	0.1521 + 02
0.4450 + 01	0.2719 + 03	0.5480 + 02	0.4960 + 01	0.1587 + 02
0.4500 + 01	0.2894 + 03	0.5731 + 02	0.5050 + 01	0.1656 + 02
0.4550 + 01	0.3080 + 03	0.5991 + 02	0.5140 + 01	0.1728 + 02
0.4600 + 01	0.3276 + 03	0.6261 + 02	0.5232 + 01	0.1802 + 02
0.4650 + 01	0.3483 + 03	0.6542 + 02	0.5324 + 01	0.1879 + 02
0.4700 + 01	0.3702 + 03	0.6833 + 02	0.5418 + 01	0.1958 + 02
0.4750 + 01	0.3933 + 03	0.7135 + 02	0.5512 + 01	0.2041 + 02
0.4800 + 01	0.4177 + 03	0.7448 + 02	0.5608 + 01	0.2126 + 02
0.4850 + 01	0.4434 + 03	0.7772 + 02	0.5704 + 01	0.2215 + 02
0.4900 + 01	0.4705 + 03	0.8109 + 02	0.5802 + 01	0.2307 + 02
0.4950 + 01	0.4990 + 03	0.8457 + 02	0.5900 + 01	0.2402 + 02
0.5000 + 01	0.5291 + 03	0.8818 + 02	0.6000 + 01	0.2500 + 02
0.5100 + 01	0.5941 + 03	0.9579 + 02	0.6202 + 01	0.2707 + 02
0.5200 + 01	0.6661 + 03	0.1039 + 03	0.6408 + 01	0.2928 + 02
0.5300 + 01	0.7457 + 03	0.1127 + 03	0.6618 + 01	0.3165 + 02
0.5400 + 01	0.8335 + 03	0.1220 + 03	0.6832 + 01	0.3417 + 02
0.5500 + 01	0.9304 + 03	0.1320 + 03	0.7050 + 01	0.3687 + 02
0.5600 + 01	0.1037 + 04	0.1426 + 03	0.7272 + 01	0.3974 + 02
0.5700 + 01	0.1154 + 04	0.1539 + 03	0.7498 + 01	0.4280 + 02
0.5800 + 01	0.1283 + 04	0.1660 + 03	0.7728 + 01	0.4605 + 02
0.5900 + 01	0.1424 + 04	0.1789 + 03	0.7962 + 01	0.4951 + 02
0.6000 + 01	0.1579 + 04	0.1925 + 03	0.8200 + 01	0.5318 + 02
0.6100 + 01	0.1748 + 04	0.2071 + 03	0.8442 + 01	0.5708 + 02
0.6200 + 01	0.1933 + 04	0.2225 + 03	0.8688 + 01	0.6121 + 02
0.6300 + 01	0.2135 + 04	0.2388 + 03	0.8938 + 01	0.6559 + 02
0.6400 + 01	0.2355 + 04	0.2562 + 03	0.9192 + 01	0.7023 + 02
0.6500 + 01	0.2594 + 04	0.2745 + 03	0.9450 + 01	0.7513 + 02
0.6600 + 01	0.2855 + 04	0.2939 + 03	0.9712 + 01	0.8032 + 02
0.6700 + 01	0.3138 + 04	0.3145 + 03	0.9978 + 01	0.8580 + 02
0.6800 + 01	0.3445 + 04	0.3362 + 03	0.1025 + 02	0.9159 + 02
0.6900 + 01	0.3779 + 04	0.3591 + 03	0.1052 + 02	0.9770 + 02
0.7000 + 01	0.4140 + 04	0.3833 + 03	0.1080 + 02	0.1041 + 03

M	$\dfrac{p_0}{p}$	$\dfrac{\rho_0}{\rho}$	$\dfrac{T_0}{T}$	$\dfrac{A}{A^*}$
0.7100 + 01	0.4531 + 04	0.4088 + 03	0.1108 + 02	0.1109 + 03
0.7200 + 01	0.4953 + 04	0.4357 + 03	0.1137 + 02	0.1181 + 03
0.7300 + 01	0.5410 + 04	0.4640 + 03	0.1166 + 02	0.1256 + 03
0.7400 + 01	0.5903 + 04	0.4939 + 03	0.1195 + 02	0.1335 + 03
0.7500 + 01	0.6434 + 04	0.5252 + 03	0.1225 + 02	0.1418 + 03
0.7600 + 01	0.7006 + 04	0.5582 + 03	0.1255 + 02	0.1506 + 03
0.7700 + 01	0.7623 + 04	0.5928 + 03	0.1286 + 02	0.1598 + 03
0.7800 + 01	0.8285 + 04	0.6292 + 03	0.1317 + 02	0.1694 + 03
0.7900 + 01	0.8998 + 04	0.6674 + 03	0.1348 + 02	0.1795 + 03
0.8000 + 01	0.9763 + 04	0.7075 + 03	0.1380 + 02	0.1901 + 03
0.9000 + 01	0.2110 + 05	0.1227 + 04	0.1720 + 02	0.3272 + 03
0.1000 + 02	0.4244 + 05	0.2021 + 04	0.2100 + 02	0.5359 + 03
0.1100 + 02	0.8033 + 05	0.3188 + 04	0.2520 + 02	0.8419 + 03
0.1200 + 02	0.1445 + 06	0.4848 + 04	0.2980 + 02	0.1276 + 04
0.1300 + 02	0.2486 + 06	0.7144 + 04	0.3480 + 02	0.1876 + 04
0.1400 + 02	0.4119 + 06	0.1025 + 05	0.4020 + 02	0.2685 + 04
0.1500 + 02	0.6602 + 06	0.1435 + 05	0.4600 + 02	0.3755 + 04
0.1600 + 02	0.1028 + 07	0.1969 + 05	0.5229 + 02	0.5145 + 04
0.1700 + 02	0.1559 + 07	0.2651 + 05	0.5880 + 02	0.6921 + 04
0.1800 + 02	0.2311 + 07	0.3512 + 05	0.6580 + 02	0.9159 + 04
0.1900 + 02	0.3356 + 07	0.4584 + 05	0.7320 + 02	0.1195 + 05
0.2000 + 02	0.4783 + 07	0.5905 + 05	0.8100 + 02	0.1538 + 05
0.2200 + 02	0.9251 + 07	0.9459 + 05	0.9780 + 02	0.2461 + 05
0.2400 + 02	0.1691 + 08	0.1456 + 06	0.1162 + 03	0.3783 + 05
0.2600 + 02	0.2949 + 08	0.2165 + 06	0.1362 + 03	0.5624 + 05
0.2800 + 02	0.4936 + 08	0.3128 + 06	0.1578 + 03	0.8121 + 05
0.3000 + 02	0.7978 + 08	0.4408 + 06	0.1810 + 03	0.1144 + 06
0.3200 + 02	0.1250 + 09	0.6076 + 06	0.2058 + 03	0.1576 + 06
0.3400 + 02	0.1908 + 09	0.8216 + 06	0.2322 + 03	0.2131 + 06
0.3600 + 02	0.2842 + 09	0.1092 + 07	0.2602 + 03	0.2832 + 06
0.3800 + 02	0.4143 + 09	0.1430 + 07	0.2898 + 03	0.3707 + 06
0.4000 + 02	0.5926 + 09	0.1846 + 07	0.3210 + 03	0.4785 + 06
0.4200 + 02	0.8330 + 09	0.2354 + 07	0.3538 + 03	0.6102 + 06
0.4400 + 02	0.1153 + 10	0.2969 + 07	0.3882 + 03	0.7694 + 06
0.4600 + 02	0.1572 + 10	0.3706 + 07	0.4242 + 03	0.9603 + 06
0.4800 + 02	0.2116 + 10	0.4583 + 07	0.4618 + 03	0.1187 + 07
0.5000 + 02	0.2815 + 10	0.5618 + 07	0.5010 + 03	0.1455 + 07

APPENDIX B

Normal Shock Properties

M	$\dfrac{p_2}{p_1}$	$\dfrac{\rho_2}{\rho_1}$	$\dfrac{T_2}{T_1}$	$\dfrac{p_{0_2}}{p_{0_1}}$	$\dfrac{p_{0_2}}{p_1}$	M_2
0.1000 + 01	0.1000 + 01	0.1000 + 01	0.1000 + 01	0.1000 + 01	0.1893 + 01	0.1000 + 01
0.1020 + 01	0.1047 + 01	0.1033 + 01	0.1013 + 01	0.1000 + 01	0.1938 + 01	0.9805 + 00
0.1040 + 01	0.1095 + 01	0.1067 + 01	0.1026 + 01	0.9999 + 00	0.1984 + 01	0.9620 + 00
0.1060 + 01	0.1144 + 01	0.1101 + 01	0.1039 + 01	0.9998 + 00	0.2032 + 01	0.9444 + 00
0.1080 + 01	0.1194 + 01	0.1135 + 01	0.1052 + 01	0.9994 + 01	0.2082 + 01	0.9277 + 00
0.1100 + 01	0.1245 + 01	0.1169 + 01	0.1065 + 01	0.9989 + 00	0.2133 + 01	0.9118 + 00
0.1120 + 01	0.1297 + 01	0.1203 + 01	0.1078 + 01	0.9982 + 00	0.2185 + 01	0.8966 + 00
0.1140 + 01	0.1350 + 01	0.1238 + 01	0.1090 + 01	0.9973 + 00	0.2239 + 01	0.8820 + 00
0.1160 + 01	0.1403 + 01	0.1272 + 01	0.1103 + 01	0.9961 + 00	0.2294 + 01	0.8682 + 00
0.1180 + 01	0.1458 + 01	0.1307 + 01	0.1115 + 01	0.9946 + 00	0.2350 + 01	0.8549 + 00
0.1200 + 01	0.1513 + 01	0.1342 + 01	0.1128 + 01	0.9928 + 00	0.2408 + 01	0.8422 + 00
0.1220 + 01	0.1570 + 01	0.1376 + 01	0.1141 + 01	0.9907 + 00	0.2466 + 01	0.8300 + 00
0.1240 + 01	0.1627 + 01	0.1411 + 01	0.1153 + 01	0.9884 + 00	0.2526 + 01	0.8183 + 00
0.1260 + 01	0.1686 + 01	0.1446 + 01	0.1166 + 01	0.9857 + 00	0.2588 + 01	0.8071 + 00
0.1280 + 01	0.1745 + 01	0.1481 + 01	0.1178 + 01	0.9827 + 00	0.2650 + 01	0.7963 + 00
0.1300 + 01	0.1805 + 01	0.1516 + 01	0.1191 + 01	0.9794 + 00	0.2714 + 01	0.7860 + 00
0.1320 + 01	0.1866 + 01	0.1551 + 01	0.1204 + 01	0.9758 + 00	0.2778 + 01	0.7760 + 00
0.1340 + 01	0.1928 + 01	0.1585 + 01	0.1216 + 01	0.9718 + 00	0.2844 + 01	0.7664 + 00
0.1360 + 01	0.1991 + 01	0.1620 + 01	0.1229 + 01	0.9676 + 00	0.2912 + 01	0.7572 + 00
0.1380 + 01	0.2055 + 01	0.1655 + 01	0.1242 + 01	0.9630 + 00	0.2980 + 01	0.7483 + 00
0.1400 + 01	0.2120 + 01	0.1690 + 01	0.1255 + 01	0.9582 + 00	0.3049 + 01	0.7397 + 00
0.1420 + 01	0.2186 + 01	0.1724 + 01	0.1268 + 01	0.9531 + 00	0.3120 + 01	0.7314 + 00
0.1440 + 01	0.2253 + 01	0.1759 + 01	0.1281 + 01	0.9476 + 00	0.3191 + 01	0.7235 + 00
0.1460 + 01	0.2320 + 01	0.1793 + 01	0.1294 + 01	0.9420 + 00	0.3264 + 01	0.7157 + 00
0.1480 + 01	0.2389 + 01	0.1828 + 01	0.1307 + 01	0.9360 + 00	0.3338 + 01	0.7083 + 00
0.1500 + 01	0.2458 + 01	0.1862 + 01	0.1320 + 01	0.9298 + 00	0.3413 + 01	0.7011 + 00
0.1520 + 01	0.2529 + 01	0.1896 + 01	0.1334 + 01	0.9233 + 00	0.3489 + 01	0.6941 + 00
0.1540 + 01	0.2600 + 01	0.1930 + 01	0.1347 + 01	0.9166 + 00	0.3567 + 01	0.6874 + 00
0.1560 + 01	0.2673 + 01	0.1964 + 01	0.1361 + 01	0.9097 + 00	0.3645 + 01	0.6809 + 00
0.1580 + 01	0.2746 + 01	0.1998 + 01	0.1374 + 01	0.9026 + 00	0.3724 + 01	0.6746 + 00

M	$\dfrac{p_2}{p_1}$	$\dfrac{\rho_2}{\rho_1}$	$\dfrac{T_2}{T_1}$	$\dfrac{p_{0_2}}{p_{0_1}}$	$\dfrac{p_{0_2}}{p_1}$	M_2
0.1600 + 01	0.2820 + 01	0.2032 + 01	0.1388 + 01	0.8952 + 00	0.3805 + 01	0.6684 + 00
0.1620 + 01	0.2895 + 01	0.2065 + 01	0.1402 + 01	0.8877 + 00	0.3887 + 01	0.6625 + 00
0.1640 + 01	0.2971 + 01	0.2099 + 01	0.1416 + 01	0.8799 + 00	0.3969 + 01	0.6568 + 00
0.1660 + 01	0.3048 + 01	0.2132 + 01	0.1430 + 01	0.8720 + 00	0.4053 + 01	0.6512 + 00
0.1680 + 01	0.3126 + 01	0.2165 + 01	0.1444 + 01	0.8639 + 00	0.4138 + 01	0.6458 + 00
0.1700 + 01	0.3205 + 01	0.2198 + 01	0.1458 + 01	0.8557 + 00	0.4224 + 01	0.6405 + 00
0.1720 + 01	0.3285 + 01	0.2230 + 01	0.1473 + 01	0.8474 + 00	0.4311 + 01	0.6355 + 00
0.1740 + 01	0.3366 + 01	0.2263 + 01	0.1487 + 01	0.8389 + 00	0.4399 + 01	0.6305 + 00
0.1760 + 01	0.3447 + 01	0.2295 + 01	0.1502 + 01	0.8302 + 00	0.4488 + 01	0.6257 + 00
0.1780 + 01	0.3530 + 01	0.2327 + 01	0.1517 + 01	0.8215 + 00	0.4578 + 01	0.6210 + 00
0.1800 + 01	0.3613 + 01	0.2359 + 01	0.1532 + 01	0.8127 + 00	0.4670 + 01	0.6165 + 00
0.1820 + 01	0.3698 + 01	0.2391 + 01	0.1547 + 01	0.8038 + 00	0.4762 + 01	0.6121 + 00
0.1840 + 01	0.3783 + 01	0.2422 + 01	0.1562 + 01	0.7948 + 00	0.4855 + 01	0.6078 + 00
0.1860 + 01	0.3870 + 01	0.2454 + 01	0.1577 + 01	0.7857 + 00	0.4950 + 01	0.6036 + 00
0.1880 + 01	0.3957 + 01	0.2485 + 01	0.1592 + 01	0.7765 + 00	0.5045 + 01	0.5996 + 00
0.1900 + 01	0.4045 + 01	0.2516 + 01	0.1608 + 01	0.7674 + 00	0.5142 + 01	0.5956 + 00
0.1920 + 01	0.4134 + 01	0.2546 + 01	0.1624 + 01	0.7581 + 00	0.5239 + 01	0.5918 + 00
0.1940 + 01	0.4224 + 01	0.2577 + 01	0.1639 + 01	0.7488 + 00	0.5338 + 01	0.5880 + 00
0.1960 + 01	0.4315 + 01	0.2607 + 01	0.1655 + 01	0.7395 + 00	0.5438 + 01	0.5844 + 00
0.1980 + 01	0.4407 + 01	0.2637 + 01	0.1671 + 01	0.7302 + 00	0.5539 + 01	0.5808 + 00
0.2000 + 01	0.4500 + 01	0.2667 + 01	0.1687 + 01	0.7209 + 00	0.5640 + 01	0.5774 + 00
0.2050 + 01	0.4736 + 01	0.2740 + 01	0.1729 + 01	0.6975 + 00	0.5900 + 01	0.5691 + 00
0.2100 + 01	0.4978 + 01	0.2812 + 01	0.1770 + 01	0.6742 + 00	0.6165 + 01	0.5613 + 00
0.2150 + 01	0.5226 + 01	0.2882 + 01	0.1813 + 01	0.6511 + 00	0.6438 + 01	0.5540 + 00
0.2200 + 01	0.5480 + 01	0.2951 + 01	0.1857 + 01	0.6281 + 00	0.6716 + 01	0.5471 + 00
0.2250 + 01	0.5740 + 01	0.3019 + 01	0.1901 + 01	0.6055 + 00	0.7002 + 01	0.5406 + 00
0.2300 + 01	0.6005 + 01	0.3085 + 01	0.1947 + 01	0.5833 + 00	0.7294 + 01	0.5344 + 00
0.2350 + 01	0.6276 + 01	0.3149 + 01	0.1993 + 01	0.5615 + 00	0.7592 + 01	0.5286 + 00
0.2400 + 01	0.6553 + 01	0.3212 + 01	0.2040 + 01	0.5401 + 00	0.7897 + 01	0.5231 + 00
0.2450 + 01	0.6836 + 01	0.3273 + 01	0.2088 + 01	0.5193 + 00	0.8208 + 01	0.5179 + 00
0.2500 + 01	0.7125 + 01	0.3333 + 01	0.2137 + 01	0.4990 + 00	0.8526 + 01	0.5130 + 00
0.2550 + 01	0.7420 + 01	0.3392 + 01	0.2187 + 01	0.4793 + 00	0.8850 + 01	0.5083 + 00
0.2600 + 01	0.7720 + 01	0.3449 + 01	0.2238 + 01	0.4601 + 00	0.9181 + 01	0.5039 + 00
0.2650 + 01	0.8026 + 01	0.3505 + 01	0.2290 + 01	0.4416 + 00	0.9519 + 01	0.4996 + 00
0.2700 + 01	0.8338 + 01	0.3559 + 01	0.2343 + 01	0.4236 + 00	0.9862 + 01	0.4956 + 00
0.2750 + 01	0.8656 + 01	0.3612 + 01	0.2397 + 01	0.4062 + 00	0.1021 + 02	0.4918 + 00
0.2800 + 01	0.8980 + 01	0.3664 + 01	0.2451 + 01	0.3895 + 00	0.1057 + 02	0.4882 + 00
0.2850 + 01	0.9310 + 01	0.3714 + 01	0.2507 + 01	0.3733 + 00	0.1093 + 02	0.4847 + 00
0.2900 + 01	0.9645 + 01	0.3763 + 01	0.2563 + 01	0.3577 + 00	0.1130 + 02	0.4814 + 00
0.2950 + 01	0.9986 + 01	0.3811 + 01	0.2621 + 01	0.3428 + 00	0.1168 + 02	0.4782 + 00
0.3000 + 01	0.1033 + 02	0.3857 + 01	0.2679 + 01	0.3283 + 00	0.1206 + 02	0.4752 + 00
0.3050 + 01	0.1069 + 02	0.3902 + 01	0.2738 + 01	0.3145 + 00	0.1245 + 02	0.4723 + 00
0.3100 + 01	0.1104 + 02	0.3947 + 01	0.2799 + 01	0.3012 + 00	0.1285 + 02	0.4695 + 00
0.3150 + 01	0.1141 + 02	0.3990 + 01	0.2860 + 01	0.2885 + 00	0.1325 + 02	0.4669 + 00
0.3200 + 01	0.1178 + 02	0.4031 + 01	0.2922 + 01	0.2762 + 00	0.1366 + 02	0.4643 + 00
0.3250 + 01	0.1216 + 02	0.4072 + 01	0.2985 + 01	0.2645 + 00	0.1407 + 02	0.4619 + 00
0.3300 + 01	0.1254 + 02	0.4112 + 01	0.3049 + 01	0.2533 + 00	0.1449 + 02	0.4596 + 00
0.3350 + 01	0.1293 + 02	0.4151 + 01	0.3114 + 01	0.2425 + 00	0.1492 + 02	0.4573 + 00
0.3400 + 01	0.1332 + 02	0.4188 + 01	0.3180 + 01	0.2322 + 00	0.1535 + 02	0.4552 + 00
0.3450 + 01	0.1372 + 02	0.4225 + 01	0.3247 + 01	0.2224 + 00	0.1579 + 02	0.4531 + 00

M	$\dfrac{p_2}{p_1}$	$\dfrac{\rho_2}{\rho_1}$	$\dfrac{T_2}{T_1}$	$\dfrac{p_{0_2}}{p_{0_1}}$	$\dfrac{p_{0_2}}{p_1}$	M_2
0.3500 + 01	0.1412 + 02	0.4261 + 01	0.3315 + 01	0.2129 + 00	0.1624 + 02	0.4512 + 00
0.3550 + 01	0.1454 + 02	0.4296 + 01	0.3384 + 01	0.2039 + 00	0.1670 + 02	0.4492 + 00
0.3600 + 01	0.1495 + 02	0.4330 + 01	0.3454 + 01	0.1953 + 00	0.1716 + 02	0.4474 + 00
0.3650 + 01	0.1538 + 02	0.4363 + 01	0.3525 + 01	0.1871 + 00	0.1762 + 02	0.4456 + 00
0.3700 + 01	0.1580 + 02	0.4395 + 01	0.3596 + 01	0.1792 + 00	0.1810 + 02	0.4439 + 00
0.3750 + 01	0.1624 + 02	0.4426 + 01	0.3669 + 01	0.1717 + 00	0.1857 + 02	0.4423 + 00
0.3800 + 01	0.1668 + 02	0.4457 + 01	0.3743 + 01	0.1645 + 00	0.1906 + 02	0.4407 + 00
0.3850 + 01	0.1713 + 02	0.4487 + 01	0.3817 + 01	0.1576 + 00	0.1955 + 02	0.4392 + 00
0.3900 + 01	0.1758 + 02	0.4516 + 01	0.3893 + 01	0.1510 + 00	0.2005 + 02	0.4377 + 00
0.3950 + 01	0.1804 + 02	0.4544 + 01	0.3969 + 01	0.1448 + 00	0.2056 + 02	0.4363 + 00
0.4000 + 01	0.1850 + 02	0.4571 + 01	0.4047 + 01	0.1388 + 00	0.2107 + 02	0.4350 + 00
0.4050 + 01	0.1897 + 02	0.4598 + 01	0.4125 + 01	0.1330 + 00	0.2159 + 02	0.4336 + 00
0.4100 + 01	0.1944 + 02	0.4624 + 01	0.4205 + 01	0.1276 + 00	0.2211 + 02	0.4324 + 00
0.4150 + 01	0.1993 + 02	0.4650 + 01	0.4285 + 01	0.1223 + 00	0.2264 + 02	0.4311 + 00
0.4200 + 01	0.2041 + 02	0.4675 + 01	0.4367 + 01	0.1173 + 00	0.2318 + 02	0.4299 + 00
0.4250 + 01	0.2091 + 02	0.4699 + 01	0.4449 + 01	0.1126 + 00	0.2372 + 02	0.4288 + 00
0.4300 + 01	0.2140 + 02	0.4723 + 01	0.4532 + 01	0.1080 + 00	0.2427 + 02	0.4277 + 00
0.4350 + 01	0.2191 + 02	0.4746 + 01	0.4616 + 01	0.1036 + 00	0.2483 + 02	0.4266 + 00
0.4400 + 01	0.2242 + 02	0.4768 + 01	0.4702 + 01	0.9948 − 01	0.2539 + 02	0.4255 + 00
0.4450 + 01	0.2294 + 02	0.4790 + 01	0.4788 + 01	0.9550 − 01	0.2596 + 02	0.4245 + 00
0.4500 + 01	0.2346 + 02	0.4812 + 01	0.4875 + 01	0.9170 − 01	0.2654 + 02	0.4236 + 00
0.4550 + 01	0.2399 + 02	0.4833 + 01	0.4963 + 01	0.8806 − 01	0.2712 + 02	0.4226 + 00
0.4600 + 01	0.2452 + 02	0.4853 + 01	0.5052 + 01	0.8459 − 01	0.2771 + 02	0.4217 + 00
0.4650 + 01	0.2506 + 02	0.4873 + 01	0.5142 + 01	0.8126 − 01	0.2831 + 02	0.4208 + 00
0.4700 + 01	0.2560 + 02	0.4893 + 01	0.5233 + 01	0.7809 − 01	0.2891 + 02	0.4199 + 00
0.4750 + 01	0.2616 + 02	0.4912 + 01	0.5325 + 01	0.7505 − 01	0.2952 + 02	0.4191 + 00
0.4800 + 01	0.2671 + 02	0.4930 + 01	0.5418 + 01	0.7214 − 01	0.3013 + 02	0.4183 + 00
0.4850 + 01	0.2728 + 02	0.4948 + 01	0.5512 + 01	0.6936 − 01	0.3075 + 02	0.4175 + 00
0.4900 + 01	0.2784 + 02	0.4966 + 01	0.5607 + 01	0.6670 − 01	0.3138 + 02	0.4167 + 00
0.4950 + 01	0.2842 + 02	0.4983 + 01	0.5703 + 01	0.6415 − 01	0.3201 + 02	0.4160 + 00
0.5000 + 01	0.2900 + 02	0.5000 + 01	0.5800 + 01	0.6172 − 01	0.3265 + 02	0.4152 + 00
0.5100 + 01	0.3018 + 02	0.5033 + 01	0.5997 + 01	0.5715 − 01	0.3395 + 02	0.4138 + 00
0.5200 + 01	0.3138 + 02	0.5064 + 01	0.6197 + 01	0.5297 − 01	0.3528 + 02	0.4125 + 00
0.5300 + 01	0.3260 + 02	0.5093 + 01	0.6401 + 01	0.4913 − 01	0.3663 + 02	0.4113 + 00
0.5400 + 01	0.3385 + 02	0.5122 + 01	0.6610 + 01	0.4560 − 01	0.3801 + 02	0.4101 + 00
0.5500 + 01	0.3512 + 02	0.5149 + 01	0.6822 + 01	0.4236 − 01	0.3941 + 02	0.4090 + 00
0.5600 + 01	0.3642 + 02	0.5175 + 01	0.7038 + 01	0.3938 − 01	0.4084 + 02	0.4079 + 00
0.5700 + 01	0.3774 + 02	0.5200 + 01	0.7258 + 01	0.3664 − 01	0.4230 + 02	0.4069 + 00
0.5800 + 01	0.3908 + 02	0.5224 + 01	0.7481 + 01	0.3412 − 01	0.4378 + 02	0.4059 + 00
0.5900 + 01	0.4044 + 02	0.5246 + 01	0.7709 + 01	0.3180 − 01	0.4528 + 02	0.4050 + 00
0.6000 + 01	0.4183 + 02	0.5268 + 01	0.7941 + 01	0.2965 − 01	0.4682 + 02	0.4042 + 00
0.6100 + 01	0.4324 + 02	0.5289 + 01	0.8176 + 01	0.2767 − 01	0.4837 + 02	0.4033 + 00
0.6200 + 01	0.4468 + 02	0.5309 + 01	0.8415 + 01	0.2584 − 01	0.4996 + 02	0.4025 + 00
0.6300 + 01	0.4614 + 02	0.5329 + 01	0.8658 + 01	0.2416 − 01	0.5157 + 02	0.4018 + 00
0.6400 + 01	0.4762 + 02	0.5347 + 01	0.8905 + 01	0.2259 − 01	0.5320 + 02	0.4011 + 00
0.6500 + 01	0.4912 + 02	0.5365 + 01	0.9156 + 01	0.2115 − 01	0.5486 + 02	0.4004 + 00
0.6600 + 01	0.5065 + 02	0.5382 + 01	0.9411 + 01	0.1981 − 01	0.5655 + 02	0.3997 + 00
0.6700 + 01	0.5220 + 02	0.5399 + 01	0.9670 + 01	0.1857 − 01	0.5826 + 02	0.3991 + 00
0.6800 + 01	0.5378 + 02	0.5415 + 01	0.9933 + 01	0.1741 − 01	0.6000 + 02	0.3985 + 00
0.6900 + 01	0.5538 + 02	0.5430 + 01	0.1020 + 02	0.1635 − 01	0.6176 + 02	0.3979 + 00

M	$\dfrac{p_2}{p_1}$	$\dfrac{\rho_2}{\rho_1}$	$\dfrac{T_2}{T_1}$	$\dfrac{p_{0_2}}{p_{0_1}}$	$\dfrac{p_{0_2}}{p_1}$	M_2
0.7000 + 01	0.5700 + 02	0.5444 + 01	0.1047 + 02	0.1535 − 01	0.6355 + 02	0.3974 + 00
0.7100 + 01	0.5864 + 02	0.5459 + 01	0.1074 + 02	0.1443 − 01	0.6537 + 02	0.3968 + 00
0.7200 + 01	0.6031 + 02	0.5472 + 01	0.1102 + 02	0.1357 − 01	0.6721 + 02	0.3963 + 00
0.7300 + 01	0.6200 + 02	0.5485 + 01	0.1130 + 02	0.1277 − 01	0.6908 + 02	0.3958 + 00
0.7400 + 01	0.6372 + 02	0.5498 + 01	0.1159 + 02	0.1202 − 01	0.7097 + 02	0.3954 + 00
0.7500 + 01	0.6546 + 02	0.5510 + 01	0.1188 + 02	0.1133 − 01	0.7289 + 02	0.3949 + 00
0.7600 + 01	0.6722 + 02	0.5522 + 01	0.1217 + 02	0.1068 − 01	0.7483 + 02	0.3945 + 00
0.7700 + 01	0.6900 + 02	0.5533 + 01	0.1247 + 02	0.1008 − 01	0.7680 + 02	0.3941 + 00
0.7800 + 01	0.7081 + 02	0.5544 + 01	0.1277 + 02	0.9510 − 02	0.7880 + 02	0.3937 + 00
0.7900 + 01	0.7264 + 02	0.5555 + 01	0.1308 + 02	0.8982 − 02	0.8082 + 02	0.3933 + 00
0.8000 + 01	0.7450 + 02	0.5565 + 01	0.1339 + 02	0.8488 − 02	0.8287 + 02	0.3929 + 00
0.9000 + 01	0.9433 + 02	0.5651 + 01	0.1669 + 02	0.4964 − 02	0.1048 + 03	0.3898 + 00
0.1000 + 02	0.1165 + 03	0.5714 + 01	0.2039 + 02	0.3045 − 02	0.1292 + 03	0.3876 + 00
0.1100 + 02	0.1410 + 03	0.5762 + 01	0.2447 + 02	0.1945 − 02	0.1563 + 03	0.3859 + 00
0.1200 + 02	0.1678 + 03	0.5799 + 01	0.2894 + 02	0.1287 − 02	0.1859 + 03	0.3847 + 00
0.1300 + 02	0.1970 + 03	0.5828 + 01	0.3380 + 02	0.8771 − 03	0.2181 + 03	0.3837 + 00
0.1400 + 02	0.2285 + 03	0.5851 + 01	0.3905 + 02	0.6138 − 03	0.2528 + 03	0.3829 + 00
0.1500 + 02	0.2623 + 03	0.5870 + 01	0.4469 + 02	0.4395 − 03	0.2902 + 03	0.3823 + 00
0.1600 + 02	0.2985 + 03	0.5885 + 01	0.5072 + 02	0.3212 − 03	0.3301 + 03	0.3817 + 00
0.1700 + 02	0.3370 + 03	0.5898 + 01	0.5714 + 02	0.2390 − 03	0.3726 + 03	0.3813 + 00
0.1800 + 02	0.3778 + 03	0.5909 + 01	0.6394 + 02	0.1807 − 03	0.4176 + 03	0.3810 + 00
0.1900 + 02	0.4210 + 03	0.5918 + 01	0.7114 + 02	0.1386 − 03	0.4653 + 03	0.3806 + 00
0.2000 + 02	0.4665 + 03	0.5926 + 01	0.7872 + 02	0.1078 − 03	0.5155 + 03	0.3804 + 00
0.2200 + 02	0.5645 + 03	0.5939 + 01	0.9506 + 02	0.6741 − 04	0.6236 + 03	0.3800 + 00
0.2400 + 02	0.6718 + 03	0.5948 + 01	0.1129 + 03	0.4388 − 04	0.7421 + 03	0.3796 + 00
0.2600 + 02	0.7885 + 03	0.5956 + 01	0.1324 + 03	0.2953 − 04	0.8709 + 03	0.3794 + 00
0.2800 + 02	0.9145 + 03	0.5962 + 01	0.1534 + 03	0.2046 − 04	0.1010 + 04	0.3792 + 00
0.3000 + 02	0.1050 + 04	0.5967 + 01	0.1759 + 03	0.1453 − 04	0.1159 + 04	0.3790 + 00
0.3200 + 02	0.1194 + 04	0.5971 + 01	0.2001 + 03	0.1055 − 04	0.1319 + 04	0.3789 + 00
0.3400 + 02	0.1348 + 04	0.5974 + 01	0.2257 + 03	0.7804 − 05	0.1489 + 04	0.3788 + 00
0.3600 + 02	0.1512 + 04	0.5977 + 01	0.2529 + 03	0.5874 − 05	0.1669 + 04	0.3787 + 00
0.3800 + 02	0.1684 + 04	0.5979 + 01	0.2817 + 03	0.4488 − 05	0.1860 + 04	0.3786 + 00
0.4000 + 02	0.1866 + 04	0.5981 + 01	0.3121 + 03	0.3477 − 05	0.2061 + 04	0.3786 + 00
0.4200 + 02	0.2058 + 04	0.5983 + 01	0.3439 + 03	0.2727 − 05	0.2272 + 04	0.3785 + 00
0.4400 + 02	0.2258 + 04	0.5985 + 01	0.3774 + 03	0.2163 − 05	0.2493 + 04	0.3785 + 00
0.4600 + 02	0.2468 + 04	0.5986 + 01	0.4124 + 03	0.1733 − 05	0.2725 + 04	0.3784 + 00
0.4800 + 02	0.2688 + 04	0.5987 + 01	0.4489 + 03	0.1402 − 05	0.2967 + 04	0.3784 + 00
0.5000 + 02	0.2916 + 04	0.5988 + 01	0.4871 + 03	0.1144 − 05	0.3219 + 04	0.3784 + 00

Prandtl-Meyer Function and Mach Angle

M	ν	μ	M	ν	μ
0.1000 + 01	0.0000	0.9000 + 02	0.1600 + 01	0.1486 + 02	0.3868 + 02
0.1020 + 01	0.1257 + 00	0.7864 + 02	0.1620 + 01	0.1545 + 02	0.3812 + 02
0.1040 + 01	0.3510 + 00	0.7406 + 02	0.1640 + 01	0.1604 + 02	0.3757 + 02
0.1060 + 01	0.6367 + 00	0.7063 + 02	0.1660 + 01	0.1663 + 02	0.3704 + 02
0.1080 + 01	0.9680 + 00	0.6781 + 02	0.1680 + 01	0.1722 + 02	0.3653 + 02
0.1100 + 01	0.1336 + 01	0.6538 + 02	0.1700 + 01	0.1781 + 02	0.3603 + 02
0.1120 + 01	0.1735 + 01	0.6323 + 02	0.1720 + 01	0.1840 + 02	0.3555 + 02
0.1140 + 01	0.2160 + 01	0.6131 + 02	0.1740 + 01	0.1898 + 02	0.3508 + 02
0.1160 + 01	0.2607 + 01	0.5955 + 02	0.1760 + 01	0.1956 + 02	0.3462 + 02
0.1180 + 01	0.3074 + 01	0.5794 + 02	0.1780 + 01	0.2015 + 02	0.3418 + 02
0.1200 + 01	0.3558 + 01	0.5644 + 02	0.1800 + 01	0.2073 + 02	0.3375 + 02
0.1220 + 01	0.4057 + 01	0.5505 + 02	0.1820 + 01	0.2130 + 02	0.3333 + 02
0.1240 + 01	0.4569 + 01	0.5375 + 02	0.1840 + 01	0.2188 + 02	0.3292 + 02
0.1260 + 01	0.5093 + 01	0.5253 + 02	0.1860 + 01	0.2245 + 02	0.3252 + 02
0.1280 + 01	0.5627 + 01	0.5138 + 02	0.1880 + 01	0.2302 + 02	0.3213 + 02
0.1300 + 01	0.6170 + 01	0.5028 + 02	0.1900 + 01	0.2359 + 02	0.3176 + 02
0.1320 + 01	0.6721 + 01	0.4925 + 02	0.1920 + 01	0.2415 + 02	0.3139 + 02
0.1340 + 01	0.7279 + 01	0.4827 + 02	0.1940 + 01	0.2471 + 02	0.3103 + 02
0.1360 + 01	0.7844 + 01	0.4733 + 02	0.1960 + 01	0.2527 + 02	0.3068 + 02
0.1380 + 01	0.8413 + 01	0.4644 + 02	0.1980 + 01	0.2583 + 02	0.3033 + 02
0.1400 + 01	0.8987 + 01	0.4558 + 02	0.2000 + 01	0.2638 + 02	0.3000 + 02
0.1420 + 01	0.9565 + 01	0.4477 + 02	0.2050 + 01	0.2775 + 02	0.2920 + 02
0.1440 + 01	0.1015 + 02	0.4398 + 02	0.2100 + 01	0.2910 + 02	0.2844 + 02
0.1460 + 01	0.1073 + 02	0.4323 + 02	0.2150 + 01	0.3043 + 02	0.2772 + 02
0.1480 + 01	0.1132 + 02	0.4251 + 02	0.2200 + 01	0.3173 + 02	0.2704 + 02
0.1500 + 01	0.1191 + 02	0.4181 + 02	0.2250 + 01	0.3302 + 02	0.2639 + 02
0.1520 + 01	0.1249 + 02	0.4114 + 02	0.2300 + 01	0.3428 + 02	0.2577 + 02
0.1540 + 01	0.1309 + 02	0.4049 + 02	0.2350 + 01	0.3553 + 02	0.2518 + 02
0.1560 + 01	0.1368 + 02	0.3987 + 02	0.2400 + 01	0.3675 + 02	0.2462 + 02
0.1580 + 01	0.1427 + 02	0.3927 + 02	0.2450 + 01	0.3795 + 02	0.2409 + 02

M	ν	μ	M	ν	μ
0.2500 + 01	0.3912 + 02	0.2358 + 02	0.5000 + 01	0.7692 + 02	0.1154 + 02
0.2550 + 01	0.4028 + 02	0.2309 + 02	0.5100 + 01	0.7784 + 02	0.1131 + 02
0.2600 + 01	0.4141 + 02	0.2262 + 02	0.5200 + 01	0.7873 + 02	0.1109 + 02
0.2650 + 01	0.4253 + 02	0.2217 + 02	0.5300 + 01	0.7960 + 02	0.1088 + 02
0.2700 + 01	0.4362 + 02	0.2174 + 02	0.5400 + 01	0.8043 + 02	0.1067 + 02
0.2750 + 01	0.4469 + 02	0.2132 + 02	0.5500 + 01	0.8124 + 02	0.1048 + 02
0.2800 + 01	0.4575 + 02	0.2092 + 02	0.5600 + 01	0.8203 + 02	0.1029 + 02
0.2850 + 01	0.4678 + 02	0.2054 + 02	0.5700 + 01	0.8280 + 02	0.1010 + 02
0.2900 + 01	0.4779 + 02	0.2017 + 02	0.5800 + 01	0.8354 + 02	0.9928 + 01
0.2950 + 01	0.4878 + 02	0.1981 + 02	0.5900 + 01	0.8426 + 02	0.9758 + 01
0.3000 + 01	0.4976 + 02	0.1947 + 02	0.6000 + 01	0.8496 + 02	0.9594 + 01
0.3050 + 01	0.5071 + 02	0.1914 + 02	0.6100 + 01	0.8563 + 02	0.9435 + 01
0.3100 + 01	0.5165 + 02	0.1882 + 02	0.6200 + 01	0.8629 + 02	0.9282 + 01
0.3150 + 01	0.5257 + 02	0.1851 + 02	0.6300 + 01	0.8694 + 02	0.9133 + 01
0.3200 + 01	0.5347 + 02	0.1821 + 02	0.6400 + 01	0.8756 + 02	0.8989 + 01
0.3250 + 01	0.5435 + 02	0.1792 + 02	0.6500 + 01	0.8817 + 02	0.8850 + 01
0.3300 + 01	0.5522 + 02	0.1764 + 02	0.6600 + 01	0.8876 + 02	0.8715 + 01
0.3350 + 01	0.5607 + 02	0.1737 + 02	0.6700 + 01	0.8933 + 02	0.8584 + 01
0.3400 + 01	0.5691 + 02	0.1710 + 02	0.6800 + 01	0.8989 + 02	0.8457 + 01
0.3450 + 01	0.5773 + 02	0.1685 + 02	0.6900 + 01	0.9044 + 02	0.8333 + 01
0.3500 + 01	0.5853 + 02	0.1660 + 02	0.7000 + 01	0.9097 + 02	0.8213 + 01
0.3550 + 01	0.5932 + 02	0.1636 + 02	0.7100 + 01	0.9149 + 02	0.8097 + 01
0.3600 + 01	0.6009 + 02	0.1613 + 02	0.7200 + 01	0.9200 + 02	0.7984 + 01
0.3650 + 01	0.6085 + 02	0.1590 + 02	0.7300 + 01	0.9249 + 02	0.7873 + 01
0.3700 + 01	0.6160 + 02	0.1568 + 02	0.7400 + 01	0.9297 + 02	0.7766 + 01
0.3750 + 01	0.6233 + 02	0.1547 + 02	0.7500 + 01	0.9344 + 02	0.7662 + 01
0.3800 + 01	0.6304 + 02	0.1526 + 02	0.7600 + 01	0.9390 + 02	0.7561 + 01
0.3850 + 01	0.6375 + 02	0.1505 + 02	0.7700 + 01	0.9434 + 02	0.7462 + 01
0.3900 + 01	0.6444 + 02	0.1486 + 02	0.7800 + 01	0.9478 + 02	0.7366 + 01
0.3950 + 01	0.6512 + 02	0.1466 + 02	0.7900 + 01	0.9521 + 02	0.7272 + 01
0.4000 + 01	0.6578 + 02	0.1448 + 02	0.8000 + 01	0.9562 + 02	0.7181 + 01
0.4050 + 01	0.6644 + 02	0.1429 + 02	0.9000 + 01	0.9932 + 02	0.6379 + 01
0.4100 + 01	0.6708 + 02	0.1412 + 02	0.1000 + 02	0.1023 + 03	0.5739 + 01
0.4150 + 01	0.6771 + 02	0.1394 + 02	0.1100 + 02	0.1048 + 03	0.5216 + 01
0.4200 + 01	0.6833 + 02	0.1377 + 02	0.1200 + 02	0.1069 + 03	0.4780 + 01
0.4250 + 01	0.6894 + 02	0.1361 + 02	0.1300 + 02	0.1087 + 03	0.4412 + 01
0.4300 + 01	0.6954 + 02	0.1345 + 02	0.1400 + 02	0.1102 + 03	0.4096 + 01
0.4350 + 01	0.7013 + 02	0.1329 + 02	0.1500 + 02	0.1115 + 03	0.3823 + 01
0.4400 + 01	0.7071 + 02	0.1314 + 02	0.1600 + 02	0.1127 + 03	0.3583 + 01
0.4450 + 01	0.7127 + 02	0.1299 + 02	0.1700 + 02	0.1137 + 03	0.3372 + 01
0.4500 + 01	0.7183 + 02	0.1284 + 02	0.1800 + 02	0.1146 + 03	0.3185 + 01
0.4550 + 01	0.7238 + 02	0.1270 + 02	0.1900 + 02	0.1155 + 03	0.3017 + 01
0.4600 + 01	0.7292 + 02	0.1256 + 02	0.2000 + 02	0.1162 + 03	0.2866 + 01
0.4650 + 01	0.7345 + 02	0.1242 + 02	0.2200 + 02	0.1175 + 03	0.2605 + 01
0.4700 + 01	0.7397 + 02	0.1228 + 02	0.2400 + 02	0.1186 + 03	0.2388 + 01
0.4750 + 01	0.7448 + 02	0.1215 + 02	0.2600 + 02	0.1195 + 03	0.2204 + 01
0.4800 + 01	0.7499 + 02	0.1202 + 02	0.2800 + 02	0.1202 + 03	0.2047 + 01
0.4850 + 01	0.7548 + 02	0.1190 + 02	0.3000 + 02	0.1209 + 03	0.1910 + 01
0.4900 + 01	0.7597 + 02	0.1178 + 02	0.3200 + 02	0.1215 + 03	0.1791 + 01
0.4950 + 01	0.7645 + 02	0.1166 + 02	0.3400 + 02	0.1220 + 03	0.1685 + 01

M	ν	μ	M	ν	μ
0.3600 + 02	0.1225 + 03	0.1592 + 01	0.4400 + 02	0.1239 + 03	0.1302 + 01
0.3800 + 02	0.1229 + 03	0.1508 + 01	0.4600 + 02	0.1242 + 03	0.1246 + 01
0.4000 + 02	0.1233 + 03	0.1433 + 01	0.4800 + 02	0.1245 + 03	0.1194 + 01
0.4200 + 02	0.1236 + 03	0.1364 + 01	0.5000 + 02	0.1247 + 03	0.1146 + 01

APPENDIX D

Standard Atmosphere, SI Units

D.1 NOTE ABOUT THE STANDARD ATMOSPHERE TABLES IN APPENDICES D AND E

The following Standard Atmosphere Tables are compiled from mean experimental data for the temperature variation with altitude, combined with the laws of physics to compute the corresponding variation of pressure and density. The laws of physics used are the hydrostatic equation (Equation 1.52) and the equation of state (Equation 7.1). The construction of the Standard Atmosphere Tables is discussed in detail in Chapter 3 of Reference 2, which you should read to learn the whole story. In Appendices D and E, the temperature, pressure, and density are tabulated versus altitude. Two columns for altitude are given, the first for the geometric altitude, h_G, and the second for the geopotential altitude, h. The geometric altitude is the actual height above standard sea level, and the geopotential altitude is a related altitude based on the assumption of a constant value of the acceleration of gravity used for the calculations. See Reference 2 for an explanation of the difference. In this book, whenever a reference is made to a certain value of the standard altitude, it means the value of h_G, the first column in the tables.

Appendix D gives the Standard Atmosphere in SI units, and Appendix E gives the Standard Atmosphere in English Engineering units. There exist tables of the standard atmosphere compiled by various organizations over the years. The values tabulated in Appendices D and E are taken from the 1959 ARDC model atmosphere compiled by the U.S. Air Force.

Altitude		Temperature T, K	Pressure p, N/m^2	Density ρ, kg/m^3
h_G, m	h, m			
−5,000	−5,004	320.69	1.7761 + 5	1.9296 + 0
−4,900	−4,904	320.03	1.7587	1.9145
−4,800	−4,804	319.38	1.7400	1.8980
−4,700	−4,703	318.73	1.7215	1.8816
−4,600	−4,603	318.08	1.7031	1.8653
−4,500	−4,503	317.43	1.6848	1.8491
−4,400	−4,403	316.78	1.6667	1.8330
−4,300	−4,303	316.13	1.6488	1.8171
−4,200	−4,203	315.48	1.6311	1.8012
−4,100	−4,103	314.83	1.6134	1.7854
−4,000	−4,003	314.18	1.5960 + 5	1.7698 + 0
−3,900	−3,902	313.53	1.5787	1.7542
−3,800	−3,802	312.87	1.5615	1.7388
−3,700	−3,702	212.22	1.5445	1.7234
−3,600	−3,602	311.57	1.5277	1.7082
−3,500	−3,502	310.92	1.5110	1.6931
−3,400	−3,402	310.27	1.4945	1.6780
−3,300	−3,302	309.62	1.4781	1.6631
−3,200	−3,202	308.97	1.4618	1.6483
−3,100	−3,102	308.32	1.4457	1.6336
−3,000	−3,001	307.67	1.4297 + 5	1.6189 + 0
−2,900	−2,901	307.02	1.4139	1.6044
−2,800	−2,801	306.37	1.3982	1.5900
−2,700	−2,701	305.72	1.3827	1.5757
−2,600	−2,601	305.07	1.3673	1.5615
−2,500	−2,501	304.42	1.3521	1.5473
−2,400	−2,401	303.77	1.3369	1.5333
−2,300	−2,301	303.12	1.3220	1.5194
−2,200	−2,201	302.46	1.3071	1.5056
−2,100	−2,101	301.81	1.2924	1.4918
−2,000	−2,001	301.16	1.2778 + 5	1.4782 + 0
−1,900	−1,901	300.51	1.2634	1.4646
−1,800	−1,801	299.86	1.2491	1.4512
−1,700	−1,701	299.21	1.2349	1.4379
−1,600	−1,600	298.56	1.2209	1.4246
−1,500	−1,500	297.91	1.2070	1.4114
−1,400	−1,400	297.26	1.1932	1.3984
−1,300	−1,300	296.61	1.1795	1.3854
−1,200	−1,200	295.96	1.1660	1.3725
−1,100	−1,100	295.31	1.1526	1.3597
−1,000	−1,000	294.66	1.1393 + 5	1.3470 + 0
−900	−900	294.01	1.1262	1.3344
−800	−800	293.36	1.1131	1.3219
−700	−700	292.71	1.1002	1.3095
−600	−600	292.06	1.0874	1.2972
−500	−500	291.41	1.0748	1.2849
−400	−400	290.76	1.0622	1.2728
−300	−300	290.11	1.0498	1.2607
−200	−200	289.46	1.0375	1.2487
−100	−100	288.81	1.0253	1.2368

Altitude		Temperature T, K	Pressure p, N/m^2	Density ρ, kg/m^3
h_G, m	h, m			
0	0	288.16	1.01325 + 5	1.2250 + 0
100	100	287.51	1.0013	1.2133
200	200	286.86	9.8945 + 4	1.2071
300	300	286.21	9.7773	1.1901
400	400	285.56	9.6611	1.1787
500	500	284.91	9.5461	1.1673
600	600	284.26	9.4322	1.1560
700	700	283.61	9.3194	1.1448
800	800	282.96	9.2077	1.1337
900	900	282.31	9.0971	1.1226
1,000	1,000	281.66	8.9876 + 4	1.1117 + 0
1,100	1,100	281.01	8.8792	1.1008
1,200	1,200	280.36	8.7718	1.0900
1,300	1,300	279.71	8.6655	1.0793
1,400	1,400	279.06	8.5602	1.0687
1,500	1,500	278.41	8.4560	1.0581
1,600	1,600	277.76	8.3527	1.0476
1,700	1,700	277.11	8.2506	1.0373
1,800	1,799	276.46	8.1494	1.0269
1,900	1,899	275.81	8.0493	1.0167
2,000	1,999	275.16	7.9501 + 4	1.0066 + 0
2,100	2,099	274.51	7.8520	9.9649 − 1
2,200	2,199	273.86	7.7548	9.8649
2,300	2,299	273.22	7.6586	9.7657
2,400	2,399	272.57	7.5634	9.6673
2,500	2,499	271.92	7.4692	9.5696
2,600	2,599	271.27	7.3759	9.4727
2,700	2,699	270.62	7.2835	9.3765
2,800	2,799	269.97	7.1921	9.2811
2,900	2,899	269.32	7.1016	9.1865
3,000	2,999	268.67	7.0121 + 4	9.0926 − 1
3,100	3,098	268.02	6.9235	8.9994
3,200	3,198	267.37	6.8357	8.9070
3,300	3,298	266.72	6.7489	8.8153
3,400	3,398	266.07	6.6630	8.7243
3,500	3,498	265.42	6.5780	8.6341
3,600	3,598	264.77	6.4939	8.5445
3,700	3,698	264.12	6.4106	8.4557
3,800	3,798	263.47	6.3282	8.3676
3,900	3,898	262.83	6.2467	8.2802
4,000	3,997	262.18	6.1660 + 4	8.1935 − 1
4,100	4,097	261.53	6.0862	8.1075
4,200	4,197	260.88	6.0072	8.0222
4,300	4,297	260.23	5.9290	7.9376
4,400	4,397	259.58	5.8517	7.8536
4,500	4,497	258.93	5.7752	7.7704
4,600	4,597	258.28	5.6995	7.6878
4,700	4,697	257.63	5.6247	7.6059
4,800	4,796	256.98	5.5506	7.5247
4,900	4,896	256.33	5.4773	7.4442

Altitude		Temperature T, K	Pressure p, N/m^2	Density ρ, kg/m^3
h_G, m	h, m			
5,000	4,996	255.69	5.4048 + 4	7.3643 − 1
5,100	5,096	255.04	5.3331	7.2851
5,200	5,196	254.39	5.2621	7.2065
5,400	5,395	253.09	5.1226	7.0513
5,500	5,495	252.44	5.0539	6.9747
5,600	5,595	251.79	4.9860	6.8987
5,700	5,695	251.14	4.9188	6.8234
5,800	5,795	250.49	4.8524	6.7486
5,900	5,895	249.85	4.7867	6.6746
6,000	5,994	249.20	4.7217 + 4	6.6011 − 1
6,100	6,094	248.55	4.6575	6.5283
6,200	6,194	247.90	4.5939	6.4561
6,300	6,294	247.25	4.5311	6.3845
6,400	6,394	246.60	4.4690	6.3135
6,500	6,493	245.95	4.4075	6.2431
6,600	6,593	245.30	4.3468	6.1733
6,700	6,693	244.66	4.2867	6.1041
6,800	6,793	244.01	4.2273	6.0356
6,900	6,893	243.36	4.1686	5.9676
7,000	6,992	242.71	4.1105 + 4	5.9002 − 1
7,100	7,092	242.06	4.0531	5.8334
7,200	7,192	241.41	3.9963	5.7671
7,300	7,292	240.76	3.9402	5.7015
7,400	7,391	240.12	3.8848	5.6364
7,500	7,491	239.47	3.8299	5.5719
7,600	7,591	238.82	3.7757	5.5080
7,700	7,691	238.17	3.7222	5.4446
7,800	7,790	237.52	3.6692	5.3818
7,900	7,890	236.87	3.6169	5.3195
8,000	7,990	236.23	3.5651 + 4	5.2578 − 1
8,100	8,090	235.58	3.5140	5.1967
8,200	8,189	234.93	3.4635	5.1361
8,300	8,289	234.28	3.4135	5.0760
8,400	8,389	233.63	3.3642	5.0165
8,500	8,489	232.98	3.3154	4.9575
8,600	8,588	232.34	3.2672	4.8991
8,700	8,688	231.69	3.2196	4.8412
8,800	8,788	231.04	3.1725	4.7838
8,900	8,888	230.39	3.1260	4.7269
9,000	8,987	229.74	3.0800 + 4	4.6706 − 1
9,100	9,087	229.09	3.0346	4.6148
9,200	9,187	228.45	2.9898	4.5595
9,300	9,286	227.80	2.9455	4.5047
9,400	9,386	227.15	2.9017	4.4504
9,500	9,486	226.50	2.8584	4.3966
9,600	9,586	225.85	2.8157	4.3433
9,700	9,685	225.21	2.7735	4.2905
9,800	9,785	224.56	2.7318	4.2382
9,900	9,885	223.91	2.6906	4.1864

Altitude		Temperature T, K	Pressure p, N/m^2	Density ρ, kg/m^3
h_G, m	h, m			
10,000	9,984	223.26	2.6500 + 4	4.1351 − 1
10,100	10,084	222.61	2.6098	4.0842
10,200	10,184	221.97	2.5701	4.0339
10,300	10,283	221.32	2.5309	3.9840
10,400	10,383	220.67	2.4922	3.9346
10,500	10,483	220.02	2.4540	3.8857
10,600	10,582	219.37	2.4163	3.8372
10,700	10,682	218.73	2.3790	3.7892
10,800	10,782	218.08	2.3422	3.7417
10,900	10,881	217.43	2.3059	3.6946
11,000	10,981	216.78	2.2700 + 4	3.6480 − 1
11,100	11,081	216.66	2.2346	3.5932
11,200	11,180	216.66	2.1997	3.5371
11,300	11,280	216.66	2.1654	3.4820
11,400	11,380	216.66	2.1317	3.4277
11,500	11,479	216.66	2.0985	3.3743
11,600	11,579	216.66	2.0657	3.3217
11,700	11,679	216.66	2.0335	3.2699
11,800	11,778	216.66	2.0018	3.2189
11,900	11,878	216.66	1.9706	3.1687
12,000	11,977	216.66	1.9399 + 4	3.1194 − 1
12,100	12,077	216.66	1.9097	3.0707
12,200	12,177	216.66	1.8799	3.0229
12,300	12,276	216.66	1.8506	2.9758
12,400	12,376	216.66	1.8218	2.9294
12,500	12,475	216.66	1.7934	2.8837
12,600	12,575	216.66	1.7654	2.8388
12,700	12,675	216.66	1.7379	2.7945
12,800	12,774	216.66	1.7108	2.7510
12,900	12,874	216.66	1.6842	2.7081
13,000	12,973	216.66	1.6579 + 4	2.6659 − 1
13,100	13,073	216.66	1.6321	2.6244
13,200	13,173	216.66	1.6067	2.5835
13,300	13,272	216.66	1.5816	2.5433
13,400	13,372	216.66	1.5570	2.5036
13,500	13,471	216.66	1.5327	2.4646
13,600	13,571	216.66	1.5089	2.4262
13,700	13,671	216.66	1.4854	2.3884
13,800	13,770	216.66	1.4622	2.3512
13,900	13,870	216.66	1.4394	2.3146
14,000	13,969	216.66	1.4170 + 4	2.2785 − 1
14,100	14,069	216.66	1.3950	2.2430
14,200	14,168	216.66	1.3732	2.2081
14,300	14,268	216.66	1.3518	2.1737
14,400	14,367	216.66	1.3308	2.1399
14,500	14,467	216.66	1.3101	2.1065
14,600	14,567	216.66	1.2896	2.0737
14,700	14,666	216.66	1.2696	2.0414
14,800	14,766	216.66	1.2498	2.0096
14,900	14,865	216.66	1.2303	1.9783

Altitude		Temperature T, K	Pressure p, N/m^2	Density ρ, kg/m^3
h_G, m	h, m			
15,000	14,965	216.66	1.2112 + 4	1.9475 − 1
15,100	15,064	216.66	1.1923	1.9172
15,200	15,164	216.66	1.1737	1.8874
15,300	15,263	216.66	1.1555	1.8580
15,400	15,363	216.66	1.1375	1.8290
15,500	15,462	216.66	1.1198	1.8006
15,600	15,562	216.66	1.1023	1.7725
15,700	15,661	216.66	1.0852	1.7449
15,800	15,761	216.66	1.0683	1.7178
15,900	15,860	216.66	1.0516	1.6910
16,000	15,960	216.66	1.0353 + 4	1.6647 − 1
16,100	16,059	216.66	1.0192	1.6388
16,200	16,159	216.66	1.0033	1.6133
16,300	16,258	216.66	9.8767 + 3	1.5882
16,400	16,358	216.66	9.7230	1.5634
16,500	16,457	216.66	9.5717	1.5391
16,600	16,557	216.66	9.4227	1.5151
16,700	16,656	216.66	9.2760	1.4916
16,800	16,756	216.66	9.1317	1.4683
16,900	16,855	216.66	8.9895	1.4455
17,000	16,955	216.66	8.8496 + 3	1.4230 − 1
17,100	17,054	216.66	8.7119	1.4009
17,200	17,154	216.66	8.5763	1.3791
17,300	17,253	216.66	8.4429	1.3576
17,400	17,353	216.66	8.3115	1.3365
17,500	17,452	216.66	8.1822	1.3157
17,600	17,551	216.66	8.0549	1.2952
17,700	17,651	216.66	7.9295	1.2751
17,800	17,750	216.66	7.8062	1.2552
17,900	17,850	216.66	7.6847	1.2357
18,000	17,949	216.66	7.5652 + 3	1.2165 − 1
18,100	18,049	216.66	7.4475	1.1975
18,200	18,148	216.66	7.3316	1.1789
18,300	18,247	216.66	7.2175	1.1606
18,400	18,347	216.66	7.1053	1.1425
18,500	18,446	216.66	6.9947	1.1247
18,600	18,546	216.66	6.8859	1.1072
18,700	18,645	216.66	6.7788	1.0900
18,800	18,745	216.66	6.6734	1.0731
18,900	18,844	216.66	6.5696	1.0564
19,000	18,943	216.66	6.4674 + 3	1.0399 − 1
19,100	19,043	216.66	6.3668	1.0238
19,200	19,142	216.66	6.2678	1.0079
19,300	19,242	216.66	6.1703	9.9218 − 2
19,400	19,341	216.66	6.0744	9.7675
19,500	19,440	216.66	5.9799	9.6156
19,600	19,540	216.66	5.8869	9.4661
19,700	19,639	216.66	5.7954	9.3189
19,800	19,739	216.66	5.7053	9.1740
19,900	19,838	216.66	5.6166	9.0313

Altitude		Temperature T, K	Pressure p, N/m^2	Density ρ, kg/m^3
h_G, m	h, m			
20,000	19,937	216.66	5.5293 + 3	8.8909 − 2
20,200	20,136	216.66	5.3587	8.6166
20,400	20,335	216.66	5.1933	8.3508
20,600	20,533	216.66	5.0331	8.0931
20,800	20,732	216.66	4.8779	7.8435
21,000	20,931	216.66	4.7274	7.6015
21,200	21,130	216.66	4.5816	7.3671
21,400	21,328	216.66	4.4403	7.1399
21,600	21,527	216.66	4.3034	6.9197
21,800	21,725	216.66	4.1706	6.7063
22,000	21,924	216.66	4.0420 + 3	6.4995 − 2
22,200	22,123	216.66	3.9174	6.2991
22,400	22,321	216.66	3.7966	6.1049
22,600	22,520	216.66	3.6796	5.9167
22,800	22,719	216.66	3.5661	5.7343
23,000	22,917	216.66	3.4562	5.5575
23,200	23,116	216.66	3.3497	5.3862
23,400	23,314	216.66	3.2464	5.2202
23,600	23,513	216.66	3.1464	5.0593
23,800	23,711	216.66	3.0494	4.9034
24,000	23,910	216.66	2.9554 + 3	4.7522 − 2
24,200	24,108	216.66	2.8644	4.6058
24,400	24,307	216.66	2.7761	4.4639
24,600	24,505	216.66	2.6906	4.3263
24,800	24,704	216.66	2.6077	4.1931
25,000	24,902	216.66	2.5273	4.0639
25,200	25,100	216.96	2.4495	3.9333
25,400	25,299	217.56	2.3742	3.8020
25,600	25,497	218.15	2.3015	3.6755
25,800	25,696	218.75	2.2312	3.5535
26,000	25,894	219.34	2.1632 + 3	3.4359 − 2
26,200	26,092	219.94	2.0975	3.3225
26,400	26,291	220.53	2.0339	3.2131
26,600	26,489	221.13	1.9725	3.1076
26,800	26,687	221.72	1.9130	3.0059
27,000	26,886	222.32	1.8555	2.9077
27,200	27,084	222.91	1.7999	2.8130
27,400	27,282	223.51	1.7461	2.7217
27,600	27,481	224.10	1.6940	2.6335
27,800	27,679	224.70	1.6437	2.5484
28,000	27,877	225.29	1.5949 + 3	2.4663 − 2
28,200	28,075	225.89	1.5477	2.3871
28,400	28,274	226.48	1.5021	2.3106
28,600	28,472	227.08	1.4579	2.2367
28,800	28,670	227.67	1.4151	2.1654
29,000	28,868	228.26	1.3737	2.0966
29,200	29,066	228.86	1.3336	2.0301
29,400	29,265	229.45	1.2948	1.9659
29,600	29,463	230.05	1.2572	1.9039
29,800	29,661	230.64	1.2208	1.8440

Altitude		Temperature T, K	Pressure p, N/m^2	Density ρ, kg/m^3
h_G, m	h, m			
30,000	29,859	231.24	1.1855 + 3	1.7861 − 2
30,200	30,057	231.83	1.1514	1.7302
30,400	30,255	232.43	1.1183	1.6762
30,600	30,453	233.02	1.0862	1.6240
30,800	30,651	233.61	1.0552	1.5734
31,000	30,850	234.21	1.0251	1.5278
31,200	31,048	234.80	9.9592 + 2	1.4777
31,400	31,246	235.40	9.6766	1.4321
31,600	31,444	235.99	9.4028	1.3881
31,800	31,642	236.59	9.1374	1.3455
32,000	31,840	237.18	8.8802 + 2	1.3044 − 2
32,200	32,038	237.77	8.6308	1.2646
32,400	32,236	238.78	8.3890	1.2261
32,600	32,434	238.96	8.1546	1.1889
32,800	32,632	239.55	7.9273	1.1529
33,000	32,830	240.15	7.7069	1.1180
33,200	33,028	240.74	7.4932	1.0844
33,400	33,225	214.34	7.2859	1.0518
33,600	33,423	241.93	7.0849	1.0202
33,800	33,621	242.52	6.8898	9.8972 − 3
34,000	33,819	243.12	6.7007 + 2	9.6020 − 3
34,200	34,017	243.71	6.5171	9.3162
34,400	34,215	244.30	6.3391	9.0396
34,600	34,413	244.90	6.1663	8.7720
34,800	34,611	245.49	5.9986	8.5128
35,000	34,808	246.09	5.8359	8.2620
35,200	35,006	246.68	5.6780	8.0191
35,400	35,204	247.27	5.5248	7.7839
35,600	35,402	247.87	5.3760	7.5562
35,800	35,600	248.46	5.2316	7.3357
36,000	35,797	249.05	5.0914 + 2	7.1221 − 3
36,200	35,995	249.65	4.9553	6.9152
36,400	36,193	250.24	4.8232	6.7149
36,600	36,390	250.83	4.6949	6.5208
36,800	36,588	251.42	4.5703	6.3328
37,000	36,786	252.02	4.4493	6.1506
37,200	36,984	252.61	4.3318	5.9741
37,400	37,181	253.20	4.2176	5.8030
37,600	37,379	253.80	4.1067	5.6373
37,800	37,577	254.39	3.9990	5.4767
38,000	37,774	254.98	3.8944 + 2	5.3210 − 3
38,200	37,972	255.58	3.7928	5.1701
38,400	38,169	256.17	3.6940	5.0238
38,600	38,367	256.76	3.5980	4.8820
38,800	38,565	257.35	3.5048	4.7445
39,000	38,762	257.95	3.4141	4.6112
39,200	38,960	258.54	3.3261	4.4819
39,400	39,157	259.13	3.2405	4.3566
39,600	39,355	259.72	3.1572	4.2350
39,800	39,552	260.32	3.0764	4.1171

Altitude		Temperature T, K	Pressure p, N/m^2	Density ρ, kg/m^3
h_G, m	h, m			
40,000	39,750	260.91	2.9977 + 2	4.0028 − 3
40,200	39,947	261.50	2.9213	3.8919
40,400	40,145	262.09	2.8470	3.7843
40,600	40,342	262.69	2.7747	3.6799
40,800	40,540	263.28	2.7044	3.5786
41,000	40,737	263.87	2.6361	3.4804
41,200	40,935	264.46	2.5696	3.3850
41,400	41,132	265.06	2.5050	3.2925
41,600	41,300	265.65	2.4421	3.2027
41,800	41,527	266.24	2.3810	3.1156
42,000	41,724	266.83	2.3215 + 2	3.0310 − 3
42,400	41,922	267.43	2.2636	2.9489
42,400	42,119	268.02	2.2073	2.8692
42,600	42,316	268.61	2.1525	2.7918
42,800	42,514	269.20	2.0992	2.7167
43,000	42,711	269.79	2.0474	2.6438
43,200	42,908	270.39	1.9969	2.5730
43,400	43,106	270.98	1.9478	2.5042
43,600	43,303	271.57	1.9000	2.4374
43,800	43,500	272.16	1.8535	2.3726
44,000	43,698	272.75	1.8082 + 2	2.3096 − 3
44,200	43,895	273.34	1.7641	2.2484
44,400	44,092	273.94	1.7212	2.1889
44,600	44,289	274.53	1.6794	2.1312
44,800	44,486	275.12	1.6387	2.0751
45,000	44,684	275.71	1.5991	2.0206
45,200	44,881	276.30	1.5606	1.9677
45,400	45,078	276.89	1.5230	1.9162
45,600	45,275	277.49	1.4865	1.8662
45,800	45,472	278.08	1.4508	1.8177
46,000	45,670	278.67	1.4162 + 2	1.7704 − 3
46,200	45,867	279.26	1.3824	1.7246
46,400	46,064	279.85	1.3495	1.6799
46,600	46,261	280.44	1.3174	1.6366
46,800	46,458	281.03	1.2862	1.5944
47,000	46,655	281.63	1.2558	1.5535
47,200	46,852	282.22	1.2261	1.5136
47,400	47,049	282.66	1.1973	1.4757
47,600	47,246	282.66	1.1691	1.4409
47,800	47,443	282.66	1.1416	1.4070
48,000	47,640	282.66	1.1147 + 2	1.3739 − 3
48,200	47,837	282.66	1.0885	1.3416
48,400	48,034	282.66	1.0629	1.3100
48,600	48,231	282.66	1.0379	1.2792
48,800	48,428	282.66	1.0135	1.2491
49,000	48,625	282.66	9.8961 + 1	1.2197
49,200	48,822	282.66	9.6633	1.1910
49,400	49,019	282.66	9.4360	1.1630
49,600	49,216	282.66	9.2141	1.1357
49,800	49,413	282.66	8.9974	1.1089

| Altitude | | Temperature T, K | Pressure p, N/m² | Density ρ, kg/m³ |
h_G, m	h, m			
50,000	49,610	282.66	8.7858 + 1	1.0829 − 3
50,500	50,102	282.66	8.2783	1.0203
51,000	50,594	282.66	7.8003	9.6140 − 4
51,500	51,086	282.66	7.3499	9.0589
52,000	51,578	282.66	6.9256	8.5360
52,500	52,070	282.66	6.5259	8.0433
53,000	52,562	282.66	6.1493	7.5791
53,500	53,053	282.42	5.7944	7.1478
54,000	53,545	280.21	5.4586	6.7867
54,500	54,037	277.99	5.1398	6.4412
55,000	54,528	275.78	4.8373 + 1	6.1108 − 4
55,500	55,020	273.57	4.5505	5.7949
56,000	55,511	271.36	4.2786	5.4931
56,500	56,002	269.15	4.0210	5.2047
57,000	56,493	266.94	3.7770	4.9293
57,500	56,985	264.73	3.5459	4.6664
58,000	57,476	262.52	3.3273	4.4156
58,500	57,967	260.31	3.1205	4.1763
59,000	58,457	258.10	2.9250	3.9482
59,500	58,948	255.89	2.7403	3.7307

Standard Atmosphere, English Engineering Units

Altitude		Temperature T, °R	Pressure p, lb/ft²	Density ρ, slugs/ft³
h_G, ft	h, ft			
−16,500	−16,513	577.58	3.6588 + 3	3.6905 − 3
−16,000	−16,012	575.79	3.6641	3.7074
−15,500	−15,512	574.00	3.6048	3.6587
−15,000	−15,011	572.22	3.5462	3.6105
−14,500	−14,510	570.43	3.4884	3.5628
−14,000	−14,009	568.65	3.4314	3.5155
−13,500	−13,509	566.86	3.3752	3.4688
−13,000	−13,008	565.08	3.3197	3.4225
−12,500	−12,507	563.29	3.2649	3.3768
−12,000	−12,007	561.51	3.2109	3.3314
−11,500	−11,506	559.72	3.1576 + 3	3.2866 − 3
−11,000	−11,006	557.94	3.1050	3.2422
−10,500	−10,505	556.15	3.0532	3.1983
−10,000	−10,005	554.37	3.0020	3.1548
−9,500	−9,504	552.58	2.9516	3.1118
−9,000	−9,004	550.80	2.9018	3.0693
−8,500	−8,503	549.01	2.8527	3.0272
−8,000	−8,003	547.23	2.8043	2.9855
−7,500	−7,503	545.44	2.7566	2.9443
−7,000	−7,002	543.66	2.7095	2.9035
−6,500	−6,502	541.88	2.6631 + 3	2.8632 − 3
−6,000	−6,002	540.09	2.6174	2.8233
−5,500	−5,501	538.31	2.5722	2.7838
−5,000	−5,001	536.52	2.5277	2.7448
−4,500	−4,501	534.74	2.4839	2.7061
−4,000	−4,001	532.96	2.4406	2.6679
−3,500	−3,501	531.17	2.3980	2.6301
−3,000	−3,000	529.39	2.3560	2.5927

Altitude		Temperature T, °R	Pressure p, lb/ft^2	Density ρ, slugs/ft^3
h_G, ft	h, ft			
−2,500	−2,500	527.60	2.3146	2.5558
−2,000	−2,000	525.82	2.2737	2.5192
−1,500	−1,500	524.04	2.2335 + 3	2.4830 − 3
−1,000	−1,000	522.25	2.1938	2.4473
−500	−500	520.47	2.1547	2.4119
0	0	518.69	2.1162	2.3769
500	500	516.90	2.0783	2.3423
1,000	1,000	515.12	2.0409	2.3081
1,500	1,500	513.34	2.0040	2.2743
2,000	2,000	511.56	1.9677	2.2409
2,500	2,500	509.77	1.9319	2.2079
3,000	3,000	507.99	1.8967	2.1752
3,500	3,499	506.21	1.8619 + 3	2.1429 − 3
4,000	3,999	504.43	1.8277	2.1110
4,500	4,499	502.64	1.7941	2.0794
5,000	4,999	500.86	1.7609	2.0482
5,500	5,499	499.08	1.7282	2.0174
6,000	5,998	497.30	1.6960	1.9869
6,500	6,498	495.52	1.6643	1.9567
7,000	6,998	493.73	1.6331	1.9270
7,500	7,497	491.95	1.6023	1.8975
8,000	7,997	490.17	1.5721	1.8685
8,500	8,497	488.39	1.5423 + 3	1.8397 − 3
9,000	8,996	486.61	1.5129	1.8113
9,500	9,496	484.82	1.4840	1.7833
10,000	9,995	483.04	1.4556	1.7556
10,500	10,495	481.26	1.4276	1.7282
11,000	10,994	479.48	1.4000	1.7011
11,500	11,494	477.70	1.3729	1.6744
12,000	11,993	475.92	1.3462	1.6480
12,500	12,493	474.14	1.3200	1.6219
13,000	12,992	472.36	1.2941	1.5961
13,500	13,491	470.58	1.2687 + 3	1.5707 − 3
14,000	13,991	468.80	1.2436	1.5455
14,500	14,490	467.01	1.2190	1.5207
15,000	14,989	465.23	1.1948	1.4962
15,500	15,488	463.45	1.1709	1.4719
16,000	15,988	461.67	1.1475	1.4480
16,500	16,487	459.89	1.1244	1.4244
17,000	16,986	458.11	1.1017	1.4011
17,500	17,485	456.33	1.0794	1.3781
18,000	17,984	454.55	1.0575	1.3553
18,500	18,484	452.77	1.0359 + 3	1.3329 − 3
19,000	18,983	450.99	1.0147	1.3107
19,500	19,482	449.21	9.9379 + 2	1.2889
20,000	19,981	447.43	9.7327	1.2673
20,500	20,480	445.65	9.5309	1.2459
21,000	20,979	443.87	9.3326	1.2249

Altitude		Temperature T, °R	Pressure p, lb/ft²	Density ρ, slugs/ft³
h_G, ft	h, ft			
21,500	21,478	442.09	9.1376	1.2041
22,000	21,977	440.32	8.9459	1.1836
22,500	22,476	438.54	8.7576	1.1634
23,000	22,975	436.76	8.5724	1.1435
23,500	23,474	434.98	8.3905 + 2	1.1238 − 3
24,000	23,972	433.20	8.2116	1.1043
24,500	24,471	431.42	8.0359	1.0852
25,000	24,970	429.64	7.8633	1.0663
25,500	25,469	427.86	7.6937	1.0476
26,000	25,968	426.08	7.5271	1.0292
26,500	26,466	424.30	7.3634	1.0110
27,000	26,965	422.53	7.2026	9.9311 − 4
27,500	27,464	420.75	7.0447	9.7544
28,000	27,962	418.97	6.8896	9.5801
28,500	28,461	417.19	6.7373 + 2	9.4082 − 4
29,000	28,960	415.41	6.5877	9.2387
29,500	29,458	413.63	6.4408	9.0716
30,000	29,957	411.86	6.2966	8.9068
30,500	30,455	410.08	6.1551	8.7443
31,000	30,954	408.30	6.0161	8.5841
31,500	31,452	406.52	5.8797	8.4261
32,000	31,951	404.75	5.7458	8.2704
32,500	32,449	402.97	5.6144	8.1169
33,000	32,948	401.19	5.4854	7.9656
33,500	33,446	399.41	5.3589 + 2	7.8165 − 4
34,000	33,945	397.64	5.2347	7.6696
34,500	34,443	395.86	5.1129	7.5247
35,000	34,941	394.08	4.9934	7.3820
35,500	35,440	392.30	4.8762	7.2413
36,000	35,938	390.53	4.7612	7.1028
36,500	36,436	389.99	4.6486	6.9443
37,000	36,934	389.99	4.5386	6.7800
37,500	37,433	389.99	4.4312	6.6196
38,000	37,931	389.99	4.3263	6.4629
38,500	38,429	389.99	4.2240 + 2	6.3100 − 4
39,000	38,927	389.99	4.1241	6.1608
39,500	39,425	389.99	4.0265	6.0150
40,000	39,923	389.99	3.9312	5.8727
40,500	40,422	389.99	3.8382	5.7338
41,000	40,920	389.99	3.7475	5.5982
41,500	41,418	389.99	3.6588	5.4658
42,000	41,916	389.99	3.5723	5.3365
42,500	42,414	389.99	3.4878	5.2103
43,000	42,912	389.99	3.4053	5.0871
43,500	43,409	389.99	3.3248 + 2	4.9668 − 4
44,000	43,907	389.99	3.2462	4.8493
44,500	44,405	389.99	3.1694	4.7346
45,000	44,903	389.99	3.0945	4.6227
45,500	45,401	389.99	3.0213	4.5134

Altitude		Temperature T, °R	Pressure p, lb/ft²	Density ρ, slugs/ft³
h_G, ft	h, ft			
46,000	45,899	389.99	2.9499	4.4067
46,500	46,397	389.99	2.8801	4.3025
47,000	46,894	389.99	2.8120	4.2008
47,500	47,392	389.99	2.7456	4.1015
48,000	47,890	389.99	2.6807	4.0045
48,500	48,387	389.99	$2.2173 + 2$	$3.9099 - 4$
49,000	48,885	389.99	2.5554	3.8175
49,500	49,383	389.99	2.4950	3.7272
50,000	49,880	389.99	2.4361	3.6391
50,500	50,378	389.99	2.3785	3.5531
51,000	50,876	389.99	2.3223	3.4692
51,500	51,373	389.99	2.2674	3.3872
52,000	51,871	389.99	2.2138	3.3072
52,500	52,368	389.99	2.1615	3.2290
53,000	52,866	389.99	2.1105	3.1527
53,500	53,363	389.99	$2.0606 + 2$	$3.0782 - 4$
54,000	53,861	389.99	2.0119	3.0055
54,500	54,358	389.99	1.9644	2.9345
55,000	54,855	389.99	1.9180	2.8652
55,500	55,353	389.99	1.8727	2.7975
56,000	55,850	389.99	1.8284	2.7314
56,500	56,347	389.99	1.7853	2.6669
57,000	56,845	389.99	1.7431	2.6039
57,500	57,342	389.99	1.7019	2.5424
58,000	57,839	389.99	1.6617	2.4824
58,500	58,336	389.99	$1.6225 + 2$	$2.4238 - 4$
59,000	58,834	389.99	1.5842	2.3665
59,500	59,331	389.99	1.5468	2.3107
60,000	59,828	389.99	1.5103	2.2561
60,500	60,325	389.99	1.4746	2.2028
61,000	60,822	389.99	1.4398	2.1508
61,500	61,319	389.99	1.4058	2.1001
62,000	61,816	389.99	1.3726	2.0505
62,500	62,313	389.99	1.3402	2.0021
63,000	62,810	389.99	1.3086	1.9548
63,500	63,307	389.99	$1.2777 + 2$	$1.9087 - 4$
64,000	63,804	389.99	1.2475	1.8636
64,500	64,301	389.99	1.2181	1.8196
65,000	64,798	389.99	1.1893	1.7767
65,500	65,295	389.99	1.1613	1.7348
66,000	65,792	389.99	1.1339	1.6938
66,500	66,289	389.99	1.1071	1.6539
67,000	66,785	389.99	1.0810	1.6148
67,500	67,282	389.99	1.0555	1.5767
68,000	67,779	389.99	1.0306	1.5395
68,500	68,276	389.99	$1.0063 + 2$	$1.5032 - 4$
69,000	68,772	389.99	$9.8253 + 1$	1.4678
69,500	69,269	389.99	9.5935	1.4331
70,000	69,766	389.99	9.3672	1.3993

Altitude		Temperature T, °R	Pressure p, lb/ft^2	Density ρ, slugs/ft^3
h_G, ft	h, ft			
70,500	70,262	389.99	9.1462	1.3663
71,000	70,759	389.99	8.9305	1.3341
71,500	74,256	389.99	8.7199	1.3026
72,000	71,752	389.99	8.5142	1.2719
72,500	72,249	389.99	8.3134	1.2419
73,000	72,745	389.99	8.1174	1.2126
73,500	73,242	389.99	7.9259 + 1	1.1840 − 4
74,000	73,738	389.99	7.7390	1.1561
74,500	74,235	389.99	7.5566	1.1288
75,000	74,731	389.99	7.3784	1.1022
75,500	75,228	389.99	7.2044	1.0762
76,000	75,724	389.99	7.0346	1.0509
76,500	76,220	389.99	6.8687	1.0261
77,000	76,717	389.99	6.7068	1.0019
77,500	77,213	389.99	6.5487	9.7829 − 5
78,000	77,709	389.99	6.3944	9.5523
78,500	78,206	389.99	6.2437 + 1	9.3271 − 5
79,000	78,702	389.99	6.0965	9.1073
79,500	79,198	389.99	5.9528	8.8927
80,000	79,694	389.99	5.8125	8.6831
80,500	80,190	389.99	5.6755	8.4785
81,000	80,687	389.99	5.5418	8.2787
81,500	81,183	389.99	5.4112	8.0836
82,000	81,679	389.99	5.2837	7.8931
82,500	82,175	390.24	5.1592	7.7022
83,000	82,671	391.06	5.0979	7.5053
83,500	83,167	391.87	4.9196 + 1	7.3139 − 5
84,000	83,663	392.69	4.8044	7.1277
84,500	84,159	393.51	4.6921	6.9467
85,000	84,655	394.32	4.5827	6.7706
85,500	85,151	395.14	4.4760	6.5994
86,000	85,647	395.96	4.3721	6.4328
86,500	86,143	396.77	4.2707	6.2708
87,000	86,639	397.59	4.1719	6.1132
87,500	87,134	398.40	4.0757	5.9598
88,000	87,630	399.22	3.9818	5.8106
88,500	88,126	400.04	3.8902 + 1	5.6655 − 5
89,000	88,622	400.85	3.8010	5.5243
89,500	89,118	401.67	3.7140	5.3868
90,000	89,613	402.48	3.6292	5.2531
90,500	90,109	403.30	3.5464	5.1230
91,000	90,605	404.12	3.4657	4.9963
91,500	91,100	404.93	3.3870	4.8730
92,000	91,596	405.75	3.3103	4.7530
92,500	92,092	406.56	3.2354	4.6362
93,000	92,587	407.38	3.1624	4.5525
93,500	93,083	408.19	3.0912 + 1	4.4118 − 5
94,000	93,578	409.01	3.0217	4.3041
94,500	94,074	409.83	2.9539	4.1992

Altitude		Temperature T, °R	Pressure p, lb/ft²	Density ρ, slugs/ft³
h_G, ft	h, ft			
95,000	94,569	410.64	2.8878	4.0970
95,500	95,065	411.46	2.8233	3.9976
96,000	95,560	412.27	2.7604	3.9007
96,500	96,056	413.09	2.6989	3.8064
97,000	96,551	413.90	2.6390	3.7145
97,500	97,046	414.72	2.5805	3.6251
98,000	97,542	415.53	2.5234	3.5379
98,500	98,037	416.35	2.4677 + 1	3.4530 − 5
99,000	98,532	417.16	2.4134	3.3704
99,500	99,028	417.98	2.3603	3.2898
100,000	99,523	418.79	2.3085	3.2114
100,500	100,018	419.61	2.2580	3.1350
101,000	100,513	420.42	2.2086	3.0605
101,500	101,008	421.24	2.1604	2.9879
102,000	101,504	422.05	2.1134	2.9172
102,500	101,999	422.87	2.0675	2.8484
103,000	102,494	423.68	2.0226	2.7812
103,500	102,989	424.50	1.9789 + 1	2.7158 − 5
104,000	103,484	425.31	1.9361	2.6520
104,500	103,979	426.13	1.8944	2.5899
105,000	104,474	426.94	1.8536	2.5293
106,000	105,464	428.57	1.7749	2.4128
107,000	106,454	430.20	1.6999	2.3050
108,000	107,444	431.83	1.6282	2.1967
109,000	108,433	433.46	1.5599	2.0966
110,000	109,423	435.09	1.4947	2.0014
111,000	110,412	436.72	1.4324	1.9109
112,000	111,402	438.35	1.3730 + 1	1.8247 − 5
113,000	112,391	439.97	1.3162	1.7428
114,000	113,380	441.60	1.2620	1.6649
115,000	114,369	443.23	1.2102	1.5907
116,000	115,358	444.86	1.1607	1.5201
117,000	116,347	446.49	1.1134	1.4528
118,000	117,336	448.11	1.0682	1.3888
119,000	118,325	449.74	1.0250	1.3278
120,000	119,313	451.37	9.8372 + 0	1.2697
121,000	120,302	453.00	9.4422	1.2143
122,000	121,290	454.62	9.0645 + 0	1.1616 − 5
123,000	122,279	456.25	8.7032	1.1113
124,000	123,267	457.88	8.3575	1.0634
125,000	124,255	459.50	8.0267	1.0177
126,000	125,243	461.13	7.7102	9.7410 − 6
127,000	126,231	462.75	7.4072	9.3253
128,000	127,219	464.38	7.1172	8.9288
129,000	128,207	466.01	6.8395	8.5505
130,000	129,195	467.63	6.5735	8.1894
131,000	130,182	469.26	6.3188	7.8449
132,000	131,170	470.88	6.0748 + 0	7.5159 − 6
133,000	132,157	472.51	5.8411	7.2019

Altitude		Temperature T, °R	Pressure p, lb/ft²	Density ρ, slugs/ft³
h_G, ft	h, ft			
134,000	133,145	474.13	5.6171	6.9020
135,000	134,132	475.76	5.4025	6.6156
136,000	135,119	477.38	5.1967	6.3420
137,000	136,106	479.01	4.9995	6.0806
138,000	137,093	480.63	4.8104	5.8309
139,000	138,080	482.26	4.6291	5.5922
140,000	139,066	483.88	4.4552	5.3640
141,000	140,053	485.50	4.2884	5.1460
142,000	141,040	487.13	4.1284 + 0	4.9374 − 6
143,000	142,026	488.75	3.9749	4.7380
144,000	143,013	490.38	3.8276	4.5473
145,000	143,999	492.00	3.6862	4.3649
146,000	144,985	493.62	3.5505	4.1904
147,000	145,971	495.24	3.4202	4.0234
148,000	146,957	496.87	3.2951	3.8636
149,000	147,943	498.49	3.1750	3.7106
150,000	148,929	500.11	3.0597	3.5642
151,000	149,915	501.74	2.9489	3.4241
152,000	150,900	503.36	2.8424 + 0	3.2898 − 6
153,000	151,886	504.98	2.7402	3.1613
154,000	152,871	506.60	2.6419	3.0382
155,000	153,856	508.22	2.5475	2.9202
156,000	154,842	508.79	2.4566	2.8130
157,000	155,827	508.79	2.3691	2.7127
158,000	156,812	508.79	2.2846	2.6160
159,000	157,797	508.79	2.2032	2.5228
160,000	158,782	508.79	2.1247	2.4329
161,000	159,797	508.79	2.0490	2.3462

REFERENCES

1. Anderson, John D., Jr.: *Gasdynamic Lasers: An Introduction,* Academic Press, New York, 1976.

2. Anderson, John D., Jr.: *Introduction to Flight,* 8th ed., McGraw Hill Education, New York, 2016.

3. Durand, W. F. (ed): *Aerodynamic Theory,* vol. 1, Springer, Berlin, 1934.

4. Wylie, C. R.: *Advanced Engineering Mathematics,* 4th ed., McGraw Hill Book Company, New York, 1975.

5. Kreyszig, E.: *Advanced Engineering Mathematics,* John Wiley & Sons, Inc., New York, 1962.

6. Hildebrand, F. B.: *Advanced Calculus for Applications,* 2d ed., Prentice-Hall, Inc., Englewood Cliffs, N.J., 1976.

7. Anderson, John D., Jr.: *Computational Fluid Dynamics: The Basics with Applications,* McGraw Hill, New York, 1995.

8. Prandtl, L., and O. G. Tietjens: *Applied Hydro- and Aeromechanics,* United Engineering Trustees, Inc., 1934; also, Dover Publications, Inc., New York, 1957.

9. Karamcheti, K.: *Principles of Ideal Fluid Aerodynamics,* John Wiley & Sons, Inc., New York, 1966.

10. Pierpont, P. K.: "Bringing Wings of Change," *Astronaut. Aeronaut.,* vol. 13, no. 10, pp. 20–27, October 1975.

11. Abbott, I. H., and A. E. von Doenhoff: *Theory of Wing Sections,* McGraw Hill Book Company, New York, 1949; also, Dover Publications, Inc., New York, 1959.

12. Munk, Max M.: *General Theory of Thin Wing Sections,* NACA report no. 142, 1922.

13. Bertin, John J., and M. L. Smith: *Aerodynamics for Engineers,* Prentice-Hall, Inc., Englewood Cliffs, N.J., 1979.

14. Hess, J. L., and A. M. O. Smith: "Calculation of potential flow about arbitrary bodies," in *Progress in Aeronautical Sciences,* vol. 8, D. Kucheman (ed.), Pergamon Press, New York, 1967, pp. 1–138.

15. Chow, C. Y.: *An Introduction to Computational Fluid Dynamics,* John Wiley & Sons, Inc., New York, 1979.

16. McGhee, R. J., and W. D. Beasley: *Low-Speed Aerodynamic Characteristics of a 17-Percent-Thick Airfoil Section Designed for General Aviation Applications,* NASA TN D-7428, December 1973.

17. McGhee, R. J., W. D. Beasley, and R. T. Whitcomb: "NASA low- and medium-speed airfoil development," in *Advanced Technology Airfoil Research,* vol. II, NASA CP 2046, March 1980.

18. Glauert, H.: *The Elements of Aerofoil and Airscrew Theory,* Cambridge University Press, London, 1926.

19. Winkelmann, A. E., J. B. Barlow, S. Agrawal, J. K. Saini, J. D. Anderson, Jr., and E. Jones: "The Effects of Leading Edge Modifications on the Post-Stall Characteristics of Wings," AIAA paper no. 80–0199, American Institute of Aeronautics and Astronautics, New York, 1980.

20. Anderson, John D., Jr., Stephen Corda, and David M. Van Wie: "Numerical Lifting Line Theory Applied to Drooped Leading-Edge Wings Below and Above Stall," *J. Aircraft,* vol. 17, no. 12, December 1980, pp. 898–904.

21. Anderson, John D., Jr.: *Modern Compressible Flow: With Historical Perspective,* 3d ed., McGraw Hill Book Company, New York, 2003.

22. Sears, F. W.: *An Introduction to Thermodynamics. The Kinetic Theory of Gases, and Statistical Mechanics,* 2d ed., Addison-Wesley Publishing Company, Inc., Reading, Mass., 1959.

23. Van Wylen, G. J., and R. E. Sonntag: *Fundamentals of Classical Thermodynamics,* 2d ed., John Wiley & Sons, Inc., New York, 1973.

24. Reynolds, W. C., and H. C. Perkins: *Engineering Thermodynamics,* 2d ed., McGraw Hill Book Company, New York, 1977.

25. Shapiro, A. H.: *The Dynamics and Thermodynamics of Compressible Fluid Flow,* vols. 1 and 2, The Ronald Press Company, New York, 1953.

26. Leipmann, H. W., and A. Roshko: *Elements of Gasdynamics,* John Wiley & Sons, Inc., New York, 1957.

27. Tsien, H. S.: "Two-Dimensional Subsonic flow of Compressible Fluids," *J. Aeronaut. Sci.,* vol. 6, no. 10, October 1939, p. 399.

28. von Karman, T. H.: "Compressibility Effects in Aerodynamics," *J. Aeronaut. Sci.,* vol. 8, no. 9, September 1941, p. 337.

29. Laitone, E. V.: "New Compressibility Correction for Two-Dimensional Subsonic Flow," *J. Aeronaut. Sci.,* vol. 18, no. 5, May 1951, p. 350.

30. Whitcomb, R. T.: *A Study of the Zero-Lift Drag-Rise Characteristics of Wing-Body Combinations Near the Speed of Sound,* NACA report no. 1273, 1956.

31. Whitcomb, R. T., and L. R. Clark: *An Airfoil Shape for Efficient Flight at Supercritical Mach Numbers,* NASA TMX-1109, July 1965.

32. Owczarek, Jerzy A.: *Fundamental of Gas Dynamics,* International Textbook Company, Scranton, Pa., 1964.

33. Anderson, J. D., Jr., L. M. Albacete, and A. E. Winkelmann: *On Hypersonic Blunt Body Flow Fields Obtained with a Time-Dependent Technique,* Naval Ordnance Laboratory NOLTR 68-129, 1968.

34. Anderson, J. D., Jr.: "An Engineering Survey of Radiating Shock Layers," *AIAA J.,* vol. 7, no. 9, September 1969, pp. 1665–1675.

35. Cherni, G. G.: *Introduction to Hypersonic Flow,* Academic Press, New York, 1961.

36. Truitt, R. W.: *Hypersonic Aerodynamics,* The Ronald Press Company, New York, 1959.

37. Dorrance, H. W.: *Viscous Hypersonic Flow,* McGraw Hill Book Company, New York, 1962.

38. Hayes, W. D., and R. F. Probstein: *Hypersonic Flow Theory,* 2d ed., Academic Press, New York, 1966.

39. Vinh, N. X., A. Busemann, and R. D. Culp: *Hypersonic and Planetary Entry Flight Mechanics,* University of Michigan Press, Ann Arbor, 1980.

40. Schlichting, H.: *Boundary Layer Theory,* 7th ed., McGraw Hill Book Company, New York, 1979.

41. White, F. M.: *Viscous Fluid Flow,* McGraw Hill Book Company, New York, 1974.

42. Cebeci, T., and A. M. O. Smith: *Analysis of Turbulent Boundary Layers,* Academic Press, New York, 1974.

43. Bradshaw, P., T. Cebeci, and J. Whitelaw: *Engineering Calculation Methods for Turbulent Flow,* Academic Press, New York, 1981.

44. Berman, H. A., J. D. Anderson, Jr., and J. P. Drummond: "Supersonic Flow over a Rearward Facing Step with Transverse Nonreacting Hydrogen Injection," *AIAA J.,* vol. 21, no. 12, December 1983, pp. 1707–1713.

45. Loftin, Lawrence K., Jr.: *Quest for Performance: The Evolution of Modern Aircraft,* NASA SP-468, 1985.

46. von Karman, T., and Lee Edson: *The Wind and Beyond: Theodore von Karman, Pioneer in*

Aviation and Pathfinder in Space, Little, Brown and Co., Boston, 1967.

47. Nakayama, Y. (ed): *Visualized Flow,* compiled by the Japan Society of Mechanical Engineers, Pergamon Press, New York, 1988.

48. Meuller, Thomas J.: *Low Reynolds Number Vehicles,* AGARDograph no. 288, Advisory Group for Advanced Research and Development, NATO, 1985.

49. Carnahan, B., H. A. Luther, and J. O. Wilkes: *Applied Numerical Methods,* John Wiley & Sons, New York, 1969.

50. Schetz, Joseph A.: *Foundations of Boundary Layer Theory for Momentum, Heat, and Mass Transfer,* Prentice-Hall, Inc., Englewood Cliffs, N.J., 1984.

51. Anderson, Dale, John C. Tannehill, and Richard H. Pletcher: *Computational Fluid Mechanics and Heat Transfer,* 2nd ed., Taylor and Francis, Washington, DC, 1997.

52. Anderson, J. D.: *Hypersonic and High Temperature Gas Dynamics,* 2nd ed. American Institute of Aeronautics and Astronautics, Reston, Virginia, 2006.

53. Kothari, A. P., and J. D. Anderson: "Flow over Low Reynolds Number Airfoils—Compressible Navier-Stokes Solutions," AIAA paper no. 85-0107, January 1985.

54. Shang, J. S., and S. J. Scherr: "Navier-Stokes Solution for a Complete Re-Entry Configuration," *J. Aircraft,* vol. 23, no. 12, December 1986, pp. 881–888.

55. Stevens, V. P.: "Hypersonic Research Facilities at the Ames Aeronautical Laboratory," *J. Appl. Phys.,* vol. 21, 1955, pp. 1150–1155.

56. Hodges, A. J.: "The Drag Coefficient of Very High Velocity Spheres," *J. Aeronaut. Sci.,* vol. 24, 1957, pp. 755–758.

57. Charters, A. C., and R. N. Thomas: "The Aerodynamic Performance of Small Spheres from Subsonic to High Supersonic Velocities," *J. Aeronaut. Sci.,* vol. 12, 1945, pp. 468–476.

58. Anderson, John, D., Jr.: *A History of Aerodynamics and Its Impact on Flying Machines,* Cambridge University Press, New York, 1997 (hardcover), 1998 (paperback).

59. Prandtl, Ludwig: *Application of Modern Hydrodynamics to Aeronautics,* NACA Technical Report 116, 1921.

60. Anderson, John D., Jr.: *Computational Fluid Dynamics: The Basics with Applications,* McGraw Hill, New York, 1995.

61. Gad-el-Hak, Mohamed: "Basic Instruments," in *The Handbook of Fluid Dynamics,* edited by Richard W. Johnson, CRC Press, Boca Raton, 1998, ch. 33, pp. 33-1–33-22.

62. Henne, P. A. (ed): *Applied Computational Aerodynamics,* vol. 125 of Progress in Astronautics and Aeronautics, American Institute of Aeronautics and Astronautics, Reston, Virginia, 1990.

63. Katz, Joseph, and Plotkin, Allen: *Low-Speed Aerodynamics, From Wing Theory to Panel Methods,* McGraw Hill, New York, 1991.

64. Anderson, W. Kyle, and Bonhaus, Daryl L.: "Airfoil Design on Unstructured Grids for Turbulent Flows," *AIAA J.,* vol. 37, no. 2, Feb. 1999, pp. 185–191.

65. Anderson, John D., Jr.: *Aircraft Performance and Design,* McGraw Hill, Boston, 1999.

66. Kuchemann, Dietrich: *The Aerodynamic Design of Aircraft,* Pergamon Press, Oxford, 1978.

67. Faulkner, Robert F., and Weber, James W.: "Hydrocarbon Scramjet Propulsion System Development, Demonstration and Application," AIAA Paper No. 99-4922, 1999.

68. Van Wie, David M., White, Michael E., and Corpening, Griffin P.: "NASP (National Aero Space Plane) Inlet Design and Testing Issues," *Johns Hopkins Applied Physics Laboratory Technical Digest,* vol. 11, nos. 3 and 4, July–December 1990, pp. 353–362.

69. Billig, Frederick S.: "Design and Development of Single-Stage-to-Orbit Vehicles," *Johns Hopkins Applied Physics Laboratory*

Technical Digest, vol. 11, nos. 3 and 4, July–December 1990, pp. 336–352.

70. Nakahasi, K., and Deiwert, G. S.: "A Self-Adaptive Grid Method with Application to Airfoil Flow," AIAA Paper 85-1525, American Institute of Aeronautics and Astronautics, 1985.

71. Hirsch, C.: *Numerical Computation of Internal and External Flows,* vols. 1 and 2, John Wiley and Sons, Chichester, 1988.

72. Jameson, Antony: "Re-Engineering the Design Process Through Computation," *J. Aircraft,* vol. 36, no. 1, Jan.–Feb. 1999, pp. 36–50.

73. Kuester, Steven P. and Anderson, John D., Jr.: "Applicability of Newtonian and Linear Theory to Slender Hypersonic Bodies," *J. Aircraft,* vol. 32, no. 2, March–April 1995, pp. 446–449.

74. Maus, J. R., Griffith, B. J., Szema, K. Y., and Best, J. T.: "Hypersonic Mach Number and Real Gas Effects on Space Shuttle Orbiter Aerodynamics," *J. Spacecraft and Rockets,* vol. 21, no. 2, March–April 1984, pp. 136–141.

75. Van Driest, E. R.: "Investigation of Laminar Boundary Layer in Compressible Fluids Using the Crocco Method," NACA TN 2579, Jan. 1952.

76. Rubesin, M. W., and Johnson, H. A.: "A Critical Review of Skin-Friction and Heat Transfer Solutions of the Laminar Boundary Layer of a Flat Plate," *Trans. of the ASME,* vol. 71, no. 4, May 1949, pp. 383–388.

77. Eckert, E. R. G.: "Engineering Relations for Heat Transfer and Friction in High-Velocity Laminar and Turbulent Boundary-Layer Flow Over Surfaces with Constant Pressure and Temperature," *Trans. of the ASME,* vol. 78, no. 6, August 1956, p. 1273.

78. Van Driest, E. R.: "The Problem of Aerodynamic Heating," *Aeronautical Engineering Review,* Oct. 1956, pp. 26–41.

79. Koppenwallner, G.: "Fundamentals of Hypersonics: Aerodynamics and Heat Transfer," in the Short Course Notes entitled *Hypersonic Aerothermodynamics,* presented at the Von Karman Institute for Fluid Dynamics, Rhode Saint Genese, Belgium, Feb. 1984.

80. Blottner, F. G.: "Finite Difference Methods of Solution of the Boundary-Layer Equations," *AIAA J.,* vol. 8, no. 2, February 1970, pp. 193–205.

81. Marvin, Joseph G.: "Turbulence Modeling for Computational Aerodynamics," *AIAA J.,* vol. 21, no. 7, July 1983, pp. 941–955.

82. Baldwin, B. S., and Lomax, H.: "Thin Layer Approximation and Algebraic Model for Separated Turbulent Flows," AIAA Paper No. 78-257, Jan. 1978.

83. Bradshaw, P., Cebeci, T., and Whitelaw, J.: *Engineering Calculational Methods for Turbulent Flow,* Academic Press, New York, 1981.

84. Reda, D. C., and Murphy, J. D.: "Shock Wave Turbulent Boundary Layer Interactions in Rectangular Channels, Part II: The Influence of Sidewall Boundary Layers on Incipient Separation and Scale of Interaction," AIAA Paper No. 73-234, 1973.

85. Beierle, Mark T.: *Investigation of Effects of Surface Roughness on Symmetric Airfoil Lift and Lift-to-Drag Ratio,* Ph.D. Dissertation, Department of Aerospace Engineering, University of Maryland, 1998.

86. Buning, P. G., Jespersen, D. C., Pulliam, T. H., Chan, W. M., Slotnick, J. P., Krist, S. E., and Renze, K. J.: "OVERFLOW User's Manual," Version 1.7v, NASA, June, 1997.

87. Hassan, A. A., and JanakiRam, R. D.: "Effects of Zero-Mass 'Synthetic' Jets on the Aerodynamics of the NACA-0012 Airfoil," AIAA Paper No. 97-2326, 1997.

88. Lombardi, G., Salvetti, M. V., and Pinelli, D.: "Numerical Evaluation of Airfoil Friction Drag," *J. Aircraft,* vol. 37, no. 2, March–April, 2000, pp. 354–356.

89. Liebeck, R. H., "Design of Subsonic Airfoils for High Lift," *AIAA J. Aircraft,* vol. 15, no. 9, September 1978, pp. 547–561.

90. Liebeck, R. H., "Blended Wing Body Design Challenges," AIAA Paper 2003-2659, April 1, 2003.

91. Roman D., J. B. Allen, and R. H. Liebeck, "Aerodynamic Design Challenges of the Blended-Wing-Body Subsonic Transport," AIAA Paper No. 2000-4335, 18th AIAA Applied Aerodynamics Conference, August 14–17, 2000.

92. Liebeck, R. H., "Design of the Blended-Wing-Body Subsonic Transport," 2002 Wright Brothers Lecture, AIAA Paper No. 2002-0002, AIAA Aerospace Sciences Conference, January 2002.

93. Busemann, A., "Drucke and Kegelformige Spitzen bei Bewegung mit Überschallgeschwindigkeit," *Z. Angew Math. Mech.,* vol. 9, 1929, p. 496.

94. Taylor, G. I., and J. W. Maccoll, "The Air Pressure on a Cone Moving at High Speed," *Proc. Roy. Soc.* (London), ser. A, vol. 139, 1933, pp. 278–311.

95. Kopal, Z., "Tables of Supersonic Flow Around Cones." M.I.T. Center of Analysis Technical Report No. 1, U.S. Government Printing Office, Washington, D.C., 1947.

96. Sims, Joseph L., "Tables for Supersonic Flow Around Right Circular Cones at Zero Angle of Attack," NASA SP-3004, 1964.

97. Tauber, M. E., and Meneses, G. P., "Aerothermodynamics of Transatmospheric Vehicles," AIAA Paper 86-1257, June 1986.

98. Zoby, E. V., "Approximate Heating Analysis for the Windward Symmetry Plane of Shuttle-like Bodies at Angle of Attack," in *Thermodynamics of Atmospheric Entry,* T. E. Horton (ed.), Vol. 82, *Progress in Astronautics and Aeronautics,* American Institute of Aeronautics and Astronautics, 1982, pp. 229–247.

99. Nonweiler, T. R., "Aerodynamic Problems of Manned Space Vehicles," *J. Royal Aeronaut. Soc.,* vol. 63, 1959, pp. 521–528.

100. Jones, J. G., "A Method for Designing Lifting Configurations for High Supersonic Speeds Using the Flow Fields of Nonlifting Cones," Royal Aeronautical Establishment Report Aero 2624, A.R.C. 24846, England, 1963.

101. Jones, J. G., K. C. Moore, J. Pike, and P. L. Roe, "A Method for Designing Lifting Configurations for High Supersonic Speeds Using Axisymmetric Flow Fields," *Ingenieur-Archiv.,* vol. 37, Band, 1. Heft, 1968, pp. 556–572.

102. Townend, L. H., "Research and Design for Lifting Reentry," *Prog. Aerospace Sciences,* vol. 18, 1979, pp. 1–80.

103. Rasmussen, M. L., "Waverider Configurations Derived from Inclined-Circular and Elliptic Cones," *J. Spacecraft and Rockets,* vol. 17, no. 6, November–December 1980, pp. 537–545.

104. Kim, B. S., M. L. Rasmussen, and M. D. Jischke, "Optimization of Waverider Configurations Generated from Axisymmetric Conical Flows," AIAA Paper 82-1299, January 1982.

105. Broadway, R., and M. L. Rasmussen, "Aerodynamics of a Simple Cone Derived Waverider," AIAA Paper 84-0085, January 1984.

106. Bowcutt, K. G., John D. Anderson, Jr., and D. Capriotti, "Viscous Optimized Hypersonic Waveriders," AIAA Paper 87-0272, January 1987.

107. Bowcutt, Kevin G., John D. Anderson, Jr., and Diego Capriotti, "Numerical Optimization of Conical Flow Waveriders Including Detailed Viscous Effects," *Aerodynamics of Hypersonic Lifting Vehicles,* AGARD Conference Proceedings, no. 428, November 1987, pp. 27-1 to 27-23.

108. Nelder, J. A., and R. Meade, "A Simplex Method of Function Minimization," *Computer J.,* vol. 7, January 1965, pp. 308–313.

109. Bowcutt, Kevin G., "Optimization of Hypersonic Waveriders Derived from Cone Flows—Including Viscous Effects," doctoral dissertation, Department of Aerospace Engineering, University of Maryland, College Park, Md., May 1986.

110. Corda, Stephan, and John D. Anderson, Jr., "Viscous Optimized Waveriders Designed from Axisymmetic Flowfields," AIAA Paper 88-0369, January 1988.

111. Anderson, John D., Jr., *The Airplane: A History of Its Technology,* American Institute of Aeronautics and Astronautics, Reston, Va., 2002.

112. Hoerner, S. F., *Fluid Dynamic Drag,* Hoerner Fluid Dynamics, Brick Town, N.J., 1965.

113. Raymer, Daniel P., *Aircraft Design: A Conceptual Approach,* 4th ed., American Institute of Aeronautics and Astronautics, Reston, Va., 2006.

114. Rizzi, Arthur, et al., "Lessons Learned from Numerical Simulations of the F-16 XL Aircraft at Flight Conditions," *Journal of Aircraft,* vol. 46, no. 2, March–April 2009, pp. 423–441.

115. Ames Research Staff, "Equations, Tables and Charts for Compressible Flow," NACA Report 1135, 1953.

116. McLellan, Charles H., "Exploratory Wind-Tunnel Investigation of Wings and Bodies at $M = 6.9$," *J. Aeronaut. Sci.,* vol. 18, no. 10, October 1951, pp. 641–648.